NORTH EUR.

N.

S.

SF.

DK.

D.

PL.

ENT. EUR.

SU.

CS.

A.

H.

R.

SOUTH-EAST EUR.

YU

AL.

BG.

TR.

GR.

8°E 15°E

57°N

46°N

FLOWERS OF EUROPE

a field guide

Flowers of Europe

OLEG POLUNIN

With 192 pages of illustrations in colour from
photographs by the author and others and 50 pages of
line drawings by Barbara Everard

London · *Oxford University Press* · New York · Toronto · 1969

Oxford University Press, Ely House, London W.1

GLASGOW NEW YORK TORONTO MELBOURNE WELLINGTON CAPE TOWN SALISBURY IBADAN
NAIROBI LUSAKA ADDIS ABABA BOMBAY CALCUTTA MADRAS KARACHI LAHORE DACCA
KUALA LUMPUR SINGAPORE HONG KONG TOKYO

Printed and bound in Great Britain by
Jarrold and Sons Ltd Norwich

To my wife

CONTENTS

PREFACE

This, so far as I am aware, is the first attempt of its kind to describe and illustrate in a single volume the commoner and most attractive seed-bearing plants to be found throughout Europe. The unprecedented increase in travel since the 1950s has made it no longer satisfactory to describe the wildlife of any one country in isolation; the naturalist and traveller are moving much further afield. This particularly applies to Great Britain, which has rather a depauperate sample of Europe's wildlife, for the last Ice Age left an indelible mark and robbed the country of many plants and animals that at one time lived here. One hundred years ago county and country were the natural limits for the naturalist; but now, for many, the area has increased enormously.

Rapid technical changes in the last two decades have helped to make a volume such as this now possible. The arrival of cheap colour photography, coupled with the evolution of the handy portable single-lens reflex camera, has introduced a new kind of plant-hunting. The vasculum and the trowel are being superseded by the camera—to the plant's great advantage. At the same time, colour printing is now sufficiently reliable to allow for the publication of relatively inexpensive books using colour photography as a means of identification. From the purely botanical point of view there is no doubt that a good line drawing cannot be bettered for conveying precise detail, an absolute essential for true identification. But the untrained enthusiast has to make a very considerable visual and mental leap forward before he can learn to match a line drawing of a plant, for example, with the same living species in the field. Trained scientists are accustomed to bridging the gap, for diagrams are part of their stock-in-trade; the interpretation of two-dimensional drawings of a three-dimensional object presents no obstacle. Experienced botanists are often unaware of the layman's difficulties. By contrast, a coloured photographic portrait of a plant with all its nuances of posture, colour, form, and habitat may at once enable the layman to put a name to his plant, and to say with conviction that the plant 'in the hand' is the same species as the one illustrated. Final confirmation, it is true, must be made by checking each characteristic against the description in the text, but the first quick visual appraisal will have gone a very long way towards identification. Many people, who in the past may have been daunted by the great variety of plants seen on their travels, will now be able to identify some of them with reasonable certainty. Modern methods of communication are making much use of this visual approach and it is becoming more and more acceptable as a means of conveying information.

Another recent trend, which has helped to make this book possible, is the enormous expansion of informed tourism that seeks more than sun and sand. It now caters for a great variety of recreational interests which can be pursued further and further afield. Rapid movement between coast and mountain, island and lake, has made possible the collection of photographs included in this book— photographs taken during holiday periods. This in itself is an indication of the great variety of plants that the ordinary person is likely to meet during his travels.

In addition, there has quite recently been an upsurge of interest in the conservation of wildlife and a realization that man is laying waste his natural environment more rapidly than ever before. In destroying a species of plant or animal he is actually destroying, for ever, precious genetical material which has taken millions of years to evolve. There is no knowing the intrinsic value of any such genetical material in breeding and selection in the future. Our modern cereals, fodder grasses, vegetables, and ornamental plants have, in many cases, had their potentiality increased manyfold by inter-breeding with material from wild stock. But more immediate and more obvious is the destruction of such natural environments as forests and coastlands, the drainage of wet-lands, and the removal of top-soil. Forests become steppes; and pelicans and flamingoes have nowhere to breed. Once the delicate balance between organism and environment is disturbed both may be changed irrevocably. In a nuclear, molecular, and mathematical age not enough thought is given to these problems; yet it is in the study of organisms—including man—living with each other and in their environments that tomorrow's great scientific advances will undoubtedly be made. Mankind must take strenuous steps to conserve at least some of his natural wildlife heritage, not only for present-day study and recreation, but for the generations that are to follow.

A knowledge of Europe's plant life, which is the bedrock of the animal life of Europe, is of the greatest importance to Europeans. It is to further this knowledge that this volume has been written.

The Text

It is estimated that there are between 15,000 and 17,000 species of seed plants, both native and naturalized, growing in Europe today. Of these, in all probability at least 3,000–4,000 are rare and grow only in very limited localities. About 9,000 more species are thought to occur in no more than five European countries, while something like 2,000 to 2,500 species are probably widespread throughout much of Europe. This volume includes about 2,800 species and can in consequence lay claim to include the majority of the commoner European species most likely to be seen by the ordinary traveller.

The selection of species has been most difficult. The criteria used in making it have been the following: abundance, attractiveness, individuality, and fame; while well-established crop plants and some widely grown ornamental plants have also been included. A countryman of any one European country is unlikely to be wholly satisfied with the selection made. In all probability, there will be certain attractive species, well known locally, which will not be found in this volume. It may be possible to travel for mile after mile past some colourful plant in an individual country and yet not find it described here. The only solution would be an even larger and more expensive book.

Keys

These have been designed to help the reader to 'run down' the family and genera to which the plant belongs. They have been made as simple as possible, often adapted from existing floras, but inevitably they depend upon small botanical differences and experience is required before they can be used with confidence. It must, however, be emphasized that the family and genera keys are designed for the identification of only those families, genera, and species which are included in this volume. A species left out of this volume, but belonging to a particular genus which is included, cannot necessarily be traced to this genus by means of these keys. The species included have been separated into groups by cross-headings to facilitate identification, but once again these cross-headings are only valid for plants that are described. For the identification of the excluded species and genera, reference to a flora of the country in which the species grows is necessary, or to *Flora Europaea* when completed.

Description of Species

The plants selected for inclusion in this volume have been dealt with in three different ways: (*a*) numbered species, (*b*) unnumbered species, (*c*) asterisked species.

(*a*) The numbered species have been given the fullest treatment. A description of the main diagnostic characters is given; this is followed by the localities in which the species is likely to be found, its time of flowering, its presence or absence in each European country, and whether it is native or introduced. The description of each species, or rather its diagnosis, has purposely been given a somewhat unorthodox treatment. By selecting certain salient features as they strike the observer on first seeing the plant, an attempt has often been made to give a thumb-nail impression of the plant first, before proceeding to the details. The inexperienced reader often finds it very difficult to obtain a general picture of a plant which is being described organ by organ; the mass of detail obscures a conception of the whole. It seemed, therefore, worthwhile breaking away from the uniform and rather monotonous format generally used in botanical descriptions, and selecting certain distinctive features first, followed by the ordered detailed description of the remaining characters. This may give the impression of an apparent lack of order in the treatment of different species, but the reader will, I think, quickly come to appreciate the merits of this system.

The geographical distribution of the (*a*) numbered species shows the actual *presence* or *absence* of the species in the country named (given in abbreviated form, see page 7), and also whether it is native or introduced. The time of flowering given must, for obvious reasons, show considerable variation in different parts of Europe and is only an indication of the months during which the plant is likely to be found in flower.

(*b*) Unnumbered species are generally associated with a related numbered species. When referred to in the text and illustrations, unnumbered species are given the number of the related species in square brackets: thus [246]=*Ranunculus platanifolius*, while 246=*R. aconitifolius*. These unnumbered species have been given shorter treatment, usually describing only those characteristics which differentiate it from its numbered relative. In general, the characters not described in the unnumbered species are common to both species. The unnumbered species are given only a general geographical distribution (see page 7).

(*c*) Asterisked species are widespread European plants, which are distinguished from already described species by relatively inconspicuous botanical differences. They are likely to be of less interest to the amateur. The asterisk indicates that a full diagnostic description can be found in *Flora of the British Isles* by Clapham, Tutin, and Warburg, second edition, Cambridge, 1962.

Distribution

It is the aim of this volume to give the exact distribution of each of the major species (the (*a*) numbered species) in the countries of Europe. This is shown by using the International Registration Letters established by the International Convention of 1949. The presence of a species as a native is shown by the appropriate letter or letters of the country, while introduced species are also indicated. Unfortunately this method has its limitations. If, for example, a species occurs in a few small colonies in any one country, there is no means of distinguishing this limited distribution from a country in which the species is widespread. A considerably more complicated system, involving at least three categories within each country, would be required to convey this more detailed information; space in one volume does not allow this. Countries have been grouped into regions which have, in the numbered species, a precise connotation (see endpaper map). For example, West Europe includes all the following countries: Portugal, Spain, France, Ireland, Great Britain, Belgium, and Holland (abbreviated West Eur.= P.E.F.IRL.GB.B.NL.). Any exceptions to this overall distribution are given in brackets thus: (except B.NL.).

Species not numbered in the text, that is to say unnumbered species and asterisked species, have a more generalized distribution which may, at first sight, seem a little confusing. However, to distinguish these general distributions from the precise distributions of the numbered species, no abbreviations have been used. Thus, Southern Europe represents those countries lying approximately south of latitude 46°N. Similarly, Western Europe may possibly include parts of Western Germany, Switzerland, and Italy, or, alternatively, it may signify presence in most of the Western European countries. Asterisked species have similar generalized distributions.

Uses

The most important uses of European plants have been given. These include not only industrial, agricultural, and crop uses, but also their value as food and for beverages, as well as their most important medicinal and herbal uses. A great many plants have been used by mankind in the past. They have been of particular importance at religious festivals, fertility rites, and seasonal observances; they have been widely used as symbols of chastity, virtue, victory, intelligence, and so on. A great many species have been used for herbal remedies. Knowledge of their healing powers has been handed down from century to century, from prehistoric and classical times to the present day, with the result that it is difficult to distinguish the real from the imaginary effects. As so much of the treatment is linked with ancient folklore and beliefs, the healing powers of many plants may well be more a matter of faith than of real efficacy. In consequence I have endeavoured to note only those plants which have well-tried remedial properties or were particularly famous as medicinal plants in the past.

ACKNOWLEDGEMENTS

A volume of this nature has inevitably to rely on many sources of information; no one person can cover such a wide field. I have drawn freely on such well-established floras as Coste's *Flore de la France*, Hayek's *Prodomus Florae Balcanicae*, and Clapham, Tutin, and Warburg's *Flora of the British Isles*, in particular, using and adapting keys and diagnoses and obtaining a great deal of detail from them. Since commencing work on this book, the first volume of *Flora Europaea* has been published. It is inevitable and desirable that such a major work should profoundly influence a work of this kind, and I should like to acknowledge my gratitude to the authors for their far-sightedness and perseverance. Their work is bound to spur on botanical research in Europe as a whole, and to spawn many lesser volumes, such as the present one. I have drawn freely on *Flora Europaea*, volume I and from the manuscript of volume II, and have endeavoured to keep as closely as possible to its order, terminology, etc. How much lighter my task would have been if all four volumes had been published before I took up my pen! The latter part of *Flowers of Europe*, from Diapensiaceae onwards, cannot unfortunately benefit from the work that is now going on under the scrutiny of the international team of experts who are helping the authors of *Flora Europaea*. The assessment of species and their geographical distribution in particular will require adjustment when volumes III and IV of *Flora Europaea* have been published.

Many people have given very generously of their knowledge and experience in the preparation of this book and to them I am greatly indebted. The final result must remain my own responsibility, but I hope that they will feel that their efforts have been worthwhile. In particular, I would like to thank Professor D. A. Webb, who has helped in so many ways. Many of his suggestions and recommendations have been incorporated in the text and he has helped me over the difficult problem of the selection of species, and with the bibliography as well. The inclusion of a Popular Names Index has only been made possible by the help of J.-L. Terretaz (French and German) and Professor E. Maugini (Italian). They have undertaken the task of selecting the most widely used popular names when, as is often the case, many local and dialect variants occur. I am greatly indebted to them both for this feature of the book which I think is particularly worthwhile. J. E. Dandy and Dr. A. Melderis have kindly checked the nomenclature of species included and have brought the plant names up to date in the light of most recent research, while Dr. A. Melderis has also helped substantially with the bibliography. I would like to thank them both for this very real and necessary help. The Regius Keeper of the Royal Botanic Gardens, Edinburgh, has generously made regular loans of herbarium specimens which has not only made it possible for Mrs. Barbara Everard to prepare the line drawings, but has enabled me to check most of the species described with herbarium material.

To the Director of the Royal Botanic Gardens, Kew, I am indebted for the use of the Herbarium and the Library, and also for certain facilities in the gardens which have enabled me to photograph examples of the extensive collection of living European plants. I would like also to thank M. J. D'Oyly and Miss M. McCallum Webster for their help in checking some of the descriptions. J. F. M.

ACKNOWLEDGEMENTS

Cannon has kindly helped with the difficult family, the Umbelliferae, and supplied me with a key to the genera. Squadron-Leader C. O. Hook has used his skill and ingenuity in keeping my photographic equipment in sound working order. To Miss A. M. Adeley I owe a special debt of gratitude. She has been intimately concerned with all stages of this book from its conception, and she has resolutely continued through all the changes and revoked decisions that have been made during the last three years. She has also prepared the indexes and helped with the Popular Names Index. Her vigilance and precision have been invaluable. I would also like to thank the following for their suggestions, help, and criticisms: Dr. K. Ferguson, J. E. Lousley, F. P. Penfold, Professor N. V. Polunin, Professor C. N. Tavares (Portugal), C. Zahariadi (Romania), and the Librarian of the Linnean Society.

A vital part of this volume are the illustrations. Mrs. Barbara Everard has prepared 50 plates of line drawings of plants with less photogenic qualities, as well as the small drawings for the illustrated glossary. Her great experience of plants has enabled her to make first-class drawings ($\frac{1}{4}$ life-size) from herbarium specimens—drawings which have all the stamp of being made from the living plant. I cherish her close co-operation in all that she has done, and thank her most sincerely.

The generosity of British botanists is remarkable and a large number of people have offered the loan of photographs of European plants, taken in the field, for this volume. After looking through several thousand transparencies, I have selected the following and I must express my most sincere thanks to these expert photographers.

Mr. & Mrs. D. Parish: numbers 3, 156, 194, 220, 221, 254, (267), 331, 382, 398, 409, 434, 441, 448, 480, 635, 761, 782, 786, (883), 912, (912), 921, 925, 933, 941, 945, 946, 949, 953, (990), 1070, 1279, 1332, 1355, 1433, (1475), 1506, 1601, 1629, 1658, 1661, 1736, 1751, 1916, (1919), 1926. M. J. D'Oyly: numbers 34, 71, 104, (146), 197, 272, 337, 363, 372, 379, 397, 420, 473, 476, 483, 594, 622, 764, 831, 871, 917, 962, 975, 1003, 1091, 1094, 1118, 1127, 1132, 1164, 1166, 1174, 1215, 1223, 1269, 1281, 1287, 1309, 1362, 1443, 1479, 1570, 1636, 1671, 1835. B. E. Smythies: numbers 78, 157, 216, 217, 222, (378), (399), 514, (545), 796, 824, (945), (976), 1013, 1074, 1081, (1204), 1211, (1269), (1280), 1587, 1595, (1662), 1666, 1670, 1672, 1685, 1888, 1906. J. H. B. Birks: numbers 38, 85, 162, 169, (247), 333, 377, (427), 684, 698, 847, 911, (922), 923, (925), 935, 937, 965, 1032, (1220), 1280, 1307, (1923). L. E. Perrins: numbers 126, 142, (203), 223, 224, 300, 446, 457, 519, 530, 785, (967), 983, 999, 1064, 1277, (1315), 1497, 1913. A. J. Huxley: numbers 70, (309), 381, 435, 922, (956), 992, 1151, 1261, 1262, 1272, (1303), (1339), 1507, 1606, 1619, 1921. H. Crook: numbers 345, 386, 506, 754, (959), (992), (1332), 1350, (1352), 1626, 1664, (1664), (1685), 1892. Miss M. McCallum Webster: numbers 204, 267, 426, 641, 783, 892, 914, 1041, 1363, 1440, 1549. C. J. Dawkins: numbers 74, 201, 234, (309), (506), (944), 1254, 1714, 1737. Miss K. M. Firby: numbers 241, 520, 593, 1224, 1338, 1419, 1537, 1925. J. Crosland: numbers 400, (445), (919), 997, 1589, 1618. Dr. P. Smith: numbers 220, 1185, 1250. Miss B. S. Smither: numbers (1510), 1673, 1886. Miss B. A. Burrough: numbers 122, 565. H. Esslemont: numbers 248, 915. Miss M. J. Robinson: numbers 773, 777. Mrs. K. Hunt: number 1390.

I would like to thank B. E. Smythies for reading the proofs, and the staff of the Herbarium, Kew, for checking the colour proofs as far as they have been able with the material available.

In conclusion, the helping and steadying hand of the publisher has always been present and this is perhaps more valuable to the author than anything else.

Godalming, December 1968   Oleg Polunin

SIGNS AND ABBREVIATIONS

agg.	an aggregate of two or more closely related species
ann.	annual
bienn.	biennial
c. (circa)	about, used in measurements
fl. fls.	flower, flowering, flowers
fr. frs.	fruit, fruiting, fruits
incl.	including
introd.	introduced
lv. lvs.	leaf, leaves
perenn.	perennial
Pl.	plate
sp. sps.	species
subsp. subsps.	subspecies
var.	variety
×	(in text) indicates hybrid; (on colour plates) indicates degree of magnification. For example, $\times \frac{1}{3} = \frac{1}{3}$ life size
*	description of this species is found in *Flora of the British Isles* by Clapham, Tutin and Warburg, second edition, Cambridge, 1962

PRECISE GEOGRAPHICAL AREAS

West Eur.	=	Portugal, Spain, France, Ireland, Great Britain, Belgium, and Holland
North Eur.	=	Iceland, Denmark, Norway, Sweden, and Finland
Cent. Eur.	=	Germany, Switzerland, Austria, Poland, Czechoslovakia, and Hungary
Med. Eur.	=	Spain, France, Italy, Yugoslavia, Albania, Greece, and Turkey
South-East Eur.	=	Yugoslavia, Albania, Greece, Turkey, Bulgaria, and Romania

GENERAL GEOGRAPHICAL AREAS

Western Europe	=	west of longitude 8°E. and south of latitude 57°N. approximately
Northern Europe	=	north of latitude 57°N. approximately
Central Europe	=	between latitude 46°N. and 57°N. approximately
Mediterranean Europe	=	countries bordering the Mediterranean Sea
South-Eastern Europe	=	east of longitude 15°E. and south of latitude 46°N. approximately
Southern Europe	=	south of latitude 46°N. approximately

COUNTRIES

A.	Austria	F.	France	P.	Portugal
AL.	Albania	GB.	Great Britain	PL.	Poland
B.	Belgium	GR.	Greece	R.	Romania
BG.	Bulgaria	H.	Hungary	S.	Sweden
CH.	Switzerland	I.	Italy	SF.	Finland
CS.	Czechoslovakia	IRL.	Ireland	SU.	Soviet Union
D.	Germany	IS.	Iceland	TR.	Turkey
DK.	Denmark	N.	Norway	YU.	Yugoslavia
E.	Spain	NL.	Holland		

AUTHORITIES

The names of the authorities first describing the species are often abbreviated. For the abbreviated forms used see *Flora Europaea*, vol. 1.
auct. = various authorities

GENERA

The number of sps. found in Europe, whether native, naturalized, or commonly cultivated, is given at the end of the genera diagnoses.

GLOSSARY

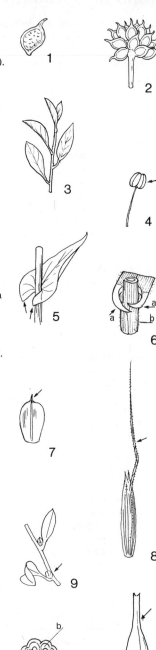

achene   A one-seeded fruit (1); there are usually many achenes on a fruiting head, as in *Ranunculus* (2).

actinomorphic   Of flowers; regular, radially symmetrical (34, 93); cf. **zygomorphic**.

adpressed   Pressed flat to a surface; commonly used of flattened hairs (56).

adventitious   Arising in an unusual position, as in adventitious roots arising from a stem (108b), or adventitious buds arising about a wound.

ala, alae   Wing; as in the flowers of Leguminosae (121b), or fruits of *Acer* (141a).

alien   Believed to have been introduced by man and now more or less naturalized.

alternate   Leaves placed singly, at different heights on a stem (3).

annual   A plant completing its life and seeding in one year or less.

anther   The part of the stamen containing the pollen grains (4).

apex, apical   Topmost point or structure; hence apical bud or flower.

apiculate   With a small broad point at the apex (80b).

apomixis   The production of fertile seed without sexual fusion. The progeny of apomictic species are genetically identical to their parents.

aril   An appendage or protuberance formed on the outside of a seed.

ascending   Rising upwards at an angle.

auricle   An ear-like flap of tissue at the base of a leaf-blade (5); curiously shaped in grasses (6a).

awl-shaped   Broad-based and tapering to a sharp point (7)=*subulate*.

awn   A long stiff bristle-like projection borne at the end or from the side of an organ (8).

axil, axillary   The angle between the leaf and stem; hence axillary flower or bud (9).

axile   Referring to the position of the ovules; those borne from a centrally placed axis which is connected by cross-walls to the outer ovary wall (10)=*axile placentation*.

bark   The outer covering of a woody stem.

beak   An elongated projection; usually referring to a slender extension of the fruit (11)=*rostellum*.

bearded   A tuft or zone of hairs (46b).

8

berry A fleshy rounded fruit, usually with hard pips or seeds (12).

biconvex With two convex surfaces; as in a lens.

biennial A plant which feeds and grows in the first year, and fruits in the second year.

bifid, bilobed Cleft in two no further than to the middle (13).

bifurcate Forked, with two equal branches (14).

bipartite Deeply divided into two to below the middle (15).

bipinnate Twice cut; of leaves cut into distinct segments which are themselves cut into distinct segments (16).

bisexual With fertile stamens and fertile ovary present in the same flower=*hermaphrodite*.

blade The flattened part of an organ, such as a leaf or petal.

bole Usually the trunk of a tree.

boss A rounded swelling raised above the surface of an organ.

bract A little leaf or scale-like structure from the axil of which a flower often arises (17, 117b).

bracteole A secondary bract, usually borne on the same axis as the flower, or in Umbelliferae one of the bract-like structures of the secondary umbels (99b).

bristle A stiff spine or fine projection; often referring to hairs.

bulb (dim. **bulblet**) A swollen underground bud-like structure, remaining dormant below ground during unfavourable growth periods (18).

bulbil A small bulb or tuber arising from the axil of the leaf or among the flowers, and reproducing the plant (19a).

calyx, calices The sepals collectively; often joined together in a tube, the calyx tube (20c).

capitulum A head of small stalkless flowers crowded together at the end of the stem, as in Compositae (21, 29).

capsule A dry fruit formed from two or more fused carpels which splits open when ripe (22).

carpel One of the units of the female part of the flower; they are either separate (2) or fused together into a fruit (23) (in transverse section (24)).

carpophore The axis to which the carpels are attached, formed as a prolongation of the receptacle, as in the Umbelliferae (25).

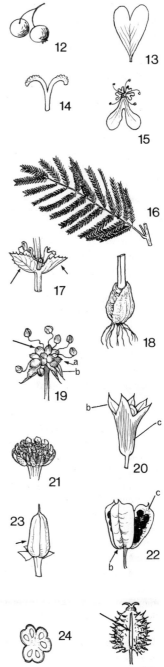

cartilaginous Tough and hard.

casual A foreign plant introduced into a country or region, which appears periodically, but cannot maintain itself for long.

catkin A crowded spike of tiny flowers, usually hanging and tassel-like (26).

cell Of a fruit, the cavity containing the seeds (10b).

ciliate Fringed with hairs along the margin (27).

cladode A green leaf-like stem, usually arising laterally from the axil of a bract (28).

claw The narrower stalk-like part of a petal (93a).

cleistogamy, cleistogamous Small flowers which never open, but which produce seed by self-pollination.

column, columnar A structure formed by the fusion of the stamens, styles, and stigmas, as in Orchidaceae.

composite A member of Compositae (29).

compound Two or more similar parts in one organ; hence a compound leaf has two or more separate leaflets (16, 96).

cone A distinct rounded or elongated structure composed of many overlapping scales which bear pollen or seeds when ripe (30).

connective The tissue or structure connecting the two anthers.

contiguous Touching each other at the edges.

converging, convergent Two or more organs separated at the base and with their tips coming together.

cordate Heart-shaped (31).

corm A swollen underground stem surrounded by scales, called the tunic, and replaced annually by a new corm (32).

corolla The petals collectively; often joined together into a tube, the corolla tube (33a, 130b).

corona Structures or appendages which stand out from the petals and together form a ring round the centre of the flower (34a).

crisped Curled and wavy, not straight.

crown See **corona** (34a).

cyme, cymose A broad, more or less flat-topped, simple or compound flower cluster, with the central flowers opening first (35).

deciduous Falling off, as with leaves in the autumn.

decurrent When the blade or stalk of the leaf continues, as a rib, down the side of the stem (36a).

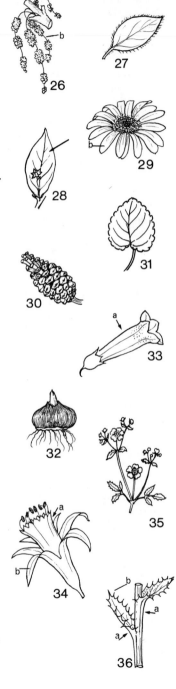

decussate Opposite; of leaves which are borne in opposite pairs with each successive pair lying at right angles to the next, and hence leaves in four ranks.

deflexed Bent sharply downwards (103)=**reflexed**.

dehiscent Splitting; as with fruits, allowing seeds to escape (37).

deltoid Shaped like the Greek letter △.

dichotomous Divided into two equal forks, and often forked again and again (38).

digitate Of leaflets or lobes of leaves, arranged as in the fingers of a hand (88)=**palmate**.

dioecious Having fertile male and female flowers on different plants; plants one-sexed.

disk A fleshy part of the receptacle which surrounds or surmounts the ovary (39).

disk-florets The tube-like flowers at the centre of the flower heads of some members of Compositae (40a).

dissected Cut or divided into parts, particularly of leaves (47, 110).

distinct, distant Of two or more organs which are separated from each other and are not fused, or do not overlap.

diverging, divergent Of two or more organs, with their tips wider apart than their bases.

dorsal The back or upper face of an organ; in flowers and fruits the parts laying nearest the main axis of the inflorescence (86b); cf. **ventral**.

downy Covered with short, weak, soft hairs.

drupe A fleshy fruit with an inner hard stone enclosing the seeds (41).

duct A slender channel carrying a fluid; as in resin canal (42).

egg-shaped With an oval outline broader towards the base than the apex, and round-ended (43)=**ovate**.

ellipsoid A solid object with an elliptic longitudinal profile.

elliptic Oval and narrowed to rounded ends in profile (44).

endemic Native of only one country or area.

entire Whole; without lobes or indentations.

epicalyx A calyx-like structure outside, but close to the true calyx (45).

epiphyte A plant growing on another plant for support but not feeding on it; usually not rooted in the soil.

evergreen Remaining green throughout the year,

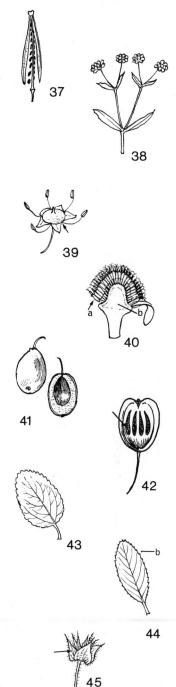

particularly of leaves that persist on the stem throughout the winter or dry season.

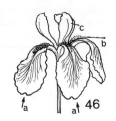

falls In *Iris*, the outer set of perianth segments which are usually turned down (46a), in contrast to the inner erect segments.

family A group of plants with many common characteristics; usually composed of many genera.

fasciculate Grouped into bundles or clusters.

46

feathery Cut into many fine segments (47).

female flowers Flowers with a fertile ovary but without fertile stamens; staminodes may be present.

ferruginous As of rusted iron; reddish-brown.

fertile Able to produce seeds capable of germination, or viable pollen.

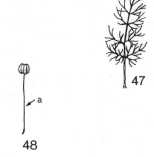

47

fetid Stinking, strong, and unpleasant-smelling.

filament The slender stalk of the stamen, which bears the anthers (48a).

flaccid Limp, floppy.

floret A small flower, usually one of a dense cluster, as in Compositae (40a, 100).

48

flower (fl. fls.) The reproductive part of certain plants, usually comprising sepals, petals, stamens, and ovary.

follicule A several- to many-seeded dry carpel which splits down one side on ripening (50).

forma, form A slight but distinctive variant of a species, often occurring sporadically.

free Not joined to other similar or different organs.

49

free-central Referring to the position of the ovules; those borne on a central column of tissue arising from the base of the ovary (51)=*free-central placentation.*

fructification The fruiting body including stalks, branches, and bracts.

fruit (fr. frs.) The ripened ovary bearing the seeds; other organs may be included (2, 22, 41, 112, 113).

50

genus, genera A classificatory term for a group of closely related species: a number of genera form a family. The generic name is the first part of the Latin binomial, as in *Ranunculus arvensis*, when *Ranunculus* is the generic name.

51

glabrous Not hairy, hence often smooth.

gland, glandular Organs of secretion usually on the tips of hairs; hence glandular-hairs (52).

glaucous Covered or whitened with a bloom which is often waxy, thus giving the organ a bluish or greyish colour.

52

globose, globular   Globe-shaped, spherical (12, 58).

glume   A chaff-like bract; in particular the bracts at the base of the spikelets of Gramineae (49a).

granular   Covered with very small grains; minutely or finely mealy.

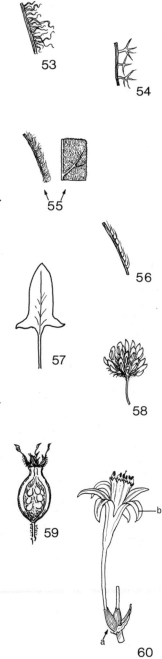

hair   A fine projection from a surface; one-celled or many-celled, simple or branched, straight or curved, etc. Special kinds of hair include glandular (52), crisped (53), and star-shaped (54) hairs. Long, soft hairs are described as woolly (55) and those pressed flat to a surface as adpressed (56).

hastate   Shaped like an arrowhead, but with the pointed basal lobes projecting outwards at an angle to the blade (57).

head   Of flowers or fruits which are crowded together at the end of a common stalk (58).

heart-shaped   Oval-shaped with a blunt, rounded apex and rounded basal lobes projecting below the point of attachment, as, for example, the leaf-blade with the leaf-stalk (31)=**cordate**.

herb   A plant which has no woody stem and is soft and leafy; also an aromatic plant used in seasoning or medicine.

herbaceous   Non-woody, soft, and leafy. Of a plant organ, having the soft texture and green colour of leaves.

hermaphrodite   With fertile stamens and ovaries present in the same flower.

hip   The fruit of the rose; the receptacle is fleshy (59).

hispid   With coarse, stiff, rough hairs.

hybrid   A plant resulting from the cross-breeding of two different species, and possessing some of the characters of each parent. Hybrids are often infertile.

hypanthium   A cup-like enlargement of the receptacle, particularly in Rosaceae (94).

indehiscent   Of fruit; not splitting to release the seeds.

inferior   Lower or below; hence inferior ovary, where the ovary is situated below the other organs of the flower (60a).

inflated   Blown-up, bladdery; separated from the organ which it encircles.

inflorescence   Flower branch, including the bracts, flower stalks, and flowers (35, 38, 98, 117).

internode   The part of the stem lying between two nodes (76b).

interrupted   Not continuous (26b).

introduced   Not native; a plant brought into a country or region from elsewhere.

investment, invested   A covering or enclosing structure.

involucel   As in Dipsacaceae; an additional calyx-like structure below the true calyx of each flower in the flower head (61).

involucre, involucral   A collection of bracts or leafy structures surrounding a flower head, groups of flowers, or a single flower; thus involucral bracts (62b).

keel   A sharp central ridge on an organ resembling the keel of a boat. Also the lower petal of the flowers of Leguminosae (63a).

labellum   Lip; in Orchidaceae the apparently lower perianth segment, usually projecting downwards, which is entire, two-, three-, or four-lobed (70, 71, 119b).

lance-shaped   Shaped like a lance with the broadest part nearer the base, with an acute apex and regularly narrowed to the base (64)=*lanceolate*.

lateral   Borne on the side of an organ.

lax   Loose and spreading; the opposite of densely clustered.

latex   A milky juice.

leaf (lv. lvs.)   A thin, expanded organ which is usually green and synthesizes food, arising from the node of a stem and commonly with a leaf-stalk or petiole and less commonly with basal stipules.

leaflet   The individual part of a compound leaf which is usually leaf-like and possesses its own stalk (65a, 128b).

legume   The fruit of the Leguminosae; comprised of one carpel which splits down two sides into two valves and is several- to many-seeded (66).

lemma   Of grasses; the lower and outer of the two fertile glumes of the floret (67a, 49b).

limb   The expanded or flat portion of an organ, particularly of calyx and corolla (20b, 60b).

ligule, ligulate   A small projection at the junction of the blade of a leaf with its sheath, found commonly in the Gramineae (68a). Ligulate-florets are those with strap-shaped corollas in the flower heads of the Compositae (100).

linear   Long and narrow, with parallel sides (69).

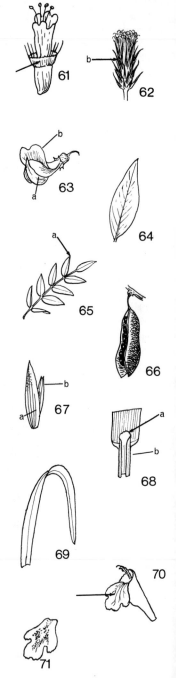

lip   A flap-like projection of a calyx or corolla (70, 71)=**labellum**.

littoral   On the sea-shore or near the sea.

lobe   A part or segment of an organ deeply divided from the rest of the organ but not separated from it (72, 20b).

72

lyrate   Shaped like a lyre; deeply cut with an enlarged terminal lobe and smaller lateral lobes (73).

male flower   A flower containing fertile stamens but no fertile ovary.

maquis   A thicket of tall shrubs and scattered trees characteristically developed in a Mediterranean climate.

membraneous   Thin, dry, flexible; parchment- or paper-like.

mericarp   A one-seeded part split off at maturity from an ovary of two or more fused carpels, as in Aceraceae (141b).

73

microspecies   A species showing small but constant differences distinguishing it from other closely related species.

moniliform   Like a string of beads; constricted at regular intervals.

74

mucilage   A viscid, slimy, gum-like substance; often produced in ducts or canals in the plant.

mucronate   With a short, narrow point or *mucro* (81b).

native   A plant naturally occurring in an area and not introduced from elsewhere.

naturalized   Thoroughly established in an area, but originally coming from another region.

75

nectary   A gland which gives out a sugary liquid and serves to attract insects; variously situated in the flower (74).

nerve   Prominent veins in a leaf or petal (75).

netted   With a net-like arrangement of veins (136b).

node   A point on the stem where one or more leaves arise (76a).

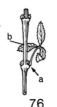

76

notch   A V-shaped indentation in an organ, as in a petal apex (87b, 121c).

nut, nutlet   A one-seeded fruit with a hard outer covering. Small nutlets, one of four, are found in the Labiatae (77a).

ob-   Inverted; thus with the broadest part of an organ near the apex and not, as is more usual, near the base.

77

Used as a prefix; thus *obconical* (78), *obcordate* (79), *obovate* (80).

oblong   An elongated but relatively wide shape, as in a leaf with parallel sides (81).

operculum   A lid or cover; as of a fruit (82).

opposite   Of two organs; arising at the same level on opposite sides of the stem (83).

orbicular   Rounded in outline, with length and breadth about the same (84).

ovary   The part of the flower containing the ovules and later the seeds, usually with one or more styles and stigmas (85).

ovate   With an outline like that of a hen's egg, with the broadest part towards the base (43).

ovoid   A solid form with an ovate outline in longitudinal profile.

ovule   A structure containing the egg, which after fertilization becomes the seed.

palate   A rounded projection on the lower lip of a two-lipped corolla, which more or less closes the opening or throat of the corolla, as in *Antirrhinum* (86a).

palea   Of grasses; the upper and inner, usually less robust of the two fertile glumes of the floret (87a, 49c, 67b).

palmate   Lobed or divided in a palm-like or hand-like manner (88)=**digitate**.

panicle   A branched inflorescence, or more precisely a branched racemose inflorescence (89).

papillae, papillate   Having small rounded or · cylindrical protuberances.

pappus   Hairs or bristles on the fruits of some Compositae which replace the calyx (90).

parasite   An organism living and feeding on another living organism; cf. **epiphyte**.

parietal   Referring to the position of the ovules; arising from the outer wall of the ovary and projecting into the central cavity.

pellucid   Clear, almost transparent.

peltate   A flat organ attached to its stalk on the underside and not to the margin (91).

pendulous   Drooping, hanging (26).

perennating   Surviving the winter, or unfavourable growth period.

perennial   Living for more than two years and usually flowering each year.

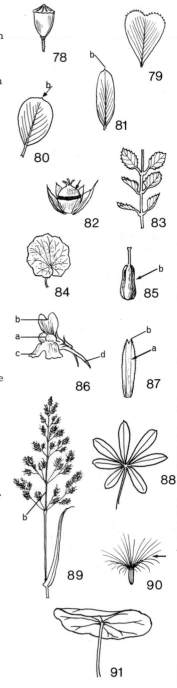

perfoliate   Of a leaf or bract with its base united round the stem (92).

perianth, perianth segments   The outer non-sexual parts of the flower, usually composed of two whorls, often with the outer sepals green (93s), the inner petals coloured (93p), or both whorls green, or both coloured (46). The perianth segments are the individual organs comprising the two whorls.

perigynium   The inflated sac enclosing the ovary in *Carex* (11).

perigynous   Of flowers; in which the receptacle forms a ring or cup surrounding the base of the ovary with the sepals, petals and anthers arising from the rim. The ovary is thus half-inferior (94).

petal   An individual member of the inner set of sterile organs surrounding the sexual parts of the flower, usually brightly coloured (93p).

phyllode   A leaf-stalk which has become flattened and leaf-like, usually replacing the true leaves (95); cf. **cladode**.

pinnate   The regular arrangement of leaflets in two rows on either side of the stalk or rachis (96).

pinnately lobed   A leaf with opposite pairs of deep lobes cut nearly to the midvein (97)=*pinnatifid*.

pit   A small depression or cavity in a surface.

placenta   The part of the ovary to which the ovules are attached (10, 51).

pollen   Small grains which contain the male reproductive cells.

pollinia   Coherent masses of pollen; as in Orchidaceae.

polyploid   A plant having a chromosome number which is a multiple of the basic number of its group.

pome   A fruit, such as an apple, in which the fleshy part is formed from the receptacle and the core from the wall of the ovary.

procumbent   Trailing or lying loosely along the surface of the ground.

proliferous   Bearing other similar structures, such as buds or offshoots from an organ.

prostrate   Lying rather closely along the surface of the ground.

raceme   A simple elongated inflorescence with the youngest flowers and branches at the apex and the oldest flowers at the base (98).

racemose   Having flowers in a raceme-like inflorescence.

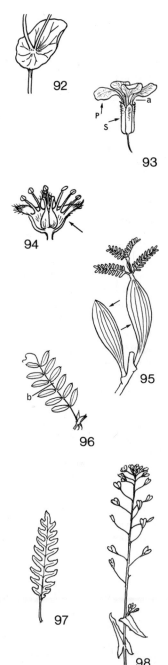

rachis   The stalk of a compound leaf (96b); the central axis bearing flowers (89b).

radical   Commonly used of leaves which arise from the rootstock (105b, 106b).

ray   One of the stalks of an umbel (99a).

ray-florets   The strap-shaped florets of many members of the Compositae (100, 29b).

receptacle   The uppermost part of the flower stalk which bears the parts of the flower (101); or the apex of the stem bearing the florets in the flower head of the Compositae (40b).

recurved   Bent backwards or downwards in a curve (102).

reflexed   Bent abruptly backwards or downwards (103)=**deflexed**.

regular   Of flowers; radially symmetrical, divisible in many planes into two equal halves (34, 93)= **actinomorphic**.

resin, resinous   A sticky substance which hardens on exposure to air and is excreted by many plants often in ducts or canals (42).

rhizome   A creeping underground stem which sends up new leaves and stems each season (104).

rhombic, rhomboidal   Diamond-shaped; approximating the shape of the diamond of playing-cards.

rootstock   A short, erect, underground stem (105).

rosette   The arrangement, usually of leaves, radiating outwards from the centre, overlapping and often spreading over the ground (106).

rostellum   A small beak (11).

runner   A trailing stem which roots at the nodes, forming new plants which eventually become detached from the parent (107).

sac   Pouch or baggy cavity=*saccate*.

samara   A dry one-seeded unsplitting carpel with a wing-like extension, as in *Acer* (141).

saprophyte, saprophytic   A plant which derives its food wholly or partially from dead organic matter.

scale   Any thin dry flap of tissue; usually a modified or degenerate leaf (108a); also see **cone**.

scorpioid   Of an inflorescence; with flowers in two ranks on one side of a coiled axis resembling a 'scorpion's tail', as in many Boraginaceae (109).

segment   One of the parts of a cut or divided leaf, calyx or corolla (110a, 34b).

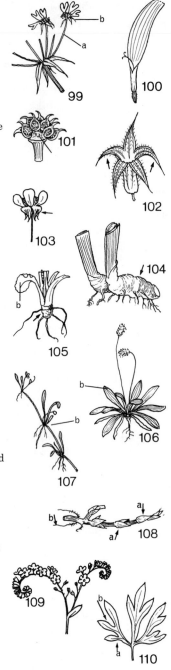

sepal One of the outer set of perianth segments, usually green and protecting in bed, less commonly coloured and petal-like (93s, 119c).

septum A partition, a cross-wall (10c).

sheath A more or less tubular structure surrounding another; as in the lower part of the leaves of the Gramineae (111, 68b).

short-shoots Lateral branches or shoots which bear leaves and have limited growth or no growth.

shrublet A small woody-stemmed plant, usually less than 30 cm. tall and often creeping over the ground.

silicula The dry splitting fruit of the Cruciferae which is not more than three times as long as broad, and more often broader than long (112).

siliqua The dry splitting fruit of the Cruciferae which is much longer than broad (113, 37).

silky Having a covering of soft fine hairs.

simple Of a leaf; not divided up into segments (116); of stems or inflorescences unbranched (98, 117).

sinus The cleft or recess between two lobes (110b).

spadix A fleshy axis bearing clusters of stalkless flowers and often ending in a swollen club-like apex, as in *Arum* (114).

spathe A large bract enclosing a flower head; it is sometimes conspicuous and coloured, as in *Arum* (115), or papery as in *Allium* (19b).

spathulate Spoon or paddle-shaped; broadest towards the apex and narrowed to the base (116).

species (sp. sps.) A group of individuals having similar characteristics, and which interbreed. Species are grouped together into genera. The specific name is the second part of the Latin binomial.

spike A slender elongated cluster of more or less stalkless flowers, the youngest flowers at the apex, the oldest at the base (117).

spikelet In Gramineae; a group of one or more florets subtended by one or two sterile bracts or glumes; a little spike (118, 49).

spine, spiny A sharp-pointed, narrow projection, often hard, woody, and piercing.

spreading Standing outwards or horizontally from an axis.

spur A hollow, more or less cylindrical projection from a petal or sepal; it usually contains nectar (118a, 86d).

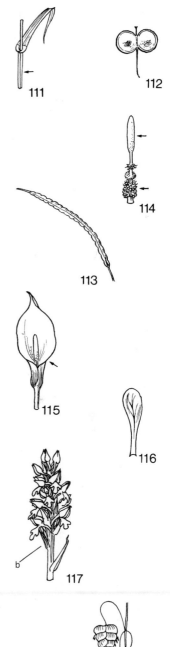

111

112

114

113

115

116

117

b

118

GLOSSARY

stamen   One of the male reproductive organs of the flower, which bears the pollen (48).

staminode   An infertile or rudimentary stamen without pollen, as in *Scrophularia* (120).

standard   The broad upper petal of the flower of Leguminosae (121a); also the inner erect perianth segments of the flower of *Iris* (46c).

stem   The main axis of a plant, which is both leaf-bearing and flower-bearing.

sterile   Lacking functional sex organs.

stigma   The part of the female organ which receives the male pollen (112a), it is normally situated at the top of the style. The stigma is sometimes lobed, hence *stigma-lobe*; less commonly radiating, thus *stigma-ray* as in *Papaver*.

stipule, stipulate   A scale-like or leaf-like appendage at the base of the leaf-stalk; usually paired (123).

stock   The short base of the stem, bearing basal leaves or scales and often roots=**rootstock**.

stolon, stoloniferous   A horizontal stem spreading above or below ground and which roots at the tip to give rise to a new plant (124).

striate   With fine longitudinal lines, grooves or ridges.

strophiole   A small hard protuberance on a seed.

style   A more or less elongated projection of the ovary which bears the stigma (125, 122b).

stylopodium   The enlarged base of the style, as in Umbelliferae (126).

subspecies (subsp. subsps.)   A group of individuals within a species which have some distinctive characteristics, and often a well-marked geographical range. The characteristics separating subspecies are more distinctive than those separating varieties.

subspontaneous   Of a species, subspecies or variety; seeding and establishing itself in the wild having been brought into a locality from elsewhere.

subtending   To stand below and close to; as of a bract bearing a flower in its axil.

subulate   Awl-shaped (7).

succulent   Fleshy, juicy, and thick.

sucker   A shoot originating and spreading below the ground and ultimately appearing above ground, often some distance from the main stem.

superior   Of an ovary; which has its origin on the receptacle above the other parts of the flower, or adjacent to the stamens (127)=*hypogynous*.

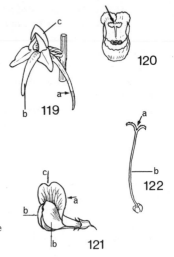

120

119

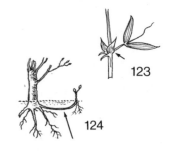

122

121

123

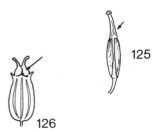

124

125

126

127

suture   The line or junction of two carpels, often where splitting takes place in a capsule (22b).

taproot   The main descending root.

tendril   A slender, clasping, twining organ, often formed from a leaf or part of a leaf (128a).

terminal   At the tip; borne at the end of the stem.

ternate   Of a compound leaf; divided into three parts which may themselves be further once or twice divided into three parts (129).

tetragonal   Square-sectioned.

thorn   A woody, spiny-pointed structure formed from a modified branch.

throat   The opening or orifice of a tubular or funnel-shaped corolla, or calyx (130a). Hence *throat-boss*, a swelling on the lower lip in the throat, as in *Antirrhinum* (86a)=*palate*.

toothed, teeth   With small triangular or rounded projections on a margin or rib (36b, 44b).

trifid   Cleft into three parts no further than the middle.

trifoliate, trefoil   Having three leaflets, as in *Trifolium* (131).

tripartite   Deeply divided to below the middle into three parts.

tube   The fused part of the calyx (132) or corolla (130b).

tuber   A swollen part of a stem or root, formed annually and usually underground (133).

twig   The youngest woody branches, usually of the present year's growth.

umbel   A cluster of flowers whose spreading stalks arise from the apex of the stem, resembling the spokes of an umbrella (134).

umbellifer   A member of Umbelliferae with flowers in umbels.

undershrub   A woody perennial, rarely as much as 1 m. tall, generally less.

undulate   Wavy in a plane at right angles to the surface (135).

unisexual   With either fertile male or fertile female organs only.

variety (var.)   A group of individual plants of a species possessing one or more distinctive charac-

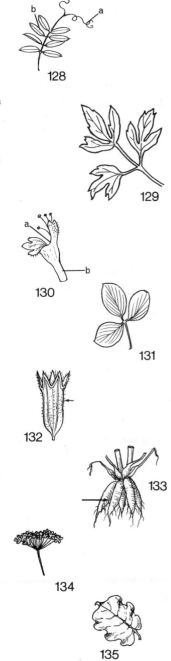

teristics, e.g. a marked colour-variation, an unusual leaf form, etc. Such variations may grade into the corresponding characteristics of closely related varieties. Varieties may have distinct geographical ranges.

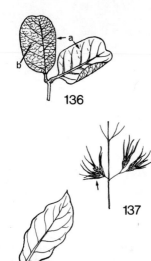

136

valve   One of the parts into which a capsule splits (22c).

valvate   Of perianth segments; with their edges in contact but not overlapping in bud.

vein   A strand of strengthening and conducting tissue running through the leaf and other organs (136a).

ventral   The front or lower face of an organ, the part furthest away from the main axis (86c); cf. **dorsal**.

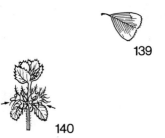

137

viscid   Sticky.

vitta, vittae   Resin canals (42).

viviparous   With flowers sprouting into leafy shoots on the parent plant and not forming seeds (137).

wavy   With regular curved indentations in the same plane as the surface (138).

138

wedge-shaped   Narrowest at the point of attachment and increasing regularly in width to the apex (139)= *cuneate*.

139

whorl   More than two organs of the same kind arising from the same level; thus whorled (140).

wing   A dry thin expansion of an organ (141a)=**ala**. Also the lateral petals of the flowers of Leguminosae (121b).

woolly   With long, soft, more or less tangled hairs (55)=*tomentose*.

140

zygomorphic   Irregular: divisible in one plane only into two equal halves (119, 130, 121); cf. **actinomorphic**.

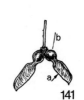

141

KEY TO FAMILIES

Adapted from *Flora Europaea*, vol. I, 1964

(Not including Tropaeolaceae, Simaroubaceae, Meliaceae, Bignoniaceae, Agavaceae, Commelinaceae)

1. Cone-bearing trees and shrubs; male and female cones separate; seeds not enclosed in an ovary. Lvs. needle-like or scale-like.   Group A
1. Flower-bearing trees, shrubs, or herbs usually with bisexual fls. composed of perianth, stamens, ovary. Seeds enclosed in an ovary. Lvs. usually broad and flattened.
  2. Plants free-floating on or below surface of water, not rooted in mud.   Group B
  2. Land plants or aquatic plants rooted in mud.
    3. Perianth of 2 (rarely more) whorls differing markedly from each other in shape, size, and colour.
      4. Petals not all united into a tube at base, very rarely cohering at apex.
        5. Ovary superior.
          6. Carpels 2 or more, free, or united at the base only.   Group C
          6. Carpels obviously united for about half their length or more, or carpel solitary.
            7. Fls. radially symmetrical.   Group D
            7. Fls. symmetrical in one plane only.   Group E
        5. Ovary inferior or partly so.   Group F
      4. Petals all united at base into a longer or shorter tube.
        8. Ovary superior.   Group G
        8. Ovary inferior.   Group H
    3. Perianth not of 2 or more markedly different whorls.
      9. Perianth entirely petal-like and brightly coloured.   Group I
      9. Perianth not petal-like, often absent, if brightly coloured then thin dry and papery.
        10. Trees or shrubs, sometimes small.   Group J
        10. Herbs.
          11. Aquatic plants; lvs. submerged or floating; fls. usually submerged, sometimes borne above the surface.   Group K
          11. Terrestrial plants or, if aquatic, with inflorescence and either stems or lvs. borne above the surface.   Group L

GROUP A. *Cone-bearing; lvs. needle-like or scale-like*

1. Lvs. all scale-like, usually brownish; internodes long.   Ephedraceae   *p.* 42
1. Lvs. usually green, needle- or scale-like; internodes short.
  2. Female fls. solitary; seeds surrounded by a fleshy aril.   Taxaceae   *p.* 42
  2. Female fls. in cones; seeds without a fleshy aril.
    3. Lvs. opposite or whorled, scale- or needle-like.   Cupressaceae   *p.* 40
    3. Lvs. alternate or 2–5, or many on short-shoots, needle-like.   Pinaceae   *p.* 38

GROUP B. *Free-floating or submerged plants, not rooted in mud*

1. Plant with small bladders on lvs. or on apparently leafless stems; lvs. divided into thread-like segments.   Lentibulariaceae   *p.* 396

1. Not as above.
  2. Plant without obvious differentiation into stems and lvs.   Lemnaceae   *p.* 557
  2. Plant with obvious stems and lvs.
    3. Floating lvs. stalkless.   Hydrocharitaceae   *p.* 479
    3. Floating lvs. with long lv. stalks.
      4. Floating lvs. orbicular, entire.   Hydrocharitaceae   *p.* 479
      4. Floating lvs. rhombic, toothed in upper two-thirds.   Trapaceae   *p.* 263

GROUP C. *Sepals and petals present; petals not fused at base; ovary superior; carpels not fused*

1. Sepals and petals 3.
  2. Carpels more than 3.
    3. Lvs. lobed.   Ranunculaceae   *p.* 95
    3. Lvs. entire.   Alismataceae   *p.* 477
  2. Carpels 3.
    4. Lvs. palmately divided; lv. stalks spiny.   Palmae   *p.* 521
    4. Lvs. simple, stalkless.   Crassulaceae   *p.* 146
1. Sepals or petals more than 3.
  5. Fls. symmetrical in 1 plane only; petals deeply divided.   Resedaceae   *p.* 144
  5. Fls. radially symmetrical; petals entire.
    6. Stamens more than twice as many as petals.
      7. Shrubs or herbs with stipules; fls. with ovary partially
        surrounded by a cup formed by the receptacle (peri-
        gynous).   Rosaceae   *p.* 157
      7. Herbs; stipules absent, though lv. bases sometimes
        sheathing; fls. with ovary not surrounded by the
        receptacular cup.
        8. Fr. a head of one-seeded carpels; sepals deciduous.   Ranunculaceae   *p.* 95
        8. Fr. of 2–5 several-seeded carpels; sepals persistent.   Paeoniaceae   *p.* 110
    6. Stamens not more than twice as many as petals.
      9. Lvs. trifoliate.   Rosaceae   *p.* 157
      9. Lvs. simple.
        10. Carpels spirally arranged on an elongated recep-
          tacle.   Ranunculaceae   *p.* 95
        10. Carpels in 1 whorl.
          11. Trees with palmately lobed lvs.; fls. in globose
            heads.   Platanaceae   *p.* 156
          11. Herbs or shrubs; lvs. not palmately lobed; fls. not
            in globose heads.
            12. Herbs or dwarf shrubs with round-sectioned
              stems; lvs. more or less succulent.   Crassulaceae   *p.* 146
            12. Shrubs with angular-sectioned stems; lvs. not
              succulent.   Coriariaceae   *p.* 231

GROUP D. *Ovary superior; carpels fused for half their length or more, or carpel 1; fls. radially symmetrical*

1. Petals more than 10.

2. Aquatic herbs with stalked floating lvs.   Nymphaeaceae *p.* 94
2. Terrestrial herbs or shrubs with stalkless or nearly stalkless lvs.
 3. Stamens 4–6.   Berberidaceae *p.* 111
 3. Stamens numerous.   Aizoaceae   *p.* 75
1. Petals fewer than 10.
 4. Stamens more than twice as many as petals.
  5. Stamens with their filaments united into a tube. Malvaceae *p.* 241
  5. Stamens free or united in bundles.
   6. Perianth segments persistent in fr., 2 large and 2 small.   Polygonaceae *p.* 61
   6. Perianth segments not as above.
    7. Ovary on a long stalk (in addition to fl. stalk). Capparidaceae *p.* 117
    7. Ovary stalkless or nearly so.
     8. Ovary surrounded by a cup formed from the receptacle; ovule 1.   Rosaceae   *p.* 157
     8. Ovary not surrounded by cup; ovules 2 or more.
      9. Carpel 1; lvs. doubly ternate, lower leaflets stalked. Ranunculaceae *p.* 95
      9. Carpels 2 or more; lvs. not as above.
       10. Large trees; inflorescence with a conspicuous bract partly fused to it. Tiliaceae   *p.* 240
       10. Not as above.
        11. Styles more than 1, free.
         12. All or most lvs. alternate; outer perianth segments petal-like. Ranunculaceae *p.* 95
         12. All lvs. opposite or in whorls; outer perianth segments sepal-like. Guttiferae   *p.* 249
        11. Style 1 or absent.
         13. Petals 4.   Papaveraceae *p.* 112
         13. Petals 5.
          14. Ovary one-celled or with a cross-wall at base only; stamens numerous. Cistaceae   *p.* 254
          14. Ovary three-celled; stamens 15. Zygophyllaceae *p.* 218
 4. Stamens not more than twice as many as petals.
  15. Trees, shrubs, or woody climbers.
   16. Fls. on tough leaf-like green stems (cladodes); lvs. scale-like, brownish. Liliaceae   *p.* 486
   16. Not as above.
    17. Lvs. small, scale-like or heather-like.
     18. Perianth segments in 2 whorls of 3; stamens 3. Empetraceae *p.* 300
     18. Perianth segments and stamens more than 3 in a whorl.
      19. Lvs. opposite.   Frankeniaceae *p.* 259
      19. Lvs. alternate.   Tamaricaceae *p.* 259
    17. Lvs. neither scale-like nor heather-like.
     20. All lvs. opposite.
      21. Fr. of 2 one-seeded winged units; lvs. usually palmately lobed. Aceraceae   *p.* 233
      21. Fr. a fleshy capsule; lvs. not palmately lobed. Celastraceae *p.* 236
     20. At least some lvs. alternate.
      22. Stamens 6.   Cruciferae   *p.* 117

22. Stamens 4, 5, or 10.
  23. Stamens 4 or 5.
    24. Stamens opposite petals.

25. Shrubs or small trees; petals shorter than sepals.	Rhamnaceae	*p.* 237
25. Woody climbers; petals longer than sepals.	Vitaceae	*p.* 240

    24. Stamens alternating with petals.

26. Bark resinous; ovule 1.	Anacardiaceae	*p.* 231
26. Bark not resinous; ovules several.	Pittosporaceae	*p.* 156

  23. Stamens 10.

27. Lvs. entire.	Ericaceae	*p.* 294
27. Lvs. pinnate.	Anacardiaceae	*p.* 231

15. Herbs, sometimes more or less woody at base.

28. Sepals 2, petals 5.	Portulacaceae	*p.* 76

 28. Sepals as many as the petals.

29. Fls. with a deep long-tubular or bell-shaped receptacle encircling ovary.	Lythraceae	*p.* 262

  29. Fls. superior, or with a flat or shallow cup-shaped receptacle encircling ovary.
    30. Stem lvs. opposite or whorled.
      31. Lvs. deeply divided, rarely only saw-toothed.

32. Petals 4.	Cruciferae	*p.* 117

        32. Petals 5.
          33. Stamens without scales on the inner side of the filaments.

	Geraniaceae	*p.* 215

          33. Stamens with scales on the inner side of the filaments.

	Zygophyllaceae	*p.* 218

      31. Lvs. simple and entire.

34. Stipules present.	Caryophyllaceae	*p.* 77

        34. Stipules absent.
          35. Sepals united to more than halfway.

36. Styles fused; placentation parietal.	Frankeniaceae	*p.* 259
36. Styles free; placentation free-central.	Caryophyllaceae	*p.* 77

          35. Sepals free or united only at base.

37. Ovary one-celled; placentation free-central.	Caryophyllaceae	*p.* 77
37. Ovary four- to five-celled; placentation axile.	Linaceae	*p.* 219

    30. Lvs. alternate or all basal.

38. Lvs. ternate.	Oxalidaceae	*p.* 214

      38. Lvs. not ternate.

39. Sepals and petals 2–3.	Polygonaceae	*p.* 61

        39. Sepals and petals 4–5.

40. Both whorls of perianth segments green.	Rosaceae	*p.* 157

          40. Inner whorl of perianth segments not green.
            41. Sepals and petals 4; stamens 4 or 6.

42. Stipules absent; stamens usually 6.	Cruciferae	*p.* 117
42. Stipules present; stamens 4.	Caryophyllaceae	*p.* 77

            41. Sepals and petals 5; stamens 5 or 10.
              43. Lvs. with conspicuous red, sticky, glandular hairs.

	Droseraceae	*p.* 145

              43. Not as above.
                44. Lvs. with numerous transparent glands, strongly scented when crushed.

	Rutaceae	*p.* 227

44. Lvs. without transparent glands.
  45. Style 1; stigma entire or shallowly lobed; anthers opening by pores.   Pyrolaceae   *p.* 293
  45. Style or stigma more than 1; anthers opening by longitudinal slits.
    46. Stigmas 5.
      47. Lvs. lobed or pinnate.   Geraniaceae   *p.* 215
      47. Lvs. entire.
        48. Sepals united; lvs. basal.   Plumbaginaceae   *p.* 308
        48. Sepals free; lvs. all from stem, not basal.   Linaceae   *p.* 219
    46. Stigmas 2–4.
      49. Fls. with conspicuous glandular, comb-like, sterile stamens.   Parnassiaceae   *p.* 155
      49. Glandular, comb-like, sterile stamens absent.
        50. Stamens 5.   Caryophyllaceae   *p.* 77
        50. Stamens 10.   Saxifragaceae   *p.* 150

GROUP E. *Ovary superior; carpels fused for half their length or more, or carpel 1; fls. symmetrical in one plane only*

1. Fls. with a pouch-like swelling or spur at the base.
  2. Sepals 2, small.   Papaveraceae   *p.* 112
  2. Sepals 3 or 5.
    3. Sepals 3, very unequal, 1 spurred; petals 3, not spurred.   Balsaminaceae   *p.* 235
    3. Sepals 5; petals 5.
      4. Lvs. alternate.   Violaceae   *p.* 251
      4. Lvs. opposite.   Geraniaceae   *p.* 215
1. Fls. without a pouch-like swelling or spur at the base.
  5. All, or all but 1, of the stamens united into a tube.
    6. Anthers opening by pores.   Polygalaceae   *p.* 230
    6. Anthers opening by slits.   Leguminosae   *p.* 175
  5. All stamens free.
    7. Trees or shrubs.
      8. Lvs. simple.
        9. Ovary on a long stalk (in addition to fl. stalk).   Capparidaceae   *p.* 117
        9. Ovary stalkless.
          10. Petals 4.   Cruciferae   *p.* 117
          10. Petals 5.   Leguminosae   *p.* 175
      8. Lvs. compound.
        11. Lvs. trifoliate or pinnate.   Leguminosae   *p.* 175
        11. Lvs. palmate with more than 3 leaflets.   Hippocastanaceae   *p.* 234
    7. Herbs.
      12. Ovary and fr. deeply five-lobed.
        13. Fls. in umbel-like clusters; fr. with a long beak.   Geraniaceae   *p.* 215
        13. Fls. in elongated spike-like clusters, lowest fl. opening first; fr. not beaked.   Rutaceae   *p.* 227
      12. Ovary and fr. not deeply five-lobed.
        14. Petals cut into narrow fringe-like segments, or lobed.   Resedaceae   *p.* 144
        14. Petals entire or notched.

15. Stamens 10.	Leguminosae	*p.* 175
15. Stamens not more than 6.		
16. Sepals free; ovary two-celled.	Cruciferae	*p.* 117
16. Sepals inserted on a cup-like flange of the receptacle; ovary one-celled.	Caryophyllaceae	*p.* 77

GROUP F. *Sepals and petals present; petals not united into a tube; ovary inferior or partly so*

1. Petals numerous.		
2. Aquatic plants; lvs. not succulent.	Nymphaeaceae	*p.* 94
2. Land plants; lvs. succulent.	Aizoaceae	*p.* 75
1. Petals 5 or fewer.		
3. Petals and sepals 3.		
4. Fls. symmetrical in 1 plane only.		
5. Style and filaments obvious.	Iridaceae	*p.* 510
5. Stigma and stamens stalkless.	Orchidaceae	*p.* 571
4. Fls. radially symmetrical.		
6. Outer whorl of perianth sepal-like.	Hydrocharitaceae	*p.* 479
6. Both whorls of perianth petal-like.		
7. Stamens 6.	Amaryllidaceae	*p.* 507
7. Stamens 3.	Iridaceae	*p.* 510
3. Petals and sepals 2, 4, or 5.		
8. Stamens numerous.		
9. Lvs. opposite, with transparent glands.	Myrtaceae	*p.* 265
9. Lvs. alternate, without transparent glands.		
10. Lvs. with finely toothed margin.	Rosaceae	*p.* 157
10. Lvs. entire; seeds covered with pulp.	Punicaceae	*p.* 265
8. Stamens 10 or fewer.		
11. Aquatic plants; lvs. pinnate, segments thread-like; fls. in spikes.	Haloragaceae	*p.* 269
11. Not as above.		
12. Trees, shrubs or woody climbers.		
13. Fls. in umbels.		
14. Climbers.	Araliaceae	*p.* 271
14. Erect shrubs.		
15. Evergreen; umbels flat-topped.	Umbelliferae	*p.* 271
15. Deciduous; umbels globose.	Cornaceae	*p.* 270
13. Fls. not in umbels.		
16. Lvs. palmately lobed.	Grossulariaceae	*p.* 155
16. Lvs. not lobed.		
17. Both whorls of perianth petal-like.	Onagraceae	*p.* 266
17. Outer whorl of perianth sepal-like.		
18. Calyx teeth very small; ovules 1 in each carpel; fr. fleshy with an inner hard wall.	Cornaceae	*p.* 270
18. Calyx teeth large; ovules numerous; fr. not fleshy, a dry splitting capsule.		
12. Herbs.		
19. Both whorls of perianth sepal-like.	Rosaceae	*p.* 157
19. Inner whorl of perianth petal-like.		
20. Petals 5.		
21. Stamens 5.	Umbelliferae	*p.* 271

21. Stamens 10.	Saxifragaceae	*p.* 150
20. Petals 4 or 2.		
22. Fls. in umbels surrounded by 4 conspicuous white bracts.	Cornaceae	*p.* 270
22. Fls. not in umbels; no conspicuous white bracts.	Onagraceae	*p.* 266

GROUP G. *Petals and sepals present; petals all united at base forming a longer or shorter tube; ovary superior*

1. Stamens at least twice as many as corolla lobes.		
2. Herbs with succulent lvs.	Crassulaceae	*p.* 146
2. Shrubs or trees.		
3. Anthers opening by pores; hairs simple or scale-like.	Ericaceae	*p.* 294
3. Anthers opening by longitudinal slits; hairs star-shaped.	Styracaceae	*p.* 310
1. Stamens as many as or fewer than corolla lobes.		
4. Plants without chlorophyll; lvs. scale-like.		
5. Fls. symmetrical in one plane only; stem stout, erect.	Orobanchaceae	*p.* 394
5. Fls. radially symmetrical; stem slender, twining.	Convolvulaceae	*p.* 327
4. Green plants.		
6. Sepals 2; fls. radially symmetrical.	Portulacaceae	*p.* 76
6. Sepals more than 2, or fls. symmetrical in one plane only.		
7. Ovary deeply four-lobed with 1 ovule in each lobe.		
8. Lvs. alternate.	Boraginaceae	*p.* 330
8. Lvs. opposite.	Labiatae	*p.* 344
7. Ovary not four-lobed.		
9. Fls. radially symmetrical or nearly so.		
10. Carpels free.		
11. Lvs. with rounded blade and stalk attached to the underside (peltate) and not to the edge; carpels 5.	Crassulaceae	*p.* 146
11. Lvs. with stalk attached to edge of blade (not peltate); carpels 2.		
12. Corolla with an additional petal-like crown or ring; styles 2, free but united by the stigma.	Asclepiadaceae	*p.* 320
12. Corolla without an additional petal-like crown or ring; styles 2, united except at the very base.	Apocynaceae	*p.* 319
10. Carpels united.		
13. Stamens fewer than corolla lobes.		
14. Herbs.	Scrophulariaceae	*p.* 374
14. Shrubs or trees.		
15. Lvs. opposite.	Oleaceae	*p.* 310
15. Lvs. alternate.		
16. Fls. yellow.	Oleaceae	*p.* 310
16. Fls. not yellow.	Scrophulariaceae	*p.* 374
13. Stamens as many as corolla lobes.		
17. Stamens opposite the corolla lobes.		
18. Styles or stigmas more than 1; ovule 1.	Plumbaginaceae	*p.* 308
18. Style 1; stigma 1; ovules numerous.	Primulaceae	*p.* 300
17. Stamens alternating with the corolla lobes.		
19. Lvs. opposite.		
20. Shrubs.		

21. Large, erect; lvs. deciduous.	Buddlejaceae	*p.* 373
21. Small, procumbent; lvs. evergreen.		
22. Lvs. elliptical or oblong; fls. pink.	Ericaceae	*p.* 294
22. Lvs. spathulate; fls. white.	Diapensiaceae	*p.* 292
20. Herbs.		
23. Land plants; lvs. stalkless.	Gentianaceae	*p.* 313
23. Aquatic plants; lvs. with lv. stalk.	Menyanthaceae	*p.* 318
19. Lvs. alternate or all basal.		
24. Sepals, petals, and stamens 4.		
25. Shrubs.	Aquifoliaceae	*p.* 236
25. Herbs.		
26. Corolla not violet-blue.	Plantaginaceae	*p.* 398
26. Corolla violet-blue.	Gesneriaceae	*p.* 394
24. Sepals, petals, and stamens 5 (rarely sepals fewer).		
27. Ovary three-celled; stigmas 3 or three-lobed.		
28. Lvs. pinnate.	Polemoniaceae	*p.* 327
28. Lvs. simple.	Diapensiaceae	*p.* 292
27. Ovary two-celled; stigmas 2 or 1.		
29. Ovules 4 or fewer.		
30. Fls. numerous in scorpioid cymes; corolla lobes distinct.	Boraginaceae	*p.* 330
30. Fls. solitary or few, not so arranged; corolla not or scarcely lobed.	Convolvulaceae	*p.* 327
29. Ovules numerous.		
31. Aquatic or bog plants; corolla with fringed lobes.	Menyanthaceae	*p.* 318
31. Land plants; corolla not fringed.		
32. Lvs. all basal.	Gesneriaceae	*p.* 394
32. Some lvs. on fl. stem.		
33. Corolla tube much shorter than lobes; stamens spreading.	Scrophulariaceae	*p.* 374
33. Corolla tube long, or anthers coming together.	Solanaceae	*p.* 368
9. Fls. conspicuously symmetrical in one plane only.		
34. Anthers opening by pores.	Ericaceae	*p.* 294
34. Anthers opening by slits.		
35. Fls. small, crowded into dense rounded heads.	Globulariaceae	*p.* 392
35. Fls. not in dense heads.		
36. Ovary one-celled; carnivorous plants.	Lentibulariaceae	*p.* 396
36. Ovary two-celled; not carnivorous plants.		
37. Ovules numerous.	Scrophulariaceae	*p.* 374
37. Ovules 4.		
38. Bracts shorter than calyx.	Verbenaceae	*p.* 342
38. Bracts or bracteoles much longer than calyx.	Acanthaceae	*p.* 393

GROUP H. *Petals and sepals present; petals all united at base forming a longer or shorter tube; ovary inferior*

1. Stamens 8–10, or 4–5 with filaments divided to base.		
2. Herbs; anthers opening by slits; lvs. ternate.	Adoxaceae	*p.* 404

2. Woody plants; anthers opening by pores; lvs. simple.	Ericaceae	*p.* 294
1. Stamens 5 or fewer; filaments not divided.		
3. Lvs. in whorls of 4 or more.	Rubiaceae	*p.* 321
3. Lvs. not in whorls.		
4. Stamens opposite corolla lobes.	Primulaceae	*p.* 300
4. Stamens alternating with corolla lobes.		
5. Lvs. opposite; stipules standing between the lvs.	Rubiaceae	*p.* 321
5. Lvs. alternate, or stipules not standing between the lvs.		
6. Fls. in dense heads of small florets surrounded by an involucre of more than 2 bracts.		
7. Anthers fused together in a ring round the style.		
8. Ovule 1; calyx, if present, represented by hairs or scales.	Compositae	*p.* 417
8. Ovules numerous; calyx lobes conspicuous, green.	Campanulaceae	*p.* 410
7. Anthers not fused together.		
9. Ovules numerous; corolla lobes longer than tube.	Campanulaceae	*p.* 410
9. Ovule 1; corolla lobes much shorter than tube.	Dipsacaceae	*p.* 407
6. Fls. not in dense heads of small florets, or if so, bracts 2.		
10. Anthers stalkless, pollen grains cohering in rounded masses.	Orchidaceae	*p.* 571
10. Stamens with filaments; pollen grains not cohering.		
11. Stamens 1–3.	Valerianaceae	*p.* 404
11. Stamens 4–5.		
12. Shrubs (sometimes small and creeping) or woody climbers.	Caprifoliaceae	*p.* 401
12. Herbs.		
13. Tendrils present.	Cucurbitaceae	*p.* 260
13. Tendrils absent.		
14. Lvs. pinnate.	Caprifoliaceae	*p.* 401
14. Lvs. not pinnate.		
15. Fls. bisexual; fr. a capsule.	Campanulaceae	*p.* 410
15. Fls. unisexual; fr. fleshy.	Cucurbitaceae	*p.* 260

GROUP I. *Perianth entirely petal-like and brightly coloured*

1. Plants without chlorophyll, parasites, or saprophytes.		
2. Filaments of stamens free.	Pyrolaceae	*p.* 293
2. Filaments united in a column.	Rafflesiaceae	*p.* 61
1. Green plants.		
3. Stems succulent, leafless, but with groups of spines.	Cactaceae	*p.* 262
3. Not as above.		
4. Stamens more than 12.	Ranunculaceae	*p.* 95
4. Stamens 12 or fewer.		
5. Fls. in ovoid dense heads without an involucre.	Rosaceae	*p.* 157
5. Fls. not so arranged or in dense heads with an involucre.		
6. Ovary superior.		
7. Perianth segments 4.		
8. Perianth tubular below.	Thymelaeaceae	*p.* 247
8. Perianth segments free.		
9. Herbs.	Liliaceae	*p.* 486

9. Shrubs.	Polygonaceae	*p.* 61
7. Perianth segments more than 4.		
10. Carpels more than 1, free or nearly so.		
11. Lvs. triangular in cross-section, all basal.	Butomaceae	*p.* 479
11. Lvs. flat, all from fl. stem.	Phytolaccaceae	*p.* 75
10. Carpel 1, or carpels obviously united.		
12. Perianth segments 6.	Liliaceae	*p.* 486
12. Perianth segments 5.		
13. Stigmas 2–3; stipules sheathing, thin, dry, stiff.	Polygonaceae	*p.* 61
13. Stigma 1; stipules absent.	Primulaceae	*p.* 300
6. Ovary inferior, or fls. male.		
14. Lvs. in whorls of 4 or more.	Rubiaceae	*p.* 321
14. Lvs. not in whorls.		
15. Fls. in dense heads surrounded by an involucre.		
16. Anthers fused into a tube round the style, or fls. unisexual.	Compositae	*p.* 417
16. Anthers not fused; fls. bisexual.	Dipsacaceae	*p.* 407
15. Fls. not in dense heads, though sometimes fls. shortly stalked in compact umbels.		
17. Ovules numerous.		
18. Perianth segments 3, or perianth tubular with a unilateral entire limb.	Aristolochiaceae	*p.* 60
18. Perianth segments 6.		
19. Stamens 6.	Amaryllidaceae	*p.* 507
19. Stamens 3.	Iridaceae	*p.* 510
17. Ovules 1 or 2.		
20. Lvs. opposite.	Valerianaceae	*p.* 404
20. Lvs. alternate.		
21. Fls. in simple cymes or solitary.	Santalaceae	*p.* 59
21. Fls. in umbels or whorls placed one above the other.	Umbelliferae	*p.* 271

GROUP J. *Perianth not petal-like, often absent, if brightly coloured then dry, thin and stiff; trees or shrubs, sometimes small*

1. Parasitic on branches of trees or shrubs.	Loranthaceae	*p.* 59
1. Not parasitic.		
2. Stems creeping, or climbing by aerial roots; evergreen.	Araliaceae	*p.* 271
2. Not as above.		
3. Fls. borne on flattened evergreen lv.-like stems. (cladodes); lvs. small, brownish, scale-like.	Liliaceae	*p.* 486
3. Not as above.		
4. Most lvs. opposite or nearly so.		
5. Stems green and fleshy or lvs. fleshy.	Chenopodiaceae	*p.* 68
5. Neither lvs. nor stems fleshy.		
6. Styles 3.	Buxaceae	*p.* 237
6. Styles 4, or 1.		
7. Fls. clustered into catkins.	Salicaceae	*p.* 43
7. Fls. not in catkins.		

8. Lvs. pinnate; stamens 2.	Oleaceae	*p.* 310
8. Lvs. simple; stamens 4 or more.		
9. Stamens 5, alternating with sepals.	Rhamnaceae	*p.* 237
9. Stamens 8; sepals 5.	Aceraceae	*p.* 233
4. All lvs. alternate.		
10. Lvs. not more than 2 mm. wide, oblong or linear.		
11. Stigma 1.	Thymelaeaceae	*p.* 247
11. Stigmas 2–9.		
12. Stamens 3; lvs. not succulent.	Empetraceae	*p.* 300
12. Stamens 5; lvs. succulent.	Chenopodiaceae	*p.* 68
10. Lvs. more than 2 mm. wide.		
13. Lvs. pinnate.		
14. Ovary inferior; styles 2; pith of stem with transverse partitions.	Juglandaceae	*p.* 47
14. Ovary superior; styles 3 or 1; pith without transverse partitions.		
15. Style 1; fr. constricted between seeds.	Leguminosae	*p.* 175
15. Styles 3; fr. a dry one-seeded drupe with hard inner wall.	Anacardiaceae	*p.* 231
13. Lvs. simple.		
16. Lv. stalk with an enlarged base enclosing the bud.	Platanaceae	*p.* 156
16. Lv. stalk base not enclosing the bud.		
17. Anthers opening by transverse valves.	Lauraceae	*p.* 112
17. Anthers opening by longitudinal slits.		
18. Fls. not in catkins or dense heads.		
19. Inflorescence of several male fls., each of 1 stamen, and a female fl., appearing as a stalked ovary, all surrounded by 4 or 5 conspicuous glands; milky sap present.	Euphorbiaceae	*p.* 222
19. Inflorescence not as above; milky sap absent.		
20. Fls. unisexual.		
21. Scale-like silvery or rust-coloured hairs present beneath the lvs. and often elsewhere; ovary one-celled; fr. fleshy.	Eleagnaceae	*p.* 249
21. Scale-like hairs absent; ovary three-celled; fr. dry.	Euphorbiaceae	*p.* 222
20. Fls. bisexual.		
22. Trees; perianth tube short, with stamens inserted near its base.	Ulmaceae	*p.* 55
22. Shrubs; perianth tube long, with stamens inserted near its apex.	Thymelaeaceae	*p.* 247
18. Fls. clustered into catkins or into dense heads.		
23. Milky sap present; fr. fleshy.	Moraceae	*p.* 56
23. Milky sap absent; fr. dry.		
24. Plants one-sexed; perianth absent.		
25. Bracts of catkins cut into slender fringe-like segments, or lobed at apex; fls. with a cup-like disk.	Salicaceae	*p.* 43
25. Bracts of catkins entire; disk absent.		
26. Lvs. without transparent glands; stamens with long filaments; ovules numerous.	Salicaceae	*p.* 43

26. Lvs. with transparent glands; stamens with short filaments; ovule 1.	Myricaceae	*p.* 47
24. Fls. unisexual, but both sexes on the same plant; perianth present in male or female fls. or both.		
27. Styles 3 or more; fls. of both sexes with perianth.	Fagaceae	*p.* 50
27. Styles 2; perianth present in fls. of 1 sex only.		
28. Male fls. 3 to each bract; perianth present.	Betulaceae	*p.* 48
28. Male fls. 1 to each bract; perianth absent.	Corylaceae	*p.* 49

GROUP K. *Perianth not petal-like; herbaceous aquatic plants with lvs. and fls. below or floating on surface, fls. rarely aerial*

1. Lvs. divided in numerous thread-like segments.		
2. Lvs. pinnately divided; fls. in a terminal spike.	Haloragaceae	*p.* 269
2. Lvs. dichotomously divided; fls. solitary, axillary.	Ceratophyllaceae	*p.* 94
1. Lvs. entire or toothed.		
3. Fls. in spikes.	Potamogetonaceae	*p.* 482
3. Fls. not spikes.		
4. Fls. solitary or few, stalkless or shortly stalked, axillary.		
5. Lvs. in whorls of 8 or more.	Hippuridaceae	*p.* 270
5. Lvs. not in whorls of 8 or more.		
6. Carpels 2 or more, free.		
7. Carpels nearly or quite stalkless in fr.	Zannichelliaceae	*p.* 485
7. Carpels in fr. with stalks several times their own length.	Potamogetonaceae	*p.* 482
6. Carpels united or solitary.		
8. Female fls. with a very long thread-like perianth tube resembling a fl. stalk.	Hydrocharitaceae	*p.* 479
8. Perianth tube short or absent.		
9. Perianth segments 4–6; stamens 4 or more; lvs. oval to obovate.		
10. Perianth segments 4; ovary inferior.	Onagraceae	*p.* 266
10. Perianth segments 6; ovary superior.	Lythraceae	*p.* 262
9. Perianth segments fewer than 4; or perianth absent; stamen 1; lvs. linear to lance-shaped.		
11. Lvs. alternate; plants of brackish water.	Zannichelliaceae	*p.* 485
11. Lvs. opposite; plants of fresh water.		
12. Lvs. entire, without sheathing base; ovary compressed, deeply four-lobed.	Callitrichaceae	*p.* 343
12. Lvs. with bristly-tipped teeth and with sheathing base; ovary rounded in section, not lobed.	Najadaceae	*p.* 486
4. Fls. in heads on long fl. stems or in compound inflorescences.		
13. Fls. bisexual; heads few-flowered.	Juncaceae	*p.* 516
13. Fls. unisexual; heads one-sexed, many-flowered.	Sparganiaceae	*p.* 558

GROUP L. *Perianth not petal-like; terrestrial plants or, if aquatic, with inflorescence and either stems or lvs. aerial*

1. Perianth absent or represented by minute scales or bristles; fls. in the axils of bracts, a number of which are also usually closely overlapping on an axis, forming a spikelet; lvs. usually linear, grass-like, sheathing below.
  2. Fls. usually with a bract above and below; lv. sheaths usually open and stems usually with hollow internodes.   Gramineae   *p.* 522
  2. Fls. with a bract below only; lv. sheaths usually closed, stems usually with solid internodes.   Cyperaceae   *p.* 559
1. Perianth present, or fls. not arranged in spikelets.
  3. Climbing plants with unisexual fls.
    4. Lvs. opposite; perianth segments 5.   Cannabaceae   *p.* 57
    4. Lvs. alternate; perianth segments 6.   Dioscoreaceae   *p.* 510
  3. Not climbing, or rarely climbers with bisexual fls.
    5. Lvs. linear.
      6. Fls. unisexual.
        7. Female fls. solitary; male fls. solitary or in short cymes.   Chenopodiaceae   *p.* 68
        7. Male and female fls. numerous, in dense heads or spikes.
          8. Male and female fls. in separate globose heads.   Sparganiaceae   *p.* 558
          8. Fls. in a dense cylindrical spike, male above, female below.   Typhaceae   *p.* 559
      6. Fls. bisexual.
        9. Plant densely hairy.   Chenopodiaceae   *p.* 68
        9. Plant hairless or sparsely hairy.
          10. Fls. in dense spikes; spikes apparently lateral on a flattened leaf-like stem.   Araceae   *p.* 555
          10. Not as above.
            11. Carpel 1.
              12. Lvs. not more or less whorled, without stipules.   Chenopodiaceae   *p.* 68
              12. Lvs. more or less whorled with minute stipules.   Caryophyllaceae   *p.* 77
            11. Carpels more than 1.
              13. Carpels free (except at base); lvs. with a conspicuous pore at apex.   Scheuchzeriaceae   *p.* 481
              13. Carpels more or less completely united; lvs. without a conspicuous pore at apex.   Juncaceae   *p.* 516
    5. Lvs. lance-shaped or wider, or sometimes small and scale-like, but never linear.
      14. Lvs. compound.
        15. Fls. in compound umbels.   Umbelliferae   *p.* 271
        15. Fls. not in compound umbels.
          16. Fls. in dense heads.
            17. Lvs. simply pinnate; style 1 or 2.   Rosaceae   *p.* 157
            17. Lvs. ternate; styles 3–5.   Adoxaceae   *p.* 404
          16. Fls. not in dense heads.
            18. Stamens numerous; epicalyx absent.   Ranunculaceae   *p.* 95
            18. Stamens 4 or 5–10; epicalyx present.   Rosaceae   *p.* 157
      14. Lvs. simple or apparently absent.

19. Fls. numerous, small, crowded on an axis (spadix), subtended and often more or less enclosed by a conspicuous bract (spathe).   Araceae   *p.* 555
19. Not as above.
  20. Inflorescence of several male fls., each of 1 stamen,
and a female fl., appearing as a stalked ovary, all
surrounded by 4 or 5 conspicuous glands; milky sap
present.   Euphorbiaceae   *p.* 222
  20. Not as above.
    21. Lvs. apparently absent; stem green and succulent.   Chenopodiaceae   *p.* 68
    21. Lvs. obvious; stem not succulent.
      22. Plant densely clothed with star-shaped hairs;
ovary three-celled with 1 ovule in each cell.   Euphorbiaceae   *p.* 222
      22. Not as above.
        23. Anns. with dense small swellings (papillae).   Aizoaceae   *p.* 75
        23. Anns. without small swellings.
          24. Lvs. whorled.   Hippuridaceae   *p.* 270
          24. Lvs. not in whorls.
            25. Lvs. alternate or all basal (rarely the lower
opposite).
              26. Stamens numerous; carpels free except some-
times at base.   Ranunculaceae   *p.* 95
              26. Stamens 12 or fewer; carpels not free, or 1 only.
                27. Carpels attached to a central axis, otherwise
free.   Phytolaccaceae   *p.* 75
                27. Carpels united, or 1 only.
                  28. Stamens 12.   Aristolochiaceae   *p.* 60
                  28. Stamens 10 or fewer.
                    29. Stipules united into a sheath.   Polygonaceae   *p.* 61
                    29. Stipules free or absent.
                      30. Epicalyx present; stipules leaf-like.   Rosaceae   *p.* 157
                      30. Epicalyx absent; stipules small or absent.
                        31. Ovary superior.
                          32. Perianth tubular below.
                            33. Ovule attached at base to ovary wall.   Chenopodiaceae   *p.* 68
                            33. Ovule attached above and hanging.   Thymelaeaceae   *p.* 247
                          32. Perianth segments free or nearly so,
rarely absent in female fls.
                            34. Perianth segments 4.
                              35. Fls. in elongated spike-like inflores-
cences without bracts.   Cruciferae   *p.* 117
                              35. Fls. in axillary clusters.   Urticaceae   *p.* 57
                            34. Perianth segments 5.
                              36. Perianth green, leaf-like.   Chenopodiaceae   *p.* 68
                              36. Perianth dry, stiff, rigid, not green.   Amaranthaceae   *p.* 74
                        31. Ovary inferior.
                          37. Lvs. kidney-shaped with heart-shaped
base.   Saxifragaceae   *p.* 150
                          37. Lvs. awl-shaped to linear-lance-shaped.   Santalaceae   *p.* 59
            25. Lvs. opposite (rarely a few upper apparently
alternate).

38. Lvs. toothed or lobed.
  39. Fls. bisexual.
    40. Ovary inferior; stigmas 2.   Saxifragaceae   *p.* 150
    40. Ovary superior; stigmas 5.   Geraniaceae   *p.* 215
  39. Fls. unisexual.
    41. Perianth segments 4 or 2; style 1.   Urticaceae   *p.* 57
    41. Perianth segments 3; styles 2.   Euphorbiaceae   *p.* 222
38. Lvs. entire.
  42. Perianth absent; ovary compressed, four-lobed.   Callitrichaceae   *p.* 343
  42. Perianth present; ovary not compressed and four-lobed.
    43. Perianth segments 3.   Polygonaceae   *p.* 61
    43. Perianth segments 4 or more.
      44. Ovary inferior.   Onagraceae   *p.* 266
      44. Ovary superior.
        45. Perianth segments 6 or 12; style and stigma 1.   Lythraceae   *p.* 262
        45. Perianth segments 4 or 5; styles or stigmas 2 or more.
          46. Lvs. without a long spiny apex; fr. unwinged.   Caryophyllaceae   *p.* 77
          46. Lvs. with a long spiny apex; fr. transversely winged.   Chenopodiaceae   *p.* 68

PINACEAE | Pine Family

Usually trees with whorled branches and narrow, needle-like, spirally arranged lvs., mostly evergreen. In addition to normal shoots in which the lvs. are spirally arranged and the axis continues growth, there are often short-shoots with the lvs. in lateral clusters of 2 or more and with the axis not elongating. Cones composed of spirally arranged scales on a central axis; male cones short-lived, bearing 2 pollen sacs on the underside of the scales; female cones woody, persistent, bearing 2 winged seeds on the upper surface of the scales. Pollen with 2 wings.

1. Lvs. all solitary; short-shoots absent.
  2. Lvs. borne on peg-like projections.   *Picea*
  2. Lvs. not borne on peg-like projections.   *Abies*
    3. Leaf-scars circular, not at all projecting; cone erect.   *Abies*
    3. Leaf-scars elliptical, slightly projecting; cone deflexed.   *Pseudotsuga*
1. Lvs. mostly in clusters of 2 or more, forming short-shoots.
  4. Lvs. evergreen.   *Pinus*
  4. Lvs. deciduous.   *Larix*

ABIES | **Fir** Lvs. single, not clustered, flattened, grooved above. Female cone erect; scales falling apart from axis when the cone is ripe. 9 sps.

1. *A. alba* Miller SILVER FIR. A pyramidal tree 20–50 m., with smooth, whitish bark and flat and spreading side branches. Lvs. 1½–3 cm., arranged in 2 ranks, thick flexible; buds not resinous. Cones erect 10–20 cm., with pointed, deflexed bracts projecting between the broad scales. △ Forming forests in mountains; rarely lower down. Apr.–May. Cent., South-East Eur. (except TR.) E.F.I.: introd. P.GB.DK.N.S. *A valuable source of turpentine which is obtained from blister-like swellings in the bark; in its raw state it is known as pitch and was used for caulking ships. The wood is light and widely used for carpentry, paper-making, and building. The bark is rich in tannin.*

A. cephalonica Loudon GREEK FIR. Distinguished by stiff, spiny-pointed lvs. arranged all round the stem, and by its resinous buds. Greece. Pl. 1.

A. grandis (D. Don) Lindley GIANT FIR. Lvs. 4–5 cm., distinctly two-ranked, dark green above, with 2 white lines beneath, apex notched. Cones cylindrical 7–10 cm. by 2½–3 cm., yellow-green. Native of North America; often planted for timber in Europe.

A. nordmanniana (Steven) Spach CAUCASIAN FIR. Lvs. 2–3 cm., forward-pointing and densely covering branchlets, glossy green above, with 2 white lines beneath, apex notched. Cones 12–15 cm. by 5 cm., reddish-brown; bracts with deflexed points longer than the scales. Native of Caucasus and Asia Minor; often planted for timber in Europe.

Pseudotsuga menziesii (Mirbel) Franco (*P. douglasii* (Lindley)·Carrière) DOUGLAS FIR. Lvs. 2–3 cm., bright green and grooved above, paler beneath, sharp-pointed. Cones 5–9 cm., pendulous, scales thin and stiff, much shorter than the narrow, three-lobed erect bracts. Native of North America; extensively planted for timber in Europe.

PICEA | **Spruce** Lvs. single, not clustered, flattened, or tetragonal in section, pointed, each attached to stem by a peg-like projection. Female cones pendulous, not breaking up on ripening. 7 sps.

2. *P. abies* (L.) Karsten NORWAY SPRUCE. A pyramidal tree with reddish scaly bark, 30–60 m. Lvs. regularly spirally arranged, diamond-shaped in section, dark green, spiny-tipped;

branches usually brown, spreading, and with ultimate branchlets usually pendulous. Cones 10–18 cm., pendulous, cylindrical, becoming russet-coloured, without bracts between the broad scales. △ Forming forests in mountains; widely planted elsewhere. May–June. North Eur. (except IS.), Cent. Eur. F.I.YU.AL.BG.R.SU.: introd. E.IRL.GB.B.NL. DK.H. *A valuable timber tree; the wood is used for many purposes, like the Silver Fir. The bark is rich in tannin; a source of resin, or Burgundy Pitch.* Pl. 1.

P. sitchensis (Bong.) Carrière SITKA SPRUCE. Lvs. 1–3 cm., light yellow-green, sharp-pointed, arranged irregularly around branchlets. Cones 6–10 cm., yellowish- or reddish-brown; scales thin and stiff with toothed or wavy-edged margin. Native of North America; often planted for timber in Europe.

LARIX | Larch Lvs. deciduous, soft, bright green, some grouped in clusters of 15–40 and forming short-shoots. Cones erect, globular; scales thin. 4 sps.

3. *L. decidua* Miller EUROPEAN LARCH. A pyramidal tree 20–35 m., with a straight trunk and greyish-brown bark which flakes in plates. Lvs. very unequal, 2–3 cm., with 2 greenish bands on the lower surface, deciduous; twigs yellowish. Fl. cones bright reddish-purple; mature cones 1½–4 cm. globular-oblong, brown, ripening at end of the first year. △ Forming forests in mountains; planted elsewhere. Apr.–June. F.D.CH.A.CS.I.: introd. GB.B.N.S. SF.YU. *A valuable timber tree used for many constructional purposes; it is supple, resistant, and durable. It yields Venice Turpentine.* Pl. 1.

*L. kaempferi (Lamb.) Carrière JAPANESE LARCH from Japan is widely planted for timber.

PINUS | Pine Lvs. evergreen, stiff needle-like, mostly in clusters of 2, 3, or 5. Female cones ripening after 2–3 years; scales thick and woody. 20 sps.

Lvs. in pairs
(a) Lvs. more than 6 cm.
4. *P. pinaster* Aiton MARITIME PINE. A tree of 20–40 m. of the Mediterranean region, with deeply fissured reddish-brown bark and very long, stiff, thick, whitish-green lvs. Lvs. 10–25 cm. by 2 mm., spiny-tipped; branches spreading upwards. Cones 8–22 cm., rounded-conical, almost stalkless, clustered, light shiny brown, with prominent pointed scale-tips. △ Sands and dunes, hills. Apr.–May. P.E.F.I.YU.: introd. GB.B.AL.GR.TR. *An important source of turpentine. Valuable for rough carpentry, boxes, paper-making, pit props, etc.* Pl. 1.

5. *P. pinea* L. STONE PINE, UMBRELLA PINE. A Mediterranean tree of 15–30 m., with a characteristically flat-topped, umbrella-shaped crown and grey-brown bark flaking to leave reddish-orange patches. Lvs. 8–20 cm., green, rather slender, somewhat stiff. Cones 10–15 cm., massive globular, shining reddish-brown, with blunt scale-tips. Seeds large, 1½–2 cm., fleshy, unwinged. △ Sands and river valleys by the sea. Apr.–May. Med. Eur. P. *The seeds are edible and much in demand; the timber is used for carpentry and furniture-making.*

6. *P. halepensis* Miller ALEPPO PINE. A tree of the Mediterranean region to 20 m., with pale silvery-grey bark when young, later becoming reddish-brown. Lvs. bright green, flexible, very narrow, 6–15 cm. by *c.* 1 mm. Cones 5–12 cm. blunt-conical, shiny brown, with thick recurved stalks. △ Rocky places, hills by the sea. Mar.–May. Med. Eur. (except AL.TR.): introd. P. *The bark is used for tanning; the timber is of mediocre quality used for coarse construction work.* Pl. 1.

7. *P. nigra* Arnold (*P. laricio* Poiret) BLACK PINE. Like 8, but upper trunk and branches dark grey. Lvs. usually dark green, commonly more than 8 cm. long. Cones 3–8 cm.,

almost stalkless, shiny light brown or yellowish-brown. A variable sp. with subsps. in Spain, Corsica, Dalmatia, and the Balkans. Tree 20–50 m. △ Mountains; extensively planted in Northern Europe. May. South-East Eur. E.A.I.: introd. IRL.GB.B.NL.DK.S.D.CH. CS. *The wood is non-durable and is used for rough carpentry and furniture; a source of turpentine.* Pl. 1.

(*b*) *Lvs. usually less than 6 cm.*

8. *P. sylvestris* L. SCOTS PINE. Readily distinguished by the upper trunk and branches which are bright orange to reddish-brown and contrasting with the dark brown lower trunk. Lvs. short, 3–7 cm. by 2 mm., paired, bluish-green, twisted. Cones 3–6 cm., sharp-pointed, dull yellowish-brown, pendulous. △ Mountains, lowlands, but widely planted. May–June. All Eur.: introd. IRL.B.NL.IS.DK.TR. *A valuable source of turpentine, rosin, pitch, creosote, and used for these products since the time of Theophrastus. A useful timber tree used for paper-making, furniture, masts, and many other kinds of constructional work.*

9. *P. mugo* Turra (*P. montana* Miller) MOUNTAIN PINE. A dense, spreading shrub up to 3½ m., with dark green lvs. 3–8 cm. Cones 2–5 cm., blunt, shiny, stalkless, horizontal, or spreading downwards; scales with glossy brown upper surface. △ Mountains. May–June. Cent. Eur. (except H.), South-East Eur. (except TR.) F.I.SU.: introd. DK.

P. contorta Loudon BEACH PINE. Lvs. 2½–7½ cm., in pairs, dark green, stiff, somewhat twisted. Cones 2–6 cm. and short-stalked, often clustered and remaining on branches for several years; scales with fragile apical prickles. Native of North America; often planted for timber in Europe.

Lvs. in fives

10. *P. cembra* L. AROLLA PINE. A tree 8–25 m., with smooth greenish-grey bark and stiff lvs. 4–8 cm., in clusters of 5. Cones 4–10 cm. by 3–6 cm. ovoid, often asymmetrical. Seeds wingless, enclosed in a hard husk, edible. △ Mountain forests, rocks. May–June. Cent. Eur. (except H.) F.I.YU.R.SU.: introd. N.S.SF. *The wood is light and used for toy-making, furniture, etc.*

P. strobus L. WEYMOUTH PINE. Lvs. 6–14 cm., in clusters of 5, triangular-sectioned, bluish-green, soft, slender. Cones 8–20 cm. by 3 cm., cylindrical, often slightly curved, stalked, pendulous, resinous. Native of North America; much planted for timber in Europe.

Lvs. in threes

P. radiata D. Don MONTEREY PINE. Lvs. 10–15 cm., in threes, bright green, slender flexible. Cones obliquely egg-shaped, 7–15 cm. by 6–9 cm., light brown, usually in clusters and remaining on branches for many years. Native of North America; often planted for timber in Europe.

CUPRESSACEAE | Cypress Family

Evergreen, resin-bearing trees or shrubs with opposite or whorled lvs. which are either short, broad, and scale-like, or long, narrow, and needle-like. Male cones short-lived, with 3–5 pollen sacs on the underside of each scale; female cones usually of a few flat-topped, dry, or fleshy scales; seeds 1–12 per scale.

1. Fr. berry-like, not separating into scales.   *Juniperus*
1. Fr. a woody cone, separating into scales.

2. Ripe cone scales flat, oblong, overlapping.   *Thuja*
2. Ripe cone scales with an enlarged, flattened, umbrella-like apex.
  3. Twigs rounded or four-angled; seeds narrowly winged, 6-20 on
    each scale.   *Cupressus*
  3. Twigs flattened; seeds broadly winged, to 5 on each scale.   *Chamaecyparis*

CUPRESSUS | **Cypress** Lvs. scale-like, opposite and four-ranked. Cones globular, formed of a few flat-topped woody scales. 4 sps.

11. *C. sempervirens* L. FUNERAL CYPRESS. A tree to 30 m., with dark green, blunt triangular, scale-like lvs. $\frac{1}{2}$-1 mm. long, closely pressed to stem. Cones globular 2$\frac{1}{2}$-4 cm.; scales 8-14, each with a rounded central boss, green, becoming yellowish-grey. There are two well-known forms: one columnar with erect fasciculate branches, the other with spreading-ascending branches forming a pyramidal crown. △ Coastal hills; widely planted, particularly the columnar form, as an ornamental. Mar.–Apr. Native of GR.: introd. Med. Eur. P.CH.BG. *The wood is hard, durable, and aromatic and was much used in ancient times for carving, shipbuilding, temples, etc. Decoctions of the leaves have been used medicinally; an essential oil is obtained from the shoots.* Pl. 2.

C. macrocarpa Hartweg MONTEREY CYPRESS. Very like 11, but distinguished by the longer scale-lvs. 1-2 mm., the smaller male cones 3-5 mm. (4-8 mm. in 11) and brown ripe female cones. A pyramidal tree when young, flat-topped in old trees. Native of North America; sometimes planted for ornament or timber in Europe.

Thuja plicata Lamb. WESTERN RED CEDAR. Lvs. variable in shape, 3-7 mm., scale-like, shiny, dark yellow-green, overlapping in pairs and pressed to opposite sides of the branchlet, with a strong resinous aroma when crushed. Cones 1-1$\frac{1}{2}$ cm. with 8-10 oval, thin, spine-tipped scales. Native of North America; widely planted for timber in Europe.

Chamaecyparis lawsoniana (A. Murray) Parl. LAWSON CYPRESS. Lvs. variable in shape, 2-7 mm., acute scale-like, bright green or glaucous green, with conspicuous glands on back, pressed close to the stem, with a strong aroma of parsley when crushed. Cones globular *c.* 2 cm., red-brown, often glaucous, with 8 flat-topped scales, each one- to five-seeded. Native of North America; widely planted for timber in Europe.

JUNIPERUS | **Juniper** Lvs. needle-like or scale-like. Cones forming a globular, somewhat fleshy, berry-like fr. 10 sps.

All lvs. sharp-pointed, needle-like

12. *J. communis* L. JUNIPER. A dense, rather columnar, silvery-grey shrub or small tree 1-6 m. Lvs. up to 2 cm., sharp-pointed, with a single whitish band on the upper surface, green and keeled beneath. Fr. globular 6-9 mm., at first green, then bluish-black when ripe in the second or third year. Subsp. *nana* Syme DWARF JUNIPER is usually a dense, spreading, mat-like shrub of mountains and northern regions. Lvs. smaller, 1-1$\frac{1}{2}$ cm., close-set, upcurved or upturned. △ Mountains, dry hills, woods, moors, heaths. May–June. All Eur. *The fruits are used for flavouring gin and sometimes as a condiment; they give Juniper Oil.* Pl. 2.

13. *J. oxycedrus* L. PRICKLY JUNIPER. Like 12, but a Mediterranean shrub or small tree with sharp-pointed lvs. with 2 white bands on upper surface. Fr. larger 6-15 mm., yellow, becoming reddish when ripe. △ Stony ground, bushy places, dry hills. May. Med. Eur. P. *Yields Oil of Cade, which is used medicinally. The wood is fine-grained and durable and was used in the past for carving.* Pl. 2.

Adult lvs. flattened, scale-like, not more than 2 mm. long

14. *J. phoenicea* L. PHOENICIAN JUNIPER. A dense, dark green shrub or small tree 4–8 m., of the Mediterranean region. Lvs. *c.* 1 mm., oval, scale-like, blunt, with a glandular furrow on back and with a papery margin, overlapping in 4–5 ranks; lvs. of young plant needle-shaped. Fr. 8–14 mm., first blackish, then greenish-yellow, and finally dark red when ripe in second year. △ Rocky places, dry hills. Feb.–Apr. Med. Eur. (except TR.) P. *The wood is very durable and used for cabinet-making, etc.; it has a disagreeable odour.* Pl. 2.

15. *J. sabina* L. SAVIN. A low, bushy, spreading shrub 2–3 m., of the mountains, with dark green, scale-like, and loosely overlapping lvs. in 4–6 ranks. Lvs. 1–1½ mm., flattened egg-shaped with elliptic gland on back and without a papery margin. Fr. 4–6 mm., bluish-black, bloomed. △ Alpine meadows and rocks. Apr. Cent. Eur. (except H.) E.F.I.YU.BG.R.SU. *Yields Oil of Savarin which is poisonous and has been used medicinally.*

16. *J. thurifera* L. SPANISH JUNIPER. Like 15, but a small pyramidal tree to 15 m. or more. Scale-like lvs. lance-shaped 1½–2 mm., bluish-green, with tip free from stem (not fused to stem as in 15). Fr. larger, 7–8 mm., dark purple when ripe. △ Limestone mountains. Feb.–Mar. E.F.

TAXACEAE | Yew Family

Evergreen trees or shrubs without resin. Lvs. linear, flattened, spirally arranged. Male cones short-lived, small axillary; female fls. not cone-forming, but ovules borne singly. Seeds partly or wholly surrounded by a brightly coloured fleshy investment.

TAXUS | **Yew**   Lvs. in 2 ranks. Male cones of 6–14 scales. 1 sp.

17. *T. baccata* L. YEW. A dense, dark green shrub or pyramidal tree 5–20 m. with stout trunk and dark, reddish-brown, scaly bark. Lvs. 1–3 cm., dark glossy green above, paler beneath, shortly stalked. Fr. with bright red, fleshy aril encircling a single dark green seed. △ Woods, rocks, screes, largely in mountains; often grown as an ornamental. Apr. All Eur. (except IS.). *All parts of the tree are poisonous to man and livestock. The poisonous substance is taxine. Livestock readily eat yew, and the result is often sudden death within a few minutes. The wood is very compact and durable, and is used for cabinet-making, bows, etc.* Pl. 2.

EPHEDRACEAE | Joint-Pine Family

Small, straggling, switch-like plants with branched stems and lvs. commonly reduced to scales. Male 'cones' in small axillary clusters; female fls. of several pairs of bracts and a terminal ovule. Fr. a fleshy berry.

EPHEDRA | **Joint-Pine**   Male 'cones' with a few two- or three-celled stamens; fr. red or yellow. 3 sps.

18. *E. fragilis* Desf. JOINT-PINE. A scrambling or spreading shrub to 5 m. of the Mediterranean

region, with greenish stems and flexible brittle branchlets. Lvs. scale-like, 2 mm., green on back. Fr. globular 8–9 mm., with red fleshy bracts almost enclosing seed. Subsp. *campylopoda* (C. A. Meyer) Ascherson and Graebner has climbing stems with long hanging branchlets; it occurs in the Eastern Mediterranean region. △ Hedges, bushy places. May–June. Med. Eur. (except F.) P.BG. Pl. 2.

E. distachya L. A low shrub with rigid stems up to ½ m. Fr. 5–7 mm., with seed distinctly emerging from the red fleshy bracts. Southern Europe.

SALICACEAE |Willow Family

Deciduous trees and shrubs with simple lvs. with stipules. Plants unisexual; fls. clustered into catkins which are composed of many stalkless fls. borne in the axils of scaly bracts. Male catkins with numerous fls., each usually with 2 conspicuous projecting stamens. Female catkins with numerous fls., each with a two-valved carpel. Seeds with long silky hairs. Most sps. hybridize with one or more sps., and intermediate forms are common, making identification of individual plants often difficult.

Buds with 1 outer scale; bracts of catkins entire.   *Salix*
Buds with several outer scales; bracts of catkins toothed or fringed.   *Populus*

SALIX |Willow   Bracts of catkins entire; stamens usually 2. Buds with 1 encircling scale. Seeds with long silky hairs. Willows occur in all kinds of damp localities such as marshes, streamsides, lakesides, fens and damp woods in the lowlands, and on damp slopes, wet rocks, and screes in the mountains. Care must be taken to select mature leafy shoots, not sucker-shoots, for identification. 70 sps.

Creeping shrubs or shrublets, or small erect shrubs usually less than 1½ m.
19. *S. reticulata* L. RETICULATE WILLOW. A lowly, creeping, woody shrublet 10–20 cm., with hairy young branches and dark green leathery lvs. Lvs. 1–4 cm., rounded-oval, wrinkled with deeply recessed veins above, much paler, almost whitish, with prominent net veins beneath, silky-haired when young. Catkins 1–3 cm., long-stalked, appearing after lvs. △ Mountains or arctic regions. May–July. Much of Eur. (except P.B.NL.IS.DK.H.AL.GR. TR.). Pl. 3.

20. *S. retusa* L. BLUNT-LEAVED WILLOW. Like 19, but lvs. smaller, 8–20 mm., oblong to oval, yellow-green, smooth, hairless, and shiny above and below, margin entire or minutely toothed. Branches hairless. Catkins 2 cm., short-stalked, greenish or yellow, appearing with lvs. △ High mountains. June–July. Cent. Eur. (except CS.H.), South-East Eur. (except TR.) E.F.I.SU. Pl. 3.

S. herbacea L. LEAST WILLOW. It has bright shining green, orbicular lvs. 1–2 cm. with prominent veins and rounded teeth. Stems woody, creeping below surface. High alpine or arctic regions. Page 45.

21. *S. myrsinites* L. agg. MYRTLE-LEAVED WILLOW. A low, spreading shrub 10–40 cm., with thick knotted stems and shiny green lvs. 1½–5 cm., netted with veins and with long straight hairs which persist mainly on the margin. The withered remains of the previous year's lvs. remain in northern forms. Catkins large, *c.* 4 cm., dark purple. △ Mountains or arctic regions. June–Aug. North Eur. (except IS.) E.F.GB.D.A.PL.CS.I.YU.R.SU. Page 45.

*S. *lapponum* L. DOWNY WILLOW. Alpine or arctic regions.

S. *helvetica* Vill. Lvs. shiny greenish on upper surface and contrastingly dense white-felted beneath. Ovary white-felted. Alpine regions of Central Europe.

22. S. *hastata* L. A compact shiny-branched shrub to 1½ m. Lvs. variable, elliptic to obovate, pale dull green above, paler beneath, becoming hairless when mature; veins netted but not prominent. Catkins appearing with lvs.; fr. hairless. Very variable, particularly in lv. shape. △ Arctic or alpine. June–July. Much of Eur. (except P.IS.IRL.GB.B.NL.H. AL.GR.TR.BG.). Page 45.

23. S. *repens* L. CREEPING WILLOW. The only widespread lowland creeping willow, 15–50 cm., which is readily distinguished by the silky hairs on the lvs. Lvs. usually 3–5 times as long as broad but very variable, older lvs. becoming more or less hairless above. Catkins appearing before lvs., small to 2½ cm.; ovary densely hairy, style short. △ Bogs, swamps, marshes. Apr.–May. Much of Eur. (except IS.H.AL.GR.TR.BG.R.). Page 45. Pl. 3.

Erect shrubs more than 1½ m., or trees
(a) Lvs. at least 3 times as long as broad
(i) Catkins appearing with lvs. on leafy side branches
24. S. *fragilis* L. CRACK WILLOW. A rather large tree 10–25 m., with wide-spreading branches in an open crown. Lvs. lance-shaped, shiny green above and with a whitish or sometimes greenish bloom beneath, soon hairless, margin coarsely toothed, tip of lv. long-pointed and often asymmetrical. Twigs olive-green, very fragile at junctions. Catkins 3–7 cm., drooping. △ Lowlands. Apr.–May. All Eur. (except IS.SF.AL.).

S. *triandra* L. ALMOND WILLOW. Distinguished by the 3 stamens in each fl. Lvs. dark green above and blue-green or paler beneath. A shrub or small tree. Widespread in Europe. Page 45.

S. *pentandra* L. BAY WILLOW. Distinguished by the 5 stamens in each fl. Lvs. dark glossy green, and sticky and aromatic when young, elliptic long-pointed, margin finely toothed. Buds sticky. Shrub or small tree. Widespread in Europe. Page 45.

25. S. *alba* L. WHITE WILLOW. A rather large tree 10–25 m., with upward ascending branches forming a narrow crown, and silvery-grey foliage. Lvs. lance-shaped, with adpressed silky hairs on both surfaces, margin finely toothed, tip of lv. more or less symmetrical. Twigs silky when young, olive-green later. Catkins 3–6 cm. △ Lowlands. Apr.–May. Most of Eur. (except North Eur.): introd. N.S.SF. *The wood is light and firm and is used for making cricket bats, baskets, etc. It yields salicine, which has been used medicinally as a substitute for quinine.* Pl. 3.

S. *babylonica* L. WEEPING WILLOW. Distinguished by its long slender drooping branches and linear-lance-shaped lvs. which are more or less hairless at maturity. Widely cultivated as an ornamental, but often as a hybrid.

(ii) Male catkins appearing before lvs.
26. S. *viminalis* L. COMMON OSIER. Usually a shrub 3–4 m., but sometimes a small tree, with flexible, greyish, hairy branches. Lvs. long, linear-lance-shaped, strongly veined, dark green and hairless above, white and silky-haired beneath, margin inrolled. Buds hairy. Catkins cylindrical-ovoid. △ Lowlands. Mar.–Apr. Cent. Eur. F.GB.B.NL.YU.GR.BG. SU.: introd. North Eur. P.E.I. *Widely cultivated for basket-making; other species such as Crack, White, and Purple Willow are also used, but are less valued for this purpose.*

27. S. *purpurea* L. PURPLE OSIER. A slender shrub 1½–5 m. with purplish shining branches which later become greenish or yellowish-grey. Lvs. very variable, 3–15 times as long as

1. *Salix herbacea* [20]
4. *S. myrsinites* 21
7. *S. aurita* [28]

2. *S. pentandra* [24]
5. *S. repens* 23
8. *S. triandra* [24]

3. *S. purpurea* 27
6. *S. hastata* 22

45

broad, very finely toothed, hairless, dull and somewhat bluish-green above, paler and often bluish-green beneath, often blackening on drying. Catkins 2–4½ cm.; stamens 2, but fused together as 1; anthers purple. △ Lowlands. Mar–Apr. Most of Eur. (except North Eur. AL.): introd. DK.N.S. *Cultivated for basket-making.* Page 45.

(*b*) *Lvs. mostly less than 3 times as long as broad and densely hairy beneath*
28. *S. cinerea* L. COMMON SALLOW. A dense, bushy shrub or small tree to 10 m., with oblong-lance-shaped lvs. which are broadest towards the apex. Lv. blades 2½–7 cm., 2–4 times as long as broad, usually greyish, woolly-haired beneath, often becoming more or less hairless above; stipules persistent. Twigs with dense grey hairs. Catkins almost stalkless, appearing before lvs. △ Lowlands. Mar.–Apr. Most of Eur. (except IS.CH.BG.).

S. aurita L. EARED SALLOW. Like 28, but a smaller shrub to 2 m. with smaller obovate lvs. 2–3 cm., with a twisted, pointed tip, rough and dull grey above, woolly-grey beneath; stipules conspicuous. Young branches brown, soon hairless. Widespread in Europe. Page 45.

29. *S. caprea* L. GREAT SALLOW, GOAT WILLOW. A small tree or shrub 3–10 m., with large lvs. 5–10 cm. broadly oval to oblong, *c.* 1½–2 times as long as broad, with a twisted tip and rounded or heart-shaped base, prominent-veined, dark green, finally hairless above, softly hairy beneath. Twigs stout, finely hairy at first, but becoming hairless. Catkins appearing on bare stems before lvs. △ Lowlands. Mar.–Apr. All Eur. (except P.IS.AL.). *The bark contains salicine which was used as a substitute for quinine; it is rich in tannin and used in curing leather.* Pl. 3.

POPULUS | Poplar   Bracts of catkins toothed or deeply divided; male catkins with 8–18 stamens. Buds with several scales. Seeds with white silky hairs. Many hybrids are cultivated for wood-pulp throughout Europe. 11 sps.

Bark smooth, grey
(*a*) *Lvs. of mature stems downy-haired beneath*
30. *P. alba* L. WHITE POPLAR. Lvs. dark green above and densely white downy-haired beneath, oval with 3–5 deep, coarsely toothed lobes. Buds and young twigs covered with dense, white, woolly hairs. A freely suckering tree with wide-spreading crown, to 30 m. △ Woods, watersides. May–Apr. Cent., South-East Eur. I.SU.: introd. P.E.F.GB.B.NL. DK.CH. *The wood of the poplars is finely textured, tough, and withstands abrasion well; it is used for floors, bottoms of trucks, etc. Also pulped for paper-making; the bark contains a yellow dye.*

31. *P. canescens* (Aiton) Sm. GREY POPLAR. Like 30, but lvs. of terminal and sucker-shoots coarsely toothed, not or shallowly lobed, with persistent grey, not white, downy hairs beneath; lvs. of short-shoots grey-haired beneath when young, but becoming nearly hairless. Buds and twigs thinly hairy. A larger tree to 35 m. △ Damp woods, riversides. Mar.–Apr. Cent. Eur. (except CS.) F.IRL.GB.B.NL.I.BG.R.SU.: introd. P.E.DK.S.GR.

(*b*) *Lvs. of mature stems hairless*
32. *P. tremula* L. ASPEN. A freely suckering tree 13–20 m., with lvs. fluttering in the wind, more or less circular in outline, with an irregularly bluntly toothed or wavy margin, hairless, paler beneath; lv. stalks strongly flattened. Lvs. of sucker-shoots with greyish silky hairs. Twigs hairless; buds slightly sticky, ciliate; young bark smooth, greyish white, later rugged dark grey. Male catkins reddish, appearing before lvs. △ Woods. Mar.–Apr. All Eur. (except P.). *The wood is used particularly in match-making and for high-quality paper.* Pl. 3.

Bark fissured, black

33. *P. nigra* L. BLACK POPLAR. A tall, long-trunked tree to 35 m. with deeply fissured, nearly black bark. Lvs. broadly oval or rhomboidal, finely toothed, and prolonged into a long, pointed, toothless tip, blade hairless, shining. Young twigs and buds sticky and hairless; branches spreading and arching downwards. Var. *italica* Duroi LOMBARDY POPLAR has a narrow pyramidal form and paler fissured bark. △ Woods, watersides. Mar.–Apr. Cent., South-East Eur. F.NL.I.SU.: introd. P.E.IRL.GB.B.DK.

*P. × canadensis Moench (*P. deltoides × nigra*) BLACK ITALIAN POPLAR has many cultivated forms, some of which are widely planted in Europe.

P. deltoides Marshall EASTERN COTTONWOOD. Distinguished by its very large lvs. 10–18 cm., which are triangular to heart-shaped, slender-pointed, irregularly coarsely toothed, bright green above, paler beneath, margin ciliate; lv. stalks flattened. A tall tree with dark grey, deeply fissured bark and shiny resinous buds. A native of North America; often planted for timber in Europe.

*P. gileadensis Rouleau BALM OF GILEAD. Sometimes planted for timber in Europe and sometimes naturalized; origin unknown.

MYRICACEAE | Bog Myrtle Family

Trees or shrubs with simple, glandular aromatic lvs. Fls. usually in one-sexed, erect catkins consisting of numerous overlapping bracts each subtending a fl.; male fls. with 2–16 stamens; female fls. with a solitary, one-seeded ovary; styles 2. Fr. a dry waxy-coated drupe. 1 genus. 3 sps.

MYRICA | Bog Myrtle

34. *M. gale* L. BOG MYRTLE, SWEET GALE. A densely branched, deciduous shrub ½–2½ m., with reddish-brown branches and buds, and very aromatic, grey-green, oblance-shaped lvs. 2–6 cm. long. Catkins ½–1½ cm., erect, in lateral clusters on previous year's stems. Plants mostly one-sexed. △ Bogs, wet heaths. Apr.–May. West, North Eur. (except IS.S.) D.PL.SU. *The bark has been used for tanning, the dried fruit as a condiment, and the leaves as an insect repellent.* Pl. 4.

JUGLANDACEAE | Walnut Family

Trees with alternate, pinnate lvs. Fls. unisexual; male fls. usually in drooping catkins, of numerous bracts each subtending 3–40 almost stalkless stamens; female spikes terminal, erect, few-flowered, each with a solitary inferior one-seeded ovary; stigmas 2. Fr. a drupe or nut.

JUGLANS | Walnut Catkins solitary; lvs. aromatic; fr. large. 3 sps.

35. *J. regia* L. WALNUT. A large deciduous tree with pale grey smooth bark, which later becomes fissured, and large pinnate lvs. which are aromatic when crushed. Leaflets

7–9, obovate-elliptic, each 6–15 cm. long. Buds blackish; twigs grey or greenish. Male catkins 5–15 cm., pendulous, appearing with young lvs. Fr. 4–5 cm. ovoid, green, glandular; nut wrinkled. △ Forests, but widely cultivated and often naturalized elsewhere. Apr.– May. South-East Eur. (except TR.): introd. P.E.F.GB.CH.A.H.I.SU. *The wood is hard and homogeneous and much valued by cabinet-makers. The nuts are important as food and also produce a valuable edible oil, which is also used by painters and in soap-making. The fleshy outer part of the fruit is very rich in vitamin C, and gives a good yellow dye.* Pl. 4.

J. nigra L. BLACK WALNUT. Leaflets 11–23, with toothed margins; bark dark brown. A native of North America; often planted for timber.

BETULACEAE | Birch Family

Deciduous trees or shrubs with simple alternate lvs. Fls. in one-sexed catkins formed of numerous overlapping fertile bracts. Male catkins drooping, with fls. in clusters of 3 in the axils of each bract and 2–4 stamens in each fl. Female catkins erect, cylindrical, with 2–3 fls. in axil of each bract; ovary one-seeded; styles 2. Fr. borne in a dense cylindrical or cone-like catkin of scales formed from the fused bracts; nutlets winged.

Stamens 2; scales of female catkin three-lobed and falling with the nutlets.　　*Betula*
Stamens 4; scales five-lobed, forming a cone-like catkin which persists after the nutlets have fallen.　　*Alnus*

BETULA | Birch　Male fls. with 2 stamens which are fused below the anthers; fr. catkins fragmenting. Fls. appearing with lvs. 4 sps.

Erect trees

36. *B. pendula* Roth (*B. verrucosa* Ehrh.) COMMON SILVER BIRCH. A small to medium-sized tree to 25 m., with slender silvery-white upper branches and a black fissured lower trunk. Male catkins 3–6 cm., drooping; female 1½–3 cm., erect. Lvs. 2½–7 cm. oval to delta-shaped, with apex narrowed to an acute point, hairless, margin double-toothed with prominent primary teeth and finer secondary teeth. Young twigs hairless, but warty, slender and pendulous. △ Woods, heaths. Apr.–May. Much of Eur. (except IS.AL.GR.TR.): introd. P. Pl. 4.

37. *B. pubescens* Ehrh. BIRCH. Very like 36, but bark grey, with lower trunk more or less smooth throughout and not deeply fissured. Lvs. acute, usually with hairs in axils of veins beneath, irregularly toothed, but without prominent primary teeth. Young twigs usually not pendulous, commonly hairy, dark grey-brown, with or without warts. Hybrids with 36 are common. △ Woods, heaths. Apr.–May. Most of Eur. (except NL.AL. GR.TR.BG.). *Both species yield Birch Tar Oil which is used for the preparation of Russia leather and gives it its peculiar smell; it is also an insect and fungal repellent. The wood is soft and light and used for making furniture and small turned articles; the twigs are used for thatching and for broom- and wattle-making; the bark for tanning. Alcoholic beverages are prepared from the sap in northern countries.*

Low shrubs

38. *B. nana* L. DWARF BIRCH. A low, spreading shrub rarely more than 1 m., of high mountains or arctic regions, with blackish-purple branches. Lvs. small ½–1½ cm., thick,

rounded-oval and margin with regular conspicuous rounded teeth. Young twigs and lvs. hairy, becoming hairless. Fr. catkins erect 5–10 mm. △ Bogs, moorlands, glacial valleys. May. North Eur. (except DK.), Cent. Eur. (except H.) F.GB.R.SU. Pl. 4.

ALNUS | Alder Male fls. with 4 stamens; female catkins woody and cone-like, not fragmenting when frs. are shed. Fls. usually appearing before lvs. 5 sps.

Catkins appearing with the lvs.
39. *A. viridis* (Chaix) DC. GREEN ALDER. A densely growing mountain shrub to 2½ m. with smooth branches and catkins borne with the lvs. Catkins stalked, on short side branches with 2–3 lvs. at the base; male and female catkins borne in separate clusters. Lvs. round-elliptic, finely toothed, sticky when young; buds stalkless. Fr. broadly winged. △ In mountains: woods, pastures. Apr.–June. Cent., South-East Eur. (except AL.GR.TR.) F.I.SU.

Catkins appearing before the lvs.
40. *A. glutinosa* (L.) Gaertner ALDER. A medium-sized tree to 20 m. or more, with dark brown fissured bark and with male and female catkins borne on the same branches and maturing before the lvs. Lvs. rounded, with a cut-off or notched apex, with 5–8 pairs of lateral veins, bright green and usually hairless, margin often double-toothed. Young twigs sticky; buds shortly stalked. △ Marshes, watersides. Feb.–Apr. All Eur. (except IS.). *The wood is light and easily worked and used for general turnery, brush-backs, etc. The charcoal is used in gunpowder-making and the bark for tanning. The leaves are used in herbal remedies.* Pl. 4.

41. *A. incana* (L.) Moench GREY ALDER. Like 40, but a smaller tree or shrub to 10 m., with smooth, pale grey bark and finely hairy lvs., greyish-green beneath, acute-tipped, with 7–12 pairs of lateral veins. Young twigs hairy, not sticky. Fr. catkins 3–5, 1–1½ cm., almost stalkless. △ Watersides. Feb.–Mar. North Eur. (except IS.), Cent., South-East Eur. (except GR.TR.) F.SU.: introd. IRL.GB.DK. Page 51. Pl. 4.

A. cordata (Loisel.) Loisel. Distinguished from 41 by its usually hairless lvs. with 6–8 pairs of lateral veins, often heart-shaped at base and with blunt or short-pointed apex and rather rounded teeth. Fr. catkins 1–3, 1½–3 cm. South-Western Europe.

CORYLACEAE | Hazel Family

Deciduous trees or shrubs with simple alternate lvs. Fls. in one-sexed catkins; male fls. numerous in pendulous catkins, each fl. solitary in the axil of a bract. Female catkins erect or drooping, with 2 fls. in axil of each bract; ovary one-seeded; styles 2. Fr. a nut surrounded by a leaf-like bract or involucre.

1. Buds ovoid, blunt; fr. large, in stalkless clusters of 1–4.   *Corylus*
1. Buds long, narrow, tapering; fr. small, several in hanging cluster.
  2. Bark grey, smooth; nut partly encircled by a lobed or toothed bract.   *Carpinus*
  2. Bark brown, rough; nut completely enclosed in a bract.   *Ostrya*

CARPINUS | Hornbeam Female fls. in drooping catkins; fr. a small nut partly encircled by a long, lobed or toothed, leaf-like bract. Buds spindle-shaped, acute. 2 sps.

42. *C. betulus* L. HORNBEAM. A medium-sized tree to 25 m., with a fluted trunk and smooth

grey bark and many acutely ascending branches. Lvs. 4–10 cm. oval, long-pointed, double-toothed, hairy on the veins beneath. Catkins appearing with the lvs.; male catkins hanging, solitary; fr. catkins 5–14 cm., stalked, hanging lax; fr. subtended by a three-lobed bract with the middle lobe much longer than the lateral lobes. △ Woods. Apr.–May. Cent., South-East Eur. F.GB.B.DK.S.I.SU. *The wood is heavy and close-grained and is resistant to abrasion and used for floors, utensils, etc.; the bark is used in tanning.* Pl. 5.

C. orientalis Miller ORIENTAL HORNBEAM. Fr. catkins smaller 3–5 cm.; bract of fr. triangular-oval, regularly toothed, not three-lobed. South-Eastern Europe. Page 51.

OSTRYA | Hop-Hornbeam Like *Carpinus*, but nuts each enclosed by a conspicuous sac-like bract. Bark rough. 1 sp.

43. *O. carpinifolia* Scop. HOP-HORNBEAM. A small tree to 15 m., with lvs. very like 42, but with rough, brown, scaly bark. Catkins appearing with the lvs.; male catkins very long to 10 cm.; fr. catkins 3–4½ cm., recalling the fr. of the hop, pendulous, cylindrical, with overlapping, pointed, whitish bracts completely enclosing the tiny nuts. △ Woods. Apr.–May. South-East Eur. (except R.) F.CH.A.I. Page 51.

CORYLUS | Hazel Female catkins small, bud-like, and with projecting styles; fr. a large nut with irregularly cut investing bracts. Buds ovoid, blunt. 3 sps.

44. *C. avellana* L. HAZEL, COB-NUT. A shrub or small tree to 6 m., with smooth brown bark and glandular-hairy twigs. Lvs. 5–12 cm. orbicular, long-pointed, hairy on both surfaces, margin double-toothed. Catkins appearing before lvs.; male catkins 2–8 cm. pendulous, in clusters of 1–4; female *c.* 5 mm., bud-like, erect. Fr. in clusters of 1–4, nuts 1½–2 cm., brown and invested by deeply lobed, irregularly toothed bracts as long as the nut. △ Woods, hedges, ravines. Jan.–Mar. All Eur. (except IS.). *The nuts are of some importance as food. The wood is used for hurdles, wattles and small articles.*

C. maxima Miller Very like 44, but bracts investing nut tubular, longer than nut, and constricted above, with a toothed apex. Balkans; widely cultivated for its fr. elsewhere in Europe.

FAGACEAE | Beech Family

Trees with spirally arranged, simple lvs. Fls. unisexual, usually in separate clusters; male fls. in catkins or tassel-like heads; female fls. solitary, or in threes in a short erect spike, or at the base of the male catkins. Ovary inferior, three- or six-celled; styles 3 or 6. Fr. a solitary nut or up to 3 nuts, surrounded or enclosed by a scaly or woody cup of fused bracts.

1. Nut triangular-sectioned, encircled by a spiny or scaly cup; male catkins pendulous. *Fagus*
1. Nut rounded-sectioned.
  2. Nut enclosed in spiny cup; male catkins erect. *Castanea*
  2. Nut half-encircled in spiny or scaly cup; male catkins pendulous. *Quercus*

FAGUS | Beach Male fls. in spherical, hanging, tassel-like heads. Fr. of 1–2 angular nuts enclosed in a woody, spiny, or scaly, four-valved cup. 2 sps.

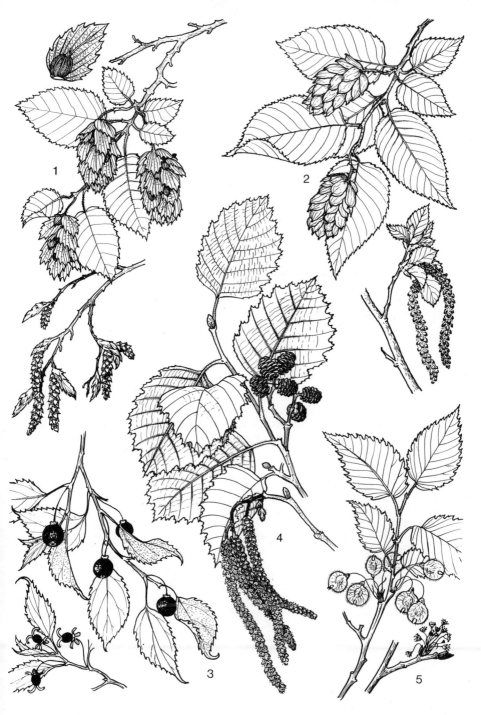

1. *Carpinus orientalis* [42]
2. *Ostrya carpinifolia* 43
3. *Celtis australis* 60
4. *Alnus incana* 41
5. *Ulmus laevis* 59

45. *F. sylvatica* L. BEECH. A tall deciduous tree to 30 m., with smooth grey trunk, smooth branches, and slender, brown, spindle-shaped buds. Lvs. 4–9 cm. oval-acute, with 5–8 pairs of conspicuous lateral veins which are silky-haired beneath, otherwise hairless, somewhat shining. Male heads long-stalked, tassel-like. Nuts 1–2 cm., brown, sharp-angled, encircled by a woody cup which is deeply divided into 4 valves which are covered outside with awl-shaped spines. △ Woods, often in mountains. Apr.–May. All Eur. (except P.IS.SF.TR.): introd. IRL. *The wood is fine-grained and used for many purposes such as furniture, flooring, turnery, etc. The nuts or 'mast' are edible and oil for cooking and light is obtained from them.*

F. orientalis Lipsky ORIENTAL BEECH. Very like 45, but lvs. with 9–12 pairs of lateral veins; cup scaly, not spiny, upper scales linear-oblong, lower spathulate. South-Eastern Europe.

CASTANEA | **Sweet Chestnut** Catkins erect, of one kind only, the male fls. numerous above, the female fls. few at the base. Fr. of 1–3 rounded nuts enclosed in a spiny cup which splits into 2–4 valves. 2 sps.

46. *C. sativa* Miller SWEET or SPANISH CHESTNUT. A tall deciduous tree to 30 m., with a stout longitudinally, often somewhat spirally furrowed trunk, and large, entire, con-spicuously veined and toothed lvs. Lvs. 10–25 cm., oblong-lance-shaped, hairless above, teeth bristly-tipped. Catkins 12–20 cm., slender erect, conspicuous, yellow. Nut smooth shiny brown, enclosed in a green densely spiny cup. △ Woods in mountains. May–July. South-East Eur. H.I.: introd. to most of Eur. (except North Eur.). *The nuts are of consider-able importance as food for humans; also used to fatten livestock. The wood is easily worked and is valuable for stakes, barrels, furniture, etc.; sometimes coppiced on a 15-year rotation for small poles, fencing, etc.* Pl. 5.

QUERCUS | **Oak** Lvs. deciduous, or persisting through winter and falling in spring and thus semi-evergreen, or evergreen. Catkins one-sexed; male catkins long pendulous; female catkins bud-like, few-flowered. Fr. with a spiny or scaly cup encircling the lower part of the nut. 27 sps.

Evergreen trees
(a) Mature lvs. hairless beneath
47. *Q. coccifera* L. KERMES OAK, HOLLY OAK. A dense shrub to 2 m. or occasionally a small tree, with rigid spiny-margined, holly-like, dark green, hairless lvs. Lvs. 1½–4 cm. oval to oblong, paler and quite hairless at maturity beneath, but hairy when young. Fr. ripening in second year; cup usually with spreading, spiny-tipped scales. △ Dry stony slopes, bushy places. Apr.–May. Med. Eur. P.BG. *The bark is rich in tannin and used for curing leather. An insect which lives on the leaves yields a red dye, which has been known since classical times, and is the 'scarlet' of the Scriptures.* Pl. 5.

(b) Mature lvs. hairy beneath
48. *Q. ilex* L. HOLM OAK. A medium-sized to large evergreen tree to 25 m., or rarely a shrub, with grey scaly bark and grey-felted buds and twigs. Lvs. 3–7 cm. oblong-lance-shaped, thick, not rigid, conspicuously grey-felted beneath and contrastingly dark green and at length hairless above, lv. margin entire, toothed or sometimes spiny-toothed, particularly on sucker-shoots; midvein of blade straight. Fr. bitter, ripening in first year; cup with short acute, but not spiny-toothed, closely adpressed scales. △ Dry stony slopes, bushy places, hills. Apr.–May. Med. Eur. P.: introd. GB.CH. *Yields a hard timber used for rough constructional work and by wheelwrights and joiners. The bark is used for tanning; produces excellent charcoal.*

Q. rotundifolia Lam. Commonly replaces 48 in the Western Iberian Peninsula. Distinguished by broadly oval to rounded, thin, papery lvs. which are greyish-glaucous above; fr. sweet.

49. *Q. suber* L. CORK OAK. Like 48, but the trunk is covered with extremely thick, rough, deeply fissured grey bark, revealing, when stripped, a pinkish or reddish-brown underbark. Lvs. 3–7 cm., dark green above, grey-felted beneath, margin entire or wavy; midvein of blade wavy. Fr. ripening in first or second year; scales of cup becoming progressively longer and more spreading from base to rim. △ Dry hills. Apr.–May, and autumn. Med. Eur. (except AL.GR.TR.) P. *The main source of commercial cork, which is cut from the trunk and lower branches every 8–10 years. The wood is of little value except for fuel.*

Semi-evergreen trees of South-Eastern Europe
50. *Q. macrolepis* Kotschy (*Q. aegilops* auct.) VALONIA OAK. A stout-trunked tree 5–15 m., with lvs. 6–10 cm., grey-felted beneath and dull dark green above, with 3–7 deep triangular, long-pointed lobes, which are often further lobed. Fr. ripening in the second year; cup large 2–4 cm. across with wide, flat, spreading, often recurved scales. △ Dry hills. Apr.–May. I.YU.AL.GR.TR. *The large acorn cups are much valued for tanning, dyeing, and ink-making. The acorns have been eaten by man in times of famine.* Pl. 5.

Q. trojana Webb MACEDONIAN OAK. Distinguished by its shining elliptic lvs., almost hairless on both surfaces and its wavy margin, with 8–14 pairs of fine-pointed teeth. Cup large 2–2½ cm. across, with lowest scales adpressed, the middle deflexed and the upper erect or incurved; nut to *c.* 3 cm. long. Balkans. Page 54.

Deciduous trees
(a) *Fr. situated on last year's stems and not among the present year's lvs.*
51. *Q. cerris* L. TURKEY OAK. Lvs. usually with 7–8 pairs of deep and narrow, more or less acute lobes, dull green and somewhat rough above, paler beneath, but very variable. Buds surrounded by narrow, long-pointed, persistent stipules. Fr. ripening in the second year; cup 1–1½ cm. across or more, with long-pointed spreading scales. A robust tree to 35 m., with dark grey fissured bark. △ Hedges, thickets, woods. Apr.–May. Cent. Eur. (except PL.), South-East Eur. F.I.: introd. GB.B. *The wood is of little use as it does not weather well; it is used for rough furniture, vine props, and heating.* Page 54.

Q. rubra L. AMERICAN RED OAK. Distinguished by its very large oval lvs. 12–20 cm. by 10–15 cm., deeply lobed to about halfway into 7–11 pairs of lobes, each with 1–3 bristle-tipped teeth. Twigs dark red, hairless. Cup 1½–2½ cm. wide, scales adpressed, finely hairy. Native of North America; often planted for timber and shelter in Europe.

(b) *Fr. situated on present year's stems and among lvs.*
52. *Q. robur* L. COMMON OAK, PEDUNCULATE OAK. A large tree to 45 m., with a stout trunk with brownish-grey, deeply fissured bark and large, widely spreading branches. Lvs. 5–12 cm., obovate, with 5–7 pairs of deep rounded lobes, and heart-shaped auricled base, hairless on both surfaces when mature. Fr. in a long-stalked cluster of 1–5; cup with oval adpressed and overlapping, finely hairy scales. △ Forests, particularly on heavier soils. Apr.–May. Much of Eur. (except IS.GR.TR.). Pl. 5.

53. *Q. petraea* (Mattuschka) Liebl. DURMAST OAK, SESSILE OAK. Like 52, but lvs. with a more or less wedge-shaped base, without auricles, and with the undersides with tufts of large, brownish, star-shaped hairs in the axils of lateral veins and midvein. Fr. 1–5 in stalkless or short-stalked clusters. Tree with upward-spreading branches, to 40 m. △ Forests, typically on lighter soils. Apr.–May. All Eur. (except P.IS.SF.GR.). *The wood of both species is hard and weathers well and is used for many purposes such as boat-building, furniture and carriage-making, houses, construction, etc. The bark is rich in tannin and is*

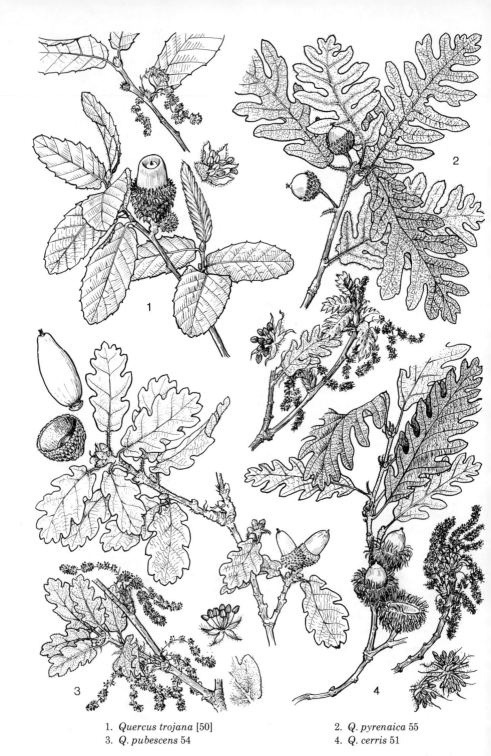

1. *Quercus trojana* [50]
2. *Q. pyrenaica* 55
3. *Q. pubescens* 54
4. *Q. cerris* 51

*used for dyeing and tanning. The acorns and leaves can be poisonous to cattle, but pigs
feed on the former. The galls contain tannin and are used in ink-making.*

54. *Q. pubescens* Willd. WHITE OAK. Like 52, but lvs. densely woolly-haired when young,
becoming hairless above but finely hairy below at maturity, and with narrower and
deeper lobes; lv. stalk 5–12 mm. Young twigs densely hairy. Fr. cluster stalkless or
shortly stalked; scales of cup lance-shaped, adpressed, softly grey-haired. A shrub or
tree to 25 m. △ Woods, dry hills. Apr.–May. Cent., South-East Eur. E.F.B.I.SU. Page 54.

55. *Q. pyrenaica* Willd. PYRENEAN OAK. A tree to 20 m., or large shrub, with large, dark-
green lvs. which are densely white-hairy beneath. Lvs. 8–20 cm., with 4–8 pairs of narrow
acute lobes; lv. stalk up to 22 mm. Young twigs woolly-haired. Fr. clusters stalkless or
short-stalked; scales of cup narrow, lance-shaped, blunt, loosely overlapping, grey-
haired. △ Mountain forests. May–June. P.E.F.I. Page 54.

ULMACEAE | Elm Family

Deciduous trees with alternate entire lvs. which are often asymmetrical at the base. Fls.
clustered on last year's twigs, bisexual or unisexual; perianth segments 4–8, somewhat
fused, often bell-shaped; stamens the same number as perianth; ovary of 2 fused carpels;
styles 2. Fr. winged and flattened, or a rounded drupe; seed 1.

Fr. dry, winged; lv. margin double-toothed.   *Ulmus*
Fr. fleshy, not winged; lv. margin single-toothed.   *Celtis*

ULMUS | Elm   Fls. appearing before lvs.; perianth four- or five-lobed. Fr. flattened and
encircled by a broad, papery wing. Bark fissured. Lvs. with very unequal basal lobes; the
angle of entry of the longer lobe to the lv. stalk is an important diagnostic character.
Hybridization and vegetative reproduction in hybrid plants makes identification difficult
in individual plants. 5 sps.

56. *U. glabra* Hudson WYCH ELM. A tall, smooth-barked tree to 40 m., often with 2–3 stout
branches arising near the base; not suckering. Lvs. medium to large 5–16 cm., broadly
oval-elliptic and long-pointed, irregularly saw-toothed, very rough on upper surface; lv.
stalk very short and more or less concealed by the longer basal lobe of the lv. Fr. orbicular,
notched, 1½–2 cm.; seed centrally placed. △ Woods, hedges, in the hills. Mar.–Apr. Most
of Eur. (except P.IS.TR.). *The wood of elms is particularly durable when under water or when
frequently wetted and consequently it is used for building boats, carts, coffins, farm buildings, etc.; it has a beautiful grain when polished and is also used for furniture. The inner
bark has been used for mats and rope-making.* Pl. 5.

57. *U. procera* Salisb. ENGLISH ELM. A tall tree to 30 m., with a stout, rough-barked trunk
and a few massive lateral branches; readily suckering. Lvs. 4–9 cm., more or less rounded
to broadly oval, sharp-toothed, rough above, hairy beneath; lv. stalk not concealed
by the longer half-heart-shaped basal lobe. Fr. orbicular, notched, 1–1½ cm.; seed placed
above middle and touching the notch. △ Roadsides, hedges. Feb.–Mar. E.F.GB.H.YU.GR.
BG.R.: introd. IRL.

58. *U. minor* Miller (*U. carpinifolia* G. Suckow) SMOOTH-LEAVED ELM. Like 57, but lvs.
narrower, obovate to oblance-shaped, shiny, usually smooth above and the longer basal
lobe of lv. joining the lv. stalk at a right angle. Fr. obovate, notched, 7–17 mm. long; seed

placed above the middle. Suckering tree to 30 m. △ Roadsides, thickets. Mar.–Apr. Most of Eur. (except IRL.IS.N SF.).

59. *U. laevis* Pallas (*U. effusa* Willd.) FLUTTERING ELM. Readily distinguished by the long-stalked fls. with stalks much longer than the perianth, and the long pendulous fr. stalks. Wings of fr. with conspicuous spreading hairs on the margin; seed centrally placed. Lvs. rounded to oval with 12–14 pairs of lateral veins, hairless or softly hairy beneath. A tree to 35 m. △ Damp ground in valleys. Mar.–Apr. Cent., South-East Eur. (except TR.) F.B.S. SF.SU. Page 51.

CELTIS Fr. a globular fleshy drupe. Fls. either male or bisexual, in leafy clusters appearing at the same time as the lvs. on young shoots. Bark smooth. 4 sps.

60. *C. australis* L. NETTLE TREE. A tree to 25 m., or a shrub, with smooth grey bark and slender, downy, flexible twigs. Lvs. 4–15 cm., oval-lance-shaped, very long-pointed, margin conspicuously and regularly sharp-toothed, rough above, soft hairy beneath. Fls. green, perianth five-lobed; anthers yellow. Fr. 9–12 mm. globular long-stalked, fleshy, sweet, brownish black, hairless. △ Hedges, banks, rocks. Apr. Med. Eur. (except GR. TR.) P.CH.BG.R. *Often planted for ornament and for its edible fruits. The bark gives a yellow dye and the wood is used for making small articles, in particular whip-handles and walking-sticks.* Page 51.

MORACEAE | Mulberry Family

Deciduous or evergreen trees or shrubs, with milky juice; lvs. simple, toothed or lobed; stipules 2. Fls. tiny, densely clustered, unisexual; perianth usually 4; male fls. with 1–4 stamens; female fls. with a one-seeded ovary, styles 3. Fr. an aggregate of fleshy carpels or an aggregate of dry carpels on a fleshy 'receptacle'.

Fr. blackberry-like, of several fleshy carpels.   *Morus*
Fr. pear-like and fleshy, containing dry carpels.   *Ficus*

MORUS | Mulberry Stipules not encircling bud. Fls. one-sexed, in separate clusters. Fr. blackberry-like with many fleshy carpels on a short axis. 2 sps.

61. *M. nigra* L. COMMON MULBERRY. A small tree to 20 m., with stout, rough, spreading branches. Lvs. 6–20 cm., oval to heart-shaped, coarsely toothed or lobed, rough above and softly hairy beneath. Fr. almost stalkless, dark red, acid until ripe. △ May. Native of Persia: introd. E.I.AL.GR.BG.R. *Ornamental, and grown for its edible fruits. The wood is used for joinery.*

M. alba L. WHITE MULBERRY. Like 61, but lvs. smooth and shiny above and almost hairless beneath except on veins. Fr. white, pink, or purplish, stalks as long as fr. Native of China; often grown in Europe as an ornamental tree, or as food for silkworms. Pl. 6.

FICUS | Fig Stipules encircling bud and leaving a circular scar. Fls. one-sexed, enclosed in a flask-like 'receptacle'. Fr. large, fleshy, pear-shaped, with the hard pip-like carpels borne on the inside. 1 sp.

62. *F. carica* L. FIG. A small tree to 3 m., or a shrub, with stout, smooth, grey branches,

yielding a milky juice when broken. Lvs. large, 10–20 cm. long and wide, usually with 3–7 deep, rounded lobes and with a heart-shaped base, dark green, rough above, hairy beneath. Fr. 5–8 cm., pear-shaped, dull green or purplish. △ Rocky places, thickets. June–Sept. Doubtfully native in Europe; much cultivated and widely naturalized in Southern Europe. *The dried fruits are an important food and a staple winter diet for some Mediterranean peoples. Figs are used medicinally as a laxative, and roast figs can be used as poultices for wounds, boils, etc. Half-ripe figs are said to be poisonous.*

CANNABACEAE | Hemp Family

Herbaceous plants, without milky juice; lvs. palmately lobed; stipules present. Plants one-sexed. Male fls. stalked, perianth 5, stamens 5; female fls. stalkless, perianth undivided and closely encircling the ovary; stigmas 2. Fr. dry, enclosed in the persistent perianth; seed 1.

Climbing perenns.   *Humulus*
Erect anns.   *Cannabis*

HUMULUS | Hop   Climbing perenns. Male inflorescences branched, pendulous; the female inflorescence stalked, dense, cone-like, with broad overlapping bracts. Fr. forming pendulous, cone-like clusters. 1 sp.

63. *H. lupulus* L. HOP. Stem green, not woody, rough, slender, climbing clockwise to 6 m. Lvs. opposite, 10–15 cm., broadly oval and mostly deeply three- to five-lobed, with toothed, long-pointed lobes. Fls. greenish; male fls. *c.* 5 mm. across; female cone-like inflorescence 1½–2 cm. and enlarging to 3–5 cm. in fr., with pale green, papery, overlapping oval bracts *c.* 1 cm. △ Hedges, damp places; widely naturalized from cultivation. June–Sept. All Eur. (except IS.). *Much cultivated for the aromatic, glandular, fruiting 'cones' which are used to give a bitter flavour to beer, and which help to preserve it. The stems produce a fibre similar to hemp; the leaves and flowers produce a brown dye.* Pl. 6.

CANNABIS | Hemp   Erect anns. with palmate lvs. Inflorescences in erect branched clusters, the male much-branched, the female spike-like. 1 sp.

64. *C. sativa* L. HEMP. A strong-smelling, glandular, stiffly erect ann. to 2½ m., with palmate lvs. and branched clusters of green fls. Lvs. stalked, with 3–9 narrow, lance-shaped, long-pointed, toothed segments. Male fls. with 5 pendulous stamens; female fls. with 2 long stigmas. Fr. globular, smooth. △ Hedges, damp places; widely cultivated and often a casual. June–Sept. SU.: introd. Cent., South-East Eur. (except AL.TR.) E.F.B.I. *An important source of fibre from which ropes, sails, etc., are made. The resin collected from the plant is narcotic and in various forms is known as bhang, hashish, or marijuana. The drug produces pleasant sensations and hallucinations in small quantities, but addiction can lead to 'stupefaction and criminal fanaticism'. The seeds yield an edible oil, which is also used in paint-making.* Pl. 6.

URTICACEAE | Nettle Family

Usually herbs with simple lvs., often with stinging hairs. Fls. small, usually unisexual, with 4–5 perianth segments. Male fls. usually with 4 stamens which spring back elastically on ripening; female fls. with a one-seeded ovary; style 1. Fr. dry, often enclosed in the persistent perianth.

1. Lvs. more than 6 mm. long; fls. clustered; stems usually erect.
  2. Lvs. opposite, with stinging hairs.   *Urtica*
  2. Lvs. alternate, without stinging hairs.   *Parietaria*
1. Lvs. less than 6 mm. long; fls. solitary; stems creeping, rooting at nodes. *Soleirolia*

URTICA | **Nettle**   Lvs. opposite, toothed, with stinging hairs. Fls. in axillary spike-like clusters; fls. one-sexed, perianth 4. 8 sps.

Male and female fl. clusters elongated
(a) *Perenns.*
65. *U. dioica* L. NETTLE. A stoloniferous perenn. with many leafy erect stems to 1½ m., with rather dull green saw-toothed lvs. with painfully stinging hairs. Plants usually one-sexed; fl. clusters green, long and tassel-like, to 10 cm., hanging or spreading. Lvs. 3–12 cm., heart-shaped long-pointed, stalked, in opposite pairs with 4 stipules at each node. A variable sp. △ Waste places, woods, hedges, tracksides. July–Sept. All Eur. *The stems produce a fibre which has been used for weaving fine textiles, etc.*

(b) *Anns.*
66. *U. urens* L. SMALL NETTLE. A branched ann. 10–60 cm., with pale green, hairy, saw-toothed lvs. with stinging hairs. Fls. green, the male and female fls. mixed, forming short clusters to 2 cm. long, spreading upwards in fr. Lvs. 1–4 cm. oval-heart-shaped, stalked; stipules 4. △ Waste ground, arable. May–Oct. All Eur.

U. dubia Forskål Distinguished by its male spikes with the fls. borne on upper side of a swollen membraneous axis. Stipules 2. Mediterranean Europe. Pl. 6.

Female fl. clusters globular
67. *U. pilulifera* L. ROMAN NETTLE. Readily distinguished by the green female fl. clusters which are globular, *c.* 1 cm. in diameter and which hang from the axils of the upper lvs. on long stalks. Male fls. in erect, branched clusters; perianth of female fls. inflated. Lvs. 2–6 cm., oval, toothed or entire, long-stalked, with stinging hairs. Erect hairy ann. or bienn. 30–100 cm. △ Waste places, waysides, Apr.–Oct. Med. Eur. P.SU.; introd. NL.D.CH. A.CS.H. Pl. 6.

PARIETARIA | **Pellitory-of-the-Wall**   Lvs. alternate, entire, without stinging hairs. Fls. in axillary branched clusters, bisexual or unisexual; perianth 4. 6 sps.

68. *P. officinalis* L. ERECT PELLITORY-OF-THE-WALL. A simple or little-branched, densely hairy, more or less erect perenn. 30–100 cm., with numerous stalked oval lvs. and green axillary fls. Fl. clusters rounded, bunched, much shorter than the lvs.; bracts free, shorter than perianth in fr. Lvs. 3–12 cm., oval to lance-shaped, long-pointed and narrowed to a short stalk. △ Walls, rocks, waste places. June–Oct. Cent., South-East Eur. F.I.SU.: introd.: B.NL.DK.PL. Pl. 7.

*P. diffusa Mert. & Koch PELLITORY-OF-THE-WALL. Western and Southern Europe.

69. *P. lusitanica* L. A delicate hairy ann. with very slender spreading stems 5–30 cm., and small, stalked, oval-pointed to rounded lvs. usually 8–12 mm., but sometimes to 4 cm. Fls. 3–5, in tiny stalkless clusters in axils of lvs.; bracts as long as or longer than perianth in fr. △ Walls, shady rocks. May–June. P.E.F.H.I.YU.GR.BG.R.SU.

*Soleirolia soleirolii (Req.) Dandy (*Helxine s.*) MIND-YOUR-OWN-BUSINESS, MOTHER OF THOUSANDS. Native of Mediterranean islands; naturalized in Western Europe.

SANTALACEAE | Sandalwood Family

Herbs or woody plants, often semi-parasitic on the roots of herbs and shrubs, with entire alternate lvs. Fls. bisexual or unisexual, tiny; perianth usually three- to five-lobed and often bell-shaped; stamens 3–5. Ovary usually inferior, one-celled; style 1. Fr. a fleshy drupe or small nut; one-seeded.

Herbaceous plants: perianth fused, tubular or bell-shaped.   *Thesium*
Woody plants: perianth not fused below.   *Osyris*

OSYRIS Shrublets with green switch-like stems and narrow evergreen lvs. Plants one-sexed; perianth 3 or 4. Fr. fleshy. 2 sps.

70. *O. alba* L. An erect, hairless, slender-stemmed shrublet 40–120 cm., with many erect, angled, lateral branches bearing numerous small lvs. and small yellowish, sweet-scented fls. Fls. 8 mm., perianth 3; male fls. in short lateral clusters, female fls. solitary. Lvs. leathery 1–1½ cm., linear to lance-shaped, one-veined. Fr. globular 5–7 mm., somewhat fleshy, at length red. △ Dry, uncultivated ground. Apr.–June. Med. Eur. P.BG. Pl. 7.

THESIUM | **Bastard Toadflax** Herbaceous anns. or perenns. with linear to lance shaped lvs. Fls. bisexual, in terminal leafy, often branched clusters. Fr. dry, green. 18 sps

71. *T. alpinum* L. ALPINE BASTARD TOADFLAX. A slender, spreading, little-branched perenn 10–20 cm., with linear lvs. and tiny greenish-yellow fls. Fl. cluster long, lax, terminal; fls. solitary, nearly stalkless, unilaterally placed in the axils of 3 linear leafy bracts. Perianth tubular and usually with 4 spreading lobes, persisting in fr. and 2–3 times as long. △ Mountain pastures and rocks. June–Aug. Much of Eur. (except North Eur. P.IRL.GB.B. NL.H.GR.TR.) S. Pl. 6.

T. pyrenaicum Pourret Distinguished from 71 in having a zigzag axis bearing several fls. on each lateral branch and forming a pyramidal, not unilateral, terminal cluster; perianth lobes usually 5. Much of Western and Central Europe.

LORANTHACEAE | Mistletoe Family

Semi-parasitic shrubs growing on the trunks and branches of trees, with opposite or whorled lvs. Fls. unisexual; perianth of 1 or 2 whorls, the inner often fleshy and united into a tube; stamens usually 4; ovary inferior, style 1 or absent. Fr. usually a berry; seed 1, without a seed coat.

Inflorescence elongated. Lvs. deciduous.   *Loranthus*
Inflorescence densely clustered. Lvs. evergreen.   *Viscum*

LORANTHUS Lvs. in our sp. deciduous, with pinnate veins. Fls. in an elongated, branched, or spike-like cluster; perianth 4–6. 1 sp.

72. *L. europaeus* Jacq. A parasitic shrub growing on the branches of oaks, sweet chestnut, and beech, forming a rounded bushy growth to *c.* 1 m. in diameter. Fls. greenish-yellow, in terminal clusters 2½–4 cm. long. Lvs. 1–5 cm. oval to obovate, short-stalked, deciduous,

leathery, dull green; branches dark brown. Fr. globular, to 1 cm., yellow, with sticky flesh. △ May–June. South-East Eur. D.A.CS.H.I.SU.

VISCUM | Mistletoe   Lvs. evergreen, with parallel veins. Fls. in a compact stalkless or almost stalkless cluster; perianth usually 4; stamens fused to perianth. 2 sps.

73. *V. album* L. MISTLETOE. An evergreen shrub growing parasitically on the branches of a considerable number of deciduous trees and less commonly on conifers. Branches greenish-yellow, mostly pendulous, much-branched, forming a rounded bushy growth to 1 m. in diameter. Fls. 3–4 in stalkless clusters in the axils of the lvs., usually unisexual. Lvs. 2–8 cm., narrowly obovate, greenish-yellow, opposite, stalkless. Fr. a white or rarely yellow berry with sticky flesh. △ Feb.–Apr. All Eur. (except IRL.IS.SF.). *The berries are somewhat poisonous and they have been used in herbal remedies and for making bird-lime.* Pl. 7.

V. cruciatum Boiss. has red berries. South-Western Spain.

ARISTOLOCHIACEAE | Birthwort Family

Herbs or woody climbers with alternate entire lvs. Fls. bisexual; perianth bell-shaped or irregularly trumpet-shaped, often lurid in colour and fetid. Stamens 6 or 12, free or fused with the style. Ovary inferior, with 6 chambers; styles usually 6, sometimes fused into a column. Fr. a capsule.

Fls. regular, bell-shaped, terminal.   *Asarum*
Fls. irregular, tubular with an oblique lip, axillary.   *Aristolochia*

ASARUM   Fls. regular, bell-shaped, three-lobed; stamens 12, free. Fr. globular, splitting irregularly. 1 sp.

74. *A. europaeum* L. ASARABACCA. A creeping perenn. with thick, hairy stems 2–10 cm. tall, with dark, shining, evergreen, broadly kidney-shaped, long-stalked lvs. 2½–10 cm. across and brown oval scales. Fls. dull brownish, solitary, terminal; corolla *c.* 1½ cm., bell-shaped, three-lobed, hairy outside. △ Open woods. Mar.–May. Much of Eur. (except IRL.IS.SF.GR.TR.): introd. NL.DK. *Used medicinally in the past as a stimulant and emetic.* Pl. 7.

ARISTOLOCHIA | Birthwort   Fls. irregular; perianth swollen at the base and with a long funnel-shaped tube with an elongated upper lip. Stamens 6, fused to the column-like style. 13 sps.

Climbing plants
75. *A. sempervirens* L. CLIMBING BIRTHWORT. A slender, twining, woody perenn. to 5 m., with glossy evergreen lvs. Fls. solitary, 2–5 cm., stalked, strongly curved, yellowish-brown with purplish stripes, yellow within. Lvs. to 10 cm. triangular-heart-shaped, stalked. Fr. 1–4 cm. △ Bushy and shady places. May–June. I.GR.: introd. F.

Non-climbing herbaceous plants
(a) Fls. clustered
76. *A. clematitis* L. BIRTHWORT. A strong-smelling stoloniferous perenn. to 1 m., bearing clusters of erect, pale yellow, irregularly trumpet-shaped fls. in the axils of the upper lvs.

Fls. 2–3 cm. long, in clusters of 2–8. Lvs. large, 6–15 cm., heart-shaped, stalked, margin finely toothed. Fr. 2–5 cm. pendulous, pear-shaped. △ Hedges, rocks, cultivated ground. May–July. Native of South-Eastern Europe; formerly cultivated and now widely naturalized in much of Eur. (except North Eur.) *Poisonous. Its medicinal use in labour has been known since the time of Theophrastus. Also used in the past as a cure for snake-bite.* Pl. 7.

(b) Fls. solitary

77. *A. rotunda* L. ROUND-LEAVED BIRTHWORT. An erect, rather weak perenn. 20–60 cm., with rounded lvs. and solitary yellowish fls. 2–5 cm., with a darker brownish strap-shaped upper lip. Lvs. 2–9 cm. oval to orbicular, with the rounded basal lobes encircling the stem, stalkless or nearly so, hairless, margin smooth. Tuber solitary, globular. △ Meadows, hedges, stony places. Apr.–June. Med. Eur. P.CH.BG. Pl. 8.

A. pistolochia L. Distinguished by its brownish fls. 2–5 cm., with a dark purple upper lip. Lvs. with rough, finely toothed margin; lv. stalks 1–5 cm.; plant hairy. Roots fibrous, numerous. Western Mediterranean region.

A. pallida Willd. Fls. 3–6 cm., green to pale brown or yellow, striped with brown or purple, upper lip as long as tube of corolla. Lv. margin not toothed. Tuber solitary, globose. Central and Eastern Mediterranean region.

A. longa L. Fls. 2–6 cm., brownish-green with a short, pointed, brownish upper lip, much shorter than the tube of the corolla; tuber cylindrical. Mediterranean region.

RAFFLESIACEAE

Parasites lacking green lvs. and living on roots of other plants; lvs. reduced to scales. Fls. unisexual, solitary or in dense heads; perianth 4–10, of 1 whorl only; stamens many, fused to fleshy column; ovary inferior, one-celled. Fr. a fleshy berry, many-seeded.

CYTINUS   Fls. in dense heads; lower fls. female, the upper male. Parasitic on roots of *Cistus* and *Halimium* sps. 2 sps.

78. *C. hypocistis* (L.) L. Fls. bright yellow, in a globular head borne on a short stem 3–7 cm. and arising direct from the soil below *Cistus* or *Halimium* bushes. Green lvs. absent, but in their place are brightly coloured, often orange-carmine, overlapping fleshy scales. Fls. 5–10, stalkless, each subtended by 2 fleshy brightly coloured bracts; perianth tubular with 4 spreading lobes; stamens fused, anthers 8. Fr. soft, fleshy. △ May–June. Med. Eur. P.BG.R.SU. *The juice is astringent and is used against dysentery, etc.—a remedy known to the Romans.* Pl. 8.

C. ruber (Fourr.) Komarov Petals white or pale pink, and scale lvs. and bracts crimson or carmine. Mediterranean region.

POLYGONACEAE | Dock Family

Usually herbs with alternate simple lvs. and with a membraneous sheath encircling the nodes of the stem in place of stipules. Fls. usually bisexual; perianth of 3–6 segments often enlarging and becoming membraneous in fr.; stamens usually 6–9. Ovary superior, one-seeded; styles 2–4. Fr. a nut, flattened and winged or three-angled.

		Nos.
1. Perianth 4; stamens 6.	*Oxyria*	90
1. Perianth 5 or 6.		
2. Climbing or twining plants.	*Bilderdykia*	87
2. Non-climbing herbs.		
3. Perianth 6, inner larger than outer in fr.	*Rumex*	91–100
3. Perianth 5, equal in fr. or outer larger.		
4. Outer perianth winged or keeled in fr.	*Reynoutria*	88
4. Outer perianth not winged or keeled in fr.		
5. Lvs. about as long as wide; fr. twice as long as perianth, or more.	*Fagopyrum*	89
5. Lvs. longer than wide; fr. usually hidden in perianth or slightly protruding.	*Polygonum*	79–86

POLYGONUM   Perianth 5, all similar, petal-like, pink or white; stamens usually 8; styles 2–3. Fr. a three-angled or flattened nut, largely encircled by the persistent perianth. 36 sps.

Fls. 1 or few in axils of lvs., and shorter than the lvs.

79. *P. maritimum* L. SEA KNOTGRASS. A prostrate perenn. of maritime sands with rather thick spreading stems 10–50 cm., and thick and leathery, lance-shaped, glaucous lvs. with inrolled margins. Fls. white or pink, 1–4 in lv. axils and much shorter than the lvs. which are $\frac{1}{2}$–$2\frac{1}{2}$ cm. Sheaths of upper lvs. conspicuous, white-papery, veined, as long as or longer than the internodes. Fr. shining dark brown, slightly projecting from perianth. △ Sea-shores, shingle. Apr.–Oct. Med., South-East Eur. P.GB.SU. Page 64.

P. oxyspermum Ledeb. Distinguished by its sheaths which are all much shorter than the internodes. Lvs. 1$\frac{1}{2}$–3 cm., elliptic to linear-lance-shaped with margins scarcely inrolled. Fr. shining, longer than the perianth. Sea-shores of much of Europe.

80. *P. aviculare* L. KNOTGRASS. A very variable prostrate or erect ann. 5–200 cm. with numerous narrow lvs. and small solitary or few fls. in the axils of the upper lvs. Fls. *c.* 3 mm. across; perianth segments green with white or pink margin. Lvs. 2–5 cm., narrow-elliptic, larger on main stem than on branches; sheaths transparent, silvery. Fr. dull brown, encircled by persistent perianth. △ Waste places, cultivated ground, tracksides, fields. May–Nov. All Eur.

Fls. in terminal leafless clusters

(a) *Fls. in narrow, elongated, interrupted spikes*

81. *P. hydropiper* L. WATER-PEPPER. Fls. greenish or pink, in slender, curved, nodding terminal spikes; perianth 3–5 mm. across, covered with yellowish glands. Lvs. 5–10 cm., lance-shaped, stalkless, ciliate; sheath with a fringe of short hairs, or without. Fr. dull brown-black, *c.* 3 mm. Slender hairless, little-branched ann. 20–80 cm., with a burning taste. △ Damp places, ditches. July–Oct. All Eur. (except IS.). *The seeds are peppery-flavoured and can be used as a condiment. A somewhat irritant plant, particularly to livestock.* Pl. 8.

P. minus Hudson SMALL WATER-PEPPER. Like 81, but perianth without glands. Nuts glossy black, 1$\frac{1}{2}$–3 mm. Widespread in Europe, but commoner in the north. Page 64.

(b) *Fls. in dense cylindrical spikes*

82. *P. persicaria* L. PERSICARIA, REDLEG. Fls. bright or pale pink in stout, cylindrical,

blunt-topped terminal spikes *c*. 2 cm. long; perianth without glands. Lvs. to 15 cm., lance-shaped, often black-blotched; sheaths fringed with short hairs; stems reddish, swollen at the nodes. An erect, nearly hairless ann. 20–80 cm. △ Damp ground, waste places, tracksides. July–Sept. All Eur.

P. lapathifolium L. (*P. nodosum* Pers.) PALE PERSICARIA. Like 82, but fls. greenish-white or dull pink and inflorescence stalks with yellow glands. Lvs. with pellucid glands, visible beneath. A very variable sp. All Europe. Pl. 8.

83. *P. amphibium* L. AMPHIBIOUS BISTORT. A floating aquatic, marsh or land perenn. with rather stout, dense, cylindrical, terminal spikes 2–4 cm. by 1 cm. of pink or red fls. The land and water forms are very dissimilar: aquatic plants have long submerged stems and oblong to oval-heart-shaped, long-stalked, hairless floating lvs. and leafless fl. spikes carried a short distance above water; terrestrial plants have erect stems 30–70 cm., with oblong-lance-shaped, rough, hairy, erect, stalkless lvs. Perianth segments without glands; stamens 5, longer. Fr. 2 mm., shining. △ Lakes, marshes, riversides, damp mud. July–Aug. All Eur. Pl. 8.

84. *P. bistorta* L. BISTORT, SNAKE-ROOT. Readily distinguished by the winged lv. stalks of the lower lvs. Fls. pink, rarely white, in a dense, stout, terminal spike 2–7 cm. by 1–1½ cm.; perianth 4–5 mm., stamens projecting. Lower lvs. broadly oblong to egg-shaped, stalked, the upper lance-shaped, stalkless, and clasping stem. Rhizome short and conspicuously contorted; stems erect, unbranched, 20–100 cm. △ Largely in mountains: damp meadows, roadsides. May–July. Much of Eur. (except North Eur. IRL.GR.TR.): introd. DK.N.S. Page 64.

85. *P. viviparum* L. VIVIPAROUS BISTORT. A slender northern or alpine perenn. with un-branched stems 6–40 cm., bearing rather slender terminal spikes of white or pink fls. with the lower fls. replaced by brownish-purple bulbils. Lvs. linear to lance-shaped, with margins inrolled, tapering at both ends. Rhizome not contorted. △ Alpine meadows, rocky places. June–Aug. Much of Eur. (except P.B.NL.DK.AL.GR.TR.BG.). *The underground stems are edible.* Pl. 9.

(c) Fls. in loose branched clusters

86. *P. alpinum* All. ALPINE KNOTWEED. A rather robust hairy perenn. 30–80 cm. with white or pale pink fls. in a leafy, branched, spreading, pyramidal cluster. Lvs. lance-shaped, long-pointed, 1–3 cm. wide, margin ciliate, paler beneath; lv. sheaths short, transparent, soon disappearing. Fr. pale glossy brown, longer than perianth. △ In mountains: damp meadows, screes. July–Aug. E.F.CH.A.I.YU.AL.GR.BG.R.SU.: introd. DK.

P. polystachyum Meissner Distinguished by its stout red stems to 120 cm., large oblong-lance-shaped lvs. 10–20 cm. by 3–8 cm. with brown persistent sheaths, and lax, leafy, branched clusters of white fls. Native of the Himalaya; sometimes naturalized in Central Europe.

BILDERDYKIA Stems usually climbing; lvs. with a heart-shaped base. Fl. clusters lax; perianth with the outer 3 segments winged; stamens 8; stigmas knob-like. 3 sps.

87. *B. convolvulus* (L.) Dumort. (*Polygonum c.*) BLACK BINDWEED. A slender climbing or spreading ann. to 1 m. with long-stalked, triangular-heart-shaped, pointed lvs. and greenish-white fls. in spike-like clusters. Outer perianth keeled or narrowly winged; fl. stalk shorter than fls., jointed above the middle and not more than 3 mm. in fr. Fr. dull black, 4–5 mm. △ Cultivated ground, waste places, hedges. June–Oct. All Eur. Page 64.

B. dumetorum (L.) Dumort. (*Polygonum d.*) COPSE BINDWEED. Like 87, but fr. stalks

1. *Polygonum maritimum* 79
3. *Polygonum minus* [81]
5. *Bilderdykia convolvulus* fr. 87
2. *Fagopyrum tataricum* [89]
4. *P. bistorta* 84
6. *B. dumetorum* [87]

5–8 mm., jointed at or below the middle, and winged above from the perianth. Fr. smooth, shining, 2½–3 mm. Widespread in Europe. Page 64.

REYNOUTRIA Stoloniferous perenns. with stout, erect, ann. stems. Fls. in small axillary, spike-like clusters; perianth with the outer 3 segments winged; stamens 8; stigmas fringed. 2 sps.

88. *R. japonica* Houtt. (*Polygonum cuspidatum* Siebold & Zucc.) JAPANESE KNOTWEED. Stems stout, 1–2 m. tall, smooth, hollow often reddish, with large, broadly oval lvs. to 12 cm. with square-cut base and narrow, long-pointed apex. Fls. white, in 2–4 slender clusters in axils of upper lvs., and shorter than lvs. Stems numerous, often forming a dense cluster. △ Waste places. Aug.–Sept. Native of Central Asia; often naturalized in Cent. Eur. P.F.IRL.GB.NL. Pl. 8.

R. sachalinensis (Friedrich Schmidt Petrop) Nakai Like 88, but more robust, to 3½ m. and lvs. larger 15–30 cm., the lower with heart-shaped bases. Fls. greenish. Native of Eastern Asia; occasionally naturalized in Western and Central Europe.

FAGOPYRUM | Buckwheat Non-climbing anns. with hollow stems and lvs. about as broad as long. Perianth 5, all similar; stamens 8; styles 3, stigmas knob-like. Fr. triangular-sectioned, longer than perianth. 2 sps.

89. *F. esculentum* Moench BUCKWHEAT. Erect ann. to 60 cm., usually with reddish stems and triangular-heart-shaped, long-pointed lvs. Fls. pink or white, in short, dense, long-stalked clusters; perianth 3–4 mm. Lvs. 7 cm. by 6 cm., dark green, the upper stalkless. Fr. smooth, dark brown, acutely three-angled, 2–3 times longer than the perianth. △ Cultivated ground, waste places. June–Aug. Native of Asia; widely cultivated and often self-seeding in much of Europe. Pl. 9.

F. tataricum (L.) Gaertner GREEN BUCKWHEAT. Like 89, but fls. greenish in lax clusters; perianth 2 mm. Stems green. Fr. rough-surfaced, with wavy toothed angles. Widely cultivated, often self-seeding. Page 64. *Both species are grown for their seeds which produce a flour suitable for baking; the seeds are also used for brewing.*

OXYRIA Perianth segments 4, enlarging in fr.; stamens 6; stigmas 2. Fr. broadly winged. 1 sp.

90. *O. digyna* (L.) Hill MOUNTAIN SORREL. A hairless, pale green, rounded-leaved, slightly fleshy, acid-tasting perenn. to 30 cm., of mountains or northern regions. Fls. green, edged with red, in a terminal, elongated, leafless, simple, or branched spike. Perianth of 2 outer spreading and 2 inner adpressed segments. Lvs. 1–3 cm., rounded kidney-shaped, long-stalked, mostly basal. Fr. 3–4 mm., wing often reddish. △ Rocky places, screes. July–Aug. Much of Eur. (except P.B.NL.DK.SF.AL.GR.TR.).

RUMEX | Dock, Sorrel Perianth segments 6, the outer 3 smaller than the inner. The inner 3 segments persist and enlarge in fr. and are sometimes called *valves*; they are often toothed and often possess a central swelling or *tubercle* on the outside. Fr. a nut surrounded by the perianth segments. Ripe fr. is necessary for identification; hybridization is frequent. 50 sps.

Fr. without swelling on perianth segments
(a) Lvs. with 2 distinct angular basal lobes
91. *R. acetosella* L. agg. SHEEP'S SORREL. An erect or procumbent, stoloniferous, hairless

perenn. 10–30 cm., with lance-shaped or narrower lvs. with spreading, forward-projecting, narrow basal lobes. Fls. reddish; perianth segments all erect c. 1½ mm., as long as the fr.; plants one-sexed. Upper lvs. short-stalked, not clasping stem; sheaths thin and transparent, cut into strips. A variable sp. △ Grasslands, heaths, sandy places. May–July. All Eur.

92. *R. scutatus* L. SHIELD DOCK. Lvs. *c.* 3–4 cm., about as long as broad, variably triangular-heart-shaped with diverging basal lobes, often rather thick, somewhat glaucous. Fls. unisexual, but on same plant, reddish in lax branched clusters. Fr. *c.* 3 mm., with orbicular to heart-shaped perianth segments 4½–6 mm. A woody-based, much-branched perenn. 25–50 cm. △ Waysides, rocks, sandy places, in mountains. May–Aug. Much of Eur. (except North Eur. IRL.TR.): introd. GB.NL.S. *Used as a pot herb.* Page 67.

93. *R. acetosa* L. SORREL. Basal lvs. large to 10 cm., oblong-oval, 2–4 times as long as broad with basal pointed lobes backward-projecting, long-stalked; upper lvs. stalkless, clasping stem. Perianth segments 3–4 mm. in fr., the outer deflexed after flowering; plants one-sexed. A variable sp. An erect, non-stoloniferous perenn. 30–100 cm. △ Meadows, hedges, woods. May–Aug. All Eur. (except GR.TR.). *Sometimes cultivated as a salad plant for its acid-tasting leaves.*

(b) *Lvs. with a heart-shaped or wedge-shaped base*

94. *R. alpinus* L. MONK'S RHUBARB. A conspicuous, rather robust perenn. ½–1 m., with very large oval lvs. 15–40 cm. and a dense compound, elongated cluster of numerous greenish fls. Lvs. oval-heart-shaped, with undulate margin and lv. stalk long, grooved above; sheath large, papery. Fr. with broadly egg-shaped perianth segments 4½–5 mm. △ In mountains: pasture, wayside halting-places, by habitation. June–Aug. Cent. Eur. (except H.), South-East Eur. (except TR.) E.F.I.SU.: introd. GB. *Formerly used medicinally like rhubarb.* Pl. 9.

95. *R. bucephalophorus* L. HORNED DOCK. A slender, usually reddish ann. 5–30 cm., often occurring abundantly in the Mediterranean region. Fls. reddish, in terminal spikes; fl. stalks recurved, usually of two kinds: either short and slender, or long, broad and flattened. Lvs. very small, lance-shaped. Fr. with thick, triangular perianth segments, with straight or curved marginal teeth. A very variable sp. △ Dry, sandy places, cultivated ground. Apr.–Sept. Med. Eur. P.

R. aquaticus L. Like 96, but without swellings on perianth segments and with basal lvs. with heart-shaped base. Widespread in Europe. Page 67.

Fr. with 1–3 distinct swellings on perianth segments or rarely without
(a) *Perianth segments with entire margins*

96. *R. hydrolapathum* Hudson GREAT WATER DOCK. A robust waterside perenn. up to 2 m. with large, broadly lance-shaped basal lvs. up to 1 m. long, tapering at both ends. Inflorescence usually much-branched with many erect branches and forming a narrow cluster. Fr. with triangular inner perianth segments 6–7 mm., each with a large swelling, and margin often with a few short teeth. Frequently hybridizes with (95). △ Borders of lakes, rivers, canals. July–Aug. Much of Eur. (except P.IS.AL.GR.TR.). Pl. 9.

97. *R. crispus* L. CURLED DOCK. Distinguished by its lower lvs. which are characteristically thick, narrow lance-shaped with undulate crisped margin. Inflorescence dense, often nearly simple or with few erect lateral branches, each subtended by a narrow crisped lv.; whorls of fls. closely bunched. Fr. with inner perianth segments 3–5½ mm., heart-shaped and usually each with a swelling, but swellings differing in size. An erect, little-branched perenn. 30–150 cm. △ Cultivated ground, grassy places, sea-shores. June–Aug. All Eur. Pl. 9.

1. *Rumex scutatus* 92
4. *R. palustris* [100]
2. *R. pulcher* 99
5. *R. maritimus* 100
3. *R. aquaticus* [95]

98. *R. conglomeratus* Murray SHARP DOCK. Like 97, but lower lvs. thin, oblong-lance-shaped with a rounded base and sometimes narrowed at the middle of the blade. Inflorescence lax, leafy, with spreading branches, whorls of fls. all distinct, not closely bunched. Fr. with inner perianth segments *c.* 3 mm., oval, and each with a large swelling often covering nearly the whole of the segment. Very variable sp. An erect, branched bienn. or perenn. 30–100 cm. △ Damp places, banks, woods. July–Sept. All Eur.

*R. sanguineus L. RED-VEINED DOCK is widespread in Europe. Pl. 9.

(b) Perianth segments with distinctly toothed or lobed margins
99. *R. pulcher* L. FIDDLE DOCK. A stiff sprawling perenn. to ½ m. with zigzag, often entangled branches and small, thick, oblong, often violin-shaped basal lvs. and lance-shaped upper lvs. Inflorescence with wide-spreading branches; whorls of fls. widely spaced and each subtended by a lv. Fr. with inner perianth segments 4½–6 mm., toothed in lower half, each with a swelling. △ Tracksides, disturbed ground. June–Aug. Much of Eur. (except North Eur. P.B.NL.D.): introd. IRL.DK.S.A. Page 67.

R. obtusifolius L. BROAD-LEAVED DOCK. Like 98, but perianth segments conspicuously toothed and 1–3 segments with swellings. Inflorescence branches erect; basal lvs. large oval. Throughout Europe. *Leaves used to reduce irritation caused by nettle stings, and to soothe burns and scalds.*

100. *R. maritimus* L. GOLDEN DOCK. A much-branched ann. or bienn. 10–50 cm., turning golden-yellow when in fr. Inflorescence of many densely bunched golden-yellow whorls of fls. subtended by strap-shaped lvs. Basal lvs. narrowly elliptical. Fr. with slender stalk and inner perianth segments with long, fine teeth and a narrow triangular apex, each with a swelling with an acute apex. △ Marshes, lakesides, sea-shores. July–Sept. Much of Eur. (except P.E.IS.AL.GR.TR.BG.). Page 67.

R. palustris Sm. MARSH DOCK. Like 100, but fr. stalks thick, rigid; inner perianth segments with short, sharp teeth and a strap-shaped blunt apex and swellings with a blunt apex. Watersides. Widespread in Europe. Page 67.

CHENOPODIACEAE | Goosefoot Family

Usually ann. or perenn. herbs with alternate, often rather fleshy and sometimes mealy lvs. Fls. bisexual or unisexual, small, greenish; perianth absent or 1–5, in female fls. often persisting and fusing together in fr.; stamens 1–5. Ovary superior or half-inferior; stigmas 2–3. Fr. usually a dry, one-seeded nut.

Nos.

1. Lvs. flattened, thin-bladed, or slightly fleshy.
  2. Fls. mostly bisexual; fr. surrounded by 2–5 persistent perianth segments.
    3. Some lvs. toothed or lobed, not more than 5 cm.   *Chenopodium*   102–105
    3. Lvs. entire, some more than 5 cm.   *Beta*   101
  2. Fls. all unisexual; fr. enclosed in 2 conspicuous enlarged flattened bracteoles.
    4. Anns.
      5. Plants with both male and female fls.   *Atriplex*   107–111
      5. Plants one-sexed.   *Spinacia*   106

4. Small shrubs.

1. Lvs. thick, fleshy, or apparently absent when stems fleshy.

  7. Lvs. alternate, spreading.

  7. Lvs. apparently absent; stems jointed.

BETA | **Beet**   Perianth 5, becoming thick, woody and fused in fr.; fls. bisexual. 5 sps.

101. *B. vulgaris* L. BEET. An erect or spreading ann., bienn., or perenn. to 1 m., with very variable large, broadly egg-shaped to rhomboid basal lvs. and narrower stem lvs., usually dark shiny green. Fls. greenish or purplish, in dense stalkless clusters forming a branched, elongated, spike-like inflorescence, the lower clusters subtended by a narrow lv. and interrupted below in fr., the upper leafless. Subsp. *vulgaris* is widely cultivated as Sugar Beet, Beetroot, Spinach Beet, Mangold, and has erect stems, large lvs., and a swollen taproot. It often occurs as a casual in cultivated ground. Subsp. *maritima* (L.) Arcangeli has spreading or erect stems, lvs. 10 cm. or less, and root usually not swollen. It is a native of the European littoral. △ June–Sept. Much of Eur. *The many cultivated forms are used as root and leaf vegetables and fodder for livestock. Sugar Beet is a very important source of sugar and is now widely cultivated throughout Europe.* Pl. 10.

CHENOPODIUM | **Goosefoot**   Fls. bisexual; perianth 2–5, rarely enlarging in fr.; stamens 1–5; stigmas usually 2. Lower lvs. toothed, lobed, or triangular. 23 sps. Many widespread sps. can only be identified by the microscopic markings on the seed coats; the following are more readily distinguished in the field.

102. *C. botrys* L. STICKY GOOSEFOOT. A sticky, glandular, unpleasantly aromatic ann., 15–60 cm., with distinctive lvs. which are oval in outline, but cut into 2–4 deep, rounded lobes on each side. Fls. in narrow elongated spike-like clusters. △ Sandy places. July–Oct. Much of Eur. (except North Eur. IRL.GB.B.NL.GR.TR.). Page 71.

103. *C. bonus-henricus* L. GOOD KING HENRY, ALL-GOOD. A rather robust perenn. to 60 cm., with large, bright green, broadly triangular lvs. to 10 cm., with spreading basal lobes and with a wavy margin. Inflorescence compound, narrow tapering, leafless; perianth and stamens 4. △ Walls, waysides, by habitation. June–Sept. All Eur. (except P.IS.TR.). *The leaves and young shoots are edible.* Pl. 10.

104. *C. foliosum* Ascherson STRAWBERRY GOOSEFOOT. Readily distinguished by its conspicuous, scarlet, fleshy frs. which are borne in dense, stalkless clusters *c.* 1 cm. across in the axils of the lvs. Lvs. narrow triangular, with several coarse, acute teeth, the uppermost narrower with 2 narrow, spreading basal lobes. A leafy erect or spreading ann. 20–100 cm. △ Waste places, waysides. June–July. Native of P.E.F.D.CH.A.; naturalized elsewhere. Pl. 10.

105. *C. vulvaria* L. STINKING GOOSEFOOT. A spreading, much-branched, leafy ann. 10–65 cm., smelling of bad fish, with distinctively small, broadly oval lvs. only $\frac{1}{2}$–$2\frac{1}{2}$ cm. long, with entire margin, grey-mealy beneath. Fl. clusters small, leafy, terminal, and axillary.

△ Waste places, walls, waysides, cultivation. July–Sept. All Eur. (except IRL.IS.N.). *The plant yields a lemon-yellow dye.* Page 71.

Widespread sps. in Europe:

Lvs. not or slightly powdery
  *C. *polyspermum* L. ALL-SEED.
  *C. *rubrum* L. RED GOOSEFOOT.
  *C. *murale* L. NETTLE-LEAVED GOOSEFOOT.
  *C. *hybridum* L. SOWBANE.
  *C. *urbicum* L. UPRIGHT GOOSEFOOT.

Lvs. powdery
  *C. *glaucum* L. GLAUCOUS GOOSEFOOT.
  *C. *album* L. FAT HEN. Pl. 10.

SPINACIA | Spinach   Like *Chenopodium*, but fls. unisexual and borne on different plants. Styles 4, very long. 1 sp.

106. *S. oleracea* L. SPINACH. An erect, branched, leafy ann. or bienn. to 1 m., with hollow, grooved stems and oval to triangular-arrow-shaped, shiny, hairless lvs. Fl. greenish, in long leafy spikes. Fr. flattened, often with 2–4 stiff, divergent apical spines. △ June–Sept. Origin unknown; widely cultivated and often escaping. *The leaves have been used as a vegetable since Greek and Roman times. The leaves give a green colouring used for liqueurs etc., and a yellow dye.* Page 73.

ATRIPLEX | Orache   Fls. unisexual, but both sexes occurring on one plant; male fls. with 5 perianth segments and 5 stamens; female fls. without perianth, but with 2 encircling bracteoles, which enlarge in fr. 19 sps.

Shrubby perenns.
107. *A. halimus* L. SHRUBBY ORACHE. An erect, silvery-white, much-branched, shrubby perenn. to 2½ m., with persistent, rather leathery, oval or angular lvs. to 4 cm. Fls. yellowish, in more or less leafless, terminal, elongated, branched spikes. Fr. with 2 rounded to kidney-shaped, shining white, leathery bracteoles. △ Coastal sands, salt-rich sands inland. Aug.–Sept. Med. Eur. (except YU.AL.TR.) P.BG.

Anns.
(a) *Lvs. green, much longer than broad*
108. *A. patula* L. COMMON ORACHE. Upper lvs. linear-lance-shaped, entire; lower lvs. rhomboid-hastate with wedge-shaped base. Fr. with bracteoles broadly rhomboid, entire or toothed, smooth or slightly warty. A very variable hairless or mealy ann. to 1½ m., with branched erect, spreading or prostrate stems. △Cultivated ground, tracksides. July–Oct. All Eur. (except AL.).

A. littoralis L. SHORE ORACHE. An ann. of sandy shores and salt-rich regions of Europe, with all lvs. linear to lance-shaped, and bracteoles strongly warty and with long-pointed apices. Page 71.

(b) *Lvs. green, usually not more than twice as long as broad*
109. *A. hortensis* L. ORACHE. Distinguished by its fr. which has conspicuous rounded, entire, shiny membraneous bracteoles ½–1½ cm., which are veined. Lvs. large, usually more than 10 cm., triangular with spreading basal lobes, somewhat glaucous, green or purplish, entire or toothed. An erect ann. to 2½ m. △ Aug. Often cultivated as a vegetable

1. *Chenopodium botrys* 102   2. *Atriplex littoralis* [108]
3. *A. hortensis* 109   4. *Chenopodium vulvaria* 105   5. *Atriplex rosea* 111

and sometimes naturalized in many European countries. *The leaves have been used as a vegetable since ancient times. It yields a blue dye similar to indigo; the fruits are purgative and emetic.* Page 71.

110. *A. hastata* L. agg. HASTATE ORACHE. Lvs. less than 10 cm., triangular, with spreading basal lobes and abruptly contracted to the lv. stalk; upper lvs. narrower, stalked. Fr. with oval to triangular-rhombic or more rounded bracteoles which are entire or toothed, and warty or smooth. A variable erect or procumbent, hairless or mealy ann. to 1 m.; some forms are very like 108. △ Coastal sands and rocks, waste places, cultivated ground. Aug.–Oct. All Eur. (except IS.).

(c) Lvs. silvery-white, about twice as long as broad

111. *A. rosea* L. A silvery-white, erect or ascending much-branched ann. to 1 m., with oval-rhombic lvs. to 6 cm. with wavy, toothed margin. Fl. clusters mostly axillary. Fr. with whitish rhombic bracteoles to 12 mm., irregularly toothed and usually with large projections on the back. △ Sandy and waste places, nitrogen-rich and salt-rich ground. Aug.–Sept. All Eur. (except North Eur. SU.): introd. GB.B.NL.CH. Page 71.

A. laciniata L. FROSTED ORACHE. Like 111, but a procumbent spreading ann. to 30 cm., with mealy, silvery stems and lvs. Lvs. 1½–2 cm. Bracteoles 6–7 mm., broadly rhombic, three-lobed, usually smooth on the back. Maritime sands of Western Europe.

HALIMIONE Like *Atriplex* and often included in the genus, but bracteoles of fr. united almost to top. 3 sps.

112. *H. portulacoides* (L.) Aellen (*Atriplex p.*) SEA PURSLANE. A small shrub with thick and fleshy, silvery-white, mealy, oblong-oval to linear lvs. and with stems rooting in the mud below and often forming dense growths; 20–80 cm. Fl. spikes terminal, little-branched, dense, slender, leafless, mealy; fls. greenish. Fr. with delta-shaped, three-lobed bracteoles 2½–5 mm. △ Salt marshes. Aug.–Sept. West, Med. Eur. DK.D.BG. Page 73.

ARTHROCNEMUM Like *Salicornia* and often included in this genus, but woody-stemmed dwarf shrubs. 3 sps.

113. *A. perenne* (Miller) Moss (*Salicornia p.*) CREEPING SHRUBBY GLASSWORT. Underground stems creeping and producing many erect, succulent, jointed, branched stems to 30 cm., the plant covering a square metre or more. Stems at first green, often becoming orange or claret-coloured. Stems 3–5 mm. broad, joints longer than broad. Seeds greenish-brown or greyish, covered with curved or hooked hairs. △ Salt marshes, coasts. July–Aug. West. Eur. (except B.NL.), Med. Eur. (except TR.).

A. fruticosum (L.) Moq. (*Salicornia f.*) SHRUBBY GLASSWORT. Like 113, but stems glaucous, usually erect to 1 m., without creeping, underground stems. Seeds covered with conical hairs. Mediterranean Europe. Pl. 10.

A. glaucum (Delile) Ung.-Sternb. Like 113, but an erect glaucous shrub to 1 m. becoming yellowish-green or reddish. Fls. of each cluster falling to leave an undivided hollow in each segment (hollow tripartite in 113 and (113)). Seeds black, hard, with swellings. Coasts of Southern Europe. Page 73.

SALICORNIA | **Glasswort** Salt marsh anns. with jointed stems. Lvs. fleshy, in opposite pairs and closely investing stems and thus appearing to be absent. Fls. 3, arranged in a triangle, enclosed in fleshy bracts; stamens 1 or 2. 7 sps.

1. *Salsola soda* [116]
2. *S. kali* 116
3. *Spinacia oleracea* 106
4. *Suaeda maritima* [115]
5. *Arthrocnemum glaucum* [113]
6. *Montia fontana* 123
7. *Halimione portulacoides* 112

114. *S. europaea* L. (*S. herbacea* (L.) L.) GLASSWORT, MARSH SAMPHIRE. A small, fleshy, very variable ann. 10–30 cm., with jointed, simple to much-branched, fleshy stems, at first bright green, later turning red or purple in the autumn. Terminal spike with 3–12 segments; lateral fls. of trio appear smaller than the central fl. △ Salt marshes. Aug.–Sept. All Eur. (except IS.CH.A.). *The ashes are rich in soda and were used like 116 in the past.*

SUAEDA | Seablite Lvs. fleshy, more or less awl-shaped, cylindrical. Fls. small, axillary; perianth segments 5, fleshy in fr.; stamens 5; stigmas 3–5. 14 sps.

115. *S. vera* J. F. Gmelin (*S. fruticosa* auct.) SHRUBBY SEABLITE. A small, densely branched bushy shrub ½–1½ m., with numerous fleshy, glaucous, semi-cylindrical, blunt-tipped or fine-pointed lvs. ½–1½ cm. Fls. green, 1–3 in axils of lvs. and shorter than the lvs.; stigmas 3; seeds smooth. △ On the seashore: rocks, sands, shingle. May–Oct. Med. Eur. (except TR.) P.GB.

S. maritima (L.) Dumort. ANNUAL SEABLITE. Like 115, but an ann. with prostrate or erect, branched stems to ½ m. or more, but very variable. Lvs. 1–5 cm., mostly acute. Stigmas 2; seeds finely netted. The coasts of Europe. Page 73.

SALSOLA | Saltwort Lvs. fleshy, cylindrical. Fls. small, axillary; bracteoles 2; perianth segments 5 and developing a transverse membraneous wing in fr.; stamens 5; stigmas 2. 25 sps.

116. *S. kali* L. PRICKLY SALTWORT. A variable, prickly-leaved, much-branched ann. to 1 m. with numerous spreading, fleshy, cylindrical, spine-tipped lvs. 1–4 cm. Fls. green, solitary, stalkless, in axils of lvs. with 2 leaf-like, spiny-tipped bracteoles. Fr. tough, with a broad transverse wing or ridge. △ Maritime sands, waste places inland. July–Aug. All Eur. (except IS.). *The plant is burnt for its ash, which is rich in soda and was formerly used for soap and glass-making.* Page 73.

S. soda L. SALTWORT. Like 116, but lvs. with a soft, terminal, non-spiny acute tip, and lv. bases partly encircling stem; bracteoles egg-shaped. Fr. with a transverse ridge or narrow wing to 1 mm. broad. Mediterranean and South-Eastern Europe. Page 73.

AMARANTHACEAE | Cockscomb Family

Herbaceous plants, commonly anns. with entire lvs.; inflorescence usually plume-like with terminal clusters of very numerous, tiny, greenish or purplish fls. Perianth dry, membraneous, usually 4–5; stamens 1–5; ovary superior, one-celled. Fr. usually opening by a lid.

Filaments of stamens fused into a tube; fr. with numerous seeds.   *Celosia*
Filaments not fused; fr. with 1 seed.   *Amaranthus*

AMARANTHUS | Amaranth Filaments of stamens unfused; fr. one-seeded. Most sps. are weeds of waste ground and most have been introduced and a number have become widespread in Europe. 12 sps.

117. *A. retroflexus* L. PIGWEED. An erect leafy ann. to 1 m., with long, dense, terminal, greenish-white fl. spikes which are branched below. Fls. 2–3 mm., unisexual; perianth

and stamens 5; bracteoles 3–6 mm., somewhat spiny. Lvs. oval, somewhat angular, pale green, white-hairy on the veins beneath; stems furrowed, hairy. △ Cultivated ground, waste places, tracksides. July–Sept. A native of North America; now naturalized in most of Eur. (except North Eur.). Pl. 10.

*A. *albus* L. WHITE PIGWEED. Widespread in Europe.
*A. *hybridus* L. Widespread in Europe.
*A. *cruentus* L. Widespread in Europe.
*A. *blitoides* S. Watson Widespread in Europe.
*A. *graecizans* L. Widespread in Europe.
*A. *lividus* L. Widespread in Europe.

CELOSIA | **Cockscomb** Like *Amaranthus* but filaments fused below into a tube; fr. many-seeded. 1 sp.

118. *C. argentea* L. COCKSCOMB. A hairless ann. to 50 cm. with a dense, cylindrical, terminal spike of very numerous tiny, white, red, or pink fls. Lvs. linear-lance-shaped. △ Summer. An ornamental tropical sp.; often found as casual in Southern Europe.

PHYTOLACCACEAE | Pokeweed Family

Herbaceous plants, shrubs, or trees, with alternate entire lvs. Fls. in spikes; perianth 5, free and persisting in fr.; stamens several; ovary of several, partially fused, one-seeded carpels. Fr. often fleshy.

PHYTOLACCA

119. *P. americana* L. (*P. decandra* L.) VIRGINIAN POKE. A tall herbaceous perenn. 1–3 m., with thick, ribbed stem, often reddish dichotomous branches, and large oval to lance-shaped lvs. 12–25 cm. Fls. stalked, greenish or pinkish, in dense, erect, cylindrical spikes to 10 cm., arising usually in the angles of the branches. Stamens and styles 10. Fr. a globular cluster of 10 fleshy, berry-like carpels, at first reddish and then purplish-black. △ Native of North America; naturalized in Southern Europe. Summer. *Widely culti-vated for ornament and for a red dye which is obtained from the fruit and used for colouring wine, confectionery, paper, cloth, etc. The roots, leaves and fruits are purgative.* Pl. 11.

AIZOACEAE | Mesembryanthemum Family

Mainly succulent herbs or low shrubs with swollen, undivided opposite lvs. Fls. solitary, often large and conspicuous with perianth with well developed calyx and corolla. Calyx tube fused to ovary and calyx lobes often fleshy; petals coloured, numerous; stamens usually very numerous; ovary usually inferior. Fr. a capsule, woody and opening when wet, or fleshy and not splitting.

Stigmas 8–20; fr. fleshy.   *Carpobrotus*
Stigmas 5; fr. a capsule.   *Mesembryanthemum*

CARPOBROTUS   Fls. with 8–20 stigmas; lvs. triangular in section. Fr. fleshy, not splitting. 3 sps.

120. *C. edulis* (L.) N.E.Br. (*Mesembryanthemum e.*) HOTTENTOT FIG. Fls. very large, 8–10 cm. across, with numerous, shining, pale lilac, yellow, or orange petals, and numerous yellow stamens; sepals 5, green. Lvs. numerous, 7–10 cm. linear, fleshy, and triangular in section, gradually tapering to a point, in opposite pairs and fused together at the base. A somewhat woody trailing perenn. spreading to 2 m. or more and forming extensive mats. △ Coastal rocks, cliffs, and sands. Apr.–July. Native of South Africa; naturalized on the littoral of P.E.F.IRL.GB.B.I.

C. acinaciformis (L.) L. Bolus (*Mesembryanthemum a.*) RED HOTTENTOT FIG. Like 120, but lvs. glaucous and when viewed from the side appear broadest towards the tip and abruptly narrowed to an acute apex. Fls. larger to *c.* 12 cm. across, brilliant red-carmine; stamens purple. Native of South Africa; naturalized in the Mediterranean region. Pl. 11.

MESEMBRYANTHEMUM   Fls. with 5 stigmas; lvs. rounded or flattened in section. Fr. a dry capsule with 5 valves. 2 sps.

121. *M. nodiflorum* L. (*Gasoul n.*) Fls. small, white or yellowish, with numerous, very narrow petals shorter than the fleshy calyx lobes, solitary and terminal or stalkless in axils of the lvs. Lvs. oblong-cylindrical, fleshy, glaucous, with some shining glossy swellings. A more or less procumbent ann. 5–30 cm. △ Sandy and rocky sea-shores, salt marshes. Apr.–July. Med. Eur. (except AL.TR.) P.

M. crystallinum L. (*Cryophytum c.*) ICE PLANT. The whole plant is covered with transparent, shiny, glossy swellings, looking like hoar-frost. Lvs. oval, flattened, fleshy. Fls. larger 2–3 cm. across, and petals longer than the sepals. Mediterranean region.

PORTULACACEAE | Purslane Family

Herbaceous anns. or perenns., usually hairless and somewhat fleshy; lvs. entire, sometimes fused in pairs at base. Sepals 2, free or fused; petals 4–6; stamens 3 to many; ovary superior or half-inferior, one-celled. Fr. a capsule opening by valves or transversely.

Stamens numerous.   *Portulaca*
Stamens 3 or 5.   *Montia*

PORTULACA | Purslane   Fls. 1 or few, often terminal; stamens numerous; ovary half-inferior. Fr. opening by a transverse lid. 1 sp.

122. *P. oleracea* L. PURSLANE. A fleshy, hairless, spreading ann. to 50 cm., usually with solitary yellow fls. towards the end of the leafy branches. Fls. 8–12 mm., shorter than lvs.; sepals *c.* 4 mm.; petals 5–6, soon falling; stamens 7–12. Lvs. 1–2 cm. oblong-oval, fleshy, shiny, crowded below fls., stems often reddish. Var. *sativa* (Haw.) Celak. is a larger, erect plant and has long been grown as a salad plant in Southern Europe. △ Cultivated ground, vineyards, waste places. May–Oct. Most of Eur. (except North Eur. IRL.). Pl. 11.

MONTIA   Fls. in terminal clusters, white or pink; stamens 3 or 5; ovary superior. Fr. opening by 3 valves. 3 sps.

123. *M. fontana* L. BLINKS. A small, bright green plant, often forming dense, low mats on damp ground, or long, slender, branching stems in water. Fls. inconspicuous, 2–3 mm. across; petals white, unequal, scarcely longer than the 2 orbicular sepals; stamens 3. Lvs. numerous, 3–20 mm. spathulate, opposite, short-stalked. A very variable ann. or perenn. 2–50 cm. △ Springs, wet pastures. Apr.–Sept. All Eur. (except AL.TR.). Page 73.

124. *M. perfoliata* (Willd.) Howell (*Claytonia p.*) PERFOLIATE CLAYTONIA. Fls. white, clustered above a green fleshy cup formed by the 2 fused uppermost lvs. Fls. *c.* 5–8 mm. across, stalked; petals little longer than sepals. Basal lvs. in rosette, long-stalked, broadly oval, fleshy. A hairless, pale green, somewhat fleshy, erect, several-stemmed ann. 10–30 cm. △ Waste ground, sandy soils. May–July. Native of North America; introd. West Eur. (except E.IRL.) DK.D.CH.H. Pl. 11.

M. sibirica (L.) Howell (*Claytonia s.*; *C. alsinoides* Sims) PINK CLAYTONIA. Distinguished by its larger pinkish fls. 1½–2 cm. across, in lax terminal clusters; petals deeply notched, about twice as long as sepals. Uppermost lvs. oval, not fused in a cup round stem. Native of North America; sometimes naturalized in Europe. Pl. 11.

CARYOPHYLLACEAE | Pink Family

Herbaceous anns. and perenns. with pairs of opposite or rarely alternate or whorled lvs. which are narrow, undivided, and often fused together at the base. Fls. usually in dichotomously branching clusters, rarely solitary. Sepals and petals 4–5, the sepals free or fused into a tube, the petals often deeply lobed; stamens 8–10. Ovary superior, one-celled with ovules arranged on a central boss, or *free-central placentation*; styles free or fused, stigmas 2–5. Fr. usually a dry capsule, splitting by teeth or valves, rarely a berry or a dry, non-splitting carpel.

1. Sepals 5–6, fused into a tube.
  2. Styles 3–5.   Group A
  2. Styles 2.   Group B
1. Sepals 4–5, free and separate.
  3. Fr. a capsule splitting by teeth or valves.   Group C
  3. Fr. not splitting, or splitting but remaining joined above.   Group D

GROUP A. *Sepals 5–6, fused; styles 3–5*

		Nos.
1. Fr. a black berry.	*Cucubalus*	177
1. Fr. a capsule splitting by teeth.		
2. Calyx much longer than petals.	*Agrostemma*	163
2. Calyx shorter than petals.		
3. Styles usually 3; fr. splitting by 6 teeth.	*Silene*	164–176
3. Styles usually 5; fr. splitting by 5 teeth.	*Lychnis*	158–162

GROUP B. *Sepals 5–6, fused; styles 2*

		Nos.
1. Calyx with epicalyx of several scales at base.		
2. Calyx with pale, papery seams between the calyx teeth.	*Petrorhagia*	183, 184
2. Calyx green, without pale seams.	*Dianthus*	185–194

CARYOPHYLLACEAE

1. Calyx without epicalyx.
 3. Petals with scales at throat.   *Saponaria* 180, 181
 3. Petals without scales at throat.
  4. Calyx angular in section and with longitudinal wings. *Vaccaria* 182
  4. Calyx circular in section, smooth, not angled.
   5. Calyx with white, papery seams. *Gypsophila* 178, 179
   5. Calyx without white seams. *Velezia* 195

GROUP C. *Sepals 4–5, free; fr. splitting by teeth or valves*

Nos.

1. Lvs. with papery stipules; lvs. often whorled or clustered.
 2. Fls. white or pink, more than $\frac{1}{2}$ cm. across.
  3. Styles 5; fr. splitting into 5 valves. *Spergula* 154
  3. Styles 3; fr. splitting into 3 valves. *Spergularia* 155–157
 2. Fls. white, less than $\frac{1}{2}$ cm. across. *Polycarpon* 153
1. Lvs. without stipules; lvs. opposite.
 4. Petals absent.
  5. Lvs. linear. *Sagina* 146, 147
  5. Lvs. not linear. *Arenaria* 125–128
 4. Petals present.
  6. Petals deeply two-lobed.
   7. Styles 4–5; fr. with 8 or 10 teeth.
    8. Fr. much longer than calyx, with twice as many teeth
    as styles. *Cerastium* 140–144
    8. Fr. ovoid, opening by 5 bilobed teeth. *Myosoton* 145
   7. Styles 3; fr. with 6 valves or teeth. *Stellaria* 134–138
  6. Petals entire or shallowly notched, jagged, or toothed.
   9. Petals jagged or toothed. *Holosteum* 139
   9. Petals entire or shallowly notched.
    10. Styles as many as sepals.
     11. Capsule opening by 8 small teeth; styles opposite
     sepals; lvs. glaucous, strap-shaped, or narrow lance-
     shaped. *Moenchia* [144]
     11. Capsule splitting to the base into 4–5 valves; styles
     alternating with sepals; lvs. not glaucous, linear. *Sagina* 146, 147
    10. Styles fewer than sepals.
     12. Fleshy maritime plants with broad lvs. *Honkenya* 133
     12. Lvs. not fleshy.
      13. Lvs. linear. *Minuartia* 131, 132
      13. Lvs. oval.
       14. Lvs. less than 1 cm. long. *Arenaria* 125–128
       14. Lvs. more than 1 cm. long. *Moehringia* 129, 130

GROUP D. *Sepals 4–5, free; fr. not splitting or, if splitting, remaining joined above*

Nos.

1. Stipules present.
 2. Bracts usually silvery and longer than fls. *Paronychia* 149, 150
 2. Bracts inconspicuous.
  3. Fls. white, sometimes red-tipped.
   4. Sepals white, spongy; lvs. opposite. *Illecebrum* 152

4. Sepals green or red in the centre, with broad white
    margin; lvs. alternate.   *Corrigiola*   [148]
3. Fls. greenish; sepals not swollen.   *Herniaria*   151
1. Stipules absent.   *Scleranthus*   148

ARENARIA | **Sandwort**   Lvs. opposite, usually oval-lance-shaped, stipules absent. Petals not notched; stamens 10; styles 3. Capsule with 6 teeth; seeds without oily appendage. 51 sps.

Lowland and hill plants

125. *A. serpyllifolia* L. THYME-LEAVED SANDWORT. A slender, much-branched, erect or spreading ann. or bienn. 3–25 cm., with tiny oval lvs. and tiny white fls. in a lax, dichotomously branched cluster. Fls. 5–8 mm.; petals shorter than oval-lance-shaped sepals. Lvs. 2½-8 mm., oval-acute, mostly stalkless, three- to five-veined, rough, hairy, and with ciliate margin. △ Bare cultivated ground, fields, walls, sands. May–Sept. All Eur. (except IS.).

*A. leptoclados (Reichenb.) Guss. is widespread in Europe.

126. *A. montana* L. A spreading, grey-green, finely hairy perenn. with erect stems 10–30 cm., bearing one or several stalked, large white fls. to *c.* 2 cm. across. Petals obovate, twice as long as the oval, one-veined sepals. Lvs. 1–2 cm. oblong to linear-lance-shaped, one-veined. △ Heaths, woods. May–July. P.E.F. Pl. 12.

Alpine plants

127. *A. ciliata* L. HAIRY SANDWORT. A low, procumbent, hairy perenn. of alpine rocks, with erect stems to 8 cm. bearing solitary or few white fls. *c.* 1 cm. across. Petals entire, 4–7 mm., and longer than the oval-lance-shaped sepals which are ciliate at the base. Lvs. oval-acute, usually ciliate-margined; stems rough. A very variable sp. △ Limestone mountains. July–Aug. E.F.IRL.N.SF.D.CH.A.PL.CS.I.YU.R.SU.

128. *A biflora* L. TWO-FLOWERED SANDWORT. Like 127, but a hairless, slender perenn. rooting at the nodes, spreading to 20 cm., with 1 or 2 white fls. on stems 2–3 cm. long. Petals only slightly longer than sepals which are generally blunt and hairless. Lvs. oval-orbicular, blunt, lv. stalk broad, ciliate. △ Alpine screes and damp places. July–Aug. South-East Eur. (except TR.) F.D.CH.A.I.SU.

A. balearica L. A delicate, densely mat-forming, slender-stemmed, creeping perenn. with tiny, hairy, broadly egg-shaped lvs. 2–4 mm. Fls. solitary on slender stems to 6 cm.; petals twice as long as blunt sepals which are *c.* 3 mm. Mountains of Mediterranean islands.

MOEHRINGIA | **Sandwort**   Like *Arenaria*, but stamens 8 or 10; styles 2 or 3. Capsule rounded, opening by 4 or 6 recurved teeth; ripe seeds with an oily appendage. 21 sps.

129. *M. trinervia* (L.) Clairv. (*Arenaria t.*) THREE-NERVED SANDWORT. Weak spreading ann. or perenn. to 40 cm., usually with conspicuously three-veined oval lvs. and tiny white fls. *c.* 6 mm. across. Petals 5, white, shorter than the sepals which are three-veined and have a wide, papery, ciliate margin. Lvs. ½–2½ cm., acute, rarely five-veined, ciliate, the lower stalked, the upper nearly stalkless. △ Woods, hedges, shady walls. May–July. All Eur. (except IS.).

130. *M. muscosa* L. MOSSY SANDWORT. A slender, weak-stemmed, hairless, bright green mountain perenn. 8–30 cm., with very narrow, linear lvs. 1–3 cm. by ½–1 mm. Fls. white,

usually with 4 entire petals 1½ times as long as the sepals which are one-veined and have a wide, papery margin; stamens 8. △ In mountains: woods, shady places, damp rocks. June–Aug. Cent., South-East Eur. (except GR.TR.) E.F.I.SU. Page 81.

MINUARTIA | Sandwort Like *Arenaria*, but lvs. usually linear, bristly-tipped or fine-pointed. Fls. white; sepals and petals 5; stamens 10; styles usually 3. Fr. usually with 3 teeth. 57 sps.

Erect anns.
131. *M. hybrida* (Vill.) Schischkin (*M. tenuifolia* (L.) Hiern) FINE-LEAVED SANDWORT. A slender, usually glandular-hairy ann. 3–12 cm., branched above and with many tiny white fls. *c.* 6 mm. across, in a lax cluster. Petals little more than half as long as the three-veined sepals; anthers yellow. Lvs. linear awl-shaped, to 12 mm. A very variable sp. △ Sandy fields, walls, rocks. May–Sept. West, South-East Eur. (except AL.) D.CH.I.SU.: introd. IRL.DK.A.

Spreading perenns. with woody base
132. *M. capillacea* (All.) Graebner A loosely tufted mountain perenn. 8–30 cm., with white fls. and petals 1½–2 times as long as the sepals. Fls. in clusters of 1–6, fl. stalks and sepals densely glandular-hairy; sepals oval-oblong, three-veined, the outer veins disappearing in upper half. Lvs. 1–2 cm., linear, stiff, sharp-pointed. △ In limestone mountains: rocks, dry places. July–Sept. F.CH.I.YU.

M. laricifolia (L.) Schinz & Thell. (*M. striata* (L.) Mattf.) Like 132, but lvs. softer, narrower, somewhat arched and fl. stalks and sepals not glandular. Sepals linear-oblong, three-veined to apex, margin usually red. Mountains, from Central Spain to the Carpathians. Page 81.

*M. stricta (Swartz) Hiern BOG SANDWORT. Northern and arctic Europe.

M. verna (L.) Hiern VERNAL SANDWORT. A very variable, usually glandular-hairy, often glaucous, loose cushion-forming perenn. Fls. white, *c.* 6 mm. across, in a lax branched inflorescence on slender, erect stems 5–15 cm.; petals ½–1½ times as long as sepals which are broadly lance-shaped, three-veined, with papery margin, and usually glandular-hairy. Mostly in mountains of Europe, except in the north.

HONKENYA Maritime plants with fleshy lvs. Styles 3; fr. globular, opening by 3 teeth. 1 sp.

133. *H. peploides* (L.) Ehrh. SEA SANDWORT, SEA PURSLANE. A spreading, mat-forming, succulent-leaved, maritime perenn. with creeping branches rooting at the nodes. Fls. unisexual, 6–10 mm. across; petals greenish-white, entire, shorter than fleshy sepals in the female fl. and equalling in the male. Lvs. fleshy, oval-acute ½–2 cm., stalkless, numerous, and overlapping; fl. stems 5–25 cm. Capsule twice as long as sepals. △ Maritime sands, shingle. May–Aug. Coasts of West, North Eur., D.PL.SU. Pl. 11.

STELLARIA | Stitchwort Petals 5, white, usually deeply bilobed; stamens usually 10; styles 3. Fr. opening by 6 long teeth. 17 sps.

Stems cylindrical; lower lvs, stalked, oval
134. *S. nemorum* L. WOOD STITCHWORT. A straggling weak-stemmed perenn. to 60 cm., with white fls. 13–18 mm. across, in lax clusters. Petals two-lobed almost to the base, twice as

1. *Moehringia muscosa* 130
3. *Sagina nodosa* 147
5. *Spergula morisonii* [154]
7. *Herniaria glabra* 151

2. *Minuartia laricifolia* [132]
4. *Cerastium uniflorum* 143
6. *Scleranthus perennis* 148
8. *Holosteum umbellatum* 139

long as sepals; fl. stalks long, glandular-hairy. Lvs. oval-acute, the lower long-stalked, the upper stalkless; stem hairy all round. Capsule up to twice as long as sepals. △ Damp woods, marshes. May–July. Much of Eur. (except P.IRL.IS.TR.).

135. *S. media* (L.) Vill. CHICKWEED. A much-branched, weak and straggling, very variable ann. to 90 cm., usually with a single line of hairs down the stem, and with numerous small white fls. 7–8 mm. across. Petals deeply bilobed, shorter or very little longer than sepals; stamens 3–10, anthers violet. Lvs. oval-acute, the lower long-stalked, the upper stalkless. Capsule little longer than the sepals. △ Cultivated and waste ground, tracksides. All the year round. All Eur.

S. neglecta Weihe GREATER CHICKWEED. Widespread in Europe.

S. pallida (Dumort.) Piré (*S. apetala* auct.) Widespread in Europe.

Stems quadrangular; lvs. all stalkless, lance-shaped
136. *S. holostea* L. GREATER STITCHWORT. A brittle-stemmed, rather glaucous, rough perenn. to 60 cm., with a lax branched cluster of rather large white fls., each 1½–3 cm. across. Petals twice as long as sepals, bilobed to about half their length; sepals 6–8 mm., three-veined, with narrow, papery margin. Bracts green, leaf-like. Lvs. 3–8 cm., narrow lance-shaped, gradually tapering to a point, very rough on margin and midvein beneath. △ Woods, hedges. Apr.–June. All Eur. (except IS.AL.). Pl. 12.

137. *S. palustris* Retz. MARSH STITCHWORT. Like 136, but fls. medium-sized, 12–18 mm. across rather few in a terminal lax cluster. Petals as long or up to twice as long as sepals, bilobed almost to their base; sepals 6–8 mm., distinctly three-veined, with a broad, papery margin. Bracts with a broad, papery margin and narrow, central, green stripe. Lvs. 1½–5 cm., often glaucous, smooth-margined. A slender, smooth-stemmed perenn. to 60 cm. △ Marshes, fens. June–July. Much of Eur. (except P.E.IRL.IS.AL.GR.TR.).

138. *S. graminea* L. LESSER STITCHWORT. Like 136, but inflorescence more branched and fls. much smaller, but variable in size, 5–12 mm. across. Petals as long as, or a little longer than, sepals. Bracts entirely papery, with ciliate margin. Lvs. bright green, smooth-margined. A delicate, slender, smooth-stemmed perenn. to 90 cm. △ Woods, heaths, grasslands. June–July. All Eur. Pl. 12.

S. alsine Grimm BOG STITCHWORT. Most of Europe.

HOLOSTEUM   Fls. in a terminal umbel-like cluster; petals toothed or jagged; styles 3. 1 sp.

139. *H. umbellatum* L. JAGGED CHICKWEED. A slender, erect ann. to 20 cm., which is glandular-sticky above and glaucous below, with an irregular terminal umbel of few long-stalked, white or pale pink fls. Fls. *c.* 8 mm. across; petals with jagged tips, twice as long as sepals. Basal lvs. oblance-shaped, acute. Young fr. stalks deflexed. △ Sandy fields, rocks, tracksides. Mar.–May. Much of Eur. (except IRL.IS.N.SF.AL.). Page 81.

CERASTIUM | **Chickweed**   Petals 5, white, usually deeply notched or bilobed; stamens 5–10; styles usually 5. Fr. cylindrical, longer than sepals, splitting with twice as many teeth as styles. 51 sps.

Fls. more than 1 cm. across; petals much longer than sepals
(a) *Lowland plants*
140. *C. tomentosum* L. DUSTY MILLER, SNOW-IN-SUMMER. A silvery-white, densely woolly-haired perenn. to *c.* 40 cm., forming loose mats with many rooting branches, and with

erect, white-hairy fl. stems bearing a lax cluster of 7–15 fls. Fls. 12–18 mm. across, petals twice as long as hairy sepals. Lvs. densely white, hairy, linear-lance-shaped. △ Widely grown as an ornamental and often excaping. May–Aug. I.

141. *C. arvense* L. FIELD MOUSE-EAR CHICKWEED. A loose, mat-forming perenn. rooting at the nodes, with green, usually downy, lvs. and erect stems 5–30 cm. with few fls. in lax clusters. Fls. 12–20 mm. across; petals 2–3 times as long as sepals. Fl. stalks and sepals glandular-hairy; bracts ciliate with papery margin. A very variable sp. △ Meadows, grassy places, glades, in mountains. Apr.–July. Much of Eur. (except P.NL.IS.TR.).

(b) Arctic or alpine plants

142. *C. alpinum* L. ALPINE MOUSE-EAR CHICKWEED. Fls. 18–25 mm. across, 1–5 in a widely spreading cluster. Petals twice as long as the acute sepals which are 7–10 mm. Bracts with narrow, papery margin. Lvs. oval-elliptic *c.* 1 cm., usually greyish-green, softly hairy. A very variable, loose, mat-forming perenn. with fl. stems to 20 cm. △ Arctic or alpine: rocks, screes, meadows. July–Aug. Much of Eur. (except P.IRL.B.NL.DK.H.TR.). Pl. 12.

143. *C. uniflorum* Clairv. Fls. usually solitary, 1½–2 cm. across, creamy-white. Petals twice as long as sepals which are 5–7 mm. Bracts leaf-like, without papery margin. Lvs. 1–2 cm. obovate-spathulate, usually bright green. A low-growing, mat-forming perenn. with fl. stem 3–10 cm. △ In mountains: rocks, screes. June–Aug. F.CH.A.PL.CS.I.YU. Page 81.

C. cerastoides (L.) Britton Distinguished from most other sps. by the 3 styles and 6 capsule teeth. Lvs. hairless, bright green; stems hairless except for a line of hairs. Arctic alpine Europe.

Fls. 8 mm. across or less; petals shorter or little longer than sepals

144. *C. fontanum* Baumg. (*C. vulgatum* auct.) COMMON MOUSE-EAR CHICKWEED. A creeping, hairy, very variable perenn. with many upright fl. stems to 45 cm. and short, creeping, flowerless shoots. Fls. in a lax cluster; sepals 5–7 mm. with papery margin and hairless tip; petals bilobed, shorter, or little longer, than the sepals. Upper bracts with narrow papery margin. Lvs. usually 1–3 cm. oblong-oval to oblance-shaped, dark greyish-green, with dense white hairs. Capsule curved. △ Grasslands, sands, waysides, waste places. Apr.–Sept. All Eur. Pl. 12.

C. glomeratum Thuill. (*C. viscosum* auct.) STICKY MOUSE-EAR CHICKWEED. A yellow-green, usually glandular ann. with fls. and fr. heads in compact clusters. Sepals hairy to tip, with narrow, papery margin; petals deeply notched to about one-quarter their length. All Europe.

C. semidecandrum L. LITTLE MOUSE-EAR CHICKWEED. Distinguished by the lax glandular, stalked inflorescence. Sepals glandular-hairy, with broad, papery margin; petals slightly notched. All Europe.

*C. pumilum Curtis Throughout Europe.

*C. diffusum Pers. (*C. tetrandrum* Curtis) Most of Europe, except in the north.

*Moenchia erecta (L.) P. Gaertner, B. Meyer & Scherb. UPRIGHT CHICKWEED. Widespread in Central and Southern Europe.

MYOSOTON Like *Stellaria*, but styles 5; capsules ovoid opening by 5 bilobed teeth. 1 sp.

145 *M. aquaticum* (L.) Moench (*Stellaria a.*) WATER CHICKWEED. A rather fragile, scrambling, glandular-hairy perenn. 30–120 cm., with white fls. 12–15 mm. across in lax clusters.

Petals white, divided almost to base, longer than blunt sepals. Lvs. 2–5 cm., oval-acute, often wavy-margined, pale green, the lower stalked, the upper stalkless. Fr. drooping, capsule ovoid, a little longer than calyx. △ Shady and wet places, marshes, ditches. June–Aug. All Eur. (except IRL.IS.TR.). Pl. 12.

SAGINA | Pearlwort Small anns. or perenns. with linear awl-shaped lvs. Sepals and petals 4 or 5; stamens 8 or 10; styles 4–5. Fr. splitting to the base into 4–5 spreading valves. 13 sps.

Petals minute and shorter than sepals or absent

146. *S. apetala* Ard. COMMON PEARLWORT. A delicate, erect, much-branched ann. 3–10 cm., with tiny, green, long-stalked, terminal fls. Fls. *c.* 1½ mm. across; sepals 4, blunt, spreading horizontally in fr.; petals minute. Lvs. linear, with fine-pointed tip. △ Damp ground, walls, tracks. May–Oct. All Eur. (except IS.N.SF.).

S. procumbens L. PROCUMBENT PEARLWORT. Distinguished by its perenn. form with many spreading stems rooting at the nodes and often mat-forming. Fls. axillary, long-stalked; sepals usually 4, at length spreading; petals usually absent, but sometimes as long as sepals. All Europe. Pl. 13.

*S. subulata (Swartz) C. Presl AWL-LEAVED PEARLWORT. Widespread in Europe.

*S. saginoides (L.) Karsten ALPINE PEARLWORT. Widespread in Europe.

*S. maritima G. Don SEA PEARLWORT. Widespread in Europe.

Petals conspicuous, longer than sepals

147. *S. nodosa* (L.) Fenzl KNOTTED PEARLWORT. A slender, spreading perenn. to 25 cm., usually with solitary terminal and axillary, white fls. ½–1 cm. across. Petals 5, 2–3 times as long as the sepals. Lvs. linear, in clusters at the nodes, decreasing in length at the upper nodes and shorter than the internodes, giving a knotted appearance. △ Damp sands, peaty and swampy places. June–Aug. Much of Eur. (except AL.GR.TR.BG.). Page 81.

SCLERANTHUS | Knawel Sepals 5; petals absent. Lvs. awl-shaped. Fr. a one-seeded nut enclosed in the persistent sepals. 3 sps.

148. *S. perennis* L. PERENNIAL KNAWEL. A small, branched, spreading perenn. 5–20 cm., with slender glaucous lvs. and rather dense clusters of green fls. Fls. *c.* 4 mm.; sepals blunt with conspicuous white papery margin, becoming incurved in fr.; petals absent. Lvs. ½–1½ cm., curved awl-shaped; stems woody below. △ Fields, turf. May–Oct. Much of Eur. (except P.IRL.IS.S.SF.). *An insect living in the roots of this plant produces swellings known as 'grains of Poland', used in the past for dyeing and in medicine.* Page 81.

S. annuus L. ANNUAL KNAWEL. An ann. or bienn. without a woody base. Sepals more or less acute with very narrow white margin, spreading or erect in fr. All Europe. Pl. 13.

*Corrigiola litoralis L. STRAPWORT. Western, Central, and Southern Europe.

PARONYCHIA Lvs. with conspicuous, usually white, papery stipules. Fls. in axillary spherical clusters, often with conspicuous silvery bracts concealing the tiny fls. Calyx 5, often awned; petals minute or absent; stamens 5; styles 2. Fr. one-seeded. 11 sps.

Calyx with papery margin, awned

149. *P. argentea* Lam. A small, matted, prostrate perenn. spreading 5–30 cm., with tiny lvs. very conspicuous silvery stipules and dense, rounded, axillary, silvery clusters of fls.

Calyx 1½-2½ mm., awned, hooded, with papery margin; petals absent; bracts silvery, oval-acute, longer than and concealing the fls. Lvs. 4-8 mm. oval-lance-shaped. △ Dry rocky places. Apr.-June. P.E.F.I.GR. Pl. 13.

Calyx with green margin, not awned

150. *P. kapela* (Hacq.) Kerner (*P. capitata* auct.) Like 149, but dense rounded clusters of fls. very conspicuous, 7-15 mm. across; bracts 3-5 mm. orbicular-oval, blunt, silvery. Lvs. linear-lance-shaped, often hairy, as long as or longer than the stipules. A spreading, much-branched, usually mat-forming or ascending perenn. 5-15 cm. △ Sandy and rocky places in the hills. May-July. Med. Eur. (except TR.) A.BG.R.

P. capitata (L.) Lam. (*P. nivea* DC.) Distinguished by the conspicuous terminal fl. cluster *c.* 10 mm. across. Calyx very unequal; bracts silvery, 6-10 mm., much longer than and concealing the fls. Lvs. greyish-green, hairy. Mediterranean region.

HERNIARIA | **Rupture Wort** Like *Paronychia*, but bracts inconspicuous. Style absent, stigmas bilobed or notched. 15 sps.

151. *H. glabra* L. GLABROUS RUPTURE-WORT. A variable branched prostrate ann. or perenn. spreading 5-20 cm., with tiny green lvs. and dense stalkless, more or less touching clusters of tiny greenish fls. ranged along the stem. Fls. *c.* 1 mm.; sepals hairless or ciliate. Lvs. 2-7 mm. oval-elliptic, stalkless, usually hairless. △ Sandy places, fallow. May-Sept. Most of Eur. (except IRL.IS.). Page 81.

H. hirsuta L. HAIRY RUPTURE-WORT. Like 151, but lvs. greyish or whitish with dense, stiff hairs. Fl. clusters somewhat distant; sepals with spreading hairs and ending in a long bristle. Widespread except in Northern Europe.

ILLECEBRUM Like *Paronychia*, but sepals white and spongy and persisting round fr. 1 sp.

152. *I. verticillatum* L. ILLECEBRUM. A small, branched, mat-forming ann. or bienn. spreading 5-20 cm., with slender, often reddish stems, rooting at the nodes, and dense clusters of tiny, shining white fls. in the axils of the lvs. Fls. 4-5 mm., 4-6 in each cluster which is much shorter than the lvs.; sepals 2 mm., shining white, thick spongy, with a fine bristle; petals 5, white. Lvs. 2-6 mm. obovate. Fr. a dry capsule enclosed by the persistent erect sepals. △ Damp sands, gravels. June-Sept. West Eur. (except IRL.), Cent. Eur. (except H.) DK.I.YU.GR.SU. Pl. 13.

POLYCARPON | **All-Seed** Lvs. whorled or opposite; stipules papery. Sepals keeled and hooded; petals shorter than sepals; stamens 1-5; style three-lobed. Fr. three-valved, many-seeded. 4 sps.

153. *P. tetraphyllum* (L.) L. FOUR-LEAVED ALL-SEED. A small, erect ann. or rarely bienn. or perenn., with slender, dichotomously branched stems 5-15 cm. and with terminal, spreading clusters of tiny whitish fls. Fls. 2-3 mm. across; sepals keeled and hooded, with broad, white, papery margin; petals white, soon falling. Lvs. 8-13 mm. obovate, the lower in apparent whorls of 4, the upper opposite. △ Fields, tracksides, sands, rocks. May-Aug. Med. Eur. (except TR.) P.GB.D.H.BG.

SPERGULA | **Spurrey** Sepals 5; petals 5, white, entire; stamens 5-10; styles 5. Lvs. in 2 bundles at each node; stipules papery, soon falling. Capsule splitting by 5 valves. 4 sps.

154. *S. arvensis* L. CORN SPURREY. A small, erect, branched, glandular-hairy ann. 5–70 cm., with linear lvs. and small white fls. in lax, dichotomously branched clusters. Fls. 4–7 mm.; petals obovate, slightly longer than the oval sepals. Lvs. 1–3 cm. or more, fleshy, grooved beneath. Fr. stalks at first deflexed, later erect; seeds with or without a very narrow wing. △ Weed of cultivation, woods. June–Sept. All Eur. (except AL.).

S. morisonii Boreau (*S. vernalis* auct.) Like 154, but stems stiff, less branched and lvs. shorter, not grooved beneath. Seeds compressed, broadly winged. Widespread in Europe. Page 81.

SPERGULARIA | Spurrey   Sepals 5; petals 5, usually pink, entire; stamens 5–10; styles 3. Lv. bundles, if present, 1 at each node; stipules papery, persistent, united round node. Capsule with 3 valves. 17 sps.

Robust plants with thick woody base; sepals more than 4 mm.

155. *S. media* (L.) C. Presl (*S. marginata* Kittel) GREATER SEA SPURREY. A fleshy-leaved, nearly hairless, spreading perenn. 5–40 cm., of the littoral, with lax heads of pale pink or white fls. Fls. 7½–12 mm.; petals as long as or longer than the blunt sepals. Lvs. 1–2½ cm. linear-acute, usually hairless, flat above and rounded beneath; stipules broadly tri-angular. Capsule 7–9 mm.; seeds usually broadly winged. △ Salt marshes, sea-shores, saline springs inland. July–Oct. All Eur. (except IS.SF.CH.AL.).

156 *S. rupicola* Le Jolis CLIFF SPURREY. Like 155, but stems densely glandular-hairy and lvs. sparsely so; stipules conspicuous, silvery, long-pointed. Fls. deep pink; petals as long as or longer than the glandular, papery-margined sepals. Capsule 4½–7 mm.; seeds winged, with minute swellings. A branched perenn. 5–35 cm. △ Coastal cliffs, rocks, walls. June–Aug. West Eur. (except B.NL.). Pl. 13.

Slender plants with slender taproot; sepals 4 mm. or less

S. marina (L.) Griseb. (*S. salina* J. & C. Presl) LESSER SEA SPURREY. Distinguished by the petals which are white at the base and pink above, and usually shorter than the sepals which are 2½–4 mm.; stamens 1–5. Stipules short, blunt. Ann. or bienn. All Europe.

157. *S. rubra* (L.) J. & C. Presl SAND SPURREY. A spreading, often matted ann. or perenn. 5–30 cm., with small pale pink fls. 3–5 mm. Petals uniformly pink and shorter than sepals which are 3–4 mm. and have a broad, papery margin. Lvs. clustered, ½–2½ cm. linear, and tapering to bristle-like apex; stipules conspicuous, silvery, long-pointed. Seeds not winged. △ Sandy fields, tracksides, cleared woods. May–Sept. All Eur. (except IS.). Pl. 13.

LYCHNIS   Sepals fused into a tubular calyx with 5 short, tooth-like lobes; petals usually reddish, with broad spreading limb and a narrow stalk-like base or *claw*, and with scales at junction of limb and claw; stamens 10; styles usually 5. Fr. a capsule with 5 teeth. 8 sps.

Plant covered in dense shaggy white hairs

158. *L. coronaria* (L.) Desr. ROSE CAMPION. A shaggy white-haired perenn. with erect stems 30–100 cm., bearing rather few large, conspicuous, red or rarely pink or whitish fls. Fls. *c.* 3 cm. across, borne on long, shaggy-haired stalks; petals notched or entire, scales lance-shaped; calyx 1½–2 cm., shaggy-haired with acute, twisted teeth. Lvs. oval-oblong. △ Rocks, glades, but often grown as an ornamental and sometimes naturalized. May–July. South-East Eur. CS.H.I.SU.: introd. P.E.F.D.CH.A.PL.

159. *L. flos-jovis* (L.) Desr. FLOWER OF JOVE. Distinguished from 158 by the short-stalked fls. which are borne in a dense terminal flat-topped cluster and the deeply bilobed petals,

which are often further divided. Fls. *c.* 2 cm. across, purplish, scarlet or rarely white; calyx 11–13 mm., teeth short, not twisted. Lvs. lance-shaped and densely shaggy-haired; perenn. 20–90 cm. △ In mountains: meadows, screes; grown as an ornamental and sometimes naturalized. June–Aug. F.CH.I.: introd. A.CS. Pl. 14.

Plants without shaggy hairs

160. *L. flos-cuculi* L. RAGGED ROBIN. Fls. rose-red, or rarely white, very jagged in appearance owing to petals being deeply cut into 4 narrow, spreading lobes. Inflorescence lax, dichotomously branched; fls. 3–4 cm. across; calyx 6–10 mm., ten-veined, often reddish. Basal lvs. stalked, spathulate, often ciliate, the upper lvs. linear-lance-shaped. A rather rough, branched perenn. 20–90 cm. △ Damp meadows, marshes, fens. May–June. All Eur. (except TR.). Pl. 15

161. *L. viscaria* L. (*Viscaria vulgaris* Bernh.) RED GERMAN CATCHFLY. Fls. reddish-purple, rarely white, rather short-stalked, forming a loose terminal interrupted, spike-like cluster on an erect stem which is conspicuously brown and sticky below the upper nodes. Fls. *c.* 2 cm. across; petals shallowly notched, with 2 conspicuous scales; calyx 6–15 mm. Lvs. linear-lance-shaped, hairless. A very variable perenn. 15–90 cm. △ Dry meadows, clearings. May–June. Much of Eur. (except IRL.IS.). Pl. 14.

162. *L. alpina* L. (*Viscaria a.*) RED ALPINE CATCHFLY. A rather small tufted alpine or northern perenn. with bright rosy-purple or rarely white fls. in a dense, terminal cluster on a short, erect stem 5–15 cm. Fls. 10–20, each 6–10 mm. across; petals deeply bilobed; calyx 4–5 mm. Basal lvs. 1½–5 cm., linear-lance-shaped, crowded, hairless but ciliate at base, stem lvs. few, broader. △ In mountains: meadows, rocks. June–Aug. North Eur. (except DK.SF.) E.F.GB.CH.A.I.SU. Pl. 14.

AGROSTEMMA   Like *Lychnis*, but calyx with long, green, leaf-like teeth. Scales on petals absent. 3 sps.

163. *A. githago* L. (*Lychnis g.*) CORN COCKLE. An erect, softly hairy ann. 30–100 cm. with large, solitary, long-stalked, reddish-purple or rarely white fls. with long, green, narrow pointed calyx spreading much beyond the petals. Fls. 3–5 cm. across; calyx 3–7 cm., hairy, the tube ten-ribbed and the teeth longer. Lvs. 5–12 cm. linear, slightly fused at base, with greyish adpressed hairs. Fr. longer than calyx tube, opening by 5 teeth. △ Cornfields. Apr.–June. Probably originally native of the Eastern Mediterranean region, but now a weed of crops throughout Europe. *The seeds are somewhat poisonous due to the presence of saponins, and if they contaminate flour in any quantity it may become dangerous to livestock and humans.* Pl. 16.

SILENE | **Campion, Catchfly**   Calyx tubular, with 10–30 veins and 5 short, tooth-like lobes; petals 5, with a narrow stalk-like base or *claw* and a broader spreading limb, with or without scales at junction of limb and claw. Stamens 10; styles usually 3, less commonly 5. Fr. opening by twice as many teeth as styles. 166 sps.

Styles 5; usually broad-leaved plants

164. *S. alba* (Miller) E. H. L. Krause (*Melandrium a.*) WHITE CAMPION. A robust, softly hairy ann. to short-lived perenn. to 1 m., with oval to broadly lance-shaped lvs. and large white fls. in a lax cluster. Fls. 2½–3 cm. across, unisexual, opening in the evening; petals deeply bilobed, scales bilobed; calyx of male fls. 15–22 mm., ten-veined, of female fls. 2–3 cm., twenty-veined. Fr. ovoid, teeth 10, erect or spreading. △ Cultivated ground, track-sides, hedges. May–July. All Eur. (except IS.).

165. *S. dioica* (L.) Clairv. (*Melandrium d.*) RED CAMPION. Like 164, but fls. red, and open during the daytime. Fls. $1\frac{1}{2}$–$2\frac{1}{2}$ cm. across, one-sexed; petals deeply bilobed; calyx 1–$1\frac{1}{2}$ cm., often reddish-tinged. Fr. globose, with 10 recurved teeth. A hairy bienn. to perenn. to 90 cm. Often hybridizes with 164 to produce pink-flowered intermediates. △ Woods, hedges, cliffs. May–Sept. Much of Eur. (except AL.GR.TR.). Pl. 15.

Styles 3; usually narrow-leaved plants
(a) Perenns. with sterile basal shoots as well as erect fl. shoots
(i) Calyx hairy

166. *S. italica* (L.) Pers. ITALIAN CATCHFLY. Fls. yellowish-white, with petals deeply bilobed, often veined reddish or violet beneath, in a lax, pyramidal inflorescence which is sticky above. Fls. erect, short-stalked; petals with scales very small or absent; calyx $1\frac{1}{2}$–2 cm., glandular-hairy, teeth blunt. Lvs. 3–5 cm., the lower linear-spathulate, all finely hairy. A slender, branched, hairy, woody-based bienn. or perenn. 20–80 cm. △ Sands, rocks, tracksides. May–Aug. Cent., South-East Eur. P.E.F.I.SU.: introd. GB. Page 90.

167. *S. nutans* L. NOTTINGHAM CATCHFLY. Like 166, but fls. drooping, white or pinkish, and bilobed petals with narrow inrolled lobes and lance-shaped scales. Inflorescence usually in a lax one-sided cluster; calyx 9–12 mm., glandular-hairy, with 10 purplish veins, teeth acute. An erect, unbranched perenn., hairy below, sticky above, 20–60 cm. △ Rocks, sands, arid ground. May–July. All Eur. (except IRL.IS.TR.). Pl. 15.

(ii) Calyx hairless

168. *S. otites* (L.) Wibel SPANISH CATCHFLY. Fls. tiny, yellowish-green, in whorl-like clusters forming a long, narrow, spike-like inflorescence which is interrupted below. Plants largely one-sexed; petals 4–5 mm. long, linear, entire; calyx 4–6 mm. bell-shaped, hairless; fl. stalks hairless. Fr. ovoid $3\frac{1}{2}$–5 mm. A viscid, hairy bienn. or short-lived perenn. 20–50 cm. △ Sands, dry places. May–Aug. Much of Eur. (except North Eur. P.IRL.) DK.: introd. B.SF. Pl. 15.

169. *S. vulgaris* (Moench) Garcke (*S. cucubalis* Wibel) BLADDER CAMPION. Fls. white, rather large, drooping, made conspicuous by the globular, inflated calyx which is often strongly veined. Fls. $1\frac{1}{2}$–$2\frac{1}{2}$ cm. across; petals deeply bilobed, scales inconspicuous; calyx with 20 netted veins. A very variable sp. Subsp. *maritima* (With.) A. & D. Love SEA CAMPION is a low-growing, mat-forming perenn. with larger and fewer fls. and wide-throated calyx, of the coasts of Western Europe. Subsp. *glareosa* (Jordan) Marsden-Jones & Turrill and subsp. *prostrata* (Gaudin) Chater & Walters ALPINE BLADDER CAMPION are cushion-forming mountain plants of Central and Southern Europe. Usually hairless, often glaucous perenn. to 60 cm. △ Cultivated ground, sands, rocks, alpine pastures. Apr.–Aug. All Eur. Pl. 14.

170. *S. acaulis* (L.) Jacq. MOSS CAMPION. A spreading, bright green, moss-like, mat-forming perenn. of arctic or alpine regions. Fls. 9–12 mm. across, rose-pink or whitish, solitary, with a short stalk 2–10 cm.; petals notched, scales bilobed; calyx 7–9 mm., bell-shaped, hairless, and often reddish. Lvs. in a rosette, linear-acute, 6–12 mm., stiffly ciliate. △ Damp rocks, crevices. June–Aug. Much of Eur. (except P.B.NL.DK.H.AL.GR.TR.). Pl. 16.

171. *S. rupestris* L. ROCK CAMPION. A slender, glaucous, hairless, erect, branched mountain perenn. 5–25 cm., with a lax, spreading cluster of small usually pink fls. on long slender stalks. Fls. c. $\frac{1}{2}$ cm. across; petals pink or white, deeply notched, with acute scales; calyx 4–6 mm. obconical, hairless, ten-veined. Basal lvs. oblance-shaped, stem lvs. lance-shaped. △ In the mountains: poor and dry pastures, rocks, sandy places. June–Sept. E.F.N.S.SF.D.CH.A.I.R.SU. Page 90.

(*b*) *Anns. or bienns. with erect fl. shoots only*
(*i*) *Calyx hairy*
172. *S. noctiflora* L. (*Melandrium n.*) NIGHT-FLOWERING CATCHFLY. Like 164, but fls. hermaphrodite, sweet-scented, and petals rosy above, yellowish beneath and rolled inwards during the day and opening in the evening. Fls. *c.* 2 cm. across; calyx 2–3 cm., woolly-haired, sticky, teeth long, slender; styles 3. Lvs. oval to oval-lance-shaped, all acute. Ann. or bienn. with erect, little-branched stems to 60 cm., sticky glandular-hairy above. △ Fields. July–Sept. Much of Eur. (except P.SF.AL.GR.): introd. IRL.IS.DK.N.S.

173. *S. viscosa* (L.) Pers. (*Melandrium v.*) STICKY CATCHFLY. A robust, densely glandular-hairy, sticky bienn. to 60 cm., with whorl-like clusters of large white fls. forming a narrow, elongated inflorescence. Fls. *c.* 2 cm. across; petals bilobed, without scales; calyx 1½–2½ cm., teeth blunt. Lvs. oval-lance-shaped with undulate margin. △ Dry grassy places. June–July. Cent. Eur. (except CH.) DK.N.SF.YU.BG.R.SU.

174. *S. gallica* L. (*S. anglica* L.) SMALL-FLOWERED CATCHFLY. Fls. small, white or pink, arranged closely along one side of the erect branches. Fls. *c.* 1 cm. across, short-stalked, erect; petals rounded or notched, little longer than calyx, scales 2; calyx 7–10 mm., rough-haired, teeth acute. Var. *quinquevulnera* (L.) Koch has petals with a conspicuous crimson spot. A very variable sp. An erect, little- or much-branched, hairy ann. 15–45 cm. △ Cultivated ground, sandy places, tracksides. May–July. Much of Eur. (except North Eur.) DK. Pl. 16.

S. colorata Poiret A hairy ann. with bright pink or white fls. 1–2 cm. across; calyx 11–13 mm. cylindrical, becoming club-shaped in fr., teeth oval blunt, densely hairy. Fr. 7–9 mm., carpophore 5–7 mm. Southern Europe. Pl. 15.

175. *S. conica* L. STRIATED CATCHFLY. A greyish hairy, erect ann. with sticky glandular stems, 10–50 cm., bearing small, bright pink fls. and conspicuous, swollen, conical-ovoid calyx in fr. Fls. 4–5 mm. across, few in a lax cluster; petals notched, scales present; calyx 8–18 mm., densely glandular-hairy tube with 30 conspicuous veins, teeth narrow, long-pointed. Lvs. linear-lance-shaped, downy. △ Inland and maritime sandy places. May–July. Much of Eur. (except North Eur. P.IRL.B.PL.): introd. DK.CH.

S. conoidea L. from the Western Mediterranean region has larger fls., and inflated calyx 22–28 mm. in fr. which is narrowed to a long apex. Lvs. greenish, sticky, less densely hairy, wider.

(*ii*) *Calyx hairless*
176. *S. armeria* L. SWEET-WILLIAM CATCHFLY. A hairless ann. or bienn. 10–40 cm., with simple, erect stems, glaucous, stalkless lvs. and a dense, flat-topped cluster of usually pink fls. Fls. numerous, 1½ cm. across, short-stalked; petals pink or white, notched, scales 2, acute; calyx 12–15 mm. cylindrical-club-shaped, teeth blunt, hairless. Lvs. oval-heart-shaped to lance-shaped, clasping the stem, decreasing in size above. △ Woods, dry places, cultivated ground; widely grown for ornament and sometimes naturalized. July–Sept. Much of Eur. (except IRL.IS.): introd. GB.B.NL.DK.N.S.SF. Page 90.

CUCUBALUS Like *Silene*, but fr. a shining berry, very loosely invested by the bell-shaped calyx. 1 sp.

177. *C. baccifer* L. BERRY CATCHFLY. A hairy, pale green, brittle-stemmed, scrambling perenn. to 2 m., with rather large greenish fls. and very conspicuous, wide, bell-shaped calyx. Inflorescence few-flowered, lax; fls. 1½–2 cm. across, drooping; petals greenish-white, widely spaced, narrow, bilobed, jagged, scales large; calyx five-toothed. Lvs.

1. *Dianthus gratianopolitanus* 192
3. *Gypsophila fastigiata* [178]
5. *Velezia rigida* 195
2. *Silene rupestris* 171
4. *Silene italica* 166
6. *Dianthus barbatus* 185   7. *Silene armeria* 176

oval-acute. Fr. 6–8 mm., globular, black, exposed from calyx. △ Hedges, woods, banks. July–Sept. All Eur. (except North Eur. IRL.): introd. GB. Pl. 15.

GYPSOPHILA Calyx tubular, five-veined, with pale, papery seams between the veins; petals without scales; styles 2. Fr. with 4 teeth. 28 sps.

Perenns. with a woody base
178. *G. repens* L. CREEPING GYPSOPHILA. A spreading, matted perenn. to 25 cm., with glaucous linear lvs. and white or pinkish fls. in a lax, somewhat flat-topped cluster. Fls. 5–8 mm. across, stalked; petals notched, without scales, twice as long as calyx or more; calyx teeth fine-pointed, with conspicuous white papery margins. Lvs. 1–3 cm., often curved, hairless. △ In mountains: rocks, screes, dry pastures. June–Sept. Cent. Eur. (except H.) E.F.I.YU. Pl. 16.

G. fastigiata L. FASTIGIATE GYPSOPHILA. Like 178, but a more robust plant with fls. in a dense, terminal, nearly flat-topped cluster, and inflorescence glandular-hairy. Petals 1½ times as long as the blunt-toothed calyx. Mainly Central and Eastern Europe. Page 90.

Slender anns.
179. *G. muralis* L. A very slender, branched ann. 5–25 cm. with many tiny, rosy, red-veined fls. in a lax, branched, terminal cluster. Fls. *c.* 4 mm. across; petals entire, twice as long as calyx; fl. stalks very slender, several times longer than calyx. Lvs. ½–2½ cm. linear, glaucous. △ Sandy places. June–Sept. Much of Eur. (except P.IRL.GB.IS.N.AL.): introd. DK.

SAPONARIA | **Soapwort** Like *Dianthus* in having a smooth, tubular calyx, but without an epicalyx; styles 2. 10 sps.

180. *S. ocymoides* L. ROCK SOAPWORT. A spreading, loosely mat-forming perenn. 10–30 cm., with bright pink fls. and conspicuous, often purplish, glandular-hairy cylindrical calyx. Inflorescence lax, fl. stalks glandular-hairy; fls. *c.* 1 cm. across; petals not notched, scales blunt; calyx 7–12 mm., teeth blunt; anthers blue. Lvs. ciliate glandular, the lower oval-elliptic, blunt, the upper narrower, acute. △ In mountains: rocks, walls, tracksides, sandy slopes. May–July. E.F.D.CH.A.CS.I.YU.: introd. GB.DK.CS. Pl. 16.

S. bellidifolia Sm. has yellow fls. in a dense globular cluster and long-projecting stamens with yellow stalks. Mainly from the Balkans.

181. *S. officinalis* L. SOAPWORT, BOUNCING BETT. A stout perenn. with creeping rhizomes and many erect fl. stems, 30–90 cm. with dense terminal clusters of large pink or flesh-coloured fls. Fls. 2½–3 cm. across, with spreading, widely spaced, shallowly notched petals, and 2 scales to each petal; calyx *c.* 2 cm., tubular, hairless, green or reddish. Lvs. 5–10 cm. broadly elliptic-oval, three-veined. △ Hedges, waste damp places, alluvium, railway-embankments. June–Sept. Much of Eur. (except NL.IS.): introd. North Eur. IRL. GB. *The plant contains saponins and is used for washing and cleaning woollen materials, tapestries, etc. Saponins are poisonous. A herbal remedy for rheumatism, gout, skin diseases, etc.* Pl. 16.

VACCARIA Like *Saponaria*, but calyx inflated and angular in section with 5 green wings; petals without scales. 1 sp.

182. *V. pyramidata* Medicus (*Saponaria vaccaria* L.) COW BASIL. An erect, dichotomously branched, glaucous-leaved hairless ann. 30–60 cm., with a lax, much-branched cluster of pink fls. with conspicious winged calyx. Fls. long-stalked, *c.* 1–1½ cm. across; petals

toothed or entire, scales absent; calyx inflated, 12–15 mm., with 5 sharp green wings, and 5 often purplish teeth; bracts papery with a green midvein. Lvs. *c*. 5 cm. oval to lance-shaped. △ Cornfields. May–July. Much of Eur. (except North Eur. IRL.GB.PL.): introd. NL.

PETRORHAGIA Like *Dianthus* with epicalyx composed of 2 or more membraneous scales at base of a single fl. or with several bracts at the base of a dense fl. head; calyx five–fifteen-veined, with whitish, papery seams; petals without scales; styles 2. 16 sps.

Fls. solitary in a lax inflorescence
183. *P. saxifraga* (L.) Link (*Tunica s.*) TUNIC FLOWER. Fls. pale rose with darker veins, or white, solitary at the ends of slender erect branches in a lax inflorescence. Petals 4½–10 mm., spreading, notched; calyx 3–6 mm., teeth blunt; epicalyx usually papery, linear acute, about half as long as calyx. Lvs. *c*. 1 cm. linear, rough-margined. A slender, erect, branched perenn. 10–45 cm. △ Rocks, sands, walls, gravel. June–Aug. Much of Eur. (except North Eur. IRL.B.PL.): introd. GB.NL.

Fls. clustered in a dense head
184. *P. prolifera* (L.) P. W. Ball & Heywood (*Tunica p.*; *Kohlrauschia p.*) PROLIFEROUS PINK. A slender, unbranched ann. 10–50 cm., with a solitary ovoid fl. head encircled by shining brown, membraneous bracts and pale pink fls. which open one at a time. Fls. small 6–8 mm. across; petals notched; bracts oval. Lvs. linear-lance-shaped, rough-margined, fused at the base into a sheath which is as broad as long. △ Dry fields, track-sides. May–Sept. Much of Eur. (except North Eur. P.IRL.) DK.S.

P. velutina (Guss.) P. W. Ball & Heywood (*Tunica v.*) Very like 184, but middle part of stem densely glandular-hairy and lv. sheath at least twice as long as broad. Outer bracts of fl. head acute. Mediterranean region.

DIANTHUS | **Pink** Calyx closely encircled by an epicalyx of 1–3 pairs of scales; calyx tubular, five-toothed; petals 5 with a broad, toothed, or entire limb and a narrow, stalk-like base or *claw*, scales absent; styles 2. Fr. opening by 4 teeth. The epicalyx closely invests the calyx and should not be confused with the bracts which subtend the fls. in an in-florescence. 121 sps.

Fls. clustered in dense heads and surrounded by bracts
185. *D. barbatus* L. SWEET WILLIAM. A rather robust hairless perenn. to 60 cm., with a dense, flat-topped, terminal cluster of numerous red fls., spotted with white. Bracts sur-rounding fl. cluster leafy, spreading, or reflexed, and about as long as fls. Fls. almost stalkless, *c*. 1½ cm. across; epicalyx green, long-pointed, longer than calyx. Lvs. lance-shaped, *c*. 1 cm. broad, with prominent midvein. △ Woods, grassy places in mountains; widely grown as an ornamental and often naturalized. June–Aug. Much of Eur. (except IRL.B.IS.CH.AL.GR.): introd. P.GB.NL.DK.N.S.SF.D. Page 90.

186. *D. armeria* L. DEPTFORD PINK. An erect hairy ann. or bienn. 30–60 cm., with a dense terminal cluster of red fls., spotted with white, surrounded by erect, leafy, hairy bracts as long as the fls. Fls. *c*. 8 mm. across; petals narrow, widely spaced; calyx woolly-haired, about as long as the pointed epicalyx scales. Basal lvs. oblong blunt, in a rosette, stem lvs. linear acute, all hairy. △ Woods, banks, sands. June–Aug. Much of Eur. (except IRL. IS.N.SF.). Pl. 17.

187. *D. carthusianorum* L. CARTHUSIAN PINK. Fls. deep pink, purple, or rarely white, usually in a dense head, surrounded by short, scaly, or leafy bracts. Fls. numerous or few; petals oval, toothed, contiguous; calyx 1–2 cm.; epicalyx obovate, fine-pointed, her-

baceous or leathery. Lvs. linear long-pointed, flat, with sheaths several times as long as wide. A very variable sp. A hairless perenn. to *c.* 60 cm. △ Dry pastures, rocks, woods. June–Oct. Much of Eur. (except North Eur. P.IRL.GB.GR.BG.). Pl. 17.

Fls. solitary or only a few, and not in compact heads
(a) Petals deeply cut into many narrow strap-shaped lobes
188. *D. superbus* L. SUPERB PINK. Fls. large, 3–6 cm. across, pink or lilac, with petals deeply cut to more than half their width into narrow, comb-like lobes. Fls. solitary or few, sweet-scented; petals with uncut part of blade oblong, green-spotted; epicalyx 2–4, oval long-pointed, one-quarter to one-third as long as calyx, which is 1½–3 cm. Lvs. soft, the lower three-veined. A hairless, erect perenn. branched above, 30–80 cm. △ Woody hills, dry meadows. June–Aug. Much of Eur. (except P.E.IRL.GB.B.IS.AL.GR.TR.). Pl. 17.

189. *D. plumarius* L. COMMON PINK. Fls. large 2–4 cm. across, solitary, very sweet-scented, white or bright pink, with petals divided to about half-way into narrow lobes. Epicalyx usually 4, obovate, short-pointed, about one-quarter as long as calyx, which is 17–30 mm. Lvs. stiff acute, *c.* 1 mm. wide. A glaucous perenn. to 40 cm. △ Mountains; widely cultivated and sometimes naturalized. Apr.–July. Cent. Eur. (except D.CH.) I.

190. *D. monspessulanus* L. FRINGED PINK. Fls. usually 2–5, 2–3½ cm. across, pink or whitish, with petals cut into narrow teeth to about half-way and with uncut part of blade oval. Fls. short-stalked, fragrant; calyx 18–25 mm., teeth narrow-pointed; epicalyx 4, oval with a long-pointed green tip, one-third to one-half as long as calyx. Lvs. long thin, soft and very acute. A lax, tufted, often glaucous perenn. 20–50 cm. △ Rocks, meadows in the mountains. May–Aug. P.E.F.CH.A.I.YU. Pl. 17.

(b) Petals with shallow teeth
(i) Petals hairy above
191. *D. seguieri* Vill. Fls. reddish-pink, often white-spotted and with a purple ring round the centre, and with toothed petals which are hairy above. Fls. one to several, 2–4 cm. across, sweet-scented; epicalyx 2–6, oval-lance-shaped and abruptly narrowed to an awl-shaped point, one-third to three-quarters as long as calyx, which is 14–20 mm. A loosely tufted, green, hairless perenn. 30–60 cm. △ Meadows, rocks in the mountains. July–Aug. E.F.D.CS.I.

192. *D. gratianopolitanus* Vill. CHEDDAR PINK. Fls. pink, sweet-scented, with toothed, hairy, unspotted petals. Fls. usually solitary, 1½–3 cm. across; epicalyx 4–6, green, obovate, very shortly pointed, lined and veined with red, one-quarter to one-third as long as calyx, which is 13–17 mm. A glaucous, hairless, densely tufted perenn. 6–25 cm. △ Rocks, grassy places in mountains. May–July. Cent. Eur. (except H.) F.GB.B.I.SU. Page 90.

193. *D. deltoides* L. MAIDEN PINK. Fls. solitary or in a branched head of 2–3, small *c.* 18 mm. across, deep pink or rarely white, with darker basal band and pale spots, scentless. Epicalyx usually 2, oval and narrowed to a long point, green with papery margin, about half as long as calyx, which is 14–18 mm. Stems and margins of lvs. rough, shortly hairy. A loosely tufted, green or glaucous perenn. 15–45 cm. △ Dry fields, woods. June–Sept. All Eur. (except IRL.IS.).

(ii) Petals without hairs
194. *D. sylvestris* Wulfen WOOD PINK. Fls. usually pink with entire or toothed, hairless petals, usually scentless. Epicalyx 2–5, broadly obovate, blunt, leathery, one-quarter as long as calyx, which is 12–29 mm. Basal lvs. very slender and less than 1 mm. wide, wiry, rough-margined, often recurved. A very variable sp. A densely tufted perenn. with fl. stems to 60 cm. △ Walls, rocks, dry places. June–Aug. Med. Eur. (except TR.) D.CH.A. Pl. 17.

D. caryophyllus L. CLOVE PINK, CARNATION. A glaucous perenn. with broad, soft, nearly flat lvs. and very fragrant, often solitary, purple, rarely pink or white fls. Epicalyx 4, broadly obovate with a broad-based point, one-quarter as long as calyx, which is 2½–3 cm. Widely cultivated and frequently naturalized. *Cultivated since the Middle Ages for 'Carnation Oil', which in its pure form is used only in the most expensive perfumes.*

VELEZIA Like *Dianthus*, but without epicalyx and usually anns. Calyx a long, cylindrical, smooth tube, with 5 teeth; styles 2. 2 sps.

195. *V. rigida* L. A stiff, much-branched, spreading or ascending ann. to 15 cm., with glandular-hairy stems and tiny stalkless pink fls. in the axils of the linear lvs. Fls. *c.* 5 mm. across; calyx tube conspicuously long and narrow, little broader than the stems of the plant, glandular-hairy; petals deeply bilobed with linear lobes. △ Dry, stony places. May–July. Med. Eur. P.BG. Page 90.

NYMPHAEACEAE | Water-Lily Family

Water or marsh plants with rounded, floating lvs., and often submerged lvs., and with stout, creeping, underwater rhizomes. Fls. solitary, generally floating, arising on long stalks directly from the rhizome. Sepals 3–6; petals and stamens numerous; ovary of many fused carpels. Fr. usually a spongy capsule which splits by the swelling of the mucilaginous inner layers.

NYMPHAEA Petals white, the outer much longer than the sepals; sepals 4, green beneath. Ovary half-inferior. 4 sps.

196. *N. alba* L. (*Castalia a.*) WHITE WATER-LILY. Fls. white, very large 10–20 cm., fragrant, floating, long-stalked, with numerous spreading petals; sepals white above. Lvs. all floating, 10–30 cm. across, blades almost circular, but with a deep narrow notch, reddish beneath. Ovary globular, stigma flat, with numerous, yellow radiating rays. △ Still or slow-flowing waters to a depth of 2 m. June–Sept. All Eur. (except IS.). *The underground stems are sometimes eaten in Northern Europe.* Pl. 18.

NUPHAR Petals yellow, much shorter than sepals; sepals 5–6, yellowish. Ovary superior. 2 sps.

197. *N. lutea* (L.) Sibth. & Sm. YELLOW WATER-LILY, BRANDY-BOTTLE. Fls. yellow, 4–6 cm. across, usually held above the water surface, smelling of alcohol; sepals arching over petals. Floating lvs. oval in outline, with a wide, deep notch, thick leathery, green beneath; submerged lvs. thin transparent, wrinkled. Ovary with 15–20 stigma-rays. △ Still and slow-flowing waters to a depth of 3 m. June–Aug. All Eur. (except IS.). Pl. 18.

*N. pumila (Timm) DC. LEAST YELLOW WATER-LILY. Like 197 but fls. smaller, 1½–2½ cm., and ovary with 8–10 stigma-rays. Often hybridizes with 197. Northern and Central Europe.

CERATOPHYLLACEAE | Horn-Wort Family

A tiny family of one genus and a few sps. of submerged aquatic plants of cosmopolitan distribution. Fls. unisexual. Perianth of 8–12 linear segments; male fls. with numerous stalkless anthers; female fls. a single, stalkless, superior ovary. Fr. a one-seeded nut.

CERATOPHYLLUM   Lvs. in whorls of 3–8. Fls. minute, axillary, rather rarely formed. 4 sps.

198. *C. demersum* L. HORN-WORT. An underwater perenn. 30–150 cm. with slender stems and numerous rather stiff whorls of dense, dark green, narrow-lobed lvs., and minute, greenish, stalkless fls. in their axils. Lvs. not collapsing out of water, once or twice forked, with toothed lobes; stems and lvs. brittle. Fr. with 2 spreading basal spines and a terminal spine as long as the fr. △ Still and slow-flowing waters. June–Aug. All Eur. (except IS.).

C. submersum L. SPINELESS HORN-WORT. Very like 198, but fr. without basal spines and with or without a short terminal spine. Lvs. bright green, less stiff, forked 3–4 times, lobes sparingly toothed. Much of Europe, except the northern regions.

RANUNCULACEAE | Buttercup Family

Herbaceous plants usually with dissected, alternate lvs. and conspicuous fls. (*Clematis* is woody and has opposite lvs.). Fls. bisexual; perianth often of 3 sets of organs: sepals, petals, and nectaries, but often when one or more is missing. The perianth may be all coloured and petal-like, or green and sepal-like, and the nectaries either funnel-shaped or petal-like. Stamens numerous. Ovary often of many. one-seeded carpels, or less commonly of fewer many-seeded carpels. This family is considered by many botanists to be one of the most primitive. The fl. is differently interpreted by botanists, the terms sepals, petals, and nectaries are here generally used in their popular sense, i.e. sepals green and protective, petals coloured and attractive, nectaries giving nectar. Many members of the family are acrid and poisonous to both man and livestock.

1. Carpels few, usually 1–7, each several-seeded and splitting down one side,
     or rarely a fleshy berry.
  2. Fls. without spur or hood, regular and symmetrical.   Group A
  2. Fls. with spur or hood, usually irregular, i.e. symmetrical in one plane
       only.   Group B
1. Carpels numerous, more than 7, each one-seeded and not splitting (except
     Trollius).
  3. Stem lvs. absent or in an apparent whorl.   Group C
  3. Stem lvs. numerous, opposite or alternate.   Group D

GROUP A. *Carpels 1–7, each several-seeded; fls. without spur or hood*
                                                                                  Nos.
1. Carpels 1; fr. a black berry.   *Actaea*   206
1. Carpels several.
  2. Fls. yellow.
    3. Sepals green; involucre absent.   *Caltha*   207
    3. A deeply lobed, leaf-like involucre present.   *Eranthis*   202
  2. Fls. white, green, or bluish.
    4. Lvs. finely divided into narrow segments; fls. usually
         blue or bluish.   *Nigella*   203
    4. Lvs. palmate or trifoliate.
      5. Petals white, soon falling; carpels flattened.   *Isopyrum*   205
      5. Petals green or rarely white, persisting; carpels
           swollen.   *Helleborus*   199–201

GROUP B. *Carpels 1–7, each many-seeded; spur or hood present*

		Nos.
1. Carpels 1; fls. with a long spur.	*Consolida*	213
1. Carpels 3–5.		
2. Hood absent; spurs projecting.		
3. Spur 1.	*Delphinium*	211–212
3. Spurs 5.	*Aquilegia*	253, 254
2. Hood present; spurs included in hood.	*Aconitum*	208–210

GROUP C. *Carpels more than 7, each one-seeded; stem lvs. absent or in a whorl*

		Nos.
1. Carpels with long feathery styles.	*Pulsatilla*	220–223
1. Carpels with very short styles.		
2. Lvs. all linear; fls. green.	*Myosurus*	252
2. Lvs. at least oval in outline and lobed; fls. not green.		
3. Basal lvs. with 3 broad, untoothed lobes.	*Hepatica*	219
3. Basal lvs. with more than 3 toothed or dissected lobes.	*Anemone*	214–218

GROUP D. *Carpels more than 7, each one-seeded (except Trollius); stem lvs. numerous*

		Nos.
1. Lvs. opposite; usually woody climbers.	*Clematis*	224–229
1. Lvs. alternate; herbaceous plants.		
2. Carpels many-seeded; fls. solitary yellow.	*Trollius*	204
2. Carpels one-seeded.		
3. Fls. inconspicuous, greenish.	*Thalictrum*	255–257
3. Fls. conspicuous yellow, white or red.		
4. Petals with a flap-like nectary; fls. yellow or white.	*Ranunculus*	232–251
4. Petals without nectary; fls. red or yellow.	*Adonis*	230, 231

HELLEBORUS | **Hellebore, Bear's Foot**   Fls. solitary or clustered, green, white, or dull purple; perianth 5; nectaries numerous, tubular; stamens numerous. Carpels 3–8, many-seeded. Lvs. palmate with toothed segments. 11 sps.

Stems and lvs. persisting through winter

199. *H. foetidus* L. SETTERWORT, STINKING HELLEBORE, BEAR'S FOOT. Fls. numerous in branched, rather one-sided, drooping clusters with rather globular-bell-shaped green perianth with a reddish-purple border, each 1–3 cm. across. Lvs. with 7–11 dark green, narrow-lance-shaped, toothed segments, contrasting with the pale green, broadly oval upper bracts; stems stout, with a rosette of over-wintering lvs. above. Carpels 2–5, usually fused slightly below. A fetid, hairless perenn. 20–80 cm. △ Woods, grassy and rocky places. Jan.–Mar. West Eur. (except IRL.NL.) D.CH.I. *All parts of the plant are poisonous to man and livestock and the poisonous glycoside persists in dried and stored plants. Used in the past as a vermifuge and against lice.* Pl. 19.

H. lividus Aiton A stout-stemmed, over-wintering plant with tough lvs. with 3 broad, leathery leaflets with spiny-toothed or entire margin, and open, pale greenish to dull-purple, cup-shaped fls. Western Mediterranean islands.

Stems dying down in winter

200. *H. viridis* L. BEAR'S FOOT, GREEN HELLEBORE. Fls. few in a spreading, branched leafy cluster, half-drooping, open cup-shaped, green, each 4–5 cm. across. Lvs. deciduous, arising mostly from the rootstock, with 7–13 lance-shaped, strongly toothed, and sometimes further lobed segments; bracts similar to lvs., stalkless. Carpels fused at base. A

hairless perenn. 20–40 cm. △ Woods, thickets. Mar.–May. E.F.GB.B.D.CH.A.I.: introd. PL.CS. *Poisonous and violently purgative.*

H. cyclophyllus Boiss. Like 200, but green fls. larger, *c.* 6 cm. across, and carpels not fused at base. Balkan Peninsula. Pl. 18.

201. *H. niger* L. CHRISTMAS ROSE. Fls. white or pinkish, large, 3–10 cm., often solitary, on a stout, almost leafless stem. Basal lvs. persistent, evergreen, with 7–9 dark green, toothed leaflets; stem lvs. entire, small oval, paler; fl. stem dying in winter. Perenn. with fl. stems 15–30 cm. △ In mountains: woods, rocky places. Jan.–Apr. D.CH.A.I.YU.: introd. F.PL.SU. *The roots are known as 'black hellebore'; they are violently purgative. Very poisonous.* Pl. 19.

ERANTHIS |Winter Aconite Fls. solitary; sepals absent, but leaf-like epicalyx present; petals usually 6; nectaries tubular. Carpels usually 6, many-seeded. 1 sp.

202. *E. hyemalis* (L.) Salisb. WINTER ACONITE. Fls. solitary, globular, golden-yellow, encircled by bright green, deeply cut, leaf-like bracts and borne on a short leafless stem arising from the rootstock. Lvs. all basal, appearing after the fls., long-stalked, with rounded blades which are deeply dissected into oblong segments resembling the bracts. Tuberous perenn. with fl. stems 5–15 cm. △ Damp woods, copses, orchards. Feb.–Mar. F.I.YU.BG.: introd. GB.B.NL.D.CH.A.CS.H.R. *Poisonous.* Pl. 18.

NIGELLA | Love-in-a-Mist Fls. solitary; sepals absent; petals, nectaries, and carpels 5. Carpels more or less fused along their inner margins, many-seeded. Lvs. much-divided into narrow segments. 12 sps.

203. *N. damascena* L. LOVE-IN-A-MIST. Fls. usually bluish, *c.* 3 cm. across, closely invested by an involucre of finely divided green lvs., resembling the stem lvs. Petals oval-oblong, bluish, claw shorter than limb. Lvs. 2–3 times cut into narrow linear segments. Fr. globular, inflated, carpels fused their whole length; styles spreading. A rather slender, erect, branched ann. 20–40 cm. △ Fields, rocky places. June–July. Med. Eur. P.BG.SU.: introd. B.NL.A.PL.CS.H.R. *The seeds are sometimes used as a condiment.* Pl. 18.

N. arvensis L. Distinguished from 203 by the long-stalked fls. which are not or rarely surrounded by an involucre of upper lvs. Carpels three-veined, fused to half their length. Fls. pale blue, often veined with green. A weed of most of Europe, except the northern regions.

N. sativa L. Like 203 with carpels fused their full length, but carpels with small swellings on their backs, not smooth. Fls. whitish, without an involucre of lvs. Frequently naturalized in Southern Europe. *The seeds are used as a condiment in cakes, curries, etc.* Pl. 18.

TROLLIUS | Globe Flower Fls. usually solitary; sepals and involucre absent; petals yellow 5–15; nectaries yellow, 5–15, strap-shaped. Carpels numerous, many-seeded. 2 sps.

204. *T. europaeus* L. GLOBE FLOWER. An erect, little-branched, hairless perenn. 10–70 cm., with large, globular, yellow or greenish-yellow, long-stalked fls. Fls. 2½–5 cm. across; petals orbicular, overlapping and incurved. Basal lvs. dark green, stalked, palmately three- to five-lobed, with lobes further lobed or toothed, the upper lvs. stalkless. △ Usually in mountains: damp pastures, thickets. May–Aug. Much of Eur. (except P.E.NL.IS.GR.TR. BG.). *Poisonous.* Pl. 19.

ISOPYRUM Fls. small, white, solitary; sepals absent; petals 5; nectaries small or absent. Carpels usually 2, many-seeded. 1 sp.

205. *I. thalictroides* L. RUE-LEAVED ISOPYRUM. A slender, sparingly branched, rather glaucous perenn. 10–30 cm., with white fls. 1–2 cm. across and with petals readily falling. Lvs. trifoliate, with leaflets further divided into three-lobed, oval segments; stipules conspicuous, papery. △ Damp, shady woods. Mar.–May. Cent. Eur. (except D.) E.F.I.YU. BG.R.SU. *Poisonous.* Page 109.

ACTAEA | **Baneberry** Fls. clustered, small, white; sepals absent; petals 3–5; nectaries 4–10, small. Fr. a fleshy berry, several-seeded. 2 sps.

206. *A. spicata* L. BANEBERRY, HERB CHRISTOPHER. Fls. white, *c.* 1 cm., numerous in a dense, terminal, long-stemmed cluster 25–50 cm. Fls. short-stalked, stalks hairy; petals soon falling; stamens white. Lower lvs. large, stalked, hairy beneath, once or twice divided into oval toothed segments, the upper much smaller, stalked. Fr. globular, *c.* 12 mm., green then shining black. An unpleasant-smelling, little-branched perenn. 30–60 cm. △ Damp woods. June–July. Much of Eur. (except P.IRL.IS.TR.). *The fruits and whole plant are poisonous; the fruit yields a black dye.* Page 109.

CALTHA | **Marsh Marigold** Lvs. entire, rounded. Fls. few, in a lax, branched cluster; perianth of only 1 whorl of 5–8 glossy yellow petals; sepals and nectaries absent. Carpels 5–13, many-seeded. 1 sp.

207. *C. palustris* L. MARSH MARIGOLD, KINGCUP. A rather fleshy, shiny, hollow-stemmed perenn. 5–30 cm., with rounded lvs. and several bright, glossy yellow fls. in a lax, leafy cluster. Fls. 1½–5 cm. across; petals 5–8, spreading, often greenish beneath. Basal lvs. stalked, more or less kidney-shaped, toothed, shiny, the upper smaller, nearly stalkless. A very variable sp. △ Fens, marshes, damp woods, ditches. Mar.–May. All Eur. (except TR.). *Poisonous to livestock. The flowers give a yellow dye.*

ACONITUM | **Monkshood** Fls. many, in a spike-like cluster; outer perianth segments 5, petal-like, the uppermost forming a large hood; inner segments 2–10, 2 of which form long-stalked spurred nectaries which are protected by the hood; stamens numerous. Carpels 2–5, many-seeded. 14 sps.

Fls. usually yellow

208. *A. vulparia* Reichenb. (*A. lycoctonum* auct.) WOLFSBANE. An erect perenn. ½–1 m., with simple or branched, elongated, leafless, spikes of erect, yellow, hooded fls. Hood elongated conical-cylindrical, about 3 times as long as broad; nectary spurs spirally curved. Lvs. deeply palmately three-lobed with main lobes *c.* 1 cm. broad, each lobe further cut into deep, narrow, toothed lobes. Carpels usually 3, hairless or nearly so. There are several closely related sps., some with violet or blue fls. △ In mountains: woods, damp meadows, rocky places. June–Aug. This and related sps. most of Eur. (except P.IRL.GB.IS. DK.TR.). *Poisonous.* Pl. 19.

209. *A. anthora* L. YELLOW MONKSHOOD. Like 208 with yellowish or rarely blue fls., but hood more or less hemispherical, as broad as long, and nectary spurs straight. Lvs. cut into very narrow strap-shaped segments, not more than 3 mm. broad. Carpels usually 5, usually hairy. △ In mountains: woods, damp meadows. July–Sept. Cent. Eur. (except D.) E.F.I.YU.BG.R.SU. *Poisonous.*

Fls. usually blue

210. *A. napellus* L. COMMON MONKSHOOD. An erect perenn. ½–1 m., usually with narrow,

dense, simple or branched spikes and blue or rarely white fls. with a hood always broader than long. Nectary spurs straight. Lvs. cut to the base into narrow linear segments which are further cut into linear lobes 1–8 mm. wide. Carpels usually 3. There are several closely related sps. △ In mountains: damp meadows, woods. June–Sept. This and related sps. much of Eur. (except North Eur. IRL.NL.AL.GR.TR.) S. *Contains aconitine, like all members of the genus, which is very poisonous and in consequence Monkshood is perhaps the most dangerous of all British plants. Used medicinally; the juice has been used as an arrow poison.* Pl. 19.

A. variegatum L. agg. BRANCHED MONKSHOOD. Like 210, but fls. in wide-spreading, branched clusters and branches with or without glandular hairs. Hood distinctly longer than broad, but not more than twice as long as broad; fls. blue, or less commonly white or variegated. Mountains of Central and South-Eastern Europe.

DELPHINIUM | **Delphinium** Fls. many, in a spike-like cluster, each with a backward-projecting spur. Outer perianth segments 5, petal-like, the uppermost long-spurred; inner segments 4, petal-like, the 2 uppermost with a spurred nectary which is inserted in the outer spur; stamens 8. Carpels usually 3. 26 sps.

211. *D. elatum* L. ALPINE LARKSPUR. A robust perenn. to 2 m., with long, slender, deep blue or bluish-violet spikes of fls. Outer perianth segments 1–2 cm. with spur 1½–3 cm. long; inner segments dark brown or blackish. Lvs. palmately divided with segments further divided into lance-shaped lobes. A very variable sp. △ In mountains: meadows, damp places, by streams. June–Aug. F.CH.A.PL.CS.YU.R.SU. Pl. 20.

D. peregrinum L. A slender ann. with blue-violet fls., usually with upward-pointing spurs, in a compact cylindrical spike. Uppermost lvs. linear-lance-shaped, entire; stem and lvs. glaucous. Carpels 3, usually hairless. Eastern Mediterranean region. Pl. 20.

212. *D. staphisagria* L. STAVESACRE, LICEBANE. Readily distinguished by its very short sac-like spur which is not more than one-third the length of the perianth segments. A robust, hairy ann. or bienn. to 1 m., with deep blue fls. in a long spike. Lvs. with 5–7 broad, lance-shaped lobes, which may be further three-lobed. Carpels 3–5, swollen, hairy. △ Field verges, stony places. May–Aug. Med. Eur. P. *Yields delphinine, which is poisonous; the seeds are particularly poisonous. Used medicinally in ancient times; now used externally as an insecticide.*

CONSOLIDA | **Larkspur** Like *Delphinium*, but always ann.; inner perianth segments only 2, petal-like, fused into a single long spur which is included in the outer spur; carpel 1. 12 sps.

213. *C. ambigua* (L.) P. W. Ball & Heywood (*Delphinium ajacis* auct.) LARKSPUR. An erect, simple- or little-branched ann. ½–1 m., usually with deep blue or rarely pink or white fls. in rather lax terminal spikes. Fls. rather large; outer perianth segments 10–14 mm. with a nearly straight spur 13–18 mm.; inner segment three-lobed, the uppermost lobe bilobed. Lvs. palmately cut into many linear acute segments, the upper lvs. stalkless, finely hairy. Carpel 1, 1½–2 cm., hairy, gradually narrowed to apex; seeds black. △ Corn-fields, tracksides, cultivated ground; often grown as an ornamental and naturalized. June–July. Med. Eur.: introd. P.B.CH.A.PL.CS.R. *Poisonous; sometimes used as an insecticide.* Pl. 20.

C. orientalis (Gay) Schrödinger Like 213, but spur shorter, not more than 12 mm., and carpel very abruptly contracted at apex; seeds reddish-brown. Southern Europe; introduced elsewhere.

C. regalis S. F. Gray Like 213 with fls. dark or light blue and spur 12–25 mm.; inflorescence in either dense and little-branched or lax and much-branched clusters, but bracts subtending fls., all linear entire, not cut. Carpel usually hairless, seeds black. Most of Europe. Pl. 20.

ANEMONE | **Anemone** Fls. usually solitary on an unbranched stem with a whorl of 3 lvs. placed some way below the fl. Perianth petal-like of only 1 whorl; sepals and nectaries absent; stamens numerous. Fr. of many one-seeded carpels. 17 sps.

Stem lvs. shortly stalked and similar to basal lvs.
(a) Carpels shortly hairy
214. *A. nemorosa* L. WOOD ANEMONE. A delicate perenn. 6–30 cm. with a small creeping rhizome and an erect stem bearing solitary, white, drooping fls. above 3 palmately lobed lvs. Fls. 2–4 cm. across; petals usually 6–7, hairless, often flushed pink or purple; anthers yellow. Basal lvs. palmately lobed and lobes further cut into oval toothed or bilobed segments similar to stem lvs. Fr. heads drooping, carpels hairy. △ Woods, shady places. Mar.–May. All Eur. (except P.IS.). *Poisonous, acid, and blistering; used medicinally in the past.* Pl. 22.

A. ranunculoides L. YELLOW WOOD ANEMONE. Like 214, but with 1, 2, or more yellow fls. 1½–2 cm. across; petals broadly oval, slightly hairy beneath. Widespread in Europe.

A. apennina L. BLUE WOOD ANEMONE. Fls. blue, with 8–14 petals which are hairy beneath. Fr. head erect. South-Eastern Europe. Pl. 22.

A. blanda Schott and Kotschy Like *A. apennina*, but petals hairless, more numerous, and fr. heads drooping. Southern Balkan Peninsula. Pl. 21.

(b) Carpels densely woolly
215. *A. sylvestris* L. SNOWDROP WINDFLOWER. Fls. large, white, 4–7 cm. across, usually solitary; petals usually 5, broadly oval, silky beneath; anthers yellow. Lvs. deeply palmately lobed, stem and basal lvs. similar. Fr. head erect, globular, carpels densely woolly-haired. A hairy non-rhizomatous perenn., spreading by root-buds; 20–50 cm. △ Woods, sunny banks. Apr.–June. Cent. Eur. F.S.I.YU.BG.R.SU.: introd. B.

Stem lvs. stalkless and dissimilar to basal lvs.
(a) Carpels densely woolly, rounded in section
216. *A. coronaria* L. CROWN ANEMONE. Fls. large, 3½–6½ cm. across, solitary, red, blue, pink, or white; petals 5–8, elliptic, overlapping; anthers blue. Stem lvs. distinctive, small, but deeply cut into numerous narrow segments. Basal lvs. 2–3 times cut into narrow segments. A tuberous perenn., with fl. stems 15–45 cm. △ Cultivated ground, fields. Feb.–Apr. Med. Eur. (except AL.) BG.: introd. P. Pl. 21.

A. hortensis L. (*A. stellata* Lam.) Like 216, but distinguished by the narrow, lance-shaped, bract-like stem lvs. which are entire or three-lobed. Petals 12–19, narrow elliptic, usually pale rose-purple. Central Mediterranean region. Pl. 21.

A. pavonina Lam. Like 216 with red, blue, or pink fls. but usually with a paler centre; petals 8–9 broader, oval. Stem lvs. narrow lance-shaped, bract-like, entire or three-lobed. Central and Eastern Mediterranean region. Pl. 21.

217. *A. palmata* L. PALMATE ANEMONE. Fls. pale yellow, 2½–3½ cm. across, usually solitary; petals 10–15, oblong, hairy beneath. Stem lvs. bract-like, fused at base, with 3–5 linear-lance-shaped segments. Basal lvs. nearly circular, but with 3–5 broad, shallow, toothed lobes. A tuberous perenn. with erect fl. stem 10–30 cm. △ Fields, hills, bushy ground. Feb.–June. P.E.F.I.GR. Pl. 21.

(b) Carpels hairless, flattened

218. *A. narcissiflora* L. NARCISSUS-FLOWERED ANEMONE. Fls. white, clustered into a terminal umbel of 3–8 long-stalked fls., and with stem lvs. immediately below umbel deeply cut into segments. Fls. 2–3 cm. across, sometimes pinkish; petals 5–6, obovate, hairless. Basal lvs. deeply palmately lobed with ultimate segments linear-lance-shaped. A rather robust, hairy perenn. 20–50 cm. △ Alpine meadows. June–July. Cent. Eur. (except H.), South-East Eur. (except GR.TR.) E.F.I.SU. Pl. 22.

HEPATICA   Like *Anemone* and often included in the genus, but fls. with 3 entire calyx-like bracts; basal lvs. with 3–5 broad, rounded, usually untoothed lobes. 2 sps.

219. *H. nobilis* Miller (*Anemone hepatica* L.) HEPATICA. Fls. blue, pinkish, or white, long-stalked, solitary, appearing to arise direct from the rootstock. Fls. 1½–3 cm. across; petals 6–9, hairless; involucre of 3 oval, green bracts placed immediately below the fls. and looking like sepals. Lvs. distinctive, with 3 broad, shallow, rounded lobes, evergreen and often purplish beneath. Perenn. 5–15 cm. △ Montane woods and copses. Mar.–May. Much of Eur. (except P.IRL.GB.IS.GR.TR.). *Used in herbal remedies for coughs and chest complaints; somewhat poisonous.* Pl. 20.

PULSATILLA | **Anemone**   Like *Anemone* and often included in the genus, but fr. with elongated feathery styles. Fls. solitary; petals 6, silky beneath. Stem lvs. usually bract-like, stalkless and fused at base. 9 sps.

Stem lvs. shortly stalked, similar to basal lvs.

220. *P. alpina* (L.) Delarbre (*Anemone a.*) ALPINE ANEMONE. Fls. solitary, white or yellow, on a short thick stem with short-stalked stem lvs. similar in appearance to the basal lvs. but smaller. Fls. large 4–6 cm.; petals oval, spreading, silky-haired beneath. Lvs. thrice-pinnate with toothed segments, densely silky-haired when young, but becoming nearly hairless. Fr. with long styles 4–5 cm., with spreading hairs. Subsp. *alpina* WHITE ALPINE ANEMONE has outer petals white or purplish and inner petals white. Subsp. *apiifolia* (Scop.) Nyman YELLOW ALPINE ANEMONE has pale yellow petals. A hairy perenn. with fl. stems 10–30 cm. △ In mountains: meadows, rocky places, streamsides. May–July. E.F.D.CH.A.I.YU. Pl. 23.

Stem lvs. stalkless, differing from basal lvs.

221. *P. vernalis* (L.) Miller (*Anemone v.*) SPRING ANEMONE. Fls. white, the outer petals flushed violet, pink, or blue, solitary, at first nodding then erect. Fls. 4–6 cm. across; petals narrowly oval, silky-haired outside. Stem lvs. bract-like, cut to the base into linear, hairy segments; basal lvs. pinnate, with oval three-lobed leaflets, evergreen and becoming hairless. A perenn. with hairy fl. stem 10–15 cm., which lengthens in fr. △ Alpine meadows. Apr.–June. North Eur. (except IS.), Cent. Eur. (except H.) E.F.I.YU.BG. Pl. 22.

222. *P. pratensis* (L.) Miller (*Anemone p.*) SMALL PASQUE FLOWER. Fls. purple, reddish-violet, greenish-yellow or white, solitary, nodding, more or less cylindrical, petals not spreading but tips of petals outcurved. Fls. 3–4 cm. across; petals less than 1½ times as long as stamens. Stem lvs. bract-like, cut into narrow lobes; basal lvs. hairy, thrice-pinnate, with narrow segments. A hairy perenn. with fl. stem 5–10 cm., which lengthens in fr. The different colour forms have distinctive geographical ranges: purple in the north, violet in the centre, and yellowish or greyish-violet in the south-east of Europe. △ Fields. May. Cent. Eur. (except CH.) DK.N.S.YU.BG.R.SU. Pl. 22.

223. *P. vulgaris* Miller (*Anemone pulsatilla* L.) PASQUE FLOWER. Like 222, but fls. larger, 5½–8½ cm. across, more or less erect with pale or dark purple, rather spreading petals

2–3 times as long as the stamens and not outcurved at the tip. Stem and basal lvs. silvery-haired when young. A very variable sp. △ Meadows, hills, glades. Mar.–May. Cent. Eur. F.GB.B.NL.DK.S.R.SU. *Poisonous; used medicinally until recently*. Pl. 22.

CLEMATIS Usually woody climbers with opposite lvs. Petals 4 or rarely more; sepals and nectaries absent; stamens numerous; carpels numerous, one-seeded. Fr. usually with long feathery styles. 10 sps.

Climbing plants
(a) Fls. many, clustered

224. *C. vitalba* L. TRAVELLER'S JOY, OLD MAN'S BEARD. A robust, woody, deciduous-leaved climbing plant to 30 m., with clusters of white fls. towards the end of the present year's branches, and conspicuous, grey, feathery fr. clusters. Fls. *c.* 2 cm., fragrant, in lax terminal and axillary clusters; petals spreading, densely hairy on both sides. Lvs. pinnate with 3–9 oval, stalked leaflets, each 3–10 cm. Fr. with numerous carpels, each with long styles with dense, spreading, white hairs. △ Woods, hedges, thickets. June–Aug. Most of Eur. (except North Eur.): introd. IRL.N.S.PL. *The stems have been used for basket-making. The young leaves are irritant and can cause ulceration and are poisonous to livestock*. Pl. 24.

225. *C. flammula* L. FRAGRANT CLEMATIS. Like 224 but lvs. twice-pinnate, with numerous small, oval, entire or three-lobed leaflets. Fls. pure white, 1½–3 cm. across, very fragrant; petals hairless on upper surface. Carpels flattened, styles shorter. A more or less woody, deciduous climber 3–5 m. △ Hedges, thickets, rocks. May–Aug. Med. Eur. P.BG.: introd. A.CS. Pl. 24.

(b) Fls. large, solitary

226. *C. cirrhosa* L. VIRGINS' BOWER. A scrambling, woody, evergreen perenn. 2–5 m., with large, solitary, nodding, cream-coloured fls. 4–7 cm. across. Fls. with a calyx-like, cup-shaped, two-lipped involucre; petals oval, hairy beneath. Lvs. somewhat leathery, clustered, very variable, either entire, lobed, or once or twice cut into lobed or toothed leaflets. △ Woods, hedges, bushy places. Mar.–Dec. Med. Eur. (except YU.AL.TR.) P.

227. *C. alpina* (L.) Miller (*Atragene a.*) ALPINE CLEMATIS. A scrambling mountain perenn. 1–2 m., with solitary, nodding, blue-violet, or less commonly yellowish-violet, bell-shaped fls. 4–6 cm. across. Petals rather spreading, narrow elliptic, hairy beneath; staminodes petal-like, about as long as the fertile stamens; involucre absent. Lvs. twice cut, usually into 9 oval to lance-shaped, coarsely toothed leaflets 2½–5 cm. △ In mountains: woods, rocky places. May–July. Cent. Eur. F.N.SF.I.YU.BG.R.SU. Pl. 23.

Herbaceous, non-climbing plants
(a) Fls. many, clustered

228. *C. recta* L. An erect herbaceous perenn. 1–1½ m., with a terminal branched cluster of numerous, fragrant, white fls. each *c.* 2 cm. across. Petals hairless, except on the margin. Lvs. large, pinnate, with 5–7 oval, entire, stalked leaflets, each 5–9 cm., somewhat glaucous beneath; stems hollow. Carpels flattened, almost hairless, with hairy styles. △ Open woods, dry hills, hedges. May–June. Cent. Eur. E.F.I.YU.BG.R.SU.: introd. N. *Poisonous.*

(b) Fls. large, solitary

229. *C. integrifolia* L. An erect herbaceous perenn. 30–70 cm., usually with a solitary, terminal, drooping, purple, rather open bell-shaped fl. Petals 3–5 cm., hairless, except near margin. Lvs. oval-acute entire, stalkless, to 9 cm. △ Meadows. June–Aug. A.CS.H.I. YU.BG.R.SU.: introd. CH. Pl. 24.

ADONIS   Herbaceous plants with finely cut lvs. Sepals usually 5; petals 3–20, glossy; nectaries absent. Fr. forming an elongated head of numerous one-seeded, wrinkled carpels with short styles. 10 sps.

Anns.

(a) Fls. usually red

230. *A. annua* L. (*A. autumnalis* L.) PHEASANT'S EYE. An erect ann. 20–40 cm., with feathery lvs. and small, glossy, scarlet fls. with a black basal patch and conspicuous blackish-purple stamens and styles. Fls. $1\frac{1}{2}$–$2\frac{1}{2}$ cm. across; sepals green, hairless, spreading. Lvs. thrice-pinnate, segments linear, fine-pointed. Fr. in a cylindrical head *c.* 2 cm. long; carpels $3\frac{1}{2}$–5 mm. with inner margin straight and without a projection. △ Cornfields. May–Aug. Med. Eur. CH.BG.: introd. GB.B.DK.A.PL.CS.H.R.SU. *Poisonous to livestock.* Pl. 24.

A. aestivalis L. Like 230, but carpels larger, 5–6 mm., with a transverse ridge round the middle and 2 projections on the inner margin. Much of Southern Europe.

A. flammea Jacq. Like 230, but fls. larger, usually deep scarlet, 2–3 cm. across, with hairy sepals pressed to the petals. Carpels with a rounded projection on the inner margin just below the style. Central and Southern Europe.

(b) Fls. usually yellow

A. microcarpa DC. Distinguished by its fls. which are usually yellow; carpels 3–4 mm., with or without a transverse ridge; upper projection from inner edge of carpel close to the style. Mediterranean region.

Perenns.; fls. yellow

231. *A. vernalis* L. YELLOW ADONIS. Fls. large, solitary, 4–8 cm. across, bright yellow, with 10–20 elliptic petals and broader hairy sepals half their length. Basal lvs. reduced to scales; stem lvs. stalkless, much-dissected into fine linear segments. Carpels hairy, style curved back. A hairless perenn. with fl. stem 10–40 cm. △ Dry pastures, rocks. Apr.–May. Cent. Eur. E.F.S.I.YU.BG.R.SU. *Poisonous; used as heart tonic.* Pl. 24.

RANUNCULUS | Buttercup, Crowfoot   Fls. yellow, white, or rarely pink; sepals usually 5; petals 5 or more. Fr. of numerous one-seeded carpels, each with a persistent style. *Most of the yellow-flowered sps. are poisonous to man and livestock when fresh; they contain an irritant poison, proto-anemonin, which is a yellow volatile oil.* 131 sps.

Fls. yellow

(a) Lvs. about as long as broad, shallowly lobed to halfway or less

(i) Carpels with spines or conspicuous swellings

232. *R. muricatus* L. SPINY-FRUITED BUTTERCUP. An almost hairless, rather shining, pale green ann. 10–30 cm., with small, pale yellow fls. 3–6 mm. across. Lvs. all circular or kidney-shaped, with shallow, rounded lobes. Carpels large, 7–8 mm., with spiny sides and a broad, sharp-edged margin; style 2–3 mm. △ Damp ground. Apr.–June. Med. Eur. BG.R. Pl. 26.

R. parviflorus L. SMALL-FLOWERED BUTTERCUP. A hairy ann. with pale yellow fls. 3–6 mm. across, and strongly deflexed sepals. Uppermost lvs. often deeply lobed. Carpels *c.* 3 mm., with hooked spines on the sides; style short, hooked. Southern Europe. Page 105.

(ii) Carpels smooth, without spines

233. *R. ficaria* L. (*Ficaria verna* Hudson) LESSER CELANDINE, PILEWORT. Readily distinguished by its triangular heart-shaped, shallowly lobed glossy lvs. and shining yellow fls., usually $1\frac{1}{2}$–3 cm. but sometimes to 5 cm. Petals spreading, rather numerous, 8–12;

sepals 3, green. A variable, weak, spreading, little-branched perenn. 5–30 cm., with spindle-shaped tubers. △ Woods, meadows, shady banks, streamsides. Mar.–May. All Eur. (except IS.). *The root tubers and mature leaves are poisonous to livestock.* Pl. 25.

234. *R. thora* L. THORA BUTTERCUP. A mountain perenn. with a slender, unbranched stem, 1 or 2 yellow fls., and a single, large, rounded, kidney-shaped, glaucous lower lv. and a small, lance-shaped upper lv. Fls. 1–2 cm. across; sepals hairless. Basal lvs. stalked, appearing after flowering, stem lvs. stalkless, toothed. △ In mountains: meadows, sunny slopes. June–July. E.F.CH.A.PL.CS.I.AL.R. *Very poisonous; used in the past as an arrow poison.* Pl. 25.

(*b*) *Lvs. about as long as broad, deeply lobed to more than halfway*
(*i*) *Carpels with spines*
235. *R. arvensis* L. CORN BUTTERCUP. A rather slender, branched ann. 15–60 cm., with almost all lvs. dissected into narrow lance-shaped or linear lobes, and with conspicuous spiny frs. Fls. 4–12 mm. across, bright lemon-yellow, in a lax, leafy cluster; sepals greenish-yellow, spreading. Carpels few, large, 6–8 mm., with conspicuous spines or swellings on the sides and a broad-edged margin; style straight, 3–4 mm. △ Cornfields. May–July. All Eur. (except IS.).

(*ii*) *Carpels smooth; sepals reflexed*
236. *R. sceleratus* L. CELERY-LEAVED BUTTERCUP. A hairless, pale green, shining-leaved, slightly fleshy ann. 10–60 cm., with many pale yellow fls. ½–1 cm. across, with reflexed sepals. Lvs. all deeply divided, the upper into narrow segments; stem much-branched, hollow, grooved. Fr. in an elongated, cylindrical head 6–10 mm. of numerous hairless carpels, each *c.* 1 mm. △ Marshes, watersides, damp mud. Apr.–Sept. All Eur. (except IS.). *Poisonous: causes blistering and inflames the mouth and gut; dangerous to livestock.*

237. *R. bulbosus* L. BULBOUS BUTTERCUP. Usually readily distinguished by the corm-like swelling at the base of the stem. Fls. shining yellow, 2–3 cm. across; sepals yellowish, strongly reflexed against fl. stalk. Basal lvs. three-lobed with lobes further divided into broad toothed segments; uppermost lvs. stalkless with linear to lance-shaped lobes. Carpels 2–4 mm.; style short, curved. A variable, hairy, little-branched perenn. 20–50 cm. △ Fields, open ground, waysides. Apr.–June. All Eur. (except IS.TR.).

R. sardous Crantz HAIRY BUTTERCUP. Like 237, but an ann. without a swollen base of stem. Fls. pale yellow; carpels 2½–3 mm. with a ring of minute swellings close to the conspicuous margin. Most of Europe. Page 105.

(*iii*) *Carpels smooth; sepals erect or spreading*
238. *R. repens* L. CREEPING BUTTERCUP. Distinguished by its long, creeping stems which readily root at the nodes, and rather few bright, glossy yellow fls. 2–3 cm. across. Sepals spreading; fl. stalks grooved; receptacle hairy. Basal lvs. three-lobed, the central lobe long-stalked, all lobes further cut into 3 toothed segments. Carpels *c.* 3 mm. flattened, bordered, hairless; style 1½ mm., curved. A very variable perenn., with fl. stems 30–50 cm. △ Damp meadows, woods, waste and shady places. May–June. All Eur. (except TR.). *Poisonous to livestock.*

R. polyanthemos L. Like 238 in having strongly bordered carpels, grooved fl. stalks and hairy receptacle, but creeping stems absent and basal lvs. usually with 5 lobes which are cut into linear to lance-shaped toothed segments. Carpels 3–5 mm.; style *c.* ½ mm. Most of Europe. Page 105.

239. *R. acris* L. MEADOW BUTTERCUP. Distinguished from 238 by its smooth fl. stalks, hairless receptacle, and absence of creeping stems. Fls. shining yellow, 1½–2½ cm. across; sepals

1. *Ranunculus polyanthemos* [238]
2. *R. parviflorus* [232]
3. *R. ophioglossifolius* [243]
4. *R. lanuginosus* 240
5. *R. sardous* [237]

spreading. Lvs. deeply divided into 3–7 stalkless, narrow oval lobes which are further toothed or lobed. Carpels 2–3½ mm.; style short, hooked. A very variable, rather hairless, erect branched perenn. 30–100 cm. △ Meadows, woods, waysides. May–July. All Eur. (except P.TR.). *Poisonous to livestock.*

240. *R. lanuginosus* L. WOOLLY BUTTERCUP. Like 239, but much more hairy, with stems and lv. stalks with long, spreading hairs and with stout, hollow stems 30–50 cm. Fls. deep orange-yellow, 2–3 cm. across. Basal lvs. generally three-lobed to three-quarters the width of the blade, lobes broadly oval and irregularly toothed. Carpels 4–5 mm., flattened, bordered; style 1½ mm., strongly curved. △ Woods in mountains. July–Aug. Cent., South-East Eur. (except TR.) F.B.DK.I.SU. Page 105.

241. *R. montanus* Willd. agg. MOUNTAIN BUTTERCUP. A mountain perenn. 5–50 cm., with one to a few large, glossy, yellow fls. 2–4 cm. across, and rounded three- to five-lobed basal lvs. contrasting with the deeply cut, narrow-lobed upper lvs. Fl. stalk smooth; sepals spreading; receptacle hairy. Carpels 2½–3½ mm., flattened and strongly keeled; style hooked. A very variable sp. with many localized microspecies. △ In mountains: pastures, woods. May–Aug. Cent. Eur. (except H), South-East Eur. (except TR.) E.F.I.SU. Pl. 25.

242. *R. auricomus* L. GOLDILOCKS. An erect, nearly hairless woodland perenn. 10–40 cm., with long-stalked, rounded, shallowly or deeply lobed lower lvs., and very dissimilar upper stem lvs. which are stalkless and dissected to the base into linear segments. Fls. yellow, 12–22 mm. across, with 0–5 fully developed petals; sepals hairy, often partly yellowish and petal-like. Carpels downy, 3½–4 mm., borne on short stalks on the receptacle. A very variable sp. △ Damp, shady woods, meadow verges. Apr.–June. All Eur. (except P.TR.).

(c) Lvs. entire, mostly much longer than broad

243. *R. flammula* L. LESSER SPEARWORT. A rather fleshy, almost hairless, hollow-stemmed perenn. 25–50 cm., rooting at the base, with narrow entire lvs. and rather few small, shiny, yellow fls. Fls. variable, 7–20 mm. across; sepals spreading, hairless. Basal lvs. lance-shaped to narrowly heart-shaped, the upper narrower. Carpels 1½–2 mm., numerous, hairless, minutely pitted. △ Damp meadows, marshes, bogs, lakesides. May–Sept. All Eur. (except TR.). *Poisonous to livestock.*

R. reptans L. Distinguished from 243 by the procumbent stems which root at each node, and the solitary yellow fls. *c.* 5 mm. across. Lvs. narrow elliptic to spathulate. Northern Europe, rarely in Central Europe.

R. ophioglossifolius Vill. SNAKESTONGUE CROWFOOT. Like 243, but basal lvs. oval or heart-shaped, to 2 cm. by 12 mm., upper lvs. lance-shaped. Fls. 6–8 mm. across, pale yellow. Carpels 1½ mm., the sides covered with small swellings. An erect ann. Widespread in Southern Europe. Page 105.

244. *R. lingua* L. GREAT SPEARWORT. A robust stoloniferous perenn. to 1½m., usually rooted in shallow water, with stout hollow stems, long spear-shaped lvs., and few large, glossy yellow fls. 2–5 cm. across. Lvs. to 20 cm., the lower shortly stalked, the upper stalkless and half-clasping the stem, usually obscurely toothed, hairless. Carpels *c.* 2½ mm., hairless, minutely pitted. △ Marshes, ditches, lakes. June–Aug. Most of Eur. (except P.IS.AL.). Pl. 25.

245. *R. gramineus* L. GRASS-LEAVED BUTTERCUP. A little-branched, glaucous perenn. 10–40 cm., with stalkless, narrow lance-shaped lvs. mostly arising from a fibrous swollen base and one to a few large, long-stalked, deep yellow fls. *c.* 3 cm. across. Sepals and receptacle hairless. Carpels 3 mm., keeled, veined. △ Dry pastures. May–June. P.E.F.CH.I.

Fls. white or pink

(a) Terrestrial plants

246. *R. aconitifolius* L. WHITE BUTTERCUP. A rather robust perenn. to ½ m., with an erect stem, and spreading branches above, bearing rather numerous white fls. 1–2 cm. across in a lax, spreading cluster. Sepals reddish-brown, hairless, soon falling. Fl. stalks 1–3 times as long as the subtending lvs., hairy above. Lvs. deeply three- to five-lobed with lobes obovate to lance-shaped, toothed, the upper lvs. stalkless. Carpels 5 mm.; style slender, hooked. △ In mountains: woods, meadows, damp places. May–Aug. E.F.D.CH.A.CS.I.YU. SU. Pl. 25.

R. platanifolius L. LARGE WHITE BUTTERCUP. Like 246, but a larger plant with five- to seven-lobed lvs. with the middle lobes not free to the base (free in 246). Fl. stalks 4–5 times as long as the subtending lvs. Central and Southern Europe.

247. *R. alpestris* L. ALPINE BUTTERCUP. A small alpine perenn. 5–10 cm., with simple or little-branched stems bearing one or a few white fls. *c.* 2 cm. across. Petals broadly oval, 5–10; sepals hairless. Lvs. mostly basal, three- to five-lobed with lobes deeply rounded-toothed, the stem lvs. 1–2, the lower three-lobed, the upper entire. Carpels globular; style slender. △ Rock crevices. June–July. E.F.D.CH.A.PL.CS.I.YU.R.

R. glacialis L. GLACIER CROWFOOT. Distinguished from 247 by the reddish-haired sepals; strongly flattened and winged fr. and rather broad style. Fls. white or pink, 1–3 cm. across; petals 5, persisting in fr. Lvs. slightly fleshy, the stem lvs. usually lobed, stalkless. Northern Europe, higher mountains of the rest of Europe. Pl. 26.

248. *R. parnassifolius* L. A small perenn. 4–20 cm. of high mountains with thick, oval-heart-shaped, entire lvs. and one to a few large white or reddish fls. Fls. 2–2½ cm. across; sepals hairy. Upper lvs, small, lance-shaped, clasping stem; stem little-branched, base fibrous. Carpels smooth, inflated. △ In mountains: rocks and screes. June–Aug. E.F.CH. A.I. Pl. 25.

R. pyrenaeus L. PYRENEAN BUTTERCUP. Distinguished from 248 by the basal lvs. which are linear to lance-shaped and never heart-shaped at the base. Fls. white, 1–2 cm. across; sepals whitish, hairless; fl. stalk hairy above. Mountains of Western and Central Europe.

(b) Aquatic plants

(i) All lvs. with broad lobes

249. *R. hederaceus* L. IVY-LEAVED CROWFOOT. A small creeping ann. or bienn. 10–40 cm., rooting regularly at the nodes, with solitary small white fls. 3–8 mm. across in the axils of the lvs. Petals not touching, little longer than the green sepals. Lvs. 1–2 cm. wide, kidney-shaped with 3–5 shallow, blunt lobes. Carpels 1–1½ mm., hairless. △ Muddy ditches, verges of ponds, riversides. Apr.–Aug. West Eur. DK.S.D.: introd. N.

(ii) Some or all lvs. with linear segments

250. *R. peltatus* Schrank POND CROWFOOT. A submerged aquatic ann. to perenn. to 1½ m., with both broad-bladed floating lvs. and finely dissected submerged lvs. usually present. Fls. large, 1½–3 cm. across; petals broad, overlapping. Floating lvs. rounded or kidney-shaped, with 3–7 shallow rounded lobes; submerged lvs. with thread-like segments shorter than the internodes. Fr. stalks long, 5 cm. or more, longer than the subtending lv. △ Still and slow-flowing waters. May–July. All Eur. (except IS.AL.). Pl. 26.

R. aquatilis L. COMMON WATER CROWFOOT. Like 250, but floating lvs. deeply lobed into broad wedge-shaped segments, often with acute teeth. Fls. smaller, 12–18 mm. across; fr. stalk less than 5 cm. and usually shorter than the subtending lv. Throughout Europe.

251. *R. trichophyllus* Chaix Distinguished by the absence of floating lvs. and its short submerged lvs. 2–4 cm. which are repeatedly trifid, the ultimate segments forked, dark green, and not lying in one plane. Fls. small, usually 8–10 mm. across; petals not touching, soon falling. Fr. stalk usually less than 4 cm.; fr. oval, more than 2 mm. △ Still waters. Apr.–June. All Eur.

R. circinatus Sibth. RIGID-LEAVED CROWFOOT. Like 251 without floating lvs., but submerged lvs. with rather stiff segments, lying in one plane and encircling the stem at each node, much shorter than the internodes, not collapsing out of water. Fls. 1–2 cm. across; petals overlapping. Widespread in Europe.

*R. fluitans Lam. Widespread in Central Europe.

*R. pseudofluitans (Syme) Baker & Foggitt Widespread in Europe. Pl. 26.

MYOSURUS   Anns. with linear lvs. Sepals 5 or more, each with a small basal spur; petals 5–7 or absent, tubular. Fr. of numerous carpels forming a much-elongated spike. 2 sps.

252. *M. minimus* L. MOUSE-TAIL. A hairless ann. with a basal rosette of linear entire lvs. and solitary, very small, pale greenish-yellow fls. on leafless stems 5–12 cm. Sepals 3–4 mm., narrowly oblong, with spur pressed to the stem; petals greenish, about as long. Carpels 1–1½ mm., one-seeded borne on a very long receptacle 2½–7 cm. △ Damp fields. Mar.–May. Most of Eur. (except P.IRL.IS.AL.). Pl. 26.

AQUILEGIA | Columbine   Fls. with 5 petal-like outer segments and 5 long-spurred petal-like inner segments; stamens numerous. Carpels usually 5, many-seeded. Lvs. compound, one- to three-ternate. 27 sps.

253. *A. vulgaris* L. COLUMBINE. A tall, slender, sometimes glaucous perenn. 30–80 cm., with very distinctive usually violet but sometimes pink or white fls. with 5 long backward-projecting spurs. Fls. nodding, 3–5 cm. long including spurs of 15–22 mm., spurs crooked, knobbed at the apex. Lower lvs. long-stalked, ternate, with 9 stalked, rounded, irregularly three-lobed toothed leaflets, the uppermost lvs. stalkless, narrowly three-lobed. Carpels glandular-hairy. A very variable sp. △ Woods, mountain pastures, rocky places. May–July. Much of Eur. (except AL.GR.TR.BG.): introd. DK.N.S.SF.SU. *Poisonous; used medicinally in the past.* Pl. 27.

254. *A. alpina* L. ALPINE COLUMBINE. An alpine perenn. 15–80 cm., with hairy stems bearing solitary or few large, bright blue, nodding fls. 3–4½ cm. long including spur. Spurs 1½–2½ cm., thick, straight, or slightly curved. Lvs. ternate with 9 two- to three-lobed, nearly hairless leaflets. Carpels 2–3 cm.; styles 6–7 mm. A variable sp. △ In mountains: rocks, thickets, grassy places. July–Aug. F.CH.A.I. Pl. 27.

A. pyrenaica DC. PYRENEAN COLUMBINE. Like 254, but petals more rounded (petals square-tipped in 254); spur slender, straight. Carpels small 13–17 mm., glandular-hairy. Pyrenees.

THALICTRUM | Meadow Rue   Fls. small, numerous, clustered; perianth 4–5, usually soon falling, nectaries absent; stamens numerous, long-stalked, conspicuous. Carpels few, often stalked. Lvs. 2–3 times divided into numerous leaflets; stipules present. 13 sps.

255. *T. aquilegifolium* L. GREAT MEADOW RUE. An erect, hairless, branched perenn. ½–1½ m., with dense, much-branched, fluffy-looking clusters of pale lilac or whitish fls., made

1. *Epimedium alpinum* 260
3. *Thalictrum minus* 256
5. *Isopyrum thalictroides* 205

2. *Actaea spicata* 206
4. *T. aquilegifolium* 255

conspicuous by the thick coloured stalks of the numerous stamens. Fls. numerous, erect; stalks of stamens as wide as the anthers. Lvs. large, two- to three-ternate; leaflets small, oval-wedge-shaped, toothed. Fr. *c.* 7 mm. long-stalked, pendulous, strongly three-angled. △ In mountains: damp meadows, bushy places, streamsides. May–July. Cent., South-East Eur. (except TR.) E.F.S.SF.I.SU. Page 109.

256. *T. minus* L. LESSER MEADOW RUE. A wiry-stemmed, wide-branched perenn. 15–150 cm., with numerous greenish-yellow fls. often tinged purple, in a lax inflorescence. Stamens very numerous, conspicuous, hanging with very slender stalks. Lvs. three- to five-ternate; leaflets rounded, irregularly lobed or toothed, hairless or glandular-hairy. Carpels 3–5 mm., stalkless, erect, weakly ribbed. A very variable sp. △ Hills, woods, rocks, screes, damp places. June–Aug. All Eur. (except IS.TR.). Page 109.

T. alpinum L. ALPINE MEADOW RUE. A delicate perenn. 5–20 cm. of northern or alpine regions, with greenish-purple fls. in an elongated, unbranched, spike-like cluster which is at first drooping, then erect. Perianth segments 4, *c.* 3 mm., spreading; stamens pendulous. Carpels 2–3, pendulous on short stalks. Lvs. almost all basal, twice-ternate. Northern regions or mountains of Central Europe.

257. *T. flavum* L. COMMON MEADOW RUE. A robust, leafy, lowland perenn. 50–125 cm., with dense, more or less cylindrical, fluffy-looking clusters of sweet-scented yellow fls. Perianth segments 4, narrow; stamens yellow, erect. Lvs. two- to three-ternate with oval-wedge-shaped leaflets, each with 3–4 terminal lobes or deep teeth, dark green above, paler beneath; stem hollow, grooved. Carpels 1½–2 mm. ovoid, six-ribbed, erect. △ Fens, damp meadows, streamsides. June–Aug. All of Eur. (except IS.GR.TR.). *The roots and leaves are purgative; poisonous to livestock.* Pl. 26.

PAEONIACEAE | Peony Family

Usually large herbaceous perenns. or rarely shrubs with very large, conspicuous, solitary fls. Sepals 5; petals 5–10; stamens numerous. Carpels borne on a fleshy disk, 2–8, each several-seeded. Included in Ranunculaceae by many botanists. 1 genus. 10 sps.

PAEONIA | Peony

258. *P. officinalis* L. PEONY. A robust herbaceous perenn. 30–60 cm., with very large red fls. 7–13 cm. across, and stalks of stamens red. Distinguished by the lower lvs. which are thrice cut into 17–30 narrow elliptic to lance-shaped segments which are hairy beneath. Carpels 2–3½ cm., 2 or 3, usually woolly. A variable sp. with distinctive subsps. occurring in different regions. △ Woods, meadows in hills. May–June. P.E.F.CH.A.H.I.YU.AL.BG.: introd. CS. *A poisonous plant which has been in long use as a medicinal plant.* Pl. 27.

P. broteroi Boiss. & Reuter WESTERN PEONY. Like 258, but lvs. with 17–20 glaucous segments, which are hairless beneath. Fls. red, 8–10 cm. across, stalks of stamens yellow. Carpels 3–4 cm., 2–4, woolly. Iberian Peninsula.

259. *P. mascula* (L.) Miller (*P. corallina* Retz.) Like 258, but lvs. mostly twice cut into 9–16 narrow, elliptic to oval segments which are usually hairless beneath. Fls. red, 8–14 cm. across. Carpels 2–4 cm., 3–5, usually hairy. A variable sp. △ Woods, thickets. May–June. South-East Eur. (except TR.) F.A. Sicily: introd. GB. Pl. 27.

P. tenuifolia L. NARROW-LEAVED PEONY. Readily distinguished by its lvs. which have numerous linear segments which are less than ½ cm. wide, hairless above, finely hairy beneath. Fls. red, 6–8 cm. across; stalks of stamens yellow. Carpels *c.* 2 cm., woolly. South-Eastern Europe.

BERBERIDACEAE | Barberry Family

Usually shrubs, less commonly herbs. Fls. usually yellow, clustered; perianth segments usually petal-like, in 3–4 whorls; nectaries 4–6; stamens 4–6, anthers opening by pores. Ovary 1; fr. a berry or capsule.

1. Herbaceous plants; fr. dry.	*Epimedium*
1. Shrubs; fr. a fleshy berry.	
2. Lvs. simple; spines present on stem.	*Berberis*
2. Lvs. pinnate; not spiny on stem.	*Mahonia*

EPIMEDIUM Herbaceous perenns. with compound lvs. Fls. with sepals, petals, nectaries, and stamens 4. Fr. a dry capsule. 2 sps.

260. *E. alpinum* L. BARREN-WORT. Fls. dull reddish-purple, in a lax, nodding, stalked, terminal cluster on a slender stem bearing a single large compound lv. Fls. 9–13 mm. across; perianth with the outer segments greyish-pink, and soon falling, middle segments dark red, inner segments light yellow. Lvs. two- to three-ternate, leaflets 5–10 oval-heart-shaped, stalked, bristly-toothed; basal lvs. also present. A stoloniferous perenn. with many erect fl. stems 15–30 cm. △ Bushy places, woods in hills. Apr.–May. A.I.YU.AL.: introd. F.GB.B.D.CH.CS. Page 109.

BERBERIS | Barberry Spiny shrubs with simple, usually deciduous, lvs. Perianth segments yellow, petal-like, in multiples of 3; stamens 6. Fr. a berry. 4 sps.

261. *B. vulgaris* L. BARBERRY. A densely branched erect shrub 1½–3 m., with entire lvs., three-pronged needle-like spines and drooping clusters of bright yellow fls. Fl. clusters 3–5 cm. long, with 15–30 fls. each 6–8 mm. across. Lvs. 2½–5½ cm. elliptic, margin spiny-toothed; wood bright yellow. Fr. cylindrical 1 cm. long, orange-red. △ Hedges, rocky banks. May–June. Most of Eur.: introd. IRL.GB.DK.N.S.SF.TR. *The alternative host to the wheat rust, Puccinia graminis. The fruit can be used as a preserve; the bark and stems give a yellow dye.* Pl. 27.

MAHONIA Like *Berberis* but stems without spines; lvs. compound, evergreen. 1 sp.

262. *M. aquifolium* (Pursh) Nutt. (*Berberis a.*) OREGON GRAPE. Distinguished by its compound lvs. with 5–9 holly-like, shiny dark green, leathery, spiny-toothed oval leaflets, each 4–8 cm. Fls. yellow, in dense, erect terminal, often three-headed clusters 5–8 cm. long Fr. a globular, black and bloomed berry. A stoloniferous, little-branched, evergreen shrub, ½–1 m. △ Bushy places. Jan.–May. Native of North America: introd. GB.NL.S.A.CS. H.GR. Pl. 27.

LAURACEAE | Laurel Family

Very aromatic trees and shrubs, which are mainly tropical. Lvs. evergreen, with shining oil glands. Fls. usually in axillary clusters, often one-sexed, small greenish or yellow. Petals 4–6; stamens many; ovary 1. Fr. a berry.

LAURUS | Laurel   Plants one-sexed; petals 4; stamens 8–12. The only European genus. 2 sps.

263. *L. nobilis* L. LAUREL, SWEET BAY. An evergreen tree or shrub 2–20 m., with black bark and dark green, leathery, lance-shaped lvs. which are very aromatic when crushed. Fls. yellowish in rounded clusters in the axils of the lvs. and much shorter than the lvs.; male fls. with 8–12 stamens, the female with 2–4 staminodes; petals 4. Fr. a black berry, 1–1½ cm. △ Thickets, rocks, hedges, old walls; often planted for ornament. Mar.–Apr. Med. Eur. P. *The leaves are used for flavouring and produce Oil of Laurel. It has been a symbol of glory since ancient times and has been used for crowning heroes and poets. It is still used in wreaths: it is the origin of the word 'baccalaureate'.*

PAPAVERACEAE | Poppy Family

Herbaceous plants with milky or watery sap. Lvs. simple or compound; stipules absent. Fls. often large; sepals usually 2, soon falling; petals 4–6; stamens usually numerous; ovary superior, of one to many carpels. Fr. a capsule, splitting by pores or valves. The genera *Corydalis, Fumaria,* and *Platycapnos* have often been placed in a separate family, Fumariaceae, owing to their distinctive small, clustered, irregular fls.

		Nos.
1. Fls. regular, i.e. radially symmetrical.		
2. Sepals fused to form a conical hood in bud; sap watery.	*Eschscholzia*	275
2. Sepals free; sap milky.		
3. Fr. less than 10 times as long as broad.		
4. Style short; stigmas distinct.	*Meconopsis*	269
4. Style absent; stigmas on disk-like top of ovary.		
5. Lvs. and stem spiny.	*Argemone*	270
5. Lvs. and stem not spiny.	*Papaver*	264–268
3. Fr. more than 10 times as long as broad.		
6. Fls. violet.	*Roemeria*	271
6. Fls. yellow or red.		
7. Fls. solitary; petals 2 cm. or more.	*Glaucium*	272, 273
7. Fls. several; petals to 1 cm.		
8. Sap orange; petals entire; fr. opening by valves.	*Chelidonium*	274
8. Sap watery; petals three-lobed; fr. breaking into sections.	*Hypecoum*	276
1. Fls. irregular, i.e. symmetrical in one plane only.		
9. Fls. slightly irregular; petals not spurred or with sac-like base.	*Hypecoum*	276
9. Fls. strongly irregular; petals with spur or sac-like base.		
10. Fr. several-seeded, splitting into 2 valves.	*Corydalis*	277–279
10. Fr. one-seeded, not splitting.		
11. Fr. globular.	*Fumaria*	280, 281
11. Fr. strongly flattened.	*Platycapnos*	282

PAPAVER | **Poppy** Fls. large; sepals 2, soon falling; petals 4, crumpled in bud; stamens numerous; ovary with 4–15 radiating stigmas. Fr. a flask-shaped capsule with a cap-like apex, opening by pores below the cap. 26 sps.

Fl. stems leafy; several-flowered
(a) Fr. hairless

264. *P. somniferum* L. OPIUM POPPY. An erect, glaucous ann. $\frac{1}{2}$–$1\frac{1}{2}$ m. with large, deeply lobed, clasping lvs. and very large white, lilac, or purple fls. Fls. to 10 cm. across; petals $3\frac{1}{2}$–$4\frac{1}{2}$ cm. rounded, with or without a dark basal blotch. Lvs. 7–12 cm., oblong, deeply lobed or toothed, hairless. Subsp. *setigerum* (DC.) Corb. has long stiff hairs on stem and lvs., and lvs. more deeply and acutely lobed. Fr. 5–9 cm. globular, smooth, hairless. △ Fields, waysides, waste ground. June–July. Probably native of Central Europe, now naturalized in most of Europe. *One of the most valuable of all medicinal plants, which produces the drug opium, and is the source of morphine and codeine. Opium is obtained by incising the fully grown, but green capsules. When smoked or chewed, opium causes intoxication and it is an addictive drug. Poppy-seed oil, extracted from the ripe seeds, known as 'Maw seed', is a fine edible oil quite free from opium, and it is also used as a drying oil in paint-making, etc.* Pl. 28.

265. *P. rhoeas* L. CORN POPPY. Distinguished by its hairless, globular to ovoid fr. 1–2 cm. which is not more than twice as long as wide. Fls. scarlet, 7–10 cm. across; petals rounded, 2–4 cm. wide, sometimes with a dark basal blotch; fl. stalks with spreading hairs. Lvs. once- to twice-pinnate, with coarsely toothed, narrow, fine-pointed segments. An erect, stiff-haired, branched, very variable ann. 25–90 cm. △ Fields, cultivated ground, waste places. May–July. All Eur. (except IS.TR.). *Poisonous to livestock. The petals have been used medicinally and to make a red ink.* Pl. 29.

P. dubium L. LONG-HEADED POPPY. Like 265, but fr. $1\frac{1}{2}$–2 cm. oblong-ovoid, 2–3 times as long as broad and tapering gradually near the top. Fls. pale scarlet, 3–7 cm. across; filaments and anthers violet; fl. stalks with adpressed hairs. Throughout Europe.

P. pinnatifidum Moris Like *P. dubium*, but anthers yellow and filaments violet. Lv. segments triangular-oval, entire or toothed. Mediterranean region.

*P. lecoqii Lamotte Western Europe.

P. orientale L. ORIENTAL POPPY. A robust rough-haired perenn. to 1 m. with very large, bright red fls. to *c.* 15 cm. across, with a dark basal blotch. Fr. to $3\frac{1}{2}$ cm., globular, glaucous, hairless. Native of Western Asia; grown as an ornamental and sometimes naturalized.

(b) Fr. with spreading bristles

266. *P. argemone* L. PRICKLY LONG-HEADED POPPY. Fr. $1\frac{1}{2}$–2 cm. oblong-club-shaped, several times as long as broad, strongly ribbed, and with a few stiff, erect, yellowish bristles. Petals $1\frac{1}{2}$–$2\frac{1}{2}$ cm. long, often not overlapping, pale scarlet and usually with a blackish basal blotch; stigma-rays 4–6. Lvs. pinnately lobed with linear to oblong segments, with stiff, bristly hairs. An erect ann. or bienn. 15–50 cm. △ Fields, sandy places, waste ground. May–July. Most of Eur. (except IS.H.AL.BG.).

P. hybridum L. PRICKLY ROUND-HEADED POPPY. Distinguished by its more or less globular fr. 10–12 mm., covered with many stiff, erect, yellow bristles. Petals 1–2 cm., crimson, with violet base; stigma-rays 4–8. Most of Europe, except Northern Europe.

Fl. stems leafless; fls. solitary

267. *P. rhaeticum* Leresche ALPINE POPPY. An alpine perenn. with a basal rosette of glaucous lvs. and leafless stems bearing solitary golden-yellow or rarely white or red fls. Petals 2–$2\frac{1}{2}$ cm.; fl. stems 5–25 cm. with adpressed hairs. Lvs. once or twice pinnately cut

into 2–4 pairs of oval blunt segments 1–6 mm. wide; lv. bases forming a fibrous investing sheath. Fr. oblong-ellipsoid, *c.* 14 mm. △ Alpine rocks and screes. July–Aug. E.F.CH.A.I. YU. Pl. 28.

P. sendtneri Hayek Like 267, but fls. usually white, smaller; petals 1½–2 cm. long. Lvs. with acute segments 1½–2½ mm. wide. The Alps. Pl. 29.

268. *P. radicatum* Rottb. ARCTIC POPPY. A northern and arctic perenn. differing from the related alpine sps. by the petals becoming blue-green on drying. Fls. very variable in size, 2–5 cm. across; petals usually yellow, rarely whitish or pink; fl. stem 5–30 cm., hairy. Lvs. usually 3–6 cm., deeply cut into 2–5 pairs of narrow or broad, toothed or entire lobes; lv. bases forming a fibrous sheath. Fr. ellipsoid, 13–17 mm., with dense, spreading or adpressed bristly hairs. A very variable sp. △ Stony places, screes. July. IS.N.SU.

P. nudicaule L. ICELAND POPPY. Like 267, but a more robust perenn. with large yellow, orange, or reddish fls. 3–6 cm. across on leafless stems to 50 cm. Lvs. all basal, broadly lobed. Native of Asia; grown as an ornamental and sometimes locally naturalized.

MECONOPSIS Like *Papaver*, but fr. with a short projecting style carrying 4–6 distinct stigmas, opening above by 4–6 valves. 1 sp.

269. *M. cambrica* (L.) Vig. WELSH POPPY. Fls. sulphur-yellow, 3–7½ cm. across, borne on long hairy stalks from leafy, branched stems. Petals 4; sepals hairy; stamens yellow. Lvs. pinnately lobed, segments oval, toothed, glaucous beneath. Fr. 2–4 cm. club-shaped, four- to six-ribbed, splitting partially into 4–6 valves. A sparsely hairy tufted perenn. with yellow latex, 30–60 cm. △ In mountains: woods, shady places. June–Aug. E.F.IRL.GB.: introd. NL.D.CH. Pl. 28.

ARGEMONE Like *Papaver*, but fr. ellipsoid, spiny, opening by 4–6 short valves; styles absent; stigmas 4. 1 sp.

270. *A. mexicana* L. PRICKLY POPPY. A rather robust glaucous ann. to 90 cm., with prickly stems and thistle-like lvs. and solitary, stalkless, pale yellow or orange fls. 5–6 cm. across. Petals 4–6, 2–3 cm. obovate; sepals 2–3. Lvs. coarsely lobed, with undulate spiny margin. Fr. *c.* 2½ cm., spiny. △ Native of America; naturalized in South and Central Europe. Summer. Pl. 29.

ROEMERIA Like *Papaver*, but fr. long and narrow, one-celled and splitting to the base into 2–4 valves. Petals 4. Sap yellow. 1 sp.

271. *R. hybrida* (L.) DC. VIOLET HORNED-POPPY. Fls. solitary, deep blue-violet with a darker basal blotch, 4–7½ cm. across. Lvs. 2–3 times pinnately cut into linear, bristle-pointed segments. Fr. 5–10 cm.; linear straight, rough-haired. An erect, branched, sparsely hairy ann. or bienn. 20–40 cm., with yellow sap. △ Fields, among crops, banks. May–June. Med. Eur. (except I.AL.) BG.: introd. GB.NL.A. *The root is poisonous.* Pl. 28.

GLAUCIUM | Horned-Poppy Like *Papaver*, but fr. long and narrow, two-celled and splitting almost to base into 2 valves and leaving the seeds embedded in the dividing wall. Fls. solitary; petals 4. Sap yellow. 3 sps.

272. *G. flavum* Crantz YELLOW HORNED-POPPY. An erect, branched, bienn. or perenn. 25–100 cm. with large yellow fls. 6–9 cm. across, glaucous lvs., and very long curved frs. Petals 3–4 cm. rounded, pale or golden-yellow; stamens yellow. Lvs. deeply pinnately

lobed with lobes further cut or toothed, the upper lvs. stalkless and encircling the hairless stem, lvs. all rough. Fr. linear, 15–30 cm., often curved, rough, hairless. △ Maritime shingle; sometimes naturalized inland. June–Aug. Most of Eur. (except IS.N.SF.): introd. CH.A.PL.CS.H. *The root is poisonous.* Pl. 29.

273. *G. corniculatum* (L.) J. H. Rudolph RED HORNED-POPPY. Like 272, but fls. scarlet or orange, smaller 2½–5 cm. across and petals often with a blackish basal blotch. Stamens with dark anthers. Lvs. deeply lobed, hairy, the upper lvs. half-clasping the hairy stem. Fr. linear, 10–20 cm., rough, stiff-haired, usually straight. Usually ann. 30–60 cm. △ Waste and cultivated places. May–June. Med., South-East Eur. (except AL.) P.H.: introd. GB.B.NL.DK.S.D.CH.A.PL.CS. Pl. 29.

CHELIDONIUM Fr. long and narrow, one-celled, opening from below by 2 valves. Fls. in an umbel; petals 4. Sap bright orange. 1 sp.

274. *C. majus* L. GREATER CELANDINE. A leafy, brittle-stemmed branched perenn. 30–90 cm. with golden-yellow fls. 2–2½ cm. across, in terminal umbels of 2–6. Petals 4, broadly obovate; sepals yellowish, hairy; stamens yellow. Lvs. pinnate with 5–7 oval, lobed, or toothed leaflets, bright green, soft; sap orange. Fr. linear, 3–5 cm., hairless; seeds black, appendage white. △ Banks, walls, hedges, waste places. May–Oct. All Eur. (except IS.): introd. IRL. *The orange sap is purgative and poisonous; formerly cultivated and used as a herbal remedy for corns, warts, and eye infections.*

ESCHSCHOLZIA Like *Papaver*, but sepals fused and forming a cap-like sheath which is shed when the fls. open; petals 4. Fr. long and narrow, ribbed, one-celled, opening by 2 valves. 1 sp.

275. *E. californica* Cham. CALIFORNIAN POPPY. Fls. orange or yellow up to 10 cm. across, solitary, long-stalked, with 4 broad, smooth, shining petals. Lvs. glaucous, several times divided into small linear segments. Fr. linear, 7–10 cm., straight, hairless. Ann. or perenn. 20–60 cm. △ Native of North America; often grown as ornamental and sometimes naturalized. Summer. Pl. 29.

HYPECOUM Fls. small; petals 4, somewhat unequal in size and fls. consequently slightly irregular, petals mostly three-lobed; stamens 4; stigmas 2. Fr. linear, divided transversely into many one-seeded sections which separate on ripening. 4 sps.

276. *H. procumbens* L. A rather delicate, glaucous, spreading ann. or bienn. 10–40 cm., with hairless, dissected lvs. and small yellow fls. in a lax, branched, leafy cluster. Fls. ½–1 cm. across; petals somewhat unequal, the outer pair broadly rhomboidal with 3 unequal, shallow terminal lobes with the middle lobe larger, the inner petals smaller, shallowly three-lobed. Lvs. 3 times dissected into linear to lance-shaped lobes. Fr. linear, 4–6 cm., curved, jointed, erect. △ Fields, cultivated ground. May–June. Med. Eur. (except AL.) P.BG.R.

H. imberbe Sibth. & Sm. (*H. grandiflorum* Bentham) Like 276, but fls. larger 1–1½ cm. across, darker, orange-yellow, with the outer petals with 3 equal lobes. Fr. scarcely jointed. More or less erect ann. or bienn. Mediterranean region. Pl. 30.

CORYDALIS Fls. irregular, distinctly two-lipped, small, and clustered; petals 4, the upper hooded and spurred, the lower keeled and boat-shaped, and the 2 lateral petals narrow; sepals 2; stamens 2. Fr. a two-valved capsule with several seeds. 14 sps.

Aerial stems with many lvs.; roots not tuberous

277. *C. claviculata* (L.) DC. CLIMBING CORYDALIS. A slender, glaucous, much-branched, climbing ann. to 1 m., with clusters of small cream-coloured fls., and with compound lvs. with tendrils. Fls. 5–6 mm. long, in dense stalked clusters of 6–8, arising opposite the lvs. Lvs. twice-pinnate with long-stalked, elliptic leaflets 5–12 mm., and terminal-branched tendrils. Fr. 6–10 mm., hairless. △ Rocks, open woods, thickets. June–Sept. West Eur. DK.S.D. Pl. 30.

278. *C. lutea* (L.) DC. YELLOW CORYDALIS. A rather glaucous, somewhat tufted branched perenn. 20–40 cm., with fern-like lvs. and dense clusters of golden-yellow fls. Fls. 12–20 mm. long, short-stalked, in clusters of 6–16; petals darker towards the tip, the upper with a short downward-curved spur. Lvs. twice-pinnate with oval, deeply three- to five-lobed stalked leaflets, hairless, glaucous beneath, without tendrils. Fr. *c.* 10 mm., drooping. △ Rocks, ravines; commonly naturalized on walls. Apr.–Sept. Native of CH.I.YU.: introd. E.F.IRL.GB.B.NL.D.A.CS. Pl. 30.

C. ochroleuca Koch Like 278, but fls. cream-coloured and petals yellow towards the tips. Lv. stalks very narrowly winged; leaflets glaucous on both surfaces. Fr. erect. Balkan Peninsula; introduced elsewhere in Europe. Pl. 30.

Aerial stems with 1 or 2 lvs.; roots tuberous

279. *C. solida* (L.) Swartz (*C. halleri* Willd.) A hairless perenn. with solitary erect stems 10–20 cm., usually with 2 dissected lvs. and a terminal, rather dense cluster of purple or whitish fls. Fls. usually 10–20, each 1½–2½ cm. long, with a nearly straight spur; bracts subtending lowest fls. always deeply lobed, the upper wedge-shaped. Lvs. usually twice-ternate with oval segments. Stem with 1–3 conspicuous oval scales below lowest lv.; tuber solid globular. Fr. 1–2½ cm., drooping when ripe. A very variable sp. △ Woods, hedges, verges, vineyards. Mar.–May. Most of Eur. (except P.IRL.IS.TR.): introd. GB.DK.N. Pl. 30.

C. bulbosa (L.) DC. (*C. cava* (L.) Schweigger & Koerte) Like 279, but bracts subtending the fls. all entire, oval-lance-shaped. Fls. white, cream, or purplish, larger, 2–3 cm. long, spur curved at apex. Scales below lvs. absent; tuber hollow. Widespread in Europe.

C. intermedia (L.) Mérat (*C. fabacea* Pers.) Like 279, but a more slender plant with smaller purple or rarely white fls. 1–1½ cm. in dense clusters of 2–8; bracts all entire, oval. Scales below lvs. present; tuber solid. Northern and Central Europe.

FUMARIA | **Fumitory** Similar to *Corydalis*, but fr. a one-seeded nutlet. Lvs. usually glaucous, 2–4 times cut into segments. Small differences in the petals and fr. distinguish a number of widespread European sps. 33 sps.

280. *F. capreolata* L. RAMPING FUMITORY. A climbing ann. to 1 m. with long-stalked axillary clusters of creamy or pinkish fls. with tips of petals dark reddish-purple. Clusters up to twenty-flowered; petals 1–1½ cm. long; sepals toothed. Lvs. compound with oblong-oval segments to 1 cm. broad; tendrils absent. Fr. stalks often rigidly reflexed; fr. smooth. △ Cultivated ground, waste places, hedges, walls. Apr.–June. Much of Eur. (except North Eur.): introd. DK.N.A.H. Pl. 30.

281. *F. officinalis* L. COMMON FUMITORY. A slender, usually erect, branched ann. 20–70 cm., with dense stalked clusters of numerous pink fls. with blackish-red wings and tips of petals. Fls. 7–9 mm. long; sepals 1½–3½ mm., toothed, more than one-quarter as long as petals. Lvs. compound with lance-shaped or linear, flat segments. Fr. stalks ascending,

longer than the subtending bracts; fr. somewhat kidney-shaped, rough when dry. A very variable sp. △ Cultivated ground, waste places. Apr.–Sept. All Eur.

F. bastardii Boreau Widespread in Europe.

F. muralis Koch Widespread in Europe.

F. densiflora DC. Widespread in Europe.

F. vaillantii Loisel. Widespread in Europe.

F. parviflora Lam. SMALL-FLOWERED FUMITORY. Distinguished by its tiny fls. 3–6 mm. long, which are whitish flushed with pink and with a blackish-red blotch on the upper petal and tips of lateral petals; sepals minute $\frac{1}{2}$–$1\frac{1}{2}$ mm. Lvs. with narrow, linear, channelled segments. Widespread in Europe, except in the north.

PLATYCAPNOS Like *Corydalis* and *Fumaria*, but fr. strongly compressed, not globular. 2 sps.

282. *P. spicata* (L.) Bernh. (*Fumaria s.*) SPIKED FUMITORY. A branched, hairless ann. 10–40 cm., with numerous tiny fls. in a very dense, rounded, clover-like head $2\frac{1}{2}$–5 cm. Fls. pink with dark blackish-purple tips, rarely whitish or yellowish. Lvs. glaucous, twice cut into narrow linear segments less than 1 mm. broad. Fr. stalk recurved in fr.; fr. oval flattened. △ Cultivated ground, tracksides. Apr.–July. P.E.F.I.

CAPPARIDACEAE | Caper Family

Herbs, trees, or shrubs with alternate, simple, or compound lvs. with stipules which are sometimes spiny. Sepals and petals 4; stamens 6 to many. Ovary long-stalked, superior, of 2 carpels; fr. a berry or capsule.

CAPPARIS Shrubs with simple lvs. usually with spiny stipules. Fls. solitary; stamens numerous. Fr. a long-stalked berry. 2 sps.

283. *C. spinosa* L. CAPER. A shrub with spreading spiny stems to $1\frac{1}{2}$ m., with conspicuous, axillary, white, or pale lilac fls. 5–7 cm. across, with numerous long-projecting stamens with violet stalks. Lvs. alternate, oval or orbicular with a blunt or notched apex, sometimes fleshy; spines short, recurved. Fr. a large berry which bursts when ripe, exposing purple seeds in pale crimson flesh. Var. *inermis* Turra has spines absent or soon falling and somewhat fleshy lvs. △ Walls and rocks. June–Sept. Med. Eur. (except TR.) P. *The flower buds when cooked and pickled produce the capers used in cooking and flavouring.* Pl. 31.

C. ovata Desf. Very like 283, but with narrow, slightly hairy lvs., with an acute or long-pointed apex; spines often strong and curved. Fls. 4–5 cm. across, conspicuously irregular. Mediterranean region.

CRUCIFERAE | Mustard Family

Herbaceous anns. or perenns. with alternate lvs., usually with stipules. Fls. usually small, numerous, in branched clusters. The floral structure is distinctive and characteristic throughout the family. Sepals 4; petals 4, usually spreading in a cross, each with a narrow

CRUCIFERAE

stalk or *claw*, and a spreading blade or *limb*; stamens usually 6; ovary of 2 fused carpels; style 1, stigma bilobed or rounded. Fr. usually opening by 2 valves, usually two-celled owing to the formation of a cross-wall or *septum*. Fr. either at least 3 times as long as wide, a *siliqua*, or less than 3 times as long as wide, a *silicula*, or less commonly splitting transversely into one-seeded sections, or not splitting. The genera can often only be distinguished by their frs. and ripe, preferably dry frs. are essential for the identification of sps.

1. Fr. winged, not jointed or splitting into valves.   Group A
1. Fr. not winged, but jointed or splitting into valves.
  2. Fr. with transverse joints often separating into sections on ripening.   Group B
  2. Fr. without transverse joints, but splitting into valves from the base.
    3. Fr. more than 3 times as long as wide, a siliqua.   Group C
    3. Fr. less than 3 times as long as wide, a silicula.
      4. Fr. not splitting into valves.   Group D
      4. Fr. splitting into 2 valves.
        5. Fr. flattened parallel to the septum lying between the valves (the septum is as wide as the widest diameter of the fr.).   Group E
        5. Fr. flattened at right angles to the septum lying between the valves (the septum is narrower than the widest diameter of the fr.).   Group F

GROUP A. *Fr. winged*

Nos.

Fr. winged, one-seeded, not splitting into valves or segments.   *Isatis*   292

GROUP B. *Fr. with transverse joints, often separating into sections when ripe*

Nos.

1. Upper segment of fr. globose or ovoid, sometimes shortly beaked (the lower segment short and usually seedless).
  2. Upper segment of fr. not beaked; style absent.   *Crambe*   368
  2. Upper segment of fr. beaked or with a distinct persistent style.
    3. Stem lvs. clasping stem with acute lobes; petals white.   *Calepina*   369
    3. Stem lvs. not clasping stem; petals yellow.   *Rapistrum*   367
1. Upper segment of fr. cylindrical or oblong-ovoid, sometimes constricted between the seeds (lower segment short, usually seedless).
    4. Upper segment of fr. not more than 3 times as long as wide, mitre-shaped, usually one-seeded.   *Cakile*   366
    4. Upper segment of fr. at least 5 times as long as wide, cylindrical and often constricted between the seeds.   *Raphanus*   370

GROUP C. *Fr. splitting from the base into 2 valves; fr. more than 3 times as long as wide, a siliqua*

Nos.

1. Plants hairless or with simple, unbranched hairs only.
  2. Sepals erect and pressed close to petals (thus calyx closed).
    3. Lvs. simple, entire, somewhat fleshy.
      4. Petals purple or violet.   *Moricandia*   356
      4. Petals yellow or yellowish-white.
        5. Stem lvs. heart-shaped, clasping stem; hairless anns.   *Conringia*   355

5. Stem lvs. not clasping stem; perenns. or bienns.	*Brassica*	359–361
3. At least some lvs. toothed, lobed, or deeply divided.		
6. Stigma deeply two-lobed.		
7. Fr. with a sword-shaped beak; petals yellowish or whitish.	*Eruca*	363
7. Fr. not beaked; petals pale purple.	*Hesperis*	296
6. Stigma club-shaped; petals yellow.		
8. Valves of fr. with 1 prominent vein.	*Brassica*	359–361
8. Valves of fr. with 3 veins, at least when young.	*Hirschfeldia*	365
2. Sepals spreading horizontally or at an angle to the petals (thus calyx open).		
9. Valves of fr. with 3–7 veins.		
10. Petals white.	*Alliaria*	289
10. Petals yellow.		
11. Fr. with a beak more than 1 cm.	*Sinapis*	362
11. Fr. not beaked, but style distinct.	*Sisymbrium*	284–287
9. Valves of fr. with 1 vein or veinless.		
12. Fr. not more than 7 times as long as wide.		
13. Seeds in 1 row in each cell.	*Arabis*	315–319
13. Seeds in 2 rows in each cell.	*Draba*	331–334
12. Fr. more than 7 times as long as wide.		
14. Valves of fr. with median vein weak, or absent.		
15. Valves flat, fr. strongly compressed.		
16. Valves coiling spirally from the base at maturity.	*Cardamine*	308–313
16. Valves not coiling spirally.	*Arabis*	315–319
15. Valves convex, fr. not strongly compressed.		
17. Petals white.	*Nasturtium*	307
17. Petals yellow.		
18. Inflorescence flat-topped; seeds in 1 row in each cell.	*Barbarea*	302
18. Inflorescence not flat-topped; seeds in 2 rows in each cell.	*Rorippa*	303–305
14. Valves of fr. with a distinct median vein.		
19. Fr. strongly compressed.		
20. Stem lvs. stalkless.	*Arabis*	315–319
20. At least lower and middle stem lvs. distinctly stalked.	*Cardaminopsis*	314
19. Fr. not strongly compressed.		
21. Petals yellow, sometimes with violet veins.		
22. Seeds in 2 rows in each cell.	*Diplotaxis*	357, 358
22. Seeds in 1 row in each cell.		
23. Valves rounded on the back; median nectaries absent; lateral nectaries prismatic.	*Brassica*	359–361
23. Valves keeled; median nectaries present or, if absent, the lateral nectaries semi-lunar or two-lobed.		
24. Median nectaries present; valves only slightly constricted between seeds.	*Barbarea*	302
24. Median nectaries absent; valves conspicuously constricted between seeds.	*Erucastrum*	364
21. Petals white, white veined with violet, or purple.		

25. Seeds in 2 rows in each cell.	*Diplotaxis*	357
25. Seeds in 1 row in each cell.		
26. Stout perenn. up to 2 m.; basal lvs. 10–40 cm.	*Brassica*	359–361
26. Slender anns. or perenns. to *c.* ½ m.; basal lvs. much less than 10 cm.		
27. Most fls. subtended by a bract.	*Sisymbrium*	284–287
27. Fls. without bracts.	*Arabidopsis*	290
1. Hairs star-shaped, branched or a mixture of branched and unbranched.		
28. Stigma deeply two-lobed, the lobes sometimes erect and coming together to form a beak to the fr.		
29. Lobes of stigma with a swelling or horn.	*Matthiola*	300, 301
29. Lobes of stigma without a swelling or horn.		
30. Petals yellow.		
31. Hairs all attached at the middle; style 2–3 mm. in fr.	*Cheiranthus*	299
31. Hairs various; style *c.* 1 mm. or less in fr.	*Hesperis*	296
30. Petals white, pink, or violet, rarely reddish.		
32. Styles short, stigma lobes free.	*Hesperis*	296
32. Styles absent, stigma lobes erect, coming together.	*Malcolmia*	297, 298
28. Stigma club-shaped, notched, or slightly two-lobed.		
33. Lvs. 2–3 times divided; petals shorter than sepals.	*Descurainia*	288
33. Lvs. entire or once divided.		
34. Petals yellow.		
35. Fr. not more than 1 cm., not more than 6 times as long as wide.	*Draba*	331–334
35. Fr. 1 cm. or more, at least 10 times as long as wide.		
36. Stem lvs. not clasping stem.	*Erysimum*	294, 295
36. Stem lvs. clasping stem with an arrow-shaped or heart-shaped base.		
37. Fr. hairless.	*Arabis*	315–319
37. Fr. finely hairy.	*Arabidopsis*	290
34. Petals white, pink, or purple.		
38. Plants densely covered with hairs attached at the middle; style usually 4–6 mm.	*Erysimum*	294, 295
38. Plants with unbranched, branched, or star-shaped hairs, sometimes with a few hairs attached at their middle.		
39. Style at least 2½ mm.	*Aubrieta*	320
39. Style not more than 2 mm.		
40. Fr. less than 10 times as long as wide.		
41. Seeds in 1 row in each cell, fr. scarcely constricted between seeds.	*Arabis*	315–319
41. Seeds in 2 rows in each cell.	*Draba*	331–334
40. Fr. at least 10 times as long as wide.		
42. Valves of fr. flat, rarely keeled.		
43. Lower and middle stem lvs. stalkless or nearly so, entire or deeply toothed.	*Arabis*	315–319
43. Lower and middle stem lvs. distinctly stalked, often deeply pinnately lobed.		
44. Valves of fr. with a distinct median vein.	*Cardaminopsis*	314

44. Valves of fr. without a median vein.	*Arabis*	315–319
42. Valves of fr. rounded or angled. Fr. less than 1 mm. wide, hairless.	*Arabidopsis*	290

GROUP D. *Fr. less than 3 times as long as wide; fr. not splitting into valves at maturity*

		Nos.
1. Fr. paired.		
2. Valves flat; petals yellow.	*Biscutella*	349
2. Valves rounded; petals white.		
3. Petals ½–1½ mm.; lvs. pinnately lobed.	*Coronopus*	354
3. Petals 3–4 mm.; lvs. entire or shallowly toothed.	*Cardaria*	353
1. Fr. not paired.		
4. Fr. with 3 cells, the upper 2 side by side, sterile, the lower one-seeded.	*Myagrum*	291
4. Fr. not as above.		
5. Fr. with 4 longitudinal wings or ridges, or covered with swellings.	*Bunias*	293
5. Fr. not as above.		
6. Fr. flattened, disk-like, pendulous.	*Clypeola*	330
6. Fr. globular, nut-like, erect.		
7. Fr. with a short, blunt beak; petals white, unequal.	*Calepina*	369
7. Fr. with a slender style; petals yellow, equal.	*Neslia*	339

GROUP E. *Fr. a silicula, flattened parallel to the septum lying between the 2 valves, rarely rounded and nut-like*

		Nos.
1. Plants hairless or with unbranched hairs.		
2. Small aquatic plants; lvs. all basal	*Subularia*	[354]
2. Not as above.		
3. Sepals erect or nearly so (thus calyx closed).		
4. Fls. violet; fr. 2–9 cm. by 1½–3½ cm.	*Lunaria*	321, 322
4. Fls. white or yellow; fr. less than 12 mm.	*Camelina*	338
3. Sepals spreading or spreading-erect (the calyx more or less open).		
5. Petals yellow.		
6. Plant with at least some lvs. pinnate, pinnately lobed or coarsely toothed; stem lvs. numerous.	*Rorippa*	303–305
6. Lvs. usually entire; stem lvs. usually absent.	*Draba*	331–334
5. Petals white.		
7. Fr. with a short, broad, blunt beak; petals unequal.	*Calepina*	369
7. Fr. not beaked, but sometimes with persistent style; petals equal.		
8. Robust plants to 1 m., with swollen roots.	*Armoracia*	306
8. Slender plants to 40 cm.; roots not swollen.		
9. Filaments straight; inflorescence with bracts.	*Cochlearia*	336, 337
9. Filaments curved; inflorescence without bracts.	*Draba*	331–334
1. At least some of the hairs branched or star-shaped.		
10. Sepals erect (thus calyx closed).		
11. Fr. stalk-like at the base.	*Alyssoides*	323
11. Fr. without a stalk-like base.		

12. Fr. flattened; petals short-clawed; perenns.	*Fibigia*	327
12. Fr. inflated, pear-shaped, obovoid, or globose; anns. or bienns.	*Camelina*	338
10. Sepals spreading or diverging (thus calyx open).		
13. Petals deeply two-lobed.		
14. Rosette lvs. present; fl. stems leafless or nearly so.	*Erophila*	335
14. Rosette lvs. absent; fl. stem leafy.		
15. Petals white; style long; seeds winged or margined.	*Berteroa*	328
15. Petals yellow.	*Alyssum*	324–326
13. Petals entire or shallowly notched.		
16. Petals yellow.		
17. Dwarf perenns. with leafless or almost leafless fl. stems; lvs. usually entire, often linear.	*Draba*	331–334
17. Anns. or perenns. with leafy stems; lvs. often toothed, wider.		
18. Inner sepals with a sac-like base; ovary with 4–8 ovules in each cell.	*Alyssoides*	323
18. Inner sepals without a sac-like base; ovary with 1–2 ovules in each cell.	*Alyssum*	324–326
16. Petals white, pink, or purplish.		
19. Hairs all attached at the middle; fr. usually finely hairy.	*Lobularia*	329
19. Hairs all star-shaped or branched.	*Draba*	331–334

GROUP F. *Fr. a silicula, flattened at right angles to the septum lying between the valves*

		Nos.
1. Fr. paired, kidney- or heart-shaped; petals $\frac{1}{2}$–$1\frac{1}{2}$ mm.	*Coronopus*	354
1. Fr. not paired or kidney- or heart-shaped.		
2. Petals unequal, the outer larger than the inner.		
3. Style distinct, stigma knob-shaped.	*Iberis*	347, 348
3. Style inconspicuous, stigma two-lobed.	*Teesdalia*	342
2. Petals equal in size, when present.		
4. Filaments with a wing or tooth-like appendage.		
5. Lvs. entire, not confined to a basal rosette; sepals erect.	*Aethionema*	346
5. Lvs. usually pinnately lobed, mostly confined to a basal rosette; sepals spreading-erect.	*Teesdalia*	342
4. Filaments without an appendage.		
6. Valves of fr. winged or strongly keeled.		
7. Seeds solitary, pendent from apex of each cell.	*Lepidium*	350–352
7. Seeds 1–8 in each cell.	*Thlaspi*	343–345
6. Valves of fr. not winged or keeled.		
8. Fr. triangular-obcordate.	*Capsella*	340
8. Fr. not as above.		
9. Lvs. simple; valves of fr. convex.	*Cochlearia*	336, 337
9. Lvs. pinnate or pinnately lobed.	*Hornungia*	341

SISYMBRIUM | **Rocket** Fls. usually yellow, rarely white. Fr. long, slender, with a strong midvein and usually 2 weaker veins on each valve; style very short. Lvs. entire or pinnately lobed; hairs unbranched. 19 sps.

Lvs. entire, toothed

284. *S. strictissimum* L. A rather tall, stiffly erect, leafy perenn. to 1½ m., distinguished by large undivided lvs. and clusters of conspicuous, sweet-scented, bright yellow fls. Fl. clusters leafless; petals 4½-10 mm., not twice as long as sepals. Fr. curving upwards, 3-8 cm. by 1-2 mm. △ Hedges, tracksides, rocks. June-Aug. Cent., South-East Eur. (except GR.TR.) F.SU.: introd. IRL.GB.B. Page 128.

Lvs. deeply lobed or compound
(a) Fr. erect and pressed to stem
285. *S. officinale* (L.) Scop. HEDGE MUSTARD. Fls. tiny 3 mm. across, pale yellow, short-stalked, at first in a more or less dense cluster, but soon elongating and appearing at the ends of long, narrow fr. spikes. Basal lvs. deeply pinnately lobed, with a larger rounded terminal lobe; upper stem lvs. with an arrow-shaped terminal lobe and 1-3 oblong lateral lobes. Fr. 1-2 cm., held stiffly against stem. Stiff, erect, bristly-haired ann. or bienn. 5-90 cm., forming an over-wintering rosette. △ Waste places, tracksides, cultivated ground. May-Sept. All Eur.

(b) Fr. held obliquely erect
286. *S. irio* L. LONDON ROCKET. Fls. yellow, inconspicuous, 3-4 mm. across and overtopped by the young frs.; petals 2½-3½ mm., little longer than the sepals. Lower lvs. deeply pinnately lobed, with a much larger terminal lobe; stem lvs. lobed or entire and hastate. Fr. slender, 2½-6½ cm. by 1 mm. An erect, branched ann. 10-60 cm., hairless or with long curved hairs. △ Waste places, waysides. Apr.-July. P.E.F.CH.A.I.YU.GR.BG.R.: introd. to rest of Eur. (except IS.SF.AL.TR.).

*S. loeselii L. Widespread in Europe.

S. austriacum Jacq. Distinguished by its golden-yellow fls. with petals 3½-7 mm., twice as long as sepals. Fr. stalks and fr. curved and twisted. Widespread in Europe, except in the north.

287. *S. altissimum* L. (*S. sinapistrum* Crantz) TALL ROCKET. A nearly hairless, erect ann. or bienn. to 1 m. with pale yellow fls. in terminal, leafless, branched clusters, and upper stalkless lvs. cut into several pairs of very narrow thread-like lobes. Fls. *c.* 11 mm. across; petals about twice as long as sepals. Lowest lvs. lance-shaped, deeply cut into triangular, toothed lobes. Fr. 5-10 cm. by 1-1½ mm., hairless, held obliquely erect. △ Waste places, tracksides, sands. May-July. Cent., South-East Eur.: introd. P.F.GB.B.NL.IS.DK.N.S.

S. orientale L. (*S. columnae* Jacq.) EASTERN ROCKET. Like 287, but uppermost lvs. shortly stalked, simple or three-lobed, with the middle lobe linear or lance-shaped. Fls. yellow, 10-12 mm. across. A greyish, hairy plant of the Mediterranean region, widely naturalized in Europe. Pl. 32.

DESCURAINIA Like *Sisymbrium*, but lvs. finely 2-3 times cut, with both simple and branched hairs present. Valves of fr. with 1 vein. 1 sp.

288. *D. sophia* (L.) Prantl (*Sisymbrium s.*) FLIXWEED. An erect, greyish-green ann, or bienn. to 1 m., with numerous very small, pale yellow fls. in a terminal leafless cluster, and all lvs. twice or thrice cut into linear segments. Fls. *c.* 3 mm. across; petals about as long as sepals. Lvs. greyish, with branched star-shaped hairs. Fr. very slender, 1-1½ cm. long, held erect on very slender spreading stalks. △ Waste places, tracksides, walls. Apr.-Sept. All Eur. (except AL.).

ALLIARIA Fls. white; fr. a four-angled siliqua, valves three-veined. Basal lvs. entire; hairs all simple. 1 sp.

289. *A. petiolata* (Bieb.) Cavara & Grande (*A. officinalis* Bieb.; *Sisymbrium alliaria* (L.) Scop.) GARLIC MUSTARD. An erect, hairless plant with terminal, leafless clusters of white fls. each *c.* 6 mm. across, and distinctive broadly heart-shaped and coarsely toothed, stalked lvs. which smell of garlic when crushed. Petals *c.* twice as long as sepals. Fr. 2–7 cm., held obliquely erect. A rosette-forming bienn., 20–120 cm. △ Hedges, shady places, woods. May–July. All Eur. (except IS.). *Used medicinally and as salad plant in the past.* Pl. 31.

ARABIDOPSIS Like *Sisymbrium*, but valves of fr. one-veined with more or less prominent midrib; seeds ovoid, in 1 row. Hairs absent or both branched and unbranched hairs present. Fls. white, 5 sps.

290. *A. thaliana* (L.) Heynh. (*Sisymbrium t.*) THALE CRESS. A slender greyish-green ann. 5–50 cm., branched above, with tiny white fls. 3 mm. across. Stem lvs. few, entire, stalkless; basal lvs. in a rosette, elliptic, toothed. Plant rough-hairy below, hairless above. Fr. 5–20 mm., hairless. △ Banks, walls, waste places, cultivated ground. Mar.–May. All Eur. (except IS.).

MYAGRUM Fr. thick corky, triangular, with 3 cells, the upper 2 seedless, the lower with 1 seed. Fls. yellow. 1 sp.

291. *M. perfoliatum* L. MITRE CRESS. A glaucous, hairless ann. to 1 m., with small yellow fls. 4–5 mm. across and upper lvs. arrow-shaped, clasping stem with acute lobes. Fls. nearly stalkless in elongated leafless clusters; petals 3–5 mm. Lowest lvs. stalked, oblance-shaped, toothed or lobed. Fr. broadly club-shaped, 5–8 mm. long, the upper 2 cells without seeds. △ Cornfield weed. May–June. Med., South-East Eur. H.: introd. to the rest of Eur. (except P.IRL.IS.DK.). Page 128.

ISATIS | Woad Fr. flattened, broadly winged, not splitting, one-seeded. Fls. yellow. 10 sps.

292. *I. tinctoria* L. WOAD. A robust, erect bienn. or perenn. 50–120 cm., with bright yellow fls. in dense, terminal, branched clusters and glaucous, clasping, arrow-shaped upper stem lvs. Fls. *c.* 4 mm. across; petals up to twice as long as sepals. Lower stem lvs. lance-shaped, stalked, often hairy above. Fr. pendulous, flattened, paddle-shaped, 1–2½ cm. by 3–7 mm. A very variable sp. △ Fields, waste places. May–June. All Eur. (except IRL.IS.). *Formerly widely cultivated for the blue dye, woad, which is obtained by the fermentation of the yellowing leaves. The plant is still used as a starter for indigo fermentation, which has now superseded woad.* Pl. 32.

BUNIAS Fr. irregularly ovoid with warty or wing-like crests, not splitting. Fls. usually yellow. 3 sps.

293. *B. erucago* L. CRESTED BUNIAS. A slender, glandular, branched ann. or bienn. 30–60 cm., with bright yellow fls. about 1 cm. across, with heart-shaped petals 8–13 mm. Lower lvs. deeply pinnately cut into triangular lobes, the upper lvs. oblong, entire, or toothed. Fr. 10–12 mm., four-angled with irregularly crested wings on the angles, and a long style 3–5 mm. △ Cornfields, tracksides. May–June. Med. Eur. P.CH.BG.: introd. GB.D.A.PL.CS.H.R. Pl. 32.

B. orientalis L. WARTED BUNIAS. Like 293, but a taller, often perenn. plant to 120 cm. Fls.

smaller, less than 1 cm. across; petals 4–8 mm., rounded. Fr. asymmetrically ovoid, covered with irregular warts; style $\frac{1}{2}$–2 mm. Eastern Europe; widely introduced elsewhere. Page 135.

ERYSIMUM | Treacle Mustard   Fr. long and narrow, four-angled, with valves conspicuously one-veined; seeds in 1 row; style distinct, stigma usually more or less two-lobed. Lvs. usually narrow, with branched flattened hairs. A very difficult genus with many closely related microspecies. 38 sps.

294. *E. helveticum* (Jacq.) DC. SWISS TREACLE MUSTARD. A tufted perenn. 10–35 cm., with lower lvs. linear or narrow lance-shaped, 1–6 mm. wide, green or greyish-green, with branched hairs. Fls. bright yellow, petals 14–22 mm.; sepals shorter, 8–11 mm., pouched at base. Fr. square in cross-section, grey-green. △ In mountains: rocks, screes. May–Aug. F.CH.A.I.YU.

E. decumbens (Willd.) Dennst. (*E. ochroleucum* DC.) Like 294, but with long creeping shoots ending in rosettes of lvs. Lower lvs. green. Fls. pale lemon-yellow, but becoming straw-coloured; petals 15–27 mm.; sepals 10–15 mm. Mountains of North Spain, Pyrenees, and South-Western Alps.

295. *E. cheiranthoides* L. TREACLE MUSTARD. A slender leafy ann. 15–100 cm., with small yellow fls. *c.* 6 mm. across and petals 3–6 mm., hairy on back, twice as long as sepals. Lower lvs. oblong-lance-shaped acute, entire or with few irregular teeth, the upper narrower, stalkless; stems with adpressed two- to three-branched hairs. Fr. stalks slender; fr. spreading upward. △ Waste places, cultivation, damp meadows. May–Oct. Most of Eur. (except P.E.AL.GR.TR.BG.).

HESPERIS   Style very short, stigma deeply two-lobed, the lobes more or less erect, free; fr. linear, valves with a distinct median vein and lateral veins. Fls. pink, purple, yellow or white. 14 sps.

296. *H. matronalis* L. DAME'S VIOLET. A robust bienn. or perenn. 40–120 cm., with terminal clusters of pale purple, lilac, or white, very sweet-scented fls. each 2–2$\frac{1}{2}$ cm. across. Lvs. numerous, lance-shaped with finely toothed margin, all stalked, becoming narrower upwards. Fr. 2$\frac{1}{2}$–10 cm., rounded in section. A variable sp. △ Hedges, clearings, damp meadows, waste places. May–Sept. E.F.A.PL.CS.H.I.YU.AL.R.SU.: introduced elsewhere. Pl. 31.

H. laciniata All. CUT-LEAVED DAME'S VIOLET. Like 296, but fls. reddish-violet or yellow, or yellow flushed with purple. Upper lvs. stalkless; lower lvs. deeply lobed; stem with stiff rough hairs. Mediterranean region.

MALCOLMIA   Fls. usually violet or pink. Styles absent, stigma deeply lobed, the lobes erect and pressed together; fr. linear, valves three-veined. 15 sps.

297. *M. littorea* (L.) R.Br. A woody-based, very leafy perenn. 10–40 cm., of the littoral, with densely white, woolly stems and lvs. covered with star-shaped hairs. Fls. purple, conspicuous; petals 14–22 mm., notched. Lvs. oblong, entire or wavy-toothed, more or less stalkless. Fr. 3–6$\frac{1}{2}$ cm., slender arched, with star-shaped hairs; style bristle-like, soon falling. △ Maritime sands. May–July. P.E.F.I.

298. *M. maritima* (L.) R.Br. VIRGINIA STOCK. A slender branched ann. 10–35 cm., with violet, pink or rarely white fls. more than 1 cm. across, and petals *c.* 6 mm. broad, notched. Lvs. oblong, entire or sparsely toothed, often greyish. Fr. 3$\frac{1}{2}$–8 cm., slender, hairy, erect or

curved, with persistent conical stigma 2–5 mm. △ Maritime sands, cultivation, waste ground; widely grown as an ornamental and often occurring as a casual. AL.GR.: introd. E.F.I. Pl. 32.

CHEIRANTHUS | **Wallflower**   Lvs. narrow entire and covered with adpressed branched hairs. Fr. oval in section, each valve one-veined; stigma with 2 distinct spreading lobes. 1 sp.

299. *C. cheiri* L. WALLFLOWER. Fls. very sweet-scented, conspicuous, 2½ cm. across, commonly yellow and often brown-veined, or brown overall, in a dense terminal cluster; petals 1½–2½ cm. Lvs. crowded, oblong-lance-shaped, entire, with forked hairs. Fr. 2½–7½ cm., more or less erect. A short-lived, woody-based perenn. 20–90 cm. △ Walls, rocks near habitation; widely grown as an ornamental and naturalized. Mar.–June. GR.: introd. to most of Eur. (except North Eur.). Pl. 32.

MATTHIOLA | **Stock**   Fr. linear, with deeply two-lobed stigmas, the lobes erect and thickened into knobs or horns. Fls. reddish, violet, yellowish, or white. Usually grey-leaved plants with branched hairs. 10 sps.

300. *M. incana* (L.) R.Br. STOCK. A robust bienn. or perenn. to 80 cm., with a stout, woody-based stem covered below with leaf-scars, and very sweet-scented, bright reddish-purple or white fls. c. 3 cm. across. Lvs. numerous above, narrow lance-shaped, entire or nearly so, white woolly or more or less hairless. Fr. flattened cylindrical in section, 4½–16 cm. long, downy but not glandular; stigma-lobes conspicuous. △ Maritime rocks, cliffs, walls; widely grown as an ornamental and sometimes escaping. Mar.–July. Med. Eur. (except AL.) P.GB. Pl. 32.

301. *M. sinuata* (L.) R.Br. SEA STOCK. A spreading, leafy, densely white-woolly, maritime bienn. 10–60 cm., with pale purple fls. and petals 17–25 mm. Lower lvs. oblong with wavy margins or cut into narrow oblong lobes, the upper lvs. entire, all woolly-haired. Fr. 5–15 cm., somewhat flattened, the valves with large conspicuous, stalked, yellow or black glands; stigmas short, inconspicuous. △ Maritime sands, rocks. May–Sept. Med. Eur. P.IRL.GB. Pl. 33.

M. tricuspidata (L.) R.Br. THREE-HORNED STOCK. Like 301, but fr. with 3 conspicuous conical horns at apex, spreading in a star. Fls. 1–1½ cm. across, violet with a white centre, unscented; petals 15–22 mm. Mediterranean region.

M. fruticulosa (L.) Maire (*M. tristis* R.Br.) SAD STOCK. Fls. variously coloured, green to reddish-purple, or rust-coloured mixed with green or violet; petals 12–28 mm. Lvs. very narrow, 1–3 mm. wide, entire or lobed, white-woolly or greenish. Fr. 2½–12 cm., with or without apical horns. Southern Europe. Pl. 33.

BARBAREA | **Winter Cress**   Fr. bluntly four-angled in section, valves with a strong midvein and netted lateral veins; stigmas slightly two-lobed. Lower lvs. stalked, deeply lobed, the upper clasping the stem. Fls. yellow. 11 sps.

302. *B. vulgaris* R.Br. WINTER CRESS. A hairless, pale green bienn. or perenn. 10–100 cm., with bright yellow fls. 7–9 mm. across in dense, leafless, terminal clusters. Petals 5–7 mm., twice as long as sepals. Lower lvs. stalked, pinnately cut into 2–5 pairs of lateral lobes and a large, oval, terminal lobe; upper lvs. stalkless with a single, oval, deeply toothed blade. Fr. linear, 1½–3 cm.; style 2–3 mm. △ Damp places, hedges, banks, meadows. Apr.–July. All Eur.: introd. IS. *Sometimes used as salad plant.*

B. stricta Andrz. SLENDER WINTER CRESS. Widespread in Europe.

B. intermedia Boreau Like 302, but distinguished by its deeply pinnately lobed upper-most lvs. and basal lvs. with 3–5 pairs of lateral lobes. Petals bright yellow, 5–6 mm., about twice as long as sepals. Fr. 1–3 cm. Central and Southern Europe, but introduced elsewhere.

RORIPPA | **Yellow-Cress**   Fr. variable in length, from spherical to linear, but with convex valves without or with a very slender and indistinct midvein; seeds usually in 2 rows. Fls. yellow. Often placed in the genus *Nasturtium*. 10 sps.

Upper lvs. toothed, not lobed

303. *R. amphibia* (L.) Besser GREAT YELLOW-CRESS. A rather fleshy, bright green, hairless perenn. 40–150 cm., rooting below, with terminal leafless clusters of bright yellow fls. with the petals about twice as long as the sepals. Upper lvs. lance-shaped, usually irregularly toothed, but not lobed, the lower lvs. variably lobed. Fr. ovoid, 3–6 mm., on spreading or somewhat reflexed stalks 5–17 mm. long. △ Margins of ponds and ditches, streamsides. June–Sept. All Eur. (except IS.AL.). Pl. 33.

R. austriaca (Crantz) Besser AUSTRIAN YELLOW-CRESS. Like 303, but all lvs. elliptic entire, with irregularly toothed margin; lower lvs. short-stalked, the upper auricled and clasping the stem. Petals yellow, slightly longer than the sepals. Fr. spherical, 1½–3 mm. on stalks 7–15 mm. Central and Eastern Europe; frequently naturalized elsewhere. Page 128.

Upper lvs. deeply lobed

304. *R. sylvestris* (L.) Besser CREEPING YELLOW-CRESS. A hairless, stoloniferous perenn. with ascending wavy stems 20–50 cm., and lvs. stalked, all pinnately lobed with oblong toothed or lobed segments. Fls. bright yellow, *c.* 5 mm. across; petals 4–5 mm., twice as long as sepals. Fr. linear, 6–18 mm. on spreading stalks. △ Damp and waste places, tracksides, salt-rich ground. June–Sept. All Eur.: introd. IS.

305. *R. islandica* (Oeder) Borbás MARSH YELLOW-CRESS. Like 304, but fls. *c.* 3 mm. across; petals shorter than or equalling the sepals. Uppermost lvs. deeply lobed or entire, more or less stalkless. Fr. swollen, oblong, 4–9 mm., on spreading stalks of similar length. Hairless or sparsely hairy ann. or bienn. to 60 cm. △ Damp places, waste ground, walls. June–Sept. All Eur. (except AL.GR.TR.).

ARMORACIA | **Horse-Radish**   Fls. white. Fr. almost spherical, with an indistinct network of veins; style short, stigma disk-like. Seeds in 2 rows in each cell. 2 sps.

306. *A. rusticana* P. Gaertner, B. Meyer & Scherb. (*Cochlearia armoracia* L.) HORSE-RADISH. A robust, hairless perenn. 40–125 cm., with large basal lvs. and much-branched fl. stems with numerous white, sweet-scented fls. Fls. 8–9 mm. across; petals 5–7 mm. Basal lvs. 30–50 cm., oval, toothed, long-stalked; lower stem lvs. often deeply lobed, the uppermost entire. Roots fleshy. Fr. globular, 4–5 mm. △ Waste places, rocks, walls; cultivated and naturalized in most of Europe. May–June. SU.: introd. to much of Eur. (except P.IS.AL.GR.BG.). *The pungent roots make a mustard-like condiment. The plant is poisonous to livestock owing to the presence of a strongly irritant volatile oil.*

NASTURTIUM | **Watercress**   Like *Rorippa* but fls. white. 2 sps.

307. *N. officinale* R.Br. (*Rorippa nasturtium-aquaticum* (L.) Hayek) WATERCRESS. A rather weak-stemmed, hairless, slightly fleshy-leaved semi-aquatic perenn. 10–60 cm., with

YELLOW-FLOWERED CRUCIFERAE

1. *Camelina sativa* 338
2. *Rorippa austriaca* [303]
3. *Sisymbrium strictissimum* 284
4. *Erucastrum gallicum* 364
5. *Myagrum perfoliatum* 291
6. *Hirschfeldia incana* 365

small white or pale purplish fls. 4–6 mm. across. Petals nearly twice as long as sepals; anthers yellow. Lvs. shining, pinnate, with 1–4 pairs of more or less oval leaflets and a larger terminal leaflet. Stems often partly submerged and rooting at the nodes. Fr. 13–18 mm., ascending, with 2 rows of seeds. △ Streams, ditches, wet places. May–Sept. All Eur. (except IS.SF.). *Cultivated as a salad plant: a good antiscorbutic.*

*N. microphyllum (Boenn.) Reichenb. (*Rorippa m.*) Mainly in Western Europe.

CARDAMINE | **Bitter-Cress, Coral-Wort** Fr. linear, strongly compressed, with valves that coil up suddenly from the base and fling out the seeds. Fls. white, purple, or rarely pale yellow. Hairs unbranched. Including the old genus *Dentaria*. 36 sps.

Fls. conspicuous, 1 cm. or more across
(a) Perenns. with subterranean, fleshy scale-covered rhizome
308. *C. bulbifera* (L.) Crantz (*Dentaria b.*) CORAL-WORT. Readily distinguished by the purplish bulbils in the axils of the upper lvs. Fls. pale lilac, rarely white, in a terminal leafless cluster; petals 12–16 mm., *c.* 3 times as long as the sepals. Lvs. more than 6, alternate, the lower pinnately lobed, the uppermost entire, all toothed. Fr. 2–3½ cm., rare. A perenn. with a rhizome with fleshy, delta-shaped scale-lvs. and fl. stems 30–70 cm. △ Shady places, woods. Apr.–June. Most of Eur. (except P.E.IRL.IS.AL.). Pl. 34.

309. *C. heptaphylla* (Vill.) O. E. Schulz (*C. pinnata* (Lam.) R.Br.) SEVEN-LEAVED CORAL-WORT. Lvs. usually 3 or more, all pinnate with 5–11 oval-lance-shaped, irregularly toothed leaflets. Fls. pale lilac, pink, or white; petals 1½–2 cm., 3 times as long as calyx. Fr. 4–8 cm. A perenn. with a creeping rhizome covered with half-moon-shaped scale-lvs. 1–2 mm. and fl. stem 30–60 cm. △ Woods in mountains. Apr.–June. E.F.D.CH.I.: introd. B.

C. pentaphyllos (L.) Crantz (*Dentaria p.*) FIVE-LEAVED CORAL-WORT. Like 309, but lvs. palmate with 3–5 leaflets. Fls. violet or pink; petals 18–22 mm. Scales of rhizome 6–10 mm. long, acute, trilobed. Mountains of Western and Central Europe. Pl. 34.

C. enneaphyllos (L.) Crantz Fls. pale yellow or white, somewhat drooping; stamens as long as or longer than petals. Lvs. 2–4 in a lax whorl, each with 3 or more oval-lance-shaped, double-toothed leaflets. Fr. 4–7½ cm. Eastern Alps, Carpathians, and Western Balkans. Pl. 33.

(b) Perenns. without a subterranean fleshy rhizome, and with few or no scale-lvs.
310. *C. pratensis* L. LADY'S SMOCK, CUCKOO FLOWER. An erect, hairless perenn. 15–60 cm., with rather large pale pink, violet, or white fls. 1½–2 cm. across in a rather dense, terminal, leafless cluster. Petals *c.* 3 times as long as sepals; anthers yellow. Lvs. pinnate, dull green or grey-green, rough-surfaced; lower lvs. with oval- to kidney-shaped leaflets, contrasting with the narrow lance-shaped leaflets of the upper lvs. A variable sp. △ Damp meadows, streamsides. Apr.–June. All Eur. (except IS.). Pl. 34.

C. amara L. LARGE BITTER-CRESS. Like 310, but fls. smaller, *c.* 1 cm. across, usually white, with petals twice as long as the sepals; anthers violet. Leaflets of upper and lower lvs. similar in shape, bright green, smooth. Stolons slender. Throughout Europe, except the extreme south-east. Pl. 34.

311. *C. trifolia* L. TRIFOLIATE BITTER-CRESS. Readily distinguished by the long-stalked trifoliate basal lvs. with 3 large, rounded, shallow-toothed leaflets which are purplish beneath. Fls. white or pink; petals 9–11 mm.; twice as long as sepals; anthers yellow. Stem lvs. much smaller than basal lvs., simple or trifoliate, or absent. A perenn. with a creeping, knobby rhizome; fl. stem 20–30 cm. △ Woods. Apr.–June. Cent. Eur. F.I.YU.SU.

Fls. inconspicuous, less than 6 mm. across

312. *C. bellidifolia* L. ALPINE BITTER-CRESS. A tiny, high alpine, rosette-forming perenn. 1–10 cm. high. Fls. white, in a dense flat-topped cluster; petals 3½–5 mm. Lvs. all thick, entire, spathulate, rarely two-lobed. △ Screes, snow valleys. July–Aug. North Eur. (except DK.) E.F.D.CH.A.I.

C. resedifolia L. Like 312, but stem lvs. with 3–7 oval or lance-shaped lobes and lv. stalk with small basal auricles; stems 3–12 cm. Fls. white; petals 5–6 mm. Seeds broadly winged. Mountains of Central and Southern Europe. Page 139.

313. *C. flexuosa* With. (*C. sylvatica* Link) WOOD BITTER-CRESS. A hairy bienn. or perenn. 10–50 cm., with a wavy, leafy stem, branched above, arising from a loose rosette of lvs. Fls. white; petals 2½–3 mm., about as long as sepals; stamens usually 6. Lvs. pinnate, leaflets entire or toothed. △ Bare ground, damp places, rocks, walls. Apr.–June. Most of Eur. (except IS.GR.TR.SU.).

C. impatiens L. NARROW-LEAVED BITTER-CRESS. Widespread in Europe.
C. hirsuta L. HAIRY BITTER-CRESS. Occurs throughout Europe.

CARDAMINOPSIS Fr. like *Cardamine*, but valves flat with a distinct median vein, not coiling on ripening. Hairs simple or forked. Fls. white, pinkish, or purplish. 5 sps.

314. *C. arenosa* (L.) Hayek (*Arabis a.*) A rough-hairy ann., bienn. or perenn. with numerous white, purplish, or lilac fls. on robust erect, usually much-branched stems to 80 cm. Petals 6–8 mm., white in northern regions and lilac in Central Europe, with a pair of small teeth on the claw. Basal lvs. in a rosette, pinnately lobed, the stem lvs. lance-shaped, toothed or pinnately cut. Fr. linear, 1–4½ cm. △ Sandy places, walls, rocks. Apr.–July. Much of Eur. (except P.E.IRL.GB.IS.AL.GR.TR.). Pl. 35.

ARABIS | **Rock-Cress** Fr. long narrow, strongly compressed; seeds in 1 row. Lvs. simple, usually hairy. 35 sps.

Fls. white, yellowish or greenish-yellow

315. *A. glabra* (L.) Bernh. (*Turritis g.*) TOWER MUSTARD. An erect, unbranched bienn. 60–120 cm., with small yellowish or greenish-white fls. in a slender spike, and many glaucous lvs. clasping the stem and a basal rosette of hairy lvs. Fls. *c.* 6 mm. across; petals less than twice as long as sepals. Rosette lvs. oblong, entire or wavy-toothed, with star-shaped hairs, dying before fls. appear; stem lvs. oval-lance-shaped with arrow-shaped clasping base, hairless. Fr. erect, 3–7 cm. by 1–1½ mm. wide, in a slender elongate cluster. △ Open woods, sandy banks, cliffs, stony places. May–July. All Eur. (except IRL.IS.).

316. *A. hirsuta* (L.) Scop. HAIRY ROCK-CRESS. Like 315, but stem lvs. and rosette lvs. all hairy. Fls. white in a dense cluster; petals 4–6½ mm. Rosette lvs. oval, stalked; stem lvs. oval to linear-lance-shaped, stalkless and clasping stem with rounded or arrow-shaped base. Fr. erect, 1½–3 cm., in a slender elongate cluster. A very variable, stiffly erect, several-stemmed perenn. or rarely bienn. 10–60 cm. △ Dry places, banks. May–July. Most of Eur. (except IS.GR.).

317. *A. turrita* L. TOWER ROCK-CRESS. Readily distinguished by the very long slender inflorescence, 10–14 cm. by 2–2½ mm., which is at first erect and then curved over to one side. Fls. pale yellow; petals 6–8 mm. Basal lvs. in a rosette, obovate, long-stalked, the upper lvs. oblong, toothed, stalkless and clasping stem. Fr. very long, 10–14 cm., at first erect, then arching. A bienn. or perenn. with greyish star-shaped hairs, 20–80 cm.

△ Woods, shady rocks, stony banks, tracksides. May–June. Cent. Eur. (except PL.), South-East Eur. (except TR.) F.B.I.SU.: introd. GB. Page 139.

318. *A. alpina* L. ALPINE ROCK-CRESS. An arctic or alpine perenn. with few sterile rosettes and erect fl. stems to 40 cm., bearing lax clusters of white fls. Petals 6–10 mm., twice as long as sepals. Rosette lvs. with 4–7 coarse teeth on each side, greenish; stem lvs. heart-shaped, clasping stem, all with dense star-shaped hairs. Fr. 2–3½ cm. △ Damp screes, springs, rivulets. Apr.–June. Most of Eur. (except P.IRL.B.NL.DK.TR.). Pl. 34.

A. caucasica Schlecht. A widely cultivated plant distinguished from 318 by the larger fls. with petals 9–18 mm., arrow-shaped stem lvs. and fr. 4–7 cm. Southern Europe; naturalized elsewhere.

Fls. blue or violet

319. *A. caerulea* (All.) Haenke BLUISH ROCK-CRESS. A tiny rosette alpine perenn. with un-branched, nearly leafless stems to 10 cm., bearing a compact cluster of few pale blue fls. Fls. 4–10; petals 4–5 mm. Rosette lvs. thick, oval, shiny, hairless or with few hairs, and with 2–5 conspicuous apical teeth; stem lvs. 1–3, stalkless, entire. Fr. 1–3 cm., erect, bluish when young. △ Rocks, screes. July–Aug. F.D.CH.A.I.YU.

A. verna (L.) R.Br. Like 319, but an ann. of the Mediterranean region with violet fls. and stem lvs. with heart-shaped base. Petals 5–8 mm., violet with a yellowish base, or white. Fl. stems several. Fr. 4½–6 cm. Pl. 35.

AUBRIETA Fls. violet or pink. Lvs. simple, with star-shaped hairs. Fr. usually linear with distinct style and rounded stigma; seeds in 2 ranks. 6 sps.

320. *A. deltoidea* (L.) DC. AUBRETIA. A perenn. of varying habit, either cushion-forming with numerous rosettes, or straggling, with ascending stems to 30 cm., with conspicuous clusters of bright reddish-purple to violet fls. Petals 12–28 mm., rarely white, twice as long as sepals or more. Lvs. narrow, spathulate to obovate, entire or with few teeth, hairy. Fr. ½–2 cm., with straight and star-shaped hairs; style 4–8 mm. △ Rocky places, walls; grown as an ornamental and sometimes naturalized. Mar.–May. I.YU.GR.: introd. E.F. GB.NL.

LUNARIA | **Honesty** Fr. flat, disk-like, elliptical or circular in outline, valves thin-walled, net-veined. Fls. purple or rarely white. 3 sps.

321. *L. rediviva* L. PERENNIAL HONESTY. An erect, leafy perenn. 25–150 cm., with a terminal cluster of very fragrant pale purple or violet fls. with petals 12–20 mm. Lvs. oval, long-pointed, finely bristle-toothed, the upper lvs. distinctly stalked. Fr. flat, 3½–9 cm. long, elliptic and usually pointed at each end and with a carpophore 2½–4 cm. △ Woods. Apr.–June. Much of Eur. (except IRL.GB.NL.IS.N.SF.GR.TR.).

322. *L. annua* L. (*L. biennis* Moench) HONESTY. Like 321, but a bienn. with upper lvs. stalk-less and coarsely and irregularly toothed. Fls. larger, reddish-purple, rarely white, sweet-scented at night; petals 1½–2½ cm. Fr. flat, 2–7 cm., broadly elliptic or often almost circular in outline, with blunt rounded ends; carpophore ½–2 cm. △ Waste ground, hedges, bushy places; widely grown for ornament and often naturalized or casual else-where. Apr.–June. I.YU.AL.GR.BG.R.: introd. to most of Eur. (except IRL.IS.DK.SF.CS.TR.). Pl. 35.

ALYSSOIDES Fr. globular, swollen; style long. Fls. yellow. 3 sps.

323. *A. utriculata* (L.) Medicus (*Vesicaria u.*) A perenn. with a woody base, basal rosettes

of lvs. and erect, leafy stems 20–40 cm. bearing rather large yellow fls. in a short, dense, terminal cluster. Petals 2 cm., with a rounded blade and long-clawed base, about twice as long as sepals. Lvs. all entire, the rosette lvs. oblong-spathulate with star-shaped hairs, the fl. stem lvs. lance-shaped, hairless. Fr. *c.* 1 cm., hairless; style long slender, 7–10 mm. A variable sp. △ Rocks, screes. Apr.–June. South-East Eur. (except TR.) F.I.

ALYSSUM   Fls. yellow. Fr. disk-like and circular in outline, valves flat; seeds 1–2 to each cell. Star-shaped hairs generally present. A difficult genus. 64 sps.

Perenns.

324. *A. saxatile* L. GOLDEN ALYSSUM. A woody-based shrubby perenn. with numerous golden-yellow fls. in rather dense, flat-topped clusters on branched stems 10–40 cm. Petals 3–6 mm., notched or bilobed, twice as long as sepals. Lvs. obovate or spathulate, entire or lobed, grey-hairy, hairs branched. Fr. hairless, flattened. △ Rocks, stony places. Apr.–June. Cent. Eur. (except CH.), South-East Eur. I.SU.

325. *A. montanum* L. A procumbent or erect perenn. to 25 cm., with green-to-whitish rosettes of lvs. and bright yellow fls. Petals 3½–6 mm., notched, spreading, to twice as long as sepals. Lvs. with star-shaped hairs, basal lvs. oblong-spathulate, the upper linear. Fr. 3–5½ mm., valves inflated but with flattened margin, in a slender elongated cluster; seeds with a narrow wing. △ Lowlands and mountains: sandy, rocky, stony places. May–July. Much of Eur. (except North Eur. P.IRL.GB.B.NL.TR.).

Anns.

326. *A. alyssoides* (L.) L. (*A. calycinum* L.) SMALL ALYSSUM. An erect, grey-hairy ann. or bienn. to 30 cm., with greyish lvs. and long or short terminal spikes of tiny, pale yellow fls. each *c.* 3 mm. across. Petals notched, little longer than sepals, becoming whitish. Lvs. 6–18 mm., spathulate, entire. Fr. 3–4 mm., almost circular, with flattened margin, finely hairy; fr. stalks spreading. △ Dry fields, rocks, walls. Apr.–June. All Eur. (except IRL. IS.). Page 135.

FIBIGIA   Fr. strongly compressed, with flat valves; style long; seeds winged 2–8 in each cell. Hairs star-shaped. Fls. yellow. 3 sps.

327. *F. clypeata* (L.) Medicus (*Farsetia c.*) An erect, densely grey-felted bienn. to 75 cm., with small yellow fls. and conspicuous, elliptic, flattened frs. Petals 8–13 mm., nearly twice as long as sepals. Lvs. elliptic, entire, grey-felted. Fr. 1½–3 cm. long, grey-felted, in an elongated cluster 10–20 cm.; style 2–3 mm. △ Rocks, walls. Apr.–June. I.YU.AL.GR.SU.: introd. F.A. Pl. 35.

BERTEROA   Fr. elliptic or orbicular, with flat or somewhat swollen valves without a conspicuous vein, style distinct; seeds 2–6 in each cell. Petals deeply bilobed, white or pale yellow. Plants with star-shaped hairs. 5 sps.

328. *B. incana* (L.) DC. (*Alyssum i.*) HOARY ALYSSUM. An erect, little-branched ann., bienn., or perenn. 20–70 cm., with greyish lvs. and small white fls. with the petals conspicuously deeply bilobed. Petals 4½–6 mm. Lvs. lance-shaped, usually entire, covered with star-shaped hairs. Fr. 4½–8 mm., elliptical, swollen, hairy, in a much-elongated cluster; style 1½–4 mm. △ Rocks, sands, dry walls, waste places. June–Oct. Cent., South-East Eur. DK.I.SU.: introd. F.GB.B.NL.N.S.SF.

LOBULARIA   Fr. circular, valves flat with a slender midvein. Lvs. narrow, untoothed, covered in hairs which are forked from the base. Fls. white or pinkish. 2 sps.

329. *L. maritima* (L.) Desv. SWEET ALISON. A small, spreading, woody-based perenn. 10–30 cm., with numerous, very sweet-scented, white or pink-flushed fls. in terminal compact heads. Petals *c.* 3 mm., rounded, not notched, spreading. Lvs. numerous, crowded, narrow lance-shaped, usually densely silvery-haired. △ Rocks, sands, dunes; widely cultivated as an ornamental annual and often naturalized or casual. Dec.–Aug. P.E.F.I.YU.SU.: introd. GB.B.NL.DK.N.A.CS.H.R. Pl. 37.

CLYPEOLA   Fr. disk-like, pendulous, one-seeded, not splitting. Fls. yellow. 2 sps.

330. *C. jonthlaspi* L. DISK CRESS. A tiny, erect ann. to 25 cm., with minute yellow fls. which later become white, and distinctive, pendulous, disk-like frs. Petals 1–2 mm. Lvs. linear to obovate. Fr. 2–5 mm., circular, flat, winged, hanging on curved stalks. △ Dry, sandy, and stony places. Apr.–June. Med. Eur. CH.BG.R.SU.

DRABA   Fr. about twice as long as broad with more or less flattened valves each with a conspicuous vein only in lower half; seeds in 2 rows in each cell. Fls. white or yellow. Lvs. simple, mostly in a basal rosette. 42 sps.

Fls. yellow

331. *D. aizoides* L. YELLOW WHITLOW-GRASS. An alpine perenn. with dense rosettes of stiff bristly lvs. and leafless fl. stems to 10 cm., bearing a rather dense cluster of 4–18 bright yellow fls. Petals 4–6 mm., equalling stamens and exceeding sepals. Lvs. linear, hairless, margin bristle-fringed. Fr. ellipsoid 6–12 mm., usually hairless, in an elongated cluster. A very variable sp. △ Alpine meadows, rocks. Apr.–June. Cent. Eur. (except H.), South-East Eur. (except GR.TR.) E.F.GB.B.I.SU. Pl. 34.

D. nemorosa L. Like 334, but fls. pale yellow; petals notched and stems almost hairless above. Stem lvs. with wedge-shaped base, not clasping; fr. hairy. Widespread in much of Europe.

Fls. white

332. *D. tomentosa* Clairv. DOWNY WHITLOW-GRASS. An alpine perenn. with soft grey rosettes of lvs. and slender, densely hairy, almost leafless fl. stems 3–20 cm., bearing small white fls. Petals 3–5 mm.; fl. stalks with dense star-shaped hairs. Lvs. oval to oblong, densely covered with star-shaped hairs. Fr. oblong-elliptic 6–14 mm. △ Rocks in mountains. July.–Aug. Cent. Eur. (except H.) E.F.I.YU.BG.

333. *D. incana* L. HOARY WHITLOW-GRASS. Like 332, but a more robust erect alpine or northern bienn. or perenn., with erect, very leafy stems 10–35 cm., and the whole plant covered in whitish branched or unbranched hairs. Fls. white; petals 4–5 mm. Lvs. lance-shaped. Fr. mostly hairless, oblance-shaped, often somewhat twisted; fr. stalks erect. A very variable sp. △ Rocks in mountains. July–Aug. North Eur. (except S.) E.F.CH.A.SU. Pl. 35.

334. *D. muralis* L. WALL WHITLOW-GRASS. A rosette ann. with an erect, simple or branched, hairy and leafy stem, 10–30 cm., bearing a cluster of many tiny white fls., each *c.* 3 mm. across. Petals *c.* 2½ mm., almost twice as long as sepals. Stem lvs. broadly oval, half-clasping, the rosette lvs. obovate, hairy. Fr. elliptic, 3–6 mm., hairless, in an elongate cluster. △ Rocks, walls, sandy places, uncultivated ground. Apr.–June. All Eur. (except IS.). Page 139.

EROPHILA   Like *Draba*, but petals deeply bilobed, white. A difficult genus with many microspecies. 2 sps.

335. *E. verna* (L.) Chevall. (*Draba v.*) COMMON WHITLOW GRASS. A tiny ann. with a rosette

of basal lvs. and slender, erect, leafless stems 3–20 cm., bearing tiny white fls. *c.* 4 mm. across with deeply lobed petals. Petals white or reddish, as long as or longer than the sepals. Lvs. spathulate to lance-shaped, entire or toothed, hairy. Fr. 3–12 mm., elliptic-globular, hairless. △ Disturbed ground, rocks, walls. Feb.–May. All Eur.

COCHLEARIA Fls. usually white or mauve. Fr. swollen, almost spherical, with a strong midvein and usually with a conspicuous network of veins. 12 sps.

336. *C. officinalis* L. SCURVY-GRASS. A somewhat fleshy, hairless bienn. or perenn. with spreading or erect branched stems 5–50 cm., bearing clusters of numerous white fls. 8–10 mm. across. Petals rounded 4–9 mm., 2–3 times as long as the sepals. Lower lvs. broadly heart-shaped, stalked, the upper clasping stem, oval or triangular, coarsely toothed, all fleshy and hairless. Fr. ovoid, 4–7 mm., valves fleshy. A variable sp. △ Maritime rocks, salt marshes, damp places in mountains. May–Aug. West Eur. (except P.) DK.N.S.D.CH.PL.

337. *C. danica* L. DANISH SCURVY-GRASS. Like 336, but a spreading ann. 2–20 cm., with small white or mauve fls. 4–5 mm. across. Lvs. mostly stalked, the lowest three- to seven-angled, the middle lvs. triangular, the upper lance-shaped. Fr. 3–6 mm. ovoid-elliptic, often narrowed at both ends, finely net-veined. △ Maritime rocks, salt marshes, disturbed ground inland. Feb.–Mar. West, North Eur. (except IS.). *Rich in ascorbic acid; it has been eaten by sailors to prevent scurvy.* Pl. 36.

*C. anglica L. LONG-LEAVED SCURVY-GRASS. North-Western Europe.

CAMELINA Fr. obovoid or pear-shaped, inflated, many-seeded; style conspicuous. Fls. usually yellow. 5 sps.

338. *C. sativa* (L.) Crantz GOLD OF PLEASURE. A more or less hairless, rather yellowish-green ann. or bienn. to 80 cm., with small yellow fls. and lance-shaped upper lvs. half-clasping the stem with acute lobes. Fls. 3 mm. across; petals to 5 mm. Lvs. 3–9 cm., the lower oblong, narrowed to a stalk. Fr. ovoid, 7–9 mm., inflated, yellowish, in an elongated rather dense cluster. △ Cultivated and waste ground; often occurring as a casual. May–July. Most of Eur. (except IS.AL.YU.TR.). *Cultivated for its seeds which yield a clear edible oil, which is also used for lighting, and softening the skin.* Page 128.

C. microcarpa DC. Like 338, but rather densely hairy. Petals pale yellow. Fr. 5–7 mm., hard and woody, in a very long cluster. Widespread in Europe.

NESLIA Fr. globular, hard woody, not splitting, one- to three-seeded; style distinct. Fls. yellow. 1 sp.

339. *N. paniculata* (L.) Desv. YELLOW BALL MUSTARD. A greyish, finely hairy, erect ann. 15–60 cm., with upper lvs. lance-shaped and clasping stem with acute lobes, and tiny yellow fls. Fls. 4–5 mm. across; petals *c.* 2 mm.; sepals yellowish-green. Lower lvs. oblong entire, narrowed to a stalk. Fr. globular, rough-netted, 1½–3 mm.; style often persisting, long slender. △ Cornfields, tracksides, waste ground. May–July. Most of Eur.: introd. in North Eur. (except IS.). Page 135.

CAPSELLA | Shepherd's Purse Fr. obcordate, the valves somewhat flattened and keeled, net-veined; seeds several in each cell. Fls. white, pink or yellowish. 5 sps.

340. *C. bursa-pastoris* (L.) Medicus SHEPHERD'S PURSE. A variable ann. or bienn., usually with a rosette of deeply lobed basal lvs., clasping entire upper lvs., and small white fls. developing into purse-like frs. Petals 2–3 mm., about twice as long as sepals. Lower lvs. mostly pinnately lobed, the upper lance-shaped clasping stem with arrow-shaped basal

YELLOW-FLOWERED CRUCIFERAE

1. *Rapistrum perenne* 367
2. *R. rugosum* [367]
3. *Neslia paniculata* 339
4. *Conringia orientalis* 355
5. *Alyssum alyssoides* 326
6. *Bunias orientalis* [293]

135

lobes. Fr. 6–9 mm., about as long as wide, triangular-obcordate, in an elongate, bractless cluster. △ Cultivated ground, waste places, tracksides. Jan.–Dec. All Eur.

C. rubella Reuter PINK SHEPHERD'S PURSE. Like 340, but petals only 1½–2 mm., usually reddish at least on the margin; sepals reddish-tipped, about as long as petals. Fr. *c.* 6 mm. Southern Europe.

HORNUNGIA Anns. Fr. elliptic to oblong-oval; seeds 1–2 in each cell. Fls. white in a bractless inflorescence. 2 sps.

341. *H. petraea* (L.) Řeichenb. (*Hutchinsia p.*) CHAMOIS CRESS. A slender ann. 3–15 cm., with pinnate lvs. and tiny white fls. in a terminal leafless cluster. Petals *c.* 1 mm., as long as or slightly longer than the sepals. Lower lvs. stalked, in a rosette, with 3–15 oval to lance-shaped acute leaflets, the stem lvs. stalkless. Fr. 2–2½ mm. △ Rocks, banks. Mar.–May. Much of Eur. (except IRL.NL.IS.SF.PL.TR.R.).

TEESDALIA Anns. Fr. flattened, rounded, and narrowly winged above. Filaments of stamens with white basal scale. Fls. white. 2 sps.

342. *T. nudicaulis* (L.) R.Br. SHEPHERD'S CRESS. A rosette ann. 8–45 cm., with pinnately lobed lvs. and tiny white fls. *c.* 2 mm. across, in a dense terminal cluster. Petals distinctly unequal. Basal lvs. in a rosette, 2–5 cm., lance-shaped, stalked, with several pairs of blunt lateral lobes and a broader, often three-lobed terminal lobe, the stem lvs. few or none, the uppermost lv. entire. Fr. 3–4 mm., broadly elliptic, flattened, narrowly winged, in an elongated bractless cluster. △ Sandy and rocky places, fallow. Apr.–June. Most of Eur. (except IS.SF.YU.AL.BG.).

THLASPI | **Pennycress** Fr. flattened, valves distinctly keeled and usually winged. Lvs. clasping stem. Fls. white or purplish, in a bractless inflorescence. 26 sps.

Fls. white or occasionally purplish
(a) Anns.

343. *T. arvense* L. FIELD PENNYCRESS. An erect, hairless, somewhat fetid ann. 10–60 cm., with white fls. 4–6 mm. across, in a terminal leafless cluster. Petals 3–4 mm.; sepals 1½–2 mm.; anthers yellow. Stem lvs. lance-shaped, entire, or toothed, clasping stem with pointed lobes, lower lvs. stalked. Fr. flattened almost circular, 10–15 mm., wings broadening upwards and forming a deep, narrow notch which includes the style, in a much-elongated cluster. △ Cultivated ground, waste places. May–Sept. All Eur. (except AL. GR.TR.). Pl. 35.

T. perfoliatum L. PERFOLIATE PENNYCRESS. Like 343, but upper stem lvs. oval with blunt rounded lobes encircling stem, glaucous. Fr. broadly obcordate 5–7 mm., with narrower wings and a wide apical notch, which includes the style. Most of Europe.

(b) Rosette-forming perenns.

344. *T. montanum* L. MOUNTAIN PENNYCRESS. A mat-forming perenn. 10–25 cm., with glaucous lvs. in rosettes and conspicuous white fls. in a terminal leafless cluster. Petals 5–7 mm., twice as long as sepals. Lvs. rather tough, entire or toothed, the rosette lvs. rounded, long-stalked, the upper lvs. oblong-oval, clasping stem with rounded or more or less acute lobes. Fr. 7–8 mm., broadly winged with a wide notch and style much longer. △ In mountains: screes, rocky places, meadows. Apr.–June. F.B.D.CH.A.CS.I.YU. Page 139.

*T. alpestre L. Widespread in Europe, except in the north.

Fls. violet

345. *T. rotundifolium* (L.) Gaudin ROUND-LEAVED PENNYCRESS. A glaucous, somewhat

fleshy, stoloniferous alpine perenn. 5–15 cm., with a terminal cluster of sweet-scented, violet fls. each 6–8 mm. across. Basal·lvs. usually in rosettes, rounded, stalked, the stem lvs. oval, clasping. Fr. oval, strongly keeled but not winged and without a notch; style conspicuous 1–2 mm. △ Loose screes in mountains. June–Aug. F.D.CH.A.I.YU. Pl. 36.

AETHIONEMA Filaments of inner 4 stamens flattened and winged and sometimes toothed above. Fr. flattened, with a broad wing, which is notched at the apex. 6 sps.

346. *A. saxatile* (L.) R.Br. BURNT CANDYTUFT. A very variable leafy ann. or woody-based perenn. 10–30 cm., with entire lvs. and violet, rose-pink or white fls. in a dense terminal cluster which elongates conspicuously in fr. Petals 2–5 mm., longer than the sepals. Lvs. oval-oblong 1–2 cm., glaucous, leathery, the lower blunt, the upper acute and often turning reddish. Fr. obovate, 5–9 mm., winged and notched at the apex; style included. △ Stony places, mainly in mountains. Apr.–Aug. Med., South-East Eur. (except TR.) D.CH.CS.H. Pl. 36.

IBERIS | Candytuft Inflorescence more or less flat-topped, with the outermost pair of petals of each fl. much longer than the inner pair. Fr. flattened with keeled valves, usually winged at apex; style long; seeds 1 in each cell. 19 sps.

Small shrubby perenns.
347. *I. sempervirens* L. EVERGREEN CANDYTUFT. A small, branched, hairless, evergreen shrublet to 25 cm., with flat-topped heads 4–5½ cm. across, of pure white fls. Lvs. 2½–5 mm. wide, narrow spathulate, flat, thick. Fr. rounded-oval, 6–7 mm., broadly winged from the base. △ In mountains: rocks, crevices, screes; grown as an ornamental and occasionally naturalized. June–Aug. Med. Eur. (except TR.): introd. GB.R.

I. saxatilis L. Like 347, but lvs. linear acute, 1–2 mm. wide, fleshy, cylindrical on non-flowering stems but flat on fl. stems, which are terminal (lateral in 347). Southern Europe.

Anns.
348. *I. amara* L. ANNUAL CANDYTUFT. An erect, little-branched ann. 10–40 cm., with white or white violet fls. in a branched, flat-topped cluster. Fls. 6–8 mm. across, the outer pair of petals 4 times as long as the inner pair. Lvs. oblong, usually with 4–8 coarse, widely spaced teeth or lobes. Fr. 3–5 mm., orbicular, with wings triangular above and with an acute apical notch, borne in an elongated cluster. △ Fields, rocks. May–Oct. West Eur. (except IRL.) D.CH.I.: introd. A.PL.CS.H.YU.R.SU. Pl. 36.

I. umbellata L. Like 348, but lvs. entire, linear-lance-shaped. Fls. larger, 7–12 mm. across, rose or purple. Fr. cluster remaining compact. Mediterranean Europe; often cultivated as an ornamental and sometimes escaping.

BISCUTELLA | Buckler Mustard Fr. with the appearance of 2 one-seeded disks placed edge to edge; style long, thread-like. Fls. yellow. A difficult genus. 41 sps.

349. *B. laevigata* L. BUCKLER MUSTARD. An extremely variable perenn. 10–50 cm., with yellow fls. in a lax or dense head. Petals 4–8 mm., abruptly narrowed into a stalk-like claw. Lvs. very variable in size, linear to oval, entire or wavy-margined, hairless, rough-hairy or finely hairy, often in a rosette; stem lvs. few. Fr. 8–14 mm. or more broad, with narrow papery margin, hairy or hairless; style 2–6 mm. △ Rocks, uncultivated places. May–Aug. Most of Eur. (except North Eur. IRL.GB.NL.AL.GR.TR.).

B. cichoriifolia Loisel. Like 349, but fls. larger with petals to 15 mm. and sepals with a long slender spur. Lvs. clasping stem. Fr. with a thin transparent margin. Mountains of Central Europe and Pyrenees.

B. didyma L. A slender southern and Mediterranean ann. with petals 4 mm., gradually narrowed to the base, not clawed. Inflorescence and fr. cluster dense; fr. 4–7 mm. long by 9–12 mm. broad, with a thick margin.

LEPIDIUM | Pepperwort Fr. flattened at right angles to the septum, strongly keeled or winged, with 1 pendent seed in each cell. Fls. usually white, small, in terminal leafless clusters. 21 sps.

Stem lvs. clasping and auricled

350. *L. campestre* (L.) R.Br. PEPPERWORT. A greyish-green, hairy, leafy ann. or bienn. 20–60 cm., branched above, with inconspicuous white fls. 2–2½ mm. across, in dense terminal spikes. Upper lvs. oblong-oval, toothed, clasping stem with narrow, pointed lobes, the lower lvs. oval, entire or slightly lobed. Fr. 5–6 mm., flattened and winged, concave, hairless, densely covered with small scale-like swellings, apex notched, with style usually included. △ Fields, tracksides, uncultivated ground. May–July. All Eur.

*L. heterophyllum Bentham SMITH'S CRESS. Western Europe.

Stem lvs. not clasping stem
(a) Lower lvs. deeply lobed

351. *L. sativum* L. GARDEN CRESS. A somewhat glaucous, hairless, strong-smelling ann. 20–50 cm., with lower lvs. very deeply cut into narrow toothed lobes and upper lvs. linear, stalkless. Fls. white or sometimes reddish; petals 2–3 mm., twice as long as sepals. Fr. hairless, 5–6 mm. elliptic flattened, narrowly winged and deeply notched, style included. △Cultivation, waste ground; often grown as a salad plant and occurring as a casual. June–July. Native of North Africa; introd. most of Eur. (except AL.IRL.IS.TR.). *Usually eaten as a seedling; the seeds produce a mustard oil.*

L. ruderale L. NARROW-LEAVED PEPPERWORT. Distinguished from 351 by the inconspicuous greenish fls. with petals often absent; fr. smaller 2–2½ mm., deeply notched, with stigma at base. Stamens usually 2. A strong-smelling ann. with basal lvs. once or twice cut into linear segments. Throughout Europe. *Used as an insecticide and medicinally in the past.* Page 139.

(b) All or most lvs. entire or toothed

352. *L. latifolium* L. DITTANDER, BROAD-LEAVED PEPPERWORT. A strong-smelling, glaucous, stoloniferous perenn. 25–150 cm., with large, leathery lower lvs. and numerous tiny white fls. 2½ mm. across in a large, branched, pyramidal cluster. Sepals with a broad white margin, petals twice as long. Lower lvs. oval, long-stalked, toothed, or sometimes deeply lobed, to 30 cm., stem lvs. oval-lance-shaped, entire. Fr. orbicular 2 mm., hairy; style short, stigma conspicuous. △ Wet places, salt marshes. June–July. All Eur. (except IS. N.SF.). *Used as an insecticide and medicinally.*

L. graminifolium L. GRASS-LEAVED PEPPERWORT. Differing from 352 in having narrow lance-shaped to linear lvs., the lower toothed or lobed. Sepals with a narrow white margin; petals 1–1½ times as long. Fr. 2½–4 mm., ovoid, with acute apex, hairless; style projecting. Widespread in Europe, except the north.

L. virginicum L. A rather robust North American ann. widely naturalized in Europe. Lowest lvs. bristly, rather deeply pinnately lobed, the middle and upper lvs. sharply toothed, ciliate. Petals white, longer than sepals. Fr. 3–4 mm., orbicular, with a broad shallow notch, style included and very short.

CARDARIA Like *Lepidium*, but fr. broadly heart-shaped, not winged or notched, not splitting. 1 sp.

WHITE-FLOWERED CRUCIFERAE

1. *Thlaspi montanum* 344
3. *Cardamine resedifolia* [312]
5. *Calepina irregularis* 369

2. *Lepidium ruderale* [351]
4. *Draba muralis* 334
6. *Arabis turrita* 317

353. *C. draba* (L.) Desv. (*Lepidium d.*) HOARY PEPPERWORT, HOARY CRESS. A hairless or sparsely hairy leafy perenn. 15–90 cm., with white fls. *c.* 5–6 mm. across, in dense flat-topped clusters. Petals *c.* 4 mm., twice as long as sepals. Lvs. oblong-oval, rather irregularly and sparsely toothed, the upper lvs. somewhat broadening towards the base and clasping stem with rounded or acute lobes. Fr. 3–4$\frac{1}{2}$ mm. by 3$\frac{1}{2}$–5 mm., heart-shaped, inflated, style prominent. Often a troublesome weed, which spreads by root-buds. △ Fields, tracksides, waste places. Apr.–July. Med., South-East Eur. P.A.H.SU.: introd. to the rest of Eur. (except IS.). *The seeds have been used as a substitute for pepper.* Pl. 36.

CORONOPUS | **Swine-Cress**   Fr. not splitting, or splitting into 2 one-seeded parts; valves of fr. hemispherical, netted or pitted. Fls. white. 3 sps.

354. *C. squamatus* (Forskål) Ascherson SWINE-CRESS, WART-CRESS. A spreading ann. to 30 cm., with dense stalkless clusters of tiny white fls. arranged opposite the lvs. Petals 1–1$\frac{1}{2}$ mm., longer than the sepals. Lvs. deeply pinnately lobed with lance-shaped toothed segments. Fr. 2–3 mm. by 3$\frac{1}{2}$–4 mm., kidney-shaped, strongly and irregularly ridged or warted, usually in densely crowded clusters. △ Waste places, tracksides. Mar.–June. All Eur.

C. didymus (L.) Sm. LESSER SWINE-CRESS. Like 354, but fr. in elongated clusters and fr. stalks longer than the fr. which is notched and covered with irregular pits, not warted. Petals whitish-yellow, *c.* $\frac{1}{2}$ mm., shorter than the sepals. Native of America; naturalized in Western and Central Europe. Pl. 36.

*Subularia aquatica L. A submerged aquatic of Northern Europe.

CONRINGIA   Glaucous hairless plants with entire lvs. and greenish or yellow fls. Fr. linear, slender, valves one- or three-veined; seeds in 1 row in each cell. 2 sps.

355. *C. orientalis* (L.) Dumort. HARE'S-EAR CABBAGE. An erect hairless, little-branched ann. or bienn. 10–70 cm., with pale yellowish or greenish-white fls. *c.* 1 cm. across and distinctive glaucous clasping stem lvs. Petals 8–13 mm. Stem lvs. alternate, elliptic-obovate with broad, rounded, clasping lobes, the basal lvs. stalked. Fr. 6–14 cm., four-angled, the valves one-veined. △ Fields, uncultivated ground. May–June. Cent., South-East Eur. I.SU.: introd. to the rest of Eur. (except P.IRL.S.). Page 135.

C. austriaca (Jacq.) Sweet Like 355, but a taller plant with smaller fls. 6–8 mm. across and lemon-yellow petals 6–10 mm. Fr. eight-angled, the valves three-veined. Central and South-Eastern Europe.

MORICANDIA   Hairless plants with simple fleshy lvs. and violet fls. Fr. linear; valves with a distinct median vein; seeds in 1 or 2 rows in each cell. 3 sps.

356. *M. arvensis* (L.) DC. VIOLET CABBAGE. A hairless glaucous ann. or perenn. 30–60 cm., with showy clusters of violet-purple fls., each *c.* 2$\frac{1}{2}$ cm. across. Lvs. glaucous, slightly fleshy, the upper clasping the stem with a broadly heart-shaped base, the lower lvs. obovate, shallowly lobed. Fr. linear 3–8 cm., four-angled. △ Fields, uncultivated ground waysides. Apr.–May. E.I.GR., Corsica. Pl. 37.

DIPLOTAXIS | **Wall Rocket**   Fls. yellow, white or pink. Fr. linear, flattened, valves prominently one-veined; seeds in 2 rows. 9 sps.

Fls. yellow
357. *D. tenuifolia* (L.) DC. PERENNIAL WALL ROCKET. A robust, hairless, woody-based

perenn. 20–80 cm., with conspicuous bright yellow fls. in lax leafless clusters. Petals 7–15 mm., twice as long as sepals. Lvs. deeply pinnately cut into 4–8 narrow lobes, glaucous, somewhat fleshy, fetid when crushed, not in a rosette. Fr. linear, 2–6 cm., borne erect on spreading fr. stalks of similar length. △ Fields, tracksides, walls, waste places. Jan.–Nov. Much of Eur. (except North Eur. P.GR.): introd. GB.DK.N.S.SF.

D. *muralis* (L.) DC. WALL ROCKET, WALL MUSTARD. Like 357, but usually an ann. with green lvs. mostly in a basal rosette. Fls. smaller, petals 4–8 mm., twice as long as sepals. Fr. 1½–4½ cm., spreading upwards with stalks shorter than fr. Widespread in Europe, except Northern Europe.

Fls. white
358. *D. erucoides* (L.) DC. WHITE WALL ROCKET. A branched ann. or over-wintering bienn. 5–50 cm., with white fls. with violet veins or becoming wholly violet, *c.* 1½ cm. across; petals 7–13 mm. Basal lvs. in a lax rosette, entire, toothed or pinnately lobed, with a large terminal lobe and with coarse, horny-tipped teeth; upper lvs. stalkless clasping stem, oblong and shallowly lobed. Fr. linear, 1–4½ cm., with a conical beak to 6 mm. △ Cultivated places. Apr.–June. E.F.I.R.

BRASSICA Fr. linear with rounded prominent one-veined valves and a conspicuous long or short beak with 0–3 seeds. Fls. yellow or white; sepals erect or spreading. 21 sps.

Some or all lvs. glaucous
359. *B. oleracea* L. (*B. sylvestris* (L.) Miller) WILD CABBAGE. A robust hairless bienn. or perenn. to 3 m., with a woody stem covered with leaf-scars and thick, glaucous and cabbage-like lvs. Fls. many in a branched cluster, yellow, 1½–2½ cm. across; petals 1½–2 cm. Basal lvs. stalked, broad, rounded, sometimes lobed, the upper lvs. oblong, stalkless. Fr. linear, 5–7 cm., with a short conical beak 4–8 mm. △ Sea cliffs; widely cultivated and often appearing as a casual. May–Aug. E.F.GB.I. *Cultivated as a vegetable or fodder crop in numerous forms such as cabbage, cauliflower, broccoli, kale, brussels sprouts, kohlrabi, which all originated largely from this species.*

360. *B. napus* L. RAPE, SWEDE. An ann. or bienn. often with swollen taproot, to 150 cm., with pale yellow fls. and hairless glaucous lvs. Petals 10–18 mm.; unopened fl. buds overtopping open fls. Lvs. not fleshy, the lowest stalked, sparsely bristly, the middle and upper lvs. oblong-lance-shaped, with heart-shaped base clasping stem. Fr. linear, 5–10 cm., narrowed to a slender beak ½–3 cm. △ Open ground, banks, streamsides, ditches; widely cultivated and often naturalized. May–July. Most of Eur. *Cultivated forms include rape, which is grown for its seeds which produce Colza oil and oil-seed cake used for cattle fodder, and swedes grown for their swollen, edible taproots. The green tops are poisonous to cattle.*

B. rapa L. (*B. campestris* L.) TURNIP. Like 360, but open fls. overtopping buds; petals 6–10 mm. Lower lvs. bright green, bristly hairy, the upper lvs. glaucous. Cultivated in much of Europe and widely naturalized; often occurring as a weed of cultivation. *The turnip is a cultivated form with a swollen taproot and is widely growr for fodder; other forms are grown for their oil-containing seeds.*

Lvs. green, not glaucous; lvs. all stalked
361. *B. nigra* (L.) Koch BLACK MUSTARD. A branched ann. to 1 m. with bright yellow fls. with darker veins; petals 7–9 mm. Lvs. all stalked, the lower rough-hairy on both sides, deeply pinnately cut into broad lobes and with a large terminal lobe, the upper lvs. entire or wavy-margined, lance-shaped, hairless. Fr. linear, 1–2 cm., narrowed to a short, slender, seedless beak and borne on short stalks pressed to the stem. △ Cultivated ground, waste

places, ditches, cliffs. June–Sept. All Eur. (except IS.AL.). *Widely cultivated for its seed which is used to make the condiment, mustard, and also the irritant mustard oil. The oil is valuable as a lubricant for precision instruments and used in medicine and in soap-making. Often grown as green manure for ploughing-in and it is a valuable fodder crop; the seedlings and green tops are used for salads.*

SINAPIS | Mustard, Charlock Fr. linear with rounded valves with 3–7 distinct veins, and with a long beak; seeds in 1 row in each cell. Fls. yellow; sepals spreading. 4 sps.

362. *S. arvensis* L. (*Brassica a.*) CHARLOCK. Usually a rough-hairy ann. 30–80 cm., distinguished by its fr. which has a straight conical beak little shorter than valves. Petals yellow, 9–12 mm. Lvs. to 20 cm., usually rough-hairy, the lower stalked, deeply pinnately lobed, with a very large, coarsely toothed terminal lobe, the upper stalkless, entire, lance-shaped. Fr. 2½–4½ cm., usually hairless, beak 1–1½ cm., fr. stalk much shorter than the valves. △ Cultivated ground, tracksides. May–Sept. All Eur. *A serious weed of cultivated ground which can be poisonous to livestock. The seeds yield a clear oil; the green tops are sometimes used as a vegetable.*

S. alba L. WHITE MUSTARD. Readily distinguished from 362 by its fr. which has a flattened, sabre-shaped beak longer than the valves, and a fr. stalk as long as the valves. Petals yellow, 1–1½ cm. Lvs. all stalked and deeply cut, sometimes twice-cut. All Europe. *Cultivated for its seeds which are used in mustard-making, also as a green manure or as a fodder crop. In excess, it can be poisonous to livestock owing to the presence of the irritant mustard oil.* Pl. 37.

ERUCA Like *Sinapis*, but sepals erect. Valves of fr. one-veined; seeds in 2 rows in each cell. 1 sp.

363. *E. vesicaria* (L.) Cav. A rough-hairy ann. or bienn. 20–100 cm., with pale yellow fls. with violet veins and which at length become whitish. Fls. 12–20 mm. across; petals 1½–2 cm. Lvs. all pinnately lobed with 2–5 pairs of narrow lateral lobes and a large, oblong, toothed terminal lobe. Fr. erect, linear, 12–25 mm., with a flattened, sabre-shaped beak about half as long as the valves. △ Cultivated and waste ground. Apr.–June. Med. Eur. P.CH.H.BG.R.SU.: introd. GB.N.D.A.PL.CS. *Cultivated as a salad plant and vegetable, and for the medicinal oil obtained from its seeds.* Pl. 37.

ERUCASTRUM Fr. linear, constricted between seeds and with strongly one-veined keeled valves; beak more or less conical with 0–3 seeds. Fls. yellow. 5 sps.

364. *E. gallicum* (Willd.) O. E. Schulz HAIRY ROCKET. A rough-hairy ann. or bienn. 20–60 cm., with pale or whitish-yellow fls. 7–8 mm. across. Petals 7–8 mm.; sepals erect, half as long as petals. Lvs. deeply pinnately cut into rather distant, parallel-sided toothed lobes. Fr. curving upwards in line with fr. stalk; beak 3–4 mm., slender, seedless. △ Tracksides, fields, banks. Apr.–Nov. E.F.NL.D.CH.A.CS.: introd. GB.DK.N.S.PL.R.SU. Page 128.

*E. nasturtiifolium (Poiret) O. E. Schultz Widespread in Europe, except the north.

HIRSCHFELDIA Like *Sinapis*, but valves of fr. three-veined when young; beak short swollen, club-shaped, containing 0–2 seeds; seeds ovoid. 1 sp.

365. *H. incana* (L.) Lagrèze-Fossat (*Sinapis i.*) HOARY MUSTARD. A hairy ann. to perenn. with branched leafy stems to 1 m. and with pale yellow fls. often with darker veins, 5–8

mm. across. Petals 6–8 mm., twice as long as sepals. Lower lvs. densely white-hairy, deeply pinnately lobed, with an oval, toothed terminal lobe, the upper lvs. stalkless, entire; stems with few stiff, white, downward-projecting hairs. Fr. 8–17 mm., erect and pressed to stem; beak swollen, blunt, about half as long as valves. △ Fields, sandy places; a weed of cultivation. June–Sept. Med. Eur. P.: introd. GB.B.NL.DK.D.CH.A. Page 128.

CAKILE | Sea Rocket Lvs. succulent, glaucous. Fr. of 2 unequal one-seeded sections, the upper mitre-shaped and larger than the top-shaped lower section. Fls. white, pink, or violet. 2 sps.

366. *C. maritima* Scop. SEA ROCKET. A fleshy glaucous branched perenn. 15–60 cm., with very sweet-scented violet, pink or white fls. in elongated leafless clusters. Petals 4–14 mm., twice as long as sepals. Lower lvs. 3–6 cm. glaucous, fleshy, deeply lobed or undivided, the upper stalkless, less divided. Fr. 1–2½ cm., the upper section up to twice as long as the lower. A very variable sp. △ Coastal sands, shingle, dunes. June–Sept. Coasts of all Eur. (except B.IS.): introd. A. Pl. 37.

RAPISTRUM Fr. constricted into 2 sections by a transverse wall, the upper section globular, one-seeded and falling at maturity, the lower more slender with 0–3 seeds; beak conical or thread-like. Fls. yellow. 2 sps.

367. *R. perenne* (L.) All. PERENNIAL BASTARD CABBAGE. A branched leafy bienn. or perenn. to 80 cm., with rough stiff, downward-pointing hairs below, but hairless above, and with bright yellow fls. with darker veins. Petals 5–7 mm., twice as long as sepals. Lower lvs. 10–15 cm., hairy, pinnate, pinnately lobed or toothed, the upper lvs. hairless, stalkless, less deeply lobed or toothed. Fr. 7–10 mm., the upper section ovoid, longitudinally ribbed and gradually narrowed into a broad conical beak ½–1 mm. long. △ Cultivated ground, waste places; often naturalized. June–July. Cent. Eur. I.YU.BG.R.SU.: introd. F.GB.B.NL. CH.PL. Page 135.

R. rugosum (L.) All. BASTARD CABBAGE. Like 367, but an ann. with pale yellow fls. and petals 6–10 mm. Fr. with upper section globose to ovoid, abruptly contracted into a slender style 1–3 mm. long. Southern Europe; introduced to much of the rest of Europe. Page 135.

CRAMBE | Seakale Fr. two-sectioned, the lower section stalk-like, seedless, the upper globular, one-seeded. Fls. white. 8 sps.

368. *C. maritima* L. SEAKALE. A stout, glaucous cabbage-like perenn. to 75 cm., with a fleshy stock and with thick, irregularly lobed basal lvs. Fls. white, rarely reddish, 8–16 mm. across, numerous in dense branched conspicuous clusters. Petals 6–10 mm., claw green; sepals spreading. Lower lvs. long-stalked, oval and pinnately lobed, glaucous and hairless, to 30 cm. the upper lvs. narrow bract-like. Fr. 8–16 mm., the lower section seedless, the upper part spherical, one-seeded. △ Maritime sands, shingle and rocks; sometimes inland. May–Aug. Coasts of Eur. (except Med. Eur. P.IS.): introd. A.CS.H.I. *Seakale is cultivated as a vegetable, particularly in Britain, for its young shoots which are forced and blanched.* Pl. 37.

CALEPINA Fr. globular to ovoid, not two-sectioned, with a short thick conical beak, netted, one-seeded, not splitting. Fls. white. 1 sp.

369. *C. irregularis* (Asso) Thell. (*C. corvini* (All.) Desv.) WHITE BALL MUSTARD. A hairless, somewhat glaucous ann. or bienn. 20–80 cm., with tiny white fls. 2–3 mm. across, with

unequal petals. Outer petals *c*. 2 mm., the inner $2\frac{1}{2}$–3 mm. Lower lvs. in a rosette, entire or deeply lobed, the upper lvs. oblong and clasping stem with acute lobes. Fr. globular-ovoid, $2\frac{1}{2}$–4 mm., shortly beaked, netted when dry. △ Bushy places, uncultivated ground, fields. Apr.–July. Med., South-East Eur. P.B.CH.H.: introd. GB.NL.D.A.CS. Page 139.

RAPHANUS | Radish   Fr. cylindrical, prolonged into a narrow seedless beak, jointed at base and often more or less constricted between the seeds. 1 sp.

370. *R. raphanistrum* L. WILD RADISH. A rough-haired, erect, branched ann. to $1\frac{1}{2}$ m., with yellow, white, lilac, or violet fls. and petals 1–2 cm., usually dark-veined, about twice as long as sepals. Lower lvs. pinnately lobed with 1–4 pairs of small lateral lobes and a large, rounded terminal lobe, the upper lvs. usually entire. Fr. narrow cylindrical, 2–10 cm. long, varying in width from $1\frac{1}{2}$ to 5 mm. or more, with 3–8 shallow constrictions; beak several times longer than the upper constriction. Subsp. *maritimus* (Sm.) Thell. SEA RADISH has frs. 5–8 mm. wide and with 1–6 deep constrictions of irregular length and beak $\frac{1}{2}$–2 cm. It is a coastal plant. △ Fields, cultivated ground, coastal rocks, and sands. May–Sept. All Eur. *Poisonous to livestock.*

R. sativus L. RADISH. Like 370, but ann. or bienn. with a swollen, often brightly coloured taproot and swollen frs. which are not constricted between the seeds. Cultivated for its edible root; often occurring as a casual in much of Europe.

RESEDACEAE | Mignonette Family

Usually herbaceous anns. or perenns. with alternate simple or divided lvs., with glands in the place of stipules. Fls. numerous, in spikes. Sepals and petals 4–8, petals at back larger and more deeply divided than those in front; stamens many, crowded in front. Ovary superior with 3–7 fused or free carpels; fr. usually a capsule.

Fr. of 1 bottle-like capsule.   *Reseda*
Fr. of 4–7 free, spreading, one-seeded carpels.   *Sesamoides*

RESEDA   Fr. a bottle-like, one-celled capsule, open above and containing several seeds. Sepals and petals 4–8; stamens 10–25. 20 sps.

Fl. spikes yellow
371. *R. luteola* L. DYER'S ROCKET, WELD. A bienn. with over-wintering rosette and stiffly erect stems $\frac{1}{2}$–$1\frac{1}{2}$ m. bearing long, slender, pale yellow fl. spikes. Fls. 4–5 mm. across; sepals and petals usually 4, the upper petals with 4–8 lobes, the lower entire. Lvs. all entire, linear to lance-shaped, usually with undulate margin. Fr. globular, 4–6 mm., with 3 pointed lobes. △ Waste ground, sands, gravels, tracksides. May–Oct. Much of Eur. (except IS.N.SF.): introd. D.CH.A.PL.H. *Once much cultivated as a dye plant. The seeds contain an oil used for illumination in the past.*

372. *R. lutea* L. WILD MIGNONETTE. Like 371, but lvs. usually pinnately cut into 1–4 pairs of narrow distant lobes which are sometimes further deeply divided, margin of lobes undulated. Fls. yellow in slender, leafless spikes; petals and sepals usually 6. Fr. oblong, 7–15 mm., with 3 lobes above. An erect, branched, bushy, hairless ann., bienn. or perenn. 20–80 cm. △ Cultivated and disturbed ground, rocks, tracksides. June–Sept. Med. Eur. P.IRL.GB.NL.A.PL.BG.R.: introd. B.D.CH.CS.H.SU. Pl. 38.

Fl. spikes whitish

373. *R. alba* L. UPRIGHT MIGNONETTE. An erect, hairless, somewhat glaucous ann., bienn., or perenn. 30–80 cm., branched above and with a long slender spike of whitish fls. Fls. *c.* 9 mm. across; sepals and petals usually 5 or 6 and petals all cut into narrow lobes. Lvs. deeply pinnately cut into 5–15 pairs of narrow lobes with undulate margins. Fr. elliptic, 8–15 mm., with 4 lobes above. △ Waste ground, on the littoral. May–Sept. Med. Eur. (except AL.TR.) P.: introd. GB.NL.D.CS.R. Pl. 38.

374. *R. phyteuma* L. RAMPION MIGNONETTE. A rather lax branched ann. or bienn. 10–50 cm., with lvs. usually all entire linear-spathulate, but sometimes with 1 pair of lateral lobes, and rather lax spikes of whitish fls. Sepals 6, oblong 3–4 mm., enlarging to 5–13 mm. in fr.; petals 6, white, 3–5 mm., deeply cut into narrow lobes. Lvs. 5–10 cm. Fr. 13–14 mm., drooping, with 3 lobes. △ Cultivated ground, rocks. June–Sept. Med. Eur. (except TR.) P.CH.PL.H.R.: introd. GB.NL.D.A.CS. Pl. 38.

R. odorata L. MIGNONETTE. Like 374 but fls. very fragrant; sepals enlarging to 4–5 mm. in fr.; petals 4 mm. Lvs. obovate, usually some with 1–2 pairs of lateral lobes. Fr. globular 9–11 mm., nodding. Native of North Africa; often grown as an ornamental and locally naturalized in Southern Europe.

SESAMOIDES Fr. of 4–7 one-seeded carpels, spreading in a star. 2 sps.

375. *S. pygmaea* (Scheele) O. Kuntze (*Reseda sesamoides* L.) A slender, spreading perenn. 5–15 cm., with white fls. in slender terminal spikes and with entire lvs. Sepals and petals usually 5, the petals cut into 5–7 narrow lobes; stamens 7–12. Lvs. linear-lance-shaped, the basal lvs. in a rosette. Carpels 4–6, spreading in a star. △ Alpine meadows, rocks, screes, alluvium. May–Sept. E.F.I.

DROSERACEAE | Sundew Family

Perennial herbaceous plants; lvs. often in rosette, lv. blades with upper surface covered with long, glistening, glandular hairs which attract and catch insects. Fls. regular, often in branched clusters; parts of fl. commonly 5; ovary superior, one-celled; styles 2–5. Fr. a capsule.

Fls. white; stamens equal in number to petals.   *Drosera*
Fls. yellow; stamens twice the number of petals.   *Drosophyllum*

DROSERA | Sundew   Lvs. all basal in a rosette; fls. with parts usually in fives; petals white; carpels usually 3. 3 sps.

376. *D. rotundifolia* L. COMMON SUNDEW. A small insectivorous perenn. with flattened rosettes of reddish, circular-bladed lvs. covered with glistening glandular hairs on the upper surface. Fl. stems slender, leafless, 6–8 cm., bearing a few white fls., each $\frac{1}{2}$ cm. across; fls. shortly stalked, usually with 6 petals. Lvs. with hairy stalks and blades 5–8 mm. across, usually spreading horizontally. △ Peat bogs, damp heaths. June–Sept. All Eur. (except AL.GR.TR.). *Used in the past in herbal remedies.*

377. *D. anglica* Hudson (*D. longifolia* L.) GREAT SUNDEW. Like 376, but blades of lvs. oblong-linear or narrowly obovate, to 3 cm. long and gradually narrowed into a long almost

hairless stalk 5–10 cm. Fls. 3–6, white; petals 6 mm.; fl. stem 10–20 cm., twice as long as the lvs. An insectivorous rosette perenn. with more or less erect lvs. △ Sphagnum bogs. July–Aug. Most of Eur. (except P.IS.H.AL.GR.TR.BG.). Pl. 38.

378. *D. intermedia* Hayne (*D. longifolia* auct.) LONG-LEAVED SUNDEW. Like 377, but a smaller plant with obovate lv. blades *c.* 7 mm. by 4 mm. which are gradually narrowed into hairless stalks. Fls. 3–7; petals white, 4–5 mm.; fl. stem 2–5 cm., scarcely longer than the erect lvs. and appearing to rise from the side of the rosette. △ Wet peat bogs. July–Aug. Most of Eur. (except IS.H.AL.GR.TR.BG.). Pl. 39.

DROSOPHYLLUM Fls. yellow; petals 5; stamens 10; styles 5.

D. lusitanicum (L.) Link YELLOW SUNDEW. Distinguished by its cluster of 5–10 bright yellow fls. each *c.* 2½ cm. across, and its slender tapering glandular, insect-catching lvs. 10–20 cm. long, coiled like a watch-spring at the apex. Portugal and Southern Spain. Pl. 38.

CRASSULACEAE | Stonecrop Family

Herbs usually with fleshy, succulent, entire lvs. without stipules. Fls. regular, usually in clusters or spikes; parts of fl. often 5, but varying from 3 to 20; stamens as many as or often double the number of the petals. Ovary superior, the carpels the same number as the petals, free or slightly fused at the base, many-seeded. Fr. a cluster of many-seeded free carpels or a capsule.

		Nos.
1. Petals fused into a tube about as long as the lobes of the corolla.	*Umbilicus*	379, 380
1. Petals free or slightly fused at base.		
2. Petals usually 4–5; plant not rosette-forming.		
3. Fls. bisexual.	*Sedum*	387–396
3. Fls. and plants unisexual.	*Rhodiola*	397
2. Petals 6–18; plants rosette-forming.		
4. Petals 6, with fringed margin.	*Jovibarba*	385
4. Petals 8–18, not fringed.		
5. A rosette plant with ann. fl. stems.	*Sempervivum*	381–384
5. A branched soft-wooded plant with perenn. fl. stems.	*Aeonium*	386

UMBILICUS | Pennywort Basal lv. with circular blades and a centrally placed stalk, or *peltate*. Fls. numerous in an elongate spike-like cluster; calyx small; petals 5, fused into a tube; stamens 10, fused to the corolla. 6 sps.

379. *U. rupestris* (Salisb.) Dandy (*U. pendulinus* DC.; *Cotyledon p.*) WALL PENNYWORT, NAVELWORT. A variable perenn. with erect, usually unbranched stems 10–40 cm., with numerous whitish-green to pinkish fls. in a long narrow leafless, many-flowered spike usually more than half the length of the stem. Fls. tubular 7–10 mm., drooping, lobes shorter than tube, with linear bracts little longer than the fl. stalk. Lower lvs. fleshy, blade rounded, 1½–7 cm. across, shallowly lobed, concave above where attached to lv. stalk; upper lvs. mostly kidney-shaped, toothed, and becoming progressively smaller above. △ Rocks, crevices, walls, May–July. Med. Eur. P.IRL.GB.BG. Pl. 39.

U. horizontalis (Guss.) DC. (*Cotyledon h.*) Like 379, but fl. stem with numerous crowded lvs. many of which are linear and fl. spike not more than half the length of the stem. Fls. 7 mm. by 3 mm., almost stalkless, horizontal. Mediterranean region.

380. *U. erectus* DC. (*Cotyledon umbilicus-veneris* L.) An erect perenn. 20–60 cm., with very numerous greenish-white or straw-coloured fls. drying to reddish-brown, in dense elongated spikes 8–25 cm. Corolla 3–13 mm. tubular, with narrow lance-shaped, long-pointed lobes, as long as the tube; fl. stalk 1–3 mm.; bracts narrow, lance-shaped. Basal lvs. rounded to heart-shaped, very large to 7 cm. across, the stem lvs. getting progressively smaller above. △ Damp shady rocks. May. I.YU.AL.GR.BG.

SEMPERVIVUM | Houseleek Lvs. all fleshy; basal lvs. in perenn. rosettes which readily form daughter rosettes; fl. stems ann. Fls. in terminal clusters; petals and carpels 8–16; stamens twice as many. Hybridization is common in this genus. 23 sps.

Fls. predominantly reddish or purplish

381. *S. arachnoideum* L. COBWEB HOUSELEEK. Rosettes small, crowded, $\frac{1}{2}$–$2\frac{1}{2}$ cm. across, greenish or whitish, with a network of fine cobweb-like hairs stretching from the tips of the lvs. Fls. rose-carmine; petals 8–10, 7–10 mm. long, broadly lance-shaped with a purple midvein; fl. stems 4–12 cm. Lvs. 7–12 mm. broadly oblance-shaped, fine-pointed, finely hairy; stem lvs. red-tipped. A very variable perenn. △ In mountains: rocks, screes, walls. July–Sept. E.F.D.CH.A.I. Pl. 39.

382. *S. montanum* L. MOUNTAIN HOUSELEEK. Rosettes variable in size 1–5 cm. across, with glandular-hairy sticky, dull-green, ciliate lvs. to *c.* 1 cm. Fl. stems 5–15 cm. bearing a cluster of 2–8 reddish-purple or rarely yellowish fls.; petals 11–13, 12–20 mm. long, 3 times as long as the sepals; filaments pale. △ Rocks in mountains. July–Aug. E.F.CH.A.PL. CS.I.R.SU.: introd. N. Pl. 39.

383. *S. tectorum* L. HOUSELEEK. Rosettes large and open, 3–8 cm. across, with stiff, sharp-pointed, often somewhat glaucous and reddish-tinged, hairless lvs. with white bristles on the margin. Fl. stem stout, 20–50 cm., with numerous fls. Fls. pale reddish 2–3 cm. across; petals 13, 9–10 mm. long, hairy beneath; filaments purple. Upper stem lvs. finely hairy. A very variable sp. △ Rocky places in mountains, rocks in lowlands; often planted on roofs and walls. July–Aug. E.F.D.CH.A.I.YU.: introd. R.SU. *Commonly grown on buildings in the belief that it is a protection against lightning, thunderbolts, and fire. A purgative plant used in herbal remedies for warts, corns, and ringworm.*

Fls. predominantly yellow

384. *S. grandiflorum* Haw. (*S. gaudinii* Christ) YELLOW HOUSELEEK. Like 382 with densely glandular-hairy, rather lax, flat rosettes 2–5 cm. across, but lvs. dark green with a reddish-brown apex and rosettes smelling of resin when bruised. Petals 12–14, yellow with a purple basal spot, 10–18 mm. long; filaments purple. Fl. stems 10–20 cm.; stolons stout 10–20 cm. △ Rocks in mountains. July–Sept. CH.I.

S. wulfenii Mert. & Koch Distinguished from 384 by its large glaucous rosettes 4–5 cm. across, and hairless lvs. except for bristles on the margins. Fls. lemon-yellow with a purple basal spot; petals 1 cm.; filaments purple. Central and Eastern Alps.

JOVIBARBA Houseleek-like plants with rosettes of fleshy lvs., but sepals and petals 6; petals pale yellow, keeled on the back and fringed with glandular hairs, petals not spreading wide, but forming a bell-shaped corolla. 5 sps.

385. *J. hirta* (L.) Opiz (*Sempervivum h.*) Fls. pale yellow in a dense, rounded, many-flowered cluster; petals 6, 15–17 mm., margin conspicuously fringed with glandular hairs: stamens 12. Rosettes 3–7 cm. across, with broadly lance-shaped, dark green, bristly-margined, spreading lvs. 1½–2 cm.; stem lvs. clasping, densely hairy. Perenn. forming lateral rosettes on slender stolons; fl. stems 20–30 cm. △ In mountains: rocks, screes. July–Aug. A.PL.CS. H.I.YU.AL.R.SU.

J. sobolifera (J. Sims) Opiz (*Sempervivum s.*) HEN-AND-CHICKENS HOUSELEEK. Like 385, but rosettes smaller, more or less globular 2½–3 cm. across, with incurved greyish or olive-green, often red-tipped lvs. and numerous smaller rosettes on slender, brittle stolons. South-Central Europe to Soviet Union.

AEONIUM Stems perenn., somewhat woody, branched, and ending in terminal rosettes of fleshy overlapping lvs. Fls. yellow, numerous in a branched cluster. 1 sp.

386. *A. arboreum* (L.) Webb & Berth. Stems stout, branched, to 1 m., light brown and with conspicuous leaf-scars and terminal, bright green, shiny rosettes *c.* 10 cm. across. Fls. bright yellow *c.* 2 cm. across, numerous, in rather dense rounded clusters; petals 9–11, narrow lance-shaped, spreading. Lvs. 6 cm. by 2 cm. oblance-shaped, flat with bristly-toothed margin, shiny, hairless. △ Walls, roofs, rock fissures. Dec.–Mar. Native of North Africa; sometimes naturalized. Introd. P.E.F.I.GR. *Used by fishermen in Portugal to harden their lines.* Pl. 39.

SEDUM | Stonecrop Lvs. fleshy, flat or cylindrical, usually on elongated stems and not forming rosettes. Petals usually 5, sometimes 6–9; stamens usually twice as many as petals; carpels as many as petals. 57 sps.

Lvs. fleshy, cylindrical, not flattened
(a) Fls. yellow or greenish-white
(i) Carpels erect; petals usually 6–9

387. *S. ochroleucum* Chaix Fl. heads erect, flat-topped in bud and forming a dense cluster 3–5 cm. across, with bright yellow or yellowish-white fls. Petals 5–8, 7–10 mm. long; sepals 5–7 mm., long-pointed, glandular-hairy. Lvs. linear-cylindrical, pointed, loosely overlapping on sterile shoots. Fr. greenish in a more or less flat-topped cluster. Perenn. 15–30 cm. △ Rocks. June–Aug. Med., South-East Eur. (except TR.) D.CH.

S. sediforme (Jacq.) Pau (*S. altissimum* Poiret) Like 387, but a more robust plant to 60 cm., with fl. clusters distinctly arched at the ends of the branches and inflorescence becoming concave in fr. Petals 4–7 mm., spreading, greenish-white or pale yellow; sepals 2 mm., oval blunt. Mediterranean region and Portugal.

388. *S. reflexum* L. (*S. rupestre* auct.) ROCK STONECROP. Like 387, but inflorescence drooping and globular in bud, and becoming concave in fr. Petals 6–7 mm., bright yellow, usually 7; sepals 3–4 mm., hairless. Lvs. evenly distributed on non-flowering shoots. Fr. yellow, erect. A perenn. with procumbent stems and erect fl. stems 15–35 cm. △ Rocks, walls. June–Aug. Much of Eur. (except P.IS.TR.R.): introd. IRL.GB. Pl. 40.

S. tenuifolium (Sibth & Sm.) Strobl (*S. amplexicaule* DC.) Distinguished by its few fls. in a lax one-sided cluster, with petals 6–10 mm., yellow with a red midvein; sepals conspicuously spurred. Non-flowering shoots with cylindrical, long-pointed, erect, and overlapping glaucous lvs. which are enlarged at the base into a broad, clasping, three-lobed base which soon withers and persists as a papery spur during the summer. Mediterranean region.

(ii) Carpels spreading; petals usually 5

389. *S. acre* L. STONECROP, WALL-PEPPER. A low, mat-forming, variable perenn. 2–10 cm., with fleshy triangular-ovoid lvs. and bright yellow star-like fls. Fls. *c.* 12 mm. across; petals 6–8 mm., 5, spreading, acute. Lvs. 3–6 mm., overlapping, flattened above, blunt-tipped, hot and acrid to the taste. △ Rocks, walls, sands. June–July. All Eur. (except TR.). *Can be eaten as an antiscorbutic; poisonous in excessive quantities.* Pl. 40.

S. alpestre Vill. ALPINE STONECROP. Like 389 but with smaller, dull yellow, oval-oblong, blunt petals 3–3½ mm.; styles very short, reflexed. Lvs. 4–6 mm. oblong, somewhat flattened, nearly parallel-sided, often streaked with red. Fr. dark red. Mountains of Central and Southern Europe.

390. *S. annuum* L. ANNUAL STONECROP. An ann. or bienn. to 15 cm., with slender stems branched from the base but without sterile shoots at time of flowering, and with yellow fls. with acute petals twice as long as the sepals. Lvs. few, *c.* 6 mm., linear-oblong, thick, flattened above and below. Fr. green or red, with spreading carpels. △ Northern regions or mountains in the south: rocks, dry pastures. June–Aug. Much of Eur. (except P.IRL.GB. B.NL.DK.PL.H.TR.).

(b) Fls. white or pink
(i) Perenns. with non-flowering shoots

391. *S. album* L. WHITE STONECROP. Fls. white in a many-flowered, flat-topped, branched cluster 2–5 cm. across. Fls. 6–9 mm. across; petals 2–4 mm., spreading, white, or sometimes pinkish. Lvs. 6–12 mm., cylindrical-ovoid, spreading, usually reddish, hairless. Fr. pink, erect. A variable perenn. with a creeping woody stem bearing numerous sterile shoots in a lax tuft; fl. stems 5–18 cm. △ Walls, rocks, roofs. June–Aug. Most of Eur. (except IS.): introd. IRL. *Considered to possess similar properties to the houseleek; sometimes eaten as a pickle or salad.* Pl. 39.

*S. anglicum Hudson ENGLISH STONECROP. Western Europe.

S. dasyphyllum L. THICK-LEAVED STONECROP. Like 391, but plant shortly and softly hairy and lvs. 3–5 mm., usually opposite, ovoid to almost globular, glandular-hairy, usually bloomed. Inflorescence few-flowered; fls. *c.* 6 mm. across; petals 3 mm., white streaked with pink. Southern Europe; naturalized further north.

(ii) Anns. or bienns. without non-flowering shoots

392. *S. atratum* L. DARK STONECROP. A much-branched ann. to 8 cm. with very blunt, club-shaped hairless lvs. 4–6 mm., at first green, but later reddish-brown. Fls. 6–12 in dense terminal clusters; petals usually 5 or 6, white or flushed with red or greenish-yellow; stamens 10–12. Fr. spreading in a star, becoming black at maturity. △ In mountains: screes, rocks. June–Aug. Cent. Eur. (except H.), South-East Eur. (except TR.) E.F.I.SU.

393. *S. hispanicum* L. Fls. white with pink midvein, numerous in one-sided clusters in a lax dichotomously branched inflorescence; petals usually 6, 5–7 mm. long, about 4 times as long as sepals. Lvs. 7–18 mm., elongate-cylindrical, fine-pointed, glaucous. Fr. spreading in a star, pink or whitish; carpels long-pointed. Hairless or glandular-hairy, usually ann.; 7–15 cm. △ Rocks, screes, walls. June–July. South-East Eur. (except GR.) CH.A.H.I.SU.: introd. S.D.

(c) Fls. blue or lilac

394. *S. villosum* L. HAIRY STONECROP. A rather delicate glandular-hairy plant with spreading heads of pink or lilac fls. with a darker central vein to each petal. Petals 4–5 mm., acute. Lvs. erect 4–7 mm., linear-oblong, yellowish-green. Usually a perenn. with offsets, but sometimes a bienn.; 5–15 cm. △ Marshes, bogs, wet fields. June–Aug. North Eur. (except DK.), Cent. Eur. (except H.) E.F.GB.I.YU.SU. Pl. 40.

S. caeruleum L. BLUE STONECROP. Distinguished by its numerous, bright blue, white-centred fls. borne in a lax cluster. Petals usually 7; stamens 14. Lvs. *c.* 1 cm. oblong-cylindrical, often turning reddish. A small branched ann., often hairy above; 5–20 cm. Corsica, Sardinia, Sicily. Pl. 40.

Lvs. flattened, fleshy

395. *S. telephium* L. ORPINE, LIVELONG. A robust perenn. to 80 cm., with numerous flat fleshy lvs. and reddish-purple, lilac or rarely whitish fls. in dense flattened heads. Petals 5, 3–5 mm.; stamens 10 as long as or longer than petals. Lvs. 2–7 cm. orbicular to narrow-oblong, flattened, toothed, sometimes glaucous. Fr. erect. A very variable sp., including the more robust subsp. *maximum* (L.) Krocker with greenish or yellowish-white fls. and oval lvs. △ Rocks, woods, hedges. June–Sept. All Eur. (except IS.TR.). Pl. 40.

396. *S. cepaea* L. A weak, slender, somewhat spreading ann. or bienn. to 30 cm., with flat lvs. in whorls of 3 or 4, or lvs. opposite, and white or pink fls. in long, lax, compound clusters. Petals 5 mm., oblong-oval, with a long fine tip and a pink hairy midvein. Lvs. obovate 1½–7 cm., flattened, hairless. △ Shady rocks, hedges, woods. June–Aug. Med., South-East Eur. (except TR.) CH.: introd. NL.D.

S. stellatum L. STARRY STONECROP. A Mediterranean ann. to 15 cm., with flattened, rounded, toothed, short-stalked, hairless lvs. 1–1½ cm. and pink fls. in a rather flat-topped outward-spreading cluster. Petals 4–5 mm., acute; sepals fused below. Fr. spreading in a star.

RHODIOLA Like *Sedum*, but plants one-sexed; sepals, petals, carpels usually 4; stamens 8. 2 sps.

397. *R. rosea* L. (*Sedum r.*) ROSEROOT, MIDSUMMER-MEN. A rather robust perenn. with several stout, erect stems to 35 cm., with numerous, overlapping, broad fleshy lvs. and compact terminal clusters of numerous yellowish-green or reddish fls. Male fls. with purple anthers and conspicuous but abortive carpels; female fls. with green carpels becoming reddish in fr. Petals 3–4 mm., sometimes absent. Lvs. obovate to lance-shaped, usually toothed, glaucous, hairless, progressively smaller from above downwards; rhizome fleshy, fragrant. A very variable sp. △ Northern regions or mountains: rocks, screes. May–Aug. E.F.IRL.GB.IS.N.S.SF.CH.A.PL.CS.I.YU.BG.R.SU. Pl. 40.

SAXIFRAGACEAE | Saxifrage Family

Herbs usually with opposite, usually simple, but sometimes deeply lobed lvs., often rosette-forming. Fls. usually in branched clusters; sepals 4 or 5; petals 4 or 5; stamens 8 or 10. Ovary distinctive, half-inferior, with 2 carpels which are fused below and are free and divergent above. Fr. a many-seeded capsule.

Petals absent; sepals mostly 4.   *Chrysosplenium*
Petals present; sepals and petals 5.   *Saxifraga*

SAXIFRAGA | **Saxifrage**   Petals 5; stamens 10. Carpels usually 2, more or less fused below. 123 sps.

Lvs. with lime-glands on margin
(*a*) *Fls. purple or pink*
398. **S. oppositifolia** L. PURPLE SAXIFRAGE. A creeping, mat-forming or trailing perenn.,
spreading 10–30 cm., with solitary, short-stalked, purple or pink fls. *c.* 1 cm. across.
Petals ½–1½ cm.; anthers bluish. Lvs. 2–6 mm., numerous, densely overlapping in 4 rows,
each thickened at tip and with 1 or more lime-glands, margin ciliate. A very variable sp.
△ Arctic or mountains: rocks, crevices, screes. May–Aug. Most of Eur. (except P.B.NL.DK.
H.GR.TR.). Pl. 41.

(*b*) *Fls. white*
399. **S. cotyledon** L. PYRAMIDAL SAXIFRAGE. Rosettes large, up to 7 cm. across, with stiff,
broad strap-shaped, finely toothed lvs. 2–6 cm. by 9–15 mm., with lime-glands on the
margin. Fl. stems 15–20 cm., branched from the base to form a large pyramidal cluster of
pure white fls. with lateral branches bearing 5–20 fls. Petals 6–10 mm.; calyx and fl. stalks
glandular-hairy. △ Northern regions or mountains: rock crevices. July–Aug. IS.N.S.D.CH.
A.

S. longifolia Lapeyr. PYRENEAN SAXIFRAGE. Distinguished by its very large, flat, solitary
rosettes 6–18 cm. across, with very numerous linear, glaucous, lime-encrusted lvs. 3–8
mm. broad; rosettes long-lived but flowering only once. Fl. stems 25–50 cm. branched
from near base to form an erect pyramidal inflorescence of numerous white fls.; petals
5–6 mm., sometimes with fine red spots. Spain and France. Pl. 41.

400. **S. paniculata** Miller (*S. aizoon* Jacq.) LIVELONG SAXIFRAGE. Like 399, but rosettes
smaller, hemispherical *c.* 1–3 cm. across; lvs. 12–40 mm. glaucous, lime-encrusted, finely
toothed. Fl. stem 12–30 cm., branched only from the upper third to form a small cluster,
with lateral-spreading, glandular-hairy branches bearing 1–3 fls. Fls. white or pale cream,
sometimes red-spotted; petals 4–6 mm. A variable sp. △ In mountains: rocks and screes.
June–Aug. Cent., South-East Eur. (except TR.) E.F.IS.N.I.SU. Pl. 42.

S. crustata Vest Distinguished by its numerous linear, glaucous lime-encrusted, entire,
not toothed lvs. 1½–6 cm. by 2–4 mm., in flat, densely tufted rosettes. Fl. stem to 30 cm.,
branched from the upper half, branches and fl. stalks glandular-hairy. Fls. white, sometimes red-spotted; petals 4–6 mm. Eastern Alps. Page 153.

Lvs. without lime-glands
(*a*) *Fls. bright yellow or orange*
401. **S. mutata** L. Recalling 399, but fls. orange-yellow in pyramidal cluster on a stout
glandular stem to 70 cm., branched from below the middle. Petals linear-lance-shaped,
5–8 mm. Rosettes large, loose; lvs. 1–7 cm. by 7–12 mm. oblance-shaped leathery, dark
shining green, margin conspicuously translucent, ciliate towards base. △ Damp rocks in
mountains. July–Aug. F.CH.A.I.R. Page 153.

402. **S. aizoides** L. YELLOW MOUNTAIN SAXIFRAGE. A rather weak-spreading, somewhat mat-
forming perenn. to 25 cm., with lax, branched clusters of yellow fls. with orange spots. Fls.
c. 1 cm. across; petals widely spaced with sepals as long and visible between. Lvs. fleshy
1–2½ cm. linear-oblong acute, entire or with few teeth, often ciliate. △ Arctic regions or
mountains: bogs, streamsides. June–Aug. North Eur. (except DK.), Cent. Eur. (except H.)
E.F.IRL.GB.I.AL.R.SU. Pl. 41.

S. hirculus L. YELLOW MARSH SAXIFRAGE. A lax tufted perenn. with erect stems bearing
1–3 yellow fls. Petals 1–1½ cm., with swellings near base; sepals reflexed in fr. Lvs. 1–2½
cm. lance-shaped, blunt, with lv. stalk with long reddish-brown hairs. Widespread,
except in Southern Europe. Page 153.

(b) Fls. white, greenish-yellow or pinkish
(i) Basal lvs. with rounded blades about as broad as long

403. *S. rotundifolia* L. ROUND-LEAVED SAXIFRAGE. An erect, rather stout-stemmed perenn. 15–40 cm., with a rather narrow branched cluster of many white fls. Petals 6–11 mm., usually spotted with yellow at base and red towards apex; sepals spreading-erect. Basal lvs. long-stalked, in a lax rosette, blade rounded kidney-shaped, deeply toothed, with a very narrow horny border, usually hairy; stem lvs. few. △ Damp and shady places in mountains. June–Aug. E.F.D.CH.PL.CS.I.YU.AL.GR.BG.R.: introd. B. Pl. 42.

404. *S. granulata* L. MEADOW SAXIFRAGE. An erect little-branched bienn. or perenn. to ½ m., with few showy white fls. in a lax cluster. Petals 1–1½ cm., hairless. Lvs. mostly basal, long-stalked, blade often somewhat fleshy, kidney-shaped with coarse rounded teeth or shallow lobes; stem lvs. smaller or absent. Bulbils brown, present only at base of stem and hidden in the soil. A very variable sp. △ Grasslands, wood margins, dry rocky places. Apr.–June. Most of Eur. (except IS.AL.GR.BG.R.).

S. bulbifera L. BULBIFEROUS SAXIFRAGE. Readily distinguished from 404 by the reddish bulbils in the axils of the stem lvs. Fls. white in an umbel-like cluster; petals 7–10 mm. Stem lvs. wedge-shaped, sharply toothed; basal lvs. kidney-shaped, rounded-toothed, glandular-hairy. Central and South-Eastern Europe. Page 153.

S. aquatica Lapeyr. PYRENEAN WATER SAXIFRAGE. A robust perenn. forming a deep cushion of foliage and stout erect stems to 60 cm., with numerous white fls. Petals 6–9 mm., twice as long as the densely glandular-hairy calyx. Lower lvs. long-stalked, semicircular in outline to 4 cm. wide, deeply divided into numerous triangular segments. Streamsides in the Pyrenees.

405. *S. umbrosa* L. WOOD SAXIFRAGE. A rosette-forming perenn. with rounded, rather leathery, spreading lvs. and erect, leafless, glandular-hairy stems to 40 cm., branched above and bearing numerous small white fls. Petals *c.* 4 mm., white with 2 yellow spots and usually some red spots; sepals reflexed. Lvs. with broad cartilaginous margin, obovate, rounded-toothed; lv. stalk broad and flat, densely ciliate, scarcely as long as blade. △ Rock crevices, shady places. May–Aug. E.F.: introd. GB.I. Page 153.

S. hirsuta L. KIDNEY SAXIFRAGE. Like 405 and often hybridizing with it. Distinguished by the more or less cylindrical lv. stalk 2–4 times as long as the blade which is hairy on both sides and often reddish beneath. Spain, France, Ireland; occasionally naturalized elsewhere. Pl. 42.

*S. spathularis Brot. ST PATRICK'S CABBAGE. Iberian Peninsula, Ireland.

406. *S. cuneifolia* L. A rosette-forming perenn. with fleshy rounded lvs. and erect, leafless, glandular-hairy fl. stems to 15 cm., bearing small white fls. Petals 2½–4 mm., yellow-spotted; sepals reflexed. Lvs. with a leathery, hairless blade, with an entire or toothed margin, and with a narrow translucent border, tapering gradually into a broad, sparsely ciliate stalk, blade rose-violet beneath. △ In mountains: woods, shady places. June–Aug. E.F.CH.A.I.YU.R.

(ii) Basal lvs. with blade twice as long as broad or more, entire or toothed

407. *S. stellaris* L. STARRY SAXIFRAGE. A rosette-forming perenn. with somewhat fleshy, shining lvs. and lax, branched clusters of white starry fls. Petals 3–7 mm., lance-shaped, spreading or turned downwards, with 2 yellow spots; sepals reflexed; anthers pink. Lvs. 1–5 cm., spathulate, variously toothed, narrowed below, sparsely hairy. A variable sp. △ Northern regions or mountains: damp places, streamsides, springs. June–Aug. Most of Eur. (except B.NL.DK.H.TR.). Pl. 41.

1. *Saxifraga bulbifera* [404] 2. *S. hypnoides* [411] 3. *S. caesia* 410
4. *S. crustata* [400] 5. *S. hirculus* [402] 6. *S. moschata* 411
7. *S. mutata* 401 8. *S. umbrosa* 405

408. *S. aspera* L. ROUGH SAXIFRAGE. A slender, spreading perenn. forming loose moss-like mats, with numerous stiff, bristly lvs. and white or cream fls. in a lax cluster on leafy stems 8–20 cm. Fls. 2–5; petals 5–8 mm., yellow towards the base. Lvs. 3–8 mm., oblong-lance-shaped, spreading, with bristly-haired margin; axillary buds conspicuous but not as long as lvs.; stem lvs. spreading 1–2 cm. △ In mountains: damp rocks, stony places, walls. July–Aug. E.F.CH.A.I.

S. bryoides L. MOSS SAXIFRAGE. Like 408, but mats denser with numerous sterile shoots and many smaller, dense, incurved linear lvs. 5–7 mm., with axillary buds as long as lvs. Fls. solitary, white with yellow base, on leafy stems 3–8 cm. and with lower stem lvs. 5 mm., erect. Mountains from the Pyrenees to the Carpathians.

409. *S. androsacea* L. A tiny hairy perenn. with soft, dark green mostly basal lvs. forming lax rosettes and fl. stem 2–8 cm., bearing 1–3 white fls. Petals white, 5–7 mm., blunt or shallowly notched, twice as long as the glandular-hairy sepals. Lvs. 1–2 cm., spathulate, entire or shortly three-lobed, margin ciliate. △ In mountains: damp pastures, scree, near snow patches. May–July. Cent. Eur. (except H.) E.F.I.YU.AL.BG.R. Pl. 41.

410. *S. caesia* L. BLUISH-GREY SAXIFRAGE. A tiny alpine perenn. forming dense cushions of small, glaucous, lime-encrusted rosettes *c.* 8 mm. across, and very slender stems to 10 cm., bearing 2–5 white fls. Petals 4–6 mm., twice as long as glandular-hairy sepals. Lvs. 3–6 mm., elliptic to spathulate, stiff, outward-curving. △ Screes, rocks. July–Aug. Cent. Eur. (except H.) E.F.I.YU. Page 153.

(iii) Basal lvs. twice as long as broad or more, mostly deeply three-lobed.

411. *S. moschata* Wulfen (*S. muscoides* Wulfen; *S. varians* Sieber) MUSKY SAXIFRAGE. A small mat-forming perenn. with small rosettes of lvs. 3–15 mm. long, and fl. stems 1–10 cm. bearing 1–7 greenish-yellow, or less commonly red or whitish fls. Petals 3–4 mm., not touching and little longer than the sepals. Lvs. very variable, stalked or stalkless, entire or usually deeply three-lobed, sometimes five-lobed, usually glandular-hairy. A very variable sp. △ In mountains: rocks, screes, walls. July–Aug. Cent. Eur. (except H.) E.F.I.YU.R. Page 153. Pl. 42.

S. exarata Vill. FURROWED SAXIFRAGE. Like 411 and sometimes hybridizing with it, but distinguished by its broader, whiter petals which often almost touch at their margins and are twice as long as the sepals. Lvs. usually densely glandular, deeply furrowed when dry; plants forming soft, dense cushions. The Alps to the Balkans.

S. hypnoides L. MOSSY SAXIFRAGE. Distinguished by its long-spreading leafy shoots with widely spaced, linear to lance-shaped lvs. which terminate in a rosette of three- to several-lobed lvs.; lobes with a fine bristle-like apex. Fl. stem 10–30 cm., slender, usually three- to seven-flowered, buds nodding; petals white, 7–10 mm. North-Western Europe. Page 153.

*S. cespitosa L. TUFTED SAXIFRAGE. Arctic and Northern Europe.

S. tridactylites L. RUE-LEAVED SAXIFRAGE. A slender, erect, branched, glandular-hairy, and often reddish ann. usually not more than 10 cm., without a basal rosette. Fls. white; petals 2–3 mm., notched, twice as long as sepals. Basal lvs. entire; stem lvs. usually three-lobed. Throughout Europe. Pl. 41.

CHRYSOSPLENIUM | **Golden Saxifrage** Fls. in terminal, leafy, flat-topped clusters; sepals 4; petals absent; stamens 8. Ovary half or quite inferior. 5 sps.

412. *C. alternifolium* L. ALTERNATE-LEAVED GOLDEN SAXIFRAGE. A delicate, slightly fleshy perenn. with many erect stems to 20 cm. arising from underground creeping stolons, and

bearing terminal flat-topped clusters of tiny golden-yellow fls. Fls. 5–6 mm. across; sepals oval, spreading; petals absent. Basal lvs. long-stalked with orbicular blade with heart-shaped base, regularly shallow-lobed, hairy above; stem lvs. alternate; bracts surrounding fls. spreading, yellowish, broadly oval, toothed. △ Shady, damp places, streamsides, springs. Mar.–June. Much of Eur. (except P.IRL.IS.AL.GR.TR.).

C. oppositifolium L. OPPOSITE-LEAVED GOLDEN SAXIFRAGE. Like 412, but stem lvs. opposite and blades of basal lvs. with a cut-off or wedge-shaped base. Basal lvs. arising from creeping, leafy, above-ground stems. Western and Central Europe. Pl. 42.

PARNASSIACEAE | Grass of Parnassus Family

Similar to Saxifragaceae where it is placed by many botanists, but differing in the presence of an additional ring of 5 sterile stamens bearing nectaries on the upper surface, alternating with the 5 fertile stamens. Ovary superior, carpels usually 4. 1 genus. 1 sp. in Eur.

PARNASSIA

413. *P. palustris* L. GRASS OF PARNASSUS. Fls. solitary, white, borne on an erect, unbranched stem 5–40 cm., with a single, stalkless, broad oval to heart-shaped lv. towards its base. Fls. 1½–3 cm. across; petals oval, conspicuously veined, spreading; nectaries branched with 9–13 long points tipped with a yellowish gland. Lvs. hairless, somewhat glaucous; basal lvs. all stalked, with oval- to heart-shaped blade 1–3 cm. A variable, sometimes tufted perenn. △ Marshes, wet meadows, damp heaths. July–Sept. All Eur. (except P.TR.). Pl. 42.

GROSSULARIACEAE | Gooseberry Family

Often included in Saxifragaceae, but differing in being shrubs with alternate, palmately three- to five-lobed lvs. Fls. in clusters; petals, sepals and stamens 5. Ovary inferior, with a conspicuous, often coloured receptacle above; carpels 2; styles 2. Fr. a berry. 1 genus. 9 sps. in Eur.

RIBES | Currant, Gooseberry

Spineless shrubs
(a) *Fr. bright red*
414. *R. rubrum* L. (*R. sylvestre* (Lam.) Mert. & Koch) RED CURRANT. An almost hairless shrub to 1½ m., not strong-smelling, with heart-shaped three- to five-lobed, toothed lvs. *c.* 6 cm., and lax clusters of yellowish-green fls. Clusters inclined or drooping, with 10–20 fls. each 5–7 mm. across; sepals rounded, spreading, petals smaller. Fr. 6–10 mm. globular, shiny red, acid, hairless. △ Damp woods, hedges; naturalized widely in Europe. Apr.– May. E.B.NL.D.I.: introd. elsewhere in Eur. *Commonly cultivated for its fruits which are particularly valued for preserves, flavouring and wine-making; the leaves produce a yellow dye and the fruits a black dye.* Pl. 43.

R. alpinum L. MOUNTAIN CURRANT. Like 414, but fl. clusters erect, one-sexed, either of 2–5 female fls., or 10–30 male fls.; axis of inflorescence glandular-hairy. Lvs. deeply three-lobed, 2–6 cm. Fr. red, insipid. Northern regions and mountains of Europe. Pl. 43.

R. spicatum Robson Woods of Northern and North-Central Europe.

(b) *Fr. black or purple-red*
415. *R. nigrum* L. BLACK CURRANT. A strong-smelling shrub to 2 m., with lax, drooping, dull reddish-green fl. clusters. Fls. *c.* 8 mm., bell-shaped; petals erect, whitish, shorter than hairy recurved sepals. Lvs. large to 10 cm. wide, deeply three- to five-lobed, with numerous aromatic glands beneath. Fr. black, to 12 mm., sweetish, aromatic. △ Woods, hedges. Apr.–May. Much of Eur. (except P.E.IS.AL.GR.TR.). *Widely cultivated for its edible fruits which are used for preserves, beverages, wine and flavouring. Used medicinally in the past; the fruits are rich in vitamin C. The leaves give a yellow dye and the fruits a blue or violet dye.*

R. petraeum Wulfen ROCK RED CURRANT. Distinguished from 415 by the lvs. which are not strong-smelling. Fl. clusters pinkish, *c.* 10 cm. long; sepals ciliate. Lvs. broader, to 15 cm. wide, without aromatic glands beneath. Fr. purple-red. Mountains of Central Europe.

Spiny shrubs
416. *R. uva-crispa* L. (*R. grossularia* L.) GOOSEBERRY. A much-branched shrub 1–1½ m. with stiff branches with stout penetrating spines, usually in threes, and small, greenish, drooping fls. in the axil of each lv. Fls. solitary or 2–3, often purple-flushed; sepals 5–7 mm., strap-shaped; petals white, smaller. Lvs. 2–5 cm. wide, rather deeply three- to five-lobed, usually hairy. Fr. large, 1–2 cm., ovoid, green or reddish, usually hairy. △ Thickets, open woods, rocks. Mar.–Apr. Most of Eur. (except North Eur. AL.TR.): introd. P.IRL.DK.N. S.SF. *Widely cultivated for its edible fruits which are used as dessert, for jams, preserves, and wine-making.* Pl. 43.

PITTOSPORACEAE | Pittosporum Family

Woody plants with alternate lvs. without stipules. Sepals 5, free or fused at the base; petals 5; stamens 5. Ovary superior with 2–5 fused carpels. Fr. a capsule or berry. 3 sps. in Eur.

PITTOSPORUM

417. *P. tobira* (Thunb.) Aiton fil. JAPANESE PITTOSPORUM. An aromatic, resinous shrub to 3 m., with very sweet-scented, creamy-white fls. borne in flat-topped clusters at the ends of short, leafy branches. Petals blunt. Lvs. oblong-obovate entire, short-stalked, leathery, dark green, shiny above. Fr. ovoid, yellowish-brown. △ Native of China and Japan; frequently grown for ornament in the Mediterranean region. Mar.

PLATANACEAE | Plane Tree Family

Trees with flaking bark and buds which are ensheathed by the base of the lv. stalk; lvs. deeply palmately lobed. Fls. clustered into one or several spherical, one-sexed heads. Perianth inconspicuous; male fls. of 4–6 stamens; female fls. with 3–4 sterile stamens and 5–9 free carpels with long slender styles. Fr. a rounded, bristly cluster of many carpels. 1 genus. 2 sps. in Eur.

PLATANUS | Plane

418. *P. orientalis* L. ORIENTAL PLANE. A tall tree to 30 m., with massive branches and greyish flaking bark. Lvs. large, about as long as broad, cut into 5–7 deep, coarsely toothed lobes,

with narrow clefts between the lobes and the middle lobe longer than broad. Fl. and fr. heads 3–6, on long, hanging stalks. △ Ravines, watersides; commonly planted in Southern Europe. May–June. YU.AL.GR.BG. *A common village shade tree. The wood takes a fine polish and is used for decorative work and by furniture makers.* Pl. 43.

P. hybrida Brot. (*P. acerifolia* (Aiton) Willd.) LONDON PLANE. Like 418, but lvs. more shallowly lobed to less than half the width of the blade, with wider clefts between the lobes and with the middle lobe about as long as broad. Fl. and fr. heads usually 2. Of unknown origin; now very widely planted as a shade and ornamental tree, particularly in towns with polluted atmospheres.

ROSACEAE | Rose Family

Trees, shrubs or herbaceous plants with alternate, simple or compound lvs., usually with stipules. Fls. radially symmetrical; sepals and petals usually 5, epicalyx sometimes present; stamens usually numerous; receptacle shallowly or deeply cup-shaped and encircling ovary, thus fls. usually half-inferior or inferior. Carpels one to many. Fr. showing great variety: carpels either solitary and one-seeded, dry or fleshy, or carpels many, dry or fleshy, either free, fused, or fused with the receptacle.

1. Trees or upright or spreading shrubs.
  2. Lvs. compound, either pinnate or palmate.   Group A
  2. Lvs. simple, entire or toothed.   Group B
1. Herbaceous plants or dwarf creeping shrublets.
  3. Dwarf creeping shrublets.   Group C
  3. Herbaceous plants.   Group D

Group A. *Trees or shrubs with compound lvs.*

		Nos.
1. Fls. in a dense head; petals absent.	*Sarcopoterium*	440
1. Fls. not in a dense head; petals 4 or 5.		
2. Carpels free, exposed over the surface of receptacle.		
3. Petals yellow; non-spiny shrubs.	*Potentilla*	445–457
3. Petals white or purple; spiny shrubs.	*Rubus*	426–429
2. Carpels enclosed in a fleshy receptacle.		
4. Spiny shrubs; carpels free within receptacle.	*Rosa*	430–435
4. Non-spiny trees; carpels fused to receptacle.	*Sorbus*	466–469

Group B. *Trees or shrubs with entire lvs.*

		Nos.
1. Carpels free, exposed over surface of the receptacle.		
2. Carpel 1, fleshy.	*Prunus*	476–484
2. Carpels 5, dry.	*Spiraea*	419
1. Carpels enclosed in a fleshy receptacle.		
3. Fls. solitary.		
4. Fls. less than 1 cm. across; fr. red or black.	*Cotoneaster*	473, 474
4. Fls. more than 1 cm. across; fr. brown or yellow.		
5. Sepals toothed, blunt, shorter than petals.	*Cydonia*	463
5. Sepals entire, long-pointed, longer than petals.	*Mespilus*	460
3. Fls. in clusters of two to many.		

6. Wall of carpels hard and stone-like in fr.
  7. Lvs. with entire margin.   *Cotoneaster*   473, 474
  7. Lvs. with toothed or lobed margin.
    8. Lvs. evergreen, toothed.   *Pyracantha*   475
    8. Lvs. deciduous, deeply lobed.   *Crataegus*   461, 462
6. Walls of carpels fleshy in fr., but seed wall sometimes
  hard.
  9. Inflorescence compound and made up of smaller
    clusters.
    10. Lvs. evergreen; cultivated plant.   *Eriobotrya*   470
    10. Lvs. deciduous.   *Sorbus*   466–469
  9. Inflorescence of a simple umbel or few-flowered
    cluster.
    11. Petals narrow, widely spaced, and not overlapping.   *Amelanchier*   471, 472
    11. Petals broad, rounded, overlapping.
      12. Styles free to base; fr. gritty.   *Pyrus*   464
      12. Styles fused below; fr. not gritty.   *Malus*   465

Group C. *Dwarf creeping shrublets*   *Nos.*

1. Petals *c.* 8, white, conspicuous.   *Dryas*   441
1. Petals 5 (or absent), minute, yellowish.   *Sibbaldia*   458

Group D. *Herbaceous plants*   *Nos.*

1. Petals absent.
  2. Lvs. simple or palmately divided; stamens usually 4
    or less.
    3. Anns.; stamens 1 or 2.   *Aphanes*   425
    3. Perenns.; stamens usually 4.   *Alchemilla*   423, 424
  2. Lvs. pinnate; stamens 4 or many.   *Sanguisorba*   438, 439
1. Petals 4 or more.
  4. Petals *c.* 8; stems woody below.   *Dryas*   441
  4. Petals 4, 5, or 6.
    5. Calyx and carpels enclosed in receptacle.   *Aremonia*   437
    5. Calyx and carpels exposed on receptacle.
      6. Sepals 4–5, with shorter lobes, the epicalyx,
        alternating with them.
        7. Lvs. pinnate.
          8. Styles elongating, persisting, and often hooked
            or feathery-haired in fr.   *Geum*   442–444
          8. Styles not elongating or hooked or feathery-haired.   *Potentilla*   445–457
        7. Lvs. palmately lobed, or with 3 leaflets.
          9. Receptacle fleshy or spongy in fr.
            10. Leaflets 5 or 7; fls. purplish.   *Potentilla*   445–457
            10. Leaflets 3; fls. white or yellowish.   *Fragaria*   459
          9. Receptacle dry in fr.
            11. Petals 1–2 mm. long; stamens 5.   *Sibbaldia*   458
            11. Petals more than 2 mm. long; stamens 10 or more.   *Potentilla*   445–457
      6. Sepals 4–6, without additional lobes, thus
        epicalyx absent.

12. Fls. yellow; fr. with hooked bristles.	*Agrimonia*	436
12. Fls. white, cream, red, or purple; fr. without hooks.		
13. Carpels 3; stipules absent.	*Aruncus*	420
13. Carpels 6 or more; stipules present.		
14. Lvs. pinnate with small leaflets alternating with larger leaflets; carpels dry.	*Filipendula*	421, 422
14. Lvs. palmate, or palmately three- to seven-lobed; carpels fleshy.	*Rubus*	426–429

SPIRAEA Deciduous shrubs with simple lvs. Sepals and petals 5; stamens 15 to numerous; carpels 5, free, several-seeded. 13 sps.

419. *S. salicifolia* L. WILLOW SPIRAEA. A suckering shrub with many erect yellowish-brown stems to 2 m., narrow toothed lvs. and dense terminal clusters of bright pink fls. Fl. clusters 2–10 cm. long, conical or cylindrical; fls. *c.* 8 mm.; stamens projecting, twice as long as petals. Lvs. 3–7 cm. oblong-elliptic, hairless. △ Hedges, bushy places. June–July. Cent. Eur. BG.R.SU.: introd. F.GB.NL.DK.N.S.D.CH.I.YU. Pl. 44.

S. hypericifolia L. Fls. small white, 5–10 in somewhat flat-topped lateral clusters forming collectively long leafy fl. spikes. Petals *c.* 3 mm.; stamens nearly as long. Lvs. 1–2 cm., obovate to spathulate, glaucous, often toothed above. Shrub ½–1 m. Parts of South-Western and South-Eastern Europe. Page 171.

ARUNCUS | Goat's-Beard Herbaceous perenns. with twice-pinnate lvs. Sepals and petals 5; carpels 3, free, several-seeded. 1 sp.

420. *A. dioicus* (Walter) Fernald (*A. sylvestris* Kostel. incl. *A. vulgaris* Rafin.; *Spiraea aruncus* L.) GOAT'S-BEARD. A tall, leafy perenn. to 2 m., with a very conspicuous, large, terminal, branched, pyramidal inflorescence of very numerous, tiny, yellowish-white fls. densely clustered along the branches. Lvs. very large to 1 m., 2–3 times cut into large, oval, long-pointed, double-toothed leaflets to 14 cm. by 7 cm. Fr. with reflexed carpels. △ Shady places and woods in mountains. June–Aug. Recalling 206 *Actaea spicata*, but differing in fr. and absence of distinctive smell. Cent. Eur. E.F.B.I.YU.AL.R.SU. *The fruit is poisonous.* Pl. 44.

FILIPENDULA | Meadow-Sweet, Dropwort Herbaceous perenns. with pinnate lvs. Sepals and petals 5; stamens 20–40; carpels 5–15, free. Fr. a group of one-seeded carpels. 2 sps.

421. *F. ulmaria* (L.) Maxim. MEADOW-SWEET. A tall leafy perenn. to 120 cm., with creamy-white fls. in compound, somewhat flat-topped terminal clusters. Sepals reflexed, hairy; petals obovate 2–5 mm., shorter than the stamens. Lvs. pinnate, basal lvs. with 8–20 pairs of large leaflets and many small leaflets, lower stem lvs. few, with 2–5 pairs of large leaflets and several pairs of smaller leaflets; leaflets all oval, toothed, 2–8 cm., usually dark green, hairless above and white-woolly beneath; stipules conspicuous. Carpels 6–10, spirally twisted, hairless. △ Damp meadows, swamps, marshes, ditches. June–Aug. All Eur. (except TR.). *The flowers have been used for flavouring beverages, the leaves as a herbal tea.* Pl. 44.

422. *F. vulgaris* Moench DROPWORT. Like 421, but basal lvs. with 5 or less pairs of large leaflets and the lower stem lvs. with 10–20 pairs of large leaflets; leaflets oblong, ½–1½ cm., pinnately lobed and deeply toothed. Fl. cluster dense, flat-topped, petals creamy-white and usually flushed reddish-purple. Fr. hairy, not spirally twisted. Roots tuberous;

fl. stem usually leafless above, 15–80 cm. △ Grasslands, clearings. May–July. Most of Eur. (except IS.). *The tubers are edible, but somewhat astringent.* Pl. 44.

ALCHEMILLA | **Lady's-Mantle** Perenns. with numerous small, greenish fls. Epicalyx 5; sepals 5; petals absent; stamens usually 4. Fr. a small, dry, one-seeded carpel. A very difficult genus with numerous microspecies. 118 sps.

Basal lvs. dissected to half width of blade or less
423. *A. vulgaris* L. agg. LADY'S-MANTLE. A hairy or almost hairless perenn. with a woody rootstock and with erect or spreading stems 5–45 cm., with tiny greenish-yellow fls. in a terminal compound, dense, or lax cluster. Fls. 3–4 mm. across; epicalyx and sepals green; petals absent. Basal lvs. rounded, long-stalked, palmately five- to eleven-lobed to half the width of the blade or less, toothed, green on both sides; upper lvs. smaller, short-stalked. An aggregate sp. with between 200 and 300 microspecies described in Europe. △ Damp meadows, open woods, rocky places. May–Sept. Most of Eur. (except P.IS.SF.AL.TR.). *Used in herbal remedies in the past.*

Basal lvs. dissected to the base of the blade or nearly so
424. *A. alpina* L. ALPINE LADY'S-MANTLE. An arctic or alpine perenn. 10–20 cm., with creeping woody rootstock, deeply lobed lvs. and tiny greenish fls. in dense clusters. Fls. *c.* 3 mm. across; calyx and fl. stalks with adpressed silky hairs. Lvs. with blade rounded in outline 2½–3½ cm. across, palmately cut to the base into 5 or 7 lance-shaped lobes which are green above with a silvery-hairy margin and densely silvery-hairy beneath; apex of lobes toothed; stipules brown papery; stem lvs. few. △ In mountains: pastures, rocky places, screes, open woods. June–Aug. E.F.IRL.GB.IS.N.S.SF.D.CH.A.I.SU. Page 161.

A. pentaphyllea L. FIVE-LEAVED LADY'S-MANTLE. A tiny, hairless, alpine perenn. 3–15 cm. with thick blackish rhizome, and lvs. *c.* 1 cm., rounded in outline and deeply cut into 5 wedge-shaped lobes which are further cut into strap-shaped segments. Fls. green, in 1 or 2 terminal whorls on slender stems. The Alps. Page 161.

APHANES Like *Alchemilla*, but anns. Fls. densely clustered in axils of lvs. Stamens 1, rarely 2. 4 sps.

425. *A. arvensis* L. (*Alchemilla a.*) PARSLEY PIERT. A small, spreading, leafy, branched ann. or bienn. 5–20 cm., with tiny green fls. *c.* 2 mm. in stalkless lateral clusters shorter than the lvs. Lvs. short-stalked, fan-shaped, with 3–5 deep lobes which are further three- to five-lobed; stipules leaf-like, deeply lobed. △ Fallow land, bare ground. May–Aug. Much of Eur. (except IS.N.SF.AL.). *Used in herbal remedies in the past.* Page 161.

*A. microcarpa (Boiss. & Reuter) Rothm. Widespread in Europe.

RUBUS | **Blackberry, Bramble** Herbs or shrub-like plants with woody stems which often die back after flowering. Fls. usually in terminal and axillary clusters; epicalyx absent; sepals and petals 5; stamens numerous. Fr. of numerous one-seeded fleshy carpels on a domed non-fleshy receptacle. 75 sps.

Lvs. simple
426. *R. chamaemorus* L. CLOUDBERRY. An arctic or alpine perenn. with creeping rhizomes and ann. leafy stems 5–20 cm., bearing solitary white fls. *c.* 2 cm. across. Fls. one-sexed; petals much longer than sepals. Lvs. heart-shaped with 5–7 shallow, rounded, toothed lobes; stipules oval, papery; stems smooth, without prickles. Fr. first red, then orange. △ Moors, bogs. June–Aug. North Eur. (except IS.) IRL.GB.D.PL.CS.SU. Pl. 44.

1. *Sibbaldia procumbens* 458
3. *A. alpina* 424
5. *P. alba* 447

2. *Alchemilla pentaphyllea* [424]
4. *Potentilla norvegica* 454
6. *P. crantzii* [455]

7. *Aphanes arvensis* 425

Lvs. compound
(a) Stems herbaceous
427. *R. saxatilis* L. ROCK BRAMBLE. A perenn. with above-ground stolons, rooting at the tips, and ann. fl. stem 8–40 cm., with 2–8 white fls. in a rather compact terminal cluster. Petals 3–5 mm., narrow, usually shorter than the sepals. Lvs. with 3 coarsely toothed, oval leaflets; stipules oval, green; stem not woody, with or without weak prickles. Carpels 2–6, scarlet, translucent. △ Northern or alpine: rocky places, thickets, woods. May–Aug. Most of Eur. (except P.TR.).

R. arcticus L. ARCTIC BRAMBLE. Like 427, but smaller, with underground creeping stems and erect stems to 15 cm. bearing solitary bright pink or red fls., with petals 7–10 mm. long, and longer than the sepals. Fr. dark red. Northern Europe. Pl. 43.

(b) Stems woody, with prickles
428. *R. idaeus* L. RASPBERRY. A suckering shrub with many stiff, erect, usually prickly, reddish stems 1–2 m., and rather dense clusters of white fls. on lateral shoots forming a compound inflorescence. Petals *c.* 5 mm., widely spaced, about as long as the sepals. Lvs. pinnate with 3–7 oval, toothed, long-pointed leaflets 5–12 cm., green above and densely white-hairy beneath. Carpels red, opaque, rarely pale yellow, hairy, on a conical receptacle. △ Woods, in mountains. May–July. All Eur. (except P.IS.TR.). *Cultivated varieties are widely grown for their edible fruits which are used for jam, preserves, and wine-making; the leaves were used in herbal remedies.*

429. *R. fruticosus* L. agg. BLACKBERRY. Stems usually angled, prickly, becoming woody and lasting two years and often rooting at the tips. Fls. white to pink, 2–3 cm. across, in compound terminal clusters; petals usually not overlapping. Lvs. usually over-wintering, usually with 5 oval, toothed leaflets 5–12 cm., with prickles on lv. stalks and midveins. Fr. black, coming away from receptacle when ripe. An extremely variable and complex aggregate of plants with *c.* 2,000 microspecies recognized. △ Woods, thickets, hedges, banks, heaths. May–Sept. All Eur. *The ripe fruits are edible and are used for jams, preserves and wine-making.*

R. caesius L. DEWBERRY. Like 429, but stems rounded, weak, rooting, with a purplish bloom; prickles scattered, very weak. Lvs. with 3 oval, coarsely toothed leaflets; stipules lance-shaped. Fls. white or pinkish, *c.* 2 cm. across. Fr. with a purplish bloom and usually with few, very fleshy carpels. Widespread in Europe. Pl. 44.

ROSA | Rose Erect or arching shrubs usually with prickles; lvs. pinnate; stipules present. Sepals and petals 5; stamens numerous. Fr. with fleshy, flask-shaped receptacle surrounding many one-seeded carpels which are borne on its inner wall. Readily hybridizing and forming a difficult complex of microspecies. 46 sps.

Leaflets 5–7
(a) Sepals usually entire; styles fused
430. *R. arvensis* Hudson FIELD ROSE. Distinguished from most other sps. by the styles which are fused together in a column which is at least as long as the shortest stamen. A trailing shrub, forming low bushes ½–1 m., with white fls. 3–5 cm. across; sepals oval, long-pointed; fl. stalks usually with stalked glands. Prickles sparse, all hooked. Lvs. deciduous, leaflets 1½–4 cm., oval-elliptic, with teeth usually simple, hairless or finely hairy. Fr. small, 1–1½ cm., red, usually globular. △ Thickets, hedges, rocks. June–July. Much of Eur. (except North Eur. P.). Pl. 45.

R. sempervirens L. Lvs. evergreen, leathery, shining, long-pointed. Fls. in flat-topped clusters of 3–7; petals white, usually 1–2 cm.; style column usually hairy. Mediterranean and South-Western Europe.

(b) Sepals with lateral lobes; styles free
431. *R. gallica* L. Fls. large, 6–9 cm. across, usually solitary, sweet-scented, bright rose or purple, with long fl. stalks with dense, stalked glands. Styles free, but included in receptacle, stigmas projecting. Prickles usually very unequal. Lvs. rather leathery, pale bluish-green, hairy and with prominent veins beneath, usually double-toothed. Sepals reflexed after flowering, falling in fr. which is globular to spindle-shaped and densely glandular-hairy. A very variable sp. A small shrub 40–80 cm. △ Hedges, woods; often grown as an ornamental. May–June. Much of Eur. (except North Eur. IRL.GB.NL.). *This and other species are cultivated in Eastern Europe to produce 'Essence of Rose of the Balkans' which is used in perfumery. The distilled essence is made into pomades and gargles and used against inflammations. Oil of Roses is obtained from the flowers of other species such as R. moschata and R. damascena. Rose hips are rich in vitamin C.*

432. *R. canina* L. DOG ROSE. A rather tall shrub 1–5 m., with clusters of 1–4 scentless white or pink fls. 4–5 cm. across on fl. stalks which are usually without stalked glands. Sepals reflexed after flowering and falling before the fr. reddens; styles free. Prickles stout, all strongly curved. Leaflets 1½–4 cm., usually hairless, and without glands on either surface. Fr. 1½–2 cm. globular, hairless, scarlet. A very variable sp. △ Woods, hedges, tracksides, thickets. May–July. All Eur. (except IS.TR.).

*R. villosa L. APPLE ROSE. Central and Southern Europe.

*R. tomentosa Sm. DOWNY ROSE. Most of Europe.

433. *R. rubiginosa* L. (*R. eglanteria* L.) SWEET BRIAR. Distinguished by the hairless or finely hairy leaflets which are covered, particularly beneath, with conspicuous russet-coloured sticky glands and which are apple-scented when crushed. Fls. bright pink, sweet-scented, *c.* 3 cm. across; fl. stalks short, with dense, stalked glands; sepals erect or spreading after flowering, persisting at least until the fr. reddens. Prickles stout, hooked, usually unequal. Fr. 1–1½ cm., globose or ovoid, scarlet, smooth or glandular hairy at base. An erect shrub ½–3 m. △ Stony dry places, hedges. June–July. Most of Eur. (except P.IS.SF.AL.).

Leaflets 9–11, rarely 7
434. *R. pimpinellifolia* L. (*R. spinosissima* auct.) BURNET ROSE. Distinguished by its stems which are densely covered with numerous unequal slender straight prickles mixed with many stiff bristles, and its small rounded-elliptic leaflets ½–1½ cm., conspicuously saw-toothed, usually sparingly hairy and without glands. Fls. solitary, creamy-white, 2½–5 cm. across; fl. stalks smooth. Fr. globular, *c.* 6 mm., blackish-purple when ripe. A low erect shrub 10–100 cm., forming large patches. △ Rocks, dunes, cliff tops, often near the sea. May–June. Most of Eur. (except P.SF.TR.). Pl. 45.

435. *R. pendulina* L. (*R. alpina* L.) ALPINE ROSE. Fls. bright carmine, 3½–5 cm. across, solitary, on reddish branches, usually without prickles; fl. stalks long, smooth or glandular-hairy, arched in fr. Leaflets oblong-oval, double-toothed, usually dark green, finely hairy, without glands above, paler and more sparsely hairy beneath. Fr. bright red, 2–2½ cm. oblong-ovoid and narrowed to a bottle-like neck, often glandular-hairy, usually pendulous, with persistent sepals which often have a leafy tip. A low, often dwarf shrub ½–2 m. △ Rocks, shady places, woods in mountains. June–July. Cent. Eur., South-East Eur. (except TR.) E.B.F.I.SU. Pl. 45.

AGRIMONIA | Agrimony   Fls. numerous, in long spikes; sepals 5, curved inwards in fr.; epicalyx absent; petals 5; stamens 10–20. Fr. of 2 carpels encircled by a hard receptacle with a ring of hooked bristles and forming a burr. 4 sps.

436.   *A. eupatoria* L. AGRIMONY. A leafy perenn. to 60 cm., with a long, slender, terminal, leafless spike of numerous small stalkless yellow fls. each 5–8 mm. across. Lvs. pinnate with 3–6 pairs of larger, oval, saw-toothed leaflets 2–6 cm., alternating with smaller leaflets; lvs. and stem with dense, spreading, non-glandular hairs; stipules leaf-like. Fr. *c.* 6 mm. obconical, deeply grooved, hairy, with spreading hooked bristles above. △ Wood verges, thickets, hedges, meadows. June–Sept. All Eur. (except IS.). *A powerfully astringent and tonic plant; it yields a yellow dye.*

*A. procera Wallr. (*A. odorata* auct.) FRAGRANT AGRIMONY. Widespread in Central Europe.

AREMONIA   Like *Agrimonia*, but fls. few in a cluster, and each fl. surrounded by a lobed involucre. Epicalyx present. 1 sp.

437.   *A. agrimonoides* (L.) DC. BASTARD AGRIMONY. A perenn. herb to 40 cm., with pinnate lvs. with 2–4 pairs of large toothed leaflets alternating with smaller leaflets and lax, few-flowered, leafy clusters of yellow fls. Fls. 7–10 mm. across; involucre of 6–10 sepal-like lobes; stamens 5–10. Fr. without spines. △ Woods in mountains. May–June. South-East Eur. A.CS.H.I.: introd. GB.CH.

SANGUISORBA | Burnet   Fls. bisexual or unisexual, in dense, globular, terminal heads; sepals 4; petals absent; stamens 4 to numerous; carpels 1–2. Sometimes included in *Poterium*. 7 sps.

438.   *S. officinalis* L. GREAT BURNET. A slender, hairless, erect perenn. 20–100 cm., with long-stemmed, leafless, dark red, ovoid fl. heads 1–2 cm. long. Fls. very densely crowded; stamens shorter or little longer than the 4 red sepals. Lvs. pinnate with 7–15 oval, toothed, long-stalked leaflets 2–4 cm., glaucous beneath. Fr. with corky receptacle without spines. △ Damp fields, marshes, valleys. June–Sept. Most of Eur. (except IS.): introd. SF. Pl. 45.

439.   *S. minor* Scop. SALAD BURNET. Like 438, but fl. heads globular *c.* 1 cm., greenish, often flushed purple on the sunny side; uppermost fls. female, with 2 feathery stigmas, the middle fls. often bisexual with 4 stamens, the lowest fls. male with 20–30 long hanging stamens. Lvs. pinnate with 4–12 pairs of oval saw-toothed, shortly stalked leaflets *c.* 1 cm. A perenn. smelling of cucumbers when crushed; fl. stems branched, 15–60 cm. △ Grasslands. May–Sept. Most of Eur. (except IS.N.). *The leaves are sometimes used for flavouring salads, soups, etc.*

SARCOPOTERIUM   Like *Sanguisorba*, but a spiny shrub. Fls. one-sexed. Fr. fleshy, with 2 nuts. 1 sp.

440.   *S. spinosum* (L.) Spach (*Poterium s.*) THORNY BURNET. A low rounded shrub, 30–60 cm., with stiff, greyish, much-interwoven spiny branches and globular heads of reddish fls. *c.* ½ cm. across. Male heads with numerous long yellow stamens; female heads with purple feathery stigmas. Lvs. pinnate with 4–7 pairs of tiny toothed leaflets *c.* 6 mm., falling in summer. Fr. swollen, fleshy, berry-like, often turning bright red. △ Dry hills, abandoned cultivation. Mar.–Apr. I.YU.AL.GR.TR. Pl. 46.

DRYAS | Mountain Avens   Shrublets with evergreen, toothed lvs. Fls. with 7–10 sepals and petals; epicalyx absent. Fr. of numerous carpels with long feathery styles. 1 sp.

441. *D. octopetala* L. MOUNTAIN AVENS. A much-branched arctic or alpine creeping shrub-let, often forming extensive mats, with large, solitary, pure-white fls. on short leafless stems 2–8 cm. and glossy oval lvs. Fls. 2½–4 cm. across, usually with 8 oblong petals; sepals with blackish glandular hairs. Lvs. ½–2 cm., margin with conspicuous rounded teeth, green above and densely white-woolly beneath. Carpels with long feathery persistent styles. △ Screes, rock ledges, rocky pastures. May–Aug. Much of Eur. (except P.B.NL.DK.H.GR.TR.SU.). Pl. 45.

GEUM | Avens   Fr. of numerous carpels, each with a long, slender, persistent, hooked or feathery style. Epicalyx present; sepals and petals 5; stamens numerous. Lvs. pinnate. 13 sps.

Fr. with hooked styles
442. *G. urbanum* L. HERB BENNET, WOOD AVENS. An erect, little-branched woodland perenn. to 60 cm., with rather few yellow, long-stalked, erect fls. in a lax inflorescence. Petals 5–9 mm. oval, usually not overlapping; sepals becoming reflexed. Basal lvs. pinnate with 2–3 pairs of oval toothed or lobed leaflets ½–1 cm., and a larger terminal lobe 5–8 cm.; uppermost stem lvs. simple; stipules similar to lv. lobes. Carpels numerous, hairy, with long crooked styles which break off near the apex leaving a hook. △ Woods, thickets, hedgerows, shady places. May–Sept. Most of Eur. (except IS.).

443. *G. rivale* L. WATER AVENS. An erect, hairy, little-branched perenn. 20–60 cm., with a few rather large, nodding, yellowish-red fls. *c.* 2 cm. long, and pinnate lvs. Sepals reddish-purple, as long as and pressed to the petals; petals 1–1½ cm. long, spathulate, notched, dull orange-pink, pressed to the stamens. Basal lvs. with 3–6 pairs of unequal oval, toothed leaflets and a larger terminal leaflet; upper lvs. simple or lobed; stipules small. Fr. with crooked styles. △ Marshes, watersides, wet fields, bushy places. May–July. Most of Eur. (except P.H.GR.TR.). Pl. 45.

Fr. with long feathery-haired styles, not hooked
444. *G. montanum* L. (*Sieversia m.*) ALPINE AVENS. A perenn. with a stout rhizome, without stolons, with solitary bright yellow fls. 2–3 cm. across, on erect unbranched, almost leafless stems 5–40 cm. Petals usually 6, slightly notched. Lvs. mostly basal, with very unequal shallow-toothed lobes, the terminal lobe rounded-heart-shaped, much larger than the lateral lobes which become progressively smaller below. Fr. with long feathery styles, recalling those of *Pulsatilla*. △ In mountains: rocky places, pastures, open woods. July–Aug. Cent. Eur. (except H.) E.F.I.YU.AL.BG.R.SU.

G. reptans L. (*Sieversia r.*) CREEPING AVENS. Like 444, but with long reddish stolons creeping above-ground to 80 cm. Lvs. more or less regularly and equally lobed from apex to base, the terminal lobe deeply three-lobed, all lobes deeply saw-toothed. Fls. larger, paler, 3–4 cm. across. Mountains of Central Europe.

POTENTILLA | Cinquefoil   Fr. of numerous dry carpels on a dry or spongy, domed-shaped receptacle; styles not elongating. Epicalyx of sepal-like lobes, alternating with the sepals; sepals and petals usually 5; stamens numerous. Usually herbaceous perenns. with palmate, pinnate, or trifoliate lvs. 76 sps.

Fls. white, pink, or reddish
(a) Basal lvs. with 3 leaflets
445. *P. sterilis* (L.) Garcke BARREN STRAWBERRY. A softly hairy perenn. with a lax rosette of lvs. and slender, sparsely leafy stems 5–15 cm. bearing 1–3 white fls. 1–1½ cm. across.

Petals obcordate, notched, widely spaced, as long as or a little longer than the hairy sepals. Lvs. trifoliate with leaflets $\frac{1}{2}$–$2\frac{1}{2}$ cm., toothed, bluish-green, hairy above, paler with silky spreading hairs beneath, the stem lvs. few, smaller; above-ground stolons present. Fr. non-fleshy, yellowish. △ Wood margins, coppices, hedges. Mar.–May. West, Cent. Eur. (except H.) DK.S.I.YU.AL.

P. micrantha DC. PINK BARREN STRAWBERRY. Like 445, but fls. pinkish, 6–8 mm. across, 1–2 on slender leafless stems much shorter than the lvs; petals shorter than sepals. Stolons absent. Much of Central and Southern Europe.

P. nitida L. PINK CINQUEFOIL. A low, tufted, woody-based perenn. 2–10 cm. of the mountains, with large, beautiful, pale or clear pink, rarely white, usually solitary fls. *c.* $2\frac{1}{2}$ cm. across. Basal lvs. usually with 3 leaflets which are silvery-haired above and below, with toothed or entire apex. Limestone mountains of the Alps. Pl. 46.

(b) Basal lvs. with 5 or more leaflets

446. *P. palustris* (L.) Scop. (*Comarum p.*) MARSH CINQUEFOIL. A perenn. with woody, creeping and rooting, reddish stems, and erect. leafy stems 15–45 cm., bearing rather few dark reddish-purple fls. *c.* 2–3 cm. across in a lax leafy cluster. Petals much smaller than the purplish, oval, long-pointed, spreading sepals; epicalyx linear, much smaller than the sepals; stamens and carpels deep purple. Lvs. stalked, palmate, with 5 or 7 oblong, coarsely toothed leaflets 3–6 cm., glaucous beneath; stipules conspicuous, papery. Fr. hairless. △ Marshes, bogs, fens. June–Aug. Much of Eur. (except P.AL.GR.TR.). Pl. 46.

447. *P. alba* L. WHITE CINQUEFOIL. A slender hairy perenn. with shortly creeping stems and slender erect stems shorter than the lvs., bearing few white fls. *c.* $1\frac{1}{2}$ cm. across. Petals notched, longer than the sepals. Lower lvs. palmate with 5 lance-shaped leaflets, green and hairless with ciliate margin above and with dense, adpressed, silvery hairs beneath; stem lvs. few, trifoliate. △ Open woods, poor pastures, rocky places. Apr.–May. Cent. Eur. F.I.YU.BG.R.SU. Page 161.

P. caulescens L. SHRUBBY WHITE CINQUEFOIL. A tufted, woody-based perenn. with lvs. with 5–7 silvery-hairy leaflets and a cluster of white fls. on weak leafy stems longer than the lvs. Petals narrow, wedge-shaped, widely spaced with calyx visible between, and slightly longer than the more or less equal calyx and epicalyx; filaments and carpels hairy. Mountains of Spain to Yugoslavia.

448. *P. rupestris* L. ROCK CINQUEFOIL. An erect, branched, somewhat glandular-hairy perenn. to 40 cm., with a lax, branched cluster of pure white fls. each 1–2 cm. across. Petals obovate 5–7 mm., longer than the calyx. Basal lvs. pinnate, with 5–7 oval, double-toothed leaflets 2–6 cm., uppermost lvs. trifoliate; bracts simple. Fr. shining, hairless. A variable sp. △ In mountains: rocks, woody slopes. June–Aug. Much of Eur. (except IRL.NL.IS.DK.SF.). Pl. 46.

Fls. yellow

(a) Lvs. palmate with 3–5–7 leaflets

(i) Lowland or hill plants

449. *P. erecta* (L.) Räuschel COMMON TORMENTIL. A slender perenn. 10–50 cm., with thick woody stock and bright yellow fls. with characteristically only 4 petals on most of the fls. Fl. clusters lax; fls. 7–11 mm. across, long-stalked; petals notched, usually longer than the calyx. Stem lvs. stalkless, with 3 obovate, coarsely toothed leaflets, 1–2 cm., the basal lvs. ternate, long-stalked; stipules leaf-like, deeply palmately lobed. △ Heaths, bogs, grasslands, mountains, woods. May–Aug. All Eur. (except GR.TR.).

450. *P. reptans* L. CREEPING CINQUEFOIL. A perenn. with a central rosette and stems

creeping extensively to 1 m. and rooting at nodes and bearing solitary, axillary, bright golden-yellow fls., 1½–2½ cm. across on long slender stalks. Petals 5, notched, to twice as long as calyx. Lvs. all stalked, usually with 5 oblong, toothed leaflets ½–3 cm.; stipules leafy, usually entire. △ Hedge banks, waste places, tracksides, damp meadows. June–July. All Eur. (except IS.).

451. *P. tabernaemontani* Ascherson (*P. verna* auct.) SPRING CINQUEFOIL. A very variable hairy perenn. 5–20 cm., with prostrate spreading stems, usually rooting and mat-forming, bearing a lax cluster, little longer than the lvs., of few, bright yellow fls. each 1–1½ cm. across. Basal lvs. palmate with 5 obovate leaflets ½–2 cm., toothed, with the terminal tooth smaller than the lateral teeth; stipules with the free part linear. △ Grasslands, rocky places, banks, waysides, sometimes alpine. Mar.–June. Cent. Eur. E.F.GB.B.DK.NL.N.S.I. BG.R.SU. Pl. 46.

452. *P. argentea* L. HOARY CINQUEFOIL. Distinguished by its dense woolly-white stems and somewhat leathery, dark green lvs. which are woolly-white beneath; leaflets 5 pinnately lobed or toothed, margin inrolled. Fls. yellow, *c.* 1–1½ cm. across, numerous in spreading clusters; petals oval, not notched, as long as the woolly-white sepals. An erect, branched variable perenn. 15–50 cm. △ Dry banks, waysides. May–Aug. Most of Eur. (except P.IRL.IS.).

453. *P. recta* L. SULPHUR CINQUEFOIL An erect hairy perenn. to 70 cm., with lower lvs. palmate with 5–7 large, narrow lance-shaped leaflets 5–10 cm. long, which are regularly and conspicuously toothed; leaflets green, with adpressed hairs on both sides. Fls. yellow, 2–2½ cm. across in many-flowered clusters; petals as long as or much longer than the densely glandular-hairy sepals. A very variable sp. △ Rocky slopes, bushy places, wood verges. June–July. Cent., South-East Eur. E.F.I.R.SU.: introd. GB.B.SF.NL.N.S.

454. *P. norvegica* L. NORWEGIAN CINQUEFOIL. Rather like 453, but fls. smaller, *c.* 1½ cm. across, and petals inconspicuous, widely spaced and shorter than the sepals. Fls. numerous; sepals 7–8 mm. long, enlarging in fr. to 1½–2 cm. Lvs. mostly with 3 coarsely toothed elliptic leaflets, green, with rough, erect hairs on both sides. A very variable, erect, rather robust ann. to perenn. 20–50 cm. △ Waste places. June–Aug. North Eur. (except IS.) D.A.CS.PL.I.R.SU.: introd. F.GB.B.NL.CH.YU. Page 161.

(ii) Alpine plants

455. *P. aurea* L. GOLDEN CINQUEFOIL. An alpine perenn. 5–25 cm., distinguished from 451 by the leaflets which are bordered with silvery silky hairs, and with the terminal tooth much smaller than the lateral teeth. Fls. 1½–2 cm. across, often darker yellow in the centre; sepals with silvery hairs. Plant not mat-forming. △ Mountain pastures. June–Aug. Cent., South-East Eur. (except TR.) E.F.I.R.SU.

P. crantzii (Crantz) Fritsch ALPINE CINQUEFOIL. Very like 455, but leaflets not bordered with silky hairs and with only 2–4 teeth on each side with the terminal tooth nearly as large as the lateral teeth. Fls. 1–2½ cm. across, very rarely darker yellow in the centre. Much of Europe. Page 161.

(b) Lvs. at least some pinnate

456. *P. anserina* L. SILVERWEED. A perenn. with a central rosette of pinnate lvs, and long creeping, often reddish, above-ground stolons to 80 cm. which root at the nodes. Fls. solitary, bright yellow, 2–3 cm. across, borne on long stalks from the axils of the lvs.; petals 5, rounded. Lvs. 5–25 cm., with 6–12 pairs of larger oval-oblong leaflets, 1–6 cm., alternating with smaller leaflets, all deeply saw-toothed and usually green above, silvery-silky-haired beneath. △ Ditches, tracksides, damp pastures, waste places, dunes. May–July. Much of Eur. (except AL.GR.TR.).

457. *P. fruticosa* L. SHRUBBY CINQUEFOIL. A small, creeping or erect, much-branched shrub to 1 m., with pinnate lvs. usually with 5 lance-shaped entire leaflets 1–2 cm. long, with inrolled margin, hairy above, densely woolly-haired beneath; stipules papery, persisting. Fls. yellow, 2–2½ cm., solitary or few in a terminal cluster; petals rounded, longer than sepals. Branches and young lvs. silvery-haired; bark peeling. △ Damp rocky ground, in mountains. June–Aug. E.F.IRL.GB.S.I.BG.SU.: introd. N. Pl. 46.

SIBBALDIA   Like *Potentilla*, but stamens 5 or rarely 4 or 10; carpels 5–12. Lvs. palmate or trifoliate. 2 sps.

458. *S. procumbens* L. A low, creeping, tufted, woody-based arctic or alpine perenn. to 5 cm., with rosettes of trifoliate lvs. and dull yellow fls. *c.* ½ cm. across in dense lateral clusters, often shorter than lvs. Petals very narrow or absent. Leaflets oval *c.* 1½ cm., stalked, three-toothed at apex, bluish-green, sparsely adpressed hairy, often purplish beneath. △ Damp rocks, screes, alpine pastures, snowy valleys. July–Aug. North Eur. (except DK.), Cent. Eur. E.F.GB.YU.BG.SU. Page 161.

FRAGARIA | **Strawberry**   Distinguished from *Potentilla* by the fr. in which the receptacle becomes enlarged, fleshy and brightly coloured and the carpels are small and dry. Herbaceous plants with above-ground stolons; lvs. trifoliate. 5 sps.

459. *F. vesca* L. WILD STRAWBERRY. A perenn. 5–30 cm., with numerous long-spreading, above-ground stolons which root at the nodes and produce new plants, and leafless stems bearing a lax cluster of white fls. each *c.* 1½ cm. across. Lvs. trifoliate with 3 oval, coarsely toothed, almost stalkless leaflets, bright green and hairy above, pale, glaucous, with adpressed silky hairs beneath. Stem about as long as the lvs.; fl. stalks with adpressed hairs. Fr. 1–2 cm., globular, bright red, with numerous small green carpels. △ Coppices, hedges, scrub. May–June. All Eur. *The edible fruits are used for jams, beverages, etc.; the roots and leaves are astringent and used as an infusion.*

F. moschata Duchesne HAUTBOIS STRAWBERRY. Like 459, but fls. larger, 1½–2½ cm. across, borne on stems longer than the lvs.; fl. stalks with spreading hairs. Leaflets shortly stalked; stolons present. Fr. reddish, or often greenish and sterile. Much of Central and South-Eastern Europe; naturalized elsewhere.

F. viridis Duchesne Like 459, but stolons short or absent; sepals pressed against fr. at maturity, not spreading or deflexed; fr. not readily detached from receptacle. Much of Europe, except in the north-west and south-east.

F. × ananassa Duchesne (*F. chiloensis × virginiana*) GARDEN STRAWBERRY. Distinguished by its large fls. 2–3½ cm. across; large fr. *c.* 3 cm. long, with carpels sunk in the fleshy receptacle. Leaflets hairless above, stalked. A garden hybrid widely cultivated for its fr. and often occurring as a casual in Europe.

MESPILUS | **Medlar**   Fls. large, solitary. Carpels 5, almost completely united. Fr. at first hard, then fleshy, crowned with persistent sepals. 1 sp.

460. *M. germanica* L. MEDLAR. A deciduous, sometimes spiny shrub or tree to 8 m., with large almost stalkless fls. 3–4 cm. across, borne at the ends of the branches. Sepals lance-shaped, long-pointed, leafy, woolly, longer than the rounded petals; anthers red. Lvs. large, 5–12 cm., oblong entire, dull green, softly hairy and very short-stalked. Fr. 2–3 cm., flattened-globular, open at top, brownish. △ Hedges, and woods. May–June. I.GR.

BG.SU.: introd. E.F.GB.B.NL.D.CH.A.CS.YU.R. *Cultivated for its fruits which are only edible when over-ripe. The wood is hard and flexible and used for whip handles.*

CRATAEGUS | Hawthorn Fls. in dense, flat-topped, umbel-like clusters; sepals and petals 5; stamens numerous. Carpels 1–5, fused below, but free above. Fr. with mealy flesh and with 1 or several hard nuts. Trees or shrubs usually with thorns; lvs. lobed or toothed. 21 sps.

461. *C. monogyna* Jacq. HAWTHORN. A spiny, much-branched shrub or small tree to 5 m., with numerous flat-topped clusters of strong-smelling white fls. with pink or purple anthers. Fls. up to *c.* 16 in a cluster; petals 4–6 mm. long; style usually 1. Lvs. oval, 1½–3½ cm., with 3–7 deep triangular lobes which are longer than broad and with entire or sparsely toothed margin; lvs. of long shoots with conspicuous leafy stipules. Fr. oval, 8–10 mm. long, deep red; nut 1. △ Thickets, hedges, woods. May–June. All Eur. (except IS.). *Commonly planted to make impenetrable thorn hedges which can readily be cut and trimmed. Used as a stock for medlar and pear grafts.*

C. laevigata (Poiret) DC. (*C. oxyacanthoides* Thuill.) Differs from 461 in having styles mostly 2–3; and fr. with 2–3 nuts. Lvs. particularly on dwarf lateral shoots distinctive, with 3–5 shallow, rounded-toothed lobes which are broader than long. Readily hybridizes with 461. Widespread in Central Europe.

462. *C. azarolus* L. AZAROLE, MEDITERRANEAN MEDLAR. A shrub or small tree to 10 m., distinguished by the white-downy hairs on the young branches, lvs., and fl. stalks and few or absent spines. Fls. in dense clusters, 5–8 cm. across; fls. white, 1½ cm. across; anthers purple. Lvs. 4–7½ cm., oval to wedge-shaped, deeply three- to five-lobed, lobes oblong entire or with a few teeth towards the apex, downy beneath. Fr. 2–2½ cm., yellow or orange-red, with 1–3 nuts. △ Woods, hedges. Apr.–May. Native of Crete: introd. E.F.I.YU. *Cultivated for its edible apple-flavoured fruits in Southern Europe.* Page 171.

CYDONIA | Quince Small trees without spines; lvs. entire. Fls. solitary; sepals and petals 5; stamens numerous; carpels 5, inferior, fused with the receptacle. Fr. hard, pear-like, with persistent sepals; seeds numerous in each carpel. 1 sp.

463. *C. oblonga* Miller QUINCE. A small tree 2–8 m., with large white or pink, solitary, and stalkless fls. 4–5 cm. across, and large oval lvs. Petals 5, orbicular; sepals leaf-like. Lvs. 5–10 cm., entire, very short-stalked, hairless above, woolly-white beneath. Fr. large, 6–10 cm. long, woolly, green becoming yellow and aromatic when fully ripe. △ Hedges; widely cultivated for its fr. May. Native of Asia; introd. to most of Eur. (except North Eur. IRL.GB.B.NL.PL.). *The fruit is only edible when cooked; it is widely used to make preserves and jams. The seeds contain much mucilage, and have been used medicinally. Used as a dwarfing stock for pears.* Pl. 47.

PYRUS | Pear Like *Sorbus*, but branches often spiny. Fls. larger, in simple umbels; anthers purple; styles free to their base. Fr. large, fleshy, and gritty. 13 sps.

464. *P. communis* L. PEAR. A somewhat pyramidal tree 12–20 m., with grey scaly bark and rather large white fls. in few-flowered umbels. Petals 1–1½ cm.; anthers purple. Lvs. variable, 8–10 cm., oval-acute, margin entire or weakly toothed, woolly when young, but becoming hairless above. Fr. 5–15 cm., pear-shaped or globular, with persistent calyx, brownish, sweet-tasting. △ Woods, hedges; widely cultivated and widely naturalized. Apr.–May. Almost throughout Europe. *Cultivated since classical times. Used for dessert,*

marmalades, as dried fruit, and for making alcoholic beverages and spirits. The wood of old trees is much in demand by cabinet- and instrument-makers.

P. pyraster Burgsd. WILD PEAR. Like 464, but usually with spiny branches and lvs. smaller, 3–6½ cm., orbicular, with a short- or long-pointed apex, margin entire except for teeth towards the apex. Fr. 1–3 cm., globular or pear-shaped, black when ripe, not sweet. Much of Europe, except the north.

MALUS | Apple   Like *Pyrus* but fr. without gritty flesh. Anthers yellow; styles fused below. 6 sps.

465. M. domestica Borkh. APPLE. A small to medium-sized tree with grey scaly bark, spineless branches, and woolly-haired twigs. Fls. white or pink, 3–5 cm. across, in woolly-stalked clusters; anthers yellow. Lvs. 4–13 cm., oval-elliptic, toothed, slightly woolly-haired above, densely so beneath. Fr. more than 5 cm., often sweet. Of hybrid origin. △ Hedges, verges of fields; cultivated almost throughout Europe and often naturalized. Apr.–May. *Cultivated since classical times. The fruit is used for dessert, preserves, jams, beverages, and is fermented as cider and distilled for liqueurs; it is rich in vitamin C and is also a source of pectin. The wood is used in cabinet-making and for carving.*

M. silvestris Miller CRAB APPLE. Like 465, but branches usually thorny, and twigs and buds hairless or loosely hairy only when young. Lvs. only hairy when young, soon quite hairless. Fl. stalks hairless or nearly so; petals 2 cm. or less. Fr. 2–2½ cm., usually green, sour. Almost throughout Europe.

SORBUS   Non-spiny trees or shrubs with deciduous, simple, or pinnate lvs. Fls. in compound umbels. Ovary of 2–5 carpels, half-embedded in the receptacle, each with 2 ovules, and with cartilaginous walls. Fr. fleshy, with 1–2 seeds in each cell. Often included in *Pyrus*. 18 sps.

Lvs. pinnate, leaflets toothed
466. S. aucuparia L. ROWAN, MOUNTAIN ASH. A slender tree to *c.* 20 m., with smooth greyish bark, pinnate lvs., and dense, flat-topped clusters of white fls. Inflorescence stalkless, woolly-haired; fls. *c.* 1 cm. across; styles 3–4. Lvs. 10–15 cm.; leaflets 9–19, oblong acute; 3–6 cm., finely saw-toothed, dark green above, paler beneath; buds hairy. Fr. globular, 6–9 mm., scarlet or yellowish-red. A variable sp. △ Woods, thickets. May–June. All Eur. (except TR.). *The fruits are used for preserves, and fermented for alcoholic beverages. The bark has been used for tanning and dyeing; the wood for turnery and carving.* Pl. 48.

S. domestica L. SERVICE TREE. Like 466, but fls. larger, 1½–2 cm. across, white or pink; styles 5. Lv. buds sticky, shiny, hairless. Fr. distinctive, *c.* 2½ cm. long, apple- or pear-shaped, green or brownish. Southern Europe; cultivated for its fr. and naturalized further north. *The fruits are only edible after being frozen or just past maturity; when green they are extremely bitter and irritant. An alcoholic beverage is made from the fruits; the wood is used in turnery, carving, and cabinet-making.* Page 171.

Lvs. entire or lobed, margin toothed
467. S. aria (L.) Crantz WHITEBEAM. A tree to 25 m. with a wide dense crown and dark grey bark, and with lvs. conspicuously white beneath and fls. in dense, white, flat-topped clusters 8–10 cm. across. Fls. *c.* 12 mm. across; petals oval; styles 2. Lvs. variable-oval, 5–12 cm., double saw-toothed, strongly veined, yellow-green, hairless or nearly so above, densely white-felted beneath. Fr. globular-ovoid, 8–15 mm., scarlet. △ Rocky places, dry woods, scrub. May–June. Much of Eur. (except North Eur. NL.GR.TR.). Pl. 47.

1. *Spiraea hypericifolia* [419]
2. *Crataegus azarolus* 462
3. *Sorbus chamaemespilus* 468
4. *S. domestica* [466]
5. *Cotoneaster nebrodensis* 474

468. *S. chamaemespilus* (L.) Crantz FALSE MEDLAR. A compact shrub 1–3 m., distinguished from 467 by the densely packed clusters of pale or dark pink fls. with erect petals, and its oblong or elliptic unequally toothed lvs. which are shiny above and green and hairless beneath, though hairy when young. Fr. orange-red, *c.* 1 cm. △ Rocky slopes, open woods. May–July. Cent., Eur. (except H.), South-East Eur. (except TR.) E.F.I. Page 171.

469. *S. torminalis* (L.) Crantz WILD SERVICE TREE. Distinguished by its conspicuously lobed, oval-heart-shaped lvs. 7–10 cm., which have 3–4 broad, triangular, long-pointed, saw-toothed lobes on each side, green above and below. Fls. white, *c.* 1 cm. across; petals almost circular; styles 2. Buds greenish, hairless; twigs woolly when young. Fr. brown, *c.* 1½ cm., with numerous corky spots. A tree with a wide crown and with dark grey, shallow-fissured bark; 10–25 m. △ Woods, rocky slopes. Apr.–May. Much of Eur. (except North Eur. IRL.NL.) DK. *The fruits are rich in vitamin C.; they are used to make conserves.*

ERIOBOTRYA | **Loquat**   Small evergreen trees with large, leathery, entire lvs. Fls. in branched terminal clusters; fr. plum-like, with 2 large seeds. 1 sp.

470. *E. japonica* (Thunb.) Lindley LOQUAT. A small evergreen tree to 10 m., with large, stiff, strongly veined, oblong-lance-shaped, toothed lvs. 15–30 cm., dark glossy green above, grey- or rusty-haired beneath. Fls. yellowish white, *c.* 1 cm. across, sweet-scented, densely clustered on rusty-haired branches. Fr. ovoid, 4 cm. long, golden-yellow. △ Native of China; widely cultivated for its edible fr. in the Mediterranean region. Nov.–Apr. Pl. 48.

AMELANCHIER   Deciduous shrubs or small trees without spines; lvs. entire, toothed. Fls. usually in clusters; sepals and petals 5, petals narrow; stamens numerous. Carpels half-encircled by the receptacle, each with 2 ovules. Fr. usually fleshy and sweet. 3 sps.

471. *A. ovalis* Medicus SNOWY MESPILUS. A shrub 1–3 m., with small, toothed, oval lvs. and erect clusters of pale yellowish, long-stalked fls. with distinctive, narrow, widely spaced petals. Fls. long-stalked, in flat-topped clusters; petals 1½–2 cm., woolly-haired beneath. Lvs. small, 2–4 cm., broadly oval, finely toothed, at first very woolly-haired beneath, later hairless; twigs downy; buds smooth and shining. Fr. *c.* 1 cm., bluish-black, sweet. △ Rocky slopes, banks, thickets, in mountains. Apr.–May. Much of Eur. (except North Eur. IRL.GB.NL.). Pl. 47.

472. *A. grandiflora* Rehder (*A. confusa* Hyl.; *A. laevis* auct.) SERVICEBERRY. A small tree to 9 m. or a shrub, with slender, spreading, or drooping clusters of white fls. with narrow, conspicuously widely spaced petals 1½–2 cm. which are hairless beneath. Lvs. 3–7 cm., oval-elliptic, finely toothed, purplish, hairy when young, but soon hairless. Fr. globular, *c.* 1½ cm., blackish purple. △ Native of North America; sometimes naturalized. Apr.–May. Introd. F.GB.B.NL.D.

COTONEASTER   Thornless shrubs with entire lvs. Fls. small, white or pinkish; sepals and petals 5; stamens 20; carpels 2–5, free on the inner side and fused outside to the receptacle. Fr. somewhat fleshy, red, purple, or black. 11 sps.

473. *C. integerrimus* Medicus A small twisted shrub 30–100 cm., with short clusters of 1–4 whitish or pink fls. in the axils of the small oval lvs. Fls. drooping after flowering; sepals hairless; petals *c.* 3 mm. Lvs. oval, 1½–4 cm., deciduous, green and hairless above, densely grey-woolly beneath. Twigs woolly when young. Fr. globular, *c.* 6 mm., pendulous, shining red, hairless. △ Rocks and screes in hills. Apr.–June. Much of Eur. (except P.IRL. NL.IS.TR.). Pl. 48.

474. *C. nebrodensis* (Guss.) C. Koch Like 473, but lvs. larger, broadly oval, blunt, hairless or finely hairy above, densely white-woolly beneath. Fls. 3–5 in small erect umbels; fl. stalks and sepals densely white-woolly. Fr. erect, bright red, hairy. Shrub 1–2 m. △ Rocks, slopes of hills. May–June. Cent., South-East Eur. F.I.: introd. DK.N.S. Page 171.

PYRACANTHA Like *Cotoneaster*, but spiny shrubs with numerous fls. in flat-topped clusters. Styles 5; fr. fleshy, containing 5 carpels. 1 sp.

475. *P. coccinea* M. J. Roemer FIRE THORN. A dense, very spiny shrub 1–2 m. with finely toothed evergreen lvs. and numerous flat-topped clusters of white or pinkish-yellow fls. Fls. *c.* 1 cm. across. Lvs. elliptic, 2–3 cm., leathery, shining, hairless; young twigs downy. Fr. globular, somewhat flattened above and below, scarlet, often persisting throughout the winter. △ Hedges, thickets; grown as an ornamental and sometimes naturalized further north. May–June. Med. Eur. P.BG.SU.

PRUNUS Spineless or spiny, deciduous or evergreen trees or shrubs with simple lvs. Fls. solitary or in spike-like or flat-topped clusters; sepals and petals 5; stamens often 20; carpel 1, style terminal; receptacle cup-shaped. Fr. fleshy and with a stony nut, one-seeded. 21 sps.

Fls. solitary or 2–3
(a) Fr. plum-like, with a waxy bloom
476. *P. spinosa* L. BLACKTHORN, SLOE. A very spiny, suckering shrub ½–4 m., with rather densely clustered, pure white fls. usually appearing before the lvs. Fls. axillary, usually solitary; fl. stalks hairless; petals 5–8 mm. Lvs. 2–4 cm., oval, toothed, finely hairy; young twigs hairy. Fr. globular 1–1½ cm., blue-black, with a whitish bloom, flesh green. △ Hedges, thickets, woods, banks. Apr.–May. All Eur. (except IS.). *The fruits are inedible and extremely astringent, but they can be fermented with the broken nuts to make a liqueur. The wood is hard and is used for walking sticks and tool handles.* Pl. 49.

477. *P. domestica* L. PLUM. A small tree to 12 m. or shrub, with large greenish-white fls. *c.* 2–3 cm. across appearing with the lvs. Fls. solitary or 2–3; petals elliptic 1–1½ cm. Lvs. 4–10 cm., dull, at first hairy on both sides, but becoming hairless above; branches usually thornless, young branches hairy. Fr. 2–4 cm., ovoid to oblong, larger in cultivated forms, varying from blue-black to greenish-yellow; stone flattened, somewhat pitted, sharply angled. Subsp. *insititia* (L.) C. K. Schneider BULLACE has twigs densely and conspicuously hairy and remaining so into the second year. Fr. globular, 2–3 cm., purple or blue-black, with waxy bloom; stone ovoid, rough. △ Hedges, thickets; often escaping from cultivation. Apr.–May. All Eur. (except IS.). *Widely cultivated for its fruits which are eaten fresh or dried, used for conserves or fermented and distilled for liqueurs. The wood is hard and compact and is used by turners and cabinet-makers. The seeds give a bitter oil.*

P. cerasifera Ehrh. CHERRY-PLUM. Distinguished from 477 by the young branches which are always hairless and its rather glossy lvs. which are hairless above. Fls. white, 1½–2½ cm. across, drooping. Fr. 2–2½ cm., globose, yellow or red; stone round, smooth. South-Eastern Europe; often planted for hedges and naturalized from cultivation in Central Europe. Pl. 48.

(b) Fr. peach-like, densely hairy
478. *P. persica* (L.) Batsch PEACH. A small deciduous tree to 6 m., with slender, hairless, greenish or pinkish, spineless twigs and usually solitary, stalkless, pink fls. 1½–4 cm. across. Lvs. 8–15 cm., narrow lance-shaped, long-pointed, finely toothed, folded longitudinally when young. Fr. globular, 5–8 cm., velvety-hairy, yellow or reddish, flesh

succulent, white or yellow; stone deeply pitted, rough. △ Possibly native of Eastern Asia; widely planted for its frs. in Southern Europe and grown as an ornamental further north. Mar.–Apr. *The fruits are eaten fresh, used for preserves, or fermented for wines and spirits. The kernels produce an oil which may be sweet or bitter and is used in perfumery.*

P. armeniaca L. APRICOT. Distinguished from 478 by the wider, oval-heart-shaped, shining, deep green, mature lvs. 6–9 cm. and young lvs. which are rolled, not folded in bud. Fls. white, flushed with pink. Fr. globular, c. 3 cm., yellow flushed with red, flesh yellow. A small spineless tree 3–6 m. Native of Asia; widely cultivated in Southern Europe. *The fruits are eaten fresh, dried, or made into conserves and alcoholic beverages. The kernels produce a sweet edible oil which is sometimes used in place of almond oil.*

479. P. dulcis (Miller) D. A. Webb (P. amygdalus Batsch) ALMOND. A small spineless tree to 10 m., distinguished from 478 by its frs. which have a hard inedible flesh, which splits to the stone on maturity. Fls. white or pink, 3½–5 cm. across. Fr. elliptical, 4–6½ cm. long, greenish, dry, velvety; nut almost smooth, with narrow fissures. △ Native of Western Asia; widely cultivated and often naturalized in Southern Europe and grown as an ornamental further north. Feb.–Apr. *The seeds are edible and are used for dessert and confectionery; they also yield Almond oil which is widely used in soaps, perfumery, and in medicine. Different varieties of Almond yield either a sweet and edible oil or a bitter and poisonous oil. The wood is used for veneering.*

Fls. numerous, clustered
(a) *Fls. in flat-topped umbel-like clusters*
480. P. avium (L.) L. GEAN, WILD CHERRY. A tree to 25 m. with shiny red-brown, horizontally peeling bark and white fls. in stalkless clusters of 2–6. Fls. 1½–2 cm. across, long-stalked; petals obovate and narrowed to the base, notched; bud scales subtending umbel mostly without leaf-like blades. Lvs. 6–15 cm., elliptic, long-pointed, toothed, dull pale green, with adpressed hairs beneath. Fr. globular, c. 1 cm., dark red or blackish, sweet or sour. △ Woods, hedges; widely cultivated for its fr. Apr.–May. All Eur. (except IS.AL.). *The edible fruit is used for conserves, alcoholic beverages, and liqueurs. The wood is used by cabinet- and instrument-makers.* Pl. 49.

481. P. cerasus L. SOUR CHERRY. Like 480, but usually a suckering shrub 1–5 m., with dark green, rather glossy, abruptly sharp-pointed lvs. 5–8 cm., almost hairless beneath. Petals white or pinkish, rounded at base, often notched at apex. Inner scales subtending umbels with leaf-like blades. Fr. bright red, or rarely black or yellowish, sour. △ Widely planted for its fr. and often naturalized. Apr.–May. Native of Asia: introd. to most of Eur.

(b) *Fls. in elongated spike-like clusters*
482. P. padus L. BIRD-CHERRY. A tree to 15 m. with sweet-scented white fls. forming long, usually drooping clusters 7–15 cm. Inflorescence borne laterally, with 1–3 lvs. at its base; petals toothed, 4–10 mm. Lvs. thin oval-elliptic, 8–12 cm., long-pointed, finely toothed, deciduous; lv. stalk 1–2 cm. Bark brown, strong-smelling. Fr. globular, 6–8 mm., becoming black. △ Woods. Apr.–June. Most of Eur. (except IS.AL.GR.TR.). *In northern countries the fruits are fermented to make an alcoholic beverage.* Pl. 49.

483. P. laurocerasus L. COMMON or CHERRY-LAUREL. An evergreen shrub or small tree to 6 m., with large, glossy, thick leathery, dark green lvs., pale green twigs, and white fls. in long, slender, erect spikes. Inflorescence 5–12 cm., dense, leafless at the base, shorter than the subtending lv.; petals c. 4 mm. Lvs. oblong-acute, 5–18 cm., entire or with few shallow teeth; lvs. stalk ½–1 cm. Fr. ovoid-conical, c. 8 mm. long, shining black-purple. △ Thickets; often grown for ornament and sometimes naturalized. Apr.–May. YU.TR.BG.:

introd. F.GB.P. *A poisonous plant; all parts contain prussic acid, hence the crushed leaves have been used by entomologists to kill insects; the leaves are used for making wreaths.* Pl. 47.

P. lusitanica L. PORTUGAL LAUREL. Like 483, but inflorescence 5–25 cm., longer than the lvs. Lvs. 6–12 cm., oval, regularly saw-toothed, very dark glossy green above; lv. stalk 2–2½ cm., without glands. Twigs and lv. stalks reddish. Native of the Iberian Peninsula; sometimes planted elsewhere for ornament. Pl. 48.

484. *P. mahaleb* L. ST LUCIE'S CHERRY. A deciduous, much-branched, spreading shrub or small tree 1–12 m., with rather few fragrant white fls. in an erect cluster 3½–5 cm. long, with several lvs. at the base. Fls. usually 4–8, long-stalked; petals 8 mm. Lvs. small, bright glossy green, 1½–6½ cm. broadly oval, acute or blunt, finely toothed, hairy beneath. Fr. oval c. ½ cm., black, bitter. △ Shrubby slopes, open woods, rocks. Apr.–May. Much of Eur. (except North Eur. IRL.GB.NL.PL.TR.): introd. N.S. *The wood is highly prized by cabinet- and toy-makers; the branches are used for cherry-wood pipes and walking sticks. The fruits are sometimes used to make a liqueur. The plant contains coumain and is used in perfumery. Used as a stock for grafting cherries.* Pl. 49.

LEGUMINOSAE | Pea Family

Trees, shrubs, or herbaceous plants usually with compound lvs. and often with tendrils; stipules present. Fls. usually with 5 sepals and 5 petals which are symmetrical in one plane only and 'butterfly-like' with the upper petal forming a broad *standard*, the 2 lateral petals forming spreading *wings*, and the 2 lower petals partly fused into a *keel* which encloses the ovary and stamens, or fls. radially symmetrical. Stamens usually 10, all fused together by their stalks to form a tube, or 9 fused and 1 free, or stamens numerous, free. Carpel 1 with marginal placentation. Fr. usually a dry carpel or *legume*, splitting longitudinally into 2 valves. Usually sub-divided into 3 sub-families: MIMOSOIDEAE with numerous clustered, radially symmetrical fls. and numerous conspicuous stamens usually much longer than the petals: CAESALPINIOIDEAE with butterfly-like fls. but stamens not fused by their stalks, and PAPILIONATAE with butterfly-like fls. and stamens fused by their stalks.

1. Stamens not fused into a tube (or if somewhat fused at base then fls. radially symmetrical); fls. usually radially symmetrical.   Group A
1. Stamens with stalks more or less fused into a tube; fls. symmetrical in one plane only, butterfly-like, with a standard, wings, and a keel which encircles the stamens and ovary.
  2. Lvs. with more than 3 leaflets and palmately or pinnately arranged.
    3. Lvs. pinnate, without a terminal leaflet, but often with a terminal tendril or spine.   Group B
    3. Lvs. pinnate with a terminal leaflet, or palmate.   Group C
  2. Lvs. with 1 or 3 leaflets, or lvs. apparently absent.   Group D

GROUP A. *Stamens not fused*

                                      Nos.

1. Stamens numerous; fls. radially symmetrical.
  2. Stamens pink.   *Albizia*   493
  2. Stamens yellow.   *Acacia*   488–492

LEGUMINOSAE

1. Stamens 10 or less.
3. Fls. green, inconspicuous, radially symmetrical.
   4. Spineless trees.   *Ceratonia*   486
   4. Spiny trees.   *Gleditsia*   487
3. Fls. coloured, conspicuous, symmetrical in 1 plane
   only.
   5. Fls. yellow; lvs. trifoliate.   *Anagyris*   495
   5. Fls. not yellow; lvs. not trifoliate.
      6. Fls. pink; lvs. not divided, entire.   *Cercis*   485
      6. Fls. creamy-white; lvs. pinnate.   *Sophora*   494

GROUP B. *Lvs. pinnate and without a terminal leaflet, but often with a tendril or spine*

		Nos.
1. Stems and lvs. glandular-hairy.	*Cicer*	541
1. Stems and lvs. not glandular-hairy.		
2. Stems winged.	*Lathyrus*	553–565
2. Stems not winged.		
3. Leaflets with parallel veins.	*Lathyrus*	553
3. Leaflets pinnately veined.		
4. Calyx teeth all equal, at least twice as long as the calyx tube.	*Lens*	552
4. All calyx teeth, or at least 2, less than twice as long as the calyx tube.		
5. Calyx teeth more or less leafy; stipules large, leafy, 2 cm. or more.	*Pisum*	566
5. Calyx teeth not leafy; stipules less than 2 cm.		
6. Style hairy all round, or on the lower side only, or hairless.	*Vicia*	542–551
6. Style hairy on the upper side only.	*Lathyrus*	553–565

GROUP C. *Lvs. pinnate with a terminal leaflet, or palmate*

		Nos.
1. Principal lateral veins of leaflets terminating at the margin, often toothed.		
2. Hairless or nearly so, without glands.	*Trifolium*	590–608
2. Hairy, glandular.		
3. Calyx swollen at base; stipules free from lv. stalk.	*Cicer*	541
3. Calyx not swollen at base; stipules fused to lv. stalk.	*Ononis*	567–572
1. Lateral veins branching and not reaching the margin.		
4. At least some fls. in terminal clusters.		
5. Lvs. palmate.	*Lupinus*	519–521
5. Lvs. pinnate.		
6. Climbing shrubs; fl. clusters pendulous.	*Wisteria*	[523]
6. Herbs or small shrubs; fls. in erect clusters.	*Anthyllis*	619–623
4. All fls. axillary, or in axillary clusters.		
7. Plant glandular, at least in part.		
8. Tree; fl. clusters pendulous.	*Robinia*	523
8. Herbs; fl. clusters erect.	*Glycyrrhiza*	536
7. Plant not glandular.		

9. Fls. in umbels or clusters with the fl. stalks arising
from approximately the same point.
  10. Fr. splitting at transverse joints.
    11. Segments of fr. horseshoe-shaped or half-moon-
shaped.   *Hippocrepis*   631, 632
    11. Segments of fr. linear or oblong, straight or
slightly curved.   *Coronilla*   624–630
  10. Fr. splitting longitudinally or not splitting.
    12. Keel prolonged into a beak.
      13. Fr. flattened, disk-like; lower lvs. simple.   *Hymenocarpus*   618
      13. Fr. linear or oblong; lvs. with 2 pairs of leaflets. *Lotus*   612–615
    12. Keel not beaked.
      14. Keel dark red or black.   *Dorycnium*   609–611
      14. Keel not dark red or black, but variously
coloured.   *Astragalus*   526–532
9. Fls. in elongate or condensed clusters with lv. stalk
arising at different levels, or fls. solitary.
  15. Leaflets with parallel veins.   *Lathyrus*   553–565
  15. Leaflets with pinnate veins.
    16. Fr. splitting at transverse joints.
      17. Fls. in clusters of 8 or more.   *Hedysarum*   634, 635
      17. Fls. solitary or few, axillary.   *Hippocrepis*   631, 632
    16. Fr. not splitting by transverse joints.
      18. Fr. not splitting, rounded, usually toothed or
spiny.   *Onobrychis*   636, 637
      18. Fr. splitting longitudinally.
        19. Fl. clusters 10 cm. or more, pendulous; trees. *Robinia*   523
        19. Fl. clusters usually less than 10 cm., shrubs or
herbaceous plants.
          20. Fr. very inflated, membraneous.
            21. Shrubs to 2 m. or more.   *Colutea*   525
            21. Stemless herbs.   *Astragalus*   526–532
          20. Fr. not, or slightly inflated, not mem-
braneous.
            22. Keel beaked.   *Lotus*   612–615
             22. Keel not beaked, but sometimes with a
fine narrow point.
              23. Keel with a fine narrow point at apex. *Oxytropis*   533–535
              23. Keel without a fine point.
                24. Stamens with stalks all united into a
tube; fls. white, pale violet or pinkish. *Galega*   524
                24. Stamens with 9 stalks fused and 1 free.
                  25. Style hairless.   *Astragalus*   526–532
                  25. Style hairy on lower side.   *Vicia*   542–551

GROUP D. *Lvs. simple or with 1 or 3 leaflets, or lvs. sometimes minute*

                                               Nos.

1. Principal lateral veins of the leaflets terminating at
the margin; leaflets often toothed.
  2. Plants glandular-hairy, at least above.

3. Stamens with stalks all fused into a tube; fr. straight or slightly curved. *Ononis*   567–572

3. Stamens with 9 stalks fused and 1 free; fr. sickle-shaped or spirally coiled. *Medicago*   581–589

2. Plant not glandular-hairy.

  4. At least some of the stalks (or claws) of petals fused to the stamen tube; corolla persisting in fr. *Trifolium*   590–608

  4. Petal stalks (or claws) not fused to the stamen tube; corolla not persisting in fr.

    5. Stalks of at least 5 stamens swollen at apex. *Trifolium*   590–608

    5. Stalks of stamens all slender.

      6. Fr. coiled in 1 or more turns of a spiral. *Medicago*   581–589

      6. Fr. straight or curved.

        7. Perenns.

          8. Fr. oval to globular. *Melilotus*   573–576

          8. Fr. oblong, sickle-shaped or kidney-shaped or curved. *Medicago*   581–589

        7. Anns. or bienns.

          9. Fls. blue. *Trigonella*   577–580

          9. Fls. white or yellow.

            10. Fr. linear or oblong, at least 3 times as long as broad. *Trigonella*   577–580

            10. Fr. oval, globular or kidney-shaped, less than 3 times as long as broad.

              11. Fr. kidney-shaped. *Medicago*   581–589

              11. Fr. oval or globular.

                12. Fr. without a beak or membraneous wing. *Melilotus*   573–576

                12. Fr. with a long curved beak or with broad membraneous wing. *Trigonella*   577–580

1. Principal lateral veins branching and not reaching margin; leaflets not toothed, sometimes very small or reduced to a spine.

13. Plant spiny.

  14. Fls. pink, red, purple, or blue-violet; lvs. *c.* ½ cm. *Erinacea*   [515]

  14. Fls. yellow.

    15. Adult lvs. reduced to persistent spines. *Ulex*   516, 517

    15. Adult lvs. often falling, not spine-tipped.

      16. Calyx tubular, with 5 short teeth, the upper portion breaking away and leaving a cup-like remnant. *Calicotome*   498

      16. Upper part of calyx not breaking away.

        17. Calyx with 5 equal teeth or two-lipped and the upper lip not deeply divided. *Anthyllis*   619–623

        17. Calyx two-lipped, the upper lip deeply divided into 2. *Genista*   508–512

13. Plant not spiny.

  18. Young stems broadly winged. *Chamaespartium*   513

  18. Young stems not broadly winged.

    19. Lvs. with stipules; lvs. trifoliate.

      20. Corolla not more than 7 mm.; plant with reddish-brown hairs. *Glycine*   540

20. Corolla 10 mm. or more; plant hairless or with
    whitish hairs.   *Phaseolus*   538, 539
19. Lvs. without stipules; lvs. simple or trifoliate.
21. Fr. and usually the calyx with prominent glan-
    dular swellings.   *Adenocarpus*   518
21. Fr. and calyx without glandular swellings.
  22. Lvs. simple or with 1 leaflet, sometimes very
      small.
    23. Anns.
      24. Lvs. linear, grass-like; fr. splitting.   *Lathyrus*   553–565
      24. Lvs. obovate or elliptic; fr. not splitting.   *Scorpiurus*   633
    23. Shrubs or woody-based perenn. herbs.
      25. Calyx falling after fls. open.   *Lygos*   514
      25. Calyx persisting.
        26. Calyx spathe-like, split to the base.   *Spartium*   515
        26. Calyx not spathe-like, not split to base.
          27. Calyx more or less tubular; fr. included
             within persistent calyx.   *Anthyllis*   619–623
          27. Calyx bell-shaped; fr. longer than calyx or
             calyx not persisting.
            28. Upper lip of calyx with short teeth.   *Cytisus*   500–505
            28. Upper lip of calyx deeply divided or
               toothed.
              29. Fr. ovoid, oblong or sickle-shaped,
                 splitting.   *Genista*   508–512
              29. Fr. globular-inflated, not or tardily
                 splitting.   *Lygos*   514
  22. At least some lvs. trifoliate.
    30. Leaflets conspicuously gland-dotted.   *Psoralea*   537
    30. Leaflets not or very minutely gland-dotted.
      31. Fr. splitting at transverse joints.   *Coronilla*   624–630
      31. Fr. not splitting transversely at the joints.
        32. Ann. or perenn. herbs.
          33. Calyx inflated, 4½–6 mm. wide in fl., up to
             12 mm. wide in fr. and enclosing fr.   *Anthyllis*   619–623
          33. Calyx less than 4½ mm. wide, not inflated.
            34. Keel very dark red or black.   *Dorycnium*   609–611
            34. Keel variously coloured, but not as above.
              35. Filaments of stamens all fused into a
                tube; stipules not fused to lv. stalk. *Argyrolobium*   522
              35. Filaments of stamens with 9 fused
                and 1 free; stipules fused to base of lv.
                stalk.
                36. Stipules inserted at the base of the lv.
                   stalk; fr. not longitudinally winged. *Lotus*   612–615
                36. Stipules inserted on the stem, and fused
                   to the base of the lv. stalk; fr. with 2–4
                   longitudinal wings.   *Tetragonolobus*   616, 617
        32. Trees or shrubs.
          37. Fr. included in the persistent calyx; calyx
             with 5 more or less equal teeth.   *Anthyllis*   619–623

37. Fr. longer than calyx, or calyx deciduous;
    calyx two-lipped.
  38. Fls. in pendulous clusters.   *Laburnum*   496, 497
  38. Fl. clusters erect.
    39. Upper lip of calyx deeply two-lobed.
      40. Calyx tube distinctly shorter than lips. *Argyrolobium*   522
      40. Calyx tube as long as or longer than the
          lips.   *Genista*   508–512
    39. Upper lip of calyx with 2 short teeth.
      41. Calyx tubular.   *Chamaecytisus*   506, 507
      41. Calyx bell-shaped.
        42. Fls. axillary, arranged in leafy clusters. *Cytisus*   500–505
        42. Fls. in leafless terminal heads or
          clusters.
          43. Twigs hairy.   *Lembotropis*   499
          43. Twigs hairless.   *Cytisus*   500–505

CERCIS | **Judas Tree** Lvs. undivided, kidney-shaped. Fls. appearing in clusters from woody branches; sepals 5, fused; petals 5, conspicuous, nearly equal; stamens 10, free. 1 sp.

485. *C. siliquastrum* L. JUDAS TREE. A small spineless tree to 10 m., with finely fissured bark, rounded kidney-shaped lvs., and clusters of bright rose-purple fls. arising abundantly from young and old branches, usually before the lvs. Fls. *c.* 2 cm., butterfly-like; calyx reddish-purple, five-toothed; petals nearly equal. Fr. to 10 cm., flattened, reddish-brown. △ Rocky hills; widely grown as ornamental. Mar.–May. F.I.YU.AL.GR.TR.BG.: introd. P.E.SU. Pl. 50.

CERATONIA | **Carob** Fls. green, in clusters arising from old branches; petals absent; stamens 5; disk swollen. Lvs. pinnate. 1 sp.

486. *C. siliqua* L. CAROB, LOCUST TREE. A spreading evergreen tree 7–10 m., with stout trunk and dark green compound lvs. with 2–5 pairs of shining, leathery, oval leaflets, 2½–7½ cm. Fls. numerous in elongated green clusters to 15 cm., arising from old branches and the trunk. Fr. very large, thick, leathery, to 20 cm. by 2 cm., occurring on old branches throughout the summer. △ Rocky places in coastal regions; widely cultivated in the Mediterranean region. Aug.–Oct. E.F.I.YU.AL.: introd. P. *The pods contain about 55 per cent of sugar and 10 per cent protein and are a source of food to humans in times of shortage; they are valuable fodder for cattle, pigs and horses. The seeds were the origin of the 'carat' weights of the goldsmiths and Locust Bean gum is extracted from them. The wood is used in cabinet-making and marquetry.* Pl. 49.

GLEDITSIA Fls. regular, green; sepals and petals 3–5, inconspicuous and more or less equal; stamens 6–10. Trees with spines; lvs. once or twice pinnate. 1 sp.

487. *G. triacanthos* L. HONEY LOCUST, THREE-THORNED ACACIA. A tree to *c.* 30 m., with inconspicuous, fragrant, greenish fls., branches with stout spines and once- or twice-pinnate lvs. Male fls. in downy, narrow clusters to 5 cm.; female clusters shorter. Lvs. 10–15 cm., usually pinnate with 18 or more dark-green, oblong-to-lance-shaped, toothed leaflets, 1–3½ cm. long. Spines often three-forked, 7–10 cm. long. Fr. leathery, dark brown, twisted, 20–40 cm. by 3–3½ cm. △ Native of North America; sometimes grown for ornament. June–July. *The fruits are used to feed sheep and the seeds are sometimes eaten by humans.*

ACACIA Trees or shrubs with twice-pinnate lvs. or with lvs. of adult plants reduced to a single flattened blade or *phyllode*. Fls. regular, numerous in dense clusters. Sepals and petals 5; stamens numerous, free, much longer than the small corolla. Fr. linear. 10 sps.

Lvs. absent; phyllodes lance-shaped, entire
488. *A. longifolia* (Andrews) Willd. WHITE SALLOW, SYDNEY GOLDEN WATTLE. Fls. bright yellow in long, cylindrical, stalkless clusters, 2–6 cm., arising from the axils of the phyllodes. Phyllodes 5–15 cm., lance-shaped, recalling willow lvs. A tall shrub or small tree to 10 m. △ Native of Australia; widely planted as an ornamental, and for stabilizing sandy soil; sometimes naturalized. Mar.–Apr. Introd. P.E.F.I. Pl. 49.

489. *A. cyanophylla* Lindley BLUE-LEAVED WATTLE. Fls. golden-yellow in shortly stalked clusters of 2–6, heads globular *c.* 1 cm. across, arising from the axils of the phyllodes and forming long leafy clusters 15–30 cm. Phyllodes bluish-green, linear to curved lance-shaped, 15–30 cm. A shrub or small tree to 6 m. △ Native of Australia; planted as an ornamental and for stabilizing sand; sometimes naturalized. Apr.–May. Introd. P.E.F.I.GR.

Lvs. present, twice pinnate
(a) Phyllodes absent
490. *A. dealbata* Link SILVER WATTLE, MIMOSA. Fls. deep yellow, fragrant, in globular heads forming simple or often compound inflorescences. Lvs. silvery-grey, twice pinnate with 13–25 pairs of primary divisions, each with 30–50 pairs of tiny, silvery-white-haired leaflets *c.* 3 mm. A tree with whitish trunk and downy branchlets. △ Native of Australia; commonly planted as an ornamental and often naturalized. Early spring. Introd. P.E.F.I.YU.R.

491. *A. farnesiana* (L.) Willd. POPINAC, OPOPANAX, CASSIE, HUISACHE. A much-branched shrub or small tree to 6 m., with zigzag spiny branches. Fl. heads deep yellow, very fragrant, in groups of 2–3 at the ends of the branches. Lvs. twice pinnate with 5–8 pairs of primary divisions and 10–25 pairs of linear hairless leaflets 3–6 mm.; spines straight, slender, 1–2½ cm. △ Native of West Indies; grown as an ornamental. Early spring. Introd. E.F.I. *The flowers produce the sweet-scented oil, Cassie, which is used in perfumery.*

(b) Phyllodes present
492. *A. melanoxylon* R.Br. BLACKWOOD ACACIA. A dense pyramidal tree to 40 m., with both compound lvs. and phyllodes present and axillary clusters of small cream-coloured fls. in rounded heads *c.* 1 cm. across. Lvs. twice-pinnate; phyllodes scimitar-shaped, 6–12 cm. Seeds scarlet. △ Native of Australia; planted for timber and locally naturalized. Early spring. Introd. P.E.F.I. *The wood is fine-grained and used for furniture and house construction.*

ALBIZIA Similar to *Acacia*, but stamens fused into a tube at the base; fr. not splitting into 2 valves. Lvs. twice pinnate. 1 sp.

493. *A. julibrissin* Durazz. PERSIAN ACACIA, PINK SIRIS. A deciduous tree to 12 m., with large, twice-pinnate lvs. and pink fls. in globular clusters crowded at the ends of the branches. Stamens conspicuous, pink, *c.* 2½ cm. long. Lvs. 22–45 cm. with 6–12 pairs of primary divisions and 20–30 pairs of oblong-acute leaflets 1½–2 cm. △ Native of Asia and tropical Africa; planted as an ornamental in the Mediterranean region. July–Aug. Pl. 50.

SOPHORA Fls. irregular, butterfly-like; calyx five-lobed; standard orbicular; stamens 10, not fused. Lvs. pinnate. 3 sps.

494. *S. japonica* L. PAGODA-TREE. A deciduous tree to 30 m., with pinnate lvs. and creamy-white fls. in large, erect, pyramidal clusters to 30 cm. Fls. *c.* 1½ cm. Lvs. 18–25 cm., with

9–17 oval leaflets each 2½–5 cm., dark glossy green above, greyish with downy adpressed hairs beneath. Fr. 5–8 cm. △ Native of China and Japan; frequently grown for ornament. July–Aug.

ANAGYRIS Fls. irregular, butterfly-like, yellow; stamens 10, not fused. Fr. not splitting; seeds separated by cross-walls. Lvs. trifoliate, fetid when crushed. 1 sp.

495. *A. foetida* L. BEAN TREFOIL. A deciduous shrub 1–3 m. with green stems, trifoliate lvs. and clusters of yellowish laburnum-like fls. Fls. 2½ cm.; standard about half as long as wings, with a blackish blotch at base; calyx bell-shaped, hairy. Leaflets 3, narrow elliptic, 2½–6 cm., silvery-haired beneath, fetid. Fr. 10–18 cm.; seeds violet or yellow. △ Rocky places, dry stony hills. Feb.–Apr. Med. Eur. P. *The whole plant is poisonous and in particular the seeds.*

LABURNUM Fls. yellow, numerous in drooping clusters; calyẍ bell-shaped, shortly two-lipped. Small trees, with smooth bark. Often included in *Cytisus*. 2 sps.

496. *L. anagyroides* Medicus (*Cytisus laburnum* L.) GOLDEN RAIN, LABURNUM. A small tree to 7 m., with smooth branches and long, hanging, leafless clusters of numerous yellow fls. streaked with brown, borne on last year's branches. Fls. *c.* 2 cm., long-stalked; calyx finely hairy. Lvs. long-stalked, trifoliate; leaflets elliptic, dark green, hairless above, paler glaucous, hairy beneath. Young branches, lv. stalks and young fr. grey-haired. Fr. 3–5 cm. △ In mountains: woods, bushy places; often grown for ornament and sometimes naturalized. May–June. F.CH.D.A.CS.H.I.YU.R.: introd. IRL.GB. *One of the most poisonous of European trees; all parts contain the poisons cytisine and laburnine and the seeds are particularly rich in these. Carrying twigs or flowers in the mouth may cause poisoning and affect respiration; treatment should be sought immediately.* Pl. 50.

497. *L. alpinum* (Miller) Berchtold and J. Presl ALPINE LABURNUM. Like 496, but lvs. hairless beneath except on midvein and margin, and inflorescence axis with lax spreading hairs. Fls. paler yellow streaked with brown, smaller *c.* 1½ cm.; calyx with lower lip larger than the undivided upper lip. Branches, lv. stalks, and fr. hairless. A small tree 3–5 m. △ Woods in mountains. May–July. F.CH.A.CS.I.YU.AL.

CALICOTOME (CALYCOTOME) Very spiny shrubs. Calyx with the upper part which is shed by a circular incision when the buds break, leaving a five-toothed, cup-shaped base. Fls. yellow. Fr. not splitting, several-seeded. 3 sps.

498. *C. villosa* (Poiret) Link SPINY BROOM. A very spiny, dense-branched shrub to 1 m. with thinly cottony-haired stems and yellow fls. usually in clusters of 5–15. Fls. 1–2 cm.; fl. stalks about as long as calyx. Lvs. trifoliate, blackening when dried; leaflets obovate 5–10 mm., softly hairy beneath; spines 1–2 cm. Fr. 2½–3 cm. with densely woolly, more or less spreading hairs. △ Sunny slopes, rocky places. Mar.–June. P.E.F.I.YU.AL.GR.TR. Pl. 51.

C. spinosa (L.) Link Like 498, but fls. usually borne singly, occasionally in clusters; fr. hairless. Spain to Yugoslavia.

LEMBOTROPIS Shrubs without spines. Fls. in terminal leafless spikes; petals yellow becoming black when dry; calyx bell-shaped, two-lipped. Seeds without a swelling or *strophiole*. 1 sp.

499. *L. nigricans* (L.) Griseb. (*Cytisus n.*) BLACK BROOM. An erect, slender shrub 1–2 m. with trifoliate lvs. and long, terminal, spike-like clusters of many short-stalked yellow fls. Petals 7–10 mm., wings shorter than keel; calyx hairy. Leaflets obovate to linear, 1–3 cm., dark green above, pale green beneath, hairy when young; stems hairy. Fr. 2–3½ cm., with adpressed hairs. △ Rocky and bushy places. June–July. Cent., South-East Eur. I.SU. Pl. 51.

CYTISUS | **Broom**   Shrubs without spines; lvs. with 1 or 3 leaflets, usually alternate. Fls. axillary and forming leafy terminal clusters; calyx often bell-shaped, two-lipped with teeth usually much shorter than the lips. Seeds usually with a swelling or *strophiole*. 21 sps.

Lvs. simple

500. *C. decumbens* (Durande) Spach (*Genista prostrata* Lam.) A spreading shrublet to 30 cm. with five-ribbed, hairy stems, simple lvs. and yellow fls. in the axils of the upper lvs. Fls. 1–1½ cm., 1–3 on stalks longer than the subtending lvs.; fl. stalks and calyx with spreading hairs. Lvs. obovate to lance-shaped 8–20 mm., almost hairless above, but margin and underside hairy; branches with crisped hairs between the ribs. Fr. 2–3 cm., usually with spreading hairs. △ Margins of woods, grassy places, banks. Apr.–July. F.CH.I.YU.AL.

Lvs. mostly trifoliate

501. *C. villosus* Pourret (*C. triflorus* L'Hér.) A stiff, erect, branched shrub 1–2 m., with rather large yellow fls. streaked with reddish-brown, solitary or 2–3 in each lv. axil and forming a lax, leafy cluster. Fls. 1½–2 cm.; petals with narrow base (or claw) distinctly longer than the hairy calyx; keel longer than the wings. Lvs. trifoliate; leaflets 1½–3 cm., elliptic, fine-pointed, with adpressed hairs; branches five-angled, hairy. Fr. 2–4½ cm., with dense adpressed hairs, but becoming hairless. △ Thickets, bushy places. Apr.–May. Med. Eur. (except TR.).

502. *C. purgans* (L.) Boiss. A rather small, densely tufted shrub 30–100 cm., with numerous short, rigid, hairless, often leafless branches bearing 1–2 pale yellow, vanilla-scented fls. at the ends of the branches. Standard 10–12 mm. rounded, often notched, hairless; calyx silvery-haired. Lvs. trifoliate or one-bladed; leaflets linear-lance-shaped, ½–1 cm., hairless above, silvery-haired beneath. Fr. 1½–3 cm., adpressed-hairy, black. △ Bushy places, uncultivated ground, rocks. June–Aug. P.E.F.

503. *C. multiflorus* (L'Hér.) Sweet (*C. albus* (Lam.) Link) WHITE SPANISH BROOM. An erect, much-branched shrub to 3 m., with long interrupted clusters of many white fls. each 1–1½ cm. long. Standard elliptic, hairless; keel blunt; fl. stalk *c.* 1 cm. Lvs. stalkless, trifoliate, or one-bladed; leaflets to 1 cm., linear-lance-shaped, silvery-haired. Fr. 1½–2½ cm., with 4–5 seeds. Recalling 514 *Lygos monosperma*, but distinguished by fr. and calyx. △ Dry fields, heaths, hills. Apr.–June. P.E.

504. *C. sessilifolius* L. A hairless shrub to 2 m., with green, very leafy branches and terminal, rather few-flowered, short, lax clusters of yellow fls. each 1–1½ cm. Standard rounded; keel beaked; calyx two-lipped. Lower lvs. long-stalked, lvs. of fl. shoots stalkless; leaflets 1–2 cm., broadly oval or rhomboidal, fine-pointed. Fr. 2½–4 cm. by 1 cm., hairless, brown. △ Hills, woods, banks. Apr.–June. E.F.I. Pl. 51.

505. *C. scoparius* (L.) Link (*Sarothamnus s.*) BROOM. A much-branched, erect shrub to 2½ m. with straight, stiff, five-angled, green branches and large golden-yellow fls. in lax, terminal, leafy clusters. Petals *c.* 2 cm., standard broad, notched; calyx hairless; style conspicuously coiled after release from keel. Lower lvs. trifoliate, the upper one-bladed;

leaflets elliptic, but soon falling and many branches leafless. Fr. 2½–5 cm. by 8–10 mm., hairy on margin, black. △ Heaths, open places in woods. Apr.–June. Much of Eur. (except IS.SF.AL.GR.TR.BG.). *A drug, obtained from the twigs, is used medicinally for heart and respiratory conditions. The plant is poisonous to livestock. The branches are used for brooms, thatching, and screens; the bark has been used to make rope and also for tanning.* Pl. 51.

CHAMAECYTISUS Shrubs usually without spines, but branches sometimes somewhat spiny; lvs. trifoliate. Fls. in axillary clusters forming leafy spikes or in terminal heads subtended by a leafy involucre. Fls. yellow, rarely white or purple; calyx tubular. Seeds with a swelling or *strophiole*. Often placed in *Cytisus* and not easily distinguished from it. 35 sps.

506. *C. hirsutus* (L.) Link HAIRY BROOM. A spreading hairy shrub to 1½ m., with few-flowered clusters of large yellow and usually darkly blotched fls. ranged along the previous year's branches. Fls. 2–2½ cm., short-stalked; calyx tubular, two-lipped, usually with long spreading hairs. Leaflets obovate, ½–2 cm., usually hairy, densely so beneath. Present year's branches and lv. stalks with dense spreading hairs. Fr. 2½–4 cm., hairy. A variable sp. △ Woods, thickets, alpine meadows. Apr.–June. Cent. Eur. (except D.), South-East Eur. (except AL.) E.F.I. Pl. 51.

C. purpureus (Scop.) Link PURPLE BROOM. Like 506, but fls. purplish-pink; calyx sparsely hairy. Stems almost hairless; leaflets glaucous, hairless. Southern and South-Eastern Alps. Pl. 51.

507. *C. supinus* (L.) Link CLUSTERED BROOM. Like 506, but fls. on present year's stems, and clustered in a terminal head surrounded by the uppermost lvs. Fls. 2–8; petals 2–2½ cm., yellow, with standard blotched with brown; calyx tubular, two-lipped, with long, dense, spreading or adpressed hairs. Leaflets large 1½–3½ cm., elliptic to obovate, hairy at least beneath; branches densely hairy. Fr. 2–3½ cm., hairy. A spreading or erect shrublet ½–1 m. △ Woods, thickets, dry banks. May–July. Cent., South-East Eur. E.F.I.SU. Page 185.

GENISTA Spiny, densely branched shrublets, or green, often leafless, rush-like shrubs without spines. Calyx shallowly two-lipped, the upper lip deeply divided, the lower lip three-toothed. Fls. yellow; standard narrow. Lvs. simple or trifoliate. Fr. splitting or not. 56 sps.

Shrubs or shrublets without spines
(a) Fls. in lax elongated clusters
508. *G. pilosa* L. HAIRY GREENWEED. A dwarf spreading shrub with ascending ribbed branches 30–100 cm., with 1–3 yellow fls. borne in the axils of the lvs. and forming lax leafy clusters. Fls. *c.* 1 cm.; petals and calyx with dense, adpressed silvery hairs. Lvs. simple, oval, 5–12 mm., hairless above and with adpressed hairs beneath. Fr. 1½–2½ cm., with dense adpressed hairs. △ Arid stony hillsides, heaths, thickets, rocks. Apr.–Oct. Much of Eur. (except P.IRL.IS.N.SF.TR.).

509. *G. tinctoria* L. DYER'S GREENWEED. Usually an erect shrub to *c.* 80 cm. with straight, green, ribbed branches and often numerous yellow fls. in long, slender, somewhat leafy, terminal clusters. Fls. 1–1½ cm.; petals hairless; calyx hairless or silvery-haired. Lvs. simple, narrow lance-shaped, 1–5 cm., hairless or silvery-haired. Fr. 2½–3 cm., hairless or silvery-haired. A very variable sp. △ Heaths, clearings, woods, grassy places, scree. Apr.–Aug. Much of Eur. (except P.E.IRL.NL.IS.SF.H.). *Known since Roman times as a dye plant; it was used especially for dyeing wool yellow or green. The seeds are purgative.*

1. *Genista germanica* 511
3. *Genista radiata* 510
5. *Coronilla coronata* 629

2. *Scorpiurus muricatus* [633]
4. *Chamaecytisus supinus* 507

185

(b) Fls. in dense rounded clusters

510. *G. radiata* (L.) Scop. (*Cytisanthus r.*) RAYED BROOM. A densely branched, tufted, apparently leafless, green, switch-like shrub to ½ m. with dense terminal clusters of 4–12 yellow fls. Petals 1–1½ cm., densely silky-haired outside; calyx hairy. Lvs. trifoliate, leaflets ½–2 cm., oblance-shaped, soon falling and leaving tough, thickened lv. stalks; branches strongly furrowed. Fr. 1 cm., hairy. △ Thickets, rocks, dry hills, and mountains. May–July. F.CH.A.I.YU.AL.GR.R. Page 185.

Spiny shrubs or shrublets

511. *G. germanica* L. A hairy-stemmed shrub to 60 cm. with slender, branched spines on the old stems, and yellow fls. in short, erect, terminal, leafless clusters. Fls. *c.* 1 cm.; petals hairy outside, standard much shorter than the keel; fl. stalks and calyx with dense spreading hairs. Lvs. simple, 8–20 mm., lance-shaped, hairy; young branches spineless. Fr. *c.* 8 mm., densely hairy. △ Thickets, poor pastures, heaths. May–Sept. Cent. Eur. F.B.NL.DK.S.I.YU.BG.R.SU. Page 185.

G. hispanica L. SPANISH GORSE. Distinguished from 511 by its 5–12 yellow fls. borne in a dense, terminal, globular head; standard hairless, as long as the hairy keel. Old branches very spiny, the young branches with few spines. Fr. hairless or slightly hairy. Spain and France. Pl. 52.

512. *G. anglica* L. NEEDLE FURZE, PETTY WHIN. Like 511, but stems, lvs., and calyx almost hairless, and young branches as well as old branches spiny. Fls. yellow, *c.* 8 mm. in short clusters; petals hairless. Lvs. simple 2–10 mm. elliptic to lance-shaped, glaucous; spines simple 1–2 cm., very slender. Fr. 1½ cm., hairless. A shrub 10–100 cm. △ Heaths, bushy places, moors, dry hills. Apr.–Aug. West Eur. (except IRL.) DK.S.D.I.

CHAMAESPARTIUM   Like *Genista* and often included in this genus. Stems with broad green wings and appearing flattened, and with alternate branches; lvs. simple or absent. 2 sps.

513. *C. sagittale* (L.) P. Gibbs (*Genista s.; Genistella s.*) WINGED BROOM. A shrublet with creeping woody stems, rooting at the nodes, and straight, erect, unbranched, almost leafless stems to 50 cm. with wide green wings which are narrowed at the nodes. Fls. yellow, 12–15 mm., numerous in a dense terminal leafless cluster; petals hairless; calyx silvery-haired; bracts linear, hairy. Lvs. few, simple, elliptic, softly hairy. Fr. 1½–2 cm., hairy. △ Dry pastures, hills, woods. May–Sept. Much of Eur. (except North Eur. P.IRL. GB.NL.TR.): introd. PL. Pl. 52.

LYGOS   Rush-like, branched, spineless shrubs with simple lvs. Fls. white or yellow; calyx two-lipped, tiny, the upper lip deeply two-toothed. Fr. ovoid or globular, usually one- or two-seeded. 3 sps.

514. *L. monosperma* (L.) Heywood (*Retama m.*) WHITE BROOM. A shrub to 3 m. with numerous, erect, ribbed, and rush-like, silky-haired, flexible branches and lax, elongated, leafy clusters of white fls. Petals 1–1½ cm.; standard erect, hairy, shorter than keel. Lvs. mostly linear-lance-shaped, silvery-haired, soon falling. Fr. oval or globular, 1–1½ cm. △ Sands by the sea, low hills. Feb.–May. P.E. Pl. 52.

SPARTIUM | Spanish Broom   Spineless shrubs with stiff, circular-sectioned, rush-like branches. Lvs. absent or simple. Calyx papery with oblique lip, becoming deeply split after flowering. 1 sp.

515. *S. junceum* L. SPANISH BROOM. A stout, rush-like, almost leafless shrub 1–3 m., with smooth, erect, glaucous-green branches and large bright yellow fls. in erect, terminal, lax, sweet-scented spikes. Fls. 2–2½ cm.; standard large, rounded, erect; keel pointed. Lvs. few, oblong-lance-shaped; twigs compressible. Fr. 5–8 cm. by 7 mm., at length hairless and black. △ Dry hills, hedges. May–Aug. Med. Eur. P.: introd. GB.R.SU. *The seeds contain cystine and are poisonous. Fibres obtained from the stems are used like flax for making rope, canvas, and coarse cloth. The young branches are used for basket-making, tying, etc.* Pl. 52.

ERINACEA Spiny shrubs usually with simple lvs. Fls. blue-violet; calyx inflated, bell-shaped, two-lipped. Fr. narrowly oblong; seeds without swelling. 1 sp.

E. anthyllis Link (*E. pungens* Boiss.) HEDGEHOG BROOM. A low, rounded, cushion-forming shrub to 30 cm. with numerous stiff, very spiny branches. Fls. blue-violet, 1½–2 cm.; calyx tubular, inflated, with 5 nearly equal teeth. Lvs. simple, hairy; young branches silvery-haired. Spain and France. Pl. 52.

ULEX | Gorse, Furze Very spiny shrubs with the adult lvs. and branchlets reduced to green, needle-like spines; juvenile lvs. trifoliate. Fls. yellow, axillary; calyx yellowish, deeply two-lipped, with 2 small yellowish bracts at its base. 7 sps.

Fls. 14 mm. or more
516. *U. europaeus* L. FURZE, GORSE, WHIN. A very spiny, densely branched shrub to 4 m., with bright yellow, sweet-scented fls. 14–18 mm. in lax clusters. Calyx yellow, with spreading hairs, two-thirds the length of the petals; bracts 3–5 mm. Spines 1½–2½ cm., deeply furrowed, very stiff and sharp-pointed. Fr. oval *c.* 1½ cm., very hairy, black. △ Grassy places, heaths. Feb.–June. West Eur. (except B.) D.CH.I. *The seeds are poisonous.*

Fls. 12 mm. or less
517. *U. minor* Roth (*U. nanus* T. F. Forster) DWARF FURZE. Differs from 516 by its smaller fls. 8–10 mm., and calyx nearly as long as the corolla, with adpressed hairs, bracts less than 1 mm. Stems with very numerous crowded, rather weak bristle-like, faintly furrowed spines to 1 cm. long. Fr. *c.* 1 cm. A spreading, spiny shrublet 20–100 cm. △ Heaths. July–Oct. P.E.F.GB. Pl. 53.

**U. gallii* Planchon Portugal to Great Britain.

U. parviflorus Pourret Distinguished from 517 by the glaucous, nearly hairless or finely hairy stems, and the long, stiff, arched, very sharp-pointed spines to 2 cm. which are widely spaced and much longer than the fls. Fls. 6–8 mm.; calyx at first hairy, later hairless and shining. Portugal to France.

ADENOCARPUS Spineless shrubs with trifoliate lvs. Fls. orange-yellow; calyx tubular, two-lipped, the upper lip divided to base, the lower three-toothed. Fr. oblong, flattened, covered all over with glandular swellings. 4 sps.

518. *A. complicatus* (L.) Gay A shrub to 4 m. with erect, silvery-haired, angular, non-spiny branches and orange-yellow fls. in a long, lax, many-flowered, leafless terminal cluster. Fls. 1–1½ cm.; standard silver-haired within; calyx usually glandular, lips unequal. Lvs. clustered, trifoliate; leaflets ½–1½ cm. oblong-lance-shaped, often folded lengthwise, adpressed hairy beneath. △ Bushy places, hills. May–Sept. Med. Eur. (except YU.AL.TR.) P. Pl. 55.

A. telonensis (Loisel.) DC. (*A. grandiflorus* Boiss.) Like 518, but fls. larger, 1½–2 cm., in a terminal umbel-like head of 2–7. Calyx densely silvery-haired, not glandular, lips equal. Portugal and Western Mediterranean.

LUPINUS | Lupin Fls. in long, conspicuous, terminal, bractless spikes; calyx two-lipped. Lvs. palmate with 5–11 leaflets. 10 sps.

Fls. yellow

519. *L. luteus* L. YELLOW LUPIN. An erect, hairy ann. 20–80 cm. with golden-yellow, violet-scented fls. in a somewhat interrupted spike 5–16 cm. long. Fls. in whorls of 6–10; petals 1–1½ cm. Lvs. with 6–8 linear to obovate leaflets, adpressed hairy on both surfaces. Fr. very broad 1–2 cm. across, shortly hairy; seeds mottled. △ Sandy pastures; widely cultivated, sometimes naturalized. Mar.–July. P.E.: introd. to most of Eur. (except IRL. GB.IS.N.SF.AL.GR.TR.). *Grown for fodder and green manure. The fresh seeds are poisonous.* Pl. 53.

Fls. blue

520. *L. angustifolius* L. NARROW-LEAVED LUPIN. An erect, hairy ann. 20–80 cm., with rather dark blue fls. in a dense, elongated spike 10–20 cm. Fls. alternate, not whorled; petals *c*. 1 cm., twice as long as the silvery-haired calyx. Lvs. distinctive, with 5–9 narrow linear to lance-shaped leaflets, hairless above, but hairy beneath. Fr. 1–1½ cm. broad, with adpressed hairs; seeds mottled. △ Sandy fields, rocks. Apr.–June. Med. Eur. (except AL.) P.BG.: introd. to Cent. Eur. R.SU. *Poisonous to livestock and in particular to sheep.* Pl. 54.

L. micranthus Guss. (*L. hirsutus* auct.) HAIRY LUPIN. Like 520, but whole plant covered with dense, spreading, often brownish hairs. Fls. blue, the lower alternate, the upper whorled. Leaflets 5–7, obovate-oblong, up to 1½ cm. wide, hairy on both sides. Fr. ½–1 cm. broad, densely shaggy-haired. Mediterranean Europe.

L. polyphyllus Lindley A robust silky-haired perenn. to 1½ m. distinguished by its rather dense spikes 15–60 cm. of whorled blue, purple, pink, or white fls. Petals 12–14 mm. Leaflets 9–17, obovate-lance-shaped 7–15 cm. long, nearly hairless. Native of North America; grown for fodder or ornament in much of Europe.

Fls. white

521. *L. albus* L. WHITE LUPIN. An erect ann. 15–100 cm. with adpressed hairs and alternately arranged white fls. tipped with blue, in a lax, few-flowered, terminal spike 5–10 cm. Petals 1–1½ cm. Leaflets 5–7, oblong-oval, hairless above. Fr. shaggy-haired, 1–1½ cm. broad; seeds flattened, white. Subsp. *graecus* (Boiss. & Spruner) Franco & P. Silva with bright-blue fls. is a plant of the Balkans and Aegean. △ Widely cultivated. May–June. South-East Eur.: introd. to much of Eur. (except North Eur. IRL.GB.PL.). *Grown as a fodder crop and as green manure. The fresh seeds are poisonous.* Pl. 54.

ARGYROLOBIUM Like *Cytisus*, but shrublets without spines; lvs. trifoliate, with stipules. Calyx deeply two-lipped, the upper lip deeply two-lobed. Fls. yellow; keel blunt. Seeds without swellings. 2 sps.

522. *A. zanonii* (Turra) P. W. Ball (*A. linnaeanum* Walpers) A silvery-grey shrublet to 25 cm., with trifoliate lvs. and a terminal, golden-yellow, long-stalked, solitary fl. or cluster of 2–3 fls. Fls. *c*. 1 cm.; petals and calyx densely silky-haired. Lvs. stalked; leaflets elliptic to lance-shaped ½–2 cm., nearly hairless above, but densely silvery-haired beneath; stipules linear-lance-shaped. Fr. 2½–3½ cm., densely silvery-haired. △ Dry places, stony ground, pine woods. May–July. P.E.F.I.YU.AL. Page 198.

ROBINIA   Fls. white, in long, hanging, axillary clusters; standard curved upwards; keel blunt. Stamens 9 fused and 1 free. Deciduous trees with pinnate lvs. 2 sps.

523. *R. pseudacacia* L. FALSE ACACIA, LOCUST. A tree to 30 m. with greyish, deeply furrowed bark, pinnate lvs. and sweet-scented hanging clusters of white or rarely pink fls. Fl. clusters to 18 cm.; fls. 1½–2½ cm.; standard erect, blotched with yellow at base. Lvs. 15–25 cm., with 7–15 oval leaflets 2–5 cm.; branches and suckering stems with long thorns. Fr. to 10 cm., compressed, reddish-brown. △ Hedges, thickets, riversides, gravel. Apr.–June. Native of North America; widely planted and commonly suckering and forming thickets. Introd. to much of Eur. (except North Eur. P.IRL.PL.H.SU.). *The wood is durable and resistant to rotting and it is used for posts, wheels, floors, and for turning. The flowers are used in perfumery; the roots are poisonous.* Pl. 53.

Wisteria sinensis (Sims) Sweet WISTERIA. Fls. mauve or deep lilac in long pendulous clusters 20–60 cm.; fls. *c.* 2½ cm. Lvs. compound, 25–30 cm., with 7–13 oval, hairy leaflets 4–10 cm. A stout woody climber. Native of China; planted for ornament in Southern Europe.

GALEGA │ Goat's Rue   Fr. linear, rounded in section. Fls. in erect clusters; calyx bell-shaped, with 5 more or less equal teeth; stamens all fused. Herbs with pinnate lvs.; stipules large. 2 sps.

524. *G. officinalis* L. GOAT'S RUE, FRENCH LILAC. An erect, leafy, hairless perenn. ½–1½ m. with pale lilac or white fls. in erect, elongate, stalked, axillary clusters longer than the lvs. Fls. 8–15 mm., short-stalked; calyx hairless, swollen at base and with 5 bristle-like teeth. Lvs. pinnate with 9–17 oblong, fine-pointed, or notched leaflets; stipules large, arrow-shaped, long-pointed. Fr. 2–3 cm. long by *c.* 4 mm., reddish-brown. △ Ditches, damp fields, streamsides; often grown as an ornamental. June–Aug. Most of Eur. (except North Eur. IRL.): introd. P.GB.B.CH. Pl. 53.

COLUTEA   Fls. yellowish in axillary clusters; standard broad, rounded; keel blunt. Fr. becoming bladder-like and papery. Deciduous shrubs with pinnate lvs. 4 sps.

525. *C. arborescens* L. BLADDER SENNA. A thornless deciduous shrub to 3 m., with green branches and erect clusters of 2–8 yellow fls., often streaked with reddish-brown. Fl. clusters shorter than the lvs.; fls. *c.* 2 cm., drooping; wings shorter than the blunt keel; calyx and inflorescence axis with short, white adpressed hairs. Lvs. pinnate with 7–15 oval leaflets each ½–1½ cm. Fr. 4–7 cm. by 2–4 cm., inflated, parchment-like. △ Bushy places, thickets; grown as an ornamental and often naturalized. Apr.–July. Med., South-East Eur. D.CH.A.CS.H.: introd. GB.B. *The leaves and seeds are purgative and have been used as a substitute for senna.* Pl. 53.

ASTRAGALUS │ Milk-Vetch   Fls. in axillary spikes or clusters. Calyx tubular with 5 short teeth; keel blunt-tipped; upper stamen free, the remaining 9 fused by their filaments. Lvs. pinnate, with a terminal leaflet. Fr. often swollen, often longitudinally two-celled. 134 sps.

Leafy stems absent and fl. stems arising from rootstock
526. *A. monspessulanus* L. MONTPELLIER MILK-VETCH. Fls. rosy-purple in rather dense ovoid clusters on stems arising from the rootstock, as long as or little longer than the lvs. Fl. clusters elongating somewhat in fr.; fls. 2 cm.; standard upturned, much longer than the wings; calyx tubular, with adpressed hairs and with linear teeth half as long as the tube. Lvs. with 21–31 oval to oblong green leaflets, adpressed-hairy beneath. Fr. cylindrical

2½–5 cm., sparsely hairy, curved. Perenn. 10–30 cm. △ Dry stony places. Apr.–Aug. Med., South-East Eur. (except TR.) P.CH.SU. Page 192.

Leafy branched stems present and fl. stems arising from them
(a) Fls. yellowish or whitish

527. *A. glycyphyllos* L. MILK-VETCH. Fls. creamy-white tinged with greenish-grey, in dense ovoid clusters on stems much shorter than the lvs. Fls. 1–1½ cm.; calyx usually hairless. Lvs. 10–20 cm., hairless, with 9–13 oval leaflets each 1½–4 cm. Fr. 2½–3½ cm. cylindrical, pointed, curved, almost hairless, not inflated. A spreading, very leafy, nearly hairless perenn. 60–100 cm. △ Grassy, bushy places, woods. May–Aug. Most of Eur. (except IS.). Pl. 54.

A. lusitanicus Lam. IBERIAN MILK-VETCH. Distinguished by its dense oblong clusters of many large white or yellow drooping fls. on stems about half as long as the lvs. Fl. clusters 5–6 cm.; fls. 2½–3 cm.; calyx hairy, becoming reddish. Fr. 5–7 cm. by 1–2 cm., hairy, blackish. Southern Iberian Peninsula and Southern Greece.

A. hamosus L. Readily distinguished by its stalked clusters of 3–12 wide-spreading but sharply upcurved, sickle-shaped, cylindrical fr. 2–6 cm. Fls. small, white, in globular, stalked clusters. Leaflets oblong 17–25. A green or greyish, hairy, spreading ann. to ½ m. Mediterranean region.

528. *A. cicer* L. WILD LENTIL. Fls. pale yellow in dense clusters on long stems about as long as the lvs.; calyx with adpressed black hairs. Lvs. hairy, with 13–25 oblong-elliptic hairy leaflets. Fr. ovoid, inflated, 1–1½ cm. by 8 mm., densely hairy. A softly hairy, spreading perenn. 30–80 cm. △ Grassy places, waysides. June–July. Cent. Eur. E.F.B.I.YU.BG.R.SU. Pl. 54.

A. massiliensis (Miller) Lam. An extremely spiny, grey-leaved, dense much-branched, low rounded, shrub-like perenn. with stalked clusters of few large white fls. Fls. in a lax globular cluster of 3–10; petals twice as long as the hairy calyx. Leaflets 13–25, soon falling; spines formed from the old lv. stalks. On the littoral; Western Mediterranean.

529. *A. frigidus* (L.) A. Gray YELLOW ALPINE MILK-VETCH. An almost hairless arctic or alpine perenn. to 40 cm. with whitish-yellow fls. in dense clusters on stems longer than the lvs. Fls. 1–1½ cm.; calyx bell-shaped and with 5 very shallow, blunt teeth tipped with black hairs. Lvs. with 7–11 oval, bluish-green leaflets, paler and sparsely hairy beneath; stipules oval, large. Fr. somewhat swollen, 1½–2½ cm. oval acute, narrowed to a distinct stalk below, densely hairy. △ Mountain meadows, damp grassy places. July–Aug. F.N.S. D.CH.A.CS.PL.I.R.SU. Page 192.

(b) Fls. blue-violet or reddish-purple

530. *A. alpinus* L. ALPINE MILK-VETCH. A small, spreading, arctic or alpine perenn. to 15 cm. with globular clusters of pale blue fls. tipped with purple. Fls. *c.* 12 mm.; calyx densely hairy. Lvs. with 17–25 lance-shaped leaflets; stipules free to their base. Fr. *c.* 1 cm. with adpressed hairs, pendulous. △ Rocks, screes and meadows. July–Aug. Cent. Eur. (except H.) E.F.GB.N.S.SF.I.YU.R.SU. Pl. 54.

531. *A. vesicarius* L. Fls. 3–10, large, violet with white wings, or yellow, in a rather lax ovoid head which later elongates, on stems twice as long as the lvs. Calyx tubular, becoming inflated after flowering, with dense short black and long white hairs, teeth very short. Lvs. with 7–19 oblong-elliptic, blunt, hairy leaflets; stipules not fused. Fr. 1–1½ cm., triangular-sectioned, white-haired, little longer than the calyx. A woody-based, hairy, ascending perenn. 10–20 cm. △ In mountains: rocks, dry pastures, river gravels. May–June. E.F.A.CS.H.I.YU.AL.GR.BG.R.SU.

532. *A. danicus* Retz. (*A. hypoglottis* auct.) PURPLE MILK-VETCH. Like 531, but a more robust, softly greyish-hairy perenn. to 35 cm. with fl. stalk usually longer than lvs. Fls. violet-purple; standard oval, little longer than the wings. Calyx tubular, not swelling after flowering, hairy, teeth about half as long as tube. Lvs. with 15–25 oblong leaflets; stipules fused at base. Fr. 1 cm., covered with crisped white hairs. △ Mountain pastures. June–Aug. F.IRL.GB.DK.S.D.A.CS.PL.I.SU.

A. onobrychis L. SAINFOIN MILK-VETCH. Distinguished from 532 by the linear-oblong blunt standard 1½–3 cm., which is much longer than the notched wings. Fls. bright violet, or sometimes white or yellowish, 10–20 in a dense oval cluster, later elongating; calyx with adpressed hairs and lance-shaped teeth 3–4 times shorter than the calyx tube. Leaflets 17–31, with adpressed greyish hairs above and below. A very variable sp. Most of Central and Southern Europe.

OXYTROPIS | **Beaked Milk-Vetch**   Very like *Astragalus*, but keel of fl. ending in a fine point. Fr. one-celled, or divided into 2 cells by a wall arising from the upper side. 22 sps.

Fl. stems leafless, arising directly from the rootstock
(*a*) *Fls. white or yellow*
533. *O. campestris* (L.) DC. MEADOW BEAKED MILK-VETCH. Fls. yellow or white, often tinged with purple, in a dense ovoid cluster of 5–15 fls. on stems arising from the thick rootstock. Lvs. with 21–31 elliptic or lance-shaped woolly-haired leaflets each 1–2 cm. Fr. oval, *c.* 1½ cm., two-celled, densely hairy. A variable, hairy perenn. 10–20 cm. △ Rocks and pastures in mountains. June–Aug. E.F.GB.N.S.SF.CH.A.PL.CS.I.YU.AL.BG.R.SU. Page 192.

(*b*) *Fls. bluish-violet or purple*
534. *O. halleri* Koch PURPLE BEAKED MILK-VETCH. Like 533, but fls. blue to purple, rarely yellowish-white, with a darker keel and fl. stem longer than the lvs. Fr. 1–2 cm., two-celled, hairy. △ Poor pastures in mountains. June–Aug. E.F.GB.CH.A.CS.PL.I.AL.R.

Fl. stems leafy at base
535. *O. jacquinii* Bunge MOUNTAIN BEAKED MILK-VETCH. Fls. bluish-violet to reddish in a globular, terminal cluster borne on short stout stems bearing 3–4 lvs. at the base. Petals 1–1½ cm. Lvs. with 14–35 oval leaflets, with adpressed whitish hairs; stipules fused to each other and to lv. stalk. Fr. 2–3 cm., one-celled, narrowed to a stalk as long as or longer than the calyx tube, more or less erect, black-haired. A spreading perenn. 5–12 cm. △ Rocks and meadows in mountains. July–Aug. F.D.CH.A.I.

O. lapponica (Wahlenb.) Gay NORTHERN BEAKED MILK-VETCH. Distinguished from 535 by the longer ascending stems to 10 cm., with 3–5 lvs. at the base. Leaflets 17–25, oblong-lance-shaped, hairy; stipules fused for at least half their length and similar to the leaflets. Fl. clusters globular; fls. blue-violet, 8–12 mm. Fr. drooping, 10–12 mm. narrowed to a stalk half as long as the calyx tube. Arctic or alpine.

GLYCYRRHIZA | **Liquorice**   Fr. short linear, straight or curved, one-celled, not splitting. Calyx bell-shaped, somewhat two-lipped; standard much longer than wings or keel. Lvs. pinnate with a terminal leaflet; stipules very small. 5 sps.

536. *G. glabra* L. LIQUORICE. A robust, erect, hairless perenn. to 1½ m., with pinnate lvs. and shorter axillary, long-stalked, spike-like clusters of bluish or violet fls. Fls. *c.* 1 cm., numerous, lax; calyx glandular-hairy; standard erect. Lvs. with 9–17 elliptic to oblong leaflets 2½–5 cm., sticky beneath; stipules minute or absent. Fr. 1½–2 cm., oblong, much flattened, hairless. △ Stony places, dry woods, ditches; often cultivated and sometimes

1. *Astragalus frigidus* 529
3. *Astragalus monspessulanus* 526
5. *Glycyrrhiza glabra* 536

2. *Cicer arietinum* 541
4. *Oxytropis campestris* 533

naturalized. June–July. South-East Eur. E.F.I.SU.: introd. P.CH.A.CS.H. *Liquorice is obtained from the creeping rhizomes; it is used in medicine, sweet-making and also in brewing and tobacco-curing.* Page 192.

G. echinata L. SPINY-FRUITED LIQUORICE. Distinguished from 536 by the oval fr. *c.* 1½ cm. covered with long, stiff, reddish-brown spines. Fls. small, lilac, in dense globular clusters on stems shorter than the lvs. Leaflets 5–13, gland-dotted beneath; stipules lance-shaped, long-pointed. Much of South-Eastern Europe.

PSORALEA Fls. bluish in dense, globular, terminal clusters. Lvs. trifoliate. Pod ovoid, one-seeded, encircled by calyx. 2 sps.

537. *P. bituminosa* L. PITCH TREFOIL. A rather slender, straggling, dark green perenn. to 1 m., with trifoliate lvs. which have a strong smell of tar when crushed, and rounded heads of blue-violet fls. Fl. heads dense, and with 2 deeply toothed, leafy bracts at base, very long-stalked; fls. 1–1½ cm. Lvs. long-stalked, the lower with rounded and the upper with lance-shaped leaflets, gland-dotted, hairy. Fr. hairy, long-beaked. △ Dry places, thickets, roadsides. Apr.–July. Med. Eur. P.BG.R.SU. Pl. 54.

PHASEOLUS Lvs. large, trifoliate, leaflets stalked, tendrils absent. Fls. clustered; calyx often two-lipped; standard broadly oval, recurved; keel with a long, spirally rolled beak. 2 sps.

538. *P. vulgaris* L. KIDNEY or HARICOT BEAN. A tall climbing or erect non-climbing ann. 30–300 cm., with white, bluish, or yellow fls. in few-flowered clusters on axillary stems shorter than the lvs. Lvs. with 3 large, oval-rhomboid, acute, entire, stalked leaflets. Fr. 10–20 cm., straight or nearly so; seeds less than 1½ cm., ellipsoid or kidney-shaped. △ Native of South America; widely cultivated and sometimes persisting. June–Sept. *Numerous varieties are cultivated in many parts of the world for human consumption. The whole pod containing the young seeds can be eaten, or the fresh or dried seeds only eaten.*

539. *P. coccineus* L. SCARLET RUNNER. A hairy twining ann. or perenn. 2–5 m., with large scarlet or white fls. 2 cm. or more in many-flowered, stalked, axillary clusters longer than the lvs. Lvs. with 3 large, oval-rhomboid, stalked leaflets. Fr. 10–30 cm., rough, somewhat curved; seeds large, 2–2½ cm., kidney-shaped, somewhat flattened. △ Native of South America; widely cultivated and sometimes persisting. June–Aug. *The green unripe pods are commonly eaten as a vegetable; the seeds are also eaten.*

GLYCINE | Soy Bean Lvs. large, trifoliate; tendrils absent. Fl. clusters very short-stalked; calyx bell-shaped; petals very unequal, the standard elongated, erect or spreading, the wings absent or very short, the keel much shorter than the standard. 1 sp.

540. *G. max* (L.) Merr. (*G. hispida* (Moench) Maxim.) SOJA or SOY BEAN. An erect, non-climbing, brown-haired, bushy ann. 40–100 cm., with inconspicuous whitish or violet fls. in small, almost stalkless clusters in the axils of the lvs. Petals little longer than the hairy calyx. Lvs. with 3 very large, oval, stalked leaflets to 9 cm. by 5 cm., usually with brown hairs. Fr. 5–7½ cm., short-stalked, usually densely hairy. △ July–Aug. Origin not known. *Cultivated for thousands of years in Eastern Asia; it has now become a very important crop plant of North America. The seeds contain 20 per cent. oil, 30–45 per cent. protein and are widely used for human consumption and cattle food. The oil is edible and also used for many industrial purposes.*

CICER | Chick-Pea Lvs. pinnate, leaflets conspicuously toothed, tendrils absent. Calyx five-toothed, the upper 2 often shorter and fused. Fr. large, globular with a conical apex and usually with 2 wrinkled, oval-conical seeds. 4 sps.

541. *C. arietinum* L. CHICK-PEA. A leafy, branched, glandular, hairy ann. to 40 cm. with solitary, bluish or white fls. arising from the axils of the lvs. on long jointed stalks. Fls. *c.* 1 cm.; petals little longer than the densely hairy calyx. Lvs. with 13–17 oval, strongly saw-toothed, glandular-hairy leaflets 1–1½ cm.; stipules egg-shaped, toothed. Fr. 2–3 cm., oval, very inflated, glandular-hairy. △ Native of Asia; often cultivated in Southern Europe and sometimes self-seeding. June–July. *Cultivated since classical times for its edible seeds; a good forage plant.* Page 192.

VICIA | Vetch Climbing or scrambling herbaceous plants with pinnate lvs. with two to many pairs of leaflets and with or without a simple or branched terminal tendril; stipules present; stems usually unwinged. Fls. solitary, axillary or in axillary clusters; tube formed by fused filaments obliquely cut-off; style hairless or equally hairy all round, or with a tuft of hairs on the lower surface only. Fr. oblong to linear, flattened. 55 sps.

Fls. often more than 6 in long-stalked clusters usually longer than lvs.
(a) *Fls. small 2–9 mm.; clusters one- to nine-flowered*
542. *V. hirsuta* (L.) S. F. Gray HAIRY TARE. A slender, nearly hairless, scrambling ann. 20–60 cm., with tiny pale violet or whitish fls. in long-stalked clusters. Fls. 1–9, each 4–5 mm. Lvs. with 4–8 pairs of linear-oblong leaflets 5–12 mm., often notched and with a fine point; tendrils usually branched. Fr. *c.* 1 cm., oblong, hairy, two-seeded. △ Hedges, fields, grassy places. May–Aug. All Eur.: introd. IS.

*V. tetrasperma (L.) Schreber SMOOTH TARE. Almost throughout Europe.

V. ervilia (L.) Willd. Distinguished from 542 by the lvs. which have 10–13 pairs of linear leaflets and a short terminal point in place of a tendril. Fls. 6–9 mm., whitish or pale pink, veined with violet. Fr. 2–3 cm., hairless, yellowish, noticeably constricted between the 3–4 seeds. Southern Europe; introduced to Central Europe.

(b) *Fls. 1–2 cm.; clusters six- to thirty-flowered*
(i) *Tendrils absent*
543. *V. orobus* DC. BITTER VETCH. An erect, stout-stemmed, non-climbing, somewhat hairy perenn. 30–60 cm., with dull white fls. tinged with purple, in clusters of 6–20 on stalks about as long as the lvs. Fls. 1–1½ cm.; calyx hairy, teeth very unequal. Lvs. with 6–15 pairs of elliptic leaflets each 1–2 cm., tendril absent; stipules conspicuous leafy, half-arrow-shaped. Fr. 2–3 cm., hairless. △ Thickets, rocky places. May–July. West Eur. (except NL.) DK.N.D.CH.

(ii) *Tendrils present*
544. *V. sylvatica* L. WOOD VETCH. A climbing perenn. 1–2 m., with white fls. with purple veins, or fls. rarely lilac, in stalked, rather lax, one-sided clusters of up to 18 fls. Fls. 1½–2 cm., drooping; calyx with bristle-like teeth, the upper teeth about half as long as the lowest tooth. Leaflets 6–10 pairs, oblong-elliptic; tendrils much-branched; stipules semi-circular, cut into narrow bristle-like teeth at base. Fr. 2½–3 cm., gradually narrowed at both ends, hairless. △ Woods in mountains. June–Aug. Most of Eur. (except P.IS.B.NL. GR.TR.BG.).

V. onobrychioides L. FALSE SAINFOIN. Distinguished by its blue-violet fls. with paler keel, and by the narrow lance-shaped or linear leaflets which are 5–8 times as long as broad. Fl. clusters lax, one-sided, with 6–12 fls. each 18–20 mm. Stipules toothed. Fr. reddish, 3–3½ cm. by ½ cm., gradually narrowing at base, hairless. Southern Europe. Page 198.

545. *V. cracca* L. TUFTED VETCH. A somewhat hairy, but variable scrambling perenn. $\frac{1}{2}$–2 m., with rather dense axillary clusters of blue-violet fls. on rather stout fl. stalks 2–10 cm. Fls. 10–40, each 10–12 mm., drooping; calyx teeth very unequal, the upper minute. Leaflets 6–20 pairs, oblong-linear, 2–5 mm. broad, somewhat hairy; tendrils branched; stipules half-arrow-shaped, entire. Fr. 1–2$\frac{1}{2}$ cm., oval, narrowed to a stalk shorter than calyx tube, hairless. △ Grassy, bushy places, hedges. June–Aug. All Eur. Pl. 55.

V. tenuifolia Roth SLENDER-LEAVED TUFTED VETCH. Like 545, but fls. larger 12–18 mm., pale blue or violet with whitish wings, numerous in conspicuous lax elongated axillary clusters longer than the lvs. Leaflets 9–10 pairs, more hairy, narrower, longer and more pointed. Fr. 2–3 cm., narrowed to a stalk equalling the calyx tube, hairless. Widespread in Europe.

V. villosa Roth Like 545, but distinguished by the calyx which has a boss-like swelling at the base; a standard with basal stalk-like part (or claw) twice as long as the terminal blade. Fls. violet, purple or blue, 12–16 mm., densely clustered, all opening at approximately the same time. Fr. hairless, narrowed to a stalk longer than the calyx tube. A very variable sp. Widespread in Europe, except in the north. Pl. 55.

V. benghalensis L. (*V. atropurpurea* Desf.) REDDISH TUFTED VETCH. Like 545, but fls. reddish with blackish-purple tips, 1–1$\frac{1}{2}$ cm., in lax clusters of 4–12, shorter than or equal to the softly hairy lvs. Stipules toothed, shaggy. Fr. hairy, 1$\frac{1}{2}$–3$\frac{1}{2}$ cm. by 1 cm., narrowed to a stalk shorter than the calyx. Mediterranean Europe. Pl. 55.

Fls. usually less than 6, stalkless or short-stalked in clusters much shorter than the lvs.
(a) *Fls. predominantly yellow or white*
546. *V. faba* L. (*Faba bona* Medicus) BROAD or HORSE BEAN. An erect, hairless, rather stout, square-stemmed ann. 30–80 cm., with large white fls. with blackish-purple blotch, in almost stalkless axillary clusters of 2–6. Fls. *c.* 2 cm. Lvs. with 1–3 pairs of large, oval, thick, glaucous leaflets, without tendrils; stipules large, toothed, blotched. Fr. very large, 5–30 cm., hairy, swollen, black; seeds large, 2–3 cm. △ Probably a native of Asia; naturalized locally in most of Europe. May–July. *Cultivated since prehistoric times as food for humans; also an important fodder plant for cattle and horses.*

547. *V. hybrida* L. HAIRY YELLOW VETCH. A spreading, branched ann. to $\frac{1}{2}$ m., usually with large, solitary, almost stalkless yellow fls. in the axils of the upper lvs. Fls. to 2$\frac{1}{2}$ cm.; standard densely hairy on the outside, often purple-veined. Leaflets 4–8 pairs, wedge-shaped, notched, with a fine point, hairy on both sides; tendril branched. Fr. 2$\frac{1}{2}$–3 cm., hairy. △ Grassy, bushy places, cultivated ground. Apr.–June. Med., South-East Eur. (except AL.) P.CH.SU.

V. lutea L. YELLOW VETCH. Like 547, but standard hairless on the outside and often reddish. Fls. 2–2$\frac{1}{2}$ cm., pale yellow. Leaflets narrower, *c.* 8 pairs, linear-lance-shaped, acute or blunt, and with a fine point. Fr. *c.* 3 cm. by 1 cm., rough-hairy with a swelling at base of each hair. Widespread in Europe, except in the north.

V. melanops Sibth. & Sm. BLACK VETCH. Distinguished from 547 by the fls. which are tricoloured with standard yellowish-green, wings dark black-purple, and keel purplish. Fls. 2 cm. in a short-stalked cluster of 2–4; calyx hairy. Fr. brown, 2$\frac{1}{2}$–3 cm. by 1 cm., hairy on the margin only. Mediterranean Europe.

V. grandiflora Scop. LARGE YELLOW VETCH. Like 547, but fls. larger, 2$\frac{1}{2}$–3$\frac{1}{2}$ cm., multi-coloured, yellow often flushed with violet, wings yellow and black-spotted, keel whitish. Leaflets 4–7 pairs, oval, notched, and with a fine point. Fr. hairless or minutely glandular-hairy. Central and South-Eastern Europe.

(b) Fls. predominantly lilac, violet or purple

548. *V. sepium* L. BUSH VETCH. A climbing or trailing, nearly hairless perenn. 30–100 cm., with almost stalkless clusters of 2–6 pale purple fls. Fls. 1–1½ cm.; calyx teeth unequal, the lower much shorter than the tube. Leaflets 4–9 pairs, 1–3 cm., oval-elliptic blunt, and with a fine point; tendrils branched; stipules usually entire. Fr. 2–2½ cm., black, hairless. △ Hedges, waysides, thickets. Apr.–July. All Eur.

549. *V. sativa* L. COMMON VETCH. A very variable ann. or bienn. distinguished from 548 by the calyx teeth which are more or less equal and as long as the calyx tube. Fls. 1 or 2, axillary, purple or violet or bicoloured. Leaflets 4–7 pairs, oval to linear; tendrils branched; stipules toothed or entire, often blotched. Fr. linear, hairy or hairless. Subsp. *sativa* is a leafy, rather hairy ann. to 1 m. or more, with larger fls. 2–3 cm. and broad, obovate leaflets. Fr. 4–8 cm. Subsp. *nigra* (L.) Ehrh. is usually a hairless, slender ann. or bienn. to 40 cm. with purple fls. 1–1½ cm.; leaflets narrow lance-shaped to linear. Fr. 3–5 cm. △ Grassy places, hedges, cultivated ground. Apr.–Sept. All Eur.: introd. IS. *Subsp. sativa is widely grown as a forage crop and as a green manure.* Pl. 55.

550. *V. narbonensis* L. PURPLE BROAD BEAN. Like 546, but fls. dull purple with darker purple wings. Leaflets thick, glaucous, oval, 2–3 cm. wide, sometimes with a toothed margin; upper lvs. with branched tendrils. Fr. 5–6 cm. by 1 cm., nearly hairless, but with small swellings on the margin, black. A rather robust, hairy, erect, square-stemmed ann. 20–50 cm. △ Thickets, damp fields, ditches. May–June. Med., South-East Eur. P.A.H.SU.: introd. D.CS.

551. *V. bithynica* (L.) L. BITHYNIAN VETCH. A hairy, climbing ann. 20–50 cm. with 1–2 axillary fls. on a short or medium-length stalk to 5 cm., with a purple standard and white or yellowish wings and keel. Fls. *c.* 2 cm. Leaflets 1–3 pairs large, elliptic-oval, 2–5 cm.; tendrils branched; stipules 1 cm. or more, toothed. Fr. 2–3½ cm. by 1 cm., hairy, yellowish. △ Fields, hedges, grassy places. May–June. Med. Eur. P.GB.BG.SU.

LENS | Lentil Very like *Vicia*, but plants not climbing and frs. one- to two-seeded; seeds disk-like, flattened. 4 sps.

552. *L. culinaris* Medicus (*Ervum lens* L.) LENTIL. A hairy, branched ann. 20–40 cm., with 1–3 white fls. with lilac veins on a long axillary fl. stalk ending in a fine point. Fls. *c.* ½ cm.; calyx hairy, teeth linear, longer than the petals. Lvs. with 5–7 pairs of narrow, oblong leaflets; tendril simple or branched; stipules entire. Fr. 1–1½ cm., rectangular, hairless. △ Origin unknown; widely cultivated and sometimes naturalized in Central and South-East Europe. May–Aug. *The seeds are very nutritious and have been an important source of food since classical times; a good forage plant for livestock.* Page 198.

L. nigricans (Bieb.) Godron (*Ervum n.*) Distinguished from 552 by the 3–5 pairs of narrower, linear-oblong leaflets and lvs. without tendrils, or only the uppermost lvs. with a simple tendril; stipules toothed. Fls. pale blue, 5 mm. Native of the Mediterranean region.

LATHYRUS | Pea, Vetch, etc. Similar to *Vicia*, but stems usually winged or at least angled and leaflets parallel-veined, usually fewer, 1–4 pairs or absent. Style flattened, with a tuft of hair on the upper side only. Fl. clusters axillary, usually long-stalked. 54 sps.

Fls. yellow

553. *L. aphaca* L. YELLOW VETCHLING. A slender, glaucous, hairless, scrambling ann. to 1 m., usually with solitary, pale yellow fls. *c.* 1 cm., on stalks as long as or longer than the apparent lvs. However, readily distinguished by the absence of true lvs. and the pre-

sence of hastate leafy stipules and by the simple or bifid tendrils which take the place of the lvs. Fr. hairless, 2–3½ cm., slightly curved. △ Fields, dry sandy places. Apr.–June. Most of Eur. (except North Eur. IRL.PL.CS.). *Famous as a botanical example in which the leaves are modified to tendrils and the stipules to leaves.* Pl. 55.

554. *L. ochrus* (L.) DC. WINGED VETCHLING. A hairless, pale glaucous ann. 30–70 cm., readily distinguished by the very broadly winged stems and lv. stalks. Fls. pale yellow, usually solitary, 1½–2 cm., on a stalk shorter than the lvs. Lower lvs. with oval to lance-shaped, flattened, leaf-like stalks ending in 3 unbranched tendrils, the uppermost lvs. with 1–2 pairs of oval leaflets and branched tendrils. Fr. with 2 membraneous wings on the back, hairless. △ Cornfields, hedges, dry places. Mar.–June. Med. Eur. P. *Sometimes grown as a fodder crop.* Pl. 55.

555. *L. pratensis* L. MEADOW VETCHLING. A stoloniferous perenn. with climbing, angled stems 30–120 cm., with yellow fls. 1½–2 cm. in clusters of 5–12 on stout stalks longer than the lvs. Lvs. with 2 lance-shaped leaflets 1–3 cm. and arrow-shaped leafy stipules 1–2½ cm.; tendrils simple or branched. Fr. 2½–3½ cm. by ½ cm., compressed, hairless or finely hairy. △ Grassy places, waysides. May–Aug. All Eur.

L. annuus L. ANNUAL YELLOW VETCHLING. A climbing ann. with 1–3 yellow fls. 1–1½ cm., on a stalk shorter than the lvs. Leaflets 2, 6–12 cm. linear-lance-shaped; stipules linear long-pointed; tendrils simple or branched; stem winged above. Pod straight, 3–8 cm. by 1 cm., hairless; seeds large, angled, with swellings. Mediterranean region and Portugal.

Fls. red, purple, pink, violet or white
(a) Lvs. simple or leaflets 2
(i) Stem with a narrow wing 1 mm. or less, or wingless
556. *L. nissolia* L. GRASS VETCHLING. Readily distinguished by its narrow, grass-like lvs. without leaflets or tendrils. Fls. crimson, *c.* 1½ cm., 1–2 on a very slender stem, shorter than or as long as the lvs. Lvs. linear-lance-shaped, up to 15 cm.; stipules minute; stem angled. Fr. straight, 3–6 cm. by 2–3 mm. A delicate grass-like, usually hairless ann. 30–90 cm. △ Dry grassy and bushy places. May–July. Much of Eur. (except North Eur. IRL.). Pl. 55.

557. *L. cicera* L. RED VETCHLING. Fls. crimson or brick-red, 1–1½ cm., solitary, on a stalk shorter than the lvs. and jointed above the middle. Calyx teeth equal, erect, 2–3 times as long as the tube. Leaflets linear to lance-shaped, 1–9 cm.; tendril branched. Fr. 2–4 cm. by ½–1 cm., hairless, grooved on upper margin; seeds 2–6. A hairless, climbing ann. 20–100 cm., with narrowly winged stems. △ Grassy places, cultivated ground. Mar.– June. Med., South-East Eur. P.CH.SU.: introd. A. *Sometimes grown for fodder.*

L. setifolius L. NARROW-LEAVED RED VETCHLING. Like 557, but fls. smaller, 8–11 mm., orange-red; calyx teeth slightly unequal, as long as or slightly longer than the tube. Leaflets narrowly linear; tendril branched; stem angled. Fr. oblong flattened, 1½–3 cm. by *c.* 1 cm., not grooved, hairy on the margin; seeds large, 2–3, with fine swellings. Southern Europe.

L. sativus L. CHICKLING PEA. Like 557, but fls. white, lilac, or blue, solitary, 1½ cm.; calyx teeth spreading, longer than the tube; fl. stalk 3–6 cm., jointed below fl. Fr. 3–3½ cm. by 12–15 mm., strongly grooved on upper margin and with 2 leafy wings. Origin not known; widely cultivated as forage crop in Central, Southern, and Eastern Europe. Pl. 56.

558. *L. hirsutus* L. HAIRY VETCHLING. Like 557, but fls. pale violet and pink, becoming bluish, *c.* 1 cm., 1–3 on slender, usually hairy stalks often 2–3 times longer than the lvs. Leaflets 2, linear-lance-shaped; stipules linear. Fr. 2–5 cm. by 5–10 mm., hairy with bases of hairs

1. *Lens culinaris* 552   2. *Ononis pusilla* 572   3. *Lathyrus hirsutus* 558
4. *Vicia onobrychioides* [544]   5. *Argyrolobium zanonii* 522   6. *Lotus creticus* 615

swollen. A sparsely hairy, climbing ann. 20–120 cm. △ Cultivated places, waysides, hedges. May–Aug. Much of Eur. (except North Eur. IRL.NL.): introd. GB. Page 198.

559. *L. tuberosus* L. EARTH-NUT PEA. Fls. crimson, sweet-scented, 1½–2 cm., 2–7 on stalks much longer than the lvs. Leaflets 2, 1½–4½ cm., elliptic to obovate, with a fine point; tendrils simple or branched; stipules to 2 cm., half-arrow-shaped. Fr. nearly cylindrical-sectioned, 2–4 cm. by 4–7 mm., brown, hairless. A hairless, climbing perenn. 30–120 cm. with tuberous roots. △ Fields, meadows, vineyards. June–Aug. Much of Eur. (except North Eur. P.IRL.TR.): introd. GB.DK.S. *The tubers are edible.* Pl. 56.

(ii) Stem broadly winged with wing more than 2 mm.
L. odoratus L. SWEET PEA. Fls. very large, 2½ cm. or more, sweet-scented, standard violet, wings and keel bluish, white, or pink. Leaflets 2, oblong-oval, blunt, but with a fine point. Fr. oblong, hairy; seeds smooth. Native of Southern Italy, but widely grown as an ornamental plant and sometimes self-seeding.

560. *L. sylvestris* L. NARROW-LEAVED EVERLASTING PEA. A rather robust, hairless, climbing perenn. 1–2 m., with rose-pink or dull pink and green fls. 14–18 mm., in clusters of 3–12. Fl. stalk 10–20 cm., longer than the lvs. Leaflets 2, 5–15 cm., narrow lance-shaped, usually three-veined; lv. stalks narrowly winged; tendrils branched; stipules not half as wide as the broadly winged stem. Fr. 4–7 cm. by 5–13 mm., hairless; seeds with small swellings. △ Hedges, thickets. June–Aug. Most of Eur. (except IRL.IS.GR.TR.).

L. latifolius L. EVERLASTING PEA. Like 560, but fls. larger, 2–3 cm., generally bright reddish-carmine. Leaflets oval, elliptic, or oblong 4–15 cm., usually five-veined; lv. stalks broadly winged; stipules more than half as wide as the broadly winged stem. Fr. 5–11 cm., brown, hairless; seeds with conspicuous swellings. Native of Southern Europe; often grown as an ornamental and escaping elsewhere. Pl. 56.

(b) Leaflets 4–12
(i) Stems winged
561. *L. clymenum* L. Fls. 1½–2 cm. with a bright reddish-purple standard and paler bluish-purple wings and keel, 1–5 on a stalk longer than the lvs. Lower lvs. simple, lance-shaped, without tendrils, the upper lvs. with 2–4 pairs of narrow lance-shaped to linear leaflets and a branched tendril; lv. stalk winged. Fr. 3–7 cm., brown, hairless. A rather glaucous climbing ann. 30–100 cm., with winged stems. △ Waysides, grassy and stony places, field verges. Apr.–June. Med. Eur. Pl. 56.

562. *L. montanus* Bernh. (*L. macrorhizus* Wimmer) BITTER VETCH. Fls. *c.* 1½ cm., crimson, becoming blue or green, in clusters of 2–6 on stalks as long as or longer than the lvs. Leaflets 2–4 pairs, 1–5 cm. linear to elliptic, glaucous beneath; tendril absent, but a fine point present; lv. stalk winged. Fr. 2½–4½ cm. by ½ cm., red-brown, hairless. A hairless perenn. with creeping rhizomes, swollen tuberous roots and erect, winged, non-climbing stems 15–30 cm. △ Bushy places, hedgerows, pastures. Apr.–July. Much of Eur. (except IS.GR.TR.BG.R.).

L. palustris L. MARSH PEA. Differs from 562 in having a climbing stem and lvs. with branched tendrils. Fls. 3–8, *c.* 18 mm., pale purplish-blue on a stem longer than the lvs. Leaflets 2–5 pairs, oblong or lance-shaped; stem and lv. stalk very narrowly winged. Fr. brown, 2½–6 cm. by 5–7 mm., flattened, hairless. Marshes. Northern and Central Europe.

(ii) Stems angled
563. *L. niger* (L.) Bernh. BLACK PEA. An erect, branched, non-climbing perenn. 30–80 cm., with violet-purple fls. 1–1½ cm., becoming blue, in clusters of 2–10 on a slender, sparsely

hairy stem equal to or longer than the lvs. Calyx with crisped hairs, teeth unequal. Lvs. more or less two-ranked, with 3–6 pairs of oval to elliptic fine-pointed leaflets, glaucous beneath; tendrils absent. Fr. *c.* 5 cm. by 5–6 mm., rough, swollen, black. Plant blackening on drying. △ Open woods. May–July. Most of Eur. Pl. 56.

564. *L. vernus* (L.) Bernh. SPRING PEA. An erect, hairless perenn. 20–40 cm., with fls. usually purple, then blue and finally brownish-purple, 1½–2 cm., in lax clusters of 3–10 on a stem longer than the lvs. Lvs. distinctive, with 2–4 pairs of broadly oval, long-pointed, soft, shining leaflets, each to 6 cm. long; tendrils absent; stipules large, longer than the lv. stalk. Fr. 4–6 cm. by ½ cm., hairless, brown or black. △ Woods, thickets. Apr.–June. Much of Eur. (except P.IRL.GB.IS.GR.TR.): introd. B.NL. Pl. 56.

565. *L. japonicus* Willd. SEA PEA. Fls. usually with purple standard and bluish wings, rather large, 1½–2½ cm., in a rather dense cluster of 2–12 on a stalk shorter than the lvs. Leaflets 3–5 pairs, elliptic, 2–4 cm.; tendril simple or branched; stipules large *c.* 2 cm., broadly triangular. Fr. 3–5 cm. by 8 mm., reddish-brown, finely hairy, but later hairless. A spreading, hairless perenn. 30–90 cm. △ Sands, dunes by the sea. July–Aug. West Eur. (except B.NL.), North Eur. D.PL. Pl. 56.

PISUM | Pea Stipules large, leafy, and rounded at base, larger than the leaflets; tendrils branched; stems not winged. Calyx lobes leafy, unequal, the upper 2 shorter and broader. Fr. cylindrical in section. 1 sp.

566. *P. sativum* L. PEA. A hairless climbing ann. to 1 m. with pink or violet fls. with black-purple or whitish wings, 1–3 on a stalk shorter than the lvs. Fls. large 1½–3½ cm. Leaflets 1–3 pairs, oval, entire or toothed, smaller than the large, leafy, oblong-oval, often blotched stipules; tendrils branched. Fr. 4–10 cm., to 1½ cm. wide. Subsp. *sativum* has white or purple fls. borne on a stem shorter than or as long as the stipules. Seeds somewhat angular, rough or smooth. Widely cultivated. Subsp. *elatius* (Bieb.) Ascherson & Graebner has a lilac standard, dark purple wings and a pink or green keel and fls. borne on a stem longer than the stipules. Seeds granular. △ Bushy places, cultivated ground; often cultivated and sometimes naturalized. Apr.–June. Med., South-East Eur. P.SU.: introd. A. *Cultivated since prehistoric times for its edible seeds which can either be eaten green, dried, or preserved. The seeds are rich in vitamins B and C.* Pl. 56.

ONONIS | Restharrow Lvs. simple or trifoliate, often both on the same plant, veins of leaflets ending in marginal teeth; tendrils absent; stipules conspicuous, fused to the lv. stalk. Calyx deeply toothed, teeth nearly equal. Fr. ovoid or oblong, longer or shorter than and encircled by the calyx. 49 sps.

Fls. pink, purple, white
(a) Anns.
567. *O. reclinata* L. SMALL RESTHARROW. A rather slender spreading ann. 5–25 cm., with sticky, densely glandular-hairy, trifoliate lvs. and pink fls. in a dense, terminal, leafy cluster. Petals 7–8 mm., as long as or shorter than the calyx. Leaflets wedge-shaped, toothed only at apex. Fr. deflexed at maturity, hairy. △ Sands and rocks by the sea. May–July. Med. Eur. P.GB.BG.

(b) Perenns.; sometimes woody
568. *O. spinosa* L. SPINY RESTHARROW. Distinguished by its strongly spiny stems, which have 1 or 2 longitudinal rows of hairs. Fls. pink, 1–2½ cm., forming a lax leafy cluster; wings shorter than keel; calyx glandular-hairy. Lvs. mostly trifoliate, but upper often simple; leaflets oblong-oval, toothed. Fr. obliquely oval, hairy, mostly one-seeded. A

glandular-hairy, erect or spreading, woody-based perenn. 30–60 cm. △ Dry, grassy places, waysides. Apr.–Sept. All Eur. (except IRL.IS.SF.).

569. *O. repens* L. RESTHARROW. Like 568, but stems spreading, stoloniferous, and rooting at the base, hairy all round, often spineless or with a few rather soft spines, but without stiff spines. Fls. pink, 1½–2½ cm.; wings equalling keel. Lvs. trifoliate or the upper simple; leaflets obovate to 2 cm. A very variable sp. △ Pastures, uncultivated ground, sandy places. June–Sept. Much of Eur. (except IS.H.AL.GR.): introd. SF.

O. fruticosa L. SHRUBBY RESTHARROW. An erect, much-branched shrub with handsome pale-pink fls. veined with purple, in terminal, leafless, elongated, branched clusters. Petals 1½–2 cm., standard hairy; bracts *c.* 5 mm., oval-acute, toothed. Lvs. trifoliate, with stalkless, leathery, spathulate leaflets with conspicuous saw-toothed margin. Fr. glandular-hairy. Iberian Peninsula and France. Pl. 57.

570. *O. rotundifolia* L. ROUND-LEAVED RESTHARROW. An erect, woody-based perenn. 30–50 cm., with 2–3 large pink fls. on long stalks up to 7 cm. in the axils of the upper lvs. Fls. 1½ cm. or more. Lvs. glandular-hairy, trifoliate with 3 almost circular, conspicuously toothed leaflets, the middle leaflet larger and long-stalked, the lateral leaflets stalkless. Fr. 2½–3 cm., pendulous, glandular-hairy. △ Rocky places, woods. May–Aug. E.F.CH.A.I. Pl. 57.

Fls. yellow

571. *O. natrix* L. LARGE YELLOW RESTHARROW. Fls. yellow, streaked with red, *c.* 1½ cm., solitary on stalks shorter than or as long as the lvs. and forming terminal leafy clusters. Lvs. mostly trifoliate, short-stalked; leaflets oblong-elliptic, toothed; stipules oval-lance-shaped, entire. Fr. ½–2 cm. by 3–4 mm., hairy, pendulous. A dense, much-branched, leafy, sticky glandular-hairy, woody-based variable perenn. 30–50 cm. △ Dry slopes. May–Aug. P.E.F.D.CH.I.YU.GR. Pl. 57.

572. *O. pusilla* L. Fls. yellow, small *c.* 8 mm., almost stalkless in a dense, terminal, leafy cluster; petals shorter or scarcely longer than the long, narrow-pointed calyx teeth. Lvs. trifoliate, the upper often simple; leaflets obovate, finely toothed. Fr. *c.* 7 mm., hairy, erect, shorter than calyx. An erect glandular-hairy perenn. 10–30 cm. △ Dry banks, gravelly places. June–Aug. Med., South-East Eur. P.CH.A.PL.CS.H.SU.: introd. B. Page 198.

MELILOTUS | Melilot Fls. small, yellow or white, numerous in narrow, elongated, leafless spikes. Lvs. trifoliate; leaflets stalked, margin toothed, apex usually blunt, but with a fine point. Fr. short, straight, thick, usually not splitting. Usually anns. or bienns. 16 sps.

Fls. white

573. *M. alba* Medicus WHITE MELILOT. An erect, slender, branched ann. or bienn. to 1½ m., readily distinguished by its slender terminal and axillary spikes of tiny white fls. much longer than the lvs. Fls. 4–5 mm.; standard longer than the wings. Fr. 3–5 mm., hairless, with netted surface, becoming blackish when ripe. △ Waysides, fields, shores; often occurring as a casual. June–Aug. Most of Eur. Pl. 57.

Fls. yellow
(a) Fr. hairless

574. *M. indica* (L.) All. (*M. parviflora* Desf.) SMALL-FLOWERED MELILOT. Distinguished by its very tiny pale yellow fls. in dense cylindrical spikes ½–2 cm. long, about as long as the lvs. Fls. 2–3 mm.; standard longer than the equal wings and keel. Fr. 1½–2½ mm., almost circular, hairless, netted, olive green when ripe. An erect, branched ann. 15–50 cm.

△ Waste places, salt-rich ground, damp sands. May–June. Med. Eur. P.: introd. Cent. Eur. (except PL.) GB.B.NL.

575. *M. officinalis* (L.) Pallas (*M. arvensis* Wallr.) COMMON MELILOT. A spreading or erect branched ann. or bienn. to 1 m., distinguished from other sps. by the keel of the fl. which is shorter than the wings and standard. Fls. yellow, 5–7 mm., in rather lax clusters 4–10 cm. long, and much longer than the lvs. Fr. 3–5 mm., hairless, transversely wrinkled, blunt with a fine tip, becoming brown when ripe. △ Waste places, tracksides, fields, rocks. May–Sept. All Eur. *Sometimes grown as a fodder crop; it has been used as an insect repellent in clothing, etc., and for flavouring cheese. It contains coumarin, giving the scent of new-mown hay on drying.*

M. sulcata Desf. FURROWED MELILOT. Like 575, but fr. globular, 3–4 mm., with numerous concentric ridges on each side. Fls. yellow, 3–5 mm., in clusters 1–1½ cm. long of 8–25 fls., as long as or longer than the lvs., but elongating to 4 cm. in fr. Mediterranean region and Portugal.

M. dentata (Waldst. & Kit.) Pers. Like [575], but fls. more numerous, 35–50, in clusters equalling the lvs. Fls. yellow, 3–3½ mm. Fr. 4–5½ mm., with a network of wrinkles, greyish to blackish. Salt-rich places of Central Europe. Page 206.

(b) Fr. hairy

576. *M. altissima* Thuill. TALL MELILOT. An erect, branched ann. to 1½ m. like 575, but distinguished by the standard, keel and wings which are all of the same length, and the hairy, netted fr. 4–6 mm. which become black when ripe. Fls. yellow, 5–7 mm., in compact elongated clusters 2–5 cm. long, twice as long as lvs. and lengthening in fr. △ Waste places, damp ground, waysides, salt-rich areas. May–Sept. Much of Eur. (except P.IS.SF. TR.BG.): introd. IRL.GB. Pl. 57.

TRIGONELLA | Fenugreek Anns. with trifoliate, toothed lvs. Fr. linear or oblong, splitting or not, straight or curved, with seeds in 2 rows. Fls. usually numerous in short clusters, rarely solitary. 23 sps.

Fls. numerous in a cluster
(a) Fls. blue

577. *T. caerulea* (L.) Ser. BLUE FENUGREEK. An erect, almost hairless, unbranched ann. 30–100 cm., with dense globular heads of lilac-blue fls. on stems longer than the lvs. Fls. 5–7 mm. Leaflets blunt, oval to oblong, finely toothed. Fr. erect, ovoid, longitudinally ribbed, little longer than the calyx. △ Fields; cultivated and rarely naturalized. June–July. Introd. to South-East Eur. *Cultivated as a fodder crop; used for flavouring cheese.* Pl. 57.

(b) Fls. yellow

578. *T. monspeliaca* L. STAR-FRUITED FENUGREEK. Fls. yellow, small 4–5 mm., 4–15 in stalk-less clusters in the axils of the much longer lvs. Calyx with equal teeth, hairy. Leaflets oval-wedge-shaped, toothed towards apex. Fr. 1–1½ cm. linear, fine-pointed, curved, hairy, carpels spreading in a star. A prostrate or spreading, finely hairy ann. 5–40 cm. △ Rocks, stony ground, dry places. Mar.–June. Much of Eur. (except North Eur. IRL.GB.NL.D.PL.).

579. *T. corniculata* (L.) L. SICKLE-FRUITED FENUGREEK. Fls. yellow, *c.* 5 mm., 8–15 in rounded heads on a long stem much longer than the lvs. Calyx with very unequal teeth, hairless. Lower lvs. with oval-wedge-shaped leaflets, the upper with narrower leaflets. Fr. drooping, 1–1½ cm., sickle-shaped, long-pointed, with transverse veins. An erect, hairless ann. 20–50 cm. △ Cultivated ground, dry banks, grassy places. Apr.–June. Med. Eur. (except TR.).

Fls. solitary or 2, whitish

580. *T. foenum-graecum* L. FENUGREEK. An erect, almost hairless ann. 15–50 cm., with stalkless solitary or paired whitish fls. in the axils of the upper lvs. Fls. 1–1½ cm.; calyx with equal linear-lance-shaped teeth, hairy. Leaflets oblong to obovate, toothed near apex. Fr. erect, hairless, linear 7–10 cm., progressively narrowing to a slender beak 3–5 cm. △ Fields; cultivated and sometimes naturalized. Apr.–May. Native of Asia: introd. to much of Eur. (except North Eur. IRL.GB.PL.) *The seeds are edible; they contain coumarin and have been used medicinally since classical times; they are used in veterinary treatments. Sometimes grown as a fodder crop.* Pl. 58.

MEDICAGO | **Medick** Usually herbaceous anns. or perenns. with trifoliate, toothed lvs. Fls. small, clustered, usually orange; petals falling early; calyx teeth more or less equal. Fr. usually coiled into a spiral, less commonly curved or sickle-shaped, often spiny, longer than the calyx. 37 sps.

Fr. sickle-shaped or bean-shaped

581. *M. lupulina* L. BLACK MEDICK. Fls. bright yellow, tiny, in dense, globular clusters 3–8 mm., on stalks longer than lvs.; like 590, but differing in fr. Fls. 2–3 mm. Leaflets 3–20 mm., oboval, finely toothed; stipules lance-shaped, long-pointed, shortly toothed at base. Fr. 2 mm., kidney-shaped, coiled in almost 1 complete turn, netted, black when ripe. An erect or spreading, branched, hairy ann. to perenn. 5–50 cm. △ Fields, grassy places, waysides. Apr.–Oct. All Eur. *Sometimes grown as a fodder crop.*

582. *M. sativa* L. subsp. *falcata* (L.) Arcangeli SICKLE MEDICK. Fls. yellow, 5–8 mm., in dense ovoid clusters up to 2½ cm. long. Leaflets narrow, oblong-wedge-shaped to 1½ cm.; stipules lance-shaped, not toothed. Fr. sickle-shaped or sometimes nearly curved in a ring, or nearly straight, often somewhat hairy. A hairy, erect, branched, woody-based perenn. 30–80 cm. *M. × varia* Martyn is the hybrid between this subsp. and 583 and shows much variation. Fls. yellow, purple, or yellow becoming dark purple. Fr. almost straight, to a spiral of 2–3 turns. △ Grassy places, waysides, banks. May–Aug. All Eur. (except IS.). Pl. 58.

Fr. in a spiral of 1 or more turns
(a) Fr. without spines

583. *M. sativa* L. subsp. *sativa* LUCERNE, ALFALFA. Fls. purple or blue, in dense, cylindrical clusters up to 4 cm. long, on stems longer than the lvs.; petals 7–11 mm. Leaflets oblong to obovate, to 3 cm.; stipules linear-lance-shaped, long-pointed, more or less toothed. Fr. an open spiral of 2–4 turns, generally hairy. A nearly hairless, erect, much-branched perenn. 30–80 cm. △ Grassy places, waste and cultivated ground; widely cultivated and often naturalized. June–Sept. Origin unknown; introd. throughout Eur. *One of the most valuable forage crops of temperate and warm-temperate regions, which thrives particularly well in semi-arid regions under irrigation.* Pl. 58.

584. *M. orbicularis* (L.) Bartal. LARGE DISK MEDICK. Distinguished by its large, flattened, disk-shaped frs. 12–18 mm. across, with a spiral of 3–5 smooth turns, closed at the centre and with conspicuous radiating veins when dry. Fls. 3 mm., 1–5 on a fl. stem shorter than the lvs. and ending in a fine point. Leaflets obovate or triangular, toothed; stipules deeply cut into slender teeth. A spreading, almost hairless ann. 20–70 cm. △ Cultivated ground, grassy places, olive grooves. Apr.–July. Med. Eur. (except TR.) P.BG.R.SU.: introd. H. Pl. 58.

585. *M. arborea* L. TREE MEDICK. A dense shrub 1–4 m. with young branches with dense silky hairs and usually entire oval leaflets which are silky-haired beneath. Fls. golden-yellow 1–2 cm., bórne in rather dense, erect clusters. Fr. hairy, flattened, net-veined, sickle-shaped, or more often in a single open spiral 1–1½ cm. across. △ Rocky places; sometimes grown as an ornamental. Apr.–Aug. E.I.AL.GR.: introd. P.F.

(b) Fr. with spines or projections

586. *M. arabica* (L.) Hudson (*M. maculata* Sibth.) SPOTTED MEDICK. A bright green, almost hairless, spreading ann. to 60 cm., distinguished by its more or less globular fr. 4–6 mm., with a spiral of 3–5 close turns and with a double row of curved hooks along the outer margins of the spirals. Fls. 4–6 mm., 1–5 on a stem shorter than the long lv. stalk. Leaflets obovate, large to 2½ cm., often spotted with brown; stipules deeply toothed. △ Waysides, fields, grassy places. May–July. Most of Eur. (except North Eur. A.PL.): introd. IRL. S.CH.CS.

587. *M. minima* (L.) Bartal. SMALL MEDICK. A hairy, often greyish, erect or spreading ann. to 20 cm., with densely hairy, globular fr. 3–5 mm., with a spiral of 4–5 close turns and with a double row of weakly hooked spines. Fls. 3–4 mm., 1–5 on a stem as long as or longer than the lv. stalk. Leaflets 3–6 mm., obovate or wedge-shaped, hairy on both sides; stipules almost entire. △ Banks, dry grassy places, waysides. Mar.–June. Most of Eur. (except IRL.IS.N.SF.).

588. *M. polymorpha* L. (*M. hispida* Gaertner) HAIRY MEDICK. Like 587, but fr. 4–9 mm. across, more disk-like, distinctly flattened from above and below, with a spiral of 1½–4 close turns, with a narrow outer margin with 2 rows of spines; fr. strongly netted on surface. Fls. 3–8, on stems about as long as the lv. stalks; keel shorter than the wings. Leaflets not blotched, hairless; stipules cut into slender segments. A very variable, spreading ann. or bienn. 5–60 cm. △ Cultivated ground, waysides. May–June. Med., South-East Eur. P.GB.CH.D.A. H.SU.: introd. B.NL.CH.CS.H. Pl. 58.

589. *M. marina* L. SEA MEDICK. A densely white-woolly spreading perenn. to ½ m., growing on the littoral. Fls. pale yellow, 5–10 in a rounded, short-stalked cluster; petals 6–8 mm. Lvs. white-woolly. Fr. densely white-woolly, in a spiral of 2–3 close turns, usually with short, spreading spines or less commonly without spines. △ Maritime sands. Apr.–June. Med. Eur. P.BG.R.SU. Pl. 58.

TRIFOLIUM | **Clover, Trefoil** Fls. usually numerous, stalkless or shortly stalked, and clustered into dense, rounded heads. Petals usually persisting in fr. Fr. small, one to several-seeded, more or less enclosed by the calyx. Anns. or herbaceous perenns. with trifoliate lvs. and stipules often fused to the lv. stalk. 99 sps.

Fls. bright yellow, or brownish-yellow

590. *T. dubium* Sibth. (*T. minus* Sm.) SUCKLING CLOVER, LESSER YELLOW TREFOIL. A slender, spreading ann. 5–25 cm., with tiny, stalked, axillary heads 5–7 mm. across, of 10–25 yellow fls. Petals *c.* 3 mm., turning dark brown; standard narrow, folded over fr. and about the same length. Leaflets to *c.* 1 cm., obovate. Fr. 2½–3 mm. △ Grassy places, waysides. May–Sept. All Eur. (except IS.SF.AL.TR.).

*T. micranthum Viv. SLENDER TREFOIL. Western and Southern Europe.

591. *T. campestre* Schreber (*T. procumbens* L.) HOP-TREFOIL. Like 590, but fls. more numerous, and fl. heads more conspicuous, larger, 10–15 mm. Petals 4–5 mm., bright yellow turning pale brown, the standard broad, conspicuously veined, strongly curved over keel, and much longer than the fr. Leaflets obovate, the terminal leaflet stalked; stipules half-oval. Fr. 2–2½ mm.; style long. A more robust, erect, or ascending ann. 10–30 cm. △ Grassy places, waysides. May–Oct. All Eur. (except IS.): introd. SF.

592. *T. aureum* Pollich (*T. agrarium* L.) LARGE HOP-TREFOIL. Like 591, but a larger, erect, branching ann. or bienn. 20–60 cm., with relatively large, oval, golden-yellow fl. heads 1½ cm. long. Petals 6 mm., turning pale brown. Leaflets oblong-oval, the terminal leaflet

LEGUMINOSAE 593–598

stalkless; stipules linear to oblong, long-pointed. Fr. usually two-seeded, with a style half as long as the ovary. △ Thickets, grassy places. June–Aug. Much of Eur. (except P.IS.AL.TR.): introd. GB.

593. *T. badium* Schreber BROWN TREFOIL. An erect mountain perenn. 10–25 cm., with golden-yellow fls. in fat, globular heads 8–20 mm., and with old fls. soon turning conspicuously leather-brown. Fls. 7–9 mm.; standard spreading, arched, strongly veined; calyx teeth very unequal, the lower with long hairs. Upper lvs. opposite; leaflets all stalkless, obovate. △ In mountains: meadows, rocky places. July–Aug. Cent. Eur. (except H.) E.F.I.YU.AL.BG.R. Pl. 60.

Fls. purple, pink, white, or yellowish-white
(a) Fl. heads elongate, at least twice as long as broad
594. *T. arvense* L. HARE'S-FOOT CLOVER. A slender, softly hairy, and often greyish, erect, or spreading, branched ann. or bienn. 5–40 cm., with ovoid-cylindrical, very softly downy heads of tiny pinkish or whitish fls. Fl. heads numerous, stalked, 1–2½ cm. long; petals white or pink, much shorter than the hairy calyx. Upper lvs. stalkless; leaflets 1–2 cm. narrowly oblong; stipules oval, fine-pointed. △ Sandy fields, dunes. May–Oct. All Eur. (except IS.). Pl. 60.

595. *T. incarnatum* L. CRIMSON CLOVER. A striking plant with long conical or cylindrical terminal heads to 6 cm. of usually crimson or rarely pink or pale cream fls. Fls. 1–1½ cm.; petals longer than the densely hairy calyx which has a strongly ribbed tube and slender teeth which spread in fr. Leaflets obovate to almost circular ½–3 cm. across; stipules papery, oval, blunt or acute, toothed. An erect, unbranched, softly hairy ann. 20–50 cm. △ Grassy places; cultivated and often escaping. May–July. All Eur. (except IRL.IS.). *Cultivated as a fodder crop.* Pl. 59.

596. *T. angustifolium* L. NARROW-LEAVED CRIMSON CLOVER. Rather like 595, but differing in the leaflets which are narrow-oblong to linear acute, and the linear acute, entire stipules. Fls. pink, all opening at about the same time; corolla smaller, about as long as calyx which has a thickened throat. An erect ann. 10–40 cm., with adpressed hairs. △ Dry hills, banks, sandy places. Apr.–July. Med. Eur. P.CS.BG.R.SU.

T. purpureum Loisel. PURPLE CLOVER. Very like 596, but fl. heads crimson-purple, more conical and fls. opening from below upwards, the lowest dying before the uppermost open. Petals much longer than the calyx which has very unequal teeth. An ann. with spreading hairs. France to Balkan Peninsula.

597. *T. rubens* L. Like 607 and 608, but distinguished by its cylindrical, reddish-purple, often paired fl. heads 5–6 cm. by 2–2½ cm., which are long-stalked and without a lv. close below the fl. head. Fls. 12–15 mm., very numerous; calyx with the lowest tooth many times longer than the upper, teeth hairy, tube hairless. Leaflets oblong 5 cm. by 1–1½ cm., somewhat leathery, strongly veined and with finely toothed margin; stipules conspicuous, narrow lance-shaped. A robust, hairless, erect perenn. 25–60 cm. △ Rocky places, woods. June–Aug. Much of Eur. (except North Eur. P.IRL.GB.NL.AL.TR.BG.). Page 206. Pl. 59.

(b) Fl. heads globular
(i) Fl. heads stalkless in axils of lvs.
598. *T. striatum* L. SOFT TREFOIL. A softly hairy, erect, or spreading ann. 5–40 cm., with pink fls. in small, almost stalkless, ovoid heads to 1½ cm. Petals little longer than the hairy calyx which is conspicuously ribbed and has bristly-pointed teeth which are erect in fr. Leaflets oval acute, hairy, lateral veins nearly straight and thin at margin. △ Dry, grassy places, sandy soils, waysides. June–Aug. All Eur. (except IS.N.SF.).

205

1. *Trifolium scabrum* [598]
3. *Trifolium rubens* 597
5. *Anthyllis hermanniae* 619

2. *Melilotus dentata* [575]
4. *Tetragonolobus maritimus* 616

T. scabrum L. ROUGH TREFOIL. Like 598, but calyx becoming stiff in fr. with spreading, curved, spiny teeth. Fls. white, in globular, stalkless heads c. 1 cm. Leaflets with lateral veins curving backwards and thickening at the margin. Western and Southern Europe. Page 206.

*T. suffocatum L. SUFFOCATED CLOVER. Western Europe and Mediterranean region.

*T. subterraneum L. SUBTERRANEAN CLOVER. Western and Southern Europe.

599. *T. tomentosum* L. WOOLLY TREFOIL. A slender, creeping ann. 5–15 cm., distinguished by its woolly-white, globular, stalkless fr. heads 1–1½ cm. across in the axils of the lvs. Fls. pink, 3 mm.; calyx soon becoming inflated, densely woolly and with a netted surface. Leaflets obovate. △ Dry, sandy places. Apr.–June. Med. Eur. P.

(ii) Fl. heads axillary or terminal, distinctly stalked and held at least 2 cm. above the attachment of the nearest lv.

600. *T. repens* L. WHITE or DUTCH CLOVER. A hairless perenn. with creeping stems rooting at the nodes, and bearing white or pinkish fls. in dense, globular heads on long leafless stalks to 30 cm. Fl. heads 1½–3 cm.; fls. 8–10 mm., drooping after flowering; calyx tube bell-shaped, white with green veins, teeth triangular, about half as long. Lvs. long-stalked; leaflets 1–3½ cm., obovate to obcordate, often whitish-blotched above. △ Meadows, waysides. May–Oct. All Eur. *An excellent forage plant for cattle; usually seeded mixed with grasses to form temporary leys, and also encouraged in permanent pastures. An important nitrogen-fixing plant like other clovers.*

601. *T. hybridum* L. ALSIKE CLOVER. Like 600, but an erect or spreading perenn. to 60 cm., not rooting at the nodes and stems leafy and usually hollow. Fls. 6–7 mm., white or bright pink, becoming brownish, in lax heads 1½–3 cm. across; fl. stalks to 3 times as long as calyx tube. Calyx tube bell-shaped, white, and with linear, long-pointed teeth nearly twice as long as the tube. Leaflets 1–3½ cm., obovate-elliptic, toothed, hairless, and without blotches. △ Waysides, fields. May–Sept. All Eur. (except IS.AL.). *Cultivated as a fodder crop.* Pl. 59.

T. montanum L. MOUNTAIN CLOVER. Very like 601, but stem and lower surface of leaflets hairy. Fls. white in very dense, globular heads c. 1½ cm. across; fl. stalks much shorter than the calyx tube. Leaflets strongly veined at the margin, finely toothed. Usually in mountains. Widespread in Europe. Pl. 59.

602. *T. stellatum* L. STAR CLOVER. A Mediterranean ann. readily distinguished by its conspicuous, globular fr. heads with stiff, often brightly coloured, pointed calyx teeth spreading in a star. Fl. heads long-stalked, solitary, terminal, pale pink; petals scarcely longer than the densely silky-haired calyx, which has teeth at first erect but soon widespreading, throat of calyx closed by hairs. Leaflets obcordate, finely toothed. An erect, softly hairy ann. 5–25 cm. △ Dry places, sands, tracksides. Apr.–June. Med. Eur. P.: introd. GB. Pl. 60.

603. *T. alpinum* L. ALPINE CLOVER. A low, creeping, woody-based perenn. of the mountains, readily distinguished by the lax head of few, very large rosy-purple, sweet-smelling fls. Fls. 3–12 on a leafless stem, 5–15 cm., arising direct from the rootstock; petals c. 2 cm., rarely white; calyx hairless. Leaflets 1–10 cm., linear-lance-shaped, strongly veined and finely toothed. △ In mountains: rocky meadows. June–Aug. E.F.CH.A.I. Pl. 60.

604. *T. fragiferum* L. STRAWBERRY CLOVER. A creeping perenn. 10–30 cm., often mistaken for 600, rooting at the nodes, with long-stalked axillary globular pink fl. heads 1–2 cm. Petals pink, 6–7 mm.; calyx very distinctive in fr., the upper lip becoming strongly

inflated and thus the whole fr. head at length globular, hairy, and strawberry-like in shape. Lvs. long-stalked; leaflets 1–1½ cm., obovate or obcordate, margin finely toothed. △ Grassy places, fields, waysides. May–Sept. All Eur. (except IS.).

(iii) Fl. heads terminal, short-stalked and with 2 lvs. close below fl. head

605. *T. pratense* L. RED CLOVER. Fls. bright pink-purple or rarely whitish, in a dense, globular head up to 3 cm. and with 2 lvs. immediately below. Head many-flowered, becoming ovoid; petals 15–18 mm.; calyx hairy, teeth unequal, throat hairless. Leaflets 1–5 cm. elliptic-obovate, often with a whitish blotch above, hairy beneath; stipules with free part triangular and bristle-tipped. A very variable, somewhat hairy, erect or spreading perenn. to 60 cm. △ Meadows, fields, waysides. May–Sept. All Eur.: introd. IS. *An excellent forage plant, rich in proteins and minerals and now commonly grown in temporary leys which can be cut three times a year. Grown in preference to lucerne in regions with a cool spring climate.*

606. *T. ochroleucon* Hudson SULPHUR CLOVER. Fls. whitish-yellow or rarely pale pink, in globular heads 2–3 cm., shortly stalked above the subtending pair of lvs. Heads becoming ovoid; petals *c.* 1½ cm.; calyx hairy, ribbed, teeth unequal, the lowest tooth 2–3 times as long as the rest. Leaflets 1–3 cm. obovate-oblong, softly hairy. An erect, hairy perenn. 10–40 cm. △ Dry, grassy places, waysides, clearings. May–Aug. Much of Eur. (except North Eur. IRL.NL.).

607. *T. medium* L. ZIGZAG CLOVER. Like 605, but leaflets narrow-elliptic; stipules with free part spreading and narrowed gradually to a long point. Fl. heads 2–4 cm., reddish-purple; petals 1–2 cm.; calyx tube hairless or nearly so, teeth and throat hairy, teeth unequal. An ascending, little-branched stoloniferous perenn. 20–50 cm. △ Grassy and cultivated places. May–July. All Eur. (except IS.AL.). *Sometimes grown as a forage crop.* Pl. 59.

608. *T. alpestre* L. MOUNTAIN ZIGZAG CLOVER. Like 607, but leaflets oblong-lance-shaped, blunt-tipped, rather stiff and leathery, strongly veined and finely toothed; stipules with free part linear. Fls. reddish-purple, in globular heads 1½–2½ cm. with 2 closely subtending lvs. with broad egg-shaped stipules. Petals 1½ cm.; calyx tube twenty-veined, tube and teeth hairy. A stoloniferous perenn. with erect, little-branched stems 10–40 cm. △ In mountains and hills: grassy places, open woods. June–Aug. Cent., South-East Eur. (except TR.) F.B.DK.I.SU.

DORYCNIUM   Fls. white with a blackish keel, rather numerous in stalked, rounded, axillary heads; wings of fls. fused. Fr. somewhat swollen, longer than calyx, splitting, one-to four-seeded. Lvs. trifoliate; stipules large, leaf-like. 4 sps.

Fls. 1 cm. long or more

609. *D. hirsutum* (L.) Ser. (*Bonjeanea h.*) A densely shaggy-haired, greyish, woody-based herbaceous perenn. or shrublet to ½ m., with lax, short-stalked, globular clusters of whitish fls. flushed with pink and with a darker keel. Fls. 5–10; petals 1½ cm.; calyx shaggy-haired, teeth awl-shaped, unequal. Leaflets and stipules ½–2½ cm., similar, narrow-egg-shaped or obovate, shaggy-haired. Fr. 6–12 mm., little longer than the calyx. △ Rocky, grassy, and sandy places. Apr.–July. Med. Eur. P. Pl. 60.

Fls. 5–8 mm.

610. *D. rectum* (L.) Ser. (*Bonjeanea r.*) Like 609, but fls. more numerous, 20–40, each 5–6 mm., white or pink with a darker or blackish keel; calyx teeth equal. Leaflets oval to obovate, glaucous, with adpressed hairs beneath; lv. stalk ½–1 cm.; stipules triangular-oval as long as the lv. stalk. Fr. 1–2 cm., much longer than the hairy calyx, valves twisting, hairless.

A hairy perenn. or small shrub to 1 m. △ Damp bushy places, ditches. May–Aug. Med. Eur. (except YU.) P. Pl. 60.

611. *D. pentaphyllum* Scop. A perenn. or dwarf shrub to 80 cm., distinguished from 610 by the leaflets which are linear to oboval-oblong; lv. stalk absent. Fls. 3–6 mm., white with dark red or black keel; calyx teeth unequal. Stems, lvs., and calyx all with adpressed silky hairs. Fr. 3–5 mm. A variable sp. △ Grassy places, dry banks, waysides. Apr.–June. Much of Eur. (except North Eur. IRL.GB.B.NL.).

LOTUS | **Birdsfoot-Trefoil** Fls. usually yellow in somewhat flat-topped clusters. Lvs. with 5 leaflets and with minute stipules (but described by many as trifoliate with 2 leafy stipules). Fr. elongated, many-seeded, usually with partitions between the seeds. 30 sps.

Anns.

612. *L. ornithopodioides* L. A hairy branched ann. 10–40 cm., distinguished by the terminal stalked clusters of 2–5 small yellow fls., with 3 broad oval bracts which are as long as or longer than the fl. cluster. In fr. the 2–5 slender curved pods hanging below the 3 spreading bracts are distinctive. Petals 5–10 mm.; calyx two-lipped, teeth very unequal. Leaflets oval-rhomboid. Fr. 2–5 cm., flattened, somewhat constricted between seeds. △ Grassy, rocky, and sandy places. Apr.–June. Med. Eur. P.SU.

L. angustissimus L. SLENDER BIRDSFOOT-TREFOIL. An ann. with lax, spreading hairs and heads of 1–3 fls. on stalks as long as or longer than the lvs. Fls. 6–12 mm.; calyx teeth nearly equal, longer than the tube. Fr. very slender, 2–3 cm. by 1–1½ mm. Southern Europe and Britain.

Perenns.

613. *L. corniculatus* L. BIRDSFOOT-TREFOIL, BACON AND EGGS. Fls. yellow, often streaked with red, *c.* 1½ cm., in a dense cluster of 2–8, on a long stalk to *c.* 8 cm. Calyx hairy, tube bell-shaped, teeth triangular and close together in bud. Upper 3 leaflets obovate 3–10 mm., the lower pair triangular-oval. Fr. to 3 cm. A very variable, more or less spreading, usually hairless perenn. 10–40 cm. △ Fields, meadows, waysides. May–Aug. All Eur.: introd. IS.

614. *L. uliginosus* Schkuhr LARGE BIRDSFOOT-TREFOIL. Like 613, but a more erect, more branched and usually more hairy perenn. ¼–1 m., with creeping stolons. Fls. 10–12 mm., deep yellow, in a dense cluster of 5–15 fls. on stalks to *c.* 15 cm.; calyx teeth spreading in bud. Leaflets obovate, usually 1½–2 cm.; stem hollow. △ Damp meadows, woods, and marshes. May–Aug. Most of Eur. (except IS.): introd. N.SF.H. Pl. 61.

*L. subbiflorus Lag. (*L. hispidus* DC.) HAIRY BIRDSFOOT-TREFOIL. Coasts of South-Western Europe.

615. *L. creticus* L. SOUTHERN BIRDSFOOT-TREFOIL. Usually distinguished by its silvery-white appearance with lvs. and stems with dense adpressed hairs, but very variable and some-times with few or no silvery hairs. Fls. yellow, in clusters of 1–8 on a stem much longer than the lvs. Petals 8–15 mm.; calyx two-lipped, the teeth unequal with the lowest tooth up to twice as long as the 2 lateral teeth. Leaflets oval, the 2 lowest longer than the common stalk bearing the 3 upper leaflets. Fr. 2–5 cm. A woody-based perenn. 10–40 cm. △ Rocks, sandy places, by the sea. Mar.–May. P.E.I.YU.GR. Page 198.

TETRAGONOLOBUS Like *Lotus*, but fr. quadrangular in section or with 4 conspicuous wings. Fls. yellow or reddish purple, 1 or 2, axillary. Lvs. trifoliate, with 2 leafy stipules. 5 sps.

616. *T. maritimus* (L.) Roth (*Lotus siliquosus* L.) WINGED PEA. A glaucous perenn. spreading and forming mats, with large and solitary pale yellow fls. 2½–3 cm. on a stalk much longer than the lvs. and with a single trifoliate bract with lance-shaped segments immediately below fl. Calyx teeth lance-shaped, about half as long as the tube. Leaflets large, obovate, hairy or hairless; stipules oval. Fr. 4–6 cm., with 4 narrow wings. △ Damp fields, marshes. May–July. Cent. Eur. E.DK.S.I.YU.BG.R.: introd. GB. Page 206.

617. *T. purpureus* Moench (*Lotus tetragonolobus* L.) ASPARAGUS PEA. A softly hairy, spreading ann. 10–40 cm., with 1 or 2 dark red-purple fls. often with an almost black keel, at first nearly stalkless, but at length on a stalk about as long as the subtending lv. Fls. *c.* 2 cm., with a trifoliate bract with obovate segments; calyx teeth as long as or longer than calyx tube. Fr. with 4 broad, undulate wings about as broad as the fr. itself, becoming black. △ Cultivated ground, waysides, banks, grassy places. Mar.–May. E.F.I.GR.SU. *Sometimes grown as a culinary plant.* Pl. 61.

HYMENOCARPUS Frs. flattened, disk-shaped; seeds kidney-shaped. Lvs. with 2–4 pairs of leaflets and a much larger terminal leaflet. 1 sp.

618. *H. circinnatus* (L.) Savi DISK TREFOIL. A softly hairy, spreading or erect ann. 10–50 cm., readily distinguished by its flattened, disk-like, almost circular, hairy fr. 1–2 cm. across, usually with finely spiny margin. Fls. yellow, *c.* 6 mm., 2–10 in a dense, long-stemmed cluster 1 cm. across. Lowest lvs. undivided, the upper lvs. with 3–5 pairs of small oval leaflets and with a terminal leaflet 2–3 times larger. △ Grassy places, cultivated ground, dry hills. Mar.–May. Med. Eur. BG.

ANTHYLLIS Fls. numerous, usually in dense heads or interrupted clusters. Calyx tubular, somewhat inflated, enclosing the fr. Lvs. pinnate with a terminal leaflet, trifoliate or simple; stipules small or none. 19 sps.

Shrubs with spines
619. *A. hermanniae* L. A dense, much-branched, spiny shrub to ½ m. with simple or trifoliate lvs. with narrow leaflets, and small, elongated, interrupted clusters of small yellow fls. Petals *c.* ½ cm.; calyx hairy; fl. clusters terminating in a spine which becomes woody. Leaflets 1–2 cm., linear-oblong blunt, with adpressed silky hairs beneath. Fr. 2–3 mm., hairless. △ Rocky places. May–Aug. I.YU.AL.GR.TR. Page 206.

Shrubs without spines
A. cytisoides L. A dense, spineless shrub with white-felted stems and young lvs. and long, interrupted spikes of pale yellow fls. Petals 5–8 mm.; calyx densely shaggy-haired; bracts leafy, entire. Lvs. somewhat fleshy, simple or trifoliate with the terminal leaflet much larger. Western Mediterranean.

620. *A. barba-jovis* L. JUPITER'S BEARD. A dense, silvery-white, leafy shrub to 1 m., with pinnate lvs. and with bright yellow fls. in compact globular heads at the end of the branches. Fl. heads with deeply cut, half-encircling silvery bracts; fls. numerous, *c.* 1 cm.; calyx woolly-haired. Lvs. with 4–9 pairs of linear-oblong leaflets and a terminal leaflet. △ Rocks by the sea. May–June. Med. Eur. (except AL.TR.).

Herbaceous plants or woody only at the base
621. *A. montana* L. MOUNTAIN KIDNEY-VETCH. A low, spreading, mat-forming perenn. 10–30 cm., with purple or pink fls. in dense solitary heads at the ends of stems arising directly from the woody base, and longer than the lvs. Fl. heads encircled by 2 deeply dissected leafy bracts; fls. *c.* 1 cm.; calyx densely woolly-haired. Lvs. with 8–15 pairs of elliptic,

hairy leaflets ½–1 cm. long and a terminal leaflet. A variable sp. △ In mountains: rocks, rocky meadows. June–July. E.F.A.CH.I.YU.AL.GR.BG.R. Pl. 61.

622. *A. vulneraria* L. KIDNEY-VETCH. Fls. yellow, reddish, or white, in dense, rounded, usually paired heads, half-encircled at the base by 2 deeply dissected leafy bracts. Fls. 12–15 mm.; calyx with dense, white-woolly hairs, distinctly contracted at the throat. Lvs. with 1–6 pairs of leaflets, the lower lvs. with the terminal leaflet much larger than the rest. A very variable sp. Subsp. *praepropera* (A. Kerner) Bornm. usually has rose, crimson, or purple fls. and occurs in the Mediterranean region; subsp. *alpestris* Ascherson & Graebner has large heads of yellow fls. and the lowest lvs. entire; it occurs in alpine regions. A hairy, branched, erect or spreading ann., bienn. or perenn. to 60 cm. △ Dry places, meadows, alpine pastures, and screes. May–Aug. All Eur. Pl. 61.

623. *A. tetraphylla* L. (*Physanthyllis t.*) BLADDER VETCH. A spreading, greyish-haired ann. 10–50 cm., distinguished in fr. by its very much inflated, often yellowish or reddish, silky-haired calyx. Fls. pale yellow, tipped with red, 2–7 in dense, stalkless, axillary clusters shorter than the lvs. Petals *c.* 1½ cm., little longer than the calyx. Lvs. with 1–2 pairs of leaflets and with a terminal oval leaflet much larger and broader than the rest. △ Grassy places, olive groves, cultivated ground. Mar.–July. Med. Eur. (except AL.TR.) P. Pl. 61.

CORONILLA   Hairless shrubs or herbs with pinnate, rarely trifoliate lvs.; stipules present. Calyx with 5 more or less equal teeth; petals narrowed to a stalk-line claw. Fr. narrow elongate, straight or curved, cylindrical-sectioned, splitting into several oblong segments. 13 sps.

Shrubs more than ½ m.

624. *C. emerus* L. SCORPION SENNA. A leafy shrub to 2 m., with slender, green, ribbed branches and clusters of yellow fls. often tipped with red. Fls. in long-stemmed clusters of 2–7; petals *c.* 2 cm. and with the claws 2–3 times as long as the cup-shaped calyx, standard abruptly upturned. Lvs. with 5–9 obovate, often notched leaflets 1–2 cm., hairless or sparsely hairy. Fr. 5–10 cm., pendulous, slender, jointed. △ Thickets, rocky places, hills. Apr.–June. Cent. Eur. (except PL.), Med Eur. N.BG.YU.SU. *The leaves are purgative.* Pl. 62.

625. *C. valentina* L. (incl. *C. glauca* L.) SHRUBBY SCORPION-VETCH. A small, glaucous, hairless tufted shrub to 1 m., with slender branches and fragrant yellow fls. in clusters of 4–12 on long stems to *c.* 7 cm. Petals 7–12 mm., claws shorter than the calyx. Leaflets 5–13, obovate, notched, to 2 cm. Fr. 1–5 cm., straight, slightly compressed, with 1–10 constrictions. △ Scrub, hills. Apr.–July. Med. Eur. (except TR.) P.: introd. GB.

626. *C. juncea* L. RUSH-LIKE SCORPION-VETCH. A hairless rush-like shrub to 1 m., with green, ribbed, sparsely branched and almost leafless, compressible branches and yellow fls. in very long-stemmed clusters of 5–8. Petals 4–6 mm., claws of petals as long as the calyx. Leaflets 3–7, linear; lvs. soon falling. Fr. 1½–4 cm., curved, four-angled, with 2–10 constrictions. △ Hills, waysides, bushy places. Apr.–June. P.E.F.I.YU. *The seeds are poisonous.*

Herbaceous plants usually less than ½ m.
(a) *Fls. white, pink or purple*
627. *C. varia* L. CROWN VETCH. A spreading, leafy, hairless perenn. to 60 cm. with dense, long-stemmed clusters of usually multi-coloured fls. with the standard lilac or pink, the wings whitish, and the keel pale pink with purple tip. Fls. 10–20, finally pendulous; calyx broadly bell-shaped. Leaflets usually 15–25, oblong-elliptic with a fine point, hairless. Fr.

$2\frac{1}{2}$–5 cm., erect, four-angled, with 3–6 constrictions and a terminal beak. △ Grassy places, fields, hills. May–Aug. Much of Eur. (except P.IRL.IS.): introd. GB.B.DK.N.S.SU. *A somewhat poisonous plant.* Pl. 62.

(b) *Fls. yellow*

628. *C. scorpioides* (L.) Koch ANNUAL SCORPION-VETCH. A glaucous, little-branched, erect, hairless ann. to 40 cm. with distinctive, stalkless, trifoliate lvs., with the middle leaflet oval and many times larger than the rounded lateral pair. Fls. pale yellow, small, 3–5 mm., in long-stemmed clusters of 2–4. Fr. to 4 cm., very slender, strongly curved, pendulous. △ Cultivated ground. Apr.–June. Med. Eur. P.BG.R.SU. *The seeds are purgative.*

629. *C. coronata* L. SCORPION-VETCH. A rather robust glaucous perenn. to 60 cm., with pinnate lvs. and long-stemmed clusters of yellow fls. Fls. 15–20, with individual fl. stalks more than twice as long as the calyx; petals *c.* 1 cm. Leaflets 7–15 obovate, to 2 cm., more or less equal. Fr. $2\frac{1}{2}$–3 cm. straight, pendulous, with 2–3 constrictions. △ Woods, thickets, grassy places, in the mountains. May–July. E.F.D.CH.A.CS.H.I.YU.AL.GR.R.SU. Page 185.

630. *C. vaginalis* Lam. SMALL SCORPION-VETCH. Like 629, but leaflets smaller, 8–10 mm., somewhat fleshy, with whitish cartilaginous margin; stipules conspicuous, 6–8 mm., white-papery, oval, fused together at base. Fls. pale yellow, *c.* 8 mm. in very long-stemmed clusters of 5–8. Fr. 2–3 cm., very slender, six-angled, pendulous, with 3–8 constrictions. A woody-based perenn. with spreading stems 10–25 cm. △ Grassy, rocky places. June–Aug. Cent. Eur. (except PL.) F.I.YU.AL.R.

HIPPOCREPIS Distinguished by its frs. which are deeply constricted into several, often horseshoe-like, one-seeded sections. Fls. yellow; calyx five-toothed with the 2 upper teeth fused to the middle. Lvs. pinnate, with a terminal leaflet. 10 sps.

Perenns.

631. *H. comosa* L. HORSESHOE VETCH. A spreading, almost hairless perenn. to 40 cm., with yellow fls. in dense, terminal, rounded, long-stemmed clusters. Fls. 5–12; petals *c.* 1 cm., petal claws longer than the hairy calyx. Lvs. with 9–31 oblong-obovate leaflets, each 4–8 mm. Fr. spreading, *c.* 3 cm. long, with 2–7 shallow horseshoe-shaped segments, smooth or with minute swellings. A very variable sp. △ Dry pastures, banks. Apr.–June. Much of Eur. (except North Eur. P.IRL.).

Anns.

632. *H. unisiliquosa* L. A spreading ann. to 20 cm. with 1–2 small, yellow, almost stalkless fls. in the axils of the upper lvs. Fls. *c.* 5 mm. Leaflets 9–15, oblong, notched. Fr. very distinctive, $2\frac{1}{2}$–4 cm., slightly curved and with 7–10 deep horseshoe-shaped constrictions, hairless. △ Dry, stony places. Apr.–May. Med. Eur. P.BG.SU.

H. multisiliquosa L. Distinguished from 632 by the larger fls. *c.* 6 mm. in a cluster of 2–6 on a stem about as long as the lvs. Fr. strongly curved to a semicircle or more, with the open ends of the horseshoe-shaped constrictions usually on the convex side, often glandular-hairy. Mediterranean region and Portugal.

SCORPIURUS Fr. elongated and strongly curved into a ring or open spiral, and with swellings or spines. Fls. yellow; petals clawed. Anns. with simple, entire long-stalked lvs. 2 sps.

633. *S. vermiculatus* L. A spreading, sparsely hairy ann. to 40 cm. with simple lvs. and usually solitary, long-stalked, orange fls. 12–13 mm. Lvs. obovate-elliptic, narrowed to a

long stalk. Fr. 6–8 mm. broad, coiled once or more in a spiral and covered with flat-topped swellings in longitudinal files, the whole recalling a caterpillar. △ Fields, cultivated ground. May–June. P.E.F.I.

S. muricatus L. (incl. *S. subvillosus* L.) Like 633, but fls. 2–3 yellow, smaller 7–9 mm. Fr. 2–2½ mm. broad, irregularly coiled, with longitudinal ridges and covered with rounded swellings or with spines. Variable, particularly the fr. Mediterranean region. Page 185.

HEDYSARUM   Fr. broad, flattened, conspicuously constricted between the seeds and breaking into one-seeded segments. Fls. pink, purple, or whitish in dense, stalked, axillary clusters. Lvs. pinnate, with a terminal leaflet. 18 sps.

Fr. spiny or with rough projections
634. *H. coronarium* L. ITALIAN SAINFOIN, FRENCH HONEYSUCKLE. A robust, leafy, branched perenn. to 1 m., with large, dense, oval clusters of very conspicuous brilliant carmine or rarely white fls. Fls. *c.* 2 cm., numerous on thick stems as long as or longer than the lvs.; calyx hairy. Leaflets 5–11, oval, 1–3½ cm., with adpressed hairs beneath. Fr. hairless, with 2–4 rounded joints and covered with sharp rough projections. △ Rich pastures, cultivated ground. Apr.–July. E.I.: introd. P.F.YU.GR. *Grown for fodder in the Mediterranean region.* Pl. 62.

H. spinosissimum L. A hairy ann. to 35 cm., with 4–10 purple or pinkish fls. in a dense long-stemmed cluster. Petals 8–11 mm.; calyx at least half as long as the petals, sparsely hairy. Leaflets 9–17, elliptic-oblong, green, hairless above, usually grey-haired beneath or nearly hairless. Fr. with 2–4 woolly-haired segments, covered with hooked spines. A variable sp. Mediterranean region.

H. glomeratum F. G. Dietrich (*H. capitatum* Desf.) Like [634], but fls. larger, 1½–2 cm., bright purple-violet; calyx 3½–4½ mm., not a quarter as long as petals. Leaflets broader, sometimes obovate. Mediterranean region and Portugal. Pl. 62.

Fr. smooth, without projections
635. *H. hedysaroides* (L.) Schinz and Thell. ALPINE SAINFOIN. An alpine perenn. to ½ m. with bright violet, reddish-purple or whitish-yellow fls. in a one-sided, elongated cluster on a stem longer than the lvs. Fls. 12–20, large 1½–2½ cm., drooping; keel longer than the standard. Lvs. with 11–22 oblong-oval, hairless leaflets. Fr. drooping, with 2–5 oval and narrowly winged constrictions, hairless or shortly hairy. △ In mountains: meadows, rocks, and sunny slopes. July–Aug. Cent. Eur. (except H.) E.F.I.YU.R.SU. Pl. 61.

ONOBRYCHIS | Sainfoin   Fr. hard, not splitting or segmenting, more or less rounded, flattened, often with a toothed margin and netted sides. Fls. usually pink, in axillary stalked clusters. Lvs. pinnate with terminal leaflet; leaflets numerous; stipules papery. 23 sps.

Robust perenns.
636. *O. viciifolia* Scop. SAINFOIN. An erect perenn. to 80 cm. with pinnate lvs. and bright pink fls. veined with purple or reddish, in dense, elongated, long-stemmed clusters of up to 50 fls. Petals 1–1½ cm.; keel and standard about equal, the wings very short and shorter than the woolly, short-tubed calyx. Lvs. with 13–25 linear to obovate leaflets, with adpressed hairs beneath. Fr. 5–8 mm., with a network of pits on both faces, shortly spined, hairy. △ Banks and grassy places; often escaping from cultivation. June–Aug. AL.A.CS. H.YU.R.: introd. P.E.F.IRL.GB.B.DK.N.S.CH.PL.I.SU. *Widely cultivated as a fodder plant.*

O. montana DC. MOUNTAIN SAINFOIN. Like 636, but fls. darker reddish; keel longer than the standard. Stems spreading, with many sterile shoots; lvs. with 7–17 leaflets. Fr. larger, 7–12 mm. Mountains of Central and South-Eastern Europe. Pl. 59.

O. arenaria (Kit.) DC. Differs from [636] by the smaller, usually pale pink fls. 7–10 mm. and smaller fr. 4–6 mm. Leaflets 7–25, linear-oblong. A variable sp. Much of Central and South-Eastern Europe.

Slender anns.

637. *O. caput-galli* (L.) Lam. COCKSCOMB SAINFOIN. A slender, greyish-haired ann. to 90 cm. with a rather dense cluster of 3–7 inconspicuous pink fls. on a stem as long as or longer than the lvs. Petals 7–8 mm.; calyx teeth as long. Lvs. with 9–15 linear to obovate leaflets. Fr. *c.* 8 mm. hard, rounded, compressed, with deep pits on the sides and slender spines on the margin of the pits and with longer spines on the circumference, finely hairy. △ Stony, dry places, waysides. Apr.–May. Med. Eur. BG.

OXALIDACEAE | Wood-Sorrel Family

Delicate herbs in our sps., often with fleshy rhizomes and usually with trifoliate lvs. Fls. regular, solitary or in umbels; calyx 5, fused at base; petals 5, free, contorted in bud; stamens 10, fused at their base; ovary usually five-celled, styles usually 5. Fr. a capsule, rarely a berry.

OXALIS | **Wood-Sorrel** Lvs. trifoliate; stamens 10; styles 5. Fr. a capsule. 11 sps.

Fls. white

638. *O. acetosella* L. WOOD-SORREL. Fls. 1–1½ cm., white and delicately veined with lilac, or rarely lilac or purple, solitary on a long, slender, hairy stalk arising direct from the rhizome. Fl. stalks 5–15 cm. with 2 small bracteoles placed near the middle. Lvs. all basal, long-stalked; leaflets 1–2 cm., obcordate, bright yellowish-green, with scattered hairs. A delicate perenn. with creeping rhizome covered by fleshy lv. bases. △ Woods, shady places. Apr.–May. All Eur. *Used medicinally in the past; the leaves are sometimes eaten as a salad plant, but they can be poisonous if eaten in considerable quantities for they contain oxalic acid.* Pl. 62.

Fls. yellow

639. *O. pes-caprae* L. (*O. cernua* Thunb.) BERMUDA BUTTERCUP. Fls. conspicuous, bright lemon-yellow, in a terminal umbel of 6–12 long-stalked fls., borne on a leafless stem longer than the lvs. Fls. drooping in bud, petals 2–2½ cm. obovate, 3–4 times as long as the calyx. Lvs. all basal, long-stalked; leaflets obcordate, *c.* 2 cm., somewhat fleshy, bright green. A perenn. to 20 cm., producing numerous bulbils at the surface and below ground, and spreading rapidly by this means. △ Cultivated ground, vineyards, olive groves. Dec.–May. Native of South Africa and now a serious weed of vineyards and olive groves in the Mediterranean region and Portugal. Pl. 63.

640. *O. corniculata* L. (*O. repens* Thunb.) PROCUMBENT YELLOW SORREL. Fls. small 8–10 mm., yellow, in long-stemmed clusters of 1–6 arising from the axils of the upper lvs. Stalks of individual fls. with adpressed hairs, deflexed after flowering. Lvs. alternate; stipules *c.* 2 mm. A spreading, branched, conspicuously hairy-stemmed ann., bienn. or perenn., rooting at the nodes and with ascending stems to 15 cm. △ Cultivated ground, waysides. Apr.–Oct. Med., South-East Eur. (except R.) P.CH.H.SU.: introd. IRL.GB.B.NL.N.S.SF.D.A. PL.CS.R. Page 220.

O. europaea Jordan (*O. stricta* auct.) UPRIGHT YELLOW SORREL. Native of North America and East Asia and widely naturalized in Europe.

GERANIACEAE | Geranium Family

Usually herbaceous plants with deeply lobed or compound lvs., usually with stipules. Fls. usually regular; sepals usually 5; petals 5; stamens 10 or 15, more or less fused at the base, often some sterile. Ovary with 3–5 fused carpels and ending in a long beak formed by the fused styles, which on maturing in fr. splits up into elastically coiled sections carrying the one-seeded carpels upwards and outwards.

1. Fls. spurred with the spur fused to the fl. stalk.   *Pelargonium*
1. Fls. not spurred.
  2. Lvs. palmate or palmately lobed; beak of fr. splitting into arched or
    coiled sections.   *Geranium*
  2. Lvs. pinnate or pinnately lobed; beak of fr. splitting into corkscrew-
    like sections.   *Erodium*

GERANIUM | **Cranesbill** Lvs. usually not longer than broad, palmate or palmately lobed, the basal lvs. long-stalked, the upper progressively shorter-stalked, hairy. Stamens 10, usually all fertile. Seeds released from the splitting carpels. 39 sps.

Perenns.; fls. large, petals usually more than 1 cm.
(a) Petals notched at apex

641. *G. sanguineum* L. BLOODY CRANESBILL. Fls. large, $2\frac{1}{2}$–3 cm., usually solitary with bright crimson-purple, notched petals, or rarely pink or white. Fl. stalks with long spreading hairs, reflexed after flowering; calyx elliptic to oval, fine-pointed, with spreading hairs. Lvs. *c.* 5 cm., rounded in outline, divided to the base into 5–7 narrow lobes which are often further three-lobed, lobes all linear-oblong. Carpels smooth, sparsely hairy. An erect or spreading hairy, bushy perenn. 10–40 cm. with a stout creeping rhizome. △ Dry hills, open woods, hedges. June–Sept. All Eur. (except IS.NL.). Pl. 63.

642. *G. pyrenaicum* Burm. fil. MOUNTAIN CRANESBILL. Fls. in pairs, *c.* $1\frac{1}{2}$ cm., rosy-purple or rarely purplish-white, rather numerous in a lax, branched cluster. Petals 7–10 mm., deeply notched, twice as long as the calyx; fl. stalks glandular-hairy, reflexed after flowering. Lower lvs. divided to about two-thirds into 5–9 rounded lobes which are further shallowly divided into 3–5 segments, the uppermost lvs. often deeply three-lobed. Carpels smooth, with adpressed hairs. A softly hairy, glandular perenn. 25–60 cm. △ Waysides, hedges, meadows, open woods. June–Oct. Most of Eur. (except IS.): introd. North Eur. B.NL.A.CS.PL.SU. Pl. 63.

G. nodosum L. BROAD-LEAVED CRANESBILL. Distinguished by its lvs. which are divided to about three-quarters into 3–5 large, entire, but strongly toothed lance-shaped lobes. Fls. large, 2–$2\frac{1}{2}$ cm., lilac with violet veins. Stems swollen at the nodes, with adpressed hairs, stoloniferous. Carpels hairy, with a transverse ridge at apex. Spain to Yugoslavia; naturalized elsewhere. Pl. 63.

G. versicolor L. (*G. striatum* L.) VEINED CRANESBILL. Like 642, but fls. white or pale lilac, with dark violet veins; petals 15–18 mm., deeply notched. Lvs. deeply five-lobed; stems and fl. stalks with long spreading hairs. Carpels smooth, hairy. Italy to Greece; sometimes naturalized elsewhere.

643. *G. tuberosum* L. TUBEROUS CRANESBILL. Fls. rosy-purple, 1–2 cm., borne in a lax cluster on nearly leafless stems 20–50 cm. arising directly from a globular, nut-like underground tuber. Petals conspicuously rounded-notched, with darker veins; sepals with long hairs. Lvs. arising mostly from the tuber, blades deeply cut almost to the middle into 5–9 narrow lobes which are further deeply cut into narrow, toothed segments; stems and fl. stalks with adpressed hairs. Carpels smooth, hairy. △ Cultivated ground, fields, vineyards, meadows. Mar.–June. South-East Eur. I.SU.: introd. F.

(*b*) *Petals rounded at apex, not notched*
644. *G. pratense* L. MEADOW CRANESBILL. Fls. large 2½–3 cm., cup-shaped, violet-blue, rarely white or lilac, in lax terminal clusters. Petals 15–18 mm.; fl. stalks reflexed after flowering, erect in fr. Basal lvs. deeply five- to seven-lobed and further deeply cut into narrow toothed segments, the upper lvs. smaller and the uppermost three-lobed and almost stalkless; stems glandular-hairy above. Carpels glandular-hairy. An erect, densely softly hairy perenn. 30–80 cm. △ Meadows, ditches, streamsides; grown as an ornamental and often naturalized. June–Sept. Most of Eur. (except P.IS.AL.TR.).

645. *G. sylvaticum* L. WOOD CRANESBILL. Distinguished from 644 by the fl. stalks which remain erect after flowering. Fls. usually smaller; petals 12–18 mm., purple-violet. Lvs. with rather broader lobes which have less deeply cut and toothed segments. Subsp. *rivulare* (Vill.) Rouy has white petals which are veined with purple and without glandular-hairy stems or fr. An erect, hairy perenn. 30–80 cm. △ Meadows and woods in mountains. June–Aug. Much of Eur. (except P.GR.TR.). Pl. 63.

646. *G. phaeum* L. DUSKY CRANESBILL, MOURNING WIDOW. Fls. blackish-purple, often with a pale centre and with reflexed or wide-spreading petals and conspicuous forward-projecting stamens and style. Fls. in lax, leafy clusters; petals *c.* 1 cm., blunt-pointed and about equalling calyx. Lvs. divided to about three-quarters into 5–7 oval irregularly toothed or cut lobes, often blotched with brown. Carpels with strong transverse ridges, hairy. An erect, hairy perenn. 30–60 cm. △ Damp meadows, hedges, open woods. June–Aug. Much of Eur. (except North Eur. P.TR.): introd. IRL.GB.B.NL.DK.S. Pl. 64.

647. *G. macrorrhizum* L. ROCK CRANESBILL. Fls. large, 2½ cm., blood-red or carmine, with reflexed or wide-spreading petals and conspicuous projecting stamens and style, in a dense cluster. Petals *c.* 1½ cm. with a hairy claw as long as the blade; sepals inflated, oval long-pointed, reddish, hairy; fl. stalks densely glandular-hairy. Lvs. divided to about two-thirds into 5–7 broad oval lobed and toothed segments. Carpels transversely ridged at apex. A hairy perenn. 10–30 cm., with thick rhizome. △ Shady rocks. July–Aug. South-East Eur. (except TR.) F.A.I.BG.R.SU.: introd. GB.B.D.SU. *Sometimes cultivated as an ornamental and medicinal plant.* Pl. 64.

Anns.; fls. small, petals usually less than 1 cm.
(*a*) *Sepals spreading or ascending*
648. *G. dissectum* L. CUT-LEAVED CRANESBILL. A hairy ann. or bienn. 10–60 cm., with lvs. deeply divided almost to the base, usually into 8 linear-oblong lobes which are further three-lobed. Fls. tiny, reddish-pink, paired, on a short stem, ½–2 cm.; individual fl. stalks ½–1½ cm.; petals notched, *c.* 5 mm., about as long as the spreading glandular-hairy calyx. Carpels smooth, hairy. △ Cultivated ground, rocks, hedges, walls. Apr.–Sept. All Eur. (except IS.).

G. columbinum L. LONG-STALKED CRANESBILL. Like 648, but stem bearing paired fls. 2–12 cm. and individual fl. stalks 2–6 cm.; petals 7–9 mm., purplish-pink, not notched. Carpels usually hairless, but sometimes sparsely hairy. Throughout Europe.

649. *G. rotundifolium* L. ROUND-LEAVED CRANESBILL. A hairy branched ann. 15–30 cm., distinguished by the rounded or kidney-shaped lower lvs. which are divided to less than half their width into blunt rounded lobes; the upper lvs. more deeply divided. Fls. pink, paired; petals 5–8 mm., wedge-shaped, rounded, or very slightly notched at apex. Carpels hairy. △ Cultivated ground, hedges, walls, waste places, waysides. Apr.–Sept. Most of Eur. (except North Eur. PL.).

G. molle L. DOVES-FOOT CRANESBILL. Like 649, but fls. bright rosy-purple; petals 3–7 mm., deeply notched. Lvs. all divided to more than half the width of the blade. Carpels hairless, usually wrinkled. Throughout Europe.

**G. pusillum* L. SMALL-FLOWERED CRANESBILL. Widespread in Europe.

(*b*) *Sepals erect and somewhat curved inwards towards apex*
650. *G. robertianum* L. HERB ROBERT. Readily distinguished by its compound lvs. which have stalked leaflets which are further pinnately cut into lobed and dissected segments. Fls. deep pink, rarely white; petals 1 cm. with the claw longer than the blade, twice as long as the glandular-hairy, somewhat incurved calyx. Carpels with a network of ridges, hairy or hairless. A strongly smelling, sparsely hairy, ferny-leaved ann. or bienn. often with reddish stems, 10–50 cm. △ Rocks, walls, hedges, shingle. Apr.–Aug. All Eur. (except IS.).

**G. purpureum* Vill. South-Western Europe.

651. *G. lucidum* L. SHINING CRANESBILL. Distinguished by its rounded shining lvs. which are divided to about half their width into broad, blunt-toothed segments, which are about as long as they are broad. Petals bright rosy-pink, 8–9 mm., with a long claw and rounded blade, not notched; calyx strongly angled and with transverse ridges between the angles, hairless. Carpels with 5 ridges, netted, hairless. A nearly hairless, brittle, bright green, shining ann., often with reddish stems, 10–40 cm. △ Shady rocks, walls, waysides. Apr.–Aug. All Eur. (except IS.NL.). Pl. 63.

ERODIUM | Storksbill Stamens 5 fertile and 5 sterile without anthers; sepals and petals 5. Beak of fr. splitting into 5 twisted corkscrew-shaped sections with the carpel attached at the base; carpel not splitting and seed retained. Lvs. mostly pinnate or pinnately lobed. 33 sps.

Some or all lvs. shallowly lobed to less than half width of blade
652. *E. malacoides* (L.) L'Hér. SOFT STORKSBILL. A spreading or erect, glandular-hairy ann. or bienn. 10–40 cm., with oval-heart-shaped lvs., rounded-toothed or shallowly cut into blunt lobes with rounded teeth. Fls. long-stalked, in a cluster of 3–8, borne on a glandular-hairy stem much longer than the lvs.; petals lilac, 4–7 mm., widely spaced, as long as or little longer than the sepals; bracts 3–6, oval. Carpels with 2 glandular pits and 2 semicircular ridges, beak 2–4 cm. △ Waysides, waste ground, poor pastures. Feb.–June. Med. Eur. P.SU. Pl. 64.

E. gruinum (L.) L'Hér. LONG-BEAKED STORKSBILL. Lower lvs. long-stalked, oval-heart-shaped, undivided, the upper lvs. divided into 3 toothed leaflets. Petals lilac, *c.* 1 cm., as long as the calyx which later enlarges in fr. to 1½ cm. Fr. rough-haired, pitted, beak very long 8–11 cm. Much of the Mediterranean region. Pl. 64.

**E. maritimum* (L.) L'Hér. SEA STORKSBILL. Atlantic and Mediterranean coasts.

Lvs. pinnate or bipinnate
653. *E. cicutarium* (L.) L'Hér. COMMON STORKSBILL. Fls. usually bright rosy-purple, often

with a dark purple spot at the base of the petals, or fls. rarely white, variable in size, 2–9 in a long-stemmed cluster. Petals often unequal, 2–4 times as long as the calyx; filaments of fertile stamens swollen at base, not toothed. Lvs. pinnate, with oval leaflets deeply once or twice cut to the midvein, but very variable; stipules and bracts papery, acute, or long-pointed. Carpel usually hairy, with a conspicuous pit, beak 1½–4 cm. A very variable, odourless, erect, or spreading ann., bienn., or perenn. to 60 cm. △ Cultivated ground, sand dunes, waysides, dry meadows. Mar.–Aug. All Eur. (except IS.).

654. *E. moschatum* (L.) L'Hér. MUSK STORKSBILL. Like 653, but with musk-smelling glandular lvs. with oval leaflets cut to less than half their width; stipules and bracts blunt. Petals rosy-purple, not blotched, little longer than the calyx; filaments of fertile stamens with a tooth on each side at the base. Carpel densely rough-hairy, with a pit, beak 2–4 cm. A variable ann. or bienn. to 60 cm. △ Waysides, grassy and waste places, by the sea. Apr.–Sept. Med. Eur. P.GB.NL.: introd. IRL.B.D.CH.A.CS.H.

E. ciconium (L.) L'Hér. Fls. lilac or blue, veined with purple, petals somewhat unequal about as long as the calyx which elongates to 1 cm. in fr. A glandular-hairy ann. with bipinnate lvs. with smaller toothed lobes between the leaflets. Beak of fr. 6–8 cm. Mediterranean and Southern Europe.

PELARGONIUM   Sepals 5, fused at base and with a spur fused to the lv. stalk; petals somewhat unequal; stamens 10. Fr. beaked and splitting into spiralling segments.

P. × hybridum (L.) L'Hér. ZONAL GERANIUM. Lvs. rounded with 5–7 shallow lobes and a deeply inset stalk, usually with dark semicircular blotch, softly hairy. Fls. numerous in a long-stemmed cluster, scarlet, pink, or white; petals 1½ cm., obovate. A native of South Africa; widely grown for ornament in numerous cultivated forms and sometimes escaping.

TROPAEOLACEAE | Nasturtium Family

Calyx two-lipped, spurred; petals 5, the upper 2 larger; stamens 8. Ovary three-lobed and separating into 3 one-seeded units in fr.

Tropaeolum majus L. NASTURTIUM. A climbing, hairless ann. or perenn. with large solitary, long-spurred, orange-red fls. arising on long stalks from the axils of the lvs. Petals to 4 cm. Lvs. circular with the stalk attached at the centre of the blade. A native of South America; commonly grown as an ornamental and quite often escaping.

ZYGOPHYLLACEAE | Caltrop Family

A largely tropical family usually with pinnate lvs., with stipules. Fls. regular; sepals and petals 4–5; stamens 8 or 10; ovary with 4–5 cells; style 1. Fr. usually a capsule.

TRIBULUS   Sepals and petals 5; stamens 10. Fr. with 5 lobes, each with stiff spreading spines. 1 sp.

655. *T. terrestris* L. MALTESE CROSS, SMALL CALTROPS. A lowly, spreading, hairy ann. 10–50 cm., with small, solitary, yellow fls. in the axils of the neatly pinnate lvs. Fls. 7–10 mm.,

on stalks shorter than the lvs. Lvs. with 5–8 pairs of elliptic, often silvery-haired leaflets *c.* 1 cm. Fr. very distinctive in the form of a Maltese cross, with 5 lobes, each with 2 long and 2 short tough, spreading spines. △ Waste ground, sandy places, abandoned cultivation. Apr.–Sept. Med., South-East Eur. P.A.CS.H.SU. *The fruits are very hard and spiny and often become lodged in the feet of animals.* Pl. 64.

LINACEAE | Flax Family

Usually herbaceous plants with simple, undivided, usually alternate lvs. without stipules. Fls. regular; sepals and petals free, 5 or rarely 4, petals contorted in bud, often falling early; stamens usually 5 and usually fused at their base, with or without alternate infertile stamens. Ovary with 3–6 cells; styles 3–5. Fr. usually a capsule.

Petals and sepals 4; petals not longer than calyx. *Radiola*
Petals and sepals 5; petals usually longer than calyx. *Linum*

RADIOLA Fls. with parts in fours; sepals two- to three-lobed. 1 sp.

656. *R. linoides* Roth ALL-SEED. A delicate, very slender-stemmed, regularly dichotomously branched ann. 2–8 cm., with tiny white fls. *c.* 2 mm. in a spreading or flat-topped cluster. Lvs. elliptic *c.* 3 mm., opposite. Fr. globular *c.* 1 mm. △ Sandy and damp places, heaths. June–Sept. Most of Eur. (except IS.SF.). Page 220.

LINUM | **Flax** Sepals, petals, and stamens 5; petals with a claw, usually soon falling; ovary with 5 cells, each two-seeded. Fr. a capsule splitting by 10 valves; seeds flat. Lvs. stalkless, linear-lance-shaped, with 1, 3, or 5 parallel veins. 36 sps.

Fls. yellow
(*a*) *Anns.*
657. *L. strictum* L. UPRIGHT YELLOW FLAX. An erect ann. 10–45 cm. with small yellow fls. *c.* 1 cm. across, in either a branched, spreading cluster or in rather dense spike-like clusters with very short lateral clusters. Petals 6–12 mm., about twice as long as the lance-shaped, long-pointed sepals which have spreading glandular hairs on the margin. Lvs. linear-lance-shaped, margin very rough. △ Dry hills, grassy places, rocks. May–July. Med. Eur. P.BG.SU.

L. nodiflorum L. (incl. *L. luteolum* Bieb.) Differs from 657 by its large almost stalkless yellow fls. in a lax, elongated, one-sided, spike-like cluster. Petals 2 cm.; sepals with a rough and minutely toothed, but not glandular-hairy margin, half as long as petals. Stem angled; lower lvs. spathulate and three- to five-veined, the upper linear, rough on margin and veins, and with 2 tiny glands at the base. Mediterranean France to Turkey. Page 220.

(*b*) *Perenns.*
658. *L. flavum* L. YELLOW FLAX. A hairless, erect, branched perenn. to 60 cm., with woody base and with large, yellow, rather numerous fls. in branched clusters. Petals obovate, shortly clawed, *c.* 2 cm.; sepals lance-shaped, long-pointed, 6–9 mm., margin sparsely or densely glandular-hairy. Lvs. 2–3½ cm. by 3–12 mm., the upper lance-shaped, three-veined, the lower spathulate. Sepals scarcely longer than the fr. A very variable sp. △ Dry grassy places. Cent. Eur. (except CH.) I.YU.AL.BG.R.SU. Page 220.

1. *Linum flavum* 658
3. *Radiola linoides* 656
5. *P. major* 699

2. *Andrachne telephioides* 685
4. *Polygala monspeliaca* 697
6. *Oxalis corniculata* 640   7. *Linum nodiflorum* [657]

L. campanulatum L. Very like 658, but upper lvs. narrow lance-shaped with a narrow whitish margin and with 2 little glands at the base. Petals 2½–3½ cm., gradually narrowed to the base, not clawed. Mediterranean Spain to Italy.

Fls. blue; sepals usually without glandular hairs
(a) Stigmas ovoid or globular, little longer than broad
659. *L. perenne* L. PERENNIAL FLAX. A slender, erect, branched perenn. 30–100 cm., with bright blue fls. *c.* 2½ cm. across in a lax, branched cluster. Petals 1½–2 cm.; sepals oval, the inner with broad papery margin, about half as long as the fr. Lvs. linear, *c.* 2 cm. by 1–3 mm., usually one-veined, very numerous. Fr. stalks erect. A variable sp. Subsp. *alpinum* (Jacq.) Ockendon is a mountain perenn. with fr. stalks slightly drooping and larger, bright blue fls. 2–3 cm. across. △ Meadows, rocky places, in mountains. May–Aug. Cent. Eur. (except CH.) F.GB.I.YU.AL.BG.SU. Pl. 65.

(b) Stigmas elongate, club-shaped
660. *L. bienne* Miller PALE FLAX. Fls. pale blue, on slender stalks 1–2 cm., forming a loose cluster; petals 8–12 mm., obovate. Sepals oval, long-pointed with papery margin, the inner sepals with glandular spreading hairs on margin; sepals about half as long as the fr. Lvs. linear-lance-shaped, 1–2½ cm., numerous, mostly three-veined. Fr. *c.* 6 mm.; seeds scarcely beaked. An ann., bienn. or perenn. 10–60 cm., with non-flowering leafy shoots. △ Grassy places, hills. June–Aug. Med. Eur. P.IRL.GB.BG.SU.

L. usitatissimum L. CULTIVATED FLAX. Very like 660, but always an ann. without non-flowering shoots and with larger three-veined lvs. 2–4 cm. by 2–4 mm. Petals 12–15 cm., bright blue; sepals ciliate or not, not glandular. Fr. 1 cm. or more; seeds with a short, blunt beak. Origin unknown; occurring as a casual from cultivation almost throughout Europe. *Flax produces linen, probably the oldest known textile, and it has been cultivated for this purpose since early Egyptian times. It is used to make cloth, ropes, sails, nets, etc., but is now largely superseded by cotton and the new man-made fibres. Linseed oil is obtained from the seeds; it is a drying oil used in varnish and printers ink; the crushed residues form oil-cake which is an important food for cattle.*

661. *L. narbonense* L. BEAUTIFUL FLAX. A glaucous, erect perenn. 20–50 cm. of the Mediterranean region, with few, very large, azure-blue fls. with petals 2½–3½ cm. Sepals lance-shaped with broad, entire, papery margin, óne-third as long as the petals and longer than the fr. Lvs. with rough margin, linear-lance-shaped, three-veined. △ Dry hills. May–July. Med. Eur. (except GR.TR.) P.: introd. CH.

Fls. pink or pale lilac; sepals glandular-hairy
662. *L. viscosum* L. STICKY FLAX. A hairy-stemmed perenn. to 60 cm. with almost stalkless pink fls. veined with violet, forming at length an elongated spike-like cluster, or fls. rarely white or blue. Petals 18–22 mm., the claw yellow; sepals lance-shaped, glandular hairy, twice as long as the fr. Lvs. oblong-lance-shaped to 8 mm. wide, with conspicuously glandular-hairy margin, three- or five-veined. △ Grassy places. May–July. E.F.D.A.I.YU.

663. *L. tenuifolium* L. Fls. pink or pale lilac, all stalked, in a lax rather flat-topped cluster. Petals distinctly blunt-tipped, not rounded, 2–2½ times as long as the sepals which are lance-shaped long-pointed, with a glandular-hairy margin, longer than fr. Lvs. numerous, narrow linear, one-veined, margin rough. A perenn. which is hairless above and with woody base below; 15–45 cm. △ Grassy and rocky places, dry hills. May–July. Cent., South-East Eur. E.F.B.I.SU.

Fls. white
(a) Lvs. opposite
664. *L. catharticum* L. PURGING FLAX. A very slender hairless ann. 5–15 cm. with opposite

lvs. and tiny white fls. borne on long slender stalks and forming a lax branched cluster. Fls. nodding in bud; petals 4–6 mm., little longer than the oval long-pointed, glandular-hairy calyx. Lvs. 5–12 mm., oblong-oval, one-veined. Fr. *c.* 3 mm., as long as the sepals. △ Grassy places, heaths, moors, dunes. June–Aug. All Eur. *The seeds are somewhat poisonous. Used in herbal remedies as a purgative and bitter.*

(*b*) *Lvs. alternate*

L. *suffruticosum* L. WHITE FLAX. A rather stiff-branched, rounded shrublet to 50 cm., with conspicuous white fls., often purplish or violet at the base, or sometimes pink-veined, in a lax branched cluster. Petals variable, 1½–3 cm., long-clawed, 3–4 times as long as sepals; fl. buds yellow. Lvs. numerous, narrow awl-shaped, rough, stiff. Spain to Italy. Pl. 65.

EUPHORBIACEAE | Spurge Family

Usually herbs in our sps., rarely shrubs. Lvs. usually alternate, and usually with stipules. Fls. one-sexed, regular; perianth 5, green and inconspicuous or absent; stamens many or 1. Ovary usually superior, three-celled, each cell usually one-seeded. Fr. usually a capsule.

1. Plants with milky juice; several male fls. and 1 female fl. encircled by an involucre and forming a compound cluster. *Euphorbia*
1. Plants without milky juice; male and female fls. in separate clusters.
   2. Lvs. palmately lobed. *Ricinus*
   2. Lvs. not palmately lobed.
      3. Lvs. opposite; plants one-sexed. *Mercurialis*
      3. Lvs. alternate; male and female fls. on same plant.
         4. Lvs. less than 1 cm., hairless. *Andrachne*
         4. Lvs. more than 1 cm., densely hairy. *Chrozophora*

CHROZOPHORA Fls. in elongated clusters with the male fls. above and the female fls. below in the same cluster. Perianth 10; male fls. with 5–10 stamens; female fls. with 1 ovary and 3 bilobed styles. 2 sps.

665. *C. tinctoria* (L.) Juss. TURN-SOLE. A densely greyish-hairy, branched ann. to 40 cm., with a short, erect cluster of yellowish fls.; the male fls. numerous, erect, almost stalkless, the female fls. 1–4 at the base of each cluster, long-stalked, drooping. Lvs. oval-rhomboid, coarsely toothed, long-stalked, with grey star-shaped hairs. Fr. pendulous, with 3 rounded segments covered with conspicuous swellings. △ Cultivated ground, waste places. June–Oct. Med. Eur. (except TR.) P.SU. *It is the source of the dye, tournesol or turn-sole; it was used as a medicinal plant in the past. The seeds are emetic.* Pl. 65.

MERCURIALIS | Mercury Male and female fls. usually on separate plants; male fls. in interrupted spikes, perianth 3, stamens 8–15; female fls. solitary or clustered, perianth 3, styles 2. Fr. with 2 cells, each one-seeded. Lvs. opposite. 7 sps.

Perenns.

666. *M. perennis* L. DOGS MERCURY. A leafy perenn. with underground stolons and erect stems 15–40 cm., with opposite lvs. and inconspicuous green fls. of one sex only. Fls. 4–5 mm., the male fls. in erect, interrupted spikes *c.* 5 cm., the female fls. in long-stemmed clusters of 1–3 fls. Lvs. 3–8 cm., elliptic to oval, saw-toothed, short-stalked, shortly hairy.

Fr. two-lobed, hairy, 6–8 mm. across. △ Woods and shady places. Mar.–June. All Eur. (except IS.). *The plant is poisonous to humans and livestock.* Pl. 65.

M. tomentosa L. HAIRY MERCURY. A woody-based, much-branched perenn. with lvs. and stems covered with dense, white-woolly hairs. Male fls. in rounded clusters forming an interrupted spike; female fls. 1–2, almost stalkless; fr. woolly-haired. Portugal to France.

Anns.

667. *M. annua* L. ANNUAL MERCURY. Like 666, but an ann. without stolons, but with branched stems to 50 cm. and almost hairless, pale yellowish-green lvs. Fls. greenish, the male fls. in clusters forming a long, interrupted, erect spike *c.* 3 cm.; female fls. almost stalkless in the axils of the lvs. Fr. rough, hairy, 3–4 mm. across. △ Fields, cultivated ground, gardens. Apr.–Nov. Most of Eur. (except IS.): introd. IRL.N.S.SF. *Poisonous.*

RICINUS Fls. in branched clusters, the female fls. above and the male below. Perianth 3–5; stamens very numerous; ovary three-celled, each one-seeded. Fr. a spiny capsule. Lvs. palmately lobed. 1 sp.

668. *R. communis* L. CASTOR OIL PLANT. A robust ann. 1–3 m., or a shrub or small tree in Southern Europe, with very large palmately lobed lvs. up to 80 cm. across, and erect, terminal, branched clusters, 30–60 cm., of reddish fls. Male fls. below, with branched filaments and numerous anthers; female fls. above, with 3 bilobed reddish styles. Fr. globular, spiny, 1–2½ cm.; seeds shining, mottled, with an aril. △ A native of the tropics; often cultivated or grown for ornament, sometimes self-seeding in waste places. Feb.–Dec. Introd. Med. Eur. (except TR.) P.R. *The seeds produce the important medicinal and industrial oil, castor oil. The seeds are poisonous and contain ricin, which remains in the residue after pressing; the pure oil used is free from this acute poison. The plant is poisonous to livestock. Castor oil is used commercially in soap-making, leather-dressing, candles, varnish, and polish. It has been cultivated since early Egyptian times, primarily as a source of illuminating oil.* Pl. 65.

EUPHORBIA | Spurge Inflorescence distinctive, composed of either simple or compound umbels, subtended by bracts. If simple, then the stalks of the umbel or *rays* bear 2–3 'flowers'; if compound, each ray bears further stalks or *secondary rays* which themselves are one- to three-'flowered'. Each 'flower' is subtended by 2 often distinctive *floral-bracts*. The 'flowers' are in reality compound; they are composed of several one-stamened male fls. and 1 stalked female fl., all encircled by a perianth-like involucre of 4–5 small teeth alternating with 4–5 conspicuous, broad, swollen glands which are either more or less oval or half-moon-shaped. Ovary three-celled; style 3, often bilobed. Fr. a capsule. Plants with milky juice; lvs. usually entire and spirally arranged. 105 sps. *The milky juice can be severely irritant to the skin and is poisonous if taken internally. The seeds have a drastic purgative action, and they have been used as a fish or arrow poison in the past.*

Woody plants, with or without spines

669. *E. dendroides* L. TREE SPURGE. A robust shrub with thick, reddish, woody stems, dichotomously branched and leafy above, forming dense rounded bushes to 2 m. Umbels with 3–10 rays and paired secondary rays; floral-bracts almost round; glands of involucre oval or half-moon-shaped. Lvs. lance-shaped, thick, somewhat glaucous, numerous. Fr. smooth, strongly three-angled. △ Cliffs and steep rocks by the sea. Apr.–June. Med. Eur. (except TR.). Pl. 66.

670. *E. spinosa* L. SPINY SPURGE. A low, twiggy, branched shrublet to 20 cm., with old branches

forming weak woody spines and intermixed with the present years leafy shoots. Umbels with 3–5 rays, each one- or two-flowered, floral-bracts obovate; glands of involucre oval. Lvs. very small, 1 cm., lance-shaped, glaucous. Fr. 6 mm., densely covered with conical swellings, weakly three-angled. △ Dry, stony places. Apr.–June. F.I.YU.AL.

E. acanthothamnos Boiss. GREEK SPINY SPURGE. Like 670, but much more branched and forming a low domed shrub of densely intertwined stiff branches like wire-netting, with branches and rays becoming sharply spiny. Umbels usually with 3 rays, each two- to three-flowered. Fr. densely covered with short, cylindrical swellings, strongly three-angled. Mediterranean; Yugoslavia to Turkey. Pl. 66.

Herbaceous plants; glands of involucre oval or rounded
(a) Fr. smooth, usually hairless
671. *E. helioscopia* L. SUN SPURGE. A hairless, usually unbranched ann. 10–50 cm., with a broad, flat-topped, often rather golden-yellow inflorescence and with obovate, finely toothed lvs. 1½–3 cm. Umbel with 5 rays little longer than the broad subtending bracts, secondary rays 2–3, each two- to three-flowered; floral-bracts oval, often yellowish-green; glands of involucre oval, green. Fr. smooth; seeds netted. △ Cultivated and waste ground. Apr.–Nov. All Eur. (except IS.). Pl. 67.

672. *E. villosa* Willd. A robust perenn. ½–1 m. with many lateral, usually hairy stems and lance-shaped lvs. 5–10 cm. by 1–2 cm. which are hairy beneath and finely toothed towards the apex. Umbels usually with 4 or more rays, each with 3 secondary rays which are two-flowered. Floral-bracts elliptic to oval, yellow, often toothed, not fused together. Fr. usually hairless and smooth; seeds smooth. △ Damp places. June–Aug. E.F.D.A.PL.CS. H.I.YU.AL.BG.R.SU. Page 225.

(b) Fr. covered with swellings
673. *E. platyphyllos* L. BROAD SPURGE. An erect, branched ann. to 90 cm., distinguished by its frs. which are covered with numerous hemispherical swellings. Umbel with 3–5 rays, each with 3 secondary rays, lateral umbels several. Floral-bracts broadly triangular-oval, finely toothed, fine-pointed, differing conspicuously from the lance-shaped lower lvs. with a heart-shaped or eared base. △ Cultivated ground, ditches, waysides. June–Sept. Cent., South-East Eur. E.F.GB.B.NL.

674. *E. palustris* L. MARSH SPURGE. A stout perenn. with creeping stolons and robust, erect stems to 1 m. or more, with many sterile branches and lateral fl. branches above. Umbels with many rays; floral-bracts obovate to rounded, yellow. Lvs. lance-shaped, often finely toothed towards apex. Fr. strongly three-angled, densely covered with short, cylindrical swellings. △ Marshes, damp valleys. Apr.–July. Most of Eur. (except P.IRL. NL.IS.TR.). Page 225.

Herbaceous plants; glands of involucre sickle-shaped or with horn-like projections
(a) Anns. or bienns. usually with 1 basal stem
675. *E. peplus* L. PETTY SPURGE. A branched, erect ann. 10–30 cm. with rather broad, oval or obovate, shortly stalked, soft green lvs. and 'flowers' with distinctive half-moon-shaped involucral glands with long slender horns. Umbel with 3 rays, each ray further 1–4 times branched; floral-bracts similar to lvs., but stalkless. Fr. with 2 narrow wings on each segment. △ Cultivated ground, fields, gardens, waste places. May–Oct. All Eur. (except IS.).

*E. exigua L. DWARF SPURGE. Almost throughout Europe.

E. falcata L. SICKLE SPURGE. A slender ann. distinguished by the rather few lance-shaped,

1. *Euphorbia villosa* 672   2. *E. pinea* [680]   3. *E. biumbellata* [684]
4. *E. falcata* [675]   5. *E. palustris* 674

three-veined lvs. which contrast strongly with the oval-rhomboid, fine-pointed floral-bracts. Glands of involucre half-moon-shaped, with short points. Fr. smooth with segments slightly keeled; seeds transversely wrinkled. Widespread in Europe, except in Northern Europe. Page 225.

676. *E. lathyris* L. CAPER SPURGE. A robust bienn. with a stout erect stem to 1 m. or more, branched dichotomously above and with 4 ranks of very regularly arranged pairs of bluish-green lvs. with prominent pale midveins. Umbels with 2–6 rays, bracts triangular-lance-shaped; floral-bracts triangular-oval, acute, heart-shaped at base; glands of involucre half-moon-shaped with blunt horns. Fr. large 8–20 mm., smooth, hairless. △ Waste places, cultivated ground. May–July. P.E.F.I.YU.GR.: introd. GB.B.NL.D.CH.A.CS. BG.R. *The fruits are not edible, even when pickled, for as in all spurges they are acrid and poisonous.* Pl. 67.

(b) Perenns. usually with several stems arising from base
(i) Floral-bracts fused in pairs
677. *E. amygdaloides* L. WOOD SPURGE. A robust, hairy woodland perenn. to 80 cm., distinguished by its lax rosette-like cluster of darker over-wintering lvs. placed about one-third of the way up the persistent stem, above which is borne the present year's fl. stem with numerous thinner, pale green lvs. Terminal umbel with 5–10 rays, several smaller lateral umbels usually present. Floral-bracts yellowish, kidney-shaped, fused in pairs to about one-third their width; glands of involucre half-moon-shaped with conspicuous, converging horns. Fr. hairless, granular. △ Woods, thickets. Apr.–June. Most of Eur. (except North Eur.). Pl. 66.

678. *E. characias* L. LARGE MEDITERRANEAN SPURGE. A stout, rather woody Mediterranean perenn. to 1 m. or more, usually distinguished by its conspicuous, dark-reddish-purple involucral glands with short, thick, rounded horns. Inflorescence cylindrical, composed of a large, many-rayed terminal umbel and numerous lateral umbels. Floral-bracts fused to half their width into a cup-like involucre. Lvs. oblong to lance-shaped, leathery, very numerous above; stem leafless below, but with conspicuous leaf-scars. Fr. densely woolly-haired. Subsp. *wulfenii* (Koch) A. R. Sm. has very robust stems up to 180 cm.; umbels up to 15 cm. across; glands yellowish, with long horns. △ Dry places. Apr.–July. Med. Eur. (except TR.) P. Pl. 66.

(ii) Floral-bracts not fused in pairs. Lvs. thin
679. *E. serrata* L. SAW-LEAVED SPURGE. Readily distinguished by its linear to lance-shaped lvs. with conspicuously saw-toothed margin, contrasting with the broad, heart-shaped, saw-toothed bracts and floral-bracts. Inflorescence golden-yellow; umbel with 3–5 rays; glands of involucre brown, with very short blunt horns, or almost oval. Fr. hairless, smooth. A hairless, glaucous, erect perenn. 20–60 cm. △ Cultivated and undisturbed ground. May–July. P.E.F.I.

680. *E. cyparissias* L. CYPRESS SPURGE. Readily distinguished by its linear lvs. which on the non-flowering stems are very numerous and form brush-like, bright greenish-yellow shoots, recalling young fir-trees; lvs. 1–3 cm. by 1–3 mm. Umbel with 9–15 rays, lateral umbels often present. Floral-bracts kidney-shaped, yellowish and often turning bright red; glands of involucre with short horns. Fr. hairless. A hairless perenn. with underground stolons and numerous erect shoots 20–50 cm. △ Cultivated ground, grassy places, waysides. Apr.–June. Much of Eur. (except P.IRL.TR.): introd. GB.DK.N.S.SF.

E. pinea L. Stems thick, erect, woody-based, often with numerous leaf-scars below and with numerous overlapping glaucous linear lvs. 2–3 mm. broad, and leafy sterile shoots. Umbels with 5–7 rays; floral-bracts contrasting with lvs., heart- or kidney-shaped, fine-pointed; glands of involucre with fine pointed horns. Fr. hairless; seeds with shallow depressions. Much of the Mediterranean littoral. Page 225.

681. *E. esula* L. A rather robust, dark green perenn. 30–80 cm. with creeping underground stolons and numerous erect fl. and non-flowering stems. Umbels with 6–20 rays, lateral umbels present below. Floral-bracts oval-triangular, yellowish or reddish, contrasting with the linear to oblance-shaped, fine-pointed lvs. Glands of involucre with rather short horns. Fr. hairless, slightly rough. △ Sandy and stony places, waysides, copses. May–Aug. Much of Eur. (except IRL.IS.AL.): introd. GB.DK.N.S.SF.CH.

(iii) Floral-bracts not fused in pairs. Lvs. fleshy
682. *E. myrsinites* L. BROAD-LEAVED GLAUCOUS SPURGE. A robust perenn. to 30 cm., distin-guished by its numerous overlapping, broad, fleshy, glaucous lvs. borne on stout, rather fleshy, spreading stems. Umbels with 7–12 rays; floral-bracts kidney-shaped; involucral glands with long club-shaped horns, often bright reddish-brown. Lvs. obovate, abruptly pointed. Fr. hairless, smooth; seeds rough. △ Rocky and grassy places. Mar.–June. South-East Eur. I.SU.: introd. CS.

683. *E. rigida* Bieb. (*E. biglandulosa* Desf.) NARROW-LEAVED GLAUCOUS SPURGE. Like 682, but stems erect to ½ m., and fleshy lvs. narrow-lance-shaped, gradually narrowed to a long point. Umbels with 9–15 rays; floral-bracts broadly rhomboid with a fine point; involucral glands half-moon-shaped, with short, blunt, rounded horns. Fr. *c.* 8 mm., hairless, granular; seeds smooth. △ Dry stony places, rocks. Feb.–May. P.I.GR.TR.SU.

684. *E. paralias* L. SEA SPURGE. A glaucous, stiff, erect perenn. 30–60 cm., of the littoral, with numerous, often closely overlapping thick, fleshy acute lvs. and broader fleshy floral-bracts. Umbels with 3–6 rays; floral-bracts rounded-rhomboid with a fine point; glands of involucre with short horns. Lvs. ½–2 cm. oblong to obovate, with broad, fleshy bases leaving conspicuous scars on the stem below. Fr. hairless, granular. △ Sands and dunes by the sea. May–Sept. West, South-East Eur. I. Pl. 66.

E. biumbellata Poiret WHORLED SPURGE. A robust perenn. of the Mediterranean littoral, often distinguished by the 2 umbels, or rarely 3, superimposed one above the other, with a more or less naked stem between. Umbels with numerous rays; floral-bracts half-orbicular, blunt; glands with 2 long club-shaped horns. Lvs. linear-oblong with a fine narrow point. Fr. hairless, finely granular; seeds netted. Spain to Italy. Page 225.

ANDRACHNE Fls. one-sexed, axillary; male fls. with 5–6 sepals and 5–6 longer, lance-shaped petals, and stamens 5–6; female fls. with 5–6 sepals and 5–6 shorter petals, or petals absent. Ovary three-celled; seeds rough. 1 sp.

685. *A. telephioides* L. A slender, glaucous, usually spreading perenn. 8–15 cm., branching from a woody base, with numerous tiny alternate, short-stalked oval lvs. 3–7 mm. Fls. axillary, solitary, short-stalked, often shorter than the lvs.; the male fls. smaller than the female fls. Fr. globular, hairless, on reflexed stalks. △ Waste places, waysides. Apr.–Oct. E.I.YU.GR.SU.: introd. F. Page 220.

RUTACEAE | Rue Family

Mostly trees and shrubs, usually with evergreen simple or pinnate lvs. with translucent oil-cavities containing volatile oils. Fls. usually regular; sepals and petals 4–5; stamens often twice as many; ovary superior with 4–5 cells, on a raised receptacle or disk; style 1. Fr. a fleshy berry or capsule.

1. Trees with evergreen lvs.; fr. large, fleshy.   *Citrus*
1. Herbaceous plants or small shrubs; fr. a dry capsule.
2. Fls. white, pink, or lilac; lvs. pinnate.   *Dictamnus*
2. Fls. yellow; lvs. 2–3 times cut.   *Ruta*

RUTA | Rue Fls. yellow; sepals and petals 4 or 5, the petals with a finely toothed or fringed margin. Capsule four- to five-lobed. Lvs. 2–3 times cut, usually strong-smelling. 5 sps.

Petals with finely toothed or entire margin
686. *R. graveolens* L. COMMON RUE. A very strongly smelling, usually glaucous, woody-based shrub-like perenn. to 80 cm., with a lax branched cluster of yellow fls. with widely spaced petals. Fls. 1½–2 cm.; petals with entire or finely toothed margin. Lvs. 2–3 times cut into oval to spathulate segments 3–12 mm., usually with aromatic glands, but variable, sometimes only slightly aromatic and yellowish-green. Fr. with blunt spreading lobes. △ Dry hills, rocks, old walls. May–July. Med. Eur. (except GR.) BG.SU.: introd. F.D.CH.A. CS.H.R. *Of very considerable medical repute in the past, and still used in veterinary medicine. A poisonous plant.* Pl. 67.

R. montana (L.) L. Like 686, but more woody and shrubby and lvs. with narrower linear-lance-shaped glaucous segments; branches glandular above. Fls. smaller *c.* 1 cm., in dense terminal clusters; petals with finely toothed, undulate margin. Mediterranean region and Portugal.

Petals with a conspicuously fringed margin
687. *R. chalepensis* L. FRINGED RUE. Like 686, but petals fringed with long teeth usually longer than the width of the petal; fls. *c.* 2 cm. borne in a dense cluster. Lvs. yellowish-green, strong-smelling, once- or twice-pinnate with oblong-elliptic, entire or blunt-lobed segments 6–20 mm. Fr. with long-pointed, erect lobes. A woody-based, shrubby perenn. to 80 cm. △ Rocky ground, dry banks, thickets. Apr.–June. Med. Eur. (except TR.) P.

DICTAMNUS Fls. somewhat irregular, in a terminal spike-like cluster. Sepals and petals 5; stamens 10. Ovary deeply five-lobed, on a thick disk. Fr. a five-lobed capsule. 1 sp.

688. *D. albus* L. BURNING BUSH. A strongly smelling, very glandular, erect perenn. to 1 m. with large, irregular, white or pink fls. veined with violet, in a long, lax, leafless, spike-like cluster. Petals lance-shaped 2–2½ cm., unequal, the upper 4 erect, the lowest reflexed; stamens iong-projecting, filaments purple, anthers green; fl. stalks and calyx glandular. Lvs. pinnate with 5–9 oval, finely toothed, leathery leaflets. Fr. deeply five-lobed. △ Woods, dry bushy places in the hills, among rocks. May–June. Med. Eur. (except F.TR.) D.CH.A.CS.PL.H.BG.R.SU. *Considered in the past as a cure-all and widely used in herbal remedies. The aromatic oil exuding from the plant is said in hot weather to burst into flames when ignited, hence its name.* Pl. 67.

CITRUS Fls. solitary or few in the axils of the lvs. Calyx with 3–5 teeth; petals 4–8, thick, strap-shaped, white or pink; stamens numerous, often 20–60 and united into bundles. Ovary superior with 5 to many cells. Fr. a large globular berry with fleshy hairs and a thick rind with translucent oil-cavities. Small evergreen trees or shrubs. Natives of Asia; introd. to Europe. 8 sps. *Citrus fruits are widely cultivated in the Mediterranean region; the flowers are used in perfumery and are candied; the fruits are eaten fresh or used for beverages, liqueurs, oils, jams, and as a source of citric acid. The juice, rind, flowers, and leaves have been used medicinally.*

Fr. globular, without a nipple-like swelling

689. *C. aurantium* L. SEVILLE ORANGE. Lvs. oblong-oval acute 5–8 cm., with shallowly indented margin and with broadly winged lv. stalk. Fls. 2–2½ cm. long, white, very sweet-scented. Fr. 7–8 cm. globular, reddish-orange, with bitter rind and sour flesh.

690. *C. sinensis* (L.) Osbeck SWEET ORANGE. Like 689, but lvs. with a narrowly winged lv. stalk. Fr. variable in size, yellow to orange. Flesh sweet, rind only slightly bitter. Pl. 67.

691. *C. deliciosa* Ten. TANGERINE. Lvs. oblong to elliptic, wavy-margined, lv. stalk narrowly winged. Fls. small, 1–1½ cm. long, white, sweet-scented. Fr. 5–7 cm. globular, strongly flattened at each end and with an easily detachable rind. Flesh very sweet, rind sweet and very aromatic.

Fr. oval, with a terminal, nipple-like swelling

692. *C. medica* L. CITRON. Lvs. oblong, acute or blunt, margin finely toothed and lv. stalk not winged. Fls. 2½–3 cm. long, white, but often flushed with pink outside, sweet-scented. Fr. very large, 13–25 cm. long, pale yellow, with a rough, fragrant rind and mildly acid flesh. Young branches reddish, spiny.

693. *C. limon* (L.) Burm. fil. LEMON. Lvs. oblong-oval acute, often irregularly toothed and lv. stalk very narrowly winged. Fls. 2 cm. long, flushed with red outside, sweet-scented. Fr. 5–10 cm. long, oval, pale yellow, smooth, but with slight swellings over the surface, flesh sour. Pl. 67.

SIMAROUBACEAE | Quassia Family

Trees or shrubs with alternate pinnate lvs. and usually with bitter bark. Fls. small, often one-sexed with 3–7 sepals and petals; stamens often twice as many. Ovary superior, surrounded by a prominent disk, with 2–5 fused or free carpels; fr. various.

AILANTHUS   Sepals 5–6, fused to the middle; petals 5–6; stamens in male fls. 10, fewer or absent in female fls. Carpels 5–6, free, but styles fused. Fr. winged. 1 sp.

694. *A. altissima* (Miller) Swingle TREE OF HEAVEN. A rapid-growing tree to 20 m., with smooth striped bark, very large compound lvs., and terminal branched clusters of numerous, small, strong-smelling, greenish-yellow fls. each 7–8 mm. Lvs. pinnate ½–1 m. long, with 13–41 oval-lance-shaped stalked leaflets up to 13 cm. long, each with 2–4 coarse glandular teeth near the base. Fr. usually in clusters of 3 winged carpels, reddish-brown. △ Native of China; commonly planted as an ornamental, sometimes naturalizing from suckers. July–Aug. Introd. to much of Eur. (except North Eur. IRL.NL.PL.TR.). Pl. 68.

MELIACEAE | Mahogany Family

Trees and shrubs of tropical and subtropical regions, with pinnate lvs. and often with a hard, scented wood. Fls. small, bisexual, in loose branched clusters; sepals and petals 4–5; stamens 8–10, usually with their stalks fused into a tube surrounding the knobbled style, and often fused to the petals. Ovary superior, with 2–5 cells; fr. a berry or capsule.

MELIA   Calyx 5; petals 5; anthers attached to the inside of the stamen tube. Fr. a berry with 5 cells, each one-seeded. 1 sp.

695. *M. azedarach* L. PERSIAN LILAC, INDIAN BEAD TREE. A small ornamental tree to 15 m., distinguished by its large, twice-pinnate lvs. and its lax branched clusters of small, fragrant, pale bluish-lilac fls. Fls. 12–15 mm. across, petals spreading in a star; stamen tube violet, anthers yellow. Lvs. 20–50 cm.; leaflets 3–5 cm., oval to lance-shaped, sharp-pointed, toothed or lobed, hairless. Fr. globular *c.* 8 mm., long-stalked, green becoming yellow. △ Native of China; often grown as an ornamental and sometimes naturalized. May. *The seeds are often used for beads, particularly in rosaries. The leaves and fruits have been used medicinally; the fruits are rich in oil and have been used as a vermifuge and antiseptic; they are somewhat poisonous.*

POLYGALACEAE | Milkwort Family

Herbs or less commonly small shrubs in our sps. usually with alternate simple lvs. and without stipules. Fls. bisexual, irregular and symmetrical in one plane only. Sepals 5, overlapping, the 2 inner often brightly coloured and petal-like (often called *'wings'* or *alae*) and much larger than the outer 3; petals 3–5. Stamens 8, usually fused into a tube below and the tube often fused to the petals; anthers opening by an apical pore. Ovary superior, with 2 one-seeded cells.

POLYGALA | Milkwort   Sepals very unequal, the 2 inner large and coloured, and often encircling the petals and stamens. Petals 3, the lowest larger, boat-shaped and often with a terminal brush-like crest, the lateral petals smaller and fused with the staminal tube. Fr. a compressed, two-lobed capsule. 33 sps.

Shrublets with a woody base
696. *P. chamaebuxus* L. BOX-LEAVED or SHRUBBY MILKWORT. A low, creeping, evergreen shrublet to 20 cm. with conspicuous yellow fls. flushed with purple at the tips, or less commonly with pink, white, or red fls. Fls. large, 1½–2 cm. long, axillary; inner sepals yellow, oval, nearly as long as the petals; lower petal three-lobed, not brush-like. Lvs. 1½–3 cm. oval-oblong, fine-pointed, leathery. Fr. *c.* 7 mm. ovoid, winged. △ Open woods, rocky places. Apr.–Sept. Cent. Eur. (except PL.) F.I.YU.R. Pl. 68.

Stems all herbaceous
(a) Anns.
697. *P. monspeliaca* L. MONTPELLIER MILKWORT. A slender, little-branched ann. 10–30 cm. with rather large whitish or greenish-white fls. in a lax terminal spike about half the length of the stem. Large sepals white, conspicuously three-veined, 6–7 mm. and much longer than the petals; bracts shorter than the fl. stalks. Lvs. linear-lance-shaped, the upper fine-pointed. Fr. broadly winged. △ Rocky places, open woods. Apr.–June. Med. Eur. P.BG. Page 220.

(b) Perenns.
(i) Lower lvs. smaller than the upper and not forming a rosette
698. *P. vulgaris* L. (incl. *P. oxyptera* Reichenb.) COMMON MILKWORT. Fls. 6–8 mm. long, pink, white, or blue in a many-flowered, lax, terminal spike. Outer sepals *c.* 3 mm., greenish, inner large sepals *c.* 6 mm., oval, brightly coloured like the petals, with branched veins which rejoin near the margin. Lvs. linear-lance-shaped, scattered, alternate. Fr. *c.* 5 mm.

long. A very variable, much-branched erect or spreading perenn. 10–30 cm. △ Grassy places, heaths, dunes. May–July. All Eur. (except IS.). Pl. 68.

P. serpyllifolia J. A. C. Hose Very like 698, but at least the lower lvs. and those of sterile shoots opposite. Fls. fewer 3–12, usually blue, but sometimes whitish or pink; large sepals 4½–5½ mm. Western and Central Europe.

699. *P. major* Jacq. LARGE MILKWORT. Fls. large, rosy- purple or rarely blue or white, in dense, erect, terminal spikes which later elongate. Large sepals 10–12 mm., elliptic, with 3–7 veins netted at the margin, shorter than the petals. Fr. shorter and narrower than large sepals, long-stalked. A woody-based perenn. to *c.* 30 cm. △ Meadows. May–June. South-East Eur. A.CS.H.I.SU. Page 220.

700. *P. comosa* Schkuhr TUFTED MILKWORT. Fls. pink, rarely blue or white in a dense spike which is hairy in bud and which soon elongates; bracts as long as or longer than the fr. stalks. Large sepals 5–6 mm. gradually narrowed to the base, with slightly netted veins, shorter and often broader than the fr. Usually an erect-stemmed perenn. to 20 cm. △ Dry pastures, hills, and open woods. May–Aug. Most of Eur. (except P.IRL.GB.IS.DK.N.).

P. nicaeensis Koch Fls. large, pink, blue, or white in a lax, soon-elongating spike. Large sepals 8–10 mm., elliptic to obovate and abruptly contracted to the base, with netted veins. Bracts as long as or longer than the fl. stalks. A variable sp. Much of Europe, except the west and north. Pl. 68.

(*ii*) *Lower lvs. forming a loose rosette and longer than the upper lvs.*
701. *P. calcarea* F. W. Schultz CHALK MILKWORT. Distinguished by the lax rosettes of obovate lvs. each ½–2 cm. which appear near the middle of the stem and a slender, almost leafless, stolon-like stem below; central axis of rosette ending in several fl. stems. Fls. 6–7 mm., usually bright blue or rarely pink or white; large sepals 5 mm. with 3–5 veins, somewhat netted near the margin, longer and nearly as broad as the fr. A spreading perenn. 10–20 cm. △ Dry grassland, rocky, sandy, gravelly places. Apr.–June. E.F.GB.B.D.CH.

CORIARIACEAE | Coriaria Family

A family of only one genus, distinguished by the petals which become fleshy and enclose the 5 separate carpels. Sepals and petals 5; stamens 10; styles 5, slender. Only 1 sp. in Europe.

CORIARIA

702. *C. myrtifolia* L. MEDITERRANEAN CORIARIA. An erect shrub with numerous suckering stems, 1–3 m., with greyish quadrangular twigs and oval-lance-shaped, conspicuously three-veined, somewhat leathery lvs. Fls. greenish, *c.* 4 mm., in short lateral clusters; petals shorter than sepals; styles 5 conspicuous, reddish. Lvs. 1½–5 cm., opposite or in threes, hairless. Fr. *c.* 6 mm., reddish-purple becoming shining black, conspicuously five-lobed, lobes shallowly ribbed. △ Thickets, rocky banks, dry hills. Mar.–July. E.F.I.GR. *A poisonous plant, particularly the fruits; the foliage is poisonous to livestock. The bark is rich in tannin and has been used for tanning and dyeing.* Pl. 68.

ANACARDIACEAE | Cashew Family

Trees and shrubs with resinous or acrid milky juice and resinous bark. Lvs. simple or compound. Fls. small, in clusters; sepals and petals usually 5; stamens 5 or 10. Ovary superior of 5 carpels usually one-celled and one-seeded; styles 1–3, widely separated. Fr. usually a drupe.

1. Lvs. simple.   *Cotinus*
1. Lvs. pinnate.
  2. Perianth of 1 whorl.   *Pistacia*
  2. Perianth of 2 whorls with petals and sepals.
    3. Stamens 10; tree with slender pendulous branches.   *Schinus*
    3. Stamens 5; shrub with stout, erect branches.   *Rhus*

PISTACIA   Fls. one-sexed on different plants, in branched axillary clusters. Male fls. with 2 bracts, calyx five-lobed and stamens 3–5; female fls. with 2 bracts, calyx two- to five-lobed, ovary 1 and style terminal, shortly three-lobed. 4 sps.

703. *P. lentiscus* L. MASTIC TREE, LENTISC. A dense, much-branched, very aromatic, resinous, evergreen shrub 1–3 m., rarely a small tree, with even-pinnate lvs. with a winged lv. stalk and bearing either male or female fls. Male fls. numerous, densely clustered; anthers at first bright red. Female fls. brownish. Leaflets 3–6 pairs, elliptic fine-pointed, 2–3 cm., leathery, dark green, shining above. Fr. globular *c.* 8 mm., red then black. △ Dry stony slopes, thickets. Apr.–June. Med. Eur. (except TR.) P. *The sap from the incised stems produces a sweet-smelling gum known as mastic, which is used in medicine, dentistry, and in varnishes. It is also used in liqueurs and cordials, and was used as a chewing-gum in classical times.* Pl. 68.

704. *P. terebinthus* L. TURPENTINE TREE, TEREBINTH. Distinguished from 703 by the much larger deciduous lvs. which are odd-pinnate with 2–5 pairs of oval leaflets to 6 cm. and lv. stalk not winged. Fls. bright reddish-purple, in compound, spreading, branched clusters, appearing on last year's branches and at the same time as the young lvs. Fr. *c.* 8 mm., red, then brown when ripe. A grey-stemmed, strongly resinous shrub, 2–5 m. △ Bushy places, rocks, in the hills. Apr.–July. Med. Eur. (except TR.) P.BG. *The sap produces a sweet-smelling gum known as Chian turpentine, which has recently become important in a treatment for cancer. The fruits are edible; the seeds produce an oil. Large horn-like galls formed by an aphis are commonly found on the leaves and these are used medicinally and produce a red dye.* Pl. 69.

SCHINUS   Resinous trees with odd-pinnate lvs. and stalkless leaflets. Fls. in axillary or terminal clusters, one-sexed, yellowish or greenish. Calyx five-lobed; petals 5; stamens 10; styles 3. Fr. a drupe. 2 sps.

705. *S. molle* L. CALIFORNIAN PEPPER-TREE, PERUVIAN MASTIC-TREE. A small tree to 10 m. with graceful pendulous branches bearing pale green, very aromatic, compound lvs. to 30 cm. and inconspicuous, greenish-yellow fls. in much-branched clusters. Leaflets 15–20 pairs, narrow-lance-shaped, entire or toothed, each 4–9 cm. long. Fr. red, globular, *c.* 5 mm., in branched pendulous clusters. △ Native of Central and South America; commonly planted as an ornamental in the Mediterranean region and sometimes self-seeding. July–Aug. Introd. P.E.I.

COTINUS   Shrubs with simple lvs. Fls. small, bisexual, in branched clusters; sepals, petals, and stamens 5; ovary 1, with 3 styles placed laterally. Fr. a drupe. 1 sp.

706. *C. coggygria* Scop. (*Rhus cotinus* L.) WIG TREE, SMOKE-TREE. A dense spreading shrub to 3 m., with neat, aromatic, rounded-to-obovate, stalked lvs. 4–7 cm., glaucous beneath.

Fls. yellowish, in terminal, lax, spreading, pyramidal clusters with numerous sterile fls.; the whole inflorescence later lengthens and becomes plume-like with spreading hairs on the fr. stalks, recalling a grey wig when seen from a distance. Fr. 3 mm., obovate, brown, shining, hairless. △ Dry hills, rocky places, open woods. May–July. Med. Eur. (except E.) H.BG.R.SU. *The wood, known as 'young fustic', is used for tanning and produces an orange dye.* Pl. 69.

RHUS | Sumach Shrubs or small trees with odd-pinnate lvs. Fls. small, clustered, bisexual; sepals, petals and stamens 5; ovary 1, styles 3, terminal. Fr. a drupe. 4 sps.

707. *R. coriaria* L. SUMACH. A softly hairy shrub 1–3 m., with milky juice, pinnate lvs., and whitish fls. in dense, erect, elongated clusters. Leaflets 7–15, broadly lance-shaped, toothed, hairy, the common lv. stalk narrowly winged above. Fr. globular, woolly-haired, brownish-purple when ripe. △ Dry, stony places, thickets. May–June. Med. Eur. P.BG.SU. *The young branches are much in demand for tanning moroccan leather; the bark produces yellow and red dyes. The fresh fruit, sap and leaves are poisonous.*

R. typhina L. (*R. hirta* (L.) Sudworth) STAGHORN SUMACH, VINEGAR TREE. Like 707, but common lv. stalk rounded, not narrowly winged above and leaflets more numerous 11–31, long-pointed; branches thick, densely brown-haired. Fr. with crimson hairs. Native of North America; grown as an ornamental and sometimes naturalized in Southern Europe.

ACERACEAE | Maple Family

Trees and shrubs usually with deciduous lvs. which are usually palmate-lobed or pinnate; stipules absent. Fls. small, greenish-yellow in clusters; sepals and petals 4–5, or petals absent; stamens usually 8, but sometimes 4–10, inserted on or inside the disk of the receptacle. Ovary two-celled, splitting in fr. into 2 winged carpels or *samaras*.

ACER | Maple Fr. with each carpel with a wing on one side. 15 sps.

Lvs. palmately lobed or entire
(*a*) *Fls. in pendulous clusters*
708. *A. pseudoplatanus* L. SYCAMORE. A tree to 30 m. with smooth grey bark which later becomes scaly, dark green palmately lobed lvs. and long-stemmed, pendulous, elongate clusters of yellowish-green fls. Fls. *c.* 6 mm., stalked; filaments hairy. Lv. blades 7–16 cm., divided to about half their width into 5 broad, coarse-toothed, acute lobes, pale and almost glaucous, and soon hairless beneath; twigs hairless. Fr. with wings widely diverging. △ Woods, hedges, streamsides, in the mountains. Apr.–May. Much of Eur. (except North Eur.): introd. IRL.GB.DK.S. *The wood is used in turnery for handles, bowls and other domestic utensils.* Pl. 69.

709. *A. opalus* Miller (*A. opulifolium* Chaix) Lvs. rather large, leathery, green above and rather pale glaucous and often hairy beneath, divided to less than half their width into 5–7 broad, shallow, blunt, weak-toothed lobes, but lvs. very variable. Fls. greenish-yellow, long-stalked, in a lax ultimately pendulous, almost stemless cluster. Fr. with wings diverging at an acute angle. A tree with smooth bark; 8–20 m. △ Open woods in hills. Apr.–May. E.F.D.CH.I. Page 239.

(*b*) *Fls. in erect clusters*
710. *A. platanoides* L. NORWAY MAPLE. Distinguished from 708 by its somewhat shining,

bright green lvs. 5–15 cm., with 5 triangular, parallel-sided, long-pointed lobes which have a few large, long-pointed teeth. Fls. bright greenish-yellow, *c.* 8 mm., in erect, branched, rather rounded clusters, appearing before the fully expanded lvs. Twigs with milky juice. Wings of fr. widely diverging. A tree to 30 m. with dark grey fissured bark, not scaling. △ Woods. Apr.–May. Much of Eur. (except P.IRL.IS.): introd. GB.NL. *The wood is used for turnery.* Pl. 69.

711. *A campestre* L. COMMON MAPLE. A small tree 9–25 m. with light grey fissured bark and small lvs. 4–7 cm. divided to about half their width into 5 blunt entire lobes or lobes with a few broad shallow lobes, but not toothed. Fls. pale green, *c.* 6 mm., in erect, branched clusters appearing with the lvs. Twigs brown, hairy. Fr. usually hairy, wings spreading in a straight line. △ Open woods, hedges. Apr.–May. Most of Eur. (except IS.SF.): introd. IRL.

712. *A. monspessulanum* L. MONTPELLIER MAPLE. Like 711, but lvs. with 3 blunt entire lobes, diverging almost at right angles, rather tough, shining above, glaucous beneath. Fls. yellowish-green in a lax, erect cluster, often appearing before the lvs. Fr. with wings not diverging but almost parallel to each other, or converging and overlapping. A shrub and small tree 5–6 m. △ Dry hills, open woods, thickets, rocky places. Apr.–May. Med., South-East Eur. P.D.CH.: introd. A.

A. tataricum L. Lvs. oval, irregularly double-toothed or very shallowly lobed, hairless. Fls. white, in erect clusters 3–8 cm. Fr. with wings nearly parallel, at length red. South-Eastern Europe.

Lvs. compound, with 3–7 leaflets
A. negundo L. BOX-ELDER. A small tree 6–20 m. with very variable pale green lvs. with 3–7 elliptic to lance-shaped leaflets which are coarsely toothed or three-lobed. Fls. one-sexed, pendulous in narrow clusters, petals absent; male fls. with 4–6 stamens, at first stalkless but later with long slender stalks; female fls. with an ovary only. Fr. in long clusters to 25 cm., wings diverging at an acute angle. Native of North America; widely planted as an ornamental, and occasionally naturalized in Southern Europe.

HIPPOCASTANACEAE | Horse-Chestnut Family

Trees or shrubs with palmate lvs., without stipules. Fls. in large, terminal, spike-like clusters; corolla irregular and symmetrical in one plane only with 4–5 unequal petals; sepals 5; stamens 5–8. Ovary superior, three-celled; fr. a tough, thick-skinned capsule opening by 3 valves; seeds large, reddish-brown, shiny.

AESCULUS | Horse-Chestnut Deciduous trees; sepals forming a tubular or bell-shaped calyx. 2 sps.

713. *A. hippocastanum* L. HORSE-CHESTNUT. A robust tree 15–30 m. with dark greyish-brown trunk, large, long-stalked, palmate lvs. and large pyramidal clusters of white fls. Fl. clusters 20–30 cm.; fls. mostly male, *c.* 2 cm. across; corolla white with yellow spots, turning pinkish; stamens projecting. Leaflets usually 7, oblong-oblance-shaped long-pointed, irregularly toothed, 8–25 cm., dark green; buds stout, sticky. Fr. *c.* 6 cm., green, with stout prickles; seeds reddish-brown, shiny, *c.* 3 cm. △ Native of South-Eastern Europe; extensively planted elsewhere. May. AL.GR.BG. *The large seeds are not edible, although they have sometimes been used to feed livestock. They contain 5 per cent*

of oil, and starch which has been used for stiffening. The wood is used for turnery, kitchen utensils, and the backs of brushes. Pl. 69.

A. *carnea* Hayne RED HORSE-CHESTNUT. Like 713, but fls. pink or red; leaflets usually 5. Fr. smaller, 3–4 cm., with few prickles. Possibly a hybrid between 713 and *A. pavia* L., an American sp. Grown as an ornamental tree. Pl. 69.

BALSAMINACEAE | Balsam Family

Herbaceous plants with somewhat translucent stems, simple lvs., and stipules absent or replaced by glands. Fls. irregular and symmetrical in one plane only; sepals 3 or 5, often coloured, the lowest much larger and with a sac-like spur; petals 5, partly fused; stamens 5, anthers fused round ovary. Ovary with 5 cells; fr. splitting by elastically coiling, fleshy valves which shoot out the seeds. Ripe frs. are often formed without the development of normal fls. (*cleistogamy*).

IMPATIENS | Balsam Sepals usually 3, very unequal, the lowest spurred. Corolla two-lipped with 1 large, upper, hood-like petal and 2 pairs of united lateral petals, forming a divided lower lip. Erect anns. 6 sps.

Fls. yellow or orange
714. *I. noli-tangere* L. TOUCH-ME-NOT. Fls. *c.* 3½ cm. long, bright yellow with small brown spots on lower lip, in few-flowered axillary clusters. Lower sepal yellow, large, *c.* 2½ cm., conical and tapering into a slender spur which curves gradually downwards or upwards. Lvs. stalked, alternate, oval-oblong, blunt or acute, margin with 10–15 coarse teeth on each side. An erect, hairless ann. with smooth stems which are swollen at the nodes; 20–100 cm. △ Damp woods, shady ravines. June–Aug. Most of Eur. (except P.IRL.IS.GR.TR.). Pl. 70.

I. capensis Meerb. (*I. biflora* Walter) ORANGE BALSAM. Like 714, but fls. orange, strongly blotched with reddish-brown within, 2–3 cm. long; lower sepal rather suddenly contracted into a narrow spur which is abruptly bent downwards or upwards, or recurved against itself. Lvs. glaucous beneath, margin with 10 or less teeth on each side. Native of North America; naturalized by rivers in France and Britain. Pl. 70.

715. *I. parviflora* DC. SMALL BALSAM. Distinguished by its small, pale yellow, unspotted fls. ½–1½ cm. long and its straight conical spur of varying length. Lvs. oval long-pointed, tapering below into a winged lv. stalk, with numerous forward-directed teeth on margin. An erect, hairless ann. 30–100 cm. △ Cultivated ground, waysides, river gravels, woods. Apr.–Oct. Native of Central Asia: introd. to much of Eur. (except P.E.IRL.NL.IS.AL.GR.TR.BG.). Pl. 70.

Fls. red, pink, or white
716. *I. glandulifera* Royle POLICEMAN'S HELMET. A very robust, thick-stemmed, fetid ann. 1½–2 m., with large pinkish-purple, or rarely white, fls. 2½–4 cm. long in long-stemmed clusters arising from the upper lvs. Lower sepal thimble-shaped with a very narrow, recurved, tail-like spur. Lvs. opposite or in whorls of 3, large, 6–15 cm., elliptic-lance-shaped, with numerous acute teeth on margin. △ Damp ground, river banks, waste places. July–Sept. Native of the Himalaya. Introd. F.IRL.GB.NL.B.DK.N.S.SF.D.CH.A.PL.CS.I.H.YU.R.SU. Pl. 70.

I. balfourii Hooker fil. Fls. pink and yellow, or white, 2½–4 cm.; lower sepal narrowed gradually to a long, slender, straight or slightly arched spur. Native of the Himalaya. Introduced to Central and Southern Europe.

AQUIFOLIACEAE | Holly Family

Trees or shrubs with simple lvs., without stipules. Fls. regular in few-flowered axillary clusters, bisexual or one-sexed and occurring on different plants. Sepals and petals 4–5; stamens 4–5; ovary three- to many-celled, stigma stalkless. Fr. a drupe with 3 or more stones.

ILEX | Holly Plants one-sexed, or rarely with bisexual fls. Sepals, petals, and stamens 4; petals united below. 3 sps.

717. *I. aquifolium* L. HOLLY. A small tree or dense shrub 3–15 m. with smooth grey bark and spiny, dark, shining, evergreen lvs. Fls. white *c.* 6 mm., in axillary clusters. Lvs. oval-elliptic, leathery with a cartilaginous margin and with several sharp stiff spines, or spineless. Fr. scarlet, 7–12 mm. △ Woods, thickets, hedges. Apr.–May. Most of Eur. (except IS.S.SF.PL.CS.H.). *The fruits are poisonous; the bark produces a sticky substance used to make bird lime. The wood is dense and is used as a substitute for box wood for printing blocks and wood engraving; it is used in turnery for small objects.* Pl. 71.

CELASTRACEAE | Spindle-Tree Family

Trees, shrubs, or woody climbers with simple lvs., with or without stipules. Fls. bisexual or one-sexed, usually small, greenish, in flat-topped clusters. Sepals and petals 4–5; stamens 4–5, rarely 8–10. Ovary with 2–5 cells, surrounded by the disk. Fr. a capsule, drupe or berry; seeds usually enveloped in a bright-pink or red pulpy aril.

EUONYMUS Trees or shrubs with opposite lvs. Sepals, petals, and stamens 4–5, the stamens attached to a wide fleshy disk. Fr. rather fleshy, forming a brightly coloured capsule. 5 sps.

718. *E. europaeus* L. SPINDLE-TREE. A deciduous shrub or small tree 2–6 m. with green, square-sectioned twigs and smooth grey bark. Fls. greenish, 8–10 mm. in lax, stalked, axillary clusters; petals 4, widely spaced, twice as long as broad. Lvs. 3–13 cm., oval to elliptic, finely toothed, stalked, turning red in autumn. Fr. deep pink, of 4 rounded segments, each splitting to reveal a bright orange aril encircling the seed. △ Woods, bushy places. Apr.–June. All Eur. (except IS.SF.). *The wood is easily split into fine strips and is used for skewers, pegs, spindles, etc. The fruits, leaves and bark are poisonous and emetic.* Pl. 71.

E. latifolius (L.) Miller Like 718, but young branches greyish-brown, rounded in section, and middle stem lvs. larger 7–15 cm. by 6 cm. Sepals, petals and stamens usually 5; petals greenish-brown, as long as broad. Fr. with 5 winged segments. South-Central and South-Eastern Europe. Page 239.

BUXACEAE | Box Family

Usually evergreen trees or shrubs with simple lvs., without stipules. Fls. one-sexed, small, in spikes or lateral clusters. Calyx usually 4 or sometimes absent, or more in female fls.; petals absent; stamens 4 or more. Ovary superior, usually three-celled. Fr. a capsule or drupe.

BUXUS | Box   Fls. in axillary clusters, with the upper fls. female and the lower male in each cluster. Male fls. with 4 sepals and 4 stamens; female fls. with several sepals and a three-celled ovary. Fr. a capsule. 2 sps.

719. *B. sempervirens* L. BOX. A shrub or small tree 2–5 m., with shiny, dark green, leathery, elliptic lvs. 1–2½ cm., with acute, blunt, or notched apex. Fls. whitish-green, 2 mm., in dense, rounded, stalkless clusters much shorter than the lvs. Fr. ovoid, *c.* 8 mm., three-horned, rough-netted. △ Dry hills. Mar.–Apr. Med. Eur. P.GB.B.D.CH.A.: introd. R. *The wood is very hard and close-grained and used for making mathematical instruments and for wood engraving and turnery. The whole plant, and in particular the leaves and seeds, is poisonous to livestock; the leaves are used medicinally.*

RHAMNACEAE | Buckthorn Family

Often spiny trees or shrubs with simple lvs. and with or without small deciduous stipules. Fls. small, usually greenish, in axillary clusters. Sepals, petals, and stamens usually 5 and placed on a cup-shaped disk, with the ovary at the base or half-sunk in the disk. Fr. a drupe or capsule.

1. Shrubs or small trees without spines.
   2. Buds naked and without scales.   *Frangula*
   2. Buds with scales.   *Rhamnus*
1. Shrubs or small trees with spines.
   3. Fr. dry, with a disk-like flange.   *Paliurus*
   3. Fr. fleshy, rounded.
      4. Stipules forming spines; fr. reddish.   *Ziziphus*
      4. Stipules not spiny, but branches forming spines; fr. small, black.   *Rhamnus*

RHAMNUS | Buckthorn   Spiny or spineless shrubs with herbaceous, deciduous stipules. Calyx tubular or bell-shaped and persisting, with 4–5 lobes which fall off; petals 4–5 or absent; style branched. Fr. a black drupe. 13 sps. *The fruits of most species are purgative; they have also been used as a colourant.*

Branches alternate, not spiny; lvs. with 4 or more pairs of lateral veins
(a) Sepals and stamens 5
720. *R. alaternus* L. MEDITERRANEAN BUCKTHORN. An erect evergreen shrub 1–5 m. with small oval to lance-shaped, shining, leathery lvs. and branched axillary clusters of small yellowish fls. Fls. one-sexed; petals absent; sepals reflexed in male fls. and erect in female fls. Lvs. variable, 2–5 cm., with 4–6 pairs of lateral veins and with a toothed or sometimes entire margin. Fr. globose *c.* 5 mm., red then black. △ Low thickets, uncultivated stony and rocky places. Mar.–Apr. Med. Eur. (except TR.) P.: introd. SU. Pl. 71.

237

(b) Sepals and stamens 4

721. *R. pumilus* Turra DWARF BUCKTHORN. A low, spreading, prostrate, much-branched shrub of the mountains, with deciduous elliptic to obovate lvs. and yellowish fl. clusters borne at the base of the young branches. Fls. one-sexed; sepals lance-shaped, long-pointed, and longer than calyx tube; petals narrow or absent. Lvs. with 4–12 pairs of prominent curved veins, margin toothed or entire. △ Rocks and screes of high mountains. May–June. E.F.D.CH.A.I.YU.AL.

722. *R. alpinus* L. ALPINE BUCKTHORN. Like 721, but an erect shrub 1–3½ m. with elliptic lvs. and greenish fls. Sepals triangular, shorter than calyx tube. Lvs. with a rounded, almost heart-shaped base and with 9–15 pairs of almost straight lateral veins. △ Open woods, rocky places in hills. May–June. Med. Eur. (except TR.) CH.A.BG. Page 239.

Branches opposite, spiny; lvs. with 2–4 pairs of lateral veins

723. *R. catharticus* L. BUCKTHORN. A large, thorny shrub or small tree 2–8 m. with opposite lateral branches spreading almost at right angles to the main branches, with deciduous lvs. and clusters of numerous greenish fls. arising from the old branches. Fls. *c.* 4 mm. across; sepals and petals 4. Lvs. 3–6 cm. oval-elliptic with 2–3 pairs of lateral veins, regularly toothed; lv. stalk hairy, 2–3 times shorter than the blade; old branchlets ending in a single spine. Fr. globular 6–10 mm., green, becoming black. △ Hedges, thickets, woods. May–June. Most of Eur. (except IS.TR.). *The fruits are poisonous; they have been used in veterinary treatment and in medicine. The fruits produce yellow, orange, and brownish dyes; the bark yields a yellow dye.* Pl. 71.

R. saxatilis Jacq. ROCK BUCKTHORN. A much-branched, very spiny, usually prostrate shrub ½–2 m., distinguished from 723 by the smaller elliptic to lance-shaped lvs. 1–3 cm. which are narrowed at the base to a very short lv. stalk about as long as the stipules. Old branchlets with several spines; twigs greyish. Central and Southern Europe.

FRANGULA Like *Rhamnus*, but buds naked and without scales; lvs. entire, not toothed. Fls. bisexual; sepals, petals, and stamens 5; style not branched. 3 sps.

724. *F. alnus* Miller (*Rhamnus frangula* L.) ALDER BUCKTHORN. An erect, slender-branched shrub 2–5 m., with deciduous, shiny green lvs. and greenish fls. *c.* 3 mm., in few-flowered axillary clusters borne on young wood. Sepals greenish, 5; petals 5, smaller. Lvs. 2–7 cm. obovate, with 6–12 pairs of conspicuous lateral veins, margin not toothed; twigs green, becoming brown. Fr. 6–10 mm., green, then red, and finally violet-black. △ Damp woods, hedges, bogs. Apr.–July. All of Eur. (except IS.). *The bark is purgative and used as a laxative. Dyes are obtained from the bark, leaves, and berries; the branches make the best drawing charcoal.*

ZIZIPHUS Shrubs with spiny stipules. Fr. fleshy, reddish. 2 sps.

725. *Z. jujuba* Miller COMMON JUJUBE. A shrub 2–8 m. with zigzag green branches, strongly curved stipular spines and 2 ranks of oval, finely toothed, shining, three-veined lvs., 2½–5 cm. Fls. yellowish, 4 mm., in axillary clusters of 3–5; petals minute; anthers golden-yellow. Fr. oval-oblong 2–3 cm., pendulous, reddish-purple. △ Native of Asia; cultivated in Southern Europe for its edible frs. and sometimes naturalized. Apr.–May. Introd. Med. Eur. BG.R.

PALIURUS Spiny shrubs like *Ziziphus*, but fr. dry, leathery, enclosed in receptacle and with a circular disk-like flange. 1 sp.

1. *Euonymus latifolius* [718]
3. *Tilia cordata* 731

2. *Acer opalus* 709
4. *Rhamnus alpinus* 722

726. *P. spina-christi* Miller. CHRIST'S THORN, JERUSALEM THORN. A much-branched shrub 2–3 m. with long, flexible, zigzag branches, numerous very sharp straight and recurved spines and 2 rows of obliquely oval, pointed, finely toothed, three-veined lvs., 2–4 cm. Fls. 2 mm., yellow, in small axillary clusters. Fr. *c.* 2–3 cm. across. flattened disk-shaped, ribbed and with undulated margin when dry. △ Thickets, dry uncultivated ground; sometimes grown as a hedge plant. June–Sept. Med. Eur. BG.R.SU.: introd. H.

VITACEAE | Vine Family

Mostly woody climbing plants, with branched tendrils modified from lateral branches. Lvs. simple, palmate or pinnate, often with glandular dots. Fls. small, green, in clusters arising opposite the lv. stalks. Sepals, petals and stamens usually 4–5, petals falling early. Ovary usually two-celled. Fr. a berry.

Tendril branches without terminal disks. *Vitis*
Tendril branches with terminal adhesive disks. *Parthenocissus*

VITIS | Vine Petals fused by their apices and falling off at flowering. Fr. large, fleshy. 1 sp.

727. *V. vinifera* L. COMMON VINE. A woody climbing plant to 30 m., with branched tendrils arising opposite the long-stalked, palmately lobed lvs. Fls. greenish, sweet-scented, in dense clusters opposite the lvs. Lvs. 5–15 cm., heart-shaped in outline and with 3–7 deep lobes which are further coarsely toothed or lobed. Fr. fleshy and sweet, green, red, or blue-black. Subsp. *sylvestris* (C. C. Gmelin) Hegi is one-sexed and with lvs. differing in each sex; fr. small, acid, bluish-black; possibly the parent of the cultivated vine. △ Probably a native of Western Asia; very widely cultivated in South Europe and often self-seeding. May–June. Introd. to much of Eur. (except North Eur. IRL.GB.NL.). *Cultivated in antiquity and thence spreading westwards throughout Southern and Central Europe. Wine, brandy, and numerous liqueurs are made from the fermented juice; the fruits are eaten fresh or dried.*

PARTHENOCISSUS Petals free and spreading at flowering. 3 sps.

728. *P. tricuspidata* (Siebold & Zucc.) Planchon JAPANESE IVY. A robust climber to 20 m. with variable lvs., either triangular-oval long-pointed and toothed, or deeply cut into 3 coarsely toothed lobes, shiny, hairless, turning bright red in autumn. Tendril branches with terminal adhesive disks. Fls. green, in axillary clusters. Fr. bluish, bloomed. △ Native of China and Japan; widely grown as an ornamental to cover buildings, etc. and sometimes self-seeding. Summer.

P. quinquefolia (L.) Planchon VIRGINIA CREEPER. Like 728, but lvs. all palmate with 3–5 shortly stalked, oblong-elliptic and coarsely toothed leaflets, softly hairy beneath. Native of America; grown for ornament and sometimes self-seeding.

TILIACEAE | Lime Tree Family

Usually deciduous trees or shrubs, with simple lvs.; stipules usually small and functioning as bud scales, or absent. Fls. bisexual, clustered; sepals and petals usually 5, but petals rarely absent; stamens 10 or more. Ovary superior, with 2–10 cells, style 1 with as many radiating lobes as cells. Fr. a drupe or nut, or rarely a berry.

TILIA | Lime, Linden Stipules forming bud scales which soon fall. Stamens many; ovary with 5 cells, stigma five-lobed. Fr. compound, of several rounded nuts, on branched fr. stalks with a broad, oblong, wing-like bract at the base, the whole detaching as a unit and spinning on descending. 6 sps.

Lvs. with simple hairs beneath
729. *T. platyphyllos* Scop. LARGE-LEAVED LIME. A tall tree to 30 m. with dark smooth bark, softly hairy twigs and dark green, heart-shaped lvs. Fls. yellowish-white, fragrant, in lax pendulous clusters of 2–5; wing-like bract 5–12 cm. Lv. blade 6–12 cm. toothed, long-pointed, paler green and softly hairy beneath; lv. stalk 1½–5 cm. Fr. 8–10 mm., densely hairy, three- to five-ribbed. △ Woods; often planted. June–July. Much of Eur. (except P.IRL.NL.IS.N.SF.). *The wood of this and other species is used for carving and turnery. The inner bark produces fibres used in basketry and rope-making in Eastern Europe. The dried flowers make a delicate medicinal tea.*

Lvs. with star-shaped hairs beneath
730. *T. tomentosa* Moench (*T. argentea* DC.) SILVER LIME. Readily distinguished by its heart-shaped lvs. which are dark green above and densely white-woolly beneath with adpressed star-shaped hairs. Fls. with groups of sterile stamens forming a corona. Lv. stalk less than half the length of the blade; twigs and buds covered with dense white hairs. Fr. five-angled. A large tree to 30 m. △ Woods; often planted. July–Aug. South-East Eur. CS.H.SU.

Lvs. hairless beneath
731. *T. cordata* Miller SMALL-LEAVED LIME. A large tree with smooth bark, to 25 m., distinguished by its smaller lvs. which are shining green above and glaucous hairless beneath except for tufts of rust-coloured hairs in the vein-angles. Fl. clusters spreading or erect, with 4–10 fls.; wing-like bract 3½–8 cm. Lv. blade 3–8 cm., toothed, more or less rounded in outline, often broader than long, abruptly long-pointed; lv. stalk 1½–4 cm.; twigs soon hairless. Fr. woolly-haired, not or hardly ribbed. △ Woods; often planted. June–July. Most of Eur. (except P.IRL.IS.). Page 239.

732. *T. × vulgaris* Hayne (*T. cordata × platyphyllos*; *T. europaea* auct.) COMMON LIME, LINDEN. A hybrid between 729 and 731 with intermediate characters. Lvs. 6–10 cm., oval, long-pointed, light green, hairless or nearly so beneath, except for the tufts of white hairs in the vein-angles; lv. stalk 3–5 cm. Fls. in pendulous clusters of 5–10; fr. slightly ribbed. △ Woods; often planted. July. E.F.S.D.CH.A.CS.PL.H.I.GR.R.SU.: introd. GB.

MALVACEAE | Mallow Family

Shrubs or herbaceous anns. or perenns., with mucilage canals, fibrous stems and usually star-shaped hairs. Lvs. often palmately lobed; stipules small, often deciduous. Fls. large, regular, solitary or in clusters, usually bisexual. Epicalyx usually present and situated immediately below the calyx and with 3 to several sepal-like lobes; calyx usually 5; petals 5, rolled together in bud. Stamens usually numerous and fused below into a tube encircling the styles. Ovary with 2 to many cells; fr. a capsule, or separating into several one-seeded nutlets.

		Nos.
1. Epicalyx absent.	*Abutilon*	749
1. Epicalyx present.		
2. Style undivided or with 5 lobes; fr. a capsule with 5 cells.		
3. Apex of style club-shaped with 5 furrows; seeds with hairs.	*Gossypium*	750
3. Apex of style with 5 lobes; seeds hairless.	*Hibiscus*	751, 752
2. Style with many branches; fr. separating into several nutlets.		
4. Epicalyx with 2–3 lobes.		
5. Epicalyx with lobes (usually) fused at base.	*Lavatera*	740–744
5. Epicalyx with lobes not fused at base.		
6. Epicalyx linear to narrowly oval.	*Malva*	734–739
6. Epicalyx broadly heart-shaped.	*Malope*	733
4. Epicalyx with 6–9 lobes.		
7. Fls. less than 6 cm., at least some stalked.	*Althaea*	745–747
7. Fls. at least 6 cm., almost stalkless in a spike.	*Alcea*	748

MALOPE Epicalyx with 3 heart-shaped lobes. Fr. strawberry-like with many carpels arranged in several rows round a central axis. 2 sps.

733. *M. malacoides* L. An erect or ascending, hairy-stemmed perenn. 10–40 cm. with deep rose fls. veined with purple. Fls. to 6 cm. across, solitary, long-stalked, borne from the axils of the upper lvs.; petals 2–3 times as long as the calyx; epicalyx half as long as calyx. Lvs. oval to 4 cm., all similar, irregularly toothed or lobed, or rarely three-lobed, sparsely rough-hairy. Fr. in a globular cluster, hairless. △ Waste places, thickets, olive groves. June–July. Med. Eur. (except YU.).

MALVA | Mallow Epicalyx of 2–3 free segments; calyx 5, fused below; petals 5, usually notched, purple, pink, or white. Fr. a whorl of nutlets arising from a central boss. 13 sps.

Fls. solitary in axils of lvs
(a) Perenns.

734. *M. moschata* L. MUSK MALLOW. An erect, branched perenn. to 1 m., usually with solitary, stalked, rosy-pink or white axillary fls. 3–6 cm. across, with deeply notched petals about 3 times as long as the calyx. Calyx and epicalyx with only simple hairs; epicalyx linear-lance-shaped, half as long as the oval calyx segments. Upper and lower lvs. differing, but very variable, the upper rounded, usually deeply cut into 3–7 narrow lobes, which are divided into linear segments, the basal lvs. kidney-shaped, with broad shallow lobes. Fr. not ribbed, rough-hairy, blackish. △ Grassy and bushy places. June–Sept. Most of Eur. (except P.IS.TR.): introd. DK.N.S.SF.H.

735. *M. alcea* L. Like 734, but calyx and epicalyx densely covered with star-shaped hairs; epicalyx with oval segments. Fls. pale pink; petals 2½–3½ cm., notched, variable in size. Upper lvs. variable, but usually less deeply and less narrowly dissected than in 734; stem with dense star-shaped hairs. Fr. finely ribbed, hairless or nearly so. Perenn. 50–120 cm. △ Meadows, hedges. June–Sept. Much of Eur. (except IRL.GB.IS.AL.GR.TR.): introd. N.SF. Pl. 72.

(b) Anns.

736. *M. cretica* Cav. An erect or spreading, hairy ann. 10–50 cm. with small pink fls. 1½ cm. across, with petals shorter than the calyx and fl. stalks as long as or longer than the lvs.

Epicalyx linear-lance-shaped, calyx linear twice as long, with bristly hairs. Upper lvs. divided almost to the base into 3-5 lance-shaped, toothed lobes, the lower lvs. nearly entire, rounded, or shallowly lobed; stem with spreading hairs. Fr. ribbed, hairless, reddish. △ Cultivated ground, stony places. Apr.-June. E.I.GR.: introd. F.

Fls. clustered in axils of lvs.
(a) *Petals 2-4 times as long as calyx*
737. *M. sylvestris* L. COMMON MALLOW. A very variable ann., bienn., or perenn. 45-120 cm. with rather large rosy-purple fls. with darker purple veins borne on stalks of unequal length, in clusters in the upper lv. axils. Fls. 2½-4 cm. across; petals 2-4 times as long as the calyx, deeply notched; epicalyx lobes oblong-lance-shaped, shorter than the oval sepals. Stem lvs. with 5-7 rather deeply toothed lobes. Fr. netted with wrinkles, usually hairy. △ Waste ground, waysides. May-Aug. All Eur. (except IS.). *It has been used as a vegetable.*

(b) *Petals less than twice as long as calyx, often considerably so*
738. *M. nicaeensis* All. (*M. montana* auct.) An erect or ascending, sparsely hairy ann. 20-50 cm. with pale lilac or whitish fls. 1-1½ cm. across, in clusters of 3-6 in the upper lv. axils. Petals nearly twice as long as the calyx; epicalyx lobes oblong, nearly as long as the hairy calyx; fr. stalks remaining erect. Lvs. rounded-heart-shaped, the upper with 3-5 acute lobes. Fr. strongly netted, not toothed on the back, usually hairless. △ Waste places, waysides; occurring as a casual elsewhere. May-July. Med. Eur. (except AL.) P.SU. Pl. 73.

M. neglecta Wallr. (*M. rotundifolia* auct.) DWARF MALLOW. Very like 738, but fr. stalks recurved; fr. with rounded teeth on the back, not wrinkled, hairy. Fls. pale pink. Almost throughout Europe.

M. pusilla Sm. (*M. rotundifolia* L.) SMALL-FLOWERED MALLOW. Distinguished by the very small pale pink to whitish fls. *c.* 5 mm. across, with the petals very little longer than the sepals. Fr. cluster 7-9 mm. across, nutlets hairy, with a network of sharp ridges on their backs; calyx scarcely enlarging in fr. Widespread in Europe, except in the west.

739. *M. verticillata* L. VERTICILLATE MALLOW. Fls. 1-1½ cm. across, whitish or pink, very short-stalked and forming dense whorls in the axils of the lvs. Petals as long or up to nearly twice as long as the calyx which enlarges, in fr.; epicalyx linear-oblong. Lvs. glaucous, large, nearly round, flat or sometimes crisped on the margin, long-stalked. Fr. slightly netted, hairless, not winged or toothed. An erect bienn. to 80 cm. △ Waste ground. Native of Asia; often escaping from cultivation. July-Sept. Introd. F.GB.NL.D.A.PL.CS. I.YU.GR.R.SU. *Grown as a salad and medicinal plant in Southern Europe.*

LAVATERA Like *Malva*, but the three-lobed epicalyx usually fused, at least in bud, at the base into a cup and encircling the calyx. 11 sps.

Fls. clustered in axils of lvs.
740. *L. arborea* L. TREE MALLOW. A very robust ann. to perenn. with a thick woody stem to 3 m. and large rosy-purple, strongly purple-veined fls. 3-4 cm. across. Petals notched, about 3 times as long as the calyx; epicalyx lobes oval, much longer than calyx and much enlarging in fr. Lvs. large to 20 cm., rounded with 5-7 shallow lobes, often somewhat folded fanwise. Fr. hairy or hairless, transversely ridged. △ Rocks by the sea, waste places; grown for ornament and often escaping from gardens. Apr.-Sept. Med. Eur. (except TR.) P.IRL.GB. Pl. 72.

741. *L. cretica* L. An ann. or bienn. ½-1½ m. with rather dense clusters of 2-8 pale violet fls., much shorter than the lvs. Petals 1-2 cm., 2-3 times as long as the calyx; epicalyx lobes

oval, only slightly fused at base, long-haired, shorter than the triangular-oval, long-pointed calyx. Lower lvs. rounded, heart-shaped, shallowly three- to seven-lobed, the upper lvs. deeply five-lobed, toothed, finely hairy above and with dense star-shaped hairs beneath. Fr. pale, hairless or hairy. △ Waysides, waste ground, fields. Apr.–June. Med. Eur. P.GB.

Fls. solitary or rarely 2 in axils of lvs.
(a) Anns. or bienns. with herbaceous stems
742. *L. trimestris* L. A bristly-haired, erect ann. to 1 m. with large solitary bright pink fls. 5–7 cm. across and petals 4–5 times as long as the calyx. Epicalyx lobes broadly triangular, spreading, shorter than the lance-shaped calyx, both enlarging in fr. Lvs. large, rounded, the upper shallowly three- to seven-lobed, bright green. Fr. hairless, nutlets completely covered by a disk-like extension of the central axis. △ Fields, rocks; often grown for ornament and sometimes naturalized. Apr.–July. Med. Eur. (except AL.TR.) P.: introd. SU. Pl. 72.

(b) Shrubs or perenns. with a woody base
743. *L. maritima* Gouan A shrub to 120 cm., with pale pink fls. with a purple centre, on stalks as long as or longer than the subtending lv. stalk. Epicalyx shorter than the calyx. Lvs. rounded, rough-hairy, greyish at least below; old branches grey, hairless, younger branches with dense white star-shaped hairs. Fr. hairless, nutlets black, strongly ridged, with concave dorsal face and very sharp angles. △ Rocky places, on the littoral. Feb.–Mar. E.F.I.

L. olbia L. Like 743, but fls. purple-violet, nearly stalkless, solitary in each lv. axil forming a long terminal cluster. Upper lvs. oval to lance-shaped, very rough-hairy, green. Nutlets yellow, not ridged, dorsal face and angles rounded. Portugal to Sicily.

744. *L. thuringiaca* L. A softly hairy, greyish-green, herbaceous perenn. to 2 m. with large, pale pink, darker-veined fls. 5–8 cm. across, long-stalked and forming a rather lax terminal cluster. Epicalyx and calyx lobes heart-shaped, densely covered with star-shaped hairs, the epicalyx shortly fine-pointed. Lower lvs. rounded-heart-shaped, five-lobed, the upper lvs. three-lobed, all irregularly toothed and covered with dense star-shaped hairs. △ Hills, thickets, waysides; sometimes grown as ornamental. July–Sept. Cent. Eur. (except CH.), South -East Eur.: introd. F.S.SF. Page 246.

ALTHAEA Epicalyx with 6–9 lobes fused together at the base and shorter than the calyx. Fr. disk-like with many nutlets surrounding and overtopping a central axis. 5 sps.

Anns. or bienns. with simple, spreading hairs
745. *A. hirsuta* L. HAIRY MALLOW. A slender, erect or spreading, bristly-haired ann. or bienn. to 60 cm., with solitary pale pink fls. 1½–2½ cm. across on hairy stalks longer than the subtending lvs. Petals turning bluish, little longer than the bristly-hairy calyx; epicalyx lance-shaped, shorter than the calyx. Uppermost lvs. deeply five-lobed, the lower lvs. rounded, shallowly lobed, all toothed. Fr. closely invested by the calyx; nutlets hairless, transversely ribbed. △ Fields, grassy places. May–July. Med., South-East Eur. P.CH. H.SU.

Perenns. with dense star-shaped hairs
746. *A. cannabina* L. A tall, erect, hairy perenn. to 2 m. with large rosy-purple fls. 3–6 cm. across, often clustered and often forming a lax, branched inflorescence. Petals more than twice as long as the calyx; epicalyx 7–9, shorter and narrower than the oval, long-pointed calyx segments. Lvs. green, all deeply five-lobed, with oblong lance-shaped

irregularly double-toothed lobes, densely covered with flattened star-shaped hairs. Nutlets hairless, transversely ribbed on the back. △ Cultivated ground, waste places. June–Aug. Med., South-East Eur. P.CS.H.SU. Page 246.

747. *A. officinalis* L. MARSH MALLOW. An erect, densely velvety-haired, greyish-white perenn. to 2 m. with pale pink fls. 2½–5 cm. across, in a branched or unbranched inflorescence. Fl. stalks much shorter than the subtending lvs.; epicalyx segments 7–9, linear-lance-shaped and shorter than oval pointed calyx, all velvety-haired. Upper lvs. triangular-oval with 3–5 shallow lobes, the lower rounded. Nutlets densely covered with star-shaped hairs. △ Damp, salt-rich places, by the sea, thickets, uncultivated ground. June–Sept. Most of Eur. (except North Eur.) DK.: introd. CH. *Used as a vegetable in Roman times and as a medicinal plant in the past.* Pl. 72.

ALCEA Like *Althaea*, but fls. very large in wand-like spikes; epicalyx usually 6. Nutlets 18–40, two-celled with the upper cell empty and the lower cell one-seeded. 6 sps.

748. *A. pallida* (Willd.) Waldst. & Kit. (*Althaea p.*) EASTERN HOLLYHOCK. A tall perenn. to 2½ m. with stiff, erect, leafless stems bearing large, pink, stalkless fls., 6–9 cm. across, irregularly along much of its length. Petals pale pink, usually with yellow at the base, not touching at their margins; epicalyx segments triangular acute, as long as or slightly shorter than the calyx. Lvs. greyish, woolly-haired, rounded-heart-shaped, undivided or divided to one-third into 3–5 blunt, toothed lobes. Nutlets rather deeply furrowed, blackish or pale brown. △ Fields, disturbed ground. Apr.–July. South-East Eur. A.CS.H. SU. Pl. 72.

A. rosea L. (*Althaea r.*) HOLLYHOCK. Like 748, but petals touching at the margins, usually pink but sometimes white or violet; epicalyx segments delta- or triangular-lance-shaped, one-half to two-thirds as long as the calyx. Stems becoming more or less hairless. Nutlets winged. Origin not known, but probably a hybrid; grown as ornamental throughout Europe and widely naturalized.

ABUTILON Epicalyx absent; calyx 5. Fr. a whorl of 5–30 carpels, fused at the base and each splitting into 2 valves. 1 sp.

749. *A. theophrasti* Medicus A tall, erect, softly hairy ann. to 2 m. with rather small, solitary, yellow fls. *c.* 1 cm. across in the axils of the upper lvs. Lvs. long-stalked, blade to *c.* 12 cm., heart-shaped long-pointed, blunt-toothed, with dense adpressed, star-shaped hairs on both sides. Fr. with 12–15 densely hairy carpels, each two-horned. △ Cultivated ground, ditches, waste places. July–Oct. H.I.YU.AL.GR.BG.R.SU.: introd. P.E.F.CS. *Cultivated in China for its fibre, known as China jute, which is used in textile manufacture.* Page 246.

GOSSYPIUM | Cotton Epicalyx of 3 large, heart-shaped lobes which are toothed or fringed; calyx cut-off or shortly five-lobed. Fr. with 4–5 cells; seeds covered with dense cottony hairs. 2 sps.

750. *G. herbaceum* L. LEVANT COTTON. A branched ann. with large, solitary, showy, pale-yellow fls. with purple centres *c.* 7½ cm. across. Lvs. heart-shaped, deeply three- to seven-lobed to one-third or one-half their width, the lobes oval, pointed and narrowed at the base, sparsely hairy or hairless. Fr. with 4–5 cells, opening into a large 'bole' with dense, greyish-white, cottony hairs surrounding the seeds. △ Probably native of Western Asia; widely grown in Southern Europe and locally naturalized on waste ground. Sept. *One of several species grown to produce cotton, the most important textile fibre in the world and*

1. *Hypericum coris* 771
3. *Hypericum richeri* 772
5. *Lavatera thuringiaca* 744

2. *Abutilon theophrasti* 749
4. *Althaea cannabina* 746

which only recently is being matched by synthetic fibre production. The seeds are rich in oil and produce cottonseed oil which is used in cosmetics, pharmacy, and as a cooking oil. Pl. 73.

HIBISCUS Epicalyx usually three-lobed; sepals 5, often fused and persisting in fr. Stamens numerous, fused into a tube. Fr. with 5 cells; seeds kidney-shaped. 4 sps.

Shrubs
751. *H. syriacus* L. SYRIAN KETMIA. A hairless, much-branched, deciduous shrub to 3 m., with large, solitary, bluish-purple or white, open-bell-shaped fls. 5–8 cm. across. Lvs. triangular-oval 5–8 cm., strongly three-veined, entire or lower lvs. mostly three-lobed and deeply toothed. Fr. oblong-ovoid *c.* 2½ cm., abruptly beaked. △ Native of Asia; often grown for ornament and sometimes self-seeding in Southern Europe. Summer.

Anns.
752. *H. trionum* L. BLADDER KETMIA. An erect, branched ann. 15–60 cm. with rather large, solitary, pale yellow and often purple-veined fls. with blackish-purple centres. Fls. 2–4 cm. across; epicalyx with 12 linear teeth covered with stiff hairs; calyx hairy, prominently veined, becoming inflated and bladder-like and enclosing fr. Lvs. entire or deeply divided into 3–5 lobes which are further toothed or lobed, nearly hairless. △ Cultivated ground, waste places. June–Sept. South-East Eur. (except TR.) SU.: introd. P.E.F.A.PL.I. Pl. 73.

THYMELAEACEAE | Daphne Family

Usually shrubs with simple, mostly stalkless lvs. without stipules. Fls. regular; calyx more or less tubular and with 4–5 spreading, petal-like lobes; petals absent or scale-like; stamens usually 8 or 10, attached to the inside of the calyx tube. Ovary superior, usually one-celled; stigma stalkless. Fr. a nutlet, drupe or berry.

Fr. dry, enclosed in calyx; fls. inconspicuous, yellowish or greenish. *Thymelaea*
Fr. fleshy, not enclosed; fls. usually conspicuous, pink, white or green. *Daphne*

THYMELAEA Calyx urn-shaped or cup-shaped with 4 petal-like lobes; petals absent; stamens 8; style laterally placed. Fr. dry, enclosed within the persistent calyx. 17 sps.

Shrubs or shrublets
753. *T. hirsuta* (L.) Endl. A much-branched shrub to 1½ m. with white-woolly branches and very small, fleshy, green, overlapping, scale-like lvs. closely pressed to the stem. Fls. small, yellowish, 4–5 mm. long, in short clusters of 2–5, one-sexed or bisexual; calyx hairy on the outside. Lvs. oval 4–6 mm., dark green above, white-cottony beneath. △ Sunny hills, rocks by the sea. Oct.–May. Med. Eur. (except AL.) P. Pl. 73.

754. *T. tartonraira* (L.) All. A branched, hairy shrublet to ½ m. with numerous flat, not fleshy, spreading lvs. and clusters of small yellowish fls. in the axils of the upper lvs. Calyx silky-haired on outside, 4–5 mm. long. Lvs. spathulate or obovate, 1–2 cm., densely covered with silky hairs. △ Rocks and sands near the sea, woods. Apr.–May. E.F.I.GR.TR. Pl. 73.

Anns.
755. *T. passerina* (L.) Cosson & Germ. A rigid, erect, hairless ann. 20–50 cm., branched only above with narrow flax-like lvs. and tiny, greenish, stalkless, axillary fls. forming a

slender, lax spike more than half the length of the plant. Fls. 2–3 mm., calyx finely adpressed hairy. Lvs. linear-lance-shaped acute, hairless. △ Dry, uncultivated places. July–Oct. Much of Eur. (except North Eur. IRL.GB.NL.TR.).

DAPHNE Like *Thymelaea*, but calyx-tube cylindrical, often swollen at the base, and with 4 spreading petal-like lobes. Style terminal. Fr. a fleshy drupe, not enclosed in the persistent calyx. 17 sps.

Shrublets usually less than 25 cm.
(a) *Fls. red or pink*

756. *D. cneorum* L. GARLAND FLOWER. A low, spreading, evergreen, alpine shrublet 10–50 cm. with very sweet-scented, red or pink fls. in rounded stalkless clusters at the ends of the branches. Fls. in clusters of 6–12; calyx *c.* 1 cm. across, greyish-haired outside. Lvs. 1–2 cm. oblong or linear-spathulate, hairless, numerous towards the ends of the brown hairy branchlets. Fr. orange-yellow. △ Alpine rocks and meadows. Apr.–Aug. Cent. Eur. E.F.I.YU.AL.BG.R.SU. Pl. 74.

D. striata Tratt. Like 756, but branchlets hairless, stouter and more erect. Fls. pink, very sweet-scented; calyx hairless or nearly so outside. Lvs. strap-shaped 2–3 cm. Fr. reddish. Mountains of Central Europe.

(b) *Fls. yellow, white, or greenish-purple*

757. *D. alpina* L. ALPINE MEZEREON. A deciduous alpine shrublet with twisted, knotted, fragile branches and white, sweet-scented fls. ½–1 cm. across in terminal clusters of 3–8. Calyx densely hairy outside. Lvs. 2–3 cm. oblong-spathulate, thin soft, at first silky-haired then hairless. Fr. reddish. △ Rocks and screes in mountains. Apr.–May. E.F.CH. A.I.YU. Page 257.

D. oleoides Schreber Like 757, but lvs. evergreen, thick, leathery and shining above, white-spotted, hairy beneath. Fls. greenish-purple or white, in clusters of 3–5. Branches straight, not twisted. Mountains of Southern Europe.

D. blagayana Freyer Distinguished by its white or cream-coloured, very sweet-scented fls. in rounded clusters of 10–15 or more, at the ends of the branches. A low, spreading shrublet with clusters of obovate evergreen lvs. 3–4 cm. at the ends of the branches. Fr. white. Mountains of South-Eastern Europe.

Shrubs usually ½ m. or more
(a) *Fls. in terminal clusters*

758. *D. gnidium* L. MEDITERRANEAN MEZEREON. An erect, little-branched, evergreen shrub ½–2 m., of the Mediterranean region, with small, sweet-scented, white fls. borne in erect, usually branched, terminal clusters. Fls. white within and brownish outside; inflorescence branches and calyx densely covered with silky white hairs. Lvs. numerous, 2–4 cm., linear-lance-shaped fine-tipped, leathery glaucous, and hairless. Fr. fleshy, red or black. △ Bushy places, uncultivated and rocky ground. Mar.–Sept. Med. Eur. (except TR.) P. *The fruit and leaves are poisonous. The plant is used externally as a wound cleanser.*

(b) *Fls. in lateral clusters*

759. *D. mezereum* L. MEZEREON. An erect, little-branched, deciduous shrub ½–1 m., with very sweet-scented, purple fls. in dense lateral clusters appearing before the lvs., on last year's stems. Fls. 8–12 mm. across, hairy outside, rarely white. Lvs. 4–10 cm., oblong-lance-shaped, short-stalked, bright green above and rather glaucous beneath; stems greyish. Fr. scarlet, ovoid, fleshy. △ Woods in the hills. Feb.–May. Most of Eur. (except P.IRL.IS.TR.): introd. DK. *The leaves, fruit, and bark are emetic and poisonous.* Pl. 74.

760. *D. laureola* L. SPURGE LAUREL. An evergreen shrub ½–1 m. with stalked, drooping, lateral clusters of small, greenish-yellow, unscented fls. arising from the axils of the uppermost lvs. Clusters with 5–10 fls.; calyx hairless outside. Lvs. 6–12 cm. oblance-shaped, thick, leathery, dark glossy green; stems greyish. Fr. black, ovoid, fleshy. △ Open woods. Feb.–Apr. Much of Eur. (except North Eur. P.IRL.NL.PL.CS.): introd. DK. SU. *The fruit and bark are poisonous.* Pl. 74.

ELAEAGNACEAË | Oleaster Family

Trees or shrubs densely covered with silvery-brown to golden-brown, scale-like or star-shaped hairs; lvs. entire, without stipules. Sepals 2 or 4, spreading; petals absent; stamens 2, 4, or 8. Ovary superior one-celled; style long. Fr. drupe-like. Very like Thymelaeaceae, but distinguished by its distinctive hairs.

Only male or female fls. present on each plant. *Hippophaë*
Bisexual fls. and male fls. present on the same plant. *Elaeagnus*

HIPPOPHAË Plants one-sexed with either male or female fls. only. Male fl. with 2 sepals and 4 stamens, female fl. with 2 minute sepals at the end of the conspicuous, elongated calyx tube; ovary 1 and style slender. 1 sp.

761. *H. rhamnoides* L. SEA BUCKTHORN. A grey-leaved, deciduous, much-branched, spiny, suckering shrub to 3 m. with very small greenish fls. appearing before the lvs. on the old wood. Fls. covered with rust-coloured scaly hairs, female fls. solitary, male fls. in short lateral clusters. Lvs. 1–8 cm. linear-lance-shaped, covered with silvery scales, but often becoming nearly hairless, dull green above. Fr. 6–8 mm., rounded to ovoid, orange. △ Dunes, sea-shores. Mar.–May. Much of Eur. (except P.IS.YU.AL.GR.TR.): introd. IRL. *The fruit is acid and poisonous. Used in the past in herbal remedies, particularly as a vermifuge.* Pl. 74.

ELAEAGNUS Fls. bisexual, or with male and bisexual fls. on the same plant; male fls. with 4 sepals and stamens. 1 sp.

762. *E. angustifolia* L. OLEASTER. A shrub or small tree 2–12 m. with dark brown branches, silvery twigs and lvs., and very sweet-scented yellowish fls. which are silvery-haired on the outside. Fls. solitary or clustered in lv. axils, bisexual or male. Lvs. oval to lance-shaped, 6–9 cm., greyish-green above, silvery-scaly beneath; branches spiny or not. Fr. oval *c.* 2 cm., yellowish or brown, silvery-scaly. △ Bushy places by streams, damp places, by the sea; often cultivated and sometimes self-seeding. May–July. Native of Asia; introd. to much of Southern Europe. *The fruits are dried and are known as 'Trebizond grapes'; they are used in cake-making. The wood is hard and fine-grained.* Page 264.

GUTTIFERAE (HYPERICACEAE) | St John's Wort Family

Small shrubs or herbaceous perenns. with simple, opposite or whorled, often gland-dotted lvs.; stipules absent. Fls. yellow in terminal clusters; sepals and petals 5; stamens numerous, grouped into 3 or 5 bundles, or irregularly grouped. Ovary of 3 or 5 carpels, one-celled, or with 3 or 5 cells; styles 3 or 5. Fr. a capsule or rarely a fleshy berry.

HYPERICUM | St John's Wort   Fls. yellow; lvs. opposite, entire, stalkless or nearly so. 62 sps.

Shrubby plants or at least woody at base; stamens in 5 bundles

763. *H. androsaemum* L. TUTSAN. A small, shrubby, half-evergreen perenn. $\frac{1}{2}$–1 m., with opposite stalkless lvs. and terminal, rather flat-topped clusters of a few yellow fls. each *c.* 2 cm. across. Sepals oval, very unequal, the larger about as long as the petals, persisting in fr.; stamens numerous, grouped in 5 bundles. Lvs. 5–10 cm., oval-heart-shaped, with translucent glands beneath; stem with 2 raised lines. Fr. a fleshy berry, at first red then turning blackish-purple. △ Damp woods, shady places. June–July. West Eur. (except NL.) CH.I.YU.TR.BG.: introd. A. Pl. 74.

H. hircinum L. (*Androsaemum h.*) STINKING TUTSAN. Like 763, but branches quadrangular, fls. large *c.* 3 cm. across; sepals lance-shaped, falling in fr. Lvs. 2$\frac{1}{2}$–6 cm., smelling strongly of goats. Fr. a capsule splitting into 3 valves. Mediterranean Europe; naturalized in Western Europe.

H. calycinum L. ROSE OF SHARON. A low, creeping, evergreen shrub with ascending stems bearing very large, solitary, terminal yellow fls. 7–8 cm. across, with very numerous projecting stamens; styles 5, shorter. Lvs. numerous, rather leathery, oblong-oval, 5–10 cm.; stems quadrangular. Native of South-Eastern Europe; naturalized elsewhere particularly in the west. Pl. 75.

Usually herbaceous perenns.; stamens in 3 bundles
(a) Hairy plants, or hairs at least on lower surface of lvs.

764. *H. montanum* L. MOUNTAIN ST JOHN'S WORT. Distinguished by its dense, almost globular head of fls. borne terminally on a long stem which is almost leafless above. Fls. 1–1$\frac{1}{2}$ cm. across, pale yellow, fragrant; sepals half as long as the petals, lance-shaped with glandular teeth on margin. Lvs. 3–5 cm., stalkless, in 2 ranks, oval to elliptic, half-clasping, hairless above and minutely downy beneath and with a row of black glands on the margin; stems rounded, unbranched. An erect perenn. 40–80 cm. △ Woods, hedges, grassy places. June–July. Most of Eur. (except IRL.IS.AL.TR.). Pl. 75.

765. *H. hirsutum* L. HAIRY ST JOHN'S WORT. Distinguished by its downy, oval lvs. 2–5 cm., which have translucent glands on the blade, but without black marginal glands. Fls. pale yellow, 1$\frac{1}{2}$–2 cm. across, numerous in lax, elongated, branched clusters; sepals about half as long as the petals, oblong-lance-shaped, with short-stalked black glands on margin. Stems hairy, rounded, little-branched, 40–100 cm. △ Woods, hedges, thickets, shady places. June–Aug. Most of Eur. (except P.IS.TR.). Pl. 75.

766. *H. elodes* L. MARSH ST JOHN'S WORT. A softly hairy, weak spreading perenn. growing in bogs, rooting below, and with erect stems 10–30 cm. usually bearing a single lateral cluster of a few pale yellow fls. Fls. *c.* 1$\frac{1}{2}$ cm. across; sepals elliptic, with fine, red or purplish glandular teeth. Lvs. 1–2 cm., heart-shaped or rounded, half-clasping stem; lvs. and stems densely downy. △ Bogs, swamps, wet heaths. June–Sept. West Eur. D.I. Pl. 75.

(b) Hairless plants
(i) Stems distinctly quadrangular

767. *H. tetrapterum* Fries (*H. acutum* Moench) SQUARE-STEMMED ST JOHN'S WORT. An erect, branched perenn. 30–70 cm. with distinctly winged quadrangular stems and oval lvs. covered with small translucent glands, half-clasping the stem. Fls. pale yellow, *c.* 1 cm.

across, in compact terminal clusters; sepals lance-shaped acute, and about two-thirds as long as the petals, without stalked glands. △ Damp meadows, marshes, by ponds and streams. June–Sept. Most of Eur. (except P.IS.N.SF.).

H. maculatum Crantz (*H. quadrangulum* auct.) IMPERFORATE ST JOHN'S WORT. Very like 767, but lvs. abruptly narrowed to the base, not half-clasping, and without or with few glandular dots; stems quadrangular, not winged. Fls. larger *c.* 2 cm., golden-yellow; sepals blunt. Widespread in Europe.

(ii) Stems rounded or with 2 raised lines
768. *H. perforatum* L. COMMON ST JOHN'S WORT. Distinguished by its hairless lvs. which have numerous, translucent, glandular dots, and its hairless stem with 2 raised lines. Fls. 2–3½ cm., many in branched clusters; petals golden-yellow, with minute black spots; sepals lance-shaped, glandular, but not fringed with glands. An erect, branched perenn. 20–100 cm. △ Open woods, thickets, grassy banks, dry fields. May–Aug. All Eur. (except IS.). *This and other species are said to speed the healing of wounds. It contains a red pigment causing photo-sensitization of white or pale-skinned animals, resulting in loss of condition.* Pl. 75.

769. *H. pulchrum* L. SLENDER ST JOHN'S WORT. Distinguished, particularly in bud, by its bright yellow petals which are flushed with red at the tips. Fls. *c.* 1½ cm. across, in a lax, leafless, terminal cluster; petals with a row of black glands near margin; sepals oval, blunt, with short-stalked black glands on margin. Lvs. ½–1 cm., oval-heart-shaped, half-clasping stem, with translucent glandular dots. A slender, erect, or ascending perenn. 20–60 cm. △ Heaths, open woods. July–Sept. Most of Eur. (except IS.SF.AL.GR.TR.BG.).

770. *H. humifusum* L. TRAILING ST JOHN'S WORT. A slender trailing perenn. 5–20 cm., with small elliptic to oblong lvs., very slender stems with 2 raised lines and few small yellow fls. *c.* 1 cm. Petals little longer than the sepals; sepals unequal, blunt, entire or toothed, teeth sometimes with glands. Lvs. *c.* 1 cm., stalkless, with translucent glandular dots. △ Heaths, moors, sandy places. June–Sept. Much of Eur. (except IS.N.SF.AL.GR.TR.BG.).

771. *H. coris* L. YELLOW CORIS. A small shrublet to 40 cm. distinguished by its linear, gland-dotted lvs. with inrolled margin, which are in whorls of 3–4 and yellow fls. streaked with red. Fl. clusters rather lax, somewhat flat-topped; sepals linear-oblong, with conspicuous glandular teeth, about one-quarter as long as the petals. △ Dry banks, uncultivated places. June–July. F.CH.I. Page 246.

772. *H. richeri* Vill. ALPINE ST JOHN'S WORT. A mountain plant with elliptic lvs. half-clasping the stem, glaucous beneath, without or with very few transparent glandular dots, but black-dotted on margin. Fls. large 2–3 cm., bright yellow, in a rather dense, few-flowered cluster; petals black-dotted, much longer than the lance-shaped long-pointed sepals, which are sparingly black-dotted and bordered with a fringe of long glandular hairs. Fr. black-dotted. An erect unbranched perenn. 15–40 cm. △ In the mountains: screes, meadows, woods. June–Aug. South-East Eur. (except TR.) E.F.CH.I.SU. Page 246.

VIOLACEAE | Violet Family

Herbs or rarely shrubs with alternate, simple lvs. with stipules. Fls. usually solitary, irregular and symmetrical in one plane only and with a spur in our sps. Sepals 5; petals 5; stamens 5, the 2 lower spurred; anthers closely pressed round the ovary. Ovary one-celled; style simple, often curved back and thickened above. Fr. usually a capsule.

VIOLA | Violet, Pansy   Fls. solitary or 2. Sepals with a flap-like projection, or appendage, below their point of insertion. Lower petal with a sac-like spur and the 2 lower stamens each with an elongated spur which is inserted into the petal spur. Fr. a three-valved capsule with elastic valves which shoot out the seeds. Cleistogamy, i.e. the formation of ripening frs. without the normal development of the fl., is common towards the end of the flowering season. Hybridization is common. 92 sps. *Most species are purgative and vomitive when eaten.*

Fls. with 2 erect petals, 2 lateral petals spreading and directed downwards, and a basal petal, all similar in size and colour (VIOLETS)
(*a*) *Fls. arising directly from the rootstock; sepals blunt*
773. *V. palustris* L. BOG VIOLET. Distinguished by the hairless, somewhat glaucous, broad heart-shaped to kidney-shaped lvs. and by the delicately veined pale lilac fls. Fls. 1–1½ cm. across, long-stalked, arising directly from the rootstock; sepals oval; petal spur longer than the calyx appendages; style straight. Lvs. 2–6 cm. wide, broader than long, weakly toothed. A perenn. *c.* 5 cm. with slender, whitish, creeping underground stolons. △ Bogs, marshes, wet heaths. Apr.–June. Much of Eur. (except AL.GR.TR.). Pl. 76.

V. pinnata L. FINGER-LEAVED VIOLET. Readily distinguished by its rosette-forming lvs. which are deeply palmately lobed, and its pale violet, fragrant fls. 1–2 cm. across. Spur blunt, about twice as long as the calyx appendages; stipules entire, whitish. The Alps.

774. *V. odorata* L. SWEET VIOLET. Fls. sweet-scented, deep violet or white with a pale violet spur, rarely purple or pink, *c.* 1½ cm. across. Calyx oval, appendages spreading, shorter than the spur; style hooked. Lvs. 1–6 cm. rounded-kidney-shaped with heart-shaped base, blunt or acute, shallowly toothed, sparsely hairy; lv. stalks with short reflexed hairs. Fr. hairy. A perenn. with a leafy rosette and long, creeping, above-ground stolons which root at their tips. △ Hedges, woods, thickets. Mar.–May. All Eur. (except IS.SF.TR.). *An oil obtained from the distillation of the flowers is used in perfumery; syrup of violets is used medicinally. The violet colour changes to green in alkali and can thus be used as a litmus.* Pl. 76.

775. *V. hirta* L. HAIRY VIOLET. Like 774, but stolons absent; lvs. narrower, triangular-oval, hairy on both sides; lv. stalks with spreading hairs. Fls. scentless, *c.* 1½ cm. across; petals paler, bright blue-violet, rarely white; spur dark blue-violet, longer than the calyx appendages. △ Grassy places, banks, open woods. Mar.–May. Most of Eur. (except P.IS.SF.TR.): introd. SF.

(*b*) *Fls. arising from short stems and not directly from the rootstock; sepals acute*
776. *V. canina* L. HEATH DOG VIOLET. A very variable sp. with bright blue or pale blue fls. 1½–2½ cm. across, with a yellowish or rarely greenish-white spur which is thick, straight, and blunt and about twice as long as the calyx appendages. Lvs. oval-oblong, usually with heart-shaped base, blunt; fl. stalks and fr. hairless. A hairless or shortly hairy perenn. 10–40 cm., with main stem ending in a fl. and not in a rosette. △ Heaths, open woods, fens. Apr.–June. All Eur. (except AL.GR.).

777. *V. riviniana* Reichenb. COMMON DOG VIOLET. Like 776, but main axis ending in a rosette of lvs. and fls. borne on short lateral branches, but very variable in habit. Fls. blue-violet, 1½–2½ cm. across, spur thick, whitish, often upcurved, furrowed, or notched; appendages of calyx large 2–3 mm. and enlarging in fr. Lvs. oval-orbicular with deeply heart-shaped base, hairless or somewhat hairy, variable in size; stipules of stem lvs. with fine spreading teeth. Fr. hairless. A spreading ascending perenn. 2–20 cm., enlarging to *c.* 40 cm. in fr. △ Woods, hedges, heaths, pastures. Apr.–May. All Eur. (except TR.). Pl. 76.

**V. reichenbachiana* Boreau (*V. sylvestris* auct.) PALE WOOD VIOLET. Widespread in Europe.

778. *V. rupestris* F. W. Schmidt A tiny, finely hairy plant with a rosette of small, broadly heart-shaped lvs., short stems and reddish-violet, blue-violet or white fls. 1–1½ cm. across, with a thick, furrowed, violet spur. Lv. blades ½–1 cm.; lv. stalks hairy; stipules large, fringed; fl. stalks longer than lvs. Fr. hairy. △ Rocky pasture and sandy places in mountains. Apr.–June. Much of Eur. (except P.IRL.IS.DK.AL.GR.TR.).

779. *V. persicifolia* Schreber (*V. stagnina* Kit.) FEN VIOLET. Fls. white or bluish-white, 1–1½ cm. across, with a blunt, conical, greenish spur, not or scarcely longer than the calyx appendages. Lvs. pale green, 2–4 cm., triangular-lance-shaped with a cut-off base; lv. stalk winged above; stipules *c.* 1 cm., lance-shaped, entire or toothed. An almost hairless perenn. with creeping, underground stems and erect aerial stems 10–25 cm. △ Fens, marshes. May–July. Much of Eur. (except P.IS.AL.GR.TR.BG.).

780. *V. elatior* Fries TALL VIOLET. A rather robust, erect, little-branched perenn. 20–50 cm. with large, solitary, pale blue fls. with a whitish centre. Fls. 2–2½ cm. across, on long stalks, much longer than the lvs. Lvs. 3–9 cm. lance-shaped; stipules of middle lvs. large 2–5 cm., leafy, oval-lance-shaped and deeply toothed below, longer than the lv. stalk. △ Wet fields, woods in valleys. May–June. Cent., South-East Eur. (except AL.TR.) F.SU.: introd. B. Page 257.

781. *V. arborescens* L. SHRUBBY VIOLET. A hairy shrublet with greyish woody and corky stems 10–20 cm., and small, whitish or pale violet fls. 1–1½ cm. across, with a curved, blunt spur *c.* 4 mm. Lvs. linear to lance-shaped, entire or toothed; stipules linear-lance-shaped, deeply lobed, about one-third as long as the lvs. △ Thickets, rocks, fissures, by the sea. Feb.–Mar. P.E.F., Sardinia.

Fls. with 2 erect petals, with 2 lateral petals directed upwards and a broader basal petal; petals unequal in size, often multi-coloured (PANSIES)

(*a*) *Spur usually not more than twice as long as calyx appendages*

782. *V. biflora* L. YELLOW WOOD VIOLET. A slender perenn. with creeping stems and usually paired golden-yellow fls. streaked with brown, 1½ cm. across and with spur little longer than the calyx appendages. Sepals acute. Lvs. kidney-shaped, broader than long, with regular rounded teeth; stipules oval 3–4 mm., entire. Fl. stem ascending to 20 cm. △ In hills and mountains: shady places, damp rocks. June–Aug. Much of Eur. (except P.IRL. GB.B.NL.IS.DK.AL.GR.TR.). Pl. 76.

783. *V. tricolor* L. WILD PANSY, HEARTSEASE. Fls. mostly tricoloured, usually predominantly violet with varying amounts of yellow and white, very variable in size, 1½–3 cm. across vertically. Petals usually longer than the sepals; spur short, usually little longer than the calyx appendages, but sometimes up to twice as long; style enlarged at the apex with a rounded head with a hollow at one side. Lvs. variable, heart-shaped to lance-shaped; stipules variable, but often palmately lobed with a larger, leafy, lance-shaped entire middle lobe. An ascending or erect, branched ann., bienn., or perenn. to 50 cm. △ Cultivated ground, meadows. Mar.–Sept. All Eur. (except P.). Pl. 75.

784. *V. arvensis* Murray FIELD PANSY. Like 783, but fls. usually smaller ½–2 cm. across vertically, predominantly cream-coloured, but tinged with violet, or if violet then the fls. very small; petals usually shorter than the sepals. Lvs. oblong-spathulate; stipules pinnately lobed with a larger, leafy, oval-lance-shaped and toothed middle lobe. A very variable, usually erect ann. 15–45 cm. △ Fields, cultivated ground. Mar.–Sept. All Eur. (except IS.).

(*b*) *Spur more than twice as long as calyx appendages*

785. *V. lutea* Hudson MOUNTAIN PANSY. Fls. yellow or violet or variegated, large, 1½–3 cm. across vertically, with a slender spur 3–6 mm., not more than half as long as the petals.

Lower lvs. oval, the upper oblong-lance-shaped, toothed, hairless or finely hairy; stipules palmately or pinnately cut into 3–5 lobes with the middle lobe larger. A very variable perenn. with creeping rhizomes and slender, unbranched stems 10–25 cm. △ Mountain pastures. July–Aug. West Eur. (except P.), Cent. Eur. (except H.). Pl. 76.

786. *V. calcarata* L. LONG-SPURRED PANSY. Like 785, but spur long and slender 8–15 mm., as long as the very wide petals; fls. violet or yellow, larger, 2–4 cm. across vertically, and to 3 cm. wide. Lvs. oval-oblong, mostly in a rosette; stipules oblong, entire or toothed, sometimes deeply lobed. A low-growing perenn. to 10 cm. △ Alpine and subalpine meadows. June–July. F.D.CH.A.I.AL.YU.GR. Pl. 76.

V. cenisia L. MT. CENIS PANSY. A hairless, tufted perenn. with numerous spreading branches to 5 cm. and large bright violet or lilac fls. 2–2½ cm. across, and entire oval lvs. *c.* 1 cm. Spur slender, 5–8 mm., about as long as the sepals. Western Alps.

CISTACEAE | Rockrose Family

Usually shrubs, or herbs with simple, often opposite lvs., with or without stipules. Fls. solitary or in loose clusters. Sepals 3 or 5; petals 5, often falling early in the day; stamens numerous, unfused. Ovary superior, usually of 3–5 fused carpels, rarely 10, one- or several-celled; style 1. Fr. a capsule, opening by as many valves as carpels and with a persistent calyx.

1. Carpels 5 or 10; inner sepals not much longer than outer.	*Cistus*
1. Carpels 3; inner sepals much longer than outer.	
2. Styles elongate, slender, bent or abruptly curved; sepals with prominent veins.	
3. All stamens fertile; lvs. all opposite.	*Helianthemum*
3. Outer stamens sterile; upper lvs. usually alternate.	*Fumana*
2. Style absent, or short and straight; sepals with inconspicuous veins.	
4. Dwarf shrubs.	*Halimium*
4. Anns. or perenns. with broad basal lvs. (often withered) and narrower stem lvs.	*Tuberaria*

CISTUS | **Cistus** Shrubs with opposite entire lvs., without stipules. Sepals 3 or 5; petals 5, pink, rosy-purple or white; stamens numerous, all fertile. Carpels 5 or rarely up to 10. Stigma stalkless or with a straight style. Fr. a capsule opening by valves. Hybridization is not uncommon. 16 sps.

Sepals 5
(a) Fls. pink or rosy-purple
787. *C. incanus* L. (*C. villosus* auct.) Fls. pink, large, 4–6 cm., in a lax terminal cluster of 1–6. Lvs. oval to elliptic, 2–5 cm. by 8–30 mm., narrowed to a short stalk, margin more or less undulate, greenish or greyish above, paler beneath and with 1 main vein, hairy and with or without glandular hairs. Fr. with dense, adpressed hairs. A variable densely branched, erect shrub 30–150 cm., with hairy branches. △ Sunny stony places, uncultivated ground, low thickets. Apr.–June. Med. Eur. (except E.F.) BG.SU. Pl. 77.

788. *C. albidus* L. GREY-LEAVED CISTUS. Distinguished from 787 by its pale greyish, densely hairy, quite stalkless, oval to lance-shaped lvs., with flat, not undulate margin, and covered with star-shaped hairs. Fls. large, pink, 4–6 cm., in terminal clusters of 1–5. Fr.

with shaggy hairs. An erect, greyish-leaved, scarcely aromatic shrub 30–170 cm. △ Stony, uncultivated ground, low thickets. Apr.–June. P.E.F.I. Pl. 77.

789. *C. crispus* L. Distinguished from 787 by the rather pale greyish-green 3-veined stalkless lvs. 1–4 cm. by 4–15 mm., fused at the base, with undulate-crisped margin and numerous, long, shaggy, adpressed hairs. Fls. rosy-purple, medium-sized, 3–4 cm., almost stalkless in a rather dense terminal cluster. Fr. shaggy-haired. A very aromatic, erect or spreading shrub 10–50 cm., with branches with long spreading hairs. △ Dry, stony ground, pine woods. Apr.–June. P.E.F.I.

(b) *Fls. white*

790. *C. salvifolius* L. SAGE-LEAVED CISTUS. Lvs. stalked sage-like, oval-oblong, wrinkled, not sticky and hardly aromatic, greenish above and whitish-hairy beneath. Fls. 2–4 cm., white and usually with an orange centre, long-stalked, in lax lateral clusters of 1–3; buds drooping. Outer sepals broadly heart-shaped, and investing the narrower inner sepals. Fr. somewhat hairy. A spreading or erect, branched shrub 20–100 cm. △ Dry, stony places, pine woods, low thickets. Apr.–June. Med. Eur. P.CH. Pl. 77.

C. psilosepalus Sweet (*C. hirsutus* Lam.) Distinguished by its flat, bright green, stalkless, oval to lance-shaped lvs. which are hairy on both sides. Fls. white, to 4 cm., 1–5 in terminal flat-topped clusters, with bracts with long spreading hairs; outer sepals much larger, heart-shaped with inrolled margin, enlarging in fr. Fr. hairy. Iberian Peninsula and France.

791. *C. monspeliensis* L. NARROW-LEAVED CISTUS. Distinguished from 790 by its stalkless linear to narrow-lance-shaped, very sticky glandular lvs. which are dark green and shining above and grey-haired beneath, and with conspicuously inrolled margins. Fls. small, 2–3 cm., white and often blotched with yellow, in a one-sided bractless cluster of 2–10. Fr. nearly hairless. A very aromatic, erect, branched shrub 30–100 cm. △ Dry, stony slopes and hills, pine woods. Apr.–June. Med. Eur. (except TR.) P. Pl. 77.

792. *C. populifolius* L. POPLAR-LEAVED CISTUS. Readily distinguished by its large, triangular-heart-shaped, stalked lvs. which are hairless and wrinkled and rough beneath. Fls. large, 6–8 cm., white usually with a yellow centre, in long-stemmed clusters of 1–8; buds often crimson, drooping; fl. stalks hairy. Fr. hairless or nearly so, with reddish calyx. An erect, aromatic shrub 80–150 cm. △ Hills, rocks, dry uncultivated places. May–June. P.E.F.

Sepals 3; fls. white

793. *C. ladanifer* L. GUM CISTUS. Fls. very large, 5–8 cm., white and usually with a brownish-purple blotch at the base of each petal, usually solitary, short-stalked. Sepals rounded, hairy especially on the margin and covered with small swellings. Lvs. 4–10 cm., usually linear-lance-shaped, stalkless, green, very sticky and hairless above, grey-hairy beneath. Fr. hairy, ten-valved. A rather tall, erect shrub 1–3 m. with much-branched sticky twigs. △ Pine woods, dry hills. May–June. P.E.F. *From this and other species a resinous gum, known as laudanum, is obtained from the leaves; it is used in perfumery and for fumigation, and was used in the past medicinally.* Pl. 77.

794. *C. laurifolius* L. LAUREL-LEAVED CISTUS. Distinguished from 793 by its oval to lance-shaped, stalked lvs. 2½–7 cm. which are very sticky above, grey woolly-haired beneath. Fls. long-stalked, in clusters of 3–8, each 5–8 cm., white, with a yellow basal blotch. Sepals oval-acute, with star-shaped and simple hairs. Fr. hairy, five-valved. A tall, erect, branched shrub 1–3 m. with sticky twigs. △ Dry hills, thickets, pine woods. June–July. P.E.F.I. Pl. 77.

HALIMIUM   Herbaceous perenns. or dwarf shrubs. Sepals usually 5, the 2 outer much smaller than the 3 inner, or all equal. Style short, straight, or absent. Fr. with 3 valves. Often placed in *Helianthemum*. 9 sps.

Fls. white

795. *H. umbellatum* (L.) Spach A sticky, spreading, or erect shrublet 20–40 cm., with densely white-felted twigs and white fls. usually in a terminal umbel of 3–6, but sometimes with a secondary cluster below. Fls. 2–2½ cm.; fl. stems and calyx with silky hairs. Lvs. linear-elliptic 1–1½ cm. by 1½–2 mm., dark green above, densely white-felted beneath, margin inrolled. Capsule ovoid, long-pointed, finely hairy. △ Bushy places, pine woods, hills. Mar.–May. P.E.F.GR. Page 257.

Fls. yellow

796. *H. commutatum* Pau (*H. libanotis* auct.) A much-branched, erect shrublet 10–40 cm., with 1–3 pale yellow fls., each *c.* 1 cm., in terminal and axillary clusters. Calyx oval, hairless. Lvs. linear or narrow oblong 1–3½ cm. by 1½–3 mm., shining green and hairless above, with white woolly hairs beneath, margin inrolled. Capsule more or less globular, with star-shaped hairs. △ Sandy places by the sea. Feb.–May. P.E. Pl. 78.

797. *H. halimifolium* (L.) Willk. Distinguished by its numerous, bright yellow fls. with a dark blotch towards the base of each petal, or rarely unblotched, in terminal, branched and somewhat lax pyramidal clusters. Fls. 2–3 cm., long-stalked; calyx oval acute, densely covered with yellow star-like scales, or star-like scales absent; epicalyx 1 or 3. Lvs. oblong or elliptic 1–4 cm. by ½–2 cm., with dense, silvery-white, velvety hairs on both faces. An erect, much-branched shrub to 1 m. with white-felted branches. △ Sandy places near the sea, sunny hills inland. Apr.–June. P.E.I.

TUBERARIA   Anns. or herbaceous perenns. with lower lvs. in a basal rosette, mostly three-veined. Sepals 5, the 2 outer narrower and smaller than the 3 inner; stamens all fertile; style absent. Ovary partially three-celled. Often placed in *Helianthemum*. 10 sps.

Anns.

798. *T. guttata* (L.) Fourr. SPOTTED ROCKROSE. An erect, hairy ann. 5–50 cm., usually with numerous, yellow, dark-blotched fls. in a lax, leafless, terminal cluster. Fls. very variable in size, petals yellow with a purple-brown blotch often covering half the petal which is longer or shorter than the hairy calyx; fl. stalks very slender, thread-like, reflexed in fr. Basal lvs. elliptic, often in a rosette and contrasting with the much narrower, one-veined upper lvs. which have stipules. A very variable sp. △ Sandy places, pine woods, heaths. Apr.–June. Med. Eur. P.IRL.GB.NL.BG. Pl. 78.

Perenns.

799. *T. lignosa* (Sweet) Samp. (*T. vulgaris* Willk.) Distinguished from 798 by the large basal oval to lance-shaped, conspicuously three-veined lvs., which are green above and white woolly-haired beneath and arranged in a plantain-like rosette. Fls. pale yellow, not blotched, 2–3 cm., numerous on an almost leafless stem. Fl. stalks nodding before flowering; epicalyx linear-lance-shaped; sepals oval, hairless. Fr. with star-shaped hairs. Usually an erect perenn. 20–30 cm., distinctly hairy below, but hairless and turning reddish above. △ Pine woods, stony and sandy places. Mar.–July. P.E.F.I.

HELIANTHEMUM | **Rockrose**   Calyx with 2 outer narrower sepals, much smaller than the 3 prominently three- to five-veined inner sepals; stamens all fertile; styles curved at base. Carpels 3. Anns., perenns., or shrublets. 31 sps.

1. *Fumana procumbens* 805
2. *Halimium umbellatum* 795
3. *Daphne alpina* 757
4. *Viola elatior* 780

Anns.

800. *H. salicifolium* (L.) Miller A small ann. to 30 cm. with yellow unspotted fls. in a rather lax, simple or branched, terminal cluster and with stout fl. stalks spreading at right angles and shorter or a little longer than the calyx. Petals 8–12 mm., longer or shorter than the sepals; inner sepals acute; bracts small, scarcely as long as the fl. stalk. Lvs. oblong-obovate $\frac{1}{2}$–3 cm. by $\frac{1}{2}$–1 cm., densely greyish-hairy. Capsule hairless, $3\frac{1}{2}$–6 mm. △ Dry sandy places, tracksides. Mar.–June. Med. Eur. P.CH.BG.R.SU.

H. ledifolium (L.) Miller Like 800, but leafy bracts as long as or longer than the yellow fls. which have a golden blotch at the base of each petal. A coarser ann. with thick, erect fl. stalks shorter than the calyx. Lvs. oblong, 1–5 cm., woolly-haired on both faces, greyish beneath. Mediterranean region.

Perenns.
(a) *Lvs. without stipules*
801. *H. canum* (L.) Baumg. HOARY ROCKROSE. A spreading and often mat-forming, or an erect shrublet 10–40 cm., with bright yellow fls. 1–1$\frac{1}{2}$ cm. in dense or rather lax, elongated, one-sided terminal clusters. Bracts small, shorter than the fl. stalks. Lvs. variable, linear-to-oval, either green or greyish above, densely woolly beneath, with or without star-shaped hairs. A very variable sp. △ Dry, stony places. Mar.–June. Much of Eur. (except North Eur. P.B.NL.) S.

(b) *Lvs. with stipules*
802. *H. nummularium* (L.) Miller COMMON ROCKROSE. Distinguished from 801 by all lvs. having linear to lance-shaped stipules which are about twice as long as the lv. stalks. Fls. varying in size, 2 cm. or more, in one-sided clusters, yellow, rarely white, pink or orange; petals 6–15 mm. Lvs. $\frac{1}{2}$–5 cm. narrow-elliptic to oblong, green or grey-hairy on both surfaces or grey-hairy below only; stems green or grey-hairy. A very variable sp. A straggling, lax shrublet 10–50 cm. △ Rocks, mountain pastures, grassy places, dry hills. May–Aug. All Eur. (except IS.N.). Pl. 78.

803. *H. apenninum* (L.) Miller (*H. pulverulentum* auct.) WHITE ROCKROSE. Distinguished from 802 by its white fls. *c.* 2 cm. with yellow centres, or rarely pink fls. and thread-like stipules which are scarcely longer than the lower lv. stalks. Lvs. $\frac{1}{2}$–2 cm. variable, grey-woolly on both sides or green above with dense, star-shaped hairs beneath, margin usually inrolled. Sepals grey, densely woolly-haired. A lax, spreading shrublet to 50 cm. △ Sunny rocks, open woods. May–July. P.E.F.GB.B.D.CH.I.AL.GR. Pl. 78.

FUMANA Like *Helianthemum* and often placed in this genus, but outer stamens sterile and with their filament looking like a string of beads or *moniliform*, the inner stamens fertile. Fls. yellow. Styles slender, curved at base. Carpels 3, valves of fr. usually spreading. 9 sps.

Stipules present
804. *F. thymifolia* (L.) Webb A variably hairy or sticky shrublet to 20 cm. with small linear lvs. with strongly inrolled margin and yellow fls. *c.* 1 cm. in a lax, terminal cluster. Fls. 3–9; fl. stalks slender, glandular-hairy, much longer than the bracts. Lvs. 5–11 mm. opposite, or the upper alternate; stipules bristle-tipped. △ Stony and rocky places. Mar.–June. Med. Eur. P.

Stipules absent
805. *F. procumbens* (Dunal) Gren. & Godron. Like 804, but fls. solitary or few and placed laterally and with robust arched fr. stalks about as long as the lvs. Fls. yellow, petals wedge-shaped, usually with a dark golden-yellow spot at the base. Lvs. 1–1$\frac{1}{2}$ cm., linear,

all alternate; stipules absent. A low, spreading, much-branched, finely hairy-stemmed shrublet 10–20 cm. △ Rocky, sandy places. May. Much of Eur. (except North Eur. IRL.GB. NL.PL.) S. Page 257.

F. ericoides (Cav.) Gand. An erect shrublet with slender spreading fr. stalks downward-curved at the apex, 2–3 times as long as the lvs. Fls. yellow, usually 1–5 terminal; petals obovate. Mediterranean region.

TAMARICACEAE | Tamarisk Family

Small trees and shrubs with tiny, alternate, needle-like or scale-like lvs. without stipules. Fls. usually small and numerous in dense elongated clusters. Sepals and petals 4 or 5; stamens as many or twice as many as the petals; ovary one-celled; styles 3–5. Fr. a capsule; seeds with long silky hairs.

Stamens 5, free; styles 3.   *Tamarix*
Stamens 10, fused into a tube below; style absent.   *Myricaria*

TAMARIX | Tamarisk   Feathery-looking deciduous shrubs with long slender branches, the smallest twigs falling with the lvs. Fls. in long, slender, cylindrical clusters, with bracts shorter than the fls. Stamens 5, inserted on a glandular disk; styles 3. Fr. a capsule; seeds with a tuft of hairs. 14 sps.

806. *T. gallica* L. (*T. anglica* Webb) TAMARISK. A slender, much-branched, glaucous shrub to 10 m. with very numerous, tiny pink or white fls. in long, slender, catkin-like clusters borne laterally towards the ends of the present year's leafy branches. Fl. clusters 1–3 cm. by 3–5 mm. wide; fls. *c.* 2 mm.; petals and sepals 5; anthers with a fine point. Lvs. glaucous, *c.* 2 mm., adpressed to and clasping the stem at the base. △ Riversides, marshes, by the sea; often planted as an ornamental shrub. Apr.–Sept. E.F.CH.I.: introd. GB.

807. *T. africana* Poiret Like 806, but fl. clusters almost stalkless and borne laterally from last year's woody branches and appearing before or with the lvs. Fls. larger to *c.* 3 mm., and fl. clusters broader 6–8 mm. across; anthers blunt. A slender, branched shrub 2–3 m. △ By the sea; often planted as an ornamental shrub. Mar.–June. P.E.F.I.: introd. GB. Pl. 78.

MYRICARIA   Like *Tamarix*, but differing in having 10 stamens which are fused into a tube below. Bracts longer than the fls. Style absent; seeds with a stalked tuft of hairs. 1 sp.

808. *M. germanica* (L.) Desv. GERMAN TAMARISK. An erect, glaucous, heather-like shrub 1–2 m., with long, slender, terminal and lateral clusters of pale pink fls. Clusters lax below, dense above; bracts subtending fls. long, papery; stamens 5 long, 5 short. Lvs. numerous, scale-like, 2–4 mm., linear-lance-shaped borne on deciduous lateral branches, which are subtended by papery bracts. Seeds with a plume of hairs. △ River gravels, streamsides, waysides. May–Aug. Cent. Eur. E.F.N.S.SF.I.YU.R.SU. Page 264.

FRANKENIACEAE | Sea Heath Family

Herbs or small, often heather-like shrubs mostly of salt-rich marshes and semi-deserts. Lvs. awl-shaped, opposite, entire. Sepals 4–7, fused to the middle into a tube, persisting;

petals 4–7, usually clawed, and with a scale-like appendage; stamens in 2 whorls, 4 or 6. Ovary one-celled; fr. a capsule opening by valves, enclosed in calyx.

FRANKENIA   Petals and sepals usually 5; stamens 4 or 6 in 2 whorls. Lvs. awl-shaped; stipules papery. 6 sps.

Anns.

809. *F. pulverulenta* L. ANNUAL SEA HEATH. An ann. with numerous prostrate branches usually spreading in a circle, and small solitary or lax clusters of stalkless, pink or pale violet fls. in the axils of the branches and the upper lvs. Fls. 3–5½ mm.; petals often notched, shorter than the hairless calyx. Lvs. broadly oval or obovate, hairless above, crisply hairy beneath, blunt or notched at apex, often becoming reddish. △ Damp sandy, salt-rich places, by the sea. May–Aug. Med. Eur. (except TR.) P.BG.SU.

Perenns.

810. *F. laevis* L. SEA HEATH. Differing from 809 by its linear heather-like lvs. 2–4 mm., with inrolled margins, which are mostly densely crowded on short lateral branches. Fls. *c.* 5 mm. across, pink or violet; petals weakly toothed at apex; calyx somewhat fleshy. A very variable procumbent perenn. with a woody base, spreading to 15 cm. or more. △ Salt marshes. June–Aug. P.E.F.GB.I.

CUCURBITACEAE | Gourd Family

Usually climbing anns. with spirally coiled tendrils arising from the side of the lv. axil; less commonly non-climbing perenns. without tendrils. Lvs. often palmately lobed. Fls. usually one-sexed; sepals 5, free; petals 5, usually fused below; stamens usually 5 and fused into a column with anthers more or less free. Ovary usually inferior, commonly three-celled; stigmas 3. Fr. usually a berry which,sometimes splits or is explosive.

1. Plants without tendrils.   *Ecballium*
1. Plants with tendrils.
   2. Tendrils branched.
      3. Corolla flat, with widely spreading petals.   *Citrullus*
      3. Corolla bell-shaped, with erect or recurved petals.   *Cucurbita*
   2. Tendrils unbranched.
      4. Fls. greenish-white, clustered; fr. small.   *Bryonia*
      4. Fls. yellow, solitary; fr. large.   *Cucumis*

ECBALLIUM   Herbaceous perenns. without tendrils. Fls. one-sexed; male fls. in axillary clusters, the female fls. solitary. Sepals, petals, and stamens 5. Fr. swollen, turgid, and ejecting seed explosively when detached from fr. stalk. 1 sp.

811. *E. elaterium* (L.) A. Richard SQUIRTING CUCUMBER. A rough, rather fleshy, procumbent, thick-stemmed perenn. spreading to 60 cm., with yellow, somewhat bell-shaped fls. *c.* 2½ cm. across, and distinctive frs. Lvs. triangular-heart-shaped, thick rough, coarsely toothed, white-haired beneath. Fr. ovoid-cylindrical 4–5 cm., long-stalked, at length inclined and when ripe explosively ejecting its seeds for several metres. △ Waste places, cultivated ground, waysides. Apr.–Sept. Med. Eur. P.BG.R.SU.: introd. GB.CS.H. *The fruit is poisonous and was used in the past medicinally.* Pl. 79.

CITRULLUS Herbaceous climbing plants with branched tendrils. Plants one-sexed; calyx bell-shaped, five-lobed; corolla with spreading lobes; stamens 5. Fr. fleshy or dry. 2 sps.

812. *C. lanatus* (Thunb.) Mansfeld (*C. vulgaris* Schrader; *Colocynthis citrullus* (L.) O. Kuntze) WATER MELON. Readily distinguished by its huge globular or ellipsoid, dark green, smooth fr. 30–80 cm. long, which has sweet, pink or yellow flesh, usually with black seeds. Fls. yellow, *c.* 4 cm. across. Lvs. deeply cut into oblong segments which may be further lobed, hairless above, sparsely hairy beneath. A spreading ann. with tendrils. △ Native of South Africa; often cultivated in the Mediterranean region and sometimes self-seeding. June–Aug.

C. colocynthis (L.) Schrader BITTER APPLE, BITTER CUCUMBER. A very rough-haired climbing or spreading perenn. with solitary, small, greenish-yellow fls. and globular, rough-haired, yellow and green fr. to *c.* 8 cm. Lvs. 5–12 cm., deeply pinnately cut into oblong, wavy-margined, bristly-haired segments. Mediterranean region; but cultivated and sometimes naturalized. *Grown for its fruits which are bitter and purgative and are used medicinally.* Pl. 78.

CUCURBITA Tendrils branched; lvs. simple or lobed. Plants one-sexed; corolla large, bell-shaped, five-lobed; anthers all fused together; stigmas 3–5. Fr. large or small, fleshy or dry. 3 sps.

813. *C. maxima* Duchesne PUMPKIN. Distinguished by its enormous frs. weighing to 100 kg., which are usually bright yellow and shallowly segmented, but white, green and white-spotted, and irregularly swollen, turban-like frs. are also grown. Fls. golden-yellow, 7–10 cm. across. Lvs. not rough to the touch, rounded and entire or very shallowly lobed. Stalks of female fls. rounded, not angular. △ Native of Central America; cultivated in gardens and sometimes as a field crop in Southern Europe. July–Aug.

814. *C. pepo* L. MARROW, ORNAMENTAL GOURD. Distinguished from 813 by its lvs. which have deeper, more pronounced acute lobes, and stalks of fls. five-angled. Fls. golden-yellow, 7–10 cm. across. Fr. very variable: large or small, globular, cylindrical or lobed, fleshy or dry, variously coloured. △ Native of America; widely cultivated as a vegetable, while dry-fruited forms are sometimes cultivated for ornament; occasionally self-seeding. June–July.

BRYONIA | Bryony Climbing plants with long, spiralling, unbranched tendrils. Fls. one-sexed, in axillary clusters; sepals 5; petals 5, free or shortly fused below; stamens 5; styles 3, bilobed. Fr. a small berry. 2 sps.

815. *B. cretica* L. WHITE BRYONY. A slender, rough, brittle-stemmed herbaceous climber to 3 m. with clusters of greenish fls. in the axils of the lvs. Male and female fls. on separate plants; the male in stalked clusters, the female clusters almost stalkless. Fls. 10–18 mm.; petals veined. Lvs. usually with 5 coarsely toothed lobes; tendril unbranched, twisting in a spring-like coil when attached to a support. Fr. globular, 5–8 mm., red when ripe. Root massive, fleshy. △ Thickets, hedges, rocky places. May–Aug. Much of Eur. (except North Eur. IRL.NL.BG.): introd. DK.N.S. *Formerly cultivated as a medicinal plant. The plant contains bryonin, a poisonous alkaloid; the roots and fruits are particularly poisonous.* Pl. 79.

B. alba L. Like 815, but plants bearing both male and female fls. and fr. black when ripe. Lvs. five-angled or five-lobed, lobes oval or triangular-acute, irregularly toothed. Central and Southern Europe; but naturalized elsewhere from cultivation.

CUCUMIS Tendrils simple; lvs. entire or dissected. Fls. one-sexed; calyx and corolla bell-shaped, five-lobed; stamens apparently 3, with 2 anthers doubled. Fr. large, fleshy, smooth or rough. 3 sps.

816. *C. sativus* L. CUCUMBER. Fr. oblong cylindrical, with rough often prickly swellings, green or yellowish rind and white flesh; seeds white. Fls. golden-yellow, 2½–4 cm. Lvs. bristly-haired, heart-shaped, palmately five-lobed, with acute, toothed lobes. A spreading ann. 1–4 m. △ Native of India; often cultivated as vegetable in Southern Europe. June–July.

817. *C. melo* L. MELON, CANTALOUPE. Fr. large, very variable globular to cylindrical, green, yellow, orange or spotted, with smooth, rough or netted rind, and white, orange or greenish flesh. Fls. golden-yellow, *c.* 2½ cm. Lvs. rough, rounded and shallowly five-lobed, with rounded and toothed lobes, the middle lobe broadest. A spreading ann. 1–1½ m. △ Native of Asia and Africa; widely cultivated in Southern Europe. May–Aug.

CACTACEAE | Cactus Family

Spiny succulent plants with swollen cylindrical stems, and lvs. reduced to spines or scales. Fls. large, bright-coloured; perianth segments numerous, fused at their base into a tube, often grading from sepal-like outer segments to petal-like inner segments. Stamens very numerous, attached to the perianth tube; ovary inferior, one-celled. Fr. a dry or fleshy berry.

OPUNTIA Lvs. cylindrical, soon falling; spines always with a tuft of barbed bristles at the base. Fls. with outer segments green and inner coloured. Fr. fleshy. 6 sps.

818. *O. ficus-indica* (L.) Miller PRICKLY PEAR, BARBARY FIG. A massive, much-branched, tree-like plant to 3 m., formed of flattened, racket-shaped, spiny green joints placed one above the other. Fls. bright yellow, 6–7 cm. across, borne on the margin of the upper joints. Joints oval 20–50 cm. by 10–30 cm. broad. Fr. large, red, yellow, or purple, edible. △ Native of tropical America; often naturalized. June–July. Introd. Med. Eur. and P. *The fruit is edible, but care must be taken to leave the fine spines of the skin. Grown as a boundary hedge in dry regions.* Pl. 79.

O. vulgaris Miller (*O. humifusa* Rafin.) PROSTRATE CACTUS. A low, spreading perenn. to ½ m. with oblong green joints 7–12 cm. by 4–5 cm. Fls. golden-yellow, 5–6 cm. across. Fr. red 2–4 cm. Native of North America; naturalized in Southern Europe. *The fruit, when attacked by the cochineal insect, produces a red dye.*

LYTHRACEAE | Loosestrife Family

Usually herbs or small shrubs with opposite or whorled entire lvs.; stipules absent or minute. Fls. solitary, or several in the axils of the lvs. and forming a spike-like cluster. Calyx tubular, usually with 5 teeth placed edge to edge in bud, or *valvate*, not overlapping, often with alternating scales. Petals crumpled in bud, 4–6 or absent, arising from the top of the calyx tube; stamens 4 or 8 or more. Ovary superior, two- or six-celled; style 1. Fr. a capsule.

LYTHRUM | Loosestrife   Calyx tubular or bell-shaped, with 4–6 teeth; petals 4–6, sometimes short-lived. Ovary two-celled. Stems usually four-angled. 13 sps.

Creeping anns. rooting at the nodes

819. *L. portula* (L.) D. A. Webb (*Peplis p.*) WATER PURSLANE. A creeping, hairless ann. 4–25 cm., with branched, four-angled stems rooting at the nodes, and opposite, broad-spathulate and shortly stalked lvs. 1–2 cm. Fls. *c.* 1 mm. across, greenish, solitary and almost stalkless in the axils of most lvs. and much shorter than the lvs.; petals minute, lilac, 6 or absent. Fr. globular, 1½ mm. △ Muddy places, ditches, bare damp ground. May–Oct. All Eur. (except IS.). Pl. 81.

Erect plants

(a) Fls. whorled, rosy-purple, conspicuous

820. *L. salicaria* L. PURPLE LOOSESTRIFE. An erect, hairy perenn. ½–2 m. with bright rosy-purple fls. in a long, dense, cylindrical, leafy, terminal spike, often somewhat interrupted below. Fls. 1–1½ cm. across; calyx tubular, hairy, ribbed, teeth unequal; petals oblong; stamens 12. Lvs. stalkless, half-clasping stem, lance-shaped, mostly opposite or in whorls of 3. Fr. 3–4 mm., oblong-ovoid. △ By streams, lakes, ponds, and marshes. June–Sept. All Eur. (except IS.). Pl. 79.

821. *L. virgatum* L. SLENDER LOOSESTRIFE. Like 820, but less robust and stems and lvs. quite hairless. Fls. rosy-purple, few in each whorl; calyx teeth equal. Lvs. opposite, narrowed to the base. Perenn. 50–120 cm. △ Meadows, marshes, damp woods. June–Aug. A.PL.CS. H.I.YU.AL.GR.BG.R.SU.: introd. F.D. Pl. 80.

(b) Fls. solitary, axillary, pinkish, small

822. *L. hyssopifolia* L. GRASS POLY. An erect, simple or branched, hairless ann. 10–60 cm. with solitary pale pinkish-lilac fls. in the axils of most of the lvs. and shorter than the lvs. Fls. ½–1 cm. across; calyx tube hairless, teeth 8–12; petals 5–6; stamens 5–6, included in the calyx tube. Lower lvs. oblong, opposite, the upper linear-lance-shaped, alternate, rough-margined. △ Damp fields, recently flooded ground. May–Sept. Most of Eur. (except IRL.NL.). Page 264.

TRAPACEAE | Water Chestnut Family

Aquatic anns. with floating lvs. in a rosette and inflated lv. stalks. Sepals, petals and stamens 4; ovary half-inferior, two-celled. Fr. top-shaped with a hard, stony inner wall and with 2–4 stout, horn-like spines, one-seeded. Often placed in Onagraceae. 1 genus. 1 sp. in Europe.

TRAPA

823. *T. natans* L. WATER CHESTNUT. A submerged aquatic ann. or perenn. with lv. blades forming a floating rosette and with solitary white fls. 1–2 cm. across, borne centrally on short stalks above the surface of the water. Lv. blades rhomboid, toothed towards apex; lv. stalks long, with a spindle-shaped hollow swelling below the blade; submerged lvs. with narrow blades. Roots abundant, much-branched. Fr. top-shaped *c.* 3 cm., with 2–4 robust spines spreading horizontally. △ Lakes and stagnant waters. June–July. Cent., South-East Eur. (except TR.) E.F.I.SU. *The fruits are edible; they form an important source of starch for some lake-dwelling communities, as in Kashmir. The plant is rich in iron.* Pl. 79.

1. *Myricaria germanica* 808
3. *E. roseum* 841
5. *Elaeagnus angustifolia* 762

2. *Epilobium fleischeri* [836]
4. *Lythrum hyssopifolia* 822

MYRTACEAE | Myrtle Family

Trees or shrubs usually with opposite and entire evergreen lvs., which are gland-dotted and aromatic when crushed. Sepals and petals 4–5; stamens numerous, often in bunches. Ovary inferior; fr. a berry or dry capsule.

Sepals and petals free; fr. a berry. *Myrtus*
Sepals and petals fused into a cap which is shed as the fl. opens;
fr. a capsule. *Eucalyptus*

MYRTUS | Myrtle Fls. solitary or few in the axils of the lvs; sepals and petals 5, free; stamens numerous. Ovary with 2–3 cells; fr. a berry. 1 sp.

824. *M. communis* L. MYRTLE. A dense evergreen shrub 2–3 m., with stiff, dark shining green, very aromatic lvs. and solitary, axillary, sweet-scented white fls. Fls. 2–3 cm. across, long-stalked; stamens numerous, conspicuous. Lvs. 2–3 cm., oval-lance-shaped, gland-dotted. Fr. a bluish-black berry. △ Evergreen thickets, stony ground, pine woods; often grown for ornament. May–July. Med. Eur. (except TR.) P. *Grown since ancient times for its very aromatic flowers, leaves, and bark; a fragrant oil, used in perfumery, is obtained from them. The fruit is sometimes used as a condiment. The bark and root are used for tanning.* Pl. 81.

EUCALYPTUS | Gum Fls. solitary or clustered; petals and sepals fused forming a cap or *operculum* over the bud which falls off when the fls. open; stamens numerous. Ovary with 2–8 cells; style elongated. Fr. a capsule. Lvs. of two kinds may be present: the juvenile lvs. are opposite and oblong, the adult lvs. are alternate and lance-shaped. 11 sps.

825. *E. globulus* Labill. BLUE GUM. A tall graceful tree 25–35 m. with dark shining, sickle-shaped lvs. and large, almost stalkless, solitary or 2–3 whitish fls. *c.* 4 cm. across. Young lvs. glaucous, clasping stem; adult lvs. green, 15–30 cm. by 2½–4½ cm. Fl. bud hard, warty, quadrangular; cap short conical, rough and waxy. Bark peeling in long strips and showing smooth greyish underbark. △ Native of Tasmania; widely planted for timber and ornament in Southern Europe. Spring. *Oil of Eucalyptus is obtained from the leaves; it is a valuable medicinal antiseptic oil and used in perfumery.*

826. *E. amygdalinus* Labill. Distinguished from 825 by its very small fls. *c.* 6 mm. across, which occur in stalked clusters of 5–12; fl. buds not quadrangular. Lvs. of fl. shoots elongate-lance-shaped 7½–12½ cm. by 6–15 mm., glaucous, and, in consequence, the tree at a distance has the appearance of possessing a head of hair. △ Native of Australia; often planted in South-Western Europe. *It is planted for timber, as an anti-malarial and to restrict the spread of the gall-forming aphis, Phylloxera.*

E. viminalis Labill. Distinguished by its small white fls. 1½ cm. across, which usually occur in threes in short-stalked axillary clusters; cap hemispherical-conical. Mature lvs. pale green, lance-shaped, long-pointed, 11–18 cm. by ½–2 cm. Tall tree to 50 m. with pendulous branches and white bark, often hanging in ribbons from the branches. Native of Australia; often planted in Western and Southern Europe.

PUNICACEAE | Pomegranate Family

A family of 1 genus and 1 sp. which is distinguished by its unusual fr. which is crowned with a persistent hard calyx tube. Fls. large, solitary; calyx bell-shaped, with 5–7 triangular teeth; petals 5–7, free; stamens numerous. Ovary inferior, many-celled; style 1. Fr. a large fleshy berry.

PUNICA | Pomegranate

827. *P. granatum* L. POMEGRANATE. A deciduous, much-branched spiny shrub or small tree 2–5 m. with oblong lance-shaped, shining lvs. and large scarlet fls. *c.* 4 cm. across. Petals crumpled; calyx fleshy, red; stamens *c.* 20. Lvs. 2–9 cm., hairless. Fr. large to 9 cm. across, with a leathery brownish-yellow rind and a crown-like calyx, and with pinkish pulpy flesh with numerous seeds. △ Hedges, rocky places; often cultivated in Southern Europe. May–Sept. Med. Eur. P. *The fruits are edible and used to make cooling drinks. The rind is used medicinally as an astringent and vermifuge and also for tanning Morocco leather; the flowers give a red dye.* Pl. 80.

ONAGRACEAE | Willow-Herb Family

Usually ann. or perenn. herbs; lvs. simple, variously arranged; stipules absent. Fls. usually regular; sepals, petals and stamens usually 4, less commonly 2 or 5, sepals free, placed edge to edge in bud or *valvate*; petals free; stamens often 8. Ovary inferior, two- to six-celled; style simple. Fr. usually a capsule, splitting longitudinally; seeds often with a plume of hairs. This family is sometimes called the Oenotheraceae.

1. Petals absent; stamens 4. *Ludwigia*
1. Petals present; stamens 2 or 8.
  2. Petals 2; stamens 2. *Circaea*
  2. Petals 4; stamens 8.
    3. Petals usually yellow; seeds without a plume of hairs. *Oenothera*
    3. Petals pink, rarely white; seeds with a plume of hairs. *Epilobium*

LUDWIGIA Fls. solitary, axillary; sepals 4; petals absent; stamens 4. Largely submerged aquatics. 3 sps.

828. *L. palustris* (L.) Elliott A slender, leafy, reddish-stemmed, submerged, aquatic perenn. 5–60 cm., rooting at the nodes and with floating lvs. above. Fls. 3 mm., greenish, solitary and stalkless in axils of the lvs.; calyx 4, often with reddish margin; petals absent. Lvs. opposite, in two ranks, 1½–5 cm., broadly oval with a fine point and narrowed abruptly to a short stalk. △ Still and flowing waters. June–Aug. Most of Eur. (except North Eur. IRL.).

CIRCAEA Fls. white or pinkish in terminal clusters; sepals 2, petals 2; stamens 2. Fr. covered with hooked bristles, not splitting. 3 sps.

829. *C. lutetiana* L. ENCHANTER'S NIGHTSHADE. A slender perenn. 20–70 cm., with creeping underground stems, opposite lvs. and a terminal, leafless, spike-like cluster of small white or pinkish fls. Fls. soon becoming well spaced on inflorescence; fl. stalks glandular-hairy, reflexed in fr.; petals 2–4 mm., deeply notched. Lvs. oval with a shallowly heart-shaped base and gradually narrowed to an acute tip, margin weakly toothed; lv. stalk hairy, not winged. Fr. ovoid *c.* 3 mm. across, densely covered with stiff, hooked bristles. △ Woods, shady ravines. June–Aug. All Eur. (except IS.SF.). Pl. 80.

830. *C. alpina* L. ALPINE ENCHANTER'S NIGHTSHADE. Like 829, but a much more delicate perenn. 5–15 cm., with fl. clusters dense and not elongating until after the petals fall; petals ½–2 mm., shallowly notched. Lvs. oval with a deeply heart-shaped base, margin strongly toothed; lv. stalks winged, nearly hairless. Fr. oblong 1–1½ mm. across, covered with soft, hooked bristles. △ Damp woods in the mountains. June–Aug. North Eur. (except IS.), Cent. Eur. (except H.) F.GB.B.NL.I.YU.R.SU.

*C. × intermedia Ehrh., the hybrid between 829 and 830, is widespread in Europe.

OENOTHERA | **Evening Primrose**   Fls. usually large, yellow or rarely pink, solitary or several in the axils of the upper lvs., and often forming elongated spike-like clusters. Sepals and petals 4, both fused into a long tube below; stamens 8; stigmas entire or four-lobed. Fr. an elongated capsule; seeds without hairs. 13 sps.

Fls. yellow
831. *O. biennis* L. EVENING PRIMROSE. An erect, robust, usually unbranched bienn. ½–1 m. with a slender spike of large yellow fls. which remain yellow as the fls. mature. Petals 2–3 cm., obcordate, corolla tube 1½–4½ cm.; styles ½–1½ cm. Rosette lvs. narrowly oblance-shaped, long-stalked, hairy; stem lvs. stalkless, broader and with finely toothed margin. Fr. 3–3½ cm. long, cylindrical, hairy but without red-based hairs. △ Disturbed ground, stony places, dunes. June–Sept. Native of America; introd. to most of Eur. (except IS. SF.). Pl. 81.

O. erythrosepala Borbás Like 831, but petals larger 2–6 cm. Lvs. with crinkled margin; stems red-spotted and with hairs with swollen red bases. Fr. with red-based hairs. Possibly of garden origin; introduced to Western and Central Europe.

832. *O. parviflora* L. SMALL-FLOWERED EVENING PRIMROSE. An unbranched or little-branched bienn. or perenn. 10–80 cm., with red-spotted stems and with red-based hairs below, and small yellow fls. in a spike. Petals 1–1½ cm., remaining yellow; sepals hairy, green and becoming red-streaked, their tips spreading. Stem lvs. narrowly lance-shaped, ascending, rather fleshy and thick, almost hairless beneath, veins reddish. △ Disturbed ground. May–Aug. Native of America; introd. F.NL.D.CS.H.PL.I.

833. *O. stricta* Link Distinguished by its petals, 3–4½ cm., which are at first yellow and then turn red; tube of corolla 1½–3 cm. Basal lvs. linear-lance-shaped and narrowed into a stalk-like base, the upper broader, stalkless, all with wavy, ciliate, and sparsely toothed margin. Fr. 1½–2½ cm., club-shaped, with silky and glandular hairs. Ann. or bienn. ½–1 m. △ Dunes, waste places. May–Aug. Introd. P.E.F.D.CH.I.SU.

Fls. pink
834. *O. rosea* Aiton PINK EVENING PRIMROSE. A rather slender, branched shrubby perenn. to 1 m. with small, solitary, pink fls. in the axils of the upper lvs. Fls. *c.* 1½ cm. across; petals 4–10 mm.; fl. stalks hollow, ribbed. Lvs. oval-lance-shaped, all stalked, entire or more or less deeply lobed. Fr. long-stalked, distinctly club-shaped, eight-angled. △ Waste places. June–July. Native of America; introd. P.E.F.GB.B.NL.I. Pl. 81.

EPILOBIUM | **Willow-Herb**   Fls. pink, purple, or rarely white; calyx 4; petals 4, usually notched; stamens 8. Ovary with 4 cells; stigma club-shaped or four-lobed. Fr. a long narrow capsule splitting into 4 valves; seeds with a long plume of hairs. Including *Chamaenerion*, which is often distinguished as a separate genus. Hybridization is frequent. 27 sps.

Lvs. all spirally arranged; fls. held horizontally

835. *E. angustifolium* L. (*Chamaenerion a.*) ROSEBAY WILLOW-HERB, FIREWEED. An erect, unbranched perenn. 20–120 cm., with numerous narrow, spirally arranged lvs. and rosy-purple fls. in a long, cylindrical, leafless, terminal spike. Fls. numerous, 2–3 cm. across; sepals dark purple; petals obovate, narrowed to a claw, unequal with the 2 upper broader than the lower 2; stigma four-lobed, style at length longer than the stamens. Lvs. 5–15 cm., narrowly oblong-lance-shaped, glaucous and net-veined beneath, often with wavy margin. Fr. 2½–8 cm. four-angled, hairy. △ Thickets, clearings in woods, screes. June–Sept. Most of Eur. (except P.). Pl. 80.

836. *E. dodonaei* Vill. (*Chamaenerion angustissimum* (Weber) D. Sosn.) Distinguished from 835 by the narrower, thick, linear to linear-lance-shaped lvs. which are green on both sides and without a distinct network of veins. Fls. rosy-purple in a short cluster which is leafy to the apex; petals not clawed; styles slender, at length about as long as the stamens. An erect perenn. 20–100 cm. △ Sand, gravels, dry slopes. June–Aug. Cent., South-East Eur. (except TR.) F.I.SU.

E. fleischeri Hochst. ALPINE FIREWEED. Like 836, but stems spreading or semi-prostrate and lvs. narrow-lance-shaped. Style thick, hairy on lower half and half as long as the stamens. Mountains of Central Europe. Page 264.

Lower lvs. opposite or whorled
(a) Stigma of 4 lobes spreading in a cross

837. *E. hirsutum* L. GREAT HAIRY WILLOW-HERB, CODLINS AND CREAM. An erect, hairy, herbaceous perenn. 80–150 cm., with terminal leafy clusters of conspicuous, relatively large, deep rosy-purple fls. 1½–2½ cm. across. Petals obovate, shallowly notched; style longer than the stamens and stigma lobes spreading and curled downwards; fl. buds erect. Lvs. large 6–12 cm., oblong-lance-shaped, half-clasping the stem, with margin of blade running a short distance down the stem; plant glandular hairy above. Fr. 5–8 cm., downy. △ Riversides, lakesides, marshes, ditches. July–Aug. All Eur. (except IS.). Pl. 81.

838. *E. parviflorum* Schreber SMALL-FLOWERED HAIRY WILLOW-HERB. Like 837, but fls. smaller 6–9 mm. across and petals pale rosy-purple, deeply notched. Style about as long as the stamens, stigma lobes spreading. Lvs. 3–7 cm. with rounded base, not half-clasping and blade not running down stem. Fr. 3½–6½ cm., nearly hairless or downy. A perenn. with soft, spreading hairs and erect stems 20–60 cm. △ Streamsides, marshes, ditches. July–Sept. All Eur. (except IS.).

839. *E. montanum* L. BROAD-LEAVED WILLOW-HERB. A small-flowered sp. distinguished by its lvs. which are mostly opposite, shortly stalked to *c.* 6 mm., with an oval to lance-shaped blade with a rounded base. Fls. pale rose, 6–9 mm. across in leafy terminal clusters; petals deeply notched; style shorter than the longer stamens, stigma of 4 short spreading lobes; fl. buds drooping. Fr. 4–8 cm., downy. An erect, little-branched, sparsely hairy perenn. 20–60 cm. △ Woods, hedges, waste places. All Eur. (except IS.TR.). Pl. 81.

*E. *lanceolatum* Sebastiani & Mauri SPEAR-LEAVED WILLOW-HERB. Widespread in Europe.

(b) Stigma club-shaped

840. *E. palustre* L. MARSH WILLOW-HERB. Distinguished by its rounded stems without raised lines or ridges, but often with 2 rows of crisped hairs running down the stem. Fls. 4–6 mm. across, buds at first erect but soon drooping and hanging to one side; petals rose, lilac, or rarely white, shallowly notched. Lvs. 2–7 cm., linear-lance-shaped to lance-shaped with a wedge-shaped base, stalkless. A perenn. with slender, underground, creeping stolons

ending in bulbil-like buds in autumn; fl. stems 15–60 cm. △ Marshes, fens, damp places. July–Sept. All Eur. (except TR.).

841. *E. roseum* Schreber SMALL-FLOWERED WILLOW-HERB. Distinguished from 840 and other sps. with club-shaped stigmas by the distinctly stalked lvs. with lv. stalk 3–20 mm., and hairless, oval-elliptic to elliptic-lance-shaped, toothed blades. Fls. 4–6 mm. across, drooping in bud; petals at first white, then streaked rose-pink. Perenn. with stem with 2 raised lines, glandular-hairy above, but hairless below; 25–80 cm. △ Damp places, woods, thickets. June–Sept. All Eur. (except IS.). Page 264.

*E. adenocaulon Hausskn. Native of North America; now widespread mainly in North-Western Europe.

*E. tetragonum L. SQUARE-STEMMED WILLOW-HERB. Widespread in Europe.

*E. obscurum Schreber Widespread in Europe.

842. *E. alsinifolium* Vill. CHICKWEED WILLOW-HERB. A slender alpine perenn. 5–30 cm., with underground creeping stolons bearing yellowish scale-lvs., and erect stems with 2–5 mostly drooping fls. 8–9 mm. across. Petals rose-red, notched; sepals 4–6 mm. Lvs. shining bluish-green, 1½–4 cm., oval-lance-shaped, shallowly toothed, shortly stalked. Fr. 3–5 cm., nearly hairless. △ Arctic or alpine: springs, streams, marshes. July–Sept. Most of Eur. (except P.B.NL.DK.AL.TR.).

E. anagallidifolium Lam. (*E. alpinum* auct.) ALPINE WILLOW-HERB. Like 842, but stolons running above ground and with pairs of small green lvs. Fls. 1–3, 4–5 mm. across on very slender stems; petals rose-red; sepals 2–4 mm. Stem lvs. yellowish-green, lance-shaped 1–2 cm. Widely distributed in arctic and alpine Europe.

HALORAGACEAE | Water-Milfoil Family

Usually submerged aquatic plants with whorled lvs. in our sps. Fls. inconspicuous, usually in spikes, often one-sexed. Sepals 2, 4, or absent; petals as many as the sepals; stamens 2, 4, or 8. Ovary usually four-celled, stigmas 1–4, feathery; fr. nut-like or a drupe.

MYRIOPHYLLUM | Water-Milfoil   Lvs. in whorls of 3–6, pinnately divided into fine linear segments. Fls. in spikes or whorls, the upper usually male, the lower female, each whorl subtended by entire or dissected bracts. Sepals and petals minute, 4 or absent; stamens usually 8; stigmas 4. Fr. usually of 4 nutlets. 5 sps.

843. *M. verticillatum* L. WHORLED WATER-MILFOIL. A submerged aquatic plant ½–3 m., branched above and rooting below, with flexible stems and dissected lvs. usually in overlapping whorls of 5. Lvs. 1½–4½ cm., much longer than the internodes, pinnate with 25–35 narrow segments. Spikes of fls. usually greenish, carried above surface; fls. in whorls of 5 all subtended by pinnate or comb-like bracts of variable length. Over-wintering buds present. △ Still or slow-flowing waters. June–Aug. All Eur.

M. spicatum L. SPIKED WATER-MILFOIL. Very like 843, but lvs. usually in whorls of 4, about equalling the internodes; lvs. with 13–35 segments. Spikes of fls. reddish, in whorls of 4, all subtended by entire bracts shorter than the fls., except the lowest whorl which has dissected bracts as long as the fls. Throughout Europe. Pl. 82.

*M. alterniflorum DC. Mainly in Western and Northern Europe.

HIPPURIDACEAE | Mare's-Tail Family

An aquatic family containing 1 genus and 1 sp. Lvs. linear, numerous in each whorl. Fls. solitary in the axils of the lvs. and thus whorled, bisexual or one-sexed; perianth absent; stamen 1; ovary inferior, one-celled; style 1, long and slender. Fr. one-seeded.

HIPPURIS | Mare's-Tail

844. *H. vulgaris* L. MARE's-TAIL. Submerged stems rather flaccid with regular whorls of long, thin, translucent lvs.; emergent stems appearing jointed and recalling a horsetail, 7–40 cm., erect, rather stiff, hollow, with whorls of 6–12 dark green, linear lvs. 1–7½ cm. with stiff acute tips. Fls. tiny, pinkish or greenish, stalkless, borne at the base of the lvs. of the upper whorls of the emergent stem. △ Still and slow-flowing waters. May–Aug. Most of Eur. (except GR.TR.).

CORNACEAE | Dogwood Family

Usually trees or shrubs with simple lvs. Fls. small, numerous in umbel-like clusters. Sepals small, 4; petals 4 or sometimes absent; stamens 4. Ovary inferior, usually two-celled; fr. a drupe or berry.

Deciduous shrubs or herbaceous perenns.   *Cornus*
Evergreen shrubs.   *Aucuba*

CORNUS   Usually trees or shrubs with opposite, entire lvs. Fls. numerous, usually in dense, flattened umbels and often with involucral bracts at base of umbel; sepals inconspicuous; petals and stamens 4. Ovary two-celled; style 1; fr. a drupe. 5 sps.

Fls. yellow, appearing before the lvs.

845. *C. mas* L. CORNELIAN CHERRY. A deciduous shrub or small tree 2–5 m., with greyish branches and small umbels of yellow fls. appearing on the previous year's branches before the lvs. Umbels *c.* 1 cm. across of 6–10 fls. and with 4 yellowish, oval bracts; fls. *c.* 4 mm. across. Lvs. 4–10 cm. oval-long-pointed, dull green, paler beneath. Fr. oblong-elliptic *c.* 1 cm. long, scarlet, acid-flavoured. △ Woods, hedges, rocky ground; sometimes grown for ornament and for its edible fr. Mar. Cent. Eur. (except PL.), South-East Eur. F.I.SU.: introd. GB. Pl. 82.

Fls. white, appearing with the lvs.

846. *C. sanguinea* L. DOGWOOD. An erect shrub 1–4 m. with slender reddish twigs, opposite oval lvs. and long-stalked, flat-topped clusters of numerous white fls. Clusters *c.* 5 cm. across, without bracts; petals 4–6 mm., oblong-lance-shaped, spreading, with adpressed hairs beneath. Lvs. 4–8 cm., stalked, with 3–4 pairs of conspicuous lateral veins, paler beneath, turning reddish in autumn. Fr. black, globular, 6–8 mm. △ Hedges, thickets, old grasslands. May–June. All Eur. (except IS.): introd. SF. *The fruits yield an oil sometimes used for illumination and soap-making; the flexible twigs can be used for basket-making.* Pl. 82.

847. *C. suecica* L. DWARF CORNEL. An arctic or alpine dwarf herbaceous perenn. 6–20 cm., usually with several simple, erect stems bearing a solitary terminal inflorescence composed of 4 conspicuous, spreading, white bracts and a compact cluster of blackish-purple

fls., the whole resembling a fl. Fls. *c.* 2 mm. across, 8–25 in each cluster; bracts 5–8 mm., oval. Lvs. 1–3 cm., oval, stalkless, three- to five-veined, somewhat bluish-green beneath. Fr. red, globular, *c.* ½ cm. △ Moors, heaths, in mountains. July–Aug. North Eur. GB.NL. D.PL.SU. Pl. 82.

Aucuba japonica Thunb. SPOTTED LAUREL. A hairless evergreen shrub with large, laurel-like, dark, shining green and often yellow-spotted, oval lvs. 8–20 cm. Fls. purplish, *c.* ½ cm. across, in erect, downy, branched clusters 5–10 cm. long. Fr. scarlet, ovoid 1½–2 cm. Native of Japan; often grown as an ornamental. Pl. 83.

ARALIACEAE | Ivy Family

Usually woody plants with alternate, simple, or compound lvs., usually with stipules. Fls. small, in umbels, or clustered into heads or spikes. Sepals small or absent; petals often 5, free or united; stamens usually as many as the petals. Ovary inferior, usually five-celled and surmounted by a disk which may be nectar-giving. Fr. a drupe or berry.

HEDERA | Ivy   Woody climbers which climb by means of rootlets. Fls. in terminal umbels; parts of fls. in fives. Fr. berry-like. 2 sps.

848. *H. helix* L. IVY. A woody plant of great plasticity with glossy, dark, evergreen lvs. either climbing to 30 m. by means of dense brown rootlets formed on the stem, or spreading over the ground in extensive carpets. Fls. yellowish-green in terminal globular umbels which are only formed on branches in full sun. Lvs. very variable, mostly with 3–5 shallow, triangular lobes or upper lvs. in full sun often entire, oval, or rhombic. Fr. globular 6–8 mm., black. △ Woods, walls, rocks, shady places. Sept.–Nov. All Eur. (except IS.SF.). *Valued as a medicinal plant in the past. The fruits and leaves are irritant and purgative.* Pl. 82.

UMBELLIFERAE | Umbellifer Family

Usually herbs, with furrowed hollow stems and compound lvs. with sheathing lv. bases. Members of this family are usually readily distinguished by their inflorescences which are composed of numerous small fls. arranged in a more or less flat-topped or umbrella-shaped umbel. Each umbel is usually compound; it is composed of several branches or *rays*, forming the *primary umbel*, and each ray usually ends in a smaller umbel of several rays called the *secondary umbel*. The rays of the secondary umbels bear fls. Primary umbels commonly have *bracts* at their base and secondary umbels *bracteoles*. Less commonly, the fls. are arranged in compact heads and partially surrounded by involucral bracts giving the inflorescence a very different appearance. Fls. with sepals absent or with 5 inconspicuous lobes; petals 5, often notched and sometimes very unequal, particularly on the outermost fls. of the umbels; stamens 5. Ovary inferior, two-celled; styles 2. Fr. composed of 2 one-seeded units which usually separate from a narrow central 'stalk' or *carpophore*, each unit or *mericarp* is often five- or nine-ribbed and often possesses resin canals or *vittae*, usually lying between the ribs. The dorsal and lateral ribs may be extended into a wing. Ripe fr. is important for the identification of genera and sps. in the Umbelliferae. In the descriptions that follow the first, second, and third divisions of pinnate, twice-pinnate, and three-pinnate lvs. are described as *segments*, and further divisions as *lobes*; lobes may be further lobed or toothed.

UMBELLIFERAE

1. Lvs. simple and entire, or lvs. lobed and spiny.   Group A
1. Lvs. deeply or shallowly lobed, or pinnately or ternately divided.
  2. Lvs. deeply or shallowly lobed or toothed; lvs. roughly rounded in
    outline.   Group B
  2. Most lvs. once, twice, or more pinnately or ternately divided; lvs.
    usually distinctly longer than wide.
    3. Lateral ribs of fr. forming distinct (but sometimes narrow) wings.
      4. Fls. yellow.   Group C
      4. Fls. white, pinkish, greenish, or cream.   Group D
    3. Fr. without wings, ribs sometimes conspicuous, but then wedge-shaped
      in section.
      5. Fls. yellow or yellowish-green.   Group E
      5. Fls. white, pink, greenish-white, or cream.
        6. Fr. at least 3 times as long as wide.   Group F
        6. Fr. less than 3 times as long as wide.
          7. Fr. hairless, not covered with prickles.   Group G
          7. Fr. finely hairy, rough-haired or with prickles. Group H

Group A. *Lvs. simple and entire, or lvs. lobed and spiny*

		Nos.
1. Lvs. simple and entire, smooth.	*Bupleurum*	881–884
1. Lvs. lobed, spiny.		
2. Male fls. clustered tightly round a female or bisexual fl., their stalks fused to each other or attached to the ovary.	*Echinophora*	857
2. Not as above; fls. bisexual, clustered into rounded heads and with spiny involucral bracts.	*Eryngium*	853–856

Group B. *Lvs. deeply or shallowly lobed or toothed; roughly rounded in outline; not pinnately or ternately divided*

		Nos.
1. Lvs. deeply lobed.		
2. Bracts and bracteoles inconspicuous; fls. white or pale pink.	*Sanicula*	850
2. Bracts large conspicuous, coloured.		
3. Fls. yellow; bracts oval.	*Hacquetia*	851
3. Fls. white, pinkish, or greenish; bracts lance-shaped.	*Astrantia*	852
1. Lvs. shallowly lobed.		
4. All lvs. with long stalks; fls. whorled, on axillary stems shorter than the lvs.	*Hydrocotyle*	849
4. Stem lvs. stalkless; fls. in compound umbels.	*Smyrnium*	865, 866

Group C. *Lateral ribs of fr. distinctly winged; fls. yellow*

		Nos.
1. Bracteoles joined, at least at the base.	*Levisticum*	896
1. Bracteoles free or none.		
2. Lvs. ternate.	*Peucedanum*	899
2. Lvs. pinnate or pinnately cut.		

3. Lvs. once-pinnate.	*Pastinaca*	900
3. Lvs. twice or more times pinnate.		
4. Slender anns.	*Anethum*	876
4. Stout perenns.		
5. Bracts numerous.		
6. Lv. lobes oval to oval-oblong.	*Peucedanum*	899
6. Lv. lobes linear to linear-oblong.	*Ferulago*	898
5. Bracts absent.		
7. Wings of fr. at least as wide as the seed-containing		
part, divergent and usually shining.	*Angelica*	894, 895
7. Wings of fr. narrow and not shining.		
8. Bracteoles absent or soon falling; fr. flat.	*Ferula*	897
8. Bracteoles several, persistent; fr. convex on back.	*Peucedanum*	899

Group D. *Lateral ribs of fr. distinctly winged; fls. white, pinkish, greenish, or cream*

		Nos.
1. Wing of fr. conspicuously thickened and corky.	*Tordylium*	903, 904
1. Wing of fr. thin and often papery.		
2. Dorsal as well as lateral ribs of fr. winged.		
3. Bracts several, persistent.		
4. Ribs of fr. all equal.		
5. Lvs. hairy on margin and on veins beneath.	*Pleurospermum*	880
5. Lvs. hairless.	*Ligusticum*	892, 893
4. Ribs of fr. unequal, the lateral wings wider.	*Laserpitium*	905, 906
3. Bracts absent or rarely several, but soon falling.		
6. Ribs of fr. all equal.	*Ligusticum*	892, 893
6. Ribs of fr. unequal, the lateral wings wider.		
7. Lv. lobes broadly oval to rounded.	*Laserpitium*	905, 906
7. Lv. lobes linear to oblong-lance-shaped.		
8. Lv. lobes *c.* 1 cm. long.	*Selinum*	891
8. Lv. lobes *c.* 3 cm. long or more.	*Angelica*	894, 895
2. Dorsal ribs of the fr. without wings.		
9. Fr. not obviously flattened from back to front.		
10. Perenns.; bracteoles not deflexed; petals cream.	*Silaum*	877
10. Anns.; bracteoles deflexed; petals white.	*Aethusa*	874
9. Fr. obviously flattened from back to front.		
11. Fr. narrowly winged; resin canals obvious.	*Heracleum*	901, 902
11. Fr. broadly winged; resin canals not obvious.		
12. Lateral wings separate and divergent.	*Angelica*	894, 895
12. Lateral wings adpressed to one another.	*Peucedanum*	899

Group E. *Fr. without wings; ribs sometimes conspicuous; fls. yellow or yellowish-green*

		Nos.
1. Bracts and bracteoles absent or few.		
2. Lv. lobes thread-like.	*Foeniculum*	875
2. Lv. lobes oval to rounded in outline.	*Smyrnium*	865, 866
1. Bracts and bracteoles numerous.		
3. Lv. lobes lance-shaped to oval.	*Petroselinum*	887
3. Lv. lobes thread-like to linear-obovate, fleshy.	*Crithmum*	871

Group F. *Fr. without wings; fls. white, pink, greenish-white, or cream; fr. at least 3 times as long as wide*

		Nos.
1. Beak well developed.		
2. Fr. ribbed.	*Scandix*	862
2. Fr. not ribbed except on the beak.	*Anthriscus*	860, 861
1. Beak very short or absent; fr. ribbed.		
3. Bracts 4–15.	*Falcaria*	889
3. Bracts normally absent.		
4. Fr. and ovary finely hairy; rays 1–3.	*Physocaulis*	858
4. Fr. and ovary hairless; rays 4–24.		
5. Fr. up to 12 mm. long; ribs inconspicuous.	*Chaerophyllum*	859
5. Fr. 2–2½ cm. long; ribs very conspicuous.	*Myrrhis*	863

GROUP G. *Fr. without wings; fls. white, pink, greenish-white, or cream; fr. less than 3 times as long as wide; fr. hairless, not covered with prickles*

		Nos.
1. Fl. stem with an underground portion arising from a tuber.	*Conopodium*	867
1. Fl. stem without an underground portion, and with all the lvs. arising from it at or above the level of the ground.		
2. Fr. globular, not separating into one-seeded units when ripe.	*Coriandrum*	864
2. Fr. usually ovoid, separating into one-seeded units when ripe.		
3. Lowest lvs. once-pinnate or simple.		
4. Stems creeping and often rooting at the nodes.		
5. Bracts absent or few and entire.	*Apium*	885, 886
5. Bracts numerous, large, and pinnately cut.	*Berula*	[870]
4. Stems more or less erect, not rooting at the nodes.		
6. Upper lvs. with thread-like to narrowly oblong, parallel-sided lobes.		
7. Bracteoles absent.	*Pimpinella*	868
7. Bracteoles several.		
8. Bracteoles more than half the length of the *longest* fl. stalk.	*Oenanthe*	872, 873
8. Bracteoles not more than a quarter the length of the longest fl. stalk.		
9. Smell strong and unpleasant, fr. 3 mm. long, nearly globular.	*Sison*	[887]
9. Smell like parsley; fr. 3–4 mm. long, ovoid.	*Petroselinum*	887
6. Upper lvs. with lance-shaped to oval or obovate lobes with curved sides.		
10. Bracts absent.	*Pimpinella*	868
10. Bracts numerous.		
11. The enlarged base of the styles or stylopodium conical.	*Berula*	[870]
11. Stylopodium nearly flat.	*Sium*	870
3. Lowest lvs. at least twice-pinnate or twice-ternate.		
12. Aquatic plants with finely divided submerged lvs. present at flowering time.	*Oenanthe*	872, 873

12. No finely divided submerged lvs. present at flowering time.

13. Larger bracts at least half as long as rays, often divided.

14. Fr. 4–12 mm. long; plant 1–2 m. — *Pleurospermum* — 880

14. Fr. not more than 2½ mm. long; plant up to 1 m. — *Ammi* — 888

13. All bracts much less than half as long as rays, sometimes absent.

15. Stem lvs. absent or very small. — *Carum* — 890

15. Stem lvs. well developed.

16. Roots usually tuberous; fr. stalks mostly shorter than fr. and often becoming thickened. — *Oenanthe* — 872, 873

16. Roots not tuberous; fr. stalks mostly longer than fr. and not becoming thickened.

17. Bracts absent, rarely 1–3.

18. Basal lvs. with oval lobes.

19. Rhizomes far-creeping; lower lvs. twice-ternate, umbels long-stalked. — *Aegopodium* — 869

19. Rhizomes absent; lower lvs. twice-pinnate; umbels mostly stalkless or nearly so. — *Apium* — 885, 886

18. Basal lvs. with linear to linear-lance-shaped lobes.

20. Lv. lobes ½–1 cm. wide, deeply saw-toothed. — *Cicuta* — [887]

20. Lv. lobes 1–2 mm. wide, entire or pinnately cut. — *Carum* — 890

17. Bracts several.

21. Lv. lobes thread-like.

22. Lvs. linear-oblong in outline. — *Carum* — 890

22. Lvs. oval to triangular in outline. — *Meum* — 878

21. Lv. lobes linear-lance-shaped to oval.

23. Stem purple-spotted; ribs of fr. undulate. — *Conium* — 879

23. Stem not purple-spotted; ribs of fr. smooth. — *Carum* — 890

GROUP H. *Fr. without wings; fls. white, pink, greenish-white, or cream; fr. less than 3 times as long as wide; fr. finely hairy, rough-haired, or with prickles*

Nos.

1. Outer one-seeded unit of each fr. with prickles, the inner with small swellings. — *Torilis* — 907

1. Both one-seeded units similar.

2. Fr. with broad-based prickles arranged in 1–3 rows on the ribs.

3. Bracts ternate or pinnately cut. — *Daucus* — 910

3. Bracts simple or absent.

4. Bracts absent or 1–2, small and inconspicuous if present. — *Caucalis* — [908]

4. Bracts 2–5, conspicuous.

5. Bracts at least half as long as the smooth rays. — *Orlaya* — 909

5. Bracts not more than a quarter as long as the bristly-haired rays. — *Turgenia* — 908

2. Fr. without broad-based prickles arranged in 1–3 rows on the ribs.

6. Most umbels shortly stalked and leaf-opposed; fr. with a distinct, hairless beak — *Anthriscus* — 860, 861

6. Most umbels long-stalked, not leaf-opposed; fr. without a distinct hairless beak. — *Torilis* — 907

HYDROCOTYLE | Pennywort   Fls. whorled and borne in stalked axillary clusters; fr. with inner wall of ovary woody. Often creeping herbaceous perenns. with simple lvs. 5 sps.

849. *H. vulgaris* L. PENNYWORT. A slender, creeping perenn. 15–50 cm., rooting at the nodes, with long-stalked almost circular, shining green peltate lvs. Fls. inconspicuous, pinkish-green, in small whorls 2–3 mm. across borne on axillary stems shorter than the lvs. Lvs. 1–3 cm. across, margin with shallow rounded lobes; lv. stalk attached to the centre of the blade. Fr. rounded, *c.* 2 mm. △ Marshes, bogs, damp meadows. June–Sept. Much of Eur. (except SF.GR.TR.BG.R.).

SANICULA | Sanicle   Secondary umbels globular; fr. with hooked bristles. Lvs. palmately lobed. 2 sps.

850. *S. europaea* L. SANICLE. A hairless, little-branched perenn. 20–60 cm., with white or pale-pink fls. in small, dense, globular umbels 4–7 mm. and glossy rounded, deeply lobed lvs. Primary umbels four-rayed, bracts simple or deeply lobed; secondary umbels with outer male fls. and inner bisexual fls., bracteoles simple. Lvs. mostly from the rootstock in a lax rosette, long-stalked; blades 2–6 cm. across, rounded in outline, deeply cut into 5–7 wedge-shaped and coarsely toothed lobes. Fr. ovoid, *c.* 3 mm., with numerous hooked bristles. △ Woods. May–July. All Eur. (except IS.).

HACQUETIA   Fls. in a simple umbel surrounded by large, leaf-like bracts. Sepals conspicuous. Fr. ovoid with prominent stout ribs. 1 sp.

851. *H. epipactis* (Scop.) DC. A hairless, tufted perenn. with short stems 8–25 cm., arising direct from the rootstock, each bearing a dense rounded head *c.* 1½ cm., of yellow fls. closely surrounded by 5–6 leaf-like, oval bracts 1–2 cm. long. Lvs. bright green, long-stalked, all basal; blade palmately divided into wedge-shaped leaflets 2–4 cm., which are further lobed and toothed towards the apex. △ Woods in mountains. Apr. A.CS.PL.I.YU.R. Pl. 83.

ASTRANTIA | Masterwort   Fls. usually in compact, rounded, primary umbels, surrounded by large, conspicuously coloured bracts; both male and bisexual fls. present. Fr. oblong-ovoid, with wrinkled or toothed ribs. 5 sps.

852. *A. major* L. GREAT MASTERWORT, MOUNTAIN SANICLE. An erect, little-branched, hairless perenn. 30–100 cm., with white, pinkish, or greenish fls. in a dense round-topped umbel surrounded by numerous lance-shaped pointed bracts, often as long as the umbel. Umbels 1½–5 cm. across; bracts 1–2 cm., spreading, whitish beneath, greenish-purple above; fls. with slender stalks. Basal lvs. with blades 8–15 cm., rounded in outline and deeply divided into 3–7 oval, coarsely toothed lobes, long-stalked, dark green shining. Fr. 6–8 mm., oblong, with blunt white scales. △ Alpine meadows, open woods. June–Sept. Much of Eur. (except P.IRL.IS.N.S.NL.GR.TR.): introd. GB.DK.SF. Pl. 83.

A. minor L. LESSER MASTERWORT. Like 852, but a smaller and more delicate pale green perenn. to 35 cm. with basal lvs. divided to their base into narrower lance-shaped, coarsely toothed lobes. Umbels *c.* 1 cm. across; fls. white; bracts 5–11 mm. Pyrenees and Apennines.

ERYNGIUM   Stiff, spiny perenns. with tough, thistle-like lvs. Fls. numerous, in compact rounded heads surrounded by rigid, spiny-toothed, involucral bracts; calyx usually with spiny teeth; petals not spreading, narrow and often notched. 26 sps.

Involucral bracts oval in outline, toothed or deeply lobed
(a) Maritime perenns.

853. *E. maritimum* L. SEA HOLLY. A stiff, branched, very spiny maritime perenn. 30–60 cm. with very glaucous, stiff, spiny-toothed, holly-like lvs. and dense globular heads of bluish fls. surrounded by broad spiny bracts similar to the lvs. Fl. heads 1½–2½ cm.; fls. *c.* 8 mm.; bracteoles narrow, three-spined, often bluish. Basal lvs. with rounded, three-lobed blades 5–12 cm.; stem lvs. palmately lobed; lvs. all spiny. Fr. covered with hooks. △ Sand and shingle beaches of the littoral. June–Sept. Much of Eur. (except IS.SF.CH.A.CS.H.). *The roots can be eaten as a vegetable or candied; also used medicinally in the past.* Pl. 83.

(b) Alpine perenns.

854. *E. alpinum* L. ALPINE ERYNGO, QUEEN OF THE ALPS. A stiff, erect, blue-topped alpine perenn. 30–60 cm., with 1–3 dense oblong-cylindrical heads of whitish fls. surrounded by conspicuous, bright blue, stiff, comb-like involucral bracts. Bracts 10–20, a little longer than fl. head, deeply cut into numerous pairs of linear spiny-tipped lobes. Lower lvs. long-stalked, triangular-heart-shaped, conspicuously spiny-toothed, the upper lvs. palmately lobed with narrow spiny-toothed lobes, the uppermost often bluish like the bracts. Fr. covered with blunt scales. △ Alpine meadows, rocky places. June–Sept. F.CH.A.CS.I.YU.

Involucral bracts lance-shaped or linear in outline, spiny-toothed or entire
(a) Basal lvs. pinnately lobed

855. *E. campestre* L. FIELD ERYNGO. A pale yellowish-green, spiny-leaved, densely branched perenn. to 60 cm., with small, ovoid, whitish-green or pale blue fl. heads 1–1½ cm. across, surrounded by 3–6 spreading, linear bracts 2–3 times as long. Fls. *c.* 6 mm. long; bracts entire or spiny-toothed. Lower lvs. with a long unwinged stalk, thrice cut into narrow spiny-toothed segments; the upper lvs. clasping the stem. Fr. oboval, covered with pointed scales. △ Dry banks, waysides, stony places. June–Sept. Much of Eur. (except North Eur. IRL.): introd. DK. Pl. 83.

856. *E. amethystinum* L. BLUE ERYNGO. Like 855, but the whole inflorescence, including the branched stems and bracts, bright bluish-violet and fl. heads larger. Involucral bracts 7–8, blue, linear-lance-shaped, entire or very sparsely spiny-toothed, 2–5 times as long as the blue fl. head. Stalks of lower lvs. with spiny-toothed wings and blades twice cut into lance-shaped spiny-toothed lobes. A much-branched stiff perenn. 50–80 cm. △ Stony ground. July–Aug. I.YU.AL.GR.BG. Pl. 83.

(b) Basal lvs. undivided or three-lobed.

E. creticum Lam. SMALL-HEADED BLUE ERYNGO. Like 856, but basal and lower stem lvs. oval-heart-shaped, entire or three-lobed, and withered at time of flowering, when the whole plant becomes blue. Fl. heads much smaller; involucral bracts 5, linear-lance-shaped, spiny at base, 2–3 times as long as fl. head. Balkan Peninsula and Bulgaria.

E. planum L. Like [856], but basal lvs. elliptic and shortly spiny-toothed, the lower stem lvs. shortly stalked or stalkless, oval or shallowly three-lobed, the uppermost lvs. palmately five-lobed, with lobes lance-shaped and spiny-toothed. Whole inflorescence often blue; fl. heads small, *c.* 1 cm. across; the involucral bracts 6–8, as long as or a little longer than the fl. head. Much of Central and South-Eastern Europe.

ECHINOPHORA   Plants with stiff spiny lvs., bracts and bracteoles. Umbels with 5–8 rays; inner fl. of umbel fertile, the outer fls. male and clustered tightly round the inner, their stalks fused to each other and attached to the ovary. 2 sps.

857. *E. spinosa* L. A very spiny, glaucous, much-branched perenn. to 50 cm. with fleshy lvs.

twice cut into stiff, linear, spiny-tipped segments. Umbels white *c.* 3 cm. across, short-stalked, with 5–8 rays; calyx spiny; bracts and bracteoles 5–10, linear-lance-shaped, spiny-tipped, about as long as the umbels. Fr. oblong with a short beak, each unit five-angled. △ Sands by the sea. July–Aug. Med. Eur. (except TR.). *The root is edible.* Page 280.

PHYSOCAULIS Fls. white; sepals absent. Fr. more or less cylindrical, scarcely beaked. Ribs 5, blunt; stigmas stalkless. 1 sp.

858. *P. nodosus* (L.) Koch (*Chaerophyllum n.*) Like *Torilis* and *Chaerophyllum*, but distinguished by its linear-lance-shaped fr. 4–10 mm., which is covered with curved, upward-pointing white bristles, often with swollen bases; styles absent. Umbels with 2–3 rays; bracts absent; bracteoles 5–7, linear; petals white, oval not notched, somewhat hairy. Lvs. twice-pinnate, lobes oval-oblong, toothed or cut. A stiff, erect, rough-haired ann. to 1 m., conspicuously swollen below the nodes, branched above. △ Woods, thickets, in mountains. May–June. Med. Eur. (except F.) P.H.BG.

CHAEROPHYLLUM Fr. cylindrical, slightly compressed, very shortly beaked; ribs of fr. broad and rounded. Bracts few or absent; bracteoles several. 10 sps.

859. *C. temulentum* L. ROUGH CHERVIL. Fls. white, umbels 3–6 cm. across, nodding in bud; rays 6–12; bracts absent; bracteoles 5–8, hairy, deflexed in fr. Petals hairless. Lvs. 2–3 times pinnate; segments oval and further shallowly lobed, hairy on both faces; stem solid, swollen below nodes, purple or purple-spotted, with short stiff hairs. Fr. 5–7 mm., usually hairless; styles diverging, recurved. An erect, branched, hairy bienn. 30–100 cm. △ Hedgebanks, open woods, grassy places, waysides. May–Aug. All Eur. (except IS.SF.N.). *A poisonous plant.*

C. hirsutum L. HAIRY CHERVIL. Like 859, but petals white or pink, bordered with hairs. Styles erect or only slightly diverging. A variable sp. Mainly in mountains of Central and Southern Europe.

*C. aureum L. Central and Southern Europe.

ANTHRISCUS Fr. cylindrical, distinctly beaked, not ribbed except on the beak. Bracts absent or rarely 1; bracteoles several. 7 sps.

860. *A. sylvestris* (L.) Hoffm. COW PARSLEY. Fls. white, umbels 2–6 cm. across, with 4–15 rays; bracts absent; bracteoles 2–5 mm., oval fine-pointed, fringed with hairs, spreading or turned down. Lvs. 2–3 times pinnate; segments oval and further lobed and toothed; stems hollow, grooved, with downy hairs below, hairless above. Fr. 5–10 mm., black or brown, smooth, styles spreading. An erect, hairy, branched bienn. or perenn. to 1½ m. △ Hedges, open woods. May–July. Most of Eur. (except IS.AL.TR.).

861. *A. cerefolium* (L.) Hoffm. CHERVIL. Like 860, but a smaller aromatic ann. to *c.* ½ m., with stems hairy above the nodes and lv. segments further once or twice cut into narrow oblong lobes. Rays of primary and secondary umbels hairy; bracteoles few, lance-shaped ciliate. Fr. linear, 1 cm., smooth, styles erect. △ Bushy places, woods; cultivated and sometimes escaping. May–Aug. South-East Eur. (except AL.TR.) A.PL.CS.H.SU.: introd. elsewhere. *Cultivated as an aromatic pot-herb.*

*A. caucalis Bieb. (*A. vulgaris* Pers.) BUR CHERVIL. Widespread in Europe, except in the north.

SCANDIX Fr. cylindrical, ribbed and with beak much longer than the lower seed-bearing part. Umbels simple or compound; bracts 1 or absent; bracteoles several. 4 sps.

862. *S. pecten-veneris* L. SHEPHERD'S NEEDLE. An erect, branched, almost hairless ann. 15–50 cm., with simple umbels, or with 2 stout primary rays and with almost stalkless white fls. 1 mm. across; fl. stalks conspicuously elongating and thickening in fr. Bracteoles ½–1 cm., usually two-lobed or pinnate, margin bristly-haired. Lvs. 2–3 times pinnate, ultimate lobes oblong. Fr. 1½–8 cm., including the long, strongly flattened, rough-margined beak which is usually 3–4 times as long as the seed-bearing part. A variable sp. △ Cultivated ground, arable. May–Aug. All Eur. (except IS.).

S. australis L. SOUTHERN SHEPHERD'S NEEDLE. Like 862, but fr. shorter, 1½–4 cm., and beak only about twice as long as the seed-bearing part. Bracteoles oblong-oval, entire, margin papery. Lvs. with ultimate lobes narrowly linear. A very variable sp. Southern Europe.

MYRRHIS Fr. linear-oblong, with prominent ribs. Umbels compound; bracts few or absent; bracteoles several, papery. 1 sp.

863. *M. odorata* (L.) Scop. SWEET CICELY. A hairy, strongly aniseed-smelling, erect perenn. 60–120 cm. with umbels of white fls. 1–5 cm. across. Primary rays 5–10; secondary rays either stout and bearing bisexual fls., or more slender and bearing only male fls.; bracteoles *c.* ½ cm., lance-shaped, fine-pointed, reflexed, hairy. Lvs. 2–3 times pinnate, ultimate segments oblong-oval, further deeply lobed and toothed. Fr. 2–2½ cm., with sharp ribs, shining black when ripe. △ Fields and woods in the hills; sometimes cultivated and often self-seeding near habitation. Apr.–Aug. Much of Eur. (except P.H.GR.BG.R.). *Sometimes used for seasoning.* Pl. 84.

CORIANDRUM | Coriander Fr. ovoid, nut-like, smooth; ribs visible only on drying when the secondary ribs are broader than the primary ribs. Umbels usually compound; bracts absent; bracteoles few. 1 sp.

864. *C. sativum* L. CORIANDER. An erect, hairless, shining ann. 20–70 cm., smelling un-pleasantly and strongly of bed bugs, with umbels of white fls. 1–3 cm. across. Primary rays 3–8, bracts absent; bracteoles linear-lance-shaped, one-sided, reflexed. Lvs. 2–3 times pinnate, the lower lvs. with broader ultimate lobes, contrasting with the narrower lobes of the upper lvs. Fr. 3–4 mm., ovoid, red-brown. △ Native of Western Asia and North Africa; often cultivated and naturalized in much of Southern Europe. June–Aug. *Grown for its seeds which are very aromatic and used for flavouring in confectionery, liqueurs, wines, and curries.* Page 280.

SMYRNIUM Fls. yellowish-green. Fr. broadly ovoid, rounded or with 3 prominent ribs on each carpel. Umbels compound; bracts and bracteoles small, few, or absent. 5 sps.

865. *S. olusatrum* L. ALEXANDERS. A stout, strong-smelling, erect bienn. ½–1½ m. with yellowish-green fls. in rather dense rounded umbels 3–8 cm. across. Primary rays 7–15, bracts absent. Lvs. large, 3 times ternate, shining dark green; the lower lvs. stalked, segments broadly oval and bluntly toothed or lobed, the upper lvs. stalkless, sheathing at base, ternate. Fr. 6–8 mm., broadly ovoid, with 3 sharp ribs on each side, becoming black. △ Hedges, waste places, cliffs near sea. Mar.–July. Med. Eur. P.: introd. IRL.GB.NL. *Grown in the past as a vegetable like celery, which has now completely superseded it.*

866. *S. perfoliatum* L. PERFOLIATE ALEXANDERS. A yellowish-green perenn. ½–1 m. with

1. *Echinophora spinosa* 857
3. *Pleurospermum austriacum* 880

2. *Peucedanum ostruthium* 899
4. *Coriandrum sativum* 864

angled, winged stems, and stalkless upper lvs. with oval or rounded and finely toothed blades. Lower lvs. 2–3 times pinnate with oval, toothed lobes, stalked and with a large inflated lv. base. Primary rays 5–10, unequal, bracts absent; fls. yellow. Fr. very small *c.* 3 mm., black. △ Woods, rocky places. Apr.–June. Med., South-East Eur. P.CS.H.SU.: introd. GB.A. Pl. 84.

S. rotundifolium Miller Like 866, but upper lvs. entire, not toothed, and often golden-yellow like the fls. Stem ribbed, not winged, and lvs. glaucous. Corsica, Italy, Yugoslavia, Greece.

CONOPODIUM Fr. ovoid, ribs inconspicuous. Umbel compound, bracts and bracteoles few and papery or absent; petals deeply notched, tip turned inwards. Root a rounded tuber. 7 sps.

867. *C. majus* (Gouan) Loret PIGNUT, EARTHNUT. A slender, little-branched, hairless perenn. with smooth, finely grooved stems 10–60 cm. and umbels of white fls. which are nodding in bud. Umbels 3–5 cm. across; primary rays 6–12; fls. 1–3 mm. across; bracts and bracteoles 5 to absent. Stem lvs. few, 2–3 times pinnate, ultimate lobes narrow linear; basal lvs. with ultimate lobes linear-lance-shaped, soon withering. Fr. 4 mm., narrowly ovoid, styles erect. △ Woods, shady meadows. May–July. E.F.IRL.GB.N.I. *The tubers are edible.*

PIMPINELLA | Burnet Saxifrage Fr. ovoid to oblong, with 5 slender ribs. Umbels compound; bracts absent; bracteoles few or absent. Petals with long inturned points. Root not tuberous. 16 sps.

868. *P. major* (L.) Hudson GREATER BURNET SAXIFRAGE. A stout, almost hairless perenn. with prominently ribbed or angled, brittle stems $\frac{1}{2}$–1 m. Umbels flat-topped 3–6 cm., with 10–25 primary rays; fls. white or pinkish, 3 mm. across; bracts and bracteoles absent. Lvs. all once-pinnate; segments of basal lvs. 2–10 cm., oval, toothed and shortly stalked, stem lvs. with narrower, stalkless segments, the uppermost lvs. small, often three-lobed. △ Damp pastures, wood verges, rocky places. May–July. Most of Eur. (except P.IS.AL. GR.TR.): introd. SF.

**P. saxifraga* L. BURNET SAXIFRAGE. Almost throughout Europe.

AEGOPODIUM Fr. ovoid with 5 slender ribs. Umbels compound; bracts and bracteoles few or absent. Lvs. once or twice ternately divided into broad segments. Underground stems far-creeping. 1 sp.

869. *A. podagraria* L. GOUTWEED, GROUND ELDER. A hairless perenn. with creeping underground stems bearing numerous leafy shoots and often forming large patches, and bearing erect, hollow, grooved stems $\frac{1}{2}$–1 m. Umbels 2–6 cm. across, with 12–20 primary rays; fls. white, *c.* 3 mm. across; bracts and bracteoles usually absent. Lvs. ternate, usually once cut into 3 oval-lance-shaped, toothed, stalked or stalkless leaflets 4–8 cm. Fr. oval 3–4 mm., styles reflexed. △ Hedges, banks, gardens, shady places, waysides. May–Aug. Much of Eur. (except P.E.IS.TR.). *It has been used in herbal remedies; eaten as a salad or vegetable in some countries.*

SIUM Fr. with slender or thickened ribs and with 1–3 resin canals in each furrow. Umbels compound; bracts and bracteoles numerous; calyx teeth distinct. Submerged lvs. 2–3 times pinnate, with linear lobes; aerial lvs. once-pinnate with broad lobes. 2 sps.

870. *S. latifolium* L. WATER PARSNIP. A stout, erect, hairless perenn. to 2 m., with grooved stems and with large umbels of white fls. 6–10 cm. across. Fls. white, *c.* 4 mm.; primary rays 20–35; bracts and bracteoles variable, often leafy. Submerged lvs. present only in spring, ultimate lobes linear; stem lvs. once-pinnate, with 4–8 opposite pairs of stalkless leaflets 2–15 cm., with oval-lance-shaped, irregularly saw-toothed blades. Fr. 3 mm., ovoid. △ Fens, marshes. July–Sept. Most of Eur. (except P.IS.N.GR.TR.). *Poisonous to livestock.*

*Berula erecta (Hudson) Coville (*Sium e.*) NARROW-LEAVED WATER PARSNIP. Widespread in Europe.

CRITHMUM   Fr. ovoid, with prominent thick ribs; calyx absent. Lvs. compound, fleshy. Umbels compound; bracts and bracteoles numerous. 1 sp.

871. *C. maritimum* L. ROCK SAMPHIRE. A fleshy, narrow-leaved, somewhat glaucous, densely branched maritime perenn. 15–30 cm., with yellowish umbels 3–6 cm. across. Primary rays rather stout 8–20; fls. *c.* 2 mm. across. Lvs. 2–3 times pinnate with linear acute, fleshy, circular-sectioned segments; lv. sheaths long, papery, encircling the solid ribbed stem. △ Rocks, cliffs and sands on the littoral. July–Oct. West Eur. (except IRL.B.), Med. Eur. BG.SU. *The fleshy leaves are sometimes pickled.* Pl. 84.

OENANTHE | Dropwort   Fr. rounded on back, with shallow grooved or thickened ribs. Umbels compound; bracts several, few or absent; bracteoles usually many; calyx teeth acute. 13 sps.

872. *O. crocata* L. HEMLOCK WATER DROPWORT. A robust, shining, hairless, leafy perenn. $\frac{1}{2}$–$1\frac{1}{2}$ m. with stout, grooved, hollow stems and thick carrot-like root tubers with yellow juice. Umbels flat-topped, 5–10 cm. across, with 12–40 rays; fls. white, *c.* 2 mm.; bracts and bracteoles linear-lance-shaped, many but soon falling. Lvs. large, shining, 3–4 times pinnate with broad oval segments which are further 2–3 times lobed and toothed; lv. stalks sheathing. Fr. 4–6 mm., oblong-cylindrical; styles erect. △ Watersides, marshes, damp meadows. June–July. West Eur. (except NL.) I. *A very poisonous plant. The tubers known as 'Dead Men's Fingers' are aptly named and the active principle 'oenanthetoxin', is not destroyed by drying or storage.* Pl. 84.

873. *O. fistulosa* L. WATER DROPWORT. A glaucous, little-branched perenn. 30–60 cm., with soft hollow stems, somewhat constricted at the nodes, and with similar-looking hollow lv. stalks bearing a few scattered pairs of linear-lance-shaped leaflets. Primary umbels of 2–4 stout rays, the secondary umbels dense, flat-topped, *c.* 1 cm. across, becoming spherical in fr.; fls. white, *c.* 3 mm. across. Fr. angular, 3–4 mm., styles longer, spreading. △ Marshes, watersides, ditches. June–Sept. Most of Eur. (except IS.N.SF.TR.). Pl. 84.

*O. lachenalii C. C. Gmelin PARSLEY WATER DROPWORT. Western and much of Central Europe.

*O. pimpinelloides L. West and Southern Europe.

*O. aquatica (L.) Poiret (*O. phellandrium* Lam.) FINE-LEAVED WATER DROPWORT. Throughout Europe.

AETHUSA   Fr. ovoid and with very broad ribs; resin canals solitary in the furrows. Umbels compound; bracts 1 or absent; bracteoles 1–5; calyx teeth small or absent. 1 sp.

874. *A. cynapium* L. FOOL'S PARSLEY. A hairless, glaucous-stemmed ann. 5–120 cm., distin-

guished by its long, slender, unequal, reflexed bracteoles borne on the outer sides of the secondary umbels. Umbels 2–6 cm. across, with 10–20 primary rays; fls. white, *c.* 2 mm.; bracts usually absent. Lvs. rather dark green, twice-pinnate, with deeply lobed or toothed oval segments; lv. stalk broad, sheathing. Fr. ovoid, 3–4 mm., with conspicuous thick ribs. △ Cultivated ground, waste places, fields. June–Oct. Most of Eur. (except P.IS.AL.GR. TR.).

FOENICULUM | Fennel Fls. yellow; petals with a blunt incurved point; calyx absent. Fr. with rounded carpels; ribs stout. Umbels compound; bracts and bracteoles few or absent. 1 sp.

875. *F. vulgare* Miller (*F. officinale* All.) FENNEL. A hairless, very strong-smelling, somewhat glaucous perenn. ½–1½ m. with much-divided feathery lvs. and umbels of yellow fls. 4–8 cm. across. Primary rays 10–30, glaucous; fls. 1–2 mm.; bracts and bracteoles usually absent. Lvs. rather dark green, with numerous, narrow thread-like, ultimate lobes 1–5 cm. long. Fr. 4–6 mm., ovoid, with broad ribs. Var. *azoricum* (Miller) Thell. is cultivated for its swollen lv. bases. △ Cultivated ground, waste places, by the sea. July–Sept. Med. Eur. P.GB.BG.: introd. elsewhere in Eur. *The aromatic leaves have been used for flavouring since classical times. The fruits have a characteristic aniseed flavour and are used in baking and for flavouring cordials and liqueurs.*

ANETHUM Fls. yellow; petals equal, entire. Fr. ellipsoid, strongly compressed, with swollen lateral ribs and narrow slender dorsal ribs. 1 sp.

876. *A. graveolens* L. DILL, FALSE FENNEL. A rather dark green, feathery-leaved, aromatic, yellow-flowered erect ann. 20–100 cm., like 875, but distinguished by its strongly flattened fr. with a thickness of less than half its width. Umbels *c.* 15 cm. across, with primary rays 15–40; petals yellow, entire, more or less rounded; bracts and bracteoles absent. Lvs. with linear ultimate lobes. △ Fields, waste places. Apr.–July. Native of Asia: introd. Med. Eur. (except TR.) P.GB.B.NL.CH.A.CS.H.BG.R.SU.

SILAUM Petals oval, yellowish; calyx teeth minute. Fr. with rounded carpels, and prominent slender ribs with the lateral pair winged; resin canals numerous, irregular, and inconspicuous. 2 sps.

877. *S. silaus* (L.) Schinz & Thell. (*S. flavescens* (Bernh.) Hayek) PEPPER SAXIFRAGE. An erect, branched, hairless perenn. 30–100 cm., with long-stalked yellowish umbels 2–6 cm. across. Primary rays 5–15, rather unequal; fls. yellowish, 1½ mm. across; bracts absent or few; bracteoles numerous, linear-lance-shaped, margin papery. Lvs. 2–3 times pinnate, segments small ½–2 cm., linear-lance-shaped, simple or lobed, with very finely toothed margin; upper lvs. few and much reduced. Fr. 4–4½ mm., oblong-ovoid, with slender ribs. △ Meadows, heaths, roadsides. June–Aug. Much of Eur. (except North Eur. P.IRL.GR.TR. BG.) S. Page 287.

MEUM Fr. with rounded carpels with slender ribs; resin canals 3–5 in each furrow. Umbels compound; bracts few or absent; bracteoles 5–8; calyx absent; petals acute. 1 sp.

878. *M. athamanticum* Jacq. SPIGNEL, BALDMONEY, MEU. A very aromatic mountain perenn. 20–50 cm., with white or purplish umbels and feathery lvs. several times dissected into numerous, slender thread-like lobes. Umbels 2–6 cm. across; primary rays 6–15, very unequal; bracts absent or several; bracteoles 3–8, thread-like. Lvs. mostly basal, ultimate

lobes more or less whorled, *c.* ½ cm. long; base of stem covered with the fibrous remains of old lvs. Fr. ovoid 6–10 mm.; ribs very prominent. △ Alpine meadows. June–Aug. Much of Eur. (except North Eur. P.IRL.NL.H.R.GR.TR.) N. Page 287.

CONIUM | Hemlock Fr. broadly ovoid to rounded with 5 prominent ribs; resin canals absent. Umbels compound; bracts and bracteoles few. Calyx absent. 1 sp.

879. *C. maculatum* L. HEMLOCK. A tall, erect, branched bienn. ½–2½ m., smelling strongly of mice and distinguished by its smooth, furrowed, glaucous-green stems which are spotted with purple along much of their length. Umbels 2–5 cm. across; fls. white, 2 mm. across; primary rays 10–20; bracts few, reflexed; bracteoles smaller, usually 3, on one side only. Lvs. large, 2–3 times pinnate, ultimate lobes 1–2 cm., oblong-lance-shaped, deeply and coarsely toothed. Fr. *c.* 3 mm. rounded, with wavy ribs. △ Hedgerows, waysides, waste places. June–Aug. All Eur. (except IS.). *A very poisonous plant which has been well known and widely used since classical times. The active principle is the alkaloid coniine; the seeds are the most poisonous part of the plant. The whole plant can be dangerous to livestock when mixed with fresh foliage, but it loses its poisonous action on drying.* Pl. 85.

PLEUROSPERMUM Fr. ovoid-oblong, with 5 conspicuous winged ribs, the wings warty and wavy-margined. Petals white, rounded. 2 sps.

880. *P. austriacum* (L.) Hoffm. A robust perenn. ½–2 m. with a very stout furrowed stem, branches which are often whorled, and large white umbels 15–30 cm. across. Primary rays 20–40; bracts 8–10 reflexed, unequal, the largest pinnately lobed; bracteoles 8–10, ciliate. Lvs. large, 3 times pinnate, ultimate lobes 4–10 cm., oblong-wedge-shaped and deeply toothed. Fr. yellow, ovoid 1 cm.; ribs wavy-margined. △ In mountains: valleys, damp rocks, meadows. June–Sept. Cent. Eur. F.S.I.YU.BG.R.SU. Page 280.

BUPLEURUM Lvs. undivided. Fls. yellow; petals entire, acute. Fr. ovoid or oblong, usually with prominent ribs. 39 sps.

Anns.

881. *B. rotundifolium* L. HARE'S-EAR, THOROW-WAX. A glaucous ann. 15–75 cm., readily distinguished by its broadly elliptic upper lvs. which encircle and are fused round the stem. Fls. yellow, in compound umbels 1–3 cm. across; primary rays 3–10, bracts absent; bracteoles 5–6, yellowish-green, broadly oval and fine-pointed, shortly fused at their base and longer than the secondary umbels. Lvs. entire 2–5 cm., the lower lvs. stalked. Fr. oblong 3–3½ mm., blackish, smooth with slender ribs; fr. cluster surrounded by the paler, erect bracteoles. △ Cultivated ground, waste places, waysides. June–Aug. Much of Eur. (except North Eur. P.IRL.): introd. GB.NL.

*B. tenuissimum L. Widespread in Central and Southern Europe.

Herbaceous perenns.

882. *B. longifolium* L. A stout, erect, pale green, little-branched perenn. ½–1 m., with oval or oblong clasping upper lvs. and yellow umbels 3–4 cm. across. Primary rays 5–12; bracts 2–4, oval or rounded; bracteoles 5–8, yellowish, shortly fused at their base and equalling or longer than the secondary umbels. Lvs. to 10 cm., the upper with rounded clasping lobes, the lower elliptic and stalked. Fr. 4–5½ mm., dark brown or black; ribs slender, very prominent. △ In mountains: open woods, pastures, rocky places. June–Aug. Cent., South-East Eur. (except GR.TR.) F.SU.

883. *B. falcatum* L. SICKLE HARE'S-EAR. A very variable perenn. or rarely ann. to 1 m., with

narrow oval, usually stalked basal lvs. and linear-lance-shaped, often curved and sickle-like upper lvs. which half-clasp the stem. Umbels yellow, 1–4 cm. across; primary rays 3–15 slender; bracts 2–5, very unequal, lance-shaped, long-pointed; bracteoles 5, linear-lance-shaped, shorter than the secondary umbels. Fr. oblong 3–6 mm.; ribs very slender, winged. △ Grassy places, fallow, woods. July–Oct. Much of Eur. (except North Eur. P.IRL.NL.TR.).

B. stellatum L. STARRY HARE'S-EAR. A glaucous alpine perenn. readily distinguished by the bracteoles which are fused together to form a shallow-lobed, often yellowish, cup-like involucre, longer than the secondary umbels. Primary rays 3–6; bracts 2–4, large, leaf-like. Basal lvs. narrow lance-shaped, arising from a rootstock covered with old lv. bases. The Alps and Corsica. Pl. 84.

Woody perenns.

884. *B. fruticosum* L. SHRUBBY HARE'S-EAR. A bushy, glaucous evergreen shrub to 2½ m. with narrow, stalkless, shining lvs., slender purplish twigs and yellow umbels 7–10 cm. across. Primary rays 5–25 stout; bracts 5–6, deflexed; bracteoles 5–7, shorter than the fl. stalks, and like the bracts soon falling. Lvs. 5–8 cm., oblong-elliptic, pinnately veined. Fr. oblong 7–8 mm., ribs narrowly winged. △ Walls, rocks, stony places. June–Sept. P.E. F.I.GR.

APIUM Fr. ovoid or oblong-elliptic and laterally compressed, with 5 usually equal ribs. Calyx teeth absent; petals white, entire acute; umbels often leaf-opposed. 5 sps.

885. *A. graveolens* L. WILD CELERY. A leafy, much-branched, erect bienn. to 1 m., smelling strongly of celery, with deeply grooved stems and rather lax greenish-white umbels 3–5 cm. across. Primary rays unequal 6–12; fls. *c.* ½ mm. across; bracts and bracteoles absent. Lvs. all shining, the basal lvs. once-pinnate, with three-lobed leaflets ½–3 cm., the lower leaflets stalked, the upper stalkless, three-lobed. Fr. ovoid 1–2 mm. △ Salt marshes, salt-rich ground. July–Sept. Much of Eur. (except North Eur. TR.) DK.S.: introd. N.SF.CH.A.CS.H. *Cultivated for its swollen leaf-stalks as celery and for its swollen stems as celeriac.*

886. *A. nodiflorum* (L.) Lag. FOOL'S WATERCRESS. A weak-stemmed, spreading and rooting perenn. of wet places, with small, almost stalkless, greenish-white umbels much shorter than the lvs. Umbels arising opposite the lvs.; primary rays 2–4; fls. *c.* ½ mm.; bracts usually absent; bracteoles *c.* 5. Lvs. bright shining green, once-pinnate with 4–6 pairs of oval-lance-shaped and toothed leaflets 1–3½ cm. Fr. *c.* 2 mm., ovoid. Fl. stem to 1 m. △ Streams, ditches, marshes, lakes. June–Sept. West, South-East Eur. D.CH.CS.I. *The leaves are not edible. The plant often grows with watercress and the leaves are similar in appearance, but the two plants can easily be distinguished by their flowers and fruits.* Pl. 85.

PETROSELINUM Fr. ovoid or rounded and laterally compressed, with 5 slender ribs; resin canals solitary in the furrows. Calyx teeth absent; petals white or yellowish, scarcely notched; bracts and bracteoles several. 2 sps.

887. *P. crispum* (Miller) A. W. Hill (*P. hortense* auct.) PARSLEY. An erect, hairless perenn. ½–1 m., with a strong and distinctive aromatic smell when crushed, and with long-stemmed, flat-topped, yellowish umbels 2–5 cm. across. Fls. 2 mm.; primary rays 8–12; bracts 1–3, entire or three-lobed, white-margined; bracteoles 5–8, often white-margined. Lower lvs. 3 times pinnate, ultimate lobes shining, broadly wedge-shaped, lobed and often crisped in cultivated forms; upper lvs. three-lobed, lobes lance-shaped. Fr. 2½–3 mm., ovoid.

△ Rocks, walls, waste places. June–Sept. Origin unknown; naturalized throughout Eur. (except NL.IS.SF.TR.). *Widely cultivated for flavouring and garnishing.*

*P. segetum (L.) Koch CORN CARAWAY. Western Europe.

*Sison amomum L. STONE PARSLEY. Western and Southern Europe.

*Cicuta virosa L. COWBANE. Northern and Central Europe.

AMMI Bracts deeply lobed; bracteoles numerous. Fr. oblong-ovoid and slightly laterally compressed, with slender ribs; resin canals solitary in the furrows. Calyx teeth absent; petals white or yellowish, unequally two-lobed. 5 sps.

888. *A. majus* L. FALSE BISHOP'S WEED. A rather glaucous, erect, hairless ann. 20–100 cm., readily distinguished by the very long bracts which are divided into long, linear, leafy segments. Umbels white, 6–10 cm. across; primary rays numerous; bracts many; bracteoles linear long-pointed, about as long as the fl. stalks. Lvs. 1–2 times pinnate, the basal lvs. with oblong-oval, toothed segments 1–3 cm., the uppermost lvs. with linear segments, all segments with a fine cartilaginous tip; stem grooved. Fr. *c.* 2 mm. oblong-ovoid, with prominent ribs, borne on slender rays in spreading umbels. △ Fields, sandy places. July–Sept. Med. Eur. P.SU. Page 287.

A. visnaga (L.) Lam. Like 888, but distinguished by the very numerous primary rays, to *c.* 150, which become thickened, erect, and closely clustered together in fr. Lv. segments all linear entire; stem robust, grooved. Mediterranean region and Portugal; naturalized further north.

FALCARIA Fr. with low, blunt ribs and solitary, slender resin canals in the furrows. Calyx teeth present; petals nearly equal, notched. Bracts and bracteoles several, very slender, entire. 1 sp.

889. *F. vulgaris* Bernh. LONGLEAF. A glaucous, hairless, much-branched bienn. or perenn. 30–80 cm., readily distinguished by its lvs. which are ternate or twice-ternate with very long linear; somewhat sickle-shaped segments to 30 cm. by ½–1 cm., with regular finely toothed margin. Umbels white, *c.* 4 cm. across; primary rays slender 10–20; bracts and bracteoles narrow linear, numerous. Fr. oblong-ovoid with 5 blunt low ribs and with 5 spreading calyx teeth. △ Fields, waysides, waste ground. July–Sept. Much of Eur. (except North Eur. P.IRL.AL.): introd. GB.B.NL.DK.S. Page 287.

CARUM Fr. ovoid to oblong with 5 slender, usually acute ribs; resin canals broad, solitary in the furrows. Calyx teeth minute; petals white. Root swollen spindle-shaped or fibrous. 5 sps.

890. *C. carvi* L. CARAWAY. An erect, hairless, much-branched bienn. 25–60 cm., with white fls. 2–3 mm. in umbels 2–4 cm. across. Primary rays very unequal 5–12; bracts and bracteoles 1, bristle-like or absent. Lvs. twice-pinnate with segments further deeply cut into narrow lance-shaped or linear lobes; root spindle-shaped. Fr. 3–4 mm., oblong with low ribs, strong-smelling when crushed. △ Meadows, woods in mountains; often naturalized from cultivation. May–July. Most of Eur. (except P.GR.TR.). *Often cultivated as a pot-herb. The seeds are used for flavouring in cooking, bakeries, cheese-making and to flavour the liqueur Kümmel; also used in herbal remedies.*

*C. verticillatum (L.) Koch WHORLED CARAWAY. Western Europe.

1. *Silaum silaus* 877
2. *Falcaria vulgaris* 889
3. *Ammi majus* 888
4. *Meum athamanticum* 878

SELINUM Fr. oblong to broadly ovoid and compressed dorsally, with winged ribs, the marginal wings distinctly wider than the dorsal wings. Calyx teeth absent; petals white, deeply notched; bracts few or absent; bracteoles several. 2 sps.

891. *S. carvifolia* (L.) L. An almost hairless, erect, little-branched perenn. smelling of parsley when crushed, with a solid, sharply angled stem 30–100 cm., and with dense, white, long-stalked umbels 3–7 cm. across. Primary rays 5–33; fls. 2 mm. across; bracts few or absent; bracteoles several, linear-lance-shaped, as long as or longer than the fl. stalks. Lvs. 2–3 times pinnate, ultimate lobes small, 3–10 mm., linear to oval, entire or sometimes lobed, finely toothed. Fr. 3–4 mm., ovoid, with winged ribs. △ Damp meadows and woods. July–Sept. Much of Eur. (except IRL.IS.AL.GR.TR.).

LIGUSTICUM Fls. white or greenish-white. Fr. ovoid, not compressed, with acute or winged ribs; resin canals many. Calyx teeth small or absent; petals notched. Bracts and bracteoles many, few, or absent. 7 sps.

892. *L. scoticum* L. NORTHERN LOVAGE. A glossy, bright green, leafy perenn. 15–90 cm., smelling of celery when crushed, with ribbed stems, broad oval leaflets, and greenish-white umbels 4–6 cm. across. Primary rays 8–20; fls. 2 mm. across; bracts and bracteoles linear, 1–7. Lvs. twice-ternate with toothed leaflets 3–5 cm. Fr. 4–8 mm., oblong, ribs acute. △ Cliffs by the sea. July. North Eur. (except SF.) IRL.GB.SU. *Sometimes used as a pot-herb in Northern Europe*. Pl. 85.

893. *L. mutellinoides* (Crantz) Vill. SMALL ALPINE LOVAGE. An alpine perenn. with short, unbranched, almost leafless stems to 30 cm. arising from a thick rootstock, and dense, almost globular, white or greenish umbels becoming reddish. Primary rays short, finely hairy, 8–20; bracts 5–10, white-bordered and often divided into narrow lobes, at least as long as the rays; bracteoles similar. Lvs. 2–3 times pinnate, ultimate lobes 2–5 mm., linear to narrow-lance-shaped. Fr. 3–5 mm., ellipsoid, blackish; ribs equal. △ In mountains: grassy places, screes. July–Aug. Cent. Eur. (except H.) F.I.YU.R.SU.

L. mutellina (L.) Crantz ALPINE LOVAGE. Like 893, but rootstock with numerous coarse fibres and stems with 1 or 2 branches subtended by small lvs. Umbels usually red or purple; primary rays 7–15; bracts absent or few and then soon falling. Mountains of Central and South-Eastern Europe.

ANGELICA Fls. greenish, white or pinkish. Fr. strongly dorsally compressed and with broadly winged lateral ribs and narrower winged dorsal ribs. Calyx teeth minute or absent; petals long-pointed, tip incurved. Bracts few or absent; bracteoles many. 8 sps.

894. *A. sylvestris* L. WILD ANGELICA, JACK-JUMP-ABOUT. A robust erect bienn. or perenn., with purplish-bloomed stems ½–2 m. and white or pinkish umbels 3–15 cm. across. Primary rays 20–30, finely hairy; fls. 2 mm., petals more or less erect; bracts few or absent; bracteoles linear-acute, as long as the fl. stalks. Basal lvs. large, 30–60 cm., 2–3 times pinnate, ultimate lobes 2–8 cm. oblong-oval, toothed; lv. stalk deeply channelled above and enlarged into a rounded sheath at the base; uppermost lvs. reduced to inflated sheaths enclosing the fl. buds. Fr. 5 mm. oval, with papery-winged ribs. △ Marshes, fens, damp meadows, cliffs. July–Sept. All Eur.

895. *A. archangelica* L. ANGELICA. Like 894, but a more robust aromatic ann. or perenn. with green stems 2–3 m. and green or greenish-white fls. in rather globular umbels. Primary rays 20–40. Lv. stalks hollow, rounded in section, not grooved; lv. lobes oval-lance-shaped, stalkless and somewhat decurrent. Fr. 5–6 mm., with ribs which become

corky. △ Damp meadows, riversides. July–Aug. North Eur. NL.D.CS.I.SU.: introd. GB.F. B.CH.A.H.YU.BG.R. *Cultivated for its leaf-stalks which are candied and used for flavouring; also used in the preparation of liqueurs such as Chartreuse. The roots are aromatic and edible and produce an oil used in confectionery, perfumery, and medicine.* Pl. 85.

LEVISTICUM | **Lovage**   Petals greenish-yellow, elliptic; calyx absent. Fr. flattened with broadly winged lateral ribs and narrowly winged dorsal ribs. 1 sp.

896. *L. officinale* Koch LOVAGE. A very strong-smelling, shining, robust, erect perenn. 1–2 m., with greenish-yellow umbels 8–10 cm. across. Primary rays 8–15; petals rounded, not notched; bracts and bracteoles numerous, lance-shaped with papery margin. Lower lvs. 2–3 times pinnate, leaflets large oval, with wedge-shaped base and toothed or lobed towards the apex. Fr. 5–7 mm., ovoid, conspicuously winged. △ Hedges, meadows in mountains. June–Aug. Doubtfully native of Europe; occurring in most of Eur. *Cultivated in the past as a medicinal plant; also used for flavouring and as a salad plant.*

FERULA   Petals yellow, oblong; calyx absent. Terminal umbel of female fls. only; bracts absent; bracteoles few or absent. Fr. dorsally compressed with very slender, dorsal ribs and somewhat swollen, winged lateral ribs, the lateral ribs of each unit closely pressed together; resin canals numerous. 8 sps.

897. *F. communis* L. GIANT FENNEL. A very robust, erect perenn. 1–5 m., with stout ribbed stems bearing large, dense, yellow umbels, and with large, dark green, feathery lvs. many times divided into very narrow, thread-like linear lobes. Primary rays 20–40; terminal umbel with female fls. only and shorter than the smaller lateral umbels; bracts absent; bracteoles linear, soon falling. Upper lvs. stalkless with a large swollen sheath enclosing the young fl. heads; lower lvs. ½ m. or more, stalked. Fr. 1½–2 cm., oval or elliptic. △ Dry hills, stony slopes. June–Aug. Med. Eur. P. *The pith of the dry stems has been used as tinder since classical times. The source of gum ammonia.* Pl. 85.

FERULAGO   Like *Ferula* but bracts and bracteoles well developed and usually conspicuous. 9 sps.

898. *F. campestris* (Besser) Grec. An erect perenn. with a strongly grooved, angular stem ½–2 m., yellow umbels and dark green lvs. to 60 cm., many times dissected into slender linear lobes. Terminal umbel stalkless with 12–20 rays, the lateral umbels smaller and long-stalked; bracts and bracteoles lance-shaped, conspicuous. Fr. 1–2 cm. by 7–9 mm. with lateral wings well developed and 1½–2 mm. broad. △ Rocky places. July–Aug. F.I.YU.AL.BG.R.SU.

PEUCEDANUM   Fls. white, yellow, or rarely pinkish; calyx small or absent; petals with long incurved tips. Fr. strongly compressed dorsally; lateral ribs with a narrow or broad wing, dorsal ribs slender, all ribs equidistant; resin canals 1–3. 29 sps.

899. *P. ostruthium* (L.) Koch MASTERWORT. An erect, more or less hairy perenn. to 1 m., with grooved stems and white or pinkish umbels 5–10 cm. across. Primary rays 20–60; bracts absent; bracteoles few, bristle-like. Lower lvs. to 30 cm., 1–2 times ternate with broadly oval, lobed and coarsely toothed segments 5–10 cm. by 2–7 cm., usually somewhat hairy beneath; upper lvs. smaller, with inflated lv. stalk. Fr. 4–5 mm., rounded, with broadly winged lateral ribs. △ In mountains: meadows, woods, rocks, streamsides. June–Aug. Cent. Eur. (except H.) E.F.I.YU.R.SU.: introd. GB.B.DK.N.S. *Formerly cultivated as a*

herbal plant; the root is strongly aromatic and bitter. Used in veterinary medicine. Page 280.

P. oreoselinum (L.) Moench Like 899, but bracts present, reflexed. Umbels white or pinkish; rays 15–30. Lvs. with ultimate segments 1–3 cm., oval-wedge-shaped, deeply three-lobed; first division of the lvs. borne at right angles to the axis or reflexed and appearing as if broken. Fr. 5–8 mm., with broad thick white lateral wings. Much of Central and Southern Europe.

*P. palustre (L.) Moench MILK PARSLEY. Widespread in much of Europe.

*P. officinale L. HOG'S FENNEL. Central and Southern Europe.

PASTINACA | Parsnip   Fls. yellow; calyx small or absent; petals with a cut-off incurved tip. Fr. ovoid and strongly compressed dorsally, with broadly winged lateral ribs and slender dorsal ribs; resin canals solitary in the furrows. 4 sps.

900. *P. sativa* L. WILD PARSNIP. An erect, rather variable, strong-smelling hairy bienn. with furrowed and angled stems $\frac{1}{2}$–$1\frac{1}{2}$ m. and yellow umbels 3–10 cm. across. Primary rays 5–15; bracts and bracteoles absent or 1–2 and soon falling. Lvs. pale yellowish-green, once-pinnate with broadly oval to lance-shaped, lobed and toothed segments *c.* 5 cm. Fr. 5–8 mm., ovoid and broadly winged, with slender dorsal ribs and conspicuous resin canals. △ Grassy places, waysides, fallow land. June–Sept. All Eur. (except IS.). *Cultivated forms with sweet, fleshy, edible roots have been grown as a vegetable since Roman times; wild forms have slender, acid roots and are inedible. Sometimes grown as a forage crop for cattle and pigs.* Pl. 86.

HERACLEUM | Hogweed   Fr. strongly compressed, with lateral ribs with thickened broad wings and slender dorsal ribs; resin canals club-shaped, solitary in the furrows. Petals white or pinkish, often very unequal, notched; calyx teeth small unequal. 9 sps.

901. *H. sphondylium* L. COW PARSNIP, HOGWEED, KECK, A stout, erect, rough bienn. with ribbed, hairy stems $\frac{1}{2}$–2 m. and with large white or pinkish umbels 5–15 cm. across. Primary rays stout, 7–25; fls. 5–10 mm., the outer with very unequal petals; bracts absent; bracteoles several, bristle-like, reflexed. Lvs. large 15–30 cm., once-pinnate with 5–9 oval to lance-shaped, often pinnately lobed or toothed, stalked segments 5–15 cm. long. Fr. 7–11 mm., flattened, with broad lateral wings; resin canals conspicuous. △ Meadows, hedgerows, waysides, woods. June–Sept. All Eur. (except IS.).

902. *H. mantegazzianum* Sommier & Levier GIANT HOGWEED. An extremely robust perenn. to 4 m., with a very stout, red-spotted stem up to 10 cm. thick, lvs. to 1 m. long and white umbels to $\frac{1}{2}$ m. across. Primary rays numerous, to 10 cm. long; fls. *c.* $\frac{1}{2}$ cm., but outer petals of outer fls. to 12 mm. long. Lvs. pinnate with large oval, deeply pinnately lobed and toothed segments. Fr. ovoid to 14 mm., hairy; resin canals conspicuous. △ Native of the Caucasus; often grown for ornament and self-seeding near habitations. June–Sept. Introd. IRL.GB.F.NL.DK.N.S.D.CH.A.CS.H.I.SU. Pl. 86.

TORDYLIUM   Fr. orbicular and strongly compressed, with lateral ribs much-thickened and dorsal ribs inconspicuous; resin canals usually 1 in each furrow. Petals white or pinkish, unequal; calyx prominent; bracts and bracteoles several or absent. 5 sps.

903. *T. maximum* L. HARTWORT. A rough-haired, branched ann. or bienn. to 130 cm., with angled stems with reflexed bristly hairs, and dense umbels of white fls. in which the 2–3

outer petals of the outer fls. are larger than the remainder. Primary rays 5–15, rough-haired; bracts and bracteoles numerous, linear long-pointed. Lvs. pinnate, segments oval to linear-lance-shaped and toothed or lobed, the terminal lobe of the upper lvs. longest. Fr. bristly-haired, 5–8 mm., orbicular flattened, with a thickened whitish margin. △ Cultivated ground, hedges, waysides, waste land. June–Aug. Med., South-East Eur. P.GB.A.CS.H.SU.: introd. B.D.CH.

904. *T. apulum* L. Like 903, but a softly and sparsely hairy ann. to ½ m., with distinctive frs. which have a thick white border which appears crinkled on the inner margin. Umbels white, lax; outer petal of outer fls. very deeply lobed and very much longer than the rest; primary rays 3–8; bracts and bracteoles linear long-pointed, with spreading hairs. Lvs. softly hairy, pinnate, the lower with oval and toothed segments, the uppermost with linear segments. Fr. orbicular, flattened 5–8 mm. △ Cultivated ground, waste land, waysides. Apr.–June. Med. Eur. Pl. 86.

LASERPITIUM Fr. scarcely compressed and with 4 conspicuous broad wings, the lateral wings larger. Petals white, pinkish, yellow, or greenish, obcordate; sepals conspicuous. 13 sps.

905. *L. siler* L. (*Siler montanum* Crantz) SERMOUNTAIN. A rather robust, glaucous, hairless perenn. ½–1½ m., with large white or pink umbels 6–10 cm. across and very large basal lvs. Primary rays 25–50; fls. *c.* 4 mm. across; bracts and bracteoles numerous, spreading, linear-lance-shaped, white-margined. Lvs. to 1 m., 2–4 times pinnate, segments 1½–7 cm., stalked, entire oblong to linear-lance-shaped, rather leathery, with narrow cartilaginous margin; upper lvs. with swollen sheaths. Fr. shining, sweet-smelling, 6–12 mm. oblong, with 4 conspicuous wings which are narrower than the fr. △ In mountains: woods, rocky places. July–Aug. E.F.D.CH.A.CS.I.YU.AL.GR.BG.

906. *L. latifolium* L. BROAD-LEAVED SERMOUNTAIN. Like 905, but lvs. twice-pinnate and segments oval-heart-shaped, toothed or lobed. Umbels white, primary rays 20–50; bracts numerous, reflexed, linear long-pointed; bracteoles few. Fr. ovoid, 5–10 mm., with broad, usually undulate wings, broader than the fr. A robust, nearly hairless perenn. ½–2½ m. △ In mountains: woods, rocky places. June–Aug. Much of Eur. (except P.IRL.GB.NL.IS. AL.GR.): introd. B.

TORILIS | **Hedge-Parsley** Fr. ovoid or nearly spherical, with slender ribs and spines or swellings in the grooves between the ribs. Petals white or pinkish; calyx teeth small, persisting. 6 sps.

907. *T. japonica* (Houtt.) DC. (*T. anthriscus* (L.) C. C. Gmelin) UPRIGHT HEDGE-PARSLEY. An erect, branched, rough-haired ann. to 125 cm., distinguished by its frs. which are densely covered with hooked and curved spines. Umbels long-stalked, pinkish or whitish-purple, 1½–4 cm. across; primary rays 5–12, hairy; bracts and bracteoles 4–12, linear, unequal, hairy. Lvs. rough with adpressed hairs, 1–3 times pinnate, segments oval to lance-shaped, deeply lobed and toothed. Fr. 3–4 mm.; styles hairless, recurved in fr. △ Woods, hedges. May–Aug. All Eur. (except IS.).

T. arvensis (Hudson) Link SPREADING HEDGE-PARSLEY. Like 907, but with fewer, usually 3–5 primary rays, and bracts 1 or absent; bracteoles numerous. Umbels white or pinkish, 1–2½ cm. across. Fr. 3–5 mm., with spines curved, not hooked, but thick-tipped; styles hairy, spreading in fr. A very variable sp. Widespread in Europe, except in the north.

*T. nodosa (L.) Gaertner KNOTTED HEDGE-PARSLEY. Most of Europe, except the north.

TURGENIA Like *Torilis*, but fr. with 7 ribs on each side and each with 2–3 rows of spines. Petals obovate, unequal; calyx teeth conspicuous. 1 sp.

908. *T. latifolia* (L.) Hoffm. (*Caucalis l.*) GREATER BUR-PARSLEY. A rough, erect, branched ann. to ½ m., distinguished by its large burred frs. 6–10 mm. long, with 2–3 rows of long spreading spines on the ribs. Umbels pink or red, 4–6 cm. across; primary rays robust, 2–5, with upward-pointing hairs; bracts 3–5, broadly lance-shaped; bracteoles 5–7, oblong, with wide papery margin. Lvs. rough-haired, once-pinnate, segments oblong to lance-shaped, coarsely toothed or lobed. △ Cultivated places, waste ground. May–Aug. Much of Eur. (except North Eur. IRL.NL.PL.): introd. GB.

*Caucalis platycarpos L. SMALL BUR-PARSLEY. Central and Southern Europe.

ORLAYA Fr. oval or oblong, with 5 slender ribs and 4 secondary ribs covered with 1–2 rows of spines. Petals white or pink, obcordate, deeply notched, often very unequal; calyx teeth present. 3 sps.

909. *O. grandiflora* (L.) Hoffm. LARGE-FLOWERED ORLAYA. A nearly hairless ann. to ½ m., readily distinguished by the very large, deeply lobed outer petals of the outer fls., which are 7–8 times as long as the other petals. Umbels white; primary rays 5–12; bracts 5–8, lance-shaped, with white papery margin; bracteoles several. Lvs. 2–3 times pinnate, with linear-oblong segments, the upper lvs. stalkless, entire or deeply lobed. Fr. *c.* 8 mm. with 2–3 rows of whitish spines on the secondary ribs. △ Grassy places. June–Sept. Med., South-East Eur. B.D.CH.A.CS.SU.

DAUCUS | Carrot Fr. ovoid to oblong, with 4 stout spiny ribs and 5 slender ribs. Petals white, notched, often unequal; calyx teeth small or absent. 9 sps.

910. *D. carota* L. WILD CARROT. A hairy, erect bienn. ½–1 m., with white or pink umbels, which are usually purple at the centre, and 3–7 cm. across. Primary rays very numerous; bracts 7–13 conspicuous, deeply divided into narrow lobes with papery margin; bracteoles several, linear-pointed. Lvs. 3 times pinnate, with the segments pinnately lobed, ultimate lobes *c.* ½ cm. lance-shaped to oval. Fr. 2½–4 mm., oblong-ovoid, with spiny ribs alternating with hairy ribs. A very variable sp.; subsp. *sativa* (Hoffm.) Arcangeli is the edible carrot with a fleshy tap-root. △ Banks, fields, waysides. May–Oct. All Eur. *Cultivated since classical times as an edible root; also used in herbal remedies.* Pl. 87.

DIAPENSIACEAE | Diapensia Family

Perenn. herbs or shrublets with simple lvs. Fls. regular, often solitary. Calyx fused below, deeply five-lobed; corolla similar; stamens 5, fertile, fused to corolla, and usually with 5 infertile stamens present. Ovary superior, three-celled, deeply three-lobed; stigma three-lobed. Fr. a capsule. 1 sp. in Europe.

DIAPENSIA

911. *D. lapponica* L. A low cushion-like evergreen shrublet to 5 cm., with dense rosettes of lvs. and short-stalked, solitary white fls. 1–1½ cm. across. Petals spreading, obovate, to 1 cm.; sepals *c.* ½ cm.; infertile stamens absent. Lvs. ½–1 cm. spathulate, leathery, numerous, densely crowded on stems and forming terminal rosettes. △ Arctic and northern regions. May–June. GB.IS.N.S.SF.SU. Pl. 87.

PYROLACEAE | Wintergreen Family

Evergreen perenns. usually with creeping underground stems, or fleshy saprophytes with colourless scale-like lvs. Calyx fused, five-lobed; petals free, 5; stamens 10, not fused to the corolla; anthers opening by pores or slits. Ovary superior, five-celled; style 1, stigma club-shaped. Fr. a capsule.

1. Plants without green lvs.	*Monotropa*
1. Plants with green lvs.	
2. Fls. solitary.	*Moneses*
2. Fls. in terminal elongated clusters or umbels.	
3. Fls. in terminal elongated clusters.	
4. Fls. turned to one side, greenish-white.	*Orthilia*
4. Fls. spreading all round stem, white or pinkish.	*Pyrola*
3. Fls. in a terminal umbel of 3–6.	*Chimaphila*

PYROLA | **Wintergreen** Fls. several or numerous in a terminal spike-like cluster, not one-sided; each anther with a very short tube ending in a pore. Lvs. usually all basal in a rosette; stems with 1 or more scales. 4 sps.

Style shorter than petals and stamens

912. *P. minor* L. SMALL WINTERGREEN. A creeping perenn. with lax rosettes of oval light green lvs. and pinkish globular fls. in a rather dense oval cluster. Fls. *c.* 6 mm.; calyx lobes acute; style straight 1–2 mm., shorter than stamens and ovary. Lvs. 2½–4 cm.; lv. stalk shorter than the blade. Fl. stem leafless, 10–30 cm. △ Woods, thickets, marshes, alpine meadows. May–July. All Eur. (except P.TR.). Pl. 86.

Style much longer than petals and stamens

P. media Swartz INTERMEDIATE WINTERGREEN. Like 912, but fl. clusters elongate cylindrical and rather lax, and fls. *c.* 1 cm., globular, white tinged with pink, Style 5 mm., longer than the stamens and ovary, straight and projecting, and thickened below the stigma. Lvs. dark green, 3–5 cm. Northern regions and the mountains of most of Europe. Pl. 87.

913. *P. rotundifolia* L. ROUND-LEAVED WINTERGREEN. Like 912, but fls. pure white in a lax cluster, and petals spreading widely, thus fls. *c.* 12 mm. across, not globular. Styles long, curved down and outwards, longer than the stamens and ovary. Lvs. 2½–5 cm., rounded-oval, dark glossy green, long-stalked. Fl. stems leafless, 20–40 cm. △ Woods. June–Aug. Most of Eur. (except P.AL.GR.TR.).

ORTHILIA Like *Pyrola*, but fl. spikes one-sided; anthers without tubes. 1 sp.

914. *O. secunda* (L.) House (*Pyrola s.*) NODDING WINTERGREEN. A creeping perenn. with lax rosettes of oval light green lvs. and a dense, one-sided cluster of white fls. tinged with green. Fls. *c.* 5 mm., with petals not spreading; calyx teeth blunt; style *c.* 5 mm., longer than the petals and stamens. Lvs. 2–4 cm., finely toothed; lv. stalk shorter than the blade. Fl. stems with scales, 5–25 cm. △ Woods in mountains. June–Aug. Much of Eur. (except P.NL.TR.): introd. B. Pl. 86.

MONESES Like *Pyrola*, but fls. solitary; anthers with long tubes. 1 sp.

915. *M. uniflora* (L.) A. Gray (*Pyrola u.*) ONE-FLOWERED WINTERGREEN. A creeping rosette

perenn. with rounded light green lvs. and solitary, long-stemmed, often drooping, white fls. 1½–2½ cm. across. Fls. cup-shaped, sweet-scented; petals spreading; style straight, longer than the petals; stigma conspicuously five-lobed. Lv. blade 1–1½ cm., rounded-oval and finely toothed, lv. stalk shorter. Fl. stem usually with 1 scale, 5–20 cm. △ Damp woods, in mountains. May–July. Much of Eur. (except P.IRL.B.NL.IS.GR.TR.). Pl. 87.

CHIMAPHILA Fls. in umbels; style absent; lvs. oblong, in whorls. 1 sp.

916. *C. umbellata* (L.) Barton UMBELLATE WINTERGREEN. A creeping perenn. with strongly toothed, dark green leathery lvs. in whorls, and an erect, leafless stem 20–40 cm., with a terminal umbel of 3–6 pink fls. Fls. with spreading petals, long-stalked; stigma pressed to ovary, style absent. Lvs. oblong-wedge-shaped. △ Woods. June–July. North, Cent. Eur. F.I.R.SU.

MONOTROPA | **Bird's Nest** Saprophytic perenns. with scale-like, brown or yellowish lvs. Fls. in short clusters; sepals and petals 4–5; stamens twice this number. Ovary partially four- to five-celled or one-celled; fr. a capsule. 1 sp.

917. *M. hypopitys* L. (*Hypopitys multiflora* Scop.) YELLOW BIRD'S NEST. A yellowish or whitish, waxy-looking, saprophytic perenn. with a rather fleshy, erect stem 8–30 cm., bearing numerous scale-like lvs. and a terminal, one-sided cluster of yellowish-white fls. Fl. clusters at first drooping, erect in fr.; fls. tubular 1–1½ cm.; petals hairy or hairless within. Scale-like lvs. ½–1 cm. oval-oblong, pressed to stem. Fr. ovoid. △ Damp woods. June–Sept. Most of Eur. (except IS.AL.TR.). Pl. 86.

ERICACEAE | Heath Family

Usually small shrubs, but sometimes small trees, with simple, usually evergreen lvs. Fls. radially symmetrical; sepals 4–6; corolla usually fused into a tube and with 4–5 lobes; stamens usually twice as many as the sepals, not usually fused to the corolla; anthers with an apical pore. Ovary superior or inferior, usually four- to five-celled; style 1. Fr. a capsule or rarely a berry.

		Nos.
1. Lvs. tiny, slender, needle-like, 1–2 mm. broad.		
2. Fls. solitary, white.	*Cassiope*	[923]
2. Fls. numerous in dense or lax clusters, usually pink-purple.		
3. Stamens fused to corolla.	*Bruckenthalia*	[934]
3. Stamens not fused to corolla.		
4. Calyx green, shorter than corolla; lvs. whorled.	*Erica*	927–934
4. Calyx coloured like corolla and longer; lvs. opposite.	*Calluna*	926
1. Lvs. small or large, with an oval blade more than 3 mm. broad.		
5. Lvs. small, less than 3 cm. long; usually small shrubs or shrublets.		
6. Fr. a dry capsule.		
7. Lvs. opposite; creeping mountain shrublets.	*Loiseleuria*	921
7. Lvs. alternate.		
8. Petals free; lvs. rusty-haired beneath.	*Ledum*	918

8. Petals fused and corolla bell- or urn-shaped; lvs. not
   rusty beneath.
   9. Corolla with 4 lobes.   *Daboecia*   922
   9. Corolla with 5 lobes.
      10. Calyx and fl. stalks glandular-hairy.   *Phyllodoce*   [922]
      10. Calyx and fl. stalks hairless.   *Andromeda*   923
6. Fr. a fleshy berry or drupe.
   11. Ovary inferior.   *Vaccinium*   935–937
   11. Ovary superior.   *Arctostaphylos*   925
5. Lvs. large, 3–12 cm. long; usually robust shrubs or small
   trees.
   12. Fls. funnel-shaped with 5 spreading petals.   *Rhododendron*   919, 920
   12. Fls. bell- or urn-shaped; fr. warty, red.   *Arbutus*   924

LEDUM   Fls. in terminal, umbel-like clusters. Calyx lobes short; petals 5, not fused; stamens usually 10. Evergreen shrubs. 1 sp.

918. *L. palustre* L. LEDUM. A dense, branched evergreen shrub to 1 m., with rusty-haired stems and undersides of lvs. and with many cream-coloured fls. in dense, umbel-like clusters. Fls. 1–1½ cm. across; petals oval, spreading; fl. stalks slender, glandular. Lvs. 1–4½ cm. linear-oblong with inrolled margin, leathery, dark green above, rusty beneath. Fr. oblong *c.* ½ cm. △ Bogs. June–July. GB.N.S.SF.D.A.PL.CS.YU.R.SU.

RHODODENDRON   Fls. in terminal, umbel-like clusters. Corolla bell- or funnel-shaped with 5 spreading lobes, slightly two-lipped; stamens 5 or 10. Ovary superior; fr. a capsule. Evergreen shrubs. *c.* 6 sps.

919. *R. ferrugineum* L. ALPENROSE. A small, azalea-like evergreen shrub to 1 m., with elliptic lvs. rusty-haired beneath, and with terminal, umbel-like clusters of 3–8 bright pinkish-red, funnel-shaped fls. Fls. *c.* 2 cm. long; corolla lobes as long as the tube; calyx lobes tiny, blunt, hairless. Lvs. 2–4 cm. long, dark green, shining, and hairless above, margin inrolled; young branches hairless. △ Mountain pastures, rocks. June–Aug. E.F.D.CH.A.I.YU.AL.R.BG. *A poisonous plant used in the past in herbal remedies.* Pl. 87.

R. hirsutum L. HAIRY ALPENROSE. Like 919, but lvs. bright green on both sides, finely toothed and with long spreading hairs on the margin and spotted with russet glands beneath. Fls. bright pink; calyx lobes narrow acute. Twigs, fl. stalks and calyx rough-hairy. The Alps. Pl. 87.

920. *R. ponticum* L. RHODODENDRON. Large, dense evergreen shrub to 5 m. with dark, glossy, laurel-like lvs. and terminal, rounded clusters of dull purple fls. *c.* 12 cm. across. Corolla 5 cm. across, widely funnel-shaped with spreading lobes longer than the tube, paler in throat and with brown spots; stamens 10, prominent; style longer. Lvs. smooth, leathery, elliptic 6–12 cm., paler beneath. △ Woods, thickets; grown for ornament and for cover and often naturalized. May–June. P.E.BG.TR.; introd. West and Cent. Eur. *Poisonous to livestock; honey obtained from rhododendron flowers can be poisonous.* Pl. 88.

LOISELEURIA   Small, creeping evergreen shrublets with opposite lvs. Fls. in tiny umbel-like clusters; calyx deeply lobed; corolla widely bell-shaped; stamens 5. 1 sp.

921. *L. procumbens* (L.) Desv. CREEPING AZALEA. A lowly, creeping, much-branched ever-green shrublet spreading 10–30 cm., with tiny oval lvs. and tiny pink fls. 4–5 mm. across.

Fls. 2–5, short-stalked in terminal umbels; corolla with 5 spreading lobes. Lvs. numerous, c. ½ cm., shining, leathery, with inrolled margin. △ Arctic or alpine: meadows, rocks. June–July. Much of Eur. (except P.IRL.B.NL.DK.H.AL.GR.TR.BG.). Pl. 87.

DABOECIA   Fls. in lax, terminal, spike-like clusters; sepals and petals 4; stamens 8. 1 sp.

922. *D. cantabrica* (Hudson) C. Koch (*Menziesia polyfolia* Juss.) ST DABEOC'S HEATH. A heather-like, glandular-hairy evergreen undershrub to ½ m., with a lax, terminal, leafless cluster of 2–10 stalked, nodding, rosy-purple fls. Corolla flask-shaped, 8–12 mm. long, with short recurved teeth; calyx and lv. stalks glandular-hairy. Lvs. ½–1 cm., linear-elliptic, margin inrolled, dark shining green above, white-hairy beneath. Fr. oblong, glandular-hairy. △ Open woods, heaths. June–Oct. P.E.F.IRL. Pl. 88.

PHYLLODOCE   Like *Daboecia*, but sepals and petals 5; stamens 10. 1 sp.

P. caerulea (L.) Bab. (*Menziesia c.*) BLUE MOUNTAIN HEATH. Distinguished by its nodding, purple, flask-shaped fls. borne on slender, reddish, glandular stalks in a lax terminal cluster of 2–6 fls.; corolla 7–8 mm., five-toothed. A low, heath-like undershrub to c. 20 cm., with densely clustered, linear evergreen lvs. ½–1 cm., with a hairy furrow beneath. Northern Europe and Pyrenees. Pl. 88.

ANDROMEDA   Fls. long-stalked in terminal, few-flowered clusters. Calyx small, five-lobed; corolla flask-shaped, five-lobed; stamens 10 and anthers with awns. Fr. a capsule. 1 sp.

923. *A. polifolia* L. MARSH ANDROMEDA. A hairless evergreen undershrub to 40 cm. with creeping, underground stems and simple, erect, leafy stems bearing a terminal cluster of a few pink, nodding, long-stalked, flask-shaped fls. Fls. 2–8; corolla 5–7 mm., mouth strongly contracted; calyx c. 2 mm., hairless. Lvs. 1½–3½ cm., broadly linear acute, margin inrolled, dark green above and glaucous beneath. Fr. globular, glaucous. △ Bogs, wet heaths. May–June. Much of Eur. (except P.E.IS.H.YU.AL.GR.TR.BG.). *A poisonous plant; it gives a black dye*. Pl. 88.

CASSIOPE   Fls. solitary. Sepals and petals 4–6; stamens 8–12. Lvs. tiny, stalkless, overlapping. 2 sps.

C. hypnoides (L.) D. Don A tiny, creeping, mat-forming evergreen shrublet with numerous overlapping and spreading lvs. 3 mm., and terminal, short-stalked, solitary, pendulous white fls. Corolla rounded bell-shaped, c. 4 mm.; calyx deep pink. Arctic and northern regions.

ARBUTUS | Strawberry Tree   Evergreen trees or shrubs. Corolla flask-shaped. Fr. a rough warty berry. 2 sps.

924. *A. unedo* L. STRAWBERRY TREE. A small evergreen tree or tall shrub 1–10 m., with rough bark, dark green lvs. and terminal, drooping clusters of globular, cream-coloured fls. Clusters c. 5 cm., many-flowered; corolla c. 7 mm., contracted at mouth, lobes reflexed. Lvs. 4–10 cm., elliptic, dark shining green above, paler beneath, margin saw-toothed. Fr. distinctive, strawberry-like and red when ripe, 1½–2 cm., globular with a warty surface. △ Bushy places, rocky ground. Oct.–Jan. Med. Eur. IRL. *The fruits are edible and are used*

for wine-making, liqueurs and preserves; the branches and leaves are used for tanning. Pl. 90.

A. andrachne L. EASTERN STRAWBERRY TREE. Like 924, but lvs. grey-green, entire not toothed, except on sucker shoots; bark very smooth, red. Fls. white in erect clusters; fl. stalks glandular. Fr. *c.* 6 mm., golden-yellow, with a rough netted surface. Greece and Turkey.

ARCTOSTAPHYLOS   Evergreen undershrubs. Fls. few, in terminal clusters. Calyx 5; corolla flask-shaped, five-lobed; stamens 10 and anthers with long awns. Fr. a fleshy drupe. 2 sps.

925. *A. uva-ursi* (L.) Sprengel BEARBERRY. A prostrate, mat-forming, hairless undershrub spreading 1–2 m., with leathery evergreen lvs. and terminal clusters of globular whitish-pink fls. Fls. 5–12, short-stalked; corolla 4–6 mm., almost globular, strongly constricted at the mouth; calyx tiny. Lvs. numerous 1–2 cm., oval, dark green above, paler beneath, conspicuously veined. Fr. 6–8 mm., globular, shining red. △ Open woods, thickets, rocks, moors in mountains. May–July. All Eur. (except GR.TR.). *The leaves are used medicinally, and also for tanning and dyeing. The fruits are edible.* Pl. 88.

A. alpina (L.) Sprengel (*Arctous a.*) BLACK BEARBERRY. Like 925, but lvs. deciduous and fls. *c.* 4 mm., white with a green throat. Lvs. bright green, conspicuously veined, finely toothed. Fr. at first red, then blue-black. Northern regions or mountains of Central Europe. Pl. 88.

CALLUNA | Ling   Involucre calyx-like, green; calyx 4, coloured like petals and longer; corolla bell-shaped, deeply four-lobed; stamens 8 with anthers awned. Fr. a capsule. 1 sp.

926. *C. vulgaris* (L.) Hull LING, HEATHER. A stiff, twiggy, much-branched evergreen shrub to 1 m. with tiny, densely overlapping lvs. and rather slender spikes of pale purple fls. Fl. spikes 3–15 cm. long, often lax; fls. *c.* 4 mm.; involucre of 4 green calyx-like bracts; calyx petal-like, pale purple, lance-shaped, longer than the similar-coloured corolla. Lvs. four-ranked, linear 1–3 mm., with 2 short projections at their base; young twigs grey-haired. △ Heaths, moors, banks, open woods. July–Oct. All Eur. (except H.BG.AL.TR.). *An important plant in the economy of northern peoples; it is used for fuel, bedding, thatching, insulation, dyeing, and also as fodder for stock and a honey plant for bees. It is also used for making brushes, brooms, baskets, and rope.*

ERICA | Heath, Heather   Evergreen shrubs with tiny narrow whorled lvs. Fls. usually in terminal clusters; calyx 4, green and not petal-like; corolla cylindrical, flask-shaped or bell-shaped, with 4 short lobes; stamens 8 and anthers usually with awns. Fr. a dry splitting capsule. *c.* 17 sps.

Stamens included in corolla
(a) Tall shrubs 1 m. or more
927. *E. arborea* L. TREE HEATH. A tall, branched, feathery-looking shrub 1–4 m., with very numerous, tiny bright green lvs. and large, dense, pyramidal clusters of extremely numerous, very small white or pinkish, sweet-scented fls. Fls. long-stalked; corolla *c.* 3 mm., bell-shaped about as long as broad; anthers included. Lvs. 5 mm. narrow linear, grooved beneath, in whorls of 3–4; twigs densely white-hairy with hairs mostly branched. △ Bushy places, heaths, banks. Mar.–May. Med. Eur. P. *The roots and knotted stems are used for making briar pipes; the stems make an excellent charcoal.* Pl. 89.

E. lusitanica Rudolphi LUSITANIAN HEATH. Like 927, but fls. larger 4–5 mm., with corolla cylindrical, and 2–3 times as long as broad; fl. stalks about as long as the corolla. Lvs. 5–6 mm.; twigs greyish-hairy, hairs simple. Portugal to France. Pl. 89.

(*b*) *Small shrubs less than 1 m.*
(*i*) *Lvs. ciliate*

928. *E. tetralix* L. BOG HEATHER, CROSS-LEAVED HEATH. A small, lax, glandular-hairy undershrub to 60 cm. with dense, globular, terminal clusters of pink, nodding, flask-shaped fls. Fls. 4–12; corolla ovoid 6–7 mm., constricted at mouth; calyx glandular-hairy, *c.* 2 mm.; anthers awned. Lvs. linear 2–4 mm., spreading and conspicuously four-whorled, greyhaired and glandular-hairy above, with inrolled margin; twigs purplish, hairy and often glandular-hairy. Fr. hairy. △ Bogs, wet heaths. June–Oct. West Eur. DK.N.S.D.SU. *Used for brush- and broom-making, also for tanning.*

929. *E. ciliaris* L. DORSET HEATH. Like 928, but fls. in elongated one-sided clusters 5–12 cm. long; corolla deep pink, 8–10 mm., cylindrical-flask-shaped, and with a constricted mouth, somewhat curved above and swollen below; style projecting; anthers without awns, included. Lvs. 1–3 mm., oval, glandular-hairy, mostly in whorls of 3. Ovary hairless. △ Heaths, pine woods. Apr.–Oct. West Eur. (except B.NL.). Pl. 90.

(*ii*) *Lvs. not ciliate*

930. *E. cinerea* L. BELL HEATHER. A stiff, branched undershrub to 60 cm. with linear lvs. in whorls of 3, and with distinctive clusters of smaller lvs. in their axils. Fls. crimsonpurple in whorls, often forming a dense elongated cluster 1–7 cm. long; corolla 5–6 mm., flask-shaped with a constricted mouth; calyx usually purple, hairless; anthers awned. Lvs. 5–7 mm., hairless, dark green. Fr. hairless. △ Heaths, moors, open woods, dry banks. May–Sept. West Eur. N.D.

931. *E. scoparia* L. GREEN HEATHER. Readily distinguished by its very numerous, tiny, greenish-yellow fls. in long, narrow, spike-like clusters. Corolla 1–2 mm., broadly bellshaped with blunt lobes as long as the tube. Lvs. in whorls of 3 or 4, hairless, linear, 4–5 mm., with 2 grooves beneath. A stiff, erect, branched shrub 40–100 cm., with whitish hairless twigs. △ Heaths, pine woods, stony uncultivated places, hills. Dec.–June. P.E.F.I.YU.

Stamens longer than corolla with anthers projecting
(*a*) *Fl. stalks about as long as calyx*

932. *E. australis* L. SPANISH HEATH. An erect, much-branched shrub to 1½ m. with glandularhairy lvs. in whorls of 4 and comparatively large purplish-red fls. in stalkless clusters and forming a long inflorescence. Corolla 7–9 mm., cylindrical, lobes recurved; calyx densely ciliate; anthers as long as or slightly longer than the corolla, with pinnately divided awns. Lvs. linear *c.* 8 mm.; twigs greyish-hairy. △ Uncultivated ground, bushy places, pine woods. Feb.–Aug. P.E.

933. *E. herbacea* L. (*E. carnea* L.) SPRING HEATH. A small, spreading, irregularly branched undershrub to 60 cm., with lvs. in whorls of 4 and dark reddish, rarely white, pendulous fls. in a short, one-sided, terminal cluster. Corolla 4–8 mm., ovoid-tubular, with blunt erect lobes; calyx about two-thirds as long as the corolla; anthers awnless, blackish-purple, completely projecting. Lvs. ½–1 cm., keeled beneath; twigs hairless. △ Often in mountains: open woods, rocky places. Dec.–Mar. Cent. Eur. (except PL.H.) F.I.YU.AL. Pl. 89.

E. hibernica (Hooker & Arnott) Syme (*E. mediterranea* auct.) Like 933, but fls. pink in a dense, one-sided, leafy cluster; corolla tubular 5–7 mm.; anthers half-projecting; style long, projecting. An erect, branched shrub 1–2 m., with greyish twigs. Lowland heaths. Portugal to France, and Ireland.

ERICACEAE 934-937

(b) *Fl. stalks 1½–3 times as long as calyx*

934. *E. multiflora* L. An erect branched shrub to 1 m. with dense, rounded, terminal clusters of pale pink fls. Fl. stalks very slender, reddish, 2–3 times as long as the cylindrical, bell-shaped corolla which is c. ½ cm. long and about 3 times as long as broad; calyx half as long as the corolla; anthers completely projecting. Lvs. in whorls of 3–5, linear 8–10 mm.; twigs hairless. △ Heaths, dry hills, thickets. July–Dec. E.F.I.YU. Pl. 89.

E. manipuliflora Salisb. (*E. verticillata* auct.) Like 934, but lvs. in whorls of 3, shorter c. ½ cm. Corolla pink, bell-shaped, about twice as long as broad; fl. stalks 1½–2 times as long as the corolla; anthers slightly projecting. Eastern Mediterranean, Italy to Turkey.

E. vagans L. CORNISH HEATH. Like 934, but fl. clusters lax, cylindrical, and terminating in a leafy apex; corolla pale lilac or white, smaller 3–4 mm., widely bell-shaped, fl. stalks 3–4 times as long; calyx one-third as long as corolla. Western Atlantic: Spain, France, Ireland.

BRUCKENTHALIA Like *Erica*, but stamens fused at the base to each other and to the corolla. Calyx bell-shaped, four-lobed; corolla four-lobed; stamens 8, not projecting; anthers awned. 1 sp.

B. spiculifolia (Salisb.) Reichenb. SPIKE-HEATH. An erect, woodland, heather-like under-shrub 10–25 cm., with downy twigs, numerous sharp, linear, glandular-hairy lvs. 3–5 mm., and crowded terminal clusters of magenta-pink fls. Corolla open bell-shaped 3 mm.; calyx pink. The Balkans.

VACCINIUM Undershrubs or shrublets with oval lvs. Corolla usually bell-shaped with 4–5 lobes; stamens 8–10; anthers awned. Fr. an inferior berry with persisting calyx. 5 sps.

Fls. bell-shaped
(a) *Lvs. evergreen*

935. *V. vitis-idaea* L. COWBERRY. A spreading evergreen undershrub to 30 cm. with dark green, two-ranked lvs. and white bell-shaped fls. tinged with pink, in terminal drooping clusters. Corolla c. 6 mm., mouth not constricted, lobes recurved; calyx short, reddish. Lvs. 1–3 cm. obovate, paler and gland-dotted beneath; twigs rounded. Fr. globular, red, △ Open woods, bogs, heaths, alpine meadows. May–July. All Eur. (except P.E.AL.TR.). *The fruits are edible and have been used in jam-making and for distillation.* Pl. 89.

(b) *Lvs. deciduous*

936. *V. myrtillus* L. BILBERRY, BLAEBERRY, WORTLEBERRY. A low, dense, much-branched undershrub 20–60 cm., with green, angled stems, deciduous lvs. and pink or greenish-pink, solitary, drooping, globular fls. in the axils of the upper lvs. Corolla 2–4 mm., mouth strongly constricted, lobes reflexed; calyx absent. Lvs. 1–3 cm. oval, finely toothed, bright green. Fr. c. 8 mm., bluish-black, bloomed. △ Heaths, moors, open woods. May–July. All Eur. (except TR.). *The fruit is edible and used for preserves, to colour wine and in the preparation of spirits. The fruits give a black or blue dye and have been used for this since Roman times.* Pl. 89.

V. uliginosum L. BOG WORTLEBERRY. Like 936, but fls. pale pink in clusters of 1–4 arising from last year's shoots. Calyx lobes short, blunt. Lvs. 1–2½ cm. oval, entire not toothed, bluish-green, particularly beneath; twigs rounded, brownish. Widespread in Europe, but only in mountains in Central and Southern Europe.

Fls. with spreading reflexed petals

937. *V. oxycoccos* L. (*Oxycoccus palustris* Pers.) CRANBERRY. A very slender-stemmed shrublet, creeping to 80 cm., with tiny, scattered evergreen lvs. and 1–4 long-stalked

small pink fls. with 4 strongly reflexed petals. Petals 5–6 mm., free almost to their base; calyx ciliate; stamens and style projecting conspicuously forward; fl. stalks 1½–3 cm., minutely downy. Lvs. 4–8 mm. oval acute, dark green above, glaucous beneath. Fr. 6–8 mm., red becoming brown. △ Bogs. May–July. Much of Eur. (except P.E.I.H.YU.AL.GR.TR.BG.). Pl. 88.

EMPETRACEAE | Crowberry Family

Small heath-like evergreen undershrubs, like Ericaceae but distinguished by the absence of bright-coloured petal-like perianth. Perianth 4–6 in 2 whorls, all similar, minute, greenish-purple; stamens 2–3, long-projecting. Fr. a dry or fleshy berry containing 2–9 one-seeded stones.

EMPETRUM | **Crowberry** ⟩ Plants often one-sexed; perianth 6; stamens 3. Ovary 1, six- to nine-celled; stigmas as many as cells. 2 sps.

938. *E. nigrum* L. CROWBERRY. A low, much-branched undershrub to 30 cm. with numerous densely overlapping, heath-like lvs. and tiny, solitary, pinkish-purple fls. in the axils of the upper lvs. Fls. 1–2 mm. across, usually one-sexed. Lvs. oblong-cylindrical *c.* ½ cm., grooved beneath, at length hairless. Fr. *c.* ½ cm. globular, black. △ Moors, damp rocks, bogs, in mountains. Apr.–June. Much of Eur. (except P.H.AL.GR.TR.). Page 304.

PRIMULACEAE | Primrose Family

Usually herbaceous plants with simple whorled lvs., or lvs. in a basal rosette. Fls. radially symmetrical, solitary or in umbels. Sepals, petals, and stamens usually 5; corolla bell-shaped or funnel-shaped, with spreading lobes; stamens attached to the corolla tube. Ovary superior, one-celled, many-seeded; style 1, stigma unbranched. Fr. a capsule.

		Nos.
1. Lvs. all basal, usually in a rosette.		
2. Lvs. rounded, as long as broad and with heart-shaped base.		
3. Fls. with reflexed petals; corm present.	*Cyclamen*	958–960
3. Fls. bell-shaped, with spreading petals; corm absent.		
4. Corolla cut into numerous narrow lobes, violet.	*Soldanella*	956
4. Corolla five-lobed, rose-purple.	*Cortusa*	955
2. Lvs. longer than broad and narrowed to lv. stalk.		
5. Corolla tube shorter than calyx.	*Androsace*	948–954
5. Corolla tube longer than calyx.	*Primula*	939–946
1. Lvs. present on fl. stems.		
6. Alpine cushion- or mat-forming plants with numerous lvs.		
7. Fls. pink or white.	*Androsace*	948–954
7. Fls. yellow.	*Vitaliana*	947
6. Not alpine plants, usually not cushion- or mat-forming.		
8. Stem lvs. in whorls or opposite.		
9. Fls. white, pink, red, or blue.		
10. Lvs. pinnate: water plants.	*Hottonia*	957
10. Lvs. entire: land plants.		

PRIMULA | **Primrose, Cowslip** Fls. large, usually clustered in a terminal umbel on an erect leafless stem, rarely solitary; corolla tube long, widening at throat, lobes often spreading; calyx tubular, five-lobed. Fr. a capsule splitting by valves. *c.* 35 sps.

Fls. yellow
(*a*) *Fls. in a terminal stalked umbel*
939. *P. veris* L. COWSLIP. Fls. deep yellow with orange spots in the throat, sweet-scented, in a terminal, nodding cluster of 9–30. Fls. 1–1½ cm. across; corolla tube *c.* 1½ cm. long, lobes ascending, curved into a cup, notched and with folds in the throat; calyx tubular wide-mouthed, pale yellowish-green. Lvs. 5–15 cm. in a rosette, oblong, toothed, and abruptly narrowed to a winged stalk, hairy; fl. stems 10–30 cm. △ Meadows, pastures, open woods. Apr.–May. All Eur. (except IRL.IS.AL.TR.). *The flowers can be used to make home-made wine; an important medicinal plant in the past.*

940. *P. elatior* (L.) Hill OXLIP, PAIGLE. Like 939, but fls. larger 1½–2 cm. across, pale yellow, with a darker yellow throat, unscented, in a nodding one-sided cluster of 1–20. Corolla with flatter, spreading lobes; calyx narrowed at the throat, with green-veined angles and paler between. Lvs. 10–20 cm. oblong, with crisped hairs; fl. stem with crisped hairs, 10–30 cm. △ Woods, hedges, meadows. Mar.–May. Most of Eur. (except IRL.IS.N.TR.). Pl. 90.

941. *P. auricula* L. AURICULA, BEAR'S-EAR. A mountain perenn. with beautiful umbels of very sweet-scented, bright yellow fls. and a lax rosette of broad, usually entire, hairless and slightly fleshy lvs. Fls. 2–25; corolla 1½–3 cm. across, throat nearly white; calyx hairless, with blunt lobes, much shorter than the corolla tube. Lvs. obovate, mealy when young, with a cartilaginous margin, toothed or not. Fl. stems 5–25 cm., hairless, longer than the lvs. A variable sp. △ Limestone rocks in mountains. May–July. Cent. Eur. F.I.YU.R. Pl. 90.

(*b*) *Fls. solitary, arising directly from rootstock*
942. *P. vulgaris* Hudson PRIMROSE. Fls. numerous, arising individually on long stalks from the rootstock, in the centre of the rosette of lvs. Corolla 2–3 cm. across, pale yellow with greenish marking at the throat, or sometimes pink particularly in the South-East of Europe, with flat, notched lobes; calyx cylindrical, angular, shaggy-haired; fl. stalks shaggy-haired. Lvs. 8–15 cm., obovate and toothed, hairless above, hairy beneath, gradually narrowed to a winged lv. stalk. △ Woods, banks, meadows, orchards. Mar.–May. Most of Eur. (except B.IS.S.SF.PL.). *An important medicinal plant in the past.* Pl. 90.

Fls. pink, purple, or violet
(*a*) *Lvs. with white or yellow mealy powder beneath*
943. *P. farinosa* L. BIRD'S-EYE PRIMROSE. A delicate perenn. with a small rosette of mealy lvs. and a slender stem to 15 cm. bearing a terminal umbel of small rose-lilac fls. with pale centres. Corolla *c.* 1 cm. across, lobes spreading, deeply notched, and corolla tube 5–6 mm.; calyx mealy; bracts much shorter than the fl. stalks. Lvs. 1–5 cm., spathulate and

toothed, hairless above and with white or yellow mealy powder beneath. Fr. 5–7 mm., much longer than the calyx. △ Mostly in mountains: damp meadows, peaty places. June–July. Much of Eur. (except P.IRL.B.NL.IS.N.AL.GR.TR.). Pl. 91.

P. halleri Honckeny (*P. longifolia* All.) LONG-FLOWERED PRIMROSE. Like 943, but readily distinguished by its very long violet corolla tube 2½–3 cm., which is about 3 times as long as the calyx. Fls. rose-violet, few; bracts of umbel as long as or longer than the fl. stalks. The Alps, Carpathians and N. Balkans.

(*b*) *Lvs. not powdery beneath*
(*i*) *Lvs. hairless*

944. *P. marginata* Curtis A very beautiful perenn. with a woody rootstock and fl. stem 5–20 cm., bearing an umbel of large, rose-violet, sweet-scented fls. and a rosette of thick, smooth, shiny, conspicuously toothed, mealy-margined lvs. Fls. 3–10; corolla 1–2½ cm. across, lobes deeply notched, tube *c.* 1 cm. and more than 3 times as long as the calyx. Bracts, fl. stalks and calyx mealy. Lvs. *c.* 3 cm., oval, narrowed to the base, not mealy beneath. △ Rock crevices of high mountains. June–July. F.I. *The petals are used to make cowslip wine; held in high repute as a medicinal plant in the past.*

P. minima L. LEAST PRIMROSE. A prostrate perenn. of alpine regions, with tiny rosettes of wedge-shaped, sharply toothed, shining, hairless lvs. bearing much larger, solitary, rose-coloured fls. Fls. 1½–3 cm. across, almost stalkless; corolla lobes spreading, deeply notched, yellow-based; calyx with blunt lobes, shorter than the corolla tube. Alps, Carpathians. Pl. 91.

(*ii*) *Lvs. glandular-hairy*

945. *P. viscosa* All. (*P. latifolia* Lapeyr.; *P. hirsuta* Vill.) Like 944, but lvs. pale green, glandular-hairy and with a coarsely toothed but not mealy margin; stem, fl. stalks and calyx all glandular-hairy. Fls. 3–20, fragrant, violet-purple throughout including throat, *c.* 1½ cm. across; calyx about one-third as long as the corolla tube. Lvs. oval, toothed, narrowed to a stalk; fl. stems 6–15 cm. △ Rock crevices of high mountains. June–July. E.F.CH.A.I. Pl. 89.

P. hirsuta All. (*P. viscosa* Vill.) RED ALPINE PRIMROSE. Like 945, but umbels with 1–3 fls., on a stem shorter than the lvs.; corolla bright purple with a white eye and throat. Lvs. small, 2–5 cm., toothed, very sticky on both sides. Pyrenees and Alps. Pl. 91.

946. *P. integrifolia* L. Fls. bright rose, pink, or lilac, *c.* 2 cm. across, in an umbel of 1–3 on a short, glandular, often reddish stem to 6 cm. Corolla lobes deeply two-lobed; calyx 6–9 mm., blunt-lobed, glandular-hairy, about half as long as the corolla tube. Lvs. in a rosette 2–4 cm. across, oval entire, with ciliate margin, glandular-hairy, not sticky. △ Rocks, alpine meadows. May-Aug. E.F.D.CH.A.I. Pl. 91.

VITALIANA   Stems creeping and often forming loose mats; lvs. linear-acute in terminal rosettes. Fls. yellow. 1 sp.

947. *V. primuliflora* Bertol. (*Douglasia vitaliana* (L.) Pax) A low, creeping, mat-forming perenn. 3–6 cm., with rosettes of lvs. 1–1½ cm. across, and solitary or paired, stalkless, bright lemon-yellow fls. *c.* 1 cm. across. Corolla turning green on drying, tube much longer than the calyx, enlarged at mouth and lobes oblong spreading; calyx and lvs. with star-shaped hairs. Lvs. awl-shaped ½–1 cm. △ Alpine rocks, pastures, sandy places. May–July. E.F.CH.A. Page 304.

ANDROSACE   Fls. small, less than 1 cm. across; corolla tube very short and narrowed at the throat, lobes usually spreading. Calyx deeply five-lobed, longer than the corolla tube. Fr. a capsule with a few seeds. *c.* 20 sps.

Fls. solitary in axils of upper lvs.

948. *A. helvetica* (L.) All. A very compact alpine cushion-forming perenn. 2–4 cm. across, composed of tiny tight rosettes each bearing a solitary, stalkless, white fl. with a yellow centre, 4–6 mm. across. Rosettes *c.* ½ cm. across, building up into columns to form a hard dense cushion; lvs. 3 mm., elliptic, densely overlapping, grey-green, covered with long, unbranched hairs. △ Alpine screes, rock clefts. July–Aug. F.D.CH.A.CS.I. Page 304.

A. vandellii (Turra) Chiov. (*A. multiflora* Boretti; *A. imbricata* Lam.) Like 948, but lvs. silvery-white and covered with star-shaped hairs. Fls. 4 mm. across, several to each rosette, white, with a throat at first yellow and becoming red; calyx with star-shaped hairs. Pyrenees and Alps.

949. *A. alpina* (L.) Lam. (*A. glacialis* Hoppe) A prostrate, loosely cushion-forming, alpine perenn. with hairy rosettes and solitary, pink or white fls. with a yellow throat, on stalks little longer than the lvs. Corolla *c.* 5 mm. across; calyx hairy. Rosettes 1–1½ cm. across; lvs. lance-shaped 2–8 mm., with star-shaped hairs. △ A high alpine of rock moraines, screes. July–Aug. F.D.CH.A.I. Pl. 91.

Fls. several or numerous in stalked umbels
(a) Slender erect anns. or bienns.

950. *A. maxima* L. A small rosette-forming ann. with several erect leafless fl. stems 5–15 cm., with umbels of tiny white fls., with the petals much smaller than the leafy calyx lobes. Umbels terminal, with 3–8 fls. and subtended by large leafy and hairy involucral bracts about as long as the fl. stalks; calyx conspicuous, hairy, enlarging to 1 cm. in fr. Lvs. 1½–2 cm. oval acute, toothed. △ Lowland fields and meadows. Apr.–May. Cent., South-East Eur. (except AL.) E.F.I.SU.

951. *A. septentrionalis* L. A finely downy rosette-forming ann. or bienn. with 1 or several erect leafless stems to 20 cm., bearing umbels of 5–30 white or pink fls. Fls. *c.* ½ cm. across; corolla lobes oblong, not notched, longer than the hairless calyx; fl. stalks slender, unequal 2–3 cm., much longer than the bracts which are *c.* 3 mm.; calyx not enlarging. Lvs. 1½–3 cm., numerous, lance-shaped, toothed. △ Dry meadows, sandy slopes, fields in mountains. May–July. North Eur. (except IS.), Cent. Eur. (except H.) F.R.SU. Page 304.

(b) Small mat-forming perenns.
(i) Stem of inflorescence and fl. stalks hairless or with fine hairs

952. *A. lactea* L. A hairless mat-forming perenn. with underground runners, rosettes of linear lvs. and a long-stemmed umbel of 1–6 white fls. with yellow centres. Fls. 1 cm. across, long-stalked; petals spreading, notched. Lvs. 1–1½ cm. linear, green, and usually hairless, in rosettes 2–3 cm. across, superimposed but distinctly separated from each other by short, thick, leafless stems; fl. stems 5–15 cm. △ Rocks, screes, alpine meadows. June–Aug. Cent. Eur. (except H.) F.I.YU.R.

953. *A. carnea* L. A perenn. forming tufts of rosettes and with small dense umbels of 2–10 pink fls. with a yellow eye, or fls. whitish. Fls. ½–1 cm.; petals not notched; fl. stalks little longer than the bracts. Lvs. linear acute 1–1½ cm., in dense green, nearly hairless rosettes; fl. stem finely hairy, 4–12 cm. A variable sp. △ In mountains: damp screes, meadows. July–Aug. E.F.D.CH.A.I. Pl. 90.

(ii) Stem of inflorescence and fl. stalks with long spreading hairs

954. *A. chamaejasme* Wulfen A rosette perenn. with 3–8 white or pink fls. in small compact umbels borne on a stem with long, silky, spreading hairs and short, glandular hairs. Fls.

1. *Androsace septentrionalis* 951
3. *Samolus valerandi* 969
5. *Androsace villosa* [954]
7. *Coris monspeliensis* 970

2. *Lysimachia thyrsiflora* 964
4. *Vitaliana primuliflora* 947
6. *A. helvetica* 948
8. *Empetrum nigrum* 938

7–10 mm. across, short-stalked; petals entire, longer than the hairy calyx. Lvs. in flattish green rosettes ½–1 cm. across, oblong-lance-shaped entire, hairless except for the long ciliate hairs on the margin; fl. stems 4–10 cm.· △ Alpine meadows and rocks. June–July. Cent. Eur. (except H.) F.I.R.

A. villosa L. Like 954, but rosettes globular, much smaller, 5–8 mm. across, and lvs. with long silky hairs on both sides. Fls. white or pink, in a dense umbel with fl. stalks shorter than the bracts, petals shallowly notched. Pyrenees, Alps, Apennines, Balkans. Page 304.

CORTUSA Fls. in an umbel; corolla deeply five-lobed; anthers long-pointed. Lvs. all basal, palmately lobed. 1 sp.

955. *C. matthioli* L. ALPINE BELLS. A hairy perenn. with long-stalked, rounded, basal lvs. and leafless stems 10–40 cm., bearing an umbel of 4–12 pink, nodding fls. which later turn violet. Fls. *c.* 1 cm., long-stalked; corolla bell-shaped with 5 spreading lobes; calyx five-lobed; style long. Lv. blades rounded-kidney-shaped and with 7–9 shallow, rounded, toothed lobes; lv. stalks hairy. △ In mountains: thickets, damp ravines, wet rocks. June–July. Cent. Eur. (except H.) F.I.YU.BG.R.SU. Pl. 91.

SOLDANELLA | Snowbell Fls. in an umbel or solitary on leafless stems; corolla bell- or funnel-shaped with lobes deeply cut into a fringe of narrow segments. Lvs. all basal, rounded, leathery. *c.* 6 sps.

956. *S. alpina* L. ALPINE SNOWBELL. A delicate, hairless, alpine perenn. with stems 5–15 cm., bearing 2–5 bluish-violet, nodding fls. with the corolla cut to half its length or more into a fringe of narrow segments. Corolla 8–13 mm., rarely white; scales present in throat. Lvs. long-stalked, blades thick, leathery, rounded-kidney-shaped, 1½–6 cm. broad. △ Damp alpine meadows, rocks. Apr.–July. E.F.D.CH.A.CS.I.YU.AL.BG. *The underground parts are purgative.*

S. pusilla Baumg. DWARF SNOWBELL. Like 956, but fls. solitary, reddish-violet; corolla 1–1½ cm., narrow bell-shaped, cut at most to one-quarter its length into a narrow fringe of slender segments; scales absent. Lvs. kidney-shaped, gland-dotted beneath, *c.* 1 cm. across. Alps, Apennines, Carpathians. Pl. 91.

S. montana Mikan MOUNTAIN TASSEL-FLOWER. Like 956, but the lv. and fl. stalks with long, dense, glandular hairs. Fls. in an umbel 3–6, reddish-violet; corolla 10–17 mm., funnel-shaped, cut to above the middle into a fringe of narrow outward-curved segments. Lvs. 2–7 cm. broad, shallowly toothed, bluish beneath; fl. stem 10–20 cm. Pyrenees, Alps, Carpathians.

HOTTONIA Submerged aquatic plants with pinnate lvs. Fl. stem borne above-water; fls. whorled; calyx 5; corolla with a short tube and 5 spreading lobes. 1 sp.

957. *H. palustris* L. WATER VIOLET. A perenn. with pale green, much-dissected submerged lvs. and erect above-water fl. stems to 40 cm., bearing several whorls of pale lilac fls. with yellow centres, forming a lax pyramidal cluster. Fls. 3–8 in each whorl, 2–2½ cm. across; petals spreading, shallowly notched; fl. stalks glandular, nodding in fr. Lvs. 1–2 times pinnate, with very numerous comb-like linear segments. Fr. longer than the glandular calyx. △ Still waters, marshes, ditches. May–July. Much of Eur. (except P.E.IS.N.SF.AL.GR.). Pl. 92.

CYCLAMEN Above-ground stem absent or very short and fls. and lvs. arising from a large underground corm. Corolla tube very short, petals reflexed. *c.* 8 sps.

Late summer and autumn flowering plants

958. *C. hederifolium* Aiton (*C. europaeum* auct.; *C. neapolitanum* Ten.) SOWBREAD. Fls. appearing before the lvs.; white or pale pink with a darker blotch at the base of the reflexed petals and with a pentagonal throat formed by the outward-flaring of the petal bases. Corolla *c.* 2½ cm.; petals elliptic-oblong; calyx as long as the corolla tube; fl. stem 10–30 cm. Lvs. appearing after the fls., long-stalked, heart-shaped but variable and often shallowly five- to nine-lobed, mottled silvery-grey above, often purplish beneath. Corms large, 2–10 cm., roots arising from the upper surface only. △ Woods, bushy places. Aug.–Nov. F.CH.I.YU.AL.GR.TR. Pl. 92.

959. *C. purpurascens* Miller COMMON CYCLAMEN. Fls. appearing with the lvs. sweet-scented, rose-carmine and with a darker basal blotch at the base of the reflexed petals and with a rounded throat. Petals oblong-elliptic; calyx a little shorter than the corolla tube. Lvs. round-heart-shaped, not angled, entire or finely toothed, silver-blotched above, reddish purple beneath, long-stalked; fl. stems 5–15 cm. Corms globular with roots arising from the whole surface. △ Woods, bushy places. June–Oct. Cent. Eur. F.I.YU.BG.: introd. R. *A rather poisonous plant; it contains the glycoside, cyclamin.* Pl. 92.

C. graecum Link GREEK CYCLAMEN. Fls. deep reddish-pink with a plum-coloured blotch at base of the reflexed petals, throat of corolla pentagonal. Lvs. richly and variably marbled with silvery-grey, margin with tiny horny teeth. Corms with a fissured corky surface, and with thick roots arising only from the base. Greece. Pl. 92.

Spring and early summer flowering plants

960. *C. repandum* Sm. REPAND CYCLAMEN. Fls. appearing in spring with the lvs., sweet-scented, bright rose-coloured, rarely white. Petals reflexed, oval or oblong 2–3 cm., slightly twisted, throat rounded; style long projecting. Lvs. thin, triangular-heart-shaped in outline, but with broad, shallow, angular lobes with cartilaginous tips, green, not mottled, above and purplish beneath. Corm small, 1–3 cm., flattened above and below, rooting only from the base. △ Woods, thickets, rocky places. Mar.–May. E.F.I.YU. GR. Pl. 92.

LYSIMACHIA | Loosestrife Fls. usually yellow; calyx five-lobed; corolla five- to seven-lobed, bell-shaped or with spreading lobes; stamens 5–7. Fr. a globular capsule opening by 2–5 valves. Lvs. usually opposite or whorled. *c.* 9 sps.

Lvs. oval, less than 4 cm.; delicate perenns.

961. *L. nummularia* L. CREEPING JENNY. A creeping perenn. rooting at the nodes and spreading to 60 cm., with rounded paired lvs. and solitary, dark yellow, short-stalked fls. Corolla 1½–2½ cm. across, open bell-shaped, gland-dotted; calyx lobes oval acute 5–10 mm. Lvs. pale yellowish-green, 1½–3 cm., oval-rounded, short-stalked, gland-dotted. Fr. rare, shorter than calyx. △ Damp meadows, ditches, by water. May–Aug. All Eur. (except P.IS.). Pl. 93.

962. *L. nemorum* L. YELLOW PIMPERNEL. A delicate, spreading perenn. to 40 cm., with ascending fl. stems bearing solitary, axillary, pale yellow fls. on long slender fl. stalks, longer than the subtending lvs. Corolla 1–1½ cm. across, with spreading lobes; calyx lobes *c.* 5 mm., narrow lance-shaped. Lvs. 2–4 cm. oval acute, short-stalked. Fr. *c.* 3 mm.; calyx longer. △ Damp woods, shady hedges, springs. May–Aug. Most of Eur. (except IS.SF.AL. GR.TR.BG.). Pl. 93.

Lvs. lance-shaped, more than 4 cm.; robust erect perenns.

963. *L. vulgaris* L. YELLOW LOOSESTRIFE. An erect, softly hairy, rhizomatous perenn. $\frac{1}{2}$–$1\frac{1}{2}$ m., with opposite or three- to four-whorled lvs. and lax, branched, pyramidal clusters of golden-yellow fls. Fls. $1\frac{1}{2}$–2 cm.; corolla bell-shaped, lobes spreading, not glandular; calyx lobes lance-shaped with orange margin, ciliate. Lvs. 5–12 cm. lance-shaped, stalkless, dotted with orange or black glands. △ Marshes, swamps, lakesides. June–Aug. All Eur. (except IS.). *It has a reputation for stopping bleeding and as a febrifuge. Yields a yellow dye.*

L. punctata L. LARGE YELLOW LOOSESTRIFE. Like 963, but fls. larger *c.* $3\frac{1}{2}$ cm. across, and corolla lobes conspicuously glandular-hairy; calyx lobes glandular-hairy, not bordered with orange. Inflorescence with the upper fls. in few-flowered, stalkless whorls. Lvs. shortly stalked. South-Eastern Europe; widely naturalized elsewhere. Pl. 93.

964. *L. thyrsiflora* L. (*Naumburgia t.*) TUFTED LOOSESTRIFE. An erect perenn. to 60 cm., usually with opposite stalkless lvs. and short-stalked globular clusters of yellow fls. arising from the axils of the middle lvs. Fl. clusters 1–2 cm.; fls. *c.* 5 mm.; petals yellow, linear, with black glands towards the apex; stamens longer than the petals. Lvs. 5–10 cm. oblong-lance-shaped, gland-dotted, decreasing in length below and becoming scale-like towards base of the stem; stem and fl. stalks hairy. △ Marshes, by water. May–July. Much of Eur. (except P.E.IRL.IS.I.AL.GR.TR.). Page 304.

TRIENTALIS Fls. white; sepals, petals, and stamens 5–9. Lvs. in a rosette towards the apex of stem. Fr. a five-valved capsule. 1 sp.

965. *T. europaea* L. CHICKWEED WINTERGREEN. A delicate perenn. with a slender, erect stem 10–25 cm., bearing a terminal, lax rosette of lance-shaped lvs. and solitary or few long-stalked white fls. Fls. $1\frac{1}{2}$–2 cm. across; petals usually 7, wide-spreading, longer than the linear pointed calyx; fl. stalks very slender 2–7 cm. Lvs. 1–8 cm., varying in size in each rosette, stiff, shining, entire or finely toothed towards apex, very short-stalked. △ Pine forests, heaths. May–Aug. Much of Eur. (except P.E.IRL.YU.AL.GR.TR.BG.). Pl. 93.

ANAGALLIS | Pimpernel Fls. red, pink or blue; sepals, petals, and stamens 5; stamens inserted at base of short corolla tube. Fr. a capsule opening by a cap-like cover. *c.* 10 sps.

966. *A. tenella* (L.) L. BOG PIMPERNEL. A very delicate, hairless creeping perenn. 5–15 cm., rooting at the nodes, and with tiny paired lvs. bearing solitary, pink fls. in their axils. Fls. to $1\frac{1}{2}$ cm. across; corolla funnel-shaped, lobes much longer than the linear acute calyx; fl. stalks slender, longer than the lvs. Lvs. *c.* $\frac{1}{2}$ cm. oval-rounded, shortly stalked. △ Bogs, damp peaty places, marshy fields. May–Sept. West Eur. D.CH.A.PL.I.R.SU.

A. minima (L.) E. H. L. Krause (*Centunculus m.*) CHAFFWEED. Throughout Central Europe.

967. *A. arvensis* L. SCARLET PIMPERNEL. A slender, spreading, hairless ann. 6–30 cm., with solitary, long-stalked, scarlet, pink or blue fls. borne from the axils of the upper lvs. Fls. to $1\frac{1}{2}$ cm. across; petals spreading, rounded or finely toothed at apex, fringed with glandular hairs; calyx lobes narrow lance-shaped, nearly as long as the petals. Lvs. $1\frac{1}{2}$–$2\frac{1}{2}$ cm. oval-lance-shaped, stalkless, dotted with black glands beneath; stems quadrangular. △ Cultivated ground, tracksides. May–Oct. All Eur.: introd. IS.SF. *An important medicinal plant in the past; the plant contains a poisonous glycoside similar to saponin.*

A. foemina Miller Very like 967, but fls. with blue petals which are oval acute and without glandular hairs on the margin. Widespread in Europe.

A. linifolia L. SHRUBBY PIMPERNEL. A woody-based, branched prostrate or ascending perenn. 5–50 cm. with large blue fls. with a purple centre, 1–2 cm. across, or fls. rarely white. Petals toothed; fl. stalks slender, longer than the lvs. Lvs. linear-lance-shaped. Portugal and Spain. Pl. 93.

GLAUX   Fls. stalkless, white or pink; petals absent; calyx 5; stamens 5. Fr. a five-valved capsule. Lvs. fleshy. 1 sp.

968. *G. maritima* L. SEA MILKWORT. A creeping, rooting, spreading perenn. to 30 cm. with fleshy, overlapping, four-ranked lvs. and small, stalkless, pinkish fls. in the axils of the upper lvs. Calyx bell-shaped, *c.* $\frac{1}{2}$ cm. across, with 5 blunt, spreading, usually pink lobes with translucent margins; petals absent. Lvs. 4–12 mm. oval, acute or obtuse, stalkless. △ Salt-rich places, usually on the littoral. May–Sept. Most of Eur. (except CH.YU.AL. GR.TR.BG.). Pl. 92.

SAMOLUS   Fls. white; sepals, petals and stamens 5. Ovary half-inferior. Lvs. mostly in a basal rosette. Fr. a five-valved capsule. 1 sp.

969. *S. valerandi* L. BROOKWEED. A hairless, pale green perenn. 5–45 cm., with a basal rosette of lvs. and an erect little-branched stem bearing a lax cluster of tiny white fls. 2–3 mm. across. Corolla bell-shaped with 5 rounded lobes; fl. stalks $\frac{1}{2}$–$1\frac{1}{2}$ cm. becoming jointed in fr. with a small bract at the joint. Lvs. 1–8 cm. oval to spathulate, entire, blunt; stem lvs. few. △ Ditches, dunes, peaty places, particularly near sea. June–Aug. All Eur. (except IS.N.). Page 304.

CORIS   Fls. reddish-violet; corolla distinctly two-lipped with 3 longer upper lobes and 2 shorter lower lobes; calyx ten-lobed or more. 2 sps.

970. *C. monspeliensis* L. A branched, stiff-stemmed, thyme-like ann. or bienn. to 20 cm. with linear lvs. and with dense, terminal clusters of rose-lilac fls. Corolla *c.* 7 mm., tubular, unequally two-lipped, with 3 longer upper, notched lobes, and 2 lower notched lobes; calyx bell-shaped with 5 inner, short, triangular teeth and 5 or more long, spreading, spiny teeth. Lvs. leathery 1–2 cm. by *c.* 2 mm., spreading, hairless. △ Stony hills, maritime sands. Apr.–July. Med. Eur. (except AL.TR.). Page 304.

PLUMBAGINACEAE | Sea Lavender Family

Usually ann. or herbaceous or woody perenns. of the sea-shore or salt-rich places, with simple, usually spirally arranged lvs. often in a basal rosette. Fls. radially symmetrical with sepals, petals, and stamens 5; calyx tubular, often bright coloured and papery, often pleated above and persisting in fr.; petals almost free or fused into a long tube. Ovary superior, one-celled; styles 5 or 1 with a five-lobed stigma. Fr. dry, often opening with a lid or remaining closed.

1. Lvs. all basal in a rosette.
2. Fls. in spreading clusters.   *Limonium*
2. Fls. in dense globular heads.   *Armeria*
1. Lvs. present on fl. stems, alternate.   *Plumbago*

PLUMBAGINACEAE   971–974

PLUMBAGO Fls. violet; corolla funnel-shaped, five-lobed; calyx deeply five-lobed with a glandular-hairy, papery margin; styles 5. 1 sp.

971. *P. europaea* L. EUROPEAN PLUMBAGO. A stiff, erect, much-branched, hairless perenn. 30–120 cm., with violet or pink fls. *c.* 1½ cm. across in clusters, and conspicuously glandular–hairy calyx. Corolla tube about twice as long as the calyx, lobes oval, spreading, veined. Lvs. rough, dark green, glandular-toothed, the lower oval and stalked, the upper linear and clasping the stem. Fr. *c.* 8 mm. cylindrical, bristly with rows of stiff, gland-tipped hairs. △ Dry rocks, hills. Aug.–Oct. Med. Eur. P.BG.R. *The plant is acid, blistering, and emetic.* Page 315.

LIMONIUM Fls. clustered into spikelets with 3 scale-like bracts at the base, the spikelets are further clustered into slender spikes borne on the branched inflorescence. Calyx tubular or funnel-shaped, widening into a papery five- to ten-lobed limb; petals almost free, or fused into a tube. Lvs. all basal in a rosette. Often placed in the genus *Statice*. *c.* 50 sps.

Stems winged above
972. *L. sinuatum* (L.) Miller WINGED SEA LAVENDER. An erect, rough-hairy, rosette-forming perenn. 20–50 cm. with conspicuously winged stems, branched above into a rather flat-topped dense cluster of bright blue-mauve fls. with tiny yellow centres. Calyx papery, pleated, blue-mauve, persisting; corolla small, pale yellow, soon withering; outer bracts papery, rusty-brown, the inner green, toothed. Lvs. in a compact, flattened rosette, deeply pinnately lobed and wavy-margined. △ Rocks and sands on littoral. Apr.–Sept. P.E.F.I.GR. Pl. 94.

Stems rounded, not winged
973. *L. vulgare* Miller SEA LAVENDER. An erect perenn. 8–30 cm. with lance-shaped basal lvs. and erect, leafless stems branched above the middle into a more or less flat-topped cluster of numerous small violet fls. Fl. spikelets closely set in 2 rows on the upper side of the branches; corolla blue-purple, 8 mm. across; calyx greenish; bracts with papery margin. Lvs. 4–12 cm., strongly pinnately veined, gradually narrowed to a long stalk. △ Salt marshes, on the littoral. July–Sept. West Eur. (except IRL.) DK.S.D.CS.I.GR.TR.BG.R.

974. *L. bellidifolium* (Gouan) Dumort. MATTED SEA LAVENDER. Rather like 973, but stems very densely zigzagged and branched above into numerous slender, fragile, rough branchlets, many of which are sterile, while others bear dense clusters of violet fls. with conspicuous white papery calyx. Spikelets in 2 rows; fls. 3–4 mm. long; bracts with very broad papery margin, the outer almost entirely white-papery. Lvs. 1½–4 cm., oval-lance-shaped, long-stalked, dying before flowering is completed. △ On the littoral. May–Sept. E.F.I.YU.GR.R.SU. Page 315.

L. virgatum (Willd.) O. Kuntze Like 974, but fls. larger 8–9 mm. long, in rather few spikelets in long, lax, slender, unilateral spikes; bracts leathery, rusty-brown, papery only on margin; calyx lobes curved; corolla violet. Stems much-branched from near the base, with numerous non-brittle sterile branches. Mediterranean region and Portugal.

ARMERIA | **Thrift** Fls. in dense, globular, terminal heads on unbranched stems arising directly from the rootstock. Papery involucre present below fl. heads, the outer bracts forming a sheath round the stem immediately below the fl. head. Lvs. linear to narrow lance-shaped, in a basal rosette. *c.* 40 sps.

975. *A. maritima* (Miller) Willd. THRIFT, SEA PINK. A densely tufted, cushion-like woody-based perenn., with numerous slender linear lvs. and with dense globular heads of pink fls. borne on leafless unbranched stems 5–30 cm. Fl. heads 1½–2½ cm.; corolla 8 mm. across, pink, rarely white; calyx 6 mm. long, with hairy ribs, lobes papery, awned; outer involucral bracts green, the inner woolly-haired, papery; sheath ½–1½ cm. Lvs. 2–15 cm. by 1–2 mm., somewhat fleshy, one-veined. △ Salt marshes, maritime rocks, sometimes in mountains. May–Aug. West, North Eur. D. Pl. 94.

A. alpina (DC.) Willd. ALPINE THRIFT. Like 975, but a cushion perenn. of high mountains, distinguished by broader lvs. 2–5 mm., which are partly three-veined at the base; bracts of involucre pale brown, not green. Fls. bright pink in spherical heads 2–3 cm.; calyx 8–10 mm. long. Pyrenees, Alps, Carpathians, Balkans.

976. *A. arenaria* (Pers.) Schultes (*A. plantaginea* Willd.) JERSEY THRIFT, PLANTAIN-LEAVED THRIFT. Like 975, but a more robust rigid perenn. with stems 20–60 cm., and with fl. heads deep rose, *c.* 2½ cm. across, with conspicuous long-pointed outer involucral bracts equalling or longer than the fl. head. Corolla *c.* 1 cm. across. Lvs. *c.* 10 cm. by 3–8 mm., linear-lance-shaped with 3, 5, or 7 veins. A variable sp. △ Sandy fields, rocks. May–Sept. P.E.F.D.CH.I.YU.

A. fasciculata (Vent.) Willd. (*A. pungens* (Brot.) Hoffmanns. & Link) SPINY THRIFT. Distinguished by its rigid glaucous, grooved, spine-tipped lvs. in a dense cluster and mixed with the old recurved lvs. at the base. Involucral bracts tough, leathery, oval, brown, and usually papery-margined. Fl. heads 1½–3 cm. across, pale rose; sheath 1½–3 cm. rusty-brown; fl. stems 10–40 cm. Portugal, Spain, Corsica, Sardinia. Pl. 94.

STYRACACEAE | Storax Family

Trees or shrubs with alternate, entire, or toothed lvs., often with scale-like or star-shaped hairs. Calyx and corolla four- to five-lobed; stamens fused to the corolla and usually double the number of the corolla lobes. Ovary three- to five-celled; fr. a drupe or dry one-seeded unit. 1 sp. in Europe.

STYRAX | Storax

977. *S. officinalis* L. STORAX. A shrub or small tree 2–7 m., with white woolly twigs, oval deciduous lvs., and clusters of large white fls. 2–4 cm. across. Clusters with 3–6 short-stalked fls. borne on a short hairy stem; corolla with a short tube and with 5–7 lance-shaped lobes; calyx cup-shaped, hairy. Lvs. entire, stalked, green above, white-woolly beneath. Fr. leathery, swollen, white-woolly with persistent calyx; seeds large 1 or 2. △ Woods, thickets, streamsides. May. F.I.YU.GR.TR. *Adult plants yield an aromatic gum known as 'Gum Storax' which is used in pharmacy, perfumery, and in incense.*

OLEACEAE | Olive Family

Trees or shrubs with entire or pinnate, often opposite lvs. without stipules. Fls. in terminal or axillary clusters. Calyx bell-shaped or tubular, four-lobed; corolla bell-shaped or tubular, four-lobed or petals rarely free; stamens 2. Ovary superior, two-celled. Fr. dry or fleshy.

Nos.

1. Lvs. compound, with 3–15 leaflets.
  2. Trees; twigs grey.   *Fraxinus*   978, 979
  2. Shrubs; twigs green.   *Jasminum*   983, 984
1. Lvs. simple.
  3. Fls. lilac or white in terminal clusters.
    4. Fls. lilac.   *Syringa*   980
    4. Fls. white.   *Ligustrum*   982
  3. Fls. greenish or yellowish in axillary clusters.
    5. Lvs. grey above and silvery beneath; usually trees.   *Olea*   985
    5. Lvs. green on both faces; usually shrubs.   *Phillyrea*   981

FRAXINUS | **Ash**   Trees with compound lvs. Petals 4, free, or petals absent; calyx four-lobed or absent; stamens 2. Fr. one-seeded and forming a winged unit, or *samara*. *c.* 8 sps.

978. *F. excelsior* L. ASH. A tall, erect tree 15–30 m., with smooth silvery-grey bark, large compound lvs., and dense rounded clusters of brownish-purple fls. borne on last year's twigs, before the lvs. Clusters axillary, branched, 3–6 cm.; calyx and corolla absent. Lvs. to 30 cm. with 9–15 lance-shaped to oval pointed, toothed, stalkless leaflets *c.* 7 cm.; buds black, swollen. Fr. oblong 3–4 cm. △ Open woods, banks. Apr.–May. All Eur. (except P.IS.). *A valuable timber tree used for many purposes owing to its lightness and elastic strength. It is used for carriage-making, ladders, handles, sports equipment, walking sticks, etc. The bark is used for tanning; it is also used medicinally. The leaves and fruits are poisonous to livestock.*

979. *F. ornus* L. MANNA or FLOWERING ASH. A rather small tree 4–10 m. with compound lvs. like 978, but at least some of the leaflets shortly stalked. Fls. conspicuous, whitish, in sweet-scented pyramidal clusters, appearing at the same time as the lvs. Petals linear, to 1½ cm.; calyx deeply four-lobed; anthers with long filaments. Leaflets 5–9, paler or whitish beneath; buds brown or silvery-grey. Fr. 3–3½ cm. long. △ Thickets, open woods, hedges; sometimes planted. Apr.–May. South-East Eur. E.F.D.CH.A.H.I. *The tree yields a sappy edible exudation called 'manna' and it is grown for this in Sicily and Calabria. Manna is used as a mild laxative, particularly for children.* Pl. 94.

SYRINGA | **Lilac**   Small trees or shrubs with simple lvs. Fls. lilac; calyx bell-shaped with 5–9 lobes; corolla funnel-shaped with 4–5 spreading lobes. Fr. a capsule. *c.* 3 sps.

980. *S. vulgaris* L. LILAC. A branched, freely suckering, deciduous shrub 2–7 m., with large heart-shaped lvs. and dense pyramidal clusters of sweet-scented lilac or rarely white fls. Corolla tube 1–1½ cm., conspicuous, lilac, lobes 4, oval, spreading lilac; stamens included; calyx very short *c.* 2 mm. Lvs. 6–16 cm., entire, pointed, hairless. Fr. 1–1½ cm., smooth. △ Thickets, rocky places; grown for ornament and widely naturalized. May–July. South-East Eur. (except TR.).

PHILLYREA   Evergreen shrubs or small trees with simple lvs. Fls. greenish or yellowish, in axillary clusters; calyx cup-shaped; corolla four-lobed; stamens projecting. Fr. a fleshy drupe. 3 sps.

981. *P. media* L. A stiff, erect, grey-stemmed shrub 2–4 m., with paired, leathery, evergreen lvs. and small axillary clusters of greenish-yellow fragrant fls. Clusters rounded *c.* 1 cm.; fls. rather numerous, *c.* 3 mm. across. Lvs. 2–4½ cm. by 1–2 cm. oval-elliptic, wedge-shaped

to rounded at the base, short-stalked, usually shallowly toothed. Fr. 5–7 mm., globular, fleshy, bluish-black. △ Evergreen thickets, rocky ground, open woods. Apr.–May. Med. Eur. P. *The stems give excellent charcoal; the bark and leaves are used in herbal remedies.*

P. latifolia L. Like 981, but lower lvs. 1½–4½ cm. by ½–3 cm. broadly oval with a heart-shaped base, toothed, the upper narrower, rounded at the base. A shrub or small tree to 6–8 m. Mediterranean Europe.

P. angustifolia L. Like 981, but lvs. 2½–4 cm. by ½–1 cm., lance-shaped with entire margin. A shrub usually less than 2 m. Portugal to Yugoslavia.

LIGUSTRUM | **Privet** Deciduous shrubs with simple lvs. Fls. white, densely clustered; calyx soon falling; corolla funnel-shaped, four-lobed; stamens not projecting. Fr. a black berry with oily flesh. *c.* 3 sps.

982. *L. vulgare* L. COMMON PRIVET. A deciduous shrub to 5 m. with opposite, lance-shaped lvs. and dense terminal clusters, 3–6 cm., of strong-smelling white fls. Corolla 4–5 mm. across, tube as long as the spreading lobes and longer than the calyx. Lvs. 3–6 cm., blunt or acute, rather thick, short-stalked, hairless, falling late in the year; young branches softly hairy. Fr. 6–8 mm., black, shining. △ Hedges, forest clearings, uncultivated ground. May–July. All Eur. (except IS.SF.). *The fruits give a black dye used by hatters and glove-makers, and also to colour wine. The branches have been used for basket- and hurdle-making. The fruits are poisonous.*

L. ovalifolium Hasskn. BROAD-LEAVED PRIVET. Very like 982, but lvs. oval and remaining on stems most of winter. Young branches hairless. Corolla tube 2–3 times as long as the lobes. Var. *aureum* GOLDEN PRIVET has lvs. with a broad, golden-yellow border. Native of Japan; often planted for hedges and sometimes naturalized.

JASMINUM | **Jasmine** Shrubs or woody climbers with green branches and alternate compound lvs. or more rarely simple opposite lvs. Fls. yellow or white; corolla funnel-shaped, four- to five-lobed; calyx four- to nine-lobed; stamens included in corolla tube. Fr. a deeply lobed or paired berry. *c.* 4 sps.

983. *J. fruticans* L. WILD JASMINE. An erect, much-branched evergreen or semi-evergreen shrub 30–100 cm. with green angular stems and small clusters of unscented yellow fls. at the ends of the branches. Clusters short-stalked with 1–5 fls.; corolla 1½ cm. long and wide; calyx lobes linear, less than half as long as the corolla tube. Lvs. usually with 3 thick, shining, oblong leaflets 1–2 cm., or simple; twigs hairless, brittle. Fr. black, shining, 6 mm. △ Bushy and rocky places. May–June. Med. Eur. P.BG.R.SU. Pl. 95.

984. *J. officinale* L. COMMON JASMINE. A climbing woody perenn. 10–15 m., with pinnate lvs. and clusters of white, very sweet-scented fls. Fls. *c.* 2½ cm. across in clusters of 3–8; calyx linear, more than half as long as the corolla tube. Lvs. opposite, with 3–7 lance-shaped entire leaflets, the terminal leaflet larger. Var. *grandiflorum* (L.) L. H. Bailey (*J. grandiflorum* L.) SPANISH JASMINE has shining lvs. with the upper 3–5 leaflets joined together at the base. Fls. larger to *c.* 4 cm. across, white tinged on the outside with purple. △ Native of Asia; widely grown for ornament and sometimes self-seeding in Southern Europe. May–Sept. *The flowers are used in perfumery.* Pl. 95.

OLEA | **Olive** Small evergreen trees with opposite entire lvs. Fls. in simple or branched axillary clusters; calyx four-lobed; corolla tube short, four-lobed; stamens projecting. Fr. somewhat fleshy with a hard, one-seeded nut. 1 sp.

985. *O. europaea* L. OLIVE. Small, often gnarled and twisted trees 2–10 m. with greyish twigs and greyish-green evergreen lvs. and small erect clusters of whitish fls. Corolla with 4 spreading oval lobes ½ cm. across; stamens 2, projecting. Lvs. oval-lance-shaped 4–10 cm., glaucous-white beneath. Fr. green, at length blackish-purple. Var. *oleaster* (Hoffmanns. & Link) Fiori has much-branched, four-angled, somewhat spiny twigs, with small oval or elliptic lvs. Fr. inedible. △ Thickets, waste ground; widely grown in the Mediterranean region and often naturalized. May–June. Med. Eur. P. *Very widely cultivated since classical times. The fruit yields up to 50 per cent of a valuable edible oil, widely used for cooking, and in soap-making, pharmacy or as a lubricant; and in the past for illumination. Black olives are partly ripened fruit, while green olives are preserved in salt and oil before ripening. The wood is hard and beautifully grained and is used in turnery and cabinet-making.* Pl. 94.

GENTIANACEAE | Gentian Family

Usually hairless herbs with opposite, entire, often stalkless lvs. without stipules. Calyx four- to five-lobed; corolla tubular with 4–5 lobes, overlapping and twisted in bud; stamens 5, attached to tube. Ovary superior, usually one-celled; style 1, stigma 1 or 2. Fr. usually a capsule with persistent corolla, many-seeded.

		Nos.
1. Corolla and calyx with 6–8 lobes; stamens 6–8; fls. yellow.	*Blackstonia*	988
1. Corolla and calyx with 4–5 lobes; stamens 4–5; fls. various-coloured.		
2. Corolla deeply divided into 5 spreading lobes; fls. dull violet.	*Swertia*	1002
2. Corolla tubular or bell-shaped with 4–5 short lobes, not or rarely dull violet.		
3. Style absent or very short.		
4. Throat of corolla fringed with hairs or corolla lobes fringed with hairs.	*Gentianella*	999–1001
4. Neither throat of corolla nor corolla lobes fringed with hairs.	*Gentiana*	989–998
3. Styles distinct, long and slender.		
5. Fls. yellow; corolla lobes 4; calyx bell-shaped, not angled.	*Cicendia*	[1001]
5. Fls. usually pink, rarely yellow; corolla lobes usually 5; calyx tubular, angled.	*Centaurium*	986, 987

CENTAURIUM | Century   Fls. usually pink, rarely white or yellow. Calyx and corolla usually 5; calyx tubular with 5 linear, keeled lobes; corolla funnel-shaped with 5 spreading lobes; stamens 5, anthers spirally twisted after shedding pollen. The genus is often called *Erythraea. c.* 15 sps.

Fls. bright pink or white
986. *C. erythraea* Rafn (*C. umbellatum* auct.; *C. minus* auct.) COMMON CENTAURY. An erect, little-branched ann. 2–50 cm., with a basal rosette of lvs. and a dense, terminal, flat-topped cluster of bright pink fls. Corolla tube longer than the calyx, corolla lobes 5–6 mm., spreading; stamens attached to top of the corolla tube. Rosette lvs. 1–5 cm., oval-elliptic, prominently three- to seven-veined, the stem lvs. narrower. △ Dry grasslands,

open woods, fallow. June–Sept. All Eur. (except IS.). *Well known as a febrifuge since classical times; it yields a yellowish-green dye.* Pl. 95.

C. littorale (D. Turner) Gilmour Like 986, but lvs. all strap-shaped or linear-spathulate, the basal lvs. 5 mm., broad or less. Corolla pink, lobes 6–7 mm., spreading. Coasts of North-Western Europe, and parts of Central and Eastern Europe.

C. pulchellum Druce Like 986, but fls. all more or less stalked at the ends of short branches in a lax, much-branched or little-branched cluster. Corolla tube longer than calyx, lobes of corolla bright pink, 2–4 mm. Basal rosette of lvs. absent. Widespread in Europe.

Fls. yellow

987. *C. maritimum* (L.) Fritsch YELLOW CENTAURY. An erect, slender, little-branched, hairless ann. 5–15 cm., with few yellow fls. *c.* 2 cm. long in a lax, branched cluster, or fls. rarely solitary. Corolla *c.* 1 cm. across; calyx lobes awl-shaped, shorter than the corolla tube; style divided to the middle into 2 stigmas. Basal rosette absent, the lower lvs. oval, smaller than the few wide-spreading, 2–3 cm., narrow-elliptic upper lvs. Fr. twice as long as the calyx. △ Heaths and sandy places on the littoral. Apr.–June. Med. Eur. P.BG. Page 315.

BLACKSTONIA | Yellow-Wort Fls. yellow; calyx, corolla, and stamens 6–8; corolla with a short tube and spreading lobes. Style deeply bilobed. The genus is sometimes called *Chlora. c.* 5 sps.

988. *B. perfoliata* (L.) Hudson YELLOW-WORT. A very glaucous, slender, erect ann. 20–80 cm. with a lax head of bright yellow fls. and broad, paired lvs. fused across the base and encircling the stem. Fls. 1–1½ cm. across; corolla lobes widely spreading; calyx lobes awl-shaped. Basal lvs. in a rosette, not fused, obovate 1–2 cm., the upper lvs. triangular-oval, widely fused. △ Grasslands, shady and rocky places, sandy ground. May–Sept. Most of Eur. (except North Eur. NL.). *The plant yields a yellow dye; a bitter plant, sometimes used in herbal remedies.* Pl. 94.

GENTIANA | Gentian Fls. yellow, purple, or blue; calyx tubular or bell-shaped, usually five-lobed; corolla bell-shaped or funnel-shaped, usually four- to five-lobed, rarely ten-lobed or ten-toothed; stamens 4–10, fused to the corolla. Ovary one-celled, stigma bilobed. Fr. a capsule splitting into 2 valves and surrounded by the dead corolla. *c.* 30 sps.

Fls. blue or violet

(a) Corolla four-lobed

989. *G. cruciata* L. CROSS GENTIAN. A robust leafy perenn. to 50 cm., with broad lvs. and dense terminal and axillary clusters of few large, stalkless, dull blue fls. Fls. 2–2½ cm. long; corolla barrel-shaped with 4 short, broad, acute lobes, pleated in the throat; anthers free; stigmas recurved. Lvs. 5–10 cm. by 1–2 cm., oval to lance-shaped, with 3–5 veins, the lower narrowed to a stalk, the upper with fused sheathing bases. △ Dry meadows, woods, rocks, banks. June–Sept. Cent., South-East Eur. (except GR.TR.) E.F.B.NL.I.SU. *The roots are tonic and febrifuge.* Pl. 96.

(b) Corolla five-lobed

(i) Anns.

990. *G. nivalis* L. SNOW GENTIAN. A slender ann. 2–15 cm., usually branched from below, with tiny lvs. and small, brilliant, deep blue, terminal fls. 6–10 mm. across. Corolla lobes elliptic-acute, spreading; calyx five-angled, divided to one-third into linear lobes as long as the corolla tube. Basal lvs. oval *c.* ½ cm., in a rosette, the stem lvs. elliptic 1 cm. △ Arctic or alpine: meadows, damp places, heaths. July–Aug. North Eur. (except DK.), Cent. Eur. (except H.) E.F.I.YU.BG.R.SU.

1. *Centaurium maritimum* 987
2. *Cynanchum acutum* 1008
3. *Plumbago europaea* 971
4. *Gomphocarpus fruticosus* 1012
5. *Limonium bellidifolium* 974

G. utriculosa L. BLADDER GENTIAN. Like 990, but fls. larger 1–1½ cm. across and corolla lobes oval-blunt, brilliant blue. Calyx ovoid and angled with 5 broad wings, lobes triangular. Branched ann. 8–25 cm. Alps, Apennines, Carpathians, Balkans. Pl. 96.

(ii) Perenns.

991. *G. verna* L. SPRING GENTIAN. A lax cushion-forming perenn. with lvs. in rosettes and solitary, brilliant deep blue, short-stemmed fls. 1½–3 cm. across. Corolla tube 3–4 cm. long, the lobes oval, spreading, and with a short white scale between each lobe; calyx angled, angles often narrowly winged. Rosette lvs. ½–2 cm. elliptic to lance-shaped; fl. stem 3–12 cm. with 1–3 pairs of smaller lvs. A variable sp. with fls. occasionally pale blue, pinkish, reddish-purple, or white. △ Alpine turf, stony grassy places. Mar.–Aug. Most of Eur. (except North Eur. P.B.NL.H.TR.). Pl. 96.

G. bavarica L. BAVARIAN GENTIAN. Like 991, but lowest lvs. not in a rosette and stems loosely mat-forming; lvs. yellow-green, all about the same size ½–1½ cm., oval to spathulate, very blunt. Fls. bright blue, 1½–2 cm. across; corolla tube often twice as long as the violet-tinged calyx; fl. stems with 3–4 pairs of closely spaced lvs. Alps, Carpathians.

992. *G. kochiana* Perr. & Song. TRUMPET GENTIAN. Readily distinguished by its very large, solitary, brilliant, dark blue, trumpet-shaped fl. 4–6 cm. long, borne on a short stalk above a small rosette of lvs. Corolla tube spotted with green within, lobes spreading, acute; calyx teeth spreading, slightly narrowed at their base and joined to each other at a blunt angle. Lvs. oval to elliptic, blunt, often little longer than broad, in a lax rosette 5–15 cm. across; fl. stem 2–8 cm., usually with 2 pairs of lvs. △ In mountains: pastures, rocky places. July–Aug. Cent. Eur. (except H.) F.I.YU.AL.BG.R. Pl. 96.

G. clusii Perr. & Song. STEMLESS TRUMPET GENTIAN. Like 992, but lvs. leathery elliptic-lance-shaped, acute, much longer than broad. Calyx teeth erect, broadest at their base and joined to each other at an acute angle. Corolla 5–6 cm., bright blue not green within; fl. stem shorter. Alps, Apennines, Carpathians. Pl. 96.

993. *G. pneumonanthe* L. MARSH GENTIAN. An erect, unbranched, leafy-stemmed perenn. 10–50 cm., with linear lvs. and a rather dense cluster of 1–7 large bright blue fls. striped with green on the outside. Fls. short-stalked, arising from the axils of the upper lvs.; corolla trumpet-shaped 2½–4 cm. long, with broadly oval, ascending lobes. Lvs. 1½–4 cm., one-veined, the lowest lvs. scale-like. △ Wet heaths, bogs, marshes. July–Oct. Most of Eur. (except IRL.IS.TR.). Pl. 96.

994. *G. asclepiadea* L. WILLOW GENTIAN. A spreading or erect, very leafy-stemmed perenn. 20–50 cm. with lance-shaped lvs. and large, dark blue, usually solitary, stalkless fls. in the axils of the upper lvs. Corolla 3½–5½ cm. long, trumpet-shaped, blue and spotted with violet within and with paler stripes outside, lobes spreading, oval-acute. Lvs. 3–8 cm., usually four-ranked, but in shady places the stem is often arched and the lvs. are two-ranked and the fls. are turned to the upper side; lv. blades strongly five-veined. △ Hills and mountains: woods, shady places, damp meadows. Aug.–Sept. Much of Eur. (except North Eur. P.IRL.GB.B.). Pl. 97.

(c) Corolla with 10 lobes

995. *G. pyrenaica* L. PYRENEAN GENTIAN. A low cushion-forming perenn. 3–10 cm. with solitary, short-stalked, blue-violet fls. with 10 distinct corolla lobes. Fls. 2–3 cm. long; corolla with a slender tube twice as long as the calyx and with 5 longer and broader entire lobes alternating with 5 shorter and toothed lobes. Lvs. in dense rosettes, elliptic-acute, rough-margined, leathery; fl. stem 3–10 cm. △ Mountain pastures. June–Sept. E.F.R.

Fls. yellow or dull purple; robust plants

996. *G. lutea* L. GREAT YELLOW GENTIAN. A stout, erect, hairless perenn. ½–2 m., with large,

rounded, strongly veined lvs. and dense axillary and terminal whorls of numerous short-stalked golden-yellow fls. Fls. *c.* 2½ cm. long; corolla with 5–9 lance-shaped lobes more or less spreading in a star, not tubular; calyx papery. Lvs. 20–30 cm., broadly oval, seven-veined, bluish-green, the upper stalkless, clasping, the lower stalked; stem robust, hollow. △ In mountains: meadows, damp places, woods. June–Aug. Much of Eur. (except North Eur. IRL.GB.N.NL.TR.SU.). *The roots, after gathering, are allowed to ferment and are then used to prepare a bitter liqueur; they are also used medicinally. They are highly valued as a tonic, febrifuge, and antiseptic in herbal remedies.* Pl. 95.

997. *G. punctata* L. SPOTTED GENTIAN. A stiff, erect, rather robust perenn. 20–60 cm., with elliptic greyish-green lvs. and dense terminal and axillary clusters of large pale yellow and purple-spotted fls. Corolla widely bell-shaped *c.* 4 cm. long, throat pleated, lobes oval; calyx tubular with 5–8 unequal teeth, not split to the base. Basal lvs. 10 cm., elliptic, stalked, the upper stalkless, blade strongly five- to seven-veined. △ In mountains: meadows, open woods, rocky places. July–Sept. Cent. Eur. (except H.), South-East Eur. (except TR.) E.I. Pl. 97.

998. *G. purpurea* L. PURPLE GENTIAN. Like 997, but fls. purplish-red, yellowish at base and striped and spotted with green; calyx papery, split on one side to the base. Corolla open-bell-shaped, 3–4 cm. long, with 5–8 rounded lobes. Lvs. *c.* 20 cm. elliptic, the basal lvs. long-stalked and with an acute apex, the stem lvs. stalkless; fl. stem 10–30 cm. △ In mountains: pastures, rocky places. July–Aug. F.N.D.CH.A.I.R. Pl. 97.

G. pannonica Scop. BROWN or HUNGARIAN GENTIAN. Like 998, but calyx with 5–8 thick, unequal, reflexed lobes and not split to the base. Corolla dull violet-purple, spotted with black-purple. Lvs. elliptic to lance-shaped. Eastern Alps and Carpathians.

GENTIANELLA Like *Gentiana*, but usually anns. or bienns. Fls. purplish, whitish or blue; throat of corolla fringed with long hairs or corolla lobes with long spreading hairs on margin. Often placed in *Gentiana. c.* 8 sps.

Fls. with dense hairs in throat of corolla
999. *G. campestris* (L.) Börner FIELD GENTIAN. An erect, often little-branched ann. or bienn. 10–30 cm., bearing a terminal cluster of dull purple or rarely white fls. with a distinctive fringe of long paler hairs in the throat of the corolla. Fls. few or many in a cluster; corolla lobes 4, the tube 1½–3 cm. long, equalling or longer than calyx which is divided nearly to the base into 4 unequal lobes, the 2 larger outer lobes encircling the 2 inner; ovary stalked. Basal lvs. 1–2½ cm. oval, blunt; stem lvs. narrower, oblong acute. △ Clearings, pastures, dunes. May–Oct. Much of Eur. (except P.IS.H.AL.YU.TR.BG.R.). *The roots contain bitter properties; used in brewing in place of hops.* Pl. 96.

G. amarella (L.) Börner FELWORT. Like 999, but calyx lobes usually 4, more or less equal and not overlapping, pressed to corolla. Fls. reddish-violet; corolla lobes 4 or 5, both often found on the same plant, corolla tube 12–22 mm. by 4 mm., twice as long as calyx or less. Ovary stalkless. Widespread in Europe, except in the South-East.

G. germanica (Willd.) Börner LARGE FIELD GENTIAN. Like 999, but corolla tube larger, 2½–3½ cm. by 1 cm., bluish-violet, and twice as long as the calyx; calyx with 5 more or less equal, spreading, hairless or finely hairy lobes. Ovary stalked. A variable sp. Central Europe.

1000. *G. tenella* (Rottb.) Börner SLENDER GENTIAN. A tiny, slender, branched or unbranched ann. 3–10 cm., with very small long-stalked terminal, rather dull blue-violet fls. 4–6 mm. across. Corolla usually four-lobed with lobes oval, and almost as long as tube, corolla tube 8–12 mm. long, hairy in throat; calyx with 4–5 deep oval lobes; fl. stalks very slender, many times longer than the fls. Lower lvs. spathulate, the upper oblong with 1 vein. △ Damp pastures in mountains. July–Aug. North Eur. (except DK.SF.), Cent. Eur. (except H.) E.F.I.R.SU.

Fls. with corolla lobes fringed with hairs
1001. *G. ciliata* (L.) Borkh. FRINGED GENTIAN. Readily distinguished by the large blue fls. with 4 broad, spreading, corolla lobes with their margins conspicuously fringed with blue hairs. Fls. 4–5 cm. across, long-stalked, 2–5 in a lax cluster from the axils of the uppermost lvs.; corolla lobes oval. Lvs. 3–4 cm. linear-lance-shaped, one-veined, not in a rosette. A slender flexuous-stemmed perenn. 8–25 cm. △ Dry meadows, rocky places, woods. Aug.–Oct. Cent., South-East Eur. (except AL.TR.) E.F.B.I.SU.

*Cicendia filiformis (L.) Delarbre (*Microcala f.*) Western and Southern Europe.

SWERTIA Calyx divided to the base into 4–5 narrow lobes; corolla divided almost to the base into 4–5 spreading lobes. Ovary one-celled. 3 sps.

1002. *S. perennis* L. MARSH FELWORT. An erect, unbranched, hairless perenn. 20–60 cm., with pale greenish-yellow, lance-shaped lvs. and terminal clusters of widely opening, dark violet-purple fls. Fls. 2½–3½ cm. across; corolla lobes 5, lance-shaped, dark-spotted and with 2 dark violet nectaries at the base; calyx with 5 linear-lance-shaped, spreading lobes. Lower lvs. 5–15 cm., long-stalked, blunt, the uppermost stalkless and clasping stem, all strongly five- to seven-veined. △ In mountains: marshes, boggy meadows. July–Sept. Cent. Eur. (except H.) E.F.I.YU.BG.R.SU. Pl. 95.

MENYANTHACEAE | Bogbean Family

Differing from Gentianaceae, but often included in this family, in being aquatic plants usually with underwater creeping and rooting stems and having alternate, simple or compound, floating or emergent lvs. Corolla with a short tube and 5 spreading lobes, placed edge to edge, or *valvate*, in bud, not overlapping, and not persisting in fr. Calyx, corolla, and stamens 5; ovary 1; fr. a capsule.

Lvs. trifoliate; fls. pink or white.   *Menyanthes*
Lvs. orbicular; fls. yellow.   *Nymphoides*

MENYANTHES | Bogbean   Lvs. compound, trifoliate; fls. in a spike-like cluster. 1 sp.

1003. *M. trifoliata* L. BOGBEAN, BUCKBEAN. A submerged aquatic perenn. with thick creeping and rooting stems and with large trifoliate lvs. carried above water, or a creeping bog plant. Fl. stems erect, bearing above water a rather dense, leafless, pyramidal cluster of pinkish-white, conspicuously fringed fls. Fls. 1½ cm. across, buds pink; corolla with 5 acute spreading lobes with long white hairs on the upper surface. Leaflets 3½–7 cm. obovate entire, lv. stalks long, with a sheathing base; fl. stem 12–30 cm. Fr. globular. △ Ponds, ditches, lake verges, marshes, bogs. Apr.–June. All Eur. *The leaves are bitter and have been used in place of hops and in herbal remedies.* Pl. 98.

NYMPHOIDES   Lvs. simple, orbicular; fls. yellow, long-stalked. 1 sp.

1004. *N. peltata* (S. G. Gmelin) O. Kuntze FRINGED WATER-LILY. A submerged aquatic plant with slender, creeping and rooting underwater stems, rather small, rounded, floating lvs. and rather large golden-yellow fls. *c.* 3 cm. across, carried above the surface. Fls. long-stalked, borne in clusters of 2–5 from the axils of the upper lvs.; corolla lobes spreading, rounded, and fringed with long spreading hairs. Lv. blade 3–10 cm. rounded, but deeply and narrowly cleft at the attachment of the stalk, shining, purple-blotched above, purplish beneath. △ Still and slow-flowing waters. June–Sept. Most of Eur. (except North Eur. IRL.TR.): introd. DK.S. Pl. 97.

APOCYNACEAE | Dogbane Family

Trees, shrubs, or herbs with milky juice and usually opposite or whorled, entire lvs. Fls. radially symmetrical; calyx 5; corolla tubular at base and with 5 spreading lobes, often with additional lobes or segments, lobes contorted in bud; stamens 5, fused to the corolla tube; ovary with 2 carpels; style 1. Fr. often an elongated capsule splitting longitudinally; seeds commonly with a tuft of hairs.

Trailing perenns.; fls. usually blue, solitary. *Vinca*
Erect shrubs; fls. usually pink in flat-topped clusters. *Nerium*

VINCA | Periwinkle Fls. solitary, axillary; corolla tubular and with 5 spreading lobes; stamens with anthers terminating in triangular projections which arch over the stigma; stigma enlarged above and with a tuft of hairs at apex. Fr. of 2 spreading carpels fused only at the base; seed without hairs. 4 sps.

1005. *V. minor* L. LESSER PERIWINKLE. A trailing slender-stemmed perenn. 1–2 m., rooting at the nodes, with evergreen lvs. and solitary blue-violet, or rarely white or pink fls. Corolla 2½–3 cm. across, lobes blunt, spreading; calyx lobes lance-shaped, hairless; fl. stalk longer than the lvs. Lvs. 2½–4 cm. elliptic, narrowed to the base to a very short stalk, leathery, hairless. Fr. 2½ cm. △ Woods, hedges, rocks. Feb.–May. Most of Eur. (except North Eur. IRL.AL.). Pl. 97.

V. difformis Pourret (*V. media* Hoffmanns. & Link) INTERMEDIATE PERIWINKLE. Like 1005, but lvs. broader, oval-lance-shaped with a rounded base. Corolla blue, 3–4 cm. across, lobes pointed; fl. stalks half as long as the lvs. Sterile branches trailing, not rooting at the nodes. Portugal to Italy.

V. herbacea Waldst. & Kit. HERBACEOUS PERIWINKLE. Like 1005, but stems and lvs. soft and herbaceous, not evergreen and not rooting at the nodes. Fls. violet-blue, 2½–3 cm. across, lobes blunt; calyx lobes linear-lance-shaped, with spreading hairs or hairless. Lvs. with finely hairy margin. Bushy places. South-Eastern Europe. Pl. 97.

1006. *V. major* L. GREATER PERIWINKLE. A rather robust perenn. spreading 1–2 m., with trailing herbaceous stems rooting only at the tips, and large solitary blue fls. 4–5 cm. across. Corolla lobes blunt; calyx linear, hairy; fl. stalks shorter than lvs. Lvs. 2–4 cm. oval with a rounded or widely heart-shaped base, margin finely hairy; lv. stalk *c.* 1 cm. Fr. 4–5 cm. △ Woods, hedges, shady rocks, streamsides; grown as an ornamental and often naturalized. Mar.–June. Med. Eur. (except TR.): introd. elsewhere.

NERIUM | Oleander Fls. large, pink or white, in terminal clusters; throat of corolla with 5 broad lobes. Bush or small tree. 2 sps.

1007. *N. oleander* L. OLEANDER. A robust shrub or small tree 2–5 m., with terminal clusters of large pink or rarely white scented fls. and stiff, leathery, grey-green lvs. Corolla 3–5 cm. across, the lobes blunt, spreading and with conspicuous toothed lobes arising from the throat; calyx lobes linear-lance-shaped, very glandular within. Lvs. 10–20 cm., lance-shaped, opposite or in whorls of 3. Fr. 10–17 cm.; seeds with long brown hairs. △ Dry water-courses, ravines; often grown as an ornamental. June–Sept. Med. Eur. P. *A poisonous plant to both man and livestock; it has been used medicinally and as an insecticide.* Pl. 98.

ASCLEPIADACEAE | Milkweed Family

Herbs or shrubs often with milky juice, and with simple, opposite or whorled lvs. Fls. radially symmetrical; calyx 5; corolla with 5 lobes and with a corona which is either fused to the corolla tube or to the 5 stems or to both; stamens with anthers fused to each other round the stigma, or with anthers fused to the stigma. Fr. of 2 often widely divergent carpels, splitting longitudinally; seeds often with long silky hairs.

1. Climbing plants or plants less than 1 m.; corolla with spreading lobes.
  2. Climbing plants.
    3. Lvs. glaucous, with deeply heart-shaped bases.   *Cynanchum*
    3. Lvs. dark shiny green, oval to lance-shaped.   *Periploca*
  2. Erect perenns.   *Vincetoxicum*
1. Robust erect perenns. 1–2 m.; corolla with reflexed lobes.
  4. Fls. white in a lax umbel; lvs. acute.   *Gomphocarpus*
  4. Fls. pink in a dense umbel; lvs. blunt.   *Asclepias*

CYNANCHUM Climbing perenns. with glaucous, oval-heart-shaped lvs. Fls. in axillary clusters; corolla with 5 spreading lobes and corona with 5 long appendages. 1 sp.

1008. *C. acutum* L. STRANGLEWORT. A rather slender, glaucous, hairless, climbing perenn. ½–6 m. with triangular-oval lvs. with a deeply heart-shaped base and axillary umbels of sweet-scented, white or pink fls. Corolla tubular 5–7 mm. across, lobes blunt, throat with 10 strap-shaped scales forming a tubular-bell-shaped corona; calyx finely hairy; inflorescence stalks hairy. Lvs. *c.* 5 cm., stalked. Fr. with diverging, almost straight carpels, each 16 cm. by *c.* ½ cm.; seed with a tuft of very long hairs. △ Damp places, maritime sands, hedges. July–Sept. Med. Eur. P.BG.R.SU. *Its white juice is a violent purgative.* Page 315.

PERIPLOCA Climbing perenns. with oval to lance-shaped lvs. Fls. in axillary clusters, greenish-brown; corona of 5 slender appendages. 2 sps.

1009. *P. graeca* L. SILK-VINE. A climbing perenn. to 10 m. with oval to lance-shaped, dark, shining green lvs. and long-stalked, lax clusters of brownish fls. which are greenish on the outside. Corolla *c.* 2½ cm. across, lobes spreading, blunt, hairy; corona of 5 slender thread-like appendages. Lvs. 7–10 cm., tardily deciduous, stalked. Fr. *c.* 12 cm. △ Bushy places, hedges. July. I.YU.GR.TR.: introd. E.BG. Pl. 97.

VINCETOXICUM Low, erect, herbaceous perenn. with oval-heart-shaped lvs. Fl. clusters terminal and axillary; corolla with 5 spreading lobes, corona usually of 5 short, rounded scales. *c.* 10 sps.

1010. V. hirundinaria Medicus (*Cynanchum vincetoxicum* (L.) Pers.) COMMON VINCETOXI-CUM. An erect, rarely weakly climbing, leafy perenn. to 1 m., with oval to lance-shaped lvs. and terminal and axillary clusters of dull greenish-white or yellowish fls. Fl. clusters rather long-stalked; corolla *c.* ½ cm. across, usually hairless within. Lvs. opposite, short-stalked, oval to lance-shaped with rounded or shallow heart-shaped base, finely hairy. Fr. 4–6 cm., enlarged towards the base. A very variable sp. △ Woods, uncultivated ground, rocks. June–Sept. Most of Eur. (except IRL.GB.IS.). *A poisonous plant.* Pl. 99.

V. nigrum Moench DARK VINCETOXICUM. Like 1010, but fls. reddish becoming blackish; corolla hairy within; corona of 10 alternating small and larger lobes. Lvs. short-stalked or stalkless, oval-acute, hairy on the veins. Fr. 6–8 cm. Mediterranean region and Bulgaria.

ASCLEPIAS Corolla with 5 usually reflexed lobes and with a corona of 5 erect or spreading, hood-like projections, each bearing an incurved horn. Fr. of 1 carpel; seeds with a tuft of hairs. 1 sp.

1011. A. syriaca L. (*A. cornuti* Decne) SILKWEED. An unbranched, stout-stemmed, very leafy perenn. 1–2 m. with large oblong lvs. and large, dense, terminal and lateral umbels of numerous sweet-scented, pale purple fls. Corolla *c.* 1½ cm. across, lobes reflexed, corona horns erect. Lvs. 10–20 cm. downy beneath, alternate, stalked, conspicuously veined. Fr. oblong-oval, erect, white-woolly, and with few short spines. △ Native of North America; often grown as an ornamental and occasionally self-seeding. June–Aug. Naturalized in much of Central and Southern Europe. *The milky juice has been used medicinally for asthma.* Pl. 99.

GOMPHOCARPUS Like *Asclepias*, but corona of 5 horns only and without smaller projections. Lvs. linear-lance-shaped. 1 sp.

1012. G. fruticosus (L.) R.Br. BRISTLY-FRUITED SILKWEED. A branched shrubby perenn. 1–3 m., with linear-lance-shaped, long-pointed lvs. and white fls. in compound axillary umbels overtopped by the lvs. Fls. *c.* ½ cm. across; petals reflexed, corona horns blunt, erect. Lvs. 5–15 cm., green on both sides, margin inrolled. Fr. borne on last year's stems and present at flowering, to 6 cm., ovoid-acute and covered with numerous bristly hairs to 1 cm.; seeds netted, with silky hairs 2–3 cm. long. △ Gravelly river-beds, water-courses; grown as an ornamental and often naturalized. Apr.–Sept. A native of Africa: introd. Med. Eur. *The plant is emetic and purgative.* Page 315.

RUBIACEAE | Madder Family

Herbaceous or woody plants usually with whorled lvs.; stipules present. Fls. radially symmetrical, small, clustered. Sepals 4–5; corolla tubular with 4–5 spreading lobes; stamens 4–5, attached to the corolla; ovary inferior, usually two-celled; styles 2 or 1. Fr. a berry or a capsule, or splitting in 2 one-seeded units.

		Nos.
1. Fls. stalkless, in elongated spikes.		
2. Fls. in dense two- to four-ranked spikes; fr. smooth.	*Crucianella*	1015, 1016
2. Fls. in whorls forming a leafy spike.		
3. Fr. with 3–4 spiny horns.	*Valantia*	1030
3. Fr. smooth.	*Cruciata*	1029

1. Fls. in spreading stalked clusters or rounded heads.

4. Lvs. opposite; fls. pink.	*Putoria*	1013
4. Lvs. mostly or all whorled.		
5. Fr. a fleshy berry; corolla usually five-lobed.	*Rubia*	1031
5. Fr. dry with 1–2 nuts; corolla with 4–5 lobes.		
6. Corolla with very short tube and 4 widely spreading lobes.	*Galium*	1020–1028
6. Corolla with tube usually at least twice as long as lobes.		
7. Calyx of 4–6 distinct teeth, persistent in fr.	*Sherardia*	1014
7. Calyx with 4 very short teeth, soon falling.	*Asperula*	1017–1019

PUTORIA   A shrublet with opposite lvs. Corolla funnel-shaped, with 4–5 spreading lobes; stamens 4–5, projecting. Fr. oblong, crowned with the calyx. 1 sp.

1013. *P. calabrica* (L. fil.) DC. A strong-smelling, much-branched prostrate or spreading shrublet with leathery lvs. and dense terminal heads of pink fls. Corolla tube slender *c.* 1½ cm. long, lobes linear-lance-shaped; stamens projecting. Lvs. shining, elliptic or lance-shaped, margin inrolled, usually hairless, but branches and lvs. sometimes finely hairy. Fr. black. △ Rocky places by the sea and in mountains. May–July. E.I.YU.AL.GR. Pl. 98.

SHERARDIA   Fls. lilac in dense terminal heads surrounded by an involucre of bracts. Fr. encircled by the enlarged calyx with 4–6 teeth. 1 sp.

1014. *S. arvensis* L. FIELD MADDER. A spreading or erect ann. 5–40 cm., with small, compact, terminal heads of 4–8 pale lilac fls., encircled by leaf-like bracts longer than the fls. Corolla tube 4–5 mm., about twice as long as the calyx, lobes of corolla 4, spreading; bracts 8–10, lance-shaped. Lvs. elliptic-acute ½–2 cm., in whorls of 5–6, rough beneath. Fr. encircled by hairy persisting calyx. △ Fields, waysides, walls. Mar.–Sept. All Eur. (except IS.). Pl. 99.

CRUCIANELLA   Fls. yellowish, stalkless, overlapping, in dense two- to four-ranked spikes; bracts papery, paired. Corolla tube long, lobes 4–5, converging, not spreading. Lvs. whorled. *c.* 7 sps.

Anns.

1015. *C. angustifolia* L. A slender ann. 10–40 cm., with long and very slender four-ranked fl. spikes of numerous overlapping, rather papery bracts and very insignificant yellowish fls. only slightly longer than the bracts. Fl. spikes 3–8 cm. by 4–6 mm., long-stalked; corolla *c.* 6 mm. long; bracts lance-shaped, fused at the base, margin papery, hairy. Lower lvs. oval to oblong, the upper linear 1–1½ cm. △ Dry, stony, and rocky places. May–June. Med. Eur. P.BG.SU. Page 326.

Perenns.

1016. *C. maritima* L. A shrublet with a woody base and many stiff, erect, glaucous stems to 30 cm. with compact, rather broad, dense spikes with yellow fls. much longer than the bracts. Spikes 2–4 cm. by 1–1½ cm.; corolla tube *c.* 1 cm.; bracts oval-acute with papery margin, not fused at base. Lvs. *c.* ½ cm. lance-shaped, leathery, spiny-tipped, white-margined. △ Sandy and rocky places on the littoral. June–Sept. P.E.F.I. Page 326.

ASPERULA   Fls. whitish or bright blue in branched clusters. Corolla funnel-shaped, usually four-lobed, less commonly three- or five-lobed; stamens arising from the tube or throat of the corolla; calyx with an inconspicuous ridge. Fr. paired, with 2 nut-like units. *c.* 50 sps.

Fls. usually in lax branched clusters, without conspicuous involucre
1017. *A. cynanchica* L. SQUINANCY WORT. A slender woody-based perenn. with spreading branches to 40 cm., with linear lvs. in whorls of 4–6 and lax or rather dense, branched clusters of pale pinkish fls. Corolla tube *c.* 6 mm. long, funnel-shaped, white within, pink outside, lobes 4, spreading; bracts lance-shaped. Lvs. variable in length, 3–12 mm., fine-pointed, spreading, or recurved. Fr. 3 mm., with numerous swellings. A very variable sp. △ Dry pastures. June–Sept. Much of Eur. (except North Eur. P.). *The underground parts produce a red dye; used in herbal remedies in the past.*

A. tinctoria L. DYER'S WOODRUFF. Like 1017, but fls. white with a corolla tube 4 mm. and lobes often 3 and a little shorter than the tube; bracts oval. Lower lvs. usually in whorls of 6, the upper in whorls of 4 and the uppermost often opposite; stems erect, quadrangular, sparsely leafy above. Widespread in much of Europe. *The roots are red and produce a red dye.*

Fls. in terminal heads surrounded by a leafy involucre
(a) Anns.; fls. blue
1018. *A. arvensis* L. BLUE WOODRUFF. A slender, square-stemmed, erect ann. to 40 cm. with whorls of linear lvs. and a small terminal head of bright blue fls. surrounded by leafy involucre longer than the fls. Fls. 5–6 mm. long, corolla tube slender; bracts linear, bristly-haired. Lvs. 1–3 cm., the lower in whorls of 4, the upper in whorls of 6–8, apex blunt, margin rough hairy. Fr. smooth. △ Cornfields, cultivated ground. Apr.–June. Cent., Med., South-East Eur. P.SU.

(b) Perenns.; fls. white or pinkish
1019. *A. taurina* L. A slender, sparsely hairy, erect perenn. to 60 cm., with broad, three- to five-veined lvs. in whorls of 4 and a dense terminal head of numerous white or yellowish-pink fls. surrounded by a leafy involucre as long as the fls. Fls. ½ cm. across; corolla tube very slender 1–1½ cm., lobes strap-shaped; anthers projecting, usually violet; involucral bracts unequal, ciliate. Lvs. oval-lance-shaped, 4–5 cm. by 1–2 cm., acute, margin and veins with bristly hairs. △ Woods, thickets, rocky places in mountains. May–July. E.F.D.CH.A.H.I.YU.AL.BG.R.SU. Page 326.

GALIUM | Bedstraw Fls. white, yellow, or pinkish, usually in compound, branched, terminal and axillary clusters. Corolla with a short or very short tube and 4–5 spreading lobes; stamens 4, longer than the tube; styles 2. Lvs. in whorls of 4–10. Fr. of 2 one-seeded nuts. *c.* 100 sps.

Lvs. three-veined, in whorls of 4
1020. *G. boreale* L. NORTHERN BEDSTRAW. Erect, stiff, leafy perenn. to ½ m., with white fls. in compact, pyramidal, terminal, leafy clusters. Corolla 3 mm. across, lobes fine-pointed. Lvs. 1–4 cm. elliptic to lance-shaped, blunt, three-veined, with rough margin and under-sides of midvein, turning black when dried. Fr. 2½ mm., covered with dense, hooked bristles. △ Damp meadows, bushy places, rocks. May–Aug. Most of Eur. (except P.AL. GR.TR.).

1021. *G. rotundifolium* L. ROUND-LEAVED BEDSTRAW. A slender leafy perenn. 10–30 cm., distinguished by its broad oval or almost circular, fine-pointed lvs. in whorls of 4. Fls. *c.* 3 mm. across in few-flowered lax clusters, white or greenish-white. Lvs. 1–1½ cm., with bristly hairs; stems quadrangular, usually hairless, smooth. Fr. covered with dense hooked bristles. △ Bushy places in mountains. May–July. Cent., South-East Eur. (except AL.) P.E.F.S.SU.: introd. N. Page 326.

Lvs. one-veined, usually in whorls of more than 4
(a) Stems with recurved bristles or prickles on the angles
1022. **G. uliginosum** L. FEN BEDSTRAW. A slender, branched perenn. to $\frac{1}{2}$ m. with lvs. in whorls of 6–8 and stems with very rough angles with recurved prickles. Fls. white, in lax branched clusters; corolla $2\frac{1}{2}$–3 mm. across, lobes acute; anthers yellow. Lvs. $\frac{1}{2}$–$1\frac{1}{2}$ cm. linear to oblance-shaped, fine-pointed, margin bristly. Fr. 1 mm., rough-surfaced. △ Marshes, wet fields, waysides. May–Sept. All Eur. (except P.AL.TR.).

G. palustre L. MARSH BEDSTRAW. Like 1022, but lvs. blunt and in whorls of 4–6, blackening on drying; stems weak, with few bristles on the angles. Fls. white, 3–$4\frac{1}{2}$ mm. across in wide-spreading lax clusters; anthers red. Throughout Europe. Pl. 99.

1023. **G. aparine** L. GOOSEGRASS, CLEAVERS. A rough, prickly, weak-stemmed ann. to 1 m. with shoots easily detached and readily clinging to clothing, and few inconspicuous whitish fls. in axillary clusters. Fls. 2 mm. across in stalked clusters of 2–5, with a whorl of leaf-like bracts below. Lvs. 1–5 cm. linear-oblance-shaped, 6–8 in a whorl, margin with hooked prickles. Fr. 4–6 mm., with hooked bristles with swollen bases. △ Hedges, waste places, shingle beaches. May–Oct. All Eur. (except AL.): introd. IS.

G. tricornutum Dandy (*G. tricorne* auct.) ROUGH CORN BEDSTRAW. Most of Europe, except Northern Europe.

G. spurium L. FALSE CLEAVERS. Much of Europe.

(b) Stems with smooth angles or stems not angled
1024. **G. verum** L. LADY'S BEDSTRAW. An erect or spreading, almost hairless perenn. 15–100 cm., with dark, shiny green lvs. in whorls of 8–12 and bright yellow fls. in long, compound, terminal, leafy clusters. Fls. 2–4 mm. across; corolla lobes fine-pointed. Lvs. $\frac{1}{2}$–$2\frac{1}{2}$ cm. by $\frac{1}{2}$–2 mm. linear, fine-pointed, margin inrolled, hairy beneath; stems bluntly four-angled. Fr. 2 mm., smooth, hairless, at length black. △ Meadows, banks, waysides, hedges. May–Sept. All Eur. (except P.). *Used to curdle milk and colour cheese and also used in herbal remedies. The roots give a red dye.*

1025. **G. mollugo** L. HEDGE BEDSTRAW. A robust perenn. with straggling or erect stems 25–120 cm. and white fls. in lax, branched, terminal clusters. Fls. 3–4 mm. across, corolla lobes fine-pointed. Lvs. 1–$2\frac{1}{2}$ cm. linear to obovate, in a whorl of 6–8, green on both sides, margin with stout forward-projecting prickles; stem smooth four-angled. Fr. 1–2 mm., rough, hairless, becoming black. A variable sp. △ Meadows, hedges, waysides. June–Oct. All Eur. (except IS.). *The roots give a red dye; used medicinally in the past.*

G. sylvaticum L. WOOD BEDSTRAW. Like 1025, but stems rounded, not four-angled in section; lvs. glaucous particularly on the underside, oblong-lance-shaped, 4–10 mm. broad, blunt with a fine point. Corolla white, 3 mm. across. Fr. smooth, hairless. Widespread in Europe, except parts of Northern and Western Europe.

1026. **G. saxatile** L. HEATH BEDSTRAW. A low, spreading, lax, mat-forming perenn. 10–20 cm., with obovate lvs. in whorls of 6–8 and white fls. in dense, terminal, cylindrical clusters. Fls. 3 mm. across; corolla lobes acute. Lvs. $\frac{1}{2}$–1 cm., margin with fine forward-pointing prickles, turning black on drying. Fr. 1–2 mm., covered with acute swellings. △ Dry, sandy pastures, rocks, heaths, woods. June–Aug. Most of Eur. (except South-East Eur. IS.H.).

G. pumilum Murray Widespread in Europe.

1027. *G. odoratum* (L.) Scop. (*Asperula o.*) SWEET WOODRUFF. A perenn. with creeping underground stolons and slender erect stems 10–30 cm., conspicuous broad lvs. in whorls of 6–9 and a lax, terminal, umbel-like cluster of pure white fls. Corolla 6–7 mm. across, deeply lobed to about half its length into 4 blunt spreading lobes. Lvs. 1–4 cm. elliptic to lance-shaped, rather stiff, margin with forward-pointing prickles; stem four-angled, hairy only at the nodes. Fr. 2–3 mm., covered with hooked bristles. △ Open woods. May–June. All Eur. (except P.IS.). *The dried shoots smell of new-mown hay; they contain coumarin which is used in perfumery and for perfuming clothes and linen in store; also used to flavour wine and liqueurs.* Pl. 99.

1028. *G. glaucum* L. (*Asperula g.*) GLAUCOUS BEDSTRAW. A rather robust, hairless, some-what glaucous, erect perenn. to 40 cm., with a weakly quadrangular whitish stem, which is swollen at the nodes and branched only above, and with widely spaced whorls of 6–8 narrow-linear rigid lvs. Fls. white, in a wide-branched, somewhat flat-topped cluster; corolla 2–3 mm. long, lobes shorter than the tube. Lvs. 1–2½ cm. with inrolled margin, fine-pointed, glaucous beneath. Fr. rough. △ Sandy, grassy places. May–July. Cent. Eur. P.E.F.I.YU.GR.BG.R. Page 326.

CRUCIATA Like *Galium*, but fls. yellow, either hermaphrodite or male. Lvs. three-veined. 2 sps.

1029. *C. laevipes* Opiz (*Galium cruciata* L.) CROSSWORT. A softly hairy, weak-stemmed perenn. to 70 cm., with whorls of 4 yellowish-green lvs. and almost stalkless clusters of yellow fls. forming an elongated, interrupted spike. Fl. clusters shorter than the lvs.; fls. 2–3 mm. across, corolla lobes 4, acute; fl. stalks with bracts and spreading hairs. Lvs. 1–2½ cm. oval-elliptic, weakly three-veined. Fr. smooth, hairless, finally blackish, on recurved stalks. △ Open woods, hedges, thickets, pastures. Apr.–June. All Eur. (except North Eur. IRL.). *The roots produce a red dye; used medicinally in the past.* Pl. 99.

C. glabra (L.) Ehrend. (*Galium g.*; *G. vernum* Scop.) SLENDER CROSSWORT. Like 1029, but a more slender perenn. to 30 cm., with smooth, hairless, quadrangular stems and lvs. shortly hairy, but becoming hairless. Fl. stalks without bracts, hairless. Fr. smooth, hairless. Widespread in Europe, except in Northern Europe.

VALANTIA Fls. white or yellowish in clusters of 3 in axils of the lvs., the lateral fls. male and the central bisexual. Fr. with fused fl. stalks which harden and develop 3 horns which are further divided into spines. *c.* 3 sps.

1030. *V. hispida* L. A tiny stiffly hairy ann. 4–10 cm., with blunt slightly fleshy lvs. in whorls of 4 and whorls of minute, stalkless, yellowish fls. forming a long, interrupted, leafy spike. Corolla three- to four-lobed. Lvs. oblong, hairy, *c.* ½ cm. Fr. rough-hairy, with 3 horns covered with spines which harden and turn ivory white. △ Dry places, rocks, sands, walls. Apr.–June. Med. Eur. P. Pl. 98.

RUBIA Fr. of 1–2 black, distinctly fleshy berries. Fls. yellow; corolla lobes usually 5, spreading. *c.* 4 sps.

1031. *R. peregrina* L. WILD MADDER. An extremely rough, prickly, scrambling or climbing perenn. 30–120 cm., with stiff, persistent, whorled lvs. and axillary clusters of greenish-yellow fls. Clusters stalked, longer than the lvs.; fls. 5 mm. across; corolla lobes 5, long-pointed. Lvs. 1½–6 cm., oval to elliptic, very rough with recurved prickles on margin and underside of midvein; stems four-angled with recurved prickles on the angles.

1. *Galium rotundifolium* 1021
3. *Asperula taurina* 1019
5. *Galium glaucum* 1028

2. *Convolvulus cantabrica* 1036
4. *Crucianella angustifolia* 1015
6. *Crucianella maritima* 1016

Fr. 4–6 mm. globular, black. △ Woods, thickets, hedges, rocks. May–July. Med. Eur. P.IRL.GB. Pl. 98.

R. tinctorum L. DYER'S MADDER. Like 1031, but lvs. softer, herbaceous, distinctly net-veined beneath. Fls. bright yellow, corolla lobes acute, not long-pointed. Fr. reddish-brown. Mediterranean region, but naturalized from cultivation in Central Europe. *At one time cultivated for its roots which yield the red dye madder, which has been known since antiquity.*

POLEMONIACEAE | Phlox Family

Usually herbaceous perenns. or anns. with simple or compound lvs. Fls. radially symmetrical; calyx bell-shaped, three- to five-lobed; corolla tubular and with 5 spreading lobes which are contorted in bud; stamens 5, attached to corolla tube. Ovary superior, with 3 carpels; style 1, stigmas 3. Fr. a capsule opening by 3 valves.

POLEMONIUM Lvs. pinnate. Filaments swollen at their base and almost closing the throat of the corolla. *c.* 2 sps.

1032. *P. caeruleum* L. JACOB'S LADDER. An erect, leafy, nearly hairless perenn. 30–90 cm. with odd-pinnate lvs. and rather large bright blue fls. in a dense terminal cluster. Fls. rarely white; corolla 2–3 cm. across, tube very short, lobes wide-spreading; calyx glandular-hairy; stamens projecting. Lvs. 10–40 cm. with 7–15 oval-lance-shaped, long-pointed, shining leaflets, each 2–4 cm., the lower lvs. long-stalked, the upper stalkless. Fr. erect, included in calyx. △ Woods, damp pastures, in mountains; often cultivated and sometimes naturalized. May–Aug. North, Cent. Eur. (except H.) F.GB.I.YU.R.SU. Pl. 100.

CONVOLVULACEAE | Convolvulus Family

Commonly climbing perenns. or anns., but sometimes erect herbs or shrubs, with alternate, usually simple lvs.; or colourless parasites with scale-like lvs. Fls. usually large, radially symmetrical; sepals 5, free, persistent; corolla contorted in bud, usually deeply funnel-shaped and without distinct lobes or with 5 spreading lobes; stamens attached to the corolla. Ovary superior, style 1–2. Fr. a capsule splitting into 2–4 valves, rarely fleshy.

1. Parasitic plants without lvs.; fls. in dense clusters. *Cuscuta*
1. Non-parasitic plants with green lvs.
  2. Fls. inconspicuous, with 5 spreading lobes; styles 2. *Cressa*
  2. Fls. conspicuous, funnel-shaped, and corolla not lobed; style 1.
    3. Stigma 1, club-shaped. *Ipomoea*
    3. Stigmas 2, linear.
      4. Calyx encircled by 2 large leafy bracteoles. *Calystegia*
      4. Calyx without large, encircling, leafy bracteoles. *Convolvulus*

IPOMOEA Corolla funnel-shaped, usually unlobed; sepals 5. Stamens included within corolla; stigma club-shaped. *c.* 4 sps.

1033. *I. purpurea* Roth MORNING GLORY. A climbing ann. to 3 m., with large funnel-shaped, violet-purple, pink, blue, or white fls. 4–5 cm. across. Fls. one to several in the axils of the upper lvs., closing by mid-day; corolla circular in outline; calyx with dense spreading hairs, lobes lance-shaped. Lvs. 6–8 cm., broadly heart-shaped, long-pointed, sparsely hairy. △ Native of South America; grown as an ornamental in Southern Europe and sometimes self-seeding. July–Sept.

I. rubrocaerulea Hooker MORNING GLORY. An ann. or perenn. with climbing stems and deeply heart-shaped and slender-pointed hairless lvs. 8–13 cm. Fls. in clusters of 3–4 on thick stalks; corolla 8–10 cm. across, bluish-purple with a white tube, the bud red before opening; calyx *c.* 6 mm., papery-margined. Native of tropical America; grown for ornament in South-Eastern Europe.

I. hederacea (L.) Jacq. (*I. nil* auct.; *Pharbitis h.*) Like (1033), but lvs. commonly shallowly and widely three-lobed. Fls. 1–3, short-stalked; corolla blue or light purple; calyx 2½ cm. or more, not recurved in fr. and longer than it. Native of tropical America; grown for ornament in South-Eastern Europe. Pl. 100.

CONVOLVULUS Corolla funnel-shaped, not or shallowly lobed; sepals 5, with small distant bracteoles; stamens included; ovary two-celled; style slender, stigmas 2, linear. Fr. a four-valved capsule. *c.* 30 sps.

Fls. at least in part blue

1034. *C. siculus* L. SMALL BLUE CONVOLVULUS. A rather weak, straggling, hairy ann. to 40 cm. with oval stalked lvs. and small, solitary blue fls. in the axils of the lvs. Corolla *c.* 1 cm. across, deeply five-lobed, tube yellowish and scarcely twice as long as the hairy elliptic sepals; fl. stalks slender and shorter than the lvs.; bracteoles 2, close to calyx. Lvs. hairy 2–3 cm., acute, with a shallow heart-shaped base. Fr. globular, hairless. △ Rocky and stony places. Apr.–May. P.E.F.I.GR.

C. pentapetaloides L. Like 1034, but lower lvs. oblong-spathulate and narrowed to the lv. stalk, the upper lance-shaped, stalkless, woolly-haired. Calyx hairless, lobes rounded-oval; corolla blue, tube yellow and twice as long as calyx. Fr. hairless. Mediterranean region.

1035. *C. tricolor* L. DWARF CONVOLVULUS. A spreading hairy ann. to 80 cm. with conspicuous tricoloured fls. 2–5 cm. across, with an orange-yellow throat, white middle zone, and a broad blue marginal zone. Fls. solitary; fl. stalks longer than the lvs.; sepals oval, densely hairy. Lvs. mostly oval-lance-shaped and narrowed to the base, the upper stalkless. Fr. hairy. △ Cultivated ground, vineyards, by the sea; often cultivated and naturalized elsewhere. Apr.–June. P.E.F.I.YU.GR. Pl. 100.

Fls. pink, white or yellowish
(a) Lvs. linear or lance-shaped; non-climbing

1036. *C. cantabrica* L. PINK CONVOLVULUS. An ascending hairy perenn. to 50 cm., branching from the base, and with a lax, spreading inflorescence of pink fls. formed of long-stalked axillary clusters longer than the lvs. Clusters with 1–6 fls.; corolla *c.* 3 cm. across, hairy outside; sepals narrow lance-shaped, with long, silvery, spreading hairs. Lvs. 2–5 cm. narrow-lance-shaped to linear, green, hairy. Fr. hairy. △ Uncultivated ground, sandy fields. June–July. Med., South-East Eur. A.CS.H.SU. Page 326.

1037. *C. lineatus* L. SILVERY-LEAVED PINK CONVOLVULUS. A silvery-haired, low-growing, densely tufted perenn. with erect unbranched stems to 30 cm., with narrow silvery lvs. and pink fls. in a dense, terminal cluster. Corolla 2–3 cm. across, 2–3 times as long as the lance-shaped sepals which have adpressed silvery hairs. Lvs. linear to lance-shaped, narrowed to the stalk below, silvery-haired on both sides. Fr. hairy. △ Dry, arid places, by the sea. May–July. Med. Eur. (except TR.) P.BG.R.

C. cneorum L. SHRUBBY CONVOLVULUS. A small shrub to 1 m. with silky-haired, silvery, lance-shaped, long-pointed lvs. 2–3 cm. and terminal heads of pale pink fls. Corolla 3–5 cm. across, and 1–1½ cm. long, silky-haired outside; calyx very hairy with spreading hairs. Italy to Greece.

(b) Lvs. heart-shaped or deeply lobed; usually climbing
1038. *C. arvensis* L. FIELD BINDWEED. A climbing perenn. to 2 m. with nearly hairless, arrow-shaped lvs. and usually solitary, long-stalked, axillary, pink or white fls. Fls. 1½–3 cm. across, hairless, often with 5 purplish stripes outside; sepals oval, hairless; fl. stalk longer than the lvs., with 2 small bracteoles. Lvs. 2–5 cm. oval-oblong with blunt or acute spreading basal lobes; rhizome long-creeping. Fr. hairless. △ Fields, waste places, roadsides, gardens. May–Oct. All Eur. (except IS.).

1039. *C. althaeoides* L. MALLOW-LEAVED BINDWEED. A hairy, usually climbing perenn. to 2 m., with deep pink fls. 3–4½ cm. across and with the upper stem lvs. deeply divided into narrow segments. Fls. 1–2 borne on a long stem longer than the subtending lv.; corolla darker pink in throat, hairy outside; sepals oval, hairy; bracteoles linear. Lower lvs. triangular-heart-shaped, blunt-toothed, only the upper lvs. deeply divided into 5–9 unequal oblong-linear segments, all green, softly hairy. △ Cultivated ground, waysides, bushy places, by the sea. Apr.–Aug. Med. Eur. (except AL.TR.) P. Pl. 100.

C. elegantissimus Miller (*C. tenuissimus* Sm.) ELEGANT BINDWEED. Like 1039, but lvs. all pale shining-whitish with silvery adpressed hairs and almost all stem lvs. deeply divided into 5–9 unequal linear segments. Corolla paler in throat. South-Eastern Europe. Pl. 100.

CALYSTEGIA | **Bindweed** Like *Convolvulus* and often included in it, but calyx encircled by 2 large leafy bracteoles forming an involucre and arising directly below the fl. *c.* 3 sps.

Climbing perenns.
1040. *C. sepium* (L.) R.Br. BELLBINE, GREATER BINDWEED. A rather robust, hairless, climbing perenn. to 3 m., with large, deeply heart-shaped to arrow-shaped lvs. and large, solitary white or rarely pinkish fls. 3½–7 cm. across. Corolla hairless; bracteoles heart-shaped, flat or strongly inflated, longer than the hairless calyx. Lvs. up to 15 cm.; underground stems long-creeping. A variable sp. △ Hedges, bushy places, woods, cultivated ground. June–Sept. All Eur. (except IS.).

Prostrate perenns.
1041. *C. soldanella* (L.) R.Br. SEA BINDWEED. A creeping hairless perenn. spreading to 60 cm., with slightly fleshy kidney-shaped lvs. and solitary deep or pale pink fls. 4–5 cm. across. Corolla hairless; fl. stalks quadrangular, usually longer than the lvs.; bracteoles rounded, shorter than the calyx. Lv. stalks longer than the blade which is 1–4 cm. across; underground stems slender, long-creeping. △ Sands and shingles on the littoral. May–Oct. West, Med. Eur. DK.D.BG.R.SU. Pl. 100.

CRESSA Fls. tiny, less than ½ cm. long; corolla tube short, lobes longer, 5, spreading; stamens projecting; sepals 5; styles 2. 1 sp.

1042. *C. cretica* L. A small, densely branched, grey-leaved shrublet of salt marshes to 25 cm., with small, terminal, globular clusters of yellowish or pinkish fls. Fl. cluster ½–1 cm. across; corolla *c.* 3 mm. across, lobes lance-shaped; sepals densely silky-haired outside.

Lvs. 2–7 mm. lance-shaped, numerous, overlapping, stalkless, silky-haired. △ Salt marshes, estuaries. July–Sept. Med. Eur. (except YU.AL.TR.) P.BG.

CUSCUTA | Dodder   Pinkish, white, or yellowish parasites with only scale-lvs. and without green lvs.; stems attached to the host plant by suckers. Fls. in small, usually stalkless clusters; corolla bell-shaped and usually five-lobed; sepals and stamens usually 5. *c.* 15 sps.

Scales not closing corolla tube; styles shorter than ovary
1043. *C. europaea* L. LARGE DODDER. A greenish-yellow, twining ann. parasite, with tiny, stalkless, pinkish-white fls. in globular clusters *c.* 1 cm. Corolla *c.* 2 mm. across, the lobes blunt, spreading, about as long as the tube; scales in throat not closing throat. Stem much-branched, yellowish or reddish, to 1 mm. thick, climbing to 1 m. △ Commonly parasitic on willows, alders, nettles, hops, mints, etc. June–Aug. Most of Eur. (except P.IRL.IS.).

Scales closing corolla tube; styles longer than ovary
1044. *C. epithymum* (L.) L. COMMON DODDER. A reddish ann. parasite with dense globular head $\frac{1}{2}$–1 cm. across of pinkish scented fls. Corolla lobes long-pointed, scales coming together and almost closing the corolla tube; stamens projecting; styles longer than the ovary. Stems reddish, very slender, about $\frac{1}{10}$ mm. thick. △ Commonly parasitic on heather, gorse, clover, and many other sps. June–Oct. Most of Eur. (except IRL.IS.). Pl. 101.

C. planiflora Ten. Like 1044, but lobes of corolla much longer than the tube and spreading widely; scales in throat deeply fringed. Mediterranean region, Portugal.

BORAGINACEAE | Borage Family

Anns. or herbaceous perenns., usually rough with stiff bristly hairs, but sometimes hairless and with alternate entire lvs. Fls. often borne on the upper side of dichotomous, outward-coiled branches, and opening progressively from the base of each branch. Calyx tubular, five-lobed; corolla bell- or funnel-shaped, five-lobed, with or without scales or hairs in the throat; stamens 5, attached to the corolla. Ovary two- or four-celled; style 1. Fr. usually of 4 one-seeded nutlets.

Corolla with 5 converging scales, or with hairs in throat and often closing it.   Group A
Corolla without scales in throat, or scales or hairs rudimentary and not converging.   Group B

GROUP A. *Corolla with converging scales or hairs in throat*
                                                                                Nos.
1. Corolla tubular with lobes very short, or corolla tube very
     short and lobes widely spreading.
   2. Corolla tubular, with lobes very short.   *Symphytum*   1052, 1053
   2. Corolla tube very short, with large spreading lobes.
      3. Stamens not projecting; lvs. smooth, hairless, or nearly so.   *Omphalodes*   1047, 1048
      3. Stamens projecting in a cone; lvs. rough, hairy.
         4. Stamens hairless, with a horn-like projection.   *Borago*   1054
         4. Stamens hairy, without horns.   *Trachystemon*   [1054]
1. Corolla funnel-shaped or saucer-shaped with flat or concave
     lobes and a short or long tube.
   5. Nutlets with spines or hooks.

6. Nutlets large, covered with short spines.   *Cynoglossum*   1049
6. Nutlets small, with long hooked spines on the margin only.
  7. Fr. erect.   *Lappula*   1050
  7. Fr. reflexed.   *Hackelia*   [1050]
5. Nutlets smooth or rough, without spines or hooks.
  8. Nutlets rough; usually roughly hairy plants.
    9. Calyx bell-shaped with 5 narrow entire lobes.
      10. Corolla tube curved, corolla lop-sided.   *Lycopsis*   1059
      10. Corolla tube straight, corolla not lop-sided.
        11. Lvs. oval, long-pointed.   *Pentaglottis*   1058
        11. Lvs. linear to lance-shaped.   *Anchusa*   1055–1057
    9. Calyx enlarged in fr. into 2 broad, toothed and veined lobes.   *Asperugo*   1051
  8. Nutlets smooth, often shining; usually softly hairy plants.
    12. Alpine cushion-forming perenns.; inflorescence very short.   *Eritrichium*   1070
    12. Not cushion plants; inflorescence soon elongating.   *Myosotis*   1065–1069

GROUP B. *Corolla without scales or hairs, or scales rudimentary*

                                   Nos.

1. Inflorescence without bracts subtending the fls.; corolla small with spreading lobes.
  2. Fls. yellow.   *Amsinckia*   1046
  2. Fls. blue, white, or lilac.   *Heliotropium*   1045
1. Inflorescence with leafy bracts subtending the fls.
  3. Corolla tubular or bell-shaped, scarcely enlarged above, with an open throat and short irregular lobes.
    4. Corolla irregular, tube curved, lobes unequal   *Echium*   1080–1083
    4. Corolla regular, tube straight, lobes equal.
      5. Almost hairless, glaucous plants.   *Cerinthe*   1078, 1079
      5. Densely stiff-haired plants.   *Onosma*   1076, 1077
  3. Corolla funnel-shaped, abruptly enlarged above into a wider barrel-shaped part with a constricted throat and spreading lobes.
    6. Calyx divided to about one-third.
      7. Corolla tube with clusters of hairs in throat.   *Pulmonaria*   1063, 1064
      7. Corolla tube with 5 hooked scales.   *Nonea*   1060
    6. Calyx divided almost to base.
      8. Tube of corolla with 5 transverse swellings near the base, alternating with the stamens.
        9. Lvs. rough, hairy.   *Alkanna*   1061
        9. Lvs. smooth, hairless, glaucous.   *Mertensia*   1062
      8. Tube of corolla naked or with 5 vertical lines of hairs near the base.   *Lithospermum*   1071–1075

HELIOTROPIUM Fls. white or lilac, borne on the upper side of outward-coiled branches; inflorescence bractless. Corolla lobes separated by a longitudinal fold and often a small tooth, throat without scales. *c.* 8 sps.

1045. *H. europaeum* L. HELIOTROPE. A softly hairy, erect, branched ann. to ½ m., with white or pale lilac fls. in one-sided, leafless, forked spikes. Corolla 3–4 mm. across, lobes wide-spreading; calyx very hairy, divided almost to the base into lance-shaped persistent lobes, spreading in fr. Lvs. oval, shortly stalked, softly hairy, green or greyish. Fr. with swellings, separating from calyx. △ Cultivated ground, waste places, waysides. June–Oct. Med., South-East Eur. P.CH.A.CS.H.SU.: introd. D. Pl. 101.

H. supinum L. Like 1045, but stems spreading and lvs. green above and white-cottony beneath. Calyx flask-like with short lobes, encircling the fr. and falling with it. Fls. white, 1–2 mm. across, in single or forked spikes. Southern Europe.

AMSINCKIA Like *Heliotropium*, but fls. yellow and without folds or teeth between the corolla lobes. *c.* 4 sps.

1046. *A. angustifolia* Lehm. A slender, erect, bristly-haired ann. 20–80 cm., with small orange-yellow fls. in one-sided, leafless, terminal spikes. Corolla tube slender, much longer than the rounded lobes; calyx divided to the base into linear lobes longer than the fr. Lvs. lance-shaped *c.* 4 cm., with numerous spreading bristly hairs. Nutlets wrinkled. △ Waste ground, waysides. Apr.–June. A native of Chile; introd. to Central Europe.

OMPHALODES Fls. blue or white; corolla tube very short and throat closed by 5 blunt, forward-projecting lobes, corolla lobes wide-spreading. Lvs. soft, hairless, or nearly so. *c.* 9 sps.

1047. *O. verna* Moench BLUE-EYED MARY. A creeping perenn. with slender, erect fl. stems 10–30 cm., bearing a lax cluster of a few sky-blue fls. Corolla 1 cm. across, lobes rounded, spreading; calyx deeply divided, hairy; fl. stalks hairy. Lvs. mostly basal, long-stalked, oval with a rounded or heart-shaped base and fine-pointed apex, light shining green, somewhat hairy. Nutlets with a hairy swollen border. △ Woods; grown as an ornamental and sometimes naturalized. Mar.–May. E.F.CH.I.YU.: introd. GB.D.CS.PL. Pl. 101.

1048. *O. linifolia* (L.) Moench VENUS' NAVEL-WORT. An erect, glaucous ann. to 40 cm. branched below and with long, lax, leafless spikes of white or bluish fls. Corolla 6–10 mm. across; calyx with sparse bristly hairs; fl. stalk 1–2 cm., spreading, arched. Lvs. *c.* 4 cm., oblong to lance-shaped, blunt or acute, almost hairless, margin bristly. Fr. with a thick ribbed border encircling a smooth depression. △ Dry arid places, banks. Mar.–June. P.E.F.: introd. I. Page 341.

CYNOGLOSSUM | Hound's-Tongue Corolla tube short and throat closed with scales, lobes spreading; stamens not projecting. Nutlets 4, flattened and covered with bristles or hooks. *c.* 11 sps.

Nutlets with a thickened border, concave above
1049. *C. officinale* L. HOUND'S-TONGUE. A softly grey-haired, leafy, erect bienn. to 80 cm., with dull reddish-purple fls. 7–10 mm. across in branched, elongated clusters, which are leafy below. Corolla lobes spreading, rarely white; calyx lobes oval, 5–7 mm. in fr. Lower lvs. elliptic, stalked, the upper stalkless, lance-shaped, all with adpressed silky hairs. Nutlets 5–6 mm. with a thickened border, concave above, and covered all over with equal-length spines. △ Woods, dry places, rocks, sands by the sea. May–July. All Eur. (except P.TR.): introd. IS. *Used medicinally in the past.* Pl. 102.

C. columnae Ten. Like 1049, with dull purple fls., but inflorescence leafless. Nutlets with a thickened border, flat or concave above and covered with finely hooked spines inter-mixed with small swellings. Upper lvs. elliptic with a heart-shaped base. Italy to Greece.

C. cheirifolium L. Distinguished by its white-woolly lvs. and stem, with the upper lvs. narrow to the base, not clasping. Fls. red, turning violet or blue; inflorescence with leafy bracts subtending each fl. Nutlets with thickened border, concave above, spines hooked. Portugal to Italy.

Nutlets without a thickened border, not concave above

C. germanicum Jacq. GREEN HOUND'S-TONGUE. Like 1049, but lvs. shiny green, rough with few bristly hairs, almost hairless above. Corolla 5 mm. across, reddish or violet. Nuts without a thickened border, marginal spines longest. Widespread in Europe, except in the north.

C. creticum Miller BLUE HOUND'S-TONGUE. Distinguished by its pale blue fls. which are obscurely veined and its rounded nutlets without a thickened border and densely covered with spines. Lvs. finely and softly hairy; fl. spikes without bracts. Mediterranean and South-Eastern Europe. Pl. 101.

LAPPULA   Fr. with small flattened nutlets with long hooked spines on the margin only. Fls. blue; corolla tube very short, throat closed with scales, corolla lobes spreading or bell-shaped. Sometimes included in *Echinospermum*. *c.* 5 sps.

1050. *L. myosotis* Moench BUR FORGET-ME-NOT. A slender, stiff, rough-haired, widely branched ann. or bienn. to 40 cm., with small sky-blue fls. forming lax leafy spikes. Corolla 2–4 mm. across, little longer than calyx. Lvs. 1–3 cm. lance-shaped, stalkless. Fr. erect, sepals spreading in a star, nutlets with 2 rows of hooked spines on the margin and a central, corrugated, hairless face. △ Dry places, vineyards, waysides, dunes. June–July. Much of Eur. (except P.IRL.GB.): introd. IS.

HACKELIA   Like *Lappula*, but fr. with a single row of hooked spines. *c.* 11 sps.

H. deflexa (Wahlenb.) Opiz (*Lappula d.*) Like 1050, but fr. stalks reflexed and nutlets flattened, winged and with a single row of hooked spines. Fls. larger 5–7 mm. across, blue; calyx at length reflexed. Widespread in Europe.

ASPERUGO   Calyx at first bell-shaped, but enlarging in fr. and becoming flattened into 2 wide, toothed lobes. Corolla funnel-shaped, throat closed by hairy scales. Nutlets keeled, not spiny. 1 sp.

1051. *A. procumbens* L. MADWORT. A rough, coarse, weak-stemmed spreading ann. to 50 cm., with inconspicuous, stalkless, solitary or paired, bluish axillary fls. much shorter than the lvs. Inflorescence lax, leafy; corolla *c.* 3 mm. across. Lvs. 2–7 cm. oval-elliptic, rough bristly-haired; stem angled with downward-directed bristly hairs. Calyx enlarging in fr. into 2 kidney-shaped, conspicuously toothed and veined lobes *c.* 1 cm. across; nutlets ovoid. △ Waste places, waysides, fields. May–July, sometimes Aug.–Jan. Much of Eur. (except P.IRL.): introd. GB.B.IS.D. *The roots yield a red dye.* Page 341.

SYMPHYTUM | Comfrey   Corolla tubular, widening somewhat above, with 5 very short erect lobes; scales narrow erect, glandular-hairy, alternating with the stamens; calyx bell-shaped or tubular, five-lobed. Fr. of 4 ovoid, smooth or granular nutlets. *c.* 12 sps.

Plants ½ m. or more
(a) Stems winged
1052. *S. officinale* L. COMFREY. A robust leafy perenn. to 120 cm., branched from the base, with conspicuously winged stems and one-sided clusters of yellowish-white, pinkish or

purple fls. Inflorescence at first densely coiled, then lengthening; corolla *c.* 1½ cm. long; calyx lobes lance-shaped. Lvs. to 25 cm. lance-shaped, sparsely hairy, the lower stalked, the upper stalkless with wings running down stem. Nutlets shiny black. △ Damp meadows, watersides, marshes. May–June. Most of Eur.: introd. IS. *Used medicinally for healing wounds, etc.; the roots were used to make a kind of gum for the treatment of wool before spinning.*

S. × *uplandicum* Nyman BLUE COMFREY. Probably a hybrid between 1052 and *S. asperum.* Fls. blue or purplish. Stem lvs. with only a narrow wing running for a short way down each internode. Naturalized in Northern and Central Europe. *Sometimes grown as a fodder crop for livestock.* Pl. 102.

(*b*) *Stems not winged or slightly winged above only*
S. asperum Lepechin ROUGH COMFREY. Like 1052, but taller to 1½ m., with rough bristly-haired lvs. and stems unwinged or only slightly winged below the uppermost lvs. Fls. red, then blue, *c.* 2 cm. long; calyx *c.* 4 mm., enlarging to 1½ cm. in fr., covered with stout short bristles with swollen bases. Native of Western Asia; sometimes grown for fodder and naturalized in Northern and Central Europe.

Plants usually less than ½ m.
1053. *S. tuberosum* L. TUBEROUS COMFREY. A rough, bristly-haired, herbaceous perenn. with swollen tuberous roots and leafy fl. stems 20–50 cm., and with few-flowered clusters of pale yellow fls. Corolla 1–1½ cm. long; calyx 7–8 mm., the lobes *c.* 3 times as long as the tube. Lvs. bristly, the lower oval, stalked, the middle lvs. with longer, broadly lance-shaped blade 10–14 cm., short-stalked, the upper stalkless. △ Damp meadows, open woods, streamsides. Mar.–June. Most of Eur. (except North Eur. P.IRL.B.NL.). Pl. 102.

S. orientale L. EASTERN COMFREY. Distinguished by its white fls. and softly hairy lvs. which are abruptly narrowed to the lv. stalk. Corolla 1½–2 cm.; calyx tubular, the lobes triangular, less than half the length of the tube. Native of Turkey; naturalized elsewhere. Pl. 101.

BORAGO | **Borage** Corolla with a very short tube and widely spreading, lance-shaped lobes. Stamens projecting forwards in a cone and with additional narrow, horn-like appendages, shorter than the stamens. Nutlets 4, rough. *c.* 2 sps.

1054. *B. officinalis* L. BORAGE. A stout-branched, rough, bristly-haired ann. 20–60 cm., with lax branched clusters of bright blue fls. with a forward-projecting cone of blackish-purple anthers. Corolla 2–2½ cm. across, lobes oval-acute, wide-spreading; calyx conspicuous, lobed almost to the base, with long bristly hairs. Lvs. 10–20 cm. oval, rough, margin often undulate, the lower stalked, the upper clasping. △ Waste places, waysides; often cultivated and self-seeding near habitation. Apr.–Sept. Med. Eur. P.: introd. to Cent. and South-East Eur. *Sometimes used as a pot herb, and to flavour beverages.* Pl. 103.

TRACHYSTEMON Like *Borago*, but lobes of corolla linear and anthers much shorter than the filament and without an additional appendage. 2 sps.

T. orientalis (L.) G. Don EASTERN BORAGE. Like 1054, but fls. bluish-violet in branched, almost leafless inflorescences; corolla *c.* 1½ cm. across, lobes linear, wide-spreading or somewhat reflexed, scales short; stamens forward-projecting in a cone. Lvs. large *c.* 30 cm., oval-heart-shaped, stalked; rhizome thick, creeping. Eastern Mediterranean region; naturalized elsewhere. Pl. 103.

ANCHUSA | Alkanet Corolla long, funnel-shaped, the throat closed with oval scales or hairs, the lobes flat or bell-shaped; stamens not projecting. Nutlets netted, obliquely conical, deeply concave at base, not stalked. *c.* 20 sps.

Lvs. flat, margin not undulate

1055. *A. officinalis* L. TRUE ALKANET. An erect much-branched perenn. or bienn. 30–60 cm., with soft spreading hairs and purple fls. which are at first in a dense, coiled cluster which later elongates. Corolla *c.* 1 cm. across, throat closed with velvet-haired scales; fl. stalks shorter than the lance-shaped bracts. Lvs. oblong-lance-shaped, the upper stalkless, with rounded base, the lower narrowed to a stalk, blade flat, margin entire or weakly toothed. △ Fields, waysides, meadows. June–Aug. All Eur. (except IRL.IS.). Pl. 103.

1056. *A. azurea* Miller LARGE BLUE ALKANET. Distinguished by its conspicuous bright blue fls. 1½–2½ cm. across, and the brush-like tuft of white hairs in the throat. Corolla tube about as long as the corolla lobes; calyx divided to the base into linear-lance-shaped lobes which enlarge in fr. to *c.* 2 cm. Lvs. thick, rough, lance-shaped, the lower stalked, the upper stalkless. An erect, thick-stemmed, usually much-branched perenn. to 120 cm. with stiff, bristly, spreading hairs. △ Fields, vineyards, waysides. May–Aug. Med., South-East Eur. P.CH.A.H.SU.: introd. D.CS. Pl. 102.

1057. *A. barrelieri* (All.) Vitm. SMALL-FLOWERED ALKANET. Fls. small, blue, 5–7 mm. across, in short, dense, many-flowered clusters in a branched inflorescence; tube of corolla much shorter than the corolla lobes; calyx only 3 mm. long; bracts linear-lance-shaped. Lower lvs. linear-oblong, long-stalked, the upper stalkless and clasping stem, entire or shallowly toothed, with a single vein, all usually with soft adpressed hairs. △ Bushy places, cultivated ground. May–July. F.CS.PL.H.I.YU.BG.

Lvs. with undulate margin

A. hybrida Ten. (*A. undulata* L.) UNDULATE ALKANET. Differs from 1057 by the longer corolla tube 1–1½ cm. which is longer than the corolla lobes and calyx. Fls. purple, in elongating dichotomous inflorescences; corolla *c.* 7 mm. across; scales in throat rounded with ciliate margin. Lvs. oblong-lance-shaped, with spreading hairs, shallowly toothed with strongly undulate margin, the lower narrowed to a stalk, the upper stalkless. Mediterranean region and Portugal.

PENTAGLOTTIS Like *Anchusa*, but lvs. oval-acute; tube of corolla straight, usually shorter than the spreading lobes, throat closed by scales. Nutlets concave at base with a small stalked attachment. *c.* 1 sp.

1058. *P. sempervirens* (L.) Tausch (*Anchusa s.*) ALKANET. A rough, bristly, leafy, and rather robust perenn. to 1 m., with bright blue fls. with a white centre, 1 cm. across, in dense, leafy, branched clusters. Fl. clusters long-stemmed and subtended by a leafy bract 1½–3 cm.; corolla tube shorter than the lobes; scales white; calyx teeth linear. Lvs. oval-acute, to 30 cm. Nutlets netted. △ Woods, damp and shady places; sometimes naturalized from cultivation. Apr.–July. West Eur. I.: introd. DK.N.S. Pl. 102.

LYCOPSIS Similar to *Anchusa*, but corolla tube curved near the middle and fls. lop-sided. *c.* 3 sps.

1059. *L. arvensis* L. (*Anchusa a.*) BUGLOSS. A rather weak-stemmed, very bristly-haired ann. or bienn. to 40 cm., with small bright blue fls. with a white centre, and lance-shaped lvs. with undulate margin. Corolla tube curved 8–10 mm. long, lobes 5–7 mm. across; scales in throat, hairy, white; calyx lobes linear-lance-shaped, somewhat enlarging in fr. Lvs.

irregularly toothed, with bristly hairs with swollen bases, the upper lvs. clasping the stem. Nutlets netted. △ Cultivated ground, sandy places. May–Sept. All Eur.

NONEA Corolla funnel-shaped to tubular, five-lobed, scales hairy, not closing the throat; stamens included. Nutlets obliquely curved. *c.* 10 sps.

1060. *N. pulla* (L.) DC. An erect, greyish, glandular-hairy bienn. to 40 cm., bristly above and with a lax, leafy, branched spike of small dark brown fls. Corolla *c.* 1 cm. long, twice as long as the calyx which enlarges in fr. Lvs. oblong to lance-shaped, with fine dense hairs, but not rough, the upper clasping the stem. Nutlets oval and with a short lateral beak. △ Stony places, pastures. Apr.–May. Cent., South-East Eur. (except AL.) SU.: introd. CH. Page 341.

N. ventricosa (Sibth. & Sm.) Griseb. (*N. alba* DC.) Like 1060, but fls. white, little longer than calyx, and lvs. not glandular. Lvs. lance-shaped, blades of the upper lvs. shortly running down the stem, margin undulate. Calyx becoming swollen and globular in fr., lobes lance-shaped; bracts large, oval-lance-shaped. Nutlets kidney-shaped, attached at the middle. Mediterranean region.

ALKANNA Corolla funnel-shaped, with 5 tiny transverse swellings alternating with the stamens, scales absent; stamens included in corolla. *c.* 15 sps.

1061. *A. tinctoria* (L.) Tausch DYER'S ALKANET. A small, spreading, bristly-haired, somewhat shrubby perenn. 6–30 cm. with bright blue fls. *c.* ½ cm. across in short, leafy, and often forked spikes. Corolla not as long as or little longer than the calyx; bracts leafy, oval-lance-shaped. Upper lvs. lance-shaped and clasping the stem, the lower narrow to the base, all densely greyish-haired and bristly. Roots with reddish skin. Nutlets with irregular rough swellings. △ Maritime sands, uncultivated ground. Apr.–June. Med., South-East Eur. (except AL.) CS.H.: introd. SU. *The roots produce a strong red dye, used by pharmacists and perfumiers to colour their products; also used to stain wood and marble, and in thermometers and microscopy.* Pl. 103.

MERTENSIA Corolla tube cylindrical, lobes spreading in a bell, throat without scales but with 5 transverse folds; stamens attached near the top of the corolla tube, slightly projecting. Nutlets rather fleshy. 1 sp.

1062. *M. maritima* (L.) S. F. Gray NORTHERN SHOREWORT, OYSTER PLANT. A somewhat fleshy, glaucous, spreading perenn. to 60 cm., with oval lvs. and leafy clusters of pink fls. turning to bluish-purple. Corolla cylindrical *c.* 6 mm. across, lobes blunt, not spreading; calyx lobes oval, hairless. Lvs. 2–6 cm. in 2 ranks, the upper stalkless, all hairless and dotted with glands on the upper surface. Fl. stalks recurved in fr. Nutlets usually smooth. △ Shingle by the sea. June–Aug. North Eur. IRL.GB.SU. Pl. 103.

PULMONARIA | Lungwort Corolla funnel-shaped with 5 tufts of hairs alternating with the stamens but not closing throat, corolla lobes short; fls. pink, blue, or violet. *c.* 6 sps.

1063. *P. officinalis* L. LUNGWORT. A hairy, leafy, rather weak-stemmed perenn. to 30 cm., with pink fls., turning blue, in dense short clusters which scarcely elongate after flowering. Corolla *c.* 1 cm. across, tube longer than the cylindrical calyx which is 6–7 mm., with lobes less than one-third its length; stamens not projecting. Basal autumn and spring lvs.

oval, often heart-shaped at base, white spotted or unspotted, lv. stalk winged above, upper stem lvs. oval, clasping stem. △ Damp open woods, shady thickets. Apr.–June. Most of Eur. (except P.IRL.IS.).

1064. *P. longifolia* (Bast.) Boreau NARROW-LEAVED LUNGWORT. Like 1063, but fls. smaller 5–6 mm. across and stamens projecting; calyx 10 mm. long, with lobes half the length of the tube. Basal lvs. lance-shaped, gradually narrowed to the winged lv. stalk, blade rough; stems with spreading and glandular hairs. △ Woods, shady places. Apr.–May. Cent. Eur. P.E.F.GB.DK.S.I.YU.R. Pl. 104.

P. montana Lejeune MOUNTAIN LUNGWORT. Like 1064, but lvs. soft, shiny, densely glandular-sticky, and stems and calyx sticky. Fls. 1½–2½ cm. long and *c.* 1 cm. across, at first pink then blue. Basal lvs. oval-lance-shaped 10–15 cm., lv. stalk winged, 10–20 cm., the upper lvs. clasping, usually unspotted. Pyrenees, Alps, Carpathians.

MYOSOTIS | **Forget-Me-Not, Scorpion-Grass** Corolla tube short, straight, throat closed by 5 short notched scales, lobes rounded or notched, spreading flat or concave. Calyx five-lobed; stamens included. Nutlets small, shining. *c.* 20 sps.

Calyx with adpressed hairs, rarely hairless
1065. *M. scorpioides* L. (*M. palustris* (L.) Hill) WATER FORGET-ME-NOT. A rather weak, leafy, ascending perenn. to 45 cm., often rooting below and with lax, elongate, leafless, spike-like clusters of short-stalked, bright blue fls. ½–1 cm. across. Corolla lobes flat, notched, rarely white; calyx bell-shaped with triangular lobes less than one-third as long as the tube, hairs adpressed. Lvs. to *c.* 7 cm. oblance-shaped, stalkless, nearly hairless or with short adpressed hairs. Nutlets 1½ mm., black, shining. △ Marshes, damp meadows, watersides. May–Sept. All Eur. Pl. 103.

M. secunda A. Murray Like 1065, but corolla smaller, 4–6 mm. across, and lobes slightly notched; calyx lobes about half the length of tube; fr. stalks 3–5 times as long as the calyx. Stems with spreading hairs below. Nutlets 1 mm., dark brown, shiny. Perenn. Northern Europe.

M. caespitosa K. F. Schultz Like *M. secunda*, but corolla smaller, 2–4 mm. across, and lobes rounded, not notched, rarely white. Stems hairless or with adpressed hairs; lvs. with adpressed hairs above and almost hairless beneath, at least on the lower lvs. Ann. or bienn. Throughout Europe.

M. sicula Guss. Like *M. caespitosa*, but calyx nearly hairless and teeth about half the length of tube, not spreading in fr.; corolla 2½–3 mm. across. Stems flexuous, spreading, often wide-branched, hairless or with scattered hairs above; lower lvs. almost hairless beneath, the upper hairy on both sides. South-Western and Mediterranean Europe. Page 341.

Calyx with spreading hairs or at least some hairs crisped or hooked
(a) Fls. usually more than 5 mm. across
1066. *M. sylvatica* Hoffm. WOOD FORGET-ME-NOT. An erect, softly hairy perenn. to ½ m., with terminal, spike-like clusters of bright blue fls., each 6–10 mm. across. Corolla lobes rounded; calyx with short crisped or hooked hairs, lobes more than half the length of the tube and spreading in fr.; fr. stalks spreading, up to twice as long as the calyx. Lvs. oblong-lance-shaped, clasping the stem, with spreading hairs. Nutlets *c.* 2 mm., dark brown, shining. △ Woods, mountain pastures, damp meadows. May–July. All Eur. (except IRL.IS.).

1067. *M. alpestris* Schmidt ALPINE FORGET-ME-NOT. A low, erect, softly hairy perenn. to 20 cm., growing in high mountains, with rather short dense clusters of bright blue fls., each 4–10 mm. across. Corolla lobes flat, rounded; calyx with dense, spreading, silvery hairs and lobes more than half the length of the tube, spreading in fr.; fr. stalks ascending, about as long as the calyx. Lvs. downy, oblong-lance-shaped, the lower long-stalked, the upper stalkless. Nutlets black. △ Rocks and pastures in high mountains. Apr.–Sept. Much of Eur. (except North Eur. P.IRL.B.NL.TR.) IS. Pl. 103.

(b) Fls. usually much smaller than 5 mm. across

1068. *M. discolor* Pers. (*M. versicolor* Sm.) YELLOW AND BLUE SCORPION-GRASS. A very slender, erect, hairy ann. 8–25 cm., with tiny yellow or white fls. which usually turn blue, *c.* 2 mm. across. Corolla tube at length twice as long as the calyx; fr. stalks ascending, shorter than the calyx. Lvs. hairy, oblong-lance-shaped, stalkless. Nutlets dark brown, shining. △ Sandy places, waysides. May–June. All Eur. (except AL.BG.).

M. ramosissima Rochel (*M. collina* auct.; *M. hispida* Schlecht.) EARLY SCORPION-GRASS. Like 1068, but fls. blue or rarely white, never yellow; corolla tube shorter than the calyx which is bell-shaped in fr. Nutlets dull brown. Throughout Europe.

1069. *M. arvensis* (L.) Hill COMMON SCORPION-GRASS. A rather slender, softly hairy ann. or bienn. 15–30 cm. with blue fls. 3–4 mm., less often to 5 mm. across. Corolla lobes flat, rounded; calyx bell-shaped with dense crisped or hooked hairs, lobes more than half the length of the tube, spreading in fr.; fr. stalks spreading, up to twice as long as the calyx. Lower lvs. oval, stalked, forming a rosette, the upper oblong-lance-shaped, stalkless. Nutlets dark brown, shining. △ Fields, waysides, cultivated ground. May–July, often Sept.–Oct. All Eur.

ERITRICHIUM Like *Myosotis*, but a cushion plant of high mountains. Calyx and lvs. with long silky hairs; inflorescence short, hardly elongating in fr. Nutlets fused at their base to the central axis. 2 sps.

1070. *E. nanum* (Vill.) Schrader KING OF THE ALPS. A dense, cushion-forming perenn. 2–10 cm. across, with rosettes of lvs. covered with long, soft, silky white hairs and small dense clusters of brilliant sky-blue fls. Inflorescence few-flowered, with bracts; corolla 6–8 mm. across, scales in throat blunt, yellow; calyx very hairy, lobes much longer than the tube. Lvs. *c.* 1 cm. oblong-lance-shaped; rootstock woody, densely branched. Nutlets three-angled. △ In high mountains: rock faces, screes. July–Aug. E.CH.A.I.R. Pl. 104.

LITHOSPERMUM | Gromwell Corolla funnel-shaped or tubular, with 5 spreading lobes; throat of the corolla with 5 small hairy scales or 5 longitudinal folds; stamens included; calyx five-lobed. Nutlets hard, spineless. *c.* 17 sps.

Fls. whitish or yellow

1071. *L. arvense* L. CORN GROMWELL. An erect, rather rough-haired, little-branched ann. 10–60 cm., with adpressed hairs and with rather short clusters of tiny dull white or rarely bluish fls. Corolla 3–4 mm. across, tube little longer than the calyx; calyx lobes linear; inflorescence elongating in fr. Lvs. to 5 cm., lateral veins inconspicuous, the lower lvs. obovate, stalked, the upper lance-shaped, stalkless. Nutlets greyish-brown, warty. △ Cultivated ground. Apr.–Sept. All Eur. (except IS.AL.).

L. apulum (L.) Vahl. An ann. like 1071, but fls. yellow, in dense branched curved clusters at the ends of the branches which do not elongate in fr. Corolla *c.* 3 mm. across; calyx very bristly-haired; bracts longer than the fls. Lvs. numerous, linear, stalkless, with rough, stiff spreading hairs. Mediterranean region and Portugal, Bulgaria.

1072. *L. officinale* L. GROMWELL. An erect, leafy, much-branched perenn. 30–100 cm., with terminal, leafy, branched clusters of small cream-coloured fls. Corolla 3–4 mm. across, the tube little longer than the calyx. Lvs. softly hairy, to *c.* 7 cm. lance-shaped, stalkless, with conspicuous lateral veins beneath. Nutlets smooth, shining white. △ Margins of woods, bushy places, banks. May–July. All Eur. (except IS.). Pl. 105.

Fls. blue
(a) Herbaceous perenns.

1073. *L. purpurocaeruleum* L. BLUE GROMWELL. A hairy perenn. with thick, creeping, underground stems from which arise many erect, leafy sterile and fl. stems 30–60 cm. Fls. large, at first reddish-purple, then bright blue, in rather few-flowered terminal clusters; corolla 1½–2 cm. across, funnel-shaped, hairy outside, twice as long as the calyx. Lvs. rough hairy, to *c.* 7 cm., narrow lance-shaped, stalkless, dark green above, paler beneath. Nutlets white, somewhat shining. △ Woods, thickets, hedges. Apr.–June. Most of Eur. (except North Eur. P.IRL.B.NL.). Pl. 104.

(b) Woody perenns.

1074. *L. diffusum* Lag. (*L. prostratum* Loisel.) SCRAMBLING GROMWELL. A low, slender, spreading, scrambling undershrub to ½ m. with terminal clusters of a few bright blue fls. Corolla 1–1½ cm. long by 1 cm. across; tube with dense silky hairs outside and 3 times as long as calyx; throat of corolla closed with hairs, corolla lobes blunt, spreading. Lvs. *c.* 1 cm., narrowly elliptic blunt, with adpressed hairs; stems very slender, with spreading hairs. △ Heaths. May–July. P.E.F. Pl. 104.

1075. *L. fruticosum* L. SHRUBBY GROMWELL. An erect, much-branched undershrub to ½ m., with greyish scaly stems, linear lvs., and terminal few-flowered clusters of large, blue, funnel-shaped fls. Corolla 1–1½ cm. long to *c.* 1 cm. across, tube quite hairless, more than twice as long as the calyx, lobes spreading. Lvs. 1–1½ cm., densely rough-hairy, whitish beneath, margin inrolled. △ Dry arid slopes. Apr.–June. P.E.F.

ONOSMA | Golden Drop Corolla cylindrical with 5 very short blunt lobes, scales in throat absent; stamens included or shortly projecting; calyx five-lobed. *c.* 20 sps.

Swollen base of hairs mostly without smaller spreading hairs

1076. *O. echioides* L. GOLDEN DROP. A stiff, erect, very bristly-haired perenn. with several erect stems to 40 cm., with terminal branched clusters of pendulous, tubular, pale yellow fls. Corolla 2 cm. long, with short spreading lobes, little longer than calyx which enlarges after flowering to 2½ cm. Lower lvs. linear to *c.* 8 cm., the upper lance-shaped 2–3 cm., all covered with bristly yellow hairs with swollen bases, but without additional small basal hairs. △ Dry, sandy slopes, rocks. May–June. E.F.D.A.H.I.YU.BG.R.SU. *The root yields a dye known as 'orcanette jaune', which is also used for colouring food and drinks.* Pl. 105.

O. arenarium Waldst. & Kit. Like 1076, but fls. smaller 1–1½ cm. long, more numerous in dense pyramidal clusters. Lvs. yellowish-green, the basal lvs. oblong-spathulate, the stem lvs. oblong, stalked; bristly hairs with basal swelling but without small hairs, or on lower surface of lv. basal swellings with small hairs. A very hispid bienn. or perenn. to ½ m. Most of Central and South-Eastern Europe.

Swollen base of hairs with smaller spreading hairs

1077. *O. stellulatum* Waldst. & Kit. Distinguished by its white bristly hairs, nearly all of which have 3–10 small lateral hairs arising from the swollen base and spreading in a star. Inflorescence single or forked; fls. pale yellow *c.* 2 cm.; corolla finely hairy, twice as long or more than the calyx; fl. stalk 6–8 mm., as long as calyx. Lvs. greenish or yellowish, oblong-linear, very rough. A perenn. with several stiff, erect, unbranched stems to 30 cm. arising from a woody base. △ Dry rocky places, in mountains. June–July. South-East Eur. I.

O. tauricum Willd. Like 1077, but fls. stalkless in dense terminal, often forked clusters with the lower bracts longer than the calyx; calyx lobes densely long-haired on the margin. Lvs. greyish, linear, densely bristly above, with many small hairs spreading in a star and arising from the swollen base of the long bristly hairs; lower surface of lv. with bristly hairs only on the midvein, and with star-shaped hairs without the central bristle on the blade. South-Eastern Europe. Page 341.

CERINTHE | Honeywort Corolla cylindrical, usually with 5 short lobes, without scales in throat; calyx five-lobed. Inflorescence with large leafy bracts. Lvs. usually glaucous, hairless or nearly so, the upper clasping the stem. Nutlets erect, globular, smooth. *c.* 10 sps.

1078. *C. minor* L. LESSER HONEYWORT. A glaucous, bloomed, hairless bienn. or perenn. to ½ m. with clasping lvs. and curved clusters of small yellowish drooping fls. subtended by conspicuous oval bracts. Corolla 1–1½ cm., often purple-blotched, divided almost to the middle into 5 narrow, acute, converging lobes a little longer than the calyx. Lvs. all smooth or with tubercles, oblong-oval, often white-spotted, margin not hairy, the lowest stalked. Fr. stalks spreading or reflexed. △ Cultivated ground, pastures, woods, rocky places, in mountains. May–Aug. Cent. Eur. (except CH.), South-East Eur. F.SU. Pl. 105.

1079. *C. major* L. HONEYWORT. A hairless glaucous ann. to ½ m. with obovate, overlapping, clasping lvs. and distinctive drooping clusters of large, yellow, cylindrical fls. often with a dark chocolate zone towards the base. Corolla 2–3 cm. long and 6–8 mm. across, twice as long as the calyx, largely pale yellow with varying amounts of darker colouring, lobes of corolla very short, recurved; bracts oval, green, as long as or longer than calyx. Lvs. with rough swellings, usually blotched with white, margin finely bristly. A variable sp. △ Cultivated places, waysides, stony ground. Feb.–June. Med. Eur. (except AL.) P.CH.: introd. A. Pl. 105.

C. glabra Miller SMOOTH HONEYWORT. Like 1079, but a perenn. with smaller golden-yellow fls. spotted with purple at the base. Corolla 1–1½ cm. by ½ cm., about 1¼ times as long as calyx. Lvs. smooth, glaucous, bloomed, usually not blotched, margin largely hairless. Fr. stalks reflexed. Widespread in Central and South-Eastern Europe.

ECHIUM | Bugloss Corolla funnel-shaped, with an open, obliquely cut mouth, the upper lip longer, entire, or shallowly two-lobed, the lower lip three-lobed, scales in throat absent; stamens unequal, some or all often longer than the corolla. *c.* 20 sps.

Stamens included in corolla; fls. usually less than 12 mm. long
1080. *E. parviflorum* Moench SMALL-FLOWERED BUGLOSS. A rough, bristly-haired ascending ann. to 40 cm. with a lax leafy cluster of small blue fls. with white throats. Corolla 10–12 mm. long, widening at the mouth, hairy, little longer than calyx; calyx with oblong-lance-shaped lobes, with adpressed hairs, enlarging to *c.* 1 cm. in fr. Lower lvs. oblong and narrowed to a stalk, the upper oblong stalkless and clasping the stem. △ Sandy places and rocks on the littoral. Mar.–May. Med. Eur. (except AL.TR.) P.

E. arenarium Guss. Like 1080, but fls. violet and corolla 8–9 mm., often longer than the calyx. Calyx with linear lobes with spreading hairs, not or little enlarging in fr. Stems slender, procumbent or ascending. Mediterranean region.

1. *Omphalodes linifolia* 1048
2. *Asperugo procumbens* 1051
3. *Myosotis sicula* [1065]
4. *Onosma tauricum* [1077]
5. *Lippia nodiflora* 1086
6. *Nonea pulla* 1060

Stamens projecting beyond corolla; fls. usually more than 12 mm. long
(a) Fls. 1–1½ cm. long

1081. *E. italicum* L. PALE BUGLOSS. A stiff, erect, greyish, very bristly-haired bienn. to 80 cm., with a very long spike-like or pyramidal inflorescence of many short lateral fl. branches with small pale pink or bluish fls., or fls. sometimes yellow with pink stripes. Corolla 1–1½ cm., hairy on the outside, twice as long as the bristly-haired calyx; stamens violet, much longer than the corolla. Basal lvs. to 20 cm., narrow lance-shaped, in a dense rosette, the upper lvs. linear with dense white bristly hairs; stem robust. △ Dry, arid places, stony ground, sands. Apr.–Aug. Med., South-East Eur. A.CS.H. *The roots produce a red dye.* Pl. 105.

(b) Fls. usually more than 1½ cm. long

1082. *E. vulgare* L. VIPERS BUGLOSS. A stiff, erect, leafy, bristly-haired bienn. to 90 cm. with very rough green lvs. and a rather narrow, branched, dense pyramidal inflorescence of bright blue, or rarely white fls. Fls. pink in bud; corolla 1½–2 cm., 2–3 times as long as the calyx, mouth oblique; 4 stamens longer than the corolla, 1 included. Rosette lvs. to 15 cm., lance-shaped and with a prominent midvein, stalked, the stem lvs. oblong-lance-shaped, stalkless. △ Cultivation, waste places, dry stony ground. May–Aug. All Eur. (except AL.). *Used in herbal remedies.* Pl. 104.

1083. *E. lycopsis* L. (*E. plantagineum* L.) PURPLE VIPERS BUGLOSS. Like 1082, but lvs. much softer and only slightly rough and rosette lvs. oval, stalked, with conspicuous midvein and lateral veins, the upper lvs. stalkless with a heart-shaped base clasping the stem. Fls. larger, 2½–3 cm. long, at first reddish-purple then blue-purple; 2 stamens considerably longer than the corolla, 3 included. An erect bienn. to 60 cm. △ Dry sandy places. Apr.–July. Med. Eur. P.GB.A.BG. Pl. 106.

VERBENACEAE | Verbena Family

Herbs, trees, shrubs or woody climbers, usually with opposite or whorled lvs. Calyx tubular, with 4–5 lobes; corolla tubular with 4–5 lobes, often somewhat two-lipped; stamens usually 4. Ovary often four-lobed; style terminal, stigma 1 or several; fr. usually a drupe with 1 or more stones, rarely of 4 nutlets.

1. Lvs. palmate; leaflets 5–7.	*Vitex*
1. Lvs. entire, toothed or pinnately lobed.	
2. Lvs. pinnately lobed.	*Verbena*
2. Lvs. entire or toothed.	
3. Shrubs.	*Lantana*
3. Creeping herbaceous perenns.	*Lippia*

VERBENA Fls. in elongate, spike-like or flat-topped, umbel-like clusters. Calyx tubular, five-lobed; corolla tubular, five-lobed, shortly two-lipped; stamens 4 included in tube. Fr. splitting into 4 nutlets. Lvs. deeply once or twice cut into narrow lobes. *c.* 5 sps.

1084. *V. officinalis* L. VERVAIN. A stiff, erect, slender, square-stemmed perenn. 30–75 cm. with tiny lilac-blue fls. at first in dense terminal clusters, but soon elongating into slender spikes. Corolla 2–5 mm. across, tube hairy, about twice as long as the hairy calyx; bracts oval about half as long as calyx. Lvs. opposite, deeply pinnately cut into oblong often toothed lobes, with rough adpressed hairs; stem with wide-spreading slender, nearly leafless branches. Nutlets 4, reddish-brown. △ Waste ground, waysides, screes. June–Oct. Much of Eur. (except IS.): introd. IRL.N.S.SF. *Used in herbal remedies.*

V. supina L. PROCUMBENT VERVAIN. Like 1084, but stems procumbent, much-branched, and lvs. twice cut into oval segments. Corolla pale lilac, shorter, 3 mm. Southern Europe.

LANTANA Fls. small, stalkless in the axils of bracts forming dense heads or spikes; corolla tube slender, with 4–5 equal spreading lobes; stamens 4. Fr. a fleshy drupe with 2 nuts. *c.* 3 sps.

1085. *L. camara* L. A prickly-stemmed shrub to *c.* 1 m. with oval-oblong toothed lvs. and dense, long-stalked, hemispherical heads of numerous yellow fls. which characteristically turn orange or red. Fl. heads 2½–5 cm. across, pink in bud; corolla tube 6–12 mm. long, with 4 nearly equal spreading lobes; calyx shortly bell-shaped. Lvs. 2½–16 cm., rather thick, rough, finely hairy beneath. △ Native of tropical America; often grown as an ornamental in Southern Europe. Summer.

LIPPIA Fls. small, numerous, in dense heads or spikes; calyx tiny, two-lobed, lobes keeled; corolla funnel-shaped, four-lobed; stamens 4. Fr. dry, with 2 carpels. Lvs. entire or toothed. 2 sps.

1086. *L. nodiflora* (L.) Michx A creeping, rooting perenn. to ½ m., with tiny, stalkless, bluish or pink fls. in dense ovoid heads ½–1 cm. long, borne on long axillary stems much longer than the subtending lvs. Corolla 3 mm. long, hairy outside; bracts broadly oval, overlapping, hairy. Lvs. opposite, 1½–3 cm., elliptic and narrowed to the base, coarsely toothed towards the apex. △ Damp places on the littoral. May–Sept. Med. Eur. Page 341.

VITEX Fls. in whorls forming elongated spikes. Corolla tubular, with 5 lobes; calyx five-lobed. Lvs. palmately compound with 5–7 leaflets. Fr. fleshy. 1 sp.

1087. *V. agnus-castus* L. CHASTE TREE. An aromatic shrub 1–3 m., with white-felted branches, palmate lvs., and with long, terminal, interrupted spikes of pale lilac or rarely rose-coloured fls. Corolla 6–9 mm., almost two-lipped, hairy outside; stamens longer, projecting; calyx hairy. Lvs. long-stalked with 5–7 lance-shaped leaflets, green above and white-felted beneath. Fr. fleshy, reddish-black. △ Damp places on the littoral, by streams. June–Sept. Med. Eur. SU. *A shrub well known since classical times and associated with chastity; it is used medicinally and is a source of a yellow dye. The fruits are used for seasoning and the twigs for basket-making.* Pl. 106.

CALLITRICHACEAE | Starwort Family

Aquatic plants with weak thread-like stems and entire opposite lvs. Fls. minute, stalkless in the axils of the lvs.; petals and sepals absent; stamen 1; ovary four-lobed, styles 2. Fr. of 4 drupelets. A family of 1 genus, showing considerable plasticity of form under different conditions. Lv. shape and fr. shape are the most important diagnostic characters, but plants are often sterile. *c.* 7 sps.

CALLITRICHE | Starwort

1088. *C. stagnalis* Scop. COMMON STARWORT. Lowest lvs. elliptic or spathulate, never linear;

the uppermost floating lvs. forming a rosette with broadly elliptic or almost circular blades abruptly narrowed to the stalk. Fr. broader than long; carpels with transparent wings. Ann. or perenn. 10–100 cm. △ Still or running waters, mud. Apr.–Oct. All Eur. Pl. 105.

C. intermedia Hoffm. Distinguished by its lowest lvs. which are linear and usually abruptly widened at the apex and deeply notched, the upper lvs. narrowly oblance-shaped, notched, and the floating lvs. not forming a well-marked rosette. Fr. almost circular, carpels keeled, not winged. Widespread in Europe.

LABIATAE | Mint Family

A distinctive family, often aromatic and glandular, usually of herbaceous plants or small shrubs with quadrangular stems and simple opposite paired lvs. which are arranged successively at right angles to each other. Fls. usually clustered into short-stalked whorls in the axils of the upper lvs. or bracts, the whole often clustered together to form a spike-like inflorescence. Fls. mostly symmetrical in one plane only, two-lipped; calyx tubular, five-lobed and often two-lipped; corolla tubular, five-lobed, usually strongly two-lipped; stamens 4, rarely 2, attached to the corolla tube. Fr. of 4 one-seeded nutlets.

1. Corolla with 4 nearly equal lobes, or with 1 lower lip.   Group A
1. Corolla distinctly two-lipped.
   2. Stamens 2.   Group B
   2. Stamens 4.
     3. Stamens and style not longer than corolla tube.   Group C
     3. Stamens and style distinctly longer than corolla tube.
       4. Upper lip of corolla distinctly arched over stamens.   Group D
       4. Upper lip of corolla more or less flat.
         5. Calyx with 5 lobes, separated into 2 lips.   Group E
         5. Calyx with 5 more or less regular lobes or calyx entire.   Group F

GROUP A. *Corolla with 4 nearly equal lobes, or with 1 lip*

		Nos.
1. Corolla with 4 nearly equal lobes.		
2. Stamens 2.	*Lycopus*	1166
2. Stamens 4.	*Mentha*	1167–1172
1. Corolla with 1 lower lip, the upper lip very short or absent.		
3. Lower lip of 3 unequal lobes; corolla tube hairy within.	*Ajuga*	1089–1094
3. Lower lip of 5 unequal lobes; corolla tube hairless within.	*Teucrium*	1095–1104

GROUP B. *Corolla two-lipped; stamens 2*

		Nos.
1. Lvs. linear, inrolled beneath.	*Rosmarinus*	1105
1. Lvs. lance-shaped to oval, flat.	*Salvia*	1143–1149

GROUP C. *Corolla two-lipped; stamens 4, not longer than corolla tube*

		Nos.
1. Fls. blue or violet, in a dense spike.	*Lavandula*	1110, 1111
1. Fls. white or yellow.		
2. Calyx without bracteoles; calyx lobes 5, spiny-tipped.	*Sideritis*	1114, 1115
2. Calyx with bracteoles; calyx lobes 5–10, not spiny-tipped.	*Marrubium*	1112, 1113

GROUP D. *Corolla two-lipped, the upper lip arched over the 4 stamens*

Nos.

1. Calyx two-lipped, or of 5 lobes separated into 2 lips.
   2. Lips of calyx closed after flowering.

3. Calyx with a boss or ridge on back; calyx lobes blunt.	*Scutellaria*	1107–1109
3. Calyx without a boss or ridge; calyx lobes narrow acute.	*Prunella*	1119, 1120

   2. Lips of calyx not closed after flowering.

4. Shrubs; calyx and corolla hairless.	*Prasium*	1106

    4. Herbaceous plants.

5. Fls. *c.* 1 cm.; calyx hairy; corolla hairless outside.	*Melissa*	1150
5. Fls. *c.* 3 cm.; calyx and corolla hairy outside.	*Melittis*	1121

1. Calyx with 5 more or less equal lobes, not separated into 2 lips.
   6. Carpels flat-topped as if cut off.

7. Calyx with spiny-tipped lobes; carpels hairy above.	*Leonurus*	1133

   7. Calyx lobes soft-tipped; carpels hairless above.

8. Lower lip of corolla two-lobed.	*Lamium*	1127–1131
8. Lower lip of corolla three-lobed.	*Galeobdolon*	1132

  6. Carpels with rounded apex.

9. Corolla with a three-lobed lower lip and 2 conical projections at base of the lip.	*Galeopsis*	1125, 1126

   9. Corolla with a three-lobed lower lip without projections.

10. Corolla tube shorter than calyx; fls. large 2 cm. long or more.	*Phlomis*	1122–1124

    10. Corolla as long as or longer than calyx; fls. *c.* 1–1½ cm. long.

11. Calyx funnel-shaped, conspicuously ten-veined.	*Ballota*	1134

     11. Calyx tubular or bell-shaped, usually with 5 veins.

12. Outer stamens spreading laterally after flowering; plant without a basal rosette of lvs.	*Stachys*	1135–1140
12. Outer stamens not spreading; plant with a basal rosette of lvs.	*Betonica*	1141, 1142

GROUP E. *Corolla two-lipped; upper lip flat; stamens 4; calyx two-lipped*

Nos.

1. Lvs. small 1 cm. or less, entire, thick.

2. Calyx tube conspicuously swollen at base.	*Acinos*	1157

  2. Calyx tube not swollen at base.
   3. Fls. in rounded or oval terminal clusters; stamens diverging above.

4. Calyx oval, strongly compressed, thirteen-veined.	*Coridothymus*	1162
4. Calyx cylindrical or bell-shaped, ten-veined.	*Thymus*	1163–1165

   3. Fls. in long slender spikes formed of many whorls; stamens converging above.

	Satureja	1152

1. Lvs. 2 cm. or more, toothed, thin.

5. Stamens parallel their whole length; fls. *c.* 3 cm.	*Melittis*	1121

  5. Stamens converging above under upper lip of corolla; fls. usually less than 2 cm.
   6. Calyx with hairs in throat.

7. Calyx tube straight; fls. in lax, stalked clusters.	*Calamintha*	1154–1156

   7. Calyx tube curved; fls. in dense more or less stalkless clusters.

8. Fl. clusters many-flowered, dense; calyx not or scarcely swollen at base.	*Clinopodium*	1158
8. Fls. in clusters of 3–8; calyx swollen at base.	*Acinos*	1157
6. Calyx hairless in throat.		
9. Fls. blue-violet.	*Horminum*	1151
9. Fls. white or rose-purple.		
10. Fls. rose-purple, inflorescence leafless at apex.	*Clinopodium*	1158
10. Fls. whitish, inflorescence leafy at apex.	*Melissa*	1150

GROUP F. *Corolla two-lipped; upper lip flat; calyx equally five-lobed or entire*

		Nos.
1. Calyx entire, split to the base.	*Majorana*	1161
1. Calyx equally five-lobed.		
2. Stamens converging above under the upper lip of corolla, or strongly divergent.		
3. Lvs. entire, thick; usually woody-based shrublets.		
4. Stamens much longer than corolla; fls. blue or violet.	*Hyssopus*	1159
4. Stamens shorter or scarcely longer than corolla; fls. pink or purple.		
5. Calyx ten-veined.	*Satureja*	1152
5. Calyx thirteen- to fifteen-veined.	*Micromeria*	1153
3. Lvs. toothed, thin: not shrublets; fls. with large bracts below calyx.	*Origanum*	1160
2. Stamens parallel their whole length.		
6. Calyx with 5–10 veins and 5 usually spiny-tipped lobes.		
7. Carpels flat-topped.	*Leonurus*	1133
7. Carpels with rounded apex.		
8. Calyx funnel-shaped; calyx lobes pleated.	*Ballota*	1134
8. Calyx cylindrical; calyx lobes not pleated.	*Stachys*	1135–1140
6. Calyx with 15 veins and 5 soft-pointed lobes.		
9. Stems creeping; lvs. kidney- or heart-shaped; fls. in axillary clusters.	*Glechoma*	1118
9. Stems erect; lvs. lance-shaped or oval; fls. in terminal clusters.	*Nepeta*	1116, 1117

AJUGA | Bugle Herbaceous perenns. distinguished by the corolla which has a very short upper lip and a conspicuous three-lobed lower lip; corolla tube with a ring of hairs within; calyx tubular-bell-shaped, nearly equally five-lobed. *c.* 12 sps.

Whorls many-flowered
(a) Above-ground stolons present
1089. *A. reptans* L. COMMON BUGLE. A creeping perenn. with many above-ground rooting stolons and erect stems to 30 cm., with densely clustered whorls of blue or rarely pink or white fls. Inflorescence rather dense, cylindrical; fls. 1½–2 cm. long; calyx shorter than the corolla tube, with long spreading hairs; bracts subtending upper whorls shorter than the fls., oval, often blue-flushed. Lvs. oblong-oval entire or toothed, nearly hairless, the basal lvs. 4–7 cm. in a rosette, the upper shorter; stem hairy on opposite sides only. △ Damp places, meadows, woods. Apr.–July. All Eur. (except IS.). Pl. 106.

(b) Above-ground stolons absent
1090. *A. genevensis* L. BLUE BUGLE. Like 1089, but above-ground stolons absent and stems white-haired all round or on 2 opposite sides. Fls. bright blue, rarely pink, in a loose

elongated spike, the upper bracts strongly flushed with blue and shorter than the fls. Calyx hairy; stamens long-projecting. Basal lvs. large, 5–12 cm. long, stalked, often dying before flowering; stem lvs. little shorter, short-stalked. A rather hairy, erect perenn. 10–40 cm. △ Dry meadows, field verges, screes. May–Aug. Cent., South-East Eur. F.B.I.SU.: introd. DK.N.S.SF. Pl. 105.

1091. *A. pyramidalis* L. PYRAMIDAL BUGLE. Readily distinguished by its dense pyramidal head of broad, oval, overlapping bracts which are bright blue or violet, longer than and half hiding the small pale blue-violet fls. Corolla 1–1½ cm.; stamens projecting. Lvs. larger at base and decreasing to apex of stem, the lowest persisting at flowering; stem hairy all round; stolons absent. Usually a densely hairy, erect perenn. 5–20 cm. △ Meadows in mountains. Apr.–Aug. Most of Eur. (except NL.H.AL.GR.TR.). Pl. 106.

A. orientalis L. EASTERN BUGLE. Like 1091, but stamens included in the corolla; bracts brightly blue-flushed, but the uppermost shorter than the blue fls. Lvs. densely hairy, oblong and deeply or shallowly lobed. Greece, Bulgaria.

Whorls two-flowered, or rarely four-flowered
(a) Lvs. undivided
1092. *A. laxmannii* (Murray) Bentham An erect perenn. with spreading hairs, oblong lvs., and short-stalked yellow fls. with purple veins. Upper lip of corolla shortly two-lobed; bracts much longer than the fls. Lvs. with a rounded base half-clasping the stem, entire or toothed towards the apex. △ Limestone slopes and hills. June–July. South-East Eur. (except AL.) PL.CS.H.SU.

1093. *A. iva* (L.) Schreber A perenn. with several very leafy stems arising from a woody base, to 20 cm., with narrow silvery-haired lvs. smelling of musk, and rosy-purple fls. Fls. stalkless, *c*. 2 cm., the upper lip not lobed; calyx woolly-haired; bracts shorter than the fls. Lvs. oblong-linear with inrolled margin, entire or with a few teeth, stalkless. △ Dry, stony places, olive groves. May–Oct. Med. Eur. (except AL.TR.) P.

(b) Lvs. deeply three-lobed
1094. *A. chamaepitys* (L.) Schreber GROUND-PINE. A leafy, branched, aromatic ann. 5–20 cm., with a strong, resinous, pine-like smell when crushed, and bright yellow fls. Corolla 5–12 mm. long, lip often red-spotted; calyx hairy; bracts usually shorter than the fls. Lvs. stalked, mostly deeply divided almost to the base into 3 strap-shaped, glandular-sticky lobes 2–4 cm. Nutlets honey-combed with depressions. △ Dry banks, fields, stony places. May–Aug. Much of Eur. (except North Eur. IRL.NL.). Pl. 106.

A. chia (Poiret) Schreber Distinguished from 1094 by its larger fls. 2 cm. long, which are yellow with purple veins, or purple throughout; corolla 4 times as long as the calyx; bracts shorter than the fls. Lvs. with 3 strap-shaped lobes, white-haired. Nutlets with transverse wrinkles. Eastern Mediterranean region.

TEUCRIUM | Germander Like *Ajuga* without an upper lip, or upper lip very short, lower lip five-lobed and corolla tube hairless within; calyx tubular, equally five-lobed or somewhat two-lipped. *c*. 50 sps.

Lvs. deeply divided into narrow lobes
1095. *T. botrys* L. CUT-LEAVED GERMANDER. A glandular-hairy aromatic ann. or bienn. 10–30 cm., distinguished by its lower lvs. which are twice cut into blunt strap-shaped segments. Fls. pinkish-purple, in numerous whorls of 4–5 fls. arranged rather one-sidedly and forming a long lax cluster. Corolla *c*. 1 cm., hairy; calyx hairy, distinctly swollen at the base; bracts pinnately lobed. Lvs. 1–2½ cm. oval in outline, stalked. △ Stony fields, waysides. June–Oct. Cent. Eur. E.F.GB.B.I.YU.BG.R.SU.: introd. S.

1096. *T. pseudochamaepitys* L. GROUND-PINE GERMANDER. Distinguished from 1095 by the lvs. which are cut into 3–5 linear pointed segments with inrolled margin. Fls. white or pink, stalked, 2 in each whorl forming a lax, one-sided, terminal cluster. Corolla large 1–1½ cm.; stamens long-projecting; calyx glandular-hairy; bracts three-lobed, shorter than the fls. A shrubby hairy perenn. 20–50 cm. △ Dry hills, grassy places, poor ground. Apr.–Sept. P.E.F. Pl. 107.

Lvs. entire or toothed
(a) Fls. in dense globular heads
1097. *T. polium* L. A small branched shrublet 5–40 cm. with white or grey-felted stems and lvs. and dense globular terminal heads of pink, white or rarely yellowish fls. Fls. numerous, stalkless; corolla little longer than the calyx which is conspicuously white-felted, calyx lobes blunt. Lvs. oblong to linear, blunt, with rounded teeth, margin inrolled. A very variable sp. △ Dry places, sunny rocks. June–Aug. Med. Eur. P.BG.R.SU.

(b) Fls. not in globular heads
(i) Fls. pink, violet, purple, or blue
1098. *T. scordium* L. WATER GERMANDER. A lax, much-branched, soft-haired perenn. 20–60 cm., with purple or lilac fls. in lax, one-sided, axillary whorls of 2–6 fls. half hidden among the leafy bracts. Corolla *c.* 12 mm.; calyx lobes equal. Lvs. ½–5 cm. lance-shaped, coarsely toothed, softly hairy, stalkless lvs. and bracts similar; stolons present. △ Streamsides, damp meadows, ditches. June–Aug. Most of Eur. (except P.IS.N.SF.AL.TR.). Page 352.

1099. *T. chamaedrys* L. WALL GERMANDER. A low, shrubby, tufted perenn. 10–30 cm., rooting below, with pinkish-purple fls. in short, somewhat one-sided, terminal clusters. Corolla *c.* 1½ cm., twice as long as calyx; calyx with spreading hairs; bracts toothed, similar to the lvs. Lvs. 1–3 cm., broadly oval, deeply rounded-toothed, somewhat leathery, usually dark green above, but very variable. △ Dry places, banks, open woods. May–Sept. Most of Eur. (except North Eur. IRL.): introd. GB. Pl. 107.

1100. *T. marum* L. CAT THYME. A leafy erect undershrub 20–50 cm., with woolly-white stems and whorls of small purple fls. forming a short, dense, one-sided cluster. Fls. *c.* 1 cm., 2–4 in a whorl; calyx densely hairy; stamens and style little-projecting. Lvs. *c.* 1 cm., oval-lance-shaped, entire, margin inrolled, green above and with white adpressed hairs beneath. △ Dry hills, bushy places. May–Aug. E.F.I.YU.

1101. *T. fruticans* L. TREE GERMANDER. A small, branched evergreen shrub 1–1½ m. with white-felted stems and few large pale blue or lilac fls. in a lax, leafy cluster. Fls. 1 or 2 in each whorl, long-stalked; lip of corolla *c.* 1½ cm. long, corolla tube very short and stamens and style much longer; calyx white-felted, bell-shaped, lobes oval. Lvs. 2–4 cm., lance-shaped entire, margin inrolled, dark shining green above, densely white-felted beneath. △ Rocks, wooded hills, on the littoral; sometimes grown for ornament and occasionally naturalized. Feb.–June. P.E.F.I. Pl. 107.

(ii) Fls. yellow, white, or greenish
1102. *T. scorodonia* L. WOOD SAGE. An erect shrubby perenn. to 50 cm., with sage-like lvs. and greenish-yellow fls. forming long leafless spikes. Fls. paired; corolla *c.* 1 cm., much longer than the calyx; calyx two-lipped, the upper lobe much broader than the others; bracts small 2–5 mm., oval entire and much shorter than the fls. Lvs. stalked, 3–7 cm., oval-heart-shaped, margin rounded-toothed, blade wrinkled above, paler greyish beneath. △ Open woods, heaths, thickets. June–Sept. West, Cent. Eur. N.I.GR.SU.: introd. DK.S. *Used as a herbal tonic.*

1103. *T. montanum* L. MOUNTAIN GERMANDER, ALPINE PENNY ROYAL. A woody-based, shrubby mat-forming perenn. 5–25 cm., with many spreading stems, narrow entire lvs. and flattened or rounded clusters of small whitish-yellow fls. Corolla 12–15 mm., twice as long as the hairless calyx. Lvs. 1–2 cm., linear-lance-shaped, greenish above, and densely white-felted beneath, margin inrolled. △ In mountains: rocks, screes, dry slopes. May–Aug. Cent., South-East Eur. E.F.B.I. Pl. 107.

1104. *T. flavum* L. YELLOW GERMANDER. A shrubby perenn. 20–50 cm., with leathery toothed lvs. and a rather dense, elongate, leafless, one-sided, terminal cluster of yellow fls. Corolla *c*. 1½ cm., about 2½ times as long as the densely hairy calyx, upper lobes of corolla blunt; bracts lance-shaped entire, shorter than the fls. Lvs. stalked, 1–1½ cm., broadly oval with a cut-off or wedge-shaped base; stems with crisped hairs. △ Stony places, rocks. May–Aug. Med. Eur. (except AL.TR.).

ROSMARINUS | Rosemary Corolla two-lipped, the upper lip erect and two-lobed, the lower three-lobed; calyx two-lipped, the upper lip oval, the lower two-lobed. Stamens 2, long-projecting, anthers hinged on filament. 3 sps.

1105. *R. officinalis* L. ROSEMARY. A dense, much-branched, aromatic evergreen shrub 1–3 m., with narrow, dark green, leathery lvs. and pale lilac fls. in leafy clusters. Fls. stalkless, several in axillary clusters; corolla two-lipped, the upper lip somewhat hooded, tube longer than the calyx; stamens and style curved, much longer than the corolla. Lvs. numerous, 2–3½ cm. linear, margin inrolled, white-hairy beneath. A variable sp. △ Dry bushy places, rocks, dry hills; often cultivated and sometimes naturalized. All the year. Med. Eur. (except AL.TR.) P.SU.: introd. CH.A. *It yields 'Oil of Rosemary' which is used in perfumery, hair lotions, and soap and is an ingredient of eau-de-Cologne. Used as a pot herb, for flavouring and in liqueurs; also used medicinally.* Pl. 107.

PRASIUM Corolla two-lipped, the upper lip arched over the stamens, the lower three-lobed; calyx two-lipped, the upper lip three-lobed; stamens 4. 1 sp.

1106. *P. majus* L. A twiggy, slender-stemmed, nearly hairless undershrub ½–1 m. with shining oval, toothed lvs. and white or pinkish fls. Fls. 2 to each whorl forming leafy clusters: corolla recalling *Lamium*, strongly two-lipped with the upper lip arched; calyx conspicuous, green, bell-shaped, ten-veined, and with spiny-tipped lobes. Lvs. 1½–3 cm. oval-heart-shaped, strongly toothed, stalked. Fr. somewhat fleshy, black. △ Rocks, bushy places. Apr.–June. Med. Eur. (except AL.TR.) P. Pl. 107.

SCUTELLARIA | Skullcap Corolla tube long, two-lipped, the upper lip hooded, the lower lip three-lobed with the middle lobe notched; calyx two-lipped, the upper lip distinctive with a small transverse scale on back, lips closing after flowering. Stamens 4, protected by the hooded upper lip of the corolla. *c*. 12 sps.

Fls. in short, dense, terminal clusters
1107. *S. alpina* L. ALPINE SKULLCAP. A low, spreading, hairy woody-based perenn. 10–30 cm., with many leafy stems and large, terminal, quadrangular clusters of blue-violet, or rarely white or purple fls. subtended by conspicuous pale or purple-flushed bracts. Corolla 2–2½ cm., the lower lip white; bracts *c*. 1 cm. oval, papery, much longer than the glandular-hairy calyx. Lvs. 1½–2 cm. oval, coarsely toothed, mostly stalkless. △ Rocks and screes in mountains. June–Aug. South-East Eur. (except TR.) E.F.CH.I. Pl. 108.

S. orientalis L. EASTERN ALPINE SKULLCAP. Like 1107, but lvs. all stalked, green, hairless or hairy above and with adpressed silvery hairs beneath. Fls. yellow, lower lip of corolla becoming pinkish; bracts green, entire or with 1–2 teeth. South-Eastern Europe and Spain. Pl. 108.

Fls. in long, one-sided, leafy clusters
(a) Bracts similar to lvs.
1108. *S. galericulata* L. SKULLCAP. An erect or ascending leafy perenn. 20–100 cm., usually with paired blue-lilac fls. arising from the upper leaf-like bracts and forming a lax, elongated, one-sided inflorescence. Corolla 1½–2 cm., slightly curved, much longer than the hairless or hairy calyx; bracts longer than the fls. Lvs. short-stalked, 3–4 cm., oblong-lance-shaped with heart-shaped base, margin with 4–8 shallow rounded teeth. △ Damp places, marshes, watersides. June–Sept. All Eur. (except IS.). Pl. 108.

S. hastifolia L. SPEAR-LEAVED SKULLCAP. Like 1108, but lvs. 1–2½ cm., toothless, oblong-lance-shaped with 2 conspicuous outward-projecting basal lobes. Fls. larger, more conspicuous, 2–2½ cm., rather densely clustered into an elongated inflorescence; corolla strongly curved at the base; calyx glandular-hairy. Lower bracts similar to the lvs., the upper lance-shaped, shorter than the fls. Widespread in Europe.

*S. minor Hudson LESSER SKULLCAP. Western and West-Central Europe.

(b) Bracts conspicuously dissimilar to lvs.
1109. *S. columnae* All. LARGE SKULLCAP. An erect, hairy, little-branched perenn. 40–80 cm. with large lvs. and slender, leafless, one-sided, sticky-glandular spikes of large violet-purple fls. Corolla 2½–3 cm.; bracts c. 1 cm., oval entire, about equalling the glandular-hairy calyx. Lvs. large, oval-heart-shaped, strongly toothed, hairy, long-stalked. △ Woods, bushy places. June–July. Med., South-East Eur. (except TR.) H. Page 352.

S. altissima L. Distinguished from 1109 by the nearly hairless lvs. and less glandular-hairy inflorescence. Fls. 1½–2 cm.; corolla 3–4 times as long as calyx, the upper lip bluish-purple, the lower whitish. East-Central and South-Eastern Europe.

LAVANDULA | Lavender Fls. blue or purple, usually in a densely clustered leafless spike. Corolla two-lipped, the upper lip two-lobed, the lower three-lobed; calyx oval, thirteen- or fifteen-veined, lobes short, the upper enlarged into a cap, converging in fr. Stamens 4, not projecting. c. 8 sps.

Fl. spikes topped by coloured bracts
1110. *L. staechas* L. FRENCH LAVENDER. A small, greyish, erect, branched shrub 30–80 cm. with dense, oval, quadrangular-sectioned, dark-purple, terminal spikes topped by a cluster of large, often pale purple bracts. Corolla dark purple, little longer than the very hairy calyx; bracts subtending fls. broadly oval, papery, veined, densely overlapping, terminal coloured bracts lance-shaped. Lvs. linear, with inrolled margin, densely white-felted on both sides; stem leafy to the inflorescence. △ Dry, stony places, hills, open thickets. Apr.–June. Med. Eur. (except YU.AL.) P. Pl. 108.

L. dentata L. TOOTHED LAVENDER. Readily distinguished from 1110 by the lvs. which have their inrolled margin deeply cut into rounded teeth. Inflorescence long-stemmed, and crowned with broad purple bracts. Spain, Italy.

Fl. spikes not topped by coloured bracts
1111. *L. angustifolia* Miller (*L. officinalis* Chaix) COMMON LAVENDER. A very aromatic, greyish-leaved, branched undershrub 30–60 cm., with long-stemmed slender spikes of blue fls. Spikes often interrupted below; corolla twice as long as the calyx, hairy outside; bracts purple, broadly oval, pointed, papery, seven-veined. Lvs. 2½–4 cm. oblong to linear,

margin inrolled, the younger lvs. white-hairy, the adult lvs. greenish. △ Dry, sunny hills; grown as an ornamental and sometimes naturalized. June–Sept. P.E.F.I.YU.

L. latifolia (L.) Vill. Like 1111, but bracts linear, green, one-veined. Fls. violet in a dense spike, usually not interrupted below; fl. stem often branched below. Lvs. blunt, broader towards apex, nearly flat. Mediterranean Europe, Spain to Yugoslavia.

L. multifida L. CUT-LEAVED LAVENDER. Readily distinguished by its lvs. which are twice cut into narrow, green, sparsely hairy segments. Fl. spikes cylindrical, often in threes at apex of the stem; bracts oval-heart-shaped, papery. Iberian Peninsula, Italy. Page 352.

MARRUBIUM | Horehound Calyx hairy within, tubular, with 5 or 10 strong veins and 5 or 10 often recurved, not spiny-tipped, tooth-like lobes. Corolla tube short, two-lipped, the upper lip flat, often notched, the lower lip three-lobed. Stamens not projecting. *c.* 10 sps.

1112. *M. vulgare* L. WHITE HOREHOUND. An erect strong-smelling perenn. 30–80 cm., with white-cottony stems and numerous dense globular whorls of inconspicuous whitish fls. forming a long, interrupted, leafy spike. Fls. *c.* 1½ cm.; the upper lip of the corolla flat and deeply notched; calyx woolly-haired, with 10 small, recurved, tooth-like lobes. Lvs. 1½–5 cm., almost circular, or broadly heart-shaped, strongly toothed, stalked; blade very hairy, greenish and strongly wrinkled above, white-cottony beneath. △ Waste places, dry slopes, waysides. May–Sept. All Eur. (except IS.): introd. IRL.N.SF. *Used as a herbal remedy for coughs and colds.* Pl. 109.

1113. *M. peregrinum* L. BRANCHED HOREHOUND. A much-branched, densely white-felted perenn. 30–80 cm., with narrow toothed lvs. and rather few-flowered whorls of whitish fls. Fls. small, 6–10 in a whorl; calyx with adpressed hairs, strongly ribbed, and with 5 unequal lobes. Lvs. 2–3 cm. oblong to lance-shaped, conspicuously veined, strongly toothed, densely white-felted. △ Dry, stony places, olive groves, waysides. July–Sept. South-East Eur. F.A.PL.CS.H.SU.: introd. D. Page 352.

SIDERITIS Calyx bell-shaped with 5 spiny-tipped, equal, or very unequal lobes. Corolla tube short, two-lipped, the upper lip flat, notched or rounded, the lower lip three-lobed. Stamens and style not projecting. *c.* 30 sps.

Anns.; bracts and lvs. similar

1114. *S. romana* L. A hairy, erect, little-branched ann. 10–30 cm., with numerous whorls of small white or pinkish fls. forming a long, lax, leafy spike. Whorls six-flowered; corolla *c.* 1 cm. across; calyx hairy, conspicuously ribbed, swollen at the base, with 5 spiny-tipped lobes, the upper lobe 2–3 times as broad as the others; bracts similar to lvs., longer than the fls. Lvs. 1–1½ cm., oval to obovate, often strongly toothed, green, hairy. △ Sandy and rocky places, dry hills. May–June. Med. Eur. P.

S. montana L. Like 1114, but calyx with 5 nearly equal spiny-tipped lobes, the 3 upper forming an upper lip. Fls. yellow, becoming brownish; corolla only a little longer than the calyx. A woolly-haired ann. with oblong-lance-shaped lvs. Widespread in Europe, except in the north and extreme west. Page 352.

Shrubby perenns.; bracts differing from lvs.

1115. *S. hyssopifolia* L. A much-branched leafy shrublet 10–40 cm., with pale yellow fls. in short, dense cylindrical clusters 1½–2 cm. wide, and broad spiny-toothed bracts as long as or longer than the calyx. Fls. in clusters of 6; corolla pale yellow, often purple-spotted; calyx bell-shaped, the lobes equal, spiny-tipped. Lvs. very variable, elliptic to linear, three-veined, toothed or entire. A very variable sp. △ Rocks in mountains. July–Aug. E.F.CH.I. Pl. 109.

1. *Scutellaria columnae* 1109
2. *Marrubium peregrinum* 1113
3. *Teucrium scordium* 1098
4. *Lavendula multifida* [1111]
5. *Nepeta nuda* 1117
6. *Sideritis montana* [1114]

S. hirsuta L. Like 1115, but upper lip of corolla white, the lower lip yellow, rarely corolla all white or yellow; calyx lobes at length spreading-erect and, like the bracts, weakly spiny-tipped; whorls usually widely spaced. Branches with long spreading hairs. A very variable sp. Portugal to Italy.

NEPETA   Calyx tubular or oval with 15 veins, distinctly two-lipped, with 5 nearly equal lobes; corolla tube slender, becoming strongly curved, the upper lip erect, two-lobed, the lower lip with 3 much larger lobes, the middle lobe largest and concave. *c.* 22 sps.

1116. *N. cataria* L. CATMINT. A greyish, erect, branched, strong-smelling, leafy perenn. 40–100 cm., forming short rounded clusters of white fls. spotted with red. Corolla 1–1½ cm., hairy, the lower lip conspicuously toothed; calyx hairy, the upper lobe longest, the 2 lower shortest. Lvs. stalked, 3–7 cm., oval-heart-shaped, strongly toothed, white-woolly beneath. △ Waste ground, rocks, waysides; often grown as an ornamental and sometimes naturalized. June–Sept. All Eur. (except IS.TR.): introd. IRL.SF.B.D.CH. Pl. 109.

N. nepetella L. LESSER CATMINT. Distinguished from 1116 by the greyish lance-shaped, toothed lvs. *c.* 1½ cm. by ½ cm. Inflorescence lax, slender, one-sided; corolla white or pink, hairy outside, tube longer than the woolly-haired calyx; calyx lobes equal. Spain to Italy.

1117. *N. nuda* L. HAIRLESS CATMINT. An erect, nearly hairless, much-branched perenn. to 1 m., with lax stalked whorls of 10–20 violet fls. forming long, slender, interrupted spikes. Corolla *c.* 1 cm., the tube hairy, and longer than the calyx; calyx hairy, lobes awl-shaped. Lvs. *c.* 3 cm., oval-oblong, strongly toothed, hairless. △ Rocks, thickets, open woods, pastures. June–Sept. Cent. Eur. E.F.I.GR.R.SU.: introd. D. Page 352.

GLECHOMA   Like *Nepeta*, but anther cells at right angles, each opening by a separate slit, not divergent and opening by a common slit as in *Nepeta*. Corolla tube straight, hairy within. *c.* 3 sps.

1118. *G. hederacea* L. (*Nepeta h.*) GROUND IVY. A creeping, rooting perenn. with lax, leafy, ascending inflorescence 5–30 cm., with whorls of violet fls. with purple spots on the lower lip. Fls. 2–10 in a whorl; corolla hairy 1½–2 cm., the tube 2–3 times as long as the hairy calyx. Lvs. long-stalked, lv. blades 1–3 cm. across, kidney-shaped, margin with large rounded teeth. △ Meadows, hedges, woods and walls. Mar.–May. All Eur. (except IS.). *Used in herbal remedies for coughs and as a tonic and vermifuge.* Pl. 109.

PRUNELLA | Self-Heal   Calyx two-lipped, the upper lip broad with 3 short teeth, the lower with 2 long narrow lobes, ten-veined, flattened, becoming closed in fr. Corolla two-lipped, the upper lip strongly hooded, the lower lip three-lobed. *c.* 5 sps.

Fl. heads with a pair of lvs. immediately below; fls. less than 2 cm. long
1119. *P. vulgaris* L. COMMON SELF-HEAL. A leafy perenn. with short creeping rhizomes and erect or ascending stems 5–30 cm., bearing dense, rounded, terminal heads of bright blue-violet fls., often with purplish bracts and calyx. Fl. heads with 2 lvs. immediately below; corolla 1–1½ cm. long, hairy, rarely white or pink; bracts differing from the lvs., rounded with a leafy apex, stalkless. Lvs. 2–5 cm. oval, entire or shallowly toothed, stalked; hairs on stem and lvs. few. △ Grasslands, open woods, rocks, waste places. June–Oct. All Eur. *Reported to heal many disorders, such as fevers, wounds, etc.*

1120. *P. laciniata* (L.) L. CUT-LEAVED SELF-HEAL. Like 1119, but upper lvs. deeply pinnately cut into linear lobes, the lower entire. Fls. usually yellowish-white, rarely violet-flushed;

corolla *c.* 1½ cm. long; bracts greenish-yellow, white-haired, the lower with a long pointed apex. A more hairy perenn. 5–30 cm., with stems with dense white hairs on the angles. △ Dry meadows, open woods, rocks. June–Oct. Most of Eur. (except North Eur. IRL.NL.). Pl. 110.

Fl. heads borne on a short stem; fls. more than 2 cm. long
P. grandiflora (L.) Jacq. LARGE SELF-HEAL. Like 1119, but fls. larger, 2–2½ cm. long, violet-blue, in dense terminal heads 3–5 cm. long borne on a short stem above the uppermost pair of lvs.; bracts often purple-flushed. Lvs. 3–5 cm. by 1–2½ cm. oval to oblong, entire or coarsely toothed. Widespread in Europe, except in the extreme west and much of the north. Pl. 110.

MELITTIS Calyx open in fr., the upper lip with 2–3 small lobes, the lower with 2 rounded lobes. Corolla two-lipped, the upper lip entire, slightly concave, the lower with rather rounded, spreading lobes. 1 sp.

1121. *M. melissophyllum* L. BASTARD BALM. A softly hairy, strong-smelling, usually un-branched herbaceous perenn. 20–50 cm., with whorls of large pink or white fls. spotted with pink. Whorls with 2–6 fls.; corolla 3½–4½ cm. long, funnel-shaped, with broad, rounded, spreading lobes; calyx papery, large, bell-shaped, ten-veined. Lvs. 5–8 cm. oval-acute, coarsely rounded-toothed, stalked. △ Woods, hedges, ravines. May–July. Much of Eur. (except North Eur. IRL.NL.AL.TR.). Pl. 109.

PHLOMIS Calyx tubular with 5 often spiny-tipped lobes. Corolla hairy outside, tube shorter than the calyx, and with a ring of hairs within, strongly two-lipped, the upper lip large and hooded, the lower three-lobed. *c.* 11 sps.

Fls. purple or pink
1122. *P. herba-venti* L. A rather robust, hairy, branched perenn. to 60 cm., with 2–5 dense axillary whorls of purple fls., the uppermost whorl with a pair of small terminal lvs. Corolla *c.* 2 cm. long, the upper lip hooded, the lower three-lobed; calyx ribbed, densely bristle-haired, lobes spine-tipped; bracts stiff, bristle-like, covered with long bristly hairs with swollen bases. Lvs. to 10 cm. lance-shaped, toothed, leathery, shining green above, paler beneath. △ Dry places, rocks, waysides. May–July. P.E.F.I. Page 360.

P. tuberosa L. Like 1122, but fls. pink in numerous whorls, the upper crowded, the lower remote; upper lip of corolla erect, margin and inner side of lip with long hairs; calyx nearly hairless, lobes spiny-tipped; bracts linear, bristly-haired. Lvs. triangular-heart-shaped, strongly toothed, nearly hairless. Roots tuberous. East-Central and South-Eastern Europe. Pl. 110.

P. purpurea L. A tufted shrubby perenn. with oblong-oval, very wrinkled, rather fleshy lvs., greenish above, with star-shaped hairs and grey- or white-woolly beneath. Fls. purple; calyx grey-woolly, lobes not spiny-tipped; bracts lance-shaped, velvety. Portugal, Spain.

Fls. yellow
1123. *P. lychnitis* L. A small shrub 20–50 cm., with white-felted stems and lvs. and 4–8 whorls of yellow fls. forming a long, interrupted spike. Whorls with 6–10 fls. subtended by a pair of broadly oval, long-pointed outer bracts which are much broader than the lvs. Corolla 2–3 cm., the upper lip hooded over the lower lip, densely felted; calyx with dense woolly hairs, lobes not spiny-tipped; inner bracts awl-shaped, softly hairy. Lvs. mostly basal, linear-lance-shaped, wrinkled above, densely white-felted beneath. △ Dry, rocky places. May–July. P.E.F. Pl. 110.

1124. *P. fruticosa* L. JERUSALEM SAGE. Distinguished from 1123 by the similar oval-oblong lvs. and outer bracts. A more robust, much-branched, white-felted shrub to 1½ m. with 1–3 compact whorls of 20–30 orange-yellow fls. Upper lip broadly heart-shaped, hooded, 2 cm. across; calyx with dense, star-shaped hairs and stiff and spiny-tipped recurved lobes; inner bracts oval-lance-shaped, softly hairy. Lvs. with wedge-shaped or rounded base, finely grey-woolly above, densely so below, the upper lvs. stalkless. △ Rocky ground. May–June. Med. Eur. (except TR.) SU. Pl. 110.

GALEOPSIS | Hemp-Nettle Calyx bell-shaped or tubular, ten-veined and with 5 spiny-pointed lobes. Corolla tube longer than the calyx, two-lipped, the upper lip flattened from side to side and helmet-shaped, the lower lip three-lobed and with 2 conical projections at the base. *c.* 9 sps.

Stems softly hairy, not swollen below nodes
1125. *G. angustifolia* Hoffm. NARROW-LEAVED HEMP-NETTLE. An erect, widely branched, hairy ann. 10–80 cm., with narrow lvs. and whorls of rosy-purple fls., which are clustered above. Corolla 1½–2½ cm., hairy; calyx densely hairy, sometimes glandular, lobes spiny-tipped, about half as long as the tube. Lvs. 1½–8 cm. linear-lance-shaped, less than 1 cm. broad, margin with 1–4 small teeth on each side. △ Stony places, cultivated ground. July–Oct. Cent. Eur. F.IRL.GB.B.NL.S.YU.BG.R.SU.

G. segetum Necker DOWNY HEMP-NETTLE. Like 1125, but lvs. oval to oval-lance-shaped, velvety-hairy particularly beneath, veins prominent. Fls. large, 2–3 cm., pale yellow or pink mixed with yellow; calyx densely velvety-haired, one-half to one-third as long as the hairy corolla tube. Western and West-Central Europe. Page 360.

Stems with rough spreading hairs, swollen below nodes
1126. *G. tetrahit* L. COMMON HEMP-NETTLE. A branched, leafy ann. 10–100 cm., with pink, purple, or white fls. in few-flowered whorls which are clustered above. Corolla 1–2 cm., the lower lip with darker marking, the tube scarcely longer than the bristly-haired calyx. Lvs. stalked, 2½–10 cm. oval-lance-shaped, toothed, with adpressed hairs; stems with red-tipped glandular hairs mostly on the swellings below the nodes. △ Cultivated ground, heaths and fens. July–Oct. All Eur. (except AL.TR.). *The seeds contain an oil; fibres obtained from the stems have been used for cord-making.* Pl. 111.

G. speciosa Miller LARGE-FLOWERED HEMP-NETTLE. Distinguished from 1126 by the much larger fls. 2–4½ cm. which are pale yellow with a dark lilac blotch on the lower lip, or rarely wholly sulphur-yellow. Calyx sparsely hairy or hairless, less than half the length of the corolla tube. Widespread in Europe, except in the south-west. Pl. 111.

LAMIUM | Dead-Nettle Calyx tubular or bell-shaped, five-veined, with 5 fine-pointed lobes. Corolla tube widening above, two-lipped, the upper lip hooded over the stamens, the lower with 2 tooth-like lateral lobes and a broad, notched middle lobe. Anthers diverging, usually hairy. *c.* 17 sps.

Anns.; fls. not more than 1½ cm. long
1127. *L. amplexicaule* L. HENBIT. A slender, little-branched ann. 5–25 cm., readily distinguished by its stalkless, kidney-shaped, strongly toothed bracts which closely encircle the whorls of rosy-purple fls. Lower whorls distant; corolla *c.* 1½ cm., tube slender, *c.* 3 times as long as the very hairy calyx, but often much smaller unopened fls. are formed. Lvs. 1–2½ cm., rounded or heart-shaped, strongly rounded-toothed, long-stalked. △ Cultivated and waste ground, tracksides. Mar.–Oct. All Eur. Pl. 111.

1128. *L. purpureum* L. RED DEAD-NETTLE. Like 1127, but bracts subtending whorls oval, all stalked and similar to the lvs. and whorls densely clustered into a pyramidal leafy inflorescence. Corolla 1–1½ cm., rosy-purple, the tube with ring of hairs within, about twice as long as the calyx. Lvs. oval-heart-shaped, strongly rounded-toothed, stalked, softly hairy. A hairy, strong-smelling, often purple-flushed ann. 10–45 cm. △ Cultivated and waste ground. Mar.–Dec. All Eur.

*L. *hybridum* Vill. CUT-LEAVED DEAD-NETTLE. Western, Central, and Southern Europe.

Perenns.; fls. more than 1½ cm. long
(a) *Fls. white*
1129. *L. album* L. WHITE DEAD-NETTLE. A hairy perenn. with creeping rhizomes and erect stems 20–60 cm. and distant whorls of large white fls. Corolla 2–2½ cm. long, the tube curved and with an oblique ring of hairs near base, longer than the calyx; upper lip with long hairs, the lower lip bilobed with a central tooth and 2–3 small lateral teeth at base; anthers blackish, hairy. Lvs. 3–7 cm. oval-heart-shaped, pointed, coarsely toothed, all stalked. △ Hedges, tracksides, waste ground. Apr.–Nov. All Eur. (except AL.TR.): introd. IRL.IS.

(b) *Fls. rosy-purple*
1130. *L. maculatum* L. SPOTTED DEAD-NETTLE. Fls. large, 2½–3 cm. long, pinkish-purple or very rarely white. Corolla tube curved with transverse ring of hairs within, longer than the calyx, the upper lip shortly hairy on margin, the lower lip with 1 tooth on each side at base. Lvs. triangular-heart-shaped, coarsely toothed, sparsely hairy, often white-blotched. A very variable perenn. 30–80 cm. △ Hedges, woods, waste places. Apr.–Oct. Much of Eur. (except North Eur. IRL.GB.): introd. N.S.SF. Pl. 111.

1131. *L. garganicum* L. LARGE RED DEAD-NETTLE. Like 1130, but fls. very large 2½–4 cm., rosy-purple, or rarely white, and corolla tube straight, hairless within, much enlarged at the throat and 2–3 times as long as the calyx; upper lip of the corolla usually two-lobed. Lvs. stalked, oval-heart-shaped, rounded- or saw-toothed, variably hairy; stems and calyx often densely white-felted, or hairless. A variable erect or ascending little-branched perenn. 30–60 cm. △ Stony places in mountains. May–Aug. F.A.I.YU.AL.GR.BG.

L. orvala L. Distinguished from 1131 by the hairless anthers and the long, curved, calyx lobes which are longer than the tube. Fls. large 2½–3 cm., dull purple or pink, the upper lip with densely ciliate margin. Lvs. large, coarsely and irregularly toothed. Southern and East-Central Europe. Pl. 111.

GALEOBDOLON   Like *Lamium*, but lower lip of corolla with 3 more or less equal lobes; anthers hairless. 1 sp.

1132. *G. luteum* Hudson (*Lamium galeobdolon* (L.) L.) YELLOW ARCHANGEL. Distinguished by its bright yellow fls. with brownish markings, in several whorls forming a long, rather lax, leafy spike. Corolla *c.* 2 cm. long, hairy outside; calyx *c.* 1 cm., nearly hairless. Lvs. stalked, 4–7 cm. oval-acute, irregularly toothed. A sparsely hairy perenn. with long, creeping, above-ground stolons and erect fl. stems 20–60 cm. △ Woods, hedges. Apr.–June. All Eur. (except P.IS.) Pl. 112.

LEONURUS   Calyx bell-shaped, five-veined, with 5 spiny-pointed lobes. Corolla tube shorter than the calyx, not widened at the throat, two-lipped, the upper lip hooded, hairy, the lower three-lobed. Anthers parallel or nearly so. 2 sps.

1133. *L. cardiaca* L. MOTHERWORT. An erect leafy perenn. 60–120 cm., with small pinkish or white fls. in numerous small whorls shorter than the subtending lv. stalks and forming a very long, interrupted, leafy spike. Corolla *c.* 12 mm. long, densely hairy outside, the tube hairy within, the lower lip yellow and spotted with purple; calyx shorter than corolla tube, lobes with bristly apices. Lvs. all stalked, the upper lance-shaped, three-lobed, the lower palmately three- to seven-lobed with the lobes strongly toothed, green above, white-hairy beneath. △ Hedges, waste places, waysides. June–Sept. Much of Eur. (except P.IRL.IS.). *Used in the past for heart and nervous diseases; a stimulant and vermifuge.* Pl. 112.

L. marrubiastrum L. FALSE MOTHERWORT. Like 1133, but a bienn. with all lvs. oval to lance-shaped, the upper strongly toothed but not lobed. Fls. pale pink in whorls as long as the subtending lv. stalks; corolla scarcely longer than calyx tube, finely hairy, hairless within. Widespread, except in Western and Northern Europe.

BALLOTA | Horehound Calyx funnel-shaped, ten-veined, usually five-lobed. Corolla tube shorter than the calyx, with a ring of hairs within, two-lipped, the upper lip erect, concave, and arched, the lower three-lobed. Stamens parallel, not diverging. *c.* 7 sps.

1134. *B. nigra* L. BLACK HOREHOUND. A dark green, hairy, fetid, branched perenn. 40–100 cm., with numerous whorls of reddish-purple fls. in the axils of the upper lvs. Corolla 1–2 cm., hairy outside, tube shorter than the calyx; calyx glandular-hairy, with 5 broad, pointed, tooth-like lobes, recurved or spreading in fr. Lvs. stalked, blades 2–5 cm. oval or rounded, coarsely toothed. △ Hedges, tracksides, waste places. May–Sept. Most of Eur. (except IS.): introd. IRL.N.SF. Pl. 112.

B. acetabulosa (L.) Bentham Readily distinguished by the much-enlarged calyx which forms a wide-spreading, umbrella-like membrane. Fls. rosy-purple. Lvs. to 6 cm., rounded-heart-shaped, coarsely rounded-toothed, rather thick, very soft and white-woolly when young, becoming grey. Greece. *The dried calices are used as wicks in oil lamps in Orthodox churches.*

STACHYS | Woundwort Calyx tubular or bell-shaped, five- or ten-veined, with 5 narrow, equal, spiny-tipped lobes. Corolla two-lipped, the upper lip flat or arched, the lower lip three-lobed with the middle lobe largest. Stamens protected by the arched upper lip of corolla, at first parallel but the outer 2 diverging after flowering. *c.* 60 sps.

Fls. yellowish-white
1135. *S. annua* (L.) L. ANNUAL YELLOW WOUNDWORT. An erect, branched, finely hairy ann. to 30 cm., with numerous whorls of yellowish-white fls. forming a long, interrupted spike. Whorls with 3–6 fls.; corolla 1–1½ cm., hairy outside, tube as long as the calyx; calyx hairy, lobes awl-shaped, curved; bracteoles tiny. Lvs. stalked, 2–6 cm. oblong-lance-shaped, shallow-toothed, nearly hairless. △ Fields, waste places. June–Oct. Most of Eur. (except P.IRL.IS.NL.): introd. N.SF.

1136. *S. recta* L. PERENNIAL YELLOW WOUNDWORT. A sweet-smelling, hairy, tufted, woody-based perenn., with many erect stems 20–60 cm. and yellowish-white fls. in long, slender, leafless spikes. Whorls with 6–15 fls., numerous, the upper whorls much broader than the subtending bracts; corolla *c.* 1½ cm.; calyx with spiny-pointed lobes and spreading hairs; bracts spiny-tipped. Lvs. hairy or hairless, the lower oval to oblong and stalked, the upper lance-shaped to linear and stalkless, all toothed, green. A very variable sp. △ Rocks, banks, dry places. June–Sept. Much of Eur. (except North Eur. P.IRL.GB.NL.). Pl. 113.

S. maritima Gouan Distinguished by its greyish, shortly woolly-haired stems and lvs., and lower lvs. in a rosette; lv. blades wrinkled. Fls. yellow, in numerous whorls forming a short dense spike; corolla tube shorter than the calyx, upper lip of corolla toothed; bracts and calyx with long spreading hairs. Mediterranean Europe and Bulgaria.

Fls. purple or pinkish

(a) *Anns.*

1137. *S. arvensis* (L.) L. FIELD WOUNDWORT. A branched, slender-stemmed, rough-haired ann. 10–40 cm., with whorls of small pale purple fls. clustered above. Whorls with 4–6 fls.; corolla 6–7 mm., little longer than the hairy calyx; lobes of calyx as long as the tube, fine-pointed. Lvs. 1½–3 cm., oval-heart-shaped with rounded teeth, the lower stalked, the upper smaller, stalkless. △ Sandy, cultivated ground. Apr.–Oct. All Eur. (except IS.TR.): introd. N.

(b) *Perenns.*

1138. *S. palustris* L. MARSH WOUNDWORT. An erect, odourless perenn. to 1 m., with lance-shaped lvs. and rather dull pinkish fls. in long, nearly leafless spikes. Whorls with 4–8 fls., mostly contiguous and only the lowest whorls distant; corolla 1–1½ cm., hairy outside; calyx hairy, lobes awl-shaped, more than half as long as the tube; bracts of upper whorls shorter than the whorls; bracteoles linear, very small. Lvs. 5–12 cm., rounded at base, toothed, the lower very shortly stalked, the upper stalkless. △ Watersides, marshes, damp places. June–Sept. All Eur. (except IS.). Pl. 112.

1139. *S. sylvatica* L. HEDGE WOUNDWORT. Like 1138, but fls. darker claret-coloured and plants strong-smelling when crushed. Whorls with 3–6 fls., forming a slender, leafless, interrupted spike; calyx glandular. Lvs. oval-heart-shaped, long-pointed, toothed, all long-stalked. An erect hairy perenn. 30–100 cm. with creeping rhizomes. △ Woods, hedges, shady places. June–Sept. All Eur. (except IS.).

S. alpina L. ALPINE WOUNDWORT. Like 1139, but fls. longer, 1½–2 cm., reddish-brown, numerous in each whorl, forming an interrupted leafy spike. Bracts longer than the whorls; bracteoles subtending the fls. as long as the calyx; calyx with long spreading hairs, lobes oval, fine-pointed, shorter than the tube. Much of Europe, except the extreme west and the north.

1140. *S. germanica* L. DOWNY WOUNDWORT. A whitish, silky-haired, erect perenn. or bienn. 30–80 cm., with numerous whorls of pale rose-purple fls. in a long, slender, densely woolly spike. Whorls many-flowered; corolla densely hairy outside, twice as long as the very densely silky-haired calyx. Lvs. usually white-silky, 2–12 cm., oval to oblong-heart-shaped, toothed, the lower stalked, the upper narrower, stalkless. A variable sp. △ Stony places, uncultivated ground. June–Sept. Much of Eur. (except North Eur. IRL.NL.). Pl. 112.

BETONICA Like *Stachys*, but outer stamens not spreading laterally after flowering; upper lip of corolla scarcely arched, corolla tube long. Basal lvs. in a rosette, stem lvs. few. *c.* 6 sps.

Fls. purple

1141. *B. officinalis* L. (*Stachys o.*) BETONY. A hairy perenn. 20–60 cm., with lvs. mostly basal in a rosette and nearly leafless, erect stems bearing a compact spike of bright reddish-purple, or rarely white fls. Whorls dense, sometimes interrupted below; corolla *c.* 1½ cm., the tube longer than the calyx, the upper lip more than twice as long as the stamens; calyx hairless or hairy in throat, 7–9 mm., lobes with an awn-like apex, half as long as the tube; anthers yellow. Lvs. all stalked, the lower oval-oblong with a heart-shaped base, and margin with regular rounded teeth, the upper lvs. few, narrower. A very variable sp. △ Meadows, heaths, woods. June–Oct. All Eur. (except IS.). *Held in high repute in the past as a medicinal plant. The leaves have been used to make tea and as a herbal tobacco.*

B. *hirsuta* L. (*Stachys densiflora* Bentham; *S. danica* Schinz & Thell.) Like 1141, but calyx longer, 12–15 mm., distinctly net-veined, sparsely hairy, with lobes one-third as long as the tube. Fls. pale pink, *c.* 2 cm., the upper lip a little longer than the stamens. Lvs. very hairy. Southern Europe.

Fls. yellow
1142. *B. alopecuros* L. (*Stachys a.*) YELLOW BETONY. A softly hairy perenn. 20–60 cm., with pale yellow fls. in a dense cylindrical spike, often interrupted below, and borne on an almost leafless stem. Corolla *c.* 1½ cm., the upper lip bilobed, hairy outside, tube with a ring of hairs within; calyx net-veined, hairy, the tube 2–3 times as long as the lobes. Lower lvs. long-stalked, oval-heart-shaped, and conspicuously rounded-toothed, softly hairy, particularly beneath, the upper stalkless or short-stalked. △ In mountains: rocky meadows, thickets. June–Aug. E.F.D.CH.A.I.YU.AL.GR. Page 360.

SALVIA | Sage   Calyx tubular or bell-shaped, two-lipped, the upper lip entire or three-lobed, the lower lip two-lobed. Corolla tubular, two-lipped, the upper lip usually hooded, the lower lip three-lobed with the middle lobe longer and often notched. Stamens 2, each with the upper fertile anther and the lower sterile anther widely separated by the connective which is hinged to the filament, thus allowing a lever-like action during pollination, *c.* 45 sps.

Low shrubs
1143. *S. officinalis* L. SAGE. A very aromatic undershrub, 20–70 cm., with thick, greyish, wrinkled lvs. and simple, terminal spikes of several rather lax whorls of large violet-blue fls. Whorls with 3–6 fls.; corolla 2–3 cm. long, 2–3 times as long as the calyx, the upper lip nearly straight; calyx often violet-flushed, not glandular, with adpressed hairs; bracts papery, oval-acute. Lvs. stalked, oblong-oval or lance-shaped, and finely rounded-toothed; stems with adpressed greyish hairs. △ Dry banks, stony places; cultivated and naturalized elsewhere. May–July. Med. Eur. (except TR.) P.BG. *A useful culinary herb: used for flavouring cheese and in wine-making and to make sage tea.*

S. triloba L. fil. THREE-LOBED SAGE. Like 1143, but some lvs. with 2 small lateral lobes at the base. Stems and calyx glandular-hairy; the whole plant very aromatic. Corolla violet. Italy, Greece. *Used for making tea in Greece.* Pl. 113.

Herbaceous anns., bienns., or perenns.
(a) *Bracts subtending whorls conspicuous, coloured and partially hiding calyx*
1144. *S. sclarea* L. CLARY. A robust, sticky, strongly smelling bienn. 30–120 cm., with a long, branched inflorescence of numerous whorls of whitish fls. flushed with violet and conspicuous papery bracts flushed with violet or pink. Corolla 2½–3 cm. long, 3 times as long as the spiny-tipped, glandular-hairy calyx; bracts longer than the calyx, rounded with a pointed apex. Lvs. greyish-hairy, wrinkled, 7–18 cm., oval-heart-shaped, irregularly shallow-lobed or toothed. △ Dry banks, rocks, tracksides; frequently cultivated and naturalized. May–Sept. Med. Eur. (except TR.) CH.A.BG.R.: introd. D.CS. *Used for adulterating wines, as a condiment and medicinally.* Pl. 113.

S. aethiopis L. WOOLLY CLARY. Like 1144, but fls. white in a large, much-branched, pyramidal cluster with conspicuous broad, rounded greenish bracts about as long as the calyx. Lvs. very silvery-white-woolly, particularly when young, not strong-smelling or glandular, the lower lvs. lobed. Southern and South-Eastern Europe. Page 360.

1. *Phlomis herba-venti* 1122
2. *Betonica alopecuros* 1142
3. *Salvia aethiopis* [1144]
4. *Galeopsis segetum* [1125]

S. argentea L. SILVER SAGE. Like [1144], but inflorescence glandular-hairy above and bracts subtending whorls much smaller and about half the length of the calyx. Corolla *c.* 1½ cm., pinkish. Lvs. very silvery, with shaggy cobweb-like hairs. Mediterranean Europe, Portugal, Bulgaria. Pl. 113.

(*b*) *Bracts not conspicuous or coloured or hiding calyx*

1145. *S. verticillata* L. WHORLED CLARY. An erect, hairy, strong-smelling perenn. 40–80 cm., with several whorls of numerous blue-violet fls. forming a long, interrupted, leafless spike. Whorls with 20–30 stalked fls.; corolla 8–12 mm. long, about twice as long as the calyx; calyx purple, bristly-haired, lobes triangular; bracts small brown, reflexed. Lvs. triangular-heart-shaped, toothed, the lower lvs. stalked and with a pair of unequal leaflets, the upper lvs. stalkless, entire. △ Waysides, dry banks, bushy and stony places. May–Aug. Much of Eur. (except P.IRL.GB.B.NL.IS.TR.): introd. N.SF.D.

1146. *S. glutinosa* L. JUPITER'S DISTAFF. Readily distinguished by its large pale yellowish fls. and its very sticky, strong-smelling lvs. Whorls 2–6 forming a lax elongated spike; corolla 3–4 cm., 3 times as long as the calyx, the upper lip hooded; calyx glandular-sticky, the upper lip entire or shallowly three-lobed; bracts tiny, green. Lvs. large, oval long-pointed and with broad, spreading, basal lobes, coarsely toothed, stalked. △ Woods and shady places in mountains. June–Sept. Cent., South-East Eur. (except TR.) E.F.I.SU. Pl. 113.

1147. *S. pratensis* L. MEADOW CLARY. A hairy perenn. with erect stems 30–80 cm., glandular-hairy above, bearing whorls of bright blue-violet fls. forming a long, sticky, interrupted spike. Corolla 1½–2½ cm. long, rarely white or pink, 3 times as long as the calyx, the upper lip hooded; calyx with the upper lip with 2–3 short teeth; bracts reflexed, shorter than the calyx. Lvs. wrinkled, mostly basal, oval-heart-shaped and double-toothed or lobed, the lower stalked, the upper stalkless. △ Waysides, grassy places, meadows. May–July. All Eur. (except P.IRL.IS.AL.GR.): introd. North Eur. Pl. 114.

S. nemorosa L. Like 1147, but fls. blue-violet or rarely white, smaller 1–1½ cm. long, in numerous small whorls forming a long, slender, interrupted spike, with oval, violet or rose-coloured bracts about as long as the coloured calyx. Lvs. finely hairy, green above, greyish beneath; stems leafy. East-Central and South-Eastern Europe.

1148. *S. horminoides* Pourret WILD CLARY. An erect, little-branched, hairy perenn. 30–80 cm., somewhat glandular and purple-flushed above, with numerous well-spaced whorls of blue-violet fls. in long slender spikes. Corolla very variable in size, ½–1½ cm., the smaller fls. not opening and shorter than the calyx; corolla with upper lip curved and lower lip with 2 white spots at the base; calyx with long spreading white hairs and glandular hairs, veins prominent. Lvs. very variable, oblong-oval, toothed or deeply pinnately lobed, the lower stalked, the upper stalkless. △ Dry meadows, waysides. May–Aug. Much of Western and Mediterranean Europe.

S. verbenaca L. Like 1148, but fls. larger 12–15 mm. long, at least as long as the calyx, lilac-blue and without white spots at base of the lower lip. Calyx hairy, the upper lip with 3 short teeth, often violet. Basal lvs. oblong, usually more than twice as long as broad, coarsely toothed or lobed, the stem lvs. several, stalkless. Var. *clandestina* (L.) Briq. has a short, almost leafless stem and a rather compact terminal cluster of pale blue or whitish fls. with the corolla twice as long as the calyx. Mediterranean Europe.

1149. *S. horminum* L. RED-TOPPED SAGE. An erect hairy ann. or perenn. 10–40 cm., usually distinguished by the cluster of broad, brightly coloured, violet or pink bracts at the apex of the inflorescence, but bracts sometimes absent. Fls. violet-purple or pink, in several or many whorls forming at length a slender, interrupted spike; corolla 1–2 cm.; calyx strongly veined, glandular-hairy, enlarging in fr.; bracts broadly heart-shaped acute, about as long as the calyx, often coloured. Lvs. stalked, 1–3 cm., oval or oblong and finely toothed, wrinkled, hairy. △ Rocky dry places. Apr.–June. Med. Eur. (except TR.) A.BG.SU. Pl. 115.

MELISSA | Balm Corolla tube curved upwards and enlarged above the middle, two-lipped, the upper lip erect, notched, the lower lip three-lobed; calyx thirteen-veined, two-lipped, the upper lip three-lobed, the lower two-lobed. Stamens 4, converging under upper lip of corolla. 1 sp.

1150. *M. officinalis* L. BALM. A much-branched, hairy or nearly hairless perenn. 30–80 cm., with white or pinkish fls. in dense, one-sided, terminal and axillary whorls, forming leafy spikes. Whorls with 6–10 fls.; corolla 8–12 mm., twice as long as the hairy calyx; bracts stalked, leaf-like. Lvs. stalked, 3–7 cm., oval and deeply toothed, sweetly lemon-scented, the upper smaller. △ Hedges, thickets, woods, vineyards. June–Sept. Med., South-East Eur. P.A.H.SU.: introd. Cent. Eur. *Used externally to cure bites, stings, and wounds; it yields a sweet-smelling balsamic oil which is used medicinally. Used for flavouring beverages and liqueurs.*

HORMINUM Calyx tubular, thirteen-veined, throat hairless, two-lipped, the upper lip three-lobed, the lower lip two-lobed. Corolla tube curved upwards, with a ring of hairs within, two-lipped, the upper lip notched, the lower shortly three-lobed. 1 sp.

1151. *H. pyrenaicum* L. DRAGONMOUTH. A stout-based perenn. with lvs. mostly basal, and erect unbranched stems 10–30 cm., bearing a long, many-whorled, one-sided inflorescence of drooping blue-violet fls. Whorls with 4–6 fls.; corolla *c.* 1½ cm., the curved tube much longer than the nearly hairless calyx. Basal lvs. 3–5 cm. oval, rounded-toothed, long-stalked, the stem lvs. tiny, stalkless. △ Dry alpine pastures, open woods. June–Aug. E.F.D.CH.I.A.YU. Pl. 114.

SATUREJA | Savory Calyx bell-shaped, ten-veined with 5 nearly equal pointed lobes. Corolla tube straight, two-lipped, the upper lip erect, notched or entire, the lower lip three-lobed. Stamens 4, converging under the corolla lip. *c.* 12 sps.

1152. *S. montana* L. WINTER SAVORY. A very aromatic slender-branched shrublet 10–40 cm., with white or lilac fls. in short-stalked clusters of 3–5 forming a lax, one-sided, leafy inflorescence. Corolla *c.* 1 cm., the lower lip three-lobed and spotted with purple; calyx usually hairy, lobes shorter than the tube. Lvs. 1–2 cm. by 2–3 mm., linear-lance-shaped, thick and dotted with shining glands, margin usually ciliate. A variable sp. △ Dry banks, rocks. July–Sept. Med., South-East Eur. CH.A.SU. *Used for seasoning.* Page 365.

S. hortensis L. SUMMER SAVORY. Like 1152, but an erect branched ann. with soft, dull greyish, usually hairless lvs. shorter than the internodes. Fls. smaller, 2–6 mm., white or pink, scarcely longer than the calyx; calyx lobes longer than the tube. Mediterranean Europe: introduced elsewhere from gardens. *Used for flavouring and seasoning; a vermifuge.*

MICROMERIA Small thyme-like shrublets with corolla and stamens like *Satureja*, but calyx with 13–15 veins, lobes 5, pointed, equal. *c.* 20 sps.

1153. *M. juliana* (L.) Bentham Fls. tiny, in numerous dense, stalkless whorls forming a very slender, somewhat interrupted, elongated spike *c.* ½ cm. broad. Corolla purple, hairy, 5–8 mm. long by 1–2 mm. across; calyx hairless in throat, hairy outside, the lobes awl-shaped, erect, hairless; bracteoles as long as the calyx. Lvs. 5–8 mm. oblong to linear-lance-shaped, margin inrolled, stalkless, in clusters at the nodes and shorter than the inter-nodes. An aromatic hairy shrublet, branched from the base with many slender, stiff, erect stems 10–30 cm. △ Rocks, dry, stony places. June–Aug. Med. Eur. P. Page 365.

M. graeca (L.) Bentham Like 1153, but whorls of fls. stalked and spreading outwards from axis and forming a much laxer spike, *c.* 1–1½ cm. broad. Calyx hairy in throat and with spreading, ciliate, awl-shaped lobes; bracteoles half as long as the calyx. Mediterranean Europe and Portugal.

CALAMINTHA | Calamint Fls. in stalked axillary clusters. Calyx tubular with 13 veins, two-lipped, the upper lip three-lobed, the lower deeply two-lobed. Corolla tube straight, two-lipped, the upper lip more or less flat, the lower three-lobed. Stamens 4, diverging below and converging above. *c.* 17 sps.

Fls. more than 1½ cm. long
1154. *C. grandiflora* (L.) Moench LARGE-FLOWERED CALAMINT. Fls. conspicuous, reddish-purple, 2½–4 cm., in stalked, few-flowered whorls forming a lax, leafy inflorescence. Corolla tube up-curved; calyx 1–1½ cm., lobes awl-shaped, ciliate, nearly equal. Lvs. 4–7 cm., oval and coarsely toothed, sparsely hairy beneath, all stalked. An aromatic, sparsely hairy perenn. with weak ascending stems 20–50 cm. △ Woods in mountains. June–Sept. Med. Eur. CH.A.BG.SU.: introd. CS. Pl. 114.

1155. *C. sylvatica* Bromf. (*Satureja s.*) WOOD CALAMINT. Fls. pinkish-purple, 1½–2 cm. long, in branched axillary clusters forming an elongated, lax, spike-like inflorescence which is leafless above. Clusters with 3–9 fls.; corolla tube straight; calyx 7–10 mm., the upper 3 tooth-like lobes recurved and much shorter than the nearly straight lower lobes. Lvs. 4–5 cm. oval, with 7–10 teeth on each side, stalked. A hairy, aromatic, little-branched, erect perenn. 30–60 cm. △ Open woods, thickets, uncultivated ground. June–Oct. Most of Eur. (except North Eur. IRL.TR.). Page 365.

Fls. less than 1½ cm. long
C. ascendens Jordan (*Satureja a.*) COMMON CALAMINT. Central and Southern Europe.

1156. *C. nepeta* (L.) Savi (*Satureja n.*) LESSER CALAMINT. Fls. pale lilac or white in lax or dense axillary clusters, the upper clusters longer than the lvs. and forming a lax, elongated inflorescence. Clusters with 3–15 fls.; corolla *c.* 1 cm.; calyx 4–6 mm., the upper lobes nearly straight, the lower lobes somewhat longer, narrower, ciliate, throat with a ring of protruding hairs. Lvs. 1–2 cm. oval and obscurely toothed, greyish-haired. A strong-smelling, branched, erect perenn. 40–80 cm. △ Dry, stony and bushy places. July–Oct. Much of Eur. (except North Eur. IRL.B.NL.CS.H.). *Yields an essential oil.* Pl. 114.

ACINOS Like *Calamintha*, but fls. in almost stalkless whorls. Calyx tubular, swollen at base, hairy within, thirteen-veined; bracteoles minute. *c.* 7 sps.

1157. *A. arvensis* (Lam.) Dandy (*Calamintha a.*; *Satureja a.*) BASIL-THYME. A low, hairy ann. or sometimes a perenn. with many spreading stems 5–40 cm., oval thyme-like lvs., and pale violet fls. in rather lax elongated inflorescence. Whorls 3–8, about six-flowered, dense; corolla 7–10 mm. long, violet with white markings on lip, the tube longer than calyx; calyx swollen at base, strongly ribbed, bristly-haired, the upper lip with middle lobe much broader than the lateral lobes. Lvs. $\frac{1}{2}$–$1\frac{1}{2}$ cm., obscurely toothed, narrowed to the short lv. stalk. △ Dry banks, pastures, cultivated ground. June–Sept. All Eur. (except P.IS.): introd. IRL.

A. *alpinus* (L.) Moench ALPINE CALAMINT. Like 1157, but a perenn. of high mountains with larger bright violet, rarely pink or white fls. 12–20 mm. long; calyx with upper lip with 3 nearly equal linear-lance-shaped lobes. A very variable sp. Widespread, except in Northern Europe. Pl. 114.

CLINOPODIUM Like *Acinos*, but whorls terminal and axillary, dense, many flowered. Calyx tube curved, not swollen at the base, with few or no hairs within, thirteen-veined; bracteoles awl-shaped. *c.* 1 sp.

1158. *C. vulgare* L. WILD BASIL. Fls. rose-purple in dense many-flowered whorls, the terminal whorl globular and larger than the axillary whorls. Corolla 1$\frac{1}{2}$–2 cm., tube longer than the calyx; calyx with long hairs, two-lipped, with unequal lobes, the upper 3 broader, the lower 2 longer; bracteoles numerous, *c.* 8 mm., awl-shaped, hairy. Lvs. 2–5 cm., oblong-. oval, and shallowly toothed, soft hairy, stalked. A softly hairy, leafy perenn. 30–80 cm. △ Hedges, woods, thickets, waysides. July–Sept. All Eur. (except IRL.IS.). Pl. 115.

HYSSOPUS | Hyssop Calyx tubular-bell-shaped with 15 veins, hairless within, and with 5 nearly equal lobes. Corolla tube straight, two-lipped, the upper lip notched, the lower lip three-lobed, with the middle lobe broader and notched. Stamens 4, diverging and projecting beyond the corolla. 1 sp.

1159. *H. officinalis* L. HYSSOP. An aromatic, nearly hairless, much-branched undershrub 20–60 cm., with narrow lvs. and whorls of blue or violet fls. usually in a compact elongated spike, which is often interrupted below. Corolla 10–12 mm. long, the upper lip and middle lobe of lower lip notched; calyx hairless 6–8 mm.; stamens long-projecting. Lvs. clustered at the nodes, 1$\frac{1}{2}$–2$\frac{1}{2}$ cm., linear or oblong-lance-shaped, hairless, with sunken glands. △ Dry banks, rocks, screes; cultivated and often naturalized. June–Sept. Med. Eur. (except GR.TR.) CH.A.SU. *An aromatic pot-herb; a tea has been made from the leaves.* Pl. 115.

ORIGANUM | Marjoram Fls. clustered into spreading terminal heads. Calyx bell-shaped, hairy within, with 13 veins and 5 equal lobes. Corolla two-lipped, the upper lip erect and notched, the lower lip three-lobed. Stamens 4, diverging. *c.* 13 sps.

1160. *O. vulgare* L. MARJORAM. A very aromatic hairy perenn. 30–80 cm., with dense, rounded, terminal and axillary clusters of rose-purple or whitish fls. in a spreading, rather rounded-topped inflorescence. Corolla 6–8 mm., the tube longer than the calyx; calyx glandular, hairy within, lobes equal; bracteoles conspicuous, oval, purplish, longer than calyx; or, less commonly, fls. white and bracteoles green. Lvs. 1$\frac{1}{2}$–4$\frac{1}{2}$ cm., oval and often shallowly toothed, stalked, hairy or hairless. A variable sp. △ Dry meadows, screes, scrub. July–Sept. All Eur. (except IS.). *Used for flavouring and in perfumery, also as a tea. It yields a purple or reddish-brown dye; it has been used externally for sprains and bruises.* Pl. 115.

MAJORANA Like *Origanum*, but calyx top-shaped, splitting nearly to the base, entire or shallowly toothed. *c.* 4 sps.

1161. *M. hortensis* Moench (*Origanum majorana* L.) SWEET MARJORAM. A shrubby, very aromatic perenn. to 50 cm., with globular clusters of small purplish or rarely white fls.,

1. *Satureja montana* 1152
2. *Lycopus exaltatus* [1166]
3. *Calamintha sylvatica* 1155
4. *Majorana hortensis* 1161
5. *Micromeria juliana* 1153

on lateral branches forming a dense pyramidal inflorescence. Corolla *c.* 4 mm., little longer than the rounded, woolly-haired, gland-dotted, overlapping bracts. Lvs. *c.* ½–1½ cm., oval to lance-shaped, densely greyish-hairy. △ Native of Africa; often cultivated for its aromatic lvs. June–July. Page 365.

CORIDOTHYMUS Calyx oval, strongly compressed dorsally, thirteen-veined, two-lipped, the upper lip shallowly three-lobed, the lower deeply three-lobed, throat hairy. Corolla two-lipped, the upper two-lobed, the lower three-lobed; stamens 4, diverging from the base. *c.* 1 sp.

1162. *C. capitatus* (L.) Reichenb. fil. (*Thymus c.*) A low, compact, much-branched, stiff shrublet 5–40 cm., with terminal rounded heads of numerous pink fls. subtended by conspicuous, oval, leafy, ciliate bracts. Heads 1–2 cm. by 1 cm.; corolla *c.* ½ cm., rarely white; stamens long-projecting. Lvs. 2–5 mm. oblong, numerous, clustered, stiff, with many sunken glands and ciliate margin; young twigs woolly-haired. △ Dry hills, stony ground. July–Aug. Med. Eur. (except F.) P. *It yields an essential oil containing Thymol.*

THYMUS | Thyme Calyx cylindrical or bell-shaped, ten-veined, two-lipped, the upper lip three-lobed, the lower with narrow, bristle-like lobes, throat closed with a ring of hairs. Corolla weakly two-lipped, the upper lip entire or notched, the lower lip three-lobed; stamens 4, diverging from the base, 2 usually longer than the corolla. *c.* 60 sps.

Erect woody shrublets
1163. *T. vulgaris* L. THYME. A very aromatic compact, greyish, much-branched shrublet 10–30 cm., with velvety-white twigs, narrow lvs., and rounded or elongate clusters of pink or whitish fls. Corolla 4–6 mm., hairy outside; calyx 3–5 mm., velvety-haired. Lvs. 5–9 mm., linear-lance-shaped, blunt, gland-dotted, velvety-white beneath, margin in-rolled. △ Rocks, dry slopes, bushy places; often cultivated and sometimes naturalized in the south. Apr.–July. P.E.F.I. *Widely used for culinary purposes and flavouring. It yields an essential oil containing Thymol which has antiseptic properties and which is used in perfumery, soap-making, and medicine.* Pl. 115.

Creeping or spreading perenns.
1164. *T. serpyllum* L. WILD THYME. A faintly aromatic, mat-forming perenn. with creeping rooting branches and ascending stems to *c.* 5 cm. bearing rounded or sometimes elongated clusters of pink fls. Fls. 3–6 mm.; calyx *c.* 4 mm., lobes ciliate. Lvs. 4–6 mm. obovate, glandular, usually ciliate, otherwise hairless; fl. stems four-angled with 2 opposite sides densely hairy and 2 sides hairless or nearly so. A very variable sp. △ Dry slopes, grass-land, dunes, bushy places. Apr.–Sept. All Eur. (except IS.TR.). Pl. 115.

1165. *T. pulegioides* L. LARGER WILD THYME. Like 1164, but fl. stem sharply four-angled with hairs only on the angles. Fls. pink in somewhat elongated clusters, interrupted below; calyx 3–4 mm., lobes long-ciliate. Lvs. ½–1½ cm., oval to elliptic, ciliate at base only, otherwise hairless. A tufted, strongly aromatic shrublet with ascending branches to 25 cm. △ Grassy places, uncultivated ground. July–Sept. All Eur. (except IS.TR.).

T. mastichina L. ROUND-HEADED THYME. Distinguished by the white fls. and the woolly-haired calyx with rigid, yellowish, spiny-tipped lobes with ciliate hairs. Fl. clusters dense, globular, usually 1–2 cm. across. An erect, branched, strongly aromatic shrublet to ½ m. Iberian Peninsula.

LYCOPUS | Gipsy-Wort Like *Mentha*, but non-aromatic perenns., and stamens 4, with

2 fertile longer than corolla and 2 sterile, included. Calyx bell-shaped, thirteen-veined, with 5 equal slender lobes; corolla tube shorter than the calyx, corolla lobes nearly equal. 2 sps.

1166. *L. europaeus* L. GIPSY-WORT. An erect, unbranched perenn. 30–100 cm., with small very compact whorls of whitish fls. in the axils of the upper lvs. Corolla *c.* 3 mm., white spotted with purple, lobes more or less equal; calyx hairy, lobes spiny-pointed, longer than the tube. Lvs. to 10 cm., elliptic to lance-shaped, pinnately cut into shallow triangular-lobes or sometimes more deeply cut nearly to the midvein. △ Watersides, damp and waste places. July–Sept. All Eur. (except IS.TR.). *It yields a black dye.* Pl. 116.

L. exaltatus L. fil. CUT-LEAVED GIPSY-WORT. Like 1166, but lvs. gland-dotted and all deeply pinnately lobed, nearly to the midvein, with the lobes further toothed or lobed. Calyx teeth at most as long as the calyx tube; sterile stamens more conspicuous. Central and South-Eastern Europe. Page 365.

MENTHA | **Mint** Calyx tubular or bell-shaped, with 10–13 veins, usually regularly five-lobed. Corolla tubular, with 4 equal lobes; stamens 4 fertile, diverging, projecting or not. Hybridisation is frequent and intermediate forms are common. *c.* 20 sps.

Inflorescence terminated by a cluster of lvs.
1167. *M. pulegium* L. PENNY-ROYAL. A very strong-smelling, spreading or ascending perenn. 10–50 cm. with numerous compact rounded, widely spaced whorls of lilac or pink fls. Corolla hairy outside; calyx slightly two-lipped, lobes unequal, hairy, throat closed with hairs. Lvs. usually greyish-haired, 1–2 cm. oval or oblong, and shallowly toothed or entire, shortly stalked, the upper lvs. shorter than the whorls of fls. A very variable sp. △ Wet places, damp meadows. July–Oct. Most of Eur. (except North Eur.). *Sometimes used for flavouring and formerly used medicinally.* Pl. 116.

1168. *M. arvensis* L. CORN MINT. Fls. lilac, in many dense, rather distant whorls forming a slender leafy inflorescence terminated by a tuft of lvs. Corolla *c.* 8 mm. long, hairy outside, longer than the calyx; calyx regularly five-lobed, not hairy in throat; stamens usually projecting. Lvs. stalked, 2–6½ cm., oval to lance-shaped, and toothed, becoming progressively smaller above and usually longer than the whorls of fls. but not always so. A very variable, erect or spreading, not strong-smelling perenn. 10–60 cm. △ Fields, damp places, marshes. July–Sept. Most of Eur. (except P.IS.AL.TR.).

Inflorescence terminated by fls. not a cluster of lvs.
(a) Terminal fl. cluster globular or ovoid, little longer than broad
1169. *M. aquatica* L. WATER MINT. Readily distinguished by its dense, globular, terminal cluster *c.* 2 cm. broad, of lilac or white fls. and with or without a few separated whorls below. Corolla hairy within; calyx and fl. stalks hairy; stamens long-projecting. Lvs. variable, 2–6 cm. or more, oval, usually hairy on both sides, toothed, stalked. An erect, simple or branched, strongly peppermint-smelling perenn. 30–80 cm. with creeping rhizomes. △ Watersides, marshes, damp places. July–Sept. All Eur.: introd. IS. Pl. 116.

(b) Terminal fl. clusters cylindrical, much longer than broad
1170. *M. spicata* L. SPEARMINT. Distinguished by its nearly hairless, bright green, lance-shaped, toothed and stalkless lvs. with a strong, sweet, peppermint-like smell. Fls. pink or lilac, in long, rather slender, cylindrical spikes 3–6 cm., becoming interrupted below. Corolla hairless; calyx tube and fl. stalks hairless; stamens projecting. An erect perenn. with hairless stems, 30–90 cm. △ Damp places in mountains, tracksides, waste places; widely cultivated and often naturalized. Aug.–Oct. Much of Eur. (except IRL.IS.SF.AL. BG.). *Widely used as a culinary herb.*

M. × *piperita* L. PEPPERMINT. A hybrid between 1169 and 1170 smelling and tasting of peppermint and distinguished by the reddish-lilac fls. with stamens not projecting and usually hairless calyx and fl. stalks. Inflorescence oblong 3½–6 cm., usually interrupted below. Lvs. all stalked, lance-shaped or oval-acute, hairless or sparsely hairy. Widespread, often from cultivation, except in the north. *The tops of the plants are distilled to produce peppermint oil which is used in pharmacy, cordials, liqueurs, and for flavouring.*

1171. *M. longifolia* (L.) Hudson HORSEMINT. Distinguished from 1170 by its silvery-grey, densely silky-haired lvs. and long, compact, cylindrical, acute spikes 3–10 cm. long of rose or lilac fls. Corolla *c.* 5 mm., hairy outside, hairless within; calyx and fl. stalks densely hairy; stamens usually projecting. Lvs. stalkless, 3–8 cm., lance-shaped or elliptic, toothed, usually densely grey-hairy but variable and sometimes with the upper side green and shortly hairy. An erect, often strong-smelling perenn. 30–100 cm. △ Wet places, watersides, damp fields. July–Sept. All Eur. (except IS.N.SF.). *Widely used in the past for flavouring and in medicine.* Pl. 116.

1172. *M. rotundifolia* (L.) Hudson APPLE-SCENTED MINT. Readily distinguished by its thick, wrinkled, rounded and toothed, stalkless lvs. 2–4 cm. which are finely hairy above and grey or white woolly-haired beneath. Fls. in dense, cylindrical acute spikes 3–5 cm. long, often with lateral spikes forming a branched inflorescence; corolla pinkish-lilac or whitish, hairy outside; calyx and fl. stalks hairy; stamens usually long-projecting. An erect aromatic perenn. with densely white-haired stems 30–90 cm. △ Damp places, ditches, waysides; often naturalized from cultivation. July–Sept. Most of Eur.

SOLANACEAE | Nightshade Family

Erect or climbing herbs, shrubs, or small trees with alternate, often entire lvs. Fls. usually radially symmetrical, variously coloured; calyx usually fused below, five-lobed, persistent; corolla fused at base, five-lobed; stamens usually 5 and attached to the corolla tube. Ovary superior, with 2 carpels; style 1; fr. a capsule or berry.

Nos.

1. Corolla with wide-spreading lobes very much longer than the very short tube.

  2. Anthers projecting forwards into a cone surrounding the style.

3. Fls. white or violet.	*Solanum*	1180–1183
3. Fls. yellow.	*Lycopersicon*	1184
2. Anthers not in a cone round style.	*Capsicum*	[1179]

1. Corolla bell-shaped, tubular, or funnel-shaped with lobes, usually shorter than the tube.

  4. Herbaceous plants.

    5. Fr. enclosed, or nearly so, by the calyx.

      6. Calyx strongly inflated and becoming papery; fr. a berry.

7. Fls. white; calyx orange in fr.	*Physalis*	1178
7. Fls. blue; calyx brownish in fr.	*Nicandra*	1179

      6. Calyx not inflated; fr. a capsule.

8. Fls. in terminal branched clusters.	*Nicotiana*	1187, 1188
8. Fls. solitary in axils of lvs.	*Hyoscyamus*	1176, 1177

    5. Fr. much longer than and not enclosed by the calyx.

9. Fr. a berry.
 10. Fls arising directly from the rootstock.   *Mandragora* 1185
 10. Fls. borne on erect aerial stems.
  11. Fls. whitish; lobes of large corolla wide-spreading;
   fr. usually conical.   *Capsicum* [1179]
  11. Fls. purple; corolla bell-shaped, lobes small; fr.
   globular.   *Atropa* 1174
9. Fr. a capsule.
 12. Fls. white or pinkish, erect; fr. usually spiny. *Datura* 1186
 12. Fls. purplish or yellow; fr. not spiny.   *Scopolia* 1175
4. Shrubs.
 13. Fls. dull purple; stamens usually protruding. *Lycium* 1173
 13. Fls. yellow or white; stamens not protruding.
  14. Fls. yellow, 3–4 cm.   *Nicotiana* 1187, 1188
  14. Fls. white, 15–25 cm.   *Datura* 1186

LYCIUM Shrubs often with spiny stems and with lance-shaped lvs. Corolla funnel-shaped, five-lobed; calyx bell-shaped, persisting but not enlarging in fr. Fr. a berry. *c.* 5 sps.

1173. *L. europaeum* L. TEA-TREE. An erect shrub 1–3 m. with pale, stiff, erect, spiny branches, thick greyish-green lvs., and tubular violet or rarely white fls. Fls. 1–3 axillary, short-stalked; corolla tube 1½–2 cm., lobes oval spreading, about one-third as long as the tube; stamens not projecting; calyx very short, at first five-toothed, but becoming two-lipped, margin papery. Lvs. narrowly spathulate 2–4 cm., stalked, hairless. Fr. globose, red. △ Hedges, waysides, thickets. Apr.–Sept. Med. Eur. (except AL.) P.A.SU. Page 384.

L. barbarum L. (*L. halimifolium* Miller) DUKE OF ARGYLL'S TEA-PLANT, BOX THORN. Like 1173, but stamens projecting and fls. rose-purple turning brownish, the lobes of the corolla little shorter than or rarely as long as the corolla tube. A spineless or weakly spiny shrub with drooping, greyish-white branches and greyish-green, lance-shaped lvs. Possibly native of South-Eastern Europe, but now widespread in Europe.

ATROPA Herbaceous perenns. with solitary or paired axillary fls. Corolla bell-shaped, shortly five-lobed; stamens inserted at base of the tube, not projecting; calyx bell-shaped, deeply five-lobed. Fr. a berry. 2 sps.

1174. *A. bella-donna* L. DEADLY NIGHTSHADE. Fls. brown-purple or greenish, usually solitary, drooping, shortly stalked from axils of the lvs. or the forks of the branches. Corolla 2½–3 cm. long, bell-shaped; calyx somewhat enlarging in fr. Lvs. to *c.* 20 cm., oval-acute, and narrowed to the stalk. Fr. globular 1½–2 cm., black. A robust, leafy, branched, glandular-hairy perenn. to 1½ m. △ Woods, thickets, waste places. June–Sept. Most of Eur. (except North Eur.) DK.: introd. P. *The shining black berries, about the size of a cherry, are sweetish to the taste and are extremely poisonous and can cause death. The whole plant contains the alkaloids, hyoscyamine, hyoscine, and atropine, which are used in medicine to stimulate the nervous system, and in particular to dilate the pupil of the eye. Used in the past for blanching the skin and removing freckles, hence the Latin name of the species.* Pl. 117.

SCOPOLIA Like *Atropa*, but fr. a capsule opening by a cap; calyx enlarging in fr. 1 sp.

1175. *S. carniolica* Jacq. A leafy, branched perenn. to 60 cm. with solitary, short-stalked, drooping axillary, brownish-purple or yellowish fls. which are olive-green within. Corolla *c*. 2 cm. long; calyx shortly five-lobed. Lvs. 8–10 cm. elliptic or obovate and long-pointed, entire; base of stem with scales. Fr. a globular capsule enclosed within the calyx. △ Woods. Spring. A.PL.CS.H.YU.R.SU.: introd. D. Pl. 116.

HYOSCYAMUS | Henbane Fls. solitary, axillary, rather numerous in outwardly coiled, elongate, leafy clusters. Corolla funnel- or bell-shaped, somewhat irregular, lobes short, rounded; calyx tubular, five-lobed. Fr. a capsule opening by a cap and encircled by the calyx. *c*. 4 sps.

Upper stem lvs. stalkless

1176. *H. niger* L. HENBANE. A coarse, hairy, sticky, and fetid ann. or bienn. 20–80 cm. with 2 axillary rows of dull yellow fls., strongly netted with purple veins, forming a leafy cluster. Fls. stalkless; corolla 2–3 cm. across; calyx with triangular-pointed lobes which become spiny-tipped; bracts leafy. Lvs. 15–20 cm. oblong, with few coarse teeth or nearly entire, the lower stalked, the upper clasping the stem. Calyx encircling fr. 1½–2 cm., strongly veined, papery. △ Waste places, waysides, sandy places by sea. May–Sept. All Eur. (except IS.). *A very poisonous plant. The drug hyoscyamine is obtained from the mature leaves and it is used medicinally as a tranquillizer and sedative and also for dilating the pupil of the eye.* Pl. 117.

Stem lvs. all stalked

1177. *H. albus* L. WHITE HENBANE. Like 1176, but corolla pale yellow without darker coloured veins and with a greenish or violet throat. Upper lvs. all stalked, oval, coarsely lobed, often heart-shaped at base. A less fetid, sticky, hairy ann. or bienn. 20–50 cm. △ Waste places, walls. Apr.–Sept. Med. Eur. P.A.BG.R.SU. *A poisonous plant with similar properties to henbane.* Pl. 117.

H. aureus L. GOLDEN HENBANE. Like 1177, but corolla bright golden-yellow with a deep violet throat and fls. all shortly stalked. Lvs. all stalked, oval or rounded and with a heart-shaped base and margin with acute lobes. Greek Islands. Pl. 117.

PHYSALIS Fls. solitary, axillary. Corolla broadly funnel-shaped with 5 spreading lobes; calyx bell-shaped, green, but becoming coloured and inflated and bladder-like, and enclosing the fr. which is a berry. *c*. 3 sps.

1178. *P. alkekengi* L. BLADDER CHERRY. A finely hairy perenn. with creeping underground stems and erect leafy stems 20–60 cm., bearing solitary, axillary, dirty white, short-stalked, drooping fls. Corolla 1½–2 cm. across, open-funnel-shaped with spreading triangular lobes; calyx bell-shaped, enlarging and becoming bright reddish-orange, thus forming in fr. a strongly veined 'bag' surrounding a red berry. Anthers yellow. Lvs. 8–10 cm. oval-acute, entire, paired, stalked. △ Fields, vineyards. May–Oct. Cent., South-East Eur. E.F.I.SU. *The fruits are edible and are rich in vitamins.* Pl. 119.

P. peruviana L. CAPE GOOSEBERRY. Like 1178, but distinguished by the whitish, purple-spotted fls. and violet anthers. Calyx pale in fr.; berry purplish. Lvs. heart-shaped. Native of the tropics; sometimes cultivated in Southern Europe.

NICANDRA Corolla broadly bell-shaped, shallowly lobed; calyx much inflated, red-veined, pentagonal in section, becoming brown and papery in fr. Fr. a brown, rather dry berry. 1 sp.

1179. *N. physalodes* (L.) Gaertner APPLE OF PERU. A hairless, fetid, leafy, branched ann. 30–130 cm., with large, solitary, axillary, drooping, blue or violet bell-shaped fls. with a white centre. Corolla 3–4 cm. across, opening for a few hours only; calyx lobes oval and with 2 downward-projecting lobes at base, enlarging in fr. into a bladder-like 'bag' enclosing the brown berry. Lvs. all stalked, oval-acute, irregularly toothed or shallow-lobed. △ Waste places, tracksides, fields. June–Oct. A native of Peru; widely naturalized in South-Cent. Eur. *A very poisonous plant.* Pl. 119.

CAPSICUM | **Peppers, Chillies** Corolla whitish, with 5 wide-spreading lobes. Fr. a berry often with dry walls. 2 sps.

C. annuum L. RED PEPPER, CHILLIES, CAPSICUMS, PIMENTO, PAPRIKA. Fls. whitish, shortly stalked, erect. Lvs. stalked, oval or elliptic, entire. Fr. very variable, but often erect, oblong-conical, red. Many forms are widely cultivated in Southern Europe; sometimes self-seeding. *The fruits are used in a variety of ways: 'Sweet Peppers', either red or green, are eaten as a vegetable; a red 'Sweet Pepper' produces pimento, whilst a long, thick, non-pungent form produces paprika. Pungent spicy forms giving red peppers and tabasco are used as condiments.*

SOLANUM Corolla usually with 5 wide-spreading petal-like lobes, shortly fused at the base or rarely broadly bell-shaped; calyx with 5–10 lobes, not enlarging in fr. Anthers projecting forward into a cone, opening by apical pores. Fr. a two-celled berry. *c.* 10 sps.

Spiny plants

1180. *S. sodomeum* L. A much-branched shrub ½–1 m., with numerous, very stout, yellow spines on the stems and lvs. with violet fls. and large yellow shining berries. Fls. few in stalked clusters; corolla *c.* 2½ cm. across; calyx and fl. stalks with long spines. Lvs. deeply pinnately lobed, lobes rounded and midvein and lateral veins with long, stout, yellow spines on both sides. Fr. globular, 2–3 cm. △ Waste places, waysides, cultivated ground. May–Aug. Med. Eur. P.: introd. A.CS. Pl. 118.

Non-spiny plants

(a) Fls. purple

1181. *S. dulcamara* L. BITTERSWEET, WOODY NIGHTSHADE. A scrambling, woody-based, shrub-like perenn. 30–200 cm., with purple fls. with conspicuous yellow anthers, in lax branched clusters arising opposite the lvs. Fls. 1–2 cm. across, lobes lance-shaped, wide-spreading, rarely white; anthers yellow, in a cone. Lvs. to *c.* 8 cm. oval, stalked, the upper entire, the lower often deeply lobed at the base. Fr. red, *c.* 1 cm. △ Woods, hedges, damp places, shingle beaches, dunes. June–Sept. All Eur. (except IS.). *Well known as a herbal plant since the time of Theophrastus: used for skin diseases, whooping cough, asthma, rheumatism, and jaundice. The berries contain solanine and are poisonous.* Pl. 119.

S. melongena L. AUBERGINE, EGG-PLANT. Distinguished by its large oblong-ovoid fr. 5–30 cm. long, which is often shining purple or white or yellowish. Fls. purple, 2–4 cm. across, few or solitary; calyx prickly. Lvs. oblong-oval, greyish-hairy with star-shaped hairs, often prickly. Native of Old World tropics; cultivated in Central and Southern Europe.

(b) Fls. white

1182. *S. nigrum* L. BLACK NIGHTSHADE. A branched, leafy, erect ann. 20–60 cm., with white fls. with a yellow cone of projecting anthers, in lax clusters arising opposite the lvs. Fls. 5–8 mm. across; corolla lobes oval-acute, spreading or somewhat reflexed. Lvs. oval and narrowed to the lv. stalk, entire or shallowly lobed. Fr. *c.* 8 mm., black or less commonly green or yellow. A very variable sp. △ Arable fields, waste ground, vineyards.

June–Oct. All Eur. (except IS.). *A somewhat poisonous plant, but in some countries the young leafy shoots are eaten as a vegetable. The berries contain solanine and are poisonous.*

S. luteum Miller Very like 1182, but a greyish, softly hairy ann. with more deeply lobed lvs. Fr. orange or brownish-yellow. Widespread in Europe, except in the extreme west and north.

S. alatum Moench Like [1182] but stems and branches rough, sharply angled and with swelling on the angles; plant with a musk-like smell. Fr. vermilion. South-Eastern Europe.

1183. *S. tuberosum* L. POTATO. Readily distinguished by its large, swollen, irregularly ovoid tubers which are borne at the ends of slender underground stolons. Fls. white or pale violet in long-stalked clusters; corolla 2½–4 cm. across, lobes triangular, wide-spreading, hairy outside. Lvs. pinnately cut into 3–5 oval or heart-shaped pairs of segments. Fr. globular, green. A hairy perenn. 30–80 cm. △ Native of South America; very widely cultivated throughout Europe. Summer. *An important vegetable crop for both humans and livestock. A primary source of commercial starch used in industrial fermentation and the sizing of cotton. Green potato tops are poisonous to livestock and green potatoes are very poisonous to humans.*

LYCOPERSICON | **Tomato** Like *Solanum* with anthers in a cone round style, but opening by longitudinal splits, not by pores, and apex of anther extended into sharp or narrow sterile tip. 1 sp.

1184. *L. esculentum* Miller (*Solanum lycopersicum* L.) TOMATO, LOVE-APPLE. A glandular-hairy, strong-smelling ann. to 1 m., with pinnate lvs., yellow fls. in branched clusters and large, red, fleshy fr. Corolla *c.* 2 cm. across; stamens forward-projecting. Lvs. large, 15–45 cm., odd-pinnate, with large, oval, irregularly toothed leaflets and smaller intermediate leaflets. Fr. globular or lobed, but usually flattened above and below, shining red or rarely yellow. △ Native of Mexico; widely cultivated in Central and Southern Europe. Summer. *The tomato has recently become very important both as a fresh 'vegetable' and also for canning, as a juice or a purée.*

MANDRAGORA | **Mandrake** Fls. short-stalked and arising directly from the root-stock. Corolla bell-shaped, five-lobed; calyx with 5 somewhat leafy lobes, enlarging in fr. Fr. a berry. *c.* 2 sps.

1185. *M. officinarum* L. MANDRAKE. A perenn. with a very stout taproot, a large flattened rosette of lvs. and with erect, short-stalked, violet bell-shaped fls. arising from centre of the rosette. Corolla 3–4 cm. long, deeply five-lobed; calyx deeply five-lobed, lobes leafy. Autumn lvs. enlarging to 40 cm., broadly lance-shaped, stalked, dark green and wrinkled above, the spring lvs. smaller and blunter. Fr. a yellow or orange berry 2½–3 cm. △ Stony places, deserted cultivation. Spring and autumn. P.E.I.GR. *A plant of great renown in classical times, famous as a pain-killer; now a medicinal and historical curiosity. Many strange superstitions have been associated with it, some of which persist to this day.* Pl. 118.

DATURA | **Thorn-Apple** Corolla long funnel-shaped, with 5 short pleated lobes; calyx tubular, not persisting in fr. Fr. usually a capsule. *c.* 3 sps.

1186. *D. stramonium* L. THORN-APPLE. A strong-smelling, hairless, branched ann. 40–100 cm., with few large, erect, white or rarely violet tubular fls. and large, very spiny green frs. Fls. solitary, axillary, short-stalked; corolla 6–8 cm. tubular, lobes pointed, twisted in bud; calyx pale green, five-angled. Lvs. to 20 cm. oval-acute, coarsely lobed or sharp-toothed, stalked. Fr. erect, ovoid 4–5 cm., very spiny or rarely without spines, splitting into 4. △ Waste places, cultivation, river sands. July–Oct. All Eur. (except IRL.IS.): introd. North Eur. GB.D. *A highly poisonous plant containing atropine, hyoscyamine, and hyoscine, which act on the nervous system producing hallucinations and dilating the pupils of the eyes. This plant was probably used by oracles and wizards in classical and medieval times.* Pl. 118.

D. metel L. Like 1186, but fls. much larger, 15–20 cm. long, white often flushed with pink; fr. pendulous, spiny. A densely hairy, greyish ann. to 1½ m., with entire or shallowly lobed lvs. Native of India; sometimes naturalized in the Mediterranean region. *A very poisonous plant with similar properties to the thorn-apple.* Pl. 118.

D. arborea L. ANGEL'S TRUMPET. A tree-like shrub 3–5 m., with very large pendulous white fls. with greenish veins, 15–23 cm. long, with a musk-like smell; calyx tubular, spathe-like with a single lobe. Lvs. softly hairy, oval-oblong, stalked. Native of Peru; often grown as an ornament in Southern Europe.

NICOTIANA | Tobacco Fls. elongate funnel-shaped or bell-shaped, in branched leafless clusters; calyx five-lobed. Stamens attached at throat of corolla, not projecting. Fr. a two-valved capsule. *c.* 5 sps.

Shrubs

1187. *N. glauca* Graham SHRUB TOBACCO. An erect, branched, hairless shrub 2–3 m., with bluish-grey lvs. and terminal clusters of yellow, narrowly funnel-shaped fls. Corolla 3–4 cm. long, with 5 short blunt lobes, hairy outside. Lvs. entire, oval-elliptic acute, glaucous, hairless. △ Waste ground, rocks, walls. Apr.–Oct. Native of South America; naturalized in the Mediterranean region. *A poisonous plant.* Pl. 119.

Herbaceous plants

1188. *N. rustica* L. SMALL TOBACCO. A strong-smelling, glandular-hairy, erect ann. 30–100 cm., with greenish-yellow fls. in a dense, many-flowered, terminal cluster. Corolla tube *c.* 2 cm., 2–3 times as long as the calyx, lobes blunt. Lvs. all stalked, oval-heart-shaped, glandular-hairy, shining. △ Native of South America; sometimes cultivated in Southern Europe. June–Aug. *Nicotine is obtained from the green parts of this plant in particular; it is a very poisonous substance and is used as an insecticide.* Pl. 119.

N. tabacum L. LARGE TOBACCO. Like 1188, but fls. pink or purple; corolla 3½–5 cm., 4–5 times as long as the calyx. Lvs. all stalkless, the lowermost with the blade running down the stem. Native of South America; widely cultivated in Southern Europe. *Tobacco is obtained from the partially fermented leaves of this species in particular; it contains from one to three per cent of nicotine, and is smoked, chewed, or inhaled as snuff by most of the peoples of the world.* Pl. 119.

BUDDLEJACEAE | Buddleja Family

Usually trees or shrubs with opposite lvs. Fls. regular; calyx bell-shaped, four-lobed; corolla tubular, four-lobed; stamens 4, attached to the corolla tube. Ovary superior, of 2 carpels; style 1. Fr. a capsule or berry.

BUDDLEJA   Fls. densely clustered in long terminal spikes or rounded heads. Calyx bell-shaped; corolla with a long straight tube and spreading lobes; stamens usually included. *c.* 2 sps.

1189. *B. davidii* Franchet A tall shrub to 5 m. with white hairy twigs, large lance-shaped lvs., and long, cylindrical acute spikes of very numerous lilac or violet fls. Spikes 10–50 cm. long; corolla tube cylindrical *c.* 1 cm. long, often orange-throated, lobes short, spreading; calyx and fl. stalks with white star-shaped hairs. Lvs. 10–25 cm., long-pointed, toothed, dark shiny green above, white woolly-haired beneath. △ Native of China; often grown for ornament and sometimes naturalized. July–Aug. Pl. 121.

SCROPHULARIACEAE | Figwort Family

Usually herbaceous plants, sometimes partially parasitic on herbaceous plants; lvs. usually entire or toothed. Fls. largely symmetrical in one plane only. Calyx four- to five-lobed; corolla fused at the base, four- to five-lobed, very variable, from more or less regularly lobed to strongly two-lipped with the lobes obscure; stamens 2, 4, or 5, attached to the corolla. Ovary superior, two-celled; style simple or two-lobed; fr. a capsule or rarely a berry.

		Nos.
1. Fertile stamens 2.	Group A	
1. Stamens 4 or 5.		
2. Stamens 4.		
3. Corolla without a spur or sac-like swelling at base.		
4. Calyx 4; corolla two-lipped.	Group B	
4. Calyx 5.	Group C	
3. Corolla with a spur or sac-like swelling at base.	Group D	
2. Stamens 5 with some or all filaments densely hairy.	*Verbascum*	1190–1196

Group A. *Fertile stamens 2*

		Nos.
1. Calyx 5; corolla tubular.	*Gratiola*	1218
1. Calyx 4; corolla with wide-spreading lobes.	*Veronica*	1219–1229

Group B. *Stamens 4; calyx 4; corolla two-lipped*

		Nos.
1. Lvs. compound, alternate or whorled.	*Pedicularis*	1249–1256
1. Lvs. entire or toothed, opposite.		
2. Corolla two-lipped with 5 more or less equal lobes; calyx sometimes five-lobed.	*Tozzia*	1261
2. Corolla two-lipped, the upper lip hooded, the lower lip three-lobed.		
3. Calyx laterally flattened and inflated.	*Rhinanthus*	1247, 1248
3. Calyx not laterally flattened, usually not inflated.		
4. Lvs. entire or with 1–3 teeth on each side at base; fr. one- to four-seeded.	*Melampyrum*	1257–1260
4. Lvs. regularly toothed or shallowly lobed; fr. usually many-seeded.		
5. Upper lip of corolla with 2 spreading or reflexed lobes.	*Euphrasia*	1242–1245
5. Upper lip of corolla entire or notched.		
6. Fls. small, 4–10 mm., in a rather one-sided spike.		

7. Fls. in a one-sided spike, yellow or pinkish-purple.	*Odontites*	1240, 1241
7. Fls. in a dense cluster, usually reddish-purple.	*Parentucellia*	1238, 1239
6. Fls. larger at least 1 cm., spike not one-sided.		
8: Calyx inflated, bell-shaped, deeply divided into 2 toothed lobes.	*Bellardia*	1237
8. Calyx not inflated, cylindrical with 4 acute lobes.		
9. Seeds large, winged, or ribbed; fls. dark purple.	*Bartsia*	1236
9. Seeds minute, smooth; fls. yellow or reddish-purple.	*Parentucellia*	1238, 1239

Group C. *Stamens 4; calyx 5*

Nos.

1. Lvs. alternate or in a basal rosette.		
2. Lvs. once or twice deeply pinnately cut.		
3. Fls. cup-shaped with 5 spreading, nearly equal lobes; stamens conspicuous, hairy.	*Verbascum*	1190–1196
3. Fls. two-lipped; stamens hidden.	*Pedicularis*	1249–1256
2. Lvs. not deeply divided, entire or toothed.		
4. Fls. open cup-shaped.	*Verbascum*	1190–1196
4. Fls. tubular or funnel-shaped.		
5. Small plants up to 15 cm.; fls. erect, lobes wide-spreading, equal.	*Erinus*	1235
5. Robust plants ½ m. or more; fls. drooping, tubular, lobes erect unequal.	*Digitalis*	1230–1234
1. Lvs. opposite.		
6. Corolla tube nearly globular, lobes short.	*Scrophularia*	1212–1216
6. Corolla tube cylindrical, lobes large, spreading.	*Mimulus*	1217

Group D. *Corolla with a basal boss or spur*

Nos.

1. Mouth of corolla not closed by a throat-boss.	*Anarrhinum*	1211
1. Mouth of corolla closed by a throat-boss.		
2. Corolla with a swollen sac-like swelling at base, not spurred.		
3. Lvs. at least twice as long as broad.	*Antirrhinum*	1197–1199
3. Lvs. about as long as broad.	*Asarina*	1200
2. Corolla with an elongated spur at base.		
4. Lvs. at least twice as long as broad, usually linear or oblong, not or shortly stalked.		
5. Fls. in terminal clusters with bracts much shorter than lvs.; fr. opening by valves.	*Linaria*	1201–1208
5. Fls. axillary with bracts similar to lvs.; fr. opening by pores.	*Chaenorhinum*	[1209]
4. Lvs. less than twice as long as broad, rounded or heart-shaped at base, stalked.		
6. Fls. yellow with purple lip.	*Kickxia*	1209
6. Fls. lilac with orange spot.	*Cymbalaria*	1210

VERBASCUM | **Mullein** Fls. with a short tube and 5 spreading, nearly equal, rounded, petal-like lobes; calyx deeply five-lobed. Stamens 4 or 5, some or all with hairy filaments. Fr. a capsule. A difficult genus; many sps. hybridize freely. *c.* 90 sps. *Mulleins are poisonous to livestock and are avoided by them.*

Stamens 5, filaments with violet or purple hairs
(a) *Fls. in clusters of 2 or more*
1190. *V. nigrum* L. DARK MULLEIN. Distinguished by its stamens with dense purple hairs on the filaments and its kidney-shaped anthers. Fls. in a long slender spike, sometimes with a few erect lateral spikes; fls. in clusters of 5–10; corolla yellow with purple spots at the base, 12–22 mm. across. Lower lvs. long-stalked, the upper nearly stalkless, all dark green, thinly hairy above, paler and conspicuously covered with star-shaped hairs beneath; stem angled above. Bienn. 50–120 cm. △ Waysides, banks, dry places. July–Sept. Much of Eur. (except P.IRL.IS.AL.GR.TR.). Pl. 120.

1191. *V. sinuatum* L. Distinguished by its distinctive basal rosette lvs. which are deeply pinnately rounded-lobed, and covered with dense greyish or yellowish woolly hairs. Lower stem lvs. oblong-spathulate and stalkless, the upper oblong with blades shortly running down the stem. Inflorescence widely branched; fls. in widely spaced clusters on twiggy, hairless branches; corolla small, 1½–2 cm. across, filaments with violet hairs, anthers attached transversely. A bienn. ½–2 m. △ Dry, uncultivated ground. June–Oct. Med. Eur. P.BG.SU.

(b) *Fls. usually solitary*
1192. *V. blattaria* L. MOTH MULLEIN. Readily distinguished by the solitary fls. with fl. stalks much longer than the calyx and forming a long, unbranched, glandular-hairy spike. Corolla 2–3 cm. across, yellow or rarely whitish; stamens unequal, the lower larger, and anthers attached obliquely to the violet-haired filament. Lvs. irregularly coarsely toothed, shining, hairless; stems hairless below. A bienn. to 1 m. △ Uncultivated ground, way-sides, banks. June–Aug. Much of Eur. (except North Eur. P.IRL.). Pl. 120.

V. phoeniceum L. PURPLE MULLEIN. Like 1192, but fls. purple, 2½ cm. across, solitary on long stalks; anthers all equal. Central and South-Eastern Europe. Pl. 121.

V. virgatum Stokes TWIGGY MULLEIN. Like 1192, with a long, unbranched, slender, glandular-hairy spike of fls., but fls. very short-stalked with stalks shorter than the bracts and calyx. Fls. 1 or several from each linear-lance-shaped bract; corolla 2½–4 cm. across, usually yellow but rarely white or flushed with violet. Lvs. hairless or glandular-hairy. Western Europe, Italy: introduced elsewhere.

Stamens 5, filaments with white or yellow hairs
(a) *Anthers partially or completely elongated down filament*
1193. *V. thapsus* L. AARON'S ROD. Fls. in a long, dense, usually unbranched spike, and the upper lvs. with blades running down stem almost to the lv. below. Corolla yellow, con-cave, 1½–3 cm. across; 3 upper stamens with whitish or yellowish hairs, lower 2 more or less hairless; anthers attached obliquely and partly fused to the filament. Lower lvs. with winged stalks. A bienn. 30–200 cm. △ Sandy banks, waste places, uncultivated ground, coppices. July–Sept. All Eur. (except IS.AL.BG.TR.). *This plant has long been famous in herbal medicine. The flowers produce a yellow dye, used by Romans. The leaves have been used as wicks and poultices.*

V. thapsiforme Schrader Like 1193, but fls. large, 3–5 cm., and corolla quite flat, bright yellow or rarely white; anthers attached their whole length to the filament. Inflorescence often shortly branched below. Lvs. with yellowish-grey woolly hairs, oblong-lance-shaped, long-pointed, coarsely rounded-toothed, with blade running down stem to the lv. below. Central Europe, but naturalized elsewhere. Pl. 120.

1194. *V. phlomoides* L. Like [1193], but whole plant covered with dense, woolly, whitish or yellow hairs and upper lvs. only shortly running down stem to less than half the length of the internodes, or lvs. not running down stem and with rounded bases. Fls. bright yellow,

3–5 cm. across, in short-stalked clusters with fl. stalks shorter than the calyx; anthers obliquely attached; inflorescence a rather lax simple spike, sometimes with a few branches at the base. Basal lvs. shortly stalked. A very variable. bienn. 30–200 cm. △ Stony, uncultivated ground. June–Sept. Much of Eur. (except North Eur. IRL.GB.B.NL.).

V. longifolium Ten. Fls. in a dense spike with stalks of fls. of unequal length, either longer or shorter than the calyx. Corolla nearly flat, 3 cm. across; longest stamens with hairless filaments. Plant covered with dense, yellowish, woolly hairs, lower lvs. short-stalked, the upper somewhat clasping with a heart-shaped base and a long-pointed apex. Italy to Bulgaria.

(b) Anthers kidney-shaped, attached in the middle
1195. *V. lychnitis* L. WHITE MULLEIN. Distinguished by the lvs. which are dark green, almost hairless above and densely white powdery-haired beneath. Fls. white or yellow, small, 1½–2 cm. across, in clusters of 2–7 in the axils of the bracts; stamens with densely white-hairy filaments; inflorescence with many short, erect, lateral branches forming a slender pyramid; fl. stalks white-woolly, longer than calyx. Basal lvs. oblong-lance-shaped, short-stalked, stem lvs. oval long-pointed, stalkless; stem angled. A bienn. ½–2 m. △ Dry places, clearings, sands. June–Sept. Most of Eur. (except P.IRL.IS.AL.TR.): introd. N.DK.SF.

V. pulverulentum Vill. HOARY MULLEIN. Like 1195, but distinguished by the whole plant being thickly covered with mealy-white woolly hairs which easily rub off; stem not angled. Fls. yellow, 2–2½ cm. across, in a wide-spreading pyramidal inflorescence with fl. clusters rather widely and irregularly spaced on the lateral branches; fl. stalks longer than the calyx. Central and Southern Europe.

V. speciosum Schrader Whole plant usually densely greyish-haired or less commonly yellow-haired. Fls. yellow, 2–2½ cm. across; fls. in stalked clusters; calyx 3 mm. long; inflorescence pyramidal, densely woolly-haired. Lower lvs. large, lance-shaped, narrowed gradually to the base, not toothed, the upper oval-acute with a heart-shaped, clasping base. South-Eastern Europe.

V. undulatum Lam. Whole plant covered with white-woolly hairs. Fls. 2–3½ cm. across, clustered, stalkless; calyx 1 cm. long; inflorescence a long interrupted spike. Lower lvs. deeply pinnately lobed with undulate margin, the upper lvs. oval, almost entire, clasping stem. Balkan Peninsula. Pl. 120.

Stamens 4
1196. *V. creticum* (L.) O. Kuntze (*Celsia c.*) A robust greyish-hairy bienn. ½–1 m., with a slender spike of large yellow fls. and strongly toothed or lobed lvs. Corolla 4½ cm. across, with unequal lobes, the 2 upper with dark brownish-purple spots; calyx lobes large and leafy, oval, toothed and glandular-hairy; stamens unequal and filaments with violet hairs. Stems very leafy, the upper lvs. oval, strongly toothed and clasping the stem, the lower lvs. stalked, deeply lobed. △ Sandy, dry places. Apr.–June. E.F.I. Crete. Pl. 120.

ANTIRRHINUM | Snapdragon Corolla cylindrical with a sac-like swelling at the base, strongly two-lipped, the lower lip three-lobed and with a swelling, or throat-boss, closing the mouth of the corolla, the upper lip two-lobed; calyx five-lobed, nearly regular; stamens 4, not projecting. Capsule two-celled, opening by 3 pores. *c.* 13 sps.

Calyx much shorter than corolla
1197. *A. majus* L. SNAPDRAGON. A short-lived, woody-based, erect perenn. 30–80 cm., with large reddish-purple fls. with a yellow throat-boss in rather dense elongated spikes, or fls. rarely whitish-yellow overall. Inflorescence densely glandular-hairy above; corolla

3½–4 cm. long; calyx with oval, blunt lobes; bracts leafy. Lvs. narrow-lance-shaped, at least 2½ times as long as broad, entire, hairless. Fr. glandular-hairy. △ Rocks, old walls, dry places; often grown for ornament and naturalized. Apr.–Nov. Med., South-East Eur. P.: introd. GB.D.CH.A.CS.

1198. *A. latifolium* Miller LARGE SNAPDRAGON. Like 1197, but lvs. oval, not more than twice as long as broad, often finely hairy; stem glandular-hairy from the base. Fls. yellow, 3½–4 cm., with a paler throat-boss, or fls. rarely purple; calyx lobes oval, glandular-hairy. A glandular-hairy woody-based perenn. 20–60 cm. △ Rocks. Apr.–Nov. E.F.CH.I.YU. Pl. 121.

Calyx as long as or longer than the corolla
1199. *A. orontium* L. (*Misopates o.*) WEASEL'S SNOUT. A slender, erect, branched ann. 20–50 cm., with pinkish-purple fls. with a yellow throat-boss in a terminal, lax, leafy cluster. Inflorescence glandular-hairy, fls. nearly stalkless; corolla small, 1–1½ cm. long; calyx with unequal linear lobes as long as or longer than the corolla. Lvs. 3–5 cm. linear-lance-shaped. Young fr. densely glandular-hairy. △ Cultivated ground, sands. June–Sept. All Eur. (except IS.): introd. N.SF. *Poisonous.* Pl. 122.

ASARINA Like *Antirrhinum*, but lvs. opposite and palmately veined. Fr. opening by 2 pores. 1 sp.

1200. *A. procumbens* Miller CREEPING SNAPDRAGON. A creeping, brittle-stemmed, spreading perenn. 40–60 cm., with opposite rounded, downy sticky lvs. and large, solitary, axillary, whitish-yellow fls. streaked with pink. Corolla stout, 3–4 cm. long; calyx glandular-hairy, the lobes lance-shaped; fl. stalk flexuous, as long as the calyx. Lvs. kidney-shaped, rounded-lobed, stalked. Fr. hairless. △ Rock fissures in mountains. Apr.–Sept. Pyrenees. Pl. 121.

LINARIA | Toadflax Fls. as in *Antirrhinum*, but corolla with a short or long spur at the base. Capsule opening by 4–10 valves. *c.* 90 sps. *Toadflaxes are poisonous to livestock and are usually avoided by them.*

Fls. predominantly purple
(a) *Spur less than half length of corolla*
1201. *L. arvensis* (L.) Desf. An erect glaucous ann. 10–40 cm. with very small, usually blue-lilac fls. in a dense, leafless, glandular-hairy, terminal cluster which later elongates. Corolla 4–8 mm. including spur; throat-boss white; spur curved, conical, shorter than the corolla; calyx glandular-hairy, blunt-lobed shorter than the corolla; bracts linear, reflexed. Lvs. linear, the lower whorled in fours, the upper alternate; sterile shoots present at the base of stem. △ Fields, sandy places. May–Sept. Cent., South-East Eur. (except TR.) E.F.B.NL.I.

1202. *L. repens* (L.) Miller PALE TOADFLAX. Distinguished by the numerous pale lilac or whitish fls. striped with violet, and forming a lax elongated inflorescence. Corolla 7–14 mm., with a short, straight, blunt spur about one-third its length; throat-boss pale yellow; calyx hairless, acute; bracts linear. Lvs. linear-lance-shaped 1–4 cm., whorled below, alternate above. An erect, often glaucous, very leafy perenn. 30–80 cm. △ Stony uncultivated ground. June–Sept. E.F.GB.B.CH.I.: introd. N.S.D.CH. Pl. 122.

(b) *Spur more than half length of corolla*
1203. *L. pelisseriana* (L.) Miller Like 1201, but fls. larger, 1–2 cm., bright violet with a white throat-boss and a straight, slender, violet spur about as long as the corolla; calyx hairless, lobes linear acute, white-margined; bracts oval, very short. Lowest lvs. elliptic, in threes, the upper linear, alternate. A hairless, erect, unbranched ann. 20–40 cm., with sterile shoots at the base. △ Cultivated ground. May–July. Med. Eur. BG.

1204. *L. alpina* (L.) Miller ALPINE TOADFLAX. A prostrate alpine ann. to perenn. with spreading stems 10–30 cm., glaucous lvs. and terminal clusters of deep violet fls. usually with a conspicuous orange throat-boss. Fl. clusters rounded, few-flowered; fls. *c.* 2 cm. long, spur almost straight and as long as the corolla, throat-boss sometimes white; calyx hairless, acute. Lvs. linear-lance-shaped in whorls of 3–5, slightly fleshy, glaucous, the upper alternate. △ In mountains: rocks, screes, river gravels. June–Sept. Cent. Eur. (except H.) E.F.I.YU.AL.R. Pl. 122.

L. triornithophora (L.) Willd. Distinguished by its interrupted spikes formed of whorls of large, striped, violet-purple fls. 2½–4 cm. long, with a long, curved, very pointed purple spur, longer than the corolla and a yellow throat-boss. An erect, glaucous perenn. 30–100 cm., with oval-lance-shaped, three-veined lvs. in whorls. Portugal and Spain. Pl. 122.

Fls. predominantly yellow

1205. *L. vulgaris* Miller COMMON TOADFLAX. Fls. sulphur-yellow with an orange throat-boss, numerous in a dense, many-flowered, somewhat glandular, elongated spike. Fls. 1½–3 cm., with a more or less straight spur as long as the corolla; lobes of calyx oval or lance-shaped. Lvs. 3–8 cm. linear to lance-shaped, numerous and closely arranged on stem. Fr. more than twice as long as the calyx; seeds winged. An erect, glaucous, leafy perenn. with many erect stems 30–80 cm. △ Hedges, banks, waste places, uncultivated ground. June–Oct. All Eur. (except P.). *The flowers give a yellow dye.*

L. angustissima (Loisel.) Borbás (*L. italica* Treviranus) ITALIAN TOADFLAX. Like 1205, but fls. pale lemon-yellow with an orange throat-boss, in quite hairless, dense, slender spikes. Corolla 16–22 m.; fl. stalks hairless. Lvs. flat, somewhat fleshy, very glaucous. Fr. globular. A mountain plant. Pyrenees, Alps, Carpathians, and Balkans.

L. dalmatica (L.) Miller Very like 1205, but lvs. oval to broadly lance-shaped and clasping the stem with a shallow heart-shaped base. Fls. large, yellow with an orange-brown throat-boss; fl. stalks longer than the bracts. Seeds not winged. Balkan Peninsula. Pl. 121.

L. genistifolia (L.) Miller Like 1205, but spikes slender elongated, usually much-branched; lvs. glaucous oblong-lance-shaped, narrowed at the base, not clasping. Fls. pale yellow, throat-boss orange-brown; fl. stalks shorter than the bracts. Seeds not winged. A variable sp. Central and South-Eastern Europe.

1206. *L. supina* (L.) Chazelles A glaucous ann. to perenn. 5–20 cm., with spreading and often prostrate branches and a terminal rounded cluster of few yellow fls. with deeper yellow throat-bosses. Inflorescence glandular-hairy; corolla 1–1½ cm., spur almost straight, nearly as long as the corolla; calyx lobes linear, glandular-hairy; fl. stalks very short. Lvs. 1–3 cm. linear, few, in whorls of 3–5. △ Sandy places. June–Sept. P.E.F.I.

Fls. white or variegated

1207. *L. chalepensis* (L.) Miller WHITE TOADFLAX. Fls. pure white with very long, slender, curved spurs and forming a slender lax spike. Fls. 12–15 mm. including the spur which is twice as long as the corolla; calyx lobes linear-acute, spreading, hairless, much longer than the fr.; bracts linear, at length reflexed. Lvs. linear 2–3 cm., mostly alternate, erect. A slender, erect, unbranched, hairless ann. 15–40 cm. △ Sandy fields. Apr.–June. Med. Eur. (except TR.) BG.

1208. *L. triphylla* (L.) Miller THREE-LEAVED TOADFLAX. Readily distinguished by its compact terminal heads of rather large tricoloured fls. and oval lvs. usually in whorls of 3. Corolla *c.* 2 cm., yellowish-white with an orange throat-boss and a curved, pointed, violet spur. Lvs. glaucous and slightly fleshy, three-veined. Seeds not winged, strongly netted. A glaucous, hairless, rather thick-stemmed, erect ann. 10–30 cm. △ Cultivated ground, fields, vineyards, near the littoral. Apr.–June. P.E.F.I.YU.GR. Pl. 122.

KICKXIA | Fluellen Like *Linaria*, but lvs. all stalked, alternate and pinnately veined. Fls. axillary, bracts similar to lvs. Capsule opening by 2 lids. Often included in *Linaria*. *c.* 6 sps.

1209. *K. spuria* (L.) Dumort. (*Linaria s.*) ROUND-LEAVED FLUELLEN. A densely glandular-hairy prostrate ann. spreading 20–50 cm., with alternate oval lvs. and small yellow axillary fls. with dark brownish-purple upper lips. Fls. solitary on long hairy stalks; corolla 8–11 mm., spur curved, as long as the corolla; calyx lobes oval. Lvs. 2–4 cm., with a rounded or heart-shaped base, densely hairy. △ Dry places, cultivated ground. June–Oct. Most of Eur. (except IRL.IS.): introd. North Eur.

K. elatine (L.) Dumort. (*Linaria e.*) SHARP-LEAVED FLUELLEN. Like 1209, but upper lvs. triangular and with spreading basal lobes, and fl. stalks usually hairless. Fls. 7–9 mm., pale yellow, with the upper lip and throat violet-purple, spur straight; calyx lobes lance-shaped acute. A prostrate less hairy or glandular ann. Most of Europe, introduced to the north.

*Chaenorhinum minus (L.) Lange (*Linaria m.*) SMALL TOADFLAX. Most of Europe, introduced to the north.

CYMBALARIA Like *Kickxia* with mostly alternate palmately veined lvs., but fr. opening by 2 lateral pores, each pore three-valved. *c.* 4 sps.

1210. *C. muralis* Gaertner, Meyer & Scherb. (*Linaria cymbalaria* (L.) Miller) IVY-LEAVED TOADFLAX. A hairless, slender, trailing ann. spreading 10–80 cm., with rounded lvs. and small, axillary, lilac fls. with a white and yellow throat-boss. Corolla 8–10 mm., rarely white, spur curved, one-third as long as the corolla; fl. stalks long slender, recurved in fr. Lvs. 2½ cm. across, long-stalked, blade thick, rounded or heart-shaped with 5–7 shallow lobes, often purplish beneath. △ Damp rocks, old walls. May–Oct. Med. Eur. (except AL.TR.) P.BG.R.SU.: introd. Cent. Eur. IRL.GB.B.NL. Pl. 122.

ANARRHINUM Corolla cylindrical with a short spur at base and with an open throat without a throat-boss, two-lipped, the upper erect lip two-lobed, the lower three-lobed. Stamens 4. Fr. two-celled, opening by 2 pores. *c.* 3 sps.

1211. *A. bellidifolium* (L.) Desf. A hairless bienn. or perenn. 20–80 cm., with a basal rosette of lvs. differing markedly from the deeply cut, narrow-lobed stem lvs. and a long, slender, rather one-sided spike of numerous blue or violet fls. Corolla 3–5 mm., spur slender, shorter than the tube and curved under it; calyx lobes linear-lance-shaped. Basal lvs. obovate or linear-elliptic, irregularly toothed; stem lvs. palmately lobed. △ Dry places, pine woods, rocks, waysides, walls. Mar.–Aug. P.E.F.D.I. Pl. 123.

SCROPHULARIA | Figwort Fls. brown, yellow, or greenish with an almost globular corolla tube with 5 small, blunt, spreading lobes; calyx five-lobed. Fertile stamens 4, usually with 1 flap-like sterile stamen present. Stems quadrangular; lvs. opposite. *c.* 40 sps.

Lvs. undivided, margin entire or toothed
(a) Calyx lobes without a membraneous margin
1212. *S. peregrina* L. NETTLE-LEAVED FIGWORT. An erect, little-branched ann. 30–60 cm.,

with brownish-purple fls. in long-stalked, lax, axillary clusters forming a leafy inflorescence. Clusters with 2–5 fls.; corolla 5–8 mm.; sterile stamen-flap circular; calyx lobes acute, without a membraneous margin; fl. stalks glandular-hairy, 2–3 times as long as the calyx. Lvs. stalked, oval-heart-shaped, acute, coarsely and irregularly toothed, hairless, pale green. △ Bushy places, cultivated ground, olive groves, vineyards. Apr.–June. Med. Eur. (except TR.) P.SU.

1213. *S. vernalis* L. YELLOW FIGWORT. Differing from 1212 in having greenish-yellow fls. in long-stemmed compound axillary clusters, with fl. stalks shorter than the calyx. Corolla globular 6–9 mm., contracted at the throat; stamens at length protruding; sterile stamen-flap absent; lobes of calyx without papery margin. Lvs. oval-heart-shaped, and coarsely toothed, stalked. An erect, softly glandular-hairy bienn. or perenn. 30–80 cm. △ Woods, shady bushy places. Apr.–July. Much of Eur. (except P.IRL.IS.AL.GR.TR.BG.): introd. DK.N.S.SF.

(b) Calyx lobes with a membraneous margin

1214. *S. nodosa* L. FIGWORT. Fls. reddish-brown in lax terminal and lateral clusters forming a narrow, leafless, glandular-hairy inflorescence. Corolla 6–9 mm., tube greenish, lobes reddish-brown; sterile stamen-flap broader than long; calyx lobes oval, with very narrow membraneous margin. Lvs. hairless, oval-acute, and coarsely double-toothed; stems sharply quadrangular, not winged. An erect perenn. 40–150 cm., with rhizome with rounded swellings. △ Damp woods, hedgebanks. June–Sept. All Eur. (except P.IS.). *Used in herbal remedies; the seeds are vermifuge.*

S. auriculata L. (*S. aquatica* auct.) WATER BETONY. Like 1214, but angles of stem distinctly winged and calyx with a broad membraneous margin ½–1 mm. wide. Lvs. oblong-oval and with regular rounded teeth, lv. stalk winged. Sterile stamen-flap rounded. Western Europe. *Somewhat poisonous to livestock.*

S. umbrosa Dumort. Very like [1214], but stems more broadly winged and lvs. oval-oblong and margin with regular sharp teeth. Corolla greenish-purple; calyx lobes with a broad membraneous margin. Sterile stamen-flap with 2 spreading lobes. Widespread in Europe, except in the north.

1215. *S. scorodonia* L. BALM-LEAVED FIGWORT. Distinguished from 1214 by the whole plant being greyish and downy-haired and the lvs. wrinkled, oval-heart-shaped, and doubly sharp-toothed. Fls. in few-flowered, long-stemmed, spreading clusters forming a lax somewhat leafy inflorescence; corolla 8–11 mm.; calyx lobes rounded, with a broad membraneous margin; sterile stamen-flap rounded. An erect perenn. to 1 m. △ Shady places by the sea. May–Sept. P.E.F.GB. Pl. 123.

Lvs. once or twice cut into narrow, toothed or lobed segments

1216. *S. canina* L. Readily distinguished by its numerous tiny blackish-purple fls. which are borne in many lax lateral clusters forming a much-branched leafless and hairless pyramid inflorescence. Corolla 3–5 mm., very rarely white; stamens projecting; sterile stamen-flap linear or absent; calyx lobes with a conspicuous broad white membraneous margin. Lvs. green, once cut into narrow lance-shaped, toothed or lobed, rather widely spaced segments. A much-branched, hairless perenn. 30–80 cm. △ Dry, stony, and sandy places, river gravels. May–Aug. Med. Eur. P.D.CH.A.CS.BG.SU.

S. hoppii Koch ALPINE FIGWORT. A mountain plant differing from 1216 in having larger blackish-purple fls. 6–8 mm. long, with the lateral lobes of the corolla white and the upper lip more than half as long as the corolla tube. Calyx with a conspicuous white membraneous margin; fl. stalks mostly as long as the calyx, with long-stalked glandular hairs. Lvs. once or twice pinnately cut. Fr. 3–5 mm. Pyrenees, Alps, Apennines. Pl. 123.

S. lucida L. Like [1216], but fl. stalks much shorter than the calyx and with stalkless glandular hairs. Stamens scarcely projecting, the sterile stamen-flap large, rounded. Fr. 5–6 mm. Mediterranean region. Page 384.

S. sambucifolia L. Distinguished by its large, dark pinkish-red or pale rusty-brown fls. 12–20 mm. long; calyx lobes rounded with a broad, membraneous, undulate margin; fl. stalks glandular-hairy. Lower lvs. once cut into lance-shaped to oval, deeply cut or lobed segments, the upper lvs. entire and toothed. A robust nearly hairless perenn. to 1 m. Iberian Peninsula.

MIMULUS Corolla with a long tube, hairy within, two-lipped, the upper lip two-lobed, the lower lip three-lobed; calyx tubular, five-angled and five-lobed. Stamens 4. Fr. included in the calyx. Lvs. opposite. *c.* 3 sps.

1217. *M. guttatus* DC. MONKEY-FLOWER. A leafy erect perenn. 20–60 cm., with large yellow fls. with red spots in the throat, forming a lax, leafy, terminal cluster. Corolla tubular 2½–4½ cm., two-lipped, the lower lip much longer than the upper and with a prominent red-spotted throat-boss; calyx tubular, shortly five-lobed, glandular-hairy, enlarging in fr. Lvs. opposite, 1–7 cm., oval and irregularly toothed, the lower stalked, the upper stalkless, all often hairless. △ Watersides. June–Sept. Native of North America; widely naturalized throughout much of Europe. Pl. 123.

M. moschatus Lindley MUSK. Like 1217, but with smaller yellow fls. 1–2 cm. without red spots. Lvs. all shortly stalked. A glandular-sticky perenn., usually with a musky scent. Native of North America; grown as an ornamental and for its scent and sometimes naturalized in Central Europe. Pl. 123.

GRATIOLA Corolla with a long tube, hairless within, indistinctly two-lipped, with 5 nearly equal lobes; calyx five-lobed. Stamens 4, the 2 upper fertile, the 2 lower sterile. *c.* 2 sps.

1218. *G. officinalis* L. GRATIOLE. A hairless perenn. 20–60 cm., with numerous lance-shaped lvs. and solitary, tubular, pinkish-white fls. borne on long stalks in the axils of the upper lvs. Corolla 1½–2 cm., tube yellowish, lobes spreading and flushed with pink or purple; stamens included; calyx lobes linear; bracteoles 2. Lvs. opposite, stalkless, toothed towards the apex, three- to five-veined; creeping rhizome present. △ Wet meadows, marshes, watersides. May–Oct. Most of Eur. (except North Eur. IRL.GB.). *A bitter plant used in herbal remedies; the underground parts are emetic.* Pl. 124.

*Limosella aquatica L. MUDWORT. Throughout Europe, except in the south-east.

VERONICA | Speedwell Fls. usually blue or rarely white or pinkish; corolla with a very short tube and 4 usually widely spreading, rounded but often somewhat unequal petal-like lobes, the upper lobe larger and the lower smallest. Stamens 2. Fr. with 2 valves. *c.* 70 sps.

Fls. in a terminal spike or cluster
(a) Fls. numerous in dense, elongated, terminal spikes
1219. *V. spicata* L. SPIKED SPEEDWELL. Fls. bright blue, very numerous in a dense, long, cylindrical, terminal spike to 10 cm. or more. Fls. *c.* ½ cm. across; corolla tube *c.* 4 mm., lobes elliptic; stamens long-projecting; fl. stalks shorter than the calyx and bracts; calyx lobes oblong, hairy. Upper lvs. linear-elliptic, narrowed to the lv. stalk, finely

rounded-toothed, the lower lvs. broader. Fr. glandular-hairy, more or less rounded, equalling the calyx. An erect, finely hairy, but variable perenn. 10–60 cm. △ Dry meadows and grasslands. July–Oct. Most of Eur. (except P.IRL.B.IS.).

V. spuria L. Like 1219, but a taller and more robust perenn. to 80 cm., with lvs. with saw-toothed margins. Fl. spikes often laxer and branched below, often several; fls. blue; fl. stalks as long as or longer than the calyx and the bracts. Fr. hairless. Central and Eastern Europe.

(b) Fls. rather few, in lax terminal clusters
1220. *V. serpyllifolia* L. THYME-LEAVED SPEEDWELL. Fls. tiny, 5–6 mm. across, whitish-lilac with darker veins, often in lax, terminal, slender spikes of up to 30 fls. Bracts narrow-oblong, longer than the fl. stalks and grading gradually into the lvs. below. Lvs. 1–2 cm. oval, with a few shallow teeth, light green. A small, creeping, rooting perenn. with erect leafy fl. stems 10–30 cm. △ Grassland, heaths, waste places. May–Oct. All Eur.

V. alpina L. ALPINE SPEEDWELL. An arctic or alpine creeping perenn. with lilac-blue fls. 5–7 mm. across, in a rather dense, terminal, leafy cluster of 4–12 fls. Bracts narrow-lance-shaped. Lvs. *c.* 1 cm., oval, short-stalked; fl. stems 5–15 cm. Widespread in Europe.

V. fruticans Jacq. (*V. saxatilis* Scop.) ROCK SPEEDWELL. A woody-based spreading perenn. with thick, shining, hairless, oblong lvs. and brilliant blue fls. with a red-purple throat. Fls. few in a lax terminal cluster; corolla *c.* 1 cm. across; fl. stalks with crisped hairs, bracts lance-shaped, differing from lvs. North Europe and mountains of Southern Europe. Pl. 124.

*V. arvensis L. WALL SPEEDWELL. Throughout Europe.

1221. *V. triphyllos* L. FINGERED SPEEDWELL. Distinguished by its palmately lobed upper lvs. which are deeply divided into 3–7 oblong-oval lobes. Fls. deep blue, 6–8 mm. across, in lax, leafy, terminal clusters; corolla lobes shorter than blunt-tipped calyx lobes; uppermost bracts entire, gradually grading to the lobed lvs. below. Seeds cup-shaped. A delicate branched, very glandular-hairy, spreading or ascending ann. 5–20 cm. △ Stony places. Feb.–May. Much of Eur. (except IRL.IS.N.SF.AL.TR.). Page 384.

*V. verna L. SPRING SPEEDWELL. Widespread in Europe, except in the extreme west.

Fls. solitary or in lateral spikes borne in axils of leafy bracts
(a) Fls. solitary, axillary
1222. *V. cymbalaria* Bodard. PALE SPEEDWELL. A trailing, weak-stemmed, hairy ann. spreading 10–60 cm., with tiny long-stalked white fls. in the axils of most of the lvs. Corolla 3–4 mm. across, lobes little longer than the calyx; calyx lobes obovate with spreading hairs; fl. stalk longer than the lvs. Lvs. stalked, rounded or kidney-shaped and with 5–9 nearly equal, shallow, rounded lobes. Fr. two-lobed, hairy. △ Cultivated ground, fields. Feb.–Oct. Med. Eur. (except AL.) P.BG.R.SU.

*V. hederifolia L. IVY SPEEDWELL. Throughout Europe.

1223. *V. persica* Poiret BUXBAUM'S SPEEDWELL. Fls. bright blue with darker blue at the base of the corolla, the lower lobe often paler or whitish. Corolla 8–12 mm. across; calyx 5–6 mm., lobes oval, ciliate, enlarging and strongly diverging in fr.; fl. stalks longer than the lvs., down-curved in fr. Lvs. triangular-oval, regularly coarsely toothed, short-stalked. Fr. 6–7 mm. broad, with 2 sharply keeled and diverging lobes, ciliate. A hairy, branched, spreading ann. 10–40 cm. △ Cultivated places, fields. Mar.–Oct. Native of Western Asia; now widely naturalized throughout Eur. Pl. 124.

*V. polita Fries GREY SPEEDWELL. Throughout Europe.

1. *Odontites lutea* 1240
2. *Veronica latifolia* 1229
3. *Lycium europaeum* 1173
4. *Melampyrum cristatum* 1257
5. *Scrophularia lucida* [1216]
6. *Veronica triphyllos* 1221

*V. agrestis L. FIELD SPEEDWELL. Throughout Europe.

1224. V. filiformis Sm. CREEPING SPEEDWELL. A hairy creeping perenn. with numerous, very slender rooting stems, often forming large patches, and solitary axillary, very slender-stalked, bright blue fls. with a narrower white lower lobe. Corolla c. 8 mm. across; calyx lobes oblong blunt, ciliate; fl. stalks many times longer than the lvs. Lvs. short-stalked, blade c. ½ cm. across, kidney-shaped and with rounded teeth. Fr. rounded. △ Cultivated ground, lawns. Apr.–May. Native of Asia Minor; becoming widely naturalized in Europe. Pl. 124.

(b) Fls. numerous in axillary spikes or clusters
(i) Hairless marsh plants
1225. V. beccabunga L. BROOKLIME. A hairless, rather fleshy perenn. 20–60 cm., with hollow stems rooting below, broad glossy lvs. and long-stemmed clusters of bright blue fls. in the axils of the upper lvs. Clusters rather lax, with 10–30 fls.; corolla 7–8 mm. across; calyx elliptic, hairless; individual fl. stalks as long as the linear bracts. Lvs. 3–6 cm. oval or oblong, shortly stalked, shallowly rounded-toothed. Fr. hairless. △ Springs, streams, marshes, wet meadows. May–Sept. All Eur. (except IS.). Pl. 124.

V. anagallis-aquatica L. WATER SPEEDWELL. Like 1225, but lvs. lance-shaped, stalkless, and clasping, and fls. pale blue, in lateral spikes of 10–50 fls. Corolla 5–6 mm. across; bracts linear, shorter than or equalling the fl. stalks; inflorescence sometimes glandular-hairy. Throughout Europe.

1226. V. scutellata L. MARSH SPEEDWELL. A slender, weak-stemmed, hairless perenn. to 50 cm. with linear-lance-shaped lvs. and white or pale blue fls. with purple lines, borne in lax axillary clusters. Clusters few-flowered; corolla 6–7 mm. across; calyx lobes oval; fl. stalks much longer than the calyx and the linear bracts. Lvs. opposite, stalkless, half-clasping stem and obscurely toothed, yellowish-green and often flushed with purple. Fr. flattened, broader than long. △ Marshes, bogs, watersides. May–Sept. All Eur.

(ii) Hairy terrestrial plants
1227. V. officinalis L. COMMON SPEEDWELL. A creeping, rooting, hairy perenn. often forming mats, with small pale blue fls. in rather dense, slender, erect spikes arising from the axils of some of the lvs. Corolla c. 6 mm. across; calyx lobes lance-shaped; bracts linear, about twice as long as the fl. stalks. Lvs. 2–3 cm. oval-oblong, narrow to the base, nearly stalkless, margin toothed, hairy on both sides. Fr. longer than the calyx. △ Grasslands, heaths, woods. May–July. All Eur. A bitter and astringent plant used in herbal remedies.

*V. montana L. WOOD SPEEDWELL. Widespread in Europe, except in the north.

1228. V. chamaedrys L. GERMANDER SPEEDWELL. Readily distinguished by the 2 lines of white hairs occurring on opposite sides of the aerial stem. Fls. 10–20 in long-stemmed, lax axillary clusters. Corolla c. 1 cm. across, bright blue with a white centre; calyx lobes lance-shaped, hairy; fl. stalks hairy, about equalling the lance-shaped bracts. Lvs. 1–2½ cm. oval-heart-shaped, strongly toothed and veined, stalkless or nearly so. Fr. broader than long, shorter than the calyx. A hairy perenn. rooting at base; fl. stems 20–40 cm. △ Meadows, hedges, woods. Apr.–June. All Eur.

1229. V. latifolia L. (V. teucrium L.) LARGE SPEEDWELL. Readily distinguished by the five-lobed calyx with very unequal, linear ciliate lobes. Fls. relatively large, c. 1 cm. across, bright blue, in slender, many-flowered, stalked, axillary spikes; calyx half as long as the corolla. Lvs. oblong-lance-shaped, and very deeply toothed, stalkless. Fr. nearly hairless, longer than broad, longer than the calyx. A variable, greyish-haired, somewhat woody-based, several-stemmed perenn. 10–40 cm. △ Dry places, meadows, open woods. June–Aug. Much of Eur. (except North Eur. IRL.GB.B.NL.AL.TR.). Page 384.

DIGITALIS | **Foxglove**   Corolla with a long bell-shaped tube, two-lipped, the upper lip shorter than the lower; calyx deeply five-lobed; stamens 4, included. Tall perenns. or bienns. with long unbranched spikes, usually of nodding fls. *c.* 18 sps.

Middle lobe of lower lip almost as long as corolla tube

1230. *D. ferruginea* L. RUSTY FOXGLOVE. An erect, almost hairless perenn. to 1 m., with a long slender spike of almost globular reddish-yellow fls. veined with brown. Corolla *c.* 1½ cm., the middle lobe of the lower lip almost as long as the tube; calyx hairless, lobes oval, with papery margin; fl. stalk very short; bracts longer than the calyx. Lvs. to *c.* 18 cm. lance-shaped, hairless or veins and margin hairy beneath. △ Woods, bushy places. July–Sept. South-East Eur. I.A.H.SU. Pl. 125.

1231. *D. laevigata* Waldst. & Kit. Like 1230, but corolla broadly bell-shaped, not globular, pale yellow with dark purple veins within, more or less rusty brown outside. Corolla *c.* 2 cm.; calyx lobes lance-shaped acute, with a very narrow papery margin; bracts about as long. Lvs. hairless, oblong-lance-shaped. An erect hairless perenn. to 80 cm. △ Woods, bushy places. July. I.YU.AL.GR.BG.

Middle lobe of lower lip much shorter than corolla tube

(a) Fls. yellow or rusty brown

1232. *D. grandiflora* Miller (*D. ambigua* Murray) LARGE YELLOW FOXGLOVE. Fls. large, pale yellow with a network of brown veins within, hairy outside and borne in a long slender spike. Corolla 3–4 cm. by 1½–2 cm., throat open; calyx lobes linear-lance-shaped, with reflexed tips, glandular-hairy. Lvs. lance-shaped, finely toothed, shining hairless above, hairy on veins beneath. An erect perenn. with glandular-hairy stem 40–100 cm. △ In mountains: rocks, open woods, clearings. June–Sept. Much of Eur. (except North Eur. P.IRL.GB.NL.TR.). *A very poisonous plant.* Pl. 125.

1233. *D. lutea* L. SMALL YELLOW FOXGLOVE. Like 1232, but fls. smaller, 1½–2 cm., yellow, but without purple veins, hairless outside; calyx not glandular. Inflorescence long, slender. Lvs. narrow lance-shaped, shining, quite hairless. An erect hairless perenn. ½–1 m. △ Woods, stony hills. June–Aug. Cent. Eur. E.F.B.I.SU. *A poisonous plant, possessing similar properties to Foxglove.* Pl. 125.

D. obscura L. SPANISH RUSTY FOXGLOVE. A shrubby hairless perenn. 30–60 cm., with leathery, linear-lance-shaped, entire lvs. and a lax one-sided spike of dark rusty-brown fls., which are dull yellow and dark-spotted within. Corolla with white marginal hairs and lower lip about half as long as the tube; bracts lance-shaped, longer than the fl. stalks. Spain. Pl. 125.

(b) Fls. pinkish-purple

1234. *D. purpurea* L. FOXGLOVE. Fls. large, 4–5 cm., tubular, bright pinkish-purple with purple spots on a white ground on the lower side of the corolla tube within, or rarely white overall. Inflorescence a long slender unbranched spike; corolla swollen at base, hairy within; calyx lobes oval fine-pointed, softly hairy. Lvs. 15–30 cm. oval to lance-shaped, and narrowed to a winged lv. stalk, softly hairy, green above and densely grey-haired beneath, margin toothed. A tall, erect bienn. ½–1½ m. △ Open woods, heaths, in mountains; often grown as an ornamental and naturalized. May–Sept. West, Cent. Eur. N.S.I.SU.: introd. IS.DK.SF.R. *The important drug, digitalin, is obtained from the leaves; it slows down the rate of the heart-beat and affects the circulation. The whole plant is poisonous.* Pl. 125.

ERINUS   Corolla with a narrow slender tube and 5 more or less equal, spreading, notched, petal-like lobes; calyx deeply five-lobed. Stamens 4; fr. ovoid. Low tufted perenns. 1 sp.

1235. *E. alpinus* L. ALPINE ERINUS. A low, tufted, hairy perenn. with many erect stems 5–15 cm. bearing a short, crowded, terminal cluster of pink to purple fls. with conspicuous, notched, spreading, petal-like lobes. Corolla tube slender, hairy outside, about as long as the calyx, corolla 5–10 mm. across; calyx lobes linear, hairy. Lvs. *c.* 1½ cm. obovate or spathulate and narrowed to a short stalk, toothed, the lowermost lvs. forming a lax rosette. △ Mountain rocks and screes, walls in lowlands. May–Oct. E.F.D.CH.A.I. Pl. 127.

BARTSIA Calyx tubular or bell-shaped, regularly four-lobed. Corolla with a long curved tube, two-lipped, the upper lip forming a hood, the lower lip three-lobed. Fr. two-valved; seeds few, with strong longitudinal ribs or wings. Semi-parasitic perenns. *c.* 3 sps.

1236. *B. alpina* L. ALPINE BARTSIA. A creeping stoloniferous perenn. with erect stems 5–30 cm. bearing a dense terminal cluster of dark purple-violet fls. from among dark, violet-flushed, glandular-hairy, leafy bracts. Corolla 8–22 mm., tip of lobes often white; bracts oval, toothed. Lvs. 1–2 cm., oval and toothed, wrinkled, opposite, stalkless. Fr. nearly twice as long as the calyx. △ Arctic or alpine: meadows, rocks. June–Aug. Most of Eur. (except P.IRL.B.NL.DK.H.AL.GR.TR.). Pl. 126.

BELLARDIA Like *Bartsia*, but calyx swollen-bell-shaped, deeply divided into 2 toothed lobes. Fr. swollen; seeds minute, longitudinally ribbed. 1 sp.

1237. *B. trixago* (L.) All. (*Bartsia t.*) Fls. rather large, *c.* 2 cm. across, pinkish mixed with yellow or yellow throughout, in a short, dense, and four-sided leafy spike. Corolla with a short, upper, hooded lip, and a much broader, deeply three-lobed lower lip with 2 throat-bosses; calyx inflated, glandular-hairy. Lvs. glandular-hairy, linear-lance-shaped and with widely spaced blunt teeth. An erect, leafy, unbranched ann. 10–80 cm. △ Damp sandy fields, pine-woods. Apr.–July. Med. Eur. P.BG. Pl. 126.

PARENTUCELLIA Like *Bartsia*, but anns. and seeds many, minute, smooth. *c.* 2 sps.

1238. *P. viscosa* (L.) Caruel (*Bartsia v.*) YELLOW BARTSIA. An erect, unbranched, pale green, glandular-sticky ann. 10–50 cm., with bright yellow fls. in a short, rather dense, leafy spike which later elongates. Corolla 1½–2 cm. long, with a shorter upper lip and a longer, broader, deeply three-lobed lower lip; calyx divided to the middle into lance-shaped glandular-sticky lobes. Lvs. 1½–3 cm. oblong-lance-shaped, regularly toothed, stalkless. △ Damp sandy places, fields. May–Sept. West. Eur. (except B.NL.), Med. Eur. (except AL.). Pl. 126.

1239. *P. latifolia* (L.) Caruel (*Bartsia l.*) SOUTHERN RED BARTSIA. Fls. small *c.* 1 cm. long, reddish-purple with a white tube or rarely all white or yellow, in a short dense, leafy spike which later elongates and becomes interrupted below. Calyx divided to one-third into oblong-lance-shaped lobes; bracts pinnately lobed, nearly as broad as long, often reddish. Lvs. *c.* 1 cm. oval, deeply rounded-toothed. An erect, little-branched, often reddish, glandular-sticky ann. 5–20 cm. △ Grassy, sandy places by the sea. Mar.–June. Med. Eur. P.BG. Pl. 126.

ODONTITES Calyx tubular or bell-shaped with 4 equal lobes; corolla tube short, two-lipped, the upper lip hooded, entire or notched, the lower lip three-lobed. Fr. laterally compressed, two-valved; seeds small, few, oblong, furrowed. Semi-parasitic anns. *c.* 20 sps.

Fls. yellow
1240. *O. lutea* (L.) Clairv. An erect, much-branched, nearly hairless ann. 10–50 cm., with

bright yellow fls. in dense, slender, elongated, one-sided spikes with narrow bracts about as long as the fls. Corolla 6–7 mm., lips wide-spreading, finely hairy and margin ciliate; calyx not glandular; anthers and style long-projecting. Lvs. linear-lance-shaped, margin inrolled, shallowly toothed. △ Dry meadows, uncultivated hilly ground, cornfields, in mountains. July–Sept. Cent., South-East Eur. (except GR.TR.) E.F.I.SU. Page 384.

O. viscosa Reichenb. Like 1240, but fls. pale yellow; corolla hairless, lips spreading little with anthers and style included; calyx glandular-hairy. A sticky aromatic ann. Portugal to Italy, Switzerland.

Fls. pinkish-purple

1241. *O. verna* (Bellardi) Dumort. (*Bartsia odontites* (L.) Hudson) RED BARTSIA. A leafy, erect, branched, purple-flushed ann. 10–50 cm., with purplish-pink fls. in long, one-sided, leafy spikes. Corolla *c.* 8 mm. with spreading lips, the lower shorter than the upper, densely covered with adpressed hairs outside. Lvs. lance-shaped, strongly toothed, stalkless. △ Cultivated ground, cornfields, waste places. May–Oct. All Eur. (except IS.GR.). Pl. 126.

EUPHRASIA | Eyebright Corolla tube straight, two-lipped with an open throat, the upper lip two-lobed, slightly concave with a narrow reflexed margin, the lower lip with 3 deep spreading and notched lobes; calyx bell-shaped, four-lobed; stamens 4. Semi-parasitic anns. There are numerous microspecies and hybridization is common. *c.* 50 sps.

Stalked glands present at least on bracts

1242. *E. rostkoviana* Hayne Fls. rather large, 6–8 mm. across, white overall, with the upper lip often violet and the lower lip spotted with yellow and veined with violet. Corolla tube elongating to 1–1½ cm. during flowering. Lvs. 6–10 mm. oval-acute, densely glandular-hairy, the lower with rounded teeth, the upper with 3–6 acute teeth on each side. Stems branched below the middle, with short internodes. △ Meadows, pastures, open woods. July–Oct. Widespread in Eur. (except North Eur.). Pl. 127.

1243. *E. brevipila* Burnat & Gremli Fls. rather large, 6–10 mm. long by 7–8 mm. across, the lower lip lilac, or white with a lilac upper lip. Bracts oval with acute or bristle-tipped teeth. Lvs. 6–18 mm., usually with short glands and short bristles, the lower oblong with rounded teeth, the upper oval with nearly acute teeth; stems robust, red-tinted, with few or many branches arising from below or about the middle. △ Poor pastures. July–Sept. Widespread in Eur.

Stalked glands absent

(a) Lvs. more than twice as long as broad

1244. *E. salisburgensis* Funck Fls. 5–8 mm. long, white veined with violet, rarely bluish or reddish. Lvs. to 7 mm. oblong-lance-shaped, with 2–4 bristle-pointed teeth on each side, hairless or with minute bristles on margin; stems slender with many spreading branches. Fr. hairless or rarely with a few weak marginal bristles. △ Rocks, poor pastures. June–Aug. Mountains of Eur.

(b) Lvs. not more than twice as long as broad

E. minima Jacq. Fls. tiny 5–6 mm. long, often with the upper lip blue or purple and the lower lip yellow, but very variable in colour and sometimes all yellow or reddish-purple. Stem usually simple, hairy; lvs. and bracts oval ½–1 cm. long, with 2–4 blunt or acute teeth on each side, and with stiff hairs on the margin and on the veins. Mountains of Central Europe, Pyrenees. Pl. 127.

1245. *E. nitidula* Reuter (*E. nemorosa* auct.) Fls. 5–8 mm. long, white or bluish, veined with violet, the lower lip spotted with yellow. Lvs. usually hairless, the upper with 4–7 acute teeth on each side; bracts spreading or arched, with acute or shortly bristle-pointed teeth; stems usually branched. △ Heaths, pastures, woods. June–Oct. IRL.GB.F.D.CS.

1246. *E. stricta* Wolf. (*E. ericetorum* Jordan) Fls. 7–10 mm. long, pale violet, veined with blue or purple, the lower lip spotted with yellow. Lvs. hairless, with 2–5 bristly-pointed teeth on each side; stems erect, branched from the base, with long erect branches. △ Dry pastures, heaths. June–Oct. Most of Cent. and South-East Eur. E.F.

RHINANTHUS | Yellow-Rattle Calyx somewhat laterally flattened, oval, shortly four-lobed, enlarging and becoming inflated and papery in fr. Corolla tubular, strongly two-lipped, the upper lip laterally flattened, hooded, yellow with violet teeth, the lower shorter, three-lobed; stamens 4. Fr. rounded, flattened, encircled by calyx. There are numerous microspecies and hybridization is frequent. Sometimes placed in the genus *Alectorolophus*. Semi-parasitic anns. *c.* 20 sps. *Yellow-rattles are poisonous to stock and are usually avoided by them.*

Fls. 10–18 mm. long, corolla tube straight, lips spreading and throat open
1247. *R. minor* L. YELLOW-RATTLE. Fls. yellow or brownish, *c.* 12–15 mm. long, with teeth of upper lip usually violet or rarely white, teeth very short, rounded and not longer than broad. Bracts green or reddish, triangular-oval to lance-shaped with acute teeth; bracts and calyx hairless. Lvs. rough, oblong-lance-shaped, and margin with rounded teeth; stems slender, often black-spotted. A simple or branched, almost hairless ann. 5–50 cm. △ Meadows, cornfields. May–Sept. All Eur. (except GR.TR.). *The leaves give a bright yellow dye. Used in the past for eye troubles.* Pl. 127.

Fls. 15–25 mm., corolla tube curved and lips coming together and throat closed
1248. *R. serotinus* (Shönh.) Oborny GREATER YELLOW-RATTLE. Like 1247, but with larger yellow fls. 1–1½ cm. long and with an up-curved corolla tube; teeth of upper lip violet, *c.* 2 mm. conical, twice as long as broad. Calyx hairless; bracts oval long-pointed, strongly and unequally toothed, the lowermost teeth deeper and more acute. Lvs. 2½–7 cm., linear to lance-shaped, strongly toothed. A robust branched ann. with a black-spotted stem 20–60 cm. △ Meadows, cornfields. May–Sept. All Eur. (except P.AL.).

PEDICULARIS | Lousewort Calyx tubular or bell-shaped, usually four-lobed, but sometimes with 2–5 lobes. Corolla narrow tubular, strongly two-lipped, the upper lip hooded and laterally flattened with the apex either cut-off and entire, or narrowed into a long beak or with 2 small teeth just below the apex; lower lip three-lobed; stamens 4. Fr. flattened. Anns. or perenns.; semi-parasites, particularly on grasses. Hybridization is frequent. *c.* 50 sps. *Louseworts are poisonous to livestock and are usually avoided by them.*

Fls. yellow or red-brown
(*a*) *Upper lip of corolla with a beak or with teeth*
1249. *P. comosa* L. CRESTED LOUSEWORT. Fls. pale yellow or whitish-yellow, in a large, compact, oval, leafless cluster. Corolla 2½–3 cm. long, the upper lip conspicuously arched and with 2 small acute teeth below the blunt apex; calyx becoming inflated and membraneous, with few hairs, lobes blunt, entire; middle and upper bracts almost entire, hardly as long as the calyx. Lvs. pinnately lobed, with lobes further divided into narrow acute segments. A hairy perenn. 10–50 cm. with erect stems, which are leafy to the inflorescence. △ Alpine meadows. June–Aug. E.F.CH.I.YU.AL.BG.R.SU.

1250. *P. tuberosa* L. TUBEROUS LOUSEWORT. Fls. whitish-yellow in a short, broad, leafy cluster borne on a nearly leafless stem. Upper lip of corolla with a conspicuous long straight beak; calyx hairy, broadly bell-shaped, with leafy toothed lobes; bracts of inflorescence all deeply lobed, much longer than the calyx. Lvs. pinnately lobed, mostly in a basal rosette. An erect perenn. with several hairy stems 10–25 cm. △ Alpine meadows. June–Aug. E.F.CH.A.I. Pl. 127.

(b) *Upper lip of corolla rounded and without a beak or teeth*

1251. *P. sceptum-carolinum* L. MOOR-KING. Fls. very large 3 cm. or more, yellow or buff, the lips often closed but if open the basal lip reddish, borne in an oblong interrupted spike, often on a nearly leafless stem. Calyx brownish-green, lobes toothed; bracts oval-acute, shallowly toothed, often nearly as long as the calyx. Lvs. pinnately cut into oval, toothed lobes, mostly in a basal rosette; stems stout 30–90 cm. △ Damp meadows, marshes, thickets. July–Aug. N.S.DK.D.A.PL.CS.SU.

1252. *P. foliosa* L. LEAFY LOUSEWORT. Distinguished by its dense cluster of sulphur-yellow fls. which is leafy throughout, due to the large pinnate bracts which are longer than the fls. Corolla 1½–2 cm. long, hairy outside, the upper lip quite rounded; calyx membraneous, hairy, lobes triangular, fine-pointed. A robust perenn. 20–60 cm., with a stout rootstock and with long twice-pinnate fern-like lvs. △ Pastures in the high mountains. June–Aug. E.F.D.CH.I.YU. Pl. 127.

Fls. reddish-purple or rarely white

(a) *Upper lip of corolla not beaked or toothed*

1253. *P. verticillata* L. WHORLED LOUSEWORT. Readily distinguished by the upper lvs. which are in whorls of 3 or 4 and by its stems which have 4 lines of hairs. Fls. purple or rarely white in a dense oval head, with leafy bracts at the base, head at length becoming interrupted below. Corolla *c.* 1½ cm., upper lip blunt-tipped and without a beak or teeth; calyx with lax hairs and short entire teeth. Lvs. mostly in a rosette, lance-shaped in outline, pinnately cut into oval toothed lobes. A rather small erect perenn. 5–25 cm. △ Damp alpine pastures. June–Sept. Cent., South-East Eur. (except TR.) E.F.SF.I.SU.

1254. *P. recutita* L. TRUNCATE LOUSEWORT. Like 1253, but upper lvs. alternate and stem hairless. Fls. brownish-purple in a short, compact, cylindrical head, later elongating; calyx hairless, lobes unequal acute, nearly half as long as the tube. Lvs. lance-shaped, pinnately cut into narrow, acute, toothed segments. A rather robust erect perenn. with leafy stems 20–40 cm. △ Damp alpine pastures, marshes. July–Aug. CH.A.I.R. Pl. 128.

P. rosea Wulfen PINK LOUSEWORT. Like 1255 with bright pink fls., but upper lip of corolla blunt, not beaked; calyx teeth triangular; inflorescence woolly-haired. Alps and Carpathians.

(b) *Upper lip of corolla prolonged into a beak or with 2 teeth*

1255. *P. kerneri* Dalla Torre (*P. rhaetica* Kerner) A low, hairless, spreading perenn., with lvs. in a dense rosette and a compact terminal cluster of 2–6 deep pink fls. on short ascending stems 5–12 cm. Corolla 1½–2½ cm., the upper lip strongly recurved and ending in a long slender beak; calyx finely hairy, with leafy toothed lobes which are recurved above; fl. stalks as long as the calyx. Lvs. 3–4 cm., narrow-lance-shaped pinnately cut into oval toothed lobes. △ Damp alpine pastures. June–Aug. F.CH. Pl. 128.

1256. *P. palustris* L. RED-RATTLE. A nearly hairless bienn. usually with a single, erect, branched stem 8–60 cm. and long, leafy, lax spikes of rosy-purple fls. Corolla 2–2½ cm., the upper lip strongly arched and with a short tooth on each side of the apex and another tooth halfway down; calyx usually hairy, often reddish, with 2 leafy and irregularly cut lobes. Lvs. 2–4 cm. oblong, pinnately cut, the lobes toothed. △ Marshes, wet meadows, wet heaths. May–Aug. All Eur. (except P.IS.DK.AL.GR.TR.).

P. sylvatica L. COMMON LOUSEWORT. Like 1256, but a perenn. with spreading stems and few-flowered compact inflorescences. Corolla rosy-purple, with only 2 apical teeth on the upper lip; calyx usually hairless, five-angled, with 4 small leaf-like toothed lobes and 1 linear lobe. Throughout Europe, except in South-Eastern Europe. Pl. 128.

MELAMPYRUM | Cow-Wheat   Calyx tubular, four-lobed. Corolla tubular, two-lipped, the upper lip hooded, flattened laterally, two-lobed, the lower lip three-lobed and with a prominent throat-boss nearly closing the mouth; stamens 4. Fr. flattened, with 1–4 seeds. All sps. show considerable variation; they are semi-parasitic anns. *c.* 15 sps. *Cow-wheats are poisonous to livestock and are avoided by them.*

Fls. in dense spikes
1257. *M. cristatum* L. CRESTED COW-WHEAT. Fls. in very dense four-sided spikes, the predominantly yellow fls. contrasting with the brightly coloured, down-curved, rosy-purple bracts. Corolla 12–16 mm. long, pale yellowish, variegated with purple and with a deeper yellow throat-boss; calyx tube with 2 lines of hairs, lobes unequal. Bracts and lvs. differing markedly; bracts with a broad heart-shaped base, which is deeply and finely toothed, the lowermost bracts prolonged into a strap-shaped leafy apex. Lvs. entire, linear lance-shaped, fine-pointed. A slender variable ann. 20–50 cm. with wide-spreading branches. △ Dry pastures, rocky slopes, wood margins. May–Sept. Most of Eur. (except P.IRL.B. NL.IS.TR.). Page 384.

1258. *M. arvense* L. FIELD COW-WHEAT. Distinguished from 1257 by the long cylindrical spikes of yellow and purple fls. in 2 ranks with brightly coloured, conspicuously fringed, reddish-purple, erect bracts. Corolla 2–2½ cm., the tube and lips pinkish-purple, the throat orange-yellow; calyx lobes linear, little longer than the tube. Bracts oval-lance-shaped, deeply cut into long narrow lobes forming a fringe-like margin. Lvs. linear-lance-shaped, the lower entire, the upper toothed at the base. An erect, often little-branched hairy ann. 20–60 cm. △ Fields. May–Aug. Most of Eur. (except P.IRL.IS.N.). Pl. 128.

Fls. paired in lax leafy clusters
(a) Bracts coloured
1259. *M. nemorosum* L. Distinguished by its paired yellow fls. with an orange upper lip and throat-boss and its broad, deeply toothed, violet bracts. Corolla *c.* 2 cm.; calyx often with long spreading hairs, lobes awl-shaped, longer than the tube; bracts oval-lance-shaped, toothed, stalked. Lvs. lance-shaped, often with one to several prominent teeth on each side at their base, all stalked. A hairy, erect, branched ann. 20–60 cm. △ Woods in mountains. June–Aug. Cent. Eur. E.F.DK.N.SF.I.YU.BG.R.SU. Pl. 128.

(b) Bracts green
1260. *M. pratense* L. COMMON COW-WHEAT. Fls. paired 12–18 mm., yellow or rarely lilac or white; bracts green, oval or lance-shaped, the uppermost usually with long, narrow, spreading teeth or untoothed. Corolla with the throat closed and lower lip straight; calyx hairless, one-third to one-quarter as long as the corolla, lobes recurved. Lvs. oval-lance-shaped to linear-lance-shaped, entire. A slender, very variable, widely branched ann. 15–50 cm. △ Meadows, woods, heaths. June–Aug. All Eur. (except IS.AL.GR.TR.).

M. silvaticum L. WOOD COW-WHEAT. Like 1260, but fls. smaller 6–12 mm., brownish-yellow, with the throat open and the lower lip deflexed; upper bracts entire or with only 1–3 lobes at their base on each side. Calyx half to one-third as long as the corolla, lobes spreading. Widespread in Europe.

TOZZIA Corolla funnel-shaped, slightly two-lipped, with 5 more or less equal spreading lobes; calyx bell-shaped, four- to five-lobed; stamens 4. Semi-parasites with opposite lvs. 1 sp.

1261. *T. alpina* L. A rather delicate erect perenn. 10–30 cm. with golden-yellow fls. with a purple-spotted throat, borne singly in the axils of the leafy bracts and forming a lax cluster. Corolla 6–10 mm. long, tube much longer than the hairless calyx. Lvs. oval with a few coarse teeth, stalkless, hairless; stem quadrangular, fleshy, with 2 conspicuous lines of hairs; rootstock scaly. Fr. globular, unsplitting, with 1 seed. △ In mountains: damp shady places, streamsides. June–Aug. Cent. Eur. (except H.) E.F.I.YU.BG. Pl. 129.

GLOBULARIACEAE | Globularia Family

Shrublets or perenn. herbs with alternate undivided lvs. Fls. in rounded heads partially surrounded by an involucre of bracts. Calyx tubular, five-lobed; corolla with a narrow tube, two-lipped, the upper lip short or almost absent, the lower with 3 long lobes. Stamens 4; fr. a one-seeded nut.

GLOBULARIA Fls. blue, numerous in a dense, globular, terminal head. *c.* 14 sps.

Shrublets

1262. *G. alypum* L. SHRUBBY GLOBULARIA. A low-growing, much-branched evergreen under-shrub 30–60 cm., with brittle woody twigs, stiff leathery lvs., and globular heads of blue fls. Heads 1½–2 cm.; fls. numerous, sweet-scented; involucral bracts oval, with brown-papery margin, ciliate, overlapping; calyx with long hairs. Lvs. oblong with a spiny tip and sometimes spiny teeth, narrowed to a short stalk. △ Dry bushy places, rocks. Winter and spring. Med. Eur. (except AL.TR.) P. *A poisonous plant which is violently purgative.* Pl. 129.

1263. *G. cordifolia* L. MATTED GLOBULARIA. A low, mat-forming, woody-stemmed perenn. with spreading and rooting branches bearing lvs. in rosettes, and short, erect, almost leafless stems bearing a globular head of blue fls. Fl. heads *c.* 1 cm. across; corolla lilac-blue, the upper lip deeply lobed; involucral bracts oval or lance-shaped, ciliate. Basal lvs. fleshy, hairless, broadly spathulate with a rounded or shortly pointed apex, or apex notched or three-toothed; stem lvs. absent or 1–2, lance-shaped, ciliate; fl. stems elongating to 15 cm. △ In mountains: rocks, screes. May–July. E.F.D.CH.A.CS.H.I.YU.AL.GR.SU. Pl. 129.

Perenns. with herbaceous stems

1264. *G. vulgaris* L. COMMON GLOBULARIA. A perenn. 5–30 cm. with herbaceous stems, and very dissimilar stem and rosette lvs. and terminal rounded heads of blue fls. *c.* 1 cm. across. Corolla rarely lilac or white, the upper lip short, bilobed; calyx hairy; involucral bracts hairy. Rosette lvs. 3–4 cm. oval entire, notched or three-lobed at apex, three- to five-veined, and narrowed to a long stalk at base, the stem lvs. 1–1½ cm. lance-shaped, stalkless. Fl. stems at first short, then elongating to 30 cm. △ Dry, grassy places, stony ground. Apr.–June. P.E.F.S.D.CH.A.H.I.YU.BG.R.SU. *The leaves produce a yellow dye.* Pl. 129.

G. nudicaulis L. Like 1264, but fl. stem leafless or with 1–4 tiny scales; rosette lvs. all one-veined. Corolla with the upper lip absent; involucral bracts hairless. Pyrenees and Alps.

BIGNONIACEAE | Bignonia Family

Trees or shrubs often with compound lvs. Fls. symmetrical in one plane only; calyx five-lobed; corolla tubular, two-lipped, with 5 lobes; stamens 4, rarely 2 or 5. Ovary superior, with 2 cells; fr. a two-valved capsule. A predominantly tropical family.

CATALPA   Corolla bell-shaped, obliquely five-lobed; stamens 2. Fr. a long slender capsule; seeds hairy. 1 sp.

1265. *C. bignonioides* Walter CATALPA, INDIAN BEAN. A deciduous tree to 12 m. with very large heart-shaped lvs. and broad pyramidal clusters of large, white, bell-shaped fls. with 2 yellow streaks and with purple spots within. Corolla 3–4 cm. across, lobes spreading, crisped; calyx two-lipped. Lvs. stalked, blades *c.* 25 cm. long. Fr. to *c.* 40 cm., pendulous. △ Native of North America; often planted as an ornamental. June–July. Pl. 129.

ACANTHACEAE | Acanthus Family

Usually herbaceous perenns. or shrubs, sometimes climbers with opposite lvs. Fls. symmetrical in one plane only, often subtended by conspicuous bracts; calyx deeply four- to five-lobed; corolla tubular, more or less two-lipped with 4–5 lobes; stamens 2 or 4. Ovary 1, superior, two-celled; fr. a capsule. Differing from Scrophulariaceae by obscure botanical characters such as the hardening of the stalks of the ovules, which act as a catapult, throwing out the seeds.

ACANTHUS   Fls. in long terminal spikes. Calyx deeply four-lobed, lobes unequal; corolla tube very short, the upper lip absent, the lower three-lobed; stamens 4. Fr. four-seeded. *c.* 6 sps.

Lvs. not spiny
1266. *A. mollis* L. BEAR'S BREECH. A robust perenn. with large, dark green, deeply lobed lvs. and stout unbranched fl. spikes to 1 m., with numerous large, stalkless, white fls. arising from broad spiny bracts. Corolla 3–5 cm., the upper lip absent, the lower lip with 3 conspicuous rounded lobes often flushed with purple; calyx two-lipped, the upper lip oblong-oval and arching over the corolla and coloured violet or green, the other lobes shorter and narrower; bracts oval with long spiny teeth. Lvs. 25–60 cm., deeply pinnately lobed, lobes oblong with soft, but not spiny teeth. △ Banks, rocks, cool places. May–July. P.E.F.I.YU.TR.BG. *The leaves of this and related species were used as an inspiration in the decoration of the classical Corinthian capital.* Pl. 130.

Lvs. spiny
1267. *A. spinosus* L. SPINY BEAR'S BREECH. Like 1266, but lvs. thistle-like, stiff, once or twice cut into deep narrow lobes with numerous long, white, spiny teeth, and with spines on midveins and lv. stalks also. Fls. in a dense, cylindrical spike with numerous, often coloured bracts with long spiny teeth; corolla whitish; calyx with upper lobe finely hairy, entire or toothed at apex, the lateral lobes much narrower, papery, and lowest lobe with a fine spiny-toothed apex. Lvs. mostly basal, to 60 cm. long. A perenn. 20–60 cm. △ Bushy, stony places, abandoned cultivation. July–Aug. I.YU.AL.GR. Pl. 130.

GESNERIACEAE | Gloxinia Family

A tropical or subtropical family closely allied to the Scrophulariaceae and generally distinguished from it by the one-celled ovary with parietal placentation.

RAMONDA   Calyx with 5 equal lobes; corolla with a very short tube and 5 spreading lobes; stamens 5, erect. 1 sp.

1268. *R. myconi* (L.) Schultz (*R. pyrenaica* L. C. M. Richard) A perenn. with a flat rosette of broad, oval, greyish-hairy lvs. and slender leafless stems bearing solitary or few large blue or violet fls. Corolla 2–3 cm. across, with 5 rounded, spreading, petal-like lobes, and with a tuft of orange hairs at the base; stamens with yellow anthers projecting in a cone. Rosette *c.* 6 cm. across, lvs. wrinkled, coarsely toothed, with dense, woolly, rusty-coloured hairs beneath and on margin; fl. stems 5–15 cm. △ Crevices and damp rocks in shady valleys. June–Aug. Pyrenees. Pl. 131.

OROBANCHACEAE | Broomrape Family

Parasites without chlorophyll, growing on the roots of flowering plants. Aerial stems usually erect, with numerous scales below in place of lvs. and bearing dense, usually cylindrical spikes of numerous stalkless fls. Fls. symmetrical in one plane only; calyx with 2–5 regular shallow lobes or deeply divided into 2 lobes; corolla tubular, two-lipped, the upper lip shortly two-lobed, the lower three-lobed; stamens 4. Ovary superior, one-celled; fr. a capsule; seeds numerous.

Plants with scaly creeping rhizome; calyx equally four-lobed.   *Lathraea*
Plants without creeping rhizome; calyx laterally two-lipped.   *Orobanche*

LATHRAEA   Calyx bell-shaped, usually equally four-lobed; corolla two-lipped, the upper often strongly curved, the lower smaller, three-lobed; style projecting. Perenns. with creeping rhizomes covered with broad, fleshy, overlapping scales. 3 sps.

1269. *L. squamaria* L. TOOTHWORT. Fls. white, tinged with dull purple, in a stout, one-sided, many-flowered spike. Corolla 1½–2 cm., slightly longer than the calyx; calyx tubular, four-lobed, glandular-hairy; bracts oval. Stem white or pale pink with whitish oval scales below, at first arching then erect, 8–30 cm. △ Parasitic on hazel, poplar, elm, and other trees. Mar.–May. All Eur. (except P.). Pl. 130.

L. clandestina L. PURPLE TOOTHWORT. Distinguished from 1269 by the large bright purple fls. borne in a dense cluster at soil level, on short stalks directly from an underground scaly rhizome. Fls. 4–6 cm., erect; corolla much longer than the hairless calyx, the upper lip strongly curved. Parasitic on the roots of trees. Spain to Italy. Pl. 130.

OROBANCHE | Broomrape   Fls. in a terminal cylindrical spike and each fl. subtended by 1 or 3 bracts. Calyx with 2–5 lobes or with 2 lateral lips which are entire, toothed or with 2 slender lobes, a fifth tooth sometimes present. Corolla with curved tube, two-lipped, the upper lip erect and more or less two-lobed, the lower spreading and three-lobed. The sps. are difficult to distinguish and it is often not easy to determine the host plant. Mature fls. should be used for identification and the profile of the fls. in side view is important. *c.* 90 sps.

Fls. subtended by 3 bracts; corolla violet or bluish

1270. *O. ramosa* L. (*Phelypaea r.*) BRANCHED BROOMRAPE. Fls. pale blue or sometimes yellowish-white, in a lax, many-flowered spike which usually has several lateral spikes arising from near the base. Corolla 10–17 mm., curved in upper half, the upper lip with 2 rounded lobes; calyx hairy, tubular, with 4 regular triangular-acute lobes shorter than the tube; bracts 3, oval-lance-shaped. Stem often purple, 5–40 cm. △ Parasitic on hemp, tobacco, potato, tomato, etc. June–Oct. Most of Eur. (except North Eur. IRL.AL.TR.).

1271. *O. purpurea* Jacq. PURPLE BROOMRAPE. Distinguished by its large bluish-purple fls. flushed with yellow at the base, 18–30 mm. long and borne in a cylindrical, rounded-topped, often lax, glandular-hairy spike. Upper lip of corolla with 2 acute lobes, the lower lip with more or less acute lobes; calyx lobes usually shorter than the tube; anthers hairless; style glandular-hairy, stigma white or blue. Fl. stem stout, unbranched, bluish, with few scales, 15–60 cm. △ Parasitic on *Achillea, Artemisia, Cirsium.* June–July. Most of Eur. (except P.IRL.IS.N.SF.AL.TR.).

O. arenaria Borkh. SAND BROOMRAPE. Like 1271, but parasitic on *Artemisia* sps., and with a dense, acute-topped, robust spike. Corolla bluish-purple, 2½–3½ cm. long, the lobes of the upper lip long-pointed, the lobes of lower lip rounded; calyx lobes as long as the tube. Filaments with few hairs at base and hairless above; anthers with long hairs. Central and Southern Europe.

Fls. subtended by 1 bract; corolla yellow, whitish or reddish
(a) *Stigma lobes yellow, white, or orange*

1272. *O. rapum-genistae* Thuill. GREATER BROOMRAPE. A stout yellowish perenn. 20–80 cm., parasitic on shrubby members of the Leguminosae, particularly gorse and brooms. Fls. 2–2½ cm. long, yellowish, tinged with purple, glandular-hairy, the tube regularly curved above, the upper lip almost entire, the lower lip three-lobed with the mid-lobe larger, lobes wavy-margined. Stamens attached to the base of the corolla, filaments hairless at base, glandular-hairy above; stigma lobes pale yellow. △ May–July. P.E.F.D.CH.A.H.I.BG. SU. Pl. 131.

1273. *O. elatior* Sutton TALL BROOMRAPE. A stout, yellowish or reddish, glandular-hairy perenn. 15–70 cm., parasitic on *Centaurea* and *Echinops.* Fls. 1½–2 cm., lengthening to 2½ cm., dusky rose-coloured, but at length yellow; corolla regularly curved from base to apex. Stamens attached 4–6 mm. above the base of the corolla, filaments hairy below, glandular-hairy above. Stigma lobes yellow. △ June–July. E.F.GB.D.CH.A.CS.H.YU.GR.BG.

O. hederae Duby IVY BROOMRAPE. Parasitic on ivy. Fls. 1–2 cm., cream-coloured and strongly veined with purple; corolla with a straight back. Stamens with filaments hairy below, hairless above, attached 3–4 mm. above the base of the corolla. Stigma lobes yellow, becoming dark. Widespread in Europe, except the north.

(b) *Stigma lobes usually purple, red or brown*

1274. *O. crenata* Forskål A rather robust pale yellowish- or bluish-stemmed perenn. 20–70 cm., parasitic on members of the Leguminosae and possibly Umbelliferae in the Mediterranean region. Fls. often large 1½–3 cm., white and veined with blue-violet, and with a scent of carnations, in a showy, cylindrical, and densely clustered spike. Corolla rather open bell-shaped, with large rounded lobes; stamens attached 2–3 mm. from the base of the corolla, filaments densely hairy below and sparsely glandular-hairy above; stigma lobes violet, rarely white. △ Apr.–June. Med. Eur. (except AL.TR.) P. Pl. 131.

O. amethystea Thuill. A reddish-violet-stemmed perenn. to 50 cm., parasitic on *Eryngium.* Fls. 15–23 mm., whitish flushed with purple and purple-veined; corolla tube abruptly bent below, spreading outwards above, nearly hairless; calyx lobes equalling or longer than the corolla tube. Stamens finely hairy at their base only; stigmas reddish-violet. Western and West-Central Europe.

1275. *O. caryophyllacea* Sm. CLOVE-SCENTED BROOMRAPE. Parasitic on the Rubiaceae. Stems 15–40 cm., yellowish tinged purple, with numerous yellow or purple-brown scales and large yellowish fls. 2–3½ cm., flushed with reddish-brown or purple. Corolla regularly arched in profile, densely glandular-hairy, the lower lip with 3 more or less equal lobes, all lobes crimped and toothed. Stamens attached 1–2 mm. above the base of the corolla, filaments hairy below. Stigma lobes purple, very rarely yellow. △ June–July. Most of Eur. (except North Eur. P.IRL.TR.). Page 399. Pl. 131.

1276. *O. alba* Willd. THYME BROOMRAPE. Stems and scales reddish, 8–25 cm., but very variable, parasitic on Thyme and other members of the Labiatae. Fls. 1–3 cm., rather few in a lax spike, yellowish veined with purple, purplish-red, or white; corolla tube curved and with violet glandular hairs, the upper lip notched. Stamens attached 1–2 mm. above the base of the corolla. Stigma lobes purple or violet, or rarely orange. △ Apr.–Aug. Cent., South-East Eur. (except TR.) GB.B.S.I.SU.

1277. *O. minor* Sm. LESSER BROOMRAPE. Stems yellowish tinged red and scales brownish, 10–50 cm., parasitic on members of the Leguminosae and Compositae. Fls. small, 10–18 mm., yellowish, flushed and veined with purple; corolla with upper lip bilobed or notched. Stamens attached 2–3 mm. from the base of the corolla, filaments almost hairless throughout; stigma lobes purple, rarely yellow. △ June–Sept. All Eur. (except North Eur.). Pl. 131.

1278. *O. picridis* Koch PICRIS BROOMRAPE. Stems yellowish often tinged purple and with a few brownish scales, 10–60 cm., parasitic on *Picris* and *Crepis* sps. Fls. 1½–2 cm., whitish-yellow with upper lip often with violet veins; corolla swollen at insertion of the stamens, narrowing at the throat. Filaments very hairy below the middle, more or less hairless above; stigma lobes purple. △ June–July. Much of Eur. (except North Eur. IRL.AL.TR.).

LENTIBULARIACEAE | Butterwort Family

Insectivorous plants which trap small invertebrate animals by various means, either aquatic, terrestrial, or rarely epiphytic herbs. Fls. symmetrical in one plane only; calyx two- to five-lobed, usually two-lipped; corolla shortly tubular and prolonged into a spur or sac at the base, two-lipped, the upper lip two-lobed; the lower lip three-lobed, throat often closed by a throat-boss. Stamens 2. Carpels 2, forming a one-celled, many-seeded ovary; fr. a capsule.

Land plants with a basal rosette of entire, sticky, glandular lvs.   *Pinguicula*
Submerged aquatic plants with compound lvs., some bearing small bladders. *Utricularia*

PINGUICULA | Butterwort   Terrestrial plants with a basal rosette of lvs. clothed with sticky glandular hairs, which attract and trap insects. Fls. solitary on a leafless stalk arising directly from the rootstock. 8 sps.

Fls. excluding spur less than 1 cm. long
1279. *P. alpina* L. ALPINE BUTTERWORT. Distinguished by its whitish fls. with 1–2 yellow spots in the throat, or fls. rarely reddish- or violet-flushed. Corolla 8–10 mm. long, the

spur 2–4 mm., shortly conical, thick blunt and abruptly turned downwards. Lvs. 2–3 cm., elliptic to lance-shaped. Fl. stalks 5–12 cm. △ In mountains: springs, damp meadows, damp rocks. June–Aug. Most of Eur. (except P.IRL.B.NL.DK.AL.GR.TR.BG.). Pl. 132.

P. villosa L. DOWNY BUTTERWORT. Fls. *c.* 7 mm. long, pale violet with 2 yellow spots on the lower lip, spur straight, about as long as wide. Lvs. brownish. Scandinavia.

**P. lusitanica* L. PALE BUTTERWORT. Western Europe.

Fls. excluding spur 1–2 cm. long

1280. *P. vulgaris* L. COMMON BUTTERWORT. Fls. bright violet or lilac, 1–1½ cm.; corolla with a broad white patch in the throat, the lower lip with lobes longer than broad, wide-spreading and not overlapping; spur 4–7 mm. directed horizontally backwards or some-what downwards, slender, acute-tipped. Lvs. 2–8 cm. oblong-oval, bright yellowish-green. Fl. stalks 5–10 cm. A perenn., over-wintering as a rootless bud. △ Bogs, wet heaths, damp rocks. May–July. All Eur. (except AL.GR.TR.). *The Laplanders use the plant to curdle milk; the fresh plant causes vomiting and is purgative.* Pl. 132.

P. leptoceras Reichenb. Like 1280, but lower lip with lobes as long as or a little longer than broad; corolla 1½–2 cm., excluding spur, violet, hairy within and with a broad white patch; spur short, straight, somewhat downcurved, about a third as long as the corolla. Pyrenees, Alps, Balkans.

P. grandiflora Lam. LARGE-FLOWERED BUTTERWORT. Like 1280, but fls. larger, 1½–2 cm. long by 2½–3 cm. broad, violet-purple with a white patch in the throat; lower lip with shallow lobes broader than long, overlapping or touching; spur rather stout, straight, 1 cm. or more, and more than half as long as the corolla. Lvs. 2–8 cm. Spain, Pyrenees, Alps, Ireland. Pl. 132.

UTRICULARIA | **Bladderwort**   Submerged aquatic plants with lvs. divided into linear segments and bearing small rounded bladders which trap tiny aquatic animals by a trap-door mechanism. Calyx two-lipped; corolla strongly two-lipped, the upper lip entire, the lower larger with a short spur or sac at the base and a conspicuous rounded throat-boss more or less closing the throat. Fl. clusters borne above water. *c.* 6 sps.

Fls. 12–18 mm.; submerged stems all green, similar

1281. *U. vulgaris* L. GREATER BLADDERWORT. Fls. bright yellow in a lax cluster of 3–12 with calyx and fl. stalks reddish-brown. Corolla with lower lip with reflexed margin and throat-boss almost as long as the rounded upper lip; spur conical. Inflorescence curved, borne 10–20 cm. above water; fl. stalks ½–1½ cm. Submerged stems free-floating; lvs. all green and with numerous bladders. △ Still waters, marshes, bogs. June–Aug. All Eur. (except IS.TR.). Pl. 132.

U. neglecta Lehm. Like 1281, but lower lip of corolla with more or less flat spreading margin and throat-boss much shorter than the upper lip. Fl. stalks 1–2½ cm. Widespread in Europe.

Fls. 6–14 mm.; submerged stems of two kinds, both green and colourless

1282. *U. intermedia* Hayne INTERMEDIATE BLADDERWORT. Fls. bright yellow with reddish-brown lines, in a lax cluster of 2–4 on stems 9–16 cm. above water. Corolla 8–15 mm., the upper lip twice as long as the rounded throat-boss; spur conical 5–7 mm. Stems of two kinds: floating and bearing green lvs. with few or no bladders, or deeper submerged, often half-buried, bearing reduced colourless lvs. with many bladders each *c.* 3 mm. Lv. seg-ments toothed, with 1–2 small bristles on the teeth. △ Still waters, marshes, and bogs. June–Aug. North Eur. (except IS.), Cent. Eur. F.IRL.GB.B.NL.I.YU.R.SU.

U. minor L. LESSER BLADDERWORT. Distinguished by its much smaller pale yellow fls. 6–8 mm., and very short, blunt, sac-like spur. Both green and colourless stems present and both bearing bladders each *c.* 2 mm.; lv. segments not toothed and without bristles. Widespread in Europe.

PLANTAGINACEAE | Plantain Family

Anns. or perenns. with lvs. usually all basal and spirally arranged in a rosette. Fls. small and inconspicuous, in dense terminal heads or spikes usually borne on long, unbranched, leafless stems. Fls. radially symmetrical, subtended by scaly bracts. Calyx fused at the base and with 4 green lobes; corolla tubular four-lobed, papery; stamens 4, much longer than corolla; ovary superior, usually one- to four-celled. Fr. a capsule opening by an apical cap, or one-seeded and not opening.

Terrestrial plants; fls. bisexual, numerous in a head or spike.   *Plantago*
Aquatic plants; fls. one-sexed, male fls. solitary, female fls. few in a head.   *Littorella*

PLANTAGO | Plantain Terrestrial plants. Fls. numerous in a head or spike, mostly bisexual; fr. opening by a cap. *c.* 50 sps.

Stems leafy and often branched
(a) Anns.
1283. *P. indica* L. (*P. arenaria* Waldst. & Kit.) BRANCHED PLANTAIN. Readily distinguished by the branched, glandular-hairy, leafy stems bearing a number of ovoid heads *c.* 1 cm. long. Fls. brownish-white, *c.* 4 mm. across; bracts with crisped hairs, the lower bracts with a green, leafy apex, the upper rounded with papery margin. Lvs. to 10 cm. narrow-linear, entire or obscurely toothed, not in a rosette. A branched ann. 10–30 cm. △ Poor fields, sandy places, waysides. Apr.–July. Most of Eur. (except North Eur.) IRL.B.NL.: introd. GB.CH. Pl. 133.

(b) Shrublets
1284. *P. sempervirens* Crantz (*P. suffruticosa* Lam.; *P. cynops* auct.) SHRUBBY PLANTAIN. A much-branched, woody-based, tufted shrublet 5–40 cm. with clusters of linear awl-shaped entire lvs. ½–1 mm. wide. Fl. heads oblong-ovoid ½–1 cm.; lowermost bracts with a long green point, the upper bracts lance-shaped with a short narrow point. Calyx lobes unequal, the anterior lobes oval fine-pointed, the posterior lance-shaped, keeled, bristly. △ Dry, stony places, rocks. May–Aug. P.E.F.I.CH.A. Page 399.

Stems leafless, unbranched; lvs. all in a rosette
(a) Lvs. linear to oblong, more than 3 times as long as broad
(i) Fl. stem smooth, not grooved or ribbed
1285. *P. coronopus* L. BUCK'S-HORN PLANTAIN. Readily distinguished by its lvs. which are usually toothed or deeply once or twice cut into linear lobes, or rarely entire. Fls. yellowish, usually in a slender cylindrical spike ½–4 cm. by 3–5 mm., on hairy stems 5–40 cm.; bracts oval, often long-pointed; corolla tube hairy; stamens pale yellow. A usually hairy rosette ann. or bienn. △ Dry, sandy places, waysides, paths, on the littoral. Apr.–Oct. All Eur. (except IS.CH.H.): introd. N.SF.A.CS. Pl. 133.

1286. *P. holosteum* Scop. (*P. carinata* Schrader) A densely tufted perenn. with short, stout, woody branches, densely covered with the bases of old lvs. and bearing apical clusters of narrow-linear, fleshy, triangular-sectioned, stiff-tipped, hairless lvs. Fl. spikes dense,

1. *Valerianella coronata* 1311
3. *Plantago sempervirens* 1284
5. *Plantago bellardii* 1288

2. *Lonicera alpigena* [1303]
4. *Orobanche caryophyllacea* 1275

long-cylindrical 3–7 cm., on slender, hairy, smooth stems 3–15 cm. △ Rocky places, dry pastures. May–Sept. Med., South-East Eur. (except TR.) P.A.

1287. *P. maritima* L. SEA PLANTAIN. A glaucous, usually fleshy-leaved perenn., with a stout woody base and narrow-cylindrical fl. spikes 2–12 cm. long. Corolla brownish, lobes with a rather broad, indistinct midvein; bracts oval, keeled, as long as the calyx. Lvs. narrow-linear, flat or grooved, usually 2–6 mm. wide and faintly three- to five-veined, but sometimes up to 1½ cm. broad and seven-veined. △ Salt marshes, on the littoral. May–Sept. All Eur. (except AL.GR.TR.BG.). Pl. 132.

P. alpina L. ALPINE PLANTAIN. Very like 1287, but a perenn. of alpine meadows. Lvs. scarcely fleshy, soft flat, not glaucous, blackening when dried. Fl. spike usually oblong, 1–3 cm.; bracts oval-lance-shaped, not keeled. Portugal, Pyrenees, Alps.

1288. *P. bellardii* All. SILKY PLANTAIN. Like 1291, but lvs. covered with long, spreading, silky hairs and short dense hairs; fl. stem smooth, not grooved, as long or half as long again as the lvs. Fl. spikes oval to cylindrical 1–3 cm.; corolla russet-coloured, hairless; bracts green, oval to lance-shaped, long-pointed, with long spreading hairs. Lvs. narrow-lance-shaped, *c.* 3–5 mm. wide. An ann. 3–15 cm. △ Sandy and stony places. Apr.–June. Med. Eur. P.BG. Page 399.

1289. *P. atrata* Hoppe Fl. head ovoid 1–1½ cm., somewhat hairy, few-flowered, borne on smooth stems which are sparsely hairy or hairless and longer than the lvs. Fls. brownish-green; corolla hairless; bracts rounded and fine-pointed, hiding the fls. Lvs. narrow-lance-shaped to oval, green, hairy or hairless, with 3–5 veins. A perenn. with a stout rootstock, 8–20 cm. △ In mountains: rocky places, pastures. May–Aug. Cent., South-East Eur. (except TR.) E.F.I.SU.

(ii) Fl. stem grooved or ribbed

1290. *P. lanceolata* L. RIBWORT PLANTAIN. Distinguished by its narrow lance-shaped, strongly three- to seven-veined lvs. usually 10–15 cm. long and gradually narrowed below into a stalk, usually about half as long as the blade. Fl. spike ovoid or cylindrical, usually 1–2 cm.; fl. stem deeply grooved, much longer than the lvs. Corolla brownish, with a prominent brown midvein running to the tip of each lobe; stamens white; bracts oval long-pointed, with papery tips. Fr. two-seeded. A hairless or finely hairy perenn. 5–70 cm. △ Meadows, waysides, heaths, waste ground. Apr.–Oct. All Eur. *A mucilage obtained from the seed coat was used to stiffen fabrics.*

1291. *P. lagopus* L. HARE'S FOOT PLANTAIN. Fls. in very dense, ovoid to oblong heads, covered with long, white, silky hairs; fl. stems grooved, 2–4 times as long as the lvs. Corolla tube hairless, lobes with long hairs on back; bracts and calyx densely hairy. Lvs. lance-shaped, narrowed to a short stalk, three- to five-veined, hairy or hairless. An ann. 10–30 cm. △ Sandy places, fields, waysides, uncultivated ground. Apr.–June. Med. Eur. (except AL.) P.BG.SU.

(b) Lvs. oval or elliptic, less than 3 times as long as broad

1292. *P. media* L. HOARY PLANTAIN. Lvs. finely hairy, usually 4–6 cm. elliptic to oval, five- to nine-veined, gradually narrowed into a short stalk, or stalkless and forming a flat rosette. Fl. spikes long-cylindrical 2–8 cm.; corolla silvery-white; filaments purple, anthers lilac or white; bracts acute, with papery margin. Fl. stem hairy, not grooved, much longer than the lvs. Fr. usually four-seeded. A hairy perenn. 20–50 cm. △ Meadows, waysides. May–Aug. All Eur. (except P.IS.): introd. IRL. Pl. 133.

1293. *P. major* L. GREAT PLANTAIN. Like 1292, with lvs. usually 10–15 cm., hairless or nearly so and abruptly contracted into a stalk usually as long as the blade. Corolla yellowish-white; anthers at first lilac, later yellowish; bracts acute, brownish with a green keel. Fl. stem little longer than lvs. Fr. with 4–16 seeds. A hairy or hairless perenn. 10–60 cm. △ Waysides, cultivated ground, farmyards. June–Oct. All Eur. *Used in herbal remedies.*

LITTORELLA Submerged aquatic plants. Fls. one-sexed; male fls. with 4 stamens; female fls. with a long rigid style. 1 sp.

1294. *L. uniflora* (L.) Ascherson SHORE-WEED. A creeping, submerged, aquatic perenn. with erect rosettes of fleshy lvs. which may form extensive mats over the mud in shallow water. Lvs. erect, 2–10 cm. linear, pointed, half-cylindrical. Fls. only formed when the plant is above water; fl. stem shorter than the lvs. and bearing a solitary, long-stalked, white male fl. with stamens with very long slender filaments and 1–3 stalkless female fls. below the male fl. △ Shallow water, lakes, ponds. May–Aug. Most of Eur. (except South-East Eur.) R.

CAPRIFOLIACEAE | Honeysuckle Family

Usually shrubs with opposite lvs.; sometimes woody climbers. Fls. radially symmetrical or less commonly symmetrical in one plane only. Calyx four- to five-lobed, small; corolla fused at base often forming a tube and with 4–5 spreading lobes, sometimes two-lipped; stamens 4–5, fused to the corolla tube. Ovary inferior, of 2–5 carpels; fr. a drupe or a dry, one-seeded unit, rarely a capsule.

1. Lvs. pinnate with 5–11 leaflets.   *Sambucus*
1. Lvs. undivided, entire, toothed, or lobed.
  2. Slender spreading plants; fl. stems 6–10 cm.   *Linnea*
  2. Shrubs, trees, or woody climbers.
    3. Style very short, stigma three-lobed.   *Viburnum*
    3. Style elongate, stigma club-shaped.
      4. Corolla with 5 unequal lobes, or two-lipped; fr. black or red.   *Lonicera*
      4. Corolla regularly five-lobed; fr. white.   *Symphoricarpos*

SAMBUCUS | Elder Usually deciduous shrubs or small trees with pinnate lvs. Fls. numerous, often in an umbel-like cluster; corolla radially symmetrical with very short tube and 5 spreading lobes; stamens 5. Fr. a drupe. 3 sps.

Herbaceous perenns.
1295. *S. ebulus* L. DANEWORT. A robust, fetid perenn. with stout, erect, leafy, green stems ½–2 m., dying down at the end of the year, bearing a broad flat-topped cluster of small white fls. Inflorescence 7–10 cm. across, with 3 primary branches; fls. *c.* 5 mm., white, sometimes reddish-tinged; anthers dark purple. Lvs. once- or twice-pinnate with 7–13 oblong-lance-shaped, long-pointed, and sharply toothed leaflets each 5–15 cm.; stipules conspicuous, oval. Fr. *c.* 4 mm. globular, black. △ Roadsides, thickets, clearings. June–Aug. All Eur. (except IS.N.SF.): introd. IRL.DK.S. *The fruit, inner part of the stem and the leaves are purgative. The leaves have been used externally in the treatment of rheumatism.* Pl. 133.

Shrubs or small trees
1296. *S. nigra* L. ELDER. A shrub or small tree to 10 m. with greyish-brown, often arched branches, pinnate lvs., and broad flat-topped clusters of numerous creamy-white fls.

Clusters 10–20 cm. across, with 5 primary branches; corolla *c.* 5 mm. across; anthers cream-coloured. Leaflets 5–7, oval to lance-shaped, long-pointed, toothed, each 3–8 cm., stipules absent or very small; twigs grey, pith white. Fr. 6–8 mm., black, rarely green, white, or pink. △ Woods, hedges, waste places. June–July. All Eur. (except IS.): introd. SF. *The leaves, flowers, fruit, and bark have many country uses, e.g. as purgatives, for inhalations, lotions, as insecticides, dyes, etc. The hard wood is used for making small articles such as skewers and toys. The flowers and fruit are used for wine-making and the leaves as a tea. The pith is used for botanical sectioning.*

1297. *S. racemosa* L. ALPINE OR RED ELDER. Like 1296, but fls. greenish-white, in dense ovoid clusters and appearing soon after the lvs. Pith of twigs cinnamon-coloured. Fr. scarlet, in drooping clusters. An erect much-branched shrub 1–4 m. △ Shady woods in mountains. Apr.–May. Cent. Eur. E.F.B.I.YU.BG.R.SU.: introd. DK.N.S.SF. *The fruits can be used to make wine.* Pl. 134.

VIBURNUM Shrubs with simple toothed or lobed lvs. Fls. usually in flat-topped umbel-like clusters. Calyx teeth very small; corolla tubular or bell-shaped, usually with 5 spreading lobes; stamens 5; stigmas 3. Fr. a drupe, usually with a flattened stone. 3 sps.

Deciduous shrubs

1298. *V. opulus* L. GUELDER ROSE. A deciduous shrub 2–4 m. with lobed lvs. and conspicuous, flat-topped, lax clusters of numerous white fls. of unequal size, with a few of the outermost fls. more than twice the size of the numerous inner fls. Clusters 5–10 cm. across; inner fls. fertile *c.* 6 mm. across, the outer fls. sterile 1½–2 cm., petals unequal. Lvs. oval 5–8 cm., usually with 3 rather deep, irregularly toothed, and long-pointed lobes, hairless above; twigs hairless, greyish. Fr. red, globular *c.* 8 mm. △ Damp woods, thickets, hedges. May–June. All Eur. (except IS.AL.TR.). *The fruits are emetic when fresh, but they can be made into a preserve or used for wine and spirit-making. The bark is used in herbal remedies.* Pl. 133.

1299. *V. lantana* L. WAYFARING TREE. Readily distinguished from 1298 by the dense, flat-topped or domed umbels of white fls. which are all of the same size, and the entire wrinkled lvs. Fls. *c.* 6 mm. across; corolla with oval spreading lobes twice as long as the tube. Lvs. 5–10 cm., oval-heart-shaped, finely toothed, nearly hairless above and with dense, whitish, star-shaped hairs beneath; twigs covered with powdery hairs. Fr. *c.* 8 mm., flattened, at first red, then black. △ Open woods, scrub, and hedges. May–June. Much of Eur. (except North Eur. P.IRL.TR.): introd. N.S. Pl. 134.

Evergreen shrubs

1300. *V. tinus* L. LAURUSTINUS. A shrub 1–3 m., with dark green leathery lvs. and equal-sized white fls. in dense flat-topped clusters. Fls. *c.* 6 mm. across, buds pink. Lvs. 3½–10 cm. oval-acute entire, shining above, with conspicuous tufts of hairs in the axils of the lateral veins beneath; twigs sparsely hairy, often reddish. Fr. metallic blue-black when ripe. △ Woods, thickets, stony places. Feb.–May. Med. Eur. (except TR.) P.SU. *The fruit is very purgative.* Pl. 134.

SYMPHORICARPOS Fls. in small, terminal, spike-like clusters; corolla tubular or bell-shaped, four- to five-lobed; calyx lobes small. Fr. a berry. 1 sp.

1301. *S. rivularis* Suksdorf SNOWBERRY. A hairless, suckering, deciduous shrub 1–3 m., readily distinguished by its soft snow-white, spherical frs. Fls. pink, bell-shaped, 3–7 in small spike-like clusters, *c.* 2 cm. long; corolla 5–6 mm. across, hairy within. Lvs. broadly oval 2–4 cm., dull bluish-green, mostly entire, but lvs. of sucker shoots deeply lobed;

twigs yellowish. Fr. 1–1½ cm. △ Often planted for ornament and naturalized on banks, hedges, and rocky places. June–Sept. Native of North America: introd. Cent.‘Eur. F.GB. BG.R.SU. Pl. 135.

LONICERA | Honeysuckle Woody climbers or non-climbing shrubs. Corolla often with a long tube which is two-lipped, the upper lip four-lobed, the lower lip with 1 lobe; or corolla with a short tube and with 5 more or less equal lobes. Fr. a two-seeded berry, often paired and somewhat fused. *c.* 20 sps.

Non-climbing shrubs; corolla with a short tube
(a) Ovary in pairs, not fused or only slightly so at the base
1302. *L. xylosteum* L. FLY HONEYSUCKLE. An upright shrub 1–2 m., with greyish somewhat hairy twigs, greyish-green lvs., and stalked axillary pairs of small yellowish fls. Corolla 1–1½ cm., at first white then yellowish, often reddish-tinged, the 2 lips longer than the tube, finely hairy outside; filaments and style hairy. Lvs. 3–6 cm., elliptic to broadly oval-acute, short-stalked, softly hairy. Fr. red, paired, not fused together. △ Woods, hedges, and thickets. May–June. All Eur. (except P.IRL.NL.IS.GR.TR.). *Fruit purgative and emetic.* Pl. 135.

1303. *L. nigra* L. BLACK-BERRIED HONEYSUCKLE, ST. FRANCIS'S WOOD. Like 1302, but lvs. hairless and fls. pinkish-white, paired and borne on a hairless axillary stem. Corolla hairless, the lips shorter than the tube. Lvs. 3–5 cm. oblong-elliptic, hairless or nearly so. Fr. black, to 1 cm., paired, not fused at their base. A hairless, grey-stemmed shrub, 60–200 cm. △ Woods, thickets, stony places. May–July. Cent. Eur. (except H.) E.F.I.YU.BG. R.SU. *The fruit is purgative and emetic.*

(b) Ovary pairs fused their whole length, or nearly so
L. caerulea L. BLUE HONEYSUCKLE. Like 1302, but stem of paired fls. much shorter than the fls.; fls. yellowish, bell-shaped, lobes nearly equal, shorter or as long as the tube, hairy. Lvs. oblong-oval blunt, slightly hairy beneath. Fr. 1 cm. glaucous, bluish-black, completely fused into a single globular berry. Mountains of Northern and Central Europe. Pl. 135.

L. alpigena L. ALPINE HONEYSUCKLE, CHERRY WOODBINE. Distinguished by its pink or reddish, open funnel-shaped paired fls. *c.* 1½ cm. long, with fl. stem 3–4 times as long as the fls. Lvs. 4–6 cm., oval, long-pointed, thick, rather densely hairy below and on margin. Fr. fused into a single red ovoid berry. Pyrenees, Alps, Carpathians, Balkans. *The fruits are purgative and emetic.* Page 399.

Woody climbers; corolla with a long tube
(a) Uppermost pairs of lvs. below fls. not fused together
1304. *L. periclymenum* L. HONEYSUCKLE. A robust woody climber to 6 m., with sweet-scented, yellowish-white, often pinkish-flushed fls. in a terminal stalked cluster. Corolla tubular 4–5 cm. long, glandular-hairy outside; bracts small, shorter than the ovary. Lvs. 3–7 cm. oval to oblong, dark green above, glaucous beneath, the upper stalkless, the lower stalked. Fr. red, in a dense globular cluster. △ Hedges, thickets, woods. June–Sept. West, Cent. Eur. (except H.) N.S.SF.I.YU.GR.SU. *The fruit and leaves are used in herbal remedies.* Pl. 135.

(b) Uppermost pairs of lvs. below fls. fused together and encircling stem
1305. *L. etrusca* Santi Like 1304 in having long-stemmed fl. clusters, but the uppermost pairs of lvs. are fused by their bases and encircle the stem. The terminal fl. clusters are in groups of 3, the lateral clusters are usually solitary; corolla whitish-yellow, flushed reddish outside, hairless. Lvs. obovate, deciduous or somewhat persistent, dark green,

the lowest stalked. Climbing to 3 m. Fr. red, clustered. △ Hedges, thickets, woods. May–June. Med. Eur. P.BG. Pl. 135.

1306. *L. caprifolium* L. PERFOLIATE HONEYSUCKLE. Distinguished by the stemless fl. clusters which arise directly from the centre of the uppermost pair of lvs. and which are fused together and encircle the stem in a cup-like involucre. Fls. yellow, purplish-flushed outside, not glandular; stamens projecting; style hairless. Lvs. elliptic, deciduous, dark green above, glaucous beneath, the uppermost fused by their bases, the lower stalked. Climbing to 2 m. Fr. red. △ Hedges, roadsides, woods; sometimes escaping from cultivation. May–July. Much of Eur. (except North Eur. IRL.B.NL.).

L. implexa Aiton Like 1306 in having stemless, terminal fl. clusters arising from a cup-like involucre of fused lvs., but lvs. evergreen, tough, leathery, shining above and glaucous beneath and with a narrow transparent margin; twigs hairless. Corolla yellow, becoming reddish-flushed, hairless or minutely hairy; stamens scarcely projecting; style hairy. Southern Europe.

LINNAEA Creeping shrublets. Fls. paired; calyx five-lobed, soon falling; corolla bell-shaped, five-lobed; stamens 4. Ovary three-celled, only 1 of which is fertile. 1 sp.

1307. *L. borealis* L. TWINFLOWER. A delicate creeping perenn., with pairs of tiny rounded lvs. and slender stems bearing a pair of drooping, pinkish-white, bell-shaped fls. Fls. sweet-scented, short-stalked; corolla *c.* 8 mm. long, hairy within, lobes equal; fl. stems 8–15 cm., glandular-hairy. Lvs. 1–1½ cm., broadly oval and shallowly toothed, shortly stalked, sparingly hairy. △ Coniferous forests; shady, mossy rocks. July–Aug. North Eur. (except IS.), Cent. Eur. (except H.) F.GB.I.SU. Pl. 135.

ADOXACEAE | Moschatel Family

A family of one genus and one sp. Fls. in a compact head, the terminal fl. with a two-lobed calyx and four-lobed corolla and stamens 4, but appearing as 8; the lateral fls. with a three-lobed calyx, five-lobed corolla, and 5 stamens, appearing as 10. Ovary half-inferior, three- to five-celled, each cell one-seeded; fr. a drupe.

ADOXA

1308. *A. moschatellina* L. MOSCHATEL, TOWN HALL CLOCK, FIVE-FACED BISHOP. A delicate hairless perenn. with creeping underground stolons with fleshy white scales and erect stems 5–10 cm., bearing a pair of compound lvs. and a terminal head of tiny green fls. Fls. 4–6 in a globular four-sided head *c.* 6 mm., reminiscent of a town hall clock. Lvs. bright green, trifoliate, the leaflets further divided into oval lobed segments, the basal lvs. long-stalked. Fr. green. △ Woods, shady places, hedges. Mar.–May. Most of Eur. (except P.IRL.IS.AL. GR.TR.R.). Pl. 137.

VALERIANACEAE | Valerian Family

Herbaceous plants, often with strong-smelling underground stems and opposite, entire, or lobed lvs. Fls. usually small in dense, many-flowered, terminal clusters. Calyx variously toothed or lobed, often forming a feathery pappus in fr.; corolla funnel-shaped, often swollen or spurred at the base, with 5 often unequal lobes; stamens 1–3 often projecting from the corolla. Ovary three-celled, with only 1 cell containing a fertile ovule; fr. not splitting.

1. Fr. without pappus; anns.
  2. Corolla tiny, regularly five-lobed.   *Valerianella*
  2. Corolla with a long tube, distinctly two-lipped.   *Fedia*
1. Fr. with pappus; usually perenns.
  3. Corolla with a spur or boss at its base; stamen 1.   *Centranthus*
  3. Corolla without spur or boss; stamens 3.   *Valeriana*

VALERIANELLA Anns. Calyx forming a toothed ring, not a pappus in fr.; corolla regularly five-lobed, without a spur or sac at base of the tube; stamens 3. *c.* 22 sps.

1309. *V. locusta* (L.) Betcke (*V. olitoria* (L.) Pollich) LAMB'S LETTUCE, CORN SALAD. A slender, much-branched, brittle-stemmed ann. 7–40 cm., with oblong or spathulate lvs. and terminal, rounded or flat-topped clusters of very tiny pale lilac fls. Clusters *c.* 2 cm. across; fls. numerous, *c.* 2 mm. across. Lvs. spathulate 2–7 cm., entire or sometimes toothed. Fr. rounded, somewhat compressed, 2½ mm. by 2 mm., corky on the back. △ Cultivated ground, bare places, rocks, walls. Apr.–June. All Eur. (except IS.). *Sometimes cultivated as a salad plant.* Pl. 136.

1310. *V. echinata* (L.) DC. SPINY LAMB'S LETTUCE. Readily distinguished by its ripe fr. which has the appearance of a cluster of spines borne at the end of the conspicuously swollen fr. stalk. Fls. white or lilac, in dense flat-topped heads on slender stems; calyx enlarging in fr. into 3 stout, outward-curved, conical spines. Lvs. lance-shaped, the upper strongly toothed or lobed. Fr. spongy, of two kinds, either linear and straight, or oblong, three-angled and irregularly furrowed. A hairless, erect, rather stout-stemmed ann. 10–30 cm. △ Cornfields. Apr.–June. Med. Eur. (except AL.) SU.

1311. *V. coronata* (L.) DC. CROWNED LAMB'S LETTUCE. Distinguished by its distinctive fr. in which the calyx enlarges into a conspicuously veined, bell-shaped, membraneous crown with 5 teeth with long crooked points. Fls. lilac or bluish; heads dense, at length globular, *c.* 1 cm. across. Lower lvs. lance-shaped, the upper linear. Ovary hairy, flattened. A slender, erect, finely hairy ann. 5–40 cm., branched above. △ Dry, arid places. May–June. Much of Eur. (except North Eur. IRL.GB.NL.CH.PL.). Page 399.

V. discoidea Loisel. Like 1311, but calyx differing in fr., and having 6 wide-spreading, often bilobed spiny-tipped lobes, calyx hairy within and often outside. Fr. obconical, hairy, with a depression on the outer side. A thick-set, rough-haired ann. Mediterranean Europe, Portugal.

*V. carinata Loisel. Central and Southern Europe.

*V. rimosa Bast. Widespread in Europe.

*V. dentata (L.) Pollich Widespread in Europe; introduced to Northern Europe.

*V. eriocarpa Desv. Western and Mediterranean Europe.

FEDIA Anns. Fls. in terminal clusters on dichotomously branched stems which soon become thick and swollen. Corolla with a long tube, two-lipped, the upper lip two-lobed, the lower three-lobed; calyx minute; stamens 2. 1 sp.

1312. *F. cornucopiae* (L.) Gaertner A hairless ann. 10–30 cm., with dichotomous fl. stems soon becoming conspicuously swollen, bearing terminal clusters of numerous stalkless, long-tubed, pink fls. Corolla tube long and slender, with a sac-like swelling on one side near the base, two-lipped; stamens 2, projecting. Lvs. oval, shallowly toothed, the upper stalkless, acute. Fr. either oblong with a papery toothed crown or flattened with a very short crown. △ Cultivated ground, cornfields. Mar.–June. Med. Eur. (except AL.TR.) P. Pl. 137.

VALERIANA | Valerian Perenns. Calyx forming a pappus in fr.; corolla funnel-shaped, somewhat swollen at the base of the tube and with 5 usually unequal, spreading lobes; stamens 3. Fr. crowned with a pappus; one-seeded. *c*. 25 sps.

Basal lvs. pinnate or pinnately lobed

1313. *V. officinalis* L. VALERIAN. A rather robust perenn. $\frac{1}{2}$–2 m. with grooved stems, compound lvs., and dense, terminal, flat-topped, branched clusters of numerous pale pink fls. Corolla 4–5 mm. across, the tube 4–5 mm. long. Lvs. very variable, to *c*. 20 cm., pinnate, the lower lvs. usually stalked, the upper stalkless, leaflets large, oval or lance-shaped, entire or strongly and unequally toothed. Fr. 4 mm. oblong-oval, hairless. △ Damp places, wet meadows, woods, watersides. May–July. All Eur. (except P.). *The underground parts have been used in herbal remedies, and also as a rat bait.* Pl. 136.

Basal lvs. undivided

V. tuberosa L. TUBEROUS VALERIAN. Differing from 1313 in having fls. in dense, rounded, not flat-topped clusters. Basal lvs. oblong, entire or with shallow blunt lobes, stem lvs. with deep linear lobes; base of stem with a swollen tuber. Fr. with silvery hairs. A perenn. of dry pastures, to 40 cm. Southern Europe.

1314. *V. dioica* L. MARSH VALERIAN. A slender hairless perenn., with creeping stolons and terminal rounded heads of pinkish, one-sexed fls. Male fls. *c*. 5 mm. across, in a rather lax head; female fls. *c*. 2 mm. across, in a dense head, each borne on different plants on erect, unbranched, sparsely leafy stems 15–40 cm. Basal lvs. 2–3 cm. oval-elliptic, entire, long-stalked, the upper lvs. stalkless, deeply divided into 5–9 lobes, with the terminal lobe larger. Fr. *c*. 3 mm. elliptic, hairless. △ Marshes, fens, bogs. Apr.–June. Most of Eur. (except P.IRL.IS.SF.AL.BG.).

1315. *V. tripteris* L. THREE-LEAVED VALERIAN. A woody-based perenn. 20–50 cm., without stolons or tubers, with a terminal, somewhat flat-topped, lax cluster of pink, bisexual fls. Bracts linear, with papery margin. Lvs. grey-green, somewhat glaucous, those of the sterile shoots oval-heart-shaped, coarsely toothed, long-stalked; lvs. of fl. stems nearly stalkless, deeply three-lobed, the terminal lobe longest, toothed. Fr. *c*. 3 mm. △ In mountains, woods, shady rocks, streamsides. June–Aug. Cent. Eur. E.F.I.YU.BG.R.SU. Pl. 136.

V. montana L. MOUNTAIN VALERIAN. Like 1315, but lvs. bright shining green, those of the sterile shoots oval and narrowed to the lv. stalk, with entire or minutely toothed margin; lvs. of the fl. stem oval-lance-shaped, entire or with a few shallow teeth. Fls. pink; bracts lance-shaped, green. Mountains of West-Central and South-Eastern Europe. Pl. 136.

CENTRANTHUS | (Kentranthus) Calyx forming a pappus in fr. Corolla tube narrow, spurred or with a small conical projection at the base, five-lobed. Stamen 1. Fr. with 1 seed. *c*. 8 sps.

1316. *C. ruber* (L.) DC. RED VALERIAN. An erect, somewhat glaucous, hairless perenn. 30–100 cm. with numerous red, pink, or rarely white fls. in a rather dense, oval or somewhat branched, pyramidal cluster. Corolla tube slender 8–10 mm., with a slender backward-projecting spur twice as long as the ovary; corolla lobes *c.* ½ cm. spreading, unequal; stamen 1, projecting. Lvs. to *c.* 10 cm. oval to lance-shaped, entire or shallow-lobed, the upper stalkless. Fr. a one-seeded nut. △ Rocks, walls; often grown for ornament and naturalized. May–Sept. Med. Eur. P.CH.A.: introd. further north. Pl. 136.

C. angustifolius (Miller) DC. NARROW-LEAVED RED VALERIAN. Like 1316, but glaucous lvs. all linear or linear-lance-shaped and only 2–14 mm. broad. Fls. in a dense, rounded, flat-topped cluster, deep pinkish-red, rarely white; spur of corolla about as long as the ovary. Spain to Italy and Switzerland. Pl. 137.

1317. *C. calcitrapae* (L.) Dufresne A hairless, often reddish ann. 10–40 cm., with pinnately cut upper lvs. and tiny pinkish or white fls. arranged in two ranks in a lax, flat-topped cluster. Corolla tube 2–5 mm., spur reduced to a rounded projection. Upper lvs. stalkless, deeply cut into oblong, often toothed lobes, the basal lvs. oval, entire or lobed; stem hollow. Fr. hairy or hairless. △ Rocks, dry hills. May–July. Med. Eur. (except AL.TR.) P.SU.

DIPSACACEAE | Scabious Family

Usually herbaceous plants with opposite or whorled lvs. Fls. small, clustered in a head or *capitulum*, which is surrounded at the base by calyx-like bracts forming an involucre. Fls. usually somewhat irregular, each surrounded at the base by an additional structure or epicalyx, known as the *involucel*. Calyx small, cup-shaped or deeply cut into segments, teeth or hairs; corolla tube often curved and two-lipped and with 4–5 often unequal lobes; stamens usually 4, rarely 2, projecting. Ovary inferior; fr. one-seeded, enclosed in the involucel and crowned by the calyx. Distinguished from the Compositae by the unfused stamens and by the presence of an involucel.

1. Receptacle without scales, but with hairs.   *Knautia*
1. Receptacle with scales.
  2. Involucre of rigid, spiny-tipped bracts.
    3. Stems spiny.   *Dipsacus*
    3. Stems without spines.   *Cephalaria*
  2. Involucre of green, soft, leafy bracts.
    4. Corolla four-lobed; involucel of 4 green lobes.   *Succisa*
    4. Corolla five-lobed; involucel funnel-shaped with papery, cup-shaped
       mouth.   *Scabiosa*

DIPSACUS | Teasel Large spiny- or prickly-stemmed herbaceous plants. Involucre of linear-lance-shaped, stiff, spine-tipped bracts; scales of receptacle also stiff spine-tipped. Calyx cup-shaped, four-angled; corolla tubular, unequally four-lobed. *c.* 6 sps.

1318. *D. fullonum* L. TEASEL. A stiff, erect, sparingly branched bienn. ½–2 m., with prickly stems and midveins of lvs., and dense, conical, spiny heads of rose-purple or whitish-violet fls. Fl. heads 3–8 cm. long; fls. maturing in a narrow zone from below upwards; involucral bracts spiny-tipped, upcurved. Over-wintering lvs. lance-shaped, in a flattened rosette; stem lvs. lance-shaped, entire or toothed, their bases fused round the stem to form a 'cup' often containing water. Fr. 5 mm. Subsp. *sativus* (L.) Thell. FULLER'S TEASEL has the involucral bracts spreading horizontally and the bracts of the receptacle ending in a stiff recurved spine. It is sometimes grown as a crop. △ Waysides, streamsides, thickets, pastures. July–Aug. Most of Eur.: introd. to North Eur. *Cultivated forms were extensively grown in the past for their dry fruit heads which were used to raise the nap on woollen cloth.*

D. laciniatus L. Like 1318, but stem lvs. deeply pinnately lobed and lobes further deeply toothed or incised, blades bristly-haired. Fls. white or pale lilac. Central and Southern Europe. Pl. 137.

1319. *D. pilosus* L. (*Cephalaria p.*) SMALL TEASEL. Distinguished by its small spherical heads of white fls. which are at first drooping and later erect, and its stalked stem lvs. Heads 2–2½ cm.; corolla 6–9 mm.; anthers dark violet; bracts of involucre spiny-tipped, spreading, shorter than fl. head; scales of receptacle oval and abruptly narrowed to a spiny point and with long, spreading, bristly hairs. Lower lvs. oval, toothed, long-stalked, the upper lvs. oval-elliptic, short-stalked, often with small, narrow, basal leaflets. An erect branched, rather weakly prickly-stemmed bienn. 30–120 cm. △ Woods, hedges, bushy places, streamsides. June–Sept. Most of Eur. (except North Eur. P.IRL.AL.GR.TR.): introd. DK.S.

CEPHALARIA   Stems with bristly hairs, not spiny. Involucral bracts stiff, in several overlapping ranks and similar to the scales of the receptacle. Involucel quadrangular with 4–10 teeth; calyx quadrangular with numerous linear teeth. *c.* 15 sps.

1320. *C. transsylvanica* (L.) Schrader A slender, erect, sparsely branched ann. 30–80 cm. with pale violet fls. in long-stalked ovoid heads 1–1½ cm. across. Fls. *c.* 8 mm. across, hairy outside; anthers purple; bracts of involucre and scales of receptacle similar, papery, ciliate, oval with a narrow spiny-pointed apex; involucel with 8 short triangular-acute teeth; calyx hairy. Lvs. pinnately cut into narrow lance-shaped lobes and a larger terminal lobe, the lower lvs. with spreading hairs, the upper nearly hairless; stems hairy below, hairless above. △ Fields, uncultivated ground. Aug.–Sept. South-East Eur. E.F.CH.A.CS.H.I.SU. Page 411.

C. leucantha (L.) Schrader Fls. white in globular heads 2–3 cm.; anthers white; involucre bracts and receptacle scales similar oval-blunt, pale, papery, with a dark tip, hairy, shorter than the fls. Involucel with hairy teeth; calyx hairy. A nearly hairless perenn. with many grooved, hollow stems to 1 m. Portugal to Yugoslavia.

SUCCISA   Like *Scabiosa*, but involucel four-angled and ending in 4 erect, triangular, green lobes. Outer fls. of fl. head differing little from the inner fls.; corolla with 4 more or less equal lobes. *č.* 4 sps.

1321. *S. pratensis* Moench (*Scabiosa succisa* L.) DEVIL'S-BIT SCABIOUS. A sparsely branched erect or spreading perenn. 15–100 cm., with dark blue-purple, rarely white or pink fls. in rounded heads 1½–2½ cm. across. Heads of bisexual fls. larger than heads of only female fls.; bracts of involucre lance-shaped, ciliate, receptacle scales elliptic, purple-tipped. Basal lvs. elliptic, stalked, in a rosette, stem lvs. few, lance-shaped or bract-like; stems with adpressed hairs. Fr. ½ cm., softly hairy, with 4–5 bristle-like calyx teeth. △ Damp meadows, marshes, fens, damp woods. July–Oct. All Eur. (except AL.GR.TR.). *Used herbally against sores and skin infections.*

KNAUTIA   Scales of receptacle absent but receptacle hairy. Involucral bracts numerous, soft. Involucel four-angled, forming a shallow cup; calyx usually with 8 teeth or bristles. *c.* 25 sps.

Perenns.
(a) At least some stem lvs. deeply lobed
1322. *K. arvensis* (L.) Coulter (*Scabiosa a.*) FIELD SCABIOUS. A little-branched hairy perenn. 30–150 cm., with long-stalked, bluish-lilac, hemispherical fl. heads 3–4 cm. across, with the outermost fls. distinctly larger than the inner. Corolla lobes unequal; involucral bracts in 2 rows, oval-lance-shaped, hairy, shorter than the fls. Basal lvs. often entire and forming an over-wintering rosette; stem lvs. deeply pinnately cut into oblong lobes with an elliptic terminal lobe, some of the uppermost lvs. entire; lvs. all grey-green, hairy. Fr. 5–6 mm., densely hairy, calyx with 8 bristle-like teeth. A very variable sp. △ Dry fields, pastures, bushy places. May–Oct. All Eur. (except TR.). Pl. 138.

(b) Stem lvs. all entire
1323. *K. sylvatica* (L.) Duby WOOD SCABIOUS. Distinguished from 1322 by its bright green stem lvs. which are entire and oblong-lance-shaped, long-pointed, usually toothed. Fl. heads lilac or purplish, 2½–4 cm. across, with outermost fls. little larger than the inner; involucral bracts lance-shaped long-pointed, nearly as long as the outer fls. A variable perenn. 30–100 cm., hairy particularly below. △ In mountains: shady places, woods, scrub. June–Sept. Cent. Eur. P.E.F.I.YU.R.SU. Pl. 138.

Anns.
1324. *K. integrifolia* (L.) Bertol. Fl. heads pale pink or lilac, flat-topped 1½–2 cm. across, the outermost fls. distinctly larger than the inner. Involucral bracts 8–12, lance-shaped, glandular-hairy, shorter than the fls.; involucel with 2 clusters of unequal teeth; calyx with numerous small teeth with or without bristles; fl. stem often glandular-hairy. Basal lvs. in a rosette, mostly deeply pinnately cut into linear lobes, the upper lvs. linear-lance-shaped entire, all hairy. An erect, slender-branched, rough-hairy but variable ann. 30–50 cm. △ Fields, bushy places. May–June. Med. Eur. BG. Page 411.

SCABIOSA | Scabious Scales of receptacle linear-lance-shaped, not spiny-tipped. Involucel ending in a papery, pleated cup; calyx with 5 spreading stiff bristles. *c.* 40 sps.

Tube of involucel with 8 conspicuous grooves from base to apex
(a) Perenns.
1325. *S. columbaria* L. SMALL SCABIOUS. Fl. heads bluish-lilac, rarely pink or white, 1½–3½ cm. across, borne on slender softly hairy stems, the outer fls. much larger than the inner. Involucral bracts *c.* 10, linear-lance-shaped, shorter than the fls. Basal lvs. obovate, simple or pinnately lobed, stalked, the uppermost lvs. pinnately cut into linear segments. Involucel of fr. hairy, with 8 deep grooves on the tube and a shallow papery cup, about one-third as long as the 5 slender, blackish calyx bristles. A slender, erect, little-branched, very variable perenn. 15–70 cm. △ Dry pastures, banks, thickets, rocks. June–Oct. Much of Eur. (except North Eur. IRL.GR.TR.) DK.S.

S. gramuntia L. Like 1325, but fl. heads globular, small, ½–1 cm. across, bright blue, and calyx bristles less than twice as long as the cup of the involucel, or bristles absent. Lvs. hairy or white-haired, the upper 2–3 times cut into linear segments 1–2 mm. broad. Western and Central Europe.

S. lucida Vill. SHINING SCABIOUS. Distinguished by its almost hairless, somewhat glossy lvs. and rose-lilac, violet, or deep mauve fls. in a head 1–2 cm. across. Involucel hairy towards the base, shallowly grooved, calyx bristles blackish, distinctly keeled and with a narrow papery margin, 3–5 times as long as the cup of the involucel. Perenn. 10–30 cm. Pyrenees, Alps, Carpathians. Pl. 138.

(b) Anns. or bienns.

1326. *S. atropurpurea* L. MOURNFUL WIDOW, SWEET SCABIOUS. An erect, almost hairless ann. 25–120 cm. with lilac, red-violet, or dark purple fls. in flattened heads which later become ovoid or oblong in fr. Outer fls. spreading, larger than the inner. Upper lvs. pinnately cut into linear lobes, the lower lvs. elliptic, toothed or lobed. Fr. with an eight-grooved involucel tube and with a lobed, incurved cup; calyx with 5 long blackish or russet-coloured bristles on a short stalk. △ Sandy places, olive groves, sunny hills by the sea; grown for ornament and sometimes naturalized. June–Oct. I.YU.GR.BG. Pl. 138.

1327. *S. ochroleuca* L. Distinguished by its yellow fl. heads *c.* 2 cm. across, with the outer fls. nearly twice as large as the inner. Involucre bracts lance-shaped acute, shorter than the fls. Lvs. finely hairy, the lowermost entire and toothed or deeply lobed, the uppermost once or twice pinnately cut into linear-lance-shaped lobes which may be further divided. Fr. with calyx bristles 2–3 times longer than the cup of the involucel, which is *c.* 1 mm. long. An erect, slender, sparsely branched bienn. 10–60 cm. △ Grassy places, uncultivated ground. July–Aug. Cent. Eur. E.F.I.YU.TR.BG.R.SU. Pl. 137.

Tube of involucel rounded or eight-grooved only towards the apex
(a) Lvs. linear

1328. *S. graminifolia* L. GRASS-LEAVED SCABIOUS. Readily distinguished by its grass-like, linear acute, silky-haired lvs. *c.* 1–3 mm. wide which are mostly basal, and its large blue-lilac fl. heads 3–6 cm. across. Fl. heads globular, long-stemmed, outer fls. longer than the inner; involucral bracts narrow lance-shaped, with adpressed hairs; involucel densely silky-haired, with a many-ribbed transparent cup; calyx with 5 whitish bristles little longer than the cup. A softly hairy, tufted, somewhat woody-based perenn. 10–50 cm. △ Rocks, scree. June–Aug. E.F.CH.A.I.YU.AL. Page 411.

(b) Lvs. at least the upper pinnately cut

S. stellata L. Distinguished by its large globular fr. heads 2–5 cm. across, with their con-spicuous, broad, yellow, papery cups of the involucels, each to 2 cm. across, 7–9 mm. high and little shorter than the 5 slender calyx bristles; tube of involucel hairy below, with 8 hairless grooves above. Fl. heads bluish or white, 2–3½ cm. across, the outer fls. larger, spreading. A rough-haired ann. with pinnately cut upper lvs. Portugal to Italy.

1329. *S. ucranica* L. Fls. yellowish-white or bluish-lilac, the outer fls. longer, spreading; involucral bracts linear-lance-shaped, usually as long as or longer than the fls. Lvs. usually hairy, sometimes densely silver-haired, pinnately cut into entire or toothed linear segments. Fr. head globular 1–2 cm.; involucel hairy below, with 8 hairless grooves above, cup whitish, 2–4 mm. high; calyx bristles russet-coloured, 3–4 times as long as the cup. A stiff, erect, leafy bienn. 40–100 cm. △ Stony hills. July–Sept. Med. Eur. BG.

CAMPANULACEAE | Bellflower Family

Usually herbaceous plants often with a milky juice; lvs. simple, usually alternate. Fls. often showy, regular; calyx five-lobed, fused to the ovary at the base; corolla usually bell-shaped with 5 lobes shorter than the tube, or tube very short and lobes much longer; stamens 5. Ovary usually inferior, two- to ten-celled; stigmas 3–5; fr. a capsule, or fleshy.

		Nos.
1. Corolla distinctly two-lipped.	*Lobelia*	1356
1. Corolla with equal lobes, not two-lipped.		

1. *Knautia integrifolia* 1324
2. *Campanula patula* 1335
3. *Cephalaria transsylvanica* 1320
4. *Phyteuma hemisphaericum* 1353
5. *Scabiosa graminifolia* 1328

2. Corolla deeply five-lobed, almost to the base.
3. Fls. in a dense spike or head.

4. Anthers fused at base; stigmas short, stout.	*Jasione*	1355
4. Anthers not fused; stigmas long, slender.	*Phyteuma*	1350–1353

3. Fls. in a lax elongate spike or branched cluster.

5. Corolla lobes linear; fr. splitting by pores.	*Asyneuma*	1349
5. Corolla lobes lance-shaped to oval; fr. cylindrical, splitting by valves.	*Legousia*	1346, 1347

2. Corolla bell-shaped or funnel-shaped, with shallow lobes usually less than half the length of the corolla tube.

6. Corolla tube slender, lobes spreading.	*Trachelium*	1348

6. Corolla funnel- or bell-shaped, lobes more or less erect.

7. Style with a disk-like swelling at base.	*Adenophora*	1345

7. Style without swelling at base.

8. Fr. opening by apical or basal pores.	*Campanula*	1330–1344
8. Fr. irregularly splitting; fls. in dense clusters.	*Edraianthus*	1354

CAMPANULA |**Bellflower** Calyx fused to the ovary, ovoid or almost globular, with 5 free lobes; corolla bell-shaped or funnel-shaped with ascending lobes; style club-shaped with 3 or 5 stigmas. Fr. a capsule opening by 3 or 5 valves or pores. *c.* 100 sps.

Calyx with 5 large lobes and 5 additional smaller lobes

1330. *C. barbata* L. BEARDED BELLFLOWER. Fls. drooping, pale blue or whitish in a few-flowered, one-sided, elongate cluster and with the corolla with conspicuous long white hairs within. Corolla 2½–3 cm. long; calyx hairy, with 5 reflexed lobes between the 5 erect lobes; stigmas 3. Basal lvs. in a rosette, oblong-lance-shaped, often wavy-margined; stem lvs. 2–3, small. Fr. drooping, three-celled. An erect, hairy, unbranched perenn. 10–40 cm. △ In mountains: rocks, meadows, open woods, heaths. June–Aug. F.N.D.CH.A.PL.CS.: introd. s. Pl. 139.

1331. *C. medium* L. CANTERBURY BELL. A stout, rough-hairy, leafy bienn. 30–60 cm. with large, inflated, bell-shaped, dark violet, blue, or white fls. 4–5 cm. long, borne in a long leafy spike. Calyx bristly-haired with 5 broad, heart-shaped, reflexed lobes between the 5 erect lobes; stigmas 5. Lvs. oval-lance-shaped, shallowly toothed, the basal lvs. stalked, the upper numerous, stalkless. Fr. five-celled. △ Woods, rocks, stony places; widely grown as an ornamental and often naturalized. May–Aug. E.F.I.: introd. elsewhere.

Calyx with 5 lobes

(a) Fls. stalkless and clustered into dense heads or elongated spikes

1332. *C. thyrsoides* L. YELLOW BELLFLOWER. Readily distinguished by its numerous pale yellow fls. clustered into a stout, compact, conical spike. Corolla tube 17–22 mm., lobes recurved, hairy on both sides; style long-projecting; stigma three-lobed. Stems thick and hollow, with numerous overlapping, rough, strap-shaped lvs. A rough-haired, rosette-forming bienn. 10–30 cm. △ In mountains: meadows, screes, rocks. July–Sept. F.CH.A. I.YU.BG. Pl. 139.

C. spicata L. SPIKED BELLFLOWER. Distinguished by its long slender spike of stalkless blue, lilac, or purple fls. which occupies about two-thirds the length of the stout, unbranched stem. Corolla 1½–2 cm. funnel-shaped, hairy within, lobes lance-shaped. Lvs. mostly basal, strap-shaped acute, with undulate margin, rough-hairy. A rosette-forming bienn. Alps, Apennines. Pl. 138.

1333. *C. glomerata* L. CLUSTERED BELLFLOWER. Readily distinguished by its terminal globular clusters of stalkless, bright blue or rarely white fls. closely encircled at the base by an involucre of leafy bracts. Corolla funnel-shaped 1½–3 cm., hairy, erect, style included; calyx tube *c.* 3 mm., five-ribbed, lobes longer, triangular acute; lateral fl. clusters often present. Basal lvs. long-stalked, oval-blunt and rounded or heart-shaped at base, finely toothed, the stem lvs. narrower acute, stalked, or half-clasping the stem, softly downy. An erect, unbranched, hairy perenn. 5–80 cm. △ Grasslands, open woods, tracksides. June–Aug. Most of Eur. (except P.IRL.IS.AL.TR.). Pl. 138.

C. cervicaria L. Like 1333, but a stiff, erect, bristly-haired perenn. with lower lvs. lance-shaped and narrowed to a winged lv. stalk. Calyx with oval blunt lobes; corolla pale blue, *c.* 1½ cm. long; style projecting. Widespread in Europe.

(*b*) *Fls. all stalked, solitary or in more or less compound clusters or spikes*
(*i*) *Blades of lower lvs. parallel-sided or broadest towards apex; fr. erect*
1334. *C. rapunculus* L. RAMPION. Fls. blue, in a long, lax, narrow cluster, often with some short, erect, lateral clusters. Corolla hairless, *c.* 2 cm., divided to one-third into lance-shaped, more or less erect lobes; calyx hairless, lobes linear, half the length of the corolla; fl. stalks with bracteoles at base. Lower stem lvs. strap-shaped and gradually narrowed to the stalk, margin undulate, the basal lvs. oval, abruptly contracted to the stalk. A slender, erect, hairy bienn. 40–80 cm., with swollen, fleshy roots. △ Fields, thickets, open woods, waysides. May–Aug. Most of Eur. (except IRL.IS.N.): introd. DK.S.SF. *Formerly grown as a salad plant; the fleshy roots and shoots are edible.*

1335. *C. patula* L. SPREADING BELLFLOWER. Like 1334, but fls. blue-violet in a broad, spreading, lax cluster formed by many upward-spreading branches. Corolla 1½–2½ cm., divided to the middle into widely spreading, triangular lobes; calyx two-thirds as long as the corolla, the lobes linear and with obscure glandular teeth; fl. stalks slender with bracteoles near the middle. Lvs. all oblong to narrow lance-shaped, obscurely toothed, the lower lvs. narrowed to a winged stalk, the upper lvs. stalkless. A slender, erect bienn. 25–60 cm., without swollen roots. △ Grasslands, woods, hedges. May–July. Most of Eur. (except P.IRL.IS.TR.): introd. DK.N.S. Page 411.

1336. *C. persicifolia* L. NARROW-LEAVED BELLFLOWER. Distinguished by its large, open, funnel-shaped fls., which are as broad as long and borne in a simple spike-like cluster of 2–8 fls. Corolla blue-violet, 3–4 cm., shallowly divided into broad triangular lobes; calyx usually hairless, lobes lance-shaped, at first spreading and then erect. Lvs. hairless, shining, the lower oblong-obovate and narrowed to a stalk, the upper linear-lance-shaped, stalkless, finely toothed. An erect, usually unbranched perenn. 40–100 cm. △ Thickets, woods, hedges. May–Aug. All Eur. (except P.IRL.IS.). Pl. 139.

1337. *C. erinus* L. ANNUAL BELLFLOWER. A slender, rough-haired, spreading, branched ann. 10–30 cm., with small, pale blue, reddish, lilac or white fls. 3–5 mm. long, borne terminally or in the axils of the branches. Fls. very short-stalked; corolla tubular-bell-shaped; calyx as long as the corolla, the lobes lance-shaped and spreading in a star in fr. Lower lvs. oval-wedge-shaped, to *c.* 3 cm., the upper lvs. oval, all lvs. coarsely toothed. △ Walls, rocky and grassy places. Apr.–May. Med. Eur. P.R.

(*ii*) *Blades of lower lvs. broadest towards base. Basal and stem lvs. differing conspicuously*
1338. *C. cochleariifolia* Lam. (*C. pusilla* Haenke) FAIRY'S THIMBLE. A low, cushion-forming perenn., with rounded, long-stalked, basal lvs. forming loose rosettes, and slender fl. stems with narrow lvs. and bearing a few nodding blue, violet, or rarely white fls. Corolla 1–2 cm. bell-shaped or almost hemispherical with a rounded base, about 4 times as long as the linear calyx lobes. Upper stem lvs. linear to lance-shaped, toothed, stalkless, differing conspicuously from the heart-shaped toothed, long-stalked, basal lvs. Fl. stems

5–20 cm.; stolons present. △ In mountains: rocks, screes, and meadows. June–Aug. F.CH.A.PL.CS.YU.AL.BG.R. Pl. 140.

1339. *C. rotundifolia* L. HAREBELL, SCOTTISH BLUEBELL. An erect hairless perenn. 15–60 cm., with drooping blue or rarely white fls. in a slender, sparsely branched, and lax cluster, or fls. solitary. Corolla *c.* 1½ cm., broadly bell-shaped and narrowed to the base, divided to about halfway into oval lobes; fl. buds erect. Lvs. of sterile shoots long-stalked, heart-shaped, and toothed; fl. stem with lower lvs. lance-shaped and the upper lvs. linear. Fr. nodding. △ Poor pastures, open woods, heaths, dunes. June–Oct. All Eur. (except P.TR.).

C. scheuchzeri Vill. Like 1339, but fls. usually solitary, rarely 2–5, dark blue-violet, broadly bell-shaped, 2–3 cm. long, nodding in bud. Stem lvs. all linear-lance-shaped, ciliate on margin below; basal lvs. kidney-shaped, shallowly lobed; stem 10–20 cm. Mountains of Central and Southern Europe. Pl. 140.

(iii) Blades of lower lvs. broadest towards base. Basal lvs. grading gradually into stem lvs.
1340. *C. trachelium* L. BATS-IN-THE-BELFRY. An erect, rough-stemmed perenn. ½–1 m. with large blue fls. 3–4 cm. long and borne in a long, leafy, spike-like cluster. Fls. solitary, or 2–4, in the axils of the upper lvs.; corolla tubular-bell-shaped, lobes acute, ciliate, one-third as long as the tube; calyx hairy, lobes triangular acute, erect. Lower lvs. large and long-stalked, blade *c.* 10 cm. oval-triangular, with heart-shaped base and an acute apex, margin strongly toothed, the upper lvs. narrower and short-stalked; stems angled. △ Woods, thickets, hedges, rocky places. June–Sept. All Eur. (except IS.). Pl. 139.

1341. *C. latifolia* L. LARGE BELLFLOWER. Like 1340, but fls. larger, 4–5 cm. long, violet, solitary in the axils of the lvs.; calyx hairless. Lvs. oval-lance-shaped and narrowed abruptly to the lv. stalk, not heart-shaped at base, irregularly toothed, finely softly hairy. Stem robust, smooth, finely hairy, to 1 m. △ In mountains: valleys, streamsides, woods. June–Sept. Most of Eur. (except P.IRL.IS.AL.GR.TR.).

1342. *C. rapunculoides* L. CREEPING BELLFLOWER. Fls. blue, drooping, borne in a slender, one-sided, spike-like leafless cluster. Fls. 1–3, short-stalked, in the axils of small linear bracts; corolla funnel-shaped, 2–3 cm., lobes ciliate, about as long as the tube; calyx with stiff adpressed hairs, lobes reflexed after flowering. Lower lvs. stalked, with an oval-heart-shaped toothed blade 5–8 cm., the upper lvs. narrower, lance-shaped acute, stalkless. An erect, finely hairy or nearly hairless perenn. which spreads by adventitious buds arising from the creeping roots; fl. stems 30–100 cm. △ Fields, woods, vineyards. July–Aug. All Eur. (except P.IS.AL.TR.): introd. IRL. Pl. 140.

1343. *C. bononiensis* L. Distinguished from 1342 by its smaller blue-lilac fls. 1–2 cm., borne in a dense, often one-sided, leafless, pyramidal, spike-like cluster. Corolla lobes hairless; calyx lobes linear-lance-shaped, spreading after flowering; fls. very short-stalked. Lvs. with fine crisped hairs above, densely white-woolly beneath. Root swollen, without adventitious buds; fl. stem hairy 30–100 cm. △ In mountains: rocks, thickets, pastures, woods. June–Aug. Much of Eur. (except North Eur. P.IRL.GB.B.NL.TR.). Pl. 140.

1344. *C. rhomboidalis* L. Distinguished from 1342 by its upper stem lvs. which are stalkless, oval-lance-shaped with a broad rounded base, and only 2–3 times as long as broad. Fls. few, 2–10, in a narrow, one-sided spike; corolla 1–2 cm., lobes hairless; calyx hairless, lobes linear, spreading, but at length reflexed; bracts lance-shaped. A slender erect perenn. with angular, leafy stem 30–70 cm. △ In mountains: pastures, open woods. June–Aug. F.D.CH.A.I.: introd. CS. Pl. 140.

ADENOPHORA   Like *Campanula*, but style encircled at the base by a fleshy tubular ring or disk. Capsule splitting by 3 pores at its base. *c*. 3 sps.

1345. *A. liliifolia* (L.) Besser An erect hairless perenn. 30–120 cm., branched above and with rather lax leafless spikes of drooping pale blue or whitish fls. Corolla 1½–2 cm., funnel-shaped, lobes oval-acute; style long-projecting; calyx lobes lance-shaped, entire or toothed, erect. Lvs. thin, hairless, the lower stem lvs. lance-shaped, stalked, toothed, the upper lvs. narrower. △ Woods, damp fields. July. Cent. Eur. I.YU.R.SU.

LEGOUSIA   Like *Campanula*, but ovary and capsule elongate-cylindrical; fls. with a very short funnel-shaped tube and with longer spreading lobes. Often included in *Specularia*. 4 sps.

Corolla shorter than the calyx

1346. *L. hybrida* (L.) Delarbre A hairy ann. 5–30 cm., with erect leafy stems and tiny violet or reddish-purple fls. which scarcely open. Corolla 8–15 mm. across; calyx lobes elliptic to lance-shaped, twice as long as the corolla and about half as long as the ovary at flowering. Lvs. oblong with undulate margin, stalkless, rough-hairy. Fr. 1½–3 cm. cylindrical, with persistent calyx. △ Sandy places, fields, stony banks. Apr.–July. P.E.F.GB.B.NL.CH.D.I.YU. GR.TR.R.SU.

L. falcata (Ten.) Fritsch Like 1346, but fls. violet and borne in a very lax, leafy spike; calyx with slender acute lobes which are strongly curved outwards and about 3 times as long as the corolla. Lvs. obovate, rather broad. A hairless, almost shining, unbranched ann. Mediterranean region.

Corolla as long as or longer than the calyx

1347. *L. speculum-veneris* (L.) Druce VENUS' LOOKING-GLASS. An erect, slender, branched ann. 10–30 cm., with conspicuous, dark violet-purple, wide-opening fls. in a lax, branched, leafy cluster. Corolla lobes spreading, *c*. 2 cm. across; calyx lobes awl-shaped, little shorter than the ovary at flowering. Lvs. rough, stalkless, oblong or obovate with weakly undulate margin. Fr. 1–1½ cm. △ Cornfields, stony places, tracksides, waste ground. May–July. Much of Eur. (except North Eur. P.IRL.GB.SU.). Pl. 141.

L. pentagonia (L.) Druce LARGE VENUS' LOOKING-GLASS. Like 1347, but fls. larger, 2–3 cm. across, and lobes of corolla spreading in a pentagonal star. Calyx lobes bristly-haired, one-half to one-third as long as the ovary at flowering. Fr. 2–3 cm. Western and Eastern Mediterranean region.

TRACHELIUM   Fls. in a dense, branched, flat-topped cluster. Corolla with a slender tube and with 5 spreading lobes; style long-projecting. Capsule opening by 2–3 basal pores. 2 sps.

1348. *T. caeruleum* L. THROATWORT. A rather robust, hairless, unbranched perenn. 30–80 cm., with numerous small, long-tubed blue or lilac fls. in a much branched, leafless, umbel-like cluster. Corolla tube very slender 6–8 mm., lobes 5, tiny, spreading; style much longer than the corolla, stigmas 3. Lvs. oval-lance-shaped, stalked, conspicuously saw-toothed, the upper lvs. narrower and shorter stalked. Fr. globular. △ Walls, shady rocks. May–Sept. P.E.I.: introd. F.

ASYNEUMA   Like *Phyteuma*, but fls. solitary or many and axillary, not densely clustered into a head; corolla funnel-shaped, divided almost to the base into 5 free, linear lobes. Ovary three-celled; capsule with 3 pores. *c*. 10 sps.

1349. *A. limonifolium* (L.) Janchen An erect, often unbranched perenn. 15–70 cm. with lvs. mostly basal and a simple elongate spike of blue-lilac fls. Fls. 1–3 in the axils of tiny triangular bracts, stalkless or nearly so; corolla *c.* 5 mm. long; calyx lobes linear-lance-shaped, one-third as long as the corolla. Lvs. lance-shaped, long-stalked, margin entire or toothed. △ Rocks in the mountains. June–July. YU.AL.GR.BG.

PHYTEUMA | **Rampion** Perenns. with numerous fls. clustered into dense heads or spikes. Corolla divided nearly to the base into linear lobes which are at first attached to each other by their tips but later separate and spread; calyx tubular with 5 narrow lobes; stamens 5; style 1; stigmas 2–3, linear, projecting. Fr. globular with 2–3 valves or pores. *c.* 15 sps.

Fls. in elongate heads, twice as long as broad or more

1350. *P. spicatum* L. SPIKED RAMPION. An erect, unbranched, hairless perenn. 30–70 cm., with numerous yellowish or rarely blue fls. in a terminal, dense, cylindrical spike 3–8 cm. long, which elongates to 5–12 cm. in fr. Corolla *c.* 1 cm. long, lobes fused above, but curved and separated below; calyx lobes minute; style *c.* 1 cm.; bracts linear-awl-shaped, the lower longer than the fls. Basal lvs. with long stalks which are winged above and with oval-heart-shaped blades 3–5 cm., with double-toothed margin, the lower stem lvs. longer and narrower, the uppermost lvs. stalkless, narrow lance-shaped. Root fleshy, swollen, spindle-shaped. △ Mountains, meadows, woods. May–July. Most of Eur. (except P.IRL. IS.GR.TR.BG.): introd. N.SF. Pl. 141.

P. betonicifolium Vill. BLUE-SPIKED RAMPION. Like 1350, but fls. blue or lilac in an ovoid head to 4 cm., which later becomes oblong-cylindrical. Fls. erect after opening; stamens hairless; stigmas 3; lower bracts of fl. spike linear and shorter than the fls. Lower lvs. oblong, 2–4 times as long as broad, stem lvs. many, small, linear. Pyrenees, Alps, Apennines. Pl. 141.

1351. *P. ovatum* Honckeny (*P. halleri* All.) DARK RAMPION. Like 1350, but fls. dark blue-purple in spikes to 6 cm. long, and with the lowest bracts of the fl. spike lance-shaped, leafy, and longer than the fls. Basal lvs. heart-shaped, as long as broad, mostly strongly and irregularly toothed, the uppermost stem lvs. lance-shaped. A more robust, hairless perenn. to 1 m., leafy to the inflorescence. △ In mountains: meadows, woods, thickets. July–Aug. E.F.D.CH.A.I.YU.

Fls. in dense globular heads, as broad as or broader than long

(a) Basal lvs. lance-shaped to oval

1352. *P. orbiculare* L. ROUND-HEADED RAMPION. An erect, hairless, unbranched perenn. 20–60 cm., with blue fls. in terminal globular heads 1½–2½ cm. Bracts below fl. heads oval-lance-shaped, shorter or longer than the fls. Basal lvs. in a rosette, heart-shaped to triangular, long-stalked; lower stem lvs. narrower, short-stalked, the upper lvs. narrow-lance-shaped, stalkless, all lvs. with rounded teeth. A very variable sp. △ Dry places, rocks, thickets, meadows. May–Oct. Cent. Eur. E.F.I.YU.AL.R.SU. Pl. 141.

P. scheuchzeri All. HORNED RAMPION. Like 1352, but bracts below fl. head linear and usually much longer than the head and often recurved. Lower lvs. conspicuously saw-toothed, not rounded-toothed, of two types, either oval heart-shaped or linear-oblong. Alps, Apennines, Carpathians.

P. comosum L. DEVIL'S CLAW. A very distinctive plant with a dense globular head of pale pink to violet-red fls., each fl. much inflated at the base and abruptly narrowed to a slender darker tube above and with protruding stigma. Fl. heads with 16–20 fls., fls. 1½–2 cm. long.

A tufted perenn. with spreading leafy stems and glossy, coarsely toothed, mostly lance-shaped lvs. Italy, Austria. Pl. 140.

(b) *Basal lvs. linear*

1353. *P. hemisphaericum* L. A tufted perenn. with narrow lvs. in a lax rosette and short unbranched fl. stems 3–15 cm. bearing a dense, flattish, globular head of blue fls. 1–1½ cm. across. Bracts broadly oval, long-pointed, obscurely toothed, ciliate, about half as long as the fls. Lvs. grass-like, narrow spathulate to linear, the stem lvs. few, broader. △ In mountains: pastures, rocks. June–Aug. E.F.D.CH.A.I. Page 411.

EDRAIANTHUS   Like *Campanula*, but capsule splitting irregularly at the apex, not at the base. Fls. in a terminal head, closely surrounded by a leafy involucre. *c.* 6 sps.

1354. *E. graminifolius* (L.) DC. FALSE BELLFLOWER. A low, tufted, rosette-forming perenn. with narrow lvs. and with several erect stems 5–20 cm., bearing a globular cluster of blue, violet, or rarely white fls. Corolla 1–2 cm., funnel-shaped; calyx tube hairless, lobes softly hairy; bracts narrow, long-pointed, about as long as the fls. Rosette lvs. broadly linear 1–4 cm. by ½–4 mm., margin entire, the stem lvs. few, narrower. A variable sp. △ Rocks in mountains. May–Aug. A.I.GR.

JASIONE   Like *Phyteuma*, but globular head of fls. closely surrounded by an involucre of overlapping bracts. Corolla lobes linear, spreading, not attached at their apex; stigmas short, stout; anthers fused at their base. Fr. two-valved. *c.* 13 sps.

1355. *J. montana* L. SHEEP'S BIT. A hairy, spreading or ascending, usually bienn. 10–50 cm. with blue or rarely white fls. densely clustered into terminal, rounded, long-stalked heads ½–3½ cm. across. Fls. *c.* ½ cm. long; calyx lobes awl-shaped, about as long as the unopened corolla; bracts more or less oval, entire or toothed, shorter than the fls. Lvs. linear-lance-shaped, with undulate margin, rough-hairy. △ Sandy places, rocks, dry meadows, thickets. June–Sept. All Eur. (except IS.GR.TR.). Pl. 141.

LOBELIA   Fls. symmetrical in one plane only; corolla tube oblique and curved, distinctly two-lipped, the upper lip two-lobed, the lower three-lobed. Calyx five-lobed; stamens 5, unequal. Ovary two-celled. 2 sps.

1356. *L. urens* L. ACRID LOBELIA. A nearly hairless perenn. 20–60 cm., with acrid juice, slender, angular, leafy stems, and blue or purplish fls. in a lax elongated spike to 20 cm. Fls. erect, short-stalked; corolla 1–1½ cm., two-lipped; calyx with bristle-like spreading teeth longer than the tube; bracts linear. Lvs. to 7 cm., oblong-oval and irregularly toothed, the upper narrower. Capsule oblong-cylindrical, erect. △ Heaths, damp woods. July–Sept. P.E.F.GB.

*L. dortmanna L. WATER LOBELIA. North-Western Europe.

COMPOSITAE | Daisy Family

The largest and probably the most successful and ubiquitous family of flowering plants in the world, showing a wide range of vegetative form. Distinguished by the uniformity of the inflorescence in which many tiny fls. are clustered together to form a flower-like head or *capitulum* surrounded by calyx-like bracts or *involucral bracts*. The tiny fls. or *florets* are arranged on the apex of the stem, often called the *receptacle*, which may be conical, flat, or concave, and with or without scales subtending each floret. The florets are usually bisexual and complete, possessing calyx, corolla, stamens, and ovary, but one or more of the organs may be absent. The calyx is often reduced to a ring of simple or branched hairs, or bristles, or scales, and becomes the *pappus* in fr. The 5 stamens are fused by their bases to the corolla and the anthers are fused together to form a tube round the style; the ovary is inferior and one-seeded; stigmas 2. The fr. is a one-seeded nutlet, sometimes lengthened above into a *beak*; the fr. is usually crowned by a pappus of scales, or simple hairs, or branched feathery hairs and more than one of these types of pappus may be present on a single fr., or the pappus may be absent. The corolla is of two basic types: (a) Bell-shaped with 5 short teeth. Florets with this type of corolla are called *disk-florets* (or *tubular-florets*). (b) Strap-shaped with a long narrow limb. Florets with this type of corolla are called *ray-florets* (or *ligulate-florets*).

Fl. heads may be composed of disk-florets only (e.g. in groundsel); or of a central group of disk-florets and peripheral row of ray-florets (e.g. in daisy); or all the florets may be ray-florets (e.g. in dandelion). The family comprises about 900 genera and the characters distinguishing each genera are often based on a combination of small botanical differences, often making their recognition in the field difficult. Characters of particular importance are those relating to the involucral bracts, the scales of the receptacle, the calyx and pappus, the floret types and the fr.

1. Fl. heads one-sexed, the male with several florets, the female with 1–2 florets. Group A
1. Fl. heads with all florets bisexual, or outer florets male, female, or sterile;
   fl. heads rarely all female and then numerous.
  2. Disk-florets present, with or without ray-florets. Plants without latex.
   3. Disk-florets only present.
    4. Lvs. not spiny.
     5. Pappus absent, or fr. with a short crown or ear-like projection. Group B
     5. Pappus of scales or hairs, at least on inner frs. Group C
    4. Lvs. spiny. Group D
   3. Disk-florets and ray-florets present.
    6. Pappus of scales or hairs. Group E
    6. Pappus absent or fr. with a crown or ear-like projection. Group F
  2. Ray-florets only present. Plants with latex.
   7. Pappus absent or fr. with a narrow membraneous margin. Group G
   7. Pappus distinct, of scales or hairs.
    8. Pappus of scales. Group H
    8. Pappus of hairs.
     9. Hairs of pappus simple. Group I
     9. Hairs of pappus branched, feathery. Group J

Group A. *Fl. head with few one-sexed florets*

		Nos.
1. Involucral bracts of female florets fused together.	*Ambrosia*	1400
1. Involucral bracts of female florets free.	*Xanthium*	1401, 1402

Group B. *Disk-florets only present; lvs. not spiny; pappus absent or nearly so*

		Nos.
1. Fl. heads pink, purple, or violet.	*Centaurea*	1499–1507
1. Fl. heads yellow, white, or brownish.		
2. Involucral bracts spiny.	*Centaurea*	1499–1507
2. Involucral bracts not spiny.		
3. Lvs. entire.		

4. Fl. heads tiny, densely clustered; florets whitish or
   brownish.   *Evax*   1374
4. Fl. heads larger; florets white; or if fl. heads tiny
   then in a branched cluster.
  5. Plant covered with dense snowy-white hairs. Fl.
     heads in a flat-topped cluster.   *Otanthus*   1423
  5. Plant not snowy-white. Fl. heads solitary, axillary,
     or in widely branched clusters.
    6. Fl. heads solitary or 2, stalkless and axillary.   *Carpesium*   1394
    6. Fl. heads in branched clusters.   *Artemisia*   1434–1438
 3. Lvs. pinnately lobed.
  7. Fl. heads many.
    8. Fl. heads in branched pyramidal clusters.   *Artemisia*   1434–1438
    8. Fl. heads in a dense flat-topped cluster.   *Chrysanthemum* 1424–1429
  7. Fl. heads solitary.
    9. Undershrubs with greyish woolly-haired lvs.   *Santolina*   1408
    9. Hairless anns.
     10. Receptacle scales present.   *Anthemis*   1409–1412
     10. Receptacle without scales.
      11. Florets with four-lobed corolla.   *Cotula*   1433
      11. Florets with five-lobed corolla.   *Matricaria*   1431, 1432

Group C. *Disk-florets only present; lvs. not spiny; pappus present on some or all frs.*
                     Nos.

1. Stem lvs. opposite.
 2. Fl. heads yellow or brownish, solitary; pappus of 2–4
   stiff barbed bristles.   *Bidens*   1406
 2. Fl. heads pink or lilac, in a flat-topped cluster; pappus
   of simple hairs.   *Eupatorium*   1357
1. Stem lvs. alternate, or stems with scale lvs. only.
 3. Stem with scale lvs. only, sometimes with juvenile lvs.
   at base.
  4. Stem with a solitary long-stalked fl. head.   *Homogyne*   1443
  4. Stem with many fl. heads in a spike-like inflorescence.   *Petasites*   1440–1442
 3. Stem lvs. alternate.
  5. Pappus of scales or simple hairs.
    6. Inner involucral bracts longer than the florets,
      brightly coloured, shining and more or less spreading.   *Xeranthemum*   1464, 1465
    6. Inner involucral bracts not longer than florets or
      brightly coloured.
     7. Florets all one-sexed, fl. heads one-sexed. Stolons
       present.   *Antennaria*   1378
     7. Florets all bisexual or outer florets female.
      8. Fl. heads yellow, whitish-green, or brownish.
       9. Lvs. hairless, narrow linear. Fl. heads yellow.   *Crinitaria*   1368
       9. Lvs. hairy.
        10. Fl. heads surrounded by an involucre of upper
          lvs. spreading in a star.   *Leontopodium*   1379
        10. Fl. head not surrounded by a leafy involucre.
         11. Fl. heads tiny, $\frac{1}{2}$ cm. long, or less.
          12. Outer involucral bracts papery, shining, bright
            yellow.   *Helichrysum*   1384, 1385

12. Outer involucral bracts not papery, or with a
    papery margin only.
   13. Involucral bracts green, without papery
       margin; fl. heads in spreading branched
       clusters. *Conyza* 1373
   13. Involucral bracts with a papery margin; fl.
       heads usually in dense stalkless clusters.
      14. Receptacle with scales among the marginal
          florets; involucre angular in section, bracts
          not spreading at maturity. *Filago* 1375–1377
      14. Receptacle without scales; involucre
          rounded in section, bracts spreading at
          maturity. *Gnaphalium* 1380–1383
   11. Fl. heads at least ½ cm. long, usually more.
      15. Involucral bracts papery or leathery; fl. heads
          1 or few. *Phagnalon* 1386
      15. Involucral bracts green; fl. heads many in a
          branched cluster.
         16. Involucral bracts of many unequal rows. *Inula* 1387–1392
         16. Involucral bracts of 1 row and of a few much
             smaller outer bracts. *Senecio* 1449–1458
8. Fl. heads pink, purple, or bluish.
  17. Involucral bracts spiny-pointed and hooked. *Arctium* 1472, 1473
  17. Involucral bracts not hooked.
   18. Plant with short, above-ground stolons.
       Involucral bracts narrow, papery. *Antennaria* 1378
   18. Plant without above-ground stolons.
      19. Involucral bracts with a papery, often ciliate
          appendage, or a papery margin, or with a
          pinnate or palmate spine. *Centaurea* 1499–1507
      19. Involucral bracts without a papery appendage
          or margin, with a simple spine, or spineless.
         20. Involucral bracts with a spine-tipped apex
             which soon falls. *Centaurea* 1499–1507
         20. Involucral bracts spineless or with a simple
             persistent spine.
            21. Pappus dark purplish. Fl. head cylindrical;
                lvs. pinnate with toothed segments. *Crupina* 1496
            21. Pappus whitish.
               22. Woody-based or shrubby plants. *Staehelina* 1474
               22. Herbaceous plants.
                  23. Involucral bracts 4–8; florets few 2–15;
                      lvs. broadly heart-shaped or kidney-
                      shaped. *Adenostyles* 1444, 1445
                  23. Involucral bracts many; florets many; lvs.
                      not heart-shaped.
                     24. Lvs. woolly-haired beneath. Pappus fused
                         in a ring at the base. *Jurinea* 1476
                     24. Lvs. not woolly-haired.
                        25. Lvs. bristly-haired; fr. attached by the
                            base; pappus fused in a ring at the base. *Carduus* 1477–1480

25. Lvs. not bristly; fr. attached laterally to the receptacle; pappus not fused in a ring.	*Serratula*	1497, 1498
5. Pappus with at least some hairs branched and feathery.		
26. Involucral bracts with an appendage or a broad, papery margin.	*Centaurea*	1499–1507
26. Involucral bracts without an appendage or papery margin.		
27. Pappus hairs in 2 rows, the outer shorter, bristly, the inner longer, feathery.	*Saussurea*	1475
27. Pappus hairs all feathery.		
28. Receptacle fleshy; fr. quadrangular in section; fl. heads usually 5 cm. or more across.	*Cynara*	1491
28. Receptacle not fleshy; fr. somewhat flattened; fl. heads usually 4 cm. or less across.	*Cirsium*	1482–1490

Group D. *Disk-florets only present; lvs. spiny*		*Nos.*
1. Pappus absent, or pappus of scales or simple hairs.		
2. Fl. heads with 1 floret, but fl. heads numerous, stalkless, and clustered into a large, globular, spiny inflorescence.	*Echinops*	1461, 1462
2. Fl. heads with many florets.		
3. Receptacle without scales, pitted.	*Onopordum*	1494, 1495
3. Receptacle with scales, or bristles.		
4. Involucral bracts spiny-toothed or with a terminal pinnate spine.		
5. Fl. heads small, with few florets, in a dense flat-topped cluster.	*Cardopatium*	1463
5. Fl. heads with numerous florets, solitary or in branched clusters.		
6. Pappus in 1 row or absent.	*Carthamus*	1508
6. Pappus in 2 unequal rows.	*Cnicus*	1509
4. Involucral bracts acute or with a simple spine.		
7. Filaments not fused; lvs. green.	*Carduus*	1477–1480
7. Filaments fused at least at the base into a tube; lvs. variegated with white.	*Silybum*	1492
1. Pappus at least on some frs. with branched, feathery hairs.		
8. Inner involucral bracts long, bright-coloured or white, and spreading outwards.	*Carlina*	1466–1469
8. Inner involucral bracts not spreading.		
9. Fr. hairy.	*Atractylis*	1470, 1471
9. Fr. hairless.		
10. Pappus of inner frs. feathery, of outer simple-haired.	*Notobasis*	1481
10. Pappus of all frs. feathery-haired.		
11. Filaments fused into a tube.	*Galactites*	1493
11. Filaments free.		
12. Receptacle fleshy; fl. heads usually 5 cm. or more; involucral bracts broad.	*Cynara*	1491
12. Receptacle not fleshy; fl. heads usually less than 5 cm.; involucral bracts narrow.	*Cirsium*	1482–1490

Group E. *Disk- and ray-florets present: pappus of scales, bristles, or hairs*

Nos.

1. Lvs. opposite.
  2. Fl. head small, globular; ray-florets short, white. Lvs. entire.   *Galinsoga*   1407
  2. Fl. heads medium-sized or large; ray florets yellow.
    3. Lvs. entire; pappus of numerous hairs.   *Arnica*   1446
    3. Lvs. saw-toothed or lobed; pappus of 2–4 rough bristles.   *Bidens*   1406
1. At least the upper lvs. alternate, or lvs. all basal in a rosette.
  4. Pappus of 2–4 narrow, deciduous scales. Fl. head large; ray-florets yellow.   *Helianthus*   1404, 1405
  4. Pappus of numerous scales or hairs.
    5. Pappus of tiny scales.
      6. Lvs. three-lobed or pinnate. Receptacle conical.   *Rudbeckia*   1403
      6. Lvs. undivided.
        7. Scales of pappus not fused.   *Asteriscus*   1398, 1399
        7. Scales of pappus fused into a crown.
          8. Marginal frs. triangular in section, with narrowly winged angles; frs. of disk-florets angular; lvs. lance-shaped.   *Buphthalmum*   1396
          8. Frs. of ray- or disk-florets similar, not winged; lvs. broad, heart-shaped or rhomboid.   *Telekia*   1397
    5. Pappus of hairs.
      9. Ray-florets white, bluish, or violet.
        10. Stem with scale lvs. only. Fl. heads many in a spike-like cluster.   *Petasites*   1440–1442
        10. Stem leafy, or without lvs. or scales.
          11. Stems leafless, all lvs. in a basal rosette. Ray-florets white.   *Bellidastrum*   1362
          11. Stems leafy.
            12. Ray-florets very narrow, in 2 or several rows.   *Erigeron*   1369–1372
            12. Ray-florets broader, usually in 1 row.   *Aster*   1363–1367
      9. Ray-florets yellow.
        13. Stems with scale lvs. only; lvs. all basal.   *Tussilago*   1439
        13. Stems leafy.
          14. Pappus two-rowed, the outer of short toothed scales, the inner of hairs.   *Pulicaria*   1393
          14. Pappus all of hairs.
            15. Involucral bracts unequal, in several overlapping ranks.
              16. Fl. heads small, with few ray-florets, in a spike-like or branched cluster. Anthers not heart-shaped at base.   *Solidago*   1358, 1359
              16. Fl. heads small or medium-sized or large, in a flat-topped cluster, a branched spreading cluster, or solitary. Anthers heart-shaped at base.   *Inula*   1387–1392
            15. Involucral bracts of equal length in 1–3 ranks, not overlapping, often with small bracts at the base.
              17. Pappus with feathery hairs.   *Ligularia*   1459
              17. Pappus with simple hairs.

18. Involucral bracts in 1 rank, often with scales at
the base; fr. all with a pappus of hairs. *Senecio* 1449–1458
18. Involucral bracts in 2–3 ranks, without scales at
base; marginal frs. often without a pappus of
hairs. *Doronicum* 1447, 1448

Group F. *Disk- and ray-florets present; pappus absent or fr. with a crown or ear-like projection*
Nos.

1. Ray-florets white, whitish-yellow, pink, or purple.
  2. Fl. heads on leafless stems. Lvs. all basal in a rosette. *Bellis* 1360, 1361
  2. Fl. heads on leafy stems.
    3. Receptacle without scales.
      4. Lvs. 1–3 times pinnately cut into slender, linear or
thread-like, ultimate segments.
        5. Fr. finely three-ribbed on outer surface, apex
without oil glands. *Matricaria* 1431, 1432
        5. Fr. strongly three-ribbed on outer surface; apex with
1–2 oil glands. *Tripleurospermum* 1430
      4. Lvs. entire or pinnately cut; ultimate segments not
linear or thread-like.
        6. Fr. with 5–10 ribs; perenns. *Chrysanthemum* 1424–1429
        6. Fr. flattened, smooth; small anns. *Bellis* 1360, 1361
    3. Receptacle with scales.
      7. Fr. not flattened, quadrangular, ribbed. *Anthemis* 1409–1412
      7. Fr. strongly flattened, not ribbed.
        8. Disk-florets with base of corolla tube enlarged and
sac-like. *Chamaemelum* 1413
        8. Disk-florets with base of corolla tube not enlarged.
          9. Marginal frs. winged; disk-florets yellow with
corolla two-lipped, five-lobed. *Anacyclus* 1414
          9. Fr. not winged; disk-florets usually whitish and
corolla not two-lipped. *Achillea* 1415–1422
1. Ray-florets yellow.
  10. Receptacle without scales.
    11. Fr. strongly incurved, with swellings on the back;
lvs. entire and remotely toothed. *Calendula* 1460
    11. Fr. not or slightly curved; lvs. saw-toothed or
pinnately lobed. *Chrysanthemum* 1424–1429
  10. Receptacle with scales.
    12. Lvs. with wedge-shaped to lance-shaped lobes. Fl.
head large; disk-florets brownish. *Rudbeckia* 1403
    12. Lvs. either once or twice pinnately cut, or entire.
      13. Lvs. entire.
        14. Fl. heads with spiny involucre. *Pallenis* 1395
        14. Fl. heads without spiny involucre.
          15. Marginal frs. triangular in section, narrowly
winged on the angles; inner frs. rounded; lvs.
lance-shaped. *Buphthalmum* 1396
          15. All frs. similar, rounded; lvs. broad, heart-shaped
or rhomboidal. *Telekia* 1397
      13. Lvs. once or twice pinnately cut.

16. Fr. rounded or quadrangular in section. Fl. heads long-stalked.	*Anthemis*	1409–1412
16. Fr. flattened.		
17. Marginal frs. winged; fl. heads solitary.	*Anacyclus*	1414
17. Marginal frs. not winged; fl. heads in a dense flat-topped cluster.	*Achillea*	1415–1422

Group G. *Ray-florets only present; pappus absent or fr. with a narrow membraneous margin*

		Nos.
1. Lvs. spiny. Pappus absent or with 2–3 bristles.	*Scolymus*	1510
1. Lvs. not spiny.		
2. Stems leafy.		
3. Mature fr. oblong or oval with 20–30 ribs, erect.	*Lapsana*	1513
3. Mature fr. linear, spreading in a star.	*Rhagadiolus*	1518
2. Stems without lvs., with 1–5 fl. heads.		
4. Stems below fl. heads not swollen; fr. with 5 ribs.	*Aposeris*	1514
4. Stems below fl. heads swollen; fr. with 10 ribs.	*Arnoseris*	[1516]

Group H. *Ray-florets only present; pappus of scales*

		Nos.
1. Florets blue.		
2. Involucral bracts papery, silvery-coloured; scales of fr. long.	*Catananche*	1511
2. Involucral bracts green, leafy; scales of fr. very short.	*Cichorium*	1512
1. Florets yellow.		
3. Stems erect, leafy at base; involucral bracts not hardening in fr.	*Tolpis*	1515
3. Stems leafless or prostrate below: involucral bracts hardening in fr.		
4. Inner frs. narrowly winged.	*Hyoseris*	1516
4. Inner frs. not winged.	*Hedypnois*	1519

Group I. *Ray-florets only present; pappus of simple hairs*

		Nos.
1. Fr. of two forms: the inner smooth or fine-pointed, the outer either rough or velvety or with inner face keeled or strongly convex.		
2. Marginal frs. strongly convex, rounded in section. Lateral fr. heads stalkless.	*Zazintha*	1517
2. Marginal frs. with inner face keeled or winged or transversely wrinkled.		
3. Marginal frs. keeled or winged on inner face.	*Crepis*	1546–1549
3. Marginal frs. transversely wrinkled.	*Reichardia*	1536
1. Fr. all similar.		
4. Fr. with a beak, pappus thus appearing stalked.		
5. Beak of fr. with a tiny crown of scales at the base.	*Chondrilla*	1534
5. Beak of fr. without a crown at base.		
6. Fl. stems without lvs.; lvs. all basal.	*Taraxacum*	1535
6. Fl. stems leafy.		
7. Pappus of an inner row of long hairs and an outer row of much shorter hairs.	*Mycelis*	1540
7. Pappus of 2 rows of hairs of equal length.		

8. Fr. flattened from side to side.   *Lactuca*   1541–1544
8. Fr. cylindrical, angled, not flattened.   *Crepis*   1546–1549
4. Fr. without a beak, pappus thus appearing stalkless.
   9. Fr. flattened from side to side.
    10. Florets blue; pappus of 2 unequal rows of hairs.   *Cicerbita*   1537
    10. Florets yellow; pappus of 2 equal rows of hairs.   *Sonchus*   1538, 1539
   9. Fr. round in section or angled.
    11. Florets violet or purple. Involucre narrow cylindrical.   *Prenanthes*   1545
    11. Florets yellow, rarely orange or pale pink.
     12. Fr. with swellings at the apex. Stems without lvs.   *Taraxacum*   1535
     12. Fr. smooth.
      13. Receptacle with pits surrounded by hairs as long as
       or longer than the frs.   *Andryala*   1533
      13. Receptacle without hairs surrounding pits.
       14. Fr. gradually narrowed to the apex; involucre
        usually of bracts of equal length; pappus snowy-
        white, soft (except *C. paludosa*).   *Crepis*   1546–1549
       14. Fr. abruptly narrowed to the apex; involucre
        usually of many rows of unequal bracts. Pappus
        dirty white to pale brown, brittle.   *Hieracium*   1550–1557

Group J. *Ray-florets only present; pappus of branched, feathery hairs*   *Nos.*

1. Receptacle with narrow, linear scales. Fl. stem leafless or
  with few lvs.   *Hypochoeris*   1520–1522
1. Receptacle without scales.
  2. Involucral bracts in 1 row, not overlapping.
   3. Lvs. entire.   *Tragopogon*   1527, 1528
   3. Lvs. toothed.   *Urospermum*   1526
  2. Involucral bracts in 2 rows or overlapping.
   4. Fl. heads with an involucre of 3–5 upper lvs.   *Picris*   1525
   4. Fl. heads without an involucre of lvs.
    5. Involucral bracts hairless, triangular-oval, to broadly
     lance-shaped, regularly overlapping; branches of
     pappus hairs intertwined.
     6. Fr. swollen at base; lvs. often pinnately lobed.   *Podospermum*   1532
     6. Fr. not swollen at base; lvs. entire.   *Scorzonera*   1529–1531
    5. Involucral bracts hairless or hairy, linear; branches of
     pappus hairs not intertwined.
     7. Stems leafless or with a small lv. at junction of the
      branches.   *Leontodon*   1523, 1524
     7. Stems leafy.   *Picris*   1525

EUPATORIUM |**Hemp Agrimony**   Fl. heads many, clustered into dense flat-topped inflorescences; disk-florets only present; involucral bracts few in 2–3 rows. Lvs. three- to five-lobed or lance-shaped. Fr. with a pappus of simple hairs. 1 sp.

1357. *E. cannabinum* L. HEMP AGRIMONY. A robust perenn. 30–120 cm., with many downy, often reddish, leafy stems and with terminal, dense, flat-topped clusters of numerous reddish-purple or white oblong fl. heads. Involucral bracts *c.* 10, purple-tipped; fl. heads *c.* 1 cm. long, with only 5–6 disk-florets. Lvs. all opposite, the lower oblance-shaped, stalked, and differing from the upper stalkless lvs. which are usually deeply three-lobed with elliptic toothed lobes 5–10 cm.; lvs of side shoots oval-lance-shaped. Fr. blackish, five-angled, pappus whitish. △ Marshes, fens, streamsides, damp woods. July–Aug. All Eur. (except IS.). *Used in herbal remedies in the past.* Pl. 141.

SOLIDAGO | Golden-Rod   Fl. heads small, yellow, with both disk- and ray-florets present; involucre of several rows of overlapping bracts. Fr. many-ribbed, not compressed; pappus of 1 or 2 rows of simple hairs. *c.* 5 sps.

1358. *S. virgaurea* L. GOLDEN-ROD. An erect, little-branched, usually hairless perenn. 5–75 cm., with small yellow fl. heads in a terminal, oblong, leafy cluster formed by many erect lateral branches. Fl. heads 7–10 mm. across; involucral bracts linear, greenish-yellow, with papery margin; ray-florets 6–12, spreading. Lower lvs. obovate-elliptic, usually toothed, narrowed to a stalk, the upper lvs. lance-shaped, entire, stalkless. Fr. brown, hairy, pappus white. A very variable sp. △ Woods, thickets, grasslands, rocks. July–Sept. All Eur. (except IS.AL.TR.). *It yields a yellow dye and has been used in herbal remedies.* Pl. 143.

1359. *S. canadensis* L. CANADIAN GOLDEN-ROD. A robust clumped perenn. 1–2½ m. with many erect stems bearing a dense pyramidal inflorescence of numerous tiny golden-yellow fl. heads arranged unilaterally on spreading branches. Fl. heads cylindrical, 5–6 mm. long; ray-florets inconspicuous, about as long as the disk-florets and scarcely longer than the lance-shaped, hairless, involucral bracts. Lvs. 10–15 cm. lance-shaped and long-pointed, three-veined, toothed, usually rough-hairy; stems hairy; rhizomes present. △ Native of North America; cultivated and often naturalized in waste places and river-sides. Aug.–Sept. Introd. to most of Eur.

S. gigantea Aiton Like 1359, but stems hairless below; lvs. hairless or hairy only on the veins beneath. Ray-florets longer than the disk-florets and involucral bracts. Native of North America; sometimes escaping from gardens. Introduced to most of Europe. Pl. 142.

BELLIS | Daisy   Fl. heads solitary, long-stalked; disk-florets yellow; ray-florets white or flushed purple; receptacle domed; involucral bracts two-rowed. Fr. compressed, without pappus. *c.* 6 sps.

Stems unbranched, leafless

1360. *B. perennis* L. DAISY. A rosette-forming perenn. with leafless fl. stems bearing solitary fl. heads 1½–2½ cm. across, with bright yellow disk-florets and numerous white or pinkish-tinged, spreading ray-florets. Involucral bracts lance-shaped, hairy. Lvs. all basal, obovate-spathulate, bluntly toothed, abruptly contracted into a short broad stalk, one-veined, sparsely hairy; fl. stem 3–15 cm., hairy. Fr. obovate, hairy, without pappus. △ Meadows, grassy places, tracksides. Mar.–Nov. All Eur.

B. sylvestris Cyr. SOUTHERN DAISY. Like 1360, but usually more robust and distinguished by its oblong, three-veined, greyish-hairy lvs. which are gradually narrowed to the long lv. stalk. Fl. heads 2–3½ cm. across; involucral bracts oblong-lance-shaped, acute; fl. stem 10–30 cm. Mediterranean region, Portugal.

Stems branched, leafy below

1361. *B. annua* L. ANNUAL DAISY. Like 1360, but an ann. which hardly forms rosettes, and with stems usually branched below and bearing a few lvs. at the base. Fl. heads small, 1–1½ cm. across; involucral bracts elliptic, blunt. Lvs. softly hairy, rather suddenly narrowed to the lv. stalk; fl. stem usually 3–10 cm. △ Salt-rich ground, moist places, grasslands. Mar.–June. Med. Eur. P. Page 432.

BELLIDASTRUM Like *Bellis,* but fr. with a pappus of white hairs. 1 sp.

1362. *B. michelii* Cass. FALSE DAISY. A daisy-like perenn. with lvs. all basal and forming a lax rosette, and with long downy fl. stems 10–30 cm. bearing large fl. heads 2–4 cm. across. Disk-florets yellow; ray-florets numerous, narrow, white or pinkish; involucral bracts linear-lance-shaped, hairy. Lvs. oval or spathulate, narrowed to a long stalk, often coarsely toothed, usually hairy. △ In mountains: rocks, pastures, clearings. Apr.–Sept. Cent. Eur. (except H.) E.F.I.YU.AL. Pl. 142.

ASTER Fl. heads with yellow disk-florets, contrasting with the single row of blue, red, or white ray-florets. Involucre of many rows of bracts; receptacle flat, pitted. Fr. compressed; pappus of 2–3 rows of simple hairs. A number of showy North American sps. have become naturalized in Europe. Also garden varieties can spread rapidly by vegetative propagation; their identity is often difficult to determine. *c.* 15 sps.

Stems unbranched; fl. heads solitary

1363. *A. alpinus* L. ALPINE ASTER. A small, hairy, alpine perenn. 5–20 cm., with large solitary fl. heads 3½–5 cm. across, or rarely 3–4 heads on a branched stem. Ray-florets blue-lilac or violet; disk-florets yellow; outer involucral bracts lance-shaped, blunt-tipped, densely hairy. Lvs. mostly basal, oval-spathulate, stalked, three-veined, hairy, the upper lvs. lance-shaped, one-veined. △ In mountains: dry pastures, rocky places. July–Sept. Cent., South-East Eur. (except TR.) E.F.I.SU. Pl. 142.

Stems branched; fl. heads many

1364. *A. amellus* L. EUROPEAN MICHAELMAS DAISY. Fl. heads with bluish-lilac ray-florets and yellow disk-florets, large 3–5 cm. across, rather few in a lax flat-topped cluster, or rarely solitary. Involucral bracts blunt-tipped, often coloured and spreading outwards at the tips, hairy. Lvs. rough-hairy, usually entire, the lower oval-elliptic, stalked, the upper lance-shaped acute, stalkless. An erect, sparsely hairy, little-branched perenn. 15–60 cm. △ Meadows, rocky places, open woods. Aug.–Sept. Cent., South-East Eur. (except AL.TR.) E.F.I.SU. Pl. 142.

1365. *A. tripolium* L. SEA ASTER. A maritime perenn. with a stout, branched, hairless stem 15–100 cm., narrow, fleshy, hairless lvs. and lax, branched, terminal clusters of bluish-purple fl. heads, which are each 1–3 cm. across. Disk-florets orange-yellow; ray-florets bluish-purple or whitish or rarely absent; involucral bracts blunt, hairless, the outer with a papery tip, the inner largely papery. Lvs. 7–12 cm., oblong-lance-shaped, entire or nearly so, quite hairless. Fr. brownish, pappus brownish. △ Salt marshes, cliffs and rocks on the littoral. July–Oct. All Eur. (except IS.CH.). *The root of this plant has been used to cure eye diseases.* Pl. 142.

1366. *A. sedifolius* L. Distinguished by its linear lance-shaped lvs. which are strongly gland-dotted and with clusters of smaller lvs. in their axils. Fl. heads lilac, *c.* 1½ cm. across, numerous, in a dense cluster; disk-florets yellow; ray-florets few, lilac; fl. stalks with many leafy bracts. An erect, leafy, rough perenn. 20–50 cm., much-branched above. △ Hills, dry uncultivated ground. June–Sept. E.F.I.YU.H.R.SU.

1367. *A novi-belgii* L. MICHAELMAS DAISY. Fl. heads violet, rather large, 2½–4 cm. across, in a broad, much-branched, pyramidal cluster. Disk-florets yellow; ray-florets violet; involucral bracts unequal, the outer about half as long as the inner, green, spreading, acute. Lvs. hairless, oval-lance-shaped, the lower stalked, the upper narrower, half-clasping the stem. An erect, nearly hairless perenn. 80–100 cm. △ Native of North America; widely naturalized by rivers, damp woods, marshes. Sept.–Oct. Much of Eur.

A. salignus Willd. Fl. heads large, 2–4 cm. across; disk-florets yellow; ray-florets at first white, then violet. Bracts of involucre violet, nearly equal, acute. Lvs. lance-shaped or linear, entire or with a few widely spaced teeth, stalkless, but not clasping the stem. Native of North America; naturalized from gardens. Widespread in Central Europe.

CRINITARIA   Like *Aster*, but ray-florets absent. *c.* 1 sp.

1368. *C. linosyris* (L.) Less. (*Aster l.*) GOLDILOCKS. Fl. heads few, bright yellow, 1–1½ cm. across, in a dense, terminal, leafy, flat-topped cluster. Disk-florets with a deeply cut corolla tube; ray-florets absent; involucre of many lax, acute, hairy bracts, the outer green, the inner yellowish. Lvs. linear, 2–5 cm. by *c.* 1 mm., very numerous. Fr. brown, hairy, pappus reddish. An erect, wiry, hairless, leafy perenn. 10–50 cm. △ Woods, pastures, rocky places. Sept.–Nov. Cent., South-East Eur. (except GR.) F.B.GB.S.I.SU. Page 432.

ERIGERON | **Fleabane**   Like *Aster*, but ray-florets numerous in several rows, with very narrow, strap-shaped corolla limbs; involucral bracts many, overlapping. Fr. compressed, two-veined, usually hairy; pappus present. *c.* 10 sps.

Ray-florets erect, little longer than the disk-florets

1369. *E. acer* L. BLUE FLEABANE. Fl. heads long-stalked, one or few in a lax cluster, with very numerous, slender, pale purple ray-florets little longer than the yellow disk-florets. Fl. heads 1–1½ cm. across; involucral bracts linear-acute, numerous, hairy. Lvs. softly hairy, the lower obovate-lance-shaped, stalked, the upper linear-lance-shaped, stalkless. A slender erect, variable, leafy ann., bienn. or perenn. 8–40 cm. △ Cultivated ground, rocks, walls, waste places, in mountains. June–Sept. All Eur. (except IS.TR.). Pl. 143.

Ray-florets spreading, much longer than the disk-florets

(a) Fl. heads solitary or several

1370. *E. alpinus* L. ALPINE FLEABANE. Fl. heads usually solitary, or rarely up to 6, 2–3 cm. across, with spreading violet or pink ray-florets and yellow disk-florets. Involucral bracts green or purple, densely short-haired, not glandular; outer disk-florets female, the inner bisexual. Lvs. oblong-oval and fine-pointed, usually hairy, sometimes nearly hairless, the upper lvs. narrower, stalkless, and half-clasping the stem. An erect, usually coarsely downy perenn. 2–40 cm. △ In the mountains: pastures, open woods, stony places. July–Sept. Much of Eur. (except P.IRL.B.NL.DK.SF.H.YU.AL.TR.).

E. uniflorus L. Like 1370, but fl. heads smaller 1–1½ cm. across, and involucral bracts with dense, long, woolly hairs. Fl. heads usually solitary; ray-florets lilac, pale rose or whitish; disk-florets yellow. Lvs. blunt, ciliate, otherwise more or less hairless. Northern and Central Europe.

(b) Fl. heads at least more than 6

1371. *E. annuus* (L.) Pers. SWEET SCABIOUS, WHITE TOP. Fl. heads with very slender whitish or pale lilac ray-florets and yellow disk-florets, 1½–2 cm. across, many in a lax, somewhat flat-topped cluster. Ray-florets *c.* 1 cm. long, nearly twice as long as the almost hairless involucral bracts. Lvs. oval to broadly lance-shaped, deeply and coarsely toothed, the uppermost only entire. Fr. with an outer row of triangular scales and an inner row of hairs. An erect, branched, sparsely hairy bienn. 40–120 cm. △ Native of North America; naturalized in fields, waste places, ballast heaps. July–Aug. Cent. Eur. Page 432.

1372. *E. karvinskianus* DC. A slender, spreading, sparsely hairy, woody-based perenn. 15–30 cm., much-branched from the base and bearing rather few, long-stemmed whitish fl. heads *c.* 1½ cm. across. Ray-florets numerous, slender, spreading, at first pale purple, then white and finally pink; involucral bracts linear-acute, hairy. Lower lvs. 1–3 cm., three-lobed or coarsely three- to five-toothed at apex, the upper lvs. linear-lance-shaped, nearly entire. Fr. reddish-brown, pappus whitish, longer. △ Native of Central America; naturalized on walls and rocks. Apr.–Aug. Cent. Eur.

CONYZA Very like *Erigeron*, but ray-florets absent or very small, and outer disk-florets female and inner bisexual. *c.* 2 sps.

1373. *C. canadensis* (L.) Cronq. (*Erigeron c.*) CANADIAN FLEABANE. A stiff, erect, leafy ann. 8–100 cm., much branched above and bearing very numerous tiny whitish fl. heads in long, lax, branched clusters. Fl. heads 2–5 mm. by 3–5 mm.; ray-florets whitish or pale lavender, scarcely longer than the yellow disk-florets; involucral bracts linear, with papery margin, hairless or nearly so. Stem lvs. 1–4 cm. linear to lance-shaped, bristly-haired, the basal lvs. broader, soon dying. △ Native of North America; naturalized in waste places, cultivated ground. July–Oct. Introd. to all Eur. (except IRL.IS.). *Oil of Erigeron is obtained from this plant and it is used medicinally.*

C. ambigua DC. Distinguished from 1373 by its greyish-green lvs. and numerous some-what larger greenish fl. heads 7–8 mm. long, without ray-florets. Involucral bracts without papery margin, hairy. Lvs. lance-shaped, the lower stalked, the upper narrower, stalkless. Mediterranean region.

EVAX Fl. heads tiny, in dense stalkless clusters; disk-florets only present, the outer female, the inner bisexual. Involucral bracts in 2 indistinct rows, thin and dry, much longer than the florets. Fr. without a pappus. *c.* 10 sps.

1374. *E. pygmaea* (L.) Brot. (*Filago p.*) A tiny, whitish, woolly-leaved ann. 1–4 cm., with several short spreading branches ending in compact clusters of yellowish fl. heads surrounded by a rosette of lvs., the whole forming a cushion-like growth. Involucral bracts lance-shaped and long-pointed, yellow, spreading. Lvs. oblong-obovate, the upper forming a rosette. Fr. flattened, with small swellings. △ Dry sandy and stony places near the sea. Apr.–May. Med. Eur. P. Pl. 143.

FILAGO | **Cudweed** Fl. heads tiny, stalkless and clustered into small, dense, rounded heads; disk-florets only present, the outer female, the inner bisexual; receptacle conical; involucre five-angled, of numerous overlapping bracts, the outer green and woolly, the inner papery. Fr. with pappus. *c.* 12 sps.

Fl. heads 8–40 in each cluster
1375. *F. vulgaris* Lam. (*F. germanica* auct.; *F. canescens* Jordan) CUDWEED. A sparsely branched, erect, greyish or yellowish woolly-haired ann. 5–30 cm., with globular yellowish clusters of fl. heads borne at the ends of the main stem and on wide-spreading lateral branches. Clusters *c.* 12 mm., with 20–40 fl. heads, not over-topped by the uppermost lvs.; outer involucral bracts densely woolly-haired, the inner papery, hairless and with a yellowish awn-like point. Lvs. 1–3 cm., linear-lance-shaped, numerous, overlapping, densely woolly-haired. △ Fields, sandy places, heaths, waysides. June–Sept. All Eur. (except N.): introd. IS.SF. Pl. 143.

*F. apiculata G.E. Sm. RED-TIPPED CUDWEED. Widespread, except in Northern Europe.

Fl. heads less than 8 in each cluster
1376. *F. gallica* L. NARROW CUDWEED. Fl. clusters *c.* 4 mm. across with 2–6 fl. heads, the uppermost lvs. much longer than the clusters. Involucral bracts hairy, with yellowish,

papery, hairless tips. Lvs. linear *c.* 1 mm. broad, with greyish silky hairs. A slender, much-branched ann. 8–20 cm. △ Sandy places, uncultivated ground. July–Sept. Med. Eur. P.GB.D.CH.BG.: introd. A. Page 432.

1377. *F. arvensis* L. Like 1376, but involucral bracts densely hairy to the tip, and spreading in a star in fr. Clusters 4–5 mm. across, with 2–7 fl. heads, usually over-topped by the lvs. Lvs. lance-shaped, densely white-woolly. An erect ann. 10–40 cm., with short lateral branches above. △ Sandy places, fields. July–Aug. All Eur. (except P.IRL.IS.TR.): introd. GB.

*F. minima (Sm.) Pers. SLENDER CUDWEED. Widespread in Europe.

ANTENNARIA | Cat's-Foot Like *Gnaphalium*, but plants one-sexed, and fl. heads densely clustered. Involucral bracts not spreading in a star in fr. *c.* 3 sps.

1378. *A. dioica* (L.) Gaertner CAT'S-FOOT. A low perenn. with a woody stock, creeping rooting stems forming rosettes of lvs. which are green above and white-woolly beneath, and erect leafy fl. stems 5–20 cm. Fl. heads white or pink, in a dense terminal cluster of 2–8; involucral bracts papery, usually white or pink and broadly oval in the male fl. heads; pink or red and narrower in the female fl. heads. Rosette lvs. 1–4 cm. spathulate, spreading, the stem lvs. lance-shaped to linear, pressed to stem. △ Mainly in mountains: meadows, heaths, sunny banks. May–June. All Eur. (except P.IS.TR.). *Used in herbal remedies for throat infections.* Pl. 143.

A. carpatica (Wahlenb.) Bluff & Fingerh. CARPATHIAN CAT'S-FOOT. Like 1378, but fl. heads brown or blackish; lower lvs. oblong-lance-shaped and narrowed at each end, not spathulate, woolly on both surfaces; creeping stems absent. Mountains of Central Europe and Northern Europe.

LEONTOPODIUM | Edelweiss Like *Gnaphalium*, but the uppermost lvs. forming an involucre and spreading in a star much beyond the dense cluster of fl. heads. 1 sp.

1379. *L. alpinum* Cass. EDELWEISS. Fl. heads yellowish, in a dense flat-topped cluster and surrounded by 6–9 lance-shaped, densely white-woolly lvs. spreading in a star. Fl. heads 2–10, nearly globular, stalkless; involucral bracts white-woolly, tipped with brown at apex. Lvs. densely white-woolly but becoming greenish, the lower oblong-lance-shaped, the upper narrower, erect. An erect, little-branched perenn. 5–20 cm. △ In mountains: pasture, screes, rocks. July–Sept. Cent., South-East Eur. (except GR.TR.) E.F.I.SU. *A famous alpine plant, used as a symbol of many activities and products connected with the mountains of Central Europe.* Pl. 144.

GNAPHALIUM | Cudweed Involucre bell-shaped, of many papery coloured bracts which spread in a star in fr. Fl. heads with disk-florets only, terminal or axillary, in spikes or flat-topped clusters. Fr. with a pappus. *c.* 7 sps.

Fl. heads densely clustered at end of the stem
(a) Anns.

1380. *G. uliginosum* L. MARSH CUDWEED. A much-branched ascending ann. 4–20 cm., with narrow whitish or greenish, woolly-haired lvs. and small, ovoid, stalkless clusters of yellowish fl. heads much shorter than the lvs. Clusters with 3–10 fl. heads; fl. heads 3–4 mm. long; involucral bracts papery, pale brown and woolly below, darker and hairless towards the tip. Lvs. 1–5 cm., narrow-oblong or spathulate, blunt or acute. △ Damp pastures and fields, tracksides, banks. June–Oct. All Eur. Pl. 144.

1381. *G. luteoalbum* L. JERSEY CUDWEED. An erect, little-branched, whitish, densely woolly-haired ann. 20–40 cm., with stalkless, yellowish fl. heads in a dense terminal cluster, or in several globular clusters forming a branched flat-topped inflorescence, not over-topped by the uppermost lvs. Fl. heads globular, 4–5 mm.; stigmas red; involucral bracts shining straw-coloured, mostly hairless. Upper lvs. lance-shaped, erect, half-clasping the stem, the lower lvs. spathulate, spreading. △ Damp sands, fields, walls. July–Sept. Much of Eur. (except IRL.IS.N.AL.): introd. DK.SF.

(b) Perenns.

1382. *G. supinum* L. DWARF CUDWEED. A low, tufted, mountain or arctic perenn. with many short, non-flowering branches, and with erect fl. stems 2–12 cm. bearing a short terminal cluster of 1–7 pale brown fl. heads. Fl. heads *c.* 6 mm. long, at first clustered, later lax; involucral bracts with a woolly-haired, greenish central stripe and brown papery margin, spreading in a star after flowering. Lvs. ½–1½ cm. linear-lance-shaped, white-woolly. △ Damp pastures, rocks, moraines. July–Sept. All Eur. (except P.IRL.B.NL.DK.H.TR.). Page 432.

Fl. heads in elongated spike-like inflorescences

1383. *G. sylvaticum* L. WOOD CUDWEED. An erect perenn. 8–60 cm., branched only from the woody base, and with fl. heads in small clusters in the axils of the upper lvs. forming a long, slender, interrupted, leafy spike to more than half the length of the stem. Fl. heads pale brown, *c.* 6 mm. long, solitary or in clusters of 2–8; involucral bracts with a central green stripe, broad pale papery margin and brown transparent apex. Lvs. 2–8 cm., becoming progressively smaller above, linear-lance-shaped, one-veined, more or less hairless on the upper side, white-woolly beneath. △ Open woods, dry pastures, heaths. June–Sept. All Eur. (except P.TR.).

G. norvegicum Gunnerus HIGHLAND CUDWEED. Like 1383, but fl. heads in a shorter, more compact, oval spike to about one-quarter the length of the stem. Fl. heads 6–7 mm., solitary or 2–3 in the axils of each lv.; involucral bracts with dark brown papery margin. Lvs. broader, oblong-lance-shaped, three-veined. Widespread in Europe. Page 432.

HELICHRYSUM | **(Elichrysum)** Involucre bell-shaped, hemispherical or cylindrical, of many brightly coloured, shining, papery, overlapping bracts, not spreading in a star in fr. Ray-florets absent; outer disk-florets female, the inner bisexual; receptacle without scales. Fr. with a pappus. *c.* 20 sps.

Lvs. flat, the lower c. 5 mm. broad

1384. *H. arenarium* (L.) DC. A greyish, woolly-haired perenn. 10–30 cm., with small globular fl. heads with bright, shining, golden-yellow, involucral bracts, forming a terminal, compact, branched, flat-topped inflorescence. Fl. heads 6–7 mm. by *c.* 5 mm.; involucral bracts about 50, the outer oval, the inner oblong and 5–6 times as long. Rosette lvs. spathulate, stalked, the stem lvs. linear-oblong, blunt, stalkless, all woolly-haired. △ Sandy places, tracksides, uncultivated ground. July–Sept. Cent. Eur. F.B.NL.DK.S.YU. BG.R.SU. Page 432.

Lvs. with inrolled margin, very narrow c. 1 mm. broad

1385. *H. stoechas* (L.) DC. Distinguished from 1384 by the much narrower lvs. which are inrolled beneath. Fl. heads globular 4–6 mm., bright yellow; involucre with the inner bracts spathulate and about twice as long as the outer oval bracts. Lvs. linear, greenish above, becoming hairless, strong-smelling when crushed. A very variable woody-based perenn. with many erect, woolly fl. stems 5–50 cm. △ Dry banks, rocks, maritime sands. Apr.–July. P.E.F.CH.I.YU. Pl. 144.

H. italicum (Roth) G. Don Distinguished from 1385 by the inner oblong involucral bracts

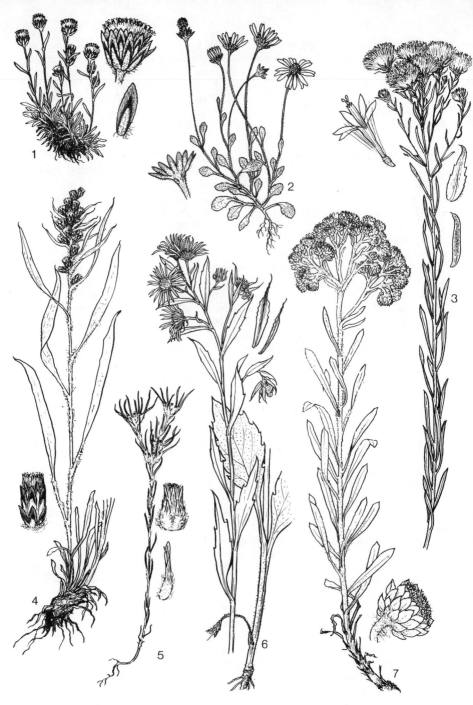

1. *Gnaphalium supinum* 1382
3. *Crinitaria linosyris* 1368
5. *Filago gallica* 1376
7. *Helichrysum arenarium* 1384

2. *Bellis annua* 1361
4. *Gnaphalium norvegicum* [1383]
6. *Erigeron annuus* 1371

which have minute reddish glands and are 5–8 times as long as the outer oval bracts. Fl. heads 4–5 mm. long and 2–4 mm. broad, clustered into dense rounded inflorescences; involucral bracts *c.* 30, straw-coloured. Mediterranean region.

PHAGNALON Like *Helichrysum*, but fl. heads usually solitary, long-stalked. Involucral bracts papery or leathery, usually brownish, not brightly coloured; disk-florets only present, outer florets female only. *c.* 5 sps.

1386. *P. rupestre* (L.) DC. An erect undershrub to 30 cm., with white-woolly stems and underside of lvs. and long-stalked, solitary, brownish-yellow, globular fl. heads *c.* 1 cm. Involucral bracts shining brown, stiff, dry, and closely pressed to fl. head, the outer oval, the inner linear. Lvs. oblong-lance-shaped, green with cobweb-like hairs above, white-woolly beneath, with undulate inrolled margin. △ Rocky and stony places. Apr.–June. Med. Eur. (except TR.) P. Pl. 145.

P. saxatile (L.) Cass. Like 1386, but outer involucral bracts of the solitary fl. heads spreading or reflexed and not closely pressed to head. Lvs. with cobweb-like hairs and green or whitish above, white-woolly beneath, the upper narrow linear with inrolled margin, the lower linear-lance-shaped, flat. Mediterranean region and Portugal.

P. sordidum (L.) DC. Like 1386, but fl. heads usually in a cluster of 2–6 at the ends of branched stems, rarely solitary; fl. heads smaller, 6–7 mm. Lvs. narrow linear, white-woolly on both sides. Spain to Italy.

INULA Fl. heads yellow, usually with both disk- and ray-florets present, solitary or rather few in a branched flat-topped cluster. Involucre hemispherical, bracts in many rows, the outer leaf-like, not papery; receptacle without scales. Fr. with a pappus. *c.* 25 sps.

Fl. heads with disk-florets only
1387. *I. conyza* DC. PLOUGHMAN'S SPIKENARD. An erect, leafy, little-branched bienn. or perenn. 20–130 cm., with numerous yellowish fl. heads *c.* 1 cm. across, in a rather dense, branched, more or less flat-topped, terminal cluster. Ray-florets absent or very small; outer involucral bracts, green, hairy, with spreading or recurved tips, the inner narrow, papery, often purple. Lower lvs. oval, stalked, the upper lvs. elliptic to lance-shaped, stalkless, all finely toothed and downy-haired especially beneath. Fr. dark brown, pappus reddish-white. △ Rocks, open woods, dry places. July–Sept. Much of Eur. (except North Eur. IRL.) DK.: introd. S. *Used in the past as an insecticide, and in herbal remedies.* Pl. 144.

Fl. heads with ray- and disk-florets
(a) Fr. hairless
1388. *I. helenium* L. ELECAMPANE. A robust perenn. 1–1½ m. with very large yellow fl. heads 6–8 cm. across and very large elliptic lvs. up to 80 cm. long. Ray-florets numerous, with a long, narrow, bright yellow, strap-shaped limb, disk-florets numerous, yellow; involucral bracts broadly oval, green and leaf-like. Lower lvs. stalked, the upper oval heart-shaped, acute, clasping the stem, all finely toothed, rough above and softly hairy beneath. Fr. hairless, strongly four-ribbed, pappus reddish. △ Damp meadows, hedges, woods. May–Sept. All Eur. (except P.IS.TR.): introd. North Eur. IRL.GB.D.CS. *Famed as a medicinal plant in the past, but now little used. The roots are used in wine-making and for flavouring spirits, and sweetmeats.* Pl. 144.

1389. *I. salicina* L. WILLOW-LEAVED INULA. A slender, unbranched perenn. 25–80 cm., with solitary or few medium-sized yellow fl. heads 2½–4 cm. across. Outermost involucral bracts lance-shaped, green, the inner bracts linear and papery, all bracts hairless but with ciliate margin. Lvs. rather stiff, hairless, often with rough margin and veins, 3–7 cm., the lower oblance-shaped and narrowed to the base, the upper elliptic with heart-shaped base half-clasping the stem and blade spreading horizontally. Rather like 1396, but readily distinguished by the absence of receptacle scales. △ Rocky and woody slopes, marshes. June–Aug. All Eur. (except IS.).

I. ensifolia L. NARROW-LEAVED INULA. Like 1389, but lvs. lance-shaped or linear with conspicuous parallel veins and lv. margin only hairy. Fl. heads often solitary, yellow, $2\frac{1}{2}$–$5\frac{1}{2}$ cm. across; ray-florets at least twice as long as the involucral bracts which are leathery and green-tipped. Central and South-Eastern Europe. Page 446.

(b) Fr. hairy

1390. *I. crithmoides* L. GOLDEN SAMPHIRE. An erect, fleshy-leaved, maritime perenn. 10–90 cm., with rather few golden-yellow fl. heads *c.* $2\frac{1}{2}$ cm. across, in a lax flat-topped cluster. Ray-florets numerous, nearly twice as long as the hairless involucral bracts; disk-florets orange-yellow. Lvs. fleshy, hairless, $2\frac{1}{2}$–6 cm., linear or oblong, with a three-toothed or entire blunt apex, the lateral lvs. in clusters. △ Salt marshes, cliffs, and shingle by the sea. Aug.–Oct. West Eur. (except B.NL.), Med. Eur. (except TR.). Pl. 145.

1391. *I. montana* L. Distinguished by its lvs. and stems which are covered with white-woolly hairs, and its usually solitary, bright yellow fl. head 4–5 cm. across. Involucral bracts very unequal, the outer green lance-shaped, woolly-haired, the inner papery, linear, ciliate. Lvs. oblong-lance-shaped, greenish above, white-woolly beneath, the upper stalkless but not clasping the stem. An erect, unbranched perenn. 10–40 cm. △ Dry rock slopes. June–July. P.E.F.CH.I.

I. britannica L. Like 1391, but lvs. and stem dark green, hairy, and stem lvs. narrow lance-shaped and half-clasping the stem. Fl. heads 1–6, 2–3 cm. across; ray-florets glandular, twice as long as the involucre; involucral bracts linear-lance-shaped, spreading or recurved, the outer green, the innermost papery. Central and Southern Europe. Pl. 145.

1392. *I. viscosa* (L.) Aiton AROMATIC INULA. A very glandular, sticky, strongly resinous-smelling, shrubby, woody-based perenn. with many erect leafy stems 40–100 cm. Fl. heads yellow, small *c.* $1\frac{1}{2}$ cm. across, short-stalked in a long, dense, leafy, pyramidal cluster; ray-florets spreading, twice as long as the involucre; involucral bracts green, with a papery margin. Lvs. lance-shaped, entire or toothed, glandular-hairy, stalkless, the upper half-clasping the stem. △ Rocky places, pine forests, olive groves. Aug.–Oct. Med. Eur. (except TR.) P.BG.

PULICARIA | Fleabane Like *Inula,* but pappus with an outer row of short scales which are free or fused into a cup, and an inner row of simple hairs. Fl. heads yellow, usually clustered, or rarely solitary. *c.* 6 sps.

1393. *P. dysenterica* (L.) Bernh. FLEABANE. An erect, leafy, branched perenn. 20–60 cm., with creeping stolons and golden-yellow fl. heads $1\frac{1}{2}$–3 cm. across in a branched, more or less flat-topped cluster. Ray-florets numerous, about twice as long as the involucre; involucral bracts green, glandular-hairy. Upper lvs. 3–8 cm. lance-shaped, clasping the stem with conspicuous basal lobes, margin undulate, toothed, lvs. all greyish and densely softly hairy. △ Damp meadows, marshes, ditches, salt-rich ground. July–Sept. All Eur. (except North Eur.) DK.: introd. N.SF. *The dried burning leaves have been used as a fumigant against insects; the plant has also been used in herbal remedies.* Pl. 145.

P. odora (L.) Reichenb. Like 1393, but stems not stoloniferous and upper lvs. oblong and clasping the stem, but without conspicuous basal lobes, lvs. loosely woolly-haired. Fl.

heads solitary or few, 2–3 cm. across, on thickened stems; involucral bracts shaggy-haired, very narrow and long-pointed, papery. Mediterranean region.

P. vulgaris Gaertner SMALL FLEABANE. Like 1393, but an ann. with smaller fl. heads *c.* 1 cm. across, and erect ray-florets little longer than the involucre. Upper lvs. softly hairy, stalkless, but not clasping the stem. Widespread, except in the north.

CARPESIUM Fl. heads solitary; disk-florets only present; receptacle without scales. Fr. elongated, pappus absent. 2 sps.

1394. *C. cernuum* L. FALSE BUR-MARIGOLD. An erect, branched, hairy perenn. 20–60 cm., with several solitary, drooping, globular, yellow fl. heads, each surrounded by an involucre of lvs., recalling *Bidens*. Fl. heads 1½–2½ cm. across, axillary and terminal and borne on thickened, recurved stems; outer involucral bracts green and leafy, rough-haired, reflexed, the inner bracts stiff, papery. Lvs. softly hairy, oval-lance-shaped, toothed, narrowed to a short stalk or stalkless. Fr. linear 4 mm. △ Wood margins, thickets, shady places. June–Sept. Cent. Eur. E.F.I.YU.BG.R.SU.

PALLENIS Fl. heads yellow; involucral bracts in 2 or 3 rows, the outermost green and spine-tipped; ray-florets in 2 rows; receptacle with scales. 1 sp.

1395. *P. spinosa* (L.) Cass. An erect, dichotomously branched, hairy ann. or bienn. 20–50 cm., with yellow fl. heads 2–3 cm. across and long, tough, spiny-tipped outer involucral bracts spreading in a star to about twice the width of the head. Ray-florets numerous, similar in colour to the disk-florets; outer involucral bracts brownish at base and green above, woolly-haired, the inner bracts leathery, hairy. Lvs. lance-shaped, hairy, the lower stalked, the upper clasping the stem with heart-shaped base, and with a spiny-tipped apex. △ Dry uncultivated places, tracksides. June–Aug. Med. Eur. P.BG.SU. Pl. 145.

BUPHTHALMUM Ray-florets in a single row; involucral bracts all similar, closely investing, not spreading; receptacle with scales. Outer fr. triangular in section, the inner smaller, quadrangular; pappus of short scales. *c.* 4 sps.

1396. *B. salicifolium* L. YELLOW OX-EYE. An erect, little-branched, hairy perenn. 30–70 cm., with large, bright yellow, solitary fl. heads 3–6 cm. across, with ray-florets 2–3 times as long as the involucre. Involucral bracts lance-shaped, green, hairy; receptacle scales linear, the inner scales all abruptly narrowed to the tip. Stem lvs. 10–12 cm., oblong-lance-shaped acute, weakly toothed, clasping the stem, with short adpressed hairs. Fr. winged. △ In hills and mountains: open woods, rocky, stony slopes. July–Aug. Cent. Eur. (except PL.) E.F.I.YU. Page 446.

TELEKIA Like *Buphthalmum*, but fl. heads large and involucral bracts herbaceous, with recurved tips; receptacle with many scales. Fr. ribbed, with a short crown-like pappus. *c.* 2 sps.

1397. *T. speciosa* (Schreber) Baumg. (*Buphthalmum s.*) LARGE YELLOW OX-EYE. A robust, erect, broad-leaved perenn. to 2 m., with large, solitary or few, orange-yellow fl. heads 5–6 cm. across. Ray-florets orange-yellow; disk-florets dark brownish-yellow. Lower lvs. very large, broadly triangular-heart-shaped, stalked, the upper lvs. rhomboid or oblong, stalkless, all lvs. strongly toothed, nearly hairless above, finely hairy below. Recalling 1388, but distinguished from it by its lvs. which are green beneath, and by the absence of a long-haired pappus and by the presence of scales on the receptacle. △ Margins of woods and streams. June–Aug. A.PL.CS.H.YU.AL.BG.R.SU.: introd. F.D. Pl. 145.

ASTERISCUS   Fl. heads solitary or clustered; ray- and disk-florets present; involucral bracts in several rows, the outer spreading, green and leafy, blunt, much longer than the inner bracts; receptacle with scales. Pappus of numerous equal scales. 3 sps.

1398. *A. maritimus* (L.) Less. (*Odontospermum m.*) A woody-based, rough-hairy, tufted perenn. 3–25 cm., with spreading stems and terminal, deep yellow fl. heads up to 4 cm. across, with 1 or 2 spreading leafy bracts immediately beneath each head. Ray-florets about 30, tips of ray-florets finely toothed; involucral bracts leathery at base and leafy-tipped, the outer spathulate, as long as the ray-florets; receptacle scales linear. Lvs. oblong or spathulate, stalked, usually rough-hairy. △ Rocks, stony banks by the sea. May–July. Med. Eur. (except YU.AL.TR.) P. Pl. 146.

1399. *A. aquaticus* (L.) Less. (*Odontospermum a.*) A regularly dichotomously branched, erect, hairy ann. 10–40 cm., with small, pale yellow, stalkless fl. heads in a flat-topped cluster. Outer involucral bracts linear-lance-shaped blunt, spreading with the upper-most lvs. in a star much beyond the small ray-florets. Upper lvs. oblong and half-clasping the stem, finely hairy, the lower stalked. △ Damp places, maritime rocks. June–Aug. Med. Eur. (except AL.) P.BG. Pl. 146.

AMBROSIA   Involucral bracts fused to form a cup; fl. heads one-sexed, globular, the male heads many, in terminal spikes, the female heads few, one-flowered and placed below the male heads and borne in the axils of the upper lvs. Fr. hard, spiny. 2 sps.

1400. *A. artemisiifolia* L. AMERICAN WORMWOOD, ROMAN RAGWEED. A branched, erect, leafy ann. with a grooved, often reddish, hairy stem 20–120 cm., compound lvs. and small, globular, greenish-yellow fl. heads in a slender, bractless, terminal spike. Fl. heads 4–5 mm., drooping; involucre cup-like; female fl. heads few, each one-flowered. Lvs. twice or thrice cut in oblong, toothed segments, green and hairy above, and with dense, white, adpressed hairs beneath. Fr. hairy, cylindrical 4–5 mm., with a whorl of 5–6 short spreading spines below the tip. △ Native of North America; widely naturalized in waste places. cultivated ground, tracksides. June–Sept. Introd. Cent. Eur. F.YU.R.SU.

XANTHIUM   Fl. heads one-sexed; male heads terminal, globular with numerous florets; female heads below male, ovoid with fused involucral bracts completely enclosing the 2 female florets, covered with recurved hooks and ending in 2 horn-like spines. *c.* 4 sps.

Stems without spines

1401. *X. strumarium* L. COCKLEBUR. A robust spineless ann. to 80 cm., with hairy grey-green lvs. and a terminal, rather dense cluster of greenish fls. Male fl. heads globular, female fl. heads ellipsoid. Lvs. triangular-heart-shaped, often three- to five-lobed, coarsely and irregularly toothed, short-stalked. Fr. ellipsoid 12–15 mm., greenish-yellow and often flushed with red, glandular-hairy and densely covered with hooked spines except near apex, and with 2 straight terminal spines. △ Probably native of America; widely naturalized in waste places, tracksides. July–Oct. Cent., Med., South-East Eur. P.SU. *Used in herbal remedies; it is poisonous to livestock and causes paralysis. The fruits are readily carried in the fur of animals and the plant may become a serious pest.* Pl. 146.

X. echinatum Murray STINKING COCKLEBUR. Like 1401, but a rough-stemmed, strongly aromatic, yellowish-green ann. without spines. Lvs. oval, entire or shallowly three-lobed, coarsely toothed, covered with yellowish glands. Fr. hairy, covered with hooked spines and with usually incurved, not straight, terminal spines. Mediterranean Europe.

Stems spiny

1402. *X. spinosum* L. SPINY COCKLEBUR. A very spiny, much-branched, nearly hairless perenn. 15–80 cm., with long yellow, three-pronged spines at the base of each lv. and ellipsoid frs. which are densely covered with dull yellowish hooked spines. Male fl. heads spherical 4–5 mm.; female fl. heads 8–12 mm., stalkless in the axils of the lvs. Lvs. dark green and with white veins above, white-felted beneath, rhomboid-entire, or with 3–5 narrow triangular lobes. Fr. 1–1½ cm., terminal spines small, unequal, straight. △ Native of America; now a cosmopolitan weed of waste places, cultivated ground, hedges, tracksides. July–Sept. Cent., Med., South-East Eur. P.SU. *It yields a yellow dye; the leaves are bitter and astringent and are used in herbal remedies.* Pl. 146.

RUDBECKIA Fl. heads large: with a conspicuous conical receptacle, with rigid acute scales and numerous brownish or purple disk-florets; ray-florets yellow, orange, or red, sterile, in a single row. Fr. four-angled; pappus absent or forming a cup. *c.* 1 sp.

1403. *R. laciniata* L. CONE FLOWER. A tall hairless perenn. ½–2½ m. with very large, long-stalked, yellow fl. heads 7–12 cm. across, with a conspicuous conical receptacle. Disk-florets brownish-black; ray-florets in a single row, few large, golden-yellow, soon reflexed. Lvs. spirally arranged, the upper three-lobed or entire. Receptacle scales as long as the hairless fr. △ Native of North America; grown as an ornamental and sometimes naturalized in damp woods, waysides. June–Oct. Cent. Eur. F.I.BG.R.SU. Pl. 146.

HELIANTHUS Fl. heads very large, receptacle flat, with scales. Disk- and ray-florets yellow; involucral bracts in several rows, with leafy tips. *c.* 3 sps.

1404. *H. annuus* L. SUNFLOWER. A stout, erect, rough-stemmed ann. to 3 m., usually with solitary and extremely large drooping fl. heads 10–30 cm. across. Disk-florets brownish, contrasting with the single row of numerous, large, golden-yellow ray-florets; involucral bracts oval, bristly-haired. Lvs. spirally arranged, broadly oval, stalked, rough-haired, the lower heart-shaped. Fr. 7–17 mm., somewhat flattened, often streaked with white and black. There are many cultivated forms. △ Native of South America; widely cultivated in Europe. July–Aug. *Introduced to Europe in the sixteenth century and now a valuable crop plant in Southern Europe. The seeds are rich in oil, which is used in soap-making, dressing wool, paints, and it is unrivalled as a lubricant. The seeds are edible and used for bread-making, as a coffee and medicinally. The fibre can be used for paper-making, and the pith is extremely light, with a specific gravity of 0·028.*

1405. *H. tuberosus* L. JERUSALEM ARTICHOKE. A leafy herbaceous perenn. with stout, erect stems 1–2½ m., and with numerous swollen, potato-like tubers produced below-ground which persist throughout the winter. Fl. heads 4–8 cm. across, solitary, erect; disk- and ray-florets yellow; involucral bracts lance-shaped, green. Lvs. mostly opposite, lance-shaped long-pointed, narrowed to a winged stalk, coarsely toothed, rough-hairy. △ Native of North America; sometimes cultivated as a vegetable and occasionally naturalized. Sept.–Oct. *Introduced to Europe in the sixteenth century and cultivated before potatoes in Britain. The edible tubers contain inulin which is a source of fructose, used by diabetics.*

BIDENS | **Bur-Marigold** Fl. heads globular, often solitary, with both disk- and ray-florets present, or ray-florets absent. Involucral bracts in 2 rows, the outer usually green and leafy, the inner papery; receptacle flat, with scales. Fr. four-angled and flattened, with 2–4 stiff, barbed terminal bristles. *c.* 4 sps.

1406. *B. cernua* L. NODDING BUR-MARIGOLD. An erect, branched, hairless ann. 10–90 cm., with solitary, somewhat drooping globular fl. heads 1½–2½ cm. across, usually without ray-florets (var. *radiata* DC. has golden-yellow ray-florets *c.* 1–1½ cm. long). Outer involucral bracts 5–8, oblong, leafy, spreading, much longer than the broadly oval, dark-streaked inner bracts. Lvs. simple, 4–15 cm. lance-shaped and long-pointed, coarsely toothed, stalkless. Fr. with 4 terminal barbed bristles. △ Pond and streamsides, damp places. July–Oct. All Eur. (except P.IS.TR.). Pl. 146.

B. tripartita L. TRIPARTITE BUR-MARIGOLD. Like 1406, but fl. heads erect; lvs. shortly stalked, usually deeply divided in 3 or 5 coarsely toothed, lance-shaped lobes. Fr. much-flattened, usually with 3 barbed bristles. Throughout Europe.

GALINSOGA Fl. heads small, with few yellow disk-florets and few short white ray-florets; involucral bracts few, oval. Fr. with pappus of several papery scales. *c.* 2 sps.

1407. *G. parviflora* Cav. GALLANT SOLDIER. A branched, nearly hairless ann. 10–75 cm., with oval-acute, toothed lvs. and tiny white and yellow fl. heads in lax terminal and axillary clusters. Fl. heads 3–5 mm. across, long-stalked; ray-florets white, usually 5, with an oval three-lobed limb little longer than the yellow disk-florets; involucral bracts oval, green and with a narrow papery margin; scales of receptacle three-lobed. Lvs. opposite, stalked. △ Native of South America; now a cosmopolitan weed of cultivated ground, tracksides, waste places. May–Oct. Most of Eur. (except P.E.IRL.IS.AL.GR.TR.). Pl. 147.

SANTOLINA Aromatic undershrubs with pinnately lobed lvs. Fl. heads more or less globular, long-stalked; disk-florets only present; receptacle with scales; pappus absent. *c.* 8 sps.

1408. *S. chamaecyparissus* L. LAVENDER COTTON. A very aromatic, much-branched, bushy evergreen undershrub 20–50 cm., with narrow, silvery-white-felted, crimped lvs. and long-stalked, globular, yellow fl. heads 1–1½ cm. across. Disk-florets only present; involucral bracts lance-shaped with a strong midvein, papery-tipped, hairy. Lvs. very crowded, linear 2–3 mm. wide, deeply cut into 2–4 rows of rounded, fleshy lobes. △ Dry banks, stony ground, rocks. July–Sept. P.E.F.I.YU.: introd. D.CS.H.SU. *Used as an insecticide and vermifuge.*

ANTHEMIS | **Chamomile** Lvs. 1–3 times cut into linear segments. Fl. heads solitary; ray- and disk-florets usually present; receptacle with scales. Fr. ribbed, not compressed, with a membraneous rim in place of the pappus. *c.* 50 sps.

Receptacle hemispherical in fr.
(a) Fr. tetragonal in section
1409. *A. tinctoria* L. YELLOW CHAMOMILE. An ascending or erect, branched, woolly-haired perenn. 30–60 cm., with solitary long-stalked fl. heads 2½–4 cm. across, with golden-yellow disk- and ray-florets, or ray-florets absent. Involucral bracts at first woolly-haired, but becoming hairless, the outer lance-shaped, the inner oblong-blunt, papery-margined. Lvs. deeply twice cut into narrow toothed segments, white-woolly beneath. Fr. head hemispherical; receptacle covered with the stiff-tipped, bristle-like scales; fr. almost

without a membraneous rim. △ Sunny slopes, rocky places, walls. June–Aug. All Eur. (except IRL.IS.): introd. GB. Pl. 147.

1410. *A. altissima* L. (*A. cota* L.) An almost hairless erect ann. 30–80 cm. with widely spreading branches and with large fl. heads 2–4 cm. across, with white ray-florets twice as long as the involucre and with yellow disk-florets. Fl. stem conspicuously thickened below fl. head at maturity. Receptacle scales abruptly narrowed into a long stiff spine, longer than the disk-florets. Lvs. 2–3 times cut into lance-shaped, fine-pointed segments. Fr. flattened, ten-ribbed. △ Cornfields, roadsides, stony places. May–July. Med. Eur. BG.

A. tomentosa L. WOOLLY CHAMOMILE. An ann. with prominent, ascending, branched stems and densely grey-haired, twice-pinnate lvs. Fl. heads 2–3 cm. across; ray-florets sterile, white, shorter than the width of the yellow disk; involucral bracts grey-haired; receptacle scales oblong and suddenly narrowed to a fine point, papery with a brownish midvein. Fr. tetragonal in section, ribbed, with or without a short apical projection. Eastern Mediterranean.

(b) Fr. cylindrical

A. chia L. CHIAN CHAMOMILE. A small, erect, little-branched ann. with almost hairless, green lvs. 3–5 times cut in narrow, acute segments. Fl. heads *c.* 3 cm. across, long-stalked, with white, often reflexed ray-florets as long as the width of the yellow disk; involucral bracts with a narrow, rusty-brown, papery margin, hairless. Fr. cylindrical, ribbed, the outer with an ear-shaped projection as long as fr., the inner with a short crown. Eastern Mediterranean.

Receptacle conical in fr.; fr. cylindrical, ribbed

1411. *A. arvensis* L. CORN CHAMOMILE. An aromatic, erect, branched ann. or perenn. 10–50 cm. with long-stalked fl. heads 2–3½ cm. across, with white spreading ray-florets as long as the width of the yellow disk. Ray-florets with styles; involucral bracts oblong, papery-tipped, with adpressed hairs; receptacle conical, scales lance-shaped and abruptly narrowed to a point. Lvs. hairy and more or less woolly beneath, usually 3 times cut into short, oblong, acute segments. Fr. strongly ribbed, smooth. △ Cultivated ground, waste places, waysides. May–Sept. All Eur.

1412. *A. cotula* L. STINKING MAYWEED. Like 1411, but a fetid much-branched ann. to 50 cm., with almost hairless lvs. Fl. heads 2–3 cm. across, long-stalked; ray-florets sterile, without styles; involucral bracts with a narrow papery margin; scales of receptacle linear. Lvs. twice cut into linear acute segments. Fr. with swellings, ten-ribbed. △ Cornfields, waste places, by habitation. May–Sept. All Eur. (except IS.): introd. IRL. *Used in herbal remedies and as an insecticide.*

CHAMAEMELUM   Like *Anthemis*, but disk-florets with the base of the corolla tube enlarged and sac-like. Fr. flattened laterally, not ribbed. *c.* 3 sps.

1413. *C. nobile* (L.) All. (*Anthemis n.*) CHAMOMILE. A pleasantly aromatic, erect, hairy perenn. 10–30 cm., with solitary long-stalked fl. heads 2–2½ cm. across, with white ray-florets as long as the width of the orange disk. Ray-florets rarely absent; involucral bracts oblong, downy, with a broad, papery, white margin; receptacle conical, the scales oblong blunt, often with a cut apex. Lvs. 1½–5 cm. green, gland-dotted, 2–3 times cut into linear fine-pointed segments, with inrolled margin. Fr. 1–1½ mm. △ Cornfields, grassy and waste places, on the littoral; often cultivated and widely naturalized. June–Sept. West Eur. (except NL.): introd. Cent. Eur. I.BG.SU. *A valuable tonic and stimulant used particularly in the form of 'Chamomile tea'; also an antiseptic and vermifuge.* Pl. 147.

ANACYCLUS Fl. heads solitary; disk-florets five-toothed and two-lipped; ray-florets usually present. Receptacle with scales. Fr. much-flattened, the outer with a broad transparent wing continued above into 2 ear-like projections, the inner without wings. *c.* 4 sps.

1414. *A. tomentosus* (Gouan) DC. (*A. clavatus* Pers.) An erect, hairy ann. 20–40 cm., with widely spreading branches and with solitary fl. heads 2½–3 cm. across, with the fl. stem conspicuously thickened below the heads. Ray-florets white; disk-florets yellow; involucral bracts with narrow papery margin, without a terminal appendage; receptacle scales obovate. Lvs. loosely hairy, twice cut into linear, fine-pointed segments. Fr. flattened, the outer broadly winged. △ On the littoral, rocks, fields, waysides. May–June. Med. Eur. P.

A. radiatus Loisel. Like 1414, but ray-florets yellow and involucral bracts terminating in a papery fringed appendage. Fr. with very broad wings. Mediterranean region.

ACHILLEA Fl. heads usually many in flat-topped clusters. Ray-florets usually broader than long, white, yellow, or reddish; disk-florets yellow; involucral bracts with a papery margin. Receptacle flat, scales narrow. Fr. strongly compressed, pappus absent. *c.* 40 sps.

Ray-florets white
(*a*) *Lvs. entire and toothed, not lobed*
1415. *A. ptarmica* L. SNEEZEWORT. Distinguished by its stalkless, linear or lance-shaped, finely saw-toothed, hairless lvs. 1½–8 cm. Fl. heads 12–18 mm. across, rather few in a lax flat-topped cluster. Ray-florets white, 8–13, oval or elliptic, as long as the involucre; disk-florets greenish-white; involucral bracts blunt, woolly-haired, green with a reddish-brown papery margin. An erect, leafy perenn. with angular hairy stems, branched above 20–60 cm. △ Damp meadows, marshes, watersides. June–Sept. Most of Eur. (except P.AL.GR.TR.BG.). *It yields an essential oil which is used medicinally; the dried and powdered leaves make a sneezing powder.*

1416. *A. erba-rotta* All. Somewhat like 1415, but lvs. with glandular depressions, aromatic when crushed and of two kinds: those of the sterile stems in rosettes, spathulate, toothed only towards the apex, and narrowed to a long stalk, those of fl. stems oblong, stalkless, with a strongly toothed margin with the basal teeth larger, linear, and comb-like, but lvs. sometimes more deeply lobed. Fl. heads few, *c.* 1½ cm. across, in a spreading flat-topped cluster; ray-florets 4–7. A creeping, woody-based perenn. with unbranched fl. stems 10–20 cm. △ In mountains: rocks, pasturage. July–Aug. F.CH.D.A.I.

(*b*) *Lvs. deeply once or twice cut into narrow lobes*
(*i*) *Large lowland perenns. 30 cm. or more*
1417. *A. millefolium* L. YARROW, MILFOIL. A strong-smelling erect perenn. 8–50 cm., with numerous fl. heads 4–6 mm. across, in dense, flat-topped, compound clusters. Ray-florets usually 5, white, pink or reddish, limb as broad as long and with a three-toothed apex; disk-florets white or cream-coloured; involucral bracts oblong, with a broad, brownish or blackish, papery margin, usually hairy. Lvs. linear to lance-shaped in outline, 2–3 times cut into short linear-lance-shaped, fine-pointed segments, densely or sparsely hairy. A variable sp. △ Grasslands, waysides, cultivated and uncultivated ground. May–Nov. All Eur. *Used widely in herbal medicine in the past, as a snuff, a tea, and as an application to wounds.*

A. nobilis L. Like 1417, but fl. heads very numerous, smaller, 2–3 mm. across; ray-florets yellowish-white, limb reflexed, broader than long and about one-third as long as the involucre; involucral bracts pale brown, hairy. Lvs. hairy, whitish-green, lv. stalk winged and toothed at the base; middle stem lvs. oval in outline. Central and South-Eastern Europe.

1418. *A. macrophylla* L. LARGE-LEAVED SNEEZEWORT. Distinguished by its large lower lvs. which are pinnately cut into 5–13 broadly lance-shaped, deeply toothed segments, contrasting with the uppermost lvs. which are entire and toothed. Fl. heads 13–15 mm. across, in a lax flat-topped cluster 3–4 cm. across; ray-florets white, usually 5, longer than the pale brown-bordered, hairy involucral bracts. A perenn. 40–100 cm. △ Shady woods in mountains. July–Aug. F.D.CH.I.

(ii) Small alpine perenns. less than 30 cm.
1419. *A. nana* L. DWARF MILFOIL. A small, very aromatic, rosette-forming, alpine perenn. 6–15 cm., with lvs. and stems thickly covered in greyish woolly hairs. Fl. heads white, 1 cm. across, in a dense rounded cluster; ray-florets dirty white, 5–8; involucral bracts woolly with dark brown margin, about as long as the ray-florets. Lvs. narrow elliptic in outline, once cut into rather broad entire or toothed lobes. △ Alpine rocks, screes and moraines. July–Aug. F.D.CH.A.I. Pl. 147.

1420. *A. atrata* L. DARK MILFOIL. Distinguished from 1419 by the hairless or nearly hairless, non-aromatic lvs. which are pinnately cut into linear segments with the upper segments further divided into 2–5 narrow lobes. Fl. heads 12–18 mm. across, in a lax spreading flat-topped cluster of 3–15; ray-florets white, 6–12; involucral bracts with a black margin, sparsely hairy. An erect or ascending perenn. 8–30 cm. △ In high mountains: damp screes, moraine streamlets. July–Sept. F.D.CH.A.I.YU.

A. moschata Wulfen MUSK MILFOIL. Very like 1420, but lvs. very aromatic, with shallow pits over their surface, bright green; lv. segments 1 mm. broad, usually entire or with 1–2 teeth. Fl. heads 1–1½ cm. across, 3–25 in a cluster; ray-florets 6–8, white. Alps.

Ray-florets yellow
1421. *A. tomentosa* L. YELLOW MILFOIL. Fl. heads bright yellow, tiny, *c.* 3 mm. across, numerous in a dense, compound, flat-topped cluster. Ray-florets rounded, golden-yellow; involucral bracts blunt, brown-margined, white-woolly. Lvs. and stem densely white-woolly; lv. linear-lance-shaped in outline, twice cut into crowded linear pointed segments. A densely woolly-haired perenn. with spreading sterile stems and erect unbranched fl. stems 8–30 cm. △ Dry slopes, rocks, walls. May–July. Med. Eur. (except TR.) D.CH.SU. Pl. 147.

1422. *A. ageratum* L. An erect, nearly hairless, woody-based perenn. 25–50 cm., with strongly toothed entire upper lvs. and a compact, flat-topped cluster of tiny yellow fl. heads. Fl. heads 2–4 mm. across; ray-florets short; involucral bracts pale. Lvs. oblong-blunt, all conspicuously saw-toothed, rough, the lower stalked and often lobed towards the base, the upper stalkless. △ Damp woods, waysides, rocky places. July–Aug. P.E.F.I.YU.GR.

OTANTHUS Disk-florets only present; corolla prolonged downwards into 2 sac-like spurs; involucral bracts with woolly hairs; receptacle conical, scales present. Fr. flattened, ribbed; pappus absent. 1 sp.

1423. *O. maritimus* (L.) Hoffmanns & Link (*Athanasia m.*; *Diotis m.*) COTTON-WEED. A somewhat woody, tufted, maritime perenn. 10–50 cm., the whole plant covered in a very dense snowy-white pile of hairs, and with rather few yellow fl. heads in a dense flat-topped cluster. Fl. heads globular, short-stalked, 6–10 mm.; ray-florets absent; involucre densely white-felted. Lvs. *c.* 1 cm., oblong or spathulate, stalkless, numerous and overlapping on the stem. △ Sands by the sea. June–Sept. Med. Eur. (except AL.) P.IRL.

CHRYSANTHEMUM Receptacle flat or convex, without scales. Fl. heads usually large, with or without ray-florets. Fr. without pappus, but often with a membraneous rim. *c.* 40 sps.

All florets yellow
(a) Ray- and disk-florets present

1424. *C. segetum* L. CORN MARIGOLD. A hairless, glaucous, erect ann. 20–60 cm., with solitary, terminal, bright golden-yellow fl. heads 3½–5½ cm. across. Limb of ray-florets oblong, toothed at apex, about as long as involucre, rarely absent; disk-florets numerous, similar-coloured to rays; involucral bracts oval glaucous, with a broad, pale brown papery margin; fl. stalks swollen below heads. Lvs. somewhat fleshy, the lower oblong, and coarsely toothed or lobed, stalked, the upper entire or toothed, and half-clasping the stem. Outermost frs. narrowly winged, inner frs. unwinged. △ Cornfields, cultivated ground. May–Oct. All Eur. (except IS.H.AL.BG.R.). *The leaves produce a yellow dye.*

C. myconis L. Like 1424, but all lvs. oblong-blunt and with a finely toothed margin, not coarsely toothed or lobed. Fl. heads yellow, 2–4 cm. across; involucral bracts oblong, almost entirely papery; receptacle convex with a central protuberance. Outer frs. with a tubular crown, central frs. with a toothed tongue. Mediterranean Europe and Portugal. Pl. 148.

1425. *C. coronarium* L. CROWN DAISY. Like 1424, but lvs. all twice cut into lance-shaped fine-pointed, entire or toothed segments. Fl. heads paler yellow, often larger, 3–6 cm. across; receptacle hemispherical. Outer fr. triangular in section, three-winged, the median wing much broader, inner frs. quadrangular. A robust, much-branched, hairless ann. 30–80 cm. △ Cultivated and waste ground; often grown as an ornamental and sometimes escaping. Apr.–Sept. Med. Eur. (except AL.). Pl. 148.

(b) Disk-florets only present

1426. *C. vulgare* (L.) Bernh. (*Tanacetum v.*) TANSY. A very aromatic perenn. with many erect stems to 1 m. arising from a creeping base, and compound flat-topped clusters of numerous, bright golden-yellow fl. heads each 7–12 mm. across. Ray-florets absent; involucral bracts pale green, papery-margined, hairless. Lvs. 15–25 cm., pinnately cut into about 12 pairs of lance-shaped leaflets each further pinnately lobed or toothed, the lower lvs. stalked, the upper lvs. half-clasping the stem. Fr. 1½ mm., greenish-white, five-ribbed. △ Pastures, waysides, hedges, screes. June–Oct. All Eur.: introd. IRL.IS. *Formerly grown as a medicinal plant and pot herb, and also used as a vermifuge and insecticide.* Pl. 147.

Ray-florets white
(a) Fl. heads solitary

1427. *C. leucanthemum* L. (*Leucanthemum vulgare* Lam.) MARGUERITE, MOON-DAISY, OX-EYE DAISY. An erect, little-branched, more or less hairless perenn. 20–70 cm., with large, solitary, long-stalked fl. heads 2½–5 cm. across, with white strap-shaped ray-florets longer than the width of the yellow disk. Involucral bracts oblong-lance-shaped, green with purplish-brown papery margin. Basal lvs. rounded or spathulate, toothed, long-stalked, the upper lvs. oblong, half-clasping the stem, deeply toothed or lobed. Fr. 2–3 mm., strongly ribbed, pale grey. A very variable sp. △ Grasslands, waysides, screes. May–Sept. All Eur. Pl. 148.

1428. *C. alpinum* L. ALPINE MOON-DAISY. A low-tufted mountain perenn. with ascending, sparsely woolly-haired stems 5–15 cm., bearing solitary fl. heads 2–4 cm. across. Ray-florets 8–12, usually white, rarely pink; disk-florets orange; involucral bracts green, with a dark brown margin. Lvs. very variable, hairless or white-woolly, nearly entire or deeply pinnately lobed. A very variable sp. △ In mountains: rocks, screes, grassy places, river gravels. July–Aug. E.F.CH.A.I.YU.R.

(b) Fl. heads in a flat-topped cluster

1429. *C. parthenium* (L.) Bernh. FEVERFEW. An erect, leafy, branched perenn. 25–60 cm., with numerous white fl. heads 1–2½ cm. across, in a lax more or less flat-topped cluster. Ray-florets with a rounded limb, as long as the width of the yellow disk or sometimes absent. Involucral bracts downy, with a pale, narrow, papery margin. Lvs. yellowish-green, strongly aromatic, 2½–8 cm., pinnate, with 3–7 oval leaflets each further divided into narrow oval, toothed or lobed segments, the lower lvs. stalked, the upper lvs. less divided, shorter stalked. △ Waysides, waste places, by habitation. June–Aug. Native of South-East Eur.: introd. to most of Eur. (except IRL.IS.). *Formerly cultivated as a medicinal plant. Also used as an insecticide and vermifuge, having similar properties to pyrethrum.* Pl. 148.

C. corymbosum L. Like 1429, but not aromatic and lvs. with 7–15 lance-shaped leaflets which are further divided into lance-shaped, toothed lobes. Fl. heads larger 2½–4 cm. across, in a flat-topped cluster; ray-florets white, limb narrow, longer than the width of the disk. Central and Southern Europe.

C. macrophyllum (Willd.) Waldst. & Kit. Distinguished by its small fl. heads only ½ cm. across, in a dense flat-topped cluster. Ray-florets white, limb much shorter than the width of the disk; disk-florets whitish; involucral bracts hairy, brown-margined. Lvs. pinnately lobed, lobes deeply double-toothed. A leafy erect perenn. South-Eastern Europe.

TRIPLEUROSPERMUM   Like *Matricaria*, but receptacle solid; fr. top-shaped with 3 ribs on one side and 2 oil-glands on the other side, and with a basal scar where it joins the receptacle; pappus absent or a short crown. 1 sp.

1430. *T. maritimum* (L.) Koch (*Matricaria m.*) SCENTLESS MAYWEED. Usually a hairless, erect, branched ann., bienn. or rarely perenn. 10–60 cm., with solitary long-stalked fl. heads 1½–4 cm. across. Ray-florets 12–30, white; disk-florets yellow; involucral bracts nearly hairless, oblong, with brown papery margin and tip. Lvs. scarcely aromatic, 2–3 times cut, with ultimate segments either long, slender, and acute, or in maritime forms fleshy and short-cylindrical, blunt. Fr. 1½–2 mm. A very variable sp. △ Cultivated ground, waste places, by habitation, on the littoral. June–Oct. All Eur.

MATRICARIA   Receptacle conical or ovoid-triangular, often hollow, without scales and hence differing from *Anthemis*. Ray-florets present or absent. Fr. ovoid, somewhat flattened, inconspicuously three- to five-ribbed, without oil-glands and with a lateral scar where it joins the receptacle. *c*. 12 sps.

1431. *M. recutita* L. (*M. chamomilla* auct.) WILD CHAMOMILE. Like 1430, but a sweetly aromatic, erect ann. 15–60 cm., with fl. heads 1–2½ cm. across, with about 15 white ray-florets which soon become strongly reflexed. Involucral bracts yellowish-green with greenish papery margin; ray-florets rarely absent; receptacle conical, hollow. Lvs. 2–3 times cut into narrow, linear, bristle-pointed segments. Fr. without oil-glands. △ Cultivated ground, waste places, tracksides. May–Oct. All Eur. *Used medicinally as a stimulant and tonic.*

1432. *M. matricarioides* (Less.) Porter (*M. discoidea* DC.) PINEAPPLE WEED, RAYLESS MAYWEED. Readily distinguished by its conical greenish-yellow fl. heads 5–8 mm. across, without ray-florets. Disk-florets numerous, borne on a hollow cone; involucral bracts hairless, oblong blunt and with a broad, transparent, papery margin. Lvs. hairless, 2–3 times cut into linear bristle-pointed segments. Fr. four-ribbed. A strongly aromatic, stiffly branched, erect leafy ann. 5–40 cm. △ Probably native of North-East Asia; spreading rapidly in Europe. May–Nov. All Eur. (except TR.). Pl. 148.

COTULA Fl. heads solitary with apparently only disk-florets present, but ray-florets very short and inconspicuous. Receptacle flat or arched, without scales. Fr. flattened, stalked; pappus absent. 1 sp.

1433. *C. coronopifolia* L. BRASS BUTTONS. An erect or creeping, hairless pale green ann. or perenn. 8–30 cm., with solitary globular yellow fl. heads 6–10 mm., which droop before and after flowering. Disk-florets white with 4 short yellow lobes; ray-florets inconspicuous but present and very short; involucral bracts hairless, in 2 rows, blunt-tipped with a papery margin. Lvs. strongly aromatic, hairless, 2–5 cm. oblance-shaped and irregularly cut into narrow lobes, and broadening below into a whitish sheathing base. Fr. rusty-brown. △ Probably native of South Africa; widely naturalized in sandy places, tracksides, on the littoral. July–Aug. Introd.: P.E.F.GB.NL.D.DK.N.S. Pl. 149.

ARTEMISIA | Wormwood Fl. heads numerous, small, in leafy clusters; ray-florets absent. Involucral bracts with papery margin; receptacle without scales. Fr. cylindrical and somewhat flattened; pappus absent. *c.* 50 sps.

Small alpine plants rarely more than 15 cm. high
1434. *A. genipi* Weber (*A. spicata* Wulfen) GENIPI. A low-tufted, silvery-haired, very aromatic alpine perenn. with erect fl. stems 5–15 cm., with a lax one-sided spike of fl. heads occupying most of its length. Fl. heads 3–5 mm., globular, short-stalked, erect; involucral bracts woolly, with a dark margin. Lower lvs. ternate, each leaflet often further three-lobed, stalked, the upper lvs. much less divided, stalkless. △ In mountains: rocks, screes, moraines. July–Sept. F.D.CH.A.I. *Genipi is the name given to a liqueur made from this plant.*

Tall lowland plants 40 cm. or more
(*a*) *Ultimate segments of lvs. at least 2 mm. broad*
1435. *A. vulgaris* L. MUGWORT. An erect, aromatic, much-branched perenn. 60–120 cm., with lvs. dark green above and contrastingly white-woolly beneath, and numerous tiny reddish-brown fl. heads in a long dense leafy cluster. Fl. heads erect, ovoid 3–4 mm. long; involucral bracts covered with dense cobweb-hairs and with a broad papery margin. Basal lvs. 5–8 cm., long-stalked, pinnately lobed, the stem lvs. stalkless and clasping the stem, once or twice pinnate, with ultimate segments lance-shaped. △ Waste places, waysides, hedgerows. July–Sept. All Eur.: introd. IS. *Used in herbal remedies as a stimulant and also in brewing. Somewhat poisonous to livestock.*

1436. *A. absinthium* L. WORMWOOD. Distinguished by its lvs. which are whitish with silky hairs on both sides and its drooping yellow, nearly globular fl. heads 3–4 mm. Fl. heads borne in lateral clusters and forming a branched pyramidal inflorescence; involucral bracts with silky hairs and a broad papery margin. Lower lvs. 3 times cut, upper lvs. twice-cut into blunt segments. A much-branched, very aromatic, silvery perenn., with grooved stems 30–100 cm. △ Rocks, screes, uncultivated ground. July–Sept. All Eur. (except IS.). *The spirit known as 'absinthe' is flavoured with this and other herbs. The plant has been used medicinally as a stimulant, vermifuge, insecticide and in brewing in place of hops.* Pl. 149.

A. dracunculus L. TARRAGON. Distinguished by its entire linear-lance-shaped, green, hairless lvs., the basal lvs. with a three-lobed apex. Fl. heads *c.* 3 mm., drooping, greenish, in a long, lax, branched cluster. A very aromatic perenn. Native of Asia; widely grown as a herb for seasoning and salads. Western and Central Europe.

(b) *Ultimate segments of lvs. 2 mm. or less broad*

1437. *A. campestris* L. FIELD SOUTHERNWOOD. An almost scentless, erect, woody-based perenn. 30–150 cm., with compound lvs. with ultimate segments linear *c.* 1 mm. wide, at first silvery-haired, but later hairless. Fl. heads green or reddish, ovoid 2½–4 mm. long, erect, stalked, in narrow clusters forming an elongated branched inflorescence; involucral bracts hairless, shining. Lower lvs. stalked, twice pinnate, the middle lvs. once-pinnate, the uppermost lvs. entire. A variable sp. △ Sandy places, rocks, uncultivated ground. Aug.–Sept. All Eur. (except IRL.IS.). Page 446.

*A. maritima L. SEA WORMWOOD. Coasts of most of Europe and salt-rich regions inland.

1438. *A. arborescens* L. SHRUBBY WORMWOOD. An aromatic, woody-based, shrubby perenn. ½–1 m., with densely silky-haired white stems and lvs.; the lower lvs. stalked, 3 times cut, the upper short-stalked, or stalkless, once or twice-cut into linear blunt segments 1–2 mm. broad. Fl. heads yellowish, globular 5–6 mm., erect during flowering, forming a broad leafy cluster; involucre white-woolly; receptacle hairy. Fr. glandular. △ Maritime rocks. May–June. P.E.F.I.YU.GR.

A. alba Turra (*A. camphorata* Vill.) CAMPHOR WORMWOOD. Like 1438, but lvs. and stems green, and smelling strongly of camphor, gland-dotted, almost hairless. Fl. heads nodding during flowering; involucral bracts whitish, finely hairy, blunt, largely papery. Ultimate segments of lvs. very narrow, thread-like; a woody-based perenn. Southern Europe.

TUSSILAGO | **Coltsfoot**   Fl. heads solitary on unbranched stems bearing only scale-like lvs. and arising direct from the rhizome. Ray-florets very numerous, in many rows; involucral bracts numerous, green, mostly in 1 row. Pappus of many rows of simple hairs. 1 sp.

1439. *T. farfara* L. COLTSFOOT. Fl. heads bright yellow, 1½–3½ cm. across, solitary, borne on thick scaly stems and appearing before the lvs. Fl. heads drooping after flowering, becoming erect in fr. Lvs. all basal, large, stalked, blade 10–30 cm. across, orbicular-heart-shaped with wide, irregular, shallow lobes, white-felted beneath. Fr. with a long white pappus. A perenn. with creeping rhizomes and erect fl. stems which lengthen in fr. to 15–35 cm. △ Waste places, banks, landslides, screes, river gravels. Feb.–Mar. All Eur. *The leaves can be smoked like tobacco and were well known as a remedy for chest complaints, coughs, and asthma.* Pl. 149.

PETASITES | **Butterbur**   Fl. heads in a dense spike-like cluster which is usually one-sexed and borne on erect scaly stems before or at the same time as the radical lvs. Ray-florets absent or inconspicuous. Fr. cylindrical; pappus of slender simple hairs. *c.* 8 sps.

Fl. stems appearing before the lvs.

1440. *P. hybridus* (L.) Gaertner, Meyer & Scherb. BUTTERBUR. Fl. stems appearing before the lvs., stout and covered with purplish lance-shaped scales and cobweb-like hairs and bearing a large dense ovoid cluster of many pale reddish-violet fl. heads. 'Male' fl. heads 7–12 mm. long, short-stalked; 'female' fl. heads smaller 3–6 mm. long, longer-stalked, and 'female' inflorescence lengthening greatly in fr. to *c.* 80 cm.; involucral bracts blunt, hairless, reddish. Lvs. all basal, long-stalked, blade rounded-heart-shaped, 10–90 cm. across, shallowly angled and irregularly toothed, at first downy, but becoming green above and greyish-downy beneath at maturity. A perenn. with stout, long-creeping,

1. *Buphthalmum salicifolium* 1396
2. *Artemisia campestris* 1437
3. *Inula ensifolia* [1389]
4. *Saussurea discolor* 1475
5. *Senecio paludosus* 1454

rhizomes and hence usually forming considerable patches. △ Riversides, damp meadows, marshes. Mar.–May. All Eur.: introd. North Eur. Pl. 149.

1441. P. albus (L.) Gaertner WHITE BUTTERBUR. Like 1440, but fl. heads whitish-yellow and scales of fl. stem greenish. Lvs. smaller, 15–30 cm. across, broadly heart-shaped, more conspicuously and irregularly toothed, white-woolly beneath at maturity. A perenn. with creeping rhizomes and stout, erect fl. stems 10–30 cm., lengthening to 70 cm. in fr. △ In hills and mountains: damp woods, streamsides, springs. Mar.–May. Much of Eur. (except P.IRL.NL.IS.SF.AL.GR.TR.): introd. GB.B. Pl. 149.

P. paradoxus (Retz.) Baumg. (*P. niveus* (Vill.) Baumg.) ALPINE BUTTERBUR. Like 1441, but lvs. broadly triangular with 2 acute, wide-spreading, basal lobes, regularly toothed and with a thick snowy-white felt of hairs beneath, hairless above. Fl. heads white or pale pink, in a dense rounded cluster; scales on fl. stem pinkish. Central Europe.

Fl. stems appearing with the lvs.
1442. P. fragrans (Vill.) C. Presl WINTER HELIOTROPE. Distinguished by its pale lilac, strongly vanilla-scented clusters of fl. heads borne on an erect stem at the same time as the lvs. Fl. heads about 10; female fl. heads with short ray-florets; scales on fl. stem green, the lower scales with rudimentary blades. Lvs. stalked and with a sheathing base, blade heart-shaped 10–20 cm. across, with regular finely toothed margin green on both sides. A perenn. with far-creeping rhizomes, often forming extensive patches; fl. stems 10–25 cm. △ Waste places, streamsides, and banks. Jan.–Mar. P.E.F.I.: introd. IRL.GB.

HOMOGYNE Fl. head solitary, borne on stems with scale lvs. and with some scales with rudimentary blades. Ray-florets absent; stigmas of disk-florets protruding. Fr. with a pappus of several rows of simple hairs. 3 sps.

1443. H. alpina (L.) Cass. ALPINE COLTSFOOT. Fl. heads erect, solitary, reddish-violet, borne on an erect woolly-haired, scaly-leaved stem 10–40 cm., the lower scales often with a rudimentary blade. Fl. heads 1–1½ cm. long; involucral bracts blunt, purplish, woolly-haired. Lvs. all basal, appearing with the fls., long-stalked, blade almost circular-heart-shaped, to *c.* 4 cm. across, broadly toothed, dark shiny green above, hairy and often purplish beneath. A stoloniferous perenn. △ In mountains: damp pastures, open woods, near springs. June–Sept. Cent., South-East Eur. (except GR.TR.) E.F.I.SU.: introd. GB. Pl. 150.

ADENOSTYLES Fl. heads small, numerous and in dense, branched, flat-topped clusters. Ray-florets absent; involucral bracts 3–8; receptacle without scales. Fr. with a pappus of several rows of hairs. *c.* 5 sps.

1444. A. glabra (Miller) DC. A robust erect perenn. 30–60 cm., with broad lvs. and small reddish-purple fl. heads clustered into a much-branched, dense, rather flat-topped inflorescence. Fl. heads usually with only 2–3 florets, narrow-cylindrical *c.* 1 cm. long; involucral bracts hairless. Basal lvs. rather tough, long-stalked, blade broadly heart-shaped to kidney-shaped, regularly and deeply toothed, grey-green and almost hairless; stem lvs. stalked, but stalks without clasping auricles at the base. △ In mountains: open woods, pastures, screes. July–Aug. Cent. Eur. E.F.I.YU.

1445. A. alliariae (Gouan) Kerner (*A. albifrons* (L. fil.) Reichenb.) Like 1444, but lvs. softer, either white- or grey-cottony beneath, or hairless and similar-coloured on both sides, margin irregularly double-toothed, the lower lvs. up to 50 cm. across. Stem lvs. stalked or stalkless, encircling the stem with rounded auricles. An erect, shortly hairy perenn. 60–100 cm. △ Streamsides, forests. July–Aug. Cent. Eur. E.F.I.YU.AL.BG. Pl. 150.

ARNICA | **Arnica** Ray-florets yellow, conspicuous, in 1 row; disk-florets similar-coloured; involucral bracts equal, in 2 rows. Stem lvs. opposite, few. Pappus with 1 row of hairs. 2 sps.

1446. *A. montana* L. ARNICA. Fl. heads orange-yellow, large 4–8 cm. across, usually solitary or rarely up to 4 on stout erect stems 20–60 cm., bearing only 1 or 2 pairs of small opposite lvs. Ray-florets wide-spreading, as long as or longer than the width of the disk; involucral bracts lance-shaped, hairy, in 2 rows. Basal lvs. in a rosette, oval-lance-shaped, glandular-hairy and aromatic, much longer than the stem lvs., lvs. all downy. Fr. cylindrical, hairy, pappus a row of rough hairs. △ In the mountains: pastures, heaths, open woods. May–July. Cent. Eur. P.E.F.B.NL.DK.N.S.I.YU.R.SU. *Often used externally for bruises, wounds, and sprains; a resinous compound, arnicine, is the active substance. Also used internally as a stimulant and febrifuge.* Pl. 151.

DORONICUM | **Leopard's-Bane** Fl. heads large; ray-florets large yellow, numerous, in 1 row; disk-florets similar-coloured. Involucral bracts equal, in 2–3 rows; receptacle flat, without scales, often hairy. Fr. usually with a pappus. Distinguished from *Arnica* by the alternate stem lvs. *c.* 12 sps.

Outer row of fr. without a pappus
(a) Fl. heads several
1447. *D. pardalianches* L. GREAT LEOPARD'S-BANE. Fl. heads usually several, yellow, 4–6 cm. across, borne on a slender, sparsely branched, woolly-haired stem 30–90 cm. Involucral bracts glandular-hairy. Basal rosette lvs. broadly oval-heart-shaped, entire, long-stalked, ciliate on both sides; middle stem lvs. narrowed below and broadened at the base into 2 rounded lobes; uppermost stem lvs. oval and clasping. A perenn. with tuberous underground stems. △ In hills: woods, shady banks. May–July. Cent., South-East Eur. (except TR.) E.F.NL.I.SU.: introd. IRL.GB. Pl. 151.

D. austriacum Jacq. AUSTRIAN LEOPARD'S-BANE. Like 1447, but stems very leafy with lvs. longer than the internodes; lower lvs. with an oval toothed blade, winged lv. stalk and rounded auricles clasping the stem; middle stem lvs. lance-shaped, suddenly narrowed . below and with clasping basal lobes. Fl. heads yellow, usually many, 5–6 cm. across; involucral bracts glandular-hairy. Central Europe.

(b) Fl. heads solitary
D. plantagineum L. LEOPARD'S-BANE. Like 1447, but basal lvs. oval-elliptic and narrowed gradually into a long stalk, the uppermost lvs. elliptic-lance-shaped, stalkless, with blades running a short distance down the stem. Fl. heads yellow, usually solitary, 5–8 cm. across; involucre ciliate; stem glandular-hairy above. South-Western Europe. *The dried leaves have been smoked like tobacco and used as snuff.*

All fr. with a pappus
1448. *D. grandiflorum* Lam. (*Aronicum scorpioides* (L.) Koch) LARGE-FLOWERED LEOPARD'S-BANE. Fl. heads yellow, usually solitary or rarely 2 or 3, 5–8 cm. across, borne on glandular-hairy stems. Involucral bracts with long glandular hairs. All lvs. sparsely glandular-hairy, deeply toothed and with numerous glandular hairs on margin; lowest lvs. stalked, oval and with a cut-off or slightly heart-shaped base, the middle lvs. often with a winged stalk and auricled base, the uppermost lvs. oval-lance-shaped stalkless and clasping the stem. A perenn. 6–50 cm. △ In mountains; screes, rocks, stony pastures. June–Sept. F.D.CH.A.I.YU.R. Pl. 150.

D. clusii (All.) Tausch (*Aronicum c.*) TUFTED LEOPARD'S-BANE. Very like 1448, but basal lvs. oval-lance-shaped and gradually narrowed at the base into a lv. stalk, blade without glandular hairs, margin entire or sparingly toothed. Fl. heads yellow, solitary, $3\frac{1}{2}$–6 cm. across; involucre rough-haired, not glandular. Pyrenees, Alps, Carpathians.

SENECIO Ray- and disk-florets yellow and similar-coloured, or ray-florets rarely absent; involucre cylindrical or bell-shaped, bracts in a single row, green and leafy, with or without a few much shorter additional bracts towards the base. Receptacle flat, without scales. Fr. cylindrical, ribbed, with a pappus of hairs borne in several rows. *c.* 65 sps. *Many species contain poisonous alkaloids which can cause death among livestock.*

Some or all lvs. deeply lobed
(a) Ray-florets absent
1449. *S. vulgaris* L. GROUNDSEL. A sparingly hairy, erect ann. or bienn. to 45 cm., branched above and with many small, yellow, cylindrical fl. heads *c.* 1 cm. long and 4 mm. across, in a branched, rather leafy cluster. Involucre with an outer row of 8–10 short additional bracts with black tips, which are about a quarter as long as the inner row of green, usually hairless acute bracts; ray-florets rarely present. Lvs. irregularly deeply cut into oblong, toothed lobes, slightly fleshy, hairless or cottony. Fr. $1\frac{1}{2}$–2 mm., hairy on ribs, pappus long, white. △ Cultivated ground, waste places, walls. All the year round. All Eur.

(b) Ray-florets inconspicuous, very narrow and soon becoming inrolled
1450. *S. viscosus* L. STINKING GROUNDSEL. A sticky-glandular fetid ann. or bienn. 20–65 cm., with long-stalked fl. heads 8 mm. across, in a lax branched cluster. Ray-florets yellow, about 13, short and inrolled from apex to base; involucre with 4–5 additional bracts nearly half as long as the inner bracts. Lvs. dark green, deeply and regularly pinnately lobed. Fr. 3–4 mm., becoming hairless, pappus white, very long. △ Clearings in woods, sands, waste places, walls. July–Sept. Most of Eur. (except P.IRL.AL.TR.): introd. North Eur.

S. sylvaticus L. WOOD GROUNDSEL. Like 1450, but not fetid and not glandular or only slightly so above. Fl. heads *c.* 5 mm. across; involucre with very small additional bracts, less than one-quarter as long as the inner bracts. Fr. hairy. Widespread in Europe.

(c) Ray-florets conspicuous, spreading
(i) Green, nearly hairless plants
1451. *S. jacobaea* L. RAGWORT. A robust, erect, branched bienn. or perenn. 30–150 cm., with numerous bright golden-yellow fl. heads $\frac{1}{2}$–$2\frac{1}{2}$ cm. across, in dense, flat-topped, compound clusters. Additional involucral bracts about one-quarter as long as the nearly hairless inner bracts and inserted on fl. stem; ray-florets sometimes absent. Basal lvs. in a rosette, disappearing at flowering; stem lvs. hairless or with cobweb hairs, once or twice deeply cut into blunt, rounded, irregularly toothed lobes, the terminal lobe scarcely longer than the lateral lobes. Stolons absent. Fr. 2 mm., the outer frs. hairless, the inner frs. hairy. △ Waysides, poor pastures, neglected cultivation, dunes. June–Oct. All Eur. (except IS.TR.). *Formerly used in herbal remedies. It contains poisonous alkaloids which are active both in fresh and dried plants, and which accumulate in the body, thus causing the death of a large number of livestock annually in Europe.*

S. erucifolius L. HOARY RAGWORT. Very like 1451, but lower stem lvs. with narrow, acute, terminal lobes and regular, parallel-sided, acute, lateral lobes. Additional involucral bracts about half as long as the inner bracts. Fr. all hairy. A perenn. with creeping stolons. Widespread, except in the north of Europe.

S. aquaticus Hill MARSH RAGWORT. Like 1451, but a marsh plant with fewer larger fl. heads 2½–3 cm. across, in an irregular, lax, nearly leafless, branched cluster. Middle and lower stem lvs. with the terminal lobes much larger and broader than the lateral lobes. Fr. hairless. Throughout Europe, except the extreme south-east.

1452. *S. vernalis* Waldst. & Kit. SPRING GROUNDSEL. An ann. 15–50 cm., with cobweb-haired stems and rather large, conspicuous fl. heads 2–3 cm. across, with about 13 spreading ray-florets, each *c.* 1 cm. long. Involucre with 6–12 short additional bracts and 21 pointed, hairless, inner bracts. Lvs. oblong in outline, pinnately lobed, at first with cobweb-hairs, later hairless, the lower lvs. stalked, the upper lvs. stalkless and broadly eared at the base. Fr. densely white hairy. △ Rocky and sandy places, uncultivated ground. Apr. South-East Eur. F.I.SU.: introd. North, Cent. Eur. Pl. 151.

(ii) Densely white-felted plants
S. incanus L. GREY ALPINE GROUNDSEL. A small, silvery-white, rosette-forming alpine perenn. with 3–10 yellow fl. heads each 1–1½ cm. across, in a dense terminal cluster. Ray-florets 3–5, limb 5–6 mm. long; involucre silvery-white. Rosette lvs. with thick white-felted blades, which are deeply cut into equal, narrow, blunt lobes, at length becoming hairless. Alps, Apennines, Carpathians. Pl. 150.

1453. *S. cineraria* DC. CINERARIA. A large, silvery-white, bushy, woody-based perenn. 30–100 cm., with rather numerous fl. heads 1–1½ cm. across in large, dense, flattish, terminal clusters. Ray-florets 10–12; involucral bracts and fl. stems densely white-felted, additional bracts minute or absent. Lvs. elliptic in outline, deeply cut into oblong lobes which are often three-lobed or further divided, greenish woolly-haired above and white-felted beneath. Fr. hairless. △ Rocks, sands, on the littoral. June–Aug. Med. Eur. (except AL. TR.) P.SU.: introd. elsewhere. *The juice of this plant has been used for eye diseases.*

Lvs. entire or toothed
(a) Involucre with an additional row of 10–20 bracts; ray-florets 10–20
1454. *S. paludosus* L. GREAT FEN RAGWORT. A stiff erect perenn. ½–2½ m., with a grooved stem which is leafy to the apex and bearing yellow fl. heads 3–4 cm. across in a dense flat-topped cluster of 12–16. Ray-florets 10–20; involucral bracts hairless or cottony, the additional bracts half as long as the inner bracts. Lvs. all stalkless, linear-lance-shaped, strongly and regularly saw-toothed, shining above and with cobweb-hairs beneath, but becoming hairless. Fr. hairless, one-third as long as the pappus. △ Watersides, marshes. June–Sept. Cent., South-East Eur. (except AL.TR.) E.F.B.NL.S.I.SU.: introd. DK. Page 446.

1455. *S. doronicum* L. CHAMOIS RAGWORT. An erect, rather greyish mountain perenn. 30–60 cm., with large, long-stalked, solitary or few golden-yellow or orange fl. heads 3½–4½ cm. across, each with 15–22 ray-florets. Additional row of involucral bracts linear, often as long as or longer than the inner row of bracts. Lvs. rather thick and leathery, with cobweb-hairs beneath, the lower lvs. elliptic and narrowed to a long stalk, the upper lvs. few, linear-lance-shaped, with a broad, rounded base half-clasping the stem. Fr. hairless, half as long as the pappus. A very variable sp. △ In mountains: pastures, rocky places, woods. June–Aug. P.E.F.D.CH.A.I.YU.AL.R. Pl. 151.

S. alpinus (L.) Scop. (*S. cordatus* Koch) ALPINE RAGWORT. Distinguished by the large, oval-heart-shaped, irregularly and coarsely toothed and stalked lower lvs. Fl. heads few, golden-yellow, 3–4 cm. across, in a flat-topped cluster. A robust leafy perenn. 30–140 cm. Alps, Balkans.

(b) Involucre with an additional row of 3–5 bracts; ray-florets 3–8
1456. *S. fuchsii* C. C. Gmelin A robust, erect, leafy, hairless perenn. 40–200 cm., with numerous yellow fl. heads 1½–2½ cm. across in a flat-topped cluster. Ray-florets few, 4–8; in-

volucre cylindrical, twice as long as broad, bracts 8–10, blunt, usually hairless. Lvs. green, lance-shaped, and narrowed to a narrowly winged stalk which is scarcely enlarged at the base, margin finely saw-toothed, hairless or sparsely glandular-haired beneath. Fr. hairless. △ In hills and mountains: damp woods, ravines, screes. July–Sept. Cent. Eur. E.F.I.YU.R.BG.SU.: introd. S.

S. nemorensis L. Like 1456, but lvs. oblong-oval to lance-shaped, the middle stem lvs. stalkless, enlarged at the base and half-clasping the stem, hairy beneath, at least on the midvein. Involucre bell-shaped, 6–8 mm., about as broad as long, sparsely hairy. Plant strong-smelling when dry. Widespread in Central and Southern Europe. Pl. 150.

*S. fluviatilis Wallr. BROAD-LEAVED RAGWORT. Widespread in Central and Southern Europe.

(c) Involucre without an additional row of bracts
1457. *S. palustris* (L.) Hooker MARSH FLEAWORT. An ann., bienn. or perenn. with a robust, grooved, hairy stem 15–100 cm. and numerous pale yellow fl. heads 2–3 cm. across, in a compact, compound cluster. Ray-florets *c.* 21; involucral bracts and fl. stems densely hairy. Lvs. numerous to the inflorescence, slightly fleshy, linear-lance-shaped and half-clasping the stem, margin undulate. Fr. hairless. △ Marshes, lakesides, dunes. May–July. B.NL.DK.S.A.PL.CS.H.R.SU.

1458. *S. integrifolius* (L.) Clairv. An erect perenn. 7–60 cm., with lvs. at first densely cobweb-haired, but later nearly hairless, and a simple, few-flowered, dense cluster of fl. heads each 1½–2½ cm. across. Ray-florets *c.* 13, bright yellow, nearly twice as long as the involucral bracts, which are cottony below. Basal lvs. in rosette, oval-spathulate, toothed or entire, and with a winged lv. stalk, the upper lvs. oblong-oval, with blade shortly running down stem, the uppermost lvs. linear. Fr. shortly hairy. A variable sp. △ Rocky pastures. June–July. Cent. Eur. E.F.GB.DK.N.S.SF.YU.R.SU.

LIGULARIA Like *Senecio*, but upper lvs. with enlarged lv.-base; involucre with 2 long opposite bracteoles arising from the base. 1 sp.

1459. *L. sibirica* (L.) Cass. A rather stout, nearly hairless, erect, unbranched perenn. 60–150 cm., with large broad lvs. and a leafless spike-like cluster of yellow fl. heads. Fl. heads numerous, *c.* 3½ cm. across; ray-florets numerous; involucre bell-shaped, bracts 8–10 lance-shaped, hairless; bracteoles linear. Lower lvs. variable, kidney-shaped, to *c.* 30 cm. across, or nearly arrow-shaped, all toothed, long-stalked and sheathing at the base, the upper lvs. much smaller, stalkless and with an enlarged sheathing base, the uppermost lvs. oval, bract-like. Pappus reddish. △ Marshes, bogs. July–Aug. E.F.A.PL.CS.H.R.SU.

CALENDULA | Marigold Ray-florets numerous in 1–3 rows; involucre broadly bell-shaped, bracts in 1 row; receptacle without scales. Fr. of three forms; pappus absent. *c.* 9 sps.

1460. *C. arvensis* L. MARIGOLD. A rough-hairy, spreading or ascending ann. 10–30 cm., with solitary orange-yellow fl. heads 1–2 cm. across. Ray-florets mostly 1–1½ cm. long, but variable and usually twice as long as the involucre; involucral bracts numerous, nearly equal, oblong-lance-shaped. Lvs. oblong-lance-shaped, shallowly toothed or entire, and half-clasping the stem. Fr. head drooping, fr. spiny or smooth, of three forms: sickle-shaped, boat-shaped, or ringed. A variable sp. △ Vineyards, fields. Apr.–Oct. Med. Eur. (except AL.) P.CH.D.SU.: introd. CS.PL.H. Pl. 151.

C. officinalis L. POT-MARIGOLD. Like 1460, but a perenn. with larger fl. heads 4–5 cm. across, usually orange-yellow. Lower lvs. blunt, spathulate. Fr. heads erect; fr. mostly boat-shaped, spiny. Probably a native of the Mediterranean region; widely grown as an ornamental and naturalized elsewhere. *Formerly used to colour butter and cheese; also used in herbal remedies for sprains and as a tonic.*

ECHINOPS | **Globe-Thistle** Lvs. thistle-like, spiny. Fl. heads grouped together in a spherical cluster; each fl. head is comprised of only one floret which is surrounded by spiny involucral bracts. *c.* 15 sps.

1461. *E. sphaerocephalus* L. PALE GLOBE-THISTLE. An erect, stiff, thistle-like perenn. $\frac{1}{2}$–2 m., with large, terminal, spherical, pale whitish-blue spiny heads 4–6 cm. across. Stamens with blue filaments; involucral bracts 16, glandular-hairy, the outer bristle-like and about half as long as the inner. Lvs. lance-shaped in outline, deeply pinnately lobed, with lobes rather weakly spiny-toothed, green and glandular above, and white-cottony beneath; stems white-woolly, glandular-hairy above. △ Dry stony places, uncultivated ground. June–Sept. Cent., South-East Eur. (except AL.) E.F.SU.: introd. DK.N.S.SF.CH.D.

E. strigosus L. ROUGH-LEAVED GLOBE-THISTLE. Like 1461, but upper surface of lvs. rough with stiff bristly hairs, white-woolly beneath; involucral bracts very unequal, not glandular-hairy, the middle bracts oblong ciliate, the inner bracts with a long-pointed and keeled apex, very long ciliate. Globular heads blue, 4–6 cm. across. Pappus hairs not fused at the base. Portugal and Spain.

1462. *E. ritro* L. GLOBE-THISTLE. Like 1461, but globular heads bright blue, 3–3$\frac{1}{2}$ cm. across. Lvs. green and shining above and densely white-woolly beneath, deeply pinnately cut into narrow, very spiny, lance-shaped lobes; fl. stems densely white-woolly but not glandular. Involucre blue, not glandular-hairy, but with numerous bristle-like hairs at the base one-quarter to one-half as long as the bracts; outer bracts triangular-spathulate, finely toothed, the inner bracts lance-shaped and long-pointed, long-ciliate. Pappus hairs fused at the base. A stiff erect perenn. 10–40 cm. △ Rocky places, uncultivated ground. July–Sept. Med. Eur. A.CS.H.BG.R.SU. Pl. 152.

CARDOPATIUM Fl. heads small and with few florets, in a dense, leafy, flat-topped cluster. Ray-florets absent; involucral bracts in several rows, leathery, with comb-like spiny teeth, all except the innermost. Pappus scaly. *c.* 3 sps.

1463. *C. corymbosum* (L.) Pers. A much-branched, erect perenn. 8–20 cm., with numerous blue fl. heads in a dense, very spiny, globular cluster surrounded by many spiny lvs. Fl. heads with 8–10 disk-florets; involucral bracts with recurved spiny apex and a spiny-haired margin. Lower lvs. pinnately cut, with segments further cut into spiny lobes, the upper lvs. smaller with narrow spiny lobes and forming a spiny wing running down the stem. △ Among vegetation by the sea. July–Aug. I.GR.TR.

XERANTHEMUM Involucral bracts all papery, the inner bracts brightly coloured and longer than the florets and sometimes spreading outwards and appearing to replace the ray-florets. Ray-florets absent. Fr. with 5–15 long, persistent, unequal bristles. *c.* 3 sps.

Outer involucral bracts hairless, acute; fr. with 5 spines
1464. *X. annuum* L. PINK EVERLASTING. An erect, slender, little-branched, whitish woolly-haired ann. 15–50 cm., with narrow lvs. and solitary, long-stalked, pinkish-purple fl. heads 3$\frac{1}{2}$–5 cm., with conspicuous, shining, papery involucral bracts. Fl. heads globular, with 100–150 purple fertile florets; outer involucral bracts oval, pale silvery-brown, the

inner bracts much longer, shining pink and spreading outwards in place of the absent ray-florets. Lvs. linear, entire. Fr. with 5 unequal bristles. △ Stony places. June–July. Med., South-East Eur. P.SU.: introd. Cent. Eur. Pl. 151.

X. inapertum (L.) Miller Like 1464, but involucre cylindrical, and bracts whitish or brownish, the inner bracts erect, not spreading and little longer than the outer. Fl. heads with 30–40 purple fertile florets. Mediterranean Europe and Portugal.

Outer involucral bracts hairy, blunt; fr. with 8–10 spines
1465. *X. foetidum* (Cass.) Moench Like 1464, but fl. heads oblong-cylindrical and with the outer involucral bracts hairy on the back, and the inner bracts shining pink, erect, hairless. Fertile florets 10–15. Lvs. lance-shaped. Fr. with 8–10 short unequal bristles. A slender, erect, little-branched ann. 30–50 cm. △ Cultivated ground, bushy places, tracksides. July–Aug. South-East Eur. F.CS.H.

CARLINA Fl. head often large; ray-florets absent; outer involucral bracts tough and leathery, spiny, the inner bracts papery, shining, with coloured apex, and longer than the florets. Receptacle with scales. Fr. cylindrical, hairy; pappus a single row of feathery hairs, soon falling. Lvs. thistle-like, spiny. *c.* 15 sps.

Fl. heads 6–14 cm. across, usually solitary
1466. *C. acanthifolia* All. Fl. heads very large 12–14 cm. across, yellow, borne stalkless at the centre of a very large rosette of spiny, spreading, thistle-like lvs. Disk-florets numerous, borne on a flat receptacle with scales; outer involucral bracts leafy, lance-shaped, spiny-toothed, the middle bracts linear, blackish, with long, comb-like, spiny teeth, the inner bracts linear, shining lemon-yellow, spreading widely when dry. Lvs. often velvety-white on both faces, or at least beneath, oblong in outline and deeply cut into spiny-toothed lobes. △ In mountains: stony places, pastures, rocks. June–Sept. South-East Eur. (except TR.) E.F.A.I. *The roots are used in herbal remedies.*

1467. *C. acaulis* L. STEMLESS CARLINE THISTLE. Like 1466, but fl. heads whitish due to the bright silvery-white spreading inner involucral bracts. Fl. heads 5–13 cm. across, solitary and stemless, or less commonly on stout stems to *c.* 30 cm., sometimes more than one-headed. Disk-florets whitish or reddish; inner involucral bracts sometimes pinkish or brownish at base. Lvs. in a large rosette, deeply pinnately spiny-lobed, hairless or with few cobweb-hairs. △ In mountains: poor pastures, rocky slopes. May–Sept. Much of Eur. (except P.IRL.GB.B.NL.IS.AL.TR.). *The dried flowers are used as a rustic hygrometer or weather-glass, the bracts closing together in damp air. The roots are used in herbal remedies.* Pl. 151.

Fl. heads 2–5 cm. across, several
1468. *C. vulgaris* L. CARLINE THISTLE. A stiff, spiny-leaved, thistle-like bienn. 5–50 cm., with few pale golden-yellow fl. heads 2–4 cm. across, in a sparsely branched spiny cluster. Outer involucral bracts green, woolly-haired, spiny, the inner bracts pale silvery-yellow, linear acute, spreading in dry weather to look like the absent ray-florets. Stem lvs. shallowly lobed, with spiny margin and half-clasping the stem at the base; rosette lvs. dying before flowering. A variable sp. △ Dry pastures, stony uncultivated ground. June–Sept. All Eur. (except P.IS.).

1469. *C. corymbosa* L. FLAT-TOPPED CARLINE THISTLE. Like 1468, but inner involucral bracts bright golden-yellow, linear-lance-shaped and broadened below apex, and spreading outwards in the place of the absent ray-florets. Fl. heads yellow, 2½–4 cm. across, short-stalked, usually several in a flat-topped cluster; outer involucral bracts green, spiny-lobed, the middle bracts woolly-haired. Lvs. oval-lance-shaped, pinnately spiny-lobed, nearly hairless, but with cobweb-hairs below; stems white-felted. An erect, stiff, thistle-like perenn. 10–50 cm. △ Stony ground. June–Aug. Med. Eur. P.BG. Pl. 152.

C. lanata L. PURPLE CARLINE THISTLE. Readily distinguished by the purple inner involucral bracts which contrast with the yellow disk-florets. Fl. heads 1½–3 cm. across, often solitary. Lvs. with white-woolly hairs, the upper stalkless, and clasping the stem. Mediterranean Europe.

ATRACTYLIS Like *Carlina*, but the outer involucral bracts leafy and deeply cut into spiny teeth, and the innermost bracts papery-tipped, not bright-coloured, little longer than the florets and not spreading. Fr. silvery-haired; pappus with 1–3 rows of feathery bristles fused at the base. 3 sps.

1470. *A. cancellata* L. A slender, erect ann. 5–25 cm., with narrow lvs. and small purple fl. heads c. 1½ cm. across, which are loosely encircled by several longer green, spiny, comb-like, outer involucral bracts, recalling a miniature Chinese lantern. Innermost involucral bracts purple, papery, lance-shaped; disk-florets only present. Lvs. stalkless, linear-lance-shaped, soft with cobweb-hairs, or white-cottony and with bristly hairs on margin; stems slender, white-felted. △ Dry, rocky places, tracksides, hills. May–July. P.E.F.I.GR.

1471. *A. gummifera* L. A spiny, thistle-like rosette perenn. bearing at the centre a large, solitary, stalkless purple fl. head 3–7 cm. across. Disk-florets only present; outer involucral bracts twice-cut into narrow spiny segments, the innermost bracts linear acute and purple-tipped. Fl. head sometimes short-stalked, and more than one. Lvs. oblong-lance-shaped, deeply pinnately cut into many spiny segments, nearly hairless. △ Dry places, field verges, waysides. Aug.–Oct. P.E.F.I.GR. *The plant produces a gum similar to mastic.*

ARCTIUM | Burdock Involucral bracts numerous, spreading, each ending in a stiff hooked spine; disk-florets only present. Fr. with a pappus of several rows of short rough hairs. Lvs. broad, entire, not spiny. Often included in *Lappa*. c. 4 sps.

Involucre densely cobweb-haired
1472. *A. tomentosum* Miller WOOLLY BURDOCK. A stiff, erect, branched bienn. 80–150 cm., with large lvs. and rather numerous globular purple fl. heads with the involucre covered with a network of fine white cobweb-hairs. Inflorescence dense, somewhat flat-topped, fl. heads 2–3 cm.; outer involucral bracts green, hooked, the inner bracts not hooked, purple-tipped, shorter than the florets. Lvs. broadly oval-heart-shaped, stalked, white-cottony beneath. △ Alluvium, waysides, habitations. June–Sept. Much of Eur. (except P.E.IRL.GB.TR.). Pl. 152.

Involucre hairless or slightly hairy
1473. *A. lappa* L. (*A. majus* Bernh.) GREAT BURDOCK. Like 1472, but involucre without cobweb-hairs. Fl. heads large, 3–4 cm. reddish-purple, usually long-stalked in a more or less flat-topped cluster. Involucral bracts green, hairless, or rarely hairy, the innermost hooked, and as long as the florets. Fl. heads opening widely above in fr. Basal lvs. with an oval-heart-shaped blade to 40 cm., grey-cottony beneath, and with a furrowed solid lv. stalk to 30 cm. A stout bienn. often with reddish and somewhat woolly stems 90–200 cm. △ Waste places, tracksides, uncultivated ground. June–Sept. All Eur. (except IS.AL.). *Used herbally for skin diseases and for purifying the blood.*

A. minus Bernh. LESSER BURDOCK. A very variable bienn. to 1½ m., differing from 1473 in often having smaller, more numerous, short-stalked fl. heads arranged in compound, spike-like clusters. Upper part of corolla equalling the lower narrower part; stalks of basal lvs. hollow. Fl. heads very variable in size, 1½–4 cm. across. Throughout Europe.

STAEHELINA Fl. heads narrow cylindrical, in clusters; ray-florets absent; involucral bracts not spiny; receptacle with scales. Fr. hairless; pappus a single row of long, smooth, simple hairs. *c.* 4 sps.

1474. *S. dubia* L. A woody-based bushy perenn. 20–40 cm., with hairy stems, narrow lvs., and usually a few purple short-stalked fl. heads in a dense cluster, or fl. heads solitary. Involucre oblong-cylindrical; bracts lance-shaped, flushed with purple, woolly-haired, the inner narrower. Lvs. linear, margin inrolled and undulate, toothed or entire, grey-haired above and white-woolly beneath. Fr. grooved, 4½–5 mm., with a conspicuous white pappus 4–5 times as long. △ Dry banks, stony places. June–July. P.E.F.I.

SAUSSUREA Involucral bracts in several ranks, entire, not spiny; ray-florets absent. Fr. cylindrical, four-ribbed; pappus with outer row of persistent rough hairs and inner row of feathery hairs which soon fall off. Lvs. not spiny. *c.* 6 sps.

1475. *S. discolor* (Willd.) DC. Fl. heads violet, sweet-scented, 3–8, almost stalkless in a dense terminal cluster. Fl. heads 1½–2 cm. long; involucral bracts oval blunt, violet-flushed, woolly-haired. Lvs. becoming hairless above, densely snowy-haired beneath, the lower lvs. with long unwinged lv. stalk, blades oval-triangular and toothed with a heart-shaped or cut-off base, the upper lvs. linear-lance-shaped, stalkless. An erect, unbranched, woolly-stemmed perenn. 5–35 cm. △ Alpine: rocky crevices and ridges. June–Aug. F.CH.A.PL.CS.I.BG.R.SU. Page 446.

S. alpina (L.) DC. Very like 1475, but stem lvs. all lance-shaped, with a rounded base or narrowed into a winged lv. stalk, greyish-woolly-haired beneath. Fl. heads bluish-lilac; involucral bracts all oval acute, with adpressed hairs. Scandinavia and mountains of Central Europe, Carpathians. Pl. 152.

JURINEA Very like *Carduus*, but fr. four- to five-angled; pappus in several rows, hairs unequal. Lvs. not spiny. *c.* 20 sps.

1476. *J. anatolica* Boiss. (*J. mollis* Ascherson) Fl. head rose-purple, large, solitary, globular, 3–5 cm., borne on a stout, white-woolly, sparsely leafy stem. Involucral bracts hairless or with cobweb-hairs, the outer bracts with long, slender, recurved, herbaceous tips, the inner bracts more or less papery, coloured, erect and long-pointed. Lvs. green above and with cobweb-hairs, white-woolly beneath, the lower lvs. pinnately cut into lance-shaped lobes with inrolled margin, the upper lvs. often entire and blade shortly running down stem. A variable perenn. with an erect, unbranched, nearly leafless stem to *c.* 60 cm. △ Dry pastures, rocks, sunny hills. Apr.–June. South-East Eur. A.PL.CS.H.I.SU.

CARDUUS | Thistle Involucral bracts many, overlapping, with spiny tips. Disk-florets only present; receptacle with scales. Fr. oval, hairless; pappus of several rows of simple hairs. Differing from *Cirsium* in possessing a pappus of simple, rough, not feathery, hairs. *c.* 45 sps.

Fl. heads small c. 1 cm. across, oblong-cylindrical
1477. *C. tenuiflorus* Curtis SLENDER THISTLE. Fl. heads pale reddish-purple, or rarely white, usually in dense terminal clusters of 3–10, borne on somewhat cottony stems with continuous, irregular, green spiny wings running their whole length. Fl. heads cylin-

drical, 1½ cm. by *c.* 8 mm.; involucral bracts spiny-tipped, glandular, the inner bracts as long as or longer than the florets. Lvs. oblance-shaped in outline, deeply pinnately spiny-lobed, more or less cottony beneath. An erect, spiny, sparsely branched ann. or bienn. 15–120 cm. △ Waste ground, waysides. May–Aug. West Eur. I.YU.GR.BG.R.SU.

*C. pycnocephalus L. Southern Europe.

Fl. heads medium or large 1½–7 cm. across, more or less globular
(a) Fl. heads usually solitary

1478. *C. nutans* L. MUSK THISTLE. Fl. heads large, solitary, reddish-purple, 3½–5 cm., more or less drooping, and borne on spiny-winged stems which are leafless below the heads. Involucral bracts cottony, the outer bracts strongly recurved and spiny-tipped, abruptly contracted below to a narrow base, often purplish, the innermost bracts papery, erect. Lvs. lance-shaped in outline, deeply pinnately cut into triangular lobes with 2–5 spiny tips and with an undulate spiny margin, usually sparsely cobweb-haired. Fr. with transverse wrinkles, and a long whitish pappus. A variable erect bienn. with cottony, interruptedly spiny-winged stems, 20–100 cm. △ Waste ground, tracksides, pastures, cultivations. June–Sept. All Eur. (except P.IS.). Pl. 152.

C. defloratus L. ALPINE THISTLE. Readily distinguished by its solitary fl. head borne on a long, leafless, unwinged stem. Fl. head oval, 1½–4½ cm. across, reddish-purple; involucral bracts erect, shortly spiny-tipped. Lvs. hairless, the lower pinnately cut into bristly-toothed lobes; stem winged below. Mountains of Central Europe.

(b) Fl. heads usually in a cluster

1479. *C. personata* (L.) Jacq. GREAT MARSH THISTLE. Fl. heads purple, oval, 1½–2 cm. across, stalkless, and usually several closely clustered together at the ends of the branches. Involucral bracts linear, fine-pointed but not spiny-tipped, hairless, the outermost bracts irregularly recurved and often blackish. Lvs. soft not spiny, the upper lvs. oval entire, bristly-toothed, and with whitish cobweb-hairs beneath, the lower lvs. deeply lobed. A perenn. to 2 m. with erect stems with narrow bristly wings up to the fl. heads. △ In mountains: woods, pastures, streamsides. July–Aug. Cent., South-East Eur. (except GR.TR.) F.I.SU. Pl. 153.

1480. *C. crispus* L. Fl. heads purple, 1–1½ cm. across, usually in a close, elongated, or flat-topped cluster of 3–5. Involucral bracts linear long-pointed, erect or spreading, weakly spiny-tipped, with some cobweb-hairs. Lvs. all deeply or shallowly pinnately lobed, never entire, margin weakly spiny, green, nearly hairless above, grey or white woolly with simple hairs beneath, the lower lvs. stalked, the upper stalkless. An erect branched bienn. 50–120 cm., with stems narrowly spiny-winged up to the fl. heads. △ Damp, shady places, hedges, waysides. July–Sept. Much of Eur. (except P.IRL.GB.IS.AL.).

*C. acanthoides L. WELTED THISTLE. Widespread in Europe.

NOTOBASIS Like *Carduus*, but fl. heads encircled by an involucre of tough, spiny, upper lvs. Fr. flattened; pappus of inner frs. with feathery hairs, pappus of outer frs. with simple hairs. 1 sp.

1481. *N. syriaca* (L.) Cass. SYRIAN THISTLE. Fl. heads purple, erect, *c.* 2 cm. across, several clustered at the ends of the stems and overtopped by the upper lvs. which form a very spiny, purple-flushed involucre longer than the fl. heads. Involucral bracts oval with a short, spreading, spiny tip, usually silvery-purple. Lvs. hairless above and strongly marked with white veins, broadly lobed and with a spiny-toothed margin, the upper lvs.

clasping the stem with broad rounded lobes. A stiff, erect, little-branched, cobweb-haired ann. 30–60 cm. △ Waste places, tracksides, uncultivated ground. Apr.–July. Med. Eur. (except AL.). Pl. 153.

CIRSIUM | **Thistle**   Like *Carduus*, but distinguished by its pappus which is composed of many rows of branched feathery hairs. Differing from *Jurinea* in often having resin glands in the outer involucral bracts and oblong, often flattened fr. Many hybrids occur. *c.* 65 sps.

Fl. heads usually whitish or yellowish
(a) Fl. heads surrounded by the upper lvs.
1482. *C. oleraceum* (L.) Scop. CABBAGE THISTLE. Fl. heads yellow or rarely reddish, densely clustered and partly concealed by the broad, yellowish, upper bract-like lvs. Fl. heads ovoid 2½–4 cm. long; involucral bracts linear-lance-shaped, spreading, the outer bracts shortly spiny. Lvs. greenish-yellow, more or less hairless, soft, but with a bristly margin, stalkless, and clasping the stem with broad rounded lobes, the lower lvs. lance-shaped and often deeply pinnately lobed, the upper lvs. entire and toothed. An erect perenn. with grooved, unwinged stems 50–120 cm. △ Marshes, damp meadows, watersides, damp woods. July–Aug. Most of Eur. (except P.E.IRL.IS.AL.GR.TR.): introd. GB. *Sometimes eaten as a vegetable.* Pl. 154.

1483. *C. spinosissimum* (L.) Scop. SPINIEST THISTLE. Differing from 1482 in having all lvs. deeply lobed and very spiny, including the uppermost, pale yellowish-green, involucre-like lvs. which spread well above the clustered fl. heads. Fl. heads whitish, *c.* 2–3 cm. across, stalkless, in a dense cluster; involucral bracts with long spiny points longer than the blade. A very leafy and very spiny erect perenn., usually 20–40 cm. △ In mountains: damp places, pastures, rocky places. July–Sept. F.D.CH.A.I. Pl. 154.

(b) Fl. heads not surrounded by the upper lvs.
1484. *C. erisithales* (Jacq.) Scop. YELLOW MELANCHOLY THISTLE. Fl. heads lemon-yellow, usually solitary and nodding, and borne on a long, leafless, wingless stem, or fl. heads 2–5 in a dense cluster. Fl. heads to 3 cm. across, rarely reddish-purple; involucral bracts hairless, spiny-tipped, and with thick resin ducts on the back. Lvs. dark green, lance-shaped in outline, deeply pinnately cut almost to the midvein into lance-shaped, finely bristly-toothed lobes, the lower lvs. stalked, the upper stalkless with heart-shaped base. An erect little-branched perenn. 80–150 cm. △ In hills and mountains: open woods, meadows, streamsides. July–Sept. Cent. Eur. F.I.YU.AL.GR.R.SU.

C. candelabrum Griseb. CANDELABRA THISTLE. A tall erect bienn. to 2 m. with many wide-spreading branches and small yellow fl. heads in branched clusters at the ends of the branches. Lvs. with very spiny lobes. Balkan Peninsula. Pl. 153.

Fl. heads purple, violet or reddish
(a) Upper surface of lvs. rough-hairy and prickly
1485. *C. eriophorum* (L.) Scop. (*Carduus e.*) WOOLLY THISTLE. Fl. heads reddish-purple, large, 4–7 cm. across, usually solitary, and with a very distinctive extremely woolly-haired involucre. Fl. heads often broader than long; involucral bracts reddish-tipped, the outer bracts spiny. Lvs. prickly-haired above, white-cottony beneath, oblong-lance-shaped in outline, and deeply pinnately cut into regular, narrow, lance-shaped, strongly spine-tipped lobes, the lower lvs. stalked, the upper stalkless. A variable, stout, branched bienn. or perenn. with furrowed, unwinged woolly stems 60–150 cm. △ Uncultivated ground, waysides, grassland. July–Aug. Cent. Eur. E.F.GB.B.NL.I.YU.R.SU. Pl. 154.

C. vulgare (Savi) Ten. (*Carduus lanceolatus* L.) SPEAR THISTLE. Like 1485, but stem with interrupted spiny wings and involucral bracts hairless or only slightly woolly-haired.

Fl. heads reddish-purple, longer than broad, 2–4 cm. across, solitary or 2–3 in a terminal cluster. Throughout Europe.

(*b*) *Upper surface of lvs. smooth, often glossy*
(*i*) *Fl. heads usually clustered*

1486. *C. palustre* (L.) Scop. (*Carduus p.*) MARSH THISTLE. Fl. heads dark reddish-purple, rarely white, 1–1½ cm. across, densely crowded into leafy clusters at the ends of the spiny stems. Fl. heads ovoid, short-stalked; involucral bracts lance–shaped, adpressed, purplish. Lvs. narrowly oblance-shaped in outline, deeply pinnately cut into narrow spiny lobes, hairy above, woolly beneath; the stem lvs. stalkless with blades running down stem in spiny wings. A slender, erect bienn. 30–250 cm., with rather few short spreading branches above. △ Damp meadows, marshes, bogs, woods. July–Sept. All Eur. (except IS.AL.GR.TR.). *The young shoots can be eaten as a salad plant.*

1487. *C. arvense* (L.) Scop. (*Carduus a.*) CREEPING THISTLE. Fl. heads pale purple or whitish, in a spreading, somewhat flat-topped cluster borne on spineless, unwinged stems. Fl. heads glandular, ovoid, 1½–2 cm. long, stalked, one-sexed; involucral bracts purplish, adpressed, with somewhat spreading spiny tips. Lvs. very variable, but usually deeply spiny-lobed, hairless on both sides or often cobweb-haired beneath. A perenn. with far-creeping roots producing many sterile lateral shoots and erect fl. stems 30–150 cm. △ Fields and cultivated ground, waste places, tracksides. July–Sept. All Eur.: introd. IS.

C. rivulare (Jacq.) All. (*C. salisburgense* (Willd.) G. Don) Distinguished by its dense terminal cluster of 2–8 stalkless, purple fl. heads borne on a long, nearly leafless, furrowed, woolly-haired stem. Fl. heads globular, 2½–3½ cm.; involucral bracts purple, with an oblong resin duct on back, not spiny-tipped. Lvs. pinnately lobed, bristly-margined, clasping the stem at the base, green on both sides, sparsely hairy above, and densely hairy beneath. Central Europe.

1488. *C. acarna* (L.) Moench (*Picnomon a.*) An extremely spiny ann. to ½ m. with white-woolly stems and lvs. and a dense cluster of purple fl. heads encircled by a longer in-volucre of very spiny upper lvs. Fl. heads 2–4; involucral bracts linear and ending in a slender, yellow, pinnate-cut, comb-like spine. Lvs. linear, spiny-pointed and with widely spaced long yellow spines on the margin; stem with broad spiny wings, much-branched above. △ Stony and rocky places. June–Aug. Med., South-East Eur. P.SU. Pl. 153.

(*ii*) *Fl. heads usually solitary*

1489. *C. heterophyllum* (L.) Hill (*Carduus h.*) MELANCHOLY THISTLE. Fl. head reddish-purple, usually solitary, 3–5 cm. across, borne on long, cottony, furrowed stems. Involucral bracts oval-lance-shaped and fine-pointed, adpressed, hairless or finely hairy, purplish-tipped. Lvs. soft, lance-shaped, and entire or toothed, green and hairless above, densely white-felted beneath, margin with soft bristles; lower lvs. long-stalked, sometimes pinnately lobed, the upper lvs. clasping the stem with enlarged, rounded, basal lobes. An erect, unbranched, leafy perenn. 60–150 cm. △ In mountains: pastures, streamsides, damp woods. July–Aug. Much of Eur. (except P.IS.AL.GR.TR.BG.).

**C. dissectum* (L.) Hill MEADOW THISTLE, MARSH PLUME THISTLE. Western Europe.

C. tuberosum (L.) All. (*Carduus t.*) TUBEROUS THISTLE. Like 1489, but differing in having swollen tuberous roots and deeply lobed lvs., with the lower lvs. stalked, and the upper lvs. stalkless but not enlarged at base; lvs. green on both sides but somewhat cottony beneath. Fl. heads dark red-purple, 2½–3 cm. across, on a very long, leafless, ribbed stem; involucral bracts cottony below. Central Europe. Pl. 153.

1490. *C. acaulon* (L.) Scop. (*Carduus a.*) STEMLESS THISTLE. Fl. heads bright reddish-purple,

borne usually stalkless at the centre of a feathered, spreading rosette of spiny lvs. Fl.
heads 1–3, 2–5 cm. across, sometimes short-stalked; involucral bracts hairless, purplish,
the outer with a short spiny tip, the inner blunt. Lvs. stiff, narrow lance-shaped in outline,
and deeply cut into undulate, very spiny-margined lobes. A rosette perenn. 2–5 cm.,
but sometimes to 30 cm. △ Dry pastures, banks, tracksides. July–Sept. Much of Eur.
(except P.IRL.IS.SF.AL.GR.TR.BG.). Pl. 154.

CYNARA | **Cardoon, Artichoke** Fl. heads very large; involucral bracts conspicuous,
stout, leathery, blunt or pointed, in many ranks. Ray-florets absent; receptacle fleshy;
scales present. c. 4 sps.

1491. *C. cardunculus* L. CARDOON. A robust perenn. to c. ½ m. with large, globular, blue fl.
heads, with conspicuous, leathery involucral bracts which are oval and prolonged into a
long, robust, yellow spine. Fl. heads 4–5 cm. across; involucral bracts glaucous, hairless,
the spines of the lower bracts spreading, the upper erect. Lvs. large white-cottony, all
deeply once or twice pinnately cut into narrow spine-tipped lobes with long, slender,
yellow spines on the margin. △ Tracksides, stony places, arid slopes. July–Aug. Med. Eur.
(except AL.TR.) P.: introd. D.CH.A.CS.SU. *The leaves and stems are blanched and eaten as a
vegetable.*

C. scolymus L. GLOBE ARTICHOKE. Closely related to 1491, but a much more robust perenn.
with very large blue fl. heads 8–15 cm. across, with conspicuous, leathery, oval, blunt
involucral bracts. Lvs. very large to 70 cm., not spiny. Unknown in the wild; widely
cultivated in Southern Europe. *The fleshy bases of the involucral bracts and the receptacle
of the young flower heads are edible and have been cultivated for this purpose since classical
times; inulin is the stored carbohydrate.* Pl. 154.

SILYBUM | **Milk-Thistle** Like *Carduus,* but differing in that the filaments of the
stamens are fused at the base to form a tube. Fl. heads large, solitary; involucral bracts
with long, stiff, recurved, spiny tips; receptacle with scales. Fr. with a pappus of simple
rough hairs. 2 sps.

1492. *S. marianum* (L.) Gaertner MILK-THISTLE, HOLY THISTLE. A robust, very spiny, simple or
branched bienn. ½–1½ m. with large, solitary, red-purple fl. heads 4–8 cm. across, with an
involucre of very sharp-pointed, spreading and recurved bracts. Fl. heads erect or drooping; involucral bracts hairless and with a cluster of short lateral spines at the base of the
long terminal green spine. Lvs. green and shiny, and usually mottled or veined with white
on the upper surface, forming conspicuous, marbled, over-wintering rosettes; lvs. all
deeply lobed, with spiny margin, the lower stem lvs. stalkless, the upper clasping the stem.
△ Waste ground, waysides, uncultivated places. June–Aug. Med., South-East Eur.
P.CH.SU.: introd. Cent. Eur. IRL.GB.R. *Grown as a salad plant in the past; its leaves and
young stems are edible. Used in herbal remedies.* Pl. 155.

GALACTITES Disk-florets only present but outer sterile florets spreading and much
larger than the inner fertile florets; involucral bracts in several rows, each ending in a long
grooved spine. Filaments fused into a tube; receptacle with scales; pappus long, feathery.
2 sps.

1493. *G. tomentosa* (L.) Moench A rather slender, thistle-like ann. or bienn. 20–60 cm., with
white-cottony stems and undersides of lvs. and few ovoid, rosy-purple, lilac, or rarely
white fl. heads in a loose cluster. Outer florets large, spreading; corollas of all florets with
narrow linear lobes; involucral bracts with cobweb-hairs, and with long, slender, erect,
and grooved spiny tips. Lvs. narrow lance-shaped in outline, stalkless with the blade

running somewhat down the stem, pinnately cut into spiny lobes, green above and often mottled with white. △ Uncultivated ground, tracksides, dry waste places. Apr.–July. Med. Eur. (except TR.) P. Pl. 154.

ONOPORDUM Fl. heads medium to large, solitary, with conspicuous, spreading or reflexed, short, spiny-tipped involucral bracts; disk-florets only present. Receptacle without scales; filaments not fused; pappus feathery or rough. *c*. 15 sps.

1494. *O. acanthium* L. SCOTCH THISTLE, COTTON THISTLE. A very robust bienn. to 1½ m. with large white-felted lvs., very broad white-felted winged stems, and a wide-branched inflorescence of solitary, pale purple fl. heads. Fl. heads more or less globular, 3–5 cm. across, broader than long; involucral bracts nearly hairless, narrow lance-shaped, gradually narrowed to a long, slender, rigid spine. Lvs. elliptic in outline, with shallow, angular, and strongly spiny lobes, stalkless with blade running down stem in a broad spiny wing. △ Waysides, waste ground, cultivated places. June–Sept. Most of Eur. (except IRL.IS.AL.TR.): introd. N.SF. Pl. 155.

O. tauricum Willd. Like 1494, but involucral bracts glandular and broadly winged stem and lvs. sticky-glandular; lvs. at first with dense woolly hairs, but becoming green and almost hairless when adult. Fl. heads large, 6–7 cm. across; florets not glandular. South-Eastern Europe and Italy.

1495. *O. illyricum* L. Like 1494, but florets with conspicuous glands on the corolla; involucral bracts purple, narrowly oval and rather abruptly ending in a short recurved spine *c*. 7 mm. long, the inner bracts much shorter than the florets. Fl. heads solitary, 3–4 cm. across, purple or rarely white, borne on white-felted, narrowly winged stems. Lvs. grey or white-felted, oblong-lance-shaped in outline, pinnately cut into lance-shaped spiny lobes. A stiff, robust, erect bienn. 30–100 cm. △ Uncultivated and waste ground, rocky hills. July–Aug. Med. Eur. (except AL.TR.) P. Pl. 155.

CRUPINA Involucral bracts rather papery, not spiny and without a papery apical appendage; fl. heads narrow cylindrical; disk-florets only present. Receptacle with scales. Fr. with shaggy hairs, flattened, with a pappus of outer rough rigid bristles and inner broad scales. Anns.; lvs. not spiny. *c*. 2 sps.

1496. *C. vulgaris* Cass. FALSE SAW-WORT. A very slender, erect, hairless, little-branched ann. 20–50 cm., with narrow spineless lvs. and narrow cylindrical fl. heads borne at the ends of the slender leafless branches. Fl. heads with 3–5 purple florets little longer than the involucre; involucral bracts lance-shaped, smooth, hairless, regularly overlapping. Lvs. hairless or sparsely cottony, pinnately cut into few linear, and sparsely toothed segments. Fr. large, with dark brown pappus. △ Dry hills, grassy places. May–July. Med., South-East Eur. (except AL.) P.CH.CS.H.SU.: introd. A.

C. crupinastrum (Moris) Vis. Like 1496, but fl. head ovoid with 9–15 purple florets. Leaflets more strongly and regularly toothed. Mediterranean region.

SERRATULA Like *Centaurea*, but involucral bracts without a papery apical appendage or spine. Fr. hairless, rounded, not flattened, with a pappus of several rows of stiff bristles. Perenns.; lvs. not spiny. *c*. 15 sps.

Fls. many
1497. *S. tinctoria* L. SAW-WORT. An erect or ascending hairless perenn. 10–90 cm., with compound spineless lvs. and with rather numerous, reddish-purple fl. heads forming a

branched flat-topped cluster, or a densely clustered head. Fl. heads narrowly ovoid, 1½–2 cm. long, the female heads larger than the male; florets rarely white; involucral bracts purplish, adpressed, the outer oval acute and with a downy margin, the inner linear, much longer. Lvs. lance-shaped in outline, entire or variously pinnately cut into narrow lobes, margin with fine bristle-tipped teeth; stems smooth, grooved. Fr. 5 mm.; pappus yellowish. A variable sp. △ Damp meadows, marshes, woods, and clearings. July–Oct. All Eur. (except IRL.IS.AL.TR.). *The juice of this plant produces a fine yellow dye.* Pl. 155.

Fls. solitary
1498. *S. lycopifolia* (Vill.) Kerner (*S. heterophylla* Desf.) Fl. head solitary, purple, 2½–4 cm. long, borne on a long, hairless, nearly leafless stem. Florets spreading; involucre globular, bracts adpressed, oval-triangular, and with a fine narrow apex. Lvs. sparsely rough-hairy, the lower lvs. long-stalked, oval and coarsely toothed or irregularly lobed, the upper lvs. stalkless, deeply pinnately cut into linear lobes and a terminal oval lobe. Pappus whitish. An erect, unbranched perenn. 30–80 cm. △ Pastures in mountains. July. F.A.CS.YU.R.SU.

CENTAUREA | Knapweed, etc. Involucral bracts either with an oval, papery appendage at the apex, which is often cut into a comb-like fringe, or involucral bracts prolonged into a terminal spine, which usually has lateral spines or at least a narrow fringed margin at its base. Florets all disk-florets, but the outer florets are often sterile and longer than the inner and spread outwards in the fl. head. Receptacle with scales. Fr. with pappus of 2 rows of hairs, the inner row with short hairs, the outer row with longer, rough, bristly hairs, but pappus sometimes absent. *c.* 200 sps.

Involucral bracts ending in a spine and usually with smaller lateral spines
(a) Fl. heads yellow
1499. *C. solstitialis* L. ST BARNABY'S THISTLE. Fl. heads pale yellow, *c.* 12 mm. long, with green involucral bracts, each ending in a long, stiff, spreading, yellow spine with 1–3 small lateral spines at its base on each side. Fl. heads solitary, short-stalked; outer florets not spreading; involucre cottony or hairless. Basal lvs. pinnately lobed, the upper linear entire and with the blade running down the stem; lvs. all woolly-haired on both sides, spineless. A white-haired, much-branched ann. 20–80 cm. △ Cultivated and waste ground. July–Sept. Med., South-East Eur. SU.: introd. Cent. Eur. *Used in herbal remedies and eaten as a vegetable.* Pl. 155.

C. melitensis L. MALTESE STAR THISTLE. Like 1499, but a greenish ann. which is rough to the touch. Fl. heads yellow, often in groups of 3–4 and surrounded by lvs.; involucral bracts ending in a brown spine with pinnately arranged lateral spines from the base to the middle. Mediterranean Europe, Portugal.

(b) Fl. heads purple
1500. *C. calcitrapa* L. STAR THISTLE. Fl. heads pale reddish-purple, oval-cylindrical 1–1½ cm. long, with green involucral bracts each ending in a stiff, robust, spreading, yellowish spine, at least as long as the width of the fl. head and with stout spines at its base. Fl. heads short-stalked, overtopped by younger heads; florets all equal, the outer not spreading. Lvs. green, rough, deeply pinnately lobed, the upper lvs. with linear acute segments with few bristle-pointed teeth. Fr. whitish, without pappus. An erect, much branched, nearly hairless bienn. or perenn. 20–50 cm. with wide-spreading branches. △ Waste places, waysides, dry stony places. Aug.–Sept. Much of Eur. (except North Eur. IRL.TR.). Pl. 156.

C. salmantica L. Distinguished by its ovoid-conical involucre with smooth, leathery, yellowish-green, oval bracts ending in a small black spine *c.* 1 mm. long, which falls off.

Fl. head purple. Lvs. nearly hairless, the lower deeply pinnately lobed, the upper linear, finely toothed. Pappus shorter than the transversely ribbed fr. Mediterranean Europe.

Involucral bracts ending in a papery fringed border which extends some way down each side
(a) *Fl. heads blue*
1501. *C. cyanus* L. CORNFLOWER, BLUEBOTTLE. Fl. heads solitary, with the tubular, outer, bright blue spreading florets much larger than the central, red-purple inner florets. Fl. heads 1½–3 cm. across, long-stalked; involucral bracts green, oval and with a narrow deeply cut fringe of pale or brownish triangular teeth. Basal lvs. with narrow distant lobes, the upper lvs. linear-lance-shaped, entire, all lvs. greyish with woolly hairs. A slender branched ann. or bienn. 30–60 cm. △ Cornfields, cultivated ground. May–Aug. All Eur.: introd. IS. *The juice of the petals has been used to make blue ink and in dyeing.* Pl. 156.

C. montana L. MOUNTAIN CORNFLOWER. Like 1501, but a mountain perenn. with underground creeping stems and lance-shaped stalkless lvs. with the blade running down the stem in narrow wings; involucral bracts with a blackish-brown fringed border. Fl. heads solitary, 6–8 cm. across, blue to blue-violet. Central Europe.

(b) *Fl. heads reddish-purple or rarely white*
1502. *C. scabiosa* L. GREATER KNAPWEED. Fl. heads solitary, large, 3–5 cm. across, reddish-purple, with an outer row of larger, spreading, tubular florets, or outer row of larger florets absent; fl. heads long-stalked, forming a lax branched inflorescence. Involucral bracts oval, green and veinless, with the upper half surrounded by a broad, blackish-brown, horse-shoe-shaped, fringed papery border. Lvs. dark green, rather rough, entire, toothed or deeply pinnately lobed, the upper stalkless and deeply pinnately lobed. A very variable, erect, branched hairy perenn. 30–90 cm. △ Dry pastures, waysides, open woods, cliffs. July–Aug. All Eur. (except P.IS.GR.TR.).

(c) *Fl. heads yellow*
1503. *C. salonitana* Vis. YELLOW KNAPWEED. An erect branched perenn. to *c.* 50 cm., with pale yellow fl. heads and an oval involucre 2–3½ cm. broad, with oval, rounded-tipped bracts with a pale brown, horseshoe-shaped, fringed papery border and with or without a short or long spine. Lvs. hairless, rough, pinnately cut into entire or toothed, lance-shaped lobes. Pappus brown. △ Grassy and rocky places. July–Aug. YU.A.GR.TR.BG. Pl. 156.

Involucral bracts distinctly enlarged into a broader papery terminal part
(a) *Terminal part fringed with fine comb-like teeth*
C. nigra L. LESSER KNAPWEED. Like 1502, but fl. heads smaller, reddish-purple, 2–4 cm. across, and larger, spreading, marginal florets usually not present. Involucral bracts with a brown or blackish, triangular, terminal part which is deeply and irregularly cut into narrow teeth. A variable perenn. Western and Central Europe.

1504. *C. nervosa* Willd. PLUME KNAPWEED. Distinguished by its large purple fl. heads with wide-spreading marginal florets, and in particular by the distinctive russet-coloured involucral bracts which have a long, slender, arched tip, with numerous brown comb-like bristles arranged along its length. Fl. heads usually solitary, 4–6 cm. across; involucre globular, 2–2½ cm. Lower lvs. oval-lance-shaped and narrowed to the stalk, the middle and upper lvs. oval or lance-shaped, usually strongly toothed, with several prominent veins beneath, stalkless with a broad, rounded, often nearly hastate, or cut-off base. An erect, little-branched, leafy perenn. 15–40 cm. △ Alpine pastures. July–Aug. E.D.CH.A.YU.AL. GR.BG. Pl. 156.

C. phrygia L. WIG KNAPWEED. Like 1504, but stems branched above, with several fl. heads; involucre oval to globose 15–18 mm. long, bracts with a black lance-shaped or oval appendage and a slender recurved apex 3–8 mm. with numerous comb-like bristles. Lower lvs. broadly oval, narrowed to the stalk; middle and upper lvs. oval to oval-lance-shaped, with a rounded or narrowed base, finely and irregularly toothed. A very variable sp. Widespread in Europe. Pl. 156.

(b) Terminal part scale-like, irregularly cut

1505. *C. jacea* L. BROWN-RAYED KNAPWEED. Distinguished from 1502 by the involucral bracts which have a broader pale brown, rounded, scaly terminal part, with a transparent whitish margin which is irregularly cut. Fl. heads reddish-purple, 1–2 cm. across, with marginal florets usually larger. Lvs. rough-hairy, the lower usually lobed and stalked, the upper lvs. narrow lance-shaped, stalkless, with 1 or 2 teeth towards the base. Fr. hairy, without a pappus. A very variable, branched perenn. 30–60 cm. △ Meadows, thickets, shady places. June–Sept. All Eur. (except P.IRL.IS.TR.): introd. GB.

1506. *C. rhapontica* L. GIANT KNAPWEED. Fl. head solitary, very large, 5–11 cm. across, reddish-purple, with a very conspicuous globular involucre, borne on a stout, nearly leafless stem. Involucral bracts broadly oval and with an irregularly cut margin, papery, rusty-brown, finely hairy. Lvs. very large, oblong-oval with heart-shaped base, toothed, greyish or white-woolly-haired beneath. A stout, unbranched perenn. with a woolly-haired stem ½–1½ m. △ Rocks in the mountains. July–Aug. D.CH.A.I. Pl. 156.

1507. *C. conifera* L. (*Leuzea c.*) Fl. head purple, large, solitary *c.* 3 cm. across, with a very conspicuous, shining, brownish involucre of numerous overlapping bracts, recalling a pine-cone. Involucral bracts with the terminal part broadly oval, papery, and often irregularly cut. Lvs. green above, white-felted beneath, usually deeply pinnately cut into narrow, widely spaced lobes and a terminal lance-shaped lobe, the lower lvs. sometimes not lobed. Pappus white, at least 5 times as long as fr. An erect, woolly-haired, usually unbranched perenn. 20–70 cm. △ Stony ground. May–Aug. P.E.F.I.R. Pl. 157.

CARTHAMUS Outer involucral bracts green and with a spiny apex and margin, similar to the upper lvs., inner bracts with a spiny tip or with a white papery appendage. Disk-florets only present; receptacle with dense scales. Fr. usually with several rows of bristles or scales. *c.* 10 sps.

1508. *C. lanatus* L. (*Kentrophyllum l.*) Fl. heads thistle-like, golden-yellow, 2–3 cm. across, solitary or few in a flat-topped cluster. Outer involucral bracts very spiny, as long as the florets, the inner bracts papery with a lance-shaped, finely fringed terminal part. Lvs. thistle-like, oval-lance-shaped, and deeply lobed with lobes all strongly spiny-toothed, leathery, sticky-glandular. Pappus greyish. A strong-smelling ann. with reddish juice and cottony, branched stems 30–60 cm. Recalling *Scolymus* which differs in having ray-florets only present and winged stems. △ Rocky places, sunny hills, uncultivated ground, tracksides. May–Aug. Med., South-East Eur. P.CH.SU.: introd. A.CS.H. Pl. 157.

C. tinctorius L. SAFFLOWER. Distinguished from 1508 by its elliptic entire lvs. with bristly-toothed margins and its smooth hairless stems. Involucral bracts greenish-yellow, with scattered spines on the margin. Fr. without a pappus. A native of Asia; sometimes cultivated in Southern Europe and sometimes escaping. *A valuable dye-plant, used particularly for dyeing silk; the dye is obtained from the fresh florets. The seeds yield a valuable oil, which is used in the Orient for culinary purposes and for illumination.*

C. caeruleus L. (*Carduncellus c.*) Like 1508, but fl. heads blue and the middle involucral bracts yellowish, with a blunt or rounded-spathulate brownish terminal part which is fringed with bristles. Fr. whitish, rough above; pappus present. Mediterranean region and Portugal.

CNICUS | Blessed Thistle Fl. heads large, surrounded by a spiny involucre formed by the upper lvs. Involucral bracts with a pinnately cut spiny tip; disk-florets only present. Pappus of 10 longer outer bristles and 10 shorter inner bristles. 1 sp.

1509. *C. benedictus* L. BLESSED THISTLE. Fl. heads yellow, solitary, each surrounded by a large involucre of oval, spiny, strongly veined lvs. arising just below the fl. head and often spreading beyond it. Involucral bracts with an oval base which is prolonged into a long spine with comb-like lateral spines on each side. Lvs. pale green, with cobweb-hairs, deeply pinnately lobed, with weakly spiny margin, prominently white-veined beneath; the lower lvs. stalked, the upper with heart-shaped base and with blade somewhat running down the stem. Fr. *c.* 8 mm., smooth, brown, strongly ribbed; pappus of stiff bristles. An erect, hairy-stemmed, branched ann. 10–40 cm. △ Sandy fields, dry ground. Apr.– July. Med. Eur. (except AL.) P.BG.: introd. Cent. Eur. SU. *Well known to herbalists as an all-healing plant and much used in the past. A bitter plant used for internal troubles.* Pl. 160.

SCOLYMUS Outer involucral bracts leaf-like, with spiny tips and grading into the similar spiny, thistle-like, upper lvs. Ray-florets only present; receptacle with scales which encircle the fr. Fr. with a pappus of 2–3 bristles or pappus absent. 3 sps.

1510. *S. hispanicus* L. SPANISH OYSTER PLANT. Fl. heads golden-yellow, *c.* 3 cm. long, borne terminally or laterally and closely encircled by spiny-toothed, thistle-like, leafy bracts longer than the florets. Ray-florets only present, hairless; anthers yellow. Lvs. stiff leathery, deeply pinnately cut into extremely spiny-tipped narrow lobes, and with their blades running down the stem forming narrow, spiny, interrupted wings. Fr. usually with 2–4 slender bristles. An erect, very spiny bienn. branched above, 20–80 cm. △ Waste and sandy places, uncultivated ground. May–Aug. Med., South-East Eur. P.SU.

S. maculatus L. Very like 1510, but stems broadly winged with wing 2–5 mm. wide at the narrowest part. Lvs., bracts and stem-wings with a continuous white cartilaginous border. Fl. heads golden-yellow, solitary or in clusters of 2–4; florets with black hairs on the outside and dark brown anthers. Fr. without bristles. Mediterranean region, Portugal, and Bulgaria. Pl. 157.

CATANANCHE Fl. heads solitary; involucral bracts papery, shining, oval and loosely overlapping. Ray-florets only present; receptacle with long hairs. Fr. five-angled; pappus with 5–7 pointed scales. 2 sps.

1511. *C. coerulea* L. CUPIDONE. Fl. heads blue or rarely white or pink, with numerous long, spreading ray-florets and an attractive inflated silvery involucre borne on long, slender, nearly leafless stems. Fl. heads *c.* 3 cm. across; involucral bracts inflated papery, with a median brown vein prolonged into a point; fl. stem with scales below involucre. Lvs. linear to narrow lance-shaped, three-veined, often with several narrow lateral lobes towards the base. An erect, sparingly branched hairy perenn. 50–80 cm. △ Dry pastures, arid ground. June–Aug. P.E.F.CH.I. Pl. 157.

CICHORIUM | Chicory Fl. heads usually blue; ray-florets only present; involucral bracts in 2 ranks, the outer shorter; receptacle without scales. Fr. angled, with 2–3 rows of very short scales. 4 sps.

1512. *C. intybus* L. CHICORY, WILD SUCCORY. Fl. heads bright blue, 2½–4 cm. across, almost stalkless and ranged along the stiff, green, grooved, and sparsely leafy stems. Ray-florets wide-spreading; involucral bracts green, glandular-hairy, the outer bracts lance-shaped, and spreading, about half as long as the erect inner bracts; florets rarely pink or white. Basal lvs. deeply pinnately lobed, the uppermost lvs. lance-shaped, entire or distantly toothed, and clasping the stem with pointed lobes. An erect perenn. 30–120 cm. with spreading branches. △ Fields, tracksides, uncultivated ground, waste places. June–Sept. All Eur. (except IS.). *Known since Roman times as a salad plant; now often blanched, when it loses its bitterness. The dried roots produce the chicory of commerce which is used for blending with coffee and for flavouring.* Pl. 158.

C. endivia L. ENDIVE. Very like 1512, but an ann. or bienn. with somewhat glaucous, less deeply lobed basal lvs. and with the upper lvs. lance-shaped or oval and clasping the stem with acute lobes. Fl. heads in clusters of 2–5, fl. stem swollen conspicuously below fl. heads; outer involucral bracts about one-third as long as the inner. Widely cultivated as a salad plant in various lv. forms; sometimes naturalized in Southern Europe.

C. spinosum L. SPINY CHICORY. Distinguished by its densely branched shrubby growth and spiny branches. Fl. heads small, usually solitary, with only 5–6 blue florets; involucral bracts hairless. Lower lvs. toothed or lobed, the upper lvs. linear. Spain, Greece, and some Mediterranean islands.

LAPSANA Fl. heads small, yellow, borne in loose clusters. Ray-florets only present. Involucre with a few tiny basal scales, bracts keeled, in one row; receptacle without scales. Fr. with many ribs, pappus absent. 4 sps.

1513. *L. communis* L. NIPPLEWORT. Fl. heads pale yellow, $1\frac{1}{2}$–2 cm. across, with 8–15 florets, borne in long-stalked lax clusters forming a rather flat-topped inflorescence. Involucral bracts 8–10, linear, blunt, becoming keeled in fr. Lvs. hairless, the lower with a terminal, oval, toothed lobe and smaller lateral lobes, the upper lvs. entire, oval-lance-shaped with a toothed or shallowly lobed margin. Fr. 3–5 mm., brownish, about half as long as the persisting involucre; pappus absent. An erect, branched, hairless ann. 20–125 cm. △ Hedges, clearings, waste ground, waysides, walls. May–Oct. All Eur. (except IS.AL.). *Used as a salad plant in the past.*

APOSERIS Fl. head solitary at end of a leafless stem; lvs. in a basal rosette. Involucre of 1 row of 8–10 bracts, with much smaller outer bracts at its base. Ray-florets only present; receptacle without scales. Fr. five-angled; pappus absent. 1 sp.

1514. *A. foetida* (L.) Less. STINKING APOSERIS. A slender rosette-forming perenn. with deeply lobed lvs., and erect, leafless and hairless stems, 5–25 cm., bearing a solitary golden-yellow fl. head which is nodding in bud. Fl. head $2\frac{1}{2}$–3 cm. across, with rather few florets; involucral bracts blackish-green, almost hairless. Lvs. oblong in outline, with deep, regularly triangular, toothed lobes. △ In mountains: woods, thickets, meadows. June–Aug. Cent. Eur. E.F.I.YU.R.SU.

TOLPIS Fl. heads yellow, long-stalked; outer involucral bracts very narrow, and thread-like. Ray-florets only present; receptacle without scales. Fr. oblong; pappus with few long bristles and with short scales at their base. 3 sps.

1515. *T. barbata* (L.) Gaertner Fl. heads yellow and often with a conspicuous dark brown or reddish-purple centre, solitary, terminal, or borne in the axils of the widely diverging branches. Fl. heads 2–3 cm. readily distinguished by the thread-like bracts which spread untidily beyond the florets; fl. stalk swollen below head. Lvs. mostly in a basal rosette, oblong-lance-shaped, toothed, hairy; stem lvs. narrower, stalked. An erect, widely branched ann. 10–40 cm. △ Arid, sandy places, uncultivated ground. May–July. P.E. F.I.GR.R. Pl. 157.

HYOSERIS Like *Aposeris*, but fr. with a pappus of unequal yellowish hairs. 2 sps.

1516. *H. radiata* L. A rosette perenn. with unbranched leafless, hairless stems 10–35 cm., which are longer than the lvs., and bear solitary yellow fl. heads. Florets twice as long as the involucral bracts which spread in a star in fr.; stem not swollen below the fl. head. Lvs. hairless or rough-haired, linear-oblong in outline, deeply and regularly triangular-lobed and coarsely toothed. △ Rocks, dry places, cultivated ground. Feb.–June. E.F.I.YU. GR.TR.

H. scabra L. Like 1516, but fl. stems procumbent, as long as the lvs., hollow and conspicuously swollen below the fl. heads; florets little longer than the involucral bracts which are erect in fr. Fl. heads yellow. Mediterranean region and Portugal.

*Arnoseris minima (L.) Schweigger & Koerte LAMB'S or SWINE'S SUCCORY. Widespread in Europe.

ZAZINTHA Involucre bell-shaped; inner bracts 8, conspicuously thickened and hardening below and encircling the fr., outer bracts 5. Ray-florets only present; receptacle with scales. Fr. with bristly-haired pappus. 1 sp.

1517. *Z. verrucosa* Gaertner A rosette ann. 10–20 cm., with spreading, curved, nearly leafless branches with rather few small yellow fl. heads with strongly swollen involucral bracts shorter than the florets. Fl. heads terminal and laterally placed, stalkless; involucral bracts swollen only at the base and with outcurved pointed apex. Rosette lvs. oblong, deeply blunt-lobed or coarsely toothed, the stem lvs. lance-shaped, and clasping the stem with acute lobes; stems becoming swollen and hollow below fr. △ Stony ground, pastures. May–June. Med. Eur. (except AL.).

RHAGADIOLUS Involucral bracts 5–8, enlarging and enclosing the outer fr. and spreading in a star, outer bracts present, tiny. Ray-florets only present; receptacle without scales. Outer fr. enlarging; pappus absent. 2 sps.

1518. *R. stellatus* DC. STAR HAWKBIT. Readily distinguished from other small yellow-flowered Compositae by the outer frs. which enlarge and spread in a persistent five-rayed star, while the inner frs. are erect and soon fall. Fl. heads yellow, *c.* 1 cm. across; involucral bracts linear-acute, enlarging to 1½ cm. in fr. Lvs. mostly basal, very variably toothed or lobed. A variable, slender, nearly hairless ann. 15–40 cm., with wide-spreading branches. △ Rocky places, cultivated and uncultivated ground. Apr.–June. Med. Eur. (except AL.) P.SU.

HEDYPNOIS Involucral bracts in 1 row, persisting, hardening, and encircling the fr.; outer fr. with a swollen, cup-shaped, and finely toothed appendage, the inner frs. with 2 rows of pappus, the outer scaly, the inner of 4–5 hairs. Ray-florets only present; receptacle without bracts. *c.* 4 sps.

1519. *H. rhagadioloides* (L.) Willd. A small rosette ann. with spreading or ascending, simple or branched stems 10–40 cm., bearing yellow fl. heads *c.* 1½ cm. across, and with stems conspicuously swollen below the fr. heads. Involucre hairless or with dense bristly hairs, hardening and thickening in fr. Lvs. bristly-haired, oblong-oval, entire to deeply lobed. Fr. heads crown-like, with stiff, incurved involucral bracts surrounding the fr. A very variable sp. △ Dry, stony ground, grassy places. May–June. Med. Eur. Pl. 158.

HYPOCHOERIS | Catsear Involucral bracts of several overlapping rows; ray-florets only present; receptacle flat, with scales. At least the inner fr. prolonged into a beak with a terminal pappus of 1 or 2 rows of feathery hairs. *c.* 8 sps.

Pappus with 2 rows of hairs, the outer simple, the inner feathery

1520. *H. radicata* L. CATSEAR. A rosette perenn. with erect, little-branched, leafless stems 25–60 cm., bearing terminal, solitary, bright yellow fl. heads 2½–4 cm. across, and with a swollen hollow stem immediately below the head. Ray-florets spreading, about twice as long as the involucral bracts which are 2–2½ cm., lance-shaped, long-pointed, bristly on the midvein. Lvs. in a flattened rosette, broadly oblong, toothed or shallowly rounded-lobed, hairless or usually with rough hairs, dull green above, somewhat glaucous beneath. Fr. orange, 1–1½ cm., beaked, covered with small swellings. △ Dry pastures, open woods, waysides. June–Sept. All Eur. (except IS.AL.). *A winter salad plant; used herbally for chest troubles.*

*H. glabra L. SMOOTH CATSEAR. Widespread, except in Northern Europe.

Pappus of hairs in 1 row, all feathery

1521. *H. uniflora* Vill. GIANT CATSEAR. A rather stout-stemmed mountain perenn. 20–40 cm., with large, solitary, golden-yellow fl. heads 3½–7 cm. across and with a conspicuously swollen hollow stem below the head. Involucral bracts blackish, oval, shaggy-haired and with a fringed comb-like margin. Lvs. mostly in a rosette, lance-shaped and usually toothed, finely hairy, the stem lvs. 1–3, smaller. △ Pastures, sunny slopes. July–Sept. Cent. Eur. (except H.) F.I.R. Pl. 158.

1522. *H. maculata* L. SPOTTED CATSEAR. A rather robust rosette perenn. 30–90 cm., with large pale yellow fl. heads 4–5 cm. across, borne singly or 2–4 at the ends of leafy branches. Fl. stems not swollen below fl. heads; involucral bracts 18–23 mm., blackish-green, woolly-haired, but without a comb-like margin. Lvs. in a large flattened rosette 10–50 cm. across, oblong, toothed, dark green and often with brownish-violet spots. △ Pastures, heaths. May–July. All Eur. (except P.IRL.IS.TR.). Page 469.

LEONTODON | Hawkbit Involucral bracts of several overlapping rows. Ray-florets only present; receptacle flat, without scales. Fr. with a short or long beak; pappus of simple or feathery hairs or a crown of scales. *c.* 25 sps.

Fr. with 1 row of feathery hairs

1523. *L. autumnalis* L. AUTUMNAL HAWKBIT. Stems usually branched, nearly leafless, bearing few or many golden-yellow fl. heads 1–3½ cm. across, on long stalks, swollen and hollow below the fl. heads, and with numerous scale-like bracts. Florets longer than the involucre, with reddish streaks on the outside of the corolla-limb. Lvs. in a lax rosette, hairless or with simple hairs, lance-shaped and toothed, or deeply pinnately cut into narrow lobes, the stem lvs. few, linear. A very variable erect perenn. 5–60 cm. △ Pastures, meadows, waysides. July–Oct. All Eur. (except P.AL.TR.).

Fr. with 2 rows of hairs, the outer simple, the inner feathery

1524. *L. hispidus* L. ROUGH HAWKBIT. Fl. heads solitary, golden-yellow, 2½–4 cm. across, on long, densely hairy unbranched stems, arising directly from the rootstock. Involucral bracts dark green or blackish, with rough whitish hairs or nearly hairless; stem scarcely swollen below the fl. head, usually with 1–2 small bracts; buds drooping. Lvs. lance-shaped and toothed, or shallow-lobed in a lax more or less erect rosette, and covered with star-shaped or forked hairs, or hairless. A very variable perenn. 10–60 cm. △ Meadows, pastures, rocky places, waysides. May–Sept. All Eur. (except IS.TR.). *Eaten as a salad plant.* Pl. 158.

*L. taraxacoides (Vill.) Mérat (*Thrincia hirta* Roth) HAIRY HAWKBIT. Widespread, except in Northern Europe.

PICRIS | **Ox-Tongue** Involucral bracts in many rows, the outer shorter and spreading and often differing from the inner. Ray-florets only present; receptacle flat, without scales. Pappus of 2 rows, the inner always feathery and deciduous. *c.* 10 sps.

1525. *P. hieracioides* L. HAWKWEED OX-TONGUE. Fl. heads yellow, 2–3½ cm. across, all stalked and borne in a lax branched terminal cluster. Involucre with short, linear, spreading outer bracts with blackish hairs, and lance-shaped inner bracts with bristly hairs and tufts of white hairs on the midvein. Middle and upper lvs. lance-shaped and with a toothed undulate margin, somewhat enlarged at base and clasping the stem, lower lvs. stalked; lvs. and stem bristly-haired. Fr. without a beak, pappus cream-coloured. A variable erect bienn. or perenn. 15–90 cm. △ Grassland, waysides, rocky places, walls, vineyards. June–Oct. All Eur. (except IS.): introd. IRL.

P. echioides L. BRISTLY OX-TONGUE. Distinguished from 1525 by its involucre which has 3–5 outer, broadly heart-shaped, bristly-haired bracts, and inner, narrow lance-shaped, long-pointed bracts. Fl. heads yellow, 2–2½ cm. across. Middle and upper lvs. lance-shaped with a more or less heart-shaped, clasping base, the lower lvs. stalked; lvs. all bristly-haired, some hairs with swollen bases. Fr. with a slender beak, pappus white. Native in the Mediterranean region and introduced to Central Europe.

UROSPERMUM Involucral bracts 7–8 in 1 row and fused below. Ray-florets only present; receptacle without scales. Fr. with a long beak which is swollen at its base and bears at its apex a pappus of feathery bristles which are fused in a ring below. 2 sps.

1526. *U. picroides* (L.) Schmidt A rather robust ann. 25–45 cm., with solitary, pale yellow fl. heads borne on a sparsely branched, leafy stem which is covered with bristly hairs, particularly below. Involucral bracts 12–20 mm. long, bristly-haired; florets about twice as long as the involucre. Lower lvs. oblong-spathulate with few teeth or irregular lobes, narrowed to a stalk, the upper lvs. oval or oblong and clasping the stem with rounded lobes; lvs. sparsely bristly-haired. Fr. with a long beak with a flask-like swelling at its base, pappus white. △ Waste places, vineyards, tracksides, stony places. May–June. Med. Eur. P. Page 469.

U. dalechampii (L.) Desf. Like 1526, but a perenn. which is softly hairy all over. Fl. head pale yellow, the outer florets often streaked with red outside; involucre 2–2½ cm., bracts softly hairy. Beak of fr. only slightly swollen at its base, pappus pale russet. Mediterranean Europe. Pl. 157.

TRAGOPOGON Involucre conical in bud, bracts in 1 row and united at their base. Ray-florets only present; receptacle without scales. Fr. with a long beak which has an inner row of numerous feathery bristles and an outer row of 5 longer simple bristles, bristles all spreading in an umbrella-like pappus. Lvs. simple, with parallel veins. *c.* 20 sps.

Fl. heads yellow

1527. *T. pratensis* L. GOATSBEARD, JACK-GO-TO-BED-AT-NOON. An erect, little-branched,

1. *Urospermum picroides* 1526
3. *Lactuca viminea* 1544

2. *Hypochoeris maculata* 1522
4. *Scorzonera austriaca* 1530

hairless ann. to perenn. 25–80 cm., with grass-like lvs. and solitary yellow fl. heads which open only for a short time in the morning. Involucre conical in bud and in fr., bracts 8–10, narrow lance-shaped, usually spreading beyond the florets but size of florets variable, hairless, or with woolly hairs at the base. Lvs. linear-lance-shaped entire, 4–10 mm. wide, parallel-veined, sheathing the stem. Fr. with a slender beak as long as the nutlet, pappus very large, spreading in a star and with interwoven feathery hairs. A very variable sp. △ Meadows, pastures, roadsides, waste places. May–Aug. All Eur. (except P.IS.). *The leaves and roots can be eaten as salad.*

Fl. heads purplish

1528. *T. porrifolius* L. SALSIFY. Like 1527, but fl. heads violet-purple and fl. stem con-spicuously swollen below the head. Involucral bracts usually 8, 3–5 cm. long and usually a little longer than the florets, but size of florets variable. Outer frs. with scaly swellings, the inner frs. smooth, beak 3–4 cm. long and pappus very large. △ Sometimes cultivated and naturalized. Apr.–June. Med. Eur. P.BG.R.: introd. Cent. Eur. *The swollen taproot is a useful vegetable, known as White Salsify; the young green shoots can also be eaten.* Pl. 158.

T. crocifolius L. Like 1528, but outer florets reddish-violet and the inner florets yellow, all much shorter than the 5–6 involucral bracts; stems not swollen below the fl. heads. Lvs. slender, 2–4 mm. wide, with sheathing base little-inflated. Fr. with beak *c.* 2½ cm. Mediter-ranean region, Portugal.

T. hybridus L. (*Geropogon glaber* L.) Fl. heads reddish-lilac, the florets much shorter than the 8 narrow involucral bracts; fl. stem soon becoming strongly swollen below head. Fr. ribbed, long-beaked, the outer frs. with a pappus of 5 rough bristles, the inner frs. with feathery branched hairs. A hairless ann. Mediterranean Europe and Portugal.

SCORZONERA   Like *Tragopogon*, but involucre of many rows of overlapping bracts; fr. linear, not beaked; pappus of several rows of unequal feathery hairs with interwoven branches, or outermost hairs simple. *c.* 25 sps.

Fl. heads pink, violet, or purple

1529. *S. purpurea* L. PURPLE VIPERGRASS. An erect, hairless, tuberous-rooted perenn. with sparsely branched stems 15–45 cm. bearing terminal, solitary, pale lilac fl. heads, with the florets longer than the involucre. Involucre 1½–2½ cm. long, hairless. Lvs. linear, to 3 mm. broad, keeled, hairless; base of stem covered with fibrous sheaths of old lvs. Fr. with smooth ribs. △ Pastures, rocky ground, screes. May–June. Cent. Eur. (except CH.) F.I. YU.GR.R.SU.

S. rosea Waldst. & Kit. PINK VIPERGRASS. Like 1529, but fl. heads pinkish-lilac, to 5 cm. across, the florets about twice as long as the involucral bracts which are 18–22 mm. long. Lvs. 3–4 mm. broad, flat. Fr. with rough, finely toothed ribs. East-Central and South-Eastern Europe.

Fl. heads yellow

(a) Base of stem with a collar of fibrous scales

1530. *S. austriaca* Willd. AUSTRIAN VIPERGRASS. An unbranched perenn. 10–40 cm., with solitary yellow fl. heads with florets longer than the involucre, borne on usually leafless stems, but with 1–2 small scale-like lvs. immediately below the fl. head. Involucral bracts narrowed above, apex blunt. Lvs. hairless, linear-lance-shaped or elliptic, acute, nar-rowed to a long stalk; base of stem with brown fibres. Fr. 8–9 mm. long, smooth or rough, pappus white. △ Dry pastures, rocks, stony places. May–July. F.D.CH.A.CS.H.I.YU.AL.BG. R.SU. Page 469.

S. mollis Bieb. Distinguished by its swollen oval rhizome and the whole plant covered with greyish cobweb-hairs; lvs. often undulate. Fl. heads yellow, with the undersides of the florets reddish, florets half as long as the involucre which is 1½–3 cm. South-Eastern Europe.

S. macrocephala DC. LARGE-HEADED VIPERGRASS. Distinguished by its very slender, glaucous lvs. usually 1 mm. broad, and long fl. heads to 5 cm. long; involucral bracts with woolly hairs, becoming hairless. Portugal, Spain.

(*b*) *Base of stem without a collar of fibrous scales*

1531. *S. hispanica* L. Like 1530, but taller 30–120 cm., usually branched, leafy, and bearing 1–5 fl. heads. Florets twice as long as the involucre which is 2–2½ cm.; bracts broad, 6–8 mm. wide, the outer triangular, the inner lance-shaped, all hairless. Stem lvs. variable, oval-lance-shaped to linear, narrowed at the base and sometimes toothed. Outer frs. densely covered with rough spines on the ribs. A variable perenn. △ Pastures, bushy places, rocks. May–Sept. Cent., South-East Eur. P.E.F.I.SU. *It has long been cultivated for its fleshy taproot, which is known as Black Salsify, and which is also used medicinally.*

S. humilis L. DWARF VIPERGRASS. Like 1531, but a shorter perenn. 10–50 cm., with unbranched stems and solitary fl. heads and lvs. nearly all basal. Florets pale yellow, twice as long as the involucre which is 2–2½ cm. Fr. with smooth ribs. Widespread in Europe.

PODOSPERMUM Like *Scorzonera*, but lvs. pinnately lobed and fr. with a basal, hollow, tubular swelling; pappus hairs all similar, feathery. *c.* 5 sps.

1532. *P. laciniatum* (L.) DC. A prostrate or ascending, very variable, leafy bienn. 20–45 cm., usually branched, with pale yellow terminal fl. heads with florets scarcely longer than the involucre. Involucral bracts in several rows, overlapping, each with a swollen apex. Lvs. deeply pinnately cut into linear or lance-shaped lobes, hairless or sparsely cobweb-haired. Fr. cylindrical, smooth, ribbed, the lower part forming a hollow swollen stalk which is nearly as long as the fertile part. △ Grassy places, waysides, cultivated and uncultivated ground. Apr.–July. Med., South-East Eur. (except AL.) P.B.D.CH.A.H.SU.

ANDRYALA Fl. heads many, clustered; involucral bracts in 1 row, with few additional bracts; ray-florets only present; receptacle covered with depressions which are bordered with hairs. Fr. with 8–10 ribs; pappus hairs simple, deciduous. *c.* 5 sps.

1533. *A. integrifolia* L. Fl. heads pale yellow, *c.* 1½ cm. across, in a dense flat-topped cluster, with involucral bracts and fl. stems conspicuously covered with dense, yellow, glandular hairs. Lvs. very soft with star-shaped hairs, oblong-lance-shaped and wavy-margined, the lower lvs. stalked, the upper toothed and pinnately lobed, enlarged at base, and half-clasping the stem. An erect, variable, leafy ann. conspicuously covered with woolly yellowish hairs, branched above, 30–60 cm. △ Sands, rocks, tracksides, heaths. Apr.–July. Med. Eur. (except AL.TR.) P. Pl. 159.

CHONDRILLA Involucre cylindrical, with 8–10 long inner bracts and a row of very short leafy outer bracts. Ray-florets only present; receptacle without scales. Fr. with an apical crown of scaly teeth from the centre of which arises a long slender beak bearing a pappus of several rows of simple, snowy-white hairs. *c.* 6 sps.

1534. *C. juncea* L. A hairless bienn. 40–100 cm., with stiff, spreading, broom-like, almost leafless, green stems and stalkless yellow fl. heads, usually in clusters of 2–5, borne along the slender stems. Fl. heads small, *c.* 1 cm. across; involucre finely downy. Lower lvs.

lance-shaped and shallowly lobed, hairless, disappearing at flowering, the upper lvs. linear or lance-shaped, entire or finely toothed; stem bristly-haired below, hairless above. △ Sandy and stony places, stony fields, waysides, ballast. July–Sept. Much of Eur. (except North Eur. IRL.GB.B.AL.). *The seeds are poisonous.*

TARAXACUM | **Dandelion** Fl. heads borne on a leafless hollow stem arising directly from the rootstock; lvs in a rosette. Involucre of 2 inner erect rows of bracts and a shorter outer spreading row of bracts; ray-florets only present; receptacle without scales. Fr. with a slender beak; pappus of many rows of simple, spreading, white hairs. *c.* 10 sps.

1535. *T. officinale* Weber DANDELION. A very variable rosette-forming perenn. with pale or dark yellow fl. heads varying greatly in size, 3–6 cm. across, borne on smooth hollow stems up to 50 cm. Ray-florets usually numerous, about twice as long as the involucre; outer involucral bracts spreading or recurved at flowering, the inner bracts erect, hairless, with or without a transparent margin. Lvs. nearly hairless, linear to oblance-shaped, varying from almost entire or toothed to deeply pinnately cut into triangular lobes, which are directed backwards towards the lv. base. A very variable sp., due in part to the lack of normal sexuality and presence of polyploidy. △ Meadows, waysides, waste places, damp and dry sands, cultivated ground. Mar.–Nov. All Eur. *The young leaves are eaten as a salad; the roots are used in place of chicory in coffee. A bitter stomachic and diuretic plant.* Pl. 159.

**T. palustre* (Lyons) DC. MARSH DANDELION. Widespread in Europe.

**T. laevigatum* (Willd.) DC. LESSER DANDELION. Widespread in Europe.

REICHARDIA Involucre urn-shaped, bracts in several rows, each with a white margin. Ray-florets only present; receptacle scales absent. Fr. thick, cylindrical, three- to five-sided, grooved and with coarse transverse swellings, the inner frs. smooth; beak absent; pappus a ring of simple hairs fused below. *c.* 4 sps.

1536. *R. picroides* (L.) Roth (*Picridium vulgare* Desf.) A hairless, often glaucous ann. with erect, branched, rather leafless stems 25–50 cm., bearing solitary, long-stemmed, pale yellow fl. heads 1½–2 cm. across. Involucre 1–2 cm., more than half as long as the florets, bracts with a white margin, the outer bracts much shorter and heart-shaped; stem with scale-lvs. immediately below the fl. head. Lvs. lance-shaped, entire or pinnately lobed, the upper clasping the stem with a heart-shaped base. A very variable sp. △ Fields, waysides, banks, rocky places. Apr.–July. Med. Eur. (except AL.) P.

CICERBITA Like *Sonchus*, but differing in its pappus which has 2 rows of very rough, simple, fragile hairs, the outer shorter; florets usually blue. *c.* 6 sps.

1537. *C. alpina* (L.) Wallr. (*Mulgedium a.*) BLUE SOW-THISTLE. A robust perenn. with a stout furrowed stem ½–2 m., with reddish glandular hairs above, and bearing a dense cylindrical cluster of short-stalked, blue-violet fl. heads each *c.* 2 cm. across. Involucre 1–1½ cm., brownish, glandular, about half as long as the florets. Lvs. large, deeply cut into triangular lobes and a much larger terminal lobe, the lower lvs. stalked, the upper clasping stem with a heart-shaped base; lvs. all glaucous, hairless beneath. Fr. whitish, *c.* 5 mm. long. △ In mountains: damp woods, bushy places, streamsides. July–Sept. Much of Eur. (except P.IRL.B.NL.IS.DK.GR.TR.). Pl. 159.

C. plumieri (L.) Kirschleger Like 1537, but involucre and fl. stems hairless and not glandular. Fr. greyish, *c.* 8 mm. long. A less robust perenn. with lvs. more regularly cut into more numerous lobes. Spain to Switzerland.

SONCHUS | **Sow-Thistle** Involucre of several rows of overlapping bracts; ray-florets only present; receptacle pitted, without scales. Fr. flattened, beak absent; pappus silvery-white of 2 equal rows of soft, simple hairs. *c.* 11 sps.

Perenns.; fr. with 5 longitudinal ribs on each face

1538. *S. arvensis* L. FIELD SOW-THISTLE. A rather robust, erect, stoloniferous perenn. 60–150 cm., with golden-yellow fl. heads 4–5 cm. across, in a lax, branched, leafless cluster. Involucre and fl. stems densely covered with conspicuous yellowish glandular hairs; ray-florets very numerous, twice as long as the bell-shaped involucre which is 1½–2 cm. long. Lvs. hairless and somewhat glaucous, oblong, and deeply pinnately triangular-lobed, the upper lvs. stalkless with a clasping base with rounded lobes. Fr. 3–4 mm., dark brown, with 5 strong ribs on each face. △ Fields, waste places, salt marshes. June–Sept. All Eur. (except P.AL.). Pl. 159.

S. palustris L. MARSH SOW-THISTLE. Very like 1538, but creeping stems absent and upper lvs. linear-lance-shaped with large, acute, clasping basal lobes. Glands on involucre usually blackish-green. Fr. *c.* 4 mm., yellowish. Widespread in Central Europe.

Anns. or bienns.; fr. with 3 longitudinal ribs on each face

1539. *S. asper* (L.) Hill PRICKLY SOW-THISTLE. An erect, usually branched, hairless ann. or bienn. 30–100 cm. with golden-yellow fl. heads 2–2½ cm. across, in an irregular umbel-like cluster. Involucre hairless when mature. Lvs. oblong, entire or deeply lobed, the upper clasping the stem with rounded, often spiny basal lobes, the lower lvs. stalked; lvs. hairless, the upper lvs. rather stiff, with a bristly, spiny margin. Fr. smooth, three-ribbed on each side. △ Cultivated ground, tracksides, waste ground. May–Sept. All Eur. Pl. 159.

S. oleraceus L. COMMON SOW-THISTLE. Like 1539, but basal lobes of upper lvs. pointed and spreading, and lvs. softer, dull, and marginal teeth less bristly. Fl. heads 2–2½ cm. across, paler yellow; involucre 1–1½ cm. long. Fr. ribbed, rough and transversely wrinkled. Widespread in Europe.

S. tenerrimus L. Like [1539] with long-pointed, clasping, basal lobes, but lvs. divided to the midvein into regular pairs of oval to lance-shaped toothed segments which are narrowed below into a stalk-like base. Fl. heads pale yellow. Fr. ribbed with fine transverse wrinkles. Mediterranean Europe.

MYCELIS Similar to *Lactuca*, but pappus of 2 rows of simple hairs, the outer row shorter; fr. beaked. *c.* 5 sps.

1540. *M. muralis* (L.) Dumort. (*Lactuca m.*) WALL LETTUCE. An erect hairless perenn. 25–100 cm., with numerous small, pale yellow fl. heads 7–8 mm. across in a compound, much-branched, lax, spreading, leafless cluster. Fl. heads usually with only 5 florets; involucre narrow cylindrical, 7–10 mm. long, often reddish, the inner bracts linear, the outer bracts much smaller and spreading. Lvs. hairless, soft, and without prickles, the lower lvs. stalked, elliptic in outline and deeply pinnately cut into broad angular lobes and a a much larger terminal lobe, the upper lvs. clasping the stem, less divided. Fr. spindle-shaped, blackish, with a short white beak, pappus white. △ Rocks, walls, damp woods, waste places. July–Sept. All Eur. (except P.IS.): introd. IRL.

LACTUCA | **Lettuce** Involucre of several rows of bracts, the outer shorter and broader; ray-florets only present; receptacle without scales. Fr. compressed, with a short or long beak; pappus of 2 equal rows of simple white hairs. Juice milky. *c.* 16 sps.

Fl. heads blue

1541. *L. perennis* L. BLUE LETTUCE. A hairless erect perenn. 20–70 cm., with blue or violet, long-stalked fl. heads 3–4 cm. across, in a loose, spreading cluster. Involucre cylindrical, 1½–2 cm. long; fl. stems with arrow-shaped bracts. Lvs. not spiny, hairless and glaucous, deeply pinnately cut into narrow oblong, nearly entire lobes, the upper lvs. stalkless and clasping the stem with rounded lobes. Fr. black, with fine transverse wrinkles and a white beak of equal length. △ Fields, banks, rocky places, vineyards. May–Aug. Cent. Eur. E.F.B.I.YU.BG.R.SU. Pl. 160.

Fl. heads yellow
(a) *Lvs. with prickles*

1542. *L. serriola* L. (*L. scariola* L.) PRICKLY LETTUCE. Readily distinguished as a 'compass' plant by its upper lvs. with their blades held vertically and orientated in a north–south direction. Fl. heads yellow, numerous, 11–13 mm. across, forming a lax, much-branched, leafless inflorescence; involucre narrow-cylindrical, 8–12 mm., glaucous, hairless. Lvs. stiff, glaucous, and with a prickly margin and underside of midvein, oblong and entire or deeply lobed and toothed, or linear-lance-shaped and spiny-toothed, the lower lvs. stalked, the upper lvs. clasping the stem with acute lobes. Fr. rough, greenish-grey, with a white beak of equal length. A stiff, erect, whitish- or reddish-stemmed bienn. 30–150 cm. △ Waysides, uncultivated ground, rocky places. July–Aug. Most of Eur. (except IRL.B. IS.AL.): introd. North Eur. *Sometimes used as salad plant.*

L. virosa L. Like 1542, but blades of stem lvs. held horizontally, and lvs. less divided, often entire or shallowly lobed. Fl. heads pale yellow, 1 cm. across. Stem more robust, usually prickly below. Fr. smooth, blackish, with white beak. Central Europe and Southern Europe. *It contains a bitter latex which is used as a sedative in respiratory complaints.*

(b) *Lvs. not prickly*

1543. *L. saligna* L. LEAST LETTUCE. A little-branched erect ann. or bienn. 30–100 cm., with small pale yellow fl. heads scattered along the slender, stiff, whitish branches and forming narrow elongated spikes. Fl. heads mostly short-stalked and borne in the axils of lance-shaped bracts with acute, basal, clasping lobes; involucre 1½ cm. long, shorter than the florets. Lower lvs. stalked, deeply cut into triangular recurved lobes, glaucous with a white midvein, upper lvs. lance-shaped, clasping the stem with acute lobes. Fr. black, half as long as the white beak. △ Poor pastures, rocky ground, waysides. June–Aug. Most of Eur. (except North Eur. IRL.AL.).

L. sativa L. GARDEN LETTUCE. Distinguished by the upper stem lvs. which are oval or rounded and with a heart-shaped clasping base and contrasting with the lower lvs. which are borne in a rosette; lvs. all smooth, hairless, usually undivided. Fl. heads pale yellow and often streaked with violet, numerous in a dense, much-branched, flat-topped cluster, with arrow-shaped bracts. Fr. with a beak equal its length. Probably a native of Asia; widely cultivated in many forms throughout Europe as a salad plant, sometimes escaping.

1544. *L. viminea* (L.) C. Presl PLIANT LETTUCE. A slender, branched, erect bienn. 20–100 cm., distinguished by the narrow stem lvs. which have basal lobes which run down the whitish stem as green ribs. Fl. heads numerous, pale yellow, in stalkless clusters of 3, forming a long slender spike; florets usually 4. Lower lvs. deeply pinnately cut into linear lobes, the uppermost lvs. linear-lance-shaped, undivided. Fr. black, with a shorter black beak. △ Dry banks, rocky places, walls. July–Sept. Med., South-East Eur. P.D.CH.A.CS.H. SU. Page 469.

PRENANTHES Involucre cylindrical, of 8–10 very unequal bracts; fl. heads with only 2–5 purple ray-florets; receptacle without scales. Fr. shallowly ribbed, beakless; pappus of 1–2 rows of simple bristles. 1 sp.

1545. *P. purpurea* L. An erect, hairless, glaucous perenn. 30–150 cm., with rather numerous small, drooping, purple or rarely white, long-stalked fl. heads in a lax pyramidal cluster. Fl. heads *c.* 2 cm. across; florets 2–5. Lvs. glaucous beneath, usually oblong-lance-shaped and often narrowed in the middle and violin-shaped, or shallowly lobed, all with clasping, rounded, basal lobes. Fr. whitish, beakless; pappus white. △ Damp woods, shady places, streamsides. July–Sept. Cent., South-East Eur. E.F.I.SU. Pl. 160.

CREPIS | Hawksbeard Involucre with unequal bracts, the outer much shorter and often spreading. Ray-florets only present; receptacle without scales, pitted, the pits with toothed or hairy margins. Fr. ribbed, narrowed above, and with or without a beak; pappus of many rows of simple hairs, usually white. *c.* 70 sps.

At least the central frs. with a slender beak
(*a*) *Fls. yellow*
1546. *C. vesicaria* L. BEAKED HAWKSBEARD. A rather robust, erect, hairy bienn. or perenn. 15–80 cm., much branched above, with numerous erect orange-yellow fl. heads 1½–2½ cm. across, in a flat-topped cluster. Involucre 8–12 mm. long, woolly-haired, and often glandular, the inner bracts hardening in fr. but not enclosing the marginal frs. and much shorter than the pappus, the outer bracts spreading, papery-margined. Lvs. mostly basal, finely hairy, oblong-lance-shaped and deeply pinnately cut into rather broad, toothed lobes, the upper lvs. often less deeply lobed, or entire and clasping the stem with acute lobes. All fr. beaked, beak as long as the nutlet. △ Meadows, waysides. May–Aug. West, Med. Eur. (except AL.) D.CH.A.H.R. Pl. 160.

*C. foetida L. STINKING HAWKSBEARD. Central and Southern Europe.

*C. setosa Haller fil. BRISTLY HAWKSBEARD. Southern and South-Eastern Europe.

(*b*) *Fls. pink*
C. rubra L. PINK HAWKSBEARD. Readily distinguished by its solitary or few pink fl. heads up to 4 cm. across, borne on long, nearly leafless stems, and drooping in bud. Involucre 11–17 mm. long; outer bracts sparsely woolly, the inner glandular-hairy. Lower lvs. deeply lobed, the uppermost lvs. lance-shaped or linear. Fr. with a long beak. Italy to Greece.

Fr. without a beak
1547. *C. biennis* L. ROUGH HAWKSBEARD. An erect, leafy bienn. 30–120 cm., with rather large golden-yellow fl. heads 2½–3½ cm. across, in flat-topped branched clusters with hairy fl. stems. Involucre 10–13 mm. long, hairy, often glandular, inner bracts downy within, the outer bracts much shorter, linear-lance-shaped, spreading. Lvs. rough with scattered hairs, lance-shaped, entire, toothed or deeply pinnately cut, the stem lvs. half-clasping the stem. Fr. 7–12 mm., reddish-brown, with 13 smooth ribs, pappus white, as long as or slightly longer than involucre in fr. △ Pastures, damp places, waysides. May–July. Most of Eur. (except P.IS.AL.): introd. IRL.N.

1548. *C. capillaris* (L.) Wallr. SMOOTH HAWKSBEARD. Distinguished from 1547 by its smaller yellow fl. heads, usually 1–1½ cm. across, and the outer involucral bracts which are adpressed and not spreading. Involucre 5–8 mm. long; bracts hairless within. Lvs. and stem almost hairless, the stem lvs. clasping with arrow-shaped basal lobes. Fr. with 10 smooth ribs. A nearly hairless, slender, much-branched ann. or bienn. 20–90 cm. △ Grasslands, heaths, waste places, waysides, walls. June–Oct. All Eur. (except IS.AL.).

C. paludosa (L.) Moench A quite hairless mountain perenn., except for the rough glandular hairs on the fl. stalks and involucre. Fl. heads yellow, clustered; involucre 12–15 mm. long, blackish, densely hairy. Upper lvs. entire, lance-shaped, long-pointed and with a heart-shaped clasping base, the lower lvs. stalked and toothed or shallowly-lobed. Fr. ten-ribbed. Central and Southern Europe.

1549. *C. aurea* (L.) Cass. GOLDEN HAWKSBEARD. An alpine rosette perenn. with conspicuous, solitary, orange or dark reddish-orange fl. heads 2–3 cm. across, borne on leafless stems 5–25 cm., or rarely with 2–3 fl. heads. Involucre covered with long, black or yellowish, flexuous hairs. Lvs. all in a basal rosette, oblance-shaped and shallowly lobed, smooth, shiny, hairless. △ Mountain pastures. June–Sept. F.CH.D.A.I. Pl. 160.

HIERACIUM Like *Crepis*, but fr. beakless and abruptly narrowed to the apex, not gradually narrowed, and pappus of brownish bristly hairs. A very difficult genus, with many sps. lacking true sexuality; numerous microspecies occur throughout the temperate regions. *c.* 200 sps.

Rosette-forming plants with creeping stolons

1550. *H. pilosella* L. MOUSE-EAR HAWKWEED. Fl. heads solitary pale yellow with the outer florets reddish beneath, borne on leafless woolly stems. Lvs. all basal in a rosette, lance-shaped entire, white-woolly beneath and with sparse, stiff white hairs above; creeping leafy stolons present. Fr. 2 mm., blackish. A very variable perenn. 5–30 cm. △ Dry, grassy places, banks. May–Sept. All Eur. *Used herbally for respiratory infections.* Pl. 160.

1551. *H. auricula* L. Distinguished by its glaucous, shining rosette of lvs. which are hairless above or with a few hairs at the base, and its long, creeping, above-ground stolons. Fl. heads pale yellow, 2–6 in a compact cluster, or rarely solitary, usually on a leafless stem 10–30 cm.; involucre 6–9 mm. long, with black glandular hairs. △ Grassy places, heaths, woods. June–Aug. Most of Eur.

1552. *H. aurantiacum* L. GRIM THE COLLIER, ORANGE HAWKWEED. Readily distinguished by the bright brick-red fl. heads borne in a dense terminal cluster of 1–6, on softly hairy, glandular, nearly leafless stems 20–50 cm. Fl. heads *c.* 1½ cm. across; involucre blackish, with long hairs. Basal lvs. in a rosette, oblong-lance-shaped, with spreading hairs on both sides; stolons mostly below ground. △ Meadows, rocks in the mountains. June–Aug. Most of Eur. (except P.IRL.IS.AL.GR.TR.).

Usually erect plants without creeping stolons
(a) Rosette lvs. absent at flowering

1553. *H. prenanthoides* Vill. PRENANTH HAWKWEED. Distinguished by its middle stem lvs. which are constricted just below the broad clasping base; all lvs. strongly net-veined and somewhat glaucous beneath. Stems rather robust, 40–100 cm., with many crowded lvs. up to the inflorescence, and with several short lateral branches bearing several fl. heads forming a rather dense, rounded-topped cluster. Involucre and fl. stems with tufts of woolly and densely glandular hairs, florets ciliate-tipped; styles dark. Fr. tawny or pale yellowish-brown. △ Woods, meadows, in mountains. July–Sept. GB.F.N.S.SF.D.CH.A.I.YU. BG.

1554. *H. umbellatum* L. Lvs. numerous, more or less crowded at least below, usually linear-lance-shaped to lance-shaped and narrowed to a stalkless base, dark green, margin recurved, rough, and with 2–3 distant teeth on each side. Fl. heads rather few, in a more or less flat-topped cluster; involucral bracts blunt, blackish-green, nearly hairless except for the recurved tips; styles yellow. △ Woods. Aug.–Sept. All Eur.

(b) Rosette lvs. present at flowering

1555. *H. alpinum* L. ALPINE HAWKWEED. Distinguished by its large, usually solitary fls. about 3 cm. across, borne on a shaggy stem, with 1–2 small bract-like lvs., and arising directly from the rosette. Rosette lvs. dark green, numerous, spathulate to oval-elliptic, narrowed to a long slender stalk, hairy and with fine yellowish glands at least on the margin. Lvs., stem and involucre covered with white-tipped, black-based hairs. A low rosette perenn. 5–20 cm. or more. △ In mountains: meadows, rocky places, screes. July–Aug. North, Cent. Eur. F.GB.I.

1556. *H. caesium* Fries Readily distinguished by its glaucous or pale green lvs. which are sometimes purple-spotted or purple beneath, hairless above and usually with soft, white, star-shaped hairs and ciliate beneath. Rosette lvs. several, toothed or pinnately cut; stem lvs. usually 1–3. Fl. heads few; involucre 10–14 mm. long, densely hairy but without glandular hairs; fl. stems with star-shaped hairs, not or scarcely glandular. △ Rocks in mountains. June–Aug. North, Cent. Eur. F.I.

1557. *H. vulgatum* Fries COMMON HAWKWEED. Rosette lvs. few, oval-lance-shaped and gradually narrowing into a stalk, lv. blade with short acute teeth, ciliate and with tufts of soft white star-shaped hairs, often violet beneath; stem lvs. usually 1–3 narrower, often with acute forward-pointing teeth. Fl. heads 1–20, in a spreading flat-topped cluster, branches with tufts of soft white star-shaped hairs, not or sparsely glandular-hairy. Involucre 8–12 mm. long, sparsely to moderately glandular-hairy; bracts incurved in bud, with paler margin with soft star-shaped hairs. △ Woods, heaths. June–Aug. North, Cent. Eur. IRL.GB.I.YU.BG.

H. maculatum Sm. SPOTTED HAWKWEED. Like 1557, but rosette lvs. elliptic-lance-shaped, long-stalked, usually spotted with dark purple blotches, toothed, often deeply so, at the base. Fl. heads 6–20, stems densely glandular and with soft white star-shaped hairs. Mountains of central Europe, Great Britain.

MONOCOTYLEDONES
ALISMATACEAE | Water-Plantain Family

Hairless aquatic or marsh plants with stalked lvs. with rounded or narrow blades. Fls. radially symmetrical; perianth of 3 outer sepal-like segments differing from the 3 inner petal-like segments; stamens usually 6. Ovary superior, carpels many, usually free; fr. a cluster or whorl of one-seeded units.

1. Fls. unisexual; stamens numerous.   *Sagittaria*
1. Fls. bisexual; stamens usually 6.
　2. Fls. solitary or in an umbel or in not more than 2 whorls.
　　3. Floating lvs. absent; lvs. strap-shaped, mostly carried above water.   *Baldellia*
　　3. Floating lvs. present, blade oval; submerged lvs. strap-shaped.   *Luronium*
　2. Fls. in more than 2 whorls.
　　4. Carpels numerous, not beaked.   *Alisma*
　　4. Carpels 6–10, long-beaked and spreading in a star.   *Damasonium*

BALDELLIA Fls. long-stalked, in umbels; lvs. all basal in a rosette. Fr. of numerous carpels in a crowded head. 1 sp.

1558. *B. ranunculoides* (L.) Parl. (*Echinodorus r.*) LESSER WATER-PLANTAIN. Fls. pale purplish, 1–1½ cm. across, borne on stalks of unequal length in a lax umbel, or sometimes with more than 1 superimposed umbel. Lvs. with a linear to narrow lance-shaped blade 2–4 cm., narrowed to a long stalk; fl. stem leafless arising directly from the rootstock. Fr. in a spherical head. An erect aquatic perenn. 5–20 cm., growing largely above water, sometimes with creeping stems rooting at the nodes. △ Watersides, ditches. May–Aug. West Eur. DK.N.S.D.CH.I.YU.GR.

LURONIUM   Stems submerged and rooting at the nodes. Fls. 1 or several, long-stalked, borne from the axils of the lvs. Fr. with 10–12 carpels in a single whorl. 1 sp.

1559. *L. natans* (L.) Rafin. (*Elisma n.*) FLOATING WATER-PLANTAIN. A delicate aquatic perenn. with submerged stems rooting at the nodes and spreading 15–50 cm., with linear, translucent, underwater lvs. and oval floating lvs. with slender lv. stalks. Fls. white with a yellow spot at centre, 1–1½ cm. across, few, axillary on very slender stalks. △ Stagnant and slow-flowing water. June–Sept. F.GB.B.NL.DK.N.S.D.PL.SU. Page 480.

ALISMA | Water-Plantain   Fls. many, borne in superimposed whorls on the main stem and on lateral branches. Carpels numerous in 1 row, strongly compressed at maturity. *c.* 3 sps.

1560. *A. plantago-aquatica* L. WATER-PLANTAIN. An erect, hairless, aquatic perenn. with lvs. and stems carried above water 20–100 cm., and with numerous whorls of pale lilac fls. forming a branched pyramidal inflorescence. Fls. *c.* 1 cm. across, long-stalked; outer perianth segments oblong, the inner rounded; anthers about twice as long as broad. Lvs. all basal, long-stalked, blade 8–20 cm., oval-acute with a rounded or heart-shaped base. Carpels with a short straight style arising from below the middle. △ Muddy places, damp ground, shallow water. June–Sept. All Eur. (except IS.).

1561. *A. lanceolatum* With. NARROW-LEAVED WATER-PLANTAIN. Very like 1560, but lv. blades lance-shaped and narrowed gradually to the long stalk. Fls. pink, the outer perianth segments oval, the inner acute; anthers as broad as long. Carpels with style arising from above the middle. △ Muddy places by water. June–Aug. Probably most of Eur.

DAMASONIUM   Carpels 6–10 in 1 row and fused at their base, each carpel prolonged above into a beak, all spreading outwards in a star in fr. *c.* 4 sps.

1562. *D. alisma* Miller THRUMWORT. A floating or submerged aquatic ann. with leafless fl. stems carried 5–20 cm. above water and bearing 1–3 simple whorls of white, stalked fls. Fls. 6–10 mm. across, about 6 in each whorl. Lvs. all basal, long-stalked, blade 3–5 cm., oval with rounded or heart-shaped base, usually floating. Fr. usually of 6 carpels, spreading in a conspicuous star. △ Still waters. June–Sept. P.E.F.GB.B.I.GR.R.SU. Page 480.

SAGITTARIA | Arrowhead   Fls. unisexual, borne in several whorls on a leafless aerial stem. Upper fls. male, with numerous stamens; female fls. with numerous carpels. Fr. a dense cluster of spirally arranged, flattened carpels. 2 sps.

1563. *S. sagittifolia* L. ARROWHEAD. Fls. large *c.* 2 cm. across, white with a dark violet basal patch, borne in several superimposed whorls of 3–5 fls., the whorls of male fls. above, the female whorls below. Lvs. all basal, usually 3 forms being present on the same plant: either submerged, when linear and translucent, or floating, when blades oval or lance-shaped, or aerial, when blade arrow-shaped. Fr. head hemispherical *c.*

1½ cm. across. Fl. stems carried above water 30–90 cm.; plant with over-wintering, sub-merged, corm-like buds. △ In still or slow-flowing shallow water. May–Aug. All Eur. (except IS.AL.TR.). *Cultivated in China for its edible tubers.*

BUTOMACEAE | Flowering Rush Family

Aquatic plants with fls. usually in an umbel. Perianth of 6 petal-like segments; stamens 6–9; carpels free, usually 6, each with many ovules—thus distinguishing it from the Alismataceae.

BUTOMUS | Flowering Rush Fls. many, in a terminal umbel, and with 2–4 papery bracts beneath the umbel; perianth persistent in fr. 1 sp.

1564. *B. umbellatus* L. FLOWERING RUSH. A handsome aquatic perenn. with a terminal umbel of many pink fls. borne on erect, leafless, rush-like, aerial stems ½–1 m. Fls. 2½–3 cm. across, flowering in succession; perianth pink with darker veins on the outer side; fl. stalks very unequal. Lvs. all aerial, rush-like, linear-acute, triangular in section, sheathing at the base; rhizome fleshy. Carpels 6–9, with erect, persistent styles. △ In still and slow-flowing shallow water. June–Aug. All Eur. (except IS.). *The rhizomes are edible.* Pl. 161.

HYDROCHARITACEAE | Frog-Bit Family

Submerged or floating aquatic plants which are usually one-sexed. Fls. encircled in bud by 1 or 2 bracts, the *spathe*. Male fls. usually numerous; perianth segments 6; stamens 3 to many. Female fls. solitary; perianth 6; ovary inferior, with many ovules.

1. Lvs. stalked, blade rounded or kidney-shaped, floating.   *Hydrocharis*
1. Lvs. stalkless, blade longer than broad, submerged.
  2. Lvs. lance-shaped or oval, usually in whorls of 3, less than 2½ cm. long. *Elodea*
  2. Lvs. linear. much longer than 2½ cm.
    3. Lvs. in a rosette, thick, spiny.   *Stratiotes*
    3. Lvs. not in a rosette, thin, smooth.   *Vallisneria*

STRATIOTES | Water Soldier Lvs. in large submerged rosettes, rising to the surface at flowering time. Fls. one-sexed, white, borne above water; stamens 12. Fr. a berry. 1 sp.

1565. *S. aloides* L. WATER SOLDIER. A submerged aquatic perenn. with large bright green rosettes 10–30 cm. across of stiff sword-shaped lvs., and white fls. 3–4 cm. across, borne above water. Male fls. several, the female solitary, borne on a stout stem 5–20 cm.; fls. with the outer perianth green and sepal-like, the inner white, petal-like, rounded; spathe green, two-lobed, 2½–3 cm. long. Lvs. linear acute, keeled, brittle, margin with spiny teeth. △ Still waters. June–Aug. Much of Eur. (except P.IS.N.AL.GR.): introd. IRL.DK.S.SF. *The plant possessed a reputation for healing wounds made by iron weapons.* Pl. 161.

HYDROCHARIS | Lvs. floating, orbicular-kidney-shaped, stalked. Fls. white, borne

1. *Luronium natans* 1559
2. *Posidonia oceanica* 1580
3. *Zannichellia palustris* 1581
4. *Triglochin palustris* 1569
5. *Damasonium alisma* 1562
6. *Vallisneria spiralis* 1567

above the surface, usually one-sexed, the male 2–3 in a spathe, the female solitary; stamens 12. Ovary six-celled with 6 bilobed styles. Fr. fleshy. 1 sp.

1566. *H. morsus-ranae* L. FROG-BIT. A submerged aquatic perenn. with rounded floating lvs. and white fls. blotched with yellow at base, *c.* 2 cm. across. Fls. with 3 crumpled white, petal-like segments, and 3 outer green sepal-like segments, long-stalked. Lvs. all floating, blades *c.* 3 cm. across, orbicular-heart-shaped, thick, long-stalked with papery stipules at the base; floating stems slender, bearing roots at each node. Plant perennating by rounded over-wintering buds. △ Still waters. June–Aug. All Eur. (except IS.). Pl. 161.

VALLISNERIA   Submerged plants with thin, ribbon-like lvs. Fls. unisexual, tiny, floating on the surface of the water. 1 sp.

1567. *V. spiralis* L. A delicate, grass-like, completely submerged perenn. with translucent, bright green lvs. 5–12 mm. broad, with a blunt apex. Fls. inconspicuous, pinkish-white, the male fls. numerous, enclosed in a spathe of 2–3 bracts, at length detaching and floating to the surface for pollination, the female fls. solitary, carried to the surface on long slender stalks which become spirally twisted in fr. △ Still and slow-flowing waters. July–Oct. South-East Eur. (except AL.) P.E.F.CH.I. Page 480.

ELODEA   Submerged plants with numerous stalkless lvs. in whorls of 3–5. Fls. one-sexed; spathe tubular and borne in the axils of the lvs.; fls. long-stalked, allowing pollination at the surface. 3 sps.

1568. *E. canadensis* Michx CANADIAN PONDWEED. A submerged, rather densely bushy, aquatic perenn. 10–30 cm., with numerous overlapping, dark green, translucent, oblong, and minutely toothed lvs. *c.* 1 cm., usually in whorls of 3. Fls. *c.* 5 mm. across, greenish-purple, carried to the surface on a long slender stalk, with a sheathing, two-lobed spathe 1–2 cm. long at its base. △ Still and flowing waters. May–Aug. Native of North America: introd. to Cent., North Eur. (except IS.) F.IRL.GB.B.NL.BG.R.SU.

SCHEUCHZERIACEAE | Arrow-Grass Family

Marsh or aquatic plants with mostly radical lvs. and numerous fls. in slender elongated spikes. Fls. tiny with 2 whorls of 3 greenish perianth segments; stamens usually 6; ovary superior with 3–6 free or fused carpels. Placed by some botanists in Juncaginaceae.

TRIGLOCHIN | **Arrow-Grass**   Fls. in a long bractless spike; perianth 6, green, deciduous. Carpels 3–6, all fertile or alternately 3 sterile. 4 sps.

1569. *T. palustris* L. MARSH ARROW-GRASS. A hairless, rush-like perenn. 15–50 cm., with a basal tuft of very slender fleshy lvs. and a long leafless spike of tiny green fls. Fls. *c.* 3 mm., erect, short-stalked; perianth segments rounded, purple-edged; fl. spike longer than the lvs. Lvs. half-cylindrical in section, deeply furrowed on the upper surface towards the base. Fr. 8–10 mm. by 1 mm., club-shaped, broader towards the apex, pressed to the stem; carpels 3 fertile, 3 sterile. △ Marshes, damp places. June–Aug. All Eur. (except GR.TR.). Page 480.

T. maritima L. SEA ARROW-GRASS. Like 1569, but a more robust maritime perenn. 15–70 cm., with slender lvs. half-cylindrical in section, but not furrowed on the upper surface.

Fr. shorter, ovoid, 3–4 mm. by 2 mm., spreading and not pressed to stem; carpels 6, all fertile. Widespread in Europe, except in the extreme south-east.

POTAMOGETONACEAE | Pondweed Family

Submerged or floating aquatic plants usually with a stipule-like sheathing scale at the base of the lvs. Fls. inconspicuous, bisexual, usually in bractless spikes; perianth segments 4; stamens 4; carpels usually 4, not fused together. Fr. fleshy or dry, one-seeded.

Nos.

1. Submerged sea plants with linear, grass-like lvs.
   2. Creeping stems slender, without fibres.
      3. Lvs. *c.* 1 mm. wide; carpels long-stalked.   *Ruppia*   1578
      3. Lvs. usually 5–8 mm. wide; carpels hidden in a sheath.   *Zostera*   1579
   2. Creeping stems stout, with numerous brown fibres.   *Posidonia*   1580
1. Submerged or floating, fresh or brackish water plants.
   4. Carpels stalkless in a dense, terminal, usually stalked cluster.
      5. Lvs. opposite; stipules absent.   *Groenlandia* 1577
      5. Lvs. alternate; stipules present.   *Potamogeton* 1570
   4. Carpels long-stalked in spreading clusters.   *Ruppia*   1578

POTAMOGETON | Pondweed Aquatic plants with alternate, floating or submerged lvs., usually with a membraneous stipule at the base which either sheathes the stem or is open and not sheathing. Fl. spikes submerged or aerial, cylindrical, of numerous tiny green fls., spike lengthening in the fr.; perianth segments 4, green; stamens 4; carpels 4. Frequent hybridization, and considerable plasticity in form combine to make the identification of individual plants often difficult. *c.* 30 sps.

Blades of lvs. lance-shaped to oval, more than ½ cm. broad
(a) Stalked floating lvs. present
1570. *P. natans* L. BROAD-LEAVED PONDWEED. Readily distinguished by its large, shining, leathery, floating lvs., with an oval blade 2½–12½ cm. long, usually shallowly heart-shaped at the base and appearing jointed at its junction with the stalk. Submerged lvs. linear, without blades; stipules large, leaf-like, 5–15 cm. Fr. spike 3–8 cm. long, carried above water on a stout but not swollen stem. Carpels obovoid, 4–5 mm. A perenn. with slender submerged stems 1–5 m. long. △ Slow-flowing and stagnant waters. June–Aug. All Eur. Pl. 161.

1571. *P. polygonifolius* Pourret (*P. oblongus* Viv.) BOG PONDWEED. Like 1570, but floating lvs. with thick elliptic blades 2–6 cm. long, which are gradually narrowed to the stalk and are not jointed at their junction with the stalk. Submerged lvs. similar to floating lvs. but blades thinner and translucent; stipules 2–4 cm., blunt. Fr. spike 1–4 cm.; carpels obovoid 2 mm. A perenn. with slender stems 20–60 cm. △ Bogs, marshes, streams. June–Aug. All Eur. (except IS.TR.R.).

P. coloratus Hornem. FEN PONDWEED. Like 1571, but all lvs. thin, translucent, and finely and distinctly net-veined, the submerged lvs. with a lance-shaped blade which is longer than the lv. stalks. Fr. spike 2½–4 cm.; carpels tiny, ovoid, 1½–2 mm. Widespread in Europe. Page 484.

P. nodosus Poiret (*P. drucei* Fryer) Readily distinguished by its contrasting submerged and floating lvs.; the submerged lvs. thin and translucent, and with finely and beautifully net-veined elliptic blades, the floating lvs. with tough leathery, opaque elliptic blades. Fr. spike 2–6 cm., stalk stout; carpels $3\frac{1}{2}$ mm. obovoid, acutely keeled. Western and Central Europe.

(b) *Floating lvs. absent and lvs. stalkless or almost so*

1572. *P. lucens* L. SHINING PONDWEED. A large, rather delicate perenn. 1–3 m., with large, thin, translucent, almost stalkless, oblong-lance-shaped lvs. 10–20 cm., with a narrow, acute or long-pointed apex. Lv. blades with 10–13 veins, margin minutely toothed; stipules conspicuous 3–8 cm., with 2 prominent ribs. Fr. spike 5–6 cm.; stalk stout and thickened above; fr. *c.* 3 mm. △ Stagnant and slow-flowing waters. June–Sept. All Eur. (except IS.TR.). Page 484.

1573. *P. perfoliatus* L. PERFOLIATE PONDWEED. Readily distinguished by its stalkless lvs. which clasp the stem with 2 broad rounded lobes. Lvs. 2–6 cm. oval, very thin and translucent, with 5–7 conspicuous veins; stipules *c.* 1 cm., soon disappearing. Fr. spike 1–3 cm.; stalk stout, not thickened above; fr. *c.* 4 mm. A variable, robust or rather delicate perenn. $\frac{1}{2}$–6 m. △ Rather deep, slow-flowing, and stagnant waters. June–Aug. All Eur.

1574. *P. crispus* L. CURLED PONDWEED. Distinguished by its dark, often reddish, thin, shining, translucent, narrowly lance-shaped, blunt-tipped, undulate lvs. Lvs. 3–9 cm. by 8–15 mm., three- to five-veined, margin with rough teeth; stipules 1–2 cm.; stem quadrangular. Fr. spike 1–2 cm., rather lax; stalk somewhat flattened. Carpels with a curved beak as long as the nut. A perenn. 30–120 cm. △ Still and slow-flowing waters. May–Sept. All. Eur. (except IS.).

Blades linear or thread-like, less than $\frac{1}{2}$ cm. broad

1575. *P. obtusifolius* Mert. & Koch GRASSY PONDWEED. Lvs. thin, translucent and dark green, flat linear 2–4 mm. wide, with a rounded apex with a fine point, narrowed to base, three-veined. Stipules open, $1\frac{1}{2}$–2 cm. long, broad blunt. Fr. spike $\frac{1}{2}$–$1\frac{1}{2}$ cm.; stalk short, slender, $\frac{1}{2}$–2 cm. A slender, delicate perenn. $\frac{1}{2}$–1 m. △ Stagnant waters. June–Aug. Probably most of Eur. Page 484.

*P. gramineus L. VARIOUS-LEAVED PONDWEED. Widespread in Europe.

1576. *P. pectinatus* L. FENNEL-LEAVED PONDWEED. Distinguished by its very narrow, thread-like lvs. $\frac{1}{4}$–2 mm. wide and its conspicuous stipules forming an open, whitish sheath 2–5 cm. long. Lvs. dark green translucent, with 3–5 minute veins and with 2 air-filled canals on each side of the midvein. Fr. spike 2–5 cm. long with 4–8 interrupted whorls borne on a very slender fl. stalk 3–15 cm. long. Fr. 3–5 mm. A very variable, much-branched perenn. 1–3 m. △ Stagnant and running waters. June–Aug. All Eur. (except IS.TR.). Page 484.

GROENLANDIA Like *Potamogeton*, but all lvs. in opposite pairs or rarely in whorls of 3; stipules absent. 1 sp.

1577. *G. densa* (L.) Fourr. (*Potamogeton d.*) OPPOSITE-LEAVED PONDWEED. Stem very leafy with 2 rows of densely overlapping, outwardly curved lvs. which are stalkless and clasp the stem. Lvs. triangular-oval to lance-shaped $1\frac{1}{2}$–$2\frac{1}{2}$ cm., translucent, often folded longitudinally, minutely toothed near the apex; floating lvs. absent. Fl. spike four-flowered, globular *c.* 3 mm. across, short-stalked; fr. *c.* 3 mm. A branched, submerged perenn. 10–30 cm. △ Fast streams, slow flowing and still waters. June–Aug. Much of Eur. (except IS.SF.AL.): introd. N.S. Page 484.

RUPPIA Submerged aquatic plants, usually of salt or brackish waters. Fls. bisexual in a stalked umbel-like cluster, at first enclosed in the sheaths of the 2 upper lvs. Perianth

1. *Groenlandia densa* 1577
3. *P. lucens* 1572
5. *Potamogeton obtusifolius* 1575

2. *Potamogeton pectinatus* 1576
4. *Najas marina* 1582
6. *P. coloratus* [1571]

absent; stamens 2; carpels 4 or more, becoming stalked in fr. Often placed in a separate family Ruppiaceae. 4 sps.

1578. *R. spiralis* Dumort. (*R. maritima* auct.) TASSEL PONDWEED. A very slender, brackish-water, grass-like perenn. 20–40 cm., with thread-like stems and lvs. *c.* 1 mm. wide. Lvs. scarcely wider than the stems, but with a broader, brownish, swollen basal sheath. Stalk of fl. cluster much-elongated after flowering to *c.* 10 cm., often spirally coiled; carpels long-stalked, ovoid with a distinct beak. △ Saline waters. May–Sept. Coasts of all Eur. (except IS.).

ZOSTERA | Eel-Grass Submerged marine perenns. with fls. borne on one side of a flattened axis and partially enclosed in a sheath. Fls. minute, one-sexed, the males and females in 2 alternating rows; pollination occurs underwater. Often placed in a separate family Posidoniaceae. 3 sps.

1579. *Z. marina* L. EEL-GRASS, GRASS-WRACK. A submerged, marine, grass-like perenn. to 60 cm. or more, with creeping rhizomes, often covering large areas of mud and sand on the sea floor. Lvs. bright green, strap-shaped, usually ½–1 cm. broad, with a rounded apex and short narrow-pointed tip, but fertile shoots with narrower lvs. Inflorescence usually 9–12 mm. long. △ Shallow salt water to *c.* 4 m. in depth. Feb.–Sept. Coastal waters of Eur. *The dried shoots have been used for stuffing and packing.*

POSIDONIA Fls. in branched, few-flowered spikes borne in the axils of long bracts, the male fls. on the upper spikes, the bisexual fls. on the lower. Fr. fleshy, the size of an olive. Often placed in a separate family Posidoniaceae. 1 sp.

1580. *P. oceanica* (L.) M. Delile Like 1579, but rhizomes much stouter and densely covered with numerous brown fibres formed from the bases of the old lvs. Lvs. strap-shaped, to *c.* 50 cm. by 6–10 mm., dark green, with a blunt, rounded tip and 10–13 longitudinal veins. Fls. greenish, inflorescence borne on a flattened stalk 15–25 cm. long, arising from the rhizome, but fls. rare. Fr. fleshy, ovoid, 1–1½ cm. A creeping perenn. covering large areas of the sea floor. △ In shallow water to a depth of *c.* 30 m. Oct.–Jan. Med. Eur. (except TR.) P. *The dried lvs. are used for stuffing, packing, and in bedding, and the thick deposits washed up on the shore used as manure.* Pl. 161. Page 480.

ZANNICHELLIACEAE | Horned Pondweed Family

Submerged aquatic plants with opposite or whorled lvs. with sheathing bases. Fls. one-sexed, very small with perianth 3 or absent; male fls. with 1–3 stamens; female fls. with 1–9 free carpels, each with a style terminating in an enlarged stigma. Fr. of 1–9 stalked or stalkless, one-seeded carpels.

ZANNICHELLIA Fls. in axillary clusters of 1 male and 2–6 female fls. borne in a colour-less, deciduous, cup-shaped spathe, the male fl. with 1 stamen, the female fls. with 1 carpel. 3 sps.

1581. *Z. palustris* L. HORNED PONDWEED. A delicate, slender, branched, submerged perenn. to 50 cm., with bright green, translucent, thread-like lvs. 1½–5 cm. or more long by ½–1 mm. broad, tapering to a point. Stipules papery, sheathing the stem, but soon falling. Fr.

stalkless in a cluster of 2–6, in the axils of the lvs.; carpels 3–5 mm. long, long-beaked and with a broad, blade-like stigma. △ Still or flowing, fresh or brackish water. May–Sept. All Eur. Page 480.

NAJADACEAE | Najas Family

Submerged aquatic plants with more or less opposite or whorled lvs. with sheathing base and 2 small scales within each sheath. Plants one-sexed; fls. small, solitary, male fls. with 1 stamen enclosed in a cup-shaped sheath; female fls. with 1 carpel with 2–4 stigmas. Fr. with 1 ovule. 1 genus. *c.* 4 sps.

NAJAS

1582. *N. marina* L. A dark green, brittle, submerged aquatic ann. 10–25 cm., with dichotomously branched, prickly stems and whorls of 3 linear-oblong lvs., with spiny-toothed margin. Lvs. *c.* 2 cm. by 2–4 mm. broad, sheathing at the base. Fls. solitary, the male in a cup-shaped spathe with 2–3 bristly teeth, the female with an ovary 4–5 mm. and a three-lobed stigma. △ Still waters. June–Sept. Most of Eur. (except IRL.IS.AL.TR.). Page 484.

LILIACEAE | Lily Family

Usually herbaceous plants, often with underground bulbs, corms, or rhizomes; lvs. entire, alternate or in whorls, often all basal. Fls. regular, with parts usually in threes. Perianth usually of 2 petal-like whorls of 3; stamens 6; ovary superior, of 3 fused carpels, three-celled with many ovules, or rarely one-celled; styles usually 1 or 3. Fr. a capsule or berry.

1. Styles 3, free or rarely fused below.   Group A
1. Styles 1 or rarely 4.
   2. Fr. forming a capsule opening by 3 valves; usually bulb- or tuber-forming   Group B
   2. Fr. a fleshy berry; not bulb-forming.   Group C

Group A. *Styles 3*

		Nos.
1. Fls. less than 1 cm. long, in a terminal, simple or branched cluster.		
2. Small plants to 30 cm.; lvs. linear.	*Tofieldia*	1583
2. Robust plants ½–1 m.; lvs. oval.	*Veratrum*	1586
1. Fls. more than 2 cm. long, arising directly from a corm or bulb.		
3. Styles fused nearly to top; fls. and lvs. appearing in spring.	*Bulbocodium*	1589
3. Styles free their whole length; fls. appearing in autumn, lvs. in the spring.		
4. Perianth fused below into a tube.	*Colchicum*	1588
4. Perianth not fused into a tube, though bases of perianth pressed together and appearing fused.	*Merendera*	1587

Group B. *Style 1; fr. a capsule; bulbous or tuberous plants*

Nos.

1. Lv. blades vertical, iris-like; stamens with hairy
filaments.   *Narthecium*   1584
1. Lvs. with more or less horizontal blades with a distinct
upper and lower surface, or all lvs. reduced to scales.
  2. Lvs. fleshy or spiny, in a rosette.   *Aloe*   1598
  2. Lvs. not fleshy or spiny.
    3. Plants with rhizome or tubers.
      4. Fls. 1–3 terminal on rush-like stems; lvs. all scaly; fls.
blue.   *Aphyllanthes*   1585
      4. Fls. several in a lax cluster; lvs. present; fls. not blue.
        5. Fls. trumpet-shaped, large, 4–10 cm. long.
          6. Fls. pure white.   *Paradisea*   1595
          6. Fls. yellow or orange-red.   *Hemerocallis*   1597
        5. Fls. smaller, less than 4 cm. long, bell-shaped, or with
wide-spreading perianth.
          7. Perianth similar-coloured on both sides.   *Anthericum*   1596
          7. Perianth distinctly darker on the outside, often with
a greenish midvein.
            8. Fl. stem leafless, or with bracts only.
              9. Filaments of stamens woolly-haired.   *Simethis*   [1596]
              9. Filaments of stamens hairless.   *Asphodelus*   1590–1592
            8. Fl. stem densely leafy.   *Asphodeline*   1593, 1594
    3. Plants with bulbs.
      10. Fls. in a terminal umbel, at first enclosed in a spathe.   *Allium*   1602–1617
      10. Fls. solitary, or in a spike-like cluster.
        11. Fls. solitary or rarely 2 or 3.
          12. Fls. large, at least 2½ cm. long.
            13. Fls. erect or spreading, cup-shaped, tubular or
globular.
              14. Style absent; anthers attached at base to filament.   *Tulipa*   1624–1627
              14. Style long; anthers attached on the back to
filament.   *Lilium*   1618–1621
            13. Fls. strongly drooping.
              15. Perianth segments turned backwards against fl.
stalk.   *Erythronium*   1628
              15. Perianth segments not turned backwards,
chequered, each with a gland at the base.   *Fritillaria*   1622, 1623
          12. Fls. small, less than 2 cm.; fls. white.   *Lloydia*   1629
        11. Fls. several or many in a spike-like cluster.
          16. Fls. large, 4 cm. long or more.   *Lilium*   1618–1621
          16. Fls. less than 3 cm. long.
            17. Perianth segments free or fused slightly at the base.
              18. Fls. blue, violet, or pink.
                19. Perianth free, spreading in a star.   *Scilla*   1631–1636
                19. Perianth segments slightly fused below, tubular
or bell-shaped.   *Endymion*   1638
              18. Fls. white, yellow, or greenish.
                20. Fls. white or greenish.
                  21. Fls. and lvs. appearing at different seasons.   *Urginea*   1630

21. Fls. and lvs. appearing together.
  22. Perianth segments with a green band on outside; plants usually more than 15 cm. tall.   *Ornithogalum*   1639–1642
  22. Perianth segments with 3 purplish veins; slender plants less than 15 cm. tall.   *Gagea*   1599–1601
20. Fls. yellow, often greenish outside.
  23. Fls. few in a lax cluster.   *Gagea*   1599–1601
  23. Fls. many in an elongated spike.   *Ornithogalum*   1639–1642
17. Perianth segments distinctly fused at base to at least a quarter their length; stamens fused to corolla.
  24. Perianth barrel-shaped, constricted at throat.   *Muscari*   1645–1648
  24. Perianth bell-shaped, not constricted at throat.
    25. Segments of perianth all spreading, 2–3 times shorter than the tube.   *Bellevalia*   1644
    25. Outer segments or all segments of perianth recurved, as long as the tube.
      26. Outer 3 segments recurved, inner 3 erect; fls. yellowish or brownish.   *Dipcadi*   1637
      26. All segments recurved; fls. not yellowish or brownish.   *Hyacinthus*   1643

Group C. *Fr. a berry; not bulbous plants*

Nos.

1. Stems unbranched, herbaceous.
  2. Stem lvs. only 2–4 or rarely more.
    3. Fls. solitary, greenish; lvs. usually 4, whorled.   *Paris*   1658
    3. Fls. in clusters, whitish; lvs. 2–3, alternate.
      4. Lvs. heart-shaped, stalked; fls. erect, star-shaped.   *Maianthemum*   1652
      4. Lvs. lance-shaped, stalkless; fls. drooping, bell-shaped.   *Convallaria*   1657
  2. Stem lvs. numerous, alternate or whorled.
    5. Fl. stalk distinctly bent; perianth segments fused only at the base.   *Streptopus*   1653
    5. Fl. stalk not bent; perianth segments fused half-way or more.   *Polygonatum*   1654–1656
1. Stems branched, herbaceous or woody.
  6. Normal lvs. absent.
    7. Branchlets needle-like, clustered.   *Asparagus*   1649, 1650
    7. Branchlets flattened and leaf-like.   *Ruscus*   1651
  6. Normal lvs. present, blade heart-shaped at base.
    8. Non-climbing herbaceous plants; fls. axillary, 1–2; fl. stalk bent.   *Streptopus*   1653
    8. Climbing woody plants with tendrils and usually spines; fls. many, clustered.   *Smilax*   1659

TOFIELDIA   Fls. tiny, borne on an almost leafless stalk arising from the rootstock. Ovary with 3 carpels fused only below; styles 3 free. 2 sps.

1583. *T. pusilla* (Michx) Pers. (*T. palustris* auct.; *T. borealis* (Wahlenb.) Wahlenb.) SCOT-TISH ASPHODEL. A delicate perenn. 5–20 cm., with tufted, grass-like lvs. and a dense cluster of tiny greenish-white fls. borne on a slender, usually leafless fl. stem. Inflorescence ½–1½ cm. long; fls. short-stalked, with a three-lobed bract, involucre absent; perianth segments blunt *c.* 2 mm.; anthers included. Basal lvs. rigid, linear 3–5 cm. by 1–2 mm., three-veined, stem lvs. sometimes present, small. △ Arctic or alpine: bogs, marshes. July–Aug. North Eur. (except DK.) GB.F.CH.A.CS.I.YU.

T. calyculata (L.) Wahlenb. Like 1583, but a more robust perenn. 15–30 cm., with yellow-ish, greenish, or rarely reddish fls. in an elongated lax or compact cluster 2–8 cm. long and borne on an erect stem with several small lvs. below the middle. Fls. with a three-lobed papery involucre and a lance-shaped bract as long as the fl. stalk; anthers longer than the perianth. Lvs. with 5–10 veins. Central Europe. Pl. 162.

NARTHECIUM | Bog Asphodel Carpels fused their whole length; style 1, stigma three-lobed. Fls. in a spike-like cluster; filaments of stamens densely hairy. 2 sps.

1584. *N. ossifragum* (L.) Hudson BOG ASPHODEL. A hairless perenn. 5–40 cm., with creeping rhizomes, spear-shaped lvs. and a terminal, cylindrical cluster of bright yellow fls. with conspicuous orange anthers. Perianth segments 6–8 mm., linear-lance-shaped, spreading in fl., erect in fr.; filaments with conspicuous, dense, yellow hairs; bracts lance-shaped. Basal lvs. 5–30 cm., rigid curved and with the blade held vertically, the stem lvs. few, much smaller. Fr. and fr. stem turning deep orange. △ Bogs, wet heaths. July–Aug. West Eur. DK.N.S.D.CS.

APHYLLANTHES Fls. blue, 1–3 in a terminal cluster surrounded by papery bracts. Stems greenish, rush-like with lvs. reduced to sheaths. 1 sp.

1585. *A. monspeliensis* L. An unusual-looking, tufted, rush-like perenn. 10–25 cm., with numerous stiff, slender, ribbed stems each ending in a terminal, oval, bud-like cluster of russet-coloured, overlapping bracts bearing one or few starry blue fls. Fls. stalkless, *c.* 2½ cm. across; perianth funnel-shaped with spreading blue lobes with a darker blue midvein, or rarely white. Lvs. reduced to sheaths which encircle the glaucous stems. Fr. a capsule. △ Dry, rocky places. Apr.–July. P.E.F.I. Pl. 162.

VERATRUM Fls. numerous, in branched clusters; perianth segments spreading, shortly fused at the base. Carpels free above, styles 3. 2 sps.

1586. *V. album* L. WHITE FALSE HELLEBORINE. A robust, erect perenn. ½–1½ m., with many broad and strongly veined lvs. in whorls of threes and a large, branched, elongated, terminal cluster of many whitish or greenish-yellow fls. Fls. 1½ cm. across, almost stalk-less; perianth white within, greenish and hairy without, or bright green or yellowish on both sides, lobes oblong, spreading in a star; bracts green, hairy. Lvs. oval, longitudin-ally pleated, hairy beneath, sheathing the finely hairy stem. △ In hills and mountains: pastures, damp grassy places. July–Aug. Much of Eur. (except IRL.GB.B.NL.IS.DK.S.TR.). *A very poisonous plant which was used in the past as an arrow poison. It can easily be mis-taken for the Yellow Gentian which is used in beverages, etc. The False Helleborine has leaves arranged in threes and they are hairy beneath, and the roots have a strong, un-pleasant smell, while the leaves of the Gentian are opposite, hairless beneath and the roots are almost odourless.* Pl. 162.

V. nigrum L. BLACK FALSE HELLEBORINE. Like 1586, but fls. blackish-purple, *c.* 1 cm. across, in a long, narrow, branched cluster; fl. stalks as long as the perianth; bracts coloured. Lvs. narrowed to a short stalk, hairless beneath. Central and South-Eastern Europe. Pl. 162.

MERENDERA Like *Colchicum*, but perianth segments not fused below, but forming a long tube with the bases of the segments closely pressed together; anthers attached to the filament near the base and similar in length. *c.* 6 sps.

1587. *M. montana* (L.) Lange (*M. bulbocodium* Ramond) A corm-forming perenn. with 1 or 2 pinkish-lilac stalkless fls. appearing in the autumn before the lvs. Perianth segments strap-shaped, acute, at length spreading in a star; anthers as long as the filament; stigmas 3. Lvs. several, usually appearing in spring, linear blunt, 4–6 mm. broad, grooved, and shorter than the fr. stems, which are 5–15 cm. Corm *c.* 2 cm. by 1 cm., with brownish scales. △ Mountain pastures. Aug.–Sept. P.E.F. Pl. 163.

COLCHICUM | **Autumn Crocus** Perianth segments fused below for much of their length into a long slender tube; anthers shorter than their filaments and attached near the middle. Lvs. appearing with or after the fls. and encircled in a sheath at the base. Styles 3. *c.* 25 sps.

1588. *C. autumnale* L. MEADOW SAFFRON, AUTUMN CROCUS, NAKED LADIES. Fls. solitary or several, pale rosy-purple or white, appearing in the autumn without lvs. from a colourless basal sheath. Perianth segments oblong, 3–4½ cm., fused below into a long, pale, stalk-like tube 5–20 cm. long; styles and stigmas strongly curved above. Lvs appearing in spring, often 3, shiny green, oblong-lance-shaped, 12–30 cm. by 1½–5 cm., lv. bases partly encircling the nearly stalkless fr.; fr. and lvs. at length carried above ground on a short stalk. Corm large, 3–5 cm. △ Damp meadows. Aug.–Sept. Most of Eur. (except North Eur.): introd. DK.N.S. *The corms produce a poisonous substance, colchicine, which is used in treatment of gout and rheumatism. Colchicine is, however, better known for its recently discovered property of upsetting the normal processes of cell division, and it is consequently used in experimental genetical work, and is a means of creating new forms and species.* Pl. 163.

C. alpinum DC. Like 1588, but whole plant smaller and perianth lobes narrower, linear-oblong, 2–3 cm. long; styles straight and with terminal pin-head stigmas. Lvs. 2–3, narrower, linear-lance-shaped, 6–12 cm. by 7–12 mm. Alps, Apennines.

BULBOCODIUM Like *Colchicum*, but spring-flowering and lvs. appearing with the fls. Styles fused for most of their length, free only at the apex. 2 sps.

1589. *B. vernum* L. (*Colchicum bulbocodium* Ker-Gawler) A crocus-like plant, with 1–3 rosy-lilac or rarely white fls. appearing in spring with the lvs. Perianth segments abruptly narrowed below, at first hooked together at the base but soon spreading wide in a star, not forming a tube; stamens 6, thus distinguishing it from *Crocus*. Lvs. 3, lance-shaped blunt, grooved, spreading, and encircled by a membraneous sheath at the base. Corm with a blackish outer scale. △ Mountain pastures. Feb.–May. E.F.CH.A.H.I.YU.R.SU. Pl. 163.

ASPHODELUS | **Asphodel** Fls. many in a simple or branched terminal cluster borne on a leafless stem. Perianth broadly funnel-shaped, divided almost to the base into 6 spreading segments, each with a deeper-coloured midvein; style slender, stigma knobbed. *c.* 8 sps.

Lvs. V-shaped in section, at least 1 cm. broad
1590. *A. albus* Miller WHITE ASPHODEL. A robust perenn. with numerous keeled and rush-like basal lvs., and a stout, leafless fl. stem to 1 m., bearing a compact, spike-like cluster of white fls. often with a few short lateral clusters below. Fls. 3–5 cm. across, borne on stalks shorter than the fls.; perianth segments to 3 cm. long, spreading, white or pinkish,

and with a brownish midvein; bracts brown, lance-shaped. Lvs. stiff, linear-acute, 1–2 cm. broad, V-shaped in section. Roots tuberous. Fr. $1\frac{1}{2}$ cm. long, with 7–8 transverse ridges. A variable sp. △ Heaths, grassy places, thickets, banks. Apr.–Aug. Med. Eur. (except TR.) P.CH.BG.

1591. *A. aestivus* Brot. (*A. microcarpus* Salzm. & Viv.) ASPHODEL. Distinguished from 1590 by its stout, much-branched, pyramidal cluster of pinkish-white fls. with the ascending lateral clusters as long or nearly as long as the terminal cluster. Fls. 3–4 cm. across; perianth segments to 2 cm. long, white with a strong reddish midvein; bracts pale brown, lance-shaped. Lvs. linear-acute, 2–4 cm. broad, V–shaped in section. Roots tuberous. Fr. 5–6 mm. obovoid with 2–5 shallow transverse ridges. △ Dry, stony places, hills, rocky ground. Apr.–June. Med. Eur. (except TR.) P. Pl. 163.

Lvs. rounded in section, less than $\frac{1}{2}$ cm. broad

1592. *A. fistulosus* L. HOLLOW-STEMMED ASPHODEL. A slender perenn. 20–60 cm., with rather few pale pinkish fls. in a simple or sparsely branched, lax inflorescence. Fls. *c.* 2 cm. across; perianth segments spreading, pale pink with a darker reddish or green midvein; bracts whitish. Lvs. linear-acute, only 1–3 mm. broad, semi-circular in section, more or less hollow; stems hollow. Roots fibrous, not swollen. Fr. globular, 4–6 mm., with 2–4 transverse ridges. △ Dry, sunny places, uncultivated ground. Apr.–May. Med. Eur. P. Pl. 163.

ASPHODELINE Like *Asphodelus*, but fl. stem leafy and perianth segments somewhat unequal, the lowest larger, narrower, and somewhat distant from the others. *c.* 4 sps.

1593. *A. lutea* (L.) Reichenb. KING'S SPEAR, YELLOW ASPHODEL. An erect perenn. 60–100 cm., with a large, dense, elongated, spike-like cluster 10–15 cm., of yellow fls., borne on a stout stem which is leafy to the inflorescence. Fls. *c.* $2\frac{1}{2}$ cm. across; perianth segments yellow with a green midvein, the lowest segment longest; bracts oval-acute, longer than the fl. stalks; spike elongating in fr. Lvs. numerous, densely clustered, linear stiff-pointed, triangular in section, smooth; fl. stem unbranched. Fr. globular, 1–$1\frac{1}{2}$ cm. △ Rocky and stony places. Apr.–May. South-East Eur. (except AL.) I.SU. Pl. 162.

1594. *A. liburnica* (Scop.) Reichenb. Distinguished from 1593 by the rough-margined, very narrow lvs. *c.* 1 mm. broad, and the leafless upper part of the fl. stem. Fls. yellow; inflorescence more lax; bracts small oval-triangular, shorter than the fl. stalks. A more slender perenn. 20–60 cm. △ Bushy places, field verges. June. South-East Eur. (except R.) I.

PARADISEA | **St Bruno's Lily** Fls. large, funnel-shaped, few, borne on leafless stems. Stamens and styles upcurved; anthers jointed and mobile on the filament. Perenns. with fibrous roots. 1 sp.

1595. *P. liliastrum* (L.) Bertol. ST BRUNO'S LILY. A slender perenn. with leafless stems bearing one or several pure white, sweet-scented, funnel-shaped fls. 4–5 cm. long, in a lax, one-sided cluster. Fls. short-stalked, often nodding; perianth segments oblong-lance-shaped, somewhat spreading above, but narrowed below into a rather slender tube; anthers and style upcurved; bracts longer than the fl. stalks. Lvs. all basal, grass-like, 20–50 cm. by 1–4 mm., about as long as the stem. Fr. ovoid acute. △ Mountain meadows, rocky places. June–Aug. P.E.F.CH.A.I. Pl. 163.

ANTHERICUM Fls. medium-sized, perianth divided almost to the base into 6 wide-spreading segments, each three-veined; anthers jointed to filament on the back and mobile. 3 sps.

1596. *A. liliago* L. ST BERNARD'S LILY. A slender perenn. with star-like, pure white fls., borne in a lax, elongated cluster on a simple or little-branched, leafless stem 30–60 cm. Fls. 3–5 cm. across; perianth segments narrow-elliptic 2 cm. long, wide-spreading up to twice as long as the stamens; bracts oval-lance-shaped, about half as long as the fl. stalks which are jointed below the middle. Lvs. all basal, in a rosette, linear-acute, 4–5 mm. broad, flat or grooved, as long as the fl. stem. Roots fibrous. Fr. oval, shallowly three-lobed. △ Hills, dry pastures. May–June. Cent. Eur. P.E.F.B.DK.S.I.YU.AL.GR.R.SU.: introd. N. Pl. 164.

A. ramosum L. Like 1596, but fls. smaller 2½ cm. across in a lax, branched, spreading cluster. Perianth about as long as the stamens; bracts about one-fifth as long as fl. stalks which are jointed almost at the base. Lvs. much shorter than the fl. stem. Widespread, except in Northern Europe.

Simethis mattiazzii (Vandelli) Saccardo (*S. planifolia* (L.) Gren. & Godron; *S. bicolor* (Desf.) Kunth). Western Europe to Italy.

HEMEROCALLIS | **Day-Lily** Fls. large, several in a somewhat flat-topped cluster. Perianth broadly funnel-shaped with segments spreading or recurved at the tips; stamens curved upwards; anthers mobile and attached to filament on their back. Fr. triangular in section. *c*. 2 sps.

1597. *H. fulva* (L.) L. DARK DAY-LILY. A rather robust perenn. with very large, dull orange-red, funnel-shaped fls. 8–10 cm. long, in an umbel-like cluster of 5–15 borne on a stout leafless stem ½–1 m. Perianth segments spreading outwards or recurved at the tips, the inner broader, with undulate margin; fls. not sweet-scented; stamens and style upcurved. Lvs. linear to 3 cm. broad, keeled, about as long as the fl. stem which has a few scale-like lvs. △ Unknown in the wild; commonly grown as an ornamental and often naturalized in ditches, damp woods, by riversides. June–Aug. Introd. Cent., South-East Eur. F.I.SU. *The buds and flowers are edible; the roots have been used in herbal remedies for the treatment of ulcers and tumours.*

H. lilioasphodelus L. (*H. flava* (L.) L.) PALE DAY-LILY. Like 1597, but fls. lemon-yellow, sweet-scented, in a cluster of 3–9; perianth 6–8 cm. long, the inner segments without undulate margin. Lvs. narrower, ½–1 cm. broad. Central and South-Eastern Europe. Pl. 164.

ALOE Stemless or stout-stemmed woody-based perenns. with fleshy, swollen, often spiny lvs. in a rosette. Fls. numerous in a spike or forming a branched inflorescence; perianth tubular, six-lobed; stamens included. Fr. a capsule or berry. *c*. 3 sps.

1598. *A. vera* L. Usually a stemless perenn. with many offsets, and with a terminal rosette of large, fleshy, spear-shaped, pale glaucous-green or pinkish lvs. with few short spiny teeth. Fls. yellow, pendulous, in a spike 60–90 cm. long, borne on a leafless stem; perianth 2½ cm. long, yellow with a greenish margin, deeply six-lobed. △ Native of Africa and Asia; naturalized in sandy and rocky places on the southern littoral. May–June. Introd. E., Sicily, GR.

GAGEA Low-growing bulbous perenns. 3–20 cm., with several small fls. in a lax terminal cluster. Perianth segments free or only slightly fused below, spreading in fl., usually yellow within and greenish without. *c*. 20 sps.

Fls. yellow
(*a*) *Basal lvs. flat or channelled*

1599. *G. lutea* (L.) Ker-Gawler YELLOW STAR-OF-BETHLEHEM. A slender perenn. 8–25 cm., with 1–7 yellow fls. in a lax terminal cluster subtended by 2 leafy bracts with ciliate margins. Perianth segments 1–1½ cm. long, bright yellow within and with a broad green band on the outside. Basal lv. solitary, hairless, broadly linear, 7–12 mm. wide, strongly three- to five-ribbed and with a hooded apex. Bulb solitary. △ Meadows, hedges, open woods. Apr.–June. Most of Eur. (except P.IRL.IS.).

1600. *G. arvensis* (Pers.) Dumort. Like 1599, but the paired leafy bracts are shaggy-haired and as long as or longer than the fls. and broader than the 2 linear, flattened and grooved, basal lvs. Fls. 1–12 in a lax cluster; perianth segments lance-shaped, 1½–2 cm. long, often hairy beneath; fl. stalks hairy, often flexuous. △ Fields, vineyards, orchards, fallow. Feb.–Apr. Most of Eur. (except P.IRL.GB.IS.N.SF.AL.).

(b) Basal lvs. half-cylindrical, hollow
1601. *G. fistulosa* (Ramond) Ker-Gawler Distinguished by its linear, usually paired basal lvs. which are *c.* 2 mm. broad, circular in section and somewhat fleshy and hollow. Fls. 1–5 in a flat-topped cluster, with hairy fl. stalks; bracts 2, unequal and up to 1 cm. broad, shorter than the fls. Bulbs 2, enclosed in a single scale. △ In mountains: pastures, alpine meadows. May–July. F.D.CH.A.I.GR.BG.R. Pl. 164.

Fls. white
G. graeca (L.) Dandy (*Lloydia g.*) GREEK LLOYDIA. Distinguished by its white fls. which are at first drooping then erect, in a lax cluster of 2–6. Perianth funnel-shaped, 1–1½ cm. long, segments white with purple veins outside, nectaries at base of segments absent, thus distinguishing it from *Lloydia*. Greece. Pl. 164.

ALLIUM | **Onion, Garlic, Leek** Fls. numerous, borne in an umbel which is enclosed before flowering in a papery sheath or *spathe* and which later splits into several segments to expose the young fls. Perianth segments free, often spreading; stamens 6, attached to the base of the perianth segments and anthers attached to filament on the back, mobile. Ovary three-celled, with a terminal style only slightly enlarged into a stigma. Fr. a capsule. Lvs. often smelling of onion or garlic when crushed; usually bulbous perenns. Placed by some botanists in the family Amaryllidaceae. *c.* 90 sps.

Lvs. linear and hollow
(a) Lvs. cylindrical in section
1602. *A. cepa* L. ONION. Fls. greenish-white, in a dense, numerous-flowered umbel 6–10 cm. across, borne on a robust stem, 60–120 cm., which is swollen below the middle. Perianth 4–5 mm. long, segments oblong-blunt; filaments of the inner stamens swollen below and prolonged above into 2 teeth on each side of the anthers; fl. stalks much longer than fls. Lvs. linear cylindrical, hollow, 6–8 mm. broad. Bulb globular, with brown, red, or white scales; bulbils absent. △ Native of Western Asia; cultivated throughout Eur. and sometimes self-seeding. June–Sept. *Cultivated since classical times and widely used as a vegetable and condiment. Onion juice is antiseptic and the roasted bulbs are used as a poultice.*

A. fistulosum L. WELSH ONION. Umbels with numerous yellowish-white fls. borne on stems which are swollen and hollow above the middle; fl. stalks about as long as fls. which are *c.* 1 cm. long. Bulb oblong, scales white then brown; bulbils present. Probably native of Asia; sometimes cultivated as vegetable.

1603. *A. schoenoprasum* L. (*A. sibiricum* L.) CHIVES. Fls. pale pink or violet-pink, in very dense globular umbels 2½–3½ cm. across, without bulbils, borne on a hollow, but not swollen stem 20–50 cm. Perianth segments 7–14 mm., spreading; stamens about half as long as the perianth; fl. stalks as long as the fls.; spathe with 1–3 lobes shorter than the umbel. Lvs. somewhat glaucous, linear-acute, cylindrical, hollow, 2–5 mm. broad. Bulbs oblong, occurring together in clumps. △ Meadows, damp rocks. June–July. Most of Eur. (except IRL.NL.TR.): introd. IS.N. *It is used for flavouring.* Pl. 165.

(b) *Lvs. half-cylindrical in section, flat or grooved above*

1604. *A. vineale* L. CROW GARLIC. Fls. pink or greenish-white in rather lax umbels with numerous stalkless bulbils, or heads entirely without fls. and with a dense cluster of bulbils only, borne on a slender stem 30–120 cm., leafy to about the middle. Fls. bell-shaped, *c.* 5 mm. long; stamens projecting, filaments with 2 projections longer than the anthers; fl. stalks much longer than the fls.; spathe 1, about as long as umbel, soon falling. Lvs. linear half-cylindrical, hollow, grooved above, *c.* 2 mm. broad. △ Fields, vineyards, waysides. June–Aug. All Eur. (except IS.).

A. sphaerocephalon L. ROUND-HEADED LEEK. Like 1604, but fls. dark reddish-purple in a dense globular umbel without bulbils; fl. stalks unequal, the outer about as long as fls., the inner longer; spathe usually 2, shorter than the umbel. Inner stamens with the 2 lateral projections shorter than the anther. Widespread, except in Northern Europe. Pl. 165.

1605. *A. oleraceum* L. FIELD GARLIC. Fls. dull pink, greenish or whitish, borne in a lax umbel with many dark stalkless bulbils; spathe 2, with very long narrow points, much longer than the umbel. Fls. bell-shaped; perianth segments 5–7 mm. blunt or with narrow points; stamens not projecting; filaments without lateral projections at apex; fl. stalks very unequal, the longest up to 3 cm. Lvs. glaucous, grooved above, 3–4 mm. broad, usually hollow, at least below. Fl. stem leafy to the middle, 30–80 cm. △ Fields, orchards, vineyards, uncultivated ground, waysides. July–Aug. Much of Eur. (except P.IRL.AL.GR.TR.): introd. IS.

Lvs. linear and solid, either cylindrical or flat or grooved

(a) *Stamens without lateral projections at the apex of the filament*

(i) *Stems leafy nearly to the middle*

1606. *A. flavum* L. YELLOW ONION. Fls. golden-yellow, numerous in a spreading umbel, without bulbils; spathe 2, long-pointed, unequal and longer than the umbel. Perianth segments blunt, 4–5 mm. long; stamens longer than the perianth, anthers yellow; fl. stalks unequal, very long. Lvs. thick, channelled, narrow linear, 1–1½ mm. broad, very smooth, hairless. Perenn. 20–50 cm. △ Dry, rocky places. June–Aug. South-East Eur. E.F.A.CS.H.I.SU. Pl. 165.

1607. *A. carinatum* L. Like 1606, but fls. purple-lilac and umbels usually with bulbils; perianth 5–7 mm. long; stamens much longer than perianth, anthers purple. Lvs. not thick, rough on the margin and on the veins beneath. △ Open woods, pastures. June–Aug. Much of Eur. (except P.IRL.IS.N.SF.): introd. GB. Pl. 165.

(ii) *Stems leafy only towards the base*

1608. *A. subhirsutum* L. A rather slender plant with white fls. in a lax, spreading umbel without bulbils, borne on a leafless, cylindrical-sectioned stem 20–50 cm. Fls. 1–1½ cm. across, perianth segments wide-spreading, at length reflexed; stamens shorter, anthers pink; spathe 3, shorter than the fl. stalks which are much longer than the fls. Lvs. 2–4, linear-acute, flat, ½–1 cm. broad, soft, with ciliate margin. △ Rocky, stony, and arid places. Apr.–May. Med. Eur. (except AL.TR.).

1609. *A. neapolitanum* Cyr. Distinguished from 1608, by its more robust three-angled fl. stem 30–60 cm., and its hairless lvs. Fls. numerous, pure white, cup-shaped, 1–2 cm. across; perianth segments blunt, spreading, shining; stamens shorter; fl. stalks longer than the fls.; spathe 1, broad, persistent, shorter than the fl. stalks. Lvs. 2–4, 1–1½ cm. broad. △ Fields, olive groves, grassy places. Mar.–May. Med. Eur. (except AL.) P. Pl. 165.

1610. *A. triquetrum* L. TRIQUETROUS GARLIC. Readily distinguished by the lax, drooping, one-sided umbel of white fls. borne on a thick, sharply three-angled stem 10–40 cm. Fls. rather few; perianth 1½–2 cm. long, bell-shaped, segments not wide-spreading, white within and with a green midvein on the outside, anthers shorter; fl. stalks longer than the fls.; spathe 2, soon falling. Lvs. 2–5, ½–1 cm. broad, strongly keeled below. △ Damp, shady places, streamsides. Mar.–May. P.E.F.I.GR. Pl. 165.

1611. *A. roseum* L. ROSE GARLIC. Fls. rose or violet, numerous in a rounded umbel with or without bulbils, borne on a cylindrical-sectioned stem 15–40 cm. Perianth bell-shaped, segments 10–12 mm. long, blunt, at length becoming papery and encircling fr.; fl. stalks longer than the fls.; spathe 2–4, persistent, shorter than the fl. stalks. Lvs. 2–4, linear-acute, 4–8 mm. broad, keeled below. Bulb with brown perforated scales; bulblets many. △ Fields, vineyards. Apr.–June. Med. Eur. P. Pl. 166.

1612. *A. nigrum* L. A robust perenn. with a large bulb and with 3–5 broad, flat, pointed lvs. 1½–6 cm. wide, with undulate margins. Fls. white or pale lilac with a green midvein, very numerous in a dense globular umbel *c.* 6 cm. across, borne on a robust cylindrical-sectioned stem to 80 cm. Perianth segments oblong-blunt, 6–8 mm. long, spreading; stamens shorter; fl. stalks 4–5 times as long as fls.; spathe 2–4, shorter than the fl. stalks. Fr. turning black. △ Cultivated ground, olive groves, vineyards. Apr.–June. Med. Eur. (except AL.) P.A.BG.

(*b*) *Stamens with 2 lateral projections at the apex of the filament, thus three-pointed*
A. sativum L. GARLIC. Fls. whitish or greenish, long-stalked, mixed with bulbils or some-times bulbils only present, forming a lax rounded umbel with 1 much longer pointed spathe. Stamens shorter than the perianth, the filaments of the inner stamens with 3 equal points. Lvs. flat, smooth, linear-acute, 4–8 mm. broad, very strong-smelling. Bulb-lets many, closely invested in white scales. Native of Asia; widely cultivated. *Cultivated since classical times and commonly used in Mediterranean countries as food and as a condiment. The juice is a mild antiseptic and is a digestive stimulant.*

1613. *A. scorodoprasum* L. SAND LEEK. Umbels lax, with rather few reddish-purple, purple-stalked fls. mixed with stalkless purple bulbils and borne on stiff stems 30–100 cm., which are leafy to the middle. Perianth bell-shaped *c.* 4 mm. long; stamens shorter than the perianth, filaments of inner stamens with lateral projections longer than the anthers; spathe 2, much shorter than the umbel. Lvs. linear-acute, 3–5 mm. wide, flat, with rough margin, very strong-smelling. Bulb with many bulblets. △ Vineyards, sandy places. June–Aug. Most of Eur. (except P.E.B.IS.AL.TR.): introd. IRL.

1614. *A. rotundum* L. Like 1613, but umbel dense, 2½–4 cm. across, without bulbils; fls. numerous, dark purple, on short stalks, the stalks of the outer fls. shorter than the fls. Perianth *c.* 5 mm. long, segments blunt and with a rough keel; stamens shorter, the fila-ments of the inner with long lateral projections. Lvs. flat, 4–6 mm. broad, margin smooth. △ Fields, vineyards. June–Aug. Cent. Eur. (except PL.), South-East Eur. (except AL.) E.F.I.SU.

1615. *A. ampeloprasum* L. WILD LEEK. Umbels large, compact, globular, 7–10 cm. across, with numerous purple or whitish fls., usually without bulbils, borne on a stout cylindrical stem to 2 m., which is leafy to the middle. Perianth 4–5 mm. long, segments blunt and keeled; stamens longer, the filaments of inner stamens with long twisted lateral points, anthers yellow; fl. stalks nearly equal, much longer than the fls.; spathe 1, soon falling.

Lvs. linear-acute, 5–7 mm. or more broad, with a rough keel and margin. Bulb with bulb-lets enclosed by scales. A variable sp. △ Dry places, hedges, banks. June–July. Med., South-East Eur. P.CH.A.H.: introd. PL.CS. Pl. 166.

A. porrum L. LEEK. Like 1615, but lvs. smooth, broad, keeled and bulb without bulblets; spathe green and persistent; stem much swollen below. Possibly a cultivated variety of 1615, but origin unknown; widely cultivated, sometimes self-seeding.

Lvs. elliptic- to lance-shaped, stalked, blade flat

1616. *A. ursinum* L. RAMSONS. Fls. white in a somewhat spreading, lax, flat-topped umbel, without bulbils, and borne on a triangular-sectioned or two-angled stem and with 2 ⟍ broad stalked lvs. near the base. Perianth 8–10 mm. long, segments spreading, lance-shaped-acute; stamens shorter; fl. stalks equal, as long as or longer than the fls.; spathe 2–3. Lvs. usually 2, bright green, oval-elliptic, 10–25 cm. by 4–7 cm. broad, with a long twisted lv. stalk ensheathing the stem at the base. Fl. stem 10–45 cm.; plant often growing in masses. △ Damp woods, shady places. Apr.–June. All Eur. (except P.IS.TR.). Pl. 166.

1617. *A. victorialis* L. ALPINE LEEK. Fls. greenish-white, becoming yellowish, in a dense globular umbel, borne on a stiff, cylindrical stem 30–60 cm., which is leafy to the middle. Perianth bell-shaped 4–6 mm. long, segments blunt, at length spreading or reflexed; stamens longer; spathe 1–2, as long as the umbel. Lvs. oblong-elliptic, short-stalked. Bulb oblong, enclosed in dense, brown, netted fibres. △ In mountains: woods, rocks. June–July. Much of Eur. (except North Eur., IRL.GB.B.NL.TR.).

LILIUM | **Lily**   Fls. large, solitary or in a cluster. Perianth funnel-shaped with free, spreading or recurved segments each with a groove-like nectary at the base; stamens 6, anthers mobile, attached to filament on the back; style long, with a three-lobed, knob-like stigma. Bulb formed by many overlapping fleshy scales and without a thin investing scale. *c.* 9 sps.

Fls. pink or purple

1618. *L. martagon* L. MARTAGON LILY. Fls. pink or pale purple, nodding, *c.* 4 cm. across, in a loose almost leafless cluster of 3–10 fls. Perianth segments strongly recurved from the middle, with dark purple raised projections on the inner face, hairy outside; stamens and style long-projecting. Lvs. oval-lance-shaped, spreading, in whorls of 4–10, the uppermost lvs. alternate, smaller. A rough, hairless perenn. with a red-spotted, erect, leafy stem 1–1½ m. △ Woods, thickets, pastures in the mountains. June–July. Much of Eur. (except North Eur.): introd. GB.DK.N.S.SF. Pl. 167.

Fls. yellow, orange or red

1619. *L. pyrenaicum* Gouan YELLOW TURK'S-CAP LILY. Fls. large, bright yellow, nodding, in a loose leafy cluster of 2–8 on a robust stem which is leafy its whole length. Perianth segments strongly recurved, spotted with blackish dots on the inner face; style stout. Lvs. numerous alternate and overlapping, linear-lance-shaped, ½–1 cm. broad, margin rough. An unpleasant-smelling perenn. 40–80 cm. △ Meadows, clearings. June–July. E.F.: introd. GB. Pl. 167.

L. pomponium L. RED LILY. Like 1619, but fls. bright red or at their palest orange-red; style slender. Lvs. linear, the upper only 1–2 mm. broad; stems almost leafless below the fl. cluster. France and Italy. Pl. 166.

L. carniolicum Bernh. Distinguished by its solitary, or rarely paired, nodding, bright vermilion, orange or golden-yellow fl. with strongly recurved perianth segments to 5 cm. long. Lvs. many, alternate, oblong-lance-shaped to 2 cm. broad, the upper gradually narrowing to 3–6 mm. broad. A variable sp. Eastern Alps.

1620. *L. bulbiferum* L. ORANGE LILY. Fls. erect, bright orange with black spots, solitary or 2–5, in a false umbel and surrounded by a whorl of 3–5 lvs. Perianth widely funnel-shaped, the segments spreading upwards but not recurved, hairy on the outside; anthers red; fl. stalks short, woolly-haired. Lvs. numerous, linear-lance-shaped, with or without bulbils in their axils. An erect, leafy, stout-stemmed perenn. 40–60 cm. △ Mountain pastures, woods, rocky places. June–July. Cent. Eur. F.I.YU.SU.: introd. R. Pl. 166.

Fls. white
1621. *L. candidum* L. MADONNA LILY. Fls. very large and funnel-shaped, very sweet-scented, brilliant white, at first erect then spreading, in a lax, few-flowered, leafless cluster. Perianth to 8 cm. long, segments spreading outwards above, but not strongly recurved; style slender, longer than the stamens; bracts small. Lvs. many, linear-lance-shaped, hairless. An erect, leafy perenn. to 1½ m. △ Rocky places in the mountains; often grown for ornament and sometimes naturalized. May–July. GR.: introd. Med. Eur. (except AL. TR.) A.H.BG.

FRITILLARIA | Fritillary Fls. solitary or few, bell-shaped, nodding, with broad perianth segments, which are often straight or incurved at the tip, often chequered, and with a large glistening nectary at base of each segment. Stamens 6, anthers attached by their back to the filament, not mobile; style entire, or two- to five-lobed. Bulbs with few swollen scales and with a thin investing scale. *c.* 25 sps.

1622. *F. meleagris* L. SNAKE'S HEAD, FRITILLARY. A strikingly beautiful bulbous perenn., usually with solitary, broad bell-shaped fls. which are chequered with dark or pale reddish-purple squares, and with a glistening 'tear-drop' gland at the base of each segment within. Perianth 3–5 cm. long, segments blunt, incurved at the tips; perianth rarely white. Lvs. 4–5 alternate, the upper linear, grooved, 2–5 mm. broad; fl. stems 20–50 cm. Fr. globose, triangular-sectioned. △ Damp meadows. Apr.–May. Most of Eur. (except P.IRL.IS.AL.GR.TR.BG.): introd. North Eur. *The bulbs are poisonous.* ·Pl. 167.

1623. *F. pyrenaica* L. PYRENEAN SNAKE'S HEAD. Like 1622, but stem lvs. more numerous 4–10, broader, 5–8 mm. across, and borne on the upper two-thirds of the stem. Fls. bell-shaped, 3–4 cm. long, brown or yellowish-purple, obscurely chequered; perianth segments unequal, the 3 inner broader above and with a short point, the 3 outer narrower oblong, all slightly outcurved at the tips. A perenn. 20–50 cm. △ Pastures in the mountains. June–July. E.F.

F. tenella Bieb. Distinguished by the very narrow lvs. only 1–4 mm. broad, which are mostly opposite or whorled. Fls. small, 2–3 cm. long, the outer perianth segments dull purple or greenish-yellow flushed with purple and often chequered, the inner segments yellow with purple chequering. Glandular 'tear-drops' oval, small. South-Eastern Europe.

TULIPA | Tulip Fls. usually solitary, erect, bell-shaped or segments spreading, without basal glands, and segments soon falling. Anthers attached to filament by their bases; style absent. Bulb with few, thick, fleshy scales, closely encircled by thin, brown, papery scales. Tulips have long been grown for ornament in Europe and some cultivated forms have become naturalized, particularly in South-East Europe, and the whereabouts of the wild parents are now no longer known. *c.* 25 sps.

Filaments of stamens hairy at their base
1624. *T. sylvestris* L. WILD TULIP. Fls. bright yellow on the inside and yellow, green, or flushed with red externally. Perianth segments opening widely in the sun, elliptic-acute 3–5 cm., hairy within at least at the base, without a black basal spot; anthers yellow,

shorter than the filaments which are hairy at their base; buds drooping. Lvs. mostly 3, linear 15–30 cm. by 5–12 mm., borne at the base of the hairless 30–60 cm. stem. A very variable sp. △ Fields, bushy places, cultivated ground, in mountains. Apr.–May. Med., South-East Eur. (except TR.) SU.: introd. North, Cent. Eur. GB.B.NL.

1625. *T. australis* Link Like 1624, but fls. smaller, bell-shaped, 2–3 cm. long; perianth segments narrow lance-shaped, pointed, yellow and hairless within and reddish outside, buds slightly nodding or erect. Lvs. linear, usually 2. A perenn. 15–30 cm. △ Uncultivated fields, meadows, rocks, in mountains. Apr.–July. Med. Eur. (except TR.) P.CH.BG. Pl. 168.

Filaments of stamens hairless at their base

1626. *T. oculus-solis* St. Amans Fls. large, 6–8 cm. long, scarlet on the inside with a black patch surrounded by a yellow band at the base of each perianth segment, yellowish outside, the inner segments elliptic-acute, the outer scarcely longer, but long-pointed. Stamens with blackish, hairless filaments. Lvs. 3–4, the lowest broadly oval-lance-shaped, the uppermost lance-shaped. Bulb with outer thin scales densely hairy within. △ Native of Asia Minor; often cultivated and sometimes naturalized in Med. Eur. Apr. Pl. 168.

1627. *T. clusiana* Vent. Fls. 3–6 cm.; perianth segments all white within, the inner segments shorter and blunt, white with a violet base outside, the outer segments longer and acute, bright pink with a white margin outside. Stamens dark violet, filaments hairless. Lvs. 4–5, linear to linear-lance-shaped, glaucous, grooved. Bulb with outer thin scales densely hairy within. △ Fields. Apr. Probably native of Western Asia: introd. P.E.F.I.GR.

ERYTHRONIUM | Dog's Tooth Violet   Fls. usually solitary, nodding; perianth segments free and soon strongly reflexed from near the base, exposing the stamens and stigma. Lvs. 2. 1 sp.

1628. *E. dens-canis* L. DOG'S TOOTH VIOLET. Fl. solitary, nodding, bright pink, 3 cm. long, with segments lance-shaped and abruptly turned back from near the base, revealing the forward-projecting bluish anthers and three-lobed stigma. Lvs. 2, nearly opposite, elliptic-lance-shaped, stalked, curiously spotted with reddish-brown or sometimes with white, borne near the base of the stem. An erect, hairless perenn. with a solitary fl. stem 10–30 cm.; bulb oblong. △ In mountains: meadows, heaths, woods. Mar.–May. Med., South-East Eur. (except GR.) P.CH.A.PL.CS.H.SU. Pl. 168.

LLOYDIA   Fls. white, tubular, erect, usually solitary; perianth segments persisting, free, blunt; anthers attached to the filament at their base. Stems leafy; bulb with brown scales, much longer than the bulb. *c.* 2 sps.

1629. *L. serotina* (L.) Reichenb. LLOYDIA. A delicate perenn. usually with a solitary, tiny, white, erect, funnel-shaped fl. *c.* 1 cm. long, borne on a slender stem 5–15 cm., with several narrow lvs. Perianth segments white with purplish veins on the outside, with a small transverse nectary at the base within. Stem lvs. 2–4, linear-lance-shaped, short, basal lvs. thread-like 15–25 cm. △ In mountains: rocky places, pastures. June–Aug. Cent. Eur. (except H.) GB.F.I.BG.R.SU. Pl. 164.

URGINEA | Sea Squill   Fls. rather small, numerous, in a long, leafless, spike-like inflorescence. Perianth narrow-bell-shaped, segments blunt; anthers attached to the filament on the back. Lvs. all basal; bulbs very large. *c.* 3 sps.

1630. *U. maritima* (L.) Baker SEA SQUILL. Fl. spikes appearing in autumn with very numerous, white, spirally arranged fls. in a cylindrical cluster 30 cm. or more long, borne on long, leafless, unbranched stems 1–1½ m., arising direct from the bare ground. Perianth *c.* 8 mm. long, segments white with green or purple midvein; anthers greenish, shorter; fl. stalks spreading, and 2–3 times as long as the linear-lance-shaped bracts. Lvs. very large, appearing only after flowering and persisting into the following summer, broadly lance-shaped, 15–30 cm. by 2½–10 cm., rather fleshy, smooth, flat, shining, and growing from extremely large, often partly exposed bulbs 10–15 cm. across. △ Dry, rocky hills, stony places. Aug.–Oct. Med. Eur. (except TR.) P. *The bulbs have been used medicinally since classical times; they are somewhat poisonous. A related North African species is a source of Red Squill rat poison.* Pl. 169.

SCILLA | Squill Fls. blue or lilac with darker midveins, in many- or few-flowered clusters; perianth segments free, usually spreading; stamens attached to the base of the perianth segments, anthers attached to the filaments on the back. Bulbous perenns. with basal lvs. and leafless stems. *c.* 22 sps.

Fls. numerous, 50–100

1631. **S. hyacinthoides** L. Fls. very numerous, blue-violet on long, spreading, blue-violet stalks and forming a conical-cylindrical, spike-like cluster borne on a stout, erect, leafless stem ½–1 m. Perianth segments 7–9 mm. long, spreading; anthers violet; fl. stalks 3–5 times longer than the fls.; bracts very short, blunt. Lvs. 10–12, linear-lance-shaped, 1½–3 cm. broad, flat with minutely ciliate margin, shorter than the fl. stem. Bulb large. △ Fields and rocky places. Apr.–May. Med. Eur. (except AL.TR.) P.

1632. **S. peruviana** L. Distinguished from 1631 by its broad hemispherical inflorescence which is at first broader than long, and its lance-shaped bracts which are nearly as long as the fl. stalks; anthers yellowish. Lvs. numerous, 4–6 cm. broad, channelled, spreading, longer than the stout, leafless fl. stem which is 20–50 cm. △ Damp, fertile ground. Mar.–May. P.E.F.I. Pl. 168.

Fls. several, to 20

(a) Spring flowering

1633. **S. verna** Hudson SPRING SQUILL. A slender perenn. with a rather dense and often somewhat flat-topped cluster of 2–12 blue-violet fls. borne on a leafless stem 5–25 cm. Perianth segments 5–8 mm. long, acute, spreading; anthers blue-violet or yellowish; lower fl. stalks much longer than the upper; bracts bluish or whitish, usually longer than the fl. stalks. Lvs. 3–6, linear-blunt, 3–6 mm. broad, grooved, shorter than the stem and appearing before the fls. Bulb with white scales. △ Heaths, grassy places, woods. Apr.–June. West Eur. (except B.NL.) N.

1634. **S. italica** L. Distinguished from 1633 by the more numerous blue fls. in a dense conical cluster; fl. stalks more or less equal in length and each subtended by 2 unequal bracts. Perianth 5–6 mm. long. Lvs. broader, 4–8 mm. wide. Bulb scales white or black. A taller perenn. 15–40 cm. △ Woods, hills, rocky places. Mar.–May. P.E.F.CH.A.I.

S. liliohyacinthus L. PYRENEAN SQUILL. Distinguished from 1634 by the large yellowish bulb, with overlapping scales and the numerous, much broader lvs. 1½–3 cm. wide. Fls. blue; perianth 1 cm. long; anthers bluish; bracts single, as long as the unequal fl. stalks; fl. stem to 40 cm. Pyrenees.

1635. **S. bifolia** L. ALPINE SQUILL. Distinguished by its opposite pair of shiny, spreading, lance-shaped lvs. with long sheathing bases, which appear to arise midway up the stem. Fls. bright blue, 2–8 in a loose cluster, on longer ascending fl. stalks; bracts absent; perianth segments oblong-elliptic, blunt, 6–9 mm., spreading; anthers shorter, violet;

fls. rarely white or pink. Lvs. rarely 3–5; fl. stem to 20 cm. △ Lowlands and in mountains: woods, thickets, meadows. Mar.–Aug. Cent., South-East Eur. E.F.B.NL.I.SU. Pl. 170.

(b) Autumn flowering

1636. *S. autumnalis* L. AUTUMN SQUILL. Fl. stems often several, appearing before the lvs. and bearing 4–20 small lilac fls. in a dense flat-topped cluster which later elongates. Perianth segments 4–6 mm. long, spreading; anthers lilac; fl. stalks as long as or longer than the fls.; bracts absent. Lvs. 4–8, narrow linear and slightly channelled. Fl. stem leafless, 12–30 cm. △ Dry, grassy, and rocky places. Aug.–Oct. Med., South-East Eur. P.GB.D.CH.H.SU. Pl. 168.

DIPCADI Like *Scilla*, but fls. yellowish; perianth funnel-shaped, segments fused below, the outer recurved, the inner straight. Bulbous perenns. with lvs. all basal. 1 sp.

1637. *D. serotinum* (L.) Medicus (*Uropetalum s.*) Fls. yellow or pale brownish, rarely orange-red, drooping, in a lax, few-flowered, and often one-sided cluster borne on an erect, leafless stem 10–40 cm. Perianth 12–15 mm. long, segments fused for about a quarter their length, the outer 3 spreading and recurved at the tip, the 3 inner straight or somewhat incurved; stamens included; fl. stalks shorter than the fls. and the lance-shaped bracts. Lvs. 3–5, all basal, narrow linear and grooved, shorter than the fl. stem. △ Sandy and rocky places, in mountains. Mar.–June. P.E.F.I.

ENDYMION | Bluebell Like *Scilla*, but fls. tubular-bell-shaped with the perianth segments united at the base and the stamens inserted about the middle of each segment; bracts subtending fls. 2. Bulb renewed annually. *c.* 5 sps.

1638. *E. non-scriptus* (L.) Garcke (*Scilla n.*) BLUEBELL, WILD HYACINTH. Fls. bright blue, at length nodding and forming a curved, one-sided, sweet-scented cluster of 4–20 fls. borne on a smooth, leafless stem 20–50 cm. Perianth $1\frac{1}{2}$–2 cm. long, tubular at the base and with tips of the segments curved outwards or reflexed, rarely white or pink; anthers yellowish; fl. stalk shorter than the perianth, elongating in fr.; bracts 2, unequal, bluish. Lvs. linear, 7–15 mm. wide, smooth, shining. △ Woods, thickets, banks. Apr.–May. West Eur.: introd. D.CH.I. Pl. 170.

E. hispanicus (Miller) Chouard (*Scilla h.*) SPANISH BLUEBELL. Like 1638, but fls. erect or ascending in a conical cluster, not one-sided; perianth segments spreading outwards from the base and thus fls. broadly bell-shaped; anthers blue. Lvs. broader, $1\frac{1}{2}$–$3\frac{1}{2}$ cm. wide. Portugal to Italy. Pl. 169.

ORNITHOGALUM Fls. white or pale yellow, usually with a green band on the outside of each segment, several or many in an elongated or more or less flat-topped terminal cluster. Perianth bell-shaped or segments spreading in a star; segments free; anthers attached on the back to the filament; style very slender. Bulbous perenns. with basal lvs. and leafless fl. stems. *c.* 23 sps.

Fls. in a lax flat-topped cluster; fr. stalks very unequal

1639. *O. umbellatum* L. STAR-OF-BETHLEHEM. Fls. white with a broad green median band on the outside of each perianth segment, 5–30 in a spreading flat-topped cluster with fl. stalks very unequal, the lowest to 10 cm. long. Perianth segments elliptic-lance-shaped $1\frac{1}{2}$–2 cm. long, spreading wide in the sun; bracts lance-shaped. Lvs. 6–9, linear, 2–8 mm. broad, grooved above and with a conspicuous white band, hairless; fl. stem 10–30 cm. Bulb white, with numerous leafy bulblets. △ Cultivated ground, grassy places, groves. Apr.–June. Most of Eur. (except IRL.IS.): introd. North Eur. Pl. 170.

Fls. in a narrow cluster or elongated spike; fl. stalks equal

1640. *O. pyrenaicum* L. BATH ASPARAGUS. Fls. numerous, small, greenish-white and becoming yellowish on drying, or yellow, in a lax, elongated, cylindrical, spike-like cluster borne on a slender leafless stem ½–1 m. Perianth segments 6–10 mm. long, oblong-linear, spreading; stamens shorter; bracts thin, whitish, long-pointed, shorter than the spreading fl. stalks and the fl. cluster when in bud. Lvs. glaucous, linear, 3–12 mm. broad, without a white band and often withered at flowering. △ Woods and meadows. May–July. Much of Eur. (except North Eur. IRL.NL.PL.TR.R.). Pl. 169.

1641. *O. pyramidale* L. Like 1640, but fls. pure white within and with a greenish band outside, not yellowing when dry; perianth segments lance-shaped 10–12 mm. long. Bracts 1–2 cm., about as long as the fl. stalk and projecting well beyond the young fl. buds. Lvs. persisting during flowering, up to 1 cm. broad; fl. stem to 80 cm. △ Grassy places, fields. May–July. Med., South-East Eur. P.A.CS.H. Page 504.

1642. *O. nutans* L. DROOPING STAR-OF-BETHLEHEM. Fls. rather large, drooping, few, whitish within and with a broad silvery-greenish band outside, 2–13 in a lax, one-sided, ovoid cluster. Perianth segments oblong-elliptic 2–3 cm., blunt or acute; stamens much shorter, filaments flattened and with 2 points at the apex; bracts long-pointed, whitish, much longer than fl. stalks which are at first erect and then arched in fr. Lvs. linear, 8–15 mm. broad, with a broad, whitish, channelled band. Fl. stem 30–60 cm. △ Fields, vineyards, grassy places, in mountains. Apr.–May. Med. Eur.: introd. elsewhere to much of Eur.

HYACINTHUS | **Hyacinth** Fls. several or numerous in a spike-like cluster; perianth tubular, fused below to one-third and with erect or recurved segments; stamens 6, attached to the base of the tube and with their filaments fused for most of their length to the segments. Bulbous perenns., with leafless stems, *c.* 3 sps.

1643. *H. orientalis* L. HYACINTH. Fls. rather large, blue, and sweet-scented, or white or pink in cultivated forms, 2–15 in a lax, one-sided, spike-like cluster borne on a stout stem to 30 cm. Perianth to 2½ cm., tubular, swollen at the base and with segments with recurved tips; fl. stalks *c.* ½ cm.; bracts tiny, triangular. Lvs. 4–6, shining, broadly linear, 1–2 cm. wide. Bulb large, outer thin scales pink, violet, or white. △ Widely grown for ornament and sometimes escaping. Mar.–May. YU.: introd. F.I.A.H.GR.BG.SU. Pl. 171.

H. amethystinus L. Distinguished from 1643 by its much more slender habit and smaller, scentless, bright blue fls. 7–10 mm. long, borne in a very lax one-sided cluster of 3–12. Bracts lance-shaped, the upper longer than the fl. stalks; lvs. thread-like. Pyrenees, Balkans.

BELLEVALIA Fls. many, bell-shaped or tubular, erect or spreading, in a lax, terminal, spike-like cluster. Perianth fused to about two-thirds its length or more, and with 6 erect or spreading lobes, not constricted at the mouth; stamens attached to the throat of the tube. Bulbous perenns. with leafless stems. *c.* 10 sps.

1644. *B. romana* (L.) Reichenb. (*Hyacinthus r.*) Fls. small, whitish or dull violet, always erect, in a lax, somewhat conical cluster which later elongates, borne on a slender leafless stem 20–50 cm. Perianth 6–10 mm. long, free part of segments nearly as long as the fused lower part; anthers violet, included; fl. stalks erect, as long as or a little longer than the fls.; bracts minute. Lvs. 4–5, linear ½–1½ cm. broad, shining, margin smooth. Fr. globular. △ Damp meadows and fields. Apr.–May. Med. Eur. (except TR.). Page 504.

B. ciliata (Cyr.) Nees (*Hyacinthus c.*) Differing from 1644 by its drooping fls. borne on very long fl. stalks, which lengthen to 8–10 cm., becoming rigid and spreading horizontally in fr. Fls. 8–10 mm. long, dull purple with green tips. Lvs. 1½–3 cm. broad, with margin densely ciliate. Fr. oblong 1½ cm. Italy to Greece.

MUSCARI | Grape-Hyacinth Fls. many, usually in a dense terminal cluster. Perianth flask-shaped, oval, or globose, constricted at the mouth and with 6 very short tooth-like lobes. Fls. short-stalked, usually spreading or nodding, the uppermost often smaller, more erect and sterile; bracts absent. A bulbous perenn. with leafless stems. *c*. 20 sps.

Fls. in a lax, spike-like cluster; terminal sterile fls. long-stalked
1645. *M. comosum* (L.) Miller TASSEL HYACINTH. A striking plant with an erect tuft of bright violet-blue, long-stalked, sterile fls. borne above a dense cluster of dark blue short-stalked buds, and at the base spreading, brownish-green mature fls. Mature fertile fls. 5–7 mm. long, tubular-bell-shaped on spreading stalks as long as or longer than the fls. Lvs. broadly linear, ½–1½ cm. wide, channelled, smooth and shining. Fl. stem 30–60 cm.; fr. cluster lengthening to 15–30 cm. △ Rocky ground, cornfields, vineyards, olive groves, cultivations. Apr.–July. Med., South-East Eur. P.CH.A.PL.CS.H.SU.: introd. DK.D. Pl. 169.

Fls. in a dense compact cluster; terminal sterile fls. short-stalked
1646. *M. commutatum* Guss. DARK GRAPE-HYACINTH. Distinguished by its dense oval cluster of dark blackish-blue, sweet-scented fls. with 'pinched-in' tooth-like lobes of the same colour. Perianth of fertile fls. 5–6 mm. long, oval-oblong, five-angled at the mouth; sterile fls. usually few, smaller, paler blue and very short-stalked. Lvs. linear, channelled, flaccid, as long as or longer than the fl. stem, which reaches 20 cm. △ Grassy places, dry hills. Mar.–Apr. I.YU.GR.R. Pl. 169.

1647. *M. atlanticum* Boiss. & Reuter (*M. racemosum* auct.) GRAPE-HYACINTH. Like 1646, but fls. dark blue with white, outcurved lobes. Perianth ovoid, 4–5 mm. long; sterile fls. paler. Lvs. linear, very narrow ½–3 mm., semi-cylindrical; fl. stems 10–30 cm. △ Cultivations, vineyards, olive groves. Mar.–May. Most of Eur. (except North Eur. IRL.NL.): introd. D.

M. neglectum Guss. Like 1647, but fls. 6–8 mm. long; lvs. broader, 3–5 mm. Southern Europe. Pl. 170.

1648. *M. botryoides* (L.) Miller SMALL GRAPE-HYACINTH. Distinguished by its almost globular fls. only 3–4 mm. long, which are pale blue or violet, with white lobes. Fl. cluster conical, 8–12 mm. broad, sterile fls. paler. Lvs. 2–4, broadening towards the apex to 7 mm., shorter than fl. stem which is 10–30 cm. △ Meadows, fields, woods. Mar.–May. Cent., South-East Eur. E.F.B.NL.I.: introd. North Eur. Pl. 170.

ASPARAGUS | Asparagus Small shrubs or herbaceous perenns. without normal lvs., but lvs. reduced to papery scales and with green, needle-like branchlets taking over the function of lvs; branchlets in clusters or solitary. Fls. often one-sexed, solitary, axillary; perianth small, bell-shaped, six-lobed; stamens 6. Fr. a berry. *c*. 10 sps.

Stems herbaceous only, not spiny; fr. red
1649. *A. officinalis* L. ASPARAGUS. A herbaceous perenn. with a smooth, erect, hairless main stem ½–1½ m., much-branched above, and with very numerous needle-like branchlets *c*. 1 cm. long, in clusters of 3–8, giving the whole plant a delicate feathery appearance. Fls. greenish or yellowish, axillary, solitary or paired, on stalks jointed near the middle; perianth bell-shaped 3–8 mm. long. Lvs. absent; scales on main stem whitish, *c*. ½ cm. long, on branchlets *c*. 1 mm. long. Fr. red, globular 5–8 mm. A variable sp.; some forms have procumbent stems. △ Hedges, waste places, dunes, cliffs; widely cultivated and

often naturalized. June–July. All Eur. (except IS.AL.). *It has been used as a vegetable since classical times. It is cultivated for its young shoots which are often forced and blanched. Used medicinally in the past.*

A. tenuifolius Lam. Like 1649, but needle-like branchlets very slender in clusters of 10–25 and young shoots much more slender. Fl. stalks with the joint much above the middle and about 1 mm. from the fl. Fr. larger *c.* 1 cm. across. Mediterranean and South-Eastern Europe. Page 504.

Stems woody, spiny; fr. black
1650. *A. acutifolius* L. Main stems climbing to 1 m. or more, with much-branched, flexuous and intertwined, whitish stems, and with a rigid cluster of 4–12 angular, sharp-pointed, glaucous branchlets spreading in a star. Scales of main branches developing into robust spines. Fls. solitary or paired, yellowish-green, globose *c.* 3 mm., sweet-scented; fl. stalks short, jointed near the middle. Fr. becoming black. △ Dry places, rocks, hedges. Apr.–June. Med. Eur. P.BG. Pl. 168.

A. aphyllus L. Stems robust, rough spiny, erect, much-branched, and branchlets in clusters of 2–6, each ½–1 cm. long and with long spiny points. Fls. solitary or clustered, greenish; fl. stalks jointed above the middle. Mediterranean region and Portugal.

A. albus L. An erect shrubby plant to 1 m. with robust white stems with straight and spreading spines. Branchlets in clusters of 8–13, soft slender, soon falling and leaving a naked spiny main stem. Fls. clustered, white, very sweet-scented, bisexual. Fr. black, with 1–2 seeds. Portugal to Italy.

RUSCUS | Butcher's Broom Small shrubs without normal lvs., but with flattened, leaf-like branchlets bearing a central cluster of tiny fls. on one face. Plants one-sexed. Perianth with 6 free segments; stamens 3, fused into a short tube. Fr. a one-seeded berry. 3 sps.

1651. *R. aculeatus* L. BUTCHER'S BROOM. A dense, dark green shrub with thick, oval, rigid, and spiny-pointed false-lvs., *c.* 2½ cm. long, borne on green, ribbed, much-branched stems 25–80 cm. Fls. greenish, *c.* 3 mm. across, 1–2, subtended by minute brown bracts and borne on the upper surface of the false-lvs. Lvs. scale-like, papery, *c.* 5 mm. Fr. a globular red berry *c.* 1½ cm. △ Woods, bushy places, dry hills. Jan.–Apr. West Eur. (except IRL.NL.), South-East Eur. CH.H.I. *The young shoots which resemble asparagus are edible; the dried branches can be used as a broom.*

R. hypoglossum L. LARGE BUTCHER'S BROOM. Like 1651, but false-lvs. leathery and flexible, and not rigid or spiny-pointed, much larger 5–9 cm. long, elliptic-lance-shaped; stems simple or little-branched, to 40 cm. Fls. 3–5, borne on the upper surface of the false-lvs. and subtended by a green leafy bract. Fr. scarlet, larger to 2 cm. Mediterranean and South-Eastern Europe. Page 504.

MAIANTHEMUM | May Lily Perenns. with fl. stems usually with 2 lvs. and with several tiny fls. in a terminal cluster; perianth and stamens 4; ovary two-celled; fr. a berry. 1 sp.

1652. *M. bifolium* (L.) Schmidt MAY LILY. A delicate perenn. with a long creeping rhizome and slender, erect, unbranched stems 8–15 cm., with a rather dense terminal cluster of 8–15 tiny white fls. Perianth segments 4, spreading, 1–2 mm. long; fls. short-stalked; bracts minute. Stem lvs. 2, stalkless or short-stalked, 3–6 cm., oval with a heart-shaped base and pointed apex; basal lv. long-stalked, usually withering before flowering; stem

1. *Streptopus amplexifolius* 1653
3. *Bellevalia romana* 1644
5. *Asparagus tenuifolius* [1649]

2. *Ruscus hypoglossum* [1651]
4. *Ornithogalum pyramidale* 1641

with stiff white hairs above. Berry red, globular, *c.* 6 mm. △ Woods. May–July. Much of Eur. (except P.E.IRL.IS.AL.GR.TR.BG.).

STREPTOPUS Herbaceous perenns. with leafy fl. stems; fls. 1 or 2 on bent stalks and borne in the axils of the upper lvs. Perianth bell-shaped, segments 6, with tips at length recurved; stamens 6, included. Fr. a many-seeded berry. 1 sp.

1653. *S. amplexifolius* (L.) DC. An erect, flexuous, simple or little-branched, herbaceous perenn. 30–80 cm., with regular alternate lvs. and small, nodding, axillary, greenish-white fls. *c.* 1 cm. long, borne on slender, jointed, and conspicuously bent fl. stalks. Lvs. many, oval-heart-shaped, long-pointed, clasping the stem, glaucous above; stem scaly below; roots fibrous. Berry red. △ In hills and mountains: woods, damp rocks. June–July. E.F.D.CH.CS.A.H.I.YU.BG.R.SU. Page 504.

POLYGONATUM | Solomon's Seal Herbaceous perenns. with leafy stems and with fls. usually in a few-flowered cluster, or solitary, borne in the axils of the alternate or whorled lvs. Perianth tubular, fused for most of its length and with 6 short tooth-like lobes; stamens included, attached to the middle of the tube. Plants with thick, white, creeping rhizomes. *c.* 6 sps.

Lvs. alternate, elliptic or oval

1654. *P. odoratum* (Miller) Druce (*P. officinale* All.) SWEET-SCENTED SOLOMON'S SEAL. Fls. white with green tips, sweet-scented, solitary or paired, borne on slender pendulous stalks from the axils of the lvs. and forming a one-sided, slightly arched cluster. Perianth cylindrical, 18–22 mm. long; filaments hairless. Lvs. oval, 5–10 cm., stalkless, alternate and more or less two-ranked, somewhat glaucous beneath; stems angled above. Berry *c.* 6 mm., bluish-black. △ Woods, rocky places. May–June. Most of Eur. (except IRL.NL. IS.TR.). Pl. 171.

1655. *P. multiflorum* (L.) All. SOLOMON'S SEAL. Like 1654, but fls. odourless and in stalked clusters of 2–5 from the axil of each lv. in a one-sided arched inflorescence. Perianth cylindrical 1–2 cm., somewhat constricted at the middle; filaments hairy. Lvs. 5–12 cm., more or less two-ranked; stem cylindrical, not angled above. △ Woods, bushy places. Apr.–June. Most of Eur. (except P.IRL.IS.). *The rhizomes were used in herbal remedies, particularly as a popular cure for black eyes.* Pl. 171.

P. latifolium (Jacq.) Desf. BROAD-LEAVED SOLOMON'S SEAL. Like 1655, but stems angled and finely downy above, to 1 m. Lvs. oval or broadly elliptic-acute, shortly stalked, green, and finely downy above. Fls. in axillary clusters of 1–4 borne on downy fl. stalks; perianth broadly tubular 12–17 mm. long; filaments hairless. East-Central and South-Eastern Europe.

Lvs. whorled, linear-lance-shaped

1656. *P. verticillatum* (L.) All. WHORLED SOLOMON'S SEAL. Readily distinguished by its narrow lvs. which are in whorls of 3–6, and its erect, angled stem to 80 cm. Fls. greenish-white in short-stalked axillary clusters of 1–4; perianth 6–10 mm. long, constricted at the middle. Lvs. 5–12 cm. linear-lance-shaped, stalkless, glaucous and hairy-veined beneath. Berry red. △ Woods in mountains. May–July. Most of Eur. (except P.IRL.NL.IS.SF.TR.).

CONVALLARIA | Lily-of-the-Valley Fls. in a terminal, leafless, spike-like cluster; perianth broadly bell-shaped and with 6 short tooth-like lobes. Stamens 6, attached to base of the perianth. Lvs. 2–4, long-stalked, the outermost sheathing the inner at the base. 1 sp.

1657. *C. majalis* L. LILY-OF-THE-VALLEY. A perenn. with a creeping rhizome and short, erect, curved stems bearing a one-sided cluster of 5–10 very sweet-scented, pure white, bell-shaped, pendulous fls. above a pair of elliptic lvs. Perianth broadly bell-shaped, 6–8 mm. long, with short, recurved, triangular lobes; fl. stalks curved, little longer than the perianth and the papery bracts. Lvs. usually 2, blade glossy, 8–20 cm., long-stalked; stem with papery sheathing bracts below. Fr. scarlet. △ Woods, thickets. Apr.–June. All Eur. (except P.IS.TR.). *A poisonous plant. Used medicinally in the past for heart disease, having a similar but less drastic action to digitalin. The leaves yield a green dye.* Pl. 171.

PARIS | Herb Paris   Fls. usually solitary, terminal; perianth segments 6–10, free; stamens 6–10; ovary with 4–5 styles. Fr. a berry or capsule. Sometimes placed in a separate family Trilliaceae. 1 sp.

1658. *P. quadrifolia* L. HERB PARIS. Readily distinguished by its whorl of 4 large stalkless lvs., above which is borne a solitary, long-stalked, greenish-yellow fl. with slender, spreading perianth segments. Perianth segments linear-lance-shaped $2\frac{1}{2}$–$3\frac{1}{2}$ cm., the outer 4 greenish, the inner 4 yellowish, narrower and shorter; stamens 8, prolonged above. Lvs. 6–12 cm. oboval and short-pointed, with 3–5 veins, lvs. sometimes more than 4; stem unbranched, 15–40 cm. Fr. globular berry-like, black. △ Damp woods. May–July. Most of Eur. (except P.IRL.GR.TR.). *A poisonous plant; much used in the past in herbal remedies.* Pl. 170.

SMILAX   Woody climbing plants with hooked spines and tendrils; plants one-sexed. Fls. in stalked clusters; perianth segments 6, spreading, at length falling; male fls. with 6 stamens, the female with 6 abortive stamens and a three-celled ovary with 3 stigmas. Fr. a berry. Placed by some botanists in a separate family the Smilacaceae. *c.* 3 sps.

1659. *S. aspera* L. Stems climbing, woody, much-branched and flexuous, often with hooked spines and with leathery, shining, evergreen lvs. with 2 tendrils arising from near the base of the lv. stalk. Fls. tiny, yellowish-green, in somewhat flat-topped, branched, terminal or axillary clusters. Lvs. 4–10 cm., very variable, lance-shaped to triangular with a heart-shaped base, with 5–9 stout veins, with or without stout hooked spines on the lv. margin and midvein beneath. Berry *c.* 7 mm., red. A variable sp. △ Bushy places. Aug.– Oct. Med. Eur. P.R. *The young asparagus-like shoots are edible.*

AGAVACEAE | Agave Family

Differing from Liliaceae in its largely woody growth, with stout, tall or short fl. stems, and narrow, often thick, fleshy, spiny-margined lvs. crowded in a rosette at the base of the stem. Fls. usually numerous, in large clusters; perianth 6; stamens 6; ovary superior or inferior, three-celled. Fr. a capsule or berry. 1 sp.

AGAVE

1660. *A. americana* L. CENTURY PLANT. Distinguished by its enormous rosettes measuring 2–4 m. across, formed of huge, spiny, spear-shaped lvs., and which produce, after 10–15 years, tree-like fl. stems up to 10 m. high. Fls. yellowish, in dense spike-like clusters and forming an enormous, branched, pyramidal inflorescence. Lvs. rigid, massive, 1–2 m. long, triangular-sectioned and gradually narrowed to a very stout spiny point, and with a spiny margin. △ Native of Mexico; widely naturalized by villages and cultivations:

rocks, waysides, coastal cliffs. June–Aug. Med. Eur. (except TR.) P. *In Mexico, a fermented beverage called 'pulque' is made from the juice obtained from the severed young flower spikes.* Pl. 171.

AMARYLLIDACEAE | Daffodil Family

Usually herbaceous perenns. with bulbs; lvs. all basal, linear, sheathing below. Fls. solitary or in umbels, enclosed in a membraneous sheath or *spathe* in bud, which splits into 1–2 papery segments at flowering. Fls. radially symmetrical; perianth of petal-like segments more or less fused at the base and sometimes with an additional ring or crown-like projection the *corona* in the throat; stamens 6. Ovary inferior, three-celled, style 1; fr. a capsule or rarely a berry.

1. Corolla with a crown-like projection or corona.
  2. Corona deeply and regularly toothed; stamens projecting.   *Pancratium*
  2. Corona entire or shallowly and irregularly toothed; stamens not projecting.   *Narcissus*
1. Corolla without a corona.
  3. Fls. erect, yellow.   *Sternbergia*
  3. Fls. drooping, white.
    4. Perianth segments all similar.   *Leucojum*
    4. Outer perianth segments twice as long and differing from inner perianth segments.   *Galanthus*

LEUCOJUM | Snowflake Fls. nodding, usually several, open bell-shaped, perianth of 6 petal-like free segments; corona absent; stamens not projecting; spathe 1 or 2; *c.* 9 sps.

1661. *L. vernum* L. SPRING SNOWFLAKE. A bulbous perenn. with usually solitary, or rarely 2, nodding, white, bell-shaped fls. with yellowish tips to the segments, borne on leafless stems 15–40 cm. Perianth segments 1½–2½ cm. long, oval and abruptly narrowed to a blunt yellowish point; spathe 1, membraneous, 3–4 cm., much longer than the fl. stalk. Lvs. all basal and arising from a large bulb, 3–4, pale green, broadly linear *c.* 1 cm. broad. △ Damp woods, meadows. Feb.–Apr. Cent. Eur. E.F.B.I.YU.R.SU.: introd. IRL.GB. Pl. 172.

1662. *L. aestivum* L. SUMMER SNOWFLAKE. Like 1661, but more robust, and fls. borne in a cluster of 2–8 on a stem up to 60 cm. Perianth segments 1–1½ cm. long, white with green tips; fl. stalks very unequal in length, the longest much longer than the spathe. Lvs. 3–5, broadly linear, *c.* 1 cm. broad. △ Damp meadows, ditches, swamps. Apr.–June. Much of Eur. (except North Eur. P.) DK. Pl. 172.

L. trichophyllum Brot. THREE-LEAVED SNOWFLAKE. A slender plant with 1–5 white or slightly pink-tinged fls. borne on a stem 10–25 cm., and with a spathe with 2 segments. Perianth *c.* 13 mm. long, the outer segments acute, the inner blunt. Lvs. 3, thread-like, about as long as the fl. stem. Portugal and Spain. Pl. 172.

GALANTHUS | Snowdrop Fl. solitary, nodding, the outer 3 perianth segments longer and at length spreading, the inner much shorter, broader, not spreading; corona absent; spathe two-lobed. Lvs. 2. *c.* 5 sps.

1663. *G. nivalis* L. SNOWDROP. Fl. nodding, 2–2½ cm. long, with 3 pure white, oval, outer perianth segments 14–17 mm. long and 3 inner, erect segments which are about half as long, deeply notched and with a green patch behind the tip; spathe arched, two-lobed, green with a white papery margin. Lvs. linear, *c.* 4 mm. broad, keeled, glaucous; fl. stem 15–25 cm. Fr. ovoid. △ Damp woods, meadows. Feb.–Mar. Most of Eur. (except North Eur. P.IRL.). *The bulbs are somewhat poisonous.* Pl. 173.

STERNBERGIA Fls. solitary or few, funnel-shaped and erect; perianth with a short, or long and slender tube, deeply divided into 6 equal segments; spathe present; fl. stalk often absent or short. *c.* 6 sps.

1664. *S. lutea* (L.) Sprengel COMMON STERNBERGIA. Fls. crocus-like, bright golden-yellow, nearly stalkless and appearing in the autumn with the young lvs. Fls. solitary or 2, 2–5 cm. long, erect and narrow funnel-shaped. Perianth tube very short, segments oblong-elliptic blunt; stamens about one-quarter as long as the perianth; spathe 1, papery, about half as long as the perianth. Lvs. broadly linear ½–1 cm. broad, shallowly channelled; fl. stem lengthening 10–30 cm. Fr. fleshy. △ Thickets, meadows. Sept.–Oct. Med. Eur. (except AL.TR.). *The bulbs are poisonous.* Pl. 172.

S. colchiciflora Waldst. & Kit. SLENDER STERNBERGIA. Distinguished from 1664 by the much more slender, paler, whitish-yellow fls. with a long slender perianth tube which is as long as or longer than the narrow segments; segments linear, 2½–3½ cm. long; fl. stalk absent and perianth tube often partly below ground. Lvs. twisted, appearing after the fls. in autumn, or in spring. The Balkans. Pl. 172.

NARCISSUS | Daffodil, Narcissus Fls. solitary or several; spathe conspicuous; fl. stem long. Perianth fused into a cylindrical tube below and with 6 spreading or reflexed segments above, with a shallow ring or broad cup-like projection, the *corona*, at the junction of the tube and the free segments. Hybridization often occurs. *c.* 40 sps. *The bulbs are poisonous and have been used medicinally in the past.*

Corona large, as long as or longer than the free perianth segments
1665. *N. pseudonarcissus* L. DAFFODIL. Fls. solitary, nodding; perianth segments pale yellow, spreading, corona deep yellow, trumpet-shaped and spreading a little at the mouth, as long as or slightly longer than the segments. Perianth 4–6 cm., with free segments 2–4 cm. long, and tube 1½–2½ cm. long; spathe to 5 cm. Lvs. glaucous, 1–2 cm. broad, linear, channelled; fl. stem somewhat flattened, 20–35 cm. A very variable sp. with numerous hybrids in cultivation. △ Meadows, woods, orchards, rocky places. Mar–May. Much of Eur. (except IRL.PL.AL.GR.): introd. DK.N.S.CS.R.

1666. *N. bulbocodium* L. HOOP PETTICOAT DAFFODIL. Fls. solitary, butter-yellow, erect or inclined, with a very conspicuous trumpet-like corona longer than the narrow triangular, similar-coloured, spreading perianth segments. Perianth variable in length, to *c.* 3 cm. long, with tube *c.* 1½ cm.; sometimes pale lemon yellow; spathe partly encircling the tube. Lvs. 2–4, very narrow, semi-cylindrical, channelled; fl. stem 5–30 cm. △ Rocky open ground, meadows in mountains, heaths. Feb.–Mar. P.E.F.

N. cantabricus DC. Like 1666, but fls. white. South West Spain. Pl. 173.

Corona medium or small, about half the length of the free segments or less
(a) Perianth segments and corona similar-coloured
1667. *N. jonquilla* L. JONQUIL. Fls. golden-yellow, very sweet-scented, 2–6 in an umbel, with the shallow cup-shaped corona similar in colour to the oval, wide-spreading perianth segments. Fls. *c.* 3 cm. across; tube of perianth greenish, 2–3 cm. long; fl. stalks long,

nearly equal. Lvs. 2–4, rush-like and grooved, more or less cylindrical in section, 3–4 mm. broad; fl. stem rounded, hollow, to 30 cm. △ Meadows. Mar.–Apr. P.E.F.I.: introd. YU. *It gives a sweet-smelling oil used in perfumery.*

N. papyraceus Ker-Gawler PAPER-WHITE NARCISSUS. Distinguished from 1669 by the similar-coloured pure white perianth segments and corona. Fls. several; perianth tube 1½–2 cm. long, white, free segments oval 12–15 mm. long, corona *c.* 4 mm. long. Lvs. glaucous, 7–15 mm. broad, grooved; fl. stem flattened. Mediterranean region.

N. × *odorus* L. Distinguished by the large, paired or sometimes solitary or 3, bright yellow fls. 4–6 cm. across, with perianth segments and corona similar-coloured. Perianth tube 2–2½ cm. long, greenish, free segments obovate-oblong, corona 10–12 mm. long, widening at the mouth, six-lobed and toothed. Lvs. 3–4, about 1 cm. broad, channelled at the base. Portugal to France.

1668. *N. requienii* Roemer (*N. juncifolius* Lag.) RUSH-LEAVED NARCISSUS. Like 1667, but a more slender plant with a shorter perianth tube 1½–2 cm. long. Fls. 1–2; corona usually more than half as long as the segments; longer fl. stalk usually included in spathe even after flowering. Lvs. very slender; 7–25 cm. △ Rocky places in mountains. Mar.–May. P.E.F. Pl. 173.

(b) Perianth segments and corona contrasting-coloured
1669. *N. tazetta* L. Distinguished by its sweet-scented bicoloured fls. 2–4 cm. across, with the free perianth segments pure white and corona golden-yellow. Fls. 3–18, in a terminal cluster; perianth tube 1½–2 cm., greenish, free segments oval-elliptic 12–15 mm., spreading or reflexed, corona about one-quarter to one-third as long. Lvs. glaucous, ½–1½ cm. broad, channelled; fl. stems robust 30–50 cm. Subsp. *aureus* P. Fourn. has perianth tube, segments, and corona all golden-yellow. △ Fields, vineyards, stony ground, damp grassy places. Jan.–Mar. Med. Eur. (except AL.TR.) P.SU.

1670. *N.* × *biflorus* Curtis Fls. usually 2, sweet-scented, bicoloured, 3–5 cm. across; perianth segments white or cream-coloured, broadly oval and with a fine point, corona bright yellow, 3–5 mm. long, with a whitish, papery, lobed and crisped margin. Lvs. 7–14 mm. broad. Probably a hybrid between 1669 and 1671. △ Meadows, hedges. Apr.–May. Naturalized in South Eur.

1671. *N. poeticus* L. PHEASANT'S-EYE NARCISSUS. Fls. large, 4–6 cm. across, solitary, nodding, sweet-scented, bicoloured; perianth segments pure white, corona yellow and with a red crisped and lobed margin. Perianth tube white, 2½–3 cm. long, free segments obovate to narrow-elliptic and fine-pointed, 2–2½ cm. long, corona 2–3 mm. long. Lvs. glaucous, 3–8 mm. broad; fl. stem flattened, to 60 cm. △ Damp meadows. Apr.–May. Med. Eur. (except TR.) BG.SU.: introd. Cent. Eur. Pl. 173.

1672. *N. serotinus* L. Fls. bicoloured, sweet-scented, usually solitary and appearing in the autumn before the lvs. Perianth tube greenish, 12–17 mm., free segments white, erect or spreading, 9–12 mm. long, elliptic fine-pointed, corona golden-yellow, ring-like, very short, 1–2 mm. Lvs. 1 or 2, thread-like, *c.* 1 mm. broad; fl. stem slender, to 20 cm. △ Rocky and stony places. Sept.–Oct. Med. Eur. (except AL.TR.) P. Pl. 173.

N. × *incomparabilis* Miller Like *N.* × *odorus*, but fls. solitary and bicoloured, with the free segments pale yellow, and the corona dark yellow. Lvs. glaucous, to 1½ cm. broad, nearly flat. Portugal to France.

PANCRATIUM   Fls. several in an umbel; perianth with a very long funnel-shaped tube which is enlarged at the throat, free segments 6, spreading, corona more or less cup-shaped, and conspicuously ten-toothed. Stamens long-projecting, attached to the margin of the corona. Bulbous perenns. with lvs. all basal. 2 sps.

1673. *P. maritimum* L. SEA DAFFODIL. Fls. white, sweet-scented, very large, 3–15 in an umbel with 2 large spathes at its base, borne on a stout flattened fl. stem to 60 cm. Fls. with a slender green tube 5–12 cm. long, which gradually widens above into the large white corona which has 12 conspicuous regular teeth, alternating with the 6 much longer stamens; free segments linear erect, to 6 cm. long. Lvs. glaucous, thick, and flat, *c.* 2 cm. broad, longer than the fl. stem. △ Maritime sands. July–Sept. Med. Eur. P.BG. *The bulbs are somewhat poisonous.* Pl. 174.

P. illyricum L. Like 1673, but fls. 7–30, smaller, 6–8 cm. long, and perianth tube widening rather abruptly at the mouth only; corona short, deeply divided into 6 narrow spreading lobes which are each deeply two-lobed, thus appearing irregularly twelve-toothed; free perianth segments lance-shaped. Lvs. 1½–3 cm. broad; stems not flattened. Spain, Corsica, Sardinia.

DIOSCOREACEAE | Yam Family

Usually slender herbaceous or woody climbers with spirally arranged, often heart-shaped and entire or lobed lvs.; roots with swollen tubers. Fls. one-sexed, small, green, borne in clusters in the axils of the lvs.; perianth bell-shaped and with 6 short spreading segments; stamens 6, with or without 3 sterile stamens, or stamens 3; ovary three-celled; style 3, or 1 with 3 stigmas. Fr. a berry or capsule.

TAMUS | Black Bryony   Plants one-sexed; herbaceous climbing perenns. Fr. a few-seeded fleshy berry. 1 sp.

1674. *T. communis* L. BLACK BRYONY. A herbaceous climber, twining to the left, to 4 m., with oval-heart-shaped, long-pointed, stalked lvs. 3–10 cm. with dark green shining blades with 3–9 conspicuous curved veins. Fls. yellowish, the male in slender spreading spike-like clusters about as long as the lvs., the female spikes shorter, pendulous, fewer-flowered. Berry *c.* 12 mm., pale shining red; tuber very large to 20 cm. or more long, blackish. △ Hedges, woods, thickets, waysides. Apr.–July. Much of Eur. (except North Eur. IRL.CS.PL.SU.). *The tubers and berries are poisonous.*

IRIDACEAE | Iris Family

Herbaceous perenns. with rhizomes, corms, or bulbs, and lvs. often sword-shaped with the blade held vertically, folded longitudinally and sheathing at the base. Fls. solitary or few in a cluster, each fl. often with a spathe which encircles the bud and splits into 1 or 2. Fls. large, bisexual, with 2 rows of brightly coloured, petal-like, spreading perianth segments which are usually fused below into a shorter or longer tube; stamens 3. Ovary inferior, three-celled; style three-lobed and sometimes brightly coloured and broad and petal-like. Fr. a capsule.

Nos.

1. Fls. regular and radially symmetrical.
  2. Fls. arising direct from the corm.   *Crocus*   1675–1680
  2. Fls. borne on aerial stems.
    3. Perianth segments more or less similar, spreading equally.
      4. Stems rounded in section.   *Romulea*   1681
      4. Stems flattened and winged in section.   *Sisyrinchium*   1682
    3. Perianth segments differing, the outer erect, the inner spreading and reflexed.
      5. Lvs. quadrangular in section; ovary one-celled.   *Hermodactylus*   1683
      5. Lvs. sword-shaped or linear, grooved. Ovary three-celled.   *Iris*   1684–1694
1. Fls. somewhat irregular and symmetrical in only one plane.
  6. Fls. purple or pink; stigmas undivided.   *Gladiolus*   1695–1697
  6. Fls. white or yellow; stigmas deeply bilobed.   *Freesia*   (1697)

CROCUS | Crocus Fls. stalkless at flowering; perianth with a slender, basal, stalk-like tube which is partly below ground at flowering, and with 6 elliptic, somewhat spreading free segments; stamens 3, shorter than the perianth. Ovary below ground; style three-lobed, with stigmas further divided or toothed; fr. a capsule borne at length above ground. Lvs. with a central white band above; corm covered with brown scales or fibres. *c.* 50 sps. *The yellow dye, saffron, is obtained from the stigmas of* C. sativus L.; *it is widely used in colouring foodstuffs, in beverages and medicinally.*

Scales of corm with parallel fibres
(a) Fls. yellow
1675. *C. flavus* Weston (*C. aureus* Sm.) DUTCH YELLOW CROCUS. Fls. 1–4, entirely golden-yellow, appearing with the lvs. in the spring. Perianth segments oblong blunt; throat hairless; stigmas orange, little-divided, shorter than the anthers. Lvs. about 6, appearing with the fls., rather broad-bladed, ciliate. Corm with rather leathery scales which later break up from below into narrow parallel fibres. △ Stony slopes. Feb.–Apr. South-East Eur. (except AL.) SU. *Often grown as an ornamental plant.*

(b) Fls. purple
1676. *C. nudiflorus* Sm. Fls. solitary, entirely purple, appearing in the autumn long before the lvs. Perianth segments *c.* 5 cm.; throat hairless; style branches divided into numerous, orange, thread-like stigmas. Lvs. 3–5, very narrow, smooth, appearing the following spring; corms forming stolons in the spring. △ Grassy places, heaths. Sept.–Nov. E.F.GB. Pl. 174.

Scales of corm with a network of fibres
1677. *C. reticulatus* Steven (*C. variegatus* Hoppe & Hornsch.) Fls. usually solitary, appearing in the spring, violet and strongly feathered with darker violet veins outside and with a violet tube. Perianth segments elliptic acute, hairless in throat; stigmas orange, little-divided and little longer than the stamens. Lvs. appearing with the fls., narrow, hairless. Corm with a coarse network of fibres. △ Grassy, stony places. Early spring. H.YU.BG.R.SU.

1678. *C. albiflorus* Kit. PURPLE CROCUS. Fls. solitary, appearing in the spring, white, usually with a violet tube, or rarely all violet. Perianth segments lance-shaped blunt, hairy in the throat; stigmas orange, with a crisped apex, shorter than the stamens. Lvs. 2–4, hairless, appearing with the fls. Corm with a fine network of fibres. △ Meadows and pastures in hills and mountains. Feb.–May. Cent. Eur. (except PL.) E.F.I.YU.R.SU. Pl. 174.

Scales of corm breaking off in concentric rings
(*a*) *Fls. golden-yellow*
1679. *C. chrysanthus* Herbich Fls. 1–3, bright golden-yellow, appearing in the spring at the same time as the lvs. Perianth segments oblong blunt, hairless in the throat; stigmas spathulate, almost entire and little longer than the anthers. Lvs. narrow, finely ciliate. △ Rocky and grassy places. Feb.–Mar. South-East Eur. (except AL.).

(*b*) *Fls. white or violet*
1680. *C. biflorus* Miller Fls. 1–3, appearing in the spring, white with 3 dull violet veins on the outside of the segments and with a yellowish, hairless throat. Stigmas golden-yellow, spathulate, entire or nearly so, longer than the anthers. Lvs. narrow, hairless, appearing with the fls. △ Sunny hills, stony and grassy places. Mar.–Apr. I.YU.GR.TR.BG.

ROMULEA Fls. long-stalked and subtended by 2 spathes. Perianth funnel-shaped, fused at base to form a very short tube and with 6 free spreading segments; stamens 6, included; style 1, three-lobed, stigmas two-lobed. Lvs. very slender. *c.* 15 sps.

1681. *R. bulbocodium* (L.) Sebastiani & Mauri A delicate, crocus-like perenn. 5–25 cm., with very narrow lvs. without a white line, and a small, slender-stalked, funnel-shaped fl., with a pair of spathes close below. Perianth segments $2\frac{1}{2}$–$3\frac{1}{2}$ cm. long, yellowish-violet and streaked with yellow or purple outside, rarely entirely white or yellow; throat of perianth and filaments hairy. Spathes 2, the inner papery, the outer green with a papery margin. Lvs. curved, rush-like, grooved, $1\frac{1}{2}$–2 mm. broad; stems recurved in fr. △ Dunes, heaths. Feb.–Apr. Med. Eur. P.BG.

R. columnae Sebastiani & Mauri (*R. parviflora* Bubani) Like 1681, but more delicate, with smaller fls. 10–12 mm. long, which are pure violet within, with a pale yellow, hairless throat, and greenish outside, or fls. whitish with darker veins. Perianth little longer than the spathes; filaments hairless. Lvs. 1 mm. broad. Western and Southern Europe.

SISYRINCHIUM Fls. several, borne at the end of a flattened, simple or branched stem; perianth tube very short, segments spreading, blue or yellow. Style branches undivided. 2 sps.

1682. *S. bermudiana* L. (*S. angustifolium* Miller) BLUE-EYED GRASS. Fls. sky-blue, *c.* $1\frac{1}{2}$ cm. across, 1–4, borne at the end of a flattened, narrowly winged, leafless stem 10–45 cm., with two conspicuous green spathes encircling the fls. in bud. Fls. very short-lived, stalked; perianth segments wide-spreading, oval blunt, and with a long, narrow-pointed tip; filaments fused most of their length. Lvs. all basal, spear-shaped, with vertical blade 1–3 mm. wide; roots fibrous. △ Native of North America; naturalized in damp meadows, copses. May–June. Introd. to much of Cent. and North. Eur.

HERMODACTYLUS Similar to *Iris*, but with swollen tuberous roots and lvs. tetragonal in section. Fls. solitary, outer perianth segments not bearded; ovary one-celled. 1 sp.

1683. *H. tuberosus* (L.) Miller (*Iris t.*) SNAKE'S HEAD IRIS, WIDOW IRIS. Fls. distinctive, greenish-yellow and with dark brownish or blackish-purple reflexed segments, fragrant, solitary at the end of a slender stem, and overtopped by a long, green, slender spathe to *c.* 20 cm. long. Perianth 4–5 cm. long, the outer 3 segments with a broad oboval base and a much smaller, oval, reflexed lobe, the inner 3 segments with an erect, lance-shaped and very deeply two-lobed blade. Lvs. rush-like, four-ribbed, 1½–3 mm. broad, longer than fl. stem; tubers oblong. △ Rocky and stony hills, bushy places. Mar.–Apr. Med. Eur. (except AL.TR.). Pl. 174.

IRIS | Iris Fls. large and showy, with 2 whorls of brightly coloured petal-like perianth segments, the outer 3, or *falls*, usually with a narrow basal part or *shaft* and broader, reflexed, terminal part; the inner 3 segments, or *standards*, usually erect and narrower. All segments are fused at the base into a long or short tube; the falls often possess a tuft of hairs on the upper surface, and are said to be *bearded*. Stamens 3; style with 3 broad petal-like arms and a bilobed tip, often curving outwards over the stamens and falls. Lvs. usually sword-shaped with the blade carried vertically; creeping rhizomes or bulbs present. *c*. 50 sps.

Plants with bulbs covered with fibrous scales

1684. *I. sisyrinchium* L. (*Gynandriris s.*) BARBARY NUT. A slender perenn. to 25 cm., with rush-like lvs. and bright blue fls. with whitish or yellowish centres, borne severally on slender stems from conspicuous, brownish, swollen, papery spathes. Fls. very variable in size, 1–3 cm. across, stalkless; the falls oval-spathulate, spreading, with a white or yellow patch at the throat, unbearded, the standards lance-shaped erect, blue; each stamen fused with a style-arm. Stem lvs. 2, 3–8 mm. broad, channelled, longer than the fl. stems. △ Hills, stony places on the littoral. Apr.–May. Med. Eur. (except YU.AL.) P. Pl. 175.

1685. *I. xiphium* L. SPANISH IRIS. A rather robust perenn. with a leafy, unbranched, hollow stem to 60 cm., bearing a large, solitary, violet-purple fl. to 10 cm. across, with a conspicuous orange-yellow patch on the spreading falls, or falls sometimes entirely yellow. Falls with a broad oval blade and a much longer, darker purple shaft, the standards lance-shaped erect; perianth tube very short. Stigmas petal-like, violet-purple, conspicuously bilobed, little shorter than the falls; spathe green and shorter than the ovary. Lvs. narrow-linear, grooved, shorter than the fl. stem. △ Damp sandy places, in mountains. May–Aug. P.E.F. Pl. 175.

I. xiphioides Ehrh. Like 1685, but fls. 2, bright blue with an orange patch on each fall, and arising from 2 much swollen equal spathes which encircle the ovary. Perianth segments unequal, the falls with an obovate, notched blade as long as the shaft and much longer than the erect bilobed standards and the bilobed, petal-like stigmas. Lvs. broadly linear, usually about as long as the fl. stem. The Pyrenees. Pl. 175.

Plants with swollen horizontal rhizomes

(a) Outer segments, or falls, beardless

1686. *I. graminea* L. GRASS-LEAVED IRIS. A slender perenn. 20–50 cm., with pairs of sweet-smelling, blue-violet fls., often strongly veined with purple. Falls spreading, with the blade much shorter than the broad shaft, the standards lance-shaped, erect, shorter than the petal-like styles and the shaft of the falls; spathes 2, green, unequal. Lvs. 5–10, grass-like, *c*. 1 cm. across, much longer than the flattened two-edged fl. stem. △ Woods, bushy places, meadows. May–June. Cent., South-East Eur. (except TR.) E.F.I.SU. Pl. 175.

1687. *I. sibirica* L. A tufted perenn. with slender stems 30–100 cm. much longer than the narrow grass-like lvs., and bearing 2–5 blue-purple or lavender fls. about 6 cm. across, and with distinctive brown spathes at flowering time. Falls rounded, *c*. 2 cm. wide, and with a longer and narrower, purplish-veined shaft; the standards erect, shorter than the falls. Lvs. *c*. 7 mm. wide. △ Pastures, heaths. May–June. Cent. Eur. F.I.YU.R.SU.

1688. *I. spuria* L. Fls. whitish-lilac, 4–5 cm. long, 1–3 in a stalkless cluster borne on a stout, rounded, little-branched stem 20–30 cm. Falls with rounded blade to *c*. $2\frac{1}{2}$ cm. wide and a shaft twice as long, whitish and veined with purple and yellow, the standards oblance-shaped, shorter than the falls, lilac; spathe inflated, enclosing the shortly stalked ovary. Lvs. unpleasant-smelling when bruised, 7–15 mm. broad. Fr. with a long point. △ Damp meadows. May–June. E.F.GB.DK.D.A.CS.H.YU.R.

1689. *I. foetidissima* L. GLADDON, STINKING IRIS. Fls. dull purplish with darker veins, or sometimes yellowish, *c*. 8 cm. across. Falls with a narrow oval blade much longer than the shaft, the standards yellowish, spathulate, shorter than the falls. Perianth tube short; style-arms yellowish; spathes with broad papery margin. Lvs. dark green, *c*. 2 cm. wide, unpleasant-smelling when crushed; fl. stem angled on one side, 50–80 cm. Fr. club-shaped, splitting to reveal large, orange-red seeds. △ Woods, shady banks, sea cliffs. May–July. West Eur. (except B.NL.) I.: introd. CH. *Poisonous to livestock.*

1690. *I. pseudacorus* L. YELLOW FLAG. Fls. large yellow, 8–10 cm. across, in terminal and lateral clusters of 2–3, with green spathes with broad papery margins. Perianth segments very unequal; the falls variable, broadly oval to lance-shaped, often marked with orange or purple, the standards very much smaller and about half as long as the yellow spathu-late styles. Lateral fl. clusters often long-stalked and overtopped by the subtending lvs. Fr. elliptic pointed; seeds brown. Stems somewhat flattened, 40–150 cm. △ Marshes, swamps, ditches, watersides. June–July. All Eur. (except IS.). *Used medicinally in the past; the leaves and roots are poisonous to livestock.* Pl. 174.

(b) Outer segments, or falls, bearded

1691. *I. pumila* L. A dwarf plant with large, solitary blue-violet or yellowish-white fls. borne on a very short stem, not as long as the basal lvs., but later elongating in fr. Falls up to 6 cm. long, broadly spathulate and with a recurved apex, and a dense beard, the standards oblong, a little longer than the falls; perianth tube 4–5 times as long as the ovary. Spathes 2, broadly papery above and with a rounded, not keeled back, not en-circling the ovary. Lvs. often curved, 7–17 mm. broad, more or less glaucous, usually shorter than the fls. △ Dry grassy, rocky places. Mar.–May. A.CS.H.YU.BG.R.SU. Pl. 176.

1692. *I. chamaeiris* Bertol. Like 1691, but with a distinct, often short, stem carried above the basal lvs. at the time of flowering; perianth tube shorter, *c*. $2\frac{1}{2}$ cm. long; spathes green. Fls. 1–2, very variable in colour; falls blue, purple, yellow or white and tinged or veined with brown, oblong-spathulate, 2 cm. broad, beard bright orange-yellow; standards erect or converging, oblong, $2\frac{1}{2}$ cm. broad, with crisped margin. Lvs. 4–6 pale green, 8–12 mm. broad. △ Dry, stony ground. Mar.–Apr. F.I. Pl. 175.

I. variegata L. Distinguished by its medium-sized yellow bearded fls. Falls *c*. 2 cm. broad, reflexed, yellow with brownish-purple veining, beard bright yellow; standards entirely yellow, or with brownish veining below; perianth tube twice as long as the ovary. Spathe inflated, fine-pointed, pale green flushed with pink. Lvs. 1–1½ cm. wide. Balkan region.

1693. *I. germanica* L. COMMON IRIS. A very robust perenn. to 1 m., with branched stems and clusters of 2–3 very large, scented, blue-violet fls. up to 10 cm. across. Falls up to 4 cm. broad, with a dark purple blade, a pale brownish-veined shaft, and a yellow beard; standards about equalling the falls, incurved, deep lilac; perianth tube longer than the ovary. Spathes 2, swollen, blunt, brown-papery in the upper half, green below. Lvs. stout, glaucous, 2–5 cm. broad. △ Rocky places; very commonly grown as an ornamental and often naturalized. Apr.–June. Med. Eur. (except TR.) P.: introd. to Cent. Eur. Pl. 175.

I. pallida Lam. PALE IRIS. Like 1693, but fls. a clear pale blue, sweet-scented, and spathes silvery-white and entirely papery at the time of flowering. Stem branched above, 1 m.

or more, longer than the pale glaucous lvs. Native of the Mediterranean region; often cultivated in Southern Europe and naturalized.

1694. *I. florentina* L. FLEUR-DE-LIS. A robust, branched perenn. to 60 cm., like 1693, but with large white, often pale blue-flushed, sweet-scented fls. Falls broader than long, their blades veined with blue, and their shafts with darker brownish veins, beard yellow. Spathes keeled, green with a papery margin. △ Widely grown as an ornamental; often semi-naturalized. Mar.–Apr. E.F.I.YU.GR. *Orris root is prepared from the dried rhizomes; it is used in perfumery and cosmetics.*

GLADIOLUS | Gladiolus Fls. showy, several or many in a one- or two-rowed spike, usually with leafy bracts. Perianth of 6 more or less equal free segments, fused below into a short tube, and arranged somewhat asymmetrically into a curved, horizontal, funnel-shaped corolla, with some or all of the segments overlapping at the margin. Corm with fibrous scales. *c.* 10 sps.

Anthers as long as or a little longer than the filament
1695. *G. segetum* Ker-Gawler FIELD GLADIOLUS. Fls. rosy-purple, 3–13 in a lax, one-sided spike, with the lower leafy bracts as long as the fls., borne on a slender leafy stem 40–100 cm. Perianth 4–5 cm. long, somewhat irregular, the uppermost segment longer and nearly twice as broad, and somewhat separated from the lateral segments, the lowest 3 segments a little longer than the lateral pair. Lvs. 7–15 mm. broad; fl. stem with 3–5 lvs. Seeds rounded, not winged. Corm scales with coarse parallel fibres. △ Cornfields, cultivated ground. Apr.–June. Med. Eur. P.BG.R.SU. Pl. 176.

G. byzantinus Miller EASTERN GLADIOLUS. Distinguished from 1695 by the 3 upper perianth segments which are more or less equal and placed close together, the uppermost segment is broadly elliptic, the lateral segments are oval-rhomboidal. Seeds flattened, winged. Central and Eastern Mediterranean region. Pl. 176.

Anthers shorter than the filament
1696. *G. communis* L. COMMON GLADIOLUS. Like 1695, but anthers distinctly shorter than filament and stigmas obovate, gradually enlarged from the base upwards. Fls. 5 cm. long, rosy-purple, and with base of the perianth reddish-brown, segments all similar, close together and forming a rather narrow funnel. Lvs. 8–15 mm. broad; stem with 3–5 lvs., to 1 m. Seeds broadly winged. △ Grassy places, fallow. May–June. Med., South-East Eur. (except TR.) A.H.SU.: introd. D.CH.

G. palustris Gaudin MARSH GLADIOLUS Like 1696, but fls. 3 cm. long, fewer 2–5, borne on a more slender stem to ½ m., with 2–3 lvs. Upper 3 perianth segments nearly equal, obovate, the lower 3 up to a third as long again. Stigmas hairy. Seeds flattened, winged. Central Europe.

1697. *G. illyricus* Koch Like 1696, but a more slender plant and fls. fewer, 3–10 in an equal, not a one-sided fl. spike. Fls. 3½–4 cm. long; perianth segments blunt, the 3 upper unequal with the uppermost longer and narrower than the oblong-rhomboidal lateral pair. Stigmas narrow below and abruptly enlarged above into oval blades. Lvs. narrow, 5–9 mm. broad. Corm scales with very fine parallel fibres. △ Meadows, bushy places, marshes, heaths. May–June. Med. Eur. (except TR.) P.GB.A.BG.

FREESIA Fls. with a short cylindrical tube and 6 short, unequal, spreading lobes, several in a one-sided spike. Stems dichotomously branched; corms with fibrous scales. 1 sp.

F. refracta (Jacq.) Klatt Fls. very sweet-scented, pale yellow or greenish-yellow, tubular and two-lipped, *c.* 3 cm. long, free segments *c.* 12 mm.; bracts papery. Lvs. linear 5–6, much shorter than the flexuous stems, which are 30–50 cm. Native of South Africa; often cultivated in various colour forms and sometimes naturalized.

JUNCACEAE | Rush Family

Often tufted herbaceous plants with narrow, rounded, channelled, or flattened grass-like lvs. which sheath the stem at the base. Fls. inconspicuous, greenish or brownish, often many and massed into dense heads or branched clusters. Perianth with 6 usually similar, small scaly segments; stamens usually 6. Ovary one- or three-celled; style 1 or absent; stigmas 3, feathery; fr. a capsule, encircled by the persisting perianth.

Lvs. hairless, usually thick, rounded, or channelled in section; fr. many-seeded   *Juncus*
Lvs. hairy, usually flattened and grass-like; fr. three-seeded.   *Luzula*

JUNCUS | Rush Stem lvs. split at the base and sheathing the stem, hairless, usually cylindrical-sectioned, seldom flattened and grass-like. Perianth brownish or greenish. Fr. many-seeded. *c.* 50 sps.

Fl. stems leafless
(a) Fl. clusters apparently laterally placed

1698. *J. inflexus* L. HARD RUSH. A densely tufted perenn. to 90 cm., with numerous, glaucous, leafless stems which are strongly ribbed, and a one-sided brownish inflorescence arising well below the sharp-pointed apex. Inflorescence of many ascending, unequal, lax branches, to *c.* 5 cm. long; perianth segments 2½–4 mm., lance-shaped with an awl-shaped apex. Lvs. absent, sheaths at base of stem shining, blackish-purple; stem with 12–18 prominent ridges, tough, hard, 1–1½ mm. broad, pith interrupted. Fr. as long as the perianth, dark, glossy chestnut-brown, narrowed gradually above to a fine point. △ Damp pastures. June–Aug. All Eur. (except IS.): introd. N.SF. Pl. 177.

1699. *J. effusus* L. SOFT RUSH. Distinguished from 1698 by its rather soft, smooth, glossy, bright yellowish-green stem 1½–4 mm. broad, with 40–90 very fine lines, pith continuous and sheaths dull, not shiny. Inflorescence lax, with many erect, spreading or reflexed branches; perianth segments 2–2½ mm. Fr. yellowish to chestnut-brown, broadly ovoid with a slightly indented apex. A glossy, tufted perenn. 30–150 cm., without true lvs. △ Wet pastures, bogs, damp woods, watersides. June–Sept. All Eur. (except IS.) Pl. 177.

1700. *J. subuliflorus* Drejer (*J. conglomeratus* auct.) CONGLOMERATE RUSH. Very like 1699, but stems not glossy, bright green to greyish-green, with numerous fine ridges which are especially prominent just below the fl. cluster. Inflorescence usually dense, rounded, compact, pressed against the side of the stem, but less commonly the cluster is stalked or branched. Fr. with a small elevated point at the apex. A tufted perenn. to 120 cm. △ Wet pastures, marshes, bogs. June–Sept. All Eur. (except IS). Pl. 177.

*J. filiformis L. Widespread, except in South-Eastern Europe.

1701. *J. jacquinii* L. BLACK ALPINE RUSH. A small alpine perenn. 15–30 cm., distinguished by its stalked, compact, laterally borne cluster of very shiny purplish-black fls. overtopped by the stem. Fls. 4–10; perianth segments lance-shaped, long-pointed. Stems very slender, closely tufted, with a fine-pointed apex and brown basal sheaths. Fr. black, nearly as long as the perianth. △ Damp pastures in mountains. July–Sept. F.D.CH.A.CS.R. Pl. 177.

J. arcticus Willd. ARCTIC RUSH. Like 1701, but fl. cluster not stalked and fls. not or scarcely shining; stems widely spaced, not closely tufted, rather thick, hollow. Fls. in a dense lateral cluster of 2–8, brownish; perianth segments brownish, unequal, the inner blunt, the outer acute. Northern Europe and mountains of Central Europe.

(b) Fl. clusters terminal

1702. *J. squarrosus* L. HEATH RUSH. Distinguished by its basal rosette of numerous stiff and tough out-spreading lvs. which are often pressed against the ground and its stiff, erect stems bearing a terminal, rather dense cluster of greenish-brown fls. Fls. in branched clusters, bract subtending the lowest clusters leaf-like, less than half the length of the inflorescence. Perianth segments 4–7 mm., oval-lance-shaped, with broad, papery margin. Lvs. deeply channelled, 8–15 cm. by 1–1½ mm., and with broader, tough, basal sheaths. △ Heaths, bogs, moors. June–Sept. Most of Eur. (except H.AL.GR.TR.BG.). Pl. 177.

1703. *J. capitatus* Weigel DWARF RUSH. A tiny tufted ann. 3–12 cm., with several stiff stems bearing a compact, usually terminal head of several fls. which is overtopped by 1 or 2 bristle-like brownish bracts. Outer perianth segments greenish, becoming reddish, with fine curved points, the inner segments shorter. Lvs. all basal, stiff, bristle-like, shorter than the fl. heads. Fr. ovoid blunt, shorter than the perianth. △ Damp, sandy places. May–Aug. Most of Eur. (except IRL.IS.N.). Page 519.

Fl. stems with 1 or more lvs.

(a) Plants of maritime and salt-rich regions

1704. *J. maritimus* Lam. SEA RUSH. A densely tufted perenn. ½–1 m., with stiff, sharp-pointed lvs. and apparently laterally placed, much-branched fl. clusters overtopped by a sharp-pointed, stem-like bract. Perianth segments 3–4½ mm., straw-coloured, lance-shaped, the outer acute, the inner blunt. Fr. straw-coloured or reddish-brown, ovoid, three-sided, pointed, about as long as the perianth; seeds with a large appendage. △ Damp sands by the sea, marshes, but sometimes found some way inland. June–Oct. The coastal countries of most of Eur. (except IS.N.SF.PL.) A.H.

1705. *J. acutus* L. SHARP-POINTED RUSH. A very robust, densely tufted perenn. 25–150 cm., with tough, rigid, very sharp-pointed lvs. and more or less rounded, compact inflorescences 4–6 cm., overtopped by the tough, sharp-pointed, stem-like bracts. Perianth segments 3–4 mm., lance-shaped long-pointed, reddish-brown with a white papery margin, very tough. Fr. c. ½ cm., reddish-brown, shining, at least twice as long as the perianth. △ Sandy places, marshes by the sea. Mar.–July. Med., South-East Eur. P.IRL.GB. *The dried stems are used for weaving mats, chair-bottoms and baskets; the pith is sometimes used as a wick.* Pl. 177.

*J. gerardii Loisel. MUD RUSH. The littoral of most of Europe.

(b) Inland plants of lowlands and mountains

1706. *J. compressus* Jacq. ROUND-FRUITED RUSH. A rather small perenn. 10–30 cm., distinguished by its narrow linear lvs. scarcely 1 mm. wide, which are flattened from above and below. Inflorescence terminal, branched, lax or compact, usually shorter than the lowest bract. Perianth segments 1½–3 mm., oval very blunt, light brown; anthers slightly shorter than the filament; style shorter than the ovary. Fr. very glossy, nearly rounded, with a short apical point, perianth one-half to one-third as long. △ Marshes, grassy places, meadows. June–Aug. Most of Eur. (except IRL.IS.). Page 519.

1707. *J. acutiflorus* Hoffm. SHARP-FLOWERED RUSH. A stiff, erect perenn. 30–100 cm., with long creeping rhizomes, distinguished by its more or less round-sectioned lvs. with conspicuous partitions in the pith. Inflorescence lax, much-branched, with numerous stalked, dense clusters of 5–12 fls.; perianth segments chestnut-brown, 3 mm. long, lance-shaped, tapering to an awl-like apex, the outer segments longer and with curved tips; style as long as the ovary. Fr. chestnut-brown, gradually tapered to an acute point; seeds *c.* 12. △ Wet meadows, moorlands, boggy places, ditches. June–Sept. Much of Eur. (except North Eur. H.GR.TR.R.SU.) DK.

J. articulatus L. JOINTED RUSH. Very like 1707, but lvs. strongly flattened and with inconspicuous transverse partitions in the pith, and creeping rhizomes short. Fls. 4–8 in dense clusters; perianth segments dark chestnut-brown to blackish, the outer longer and more acute than the papery-margined inner segments. Fr. rather abruptly narrowed to a fine point, usually blackish and shining; seeds *c.* 40. Throughout Europe. Pl. 177.

**J. subnodulosus* Schrank BLUNT-FLOWERED RUSH. Widespread, except in Northern Europe.

1708. *J. bulbosus* L. BULBOUS RUSH. A slender, very variable, grass-like perenn. to 20 cm., which may be erect and tufted, prostrate and rooting at the nodes, or very slender and submerged with thread-like, much-branched, underwater floating stems. Inflorescence simple or branched, often viviparous and producing leafy shoots in place of fls. Fls. in clusters of 2–10; perianth segments $3\frac{1}{2}$–4 mm., equal, green or pinkish, the outer acute, the inner blunt; stamens usually 3. Lvs. very slender, 1–3 mm. broad, not stiff; stem often with a bulbous swelling at the base. Fr. $2\frac{1}{2}$–3 mm., yellowish-brown, oblong, blunt, little longer than the perianth. △ Damp heaths, bogs, damp woods. June–Sept. Much of Eur. (except AL.GR.TR.BG.). Page 519.

1709. *J. trifidus* L. THREE-LEAVED RUSH. A densely tufted, grass-like, northern or alpine perenn. 10–30 cm., readily distinguished by the 2–4 very long, thread-like, leafy bracts 2–8 cm., which are much longer than the dense cluster of brownish-black fls. Fl. clusters one or several, stalkless, one- to four-flowered; perianth segments 4 mm., narrowed to a fine point. Lvs. mostly with only basal sheaths, but some with narrow thread-like blades. Fr. with a slender beak, about as long as the perianth. △ Rocky, grassy slopes in mountains. July–Sept. Much of Eur. (except P.IRL.B.NL.DK.H.TR.). Page 519.

1710. *J. tenuis* Willd. (*J. macer* S. F. Gray) SLENDER RUSH. A slender, pale green, tufted, lowland perenn. 15–30 cm., with at least 1 of the slender, curved, leafy bracts much longer than the lax branched inflorescence. Perianth segments 3–5 mm., greenish, but becoming straw-coloured, narrowly lance-shaped, long-pointed. Lvs. flat, channelled, 1 mm. broad, not stiff, with broad sheathing base and papery auricles; fl. stems slender 15–35 cm. Fr. ovoid blunt, shorter than the perianth. △ Native of America; widely naturalized in waysides, woodlands, waste ground. July–Sept. Introd.: North Eur. (except IS.) Cent. Eur. F.IRL.GB.B.NL.R.SU. Page 519.

1711. *J. bufonius* L. TOAD RUSH. A delicate, variable, erect or prostrate, often tufted, much-branched perenn. 3–25 cm., with solitary stalkless fls. ranged rather regularly along one side of the slender leafy branches. Perianth segments 3–7 mm., lance-shaped with a fine point, pale green and with a transparent margin. Lvs. thread-like, deeply channelled. Fr. oblong blunt, usually shorter than the perianth. △ Mud, damp sands, arable, tracksides. June–Sept. All Eur.

LUZULA | Woodrush Tufted grass-like perenns. with flat or channelled lvs., usually with sparse, spreading, long white hairs. Fls. in a branched or sometimes dense cluster; perianth segments papery, coloured; stamens 6. Fr. a one-celled capsule; seeds 3, usually with an appendage. *c.* 25 sps.

1. *Juncus capitatus* 1703
3. *Juncus trifidus* 1709
5. *J. tenuis* 1710

2. *Luzula lutea* [1712]
4. *J. bulbosus* 1708
6. *J. compressus* 1706

Fls. white or yellow

1712. **L. nivea** (L.) DC. SNOWY WOODRUSH. Distinguished by the comparatively large white fls., which are borne in a compound, somewhat flat-topped, compact inflorescence shorter than the uppermost lvs. Fls. in clusters of 3–8; perianth 5–6 mm. long, segments snow-white, unequal, acute; anthers brown. Lvs. 2–4 mm. broad, with long hairs. Fr. shorter than the outer perianth segments. An erect leafy perenn. 40–80 cm. △ In mountains: sunny slopes in woodlands. June–Aug. E.F.D.CH.A.I.R. Pl. 177.

L. luzuloides (Lam.) Dandy & Wilmott (*L. albida* (Hoffm.) DC.) Distinguished from 1712 by its umbel-like inflorescence which soon becomes lax and spreading and is a little shorter than the uppermost lvs. Fls. tiny, dirty white or sometimes pinkish, in clusters of 2–6; perianth segments 3 mm. long, about as long as the fr. Western and Central Europe.

L. lutea (All.) DC. YELLOW WOODRUSH. Like 1712, but fls. pale golden-yellow and borne in several, very dense, rounded, slender-stemmed clusters in a simple spreading inflorescence which is much longer than the uppermost lvs. Perianth segments equal; anthers yellow. Lvs. hairless, greenish-yellow, 4–8 cm. by 6 mm., much shorter than the fl. stem. Pyrenees, Alps, and Apennines. Page 519.

Fls. brown or green
(a) Fls. usually solitary, long-stalked

1713. **L. pilosa** (L.) Willd. HAIRY WOODRUSH. A tufted grass-like perenn. 20–40 cm., with hairy lvs. and a lax umbel-like cluster of dark chestnut-brown fls. borne on very unequal, slender spreading, and later conspicuously reflexed stalks. Fls. usually solitary, c. ½ cm. across; perianth segments 3–4 mm., acute and with a broad transparent margin and tip. Basal lvs. 3–4 mm. broad, stem lvs. 7–10 mm. broad, all with a hairy margin. Fr. a little shorter than the perianth; seeds with a long hooked appendage. △ Woods. Mar.–May. Most of Eur. (except P.IS.TR.).

*L. forsteri (Sm.) DC. Western and Southern Europe.

(b) Fls. in clusters, clusters stalked

1714. **L. campestris** (L.) DC. FIELD WOODRUSH, SWEEP'S BRUSH. Fls. 3–10, in dense oval clusters borne on stalks of very unequal length, and forming an umbel-like inflorescence. Terminal fl. clusters stalkless and erect, the longest stalked clusters often drooping, all clusters reflexed in fr. Perianth segments 3–4 mm., chestnut-brown with a transparent margin; anthers longer than the filament; bracts papery. Lvs. grass-like, 2–5 mm. broad, with long, sparse, whitish hairs. A loosely tufted perenn. 10–30 cm. △ Grassy places, meadows. Apr.–June. All Eur. (except IS.). Pl. 178.

L. spicata (L.) DC. SPIKED WOODRUSH. Distinguished by its dense, nodding, spike-like inflorescence of short-stalked clusters of chestnut-brown fls. Perianth segments 2½–3 mm. long, with a brown, fine-pointed apex. Lvs. 1–2 mm. broad. A slender tufted perenn. to 20 cm. Arctic region and mountains of Europe.

*L. multiflora (Retz.) Lej. MANY-HEADED WOODRUSH. Widespread in Europe.

1715. **L. sylvatica** (Hudson) Gaudin GREATER WOODRUSH. A robust perenn. often forming mats or tussocks, with erect stems usually more than 40 cm. and up to 80 cm., bearing a lax, spreading, compound inflorescence of numerous chestnut-brown, unequal-stalked fl. clusters. Fls. 3–4 in each cluster; perianth segments 3–3½ mm., about equalling the fr. Basal lvs. broad, glossy green, 10–30 cm. by 6–12 mm. gradually tapering to a very acute point, with sparse long hairs; stem lvs. c. 4 up to 5 cm. long. Fr. ovoid, with a fine apex, about as long as the perianth. △ Woods, moors. Apr.–July. Most of Eur. (except IS.SF.): introd. S.

COMMELINACEAE | Spiderwort Family

Anns. or herbaceous perenns. of tropical and subtropical regions. Lvs. alternate entire and clasping the stem, which is swollen at the nodes. Fls. with 3 green sepal-like segments and 3 white or violet petal-like segments; stamens usually 6; ovary superior, two- to three-celled; style 1. Fr. a capsule.

Fertile stamens 3; filaments hairless.	*Commelina*
Fertile stamens 6; filaments hairy.	*Tradescantia*

COMMELINA   Stamens 3 fertile and 3 sterile, filaments hairless. 1 sp.

1716. *C. communis* L. DAY-FLOWER. Fls. bright blue, several, clustered in the axil of a broad, oval, leafy bract resembling a sheath. Sterile stamens with 4 yellow lobes in a cross. Lvs. lance-shaped, sheathing; stems erect or creeping and rooting at the nodes. △ Native of China; grown as ornamental, sometimes naturalized. July–Sept.

TRADESCANTIA   Stamens 6 fertile, filaments hairy. 2 sps.

1717. *T. virginiana* L. COMMON SPIDERWORT. Fls. violet-purple, long-stalked in a rounded umbel-like cluster and subtended by narrow, lance-shaped, leafy bracts much longer than the fls. Fls. *c.* 2½ cm. across; sepals and fl. stalks hairy. Lvs. linear 15–40 cm. An erect, nearly hairless perenn. to 70 cm. △ Native of North America; grown as an ornamental and sometimes escaping. May–July.

T. fluminensis Vell. WANDERING JEW. Stems creeping and rooting at nodes; lvs. *c.* 4 cm., oval-elliptic, and sheathing the stem at the base, often violet beneath. Fls. white, *c.* 1 cm. across, in stalkless umbels subtended by 2 much larger unequal leafy bracts. Native of South America; often grown for ornament and sometimes escaping.

PALMAE | Palm Family

Trees, shrubs, or woody climbers widely distributed in the tropical regions, often with unbranched trunks and with very large, pinnately or palmately divided lvs. usually clustered into a crown. Plants often one-sexed. Fl. clusters at first enclosed in a large sheath or *spathe*; fls. small, numerous, one-sexed; perianth 6; stamens 3, 6, or numerous; ovary 1 or 3. Fr. a berry, drupe or nut.

Lvs. pinnate; trunk stout, elongate.	*Phoenix*
Lvs. palmately cut or fan-shaped; trunk short.	*Chamaerops*

PHOENIX   Lvs. pinnate with numerous leaflets; trunk stout. Stamens 6; spathe 1. *c.* 2 sps.

1718. *P. canariensis* Chaub. CANARY PALM. A typically palm-like tree with a stout, unbranched trunk 6–8 m., closely covered with fibrous lv. bases, or when trimmed, with horizontal scars, and with an enormous terminal rosette of spreading pinnate lvs. Lvs. to 7 m., arching upwards, with very numerous, lance-shaped, strongly keeled leaflets. Fr. smooth brown, the size of an olive, flesh dry, tasteless. △ Native of the Canary Islands; commonly planted for ornament in the Mediterranean region. Pl. 176.

P. dactylifera L. DATE PALM. Like 1718, but trunk taller to 20 m. and more slender. Lvs. grey-green, arching or somewhat down-curved; leaflets stiffer. Fr. *c.* 3 cm., reddish or yellowish-brown and borne on long, hanging, fruiting branches. Native of the Middle East; planted for fr. and ornament, naturalized in South Spain.

CHAMAEROPS   Lvs. fan-shaped, palmately cut into stiff segments; stems usually very short, suckering. Stamens 6–9; spathe 2–4. 1 sp.

1719. *C. humilis* L. DWARF FAN PALM. Lvs. with rounded blades, deeply cut to half their width or more into 12–15 stiff, lance-shaped segments, and with a stout lv. stalk with straight spines. Fls. in dense many-flowered clusters borne among the lvs. Stems suckering and often forming dense bushy patches *c.* ½ m. high, rarely forming a short trunk. The only native European palm. △ Dry, sandy, and rocky places. Mar.–June. P.E.I. *Fibres obtained from the leaves are used for brush- and rope-making.* Pl. 176.

GRAMINEAE | Grass Family

Herbaceous anns. or perenns., rarely woody, usually with hollow stems and solid nodes. Lvs. alternate, with a sheathing base and a narrow spreading blade. At the junction of the sheath and the blade there is usually a flap of tissue, or *ligule*, or sometimes a ring of hairs, and less commonly 2 thickened projections, or *auricles*, on each side at the junction. Inflorescence simple and spike-like, or branched with the branches often further branched and forming a simple or compound cluster, or *panicle*. Fls., or *florets*, tiny, usually bisexual and often densely clustered and overlapping and forming a one- to many-flowered *spikelet*. The spikelet is usually stalked and usually has 2 lower, sterile, scale-like outer bracts or *glumes*, commonly partly encircling the florets. Florets each consisting of an outer bract or *lemma* and an inner thinner bract or *palea*, 3 stamens or rarely 1 or 2, and a superior one-celled ovary with 2 feathery styles. Spikelets often with sterile or male florets as well as fertile florets. Fr. with 1 seed closely invested and fused with the ovary wall.

1. Florets all one-sexed and dissimilar, the male differing from the female.   Group A
1. Florets all similar, either bisexual or one-sexed.
  2. Spikelets stalkless or nearly so, and arising from a tooth or excavation in the axis of the inflorescence.
    3. Spikelets in a solitary, terminal, spike-like cluster.   Group B
    3. Spikelets in several, often terminal, spike-like clusters.   Group C
  2. Spikelets stalked or very short-stalked, but never from excavations and forming a compound, narrow- or broadly-branched inflorescence.
    4. Spikelets in short-stalked branched clusters, closely grouped and forming a slender spike-like or rounded inflorescence.   Group D
    4. Spikelets long-stalked, forming a branched open or spreading inflorescence.
      5. Spikelets with 1 fertile floret and often with sterile or rudimentary florets.   Group E
      5. Spikelets with 2 or more fertile florets.
        6. Glumes as long as or longer than the florets.   Group F
        6. Glumes much shorter than the florets.   Group G

Group A. *Florets one-sexed, dissimilar*

Nos.

1. Lvs. 5–12 cm. wide; male florets in slender terminal spikelets; female spikelets among lower lvs.   *Zea*   1814

1. Lvs. 2 cm. wide; male and female fls. in the same spikelet.   *Coix*   1815

Group B. *Spikelets stalkless or nearly so, in a solitary spike-like cluster*   *Nos.*

1. Spikelets enclosed in a sheath which replaces the glumes.   *Lygeum*   1799
1. Spikelets not enclosed in a sheath.
  2. Spikelets arranged unilaterally on one side of the axis.
    3. Spikelets solitary, arising from each excavation of the axis.   *Nardus*   1763
    3. Spikelets 2–3, arising from each excavation of the axis.   *Stenotaphrum*   1810
  2. Spikelets arranged on 2 sides, or all round the axis.
    4. Inflorescence spiny, little longer than broad.   *Echinaria*   1738
    4. Inflorescence not spiny, usually much longer than broad.
      5. Spikelets 2–3, arising from each tooth on the axis.
        6. Spikelets with long awns.   *Hordeum*   1734, 1735
        6. Spikelets awnless.   *Elymus*   1736
      5. Spikelets solitary, arising from each tooth or excavation of the axis.
        7. Spikelets with 1 floret only; inflorescence very slender.
          8. Spikelets awnless.   *Parapholis*   1761
          8. Spikelets with awns.   *Psilurus*   1762
        7. Spikelets with several florets.
          9. Ligule replaced by a ring of hairs; glumes as long as spikelet.   *Sieglingia*   [1763]
          9. Ligule papery.
            10. Spikelets placed edgeways to axis; glume 1.   *Lolium*   1760
            10. Spikelets placed broadside to axis; glumes 2.
              11. Lemma with a bent awn arising from the middle.   *Gaudinia*   1769
              11. Lemma with straight terminal awn, or awnless.
                12. Spikelet with 2 fertile florets and usually 1 sterile floret.   *Secale*   1733
                12. Spikelets with 3–25 florets.
                  13. Spikelets arranged on 2 sides of a simple axis.
                    14. Spikelets 2–4 cm. long.   *Brachypodium*   1726
                    14. Spikelets shorter.
                      15. Low spreading or ascending anns.   *Catapodium*   1759
                      15. Tall erect perenns.; inflorescence quadrangular in section.   *Agropyron*   1727–1729
                  13. Spikelets arranged all round axis forming a dense, cylindrical, spike-like inflorescence.
                    16. Glumes rounded on back, with 2–4 awns or teeth.   *Aegilops*   1730, 1731
                    16. Glumes keeled on back, blunt or with a short point.   *Triticum*   1732

Group C. *Spikelets in several, often terminal, slender, spike-like clusters*   *Nos.*

1. Spike-like clusters several, in a terminal whorl; spikelets often arranged on one side of the branches.
  2. Spikelets with usually 1–2 florets; spike-like clusters long and narrow.

3. Clusters paired; bracts leafy, conspicuous.	*Hyparrhenia*	1809
3. Clusters several; bracts not conspicuous.		
4. Lemma rounded on the back.		
5. Lemma with a long awn.	*Bothrichloa*	1808
5. Lemma awnless.	*Digitaria*	1802
4. Lemma keeled on the back.	*Cynodon*	1801
2. Spikelets with 3–8 florets; inflorescence oval.	*Sclerochloa*	1750
1. Spike-like clusters several, alternate or opposite, spikelets not arranged on one side of the branches.		
6. Spikelets arranged in 2 regular ranks on lateral branches; glumes swollen, boat-shaped.	*Beckmannia*	1800
6. Spikelets arranged all round the lateral branches.		
7. Spikelets plump; lemma rounded òn the back; ligule absent.	*Echinochloa*	1804
7. Spikelets flattened; lemma keeled on the back; ligule a fringe of hairs.	*Spartina*	1796

Group D. *Spikelets in short-branched clusters, closely grouped and forming a slender spike-like, or a dense rounded, inflorescence*

		Nos.
1. Spikelets of 2 kinds, sterile and fertile.		
2. Sterile spikelets with oval, blunt, awnless glumes.	*Lamarckia*	1749
2. Sterile spikelets with sharp-pointed or awned, linear-lance-shaped glumes.	*Cynosurus*	1748
1. Spikelets all similar and all fertile.		
3. Spikelets with hooked bristles on glumes.	*Tragus*	1813
3. Spikelets without hooked bristles..		
4. Glumes whitish with long silky hairs.		
5. Lvs. stiff, inrolled; spikelets awnless.	*Imperata*	1806
5. Lvs. flexible, broad, flat; spikelets awned.		
6. Lvs. hairless; ligule elongated; inflorescence elongated.	*Polypogon*	1778
6. Lvs. softly hairy; ligule short; inflorescence oval.	*Lagurus*	1779
4. Glumes hairless or finely hairy, or ciliate on keel.		
7. Lemmas with white or grey silky hairs.		
8. Spikelets awned.	*Achnatherum*	1789
8. Spikelets not awned.	*Melica*	1744, 1745
7. Lemmas without silky hairs.		
9. Spikelets surrounded by a few stiff, bristle-like hairs.	*Setaria*	1805
9. Spikelets not surrounded by hairs.		
10. Spikelets with only 1 fertile floret.		
11. Spikelets rather few, linear, 1–6 cm. Awn very long.	*Stipa*	1786–1788
11. Spikelets numerous, oval-lance-shaped, usually much less than 1 cm.		
12. Glumes with a distinctly winged keel. Inflorescence dense.	*Phalaris*	1794, 1795
12. Glumes with keel not winged.		
13. Glumes 4; the 2 outer very unequal; stamens 2.	*Anthoxanthum*	1793
13. Glumes 2; stamens 3 (except in *Crypsis*).		
14. Spikelets *c.* 1 cm.; inflorescence 10–25 cm. long.	*Ammophila*	1775

14. Spikelets 2–4 mm.; inflorescence usually less than 10 cm.		
15. Ligule replaced by hairs.	*Crypsis*	1783
15. Ligule papery.		
16. Glumes fused at base, tapering to an incurved apex; awn of lemma bent, rarely absent.	*Alopecurus*	1780–1782
16. Glumes not fused, usually abruptly contracted to an outward-curving tip; lemma not awned.	*Phleum*	1784, 1785
10. Spikelets with 2 or more fertile florets.		
17. Lemma with a long awn.		
18. Awn of lemma terminal, straight.	*Vulpia*	1758
18. Awn arising from the back of the lemma, bent.	*Trisetum*	1764
17. Lemma with a very short awn, or awn absent.		
19. Spikelets arranged on one side of branches; inflorescence one-sided.		
20. Inflorescence stiff erect, almost stalkless and scarcely free from the upper lv. sheath.	*Sclerochloa*	1750
20. Inflorescence long-stalked, carried well above the upper lv. sheath.	*Dactylis*	1747
19. Inflorescence nearly cylindrical, with spikelets arranged all round the axis.		
21. Lemma blunt, with 3–5 teeth at apex.	*Sesleria*	1737
21. Lemma acute, or two-toothed with a central short awn.	*Koeleria*	[1763]

Group E. *Spikelets long-stalked in a branched inflorescence; spikelets with 1 fertile floret*

		Nos.
1. Spikelets in clusters of 2–3, dissimilar, the fertile stalkless, the sterile stalked. Ligule replaced by hairs.		
2. Lvs. narrow, 2–4 mm.; a tuft of russet hairs present below each spike.	*Chrysopogon*	1811
2. Lvs. broad, 6–30 mm.; without tuft of hairs below each spike.		
3. Spikelets surrounded by long white silky hairs.	*Erianthus*	1807
3. Spikelets not surrounded by hairs.	*Sorghum*	1812
1. Spikelets all similar and all stalked, solitary, or clustered.		
4. Glumes absent or very small and much smaller than lemma.		
5. Inflorescence enclosed for the greater part by the upper lv. sheath; lemma 4–5 mm.	*Leersia*	1798
5. Inflorescence soon free from upper sheath; lemma 7–9 mm.	*Oryza*	1797
4. Glumes well developed and easily visible, at least equal the lemma.		
6. Spikelets always containing 1–2 conspicuous sterile male florets, as well as the fertile florets.		
7. Spikelets surrounded with long white hairs at the base.	*Erianthus*	1807
7. Spikelets without hairs at the base.		
8. Spikelets awnless.		

9. Branches of inflorescence bearing 1–3 spikelets only.	*Melica*	1744, 1745
9. Branches of inflorescence bearing more than 3 spikelets.		
10. Spikelets flattened from side to side; ligule papery.	*Hierochloe*	1792
10. Spikelets flattened from front to back; ligule replaced by hairs.	*Panicum*	1803
8. Spikelets awned, flattened from side to side.		
11. Spikelets hairless, 7–10 mm. long.	*Arrhenatherum*	1768
11. Spikelets softly hairy, 3–6 mm. long.	*Holcus*	1770
6. Spikelets without sterile florets, or, if present, then very inconspicuous.		
12. Lemma leathery, margin inrolled; awn often very long 2–30 cm.	*Stipa*	1786–1788
12. Lemma not leathery, and margin not inrolled; awn less than 2 cm.		
13. Florets with long hairs at the base and often as long as the lemma.	*Calamagrostis*	1774
13. Florets without hairs at the base.		
14. Spikelets rounded on the back.		
15. Lemma awned.	*Oryzopsis*	1790
15. Lemma not awned.	*Milium*	1791
14. Spikelet keeled on the back.		
16. Spikelets not awned; lvs. 8–16 mm. broad.	*Phalaris*	1794, 1795
16. Spikelets awned; lvs. narrow, usually 1–6 mm. broad.		
17. Awn arising near apex of lemma and more than twice as long as lemma.	*Apera*	1776
17. Awn arising near base of lemma, not twice as long as lemma.	*Agrostis*	1777

Group F. *Spikelets long-stalked with 2 or more fertile florets; glumes longer than florets*

		Nos.
1. Florets surrounded by long white hairs; inflorescence silky-haired.	*Arundo*	1739
1. Florets not surrounded by long hairs.		
2. Awns absent.	*Melica*	1744, 1745
2. Awns present and inserted on the back of the lemma.		
3. Spikelets two-flowered, tiny, 2–6 mm. long; awn projecting little.		
4. Tufted perenns.		
5. Awn with a ring of minute hairs at about the middle and with club-shaped apex.	*Corynephorus*	[1772]
5. Awn without a ring of hairs, not club-shaped. Lemmas blunt.	*Deschampsia*	1771, 1772
4. Anns. Upper glume one-veined; lemma acute, two-toothed.	*Aira*	1773
3. Spikelets with 2–6 florets, usually more than 5 mm. long; awn long, bent and strongly projecting.		
6. Spikelets small, 5–7 mm.; lemma keeled.	*Trisetum*	1764
6. Spikelets longer, 8–35 mm.; lemma rounded on back.		
7. Spikelets always erect.	*Helictotrichon*	1767
7. Spikelets at length drooping.	*Avena*	1765, 1766

Group G. *Spikelets long-stalked with 2 or more fertile florets; glumes shorter than the florets*

		Nos.
1. Florets surrounded by long white silky hairs.		
2. Lvs. narrow, less than 1 cm. wide, margin inrolled; ligule long, papery.	*Ampelodesma*	1741
2. Lvs. broad, 1–5 cm. wide; ligule replaced by hairs.		
3. Stems robust, *c.* 2 cm. or more across; axis of spikelets and lemma with long hairs.	*Arundo*	1739
3. Stems slender *c.* 1 cm. across; axis of spikelets with long hairs; lemma hairless.	*Phragmites*	1740
1. Florets not surrounded by long hairs.		
4. Spikelets with only 2 fertile florets.		
5. Ligule replaced by long hairs.	*Molinia*	1742
5. Ligule papery.		
6. Lemmas blunt and appearing cut-off.	*Catabrosa*	[1743]
6. Lemmas acute.	*Melica*	1744, 1745
4. Spikelets of 3–30 fertile florets.		
7. Spikelets nearly as broad as long, with a heart-shaped base, at length trembling on slender stalks.	*Briza*	1746
7. Spikelets distinctly longer than broad, not heart-shaped at base, not trembling.		
8. Spikelets with overlapping florets arranged in 2 conspicuous ranks, awnless; ligule replaced by hairs.	*Eragrostis*	1743
8. Spikelets not conspicuously two-ranked, or if two-ranked then awned; ligule papery.		
9. Spikelets in compact, one-sided clusters forming an asymmetrical branched inflorescence.	*Dactylis*	1747
9. Spikelets not compact and inflorescence usually not one-sided.		
10. Awn inserted about the middle, or below the middle of the lemma.	*Avena*	1765, 1766
10. Awn inserted at or very near the apex of the lemma, or absent.		
11. Lemma finely notched with awn arising from base of the notch, or awn absent.	*Bromus*	1720–1725
11. Lemma not notched, awnless, or with a terminal awn.		
12. Ligule replaced by hairs.	*Molinia*	1742
12. Ligule papery.		
13. Spikelets rounded on the back, not keeled. Hairless plants.		
14. Glumes and lemmas blunt, awnless; aquatic plants.	*Glyceria*	1755, 1756
14. Glumes and lemmas acute, often awned; usually plants of dry places.	*Festuca*	1757
13. Spikelets keeled on the back.		
15. Lemma with an awn as long as itself. Inflorescence usually very slender.	*Vulpia*	1758
15. Lemmas not awned.		
16. Spikelets on very short, rather stout stalks forming a one-sided stiff spike; low anns.	*Catapodium*	1759

16. Spikelets on slender, usually spreading stalks;
lemma often with cottony hairs at base.   *Poa*   1751–1754

BROMUS | Brome   Spikelets of many overlapping florets, somewhat flattened from side to side and usually long-stalked. Glumes unequal and often awned, the lower one- to five-veined, the upper three- to seven-veined; lemma with an awn often longer than itself, but awn sometimes absent. *c.* 35 sps.

Perenns. usually more than ½ m.
(a) Inflorescence erect
1720. *B. erectus* Hudson (*Zerna e.*) UPRIGHT BROME. A stiff, erect, tufted perenn. to 1 m., with an erect or somewhat nodding, reddish, purplish, or green, nearly simple, inflorescence 10–15 cm. Spikelets 1½–4 cm., narrowly oblong. Glumes nearly equal, keeled, awnless; lemma usually hairless, seven-veined and with a transparent margin above, about 3 times as long as its awn; anthers orange. Lower lvs. inrolled, upper lvs. flat, hairless or ciliate. △ Dry meadows, banks. May–July. Much of Eur. (except IS.AL.TR.): introd. North Eur. Pl. 178.

1721. *B. inermis* Leysser AWNLESS or HUNGARIAN BROME. An extensively creeping perenn. with stiff, erect stems to 1½ m., and a dense, erect, compound inflorescence with lance-shaped, awnless spikelets 1½–3 cm. Glumes unequal, at least the upper with a blunt transparent tip; lemma 7–9 mm., blunt or pointed, thin, transparent, hairy. Lvs. usually hairless, dark green, 5–8 mm. broad. △ Waysides, hedges, hollows. June–Sept. E.F.D. CH.A.CS.I.GR.TR.BG.

(b) Inflorescence drooping
1722. *B. ramosus* Hudson (*Zerna r.*) HAIRY BROME. A tall hairy perenn. 1–1½ m. with slender long-stalked spikelets nodding gracefully in a lax arched cluster 15–45 cm. long. Spikelets 2–4 cm., linear, awned. Glumes unequal, keeled, the lower one-veined, the upper three- to five-veined; lemma 10–14 mm., keeled, awn 4–8 mm. Lvs. dark green, to 60 cm. long and 6–16 mm. wide, drooping. △ Woods and wood-margins, roadsides, hedgerows. June–Aug. Most of Eur. (except P.NL.IS.SF.AL.). Pl. 178.

Anns. usually less than ½ m.
(a) Inflorescence for the most part erect
1723. *B. hordeaceus* L. (*B. mollis* L.) LOP-GRASS. Inflorescence erect, usually ovoid, of many densely clustered hairy spikelets with awns to 1 cm. Spikelets rather rotund, 1–2 cm., lance-shaped; glumes unequal, the lower lance-shaped, the upper oval; lemma 6–9 mm., awned, hairy, veins prominent. Lvs. greyish-green, blades flaccid, 2–7 mm. wide. A variable sp. with several closely related microspecies. A rather stout, usually softly hairy ann. or bienn. 5–80 cm. △ Cultivated and uncultivated places. May–July. All Eur.

*B. *racemosus* L. SMOOTH BROME. Widespread in Europe.

*B. *arvensis* L. Widespread in Europe.

*B. *madritensis* L. (*Anisantha m.*) COMPACT BROME. Western and Mediterranean Europe.

(b) Inflorescence drooping after flowering
1724. *B. tectorum* L. (*Anisantha t.*) DROOPING BROME. A hairy ann. 10–60 cm., with one or many erect stems and a lax or rather dense glistening inflorescence 4–18 cm., drooping to one side. Spikelets numerous, 2½–3½ cm., in clusters of up to 8 on the spreading, hairy side branches, florets 4–5 in each spikelet. Glumes unequal, thin-papery, hairy, the lower one-veined, the upper three-veined; lemma 9–13 mm., seven-veined, hairy, awn rough,

straight, 10–18 mm. Lvs. softly hairy. △ Sandy places, walls, fields, uncultivated ground, waysides. May–Aug. All Eur. (except IRL.): introd. North Eur. GB.IS.SF.

*B. sterilis L. (Anisantha s.) BARREN BROME. Almost throughout Europe.

*B. commutatus Schrader MEADOW BROME. Widespread in Europe.

1725. B. squarrosus L. An ann. to 60 cm., with a lax, one-sided green or purplish, often simple inflorescence with large oval to lance-shaped stalked spikelets 2½–4 cm., which at length become pendulous. Spikelets with awned lemmas, awns to 13 mm., twisted at the base and strongly curved outwards, but awns shorter or absent in the lower florets; glumes nearly equal, the lower five-veined, the upper seven-veined. Lvs. and stems hairy below. △ Rocky places, dry fields, vineyards. May–July. Much of Eur. (except North Eur. P.IRL.B.NL.): introd. GB.D.A.

*B. secalinus L. RYE-BROME. Widespread in Europe.

BRACHYPODIUM | False-Brome Like Bromus, but spikelets rounded, not flattened, and almost stalkless in 2 alternating rows, inserted edgeways on to the axis. c. 7 sps.

1726. B. pinnatum (L.) Beauv. HEATH FALSE-BROME. A tall, stiff, erect, usually hairless and often glaucous perenn. 30–120 cm., with a slender, usually erect, spike-like inflorescence 4–25 cm., with rather few greenish or yellowish, almost stalkless alternate spikelets. Spikelets 3–15, each 2–4 cm. long, straight or curved, with many florets. Glumes unequal, hairless, the lower three- to six-veined, the upper five- to seven-veined; lemma 6–10 mm., seven-veined, hairless, smooth, with a fine awn 1–5 mm. Lvs. variable, flat or inrolled, stiff or flaccid, usually 2–6 mm. wide. △ Dry uncultivated ground, grasslands. June–Aug. All Eur. (except IS.).

B. sylvaticum (Hudson) Beauv. SLENDER FALSE-BROME. Like 1726, but stem and lvs. with long spreading hairs; inflorescence spike-like, but rather lax and usually nodding. Awns to 12 mm., as long as or longer than the lemmas; glumes and lemmas usually hairy. A soft, yellow-green perenn. of shady places and woods. Almost all Europe.

AGROPYRON | Couch-Grass Like Brachypodium, but spikelets inserted broadside on to the axis in 2 alternating rows, spikelets quite stalkless and forming a slender, spike-like inflorescence. c. 25 sps.

Awns absent
1727. A. junceiforme (A. & D. Löve) A. & D. Löve (A. junceum auct.) SAND COUCH. A far-creeping perenn. with erect, very glaucous stems and lvs. to 60 cm., and with a stiff, spike-like inflorescence 4–20 cm., with rather few alternately arranged spikelets. Spikelets stout, 1½–3 cm., their own length or less apart, readily breaking up between each lemma at maturity. Glumes very tough, rigid, seven- to eleven-veined, two thirds as long as the spikelet; lemma thick, rigid, 1–2 cm., five-veined, blunt or notched and with a short hard point. Lvs. glaucous, flat or inrolled, minutely hairy above. △ Sandy shores, dunes. June–Aug. Coastal regions of all Eur. (except IS.). Pl. 178.

*A. pungens (Pers.) Roemer & Schultes SEA COUCH. Coasts of Western and Mediterranean Europe. Often hybridizes with 1727.

A. elongatum (Host) Beauv. Like 1727, but densely tufted and without creeping rhizomes, and lvs. green, inrolled, very rough above. Fl. spike slender; spikelets oval, widely spaced, the lower not overlapping, not readily breaking up; glumes c. 1 cm., usually nine-veined, little more than half as long as the spikelets. Mediterranean region, Portugal.

1728. *A. repens* (L.) Beauv. COUCH-GRASS, TWITCH. A creeping perenn. spreading extensively by underground stolons and often forming large patches, with stiff erect stems 30–120 cm., bearing a straight, erect, spike-like inflorescence 5–30 cm. Spikelets numerous, overlapping, 1–2 cm. long. Glumes three- to seven-veined, blunt or pointed; lemmas tough, 8–13 mm., five-veined, blunt or pointed. Lvs. dull green, usually sparsely hairy above, but a very variable sp. Hybridizes with 1727 and with *A. pungens*. △ Waysides, waste ground, rough grassland; often a noxious weed of cultivated ground. June–Sept. All Eur. *Used medicinally in the past*. Pl. 178.

Awns present

1729. *A. caninum* (L.) Beauv. BEARDED COUCH. A loosely tufted, but not creeping perenn., with erect, hairless stems to 1 m., with a long, slender, curved or nodding, awned, spike-like inflorescence 5–20 cm. Spikelets 1–2 cm., oblong-lance-shaped. Glumes rigid, sharply pointed and sometimes awned, with 2–5 rough veins; lemma 9–13 mm., rigid, five-veined, with a terminal awn 7–20 mm. Lvs. bright green, flat, loosely hairy above, or hairless. △ Shady places, woods, hedgerows, ravines. May–Aug. All Eur. (except P.TR.).

AEGILOPS Spikelets stalkless, placed broadside on to the axis, in a compact ovoid or cylindrical inflorescence. Glumes large, tough, and strongly veined, with 2–4 long awns or teeth at the apex; lemmas also with 3–5 long awns or teeth. *c*. 10 sps.

1730. *A. ovata* L. A hairy or hairless, tufted ann. to 40 cm., with erect stems bearing a short oval inflorescence 1–2 cm. long, with a few long-awned spikelets. Spikelets 2–5, the upper slender and sterile, the lower swollen, oval, with 3–4 fertile florets. Glumes thick-veined, usually with 4 long, more or less equal stiff spreading awns 2–7 cm.; lemma with 2–3 slender awns. Lvs. flat, rough. △ Dry uncultivated ground, waysides. May–July. Med. Eur. P.A.BG.R.SU.: introd. CS. Pl. 178.

A. triuncialis L. Like 1730, but inflorescence linear-lance-shaped, 4–6 cm. long excluding awns, with 4–7 oblong-cylindrical, scarcely swollen, not overlapping spikelets and with 2–3 rudimentary spikelets at the base of the inflorescence. Glumes rough or hairy, with 2–3 long stiff awns, 4–5 cm.; lemmas of lower spikelets with 3 teeth, lemma of upper spikelet with 3 long awns. Mediterranean region.

1731. *A. cylindrica* Host Inflorescence very slender, fragile, 6–12 cm. long by 3–4 mm. broad, with 5–11 alternately arranged spikelets sunk into the axis, and with only the 2 uppermost spikelets with long awns. Glumes linear, those of lower spikelets with a short awn and 1 tooth, glume of the uppermost spikelet with 2 teeth and a long awn 5–8 cm.; lemmas mostly shortly awned, only the uppermost long-awned. Lvs. 1–2 mm. broad, hairy. △ Dry, grassy places. May–July. A.H.YU.GR.BG.SU.: introd. elsewhere in Eur. Page 534.

TRITICUM | Wheat Like *Aegilops*, but glumes keeled and blunt, or with a short narrow point; lemma usually with a long awn. *c*. 10 sps.

1732. *T. turgidum* L. RIVET or CONE WHEAT. Distinguished by its dense, oblong, and angled spike, *c*. 6 cm., with numerous very long-awned spikelets, which are as broad as long. Glumes broadly oval, 1 cm., strongly keeled their whole length, apex three-pointed, the middle point longest; lemmas all with long awns. Fr. oval, little longer than broad. Ann. usually with solid stems, to 120 cm. △ Cultivated chiefly in Southern and Mediterranean Europe. June–Aug.

T. durum Desf. MACARONI WHEAT. Like 1732, but spikelets longer than broad; glumes oblong, strongly keeled and with a long fine point; lemmas with a long awn. Fr. hard, oblong, about 3 times as long as broad. Cultivated chiefly in Southern and Mediterranean Europe. Pl. 178.

T. aestivum L. BREAD WHEAT. Distinguished by the glumes which are rounded below on the back and keeled only in the upper half. Spike dense, quadrangular, and usually without long awns. Glumes oval, less than 1 cm. long; lemma with a short narrow point, rarely long-awned. Fr. ovoid. Widely cultivated in Europe. *The most important cereal crop plant, occurring in nearly 50 per cent of the total area sown with grain crops in Europe. Largely consumed by humans, but valuable as a food for livestock; the basis of starch-making and distillation. The straw is used for mats, wrappers, and thatching.*

T. polonicum L. POLISH WHEAT. Readily distinguished from 1732 by the much longer spikelets 2–3 cm., 3 times as long as broad and possessing very long awns. Glumes 2 cm. or more, lance-shaped, longer than the florets, keeled from the base and with stiff,hairs on the keel, apex two-toothed; lemma keeled, awn very long, to 10 cm. Fr. oblong. Occasionally cultivated in Southern Europe.

SECALE | **Rye** Like *Triticum*, but spikelets two-flowered, rarely with 3 or 4 fls. Glumes linear-lance-shaped, one-veined, acute or awned; lemma keeled and keel with stiff spreading hairs, long-awned. *c.* 3 sps.

1733. *S. cereale* L. RYE. A somewhat glaucous, erect, hairless ann. 1–2 m., with a long, slender, erect or somewhat nodding, dense spike-like inflorescence 8–15 cm., of numerous long-awned, stalkless spikelets. Glumes keeled, linear and narrowed to a long awn-like tip; lemma with a ciliate keel, awn to 3 cm. Lvs. soft, 3–8 mm. wide, rough. △ Widely cultivated in Europe; sometimes naturalized. May–July. *An important cereal crop in Central and Eastern Europe, used in bread-making, for alcoholic fermentation and as food for livestock. Rye bread is greyish or 'black', and was widely consumed, but wheat bread has superseded it in many countries. The straw is strong and used for thatching. It can be grown further north than any other cereal crop.*

HORDEUM | **Barley** Like *Triticum* with dense, long-awned, spike-like inflorescences, but spikelets in clusters of 3, arising from each joint of the axis, or rarely 2. Glumes narrow, one-veined, awned, placed side by side in front of the spikelets. *c.* 12 sps.

1734. *H. bulbosum* L. BULBOUS BARLEY. Distinguished by the small, bulb-like swelling at the base of the stem and the slender, linear, yellowish or purplish, dense cylindrical, spike-like inflorescence, 6–12 cm. by 1 cm. Central fertile spikelet of trio shortest, with lance-shaped glume with awns *c.* 1 cm. and lemma with awn 2–2½ cm.; lateral spikelets sterile, with the outer glume with awn *c.* 2 cm.; lemma blunt. Lvs. flat, the upper hairless, the lower with a hairy sheath. A robust erect perenn. to 1 m. △ Waste places, waysides. May–July. Med. Eur. (except AL.) BG.R.SU.

1735. *H. murinum* L. WALL BARLEY. A rather stout, often tufted ann. to 60 cm., with flat, light green, hairy lvs. and flattened, cylindrical, spike-like inflorescences 4–12 cm., with numerous long awns. Central spikelet of trio fertile, with lance-shaped, ciliate, long-awned glumes; lemma five-veined, with an awn 2–5 cm.; lateral spikelets sterile or with 2 male florets, glumes bristle-like, long-awned, minutely toothed, lemma lance-shaped, with an awn 2–4 cm. △ Waste places, waysides, dry ground. May–Aug. All Eur. (except IS.AL.): introd. IRL.SF. Pl. 178.

*H. secalinum Schreber (*H. nodosum* auct.) MEADOW BARLEY. Western and Southern Europe.

H. marinum Hudson SEA BARLEY. Maritime Western and Southern Europe.

H. vulgare L. FOUR-ROWED or SIX-ROWED BARLEY. Distinguished by the four-sided, spike-like inflorescence 6–10 cm., with spikelets arranged in 6 longitudinal rows. Spikelets all fertile; glumes and lemma with long awns to 10 cm. Widely cultivated in Europe, rarely naturalized.

H. distichon L. TWO-ROWED BARLEY. Distinguished by the spikelets which are arranged in 2 longitudinal rows; the outer 2 spikelets of each trio sterile or male. Sometimes culti-vated in Europe. *Barley is one of the oldest cereal crop plants and has been in cultivation since about 2800 B.C. A very important food crop for man and livestock, and a source of malt used largely in brewing. It produces a non-glutinous flour of less value for bread-making than wheat, the latter has now largely replaced barley for this purpose.*

ELYMUS Spikelets stalkless, in clusters of 2–3, arising from each joint of the axis and placed broadside on to it. Glumes equal, leathery, five-veined, about equalling the florets and often placed side by side in front of it; lemma leathery. *c.* 5 sps.

1736. *E. arenarius* L. LYME GRASS. A robust glaucous perenn., with creeping underground stems and numerous erect stems forming large tufts, up to 2 m. high. Inflorescence spike-like 15–35 cm. long, stout, cylindrical, dense and stiff; spikelets 2–3 cm., oblong, in stalk-less pairs arising alternately on each side of the axis. Glumes lance-shaped, keeled, hairy; lemma lance-shaped, densely hairy, seven-veined, awnless. Lvs. stiff, broad 8–20 mm., sharp-pointed, rough on the veins above, smooth beneath. △ Sands and dunes on the littoral. June–Aug. West Eur. (except P.), North Eur. D.PL.H.I.YU.R.SU. Pl. 179.

SESLERIA Inflorescence spike-like, dense, ovoid, with numerous short-stalked spike-lets, and with tiny sheathing bracts at the base. Glumes nearly equal, keeled; lemma boat-shaped and with veins prolonged into teeth. *c.* 15 sps.

1737. *S. caerulea* (L.) Ard. BLUE MOOR-GRASS. A tufted perenn. with many erect, nearly leafless stems 10–50 cm., bearing a solitary, glistening, rather bluish or purplish, dense, oval to cylindrical, spike-like inflorescence 1–3 cm. Spikelets 4½–7 mm., two- to three-flowered. Glumes 3–6 mm., oval, one-veined, thin translucent; lemmas oval, hairy, with a broad five-toothed apex. Lvs. hairless, mostly basal with persistent keeled sheaths, blades 2–6 mm. wide, keeled beneath, abruptly narrowed to a fine point. △ Rocky places, dry banks, wet fields. Mar.–Aug. Much of Eur. (except P.NL.DK.GR.TR.): introd. N. Pl. 179.

ECHINARIA Inflorescence spike-like, globular; spikelets stalkless with one to several florets. Glumes leathery, one- to three-spined; lemma leathery, with 5 or 6 spines. 1 sp.

1738. *E. capitata* (L.) Desf. PRICKLE-GRASS. A low ann. to *c.* 15 cm., of the Mediterranean region, with very distinctive, long-stalked, globular inflorescences, with numerous stiff, spreading spines. Inflorescence 1–1½ cm., with numerous spikelets. Lower glumes with 2 or 3 short spines; lemma with 5–6 unequal, spreading and curved spines. Lvs. short, blade rough and flat. △ Dry places. May–July. Med. Eur. (except AL.) BG. Page 534.

ARUNDO Inflorescence large, compound, feathery, with numerous spikelets each one-to seven-flowered. Glumes lance-shaped, papery, keeled, three-veined; lemma papery with long, dense, silvery hairs and with a three-lobed apex. 2 sps.

1739. *A. donax* L. GIANT REED, CANE. A very robust, bamboo-like perenn. with stout, very leafy woody stems, 1½–5 m. high, usually growing in dense groups from swollen, underground, creeping stems. Inflorescence dense, feathery, 40–70 cm., whitish-green or violet and at length silvery; branches very rough; spikelets very numerous, *c.* 12 mm.; lemma with hairs up to 1 cm. long. Lvs. flat, 2–5 cm. broad. △ Damp places, watersides. Sept.–Oct. Med. Eur. (except TR.) P.: introd. R. *Grown as a wind-break or shelter for crops; the dried stems are used for fishing-rods, walking-sticks, baskets, and shelters. The underground stems have been used medicinally.*

PHRAGMITES | **Reed** Inflorescence large, compound, feathery with numerous slender spikelets, three- to seven-flowered. Stalks of florets clothed with a tuft of long hairs; glumes unequal, three-veined; lemma one- to three-veined, hairless. *c.* 2 sps.

1740. *P. communis* Trin. COMMON REED. A robust, bamboo-like perenn. with stiff, tough, non-woody, smooth stems 1½–3 m. high, growing in dense groups from stout, creeping stems. Inflorescence soft and feathery, brownish or purplish, erect or finally nodding, 15–40 cm. Spikelets 1–1½ cm., with silky hairs to 1 cm. long, arising from axis below the florets; glumes papery, smooth; lemma hairless. Lvs. greyish-green, flat, 1–3 cm. wide, tapering to a slender point. △ Shallow water, swamps, fens, rivers. Aug.–Sept. All Eur. (except IS.). *The dried stems are used for thatching, shelters, mats, etc. and recently the plant has been cropped to yield cellulose.* Pl. 179.

AMPELODESMA Hairless perenns. with a much-branched, spike-like inflorescence; spikelets lance-shaped, with 2–5 florets. Glumes equal, lance-shaped long-pointed; lemma hairy below, leathery, five-veined, apex bilobed and with a short awn between. 1 sp.

1741. *A. mauritanica* (Poiret) Durrand & Schinz (*A. tenax* (Vahl) Link) A very robust, densely tufted perenn. 1–3 m., with long, much-branched, interrupted, somewhat one-sided, purplish-green, spike-like inflorescences. Spikelets very numerous 1–1½ cm., shortly awned. Glumes often purplish, 9–12 mm., long-pointed; lemmas with long silky hairs at the base and on the keel, margin papery, awn 2 mm. Lvs. very long, tough, rigid and rush-like, with inrolled margin. △ Rocks, dry sunny places. May–June. E.F.I.GR. *The leaves are used like esparto grass.* Pl. 182.

MOLINIA Spikelets awnless, rounded, not flattened, with 2–5 florets, in a branched inflorescence. Glumes papery, about half the length of the first floret, one-veined; lemma stiff, three- to five-veined. Ligule a ring of hairs. 2 sps.

1742. *M. caerulea* (L.) Moench PURPLE MOOR-GRASS. A tufted, often tussock-forming perenn. with slender, stiff, erect stems 15–120 cm., bearing a variable, dense, interrupted, spike-like or open and lax, purplish, brownish or green inflorescence 5–40 cm. Spikelets oblong, 4–9 mm., glumes oval to lance-shaped, one- to three-veined or veins absent; lemma 4–6 mm. lance-shaped, acute or blunt. Lvs. flat, rough, 3–10 mm. broad, slightly hairy above or hairless. A variable sp. △ Heaths, fens, damp woods, and meadows. July–Sept. All Eur. Pl. 179.

ERAGROSTIS Spikelets awnless, laterally flattened, many-flowered, in a branched inflorescence. Glumes more or less equal; keeled, one-veined; lemma three-veined, keeled, blunt; palea with 2 keels, ciliate. *c.* 5 sps.

1743. *E. cilianensis* (All.) Vign. Lut. (*E. major* Host) An erect ann. to 50 cm. with a dense, branched, pyramidal inflorescence of large spreading spikelets with their florets arranged

1. *Parapholis incurva* 1761
3. *Aegilops cylindrica* 1731
5. *Eragrostis cilianensis* 1743
7. *Crypsis aculeata* 1783

2. *Echinaria capitata* 1738
4. *Lamarckia aurea* 1749
6. *Polypogon monspeliensis* 1778

in 2 very regular overlapping ranks. Spikelets $\frac{1}{2}$–3 cm. by 3 mm., with 15–40 florets. Glumes *c.* 2 mm.; lemma 2 mm., blunt and notched or with a short narrow point, and with conspicuous lateral veins. Lvs. with fine swollen teeth on the margin; sheath smooth. △ Cultivated ground, sandy places, walls, waysides. June–Oct. Med., South-East Eur. P.A.CS.H.SU.: introd. D.CH. Page 534.

*Catabrosa aquatica (L.) Beauv. WATER WHORL-GRASS. Most of Europe.

MELICA | **Melick** Spikelets awnless, rounded, not flattened, with 2–4 florets, the upper floret sterile and club-shaped. Glumes thin, three- to five-veined; lemma leathery, blunt, seven- to nine-veined; palea two-veined, tough. *c.* 13 sps.

1744. *M. nutans* L. MOUNTAIN or NODDING MELICK. Inflorescence 5–15 cm., slender, un-branched, one-sided and somewhat arched. Spikelets oval, 6–7 mm., violet or reddish, short-stalked, drooping; glumes oval; lemma oblong, hairless, strongly veined. Lvs. flat, sheaths hairless, ligule blunt. A slender hairless perenn. 30–60 cm., with creeping underground stems. △ Woods, shady rocks. May–July. Most of Eur. (except P.IRL.NL. IS.TR.).

*M. uniflora Retz. WOOD MELICK. Most of Europe.

1745. *M. ciliata* L. HAIRY MELICK. Readily distinguished by its spikelets which are covered with long, silvery-white or yellowish, silky hairs. Inflorescence slender, spike-like, 8–15 cm., with short lateral branches with up to 10 spikelets. Glumes *c.* 5 mm. oval long-pointed, pale and hairless; lemma lance-shaped, with long hairs on the back; palea hairy. Lvs. stiff, narrow, margin inrolled. A variable, often glaucous perenn. 30–80 cm. △ Rocks, dry places, walls, hills. May–Aug. Much of Eur. (except IRL.GB.IS.N.): introd. DK. Pl. 179.

BRIZA | **Quaking-Grass** Spikelets rotund, ovoid to heart-shaped, somewhat flattened, awnless, usually pendulous, with 5–15 closely overlapping florets, borne in a lax, branched inflorescence. Glumes equal, rounded on back, seven- to nine-veined; lemma thin and papery, boat-shaped, blunt; palea with 2 keels. 4 sps.

1746. *B. maxima* L. LARGE QUAKING-GRASS. A very handsome ann. to 50 cm., with neat, plump, ovoid, pendulous spikelets on very slender branches, in a lax, simple, one-sided inflorescence. Spikelets many-flowered, 1–2 cm. long, silvery-white, then brownish-purple, trembling. Glumes oval-boat-shaped, thin, overlapping; lemma rounded and with a heart-shaped base, thin. Lvs. flat, ligule lance-shaped, the uppermost lv. sheath some-what inflated. △ Dry places, hills. Apr.–June. Med. Eur. P.BG.SU. Pl. 179.

B. media L. QUAKING-GRASS. Like 1746, but inflorescence erect and pyramidal, with long, slender, spreading side branches, which are often further branched and which bear drooping purplish spikelets. Spikelets small, 4–5 mm., rounded or triangular, trembling. Ligule short. A perenn. to 50 cm. Widespread in Europe.

B. minor L. SMALL QUAKING-GRASS. Like [1746], but an ann. with numerous smaller, drooping, broadly triangular, pale green spikelets 2–4 mm.; glumes larger than the lemmas. Ligule lance-shaped acute. Mediterranean Europe and Portugal; naturalized elsewhere.

DACTYLIS | **Cock's-Foot** Inflorescence compound, with spikelets crowded into dense masses at the ends of the side branches. Spikelets flattened, short-stalked, with 2–5 florets; glumes keeled, three-veined; lemma keeled, five-veined, very shortly awned. *c.* 1 sp.

1747. *D. glomerata* L. COCK'S-FOOT. A coarse, tufted, usually glaucous, hairless perenn. to 1½ m. Inflorescence erect, in a rather unequal, one-sided, pyramidal cluster 3–15 cm., with the lower branches usually long and spreading horizontally, or reflexed in fl. and erect in fr., the upper branches much shorter. Spikelets 5–7 mm., green or violet; glumes ciliate on keel; lemmas closely overlapping, ending in a rigid awn to 1½ mm. Lvs. 2–14 mm. broad, keeled, rough on the margin; non-flowering shoots distinctly flattened. There are many cultivated forms. △ Meadows, rough grassland, roadsides. June–Sept. All Eur. *A valuable pasture and hay grass.*

CYNOSURUS | Dog's-Tail Inflorescence dense and spike-like, of numerous, almost stalkless spikelets. Spikelets paired and of 2 kinds: the outer sterile and comb-like, with many spreading lemmas, the inner fertile and with 2–7 florets. Glumes equal, papery, one-veined; lemma awned, three- to five-veined. *c.* 4 sps.

1748. *C. echinatus* L. ROUGH DOG'S-TAIL. Readily distinguished by its dense, rather one-sided ovoid, shining green or purplish, plume-like inflorescence 1–8 cm., with numerous slender, rough awns. Outer spikelet of each pair comb-like with several pairs of spreading, long-awned, narrow, sterile lemmas; the inner spikelet fertile, wedge-shaped, 8–14 mm., with lance-shaped, long-pointed glumes; lemma stiff, long-awned. An erect, hairless ann. to 50 cm. with inflated upper lv. sheaths and short, pointed lv. blades. △ Dry banks, sandy fields. Apr.–July. Med., South-East Eur. P.CH.A.H.: introd. D.CS. Pl. 179.

*C. cristatus L. CRESTED DOG'S-TAIL. Throughout Europe.

LAMARCKIA Like *Cynosurus* with spikelets of 2 kinds: the upper spikelet with 1 fertile floret and 1 rudimentary floret, the lower spikelets with several pairs of over-lapping, blunt, sterile lemmas in 2 ranks. 1 sp.

1749. *L. aurea* (L.) Moench (*Cynosurus a.*) GOLDEN DOG'S-TAIL. Distinguished by its dense, somewhat one-sided, oval-oblong, awned inflorescence 2–2½ cm. broad, of many short-stalked spikelets, at first green then golden-yellow. Fertile spikelet linear, surrounded by a cluster of 3–4 sterile spikelets each 6–10 mm. long; branches immediately below spikelets hairy. Lemma of fertile spikelet with awn 6–10 mm.; lemmas of sterile spikelets 6–12 in 2 ranks, obovate, awnless, finely hairy. Lvs. soft, pale, 3–5 mm. wide. A hairless ann. to 20 cm. △ Shady places, rocky and cultivated ground. Mar.–June. Med. Eur. (except AL.TR.) P. Page 534.

SCLEROCHLOA Inflorescence dense, spike-like; spikelets awnless, overlapping in 2 rows and each with 3–5 florets. Glumes unequal, keeled, blunt, strongly veined; lemma leathery keeled, blunt, strongly veined. *c.* 6 sps.

1750. *S. dura* (L.) Beauv. A prostrate ann. with several spreading stems to 15 cm., and rather broad flat lvs., the uppermost usually longer than the dense, oval, one-sided, stiff inflorescence 2–4 cm. long. Spikelets greenish-white, 6–10 mm.; glumes blunt, with a papery margin and 3–5 strong veins; lemma blunt with a papery margin, fine-veined. Stems flattened; lvs. hairless, 2–4 mm. broad. △ Waysides, sandy places. Apr.–June. Cent. Eur. (except PL.), South-East Eur. (except AL.) E.F.I.SU.

POA | Meadow-Grass Inflorescence compound; spikelets flattened from side to side, with many florets. Glumes keeled, awnless, the lower one- to three-veined, the upper three-veined; lemma green, keeled, awnless, often with a tuft of cottony hairs at their base, three- to seven-veined. *c.* 35 sps.

Rootstock stout and clothed with persistent sheaths; often with proliferous fls.

1751. *P. alpina* L. ALPINE MEADOW-GRASS. A perenn. to 40 cm., of northern or alpine regions, with an erect or nodding, open, purplish or green oval inflorescence 3–7 cm. long. Spikelets 4–7 mm., oval, flattened; often proliferous with the upper part of each spikelet replaced by a miniature leafy plant. Glumes 2½–4 mm., acute, with a whitish papery margin; lemma with a dense fringe of white hairs along the keel and veins, apex papery. Lvs. mostly basal, short and stiff, hairless, 2–5 mm. wide; base of stem with many persisting fibrous lv. bases. △ Stony and rocky places, in mountains. June–Aug. Much of Eur. (except P.B.NL.DK.TR.). Pl. 179.

1752. *P. bulbosa* L. BULBOUS MEADOW-GRASS. Distinguished by the small, swollen, bulbous bases of the stem which remain alive during periods of dry weather after the stems have died down and can be dispersed by the wind. A tufted erect perenn. to 30 cm. with short, narrow lvs. 1–2 mm. wide, and an erect, green or violet, oval-oblong inflorescence 2–6 cm. Spikelets oval, 3–5 mm.; often proliferous with miniature leafy plants developing from the florets; glumes 2–3 mm., lance-shaped, keeled, rough; lemma *c.* 3 mm., with a fringe of hairs on the keel and veins, rough-margined. A variable sp. △ Walls, banks, dry places, coasts. Apr.–June. Most of Eur. (except IRL.IS.AL.): introd. N.SF.

Rootstock slender; not proliferous

1753. *P. pratensis* L. MEADOW-GRASS. A very variable turf-forming perenn. with creeping rhizomes and with erect stems 10–40 cm. Inflorescence 2–20 cm., compound, oval, pyramidal or oblong, erect or nodding, open or rather dense, purplish, green or greyish, with very slender lateral branches, mostly in clusters of 3–5. Spikelets 4–6 mm., flattened, oval; glumes oval, unequal, rough on the keel; lemma 3–4 mm., hairy on keel and veins. △ Meadows, pastures. May–Aug. All Eur. *An important hay and pasture grass, with numerous varieties grown in cultivation.*

**P. trivialis* L. ROUGH MEADOW-GRASS. Throughout Europe.

**P. nemoralis* L. WOOD MEADOW-GRASS. Throughout Europe.

**P. compressa* L. FLATTENED MEADOW-GRASS. Almost throughout Europe.

1754. *P. annua* L. ANNUAL MEADOW-GRASS. Usually a compactly tufted ann. 3–30 cm., but rarely a perenn., with a lax, open or somewhat dense, pale or bright green, reddish or purplish inflorescence 1–12 cm., with branches mostly solitary or paired and spreading or deflexed in fr. Spikelets 3–10 mm.; glumes acute, the lower oval to lance-shaped 1½–3 mm., the upper broader and longer; lemma 2½–4 mm., with a papery margin, hairy on the keel and veins, or hairless. Lvs. often transversely wrinkled. △ Cultivated places, bare ground, tracksides, gardens. All the year round. All Eur.

GLYCERIA | **Sweet-Grass** Aquatic perenns. with submerged creeping stolons growing over the mud of lake or river bottoms. Spikelets rounded on back, not flattened or keeled, and with many overlapping florets in 2 ranks, awnless. Glumes thin, translucent, one-veined; lemma papery, prominently seven- to nine-veined and with a translucent apex; palea tough, two-veined. *c.* 13 sps.

1755. *G. maxima* (Hartman) Holmberg REED SWEET-GRASS. A stout, erect, leafy, aquatic perenn. 1–2½ m., usually forming extensive patches around the margin of open water. Inflorescence compound, 15–45 cm. long with many slender branches, at first lax but closing up and becoming rather dense. Spikelets oblong, 5–12 mm., green and tinged yellow or purple; glumes oval; lemma elliptic, very blunt, prominently seven-veined; palea with 2 keels. Lvs. pale green, 30–60 cm. by 1–2 cm. wide, rough on the margin, lv.

sheath keeled. △ Watersides, shallow lakes and rivers. June–Aug. Most of Eur. (except P.E.IS.). Pl. 180.

1756. *G. fluitans* (L.) R.Br. FLOATING SWEET-GRASS, FLOTE-GRASS. A rather succulent perenn. forming loose, bright green, spreading or floating masses with slender fl. stems to 1 m. Inflorescence with wide-spreading branches during flowering, but soon contracted into a very narrow curved or nodding spike 10–50 cm. long. Spikelets narrow-oblong, 2–3½ cm., with many overlapping green or purplish florets. Glumes unequal, blunt; lemma 6–7½ mm., seven-veined; palea sharply two-toothed. △ Margins of lakes, ponds, rivers, and ditches. May–Aug. All Eur. *Sometimes known as Manna-grass; the seeds are used as an edible grain in some Slav countries.*

G. plicata Fries PLICATE SWEET-GRASS. Most of Europe.

G. declinata Bréb. GLAUCOUS SWEET-GRASS. Western Europe. Pl. 180.

FESTUCA | Fescue Spikelets rounded on back, not flattened or keeled, with 3 or more florets. Glumes acute, the lower one-veined, the upper one- to three-veined; lemma five- to seven-veined, usually with a short, slender, terminal awn; palea two-veined, two-toothed. *c.* 60 sps.

1757. *F. gigantea* (L.) Vill. TALL BROME. A loosely tufted, rather stout, hairless perenn. ½–1½ m., with a lax nodding inflorescence 10–50 cm., with rough and spreading, usually paired branches bearing several long-stalked awned spikelets. Spikelets lance-shaped 8–20 mm., green; glumes 4–8 mm., fine-pointed; lemma with a straight, terminal, hair-like awn 1–2 cm. Lvs. 6–18 mm. wide, smooth, glossy beneath, with hairless auricles at the junction of blade and the hairless lv. sheath. △ Woods, shady places. June–Aug. Most of Eur. (except P.IS.AL.).

F. arundinacea Schreber TALL FESCUE. Like 1757, but spikelets awnless, or lemmas with a fine tip 1–4 mm.; lvs. with minutely hairy auricles. Spikelets with short stalks less than 1 cm. long. A tall tufted perenn. 1–2 m. Widespread in Europe. Pl. 180.

F. pratensis Hudson MEADOW FESCUE. Throughout Europe.

F. rubra L. CREEPING FESCUE. Throughout Europe.

F. ovina L. agg. SHEEP'S FESCUE. Throughout Europe.

VULPIA Anns. with a slender, one-sided, wand-like inflorescence and very short-stalked spikelets each with 4–6 florets. Glumes unequal; lemma tough, margin inrolled, veins obscure, awn minutely toothed and about as long as lemma; palea thin, two-veined, two-toothed. Stamens usually 1. *c.* 15 sps.

1758. *V. ciliata* Link An erect ann. to 40 cm., of the Mediterranean region, distinguished by its long, slender, awned, erect or arched, wand-like inflorescence 6–20 cm., and its long-haired lemmas. Spikelets with very short hairy stalks; upper glume narrow linear, the lower minute or absent; lemmas of fertile florets with long spreading hairs on the back, of infertile florets with long hairs on the margin only. Lvs. linear, *c.* ½ mm. broad, sheath of uppermost lv. encircling the base of the inflorescence. △ Dry, sandy places. Apr.–July. Med. Eur. (except AL.) P.CH.BG.R.SU.

V. bromoides (L.) S. F. Gray (*Festuca b.*) SQUIRREL-TAIL FESCUE. Most of Europe, except the north.

V. myuros (L.) C. C. Gmelin (*Festuca m.*) RAT'S-TAIL FESCUE. Most of Europe, except the north.

CATAPODIUM Anns. with a rigid, simple or branched, one-sided, spike-like inflorescence. Spikelets somewhat compressed, of many florets arranged in 2 ranks; glumes nearly equal, papery, one- to three-veined; lemma blunt, leathery, indistinctly three-veined; palea thin. *c.* 4 sps.

1759. *C. rigidum* (L.) Hubbard (*Scleropoa r.*; *Desmazeria r.*) HARD POA. A small, stiff, glaucous, hairless ann. 2–30 cm., with several or many erect or spreading stems, and stiff, lax or dense inflorescences 1–8 cm., which are often branched below. Spikelets tiny, narrow-oblong, 4–7 mm. by 1–1½ mm., with 5–10 minute florets. Glumes 1–2 mm., pointed; lemmas longer, 2–3 mm., blunt. Lvs. green or purplish, fine-pointed, inrolled or flat, 1–10 cm. by ½–2 mm. △ Dry places, walls, banks, rocks. May–July. West Eur. (except NL.), Med. Eur. BG.R.SU. Pl. 180.

LOLIUM | Rye-Grass Inflorescence simple, unbranched, spike-like, with stalkless spikelets alternately arranged and placed edgeways on to the jointed axis. Spikelets flattened; 1 glume only present, five- to nine-veined; lemma five-veined, awned or awnless; palea translucent, two-veined. *c.* 10 sps.

1760. *L. perenne* L. RYE-GRASS. Inflorescence slender spike-like, with numerous alternating spikelets placed on opposite sides of the axis, their own length or less apart, and with their edges fitting into hollows in the axis. Spikelets oval, flattened, 7–20 mm.; glume usually shorter than the spikelet, oblong blunt, smooth; lemma oblong blunt, awnless. A loosely to densely tufted perenn. to 90 cm. with smooth glossy lvs. 2–6 mm. wide. △ Meadows, pastures, waysides. May–Oct. All Eur. (except IS.). *Widely cultivated in temporary leys for grazing, hay and ensilage.*

L. multiflorum Lam. ITALIAN RYE-GRASS. Like 1760, but an ann. with awned lemmas, with awns to 1 cm.; glume much shorter than the spikelet. Readily hybridizes with 1760, and both the hybrid and its parent are often cultivated as a short-term fodder crop. Native of Central and Southern Europe and introduced elsewhere. Pl. 180.

L. temulentum L. DARNEL. Like 1760, but glume large, as long as or longer than the spikelet. Spikelets large, to 2½ cm., wedge-shaped and with awns up to 2 cm., or awnless. A weed of cultivation almost throughout Europe. *The seeds can be infected by a fungus, or ergot, which probably results in the presence of a poisonous substance, and it has been known since classical times that flour infected with seeds of this plant can cause blindness when eaten.*

PARAPHOLIS Anns. with very slender, whip-like inflorescences. Spikelets alternate and placed broadside on and set into a hollow in the axis. Spikelets with 1 floret; glumes equal, three- to five-veined; lemma finely three-veined. *c.* 4 sps.

1761. *P. incurva* (L.) Hubbard (*Lepturus i.*) CURVED SEA HARD-GRASS. A distinctive ann. of the littoral, with a slender, rigid, curved and pointed inflorescence with alternately arranged spikelets deeply set into hollows in the axis, the whole appearing as a continuous stem. Spikelets 6–7 mm., a little longer than the joints of the axis; glumes equal, rigid, closing the cavity of the axis; lemma like the glumes. Lvs. flat or inrolled, 1–2 mm. wide, the upper lv. sheaths somewhat inflated; stems usually numerous to 10 cm. △ Salt marshes, cliffs. Apr.–July. Med. Eur. (except AL.) P.GB.BG. Page 534.

PSILURUS Anns. with a very slender, whip-like inflorescence. Spikelets stalkless, alternately arranged and pressed close to the stem axis. Spikelets with 1 or 2 florets; glume 1, very short; lemma finely awned. 1 sp.

1762. *P. incurvus* (Gouan) Schinz & Thell. (*P. aristatus* (L.) Duval-Jouve) WHIP-GRASS. A slender ann. of the Mediterranean littoral, with a very slender, curved, whip-like inflorescence to 20 cm. and scarcely 1 mm. broad. Spikelets widely spaced, closely pressed against the axis, awl-shaped, finely awned. Glume very short; lemma to 5 mm., with a rough awn 3–6 mm. Lvs. bristle-like, much shorter than the inflorescence. △ Dry hills, rocky places. Apr.–June. Med. Eur. P.BG.R.SU.

NARDUS | Mat-Grass Perenns. with very slender inflorescences, with stalkless spikelets arising alternately from notches in the stem axis. Spikelets with 1 floret; lower glume very small, the upper absent; lemma three-veined, with 2 or 3 keels, awned. 1 sp.

1763. *N. stricta* L. MAT-GRASS. A tough, wiry, densely tufted perenn. 10–40 cm., with very slender, one-sided, erect inflorescences 3–8 cm. long, and long, bristle-like, sharp-pointed lvs. about ½ mm. wide. Spikelets very slender, 5–9 mm. long, fine-pointed, arranged on one side of the axis and closely pressed to it after flowering. Glume 1, minute; lemma lance-shaped, with a fine awn 1–3 mm. Base of stem with many pale, persistent, tough and shining sheaths. △ Heaths, moorlands, in mountains. May–Aug. All Eur. (except TR.).

Sieglingia decumbens (L.) Bernh. HEATH GRASS. All Europe.

Koeleria cristata (L.) Pers. (*K. gracilis* Pers.) CRESTED HAIR-GRASS. Much of Europe, except the north.

TRISETUM Perenns. with a compound branched inflorescence and numerous compressed, awned spikelets with 2–4 florets. Glumes translucent, firm, keeled, one- to three-veined; lemma strongly keeled, five-veined, apex with 2 fine points, and with a jointed awn attached to the back of the lemma; palea with 2 fine points. *c.* 15 sps.

1764. *T. flavescens* (L.) Beauv. YELLOW OAT. A loosely tufted perenn. to 80 cm. with rather dense or lax, erect or nodding, usually yellowish compound inflorescences 5–17 cm., with numerous, tiny, glistening, awned spikelets. Spikelets 5–7 mm.; glumes 3–5½ mm., unequal, the upper one-veined, the lower broader, three-veined; lemma with slender awn 5–9 mm., bent below middle when dry. Lvs. softly hairy, or hairless, flat 2–4 mm. wide. △ Waysides, pastures. May–Aug. All Eur. (except IS.): introd. North Eur. Pl. 180.

AVENA | Oat Stout anns. with a compound branched inflorescence, and large, long-stalked, and at length drooping spikelets with 2–4 florets. Glumes papery, longer than the florets, the lower seven-veined, the upper nine-veined; lemma leathery, seven-veined, and with a stout, rough, bent awn arising from the back and much longer than it; palea ciliate. *c.* 8 sps.

1765. *A. sterilis* L. ANIMATED OAT. An erect ann. ½–1½ m., with a spreading, lax, one-sided, erect or nodding inflorescence with very large open spikelets, 3–3½ cm., and a long stiff awn to *c.* 7 cm. Spikelets with 3–4 florets; glumes nearly equal, 3–3½ cm., long-pointed, pale; lower lemmas with long silky hairs on the lower half, hairless and green-veined above, apex two-toothed, and with a stout awn which is spirally twisted below, strongly bent, and straight above. Axis below lower floret with a tuft of hairs. Lvs. flat, hairless. △ Cultivated and waste places. May–July. Med. Eur. (except AL.) P.A.R.SU. Pl. 180.

A. fatua L. COMMON WILD OAT. Very like 1765, but spikelets smaller 2–2½ cm., with 2–3 florets; awns 3–5 cm. Glumes green; all lemmas hairy on the lower two-thirds only. Inflorescence pyramidal. Widespread in Europe.

A. strigosa Schreber BRISTLE or SMALL OAT. Widespread in Europe.

1766. *A sativa* L. OAT. Distinguished by the absence of hairs at the base of the lemmas and by the florets which are not jointed on the axis below and in consequence the spikelets do not shatter at maturity. Inflorescence erect, lax; glumes 2–2½ cm.; lemmas hairless or sometimes sparsely hairy and with a twisted, jointed awn borne on lemma of the lower floret only and about twice as long as it. △ Widely cultivated; sometimes persisting in waste places. May–July. *Cultivated since classical times, now extensively grown in Central and Northern Europe to 60° N. An important food for livestock; it has been a staple human diet for northern peoples such as the Scottish.*

HELICTOTRICHON Like *Avena* and often included in this genus, but perenns. with erect spikelets. Glumes bluntly keeled, the lower one- to three-veined, the upper three- to five-veined; lemma five-veined. *c.* 18 sps.

1767. *H. pubescens* (Hudson) Pilger (*Avena p.*) HAIRY OAT-GRASS. A loosely tufted perenn. to 1 m., with a lax, compound, erect, greenish or purplish, glistening inflorescence 6–20 cm. Spikelets 11–17 mm., with 2–3 florets, axis with long hairs; glumes unequal, 1–1½ cm., fine-pointed; lemma with long hairs at its base, apex four-toothed, and with a spirally twisted, strongly jointed awn attached to its back and about twice as long as it. Sheaths of basal and lower stem lvs. with loose spreading or deflexed hairs, the upper sheaths hairless; blades soft, green, 2–6 mm. wide, hairy or becoming hairless. △ Damp grasslands, banks, woods. May–July. All Eur. (except IS.) A.GR.TR.

H. pratense (L.) Pilger (*Avena p.*) MEADOW OAT-GRASS. Like 1767, but lv. sheaths all hairless and lv. blades stiff, glaucous, hairless. Inflorescence erect, nearly simple with only 1–2 branches from each joint of the axis; spikelets with 3–6 florets; lemma with a two-toothed apex. Widespread in Europe.

ARRHENATHERUM | **Oat-Grass** Perenn. with erect spikelets like *Helictotrichon*, but each spikelet with 2 florets, the lower male only and awned, the upper female or bisexual, usually awnless or with a fine bristle. *c.* 4 sps.

1768. *A. elatius* (L.) Roemer & Schultes OAT-GRASS. A robust, loosely tufted perenn. ½–1½ m. with narrow, compound, erect or nodding, lax or dense, green or purplish inflorescence 10–30 cm. with many lateral branches from each joint of the axis. Spikelets 7–11 mm., erect; glumes unequal, the lower smaller, one-veined, the upper three-veined; lemmas shortly hairy at their base, the lemma of the lower floret with an awn 10–17 mm. arising from near the base, the upper lemma awnless or with a nearly terminal bristle or short-awned. Lv. blades weak, 4–10 mm. wide, rough, loosely hairy or hairless. Var. *bulbosum* (Willd.) Spenner ONION COUCH has swollen, bulbous basal stems and it is sometimes a troublesome weed. △ Meadows, woods, hedgerows, roadsides, waste ground. May–Aug. All Eur. (except IS.AL.).

GAUDINIA Spikelets stalkless, arranged alternately and overlapping on the axis and forming a slender, spike-like, awned inflorescence. Spikelets with many florets; glumes unequal; lemma two-toothed, awn bent, twisted and arising from above the middle. 1 sp.

1769. *G. fragilis* (L.) Beauv. FRENCH OAT. A softly hairy ann. with a long, slender, pale green inflorescence *c.* 10 cm., with 2 alternate rows of stalkless awned spikelets placed edgeways to the axis. Lower glume *c.* 3 mm., three-veined, upper glume 7 mm., seven- to nine-veined, both finely hairy; lemma with a two-toothed apex and a twisted, bent, outspreading awn about as long as the lemma. Lvs. with the lower sheaths with spreading hairs, upper sheaths hairless, blades flat with a hairy margin. △ Cultivated ground, sandy and grassy places by the sea. Apr.–Aug. Med. Eur. P.CH.BG.SU.: introd. D.CS.

HOLCUS   Hairy perenns. with compound, lax or dense inflorescences. Spikelets with 2 florets, the lower bisexual and awnless, the upper male and finely awned. Glumes hairy, strongly keeled, longer than the spikelets, the lower glume one-veined, the upper glume three-veined; lemma shiny, leathery, strongly keeled, weakly five-veined. *c*. 8 sps.

1770. *H. lanatus* L. YORKSHIRE FOG. A softly hairy, loosely tufted perenn. 20–100 cm., with lance-shaped to oval, very dense or rather lax, erect or nodding inflorescences 3–20 cm., which are whitish, pale green, pinkish, or purple. Spikelets 4–6 mm.; glumes stiffly hairy on the keel and veins, the upper usually with a fine awn to 1 mm.; lemma of lower floret awnless, of upper floret with awn 1–2 mm., becoming curved like a fish-hook when dry. Lvs. softly hairy, lv. sheaths with reflexed hairs. △ Meadows, open woods, waste places. May–Aug. All Eur. Pl. 181.

H. mollis L. CREEPING SOFT-GRASS. Like 1770, but awn easily visible and conspicuously projecting beyond the tips of the glumes; nodes of fl. stems with long downward-spreading hairs. A variable sp. with long creeping rhizomes and forming tufts or loose mats. Throughout Europe.

DESCHAMPSIA | **Hair-Grass**   Tufted perenns. with slender, much-branched inflorescences. Spikelets with 2 bisexual florets and an axis with silky hairs. Glumes thin and translucent, one- to three-veined, about as long as the spikelet; lemma thin, translucent, weakly five-veined, with a cut-off jagged apex and a straight or curved awn arising from the back. *c*. 10 sps.

1771. *D. cespitosa* (L.) Beauv. TUFTED HAIR-GRASS. A dense, tussock-forming perenn. with coarse, tough lvs. and stiff, smooth stems to 2 m., with a lax, spreading, erect or nodding, green, silvery, golden or purple compound inflorescence 10–50 cm. Spikelets 4–6 mm., numerous on very slender branchlets; glumes shining, the lower one-veined, the upper broader, three-veined; lemma with a papery jagged tip, awn slender, 4 mm. Lvs. 2–5 mm. wide, with strong, very rough ribs above, smooth beneath. △ Wet meadows, grassy places, moorlands. June–Aug. All Eur.

1772. *D. flexuosa* (L.) Trin. WAVY HAIR-GRASS. Usually a densely tufted perenn. 20–100 cm., with slender, bristle-like lvs. and a delicate, very lax open inflorescence 4–15 cm. by 8 cm., with flexuous hair-like branchlets bearing shining spikelets. Spikelets 4–6 mm., awned, purplish, brownish, or silvery; glumes very thin, nearly equal, the lower one-veined, the upper one- to three-veined; lemma blunt, minutely toothed at apex, awn brown, bent, 4–7 mm. Lvs. much shorter than the inflorescence, very slender *c*. ½ mm. wide, with inrolled margin. △ Dry, sandy rocks, moors, heaths, open woods. June–July. All Eur. (except IS.AL.). Pl. 180.

*Corynephorus canescens (L.) Beauv. GREY HAIR-GRASS. Much of Europe, except the north and Balkans.

AIRA | **Hair-Grass**   Slender anns. with delicate, often much-branched compound inflorescences. Spikelets very small, with 2 florets; glumes equal, one-veined, slightly keeled; lemma three- to five-veined, apex with 2 bristle-like points, awn slender, attached below the middle. *c*. 5 sps.

1773. *A. elegans* Gaudin (*A. capillaris* Host) A slender, erect, hairless ann. with a delicate, very lax, spreading, oval inflorescence with very slender, thread-like branchlets bearing numerous tiny, pale, shining spikelets 1–2 mm. long. Spikelets oblong, with thread-like stalks 2–3 times as long which are abruptly thickened below the spikelets. Glumes whitish, finely toothed; 1 or both lemmas with awn twice as long. Lvs. thread-like, blade short; stems 1 or several, 10–30 cm. △ Sandy and stony places. May–June. Med., South-East Eur. CH.CS.H.SU.

A. caryophyllea L. SILVERY HAIR-GRASS. Most of Europe, except the north.

A. praecox L. EARLY HAIR-GRASS. Western, Northern, and Central Europe.

CALAMAGROSTIS Perenns. with compound branched inflorescences; often growing in damp places. Spikelets with 1 fertile floret, with numerous long silky hairs at its base, and a rudimentary floret; glumes papery, equal, the lower one-veined, the upper three-veined; lemma thin, stiff, three- to five-veined, apex bilobed, awn short, slender, arising from the back or apex. *c.* 20 sps.

1774. *C. arundinacea* (L.) Roth A hairless, stiff, erect, leafy perenn. $\frac{1}{2}$–$1\frac{1}{2}$ m., with creeping underground stems, and with a narrow, erect, straw-coloured or greenish-violet compound inflorescence 10–30 cm., with lateral branches clustered with spikelets to the base. Spikelets with 1 fertile floret and an upper rudimentary floret; glumes 5–6 mm.; lemma apex toothed, awn conspicuous, about twice as long as glumes; hairs at base of floret about one-quarter as long as the lemma. Lvs. 5–10 mm. wide, rough, lv. sheaths smooth. △ Woods, thickets, rocks. July–Aug. Much of Eur. (except P.IRL.GB.NL.IS.AL.TR.).

C. epigejos (L.) Roth BUSHGRASS. Most of Europe.

C. canescens (Weber) Roth PURPLE SMALLREED. Much of Europe, except the south-east.

AMMOPHILA | Marram Grass Hairless perenns. with stiff slender lvs. with inrolled margins. Inflorescence spike-like, dense, cylindrical; spikelets large, with 1 floret, axis silky-haired; glumes papery, keeled, the lower one-veined, the upper three-veined; lemma papery, five- to seven-veined, apex bilobed, awn very short or absent. *c.* 2 sps.

1775. *A. arenaria* (L.) Link MARRAM GRASS. A robust maritime perenn. with long creeping stems, and erect leafy stems forming compact tufts $\frac{1}{2}$–$1\frac{1}{4}$ m. high with dense, cylindrical, tapering, whitish, spike-like inflorescences 7–22 cm. Spikelets 1–1$\frac{1}{2}$ cm., flattened, short-stalked, closely overlapping; glumes stiff; lemma blunt, with fine white hairs at its base. Lvs. greyish-green, sharp-pointed, rigid, tightly inrolled, ribbed above; lv. sheaths smooth; ligule 1–2 cm. △ Sand dunes, on the littoral. May–July. Coasts of Eur. (except R.). *Often planted to stabilize dunes; sometimes used for thatching.* Pl. 181.

APERA Anns. with compound inflorescences with very numerous, very slender rough branchlets and tiny spikelets. Spikelets one-flowered, shining, long-awned. Glumes unequal, papery, the lower one-veined, the upper three-veined; lemma papery, with a long terminal awn. *c.* 2 sps.

1776. *A. spica-venti* (L.) Beauv. SILKY APERA. A tall slender ann. 20–100 cm., with a very delicate, usually open and very lax, much-branched, green or purplish inflorescences 10–25 cm. by 3–15 cm., with numerous tiny awned spikelets. Spikelets 2$\frac{1}{2}$–3 mm., floret 1; glumes fine-pointed; lemma awned from near the apex, awn 5–10 mm., very slender, straight or wavy. Lvs. hairless, 3–10 mm. wide, flat, rough all over or smooth beneath. △ A weed of cultivated ground, sandy fields, waste places. June–Aug. Most of Eur. (except P.IRL.IS.).

AGROSTIS | Bent-Grass Tufted or creeping perenns. with slender-branched compound inflorescences. Spikelets small, one-flowered; glumes papery, one-veined; lemma oval, blunt, three- to five-veined, awn arising from near base, short and slender, or awn absent. *c.* 23 sps.

1777. *A. stolonifera* L. (*A. alba* auct.; *A. palustris* Hudson) FIORIN, CREEPING BENT-GRASS. A very variable, tufted, hairless perenn. with slender, spreading, creeping stolons, and ascending stems 10–140 cm., bearing a lax open inflorescence when flowering which later contracts into a rather dense spike-like cluster. Spikelets clustered, 2–3 mm., green, whitish or purplish; glumes rough on the keel; lemma shorter, usually awnless. Lvs. $\frac{1}{2}$–5 mm. wide, minutely rough, inrolled when young; ligule blunt. △ Grassland, open woods, waysides, salt marshes, cultivated ground. July–Aug. All Eur.

*A. *tenuis* Sibth. COMMON BENT-GRASS. Widespread in Europe.

*A. *canina* L. BROWN BENT-GRASS. All Europe.

POLYPOGON | **Beard-Grass**   Anns. or perenns. with very dense, compound, spike-like inflorescences. Spikelets with 1 floret and falling as a unit when ripe; glumes bilobed, three-veined, awned, longer than lemma; lemma obscurely veined, notched or toothed, usually shortly awned. *c.* 4 sps.

1778. *P. monspeliensis* (L.) Desf. ANNUAL BEARD-GRASS. A hairless ann. with very dense, silky, pale green or yellowish, cylindrical spike-like inflorescences $1\frac{1}{2}$–16 cm. long, formed of numerous short branches bearing tiny stalkless spikelets, with very slender awns. Spikelets 2–3 mm.; glumes blunt, rough, hairy, and with a terminal straight awn 4–7 mm.; lemma smooth, shining, awnless or with short fine awn. Lvs. flat, rough, 2–8 mm. wide; stems one or several 6–80 cm. △ Sandy places by the sea, occasionally inland. May–Aug. Med. Eur. (except AL.) P.GB.BG.SU.: introd. B.A.CS. Page 534.

LAGURUS | **Hare's-Tail**   Anns. with dense, rounded, very softly hairy, spike-like inflorescences. Spikelets with 1 floret; glumes with many fine spreading hairs, one-veined; lemma five-veined, apex with 2 long bristle-like teeth, awned on back. 1 sp.

1779. *L. ovatus* L. HARE'S-TAIL. A greyish-green, softly hairy ann. bearing dense, erect, globular or ovoid, very softly hairy, whitish inflorescences 1–7 cm. by $\frac{1}{2}$–2 cm., with very slender awns. Spikelets stalkless, densely overlapping, 8–10 mm.; glumes narrow-lance-shaped, long-pointed, with numerous long, white, spreading hairs; lemma hairy only at its base and with a slender bent and twisted awn 8–18 mm. Lv. blades short, flat, with soft velvety hairs; upper lv. sheaths somewhat inflated; stems one or several, 5–60 cm. △ Sandy places by the sea. Apr.–July. Med. Eur. P.BG.: introd. GB.A.CS. Pl. 181.

ALOPECURUS | **Fox-Tail**   Anns. or perenns. with dense, narrow, spike-like inflorescences. Spikelets stalkless, strongly compressed, with 1 floret, and falling as a unit when ripe. Glumes tough, often fused below, three-veined; lemma transparent, usually awned from the back; palea absent. *c.* 12 sps.

1780. *A. pratensis* L. MEADOW FOX-TAIL. An erect, slender, hairless perenn. 30–120 cm., with very soft, dense, long, cylindrical inflorescences 2–13 cm. by $\frac{1}{2}$–1 cm. Spikelets 4–6 mm., stalkless and arranged all round the axis; glumes parallel, with fine hairs on keel; lemma as long, oval, awn attached near the base and nearly twice as long as the lemma. Lvs. with flat, mostly rough blades 3–10 mm. wide, lv. sheaths smooth. △ Damp meadows. Apr.–July. All Eur. (except P.TR.): introd. CS. *An important forage grass, particularly early in the season.* Pl. 181.

A. arundinaceus Poiret Like 1780, but larger and with long creeping rhizomes; lv. blades 5–12 mm. wide; inflorescence 5–15 cm. by 8–15 mm. Glumes distinctly diverging above, whitish, veins green or blackish; lemma little shorter than the glumes and awn inserted at or above the middle of the lemma and little longer. Most of the Mediterranean region.

1781. *A. utriculatus* (L.) Solander A hairless ann. with many erect stems to 40 cm., with an oval or shortly cylindrical, green or violet inflorescence 1–2 cm. by ½–1 cm., and with the uppermost lv. sheath spindle-shaped and very conspicuously inflated. Spikelets 6–7 mm.; glumes fused to the middle, abruptly contracted to the apex, keel ciliate below; lemma with awn attached to the lower half and twice its length. Lvs. hairless, blade short, 1–3 mm. wide. △ Damp grassy places, marshes. May–June. Med. Eur. P.B.D.A.BG.

1782. *A. myosuroides* Hudson SLENDER FOX-TAIL, BLACK GRASS. Distinguished by the long, dense, narrow-cylindrical, pointed inflorescence 2–12 cm. by 3–6 mm. Sheaths of upper lv. not or slightly inflated, blades 3–8 mm. wide; spikelets 4–7 mm.; glumes fused to one-third their length, shortly hairy on the veins and keel; lemma with awn arising from near its base and twice its length. △ Often a serious pest of cultivated ground. Apr.–Aug. All Eur. (except IS.).

*A. geniculatus L. MARSH FOX-TAIL. All Europe, except extreme south-east.

CRYPSIS Inflorescence a dense spike or broad head, often partly enveloped by the swollen bases of the uppermost lvs. Spikelets compressed, with 1 floret; glumes one-veined, awnless; lemma like glumes, one- or two-veined. Ligule replaced by hairs. *c.* 4 sps.

1783. *C. aculeata* (L.) Aiton FLAT-TOPPED GRASS. A very distinctive, spreading, many-stemmed ann. to 30 cm., of the Mediterranean region, with very compact, flat-topped inflorescences 1–1½ cm. across, broader than long, and partially surrounded by an involucre formed from the large swollen sheaths of the uppermost lvs. Spikelets very densely clustered; glumes 2–3 mm., rough on the keel; lemma longer, awnless; stamens 2. Lvs. glaucous, short, rigid and spiny-tipped, 2–4 mm. wide, sheath inflated. △ Damp places on the littoral. July–Oct. Med., South-East Eur. P.A.H.CS.SU. Page 534.

C. schoenoides (L.) Lam. (*Heleochloa s.*) Differs from 1783 in having a dense, oval, pale green inflorescence ½–3 cm. by 6–9 mm., free from or only partly surrounded by the swollen sheath of the upper lv. Spikelets *c.* 3 mm.; glumes narrow, with a rough keel; lemma little longer, with a green keel; stamens 3. Lvs. rigid, glaucous. Much of Southern Europe.

PHLEUM | Cat's-Tail Hairless anns. or perenns. with a cylindrical or ovoid spike-like inflorescence of numerous densely arranged stalkless spikelets. Spikelets flattened, with only 1 fertile floret; glumes equal, papery, strongly keeled and with stiff spreading hairs on the keel, three-veined, long-pointed; lemma much shorter, thin, blunt, three- to five-veined. *c.* 12 sps.

1784. *P. pratense* L. TIMOTHY GRASS, CAT'S-TAIL. A hairless perenn. with erect stems ½–1½ m. with a dense, long-cylindrical, spike-like inflorescence usually 5–15 cm. by 6–10 mm., but sometimes longer, with green or purplish, tightly packed, stalkless spikelets. Spikelets flattened, 3–4 mm.; glumes with stiff, spreading white hairs on the keel, with a broad apex which is abruptly contracted into a stiff, rough, terminal spine 1–2 mm.; lemma short, very broad, papery. Lvs. flat, firm, rough, 3–9 mm. wide; ligule blunt, to 6 mm. △ Meadows, roadsides, waste places. May–Aug. All Eur. *An important fodder grass, often grown for hay and grazing; numerous strains are in cultivation.* Pl. 181.

P. phleoides (L.) Karsten PURPLE-STEM CAT'S-TAIL. Widespread in Europe.

P. alpinum L. (*P. commutatum* Gaudin) ALPINE CAT'S-TAIL. Mountains or arctic regions of Europe.

1785. P. arenarium L. SAND CAT'S-TAIL. A small stiff ann. with solitary or clustered stems 6–15 cm. or more, with very dense, cylindrical or ovoid, pale green or whitish-green inflorescences ½–5 cm. by 3–7 mm. Spikelets 3–4 mm.; glumes gradually narrowed to a sharp apex, keel hairy above the middle only; lemma about one-third as long, very blunt. Lvs. hairless, short, flat, to 4 mm. wide, the uppermost lv. sheath distinctly inflated. △ Sandy places, dunes by the sea, sometimes inland. May–July. West, Med. Eur. (except AL.) DK.N.S.

P. echinatum Host Like 1785, but inflorescence shorter, fatter, more globular, 1–2½ cm. by 1 cm.; glumes whitish, papery, with a green, stiffly ciliate keel, blunt-tipped but with the keel prolonged into a narrow, rough, rigid, three-veined awn 2–5 mm. Lvs. 2–3 mm. wide. Italy to Greece.

STIPA Anns. or perenns. usually with very slender, 'feathery' compound inflorescences with slender, erect, lateral branches, usually bearing solitary spikelets. Spikelets linear, flattened, with 1 very long-awned floret; glumes long-pointed, papery, three-veined; lemma leathery, inrolled, five-veined, and with a very long twisted and bent awn. *c.* 12 sps.

Awns with spreading plume-like hairs
1786. S. pennata L. FEATHER GRASS. A densely tufted glaucous perenn. to 60 cm., with narrow, 'feathery' compound inflorescences of few spikelets each, and with very long trailing awns which have many conspicuous, silvery-white, spreading hairs. Spikelets yellowish-green; glumes 1½–2 cm., with a hairless slender awn 2–3 times as long; lemma 1½–2½ cm. with an awn up to 35 cm. long, which is twisted below and bent and plume-like above. Lvs. narrow, bristle-like, ½–1 mm. wide, with inrolled margin. A variable sp. △ Dry rocky and stony places, dry grasslands. May–July. Cent. Eur. (except PL.) South-East Eur. E.F.S. I.SU. Pl. 182.

Awns without hairs, or very finely hairy
(a) Perenns.
1787. S. capillata L. A densely tufted perenn. to 80 cm., with rough, narrow, glaucous, inrolled lvs. ½–1 mm. wide, and lax inflorescences with the lower branches 3–4 times branched and at first encircled at the base by the uppermost lv. sheath. Spikelets greenish, 2½–3 cm.; glumes narrow lance-shaped, long-pointed; lemma 12–13 mm., and with a very long, rough, hairless, bent and spirally twisted awn 10–15 cm. △ Dry exposed places, rocks. June–Nov. Much of Eur. (except North Eur. P.IRL.GB.B.NL.).

S. gigantea Lag. A large perenn. ½–1½ m., of the Iberian Peninsula, with a slender, spike-like inflorescence with long-stalked, narrow spikelets with very long awns 15–20 cm., which are not jointed and are finely hairy their whole length. Lvs. inrolled, channelled, very slender, finely hairy above; ligule blunt, woolly-haired outside. Portugal, Spain, Italy.

(b) Anns.
1788. S. capensis Thunb. (*S. tortilis* Desf.) An erect ann. to 60 cm., with glaucous inrolled lvs. which are finely hairy above, and have a very short ligule. Inflorescence narrow and rather contracted, 3–8 cm. long, the base encircled by the upper enlarged lv. sheath. Spikelets whitish, with awns 5–10 cm., once or twice bent, conspicuously spirally twisted and shortly hairy below, rough, hairless, and straight above. Lemma 7 mm. long, awnless,

with fine spreading hairs. △ Rocks, dry sunny places near the coast. Apr.–June. Med. Eur. (except AL.TR.) P.

ACHNATHERUM   Like *Stipa*, but lemmas with long, dense, silky hairs. *c.* 1 sp.

1789. *A. calamagrostis* (L.) Beauv. (*Stipa c.*) A tufted perenn. to 1 m., distinguished by its silver, then russet-coloured, lax and much-branched compound inflorescence 12–25 cm., often nodding above. Spikelets pale, 8–10 mm.; glumes equal, papery, lance-shaped; lemmas with long silky hairs to 4 mm. long; awn slender, 1–1½ cm., bent at base. Lvs. narrow, stiff, later inrolled and rush-like. △ Rocky places in mountains. June–Aug. F.D.CH.A.I.YU.AL.GR.BG.

ORYZOPSIS   Hairless perenns. with lax, branched inflorescences and awned spikelets with 1 floret. Spikelets flattened; glumes nearly equal, acute or blunt, three- to five-veined; lemma shining, becoming leathery, with a straight awn which readily drops off after flowering. *c.* 4 sps.

1790. *O. coerulescens* (Desf.) Richter (*Piptatherum c.*) A dense, tufted perenn. with erect stems to *c.* 60 cm., and a very lax, branched inflorescence with single or paired long lateral branches bearing rather few spikelets. Spikelets 6–7 mm.; glumes thin, five-veined, fine-pointed, violet-flushed at base and with a colourless margin; lemma 3 mm., hairless, with a short awn of about the same length and not longer than the glumes; anthers hairy above. Lvs. 2–4 mm. wide, often inrolled; ligule 6 mm. △ Dry places, sunny hills. Apr.–June. Med. Eur. (except YU.AL.TR.).

MILIUM | Millet   Hairless anns. or perenns. with lax, branched inflorescences. Spikelets rounded, not flattened, with 1 floret; glumes equal, three-veined; lemma leathery, margin inrolled, five-veined, awnless. *c.* 4 sps.

1791. *M. effusum* L. WOOD MILLET. A loosely tufted perenn. 45–180 cm., with a lax, spreading, pyramidal inflorescence 10–40 cm. long to *c.* 20 cm. wide, and very slender, spreading or reflexed, curved branches bearing a few green awnless spikelets. Spikelets oval 3–4 mm.; glumes oval, greenish, papery; lemma smooth, shining, becoming very hard and tough. Lvs. flat, ½–1½ cm. wide, hairless, rough on margin. △ Woods, shady places. May–Aug. All Eur. (except P.). Pl. 181.

HIEROCHLOE   Perenns. with compound inflorescences. Spikelets flattened, with 3 florets, the uppermost bisexual, the 2 lower male; glumes equal, keeled, three-veined; lemma keeled, five-veined, awned or not. *c.* 4 sps.

1792. *H. odorata* (L.) Beauv. HOLY-GRASS. A hairless, aromatic perenn. with long-creeping, underground stems and erect stems forming tufts or patches to ½ m. high. Inflorescence oval, lax, 4–10 cm. by up to 8 cm., with smooth, spreading, unequal branches bearing rather plump, shining, golden-brown spikelets. Spikelets oval, 4–6 mm.; glumes blunt or short-pointed, papery; lemma of male florets shortly awned, rough, hairy. Lvs. 3–10 mm. wide, upper blades very short, flat, hairy or hairless above, glossy green beneath, margin rough. △ Northern pastures, rocks, wet banks, mountains in the south. May–Aug. North, Cent. Eur. F.IRL.GB.NL.YU.GR.BG.R.SU. *It contains coumarin and is used for flavouring vodka in Poland. It is called Holy-grass because it was strewn before the doors of churches in Central Europe at certain Christian festivals.*

ANTHOXANTHUM | Vernal-Grass Ann. or perenn. with dense, ovoid or oblong, spike-like inflorescences. Spikelets flattened with 2 sterile florets and an uppermost fertile floret; glumes very unequal, the lower one-veined, the upper three-veined, shortly awned; lemma of sterile floret three-veined, awned, of fertile seven-veined, awnless. Stamens 2. *c*. 6 sps.

1793. *A. odoratum* L. SWEET VERNAL-GRASS. A tufted perenn. to $\frac{1}{2}$ m., smelling of new-mown hay, with a spike-like, oval-oblong, dense or rather lax, pale green or purplish inflorescence 1–12 cm. by 6–15 mm. Spikelet 6–10 mm.; glumes hairy, finely pointed, the lower glume half the length of the upper which encloses the florets; lemma of sterile fls. brown, hairy, the lower with a straight awn 2–4 mm., the upper with bent awn 6–9 mm., fertile lemma smooth, shining, awnless; stamens 2. Lvs. flat, hairless or hairy, usually 2–5 mm. wide. A very variable sp. △ Meadows, pastures, heaths, moors, open woodlands. Apr.–July. All Eur. *Contains coumarin, which gives the scent to new-mown hay.*

PHALARIS | Canary-Grass Hairless anns. or perenns. with dense, oval or cylindrical inflorescences. Spikelets strongly flattened, with 2 rudimentary florets and a terminal fertile floret. Glumes papery, keeled, and keel often somewhat winged, three- to seven-veined; 2 sterile lemmas present at the base of the fertile five-veined lemma. *c*. 8 sps.

1794. *P. arundinacea* L. REED CANARY-GRASS. A robust perenn. to 2 m., spreading by creeping underground stems and with many erect stems often forming clumps, with a dense, cylindrical, compound, often shortly branched, whitish-green or purplish inflorescence 5–25 cm. Spikelets awnless, 5–7 mm.; glumes pointed, firm; sterile lemmas narrow, with short hairs; fertile lemma longer, broader, smooth and glossy below, and with adpressed hairs above. Lvs. 6–18 mm. wide, flat, firm, rough. △ Wet places, marshes, watersides. May–July. All Eur. (except TR.): introd. IS. *Sometimes cultivated as a forage grass in damp areas liable to flooding.*

1795. *P. canariensis* L. CANARY-GRASS. A tufted ann. to 120 cm., distinguished by its dense, oval inflorescence 1$\frac{1}{2}$–6 cm. long, with numerous broad overlapping, distinctive, whitish spikelets with green veins. Spikelets much flattened, obovate, 6–10 mm. by 6 mm.; glumes stiff, with a green keel which is broadly winged above, and with an abruptly pointed apex. Lvs. flat, rough, 3–12 mm. wide. △ Waysides, cultivated ground, waste places; widely cultivated and often escaping. May–July. Native of Africa and Canary Isles: introd. to most of Eur., except North Eur. *The grain is used for feeding cage-birds.* Pl. 181.

SPARTINA | Cord-Grass Stout, hairless perenns. of salt marshes, with compound inflorescences of several erect, spike-like branches with numerous, alternately arranged, stalkless spikelets. Spikelets awnless, with 1 floret; glumes papery, keeled, three-veined; lemma similar, three- to five-veined. *c*. 4 sps.

1796. *S.* × *townsendii* H. & J. Groves TOWNSEND'S CORD-GRASS. A stiff, robust perenn. $\frac{1}{2}$–1$\frac{1}{2}$ m., with soft fleshy rhizomes creeping in mud on the littoral, and forming clumps or extensive meadows. Inflorescence of 3–6 slender, erect, spike-like branches, which finally come together into a single, slender, dense cluster 12–35 cm. long. Spikelets stalkless, overlapping, two-rowed, and pressed to axis, each 14–20 mm. long; glumes unequal, shortly hairy; lemma minutely hairy. Lvs. stiff, ribbed above, smooth, flat or inrolled upwards, 7–15 mm. wide; ligule a dense fringe of hairs; lv. sheaths smooth, overlapping. A fertile hybrid between *S. alterniflora* Loisel. and *S. maritima* (Curt.) Fernald; first recorded in Southampton Water in 1870. △ Salt marshes, mud-flats. July–Nov. GB.: introd. F.IRL.B.NL.D.DK. *Often planted to stabilize coastal mud-flats and to reclaim land.* Pl. 181.

S. maritima (Curt.) Fernald CORD-GRASS. Like 1796, but less robust and inflorescence shorter, 6–12 cm., usually with only 2–3 branches and with the axis not prolonged beyond the spikelets as in 1796. Lvs. to 5 mm. wide; stems to ½ m. Coasts of Western Europe and Northern Adriatic.

ORYZA | Rice Anns. of shallow water, with compound lax inflorescences with many spikelets each with 1 floret. Glumes much smaller than the lemma or palea, one-veined; lemma compressed, keeled, strongly five-veined; palea as long or longer, two-veined. Stamens 6. 1 sp.

1797. *O. sativa* L. RICE. A hairless aquatic ann. with leafy, above-water stems to 130 cm., with a large, lax, erect or curved, greenish-white inflorescence, with numerous long, erect, lateral branches, bearing many spikelets. Glumes tiny, equal, 2–2½ mm.; lemma 7–9 mm., hairy above, short-pointed or with an awn up to 1 cm. long. Lvs. flat, hairless, smooth, 1–1½ cm. wide. △ Native of the tropics; cultivated in shallow water in river deltas in the Mediterranean region. Aug.

LEERSIA Perenns. with lax compound inflorescences. Spikelets with 1 floret; glumes absent; lemma flattened, tough, three-veined, awnless. 1 sp.

1798. *L. oryzoides* (L.) Swartz CUT-GRASS. A perenn. to 1 m. or more, with creeping stolons and erect stems forming patches or loose tufts, with pale, lax inflorescences 10–22 cm., partly encircled by the sheath of the upper lv. for a long time. Branches of inflorescence very slender, little-branched, wavy, each with a row of short-stalked spikelets towards the end. Spikelets pale green, elliptic, flattened, 4–5 mm., easily falling; lemma with spreading hairs. Lvs. yellowish-green, ½–1 cm. wide, flat, rough, with a spiny-toothed margin. △ Margins of lakes and rivers. July–Sept. Much of Eur. (except IRL.IS.N.AL.).

LYGEUM Spikelets 2–3, surrounded by a broad, spathe-like sheath, spikelets fused at the base into a tube which is covered with long silky hairs; tube becoming hard and enclosing the fr. 1 sp.

1799. *L. spartum* L. ALBARDINE. A creeping, rush-like perenn., with ascending tufted stems, very leafy at the base and ending in a lance-shaped terminal sheath 4–5 cm. long, enclosing the spikelets. Silky hairs on tube at base of spikelets up to 1 cm. long; palea lance-shaped, to 3 cm. long. Lvs. stiff, rush-like, cylindrical in section, 1–1½ mm. wide. △ Dry places. Mar.–July. E.I. Crete. *An Esparto-furnishing grass, used in paper-making.*

BECKMANNIA Inflorescence compound; spikelets arranged in 2 ranks on each side of short lateral branches, florets 2. Glumes 2, swollen; lemma lance-shaped, keeled, not awned. 1 sp.

1800. *B. eruciformis* (L.) Host A very distinctive perenn. to 1½ m., with a slender, one-sided inflorescence *c.* 20 cm. long, composed of many short regularly alternate, lateral branches in 2 rows pressed to the axis, and each branch with 2 ranks of closely overlapping spikelets. Spikelets 2½–3 mm., as broad as long; glumes rounded, with a swollen keel, and a white or violet margin. Lvs. flat, 4–10 mm. wide, rough above; base of stem swollen. △ Damp fields, marshes, salty ground. May–June. South-East Eur. (except R.) A.CS.PL. H.I.SU. Page 553.

CYNODON Perenns. with a compound inflorescence of a terminal whorl of several slender branches with stalkless spikelets arranged in 2 rows. Spikelets flattened, with 1 floret; glumes equal, papery, one-veined; lemma leathery, strongly flattened, obscurely three-veined, awnless. 1 sp.

1801. *C. dactylon* (L.) Pers. BERMUDA GRASS. A low greyish perenn., often forming mats, with creeping stems with numerous scaly lvs. and erect stems usually less than ½ m. Inflorescence with 3–7 narrow, spreading, equal terminal branches, each 2½–5 cm. long by *c.* 2 mm. Spikelets 2–3 mm., stalkless, pressed to stem, in 2 rows on one side of the branches. Glumes pointed, keeled; lemma longer, boat-shaped, keeled, minutely hairy. Lvs. 2–4 mm. wide, short, flat, spreading; ligule a row of hairs. △ Dry, sandy places, poor pastures. July–Sept. Much of Eur. (except North Eur. IRL.PL.).

DIGITARIA | Finger-Grass Anns. or perenns. with a compound inflorescence of 4–10 long, narrow, spreading branches arising from near the apex of the stem. Spikelets with 2 florets, the lower sterile, the upper bisexual, usually paired and in 2 rows on one side of the branches; glumes very unequal, three-veined; sterile lemma five-veined; fertile lemma three-veined, shiny. *c.* 4 sps.

1802. *D. sanguinalis* (L.) Scop. HAIRY FINGER-GRASS, CRAB GRASS. An ann. to *c.* 30 cm., with many ascending stems rooting below, and a compound inflorescence of 4–10 up-spreading, very slender branches each 4–18 cm. long, with 2 rows of spikelets pressed to the branches. Spikelets 2½–3½ mm., often purple; lower glume minute, the upper glume half the length of the spikelet; lower sterile lemma finely hairy, upper fertile lemma smooth. Lv. sheaths hairy, blade flat, 3–8 mm. wide, hairy or hairless; ligule papery. △ Cultivated ground, sandy places. July–Oct. Med. Eur.: introd. to most of Eur. (except North Eur. IRL.).

*D. ischaemum (Schreber) Muhl. RED MILLET. Central and Southern Europe.

PANICUM | Millet Inflorescence compound, lax, much-branched, with slender branches bearing several stalked spikelets and clustered into a spike-like inflorescence. Spikelets flattened, the upper floret fertile, the lower sterile. Glumes with the lowest minute and very much smaller than the similar upper glume and sterile lemma, keeled, one-veined; fertile lemma tough, smooth, blunt, awnless. *c.* 4 sps.

1803. *P. repens* L. CREEPING MILLET. A perenn. with creeping underground stems and stiff, erect stems to 80 cm., and a narrow, erect inflorescence with slender ascending branches bearing rather numerous, tiny, whitish spikelets. Spikelets 2 mm., awnless; glumes unequal, the lower one-quarter as long as the upper. Lvs. in 2 ranks, glaucous, stiff, 3–6 mm. wide, the uppermost lv. often as long as the inflorescence; lv. sheaths ciliate; stems scaly at the base. △ Damp sandy places, especially near the sea. June–Oct. Med. Eur. (except AL.TR.) P. Page 553.

P. miliaceum L. COMMON or BROOM-CORN MILLET. A robust ann. to 1 m., with a rather dense inflorescence 10–20 cm., of numerous long, slender, erect branches, and numerous plump spikelets; inflorescence later drooping. Spikelets often purplish, 3–4 mm.; glumes hairless, the lower about one-third as long as upper. Lvs. 1–2 cm. wide; sheaths with long hairs. Native of Asia; widely cultivated in Southern Europe. *A cereal crop for humans and livestock; also grown as a forage crop.* Pl. 182.

ECHINOCHLOA Inflorescence with rather few triangular-sectioned lateral branches forming a compound inflorescence. Spikelets somewhat flattened, with 2 florets, the lower sterile, the upper fertile, with or without awns, in stalkless clusters on one side of the branches. Glumes very unequal; lemma leathery, smooth. *c.* 3 sps.

1804. *E. crus-galli* (L.) Beauv. COCKSPUR GRASS. A robust, rather broad-leaved tufted ann. 30–120 cm., with an irregular inflorescence 6–20 cm. of few or many, unequal and scattered lateral branches densely covered with clusters of spikelets. Spikelets broadly oval, 3–4 mm., pointed or with stiff, rough awns; lower glume one-third the length of the upper, which is five-veined, spiny-haired, and encircles the spikelet; lemma of infertile floret pointed, or narrowed to a short or long awn to 5 cm. Lvs. hairless, rough on the thickened margin, 8–20 mm. wide. △ Damp, sandy fields, ditches. July–Oct. Most of Eur. (except IRL.IS.): introd. N.SF.CS. Pl. 182.

SETARIA | **Bristle-Grass** Anns. or perenns. with a cylindrical, spike-like inflorescence of numerous, densely clustered, stalkless spikelets, with a number of stiff bristles arising immediately below each spikelet. Spikelets with 2 florets, the lower sterile, the upper fertile; glumes very unequal, the lower three-veined, the upper five-veined; lemma of sterile floret five-veined, of fertile floret becoming firm. *c.* 5 sps.

1805. *S. viridis* (L.) Beauv. GREEN BRISTLE-GRASS. A loosely tufted ann. to ½ m., with a dense, cylindrical, very bristly, erect inflorescence 1–10 cm. by 4–10 mm. Bristles below spikelets several, flexuous, green, with forward-pointing teeth, 2–4 times as long as the blunt spikelets which are 2–3 mm. Lower glume one-third as long as the floret, the upper covering the floret; upper lemma blunt, becoming tough and rigid, very finely wrinkled. Lv. blades 4–10 mm. wide, flat, hairless; ligule a ring of hairs. △ Cultivated ground, sandy places. June–Oct. All Eur. (except IRL.IS.AL.).

S. lutescens (Weigel) Hubbard (*S. glauca* auct.) YELLOW BRISTLE-GRASS. Like 1805, but with yellowish or rusty-brown bristles 2–3 times as long as the spikelets; fertile lemma conspicuously transversely wrinkled; glumes nearly equal. Widespread, except in Northern Europe.

S. verticillata (L.) Beauv. BUR or ROUGH BRISTLE-GRASS. Distinguished from 1805 by the slender, cylindrical inflorescence which is distinctly whorled or interrupted below with the lower clusters separated from each other. Bristles 1–2, with backward-pointing teeth, green, 3–5 times as long as the spikelets. Widespread, except in Northern Europe.

S. italica (L.) Beauv. FOXTAIL or ITALIAN MILLET. Like 1805, but taller to 1 m., distinguished by its large green compound and lobed inflorescences 20–30 cm. by 2–3 cm., with short lateral branches densely covered with spikelets. Bristles 1–3, green, with forward-pointing teeth, up to twice as long as the spikelets. Lvs. 8–15 mm. wide. Fr. becoming detached from the rest of the spikelet at maturity. Native of Asia; cultivated as a crop and sometimes occurring as a casual. *Grown for its grain which is used for feeding cage-birds.* Pl. 182.

IMPERATA Robust perenns. with a dense, silvery, spike-like inflorescence of numerous spikelets each of 1 floret and surrounded by long silky hairs. Glumes 3, the lower 2 with long hairs, the uppermost and the lemma hairless. Stamens 1–2. 1 sp.

1806. *I. cylindrica* (L.) Beauv. Readily distinguished by its shining, silky-white, dense, long-cylindrical inflorescence 8–15 cm. Spikelets hidden in the long hairs which are more than twice as long as the acute glumes; lemma awnless, hairless, fringed at the apex. A robust perenn. to 1 m. with creeping underground stems, and with flat lvs. 6–8 mm. wide which are inrolled towards the apex, and with broad enlarged lv. sheaths. △ Sandy, dry river beds, banks. May–July. Med. Eur. P.BG. Page 553.

ERIANTHUS Inflorescence dense, spike-like with numerous spikelets hidden by long hairs; spikelets with 1 stalkless fertile floret and 1 stalked sterile floret. Glumes densely silvery-haired; lemma papery, with a long awn. 2 sps.

1807. *E. ravennae* (L.) Beauv. A very robust perenn. forming large tufts, with stout, leafy stems 1–2 m., bearing a silvery-white, plume-like, pyramidal or cylindrical inflorescence 30–60 cm. long, with branches at first somewhat spreading but later coming together. Spikelets very numerous, hidden by long silky hairs arising from the axis and detached with the fr.; glumes pale, long-pointed; lemma hairless with an awn twice its length. Lv. blades 1–1½ cm. wide, grooved, rough-margined, lv. sheaths hairy; ligule replaced by long silky hairs. △ Sandy places, dry riverbeds. Aug.–Oct. Med., South-East Eur. Pl. 182.

BOTHRIOCHLOA Inflorescence branched, with 2 or more slender, spike-like clusters arising from the ends of the branches. Spikelets 2 or 3 arising from each node of the branches, overlapping, the lower stalkless and fertile, the upper stalked, sterile, each with 1 floret. Glumes 3, awnless; lemma of fertile spikelet awned. *c.* 3 sps.

1808. *B. ischaemum* (L.) Keng (*Andropogon i.*) A tufted perenn. 40–80 cm., with an inflorescence of 2–10 narrow, spike-like clusters each 3–6 cm. long, arising from the apex of the stem. Spikelets paired, the lower fertile spikelet stalkless, awned, the upper sterile spikelet stalked, spikelets and stalks densely covered with conspicuous, long, spreading hairs. Glumes violet, awnless; lemma of fertile spikelet with a bent awn, to 1½ cm. Lvs. glaucous, grooved, 2–3 mm. wide; ligule a row of hairs. △ Dry and arid places. Apr.–Nov. Much of Eur. (except North Eur. IRL.GB.NL.). Pl. 182.

HYPARRHENIA Like *Bothriochloa*, but branched inflorescence with paired, slender, spike-like clusters arising from leaf-like bracts. *c.* 2 sps.

1809. *H. hirta* (L.) Stapf (*Andropogon h.*) A tufted perenn. with erect stems 40–120 cm. with a lax, branched inflorescence with broad leaf-like bracts and paired, silvery-haired, spike-like clusters 2–3 cm. long. Spikelets paired, as in 1808 densely arranged on the hairy branches; glumes hairy; lemma of fertile floret with a bent awn, 2–3 cm., which is hairy below. Lvs. flat, 2–3 mm. wide, margin rough. △ Dry hills, rocky places, cultivation, waysides. Apr.–Nov. Med. Eur. P.R. Pl. 182.

STENOTAPHRUM Inflorescence slender, spike-like; spikelets paired, stalkless, arranged in 2 rows on an erect, thickened, unbranched axis. Glumes very unequal leathery, the lower very short, the upper oval, seven-veined; lemma leathery, three- to five-veined. *c.* 1 sp.

1810. *S. dimidiatum* (L.) Brongt A tough, long-creeping perenn. rooting at the nodes, with short, blunt, leathery lvs. and slender, flattened, erect inflorescences 4–10 cm., with 2 rows of spikelets pressed close into cavities in the conspicuously thickened axis. Spikelets *c.* 4 mm., whitish, hairless, awnless; lower glume rounded, 3–4 times shorter than the oval upper glume. Lvs. with broad overlapping sheaths and short blunt blades 1 cm. wide, margin inrolled. △ Native of Africa and America; sometimes naturalized and replacing the native grasses on the Mediterranean littoral. July–Sept. Page 553.

CHRYSOPOGON Like *Bothriochloa*, but spikelets 2–3 at the ends of the slender branches which are whorled, numerous, and unbranched. *c.* 1 sp.

1. *Stenotaphrum dimidiatum* 1810
3. *Tragus racemosus* 1813
5. *Beckmannia eruciformis* 1800
2. *Imperata cylindrica* 1806
4. *Panicum repens* 1803

1811. *C. gryllus* (L.) Trin. (*Andropogon g.*) An erect, tufted perenn. ½–1 m., with a large spreading inflorescence 10–20 cm. long, of many slender, unequal, whorled branches bearing a terminal group of 2–3 violet spikelets with a conspicuous cluster of russet hairs to *c.* 1 cm. long immediately below. Fertile lower spikelet stalkless, the upper glume with a fine awn, and the lemma with a bent, hairy awn 3–4 cm.; upper sterile spikelets stalked, shortly awned. Lvs. sparsely hairy, 2–3 mm. broad; ligule a row of hairs. △ Sandy places, dry hills. May–July. Med., South-East Eur. CH.A.CS.H.SU.

SORGHUM Inflorescence large, much-branched, with terminal clusters of shining spikelets. Spikelets usually with 1 stalkless fertile floret and 1 or more stalked sterile florets. Glumes all somewhat compressed, with a three-pointed apex, becoming shining and hard at maturity; lemma with long hairs. *c.* 2 sps.

1812. *S. halepense* (L.) Pers. JOHNSON GRASS. A robust perenn. with white creeping stolons, and erect stems ½–2 m., with an elongated pyramidal inflorescence 10–30 cm., with spreading branches. Fertile floret oval, 5 mm. with shining glumes with adpressed hairs and lemma usually with an awn to 1 cm.; sterile floret lance-shaped, violet, hairy. Lvs. smooth, flat, 1–2 cm. broad; ligule a row of hairs. △ Sandy places, waysides, cultivated ground. July–Sept. Med., South-East Eur. P.SU.: introd. A.CS.

S. bicolor (L.) Moench MILLET, GUINEA CORN. An ann. to 3 m., without creeping stems, and with a variable inflorescence 10–50 cm., either oblong-oval or rather dense and flat-topped, erect or nodding. Fertile spikelets broadly oval, 5–7 mm., tawny or rusty-brown, borne on hairy branches. Native of Asia and Africa; cultivated in the Mediterranean region and sometimes occurring as a casual. *The grain and foliage are used to feed livestock; var. saccharatum Koern. is the source of a syrup called Sorghum, which is widely consumed in the U.S.A. The inflorescence stalks, when dried, are used for brushes and brooms. Known in Egypt since about 2200 B.C.*

TRAGUS Inflorescence branched, lax or dense; spikelets several on short side branches, each with 1 floret. Glumes 2, the lower minute, papery, the upper leathery and with 3–7 rows of hooked bristles on the outside; lemma smooth, leathery, awnless. 1 sp.

1813. *T. racemosus* (L.) All. Readily distinguished from all other grasses by the 5 rows of fine crooked bristles on the backs of the leathery glumes. A creeping, much-branched, spreading ann. rooting at the nodes, with erect stems to 30 cm., with a long, cylindrical, spike-like, purplish inflorescence 3–8 cm. Spikelets short-stalked, spreading, *c.* ½ cm.; lemma papery, smooth, awnless. Lvs. short flat, 2–3 mm. broad, margin ciliate. △ Dry sandy places, waysides. June–Oct. Med., South-East Eur. (except AL.) A.H.SU.: introd. D.CS. Page 553.

ZEA | Maize Spikelets in one-sexed clusters. The terminal cluster is branched and bears male spikelets only. The female spikelets are borne on axillary stems surrounded by broad leafy bracts; styles long-projecting. 1 sp.

1814. *Z. mays* L. MAIZE. A very robust ann. to 5 m., with a stout stem with many broad lvs. and a terminal plume-like inflorescence of male spikelets, and several large, lateral, bud-like inflorescences of female spikelets closely invested by overlapping leafy bracts. Male spikelets 6–8 mm. arranged closely along many slender branches; female spikelets densely arranged on swollen lateral branches and completely enclosed in leafy bracts, with only the very long stigmas 12–20 cm. projecting. Lvs. 5–12 cm. wide, margin rough and finely ciliate. Fr. hard, shining yellow, white or purple, arranged round a much swollen axis. △ Native of South America; very widely cultivated in South and Central Europe. July–Oct. *A very important crop plant which is the basis of two important industries, distillation and starch-making. The grain and foliage are very valuable as fodder for livestock, while as sweet-corn, popcorn, or baking-flour it is of some importance as food for humans.*

COIX Spikelets one-sexed and very different in appearance. The male spikelets are placed above the female and form a short cluster; the female spikelet is composed of 1 fertile floret which is enclosed in a swollen, woody sheath. 1 sp.

1815. *C. lacryma-jobi* L. JOB'S TEARS. A broad-leaved grass to 1 m., with few, very distinctive, globular female florets which become hard, shining white, and pearl-like in fr. and from the apex of which arise several male spikelets. Male spikelets 8 mm. long; fr. *c.* 1 cm. Stems leafy, branched; lvs. 1–2½ cm. wide. △ Native of India; often grown as an ornamental in South Europe. July–Oct. *The fruits are used for necklaces, chaplets, etc. They probably contain more protein than any other cereal; they are consumed by some far Eastern people.*

ARACEAE | Arum Family

Herbaceous perenns. with swollen rhizomes or tubers. Fls. small, numerous and densely crowded on to a club-like stem or *spadix*, which is usually enlarged and coloured at the apex; a large and often conspicuously coloured sheath, or *spathe*, may enclose or partially surround the fls., the whole constituting what is popularly known as the 'flower', but which is in reality an inflorescence. Fls. greenish, often one-sexed, when the cluster of male fls. are usually borne on the spadix above the female fls., with a cluster of sterile fls. sometimes between. Perianth 4–6, or absent; stamens usually 4–6 fused or free; ovary with 1 to many cells; fr. a berry.

		Nos.
1. Lvs. sword-shaped or lance-shaped.		
2. Spathe apparently absent; fls. exposed and covering spadix.	*Acorus*	1816
2. Spathe present, tubular below and enclosing fls. and with an elongated blade.	*Biarum*	1820
1. Lvs. oval, arrow-shaped, or compound. Spathe present.		
3. Lvs. compound; spadix about as long as spathe.	*Dracunculus*	1819
3. Lvs. simple; spadix usually shorter than spathe.		
4. Spathe wrapped round spadix at base and enclosing fls.		
5. Spathe fused at base forming a funnel; spadix curved.	*Arisarum*	1821
5. Spathe overlapping at base, not fused; spadix straight.	*Arum*	1818
4. Spathe flat; fls. exposed.	*Calla*	1817

ACORUS Spadix appearing to be borne laterally, without a spathe, but in reality the spathe continues the leafy blade of the stem, the fl. stems in consequence appearing similar to the lvs. Fls. all bisexual; perianth 6. 1 sp.

1816. *A. calamus* L. SWEET FLAG. Lvs. sword-shaped, with the blades held vertically, with undulate margin, ½–2 cm. broad, sweet-smelling when crushed. Fls. on a cylindrical, tapering, spike-like branch *c.* 8 cm. long, placed laterally at an angle of about 45 degrees two-thirds of the way up the flattened leaf-like stem. Fls. greenish-yellow, very numerous and completely covering the branch-like spadix. Stems and lvs. to *c.* 1 m.; rhizome creeping in mud underwater. △ Shallow water, margins of ponds, lakes, rivers. May–July. Native of Asia and America: introd. to most of Eur. (except P.IS.AL.TR.). *The whole plant has a sweet aromatic scent and is sometimes used for flavouring. It yields Oil of Calamus which is used in perfumery and medicine.* Pl. 183.

CALLA   Spathe white, oval, spreading and not encircling the fls. and spadix. Lvs. heart-shaped. 1 sp.

1817. *C. palustris* L. BOG ARUM. An aquatic perenn. with a stout, underwater rhizome, long-stalked basal lvs. carried above-water and leafless stems bearing a solitary, white, flat spathe. Spathes oval-acute, 6–7 cm., spadix stout, oval, more than half as long and covered with fls. to the apex. Fls. one-sexed, densely clustered. Lvs. broadly heart-shaped, long-pointed, stalked, sheathing below; fl. stem 15–30 cm. above-water. Fr. coral red. △ Swamps, lakes. June–Aug. North Eur. (except IS.) Cent. Eur. (except H.) F.B.NL.BG.R.SU. *The fresh rhizomes are poisonous; the dried rhizomes have been ground up for flour in northern countries.* Pl. 183.

ARUM   Spathe large and conspicuous, funnel-shaped, partly encircling the spadix above, constricted below the middle and surrounding the fls. below. Spadix with a long club-like apex shorter than the spathe, and with fls. arranged in 4 zones near the base; the uppermost zone is sterile, below this is a zone of male fls., lower still a zone of sterile female fls. and at the base fertile female fls., allowing for a very complex pollination mechanism. *c.* 8 sps.

1818. *A. maculatum* L. LORDS-AND-LADIES, CUCKOO-PINT. Spathe large, erect, pale greenish-yellow, edged and sometimes spotted with purple, 15–25 cm., about twice as long as the dull purple or rarely yellowish spadix. Lvs. appearing in the spring, long-stalked; blade 7–20 cm. triangular with blunt or acute basal lobes, shiny, with or without black spots. Fr. scarlet, fleshy, *c.* ½ cm., in a spike 3–5 cm. long. A tuber-forming, hairless perenn. 30–50 cm. △ Woods, hedges. Apr.–May. Much of Eur. (except North Eur. P.TR.) DK. *The berries are poisonous and the acrid juice is an acute irritant. The tubers have been used as food and the starch for stiffening clothing.*

A. italicum Miller ITALIAN ARUM. Like 1818, but spadix almost always yellow, about one-third as long as the spathe which is usually whitish or yellowish, but variable and up to 40 cm. long. First lvs. appearing in autumn, arrow-shaped with pointed lobes, often white-veined, the later winter lvs. longer, with rounded, overlapping basal lobes, green-veined or with pale whitish spots. Mediterranean and Western Europe. Pl. 183.

A. pictum L. fil. PAINTED ARUM. Distinguished by its dark violet spathe 12–24 cm., and dark blackish-purple spadix which is three-quarters or more as long, autumn-flowering and appearing before the lvs. Sterile fls. between male and female fls. absent. Lvs. oval-oblong with a heart-shaped base, whitish-veined, and with a large ensheathing lv. stalk base; tubers nearly globular. Balearics, Corsica, Sardinia.

A. orientalis Bieb. EASTERN ARUM. Like *A. pictum*, but spring-flowering and usually with a deep blackish-purple spathe, but sometimes greenish or purple-flushed, and spadix blackish-purple, about half as long as the spathe. Rudimentary sterile fls. between male and female fls. present. Lvs. with acute lobes; tubers disk-shaped. Greece to Soviet Union.

DRACUNCULUS   Like *Arum*, but lvs. fan-like, palmately cut into many lance-shaped leaflets and fl. stem stout, leafy. Spadix large, about as long as the spathe. 2 sps.

1819. *D. vulgaris* Schott DRAGON ARUM. A stout, spotted-sheathed, leafy perenn. to 1 m., with a very large, deep chocolate-purple spathe with a wavy margin, and palmate lvs. Spathe to 35 cm. long, green outside; spadix thick fleshy, often as long as the spathe and similarly chocolate-purple; very fetid. Lvs. with 11–15 green lance-shaped leaflets, long-stalked and with conspicuous green-mottled sheaths; stems usually unspotted. Tubers globular. △ Uncultivated ground, bushy and woody places. Apr.–June. Med. Eur. (except AL.) P. Pl. 183.

BIARUM Stemless perenns. with the spathe arising directly from the ground; spathe with an oblong-oval acute blade and fused into a stalk-like tube at the base surrounding the fls. Male and female fl. cluster separated by sterile fls. Lvs. lance-shaped. 5 sps.

1820. *B. tenuifolium* (L.) Schott Distinguished by its dark brownish-purple, lance-shaped spathe which appears leafless and stalkless from the ground, and its narrow, cylindrical, purple spadix longer than the spathe. Fls. enclosed in the pale cylindrical tube of the lower part of the spathe; rudimentary male fls. present above the zone of fertile male fls. Lvs. usually produced after the inflorescence, lance-shaped to spathulate, long-sheathed and with linear-lance-shaped basal scales; stem very short. Fr. white. A tuber-forming perenn. △ Rocky places. Spring or autumn and winter. P.E.I.YU.GR. Pl. 183.

ARISARUM Spathe fused into a wide tube which is not constricted towards the base, blade of spathe very short; spadix narrow cylindrical, curved forwards. Male and female fls. not separated by a zone of sterile fls. 2 sps.

1821. *A. vulgare* Targ.-Tozz. FRIAR'S COWL. A small tuberous perenn. 5–30 cm., with an unusual-looking, longitudinally striped green and brown, flask-shaped spathe and a slender, forward-curved spadix. Spathe *c.* 3 cm. by 1½ cm., broadly cylindrical with an oblique mouth and an over-arching, brownish-purple upper lip; spadix brown-purple, a little longer than the spathe. Lvs. oval-arrow-shaped, long-stalked; stem leafless, spotted, often longer than the lvs. △ Grassy places, hedges, orchards, olive groves. Mar.–May and Oct.–Nov. Med. Eur. (except AL.TR.) P. *The plant is somewhat poisonous.* Pl. 183.

LEMNACEAE | Duckweed Family

Cosmopolitan water plants often occurring in large numbers on the surface or below the surface of the water and unattached to the soil; without typical lvs. and stems, but with small, rounded, leaf-like structures which are continuously budding, usually with simple roots beneath. Fls. minute, one-sexed; perianth absent, the male fls. of 1–2 stamens, the female of 1 ovary.

LEMNA | **Duckweed** 'Leaves' more or less flattened; roots present. 4 sps.

'Leaves' floating on surface

1822. *L. polyrhiza* L. (*Spirodela p.*) GREAT DUCKWEED. Readily distinguished by the presence of several straight roots up to 3 cm. long, arising from the lower surface of each 'leaf'. 'Leaves' oval to orbicular, 5–10 mm. across, thick but flat on each side, shiny green above and purplish with faintly marked veins beneath. △ Still waters. May–June. Most of Eur. (except IS.AL.TR.). Pl. 184.

1823. *L. minor* L. DUCKWEED. Like 1822, but 'leaves' smaller, oval, 2–5 mm. long, and with a solitary root arising from beneath up to 15 cm. long, but usually much less. Fls. not uncommon; ovary with 1 seed. △ Still waters. Apr.–June. All Eur. (except IS.). Pl. 184.

*L. gibba L. GIBBOUS DUCKWEED. Widespread in Europe.

'Leaves' floating below surface

1824. L. trisulca L. IVY DUCKWEED. Readily distinguished by its lance-shaped, rather translucent, often shortly stalked 'leaves' which are submerged and float below, not on the surface. 'Leaves' ½–1 cm. long, the young 'leaves' arising in opposite pairs at right angles to the parent 'leaf' and remaining attached to it and forming small irregular clusters. Fls. rare, borne on 'leaves' floating on the surface. △ Still waters. May–July. All Eur. (except IS.AL.).

*Wolffia arrhiza (L.) Wimmer Widespread in Central and Southern Europe.

SPARGANIACEAE | Bur-Reed Family

Herbaceous aquatic and swamp perenns. with creeping rhizomes, simple or branched stems and narrow, erect or floating lvs. sheathing at the base. Fls. one-sexed, numerous, crowded into several globular heads, the female heads subtended by leafy bracts and placed below the naked male heads, in a simple or branched inflorescence. Perianth of 3–6 membraneous scales; the male fls. with 3 or more stamens, the female fls. with a one-celled, stalkless ovary. Fr. dry, not splitting. 1 genus. c. 8 sps.

SPARGANIUM | Bur-Reed

Lateral branches of inflorescence with both male and female fl. clusters

1825. S. erectum L. (S. ramosum Hudson) BUR-REED. A robust, erect, hairless perenn. ½–2 m., usually standing in shallow water, with erect, sword-shaped lvs. and a branched, or rarely simple inflorescence of many globular greenish or yellowish clusters of fls. Male fl. clusters yellowish, numerous towards the ends of terminal and lateral branches, the female fl. clusters fewer and much larger, 1–2 cm. across, stalkless; perianth segments black-tipped. Lvs. rather stiff, 1–1½ cm. wide, keeled and triangular in section. Fr. abruptly narrowed above; seeds longitudinally ribbed. A variable sp. △ Shallow water, lakesides, margins of slow-flowing rivers. June–Aug. All Eur. (except IS.). Pl. 184.

Lateral branches of inflorescence with only female fl. clusters

1826. S. emersum Rehman (S. simplex Hudson) UNBRANCHED BUR-REED. Like 1825, but a smaller perenn. 20–60 cm., with narrower, soft, flexible lvs. 3–12 mm. wide, and a simple unbranched inflorescence with 3–10 upper stalkless male heads and 3–6 female heads below, the lowermost heads often stalked. Floating lvs. as well as aerial lvs. usually present, the former keeled, the latter triangular in section. Fr. 4–5 mm. long, narrowed to a slender beak. △ Shallow water of rivers, ponds, and lakes. June–Sept. Most of Eur. (except IS.AL.GR.TR.).

S. angustifolium Michx (S. affine Schmizel) FLOATING BUR-REED. Like 1826, but stems largely submerged and lvs. all floating, with very long flat blades c. ½ cm. wide, and with an inflated basal sheath. Male fl. clusters usually 2–3, the female clusters 2–4, the lowest stalked. Fr. c. 8 mm. long. Widespread, except South-Eastern Europe.

*S. minimum Wallr. SMALL BUR-REED. Widespread in Europe.

TYPHACEAE | Reedmace Family

Herbaceous aquatic and swamp perenns. with stout rhizomes and linear lvs. sheathing at the base. Fls. one-sexed, numerous, in a dense cylindrical spike, the male fls. above, the female fls. below. Fls. surrounded by jointed hairs or spathulate scales; perianth absent; the male fls. usually with 2–3 stamens with a common stalk, the female with 1 stalked ovary and an elongated style. Fr. one-seeded. 1 genus. *c.* 8 sps.

TYPHA | Reedmace

1827. *T. latifolia* L. GREAT REEDMACE, CAT'S-TAIL. A robust perenn. with creeping under-water rhizomes and erect, aerial, leafy stems to 2½ m., with a very dense cylindrical spike of numerous tiny fls. at the apex. The lower female fls. form a broad cylindrical spike which turns brown in fr., and the upper male fls. form a narrower, yellow spike which later falls off leaving a slender, naked, terminal axis; male and female clusters usually contiguous and about equal. Bracteoles of female fls. absent. Lvs. thick flat, 1–2 cm. broad, little longer than the inflorescence. △ Swamps, ditches, margins of lakes and ponds. June–Aug. All Eur. (except IS.). Pl. 184.

1828. *T. angustifolia* L. LESSER REEDMACE. Like 1827, but lvs. narrower, 3–10 mm. broad, convex beneath. Male and female fl. clusters separated by a naked section of the stem 1–9 cm. long. Bracteoles of female fls. shorter than the stigmas. An erect perenn. to 3 m. △ Swamps, ditches, margins of lakes and ponds. June–Aug. All Eur. (except IS.).

T. minima Hoppe LEAST REEDMACE. Like 1828, but a more slender perenn. 30–100 cm., distinguished by the short, brown, ovoid female fl. cluster, which may or may not be separated by a naked section of stem from the narrow, cylindrical male cluster. Lvs. all basal, very narrow, 1–2 mm. broad, shorter than the inflorescence; stem lvs. reduced to sheaths. Central and Southern Europe.

CYPERACEAE | Sedge Family

Usually herbaceous perenns., often with rhizomes, usually growing in moist or watery places. Lvs. linear, long-sheathing at the base, sometimes reduced to sheaths only. Fls. bisexual or one-sexed, each subtended by a scaly bract or *glume*, and clustered into *spike-lets*. Spikelets either solitary, or grouped into a branched or spike-like inflorescence, often subtended by bracts. Perianth often absent or reduced to bristle-like hairs or scales; stamens usually 3, rarely 2; ovary with 1 ovule; stigmas 3 or 2, feathery. Fr. more or less globular, three-sided or biconvex, not splitting.

		Nos.
1. Fls. one-sexed in separate male and female spikelets, or on different parts of the same spikelet; fr. enclosed in a sac or *perigynium.*	*Carex*	1850–1871
1. Fls. bisexual; spikelets all similar; fr. not enclosed.		
2. Spikelets flattened, with fls. and glumes two-ranked. Inflorescence with leafy bracts.		
3. Spikelets of 1–4 fertile fls.; leafy bract below inflorescence 1.	*Schoenus*	1847
3. Spikelets of 20–30 fertile fls.; leafy bracts below inflorescence 2–6.	*Cyperus*	1829–1832

2. Spikelets not flattened, with fls. and glumes spirally
  arranged, or in 3 ranks.
  4. Fr. encircled by numerous white bristly hairs, much
    longer than the spikelet.   *Eriophorum*   1833–1836
  4. Fr. without hairs, or bristly hairs shorter than the
    spikelet or glumes.
    5. Spikelets few-flowered, with 5–7 glumes and with only
      the upper 1–3 fertile and longer.
      6. Tall plants over 70 cm.; lvs. and bracts ½–1 cm. broad,
        very rough on margin.   *Cladium*   1849
      6. Slender plants 10–50 cm.; lvs. slender, 1–2 mm. broad,
        smooth on margin.   *Rhynchospora*   1848
    5. Spikelets usually many-flowered, glumes usually all
      fertile and equal, or the lower glumes longer.
      7. Spikelets solitary, terminal.
        8. Uppermost lv. sheath without a green blade.   *Eleocharis*   1846
        8. Uppermost lv. sheath with a short or long green
          blade.   *Scirpus*   1837–1844
      7. Spikelets several or many.
        9. Spikelets clustered into an ovoid, two-ranked
          inflorescence.   *Blysmus*   1845
        9. Spikelets not so clustered.   *Scirpus*   1837–1844

CYPERUS Spikelets with many fls. arranged in 2 opposite ranks; spikelets either clustered into umbels or into terminal heads; inflorescence with leaf-like bracts. Perianth absent; stamens usually 3; stigmas 3. *c.* 21 sps.

Anns.; lvs. very narrow, 1–4 mm. broad

1829. *C. fuscus* L. BROWN CYPERUS. Spikelets chestnut or blackish-brown, short-stalked, in a dense head or a short dense umbel with 3–7 rays, and with leafy bracts much longer than the inflorescence. Spikelets ½–1 cm. by 1½ mm. oblong, flattened; glumes *c.* 1 mm. long, acute, spreading, with reddish-brown midvein; stamens usually 2; stigmas 3; bracts 3–5, the longest to 10 cm. Lvs. flat, 2–4 mm. broad. A small tufted ann. with numerous acutely triangular-sectioned stems 5–30 cm. △ Damp sands. July–Sept. Most of Eur. (except IRL.IS.N.SF.). Page 564.

C. flavescens L. YELLOWISH CYPERUS. Like 1829, but spikelets yellowish, broader, *c.* 3 mm. wide, and somewhat longer; glumes longer, 1½–2 mm., broadly oval, blunt. Stamens 3; stigmas 2. Widespread, except in Northern Europe.

Large perenns.; lvs. 4–10 mm. broad

1830. *C. longus* L. GALINGALE. Spikelets densely arranged in fan-like clusters of 3–8 at the ends of long, very unequal branches and forming a spreading, very unequal umbel over-topped by 1 or more long leafy bracts. Spikelets 1–2 cm. by 2 mm., pale reddish-brown borne on repeatedly branched branches 8–30 cm. long; glumes 2½ mm. oval, closely over-lapping, green-keeled, margin papery, running down stalk in a delicate papery wing; stamens 3; stigmas 3. Lvs. 4–7 mm. wide, about as long as the inflorescence. A robust perenn. with smooth triangular-sectioned stems to 1 m. △ Marshes, watersides. July–Oct. Most of Eur. (except North Eur. IRL.B.NL.PL.CS.). *Used for paper-making and basket-weaving in the U.S.A.* Page 564.

C. badius Desf. BROWN GALINGALE. Like 1830, but umbel more compact and more flat-topped, the longest branches at most 8 cm.; spikelet 8–20 mm. long, dark reddish-brown or brown, 3–5 at the ends of the branches. Lvs. 3–5 mm. wide; leafy bracts 3–5, longer than the inflorescence. Mediterranean Europe and Portugal.

C. esculentus L. EDIBLE CYPERUS. Like 1830, but distinguished by its globular underground tubers and its short, straw-coloured or golden spikelets 8–15 mm. long, arranged in an umbel, and forming a lax cluster on branches at most up to 6 cm. long. Lvs. 2–5 mm. broad; leafy bracts 2–3, longer than the inflorescence. Glumes loosely overlapping; stamens and stigmas 3. Sometimes cultivated. Mediterranean Europe and Portugal. *The tubers are edible and used to make a beverage.*

1831. *C. serotinus* Rottb. Distinguished from 1830 by the spikelets which are densely arranged in opposite pairs, not in fan-shaped clusters, at the end of the unequal branches of the umbel. Spikelets reddish-brown, 5–20 mm. by 2–3 mm., spreading; glumes broadly oval, loosely overlapping; stigmas 2. Lvs. 4–10 mm. broad, thick, keeled beneath; leafy bracts usually 3, longer than the inflorescence; stems often solitary, thick, sharply three-angled, to 1 m. △ Marshes, watersides. July–Sept. E.F.CH.I.YU.AL.BG.R.SU.

1832. *C. glomeratus* L. ROUND-HEADED CYPERUS. Distinguished by its oval or globular clusters of spikelets which are themselves either stalked and form a lax umbel, or stalkless and densely clustered. Glumes linear-lance-shaped $1\frac{1}{2}$–2 mm. by $\frac{1}{2}$ mm., reddish-brown with a green midvein; stigmas 3. Lvs. 2–7 mm. broad, keeled; leafy bracts 3–6, longer than the inflorescence; stem three-angled, to $\frac{1}{2}$ m. △ Damp, sandy ground, riversides. July–Sept. I.YU.GR.BG.R.SU.

ERIOPHORUM | **Cotton-Grass**   Spikelets solitary terminal, or several and forming an umbel-like cluster. Fls. bisexual; glumes spirally arranged; perianth of numerous bristle-like hairs which elongate and become cottony after flowering; stamens and stigmas 3. *c.* 8 sps.

Fl. clusters several, stalked

1833. *E. angustifolium* Honckeny (*E. polystachion* L.) COMMON COTTON-GRASS. A sedge-like perenn. 20–60 cm., distinguished by its several more or less nodding fruiting clusters or spikelets with a dense tuft of pure-white silky hairs 3–4 cm. long. Spikelets 2–7, brownish, 1–1$\frac{1}{2}$ cm. borne on unequal smooth stalks, at first erect; bracts several, leafy, shortly sheathing. Glumes *c.* 7 mm. lance-shaped long-pointed, brownish below, with transparent silvery margin. Lvs. 3–6 mm. wide, channelled above and narrowed to a three-angled apex; stems smooth, nearly cylindrical-sectioned above; plant not tufted. △ Bogs, marshes. Apr.–July. Most of Eur. (except GR.TR.). *The fruiting heads have been used for wicks.*

1834. *E. latifolium* Hoppe BROAD-LEAVED COTTON-GRASS. Like 1833, but a tufted perenn. 20–60 cm., with flat lvs. 3–8 mm. wide and fl. stems three-angled above. Spikelets 5–12, borne on very rough stalks, nodding in fr.; hairs of fr. shorter, 2 cm., silvery-white. Glumes 4–5 mm., blackish with very narrow papery margin. Uppermost lv. with a tight-fitting sheath, and without a ligule. △ Wet places. Apr.–June. Most of Eur. (except P.IS.TR.). Pl. 184.

Fl. cluster solitary, terminal

1835. *E. vaginatum* L. COTTON-GRASS, HARE'S-TAIL. A tussock-forming perenn. with many slender, erect, apparently leafless stems 20–60 cm., bearing an erect, terminal, ovoid cluster or spikelet of dense, white, silky hairs when in fr. Spikelets *c.* 1$\frac{1}{2}$ cm., brownish at flowering; bracts below spikelet absent; glumes *c.* 7 mm. translucent, slaty-grey; hairs to 2 cm. in fr. Stem lvs. 2–3 with strongly inflated sheathing base and a short or very short blade, the basal lvs. numerous, thread-like to 1 mm. wide, rough. Stem slender, stiff, three-angled above. △ Bogs, wet grassy places. Apr.–July. Much of Eur. (except P.IS. AL.GR.TR.). Pl. 184.

1836. *E. scheuchzeri* Hoppe Like 1835, but stems cylindrical-sectioned above and stem lvs. with sheaths not or only slightly inflated, the basal lvs. few, smooth. Spikelet globular; hairs 1½–2½ cm. long in fr. A less tufted and more creeping perenn. 10–30 cm. △ Arctic or alpine: marshes, bogs, and lakesides. June–July. North Eur. (except DK.) E.F.D.CH.A.CS. I.YU.R.

SCIRPUS Spikelets oval or cylindrical, solitary or clustered, stalkless or borne on simple or branched stems forming a lax or dense inflorescence. Fls. bisexual; perianth of 3–6 bristly hairs, or absent; stamens usually 3; stigmas 2–3. *c.* 22 sps.

Spikelet solitary, terminal, without leafy bract at base
1837. *S. caespitosus* L. (*Trichophorum c.*) DEER GRASS. A densely tufted, rush-like perenn. 5–25 cm., with numerous, very slender, leafless stems ending in a tiny pale brownish spikelet 3–7 mm. long. Fls. 3–6; outermost glume largest, blunt, about equalling or longer than the spikelet and often with a green leaf-like apex. Stems smooth, with closely investing brownish sheaths below and with the uppermost sheath prolonged into a short green blade to 1 cm. Fr. *c.* 2 mm., surrounded by brownish bristles a little longer but shorter than the glumes. △ Bogs, damp heaths. May–Aug. Most of Eur. (except H.AL.GR. TR.R.).

1838. *S. fluitans* L. (*Eleogiton f.*) FLOATING SCIRPUS. A slender, leafy, grass-like, floating perenn. 15–40 cm. long, with leafy stems and with solitary, terminal, long-stalked, pale greenish spikelets only 2–3 mm. long. Fls. 3–5; glumes *c.* 2 mm., oval, papery; bristles absent; stamens and stigmas 3. Lvs. with sheathing bases and blades to 5 cm. by *c.* 1 mm. wide; stems branched, flattened. △ Stagnant water, marshes. June–Sept. West Eur. DK.S.R. Page 564.

Spikelets many, or 1–3 and with a leafy bract at base
(a) Spikelets terminal, with leafy bracts not appearing to continue the stem
1839. *S. maritimus* L. SEA CLUB-RUSH. A tall, erect perenn. 30–120 cm., with leafy, triangular-sectioned, rough stems and a dense, terminal cluster of reddish-brown spikelets *c.* 5 cm., much overtopped by one or more leafy bracts. Spikelets 1–2 cm., ovoid, stalkless or short-stalked in clusters of 2–5; glumes rusty-brown with a green midvein, *c.* 7 mm., oval with a bilobed apex and a longer central awn in the notch. Lvs. flat, keeled beneath, 2–10 mm. broad, apex triangular. △ Marshes and ditches by the sea, rarely inland. June–Sept. All Eur. (except IS.). Pl. 185.

1840. *S. sylvaticus* L. WOOD CLUB-RUSH. Distinguished by its lax, much-branched, spreading, rounded inflorescence *c.* 15 cm. across, with leafy bracts of about the same length. Spikelets very numerous, ovoid, 3–4 mm. long, greenish, usually in small dense clusters at ends of the branchlets; glumes 1½ mm., oval with entire apex. Lvs. flat, to 2 cm. broad, margin rough; stem stout, leafy, bluntly three-angled, smooth, 30–100 cm. △ Marshes, wet woods, meadows, ditches, riversides. May–Aug. Most of Eur. (except P.IS.). Pl. 185.

(b) Spikelets apparently lateral with bract appearing to continue the stem above
1841. *S. holoschoenus* L. (*Holoschoenus vulgaris* Link) ROUND-HEADED CLUB-RUSH. Readily distinguished by its very dense, small, quite globular heads ½–1½ cm., borne laterally and subtended by a much longer bract forming a stem-like extension above the inflorescence. Heads 1–5, greyish-brown, composed of numerous rounded spikelets; heads stalkless or stalked with flattened stalks with a rough margin. Glumes *c.* 2½ mm. obovate, fine-pointed, ciliate. Stem smooth, ribbed, cylindrical-sectioned, leafless but with scaly sheaths below. A densely tufted, creeping perenn. ½–1 m. △ Damp places, watersides. July–Aug. Most of Eur. (except North Eur. IRL.NL.). Pl. 185.

1842. *S. triquetrus* L. (*Schoenoplectus t.*) TRIANGULAR SCIRPUS. Readily distinguished by its sharply triangular-sectioned, leafless stem bearing below the apex a rather dense cluster of reddish-brown spikelets *c.* 3 cm. long, and overtopped by a triangular pointed extension of the stem about twice as long. Spikelets ovoid, 5–8 mm. long, stalked; glumes 4 mm. with a brownish transparent margin, green midvein, and a shallowly notched apex. A stout, creeping perenn., with erect stems 1–1½ m. △ Marshes, riversides, by the sea. July–Sept. Most of Eur. (except North Eur. AL.GR.TR.).

1843. *S. lacustris* L. (*Schoenoplectus l.*) BULRUSH. Distinguished from 1842 by its smooth stems which are round-sectioned and up to 1½ cm. in diameter. Inflorescence lateral, forming either a densely clustered head or more often an irregular lateral umbel, over-topped by a rounded-sectioned, stem-like extension as long as or shorter than the inflorescence. Spikelets reddish-brown, cylindrical-ovoid, ½–1 cm. long; glumes 3–4 mm., broadly oval, notched, often with a short, rough, projecting awn, smooth on back; stigmas 3. A robust perenn. with underwater creeping rhizomes producing tufts of sub-merged lvs. and almost leafless, erect, above-water stems 1–3 m. △ Rivers, lakes, ponds. May–Aug. All Eur. (except IS.). *The dried stems are used for making mats, mattresses, baskets, chair-bottoms, and for thatching. The pith is used for paper-making.*

S. tabernaemontani C. C. Gmelin (*Schoenoplectus t.*) Very like 1843, but stems glaucous, not green; glumes with small brown swellings on back; stigmas 2. Almost all Europe. Pl. 185.

1844. *S. setaceus* L. (*Isolepis s.*) BRISTLE-SCIRPUS. A tiny, densely tufted, bright green, grassy perenn. 5–15 cm. with many thread-like stems usually bearing 2–4 tiny, dark purplish-brown spikelets apparently just below the tip of the stem. Spikelets ovoid, less than ½ cm. long, densely clustered, and with a longer green bract; glumes 1½ mm. oval fine-pointed, purple-brown, with a green midvein and a transparent margin. Fr. dark brown, shiny, longitudinally ribbed. △ Damp, sandy places, ditches, lakesides. June–Sept. All Eur. (except IS.TR.). Page 564.

BLYSMUS Spikelets several, in 2 ranks and forming a short spike-like cluster subtended by a leafy bract. Bristly hairs 3–6. Stigmas 2. 2 sps.

1845. *B. compressus* (L.) Link (*Scirpus planifolius* Grimm) BROAD BLYSMUS. Inflorescence simple, reddish-brown, *c.* 2 cm. long, usually of 10–12 spikelets arranged in 2 ranks, and with a slender leafy bract as long as or shorter than the inflorescence. Spikelets 5–7 mm.; glumes 3 mm. acute, reddish-brown with pale midrib and a narrow papery margin. Lvs. 1–3 mm. wide, flat, keeled, with a rough margin; a perenn. with a creeping rhizome and leafy shoots and fl. stems 10–30 cm. △ Damp meadows, marshy places. June–Aug. Much of Eur. (except P.E.IRL.IS.TR.). Page 564.

ELEOCHARIS | Spike-Rush Spikelets solitary, terminal, the lowest glume differing somewhat from the fertile glumes and usually sterile; bristly hairs usually present. Stems round-sectioned, with papery sheaths below and without leafy blades. *c.* 8 sps.

1846. *E. palustris* (L.) Roemer & Schultes COMMON SPIKE-RUSH. An aquatic perenn. with far-creeping rhizomes producing single or small tufts of leafless stems 10–60 cm. each bearing a terminal, cylindrical spikelet ½–2 cm. long. Glumes brown and with green midvein and transparent margin, the lowest 2 sterile and much shorter than the spikelet and each not more than half-encircling its base; stamens 3; stigmas 2. Stems stout or

1. *Rynchospora alba* 1848
2. *Scirpus setaceus* 1844
3. *Cyperus fuscus* 1829
4. *Blysmus compressus* 1845
5. *Cyperus longus* 1830
6. *Scirpus fluitans* 1838

slender, 1–4 mm. broad; sheaths pale brownish. Fr. biconvex. △ Marshes, margins of pools, damp meadows, ditches. May–Aug. All Eur. Pl. 185.

E. multicaulis (Sm.) Sm. MANY-STEMMED SPIKE-RUSH. Like 1846, but stems densely tufted and spikelets smaller, *c.* 1 cm. long, the lowest glume less than one-quarter as long as the spikelet and encircling its base. Spikelets often viviparous; stigmas 3. Stems slender; sheaths obliquely cut-off. Fr. three-angled. Often mistaken for 1837 with which it may grow, but distinguished from it by the absence of a green lv. blade to the uppermost lv. sheath and by the much shorter lowest glume. Western, Northern, and much of Central Europe.

*E. uniglumis (Link) Schultes Widespread in Europe.

*E. acicularis (L.) Roemer & Schultes SLENDER SPIKE-RUSH. Widespread in Europe.

SCHOENUS | Bog-Rush Spikelets several, in a dense terminal head with the bract of the lowest spikelet largely encircling the whole inflorescence. Spikelets flattened laterally, with 1–4 fls.; glumes in 2 opposite rows, the lower sterile. Bristly hairs 1–6; stamens and stigmas 3. 3 sps.

1847. *S. nigricans* L. BOG-RUSH. A densely tufted perenn. 15–75 cm., with many wiry stems each bearing a dense, blackish, ovoid head 1–1½ cm., overtopped by the lowest leafy bract. Spikelets 5–20 in a head, stalkless, each 5–8 mm. long; glumes two-rowed, acute, and keeled; bristles 3–5; bracts 3–4, blackish. Lvs. wiry, almost cylindrical in section and with inrolled margin; sheaths of lower lvs. tough, shining, blackish-brown. △ Marshes, bogs, sandy places by the sea. June–July. Most of Eur. (except IS.SF.). Pl. 185.

S. ferrugineus L. Like 1847, but a more slender perenn. with a narrower fl. head of 1–3 spikelets, to 1 cm. long; lowest bract shorter or a little longer than the fl. head. Central Europe and Balkans.

RHYNCHOSPORA | Beak-Sedge Spikelets several, in a small, compact, terminal cluster with or without stalked lateral heads below. Spikelets with 1–2 fls. and several sterile overlapping glumes; bristly hairs 5–13; stigmas usually 2. Fr. beaked, usually biconvex. 2 sps.

1848. *R. alba* (L.) Vahl WHITE BEAK-SEDGE. Distinguished by its dense, white, terminal cluster of tiny spikelets, *c.* 7 mm. long, not or slightly exceeded by a slender bract. Clusters at length turning pale reddish-brown; spikelets 4–5 mm., usually two-flowered; bristles 9–13. Upper stem lvs. with a blade about as long as the stem, the lower sheaths bladeless and often bearing bulbils in their axils. A slender, somewhat tufted perenn. 10–50 cm. △ Bogs, marshes. June–Aug. Most of Eur. (except IS.AL.GR.TR.). Page 564.

CLADIUM Spikelets clustered into dense heads and forming a branched inflorescence. Spikelets with 1–3 fls.; stamens 2–3; stigmas 2–3, style swollen at the base; bristles absent. 2 sps.

1849. *C. mariscus* (L.) Pohl CUT-SEDGE. A tall, hollow-stemmed, rough-leaved perenn. 70–300 cm., with a much-branched inflorescence with unequal branches each ending in dense, rounded, reddish-brown heads ½–1 cm. Spikelets 3–10 in each head, 3–5 mm. long with 1–3 fls.; stamens usually 2; stigmas usually 3, style base much enlarged. Lvs. 1–2 cm. broad, grey-green, with a finely toothed, very rough margin and keel, and a long three-angled tip; stems leafy, rounded, rough. Rhizome creeping. △ Fens, reed-swamps, marshes. July–Aug. Much of Eur. (except IS.TR.).

CAREX | Sedge Fls. one-sexed, each subtended by a papery bract or *glume,* and densely clustered and overlapping in a compact spike. Spikes usually several, the upper spikes usually male, the lower spike female, but sometimes the male spikes have a few female fls. at the base, or the female spikes a few male fls. at the apex. Solitary spikes have male and female fls. in the same spike, or are very rarely one-sexed. Perianth absent; male fls. with 2–3 stamens; female fls. with a single ovary closely surrounded by a sac, the *perigynium,* which is usually crowned by a beak from which the stigmas project. Ovary three-angled, with 3 stigmas, or biconvex with 2 stigmas. Fr. a one-seeded nut surrounded by the perigynium. Measurement of fr. includes perigynium and beak. *c.* 160 sps.

Spikes dissimilar, with distinctive male and female spikes
(a) *Spikes densely clustered and overlapping*
1850. *C. flava* L. YELLOW SEDGE. A densely tufted perenn. to 50 cm., with a dense terminal cluster of spikes comprising a single, pale, linear male spike 1–2 cm., and 2–4 erect, oval to globular, stalkless and overlapping female spikes ½–1½ cm. Bracts leafy, reflexed, the lowest longer than the inflorescence. Female glumes 4 mm., lance-shaped, transparent-brownish and with a green midrib. Lvs. 4–7 mm. wide, channelled, bright green, smooth, sheaths of lvs. becoming fibrous. Fr. 5–7 mm., greenish to golden-yellow, curved outwards or reflexed, beak saw-toothed or ciliate. △ Marshy places, damp fields. May–July. Most of Eur. (except GR.TR.). Page 569.

*C. lepidocarpa Tausch Widespread in Europe, except in the south.

1851. *C. extensa* Gooden. SALTMARSH SEDGE. A tufted littoral or salt marsh perenn. to 40 cm., distinguished by its densely clustered inflorescence and its 2 or 3 long, rigid, spreading or reflexed, leafy bracts which are many times longer. Male spike 1, 1–1½ cm.; female spikes 2–4, oblong-oval, 8–15 mm. long, clustered at base of the male spike, overlapping, or the lowest spike distant. Female glumes *c.* 2 mm., broadly oval fine-pointed, straw-coloured with brownish patches. Lvs. 2–3 mm. wide, nearly all basal. Fr. 3–4 mm., greenish or light brown, beak short and smooth. △ Apr.–July. All coastal countries of Eur. (except IS.AL.). Pl. 185.

*C. pilulifera L. PILL SEDGE. Widespread in Europe.

*C. caryophyllea Latourr. SPRING SEDGE. Widespread in Europe.

*C. pallescens L. PALE SEDGE. Widespread in Europe.

(b) *Female spikes remote from each other, not overlapping, or if overlapping and clustered then more or less nodding on long slender stalks*
(i) *Male spike 1, or with a second very much smaller male spike*
1852. *C. sylvatica* Hudson WOOD SEDGE. A tufted woodland perenn. to 60 cm., with a single, slender, male spike 1½–3 cm. and 3–4 long-stalked, widely spaced, drooping, slender, lax female spikes 2–5 cm. by 3–4 mm. Lower bracts leafy, but shorter than the inflorescence. Female glumes *c.* 3 mm. oval-acute, brown with transparent margin and green midvein. Lvs. 3–8 mm. broad, flat, shining, soft. Fr. 4–5 mm., green, beak long, smooth. △ Woods, shady ravines. May–July. Most of Eur. (except P.IS.).

1853. *C. pseudocyperus* L. HOP SEDGE. A handsome tufted perenn. of watersides, with very rough, sharply three-angled stems to 1 mm., and broad, rough-edged, yellowish-green lvs. 5–12 mm. wide. Male spike solitary, 3–6 cm.; the female spikes 3–5, cylindrical, 3–5 cm. by 1 cm., clustered below the male spike but stalked and pendulous, the lowest more or less remote. Bracts leaf-like, much longer than the inflorescence. Male glumes *c.* 6 mm., long-pointed, light brown; female glumes 3½–4½ mm., greenish with transparent margin and a long saw-toothed apex. Fr. spreading, 5–6 mm., green and shining, beak 2 mm., smooth or toothed. △ Marshes, watersides. May–July. Most of Eur. (except IS.AL.TR.). Pl. 185.

1854. *C. pendula* Hudson DROOPING SEDGE. One of the tallest and most graceful sedges, with slender smooth stems $\frac{1}{2}$–$1\frac{1}{2}$ m., broad lvs. 1–2 cm. wide and very long, green, pendulous female spikes. Male spike solitary 6–10 cm., brownish; female spikes 4–5, distant, stalked, 7–16 cm. by 5–7 mm. Male glumes *c.* 7 mm., long-pointed, brownish and papery; female glumes 2–2$\frac{1}{2}$ mm., oval acute, reddish-brown. Lvs. with rough margin, keeled, somewhat glaucous beneath. Fr. *c.* 3 mm., greenish-brown, beak short. △ Damp woods, shady ravines. May–July. Most of Eur. (except North Eur.) DK. Pl. 186.

C. panicea L. CARNATION SEDGE. Widespread in Europe.

C. limosa L. MUD SEDGE. Widespread in Europe.

1855. *C. distans* L. DISTANT SEDGE. A densely tufted perenn. with triangular-sectioned, smooth stems to 45 cm., and with a slender inflorescence of a single male spike and 2–3 widely spaced, shortly stalked, erect female spikes, with leafy bracts shorter than the inflorescence. Male spike 1$\frac{1}{2}$–3 cm. by 2–3 mm.; female spikes 1–2 cm. by 4–5 mm., with stalks more or less included in the bract sheath. Male glumes oval blunt, *c.* 4 mm.; female glumes oval fine-pointed, reddish-brown or greenish-brown with pale midvein, which is conspicuous in fr. Lvs. grey-green, nearly flat, 2–5 mm. wide, much shorter than stem. Fr. 3$\frac{1}{2}$–4 mm., many-ribbed, erect, beak rough. △ Marshes, damp meadows, often near sea. May–July. All Eur. (except IS.).

C. hostiana DC. TAWNY SEDGE. Widespread in Europe, except in the south-west.

C. laevigata Sm. SMOOTH SEDGE. Western Europe.

1856. *C. sempervirens* Vill. A densely tufted alpine perenn. with smooth stems to $\frac{1}{2}$ m., solitary brown male spikes and 2–3 erect, stalked, rather widely spaced, lax female spikes and with bracts often shorter than stalks of the spikes. Female glumes oval-lance-shaped, brownish with transparent margin and a pale midvein. Lvs. flat, rough, 2–5 mm. wide, shorter than the fl. stem. Fr. 5–6 mm. oblong-lance-shaped, narrowed to a long ciliate beak, becoming rusty-brown. △ Mountain pastures. June–Aug. Cent., South-East Eur. (except AL.TR.) E.F. Page 569.

(ii) Male spikes usually more than 1

1857. *C. rostrata* Stokes (*C. ampullacea* Gooden.) BOTTLE SEDGE. A greyish-green perenn. to 60 cm., usually rooting in shallow water, with glaucous inrolled lvs., 2–4 light brown male spikes, and 2–4 remote, cylindrical female spikes and bracts equal, overtopping the inflorescence. Male spikes 2–7 cm. by 1–3 mm.; female spikes 2–8 cm. by 1 cm., often with some male fls. at the apex, short-stalked, more or less erect. Female glumes 5 mm., lance-shaped, brown with a pale midvein. Lvs. 3–7 mm. wide, rough. Fr. ovoid, inflated and bottle-like, spreading, yellow-green, abruptly narrowed to a long smooth beak. △ Watersides, peaty pools. May–July. Most of Eur. (except P.TR.). Pl. 186.

C. vesicaria L. BLADDER SEDGE. Widespread in Europe, except in the extreme south-east.

1858. *C. riparia* Curtis GREAT POND-SEDGE. A robust waterside perenn. 1–1$\frac{1}{2}$ m., with rough, sharply three-angled stems, and glaucous, sharply keeled lvs. 6–20 mm. wide. Male spikes 2–5, broadly cylindrical, dark brown; female spikes 1–5, large and robust 3–9 cm. by 1–1$\frac{1}{2}$ cm., distant, the upper erect and nearly stalkless, the lower long-stalked, at length nodding; bracts leaf-like, about equalling the inflorescence. Male glumes *c.* 8 mm., dark brown, and with a paler midvein; female glumes *c.* 7 mm., brown, and with a paler midvein projecting at the apex. Fr. brown *c.* 8 mm., beak 1$\frac{1}{2}$ mm., smooth. △ Watersides. Apr.–June. All Eur. (except IS.). Pl. 186.

C. acutiformis Ehrh. LESSER POND-SEDGE. Widespread in Europe.

1859. *C. flacca* Schreber GLAUCOUS SEDGE. A glaucous perenn. commonly found in dry grasslands, with smooth, bluntly three-angled stems to 40 cm., and glaucous, rough, slightly keeled lvs. 2–4 mm. wide. Male spikes usually 2–3, rarely 1, purple-brown, 1–3 cm.; female spikes usually 2, mostly short-stalked, erect or nodding, 1½–4 cm. by 4–6 mm.; bracts usually shorter than the inflorescence. Female glumes 1½–2 mm., oval acute, purplish-brown or black, often with a pale, straw-coloured midvein and a transparent margin. Fr. 2–2½ mm., curved, variously coloured greenish-yellow, reddish, or blackish, and covered with minute but distinctive swellings, beak very short. △ Dry grassland, marshes, bogs. Apr.–June. All Eur. Pl. 186.

1860. *C. hirta* L. HAIRY SEDGE. Readily distinguished by its hairy lvs., densely shaggy-haired lv. sheaths, and its conspicuously hairy frs. Male spikes 2–3, slender; female spikes usually 2–3, erect 1–3 cm., short-stalked with the stalk often included in bract sheath, very widely spaced, the lowest often near the base of the stem; bracts leafy, all much longer than the inflorescence. Female glumes 6–8 mm., pale greenish, and tapered to a long ciliate point. Lvs. 2–4 mm. wide, shorter than the fl. stem. Fr. downy-haired, greenish, 6–7 mm., many-ribbed. A creeping perenn. with erect, hairless, shiny, three-angled stems 30–60 cm. △ Damp, sandy places, damp meadows, woods. May–July. All Eur. (except IS.). Pl. 186.

1861. *C. nigra* (L.) Reichard (*C. vulgaris* Fries) COMMON BLACK SEDGE. A variable, often tufted perenn. with slender three-angled stems 7–70 cm. and male spikes 2, usually purplish; female spikes 1–4, usually stalkless and clustered below male or somewhat distant, 1–3 cm. by 4–5 mm. Male spikes often unequal, the lower smaller, or less frequently 1; lowest bract shorter or longer than the inflorescence. Female glumes 3–3½ mm., acute or blunt, black or rarely brown, and with a pale midvein ceasing below the apex, margin narrow papery. Lvs. 2–3 mm. wide, rather rough, the lower lvs. with black fibrous bases; stem rough above. Fr. 2½–3 mm., broader than the glumes, green or purplish, beak very short. △ Wet places, watersides. May–July. All Eur. (except TR.). Pl. 186.

1862. *C. elata* All. TUFTED SEDGE. A tussock-forming perenn. with roughish, sharply three-angled stems to 1 m., and glaucous, sharply keeled lvs. 4–6 mm. wide. Male spikes usually 1–2, 3–4 cm. long, dark purplish-brown; female spikes usually 2, rather distant, stalkless, and erect, 1½–4 cm. by 5–7 mm., purple-brown with glumes arranged in conspicuous longitudinal rows, often with a few male fls. at the apex; bracts bristle-like, much shorter than the inflorescence. Female glumes 3 mm., purple-brown, and with a paler midvein ceasing below the apex, margin papery. Fr. *c.* 3 mm. oval, veined, beak very short, and not notched. △ Fens, ditches, watersides. Apr.–June. Most of Eur. (except P.IS.AL.TR.BG.). Page 569.

C. acuta L. GRACEFUL SEDGE. Widespread in Europe.

Spikes all similar with both male and female fls. present, or spikes one-sexed
(a) Spikes several or many
(i) Spikes usually stalkless and overlapping in a dense oval or cylindrical inflorescence
1863. *C. paniculata* L. TUSSOCK SEDGE. A very densely tufted perenn. often forming large tussocks *c.* 1 m. across and up to *c.* 1 m. tall, with very rough, sharply three-angled stems ½–1½ m., and very rough, stiff, dark-green, inrolled lvs. 3–7 mm. wide. Spikes ovoid, pale brownish, numerous, in a lax branched, or slender little-branched inflorescence 6–10 cm. long; bracts bristle-pointed, shorter. Female glumes *c.* 3 mm., oval-triangular, transparent, brownish, and with a yellowish midvein. Fr. *c.* 3 mm., dark brown, corky at base, beak winged and toothed or ciliate. △ Marshes. May–July. All Eur. (except IS.TR.).

1. *Carex baldensis* [1867]
2. *C. sempervirens* 1856
3. *C. elata* 1862
4. *C. dioica* 1871
5. *C. pulicaris* 1870
6. *C. flava* 1850
7. *C. capitata* [1870]

1864. *C. vulpina* L. GREATER FOX SEDGE. A stout tufted perenn. to 1 m., with robust and very sharply three-angled or almost winged stems, bearing a reddish-brown, oblong-cylindrical cluster of spikes 3–8 cm. long, often interrupted below. Spikes *c.* 1 cm. globular, stalkless, overlapping, with short, inconspicuous, bristle-like bracts. Male fls. at apex of spike; female glumes 4–5 mm., long-pointed, and with a green midvein and a brownish margin. Fr. 4–5 mm., green then brown, ribbed; beak toothed. △ Ditches, damp meadows, streamsides. May–June. All Eur. (except IS.AL.).

C. otrubae Podp. FALSE FOX SEDGE. Widespread in Europe.

1865. *C. arenaria* L. SAND SEDGE. A perenn. of sandy places distinguished by its far-creeping underground stolons bearing usually widely spaced, erect, aerial stems 10–40 cm. Inflorescence usually dense, but sometimes lobed below, oval-oblong to 4 cm., with 5–12 stalkless, overlapping, brownish spikes; bracts bristle-pointed, shorter. Glumes 5–6 mm., oval long-pointed, brownish and papery. Lvs. 1½–4 mm. broad, tough, shorter than or equalling the three-angled stems, sheaths brown. Fr. 4–5 mm., ovoid many-ribbed, the upper half and the beak broadly winged and toothed. △ Sandy places, dunes on the littoral. May–Aug. Coastal countries of Eur. (except IS.).

1866. *C. spicata* Hudson (*C. contigua* Hoppe) SPIKED SEDGE. A tufted perenn. 30–75 cm., with slender, rough, three-angled stems, bearing a slender, rather dense or somewhat lax, cylindrical, pale greenish-brown inflorescence 2–4 cm. long, with much shorter bristle-like bracts. Spikes *c.* ½ cm., stalkless and usually overlapping and crowded or the the lower spike not more than its own length below the spike above. Glumes 3–4 mm., long-pointed, pale brown and with a green midvein. Lvs. 2–3 mm. wide, channelled, basal sheaths often tinged purple. Fr. 4–5½ mm., greenish, gradually tapering to a rough beak *c.* 2 mm. long. △ Grasslands, roadsides, marshes. Apr.–June. All Eur. (except IS.).

1867. *C. ovalis* Gooden. OVAL SEDGE. A coarse, densely tufted perenn. 20–90 cm., distinguished by the dense oval-acute cluster of pale brown spikes, with a single, usually longer, bristle-like bract. Spikes 3–9, each 8–12 mm. long, oval, stalkless, and overlapping; male fls. at the base of the spikes; glumes 3–4 mm., brownish and papery. Lvs. narrow, 2–3 mm. broad, bright green, rough-margined, much shorter than the three-angled, rough stems. Fr. 4–5 mm., light brown, flattened and many-veined, and with rough flanges, beak rough, green. △ Grassy places, damp woods. May–Aug. All Eur. (except IS.AL.TR.). Pl. 186.

C. echinata Murray STAR SEDGE. All Europe.

C. baldensis Torner MONTE BALDO SEDGE. A very distinctive perenn. with several white oval spikes grouped into a dense, hemispherical, whitish head 2½ cm. across, and with 2 or more long spreading bracts to 5 cm. immediately below. Female glumes white, blunt. Fr. 3–4 mm., beakless. Eastern Alps, Italy. Page 569.

(*ii*) *Spikes stalkless but not overlapping, in a lax inflorescence; or lower spikes stalked*
C. divulsa Stokes GREY SEDGE. Widespread in Europe.

1868. *C. remota* L. REMOTE SEDGE. Distinguished by its very small, widely spaced green spikes arising stalkless from the axils of the bracts, the lower bract leafy and much longer than the inflorescence, the upper bristle-like. Spikes 4–7, ovoid, the lower 7–10 mm. long, the upper spikes smaller; glumes *c.* 2½ mm. oval-triangular, transparent, and with the midvein green below and brown above. Fr. *c.* 3 mm., greenish, beak broad, and sawtoothed. A densely tufted, grass-like perenn. with weak, slender, bluntly three-angled, rough stems to 60 cm., and narrow, bright green, channelled lvs. 2 mm. wide. △ Shady, damp places. May–July. All Eur. (except IS.AL.).

C. elongata L. GINGERBREAD SEDGE. Much of Europe, except the south.

1869. *C. atrata* L. DARK SEDGE. Readily distinguished by its 3–5 almost black, fat spikes borne in a lax cluster, the upper spikes stalkless and erect, the lowest often long-stalked and at length nodding. Spikes all similar, oval-oblong 1–2 cm., the upper spike with male fls. at its base, the lower spikes with only female fls.; lower bracts leafy, scarcely longer than the inflorescence. Female glumes *c.* 4 mm. oval-acute, blackish and with paler midvein and papery-margined towards the apex. Fr. *c.* 4½ mm., brownish, shortly beaked. A mountain perenn. with smooth, often curved, three-angled stems 20–40 cm. and keeled, somewhat glaucous flat lvs. 3–5 mm. wide. △ In mountains: damp rocks, pastures. July–Aug. Much of Eur. (except P.IRL.B.NL.DK.PL.AL.TR.). Pl. 186.

(b) Spikes solitary, terminal

1870. *C. pulicaris* L. FLEA SEDGE. Readily distinguished by its solitary slender spike, with the upper part consisting of a cluster of several male fls. and the lower part of several spreading and later drooping female fls. Spike 1–2½ cm. long; bracts absent. Male glumes *c.* ½ cm., brown; female glumes soon falling. Fr. reflexed, spindle-shaped 4–6 mm., dark brown and shiny. A shortly creeping perenn. forming dense patches, with very slender, rigid, smooth stems 10–30 cm., and very narrow, channelled, dark green lvs. *c.* 1 mm. wide. △ Grasslands, fens, damp places. May–June. Most of Eur. (except P.E.IS.AL.GR.TR.BG.). Page 569.

C. capitata L. Fl. spike terminal and globular, *c.* ½ cm.; bracts absent; female glumes oval to orbicular, persistent. Fr. 3½ mm., oval to orbicular, erect or spreading, beaked. A delicate, tufted perenn. Alps, Carpathians, and Northern Europe. Page 569.

1871. *C. dioica* L. DIOECIOUS SEDGE. Readily distinguished by its solitary spikes without bracts, which are usually one-sexed, with male and female fls. occurring on different plants. Male spikes narrow, 1–1½ cm. by 2–3 mm., pale brown, rarely with female fls. at base. Female spikes oval, dense, 1–1½ cm. by 5–7 mm.; female glumes 2½–3 mm., brown. Fr. spreading, *c.* 3½ mm., greenish-brown and with dark brown veins, tapering to a broad, saw-toothed, blackish beak. A slender-stemmed perenn. to 35 cm., with very narrow, dark green, three-veined, channelled lvs. ½–1 mm. wide. △ Fens, marshes. Apr.–July. Much of Eur. (except P.AL.GR.TR.BG.). Page 569.

ORCHIDACEAE | Orchid Family

Herbaceous terrestrial perenns. in our sps., often with tuberous roots; lvs. usually spirally arranged or in 2 rows, often with a sheathing base, sometimes spotted. In saprophytic sps. the lvs. are reduced to scales and are not green. Fls. usually in a spike-like cluster and each fl. is subtended by a leafy or scaly bract. The floral structure is unique and very highly specialized for pollination. Fls. symmetrical in one plane only, or *zygomorphic*; perianth of 2 whorls of 3 segments, the outer whorl consisting of 3 often coloured, petal-like segments, the inner whorl consists of 2 petal-like segments similar to the outer 3, and a single, often much larger, lip-like segment or *labellum*, frequently with a hollow nectar-producing *spur* at its base which projects downwards or backwards. The labellum is often three-lobed, or enlarged and convex, or sometimes concave. The 1–2 stamens are fused to the stigma to form a *column*, bearing detachable pollen masses, or *pollinia*, and stigma with 2–3 stigmatic areas. Where only 2 stigmas occur, the third is generally reduced to a beak-like process or *rostellum*. Ovary inferior, usually one-celled, often twisted; fr. a capsule opening by 3 or 6 longitudinal slits; seeds minute, very numerous.

ORCHIDACEAE

1. Plants without green lvs.		Group A
1. Plants with green lvs.		
2. Fls. with a spur, sometimes very short and sac-like.		Group B
2. Fls. without a spur.		Group C

GROUP A. *Plants without green lvs.*

		Nos.
1. Fls. spurred, violet.	*Limodorum*	1921
1. Fls. not spurred, not violet.		
2. Lip with 2 lobes; fls. yellowish or brown.	*Neottia*	1924
2. Lip undivided or three-lobed; fls. greenish.	*Corallorhiza*	1926

GROUP B. *Fls. with a spur or sac*

		Nos.
1. Lip undivided.		
2. Spur usually longer than ovary; fls. whitish or greenish.	*Platanthera*	1914, 1915
2. Spur shorter than ovary; fls. rarely whitish or greenish.		
3. Lip directed upwards; fls. red or blackish-purple.	*Nigritella*	1901
3. Lip directed downwards; fls. usually various shades of pink to purple.	*Orchis*	1884–1896
1. Lip divided into 3 or 4 lobes.		
4. Lip more than twice as long as the perianth segments.	*Himantoglossum*	1906, 1907
4. Lip less than twice as long as perianth segments.		
5. Fls. greenish-purple or greenish-white; spur 2–3 mm.		
6. Lip with 2 spreading lateral lobes.	*Leucorchis*	1913
6. Lip with 2 parallel lateral lobes.	*Coeloglossum*	1911
5. Fls. white, yellow, pink, or purple, not greenish; spur usually more than 3 mm.		
7. Spur thread-like, *c.* 1 mm. broad.		
8. Fl. spike short, pyramidal; throat with 2 bosses.	*Anacamptis*	1908
8. Fl. spike long-cylindrical; throat without bosses.	*Gymnadenia*	1912
7. Spur cylindrical, broader than 1 mm.		
9. Perianth segments crowded together in a 'helmet' above the lip.	*Orchis*	1884–1896
9. Outer perianth at least spreading.		
10. Emerging fl. spike enclosed in thin spathe-like lvs.; bracts membraneous.	*Orchis*	1884–1896
10. Emerging fl. spike not enclosed in spathe-like lvs.; bracts leafy.	*Dactylorhiza*	1897–1900

GROUP C. *Fls. without a spur*

		Nos.
1. Fls. 1–2; lip deeply concave, slipper-like.	*Cypripedium*	1872
1. Fls. several or numerous; lip not slipper-like.		
2. Lip more or less convex, velvet-haired, insect-like.	*Ophrys*	1873–1883
2. Lip more or less flat, not insect-like.		
3. Fls. small, 1 cm. across, whitish, arranged in one or more spirals on the stem.		
4. Plant with creeping stolons; lvs. conspicuously net-veined.	*Goodyera*	1925
4. Plants without stolons; lvs. not conspicuously net-veined.	*Spiranthes*	1922

3. Fls. usually more than 1 cm., green, purple, or reddish, not in spirals.
5. Lip distinctly constricted into a basal part and a lip-like terminal part.
6. Fls. reddish-brown; lip hanging down, tongue-like, longer than broad, acute. *Serapias* 1902–1904
6. Fls. usually white or purplish; lip directed forward, about as long as broad, blunt.
7. Fls. erect, stalkless; ovary twisted. *Cephalanthera* 1919, 1920
7. Fls. horizontal or drooping, stalked; ovary straight. *Epipactis* 1916–1918
5. Lip not constricted, one- to four-lobed.
8. Lvs. narrow linear, as long as or longer than stem. *Chamorchis* 1910
8. Lvs. broader, shorter than stem.
9. Lvs. 2, sometimes with 1–2 small stem lvs.
10. Lvs. opposite on stem; lip with 2 terminal lobes. *Listera* 1923
10. Lvs. basal; lip three-lobed. *Herminium* 1909
9. Lvs. several. Fls. man-like with 'arms' and 'legs'. *Aceras* 1905

CYPRIPEDIUM | Lady's Slipper Orchid Fls. with perianth of apparently 4 lance-shaped, spreading segments and a large, concave, slipper-like lip; spur absent. Column projecting forward and partly closing the opening of the lip; fertile anthers 2, stigma 1. 3 sps.

1872. *C. calceolus* L. LADY'S SLIPPER ORCHID. Fls. very striking, large, usually solitary with reddish-maroon perianth segments contrasting with the large, pale yellow, slipper-shaped lip with darker veins and reddish spots within. Perianth segments 6–9 cm. lance-shaped, arranged in a cross, the 2 lower segments fused and situated below the shorter lip. Lvs. 3–5, oblong-oval acute, sheathing below, with several strong veins. Ovary finely hairy. An erect perenn. 15–50 cm., with a creeping rhizome. △ Woods and thickets in mountains. May–July. Much of Eur. (except P.IRL.B.NL.IS.DK.TR.BG.).

OPHRYS Fls. with 5 unequal, spreading perianth segments, the outer 3 petal-like and longer than the inner 2, and a much larger, swollen, convex lip which is hairy and variously and conspicuously marked and which mimics the body of sps. of bee; spur absent. Column often surmounted by a forward-projecting beak; pollinia 2, stalked. Hybridization is not infrequent, particularly in the Mediterranean region, often resulting in distinctive colouration and markings of the lip and making the identification of individual plants difficult. *c.* 25 sps.

Lip three-lobed
(a) Lateral lobes arising from near the tip of the lip
1873. *O. fusca* Link BROWN BEE ORCHID. Fls. with greenish-yellow outer perianth segments, and with a deep chocolate-brown, velvet-haired lip with bluish or slaty reflective patches near the base; lip longer than broad, and with the median lobe notched and longer than the lateral lobes. Outer perianth segments oval-blunt, often incurved, the inner 2 about three-quarters as long, hairless, strap-shaped, greenish or purplish. Inflorescence lax; fls. 2–6; bracts longer than the ovary. Lvs. 3–4; stems 10–30 cm. △ Grassy, rocky ground, bushy places, olive groves. Feb.–May. Med. Eur. P.R. Pl. 188.

1874. *O. lutea* Cav. YELLOW BEE ORCHID. Distinguished from 1873 by the broad three-lobed lip which has a conspicuous broad or narrow yellow margin, and a brown central raised boss with 2 bluish reflective patches towards the base of the lip. Lip often broader than long, the middle lobe widening towards the apex; perianth segments yellowish-green, the outer oval-blunt, the inner about half as long, strap-shaped. Fls. 1–6, varying in size of lip. Lvs. 3–4; stems 10–25 cm. △ Grassy places, open ground, olive groves, clearings. Mar.–May. Med. Eur. P. Pl. 188.

(b) Lateral lobes arising from the base, or below middle of the lip

1875. *O. speculum* Link MIRROR ORCHID, MIRROR-OF-VENUS. Readily distinguished by its very striking fls. which have a broad rounded lip with a brilliant metallic-blue central reflective patch circled with yellow, and with a dense fringe of black or brown hairs on the down-turned margin. Lip also possessing 2 broad or narrow lateral hairy lobes arising from near its base. Outer perianth segments oblong-oval, hairy, dark purple. Fls. 2–6. Lvs. 5–8; stems 10–30 cm. △ Grassy places. Mar.–May. Med. Eur. (except YU.AL.) P.

1876. *O. scolopax* Cav. WOODCOCK ORCHID. Fls. with a large cylindrical-ovoid to elliptic lip, strongly marked with a variable pattern of white and yellow circles and lines over a reddish-purple ground, and with a yellowish knob-like apex. Lip also possessing 2 conspicuous, forward-projecting, long-pointed horns or rounded hairy 'shoulders', arising laterally from its base. Outer perianth segments white or purplish, oblong-blunt, the inner segments much smaller, usually pink, but very variable. Fls. 3–7. Lvs. 2–4; stem to 40 cm. △ Grassy, stony places, olive groves. Apr.–May. Med. Eur. P.H.BG. Pl. 188.

1877. *O. insectifera* L. (*O. muscifera* Hudson) FLY ORCHID. Fls. rather small, with a slender purplish-brown lip with a broad, shining, reflective patch at base and with 2 spreading basal lateral lobes and a narrowly oblong, central, notched lobe. Outer perianth segments spreading, oblong, greenish-yellow, the inner 2 half as long, brownish-purple, thread-like, hairy. Fls. 4–20, in a lax cluster. Lvs. 4–5; stem slender 15–60 cm. △ Field verges, woods, thickets, grassy places. May–June. Much of Eur. (except P.IS.AL.TR.BG.). Pl. 188.

1878. *O. apifera* Hudson BEE ORCHID. Fls. with perianth segments usually pink and contrasting with the globular, velvety, red-brown lip, with a variable yellow pattern enclosing a red patch and with 2 yellow spots towards the apex. Lip also possessing a pointed apex tucked under the lobe and invisible from above, and hairy basal lateral lobes, which are swollen at their base, triangular-oval, and reflexed. Perianth segments oval-oblong, rarely white, greenish outside, the inner 2 very much smaller, green and velvety. Fls. 2–7. Lvs. 4–7; stem to 50 cm. △ Grassy banks, thickets, pastures. May–July. Much of Eur. (except North Eur. PL.BG.). Pl. 188.

1879. *O. bombyliflora* Link BUMBLE BEE ORCHID. Readily distinguished by its small, broad, rounded lip which is shorter than the greenish perianth segments. Lip velvety-brown with an ill-defined central slaty-blue patch, and with 2 short basal lateral lobes which project forwards into hairy humps and turn backwards to a point. Outer 3 perianth segments broadly oval, greenish-yellow, the inner 2 about one-third as long, velvety-green. Fls. 1–4; bracts oval, shorter than the ovary. Lvs. 4–8, the lower oval; stem 10–25 cm. △ Grassy places, olive groves, damp, sandy places by the sea. Mar.–Apr. Med. Eur. (except TR.) P. Pl. 188.

Lip entire, not three-lobed but sometimes with 2 basal swellings

1880. *O. sphegodes* Miller (*O. aranifera* Hudson) EARLY SPIDER ORCHID. Fls. with a velvety dark brown or blackish-purple ovoid lip with paler, hairless, bluish markings, either in 2 parallel lines or in an H-shaped or X-shaped pattern, and lip often with 2 rounded bosses or short arms arising near the base. Outer 3 perianth segments green, pink, or white, the inner 2 to about three-quarters as long and often with a waved margin, similar-coloured or differing in colour from the outer. Fls. 2–10, in a lax cluster. Lvs. 4–7; stem to 30 cm.

A very variable sp., particularly in South-Eastern Europe. △ Grassy, rocky places, sunny hills. Apr.–June. Much of Eur. (except North Eur. P.IRL.PL.).

1881. *O. bertolonii* Moretti Distinguished by the large, almost black, hairy lip with a pale, shining blue, indented patch towards the tip, and contrasting with the pink perianth segments. Lip large, ovoid to *c.* 2 cm., curved forward, angled, often with 2 swellings or arms at the base, and with a forward-projecting knob in the notched apex. Outer 3 perianth segments oblong, pink, the inner 2 half as long, linear, purplish, hairy. Fls. 1–6. Lvs. 6–10, the lower oval; stems 10–30 cm. △ Stony, grassy places. Apr.–June. Med. Eur. (except TR.). Pl. 188.

1882. *O. tenthredinifera* Willd. SAWFLY ORCHID. In general, distinguished by its brilliant pink perianth and large yellow and brown lip. Outer 3 perianth segments oval, concave, the inner 2 much shorter, triangular, hairy. Lip somewhat wedge-shaped, shaggy-haired and with yellowish margin and dark brown centre, and a large, reddish-brown, hairless basal patch surrounded by a blue line; apex of lip broad, deeply notched and with a yellowish hairy knob; lip often with 2 swellings at its base. Fls. 2–10. Lvs. 4–9; stems to 40 cm. △ Stony, grassy places. Mar.–May. Med. Eur. (except YU.AL.) P. Pl. 188.

1883. *O. fuciflora* (Crantz) Moench (*O. arachnites* (L.) Reichard) LATE SPIDER ORCHID. A very variable plant with a dark velvety-brown, rather quadrangular lip with a bold symmetrical pattern of yellowish-green lines, often enclosing a blue area; apex of lip with a terminal green, variously shaped, flat or incurved, projecting knob; lip with 2 basal swellings. Outer 3 perianth segments white to pink, green-veined, the inner 2 one-third to one-quarter as long, similar-coloured, hairy. Fls. 2–10. Lvs. 3–6; stem 10–35 cm. △ Stony, grassy places. May–June. Much of Eur. (except North Eur. IRL.NL.PL.BG.).

ORCHIS | Orchid Fls. with 5 equal perianth segments often coming together in a 'helmet' which stands over the column, and with a conspicuous lip which is usually either three- or four-lobed; spur present. Tubers 2, egg-shaped. *c.* 22 sps.

Perianth with all 5 segments coming together in a 'helmet'
(a) Lip undivided, but toothed
1884. *O. papilionacea* L. PINK BUTTERFLY ORCHID. A very beautiful orchid of the Mediterranean region, with strongly veined rosy-pink fls. Lip large fan-shaped, conspicuously toothed, usually pale pink but sometimes violet or white, veined with dark pink or crimson; spur narrowly conical, reflexed, shorter than the ovary. Perianth segments rosy-pink or purple, coming together above, the inner 2 a little narrower. Fls. 3–10, often large to 3 cm.; bracts conspicuous, flushed rosy-purple, longer than the ovary. Lvs. lance-shaped; stem 10–40 cm. △ Dry, grassy places, thickets, olive groves. Mar.–May. Med., South-East Eur. P.CH. Pl. 187.

(b) Lip three-lobed in general outline
1885. *O. morio* L. GREEN-WINGED ORCHID. Fls. reddish-purple or sometimes white or flesh-coloured, usually many in a short, dense, ovoid cluster and distinguished by the 2 broad, lateral, outer perianth segments which have conspicuous parallel greenish veins on a purplish ground. Lip reddish-purple and spotted with deeper reddish-purple, broader than long, the median lobe much smaller than the broad lateral lobes; spur straight, thick and blunt, horizontal or ascending, about as long as the ovary; bracts linear-acute, as long as the ovary. Lvs. broadly lance-shaped, unspotted; stem 10–40 cm. △ Grassy meadows. Apr.–June. Most of Eur. (except IS.SF.).

1886. *O. coriophora* L. BUG ORCHID. Distinguished by its dull reddish-green fls. and its pointed, green-veined 'helmet' which is longer than the three-lobed lip. Fls. smelling of bed-bugs, borne in a long spike-like cluster; lip with 3 more or less equal lobes, the middle pointed, the lateral lobes blunt and toothed; spur conical, half as long as the ovary, downward-pointing; bracts purplish, as long as or longer than the ovary. Var. *fragrans* (Pollich) Fiori has sweet vanilla scented fls. which are darker red, and whitish bracts. Lvs. lance-shaped to linear-lance-shaped; stem 20–40 cm. △ Damp meadows. Apr.–June. Much of Eur. (except North Eur. IRL.GB.NL.). Pl. 187.

1887. *O. globosa* L. (*Traunsteinera g.*) ROUND-HEADED ORCHID. A mountain orchid readily distinguished by the perianth segments which have elongated tips ending in an enlarged rounded apex. Fls. small, lilac to pale pink and purple-spotted, in a very dense globular-conical head, which later somewhat elongates. Perianth in a 'helmet'; lip about as long, with 3 more or less equal-length lobes, the middle lobe rectangular with a three-toothed apex, the lateral triangular, toothed; spur slender, cylindrical, half as long as the ovary, downward-pointing; bracts greenish, often purple-flushed, as long as or longer than the ovary. Lvs. oblong-lance-shaped; stem 25–50 cm. △ Pastures in the mountains, woods. June.–Aug. Cent. Eur. E.F.I.YU.AL.BG.R.SU.

(c) *Lip four-lobed, man-like in general outline*

1888. *O. ustulata* L. DARK-WINGED ORCHID, BURNT ORCHID. Fls. in a dense conical cluster with the uppermost buds dark brownish-purple and the fls. becoming paler and finally whitish when fully open below. 'Helmet' c. $\frac{1}{2}$ cm., at first dark but becoming whitish; lip white, and strongly dotted with reddish-purple spots, four-lobed, the basal lobes oblong and spreading, the apical lobes similar and often with a short tooth between; spur very short, conical, curved downward and forwards, about one quarter as long as the ovary. Lvs. oblong-acute; stem 8–20 cm. △ Dry meadows, grassy places in the mountains. Apr.–July. Much of Eur. (except P.IRL.NL.IS.N.SF.TR.). Pl. 187.

1889. *O. tridentata* Scop. TOOTHED ORCHID. Fls. fragrant, pink or white in a dense conical head, with a distinctive purple-spotted lip with 2 oblong lateral lobes and a larger, broad, approximately wedge-shaped median lobe which is divided into 2 broad, rounded, terminal lobes, often with a central tooth between. Perianth segments pink or whitish, usually long-pointed; spur more than half as long as the ovary, downward-pointing; bracts small, papery. A variable sp. Lvs. oblong-lance-shaped; stems 15–40 cm. △ Woods, thickets, grassy places. Mar.–May. Much of Eur. (except North Eur. IRL.GB.B.NL.). Pl. 187.

1890. *O. simia* Lam. MONKEY ORCHID. Fls. distinctly monkey-like with a slender lip of 2 long, basal, strap-like lobes, the 'arms', and 2 long, narrow, apical lobes, the 'legs', and a central narrow tooth, the 'tail', lobes all curved forward, pale pink-spotted towards the base and with darker rosy-purple tips. Fls. in a dense oval cluster; perianth segments forming a 'helmet', pale rose or violet, faintly darker veined; spur short and thick, half as long as the ovary; bracts minute, rounded. Lvs. oblong-oval, flat, shining; stem 20–50 cm. △ Grassy places, groves. Apr.–June. Med. Eur. GB.B.D.CH.BG.R.SU.

O. italica Poiret (*O. longicruris* Link) Like 1890, but fls. larger, up to 2 cm. long, with the 'helmet' strongly veined with pink or purple; lip with downward-curved lateral lobes with their tips scarcely darker and with a much longer, tail-like central tooth, tips of lobes little darker than centre of lip. Lvs. with an undulate margin. Mediterranean Europe and Portugal. Pl. 187.

1891. *O. militaris* L. SOLDIER or MILITARY ORCHID. Like 1890, but lobes of lip dissimilar, the basal arm-like lobes linear and much narrower and longer than the broad, oblong-oval, apical or leg-like lobes; lobes all pink or violet, with the centre of the lip whitish and dotted with tufts of violet swellings. Fls. in a dense oval cluster; perianth segments in a 'helmet', ash-grey flushed with rose or violet; spur cylindrical, half as long as the ovary; bracts minute, scale-like. Lvs. oblong-lance-shaped, shining; stem 20–45 cm. △ Grassy places, thicket verges. Apr.–June. Much of Eur. (except North Eur. P.IRL.NL.) S. Pl. 189.

1892. *O. purpurea* Hudson LADY ORCHID. A rather stout plant distinguished by its very dark reddish-purple 'helmet' which appears blackish in bud, and its whitish lip flushed with violet or pink, particularly on the margin, and dotted with raised purple swellings. Fls. in a dense cylindrical cluster; perianth segments long-pointed, the lip with linear basal lobes and 2 much broader, rhomboidal and finely toothed apical lobes, often with a tooth between; spur cylindrical, about half as long as the ovary; bracts minute. Lvs. oblong-oval, shining; stem 20–80 cm. △ Banks, copses. Apr.–June. Much of Eur. (except North Eur. P.IRL.) DK. Pl. 187.

Perianth with outer 2 segments spreading and inner 3 coming together in a 'helmet'
(a) Fls. rose or purple
1893. *O. laxiflora* Lam. JERSEY ORCHID. Fls. large, dark purple and unspotted, in a long lax cluster, with bracts and fl. stem often purple-flushed. Perianth segments oblong-blunt, the outer 2 reflexed, the inner 3 in a shorter 'helmet'; lip somewhat two-lobed, the lobes broad and turned down to lie against each other, the median lobe very short or absent; spur horizontal or upward-pointing, swollen and notched at the tip; bracts reddish-purple, nearly as long as the ovary. Lvs. lance-shaped; stem 30–50 cm. △ Marshy meadows. Mar.–June. Much of Eur. (except North Eur. IRL.GB.NL.) S. Pl. 187.

O. palustris Jacq. BOG ORCHID. Like 1893, but fls. paler and plant more slender; lip distinctly three-lobed with a broad, deeply notched median lobe, usually longer than the 2 lateral lobes which are at first spreading and only later turned down. Spur horizontal or downward-pointing, narrowed to the tip; lower bracts longer than the ovary. Central and South-Eastern Europe.

1894. *O. mascula* (L.) L. EARLY PURPLE ORCHID. Fls. rather numerous in an oval or cylindrical cluster, dark crimson-purple, with a strongly spotted three-lobed lip, which is usually paler towards the base, and with a stout horizontal or upward-pointing spur as long as or longer than the ovary. Perianth segments purplish-crimson, the 2 outer spreading or folded back, the inner 3 forming a 'helmet'; lip with broad, rounded and toothed lateral lobes and a median notched lobe; bracts purple. Fls. smelling of tom-cats. Lvs. oblong-lance-shaped, usually spotted; stem 15–60 cm. △ Woods, copses, and pastures. May–July. Most of Eur. (except IS.). *The tubers of this and other orchids produce a starch-like substance called bassorine which is said to contain more nutritive matter than any other plant product, an ounce per day is said to be sufficient to sustain a man. A drink called Salep is made from the macerated tubers; it is still drunk in South-Eastern Europe.*

1895. *O. quadripunctata* Cyr. FOUR-SPOTTED ORCHID. A slender orchid of the Mediterranean region, with a lax cylindrical cluster of small pale pink or violet fls. with whitish centres and with 2–4 conspicuous purple spots; spur very slender downward-pointing. Fls. 8–10 mm. across; perianth segments oval-blunt, the outer 3 spreading, the inner 2 converging; lip 4–5 mm. long, rather broader than long, three-lobed, the median lobe longest; bracts shorter than the ovary. Lvs. oblong-lance-shaped; stem 10–25 cm. △ Grassy hills, bushy places. Apr.–May. I.YU.AL.GR. Pl. 189.

(b) Fls. usually yellow
1896. *O. provincialis* Balbis PROVENCE ORCHID. Fls. pale or dark yellow, in a rather dense, few or many-flowered cluster, with a three-lobed lip, often darker orange in the centre and with brownish spots. Perianth segments blunt, the outer 2 spreading or reflexed, the inner 3 coming together; lip with rounded lateral lobes and with an often smaller notched

or bilobed median lobe; spur upward-pointing, as long as or longer than the ovary; bracts pale, equalling the ovary. Lvs. oblong-lance-shaped, brown-spotted or unspotted; stems 10–30 cm. △ Bushy and grassy places. Apr.–June. Med. Eur. (except TR.) P.CH.BG.

O. *pallens* L. PALE-FLOWERED ORCHID. Like 1896, but a mountain plant with pale yellow, elder-scented fls. in a dense spike, or fls. rarely purplish. Lip darker yellow, three-lobed, unspotted; spur a little shorter than the ovary. Lvs. broadly obovate, widening above the middle, unspotted. Central and Southern Europe.

DACTYLORHIZA Like *Orchis*, but tubers always palmately lobed or divided. Perianth segments erect or spreading, not coming together in a 'helmet'; bracts always greenish and leaf-like. Readily hybridizing; intermediates are common. The genus was earlier included in *Orchis* and until very recently in *Dactylorchis*. c. 12 sps.

Stems hollow
(a) Fls. purple
1897. *D. incarnata* (L.) Soó (*Orchis i.*; *O. strictifolia* Opiz; *O. latifolia* auct.; *Dactylorchis i.*) MARSH ORCHID. Fls. in a dense cylindrical cluster, variously coloured, reddish-purple, red or flesh-coloured, and with broad lower bracts more than 3 mm. wide, and much longer than the lower fls. Perianth with outer 2 diverging and inner 3 converging segments; lip about as long as broad, shallowly three-lobed, with the median lobe triangular pointed and slightly longer, and with strongly turned down lateral lobes so that the lip appears narrow, lip paler in the centre, with darker purple spots and lines; spur usually shorter than the ovary. A very variable sp. Lvs. linear-lance-shaped, pale greenish-yellow, unspotted; stem 15–80 cm. △ Wet meadows, marshes. May–June. Most of Eur. (except IS.TR.).

1898. *D. majalis* (Reichenb.) P. F. Hunt & Summerh. (*Orchis m.*; *Dactylorchis m.*) BROAD-LEAVED MARSH ORCHID. Like 1897, but distinguished by its broader, dark green and somewhat bluish, often spreading, spotted or unspotted lvs., its hollow slender-cavity or almost solid stem, and the lateral lobes of the lip which are spreading or slightly turned down, the lip thus appearing broader. Fls. lilac-purple, in a dense oval cluster; lip usually shallowly three-lobed, with undulate margin, strongly marked with a series of darker lines or dots; spur stout, downward-pointing, shorter than the ovary; lowest bracts 4–6 mm. wide, longer than the lower fls. Stem lvs. 3–6; stem 30–80 cm. △ Damp meadows, marshes. May–July. Probably most of Eur. (except IS.N.SF.). Pl. 187.

(b) Fls. yellow
1899. *D. sambucina* (L.) Soó ELDER-FLOWERED ORCHID. A mountain plant with yellow fls., or fls. less commonly purple with a yellowish patch at the base of the lip. Fls. rather large, in a dense ovoid cluster, without a distinctive smell; lip broader than long, toothed, very shallowly three-lobed, indistinctly spotted; spur very stout, conical and blunt, downward-pointing, as long as or longer than the ovary; bracts green or reddish, longer than the fls. Lvs. pale green, shining, unspotted; stem 10–30 cm. △ Mountain meadows. Apr.–July. P.E.F.D.CH.A.CS.I.YU.GR.BG.R. Pl. 189.

Stems solid
1900. *D. maculata* (L.) Soó (*Orchis m.*; *Dactylorchis m.*) HEATH SPOTTED ORCHID. Distinguished by its solid stem, usually spotted lvs., and pale lilac or often whitish fls., with a finely spotted lip which has broad, spreading lateral lobes and a much smaller and shorter median lobe. Fls. numerous, in a dense conical cluster which later becomes cylindrical; perianth with the outer 2 segments spreading and the inner 3 coming together; spur 5–8 mm., very slender and less than 1 mm. wide in the middle; bracts green,

shorter or longer than the ovary. A very variable sp. Lvs. 6–9, nearly always spotted, the lower lance-shaped, the uppermost much narrower; stem 15–50 cm. △ Meadows, heaths, bogs, moors. May–July. All Eur.

D. fuchsii (Druce) Soó (*Orchis f.*; *Dactylorchis f.*) SPOTTED ORCHID. Like 1900, but lip with 3 more or less similar triangular lobes, the median somewhat longer than the lateral lobes; lip strongly marked with more or less continuous curved reddish lines; spur thicker, *c.* 1½ mm. wide. Basal lvs. broader, elliptic to obovoid, blunt. Widespread in Europe. Pl. 189.

NIGRITELLA Like *Orchis*, but fls. vanilla-scented and lip placed above the remaining perianth segments, and not longer than them. Ovary not twisted. 2 sps.

1901. *N. nigra* (L.) Reichenb. BLACK VANILLA ORCHID. Fls. numerous, sweetly vanilla-scented, dark blackish-purple or rarely pink, in a very dense, at first conical and later globular cluster. Perianth segments lance-shaped, the inner segments half as broad as the outer, lip triangular entire, and directed upwards; spur blunt, sac-like, much shorter than the ovary; bracts lance-shaped, as long as the fls. Lvs. numerous, linear-acute; stem angled, 10–25 cm. △ Alpine meadows. June–Aug. E.F.N.S.D.CH.A.I.YU.GR.BG.R. Pl. 189.

N. miniata (Crantz) Janchen (*N. rubra* (Wettst.) Richter) ROSY VANILLA ORCHID. Like 1901, but fls. red, in an ovoid or cylindrical cluster. Perianth segments rather blunt, broader, oblong-oval, the inner segments as broad as the outer segments. Flowering time about 2 weeks earlier than 1901. Alps, Carpathians.

SERAPIAS Perianth forming a 'helmet' of 3 fused outer segments which conceal the 2 free inner segments, and with a long, triangular, tongue-like lip; spur absent. Lip with 1 or 2 basal swellings, and with 2 small, rounded basal lobes which curve upwards and are often partly concealed by the 'helmet'. Stamen 1; column with a long beak. *c.* 6 sps.

Lip with 1 basal swelling
1902. *S. lingua* L. TONGUE ORCHID. Fls. 2–5, in a lax cluster with conspicuous, rather rotund, pale reddish, and obscurely veined bracts, and similar-coloured fls. with an erect 'helmet' and a long-hanging, reddish, tongue-like lip twice as long. 'Helmet' *c.* 1½ cm. long, with free tips; lip with a dark purple, oblong basal swelling, and with 2 erect, rounded, blackish-purple lateral lobes half-hidden by the 'helmet', and an oval to lance-shaped acute tongue-like lobe 1⅔ cm. by 6–8 mm., which is finely hairy at the base. Lvs. lance-shaped; stems to 30 cm. △ Grassy places, olive groves. Apr.–July. Med. Eur. (except TR.) P.

Lip with 2 basal swellings
(*a*) *Lip with a broad, tongue-like median lobe 6–15 mm. wide*
1903. *S. cordigera* L. HEART-FLOWERED SERAPIAS. Distinguished by its large dark wine-purple fls. 2½–4 cm. long, with an erect, reddish-violet or wine-coloured 'helmet' which is paler outside, and a large black-purple, hairy, tongue-like lip. Fls. 3–8, in a dense cluster; lip densely hairy, with 2 blackish divergent basal swellings, rectangular lateral lobes, and a very large, hanging, oval or heart-shaped acute tongue-like lobe; bracts rotund lance-shaped, pale pinkish, little shorter than the fls. Lvs. linear-lance-shaped, sheaths and lower stem spotted; to 40 cm. △ Damp heaths and sandy places, woods. Mar.–May. Med. Eur. P.R.

S. neglecta De Not. Like 1903, but lip brick-red and yellowish towards the centre, oval-acute; basal swellings parallel; 'helmet' reddish-violet; bracts flushed violet. Lv. sheaths not spotted. France, Italy, Ionian Islands. Pl. 189.

1904. *S. vomeracea* (Burm.) Briq. (*S. pseudocordigera* Moric.; *S. longipetala* Pollich) LONG-LIPPED SERAPIAS. Like 1902, but fls. smaller, 2½–3 cm. long, and bracts longer, reddish-violet. 'Helmet' violet with deeper veins; lip brick-red to reddish-brown, hairy, oblong-lance-shaped, scarcely narrowed above the lateral lobes, which are hidden by the 'helmet'; basal swellings 2, nearly parallel. Lvs. lance-shaped; stem 10–40 cm. △ Damp, sandy places, grassy places, woods. Apr.–June. Med. Eur. P.CH.BG. Pl. 189.

(*b*) *Lip with a narrow, tongue-like median lobe 2–5 mm. wide*
S. *parviflora* Parl. (*S. occultata* Gay; *S. laxiflora* Chaub.) SMALL-FLOWERED SERAPIAS. Distinguished by its small fls. 1½–2 cm. long, with the lip scarcely longer than the 'helmet' and reflexed against the ovary. 'Helmet' reddish-violet, lip rusty-red; bracts pale red and obscurely veined. Mediterranean region.

ACERAS Perianth segments forming a 'helmet' and lip 'man-like' with slender lateral lobes, the 'arms', and a median lobe divided into 2 narrow apical lobes, the 'legs'. Distinguished from certain *Orchis* sps. by the absence of a spur. 1 sp.

1905. *A. anthropophorum* (L.) Aiton fil. MAN ORCHID. Fls. greenish-yellow, often marked or edged with reddish-brown, numerous in a dense, slender, cylindrical, spike-like cluster. 'Helmet' conspicuous, greenish-yellow, strongly bent forward over the column; lip slender *c.* 12 mm. long, hanging vertically below the 'helmet', greenish-yellow and often flushed with reddish-brown, the lobes linear, and the terminal lobes often with a tooth between; bracts lance-shaped. Lvs. several, oblong-lance-shaped, grading into the bracts above, keeled, glossy on both sides; stem 20–40 cm. △ Grassy, bushy places. Apr.–June. West Eur. (except IRL.NL.) D.CH.I.YU.GR. Pl. 189.

HIMANTOGLOSSUM Perianth segments forming a 'helmet', the lip three-lobed, the median lobe deeply or shallowly two-lobed and often much elongated; spur short. 2 sps.

1906. *H. longibracteatum* (Biv.) Schlecht. GIANT ORCHID. A very robust orchid with a very dense spike of dull greenish-purple fls., often with varying amounts of pink or purple on the same inflorescence, and with similar-coloured bracts longer than the fls. Lip large, dull purple, with wavy-margined lobes, lateral lobes of lip sickle-shaped, median lobe broader and longer, deeply two-lobed at the apex and with a deeply toothed or fringed margin. Outer perianth segments spreading, the inner incurved; spur short, stout, and conical, downward-pointing. Lvs. broadly elliptic blunt, unspotted, glossy; stem stout, 20–50 cm. △ Grassy places, thickets. Feb.–Apr. Med. Eur. (except AL.TR.) Pl. 190.

1907. *H. hircinum* (L.) Sprengel LIZARD ORCHID. A very striking, robust orchid smelling strongly of billy-goats, with untidy-looking, pale greenish-purple fls. with extremely long lobes to the lip which are curled up like a watch-spring in bud and irregularly twisted in fl. Fls. numerous, in a cylindrical, spike-like cluster 10–50 cm. long; perianth forming a greenish 'helmet', lip very long and slender, 3–12 cm., whitish with purple spots and greenish towards the apex. Lateral lobes of lip narrow, ½–2 cm. long, curly, the median lobe very much longer, undulate, coiled; spur short and conical, downward-pointing. Lvs. oblong-lance-shaped; stems 20–80 cm. △ Grassy, rocky places, dry banks, wood verges. May–July. Much of Eur. (except North Eur. P.IRL.NL.PL.R.) Pl. 190.

ANACAMPTIS Like *Orchis*, but lip deeply three-lobed and bearing 2 obliquely erect swellings or 'guide-plates' at the base; spur long, slender. 1 sp.

1908. *A. pyramidalis* (L.) L. C. M. Richard PYRAMIDAL ORCHID. Fls. rosy-purple, at first in a dense conical cluster which later becomes oval, and with a foxy smell. Outer perianth segments spreading, the inner coming together; lip with 3 more or less equal spreading lobes and with raised swellings at the base; spur *c.* 12 mm. long, thread-like, downward-pointing, as long as or longer than the ovary; bracts lance-shaped. Lvs. unspotted, linear-lance-shaped acute, the upper long-pointed; stem 10–60 cm. △ Grassy places, meadows. May–July. Most of Eur. (except IS.N.SF.PL.). Pl. 190.

HERMINIUM Fls. greenish-yellow, spurless. Perianth more or less bell-shaped with nearly equal, oval-lance-shaped segments and a three-lobed lip, about as long as the other segments. 1 sp.

1909. *H. monorchis* (L.) R.Br. MUSK ORCHID. A small plant 7–20 cm. with a dense, often rather one-sided, narrow cylindrical spike of very sweet-scented, tiny greenish-yellow, somewhat bell-shaped fls. Lip inconspicuous, as long as the remaining perianth segments, with 3 triangular acute lobes, the median lobe about twice as long as the lateral lobes; bracts lance-shaped, shorter than the ovary. Lvs. oval, basal, usually 2 present at flowering; tuber 1 at flowering. △ Dry and damp meadows, by springs. May–July. Much of Eur. (except P.E.IRL.NL.IS.DK.AL.GR.TR.). Pl. 190.

CHAMORCHIS Like *Herminium*, but lvs. linear, grooved, and grass-like; tubers 2 at flowering. 1 sp.

1910. *C. alpina* (L.) L. C. M. Richard FALSE MUSK ORCHID. Distinguished from 1909 by the numerous linear basal lvs. as long as or longer than the fl. spike. Fls. rather few in a short lax spike, greenish-yellow and brownish-violet outside. Perianth segments in a 'helmet'; lip scarcely lobed or with 2 short, tooth-like lobes near the base, reflexed and a little longer than the 'helmet'; bracts longer than the fls. Stem leafless, 6–12 cm. △ Mountain pastures. July–Aug. Cent., North Eur. (except IS.) F.I.R.SU.

COELOGLOSSUM Fls. greenish-purple; perianth segments forming a 'helmet'; lip narrow oblong; spur very short. *c.* 1 sp.

1911. *C. viride* (L.) Hartman FROG ORCHID. A small plant 6–35 cm., with small greenish fls. with varying amounts of brownish-purple, in a rather lax spike, with bracts about as long as the fls. Outer perianth segments oval, forming a 'helmet' and hiding the inner, greenish or greenish-purple segments; lip greenish to brownish, hanging down, narrowly oblong and three-lobed at the apex, with the lateral lobes parallel, not spreading, and longer than the median lobe; spur very short, sac-like. Lvs. usually 5, oval to oblong, unspotted, the lower blunt. △ Meadows, grassy places, in mountains. May–June. Most of Eur. (except P.NL.TR.).

GYMNADENIA Like *Orchis*, but differing in minute characters of the column and the pollinia. Fls. fragrant; the outer perianth segments spreading, the inner forming a 'helmet'; lip entire or three-lobed; spur usually long, often slender. *c.* 5 sps.

1912. *G. conopsea* (L.) R.Br. FRAGRANT ORCHID. Fls. rose-violet, very fragrant, in a slender cylindrical spike up to 15 cm. long, with bracts about as long as the ovary. Fls. small *c.* 1 cm. across, the lateral perianth segments rather downward-curved, the upper in a 'helmet'; the lip wedge-shaped, broader than long, and with 3 blunt, nearly equal lobes; spur very slender, nearly twice as long as the ovary. Lvs. 6–9, unspotted, the lower oblong-lance-shaped, keeled and folded, the upper narrower; stem 15–60 cm. △ Grassy places, marshes, woods. May–July. All Eur. (except IS.TR.). Pl. 190.

G. odoratissima (L.) L. C. M. Richard SHORT-SPURRED FRAGRANT ORCHID. Like 1912, but spur thicker and shorter than the ovary; lateral perianth segments spreading horizontally, and lip somewhat longer than the 'helmet'. Fls. rose-violet or yellowish-white, with a strong vanilla scent, very small, 5–7 mm. across, in a spike rarely more than 6 cm. long. Lvs. 4–5. Central Europe.

LEUCORCHIS   Like *Gymnadenia* and often included in it, but outer perianth coming together with inner to form a 'helmet' and the lip somewhat pressed to the 'helmet' so that the fl. appears more or less bell-shaped; spur short. 1 sp.

1913. *L. albida* (L.) E. H. F. Meyer (*Gymnadenia a.*) SMALL WHITE ORCHID. Fls. numerous, greenish-white, very small, half-drooping, in a dense, narrow cylindrical, one-sided spike 3–6 cm. with bracts as long as the ovary. Fls. 2–2½ mm. across, faintly scented; perianth segments all coming together, the lip scarcely longer, three-lobed with a downward-curved central lobe longer than the lateral lobes which are closely pressed to the perianth; spur short thick, downward-curved, less than half as long as the ovary. Lvs. 4–6 oblong, keeled, glossy and unspotted; stem 10–35 cm. △ Pastures in mountains. June–Aug. Much of Eur. (except P.B.NL.IS.SF.PL.TR.). Pl. 192.

PLATANTHERA | **Butterfly Orchid**   Lip undivided, elongated strap-shaped; spur usually long and slender; outer 2 perianth segments spreading, the inner 3 forming a 'helmet'. *c.* 4 sps.

1914. *P. bifolia* (L.) L. C. M. Richard LESSER BUTTERFLY ORCHID. Fls. whitish, with a slender strap-shaped lip and a long curved spur placed almost horizontally, numerous, sweet-scented, in a lax, cylindrical spike. Fls. 11–18 mm. across; lip ½–1 cm. long; spur thread-like, greenish, 1½–2 cm. long, from half to twice as long as the ovary; yellowish anthers lying parallel to each other. Lower lvs. usually 2, large 3–9 cm., oval to elliptic, the upper lvs. much smaller and bract-like, all unspotted; stem 15–45 cm. △ Woods, grassy places, marshes. May–July. All Eur. (except IS.). Pl. 191.

1915. *P. chlorantha* (Custer) Reichenb. GREATER BUTTERFLY ORCHID. Like 1914, but fls. greenish-white, larger, 18–23 mm. across. Lip 1–1½ cm. long; spur 2–3 cm. long, curved downwards and forwards and thickened towards the apex. Inflorescence more or less pyramidal, lax; yellow anthers widely diverging downwards. Lower lvs. usually 2, large, 5–15 cm., the upper smaller and bract-like; stem 30–60 cm. △ Woods, grassy places, heaths. May–July. Almost throughout Eur. Pl. 191.

EPIPACTIS | **Helleborine**   Fls. distinctly stalked and held horizontally or drooping in a lax spike; perianth segments reddish or greenish, spreading or coming together in a 'helmet'; spur absent. Lip composed of 2 parts, the basal part forming a cup, the apical part enlarging into a triangular or heart-shaped, downward-pointing lobe. Ovary not spirally twisted. *c.* 4 sps.

Plants of damp places

1916. *E. palustris* (L.) Crantz MARSH HELLEBORINE. Fls. brownish or greenish-purple, 7–14 in a more or less one-sided, lax, hairy-stemmed spike with the lower bracts leafy and about equalling the fls. Perianth segments hairy outside, the outer 3 greenish or brownish, the inner 2 whitish with purple veins; lip white, veined with red, about twice as long as the perianth, the terminal lobe broadly oval with waved and frilled margin and narrowly hinged to the basal part. Lower lvs. elliptic, the upper lance-shaped; stem 20–50 cm.

Mature ovary hairy. △ Marshes, boggy fields, fens. June–Aug. Most of Eur. (except IS.TR.). Pl. 191.

Plants of woodlands or dry places

1917. *E. helleborine* (L.) Crantz (*E. latifolia* (L.) All.) BROAD HELLEBORINE. Fls. greenish or dull purple, numerous 15–50, drooping, scentless, in a long one-sided spike to 30 cm., with the lowest bracts leafy and equalling the fls. Fls. *c.* 1 cm. across; perianth segments *c.* 1 cm. long, green to dull purple; lip shorter than the perianth and with the apical part broader than long, purplish, rose, or greenish-white and with 2 more or less smooth bosses. Lvs. numerous, spirally arranged, dull green and with prominent veins, the lower broadly oval to almost rounded, the upper lance-shaped and grading into the linear-acute bracts; stem often violet-flushed below, 30–100 cm. △ Woods, rocky and grassy places. July–Sept. All Eur. (except IS.).

1918. *E. atrorubens* (Hoffm.) Schultes (*E. atropurpurea* Rafin.) DARK RED HELLEBORINE. Like 1917, but fls. entirely dull reddish-purple, fewer 8–18, and vanilla-scented. Perianth segments 5–8 mm. long, dark reddish-purple, finely hairy outside; lip shorter than the perianth, apical part deep reddish-violet with a small acute tip and with 2 rough basal bosses. Lvs. elliptic to broadly lance-shaped, in 2 opposite ranks; stems with crisp hairs above, violet below, 20–50 cm. △ Woods, grassy places, dunes. June–Aug. All Eur. (except NL.IS.).

CEPHALANTHERA Like *Epipactis*, but fls. all stalkless, erect; ovary stalk-like, spirally twisted. Fls. white or pink, not opening widely. 4 sps.

Fls. white; ovary hairless

1919. *C. damasonium* (Miller) Druce (*C. latifolia* Janchen; *C. pallens* L. C. M. Richard) WHITE HELLEBORINE. Fls. tubular, erect, creamy-white, 3–12 in a lax, leafy inflorescence, the bracts longer than the ovary. Fls. *c.* 2 cm. long, the outer perianth segments oblong-blunt, the inner shorter, all closely pressed together and not opening wide; lip shorter, with an orange blotch on the basal lobe and orange ridges on heart-shaped terminal part. Lvs. 5–10 cm., oval-lance-shaped, gradually narrowing above to the bracts; stem 15–60 cm. △ Woods, shady, bushy places. Apr.–June. Most of Eur. (except P.IRL.IS.N.SF.PL.).

C. longifolia (L.) Fritsch (*C. ensifolia* (Schmidt) L. C. M. Richard) LONG-LEAVED HELLE-BORINE. Like 1919, but lvs. mostly lance-shaped, the uppermost linear. Outer perianth segments acute; bracts shorter than the ovary. Fls. pure white, smaller, *c.* 1½ cm. long. Almost throughout Europe. Pl. 191.

Fls. pink; ovary hairy

1920. *C. rubra* (L.) L. C. M. Richard RED HELLEBORINE. Fls. bright pink, 3–15 in a lax cluster, bracts mostly shorter than the fls. but longer than the ovary. Perianth segments lance-shaped acute, bright pink, coming together, the lip shorter erect, with the terminal part with a red-violet margin and tip, and with narrow, crested, yellowish ridges. Lower lvs. oblong, the upper lance-shaped, all very acute; stem glandular-hairy above, 20–60 cm. Fr. hairy. △ Woods, bushy places. May–July. Most of Eur. (except P.IRL.IS.). Pl. 191.

LIMODORUM Saprophytic plants with stout, cane-like stems and scale-like lvs.; fls. violet, spur present. Placed in the past in the genera *Epipactis*, *Ionorchis*, *Centrosis*, and *Orchis*. 1 sp.

1921. *L. abortivum* (L.) Swartz LIMODORE. Fls. large to *c.* 4 cm. across, violet, several, rather widely spaced in a long slender spike borne on a stiff violet stem with scattered, violet-flushed, scale-like lvs. Perianth segments coming together in a 'helmet', violet and veined with darker purple; lip a little shorter, violet with yellowish shading, variously shaped but usually oval and with a toothed margin; spur downward-pointing, as long as the ovary. Lvs. reduced to sheathing scales; stem 30–80 cm.; roots tuberous. △ Woods, bushy places in the hills. May–July. Much of Eur. (except North Eur. IRL.GB.B.NL.PL.TR.). Pl. 191.

SPIRANTHES | Lady's Tresses Fls. tiny, white or pink, arranged in spirally twisted rows up the stem; lip embracing column at its base, little longer than the perianth segments, lip-margin frilled; spur absent. 3 sps.

1922. *S. spiralis* (L.) Chevall. (*S. autumnalis* L. C. M. Richard) AUTUMN LADY'S TRESSES. Fls. tiny, white, stalkless, arranged in a slender, spirally twisted, elongated spike and appearing in the autumn. Fls. 4–5 mm. across, sweet-scented, somewhat tubular; lip green with a broad, white, frilled margin, little longer than the white perianth segments; bracts sheathing, little shorter than the fls. Fl. stem with pale green overlapping scales only, and with the basal rosette of lvs. placed at the side of the stem; stem 10–30 cm. Tubers usually 2. △ Dry grassy places, meadows, heaths. Aug.–Oct. Much of Eur. (except North Eur.) DK. Pl. 192.

S. aestivalis (Poiret) L. C. M. Richard SUMMER LADY'S TRESSES. Like 1922, but summer-flowering and fl. stem with glossy linear-lance-shaped lvs. 5–12 cm.; tubers 3–6, long cylindrical. Fls. somewhat larger, pure white, slightly scented. Central and Southern Europe.

LISTERA | Twayblade Lvs. usually 2, opposite and placed towards the base of the stem. Perianth segments coming together; lip linear and two-lobed at its apex, often with central tooth; spur absent. 2 sps.

1923. *L. ovata* (L.) R.Br. TWAYBLADE. Fls. greenish-yellow, numerous, and borne in a long slender spike on a long hairy stem, with 2 opposite rounded lvs. towards the base. Fls. *c.* 5 mm. across, short-stalked; perianth segments green, coming together; lip green, linear, 1–1½ cm., hanging downwards and divided into 2 deep lobes for nearly half its length; bracts minute, shorter than the fl. stalks. Lvs. 5–20 cm., broadly oval-elliptic, stalkless, with 3–5 prominent veins, and 1–2 tiny bract-like lvs. on the stem above. Rhizome creeping; fl. stem 20–60 cm. △ Woods, thickets, pastures. May–July. All Eur. (except P.).

L. cordata (L.) R.Br. LESSER TWAYBLADE. A very slender plant 6–20 cm. recalling 1923, with a short spike of few tiny reddish-green fls. and a pair of small, opposite, rounded-heart-shaped lvs. 1–2½ cm. long, near the base of the stem. Fls. 3–4 mm. across; lip 3½–4 mm. long with 2 tiny narrow lobes at its base and a central lobe, forked to about halfway. Widespread in Europe. Pl. 192.

NEOTTIA | Bird's-Nest Orchid Saprophytic plants without green lvs. and with stems covered with brown scales. Perianth segments coming together; lip two-lobed; spur absent. 1 sp.

1924. *N. nidus-avis* (L.) L. C. M. Richard BIRD'S-NEST ORCHID. Readily distinguished by the dull yellowish-brown colour of the whole plant which is without green lvs., and its rather numerous brownish fls. in a dense oval or cylindrical spike. Perianth segments oval-oblong blunt, coming together; lip *c.* 12 mm., about twice as long, divided to nearly half its length into 2 diverging blunt lobes; bracts lance-shaped, papery, shorter than the ovary. Scales on stem brownish, lance-shaped, sheathing; stem brownish, 20–50 cm. Roots tuberous and densely massed into the semblance of a 'bird's nest'. Recalling a Broomrape, but distinguished from it by the fertile parts and the presence of a lip. △ Shady woods, particularly in beech woods. May–July. All Eur. (except IS.). Pl. 192.

GOODYERA   Fls. tiny, whitish, in a twisted, one-sided spike; perianth segments unequal, in a 'helmet'; lip shorter, with a basal, pocket-like part and an apical, spout-like part. Lvs. green; stolons creeping. 1 sp.

1925. *G. repens* (L.) R.Br. CREEPING LADY'S TRESSES. A delicate plant with creeping stolons, rosettes of lvs., and creamy-white, fragrant fls. in a slender cylindrical cluster. Fls. *c.* 4 mm. long; perianth segments blunt, glandular-hairy; lip shorter, the apical part of lip narrow, furrowed, and recurved; bracts longer than the ovary. Lvs. all basal, 1½–2½ cm., in a lax rosette, oval, narrowed to a stalk, dark green and often mottled with lighter green, and with a conspicuous network of veins. Fl. stems erect, glandular-hairy, 10–25 cm., with whitish, sheathing scale lvs. △ Mossy woods, particularly in mountains. July–Aug. Much of Eur. (except P.IRL.B.IS.AL.TR.). Pl. 192.

CORALLORHIZA | **Coral-Root**   Brown saprophytic plants without green lvs. and with reddish, much-branched, and coral-like rhizomes. Perianth in a 'helmet'; lip shorter, entire or three-lobed; spur short and fused to the ovary, or absent. 1 sp.

1926. *C. trifida* Chatel. CORAL-ROOT ORCHID. Fls. rather few, short-stalked, spreading, yellowish-green, often with reddish spots or a reddish margin, forming a lax spike. Fls. *c.* 4 mm. long; the outer perianth segments curved downwards close to the lip, the inner coming together above in a 'helmet'; lip whitish with crimson lines or spots, usually three-lobed with smaller lateral lobes. Stem pale green, 10–30 cm. with a few sheathing, scale-like bracts. △ Woods in the mountains. June–Aug. Much of Eur. (except P.E.IRL. B.NL.TR.). Pl. 192.

BIBLIOGRAPHY

FLORAS AND POPULAR GUIDES TO THE SEED PLANTS OF EUROPE

R   Larger reference floras of regions
F   Flora of country or region, with descriptions of plants
E   Excursion flora, portable and for field identification
P   Popular selection of plants of a country or region
C   Lists of plants without descriptions, often with distributions and notes
A   Atlas, showing distribution of plants
I   Well illustrated by line drawings or colour photographs
i   With many line drawings of diagnostic features or a limited selection of species drawn

Regional Floras (reference)

COSTE, H., *Flore descriptive et illustrée de la France, de la Corse et des Contrées limitrophes*, Paris, 1900–1906.   F I R

HAYEK, A. VON, *Prodromus Florae Peninsulae Balcanicae* (In *Feddes Repert* (*Beih.*), 30), Berlin-Dahlem, 1924–1933.   F R

HEGI, G., *Illustrietre Flora von Mittel-Europa*, München, 1906–1931, 2nd ed., München, 1936–   F I R

HYLANDER, N., *Forteckning over Nordens växter*, 1 Kärlväxter Lund, 1955. Tilläggoch Rättelser (in *Bot. Nat.* 112) Lund, 1959.   C

HYLANDER, N., *Nordisk Kärlväxtflora*, Stockholm, 1953–   F

KOMAROV, V. L., *et al.*, eds., *Flora URSS*, Leningrad and Moskva, 1934–1964.   F I R

TUTIN, T. G., *et al.*, eds., *Flora Europaea*, Cambridge, 1964, 1968.   F

Useful, more or less portable (rucksack or pocket) floras, with descriptions, covering most of Europe except Spain, Albania, Greece, Turkey, Romania

AUSTRIA (A) Fritsch, K., *Exkursionsflora für Österreich und die ehemals österreichischen Nachbargebiete*, 3rd ed., 1922.

BELGIUM (B) Mullenders, E., ed., *Flore de la Belgique, du Nord de la France et des Régions Voisines*, 1967.

BULGARIA (BG) Stojanov, N., & Stefanov, B., *Flora na Bălgarija*, 3rd ed., 1948.

CZECHOSLOVAKIA (CS) Dostál, J., *Květena ČSR*, 1950.

DENMARK (DK) Rostrup, F. G. E., *Den danske Flora*, 19th ed., 1961.

FINLAND (SF) Alcenius, O., *Finlands Kärlväxter*, 3rd ed., 1958.

FRANCE (F) Fournier, P., *Les quatre Flores de la France, Corse comprise*, 1961.

GERMANY (D) Rothmaler, W., *Exkursionsflora von Deutschland*, 1963.

GREAT BRITAIN (GB) Clapham, A. R., *et al.*, *Flora of the British Isles*, 2nd ed., 1962.

HOLLAND (NL) Heukels, H., *Flora van Nederland*, 15th ed., 1962.

HUNGARY (H) Soó, R. de, & Jávorka, S., *A Magyar Növényvilág Kézikönyve*, 1951.

ICELAND (IS) Ostenfeld, C. E. H., & Gröntved, J., *The Flora of Iceland and the Faeroes*, 1934.

IRELAND (IRL) Webb, D. A., *An Irish Flora*, 5th ed., 1967.

ITALY (I) Arcangeli, G., *Compendio della Flora italiana*, 2nd ed., 1894.

—— Baroni, E., *Guida Botanica d'Italia*, 1955.

NORWAY (N) Lid, J., *Norsk og svensk Flora*, 1963.

POLAND (PL) Szafer, W., *et al.*, *Rósliny Polskie*, 1953.

PORTUGAL (P) Coutinho, A. X. Pereira, *Flora de Portugal*, 2nd ed., 1939.

SOVIET UNION (SU) Stankov, S. S., & Taliev, V. I., *Opredelitel vysših Rastenij Evropejskoj Casti SSSR*, 2nd ed., 1957.

SWEDEN (S) Lid, J., *Norsk og svensk Flora*, 1963.
SWITZERLAND (CH) Binz, A., & Thommen, E., *Flore de la Suisse*, 2nd ed., 1953.
—— Binz, A., *Schul- und Exkursionsflora für die Schweiz*, 11th ed., 1964.
YUGOSLAVIA (YU) Domac, R., *Flora za odredivanje i upoznavanje Bilja*, 1950.

Regional guides, popular accounts, atlases, groups of plants

BARNEBY, T., *European Alpine Flowers in Colour*, London, 1967.   P I
BONNIER, G., *Flore complète illustrée en couleurs de France, Suisse et Belgique*,
   Neuchâtel, Paris and Bruxelles, 1912–1934.   F I
DUPERREX, A., *Orchids of Europe*, London, 1961.   P I
GAMS, H., *The Natural History of Europe*, rev. and edit. by A. Melderis and Joyce Pope,
   London, 1967.   P I
GJAERVOLL, O., & JÖRGENSEN, R., *Mountain Flowers of Scandinavia*, Trondheim, 1963. P I
HEGI, G., *Alpenflora*, 18th ed. by H. Merxmüller, München, 1963.   P I
HUBBARD, C. E., *Grasses*, London, 1954.   E I
HULTÉN, E., *Atlas of the Distribution of Vascular Plants in N.W. Europe*, Stockholm,
   1950.   A
HUXLEY, A., *Mountain Flowers*, London, 1967.   P I
KOHLHAUPT, PAULA, *Alpenblumen, Farbige Wunder*, Stuttgart, 1963–1964.   P I
LANDOLT, E., *Unsere Alpenflora*, Zürich, 1960.   P I
LID, J., *Norsk og Svensk Flora*, Oslo, 1963.   F I
MEUSEL, H., JÄGER, E., & WEINERT, E., *Vergleichende Chorológie der Zentraleuropäischen*
   Flora, Jena, 1965.   A
PHITOS, D., *Wild Flowers of Greece*, Athens, 1965.   P I
POLUNIN, O., & HUXLEY, A., *Flowers of the Mediterranean*, London, 1965.   P I
TURRILL, W. B., *The Plant-Life of the Balkan Peninsula*, Oxford, 1929.   C

Floras, guides, etc. to the countries of Europe

AUSTRIA (A)
FRITSCH, K., *Exkursionsflora für Österreich und die ehemals österreichischen Nach-*
   bargebiete, 3rd ed., Wien and Leipzig, 1922.   E
JANCHEN, E., *Catalogus Florae Austriae*, Wien, 1956–1960. Supplements, Wien, 1963–
   1966.   C
RECHINGER, K. H., & GÖTTING, F., *Pflanzenbilder aus den Ostalpen*, Wien, 1947.   P I

BELGIUM (B)
CRÉPIN, Fr., *Manuel de la Flore de Belgique*, 5th ed., Bruxelles, 1884.   E
GOFFART, J., *Nouveau Manuel de la Flore de Belgique*, 3rd ed., Liège, 1945.   E i
MULLENDERS, E., *Flore de la Belgique, du Nord de la France et des Régions Voisines*,
   Liège, 1967.   F i
ROBYNS, W., *Flore générale de Belgique*, Bruxelles, 1952.   F

BULGARIA (BG)
JORDANOV, D., *Flora na Narodna Republika Bălgarija*, Sofija, 1963–   F I
STOJANOV, N., & STEFANOV, B., *Flora na Bălgarija*, 3rd ed., Sofija, 1948.   F I

CZECHOSLOVAKIA (CS)
DOSTÁL, J., *Květena ČSR*, Praha, 1948–1950.   F i
DOSTÁL, J., *Klíč k úplné květeně ČSR*, Praha, 1958.   E i
PILÁT, A., *Atlas Roslin*, Praha, 1953.   P I

DENMARK (DK) and the FAEROES

HAGERUP, O., & PETERSON, V., *Botanisk Atlas*, København, 1956–1960. I

RASMUSSEN, R., *Föroya Flora*, 2nd ed., Tórshavn, 1952. F

RAUNKIAER, C., *Dansk Ekskursions—Flora*, 7th ed. by K. Wiinstedt, København, 1950. E

ROSTRUP, F. G. E., *Den Danske Flora*, 19th ed. by C. A. Jørgensen, København, 1961. E I

WARMING, E., *et al., Botany of the Faeroes*. London, 1901–1908. F

FINLAND (SF)

ALCENIUS, O., *Finlands Kärlväxter*, 3rd ed., Helsinki, 1958. E

HIITONEN, H. I. A., *Suomen Kasvio*, Helsinki, 1933. E i

FRANCE (F)

BRIQUET, J., *Prodrome de la Flore corse*, Genève, Bâle, Lyon, and Paris, 1910–1955. C

COSTE, H., *Flore descriptive et illustrée de la France, de la Corse et des Contrées limitrophes*, Paris, 1900–1906. F I R

FOURNIER, P., *Les Quatre Flores de la France, Corse comprise*, Poinson-les-Grancey, 1934–1940. Reprint with additions and corrections, Paris, 1961. E i

ROUY, G. C. C., *et al., Flore de la France*, Asnières, Paris and Rochefort, 1893–1913. F R

ROUY, G. C. C., *Conspectus de la Flore de France*, Paris, 1927. F

GERMANY (D)

GARCKE, A., *Illustrierte Flora von Deutschland*, Berlin, 1898. E I

POTONIE, H., *Taschenatlas zur Flora von Nord- und Mitteldeutschland*, 7th ed., Jena, 1923. I

SCHMEIL, O., & FITSCHEN, J., *Flora von Deutschland*, 78th ed. by Dr. Werner Rauh, Heidelberg, 1965. E i

ROTHMALER, W., *Exkursionsflora von Deutschland*, 2: *Gefässpflanzen*, Berlin, 1962. 4: *Kritischer Ergänzungsband Gefässpflanzen*, Berlin, 1963. E

GREAT BRITAIN (GB)

BUTCHER, R. W., *A New Illustrated British Flora*, London, 1961. F I

CLAPHAM, A. R., *et al., Flora of the British Isles*, 2nd ed., Cambridge, 1962. F

CLAPHAM, A. R., *et al., The Excursion Flora of the British Isles*, Cambridge, 1959. E

CLAPHAM, A. R., *et al., Illustrations to the Flora of the British Isles*, Cambridge, 1957–1965. I

DANDY, J. E., *List of British Vascular Plants*, London, 1958. C

KEBLE MARTIN, W., *The Concise British Flora in Colour*, London, 1965. P I

LEWIS, PATRICIA, *British Wild Flowers*, London, 1958. P I

MCCLINTOCK, D., & FITTER, R. S. R., *The Pocket Guide to Wild Flowers*, London, 1961. P I

MCCLINTOCK, D., *Supplement to the Pocket Guide to Wild Flowers*, Platt, 1957. P

MEIKLE, R. D., *British Trees and Shrubs*, London, 1958. P I

MELDERIS, A., & BANGERTER, E. B., *British Flowering Plants*, London and Melbourne, 1955. P I

NICHOLSON, B. E., *et al., The Oxford Book of Wild Flowers*, London, 1960. P I

PERRING, F. H., & WALTERS, S. M., *Atlas of the British Flora*, London, 1962. A

ROSS-CRAIG, STELLA, *Drawings of British Plants*, London, 1950– I

GREECE (GR)

ATCHLEY, S. C., *Wild Flowers of Attica*, Oxford, 1938. P I

DIAPOULIS, K. A., *Ellenike Khloris*, Athenai, 1939–1949. F

HALACSY, E. VON, *Conspectus Florae Graecae*, Leipzig, 1900–1904. Supplements: Leipzig, 1908; Budapest, 1912. F R

RECHINGER, K. H., *Flora Aegaea*, Wien, 1943. Supplement in Phyton (Austria), 1949. C

RECHINGER, K. H., *Neue Beiträge zur Flora von Kreta*, (*Denkschr. Akad. Wissensch.* Wien, 105, 2, 2), Wien, 1943. F

HOLLAND (NL)

HEIMANS, E., *et al.*, *Geillustreerde Flora van Nederland*, Amsterdam, 1948.   E I
HEUKELS—VAN OOSTSTROOM, S. J., *Flora van Nederland*, 15th ed. by S. J. van Oost-
stroom, Groningen, 1962.   F i
WEEVERS, T., *et al.*, *Flora Neerlandica*, Amsterdam, 1948–   F

HUNGARY (H)

DEGEN, A. VON, *Flora vilebitica*, Budapest, 1936–1940.   F
SANDOR, VERA, *Erdo-Mezo Viragai*, Budapest, 1965.   P I
SOÓ, R. de, & JÁVORKA, S. A., *Magyar Növényvilág Kézikönyve*, Budapest, 1951.   E
SOÓ, R. de, *A magyr flóra és vegetació rendrzertaninövényföldrajzi kézikönyve*,
Budapest, 1966.   C

ICELAND (IS)

LÖVE, Á., *Íslenzkar Jurtir*, Kφbenhavn, 1945.   F I
GRÖNTVED, J., *The Pteridophyta and Spermatophyta of Iceland* (*Botany of Iceland*, ed.
by Gröntved, J., Paulsen, O., & Sφrensen, T.), Kφbenhavn and London, 1942.   F
OSTENFELD, C. E. H., & GRÖNTVED, J., *The Flora of Iceland and the Faeroes*, Kφbenhavn
and London, 1934.   F
STEFÁNSSON, S., *Flóra Islands*, 3rd ed. by S. Steindórsson, Akureyri, 1948.   E

IRELAND (IRL)

WEBB, D. A., *An Irish Flora*, 5th ed., Dundalk, 1967.   E

ITALY (I)

ARCANGELI, G., *Compendio della Flora italiana*, 2nd ed., Torino, 1894.   F
BARONI, E., *Guida Botanica d'Italia*, Rome, 1955.   E i
DALLA FIOR, G., *La Nostra Flora*, 2nd ed., Trento, 1963.   F i
FENAROLI, L., *Flore della Alpi*, Milano, 1955.   F I
FIORI, A., *Nuova Flora Analitica d'Italia*, Firenze, 1923–1929.   F R
FIORI, A., & PAOLETTI, G., *Iconographia Florae Italicae*, 3rd ed., San Casciano Val di
Pesa, 1933.   I
GIACOMINI, V., & FENAROLI, L., *Conosci l'Italia: II La Flora*, Milano Touring Club
Italiano, 1958.   P I
MEZZENA, R., *Flora del Carso*, Trieste, 1965.   P I
PITSCHMANN, REISIGL, & SCHIECHTL, *Flora de Südalpen von Gardasee zum Comersee*,
Stuttgart, 1965.   E I
ZENARI, S., *Flora Escursionistica*, Padova, 1958.   E

MALTA

BORG, J., *Descriptive Flora of the Maltese Islands*, Malta, 1927.   F
LANFRANCO, GUIDO G., *Guide to the Flora of Malta*, Malta, 1955.   E i

NORWAY (N) and SPITSBERGEN

FAEGRI, K., *A Botanical Atlas. The Distribution of Coast Plants*, Oslo, 1960.   A
LID, J., *Norsk Flora*, 2nd ed., Oslo, 1952.   F I
NORDHAGEN, R., *Norsk Flora*, Oslo, 1940.   I
RESVOLL-HOLMSEN, H., *Svalbards Flora*, Oslo, 1927.   I

POLAND (PL)

KULCZYNSKI, st., *Florae Polonicae Iconographia*, Krakow, 1930–1936.   I
MADALSKI, J., *Florae Polonicae Terrarumque Adiacentium Iconographia*, Warczawa
and Wroclaw, 1954–   I
PAWLOWSKI, B., *Flora Tatr*, Varsoviae, 1956–   F

BIBLIOGRAPHY

RACIBORSKI, M., SZAFER, W., & PAWLOWSKI, B., *Flora Polska*, Kraków and Warszawa,
1919– .   F R
SZAFER, W., KULCZYŃSKI, S., & PAWLOWSKI, B., *Rósliny Polskie*, Warszawa, 1953.   E i

PORTUGAL (P)
COUTINHO, A. X. P., *Flora de Portugal*, 2nd ed. by R. T. Palhinha, Lisboa, 1939.   F
SAMPAIO, G. A. de SILVA FERREIRA, *Flora portuguesa*, 2nd ed. by A. Pires de Lima,
Porto, 1947.   F

ROMANIA (R)
BICHICEANU, R. R. & M., *Flowers of Rumania*, Bucureşti, 1964.   P I
BORZA, A., *Conspectus Florae Romaniae*, Cluj, 1947–1949.   C
SĂVULESCU, T., *Flora Republicii Populare Române*, Bucureşti, 1952– .   F I

SOVIET UNION (SU)
FOMIN, A. V., *et al.*, *Flora RSS Ucr.*, Kijiv, 1936–1965. 2nd ed., Kijiv, 1938.   F i R
GALENIEKS, P., *Latvijas, PSR flora*, Riga, 1953– .   F i
GEIDEMAN, T. S., *Opredelitel Rastenij Moldavskoj SSR*, Moskva and Leningrad, 1954.   E
GORODKOV, B. N., & POJARKOVA, A. I., *Flora Murmanskoj Oblasti*, Moskva and Lenin-
grad, 1953– .   F I
KASK, M., & VAGA, A., *Eesti Taimede Määraja*, Talinn, 1966.   E i
KOLAKOVSKII, A. A., *Rastitel' hyi mir Kolkhidy*, Moskva, 1961.   F I
MAEVSKIJ, P. F., *Flora srednej Polosy evropejskoj Časti SSR*, 9th ed. by B. K. Schischkin,
Leningrad, 1964.   F i
NATKEVIAITE-IVANAUSKIENE, M., *Lietuvos T.S.R. flora*, Vilnius, 1959– .   F i
NEISHTADT, M. I., *Opredelitel' rastenii srednei polosy Evropejskoj Časti SSSR*, Moskva,
1963.   E i
PETERSONE, A., & BIRKMANE, K., *Latvijas P.S.R. augu noteicejs*, Riga, 1958.   E i
RAMENSKAYA, M. L., *Opredelitel' vysshikh rastenii Karelii*, Sortavala, 1960.
SHISHKIN, B. K., *Botanicheskii atlas*, Moskva and Leningrad, 1963.   P I
SHISHKIN, B. K., *Flora Leningradskoi Oblasti*, Leningrad, 1955– .   F
SHISHKIN, B. K., *Flora B.S.S.R.*, Moskva, 1949– .   F I
STANKOV, S. S., & TALIEV, V. I., *Opredelitel vysših Rastenij Evropejskoj Časti SSSR*, 2nd
ed., Moskva, 1957.   F i
WULF, E. V., *Flora Kryma*, Yalta, Leningrad, and Moskva, 1927.   F
ZEROV, D. K., *et al.*, *Viznacnik Roslin Ukrajini*, 2nd ed., Kijiv, 1965.   E i

SPAIN (E) and GIBRALTAR
BARCELÓ Y COMBIS, F., *Flora de las Islas Baleares*, Palma de Mallorca, 1879–1881.   F
CABALLERO, A., *Flora Analitica de Espana*, Madrid, 1940.   E i
CADEVALL I. DIARS, J., *Flora de Catalunya*, Barcelona, 1913–1937.   F i R
DE BOLÓS Ÿ VAYREDA, A., *Vegetación de las Comarcas Barcelonesas*, Barcelona, 1950.   F I
KNOCHE, H., *Flora Balearica*, Montpellier, 1921–1923.   C
LÓPEZ, E. G., *Vizcaya y su paisaje Vegetal*, Bilbao, 1949.   F I
MERINO Y ROMÁN, B., *Flora descriptiva é illustrada de Galicia*, Santiago de Com-
postela, 1905–1909.   F i
WILLKOMM, H. M., & LANGE, J., *Prodromus Florae Hispanicae*, Stuttgart, 1861–1880.   F R
WILLKOMM, H. M., *Supplementum Prodromi Florae Hispanicae*, Stuttgart, 1893.   F R
WOLLEY-DOD, A. H., *A Flora of Gibraltar and the Neighbourhood* (Suppl. to *Journal of
Botany*, 1914), London, 1914.   F

SWEDEN (S)
BOLIN, L., & POST, VON L. O. A., *Floran i färg*, Stockholm, 1957.   P I
LAGERBERG, A. V., *Svenska Fjällblommor*, Stockholm, 1952.   P I

BIBLIOGRAPHY

LINDMAN, C. A. M., *Svensk Fanerogamflora*, 2nd ed., Stockholm, 1926.   E i
SÖDERBERG, E., *Våra Vilda Växter*, Stockholm, 1943.   P I
URSING, B., *Svenska Växter*, Stockholm, 1945.   E I
WEIMARCK, H., *Skånes Flora*, Lund, 1963.   F

SWITZERLAND (CH)

BINZ, A., & THOMMEN, E., *Flore de la Suisse*, 2nd ed., Lausanne, 1953.   E i
BINZ, A., *Schul- und Exkursionsflora für die Schweiz*, 11th ed. by A. Becherer, Bâle, 1964.   E i
LANDOLT, E., *Unsere Alpenflora*, Zürich, 1960.   P I
SCHINZ, H., & KELLER, R., *Flora der Schweiz*, 4th ed., Zürich, 1923.   F
THOMMEN, E., *Taschenatlas der Schweizer Flora*, 2nd ed., Bâle, 1951.   I

TURKEY (TR)

DAVIS, P. H., *Flora of Turkey*, Edinburgh, 1965– .   F i
WEBB, D. A., *Flora of European Turkey* (*Proceedings of the Royal Irish Academy*, vol. 65, Sec. B, 1), Dublin, 1966.   C

YUGOSLAVIA (YU)

DOMAC, R., *Flora za odredivanje i upoznavanje Bilja*, Zagreb, 1950.   E
HORVATIC, S., *Illustrirani Bilinar*, Zagreb, 1954.   E i
MAYER, E., *Seznam praprotnic in Cvetnic Slovenskega Ozemlja*, Ljubljana, 1952   C
PISKERNIK, ANGELA, *Kljuc za dolocanje cvetnic in praprotnic*, Ljubljana, 1951.   E i
ROHLENA, J., *Conspectus Florae Montenegrinae*, Praha, 1942.   C

INDEX OF POPULAR NAMES

	English	Français	Deutsch	Italiano
Aira	Hair-grass	Aira	Nelkenhafer	Nebbia
Ajuga	Bugle	Bugle	Günsel	Camepizio
chamaepitys	Ground-Pine	B.-petit pin	Gelber G.	
Alcea	Hollyhock	Rose Trémière	Stockrose	Malvarosa, Malvoni
Alchemilla				
alpina	Alpine Lady's-mantle	Alchémille argentée	Silbermantel	
vulgaris	Lady's-mantle	A. vulgaire	Frauenmantel	Erba ventaglina
Alisma	Water-plantain	Plantain-d'eau	Froschlöffel	Mestola, Alismante
Alkanna	Alkanna, Dyer's A.	Orcanette	Alkanna	Alcanna
Alliaria	Garlic Mustard	Alliaire, Herbe-à-l'ail	Knoblauchsrauke	Erba alliaria
Allium		Ail	Lauch	Aglio
ascalonicum	Shallot	Echalote	Schalotte	Scalogno
cepa	Onion	Oignon	Zwiebel	Cipolla
fistulosum	Welsh Onion	Ciboule	Winter-Z.	C. d'inverno
porrum	Leek	Poireau	Lauch	Porro
sativum	Garlic	Ail	Knoblauch	Aglio
schoenoprasum	Chives	Ciboulette	Schnittlauch	A. di serpe
ursinum	Ramsons	Ail-des-ours	Bären-Lauch	A. orsino
Alnus	Alder	Verne, Aune	Erle	Ontano
Alopecurus	Fox-tail	Vulpin	Fuchsschwanzgras	Borsette
Althaea	Marsh Mallow	Guimauve	Eibisch	Altea
Alyssum		Alysson	Steinkraut	
saxatile	Golden Alyssum	Corbeille-d'or		Alisso dorato
Amaranthus	Amaranth	Amarante	Amarant	Amaranto
Ambrosia	American Wormwood	Ambroisie	Ambrosie	Ambrosie
Amelanchier	Snowy Mespilus	Amélanchier	Felsenmispel	Pero corvino
Ammi	False Bishop's Weed	Ammi	Knorpelmöhre	Ammi
Ammophila	Marram Grass	Roseau-des-sables	Strandhafer	Sparto pungente, Ammofila
Amorpha	False Indigo	Faux indigo	Falscher Indigo	Indaco bastardo
Ampelodesma				Saracchio
Anacamptis	Pyramidal Orchid	Orchis pyramidal	Spitzorchis	Giglione

GENUS	ENGLISH	FRENCH	GERMAN	ITALIAN
Anacyclus		Pyrèthre	Bertram	Piretro
Anagallis				
arvensis	Scarlet Pimpernel	Mouron-des-champs	Acker Gauchheil	Anagallide
tenella	Bog Pimpernel	Mouron délicat	Zarter G.	
Anagyris	Bean Trefoil	Bois puant		Anagiride
Anarrhinum		Anarrhinum	Lochschlund	Muffolino frastagliato
Anchusa	Alkanet	Buglosse	Ochsenzunge	Buglossa, Ancusa
arvensis	Bugloss		Wolfsauge	Lingua di bove
Andromeda	Marsh Andromeda	Andromède	Rosmarinheide	
Andropogon		Barbon	Bartgras	Sanguinella
Androsace		Androsace	Mannsschild	
Anemone	Anemone	Anémone	Windröschen	Anemone
Anethum	Dill, False Fennel	Aneth	Dill	Aneto
Angelica	Angelica	Angélique	Engelwurz	Angelica
Antennaria	Cat's-foot	Pied-de-chat	Katzenpfötchen	Bambagia selvatica
Anthemis	Chamomile	Camomille	Hundskamille	Camomilla
Anthericum	St. Bernard's Lily	Petit lis	Graslilie	Giglio, Anterico
Anthoxanthum	Vernal-grass	Flouve odorante	Ruchgras	Paleina odorosa
Anthriscus	Cow Parsley	Persil sauvage	Kerbel	Lappola
cerefolium	Chervil	Cerfeuil	Garten-K.	
Anthyllis		Anthyllide		
vulneraria	Kidney-vetch	Vulnéraire	Wundklee	Vulneraria, Antillide
Antirrhinum	Snapdragon	Muflier	Löwenmaul	Bocca di leone
Aphanes	Parsley Piert	Alchémille	Frauenmantel	Erba Ventaglina
Aphyllanthes		Aphyllanthe	Blausternbinse	Giunchetta turchina di Monpellieri
Apium				
graveolens	Wild Celery	Céleri	Sellerie	Sedano
nodiflorum	Fool's Watercress	Ache nodiflore	Knotenblutiger Sellerie	Sedanina d'acqua
Aquilegia	Columbine	Ancolie	Akelei	Amor nascosto

Arabidopsis	Thale Cress	Fausse arabette	Schotenkresse	Pelosella
Arabis	Rock-cress	Arabette	Gänsekresse	
glabra	Tower Mustard	Tourette	Turmkraut	
Arachis	Ground-nut, Pea-nut	Arachide	Erdnuss	Arachide
Arbutus	Arbutus, Strawberry Tree	Arbousier	Sandbeerbaum	Albatro, Corbezzolo
Arctium	Burdock	Bardane	Klette	Bardana, Lappola
Arctostaphylos	Bearberry	Raisin-d'ours	Bärentraube	Uva orsina
Arenaria	Sandwort	Sabline	Sandkraut	Erba pondina
Argyrolobium		Argyrolobe	Silberginster	
Arisarum	Friar's Cowl			Arisaro
Aristolochia	Birthwort	Aristoloche	Osterluzei	Aristolochia
Armeria	Thrift	Arméria	Grasnelke	Armeria
arenaria	Jersey Thrift	A.-faux-plantain	Widerstoss	
maritima	Sea Pink	Gazon d'Espagne		
Armoracia	Horse-radish	Raifort	Meerrettich	Barba forte, Cren
Arnica	Arnica	Arnica	Arnika	Arnica
Arrhenatherum	Oat-grass	Fromental	Glatthafer	Perlaria
Artemisia				
abrotanum	Southernwood	Aurone	Eberreis	Abrotano
absinthium	Wormwood	Absinthe	Wermut	Assenzio
campestris	Field Southernwood	Armoise-des-champs	Feld-Beifuss	Abrotano dei campi
dracunculus	Tarragon	Estragon	Estragon	Dragoncello
vulgaris	Mugwort	Armoise	Beifuss	Amarella
Arum	Lords-and-Ladies, Cuckoo-pint	Gouet	Aronstab	Gigaro, Aro
Aruncus	Goat's-beard	Barbe-de-bouc	Geissbart	Barba di capra
Arundo	Reed	Grand roseau	Flöte, Pfahlrohr, Riesenschilf	Canna
Asarum	Asarabacca	Asaret	Haselwurz	Asaro
Asclepias	Silk Weed	Herbe-à-la-ouate	Seidenpflanze	Albero della seta
Asparagus	Asparagus	Asperge	Spargel	Asparago

GENUS	ENGLISH	FRENCH	GERMAN	ITALIAN
Asperugo	Madwort	Râpette	Scharfkraut	Lingua di bue
Asperula	Squinancy Wort	Aspérule	Waldmeister	Stellina
Asphodelus	Asphodel	Asphodèle	Affodill	Asfodelo
Aster				
alpinus	Alpine Aster	Aster des Alpes	Alpen-Aster	Amello, Astro
amellus			Berg-Aster	
linosyris	Goldilocks	Chevelure-dorée	Gold-Aster	
tripolium	Sea Aster		Salz-Aster	
Astragalus	Milk-vetch	Astragale	Tragant	Dragante
Astrantia	Masterwort	Astrance	Sterndolde	Astranzia
Atractylis	Distaff Thistle			Masticogna
Atriplex	Orache	Arroche	Melde	Atriplice
Atropa	Deadly Nightshade	Belladone	Tollkirsche	Belladonna
Avena	Oat	Avoine	Hafer	Avena
Ballota	Black Horehound	Marrube noir	Schwarznessel	Ballota
Barbarea	Winter Cress	Barbarée	Barbenkraut	Barbarea
Bartsia	Alpine Bartsia	Bartsie	Alpenhelm	Sporchia del grano
Bellidastrum	False Daisy	Fausse pâquerette	Sternlieb	Margarita d'alpe
Bellis	Daisy	Pâquerette	Massliebchen, Gänseblümchen	Margheritina, Pratolina
Berberis	Barberry	Epine-vinette	Sauerdorn	Crespino
Berteroa	Hoary Alyssum		Graukresse	
Beta	Beet	Bette	Runkelrübe	
vulgaris		Betterave	Mangold	Barbabietola
Betonica	Betony	Bétoine	Ziest	Betonica
Betula	Birch	Bouleau	Birke	Betulla
Bidens	Bur-Marigold	Bident	Zweizahn	
Bilderdykia	Black Bindweed	Petite vrillée		Erba leprina
Biscutella	Buckler Mustard	Lunetière	Brillenschötchen	Occhi di S. Lucia
Blackstonia	Yellow-wort	Centaurée jaune	Bitterling	Centaurea gialla
Borago	Borage	Bourrache	Boretsch	Borrana, Borragine

	English	French	German	Italian
Brachypodium	False-brome	Brachypode	Zwenke	Paleo
Brassica		Chou	Kohl	
napus	Rape, Cole, Swede	Colza	Raps	Cavolo rapone, Navone
nigra	Black Mustard	Moutarde noire	Schwarzer Senf	Senapa nera
oleracea	Wild Cabbage	Chou, Choufleur	Gemüsekohl	Cavolo
rapa	Turnip	Rave, Navet	Rübenkohl	Rapa
Briza	Quaking Grass	Amourette	Zittergras	Tamburini
Bromus	Brome	Brome	Trespe	Forasacco peloso
Bryonia	Bryony	Bryone	Zaunrübe	Brionia, Barbone
Buddleja	Buddleja	Buddléa	Buddleja	
Bulbocodium		Bulbocode	Lichtblume	Colchico di spagna
Bunias	Bunias	Bunias	Zackenschötchen	Bunio
Buphthalmum	Yellow Ox-eye		Rindsauge	Astri gialli
Bupleurum	Hare's-Ear	Buplèvre	Hasenohr	Perfoliata
Butomus	Flowering Rush	Jonc fleuri	Schwanenblume	Giunco fiorito, Biodo
Buxus	Box	Buis	Buchsbaum	Bosso, Bossolo
Cakile	Sea Rocket	Roquette-de-mer	Meersenf	Radicetta marina
Calamagrostis	Bushgrass, Smallreed	Calamagrostide	Reitgras	Cannuccia, Cannella
Calamintha	Calamint	Sarriette	Kalaminthe	Nepitella
Calendula	Pot-marigold	Souci	Ringelblume	Cappuccina dei campi
Calepina	White Ball Mustard	Calépine	Wendich	
Calicotome	Spiny Broom	Cytise épineux		Ginestra spinosa
Calla		Calla-des-marais	Calla	Dragontea acquatica
Callitriche	Starwort	Étoile-d'eau	Wasserstern	Erba gamberaia
Calluna	Ling, Heather	Bruyère	Besenheide	Grecchia, Bréntoli
Caltha	Kingcup, Marsh Marigold	Populage	Dotterblume	Calta palustre
Calystegia	Large Bindweed	Liseron	Winde	Campanelle
soldanella	Sea Bindweed	Chou-marin	Strand-Winde	Soldanella, Cavolo di mare

GENUS	ENGLISH	FRENCH	GERMAN	ITALIAN
Camelina	Gold of Pleasure	Caméline	Leindotter	Camarina, Miagro
Campanula	Bellflower, Campanula	Campanule	Glockenblume	
rapunculus	Rampion	C. gantelée	Rapunzel-G.	Raperonzolo
trachelium	Bats-in-the-Belfry	Gant de Notre-Dame		Imbutini
Cannabis	Hemp	Chanvre	Hanf	Canapa
Capparis	Caper	Câprier	Kapernstrauch	Cappero
Capsella	Shepherd's Purse	Bourse-à-pasteur	Hirtentäschelkraut	Borsa pastore
Capsicum	Chili, Peppers	Piment	Spanischer Pfeffer	Peperone, Pepe rosso
Cardamine	Cuckoo Flower, Bittercress	Cardamine	Schaumkraut	Billeri
bulbifera	Coral-wort	Dentaire	Zahnwurz	Dentaria
Cardaria	Hoary Cress, Hoary Pepperwort	Passerage Drave	Pfeil-Kresse	Cocola, Lattona
Carduus	Thistle	Chardon	Distel	Cardo
Carex	Sedge	Laîche	Segge	Carice
Carlina	Carline Thistle	Carline	Eberwurz	Carlina bianca
acaulis		Chardon argenté	Silberdistel	
Carpinus	Hornbeam	Charme, Charmille	Hainbuche	Carpino
Carpobrotus	Hottentot Fig	Mésembrianthème		Fico degli Ottentotti
Carthamus				
lanatus		Chardon béni des Parisiens		
tinctorius	Safflower	Faux-safran	Saflor	Cartamo
Carum	Caraway	Carvi, Cumin-des-prés	Kümmel	Cumino Tedesco
Castanea	Sweet Chestnut, Spanish Chestnut	Châtaignier	Edelkastanie	Castagno
Catabrosa	Whorl-grass	Catabrosa	Quellgras	
Catalpa	Catalpa, Indian Bean	Catalpa	Catalpa	Catalpa

Catananche	Cupidone	Cupidone	Rasselblume	Madre d'amore, Turchina
Catapodium	Hard Poa	Scléropoa	Hartgras, Steifgras	
Caucalis	Bur-parsley	Caucalis	Haftdolde	Lappola
Celtis	Nettle Tree	Micocoulier	Zürgelbaum	Bagolaro, Spaccasassi, Arcidiavolo
Centaurea	Knapweed	Centaurée	Flockenblume	Centaurea
calcitrapa	Star Thistle	Chausse-trape	Fussangel-F.	Calcatreppola
cyanus	Cornflower	Bluet	Kornblume	Fiordaliso, Ciano, Ciano maggiore
montana		Grand bluet		
solstitialis	St. Barnaby's Thistle			Spino giallo
Centaurium	Centaury	Petite centaurée	Tausendguldenkraut	Biondella
Centranthus	Red Valerian	Centranthe	Spornblume	Valeriana rossa
Cephalanthera	Helleborine	Céphalanthère	Waldvöglein	Elleborina
Cephalaria		Verge-à-pasteur	Kopfblume	Vedovina
Cerastium	Chickweed	Céraiste	Hornkraut	Orecchio di topo
Ceratonia	Carob, Locust Tree	Caroubier	Johannsbröt	Carrubo
Ceratophyllum	Horn-wort	Hydre	Hornblatt	Coda di volpe
Cercis	Judas Tree	Arbre de Judée	Judasbaum	Albero di guida, Siliquastro
Cerinthe	Honeywort	Mélinet	Wachsblume	Cerinta, Scarlattina
Chaerophyllum	Rough Chervil	Cerfeuil sauvage	Kälberkropf	Anacio selvatico
Chamaemelum	Chamomile	Camomille romaine	Römische Kamille	Camomilla bastarda
Chamaerops	Fan Palm			Palma nana
Chamorchis	False Orchid	Chamorchis	Zwergorchis	
Cheiranthus	Wallflower	Giroflée, Violier	Goldlack	Violaciocca gialla
Chelidonium	Greater Celandine	Chélidoine	Schöllkraut	Celidonia
Chenopodium	Goosefoot	Chénopode	Gänsefuss	
bonus-henricus	Good King Henry	Epinard sauvage	Guter Heinrich	Bono-enrico
Chondrilla			Knorpellatich	Condrilla
Chrozophora		Tournesol		Girasole piccolo

GENUS	ENGLISH	FRENCH	GERMAN	ITALIAN
Chrysanthemum				
coronarium	Crown Daisy	Chrysanthème	Wucherblume	Bambagella
leucanthemum	Marguerite	Grande-marguerite		Margherita
parthenium	Feverfew	Grande camomille	Mutterkraut	Matricale
segetum	Corn Marigold	Chrysanthème-des-blés	Saat-Wucherblume	Ingrassabue
vulgare	Tansy	Tanaisie	Rainfarn	Tanaceto
Chrysosplenium	Golden Saxifrage	Dorine	Milzkraut	
Cicer	Chick-pea	Pois-chiche	Zichererbse	Cece
Cicerbita	Sow-thistle	Laitue	Lattich	
Cichorium	Chicory	Chicorée	Wegwarte	Cicoria, Radicchio
Cicuta	Cowbane	Ciguë aquatique	Wasserschierling	Cicuta
Circaea	Enchanter's Nightshade	Herbe-aux-sorciers	Hexenkraut	Circea
Cirsium	Thistle	Cirse, Chardon	Kratzdistel	Cardo, Cirsio
arvense	Creeping Thistle			Scardaccione
eriophorum	Woolly Thistle	Chardon-des-ânes		Cardo scardaccio
Cistus	Cistus	Ciste	Zistrose, Sandröschen, Sonnenröschen	Cisto
Citrullus				
colocynthis	Bitter Apple	Coloquinte	Koloquinte	Coloquintide
lanatus	Water Melon	Pastèque	Wassermelone	Cocomero, Anguria
Citrus				
aurantium	Seville Orange	Bigaradier	Pomeranzenbaum	Arancia amara
deliciosa	Tangerine	Mandarine	Mandarine	Mandarino
limon	Lemon	Citronnier	Zitronenbaum	Limone
medica	Citron	Cédratier	Limone	Cédro
sinensis	Orange	Oranger doux	Apfelsinenbaum	Arancia
Cladium	Sedge	Marisque	Schneide	Scarzone
Clematis	Clematis	Clématite	Waldrebe	Clematide
vitalba	Travellers' Joy, Old Man's Beard	Vigne blanche		Vitalba
Clinopodium	Wild Basil	Rouette		Menta dei greppi
Clypeola	Disk Cress	Clypéole	Schildkraut	Filigrano

Cnicus	Blessed Thistle	Benediktenkraut	Chardon béni	Cardo santo
Cochlearia	Scurvy-grass	Löffelkraut	Cranson	Coclearia
Coeloglossum	Frog Orchid	Hohlzunge	Orchis	Testicolo di volpe
Coix	Job's tears	Hiobsträne	Larme de Job	Lacrime di Giobbe
Colchicum autumnale	Meadow Saffron, Autumn Crocus	Herbst-Zeitlose	Colchique	Zafferano bastardo
Colutea	Bladder Senna	Blasenstrauch	Baguenaudier	Vescicaria
Conium	Hemlock	Schierling	Grande ciguë	Cicuta maggiore
Conopodium	Pignut, Earthnut		Janotte, Génotte	
Conringia	Hare's-ear Cabbage	Ackerkohl	Roquette d'Orient	Perfoliata orientali
Consolida	Larkspur		Delphinette	Consolida regale
Convallaria	Lily-of-the-Valley	Maiblume	Muguet	Mughetto
Convolvulus	Bindweed, Cornbine	Winde	Liseron	Vilucchio, Erba leporina
Corallorhiza	Coral-root	Korallenwurz	Racine-de-corail	
Coriandrum	Coriander	Koriander	Coriandre	Coriandolo
Coriaria			Corroyère	Coriaria
Cornus	Dogwood	Roter Hartriegel	Sanguine	Sanguinella
mas	Cornelian Cherry	Kornelkirsche	Cormier	Corniolo
Coronilla	Crown Vetch	Cronwicke	Coronille	Vecciarini
Coronopus	Swine Cress, Wart Cress	Krahenfuss	Coronope	Lappolino
Cortaderia	Pampas Grass		Herbe des Pampas	
Cortusa	Alpine Bells	Heilglöckchen	Cortusa	
Corydalis	Corydalis	Lerchensporn	Corydale	Fumaria bulbosa
Corylus	Hazel, Cob-nut	Hasel	Noisetier	Nicciolo, Avellano
Cotinus	Wig Tree, Smoke-tree	Perückenbaum	Fustet, Arbre-à-perruque	Cotino
Cotoneaster		Zwergmispel	Cotonéaster	Cotognastro
Crambe	Seakale	Meerkohl	Chou-marin	Cavolo di mare
Crataegus	Hawthorn	Weissdorn	Aubépine, Epine blanche	Biancospino
azarolus	Mediterranean Medlar		Azerolier	Azarolo
Crepis	Hawk's-beard	Pippau	Crépide	Crepide
Crithmum	Rock Samphire		Criste-marine	Finocchio marino
Crocus	Crocus	Safran	Crocus	Zafferano, Croco

GENUS	ENGLISH	FRENCH	GERMAN	ITALIAN
Crucianella		Crucianelle	Kreuzkraut	Spigenio
Cucubalus	Berry Catchfly	Cucubale à baies	Taubenkropf	Cucubalo
Cucumis				
melo	Melon	Melon	Melone	Melone, Popone
sativus	Cucumber	Concombre, Cornichon	Gurke	Cetriolo
Cucurbita				
maxima	Pumpkin	Courge	Kürbis	Zucca
pepo	Marrow, Ornamental Gourd	Citrouille	Kürbis	Zucca, Cocuzza
Cupressus	Cypress	Cyprès	Zypresse	Cipresso
Cuscuta	Dodder	Cuscute	Seide	Cuscuta
Cyclamen	Sowbread	Cyclamen	Alpenveilchen	Ciclamino
Cydonia	Quince	Cognassier	Quittenbaum	Cotogno
Cymbalaria	Ivy-leaved Toadflax	Ruine de Rome	Zimbelkraut	Cimbalaria
Cynanchum	Stranglewort	Scammonée de Montpellier	Schwalbenwurz	Vincetossico
Cynara				
cardunculus	Cardoon	Cardon	Kardone	Carduccio, Caglio
scolymus	Artichoke	Artichaut	Artischocke	Carciofo
Cynodoi	Bermuda-grass	Pied-de-poule	Hundszahngras	Gramigna
Cynoglossum	Hound's-tongue	Langue-de-chien	Hundszunge	Cinoglosso
Cynosurus	Dog's-tail Grass	Cynosure	Kammgras	Covetta, Ventolana
Cyperus	Galingale	Souchet	Zypergras	Cipero
Cypripedium	Lady's Slipper	Sabot de Vénus	Frauenschuh	Pianella della Madonna
Cytinus		Cytinet		Ipocisto
Cytisus	Broom	Cytise	Geissklee	Maggio ciondolo
Daboecia	St Dabeoc's Heath	Bruyère de Saint-Daboec		
Dactylis	Cock's-foot	Dactyle pelotonné	Knäuelgras	Erba mazzolina
Dahlia	Dahlia	Dahlia	Dahlie	Dahlia, Giorgina
Damasonium	Thrumwort	Damasonium	Stern-Froschloffel	
Daphne				
alpina	Alpine Mezereon	Daphné des Alpes	Alpen-Seidelbast	Olivella

Daphne				
cneorum	Garland Flower	Camélée	Fluhröschen	Cneoro
cneorum	Garland Flower	Camélée	Fluhröschen	Cneoro
gnidium		Garou, St.-Bois		Cocco cnidio
laureola	Spurge Laurel	Lauréole	Lorbeer-Seidelbast	Laureola
mezereum	Mezereon	Bois gentil	Ziland	Camelea, Mezzereo
striata		Daphné strié	Steinröschen	
Datura	Thorn Apple	Datura stramoine, Pomme épineuse	Stechapfel	Stramonio, Noce spinosa
Daucus	Carrot	Carotte	Möhre, Gelbe Rübe	Carota selvatica, Gallinacci
Delphinium	Delphinium	Pied-d'alouette	Rittersporn	
staphisagria	Stavesacre, Licebane	Herbe-aux-poux		Stafisagria
Deschampsia	Hair-grass	Canche	Schmiele	
Descurainia	Flixweed	Sisymbre, Sagesse	Sophienkraut, Basenrauke	Nebbia
Dianthus	Pink	Oeillet	Nelke	Garofano
carthusianorum	Carthusian Pink	O.-des-chartreux	Kartäusernelke	Garofano selvatico
caryophyllus	Carnation	O.-des-fleuristes	Garten-Nelke	Garofano
plumarius	Common Pink	Mignardise		Garofanino
Dictamnus	Burning Bush	Fraxinelle	Diptam	Dittamo
Digitalis	Foxglove	Digitale	Fingerhut	Digitale
Digitaria	Crab-grass	Panic	Fingerhirse	Sanguinaria
Diplotaxis	Wall Rocket, Wall Mustard	Diplotaxis	Doppelsame	Ruchetta selvetica
Dipsacus	Teasel	Cardère	Karde	Cardo dei lanaioli
Doronicum	Leopard's-bane	Doronic	Gemswurz	Doronico
Doryenium		Badasse	Backenklee	Trifoglino
Draba	Whitlow Grass	Drave	Felsenblümchen	Pelosella
Dracunculus	Dragon Arum	Serpentaire	Schlangenwurz	Serpentaria, Dragontea
Drosera	Sundew	Rossolis	Sonnentau	Rosolida
Dryas	Mountain Avens	Chénette	Silberwurz	Camedrio cervino
Ecballium	Squirting Cucumber	Momordique, Giclet	Spritzgurke	Cocomero asinino, Schizzetti, Sputaveleno

GENUS	ENGLISH	FRENCH	GERMAN	ITALIAN
Echinochloa	Cockspur	Pied-de-coq	Hühnerhirse	Giavone
Echinops	Globe Thistle	Echinope	Kugeldistel	Cardo pallotola
Echium	Viper's Bugloss	Vipérine	Natterkopf	Echio, Lingua di cane
Elaeagnus	Oleaster	Chalef	Ölweide	Olivagno
Eleocharis	Spike-rush	Héléocharis	Nadelbinse	Giunco tondo
Elodea	Canadian Pondweed	Elodea du Canada, Peste d'eau	Wasserpest	Peste d'acqua
Elymus	Lyme-grass	Elyme	Haargras	
Empetrum	Crowberry	Camarine	Krähenbeere	
Endymion	Bluebell	Jacinthe sauvage	Wildhyazinthe	Giacinto a campanelle
Ephedra	Joint-Pine	Uvette	Meerträubchen	Uva marina
Epilobium	Willow-herb	Epilobe	Weidenröschen	Epilobio
angustifolium	Rosebay Willow-herb, Fireweed	Nériette	Wald-Weidenröschen	
Epimedium	Barren-wort	Chapeau-d'évêque	Sockenblume	Cappello di vescovo
Epipactis	Helleborine	Helléborine	Sumpfwurz	Mughetto pendolino
Eragrostis			Liebesgras	Gramignone
Eranthis	Winter Aconite	Hellébore-d'hiver	Winterling	Piè di gallo
Erianthus		Érianthe		
Erica	Heather	Bruyère	Erika	Scopa
Erigeron	Fleabane	Vergerette	Berufkraut, Feinstrahl	Impia
Erinus		Erine	Leberbalsam	
Eriobotrya	Loquat	Bibacier, Néflier du Japon		Nespolo del Giappone
Eriophorum	Cotton-grass	Linaigrette	Wollgras	Pennacchio
Erodium	Storksbill	Bec-de-grue	Reiherschnabel	Geranio selvatico
Erophila	Whitlow Grass	Drave	Hungerblümchen	
Eruca		Roquette	Ruke	Ruchetta, Eruca
Erucastrum	Hairy Rocket	Fausse roquette	Rampe	Arosseno
Eryngium	Sea Holly	Panicaut	Mannstreu	Calcatreppolo
alpinum		Chardon bleu	Alpen-M.	

Erysimum cheiranthoides	Treacle Mustard	Vélar-giroflée, Fausse giroflée	Acker-Schoterich	Violacciocca selvatica
Erythronium	Dog's Tooth Violet	Dent-de-chien	Hundszahn	Dente di cane
Eschscholzia	Californian Poppy	Pavot de Californie	Eschscholtzie	
Eucalyptus	Gum, Eucalyptus	Eucalyptus	Eucalyptus	Eucalitto
Euonymus	Spindle-tree	Fusain	Spindelstrauch	Beretta da prete
Eupatorium cannabinum	Hemp Agrimony	Eupatoire-chanvrine	Wasserdost	Canapa selvatica
Euphorbia	Spurge	Euphorbe	Wolfsmilch	Euforbia
cyparissias	Cypress Spurge	Petit-cyprés	Zypressen-W.	Erba cipressina
helioscopia	Sun Spurge	Réveil-matin	Sonnenwend-W.	Erba calenzola
lathyris	Caper Spurge	Epurge		Catapuzza
Euphrasia	Eyebright	Casse-lunettes, Euphraise	Augentrost	Eufragia
Fagopyrum	Buckwheat	Sarrasin, Blé noir	Buchweizen	Grano saraceno
Fagus	Beech	Hêtre, Fayard	Buche, Rotbuche	Faggio
Falcaria	Longleaf	Falcaire commune	Sicheldolde	
Ferula	Giant Fennel	Férule		Ferola
Festuca ovina	Fescue	Fétuque, Pois-de-chien	Schwingel	Paleo, Setaiola
Ficus	Fig	Figuier	Feigenbaum	Fico
Filago	Cudweed	Cotonnière	Fadenkraut, Filzkraut	Fignamichino
Filipendula ulmaria	Meadow-sweet	Reine-des-prés	Spierstaude	Regina dei prati
vulgaris	Dropwort	Spirée	Mädesüss	
Foeniculum	Fennel	Fenouil	Fenchel	Finnocchio
Fragaria	Wild Strawberry	Fraisier	Erdbeere	Fragola
Frangula alnus	Alder Buckthorn	Bourdaine	Brechwegdorn, Pulverholz	Frangola
Frankenia	Sea Heath			Erba franca
Fraxinus	Ash	Frêne	Esche	Frassino
ornus	Manna Ash	F.-à-fleurs	Manna-E.	Orno, Orniello
Fritillaria	Snake's Head, Fritillary	Fritillaire	Schachblume	Fritillaria

GENUS	ENGLISH	FRENCH	GERMAN	ITALIAN
Fumaria	Fumitory	Fumeterre	Erdrauch	Fumoterra, Fumosterno
Gagea	Yellow Star-of-Bethlehem	Etoile jaune	Gelbstern	Cipollaccia, Gialla dei campi, Scarline
Galactites				
Galanthus	Snowdrop	Perce-neige	Schneeglöckchen	Foraneve, Bucaneve
Galega	Goat's Rue, French Lilac	Galéga	Geissraute	Capraggine, Avanese
Galeobdolon	Yellow Archangel	Lamier Galeobdolon, Ortie jaune	Goldnessel	Ortica mora
Galeopsis	Hemp-nettle	Gáléopsis	Hohlzahn	Gallinaccia
Galinsoga	Gallant Soldier		Knopfkraut	
Galium	Bedstraw	Caille-lait, Gaillet	Labkraut	Caglio
odoratum	Sweet Woodruff	Aspérule odorante	Waldmeister	
Gastridium	Nitgrass	Gastridie	Nissegras	
Genista	Greenweed, Needle Furze, Petty Whin	Genêt	Ginster	Ginestrella
Gentiana	Gentian	Gentiane	Enzian	Genziana
asclepiadea	Willow Gentian	G.-à-feuilles d'asclépiade	Schwalbenwurz-E.	
cruciata	Cross Gentian	Croisette	Kreuzenzian	
pneumonanthe	Marsh Gentian	Pneumonanthe	Lungen-E.	Genziana minore
Geranium	Cranesbill	Géranium	Storchschnabel	Mettimborsa, Geranio
robertianum	Herb Robert	Herbe à Robert	Ruprechtskraut	Erba roberta
Geum	Herb Bennet, Avens	Benoîte	Nelkenwurz	Ambretta, Garofanaia
Gladiolus	Gladiolus	Glaieul	Siegwurz	Gladiolo
Glaucium	Horned-poppy	Pavot cornu	Hormohn	Glaucio, Papavero cornuto
Glaux	Sea Milkwort		Milchkraut	
Glechoma	Ground Ivy	Lierre terrestre	Gundelrebe	Edera terrestre
Gleditsia	Honey Locust	Févier	Dornbaum, Christusdorn	Spino di giuda
Globularia	Globularia	Globulaire	Kugelblume	Globularia
alypum		Séné des Provençaux		Erba dei frati
Glyceria	Flote-grass, Reedgrass	Brouille	Schwaden	Gramigna da padule
Glycine	Soja Bean	Soja	Soja	Glicine

Glycyrrhiza	Liquorice	Réglisse	Süssholz	Liquirizia, Regolizia
Gnaphalium	Cudweed	Gnaphale	Ruhrkraut	Impia
Goodyera	Creeping Lady's Tresses	Goodyère	Netzblatt	
Gossypium	Cotton	Cotonnier	Baumwolle	Cotone
Gratiola		Gratiole	Gnadenkraut	Graziola
Gymnadenia	Fragrant Orchid	Gymnadénia	Händelwurz	Orchide odorosa
Gypsophila	Gypsophila	Gypsophile	Gipskraut	Radice saponaria
Halimione	Sea Purslane		Salzmelde	
Hedera	Ivy	Lierre	Efeu	Edera
Hedysarum		Sainfoin		
coronarium	Italian Sainfoin	Sulla	Süssklee	Sulla
hedysaroides	Alpine Sainfoin	Sainfoin des Alpes		
Helianthemum	Rockrose	Hélianthème	Sonnenröschen	Eliantemo
Helianthus				
annuus	Sunflower	Soleil, Tournesol	Sonnenblume	Girasole
tuberosus	Jerusalem Artichoke	Topinambour	Topinambur	Patata americana
Helichrysum	Everlasting	Immortelle	Strohblume	Elicriso, Ambrenti
Heliotropium	Heliotrope	Héliotrope	Sonnenwende	Eliotropio
europaeum				Erba porraia
Helleborus	Hellebore	Ellébore	Nieswurz	
foetidus	Bear's-foot, Stinking Hellebore	Pied-de-griffon	Stinkende N.	Elabro puzzolente, Elabro nero
niger	Christmas Rose	Rose de Noël	Christrose	Elleboro nero
viridis	Green Hellebore	Ellébore vert	Grüne Nieswurz	Elleboro verde
Hemerocallis	Day-lily			Emerocalle
fulva		Lis rouge	Rote Taglilie	Giglio turco
lilioasphodelus		Lis jaune	Gelbe T.	Giglio dorato
Hepatica	Hepatica	Hépatique	Leberblümchen	Erba trinita, Trifoglio epatico
Heracleum	Cow Parsnip, Hogweed	Berce	Bärenklau	Panace
Herminium	Musk Orchid	Herminium	Einknolle	
Hermodactylus	Snake's Head Iris			Ermodattilo vero
Herniaria	Rupture-wort	Herniaire	Bruchkraut	Erniaria

GENUS	ENGLISH	FRENCH	GERMAN	ITALIAN
Hesperis	Dame's Violet	Julienne	Nachtviole	Esperide, Antoniana, Viola matronale
Hibiscus	Hibiscus	Ketmie	Roseneibisch	Ibisco
Hieracium	Hawkweed	Epervière	Habichtskraut	
pilosella	Mouse-ear Hawkweed	Piloselle	Kleines H.	Pelosella
Hierochloë	Holy-grass	Hiérochloé	Mariengras	Gramigna odorosa
Himantoglossum	Lizard Orchid	Orchis-bouc	Bocksorchis	Fior del cucullo
Hippocrepis	Horse-shoe Vetch	Fer-à-cheval	Hufeisenklee	Sferra cavallo
Hippophaë	Sea Buckthorn	Argousier	Sanddorn	Olivello
Hippuris	Mare's-tail	Pesse	Tannenwedel	Coda di cavallo
Hirschfeldia	Hoary Mustard	Hirschfeldie gusâtre	Graukohl	
Holcus	Yorkshire Fog	Houque	Honiggras	Bambagiona
Holoschoenus	Round-headed Club-rush	Holoschoenus de Rome	Kugelbinse	Giunco
Holosteum	Jagged Chickweed	Holostée	Spurre	Garofolino
Homogyne		Homogyne	Alpenlattich	
Honkenya	Sea Sandwort	Pourpier-de-mer	Salzmiere	
Hordeum	Barley	Orge	Gerste	Orzo
Hottonia	Water Violet	Millefeuille-aquatique	Wasserfeder	Erba scoparia
Humulus	Hop	Houblon	Hopfen	Luppolo
Hyacinthus	Hyacinth	Jacinthe	Hyazinthe	Giacinto
Hydrocharis	Frog-bit	Morrène	Froschbiss	Morso di rana
Hydrocotyle	Pennywort	Ecuelle-d'eau	Wassernabel	Soldinella acquatica
Hyoscyamus	Henbane	Jusquiame	Bilsenkraut	Giusquiamo
Hypecoum			Gelbäugelchen	Cornacchina
Hypericum	St John's Wort	Millepertuis	Johanniskraut	Iperico
Hypochoeris radicata	Cat's Ear	Porcelle	Ferkelkraut	Costole d'asino
Hyssopus	Hyssop	Hysope	Ysop	Isopo
Iberis	Candytuft	Ibéride	Bauernsenf	
sempervirens		Corbeille-d'argent	Immergrüner B.	Porcellana minore
Ilex	Holly	Houx	Stechpalme	Agrifoglio
Illecebrum	Illecebrum	Illécèbre	Knorpelblume	Arsinedda

Impatiens	Touch-me-not, Balsam	Balsamine, Impatiente	Springkraut	Balsamina, Erba impaziente
Inula		Aunée	Alant	
conyza	Ploughman's Spikenard	Œil-de-cheval	Gemeine Dürrwurz	Enula bacicci
crithmoides	Golden Samphire	Inule-faux, crithmum		
helenium	Elecampane	Grande aunée	Echter Alant	Enula
Iris	Iris, Flag	Iris	Schwertlilie	Giaggiolo
foetidissima	Gladdon	Iris puant		Ricottaria
pseudacorus	Yellow Flag	Iris jaune	Sumpf-S.	Acoro, Giglio giallo
Isatis	Woad	Pastel	Waid	Guado
Isolepis	Bristle Scirpus	Isolepis sétacé	Moorbinse	
Isopyrum	Isopyrum	Isopyre	Muschelblümchen	Ellèbanu
Jasione	Sheep's-bit	Jasione	Sandglöckchen	Vedovelle celesti
Jasminum	Jasmine	Jasmin	Jasmin	Gelsomino
Juglans	Walnut	Noyer	Walnuss	Noce
Juncus	Rush	Jonc	Binse	Giunco
Juniperus	Juniper	Genêvrier	Wacholder, Reckolder	Ginepro
Jurinea		Jurinée		
Knautia	Field Scabious	Scabieuse	Witwenblume	Vedovina, Scabiosa
Koeleria	Crested Hair-grass	Koelérie	Schillergras Kammschmiele	Paleo argenteo gentile
Laburnum	Laburnum	Cytise, Aubour	Goldregen	Avorniello, Laburno, Maggio ciondolino
Lactuca	Lettuce	Laitue	Lattich	Lattuga
Lagurus	Hair's-tail	Lagure		Piumino, Coda di lepre
Lamium	Dead-nettle	Ortie morte, Lamier	Taubnessel	Ortica che non punge, Lamio
Lappula			Igelsame	Lappolini
Lapsana	Nipplewort	Grageline	Rainkohl	Lassana
Larix	Larch	Mélèze	Lärche	Larice
Laserpitium		Laser	Laserkraut	Lasero
Lathraea	Toothwort	Lathrée	Schuppenwurz	Fuoco dei boschi, Fior galletto,
Lathyrus	Vetchling, Pea Vetch	Gesse	Platterbse	Veccia bastarda

GENUS	ENGLISH	FRENCH	GERMAN	ITALIAN
Laurus	Laurel	Laurier	Lorbeer	Alloro, Lauro
Lavandula	Lavender	Lavande	Lavendel	Lavanda, Spico
Lavatera	Tree Mallow	Mauve royale	Strauchpappel	Malva arborea
Leersia	Cut-grass	Faux riz	Reisquecke	
Legousia	Venus' Looking-Glass	Miroir de Vénus	Frauenspiegel	Specchio di Venere
Lemna	Duckweed	Lenticule-d'eau	Wasserlinse	Lente di padule
Lens	Lentil	Lentille	Linse	Lenticchia
Leonurus	Motherwort	Agripaume	Herzgespann	Cardiaca
Leontodon	Hawkbit	Liondent	Löwenzahn	Radicchiello
Leontopodium	Edelweiss	Edelweiss	Edelweiss	Stella alpina, Edelweiss
Lepidium	Pepperwort	Passerage	Kresse	Mostardina
Leucojum	Snowflake	Nivéole	Knotenblume	Campanelle
Leucorchis	Small White Orchid	Gymnadénia	Weisszüngel, Handwurz	
Levisticum	Lovage	Livèche	Liebstöckel	Ligustio, Levistico
Ligusticum	Lovage	Ligustique	Mutterwurz, Liebstock	Ligustio
Ligustrum	Privet	Troène	Rainweide	Ligustro
Lilium	Lily	Lis	Lilie	Giglio
martagon	Martagon Lily	Martagon	Türkenbundlilie	G. martagone
Limodorum	Limodore	Limodorum	Dingelorchis	Limodoro
Limonium	Sea Lavender	Lavande-de-mer	Widerstoss	Statice
Linaria	Toadflax	Linaire	Leinkraut	Linaria
Linnaea	Linnaea	Linnée boréale	Moosglöckchen	Linnea
Linum	Flax	Lin	Lein	Lino
Lippia	Verveine	Verveine	Verveine	Erba limonaria
Listera	Twayblade	Listéra	Zweiblatt	Giglio verde
Lithospermum	Gromwell	Grémil	Steinsame	Lithospermo, Migliarino
Littorella	Shore-weed	Littorelle	Strandling	Gramignola d'acqua
Lloydia	Lloydia	Lloïdie	Faltenlilie	
Lobularia	Sweet Alison			
Loiseleuria	Loiseleuria	Azalée couchée	Alpenheide	
Lolium	Rye-grass	Ivraie	Lolch	Loglio, Gioglio

Lonicera	Honeysuckle	Chèvrefeuille	Geissblatt	Caprifoglio, Madreselva
Lotus	Birdsfoot-trefoil	Lotier	Hornklee	Trifoglina, Ginestrina
Ludwigia		Isnardie	Heusenkraut	
Lunaria	Honesty	Monnaie-du-pape	Silberblatt	Lunaria
Lupinus	Lupin	Lupin	Lupine	Lupino
Luzula	Woodrush	Luzule	Heinsimse	Erba lucciola
Lychnis				
coronaria	Rose Campion	Coquelourde	Kranzrade	Fior del cucolo
flos-cuculi	Ragged Robin	Fleur-de-coucou	Lichtnelke	Fior di giove
flos-jovis	Flower of Jove	Fleur de Jupiter	Jupiternelke	Ogio de criste
viscaria	Red German Catchfly	Œillet-des-prés	Schlampeter-Mädli	Spina-Cristi, Agutoli
Lycium	Duke of Argyll's Tea-plant	Lyciet	Bocksdorn	Marrobio acquatico
Lycopus	Gipsy-wort	Chanvre-d'eau	Wolfstrapp	Erba quattrina dei boschi
Lysimachia		Lysimaque	Lysimachie	
nemorum	Yellow Pimpernel	L.-des-bois	Hain-Friedlos	Borissa
nummularia	Creeping Jenny	Herbe-aux-écus	Pfennigkraut	Mazza d'oro,
vulgaris	Yellow Loosestrife	Lysimaque	Gewöhnlicher Lysimachie	Lisimachia
Lythrum	Purple Loosestrife	Salicaire	Weiderich	Salicaria, Salcerella
portula	Water Purslane	Pourpier-des-marais	Sumpfquendel	
Maianthemum	May Lily	Petit-muguet	Schattenblume	Mughetto
Malcolmia	Virginia Stock			Violacciocca
Malus	Apple	Pommier	Apfelbaum	Melo
Malva	Mallow	Mauve	Malve	Malva
Mandragora	Mandrake	Mandragore	Alraun	Mandragora
Marrubium	White Horehound	Marrube	Andorn	Erba apiola, Marrubio
Matricaria		Matricaire	Kamille	
recutita	Wild Chamomile	Camomille vulgaire, C. allemande	Feldkamille	Camomilla

GENUS	ENGLISH	FRENCH	GERMAN	ITALIAN
Matthiola	Stock	Violier	Levkoje	Violacciocca
Meconopsis	Poppy	Pavot	Mohn	
Medicago	Medick	Luzerne	Schneckenklee	
lupulina	Black Medick	Minette	Hopfenklee	Trifogliolino selvatico
sativa	Lucerne, Alfalfa	Luzerne cultivée	Luzerne	Erba medica
Melampyrum	Cow-wheat	Mélampyre	Wachtelweizen	Fiamma, Melampiro
Melia	Indian Bead Tree, Persian Lilac	Margousier		
Melica	Melick	Mélique	Perlgras	
Melilotus	Melilot	Mélilot	Steinklee	Meliloto
Melissa	Balm	Mélisse	Melisse	Melissa, Appiastro
Melittis	Bastard Balm	Mélitte	Immenblatt	Bocca di lupo
Mentha	Mint	Menthe	Minze	Mentastro
pulegium	Penny-royal	Pouillot	Polei	Puleggio
Menyanthes	Bogbean, Buckbean	Trèfle-d'eau	Fieberklee	Trifoglio fibrino
Mercurialis	Mercury	Mercuriale	Bingelkraut	Mercorella
Mesembryanthemum		Ficoïde glaciale		Erba cristallina
Mespilis	Medlar	Néflier	Mispel	Nespolo
Meum	Spignel, Baldmoney	Cistre	Bärwurz	Meo, Finocchiella
Milium	Millet	Millet	Flattergras	Miglio
Mimulus	Monkey-flower, Musk	Mimule	Gauklerblume	Momolo
Minuartia	Sandwort	Minuartie	Miere	
Mirabilis	Four-o'clock	Belle-de-nuit		Bella di notte
Moehringia	Three-nerved Sandwort	Moehringie	Nabelmiere	Centonchio minore
Molinia	Purple Moor-grass	Canche bleue	Pfeifengras	Gramigna liscia
Monotropa	Yellow Bird's Nest	Sucepin	Fichtenspargel	
Montia	Blinks	Montie	Quellkraut, Claytonie	Centonchiello

Morus	Mulberry	Mûrier	Maulbeerbaum	Gelso, Moro
Muscari	Grape-hyacinth	Muscari	Traubelhyazinthe	Giacinto
comosum	Tassel Hyacinth	M.-à-houppe	Schopfartige Bisamhyazinthe	Muschini
Myagrum	Mitre Cress	Laitue-des-murs	Hohldotter	
Mycelis	Wall Lettuce		Mauerlattich	
Myosotis	Forget-me-not, Scorpion Grass	Myosotis	Vergissmeinnicht	Non ti scordar di mi
Myrica	Bog Myrtle, Sweet Gale	Piment royal	Gagel	
Myricaria		Tamarin	Rispelstrauch, Tamariske	Tamerice, Scopa marina
Myriophyllum	Water-milfoil	Myriophylle	Tausendblatt	Millefoglio d'acqua
Myrrhis	Sweet Cicely	Cerfeuil musqué	Sussdolde	Finocchiella
Myrtus	Myrtle	Myrte	Myrte	Mirto, Mortella
Najas	Naiad	Naiade	Nixenkraut	Spini da ranocchi
Narcissus		Narcisse	Narcisse	Narciso
poeticus	Pheasant's Eye	N.-des-poetes	Weisse N.	Fior-maggi
pseudonarcissus	Wild Daffodil	Jonquille	Gelbe N.	Trombone
Nardus	Mat-grass	Nard	Borstgras	Cervino
Narthecium	Bog Asphodel	Ossifrage	Beinbrech	Ossifrago
Nasturtium	Watercress	Cresson-de-fontaine	Brunnenkresse	Crescione
Neottia	Bird's-nest Orchid	Neottie	Nestwurz	Nido d'uccello
Nepeta	Cat-mint	Herbe-aux-chats	Katzenminze	Erba gatta
Nerium	Oleander	Laurier rose		Oleandro, Leandro
Nicandra	Apple of Peru		Giftbeere	
Nicotiana	Tobacco	Tabac	Tabak	Tabacco
Nigella	Love-in-a-Mist	Nigelle	Schwarzkümmel	Nigella, Fanciullaccia
Nigritella	Vanilla Orchid	Orchis vanillé	Männertreu Mönchskraut	Morettina
Nonea				
Nuphar	Yellow Water-lily	Nénuphar jaune	Teichrose	Ninfea gialla
Nymphaea	Water-lily	Nénuphar	Seerose	Ninfea bianca
Nymphoides	Fringed Water-lily	Petit nénuphar	Seekanne	

GENUS	ENGLISH	FRENCH	GERMAN	ITALIAN
Odontites	Red Bartsia	Euphraise	Zahntrost	Perlina
Oenanthe	Water Dropwort	Oenanthe	Rebendolde	
aquatica	Fine-leaved Water D.	O. aquatique	Wasserfenchel	
Oenothera	Evening Primrose	Onagre	Nachtkerze	Enagra
Olea	Olive	Olivier	Olive, Ölbaum	Olivo
Omphalodes	Blue-eyed Mary	Nombril de Vénus	Nabelnuss	Bucalosso
Onobrychis				
caput-galli		Tête-de-coq		Cresta di gallo
viciifolia	Sainfoin	Sainfoin	Esparsette	Lupinella
		Esparcette		
Ononis	Restharrow	Bugrane	Hauhechel	Anonide,
spinosa	Spiny Restharrow	Arrête-bœuf		Arrestabue, Bonaga
Onopordum	Scotch Thistle	Onoporde	Eselsdistel	Cardo asinino
Onosma	Golden Drop	Orcanette jaune	Lotwurz	Ancusa gialla
Ophrys	Orchid	Ophrys	Ragwurz,	Fior ragno
			Kerfstendel	
Opuntia	Prickly Pear,	Figuier de Barbarie	Feigenkaktus	Fico d'India
	Barbary Fig			
Orchis	Orchid	Orchis	Knabenkraut	
Origanum	Marjoram	Origan	Dost	Origano
Orlaya	Orlaya	Orlaya	Breitsame	
Ornithogalum				
umbellatum	Star-of-Bethlehem	Ornithogale	Milchstern	Latte di gallina
		Dame-d'onze-heures	Doldiger M.	Aglio selvatico
Ornithopus	Birdsfoot	Pied-d'oiseau	Vogelfuss	Serradella
Orobanche	Broomrape	Orobanche	Sommerwurz	Succiamele,
				Erba lupa, Fiamma
Ostrya	Hop-Hornbeam	Charme-houblon	Hopfenbuche	Carpinella
Osyris		Rouvet	Harnstrauch	Casia poetica,
				Osiride
Otanthus	Cotton-weed			Gnafalio marittimo
Oxalis	Sorrel	Surelle	Sauerklee	Acetosella
Oxyria	Mountain Sorrel	Oxyria	Säuerling	

Oxytropis	Beaked Milk-vetch	Oxytropis	Spitzkiel	
Paeonia	Peony	Pivoine	Pfingstrose	Peonia
Paliurus	Christ's Thorn	Épine du Christ		Paliuro, Marruca
Pancratium	Sea Daffodil	Lis-mathiole		Pancrazio, Narciso marino
Panicum	Cockspur	Millet	Hirse	Miglio
Papaver	Poppy	Pavot, Coquelicot	Mohn	Papavero, Rosolaccio
Paradisea	St Bruno's Lily	Faux-lis	Trichterlilie	
Parapholis	Hard-grass	Lepture	Dünnschwanz	
Parietaria	Pellitory-of-the-Wall	Pariétaire	Glaskraut	Parietaria, Spaccapietra
Paris	Herb Paris	Parisette	Einbeere, Wolfsbeere	Erba paris, Erba crociola
Parnassia	Grass of Parnassus	Parnassie	Herzblatt, Studentenröschen	Parnassia
Paronychia		Panarine		
Parthenocissus	Virginia Creeper	Vigne vierge	Jungferurebe	Vite del Canadà
Pastinaca	Wild Parsnip	Panais	Pastinak	Pastinaca
Pedicularis	Lousewort, Red-rattle	Pédiculaire	Läusekraut	Pediculare
Pelargonium	Zonal Geranium	Géranium	Geranium	Geranio
Petasites		Pétasite		Petasite, Farfaraccio
fragrans	Winter Heliotrope	Héliotrope-d'hiver		
hybridus	Butterbur	Herbe-aux-teigneux	Pestwurz	
Petrorhagia	Proliferous Pink	Œillet	Felsennelke	
Petroselinum	Parsley, Corn Caraway	Persil	Petersilie	Prezzemolo
Peucedanum	Hog's Fennel, Milk Parsley, Master-wort	Peucédan	Haarstrang	Finocchio di porco
Phalaris	Reed-grass, Canary Grass	Alpiste	Glanzgras	Scagliola, Canaria

GENUS	ENGLISH	FRENCH	GERMAN	ITALIAN
Pharbitis	Morning Glory	Volubilis		
Phaseolus	Bean	Haricot	Gartenbohne	Fagiuolo
Phillyrea			Steinlinde	Fillirea, Olivastro
Phleum	Timothy	Fléole	Lieschgras	Codolina, Coda di topo
Phlomis	Jerusalem Sage			Flomide
fruticosa				
herba-venti		Herbe-au-vent		
Phragmites	Reed	Roseau-à-balais	Schilf	Canna palustre
Physalis	Winter Cherry	Coqueret	Judenkirsche	Alchechengi
Phyteuma	Rampion	Raiponce	Teufelskralle	
Phytolacca	Virginian Pokeweed	Raisin d'Amérique	Kermesbeere	Amaranto del Perù
Picea	Spruce	Epicéa, Pesse	Fichte	Abete rosso
Picris	Ox-tongue	Picride	Bitterkraut	Lattaiola pungente
Pimpinella	Burnet Saxifrage	Boucage Pimprenelle	Bibernelle	Anice selvatico
Pinguicula	Butterwort	Grassette	Fettkraut	Erba grassa
Pinus	Pine	Pin	Kiefer	Pino
Pistacia				
lentiscus	Mastic Tree, Lentisc	Lentisque, Arbre-au-mastic	Mastixbaum	Lentisco
terebinthus	Turpentine Tree, Terebinth	Térébinthe	Pistazie, Terebinthe	Terebinto
Pisum	Pea	Pois	Erbse	Pisello
Plantago	Plantain	Plantain	Wegerich	Piantaggine
Platanthera	Butterfly Orchid	Platanthère	Breitkolbohen	Testicolo di cane
Platanus	Plane	Platane	Platane	Platano
Pleurospermum			Rippensame	Cicutaria
Plumbago	Leadwort	Dentelaire		Piombaggine
Poa	Poa	Paturin	Rispengras	Fienarola
Polemonium	Jacob's Ladder	Polémoine, Valériane grecque	Himmelsleiter	
Polycarpon	Four-leaved All-seed	Polycarpon	Nagelkraut	Erba migliarina

Polygala	Milkwort	Polygala	Kreuzblume	
chamaebuxus	Box-leaved M.	Faux buis	Buchs-K.	
Polygonatum	Solomon's Seal	Sceau de Salomon	Salomonsiegel	Sigillo di Salomone
Polygonum	Knotgrass	Renouée	Knöterich	
Polypogon	Beardgrass			Coda di topo
Populus	Poplar	Peuplier	Pappel	Pioppo
tremula	Aspen	Tremble	Zitter-P.	Tremolo
Portulaca	Purslane	Pourpier	Portulak	Porcellana
Posidonia		Paille-de-mer		Alga
Potamogeton	Pondweed	Potamot	Laichkraut	Lattuga ranina
Potentilla	Cinquefoil	Potentille	Fingerkraut	
erecta	Common Tormentil	Tormentille	Tormentill	Tormentilla
palustris	Marsh Cinquefoil	Comaret-des-marais	Blutange	
reptans	Creeping Cinquefoil	Quintefeuille	Kriechendes Fingerkraut	Cinquefoglio, Pentafillo
sterilis	Barren Strawberry	Potentille-faux fraisier	Erdbeer-F.	
Prenanthes		Prénanthe	Hasenlattich	Lattuga montana
Primula	Primrose, Cowslip	Primevère	Primel, Schlüsselblume	Primula, Primavera
elatior	Oxlip	P. élevée	Wald-Primel	
veris	Cowslip	P. officinale	Himmelschlüsselblume	
vulgaris	Primrose	Primevère	Primel	
Prunella	Self-Heal	Brunelle	Brunelle	Brunella
Prunus	Cherry	Prunier		Pruno
armeniaca	Apricot	Abricotier	Aprikosenbaum	Albicocco
avium	Wild Cherry	Merisier	Süsskirschbaum	Ciliegio
cerasifera	Cherry-plum	Prunier-cerise	Kirschpflaumenbaum	Ciliegia susina
dulcis	Almond	Amandier	Mandelbaum	Mandorlo
padus	Bird-cherry	Merisier-à-grappes	Traubenkirschenbaum	Pado
persica	Peach	Pêcher	Phrsichbaum	Pesco
spinosa	Blackthorn, Sloe	Prunellier	Schlehdorn	Pruno selvatico
Psoralea	Pitch Trefoil	Herbe-au-bitume		Trifoglio bituminoso
Pulicaria	Fleabane	Pulicaire	Flohkraut	Incensaria

GENUS	ENGLISH	FRENCH	GERMAN	ITALIAN
Pulmonaria	Lung-wort	Pulmonaire	Lungenkraut	Pulmonaria
Pulsatilla	Pasque Flower	Pulsatille, Anémone	Küchenschelle	Pulsatilla
Punica	Pomegranate	Grenadier	Granatbaum	Melograno, Granato
Pyracantha	Pyracantha	Buisson ardent		
Pyrola	Wintergreen	Pyrole	Wintergrün	Pirolatta soldanina
Pyrus				
communis	Pear	Poirier	Birnbaum	Pero
Quercus	Oak	Chêne	Eiche	Querce
cerris	Turkey Oak	C. chevelu	Zerr-E.	Cerro
coccifera	Kermes Oak	C.-kermès		Q. spinosa
ilex	Holm Oak, Evergreen Oak	Yeuse, C. vert	Stein-E.	Leccio
petraea	Durmast Oak	C. noir	Trauben-E.	
robur	Common Oak	C. rouvre	Stiel-E.	Q. comune, Rovere
suber	Cork Oak	C.-liège		Sughera
Radiola	All-seed	Radiole	Zwergflachs	
Ranunculus	Buttercup, Crowfoot	Renoncule	Hahnenfuss	Ranuncolo
ficaria	Lesser Celandine, Pilewort	Ficaire	Scharbockskraut	
Raphanus	Radish	Radis	Rettich	Radice, Rafano
Rapistrum	Bastard Cabbage		Rapsdotter	
Reseda	Dyer's Rocket, Mignonette	Réséda	Wau, Reseda	Reseda, Amorino
Rhamnus				
alaternus		Alaterne		Alaterno, Legno puzzo
catharticus	Buckthorn	Nerprun	Kreuzdorn	Spino cervino
Rhinanthus	Yellow Rattle	Cocriste, Rhinante	Klappertopf	Cresta di gallo
Rhodiola	Midsummer-men, Rose-root	Orpin rose	Rosenwurz	
Rhododendron	Rhododendron	Rhododendron	Alpenrose	Rododendro
Rhus	Sumach	Sumac	Sumach	Sommacco

Rhynchospora	Beak-sedge	Rhynchospora	Schnabelbinse	Giunco tenero
Ribes	Currant	Groseillier	Johannisbeere	
nigrum	Black Currant	Cassis	Schwarze J.	Ribes nero
rubrum	Red Currant	G. à grappes	Rote J.	Ribes volgare
uva-crispa	Gooseberry	G. épineux	Stachelbeerstrauch	Uva crispa
Ricinus	Castor Oil Plant	Ricin	Rizinus	Ricino, Palma-Christi
Robinia	False Acacia	Robinier	Robinie	Acacia, Robinia
Roemeria	Violet Horned-poppy			
Rorippa	Yellow-cress	Cresson	Sumpfkresse	
Rosa	Rose	Rosier, Eglantier	Rose, Hagrose	Rosa
Rosmarinus	Rosemary	Romarin	Rosmarin	Rosmarino
Rubia	Wild Madder	Garance	Färberröte	Robbia
Rubus	Blackberry, Bramble	Ronce	Brombeere	Rovo
caesius	Dewberry	R. bleuâtre	Taubenkropf	
idaeus	Raspberry	Framboisier	Himbeere	Lampone
saxatilis	Rock Bramble	R.-des-rochers	Steinbeere	Rogo erbaiolo
Rumex	Dock	Patience, Rumex	Ampfer	Romice
acetosa	Sorrel	Oseille	Sauerampfer	Acetosa
alpinus	Monk's Rhubarb	Lampé	Blacke	Rabarbaro alpino
Ruscus	Butcher's Broom	Fragon	Mäusedorn	Pungitopo
Ruta	Rue	Rue	Raute	Ruta
Sagina	Pearlwort	Sagine	Mastkraut	Burinella
Sagittaria	Arrow-head	Sagittaire, Flèche-d'eau	Pfeilkraut	Erba saetta
Salicornia	Glasswort, Marsh Samphire			Salicornia
Salix	Willow	Saule	Weide	Salice
Salsola	Saltwort	Soude	Salzkraut	Soda, Erba-kali
Salvia	Clary, Sage	Sauge	Salbei	Salvia
Sambucus				
ebulus	Danewort	Hièble	Attich	Ebbio
nigra	Elder	Sureau	Holunder	Sambuco

GENUS	ENGLISH	FRENCH	GERMAN	ITALIAN
Samolus	Brookweed	Mouron-d'eau	Bunge	Samolo
Sanguisorba				
minor	Salad Burnet	Petite sanguisorbe		Pimpinella
officinalis	Great Burnet	Pimprenelle	Wiesenknopf	Sanguisorba
Sanicula	Sanicle	Sanicle	Sanikel	Sanicola, Erba frangolina
Santolina				
chamaecyparissus	Lavender Cotton	Petit cyprès		Santolina
Saponaria	Soapwort, Bouncing Bet	Saponaire	Seifenkraut	Saponaria
Sarcopoterium	Thorny Burnet			Spina porci
Sarothamnus	Broom	Genêt à balais	Besenginster	Ginestra dei carbonai
Satureja	Savory	Sarriette	Bergminze	Santoreggia
Saussurea	Alpine Saussurea	Saussurée	Alpenscharte	
Saxifraga	Saxifrage	Saxifrage	Steinbrech	Sassifraga
granulata	Meadow Saxifrage	Casse-pierre	Knollen-S.	
Scabiosa	Scabious	Scabieuse	Skabiose	Scabiosa
Scandix	Shepherd's Needle	Peigne de Vénus	Nadelkerbel	Pettine di Venere, Acicula
Schinus	Californian Pepper-tree, Peruvian Mastic-tree	Faux poivrier		Pepe, Falsopepe
Schoenus	Bog-rush	Choin	Kopfried	Giunco nero
Scilla	Squill	Scille	Blaustern	Scilla
Scirpus	Club-rush	Souchet	Simse	Giunco
Scleranthus	Knawel	Gnavelle	Knäuel	Centigrani
Scolymus	Spanish Oyster Plant	Épine-jaune		Cardo scolimo, Scardiccione
Scorpiurus		Chenille		Coda di scorpione
Scorzonera	Dwarf Scorzonera	Scorsonère	Schwarzwurzel	Scorzonera
Scrophularia	Figwort, Water Betony	Scrofulaire	Braunwurz	Scrofolaria
Scutellaria	Skull-cap	Scutellaire	Helmkraut	Scutellaria
Secale	Rye	Seigle	Roggen	Segala
Sedum	Stonecrop	Orpin	Fettkraut, Fetthenne	
acre	Wall-pepper	Poivre-de-muraille	Scharfer Mauerpfeffer	Borraccina
album	White Stonecrop	Trique-madame	Weisses Fettkraut	Pinocchiella

Sedum telephium	Orpine, Livelong	Reprise	Breitblattriges, Fettkraut	Erba di S. Giovanni
Selinum		Sélin	Silge	Carvifolio
Sempervivum tectorum	Houseleek	Joubarbe / Artichaut-de-murailles	Hauswurz / Echte H.	Semprevivo
Senecio		Séneçon	Kreuzkraut	
cineraria	Cineraria	Cinéraire		Cineraria
jacobaea	Ragwort	Herbe de St.-Jacques	Jakobskraut	Erba chitarra, Jacobea
vulgaris	Groundsel	Séneçon vulgaire	Gemeines Kreuzkraut	Erba calderugia
Serapias	Helleborine	Sérapias	Stendelwurz	Limodoro
Serratula	Saw-wort	Sarrette	Scharte	Serratola
Sesamum	Sesamum	Sésame	Sesam	Sesamo
Sesleria	Blue Sesleria	Seslérie	Kopfgras	Gramigna liscia
Setaria	Bristle-grass	Sétaire	Borstenhirse	Panico
Sherardia	Field Madder	Rubéole	Ackerröte	Toccamano
Sibbaldia		Sibbaldie	Gelbling	
Sideritis		Crapaudine	Gliedkraut	
Silaum	Pepper Saxifrage	Silaum	Rosskummel	
Silene	Campion	Silène	Leimkraut	Silene
acaulis	Moss Campion	Gazon rose		
dioica	Red Campion	Compagnon rouge		
Silybum	Milk-thistle	Chardon Marie	Mariendistel	Cardo mariano
Sinapis	Mustard	Moutarde	Senf	Senape
Sison	Stone Parsley	Sison	Gewürzdolde	Sisero, Ammone
Sisymbrium	Hedge Mustard, London Rocket	Vélar	Rauke	
Sisyrinchium	Blue-eyed Grass	Sisyrinchium	Grasschwertel	
Sium	Water Parsnip	Berle	Merk	Erba cannella
Smilax		Liseron épineux		Smilace, Salsapariglia nostrale
Smyrnium	Alexanders	Maceron	Gelbolde	Smirnio
Solanum dulcamara	Bittersweet, Woody Nightshade	Douce-amère	Bittersüss	Dulcamara

GENUS	ENGLISH	FRENCH	GERMAN	ITALIAN
Solanum				
lycopersicum	Tomato	Tomate	Tomate	Pomodoro
melongena	Aubergine	Aubergine	Aubergine	Melanzana
nigrum	Black Nightshade	Morelle noire	Schwarzer Nachtschatten	Morella
sodomaeum				Pomo di Sodoma
tuberosum	Potato	Pomme-de-terre	Kartoffel	Patata
Soldanella	Soldanella	Soldanelle	Troddelblume	Soldanella
Solidago	Golden-rod	Verge-d'or	Goldrute	Verga d'oro
Sonchus	Sow-thistle	Laiteron	Gänsedistel	Cicerbita
Sorbus		Sorbier		Sorbo
aria	White Beam	Alisier	Mehlbeerbaum	Sorbo montano
aucuparia	Rowan, Mountain Ash	Sorbier-des-oiseleurs	Vogelbeerbaum	Sorbo rosso
domestica	Service Tree	Cormier	Speierling	Sorbo domestico
torminalis	Wild Service Tree	Alisier	Elsbeerbaum	Bacarello
Sorghum	Millet	Sorgho	Sorgho	Saggina
Sparganium	Bur-reed	Rubanier	Igelkolben	Sala, Biodo
Spartina	Cord-grass		Reisgras	
Spartium	Spanish Broom	Genêt d'Espagne	Spanischer Ginster	Ginestra
Spergula	Corn Spurrey	Espargoutte	Spark	Spergola
Spergularia	Sand Spurrey	Spergulaire	Spärkling	Lupinaia
Spinacia	Spinach	Épinard	Spinat	Spinacio
Spiraea	Willow Spiraea	Spirée	Spierstrauch	Spirea
Spiranthes	Lady's Tresses	Spiranthe	Wendelähre	Testicolo odoroso
Stachys	Woundwort	Bétoine, Epiaire	Ziest	
Staphylea	Bladder-nut	Staphylier	Pimpernuss	Pistacchio falso
Stellaria	Stitchwort	Stellaire	Sternmiere	
media	Chickweed	Mouron-des-oiseaux, Morgeline	Hühnerdarm	Morso di gallina
Stipa		Plumet	Federgras	
Stratiotes	Water Soldier	Faux-aloès	Krebsschere	Stilletti d'acqua
Streptopus			Knotenfuss	Lauro Alessandrino
Styrax	Storax	Aliboufier		Storace

Suaeda	Seablite	Soude	Soda	Sopravvivolo legnoso
Succisa	Devil's-bit Scabious	Succise	Teufelsabbiss	Succisa
Swertia		Swertie	Moorenzian	
Symphoricarpos	Snowberry	Symphorine	Schneebeere	Pianta delle perle
Symphytum	Comfrey	Consoude	Beinwell	Consolida
Syringa	Lilac	Lilas	Flieder	Lillà
Tamarix	Tamarisk	Tamarin		Tamerice
Tamus	Black Byrony	Tamier	Schmerwurz	Tamaro, Vite nera
Taraxacum	Dandelion	Dent-de-lion, Pissenlit	Löwenzahn, Kuhblume	Tarassaco Piscialletto, Dente di leone
Taxus	Yew	If	Eibe	Tasso
Teesdalia	Shepherd's Cress	Teesdalie	Bauernsenf, Tisdalie	
Telekia		Télékie	Telekie	
Tetragonolobus purpureus	Winged Pea	Lotier rouge	Spargelbohne	Loto rosso
	Asparagus Pea			
Teucrium	Germander	Germandrée	Gamander	Camedrio
chamaedrys	Wall Germander	Petit chéne		Polio
polium				
scorodonia	Wood Sage	Sauge-des-bois	Wald-G.	Scorodonia
Thalictrum	Meadow Rue	Pigamon	Wiesenraute	Pigamo
Thesium	Bastard Toadflax	Thésion	Leinblatt	Mbriaca-voi
Thlaspi	Penny Cress	Tabouret	Hellerkraut	Erba storna
Thymelaea		Passerine	Vogelkopf, Spatzenzunge	Barbosa, Spazzaforno
Thymus	Thyme	Thym	Thymian	Timo
serpyllum		Serpolet	Quendel	Pepolino
Tilia	Linden, Lime	Tilleul	Linde	Tiglio
Tofieldia	Scottish Asphodel	Tofieldie	Simsenlilie	Tajola
Tolpis		Œil du Christ		Barbatella
Tordylium	Hartwort	Tordyle	Zirmet, Drehkraut	Tordilio
Torilis	Hedge-parsley	Torilis	Klettenkerbel	Lappolina
Tozzia		Tozzie	Tozzie, Alpenrachen	
Tradescantia	Spiderwort			
Tragopogon	Goats-beard	Barbe-de-bouc	Bocksbart	

GENUS	ENGLISH	FRENCH	GERMAN	ITALIAN
Tragopogon				
porrifolius	Salsify	Salsifis	Salsifis	Barba di becco, Tragopogono
Tragus		Bardanette	Klettengras	
Trapa	Water Chestnut	Mâcre	Wassernuss	Castagna d'acqua
Tribulus	Maltese Cross, Small Caltrops	Croix de Malte		Tribolo
Trichophorum	Deer-Grass	Trichophorum	Haarbinse	
alpinum		T. des Alpes	Alpen-Haarbinse	
caespitosum		T. gazonnant	Rasenbinse	
Trientalis	Chickweed, Wintergreen	Trientalis	Siebenstern	Trientale stellata
Trifolium	Clover, Trefoil	Trèfle	Klee	Trifoglio
arvense	Hare's-foot	Pied-de-lièvre	Hasenklee	Moscino
fragiferum	Strawberry Clover	Porte-fraise	Erdbeerklee	T. fragolino
Triglochin	Arrow-grass	Troscart	Dreizack	Giuncastrello
Trigonella		Trigonelle	Hornklee	
foenum-graecum	Fenogreek	Fenugrec	Bockshornklee	Fieno greco
Trisetum	Yellow Oat	Avoine dorée	Goldhafer	Gramigna bionda
Triticum	Wheat	Blé	Weizen	Grano, Frumento
Trollius	Globe Flower	Trolle	Trollblume	Luparia, Paparia
Tropaeolum	Nasturtium	Capucine	Kapuzinerkresse	Nasturzio, Cappuccina
Tulipa	Tulip	Tulipe	Tulpe	Tulipano
Tussilago	Coltsfoot	Taconnet, Tussilage	Huflattich	Tossilaggine, Farfara
Typha	Reedmace, Bulrush, Cat's-tail	Massette	Rohrkolben	Mazza-sorda, Stiancia
Ulex	Furze, Gorse, Whin	Ajonc	Stechginster	Ginestrone
Ulmus	Elm	Orme	Ulme	Olmo
Umbilicus	Pennywort, Navelwort	Gobelet, Écuelle	Venusnabel	Ombelico di Venere, Cappellini
Urginea	Sea Squill	Scille	Meerzwiebel	Scilla
Urtica	Nettle	Ortie	Brennessel	Ortica
Utricularia	Bladderwort	Utriculaire	Wasserschlauch	Erba vescica
Vaccaria	Cow Basil	Vaccaire	Kuhkraut	Cetino, Mezzettino

Vaccinium myrtillus	Bilberry, Whortleberry, Blaeberry, Huckleberry	Myrtille	Heidelbeere	Mirtillo
oxycoccos	Cranberry	Canneberge	Moosbeere	Mortellina di palule
uliginosum	Bog Whortleberry	Airelle-des-marais	Moorbeere, Rauschbeere	
vitis-idaea	Cowberry, Red Whortleberry	Airelle rouge	Preiselbeere	Mortella rossa
Vaillantia				Erba croce dei muri
Valeriana	Valerian	Valériane	Baldrian	Valeriana
Valerianella	Lamb's Lettuce, Corn Salad	Mâche, Rampon, Doucette	Nüsslisalat, Feldsalat	Agnellino, Locusta
Vallisneria		Vallisnérie	Wasserschraube	Alga di Chiana
Veratrum	White Hellebore	Vératre	Germer	Veratro
Verbascum	Aaron's Rod, Mullein	Molène, Bouillon blanc, Bonhomme	Königskerze	Verbasco
Verbena	Vervain	Verveine	Eisenkraut	Verbena
Veronica	Speedwell	Véronique	Ehrenpreis	Veronica
beccabunga	Brooklime	Cresson-de-cheval	Bachbungen-E.	Beccabunga
Viburnum		Viorne	Schneeball	Viburno
lantana	Wayfaring Tree	Mancienne, V.-lantane	Wolliger S.	Lantana
opulus	Guelder Rose	V.-obier, Boule-de-neige	Bach-Holder	Pallon di maggio
tinus	Laurustinus	Laurier-tin		Lauro tino
Vicia	Vetch, Tare	Vesce	Wicke	Veccia
faba	Broad Bean	Fève	Saubohne	Fava
sativa	Common Vetch	Poisette		Veccia buona
Vigna	Cherry-bean, Cow-pea	Mongette, Banette, Cornille		Fasulèin zal
Vinca	Periwinkle	Pervenche	Immergrün	Pervinca
Vincetoxicum		Dompte-venin	Schwalbenwurz	Vincetossico, Bozzoli
Viola	Violet, Pansy	Violette, Pensée	Veilchen, Stiefmütterchen	Viola
Viscum	Mistletoe	Gui	Mistel	Vischio
Vitaliana		Douglasia, Grégorie	Douglasie, Goldprimel, Rasenprimel	

GENUS	ENGLISH	FRENCH	GERMAN	ITALIAN
Vitex	Chaste Tree	Gattilier	Mönchspfeffer	Vitice, Agno casto
Vitis	Vine	Vigne	Weinrebe	Vite
Vulpia	Fescue	Vulpie	Fuchsschwingel	Paleo forasacco
Wahlenbergia	Ivy Campanula		Moorglocke	
Wolffia			Zwerglinse	
Xanthium		Lampourde, Petite bardane	Spitzklette	Lappola
strumarium	Cocklebur	Herbe-aux-écrouelles, Grappille		Lappola, Bardane minore
Xeranthemum		Immortelle	Spreublume	Perpetuini
Zannichellia	Horned Pondweed	Zannichellie	Teichfaden	Alga di fiume
Zea	Maize	Maïs, Blé de Turquie	Mais	Granoturco, Mais
Ziziphus	Jujube	Jujubier	Brustbeer	Giuggiolo, Zizzolo
Zostera	Eel-grass, Grass-wrack	Varech, Goémon	Seegras	Alga

COLOUR PLATES

(1) *Abies cephalonica* × $\frac{1}{5}$

2 *Picea abies* × $\frac{1}{2}$

4 *Pinus pinaster* × $\frac{1}{4}$

3 *Larix decidua* × $1\frac{1}{2}$

1

7 *Pinus nigra* × $\frac{1}{2}$ (*short-leaved form*)

6 *Pinus halepensis* × $\frac{1}{2}$

11 *Cupressus sempervirens* × ⅔

14 *Juniperus phoenicea* × ½

13 *Juniperus oxycedrus* × ⅘

12 *Juniperus communis* × ¾

2

17 *Taxus baccata* × ½

18 *Ephedra fragilis* subsp.
campylopoda × ¼

19 *Salix reticulata* $\times \frac{3}{4}$

20 *Salix retusa* $\times \frac{1}{2}$

22 *Salix hastata* ? $\times \frac{1}{3}$

25 *Salix alba* $\times \frac{4}{5}$

3

29 *Salix caprea* $\times \frac{1}{2}$

32 *Populus tremula* $\times \frac{2}{3}$

34 *Myrica gale* × ⁴⁄₅

35 *Juglans regia* × ⅓

36 *Betula pendula* × ½

38 *Betula nana* × ⅔

4

41 *Alnus incana* × ⅓

40 *Alnus glutinosa* × ⁴⁄₅

50 *Quercus macrolepis* × $\frac{1}{3}$

42 *Carpinus betulus* × $\frac{4}{5}$

52 *Quercus robur* × $\frac{1}{2}$

46 *Castanea sativa* × $\frac{1}{2}$

5

56 *Ulmus glabra* × $\frac{2}{3}$

47 *Quercus coccifera* × $\frac{1}{2}$

67 *Urtica pilulifera* × $\frac{4}{5}$

(66) *Urtica dubia* × $\frac{4}{5}$

64 *Cannabis sativa* × 1

6

(61) *Morus alba* × $\frac{1}{3}$

63 *Humulus lupulus* × $\frac{1}{3}$

71 *Thesium alpinum* × $\frac{2}{3}$

74 *Asarum europaeum* × ⅓

68 *Parietaria officinalis* × ⅓

70 *Osyris alba* × ⅘

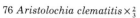

73 *Viscum album* × ⅓

7

76 *Aristolochia clematitis* × ⅔

78 *Cytinus hypocistis* × ½

77 *Aristolochia rotunda* × ⅘

83 *Polygonum amphibium* × ½

81 *Polygonum hydropiper* × ⅔

8

88 *Reynoutria japonica* × ⅔

(82) *Polygonum lapathifolium* × ½

85 *Polygonum viviparum* × 1½

97 *Rumex crispus* × ⅔

(98) *Rumex sanguineus* × ½

9

89 *Fagopyrum esculentum* × ½

96 *Rumex hydrolapathum* × ¼

94 *Rumex alpinus* × ⅒

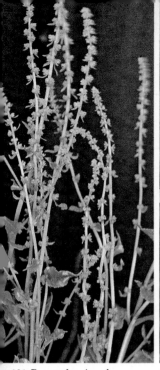

101 *Beta vulgaris subsp. maritima* × $\frac{1}{3}$

(105) *Chenopodium album* × $\frac{1}{2}$

103 *Chenopodium bonus-henricus* × $\frac{4}{5}$

10

(113) *Arthrocnemum fruticosum* × $\frac{1}{2}$

117 *Amaranthus retroflexus* ? × $\frac{1}{2}$

104 *Chenopodium foliosum* × 1

(124) *Montia sibirica* $\times \frac{2}{5}$ (120) *Carpobrotus acinaciformis* $\times \frac{2}{5}$

122 *Portulaca oleracea* $\times \frac{1}{2}$ 124 *Montia perfoliata* $\times 1$

11

(119) *Phytolacca sp.* $\times \frac{1}{2}$ 133 *Honkenya peploides* $\times \frac{2}{3}$

126 *Arenaria montana* × 1

136 *Stellaria holostea* × ½

138 *Stellaria graminea* × 1

142 *Cerastium alpinum* × 1

12

144 *Cerastium fontanum* × 1

145 *Myosoton aquaticum* × ½

(148) *Scleranthus annuus* × 1

(146) *Sagina procumbens* × 1

149 *Paronychia argentea* × ½

152 *Illecebrum verticillatum* × ⅓

13

157 *Spergularia rubra* × 1½

156 *Spergularia rupicola* × 1

159 *Lychnis flos-jovis* × ¾   162 *Lychnis alpina* × ½   161 *Lychnis viscaria* × ⅖

14

169 *Silene vulgaris subsp. maritima* × ½

160 *Lychnis flos-cuculi* × ½

167 *Silene nutans* × ⅘

(174) *Silene colorata* × 1

15

165 *Silene dioica* × ⅖

168 *Silene otites* × ½

177 *Cucubalus baccifer* × ⅗

170 *Silene acaulis* × $\frac{2}{3}$

163 *Agrostemma githago* × $\frac{4}{5}$

174 *Silene gallica var. quinquevulnera* × $\frac{2}{3}$

181 *Saponaria officinalis* × $\frac{1}{4}$

16

180 *Saponaria ocymoides* × $\frac{2}{3}$

178 *Gypsophila repens* × $\frac{4}{5}$

187 *Dianthus carthusianorum* × 1

194 *Dianthus sylvestris* × $\frac{1}{3}$

17

188 *Dianthus superbus* × $\frac{2}{3}$

190 *Dianthus monspessulanus* × $\frac{3}{4}$

186 *Dianthus armeria* × $1\frac{1}{2}$

196 *Nymphaea alba* × ⅓

(203) *Nigella sativa* × 1

197 *Nuphar lutea* × 1/10

202 *Eranthis hyemalis* × ⅔

18

203 *Nigella damascena* × 1¼

(200) *Helleborus cyclophyllus* × ⅙

201 *Helleborus niger* $\times \frac{1}{3}$

199 *Helleborus foetidus* $\times \frac{4}{5}$

19

204 *Trollius europaeus* $\times \frac{1}{10}$

210 *Aconitum napellus* $\times 1$

208 *Aconitum vulparia* $\times 1\frac{1}{4}$

(211) *Delphinium peregrinum* × 1

213 *Consolida ambigua* × 1⅕

211 *Delphinium elatum* × ⅘

20

(213) *Consolida regalis* × 1⅓

219 *Hepatica nobilis* × ⅘

217 *Anemone palmata* $\times \frac{2}{5}$ (216) *Anemone hortensis* $\times \frac{1}{2}$

(216) *Anemone pavonina* $\times \frac{1}{2}$ (214) *Anemone blanda* $\times 1$

21

216 *Anemone coronaria* $\times \frac{2}{3}$ 216 *Anemone coronaria* $\times \frac{1}{2}$

218 *Anemone narcissiflora* × ½

222 *Pulsatilla pratensis* × ⅔

214 *Anemone nemorosa* × ⅔

(214) *Anemone apennina* × ⅔

22

223 *Pulsatilla vulgaris* × ½

221 *Pulsatilla vernalis* × ½

227 *Clematis alpina* × $\frac{4}{5}$

220 *Pulsatilla alpina* subsp. *apiifolia* × $\frac{3}{4}$

220 *Pulsatilla alpina* × $\frac{1}{2}$

225 *Clematis flammula* × 1   229 *Clematis integrifolia* × ½   230 *Adonis annua* × $\frac{4}{5}$

24

224 *Clematis vitalba* × ½   231 *Adonis vernalis* × 1

233 *Ranunculus ficaria* × $\frac{4}{5}$

234 *Ranunculus thora* × $\frac{1}{2}$

241 *Ranunculus montanus* × $\frac{1}{2}$

244 *Ranunculus lingua* × $\frac{1}{2}$

25

246 *Ranunculus aconitifolius* × $\frac{2}{3}$

248 *Ranunculus parnassifolius* × 1

232 *Ranunculus muricatus* × 1

257 *Thalictrum flavum* × ½

250 *Ranunculus peltatus* × ½

(251) *Ranunculus pseudofluitans* × ½

26

(247) *Ranunculus glacialis* × ⅘

252 *Myosurus minimus* × 1¼

258 *Paeonia officinalis* × $\frac{2}{5}$

259 *Paeonia mascula* × $\frac{1}{4}$

254 *Aquilegia alpina* × $\frac{1}{2}$

253 *Aquilegia vulgaris* × $\frac{2}{3}$

27

262 *Mahonia aquifolium* × $\frac{1}{2}$

261 *Berberis vulgaris* × $\frac{4}{5}$

264 *Papaver somniferum* × ¼   269 *Meconopsis cambrica* × ¼   271 *Roemeria hybrida* × ½

28

267 *Papaver rhaeticum* × ⅕

(267) *Papaver sendtneri* × $\frac{1}{2}$

265 *Papaver rhoeas* × $\frac{1}{3}$

272 *Glaucium flavum* × $\frac{1}{6}$

273 *Glaucium corniculatum* × $\frac{1}{2}$

29

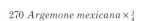

270 *Argemone mexicana* × $\frac{3}{4}$

275 *Eschscholzia californica* × $\frac{1}{3}$

(276) *Hypecoum imberbe* × $\frac{3}{4}$ 277 *Corydalis claviculata* × $\frac{5}{6}$

278 *Corydalis lutea* × $\frac{1}{2}$ (278) *Corydalis ochroleuca* × $\frac{2}{3}$

30

280 *Fumaria capreolata* × $\frac{1}{2}$ 279 *Corydalis solida* × $\frac{1}{2}$

289 *Alliaria petiolata* × 1

296 *Hesperis matronalis* × $\frac{1}{4}$

31

283 *Capparis spinosa* × 1

(287) *Sisymbrium orientale* × $\frac{1}{4}$

292 *Isatis tinctoria* × $\frac{1}{10}$

299 *Cheiranthus cheiri* × $\frac{1}{10}$

293 *Bunias erucago* × $\frac{4}{5}$

32

298 *Malcolmia maritima* × $1\frac{3}{4}$

300 *Matthiola incana* × $\frac{4}{5}$

(301) *Matthiola fruticulosa* $\times \frac{1}{3}$

303 *Rorippa amphibia* $\times \frac{4}{5}$

33

(309) *Cardamine enneaphyllos* $\times \frac{1}{2}$

301 *Matthiola sinuata* $\times \frac{2}{3}$

308 *Cardamine bulbifera* × ½

(309) *Cardamine pentaphyllos* × $\frac{1}{10}$

310 *Cardamine pratensis* × $1\frac{1}{3}$

(310) *Cardamine amara* × ½

34

331 *Draba aizoides* × $\frac{2}{3}$

318 *Arabis alpina* × $\frac{1}{3}$

314 *Cardaminopsis arenosa* × ⅓

(319) *Arabis verna* × 1

322 *Lunaria annua* × ⅓

327 *Fibigia clypeata* × ⅓

333 *Draba incana* × ⅔

343 *Thlaspi arvense* × ½

337 *Cochlearia danica* × 1

345 *Thlaspi rotundifolium* × $\frac{3}{4}$

346 *Aethionema saxatile* × $\frac{1}{2}$

348 *Iberis amara* × $1\frac{1}{4}$

36

353 *Cardaria draba* × $\frac{1}{2}$

354 *Coronopus Didymus* × $\frac{2}{3}$

329 *Lobularia maritima* × ¼

356 *Moricandia arvensis* × ¼

(362) *Sinapis alba* × ⅓

366 *Cakile maritima* × ⅓

37

363 *Eruca vesicaria* × 2

368 *Crambe maritima* × ⅛

374 *Reseda phyteuma* × ½ 　　372 Reseda lutea × ⅛ 　　　373 Reseda alba × 1½

38

(378) *Drosophyllum lusitanicum* × ¾ 　　　　377 *Drosera anglica* × ¾

378 *Drosera intermedia* × 1

382 *Sempervivum montanum* × 1

381 *Sempervivum arachnoideum* × $\frac{1}{3}$

386 *Aeonium arboreum* × $\frac{1}{4}$

39

379 *Umbilicus rupestris* × $\frac{1}{10}$

391 *Sedum album* × $\frac{1}{3}$

388 *Sedum reflexum* × $\frac{1}{4}$

389 *Sedum acre* × $\frac{3}{4}$

(394) *Sedum caeruleum* × $\frac{4}{5}$

394 *Sedum villosum* × 1

40

395 *Sedum telephium* subsp. *maximum* × 1

397 *Rhodiola rosea* × $\frac{1}{2}$

398 *Saxifraga oppositifolia* $\times \frac{1}{3}$

(399) *Saxifraga longifolia* $\times \frac{1}{4}$

402 *Saxifraga aizoides* $\times \frac{1}{3}$

(411) *Saxifraga tridactylites* $\times 1$

41

409 *Saxifraga androsacea* $\times 1$

407 *Saxifraga stellaris* $\times \frac{1}{2}$

400 *Saxifraga paniculata* × ⅓

403 *Saxifraga rotundifolia* × ⅓

411 *Saxifraga moschata* × 1

(405) *Saxifraga hirsuta* × ⅓

42

(412) *Chrysosplenium oppositifolium* × ½

413 *Parnassia palustris* × ½

(414) *Ribes alpinum* × ½

414 *Ribes rubrum* × ½

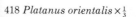
416 *Ribes uva-crispa* × 1

43

418 *Platanus orientalis* × ⅓

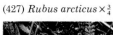
(427) *Rubus arcticus* × ¾

426 *Rubus chamaemorus* $\times \frac{1}{2}$

(429) *Rubus caesius* $\times \frac{2}{3}$

421 *Filipendula ulmaria* $\times \frac{1}{6}$

422 *Filipendula vulgaris* $\times \frac{1}{2}$

44

419 *Spiraea salicifolia* $\times \frac{1}{2}$

420 *Aruncus dioicus* $\times \frac{1}{10}$

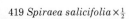

430 *Rosa arvensis* × ½

434 *Rosa pimpinellifolia* × ⅔

435 *Rosa pendulina* × ⅓

441 *Dryas octopetala* × ⅔

45

438 *Sanguisorba officinalis* × ½

443 *Geum rivale* × ⅓

(445) *Potentilla nitida* × ⅔

440 *Sarcopoterium spinosum* × ⅔

446 *Potentilla palustris* × 1¼

448 *Potentilla rupestris* × ⅔

46

457 *Potentilla fruticosa* × 1

451 *Potentilla tabernaemontani* × ⅔

467 *Sorbus aria* × 1

463 *Cydonia oblonga* × $\frac{1}{2}$

47

483 *Prunus laurocerasus* × $\frac{4}{5}$

471 *Amelanchier ovalis* × $\frac{1}{3}$

(483) *Prunus lusitanica* × ½

48

(477) *Prunus cerasifera* × 1

466 *Sorbus aucuparia* × ⅓

470 *Eriobotrya japonica* × ¼

473 *Cotoneaster integerrimus* × ¾

482 *Prunus padus* × ⅓

480 *Prunus avium* × 1⅓

476 *Prunus spinosa* × ⅘

484 *Prunus mahaleb* × ⅘

49

486 *Ceratonia siliqua* × ⅓

488 *Acacia longifolia* × ⅓

496 *Laburnum anagyroides* × 1

493 *Albizia julibrissin* × $\frac{1}{2}$

50

485 *Cercis siliquastrum* × $\frac{4}{5}$

498 *Calicotome villosa* × 1

504 *Cytisus sessilifolius* × $\frac{1}{2}$

505 *Cytisus scoparius* × $\frac{1}{2}$

499 *Lembotropis nigricans* × $\frac{1}{2}$

51

506 *Chamaecytisus hirsutus* × $\frac{1}{2}$

(506) *Chamaecytisus purpureus* × $\frac{1}{3}$

(511) *Genista hispanica* × ¾

514 *Lygos monosperma* × ¼

52

515 *Spartium junceum* × ⅘

513 *Chamaespartium sagittale* × ½

(515) *Erinacea anthyllis* × ½

517 *Ulex minor* × ½   519 *Lupinus luteus* × ⅔   525 *Colutea arborescens* × ⅓

53

523 *Robinia pseudacacia* × ½   524 *Galega officinalis* × ⅓

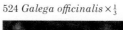

527 *Astragalus glycyphyllos* × ½

520 *Lupinus angustifolius* × ⅖

530 *Astragalus alpinus* × ¾

521 *Lupinus albus* × ⅓

54

537 *Psoralea bituminosa* × ⅖

528 *Astragalus cicer* × ⅔

(545) *Vicia benghalensis* × 1

549 *Vicia sativa* × $\frac{2}{3}$

(545) *Vicia villosa* × $\frac{2}{3}$

55

518 *Adenocarpus complicatus* × $\frac{4}{5}$

545 *Vicia cracca* × $\frac{1}{3}$

553 *Lathyrus aphaca* × $\frac{3}{4}$

554 *Lathyrus ochrus* × $\frac{1}{3}$

556 *Lathyrus nissolia* × $\frac{2}{5}$

559 *Lathyrus tuberosus* × ⅔

565 *Lathyrus japonicus* × ½

561 *Lathyrus clymenum* × ½

56

563 *Lathyrus niger* × ⅓

(560) *Lathyrus latifolius* × ½

(557) *Lathyrus sativus* × ¾

564 *Lathyrus vernus* × ⅘

566 *Pisum sativum* × ⅓

571 *Ononis natrix* × ⅔

(569) *Ononis fruticosa* × ⅗

570 *Ononis rotundifolia* × ⅗

573 *Melilotus alba* × 1/10

57

576 *Melilotus altissima* × ½

577 *Trigonella caerulea* × ⅓

580 *Trigonella foenum-graecum* × ½

589 *Medicago marina* × ⅘

583 *Medicago sativa* subsp. *sativa* × ⅓

582 *Medicago sativa* subsp. *falcata* × ⅖

58

588 *Medicago polymorpha* × 1⅓

584 *Medicago orbicularis* × ⅘

597 *Trifolium rubens* × 1

601 *Trifolium hybridum* × ½

(601) *Trifolium montanum* × 1¼

59

595 *Trifolium incarnatum* × ⅓

607 *Trifolium medium* × ⅔

(636) *Onobrychis montana* × ⅘

594 *Trifolium arvense* × ⅔

593 *Trifolium badium* × ¾

602 *Trifolium stellatum* × ¾

603 *Trifolium alpinum* × 1

60

610 *Dorycnium rectum* × 1

609 *Dorycnium hirsutum* × ¾

614 *Lotus uliginosus* × ½

621 *Anthyllis montana* × ⅔

617 *Tetragonolobus purpureus* × 1

622 *Anthyllis vulneraria* × ⅓

61

635 *Hedysarum hedysaroides* × ½   623 *Anthyllis tetraphylla* × 1

624 *Coronilla emerus* $\times \frac{1}{4}$

62

634 *Hedysarum coronarium* $\times \frac{1}{3}$

627 *Coronilla varia* $\times \frac{1}{2}$

(634) *Hedysarum glomeratum* $\times \frac{3}{5}$

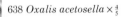

638 *Oxalis acetosella* $\times \frac{4}{5}$

639 *Oxalis pes-caprae* $\times \frac{1}{3}$

641 *Geranium sanguineum* $\times \frac{3}{4}$

642 *Geranium pyrenaicum* $\times \frac{2}{3}$

(642) *Geranium nodosum* $\times \frac{4}{5}$

63

645 *Geranium sylvaticum* $\times \frac{2}{3}$

651 *Geranium lucidum* $\times \frac{1}{2}$

646 *Geranium phaeum* × 1¼

647 *Geranium macrorrhizum* × 1⅓

64

(652) *Erodium gruinum* × 1¼

652 *Erodium malacoides* × ⅘

655 *Tribulus terrestris* × ⅓

(664) *Linum suffruticosum* × $1\frac{1}{4}$

659 *Linum perenne* × $\frac{2}{3}$

666 *Mercurialis perennis* × $\frac{1}{2}$

65

668 *Ricinus communis* × $\frac{1}{2}$

665 *Chrozophora tinctoria* × $\frac{1}{2}$

678 *Euphorbia characias* subsp. *wulfenii* $\times \frac{1}{15}$

678 *Euphorbia characias* $\times \frac{1}{2}$

(670) *Euphorbia acanthothamnos* $\times \frac{1}{6}$

669 *Euphorbia dendroides* $\times \frac{1}{30}$

66

684 *Euphorbia paralias* $\times 1\frac{1}{2}$

677 *Euphorbia amygdaloides* $\times \frac{1}{2}$

693 *Citrus İimon* × $\frac{1}{3}$

671 *Euphorbia helioscopia* × $\frac{2}{3}$

686 *Ruta graveolens* × $\frac{2}{5}$

690 *Citrus sinensis* × $\frac{1}{3}$

67

688 *Dictamnus albus* × $\frac{1}{6}$

676 *Euphorbia lathyris* × $\frac{1}{2}$

694 *Ailanthus altissima* × ¼

702 *Coriaria myrtifolia* × ⅓

696 *Polygala chamaebuxus* × ⅘

(700) *Polygala nicaeensis* × ¾

68

698 *Polygala vulgaris* × 1

703 *Pistacia lentiscus* × ⅘

706 *Cotinus coggygria* $\times \frac{1}{2}$

704 *Pistacia terebinthus* $\times \frac{1}{3}$

713 *Aesculus hippocastanum* $\times \frac{1}{5}$

710 *Acer platanoides* $\times \frac{1}{2}$

69

(713) *Aesculus carnea* $\times \frac{1}{3}$

708 *Acer pseudoplatanus* $\times \frac{4}{5}$

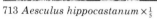

716 *Impatiens glandulifera* × ⅓

714 *Impatiens noli-tangere* × ⅘

70

715 *Impatiens parviflora* × ½

(714) *Impatiens capensis* × 1¼

720 *Rhamnus alaternus* $\times \frac{1}{6}$

723 *Rhamnus catharticus* $\times \frac{4}{5}$

717 *Ilex aquifolium* $\times \frac{2}{3}$

718 *Euonymus europaeus* $\times \frac{3}{4}$

71

718 *Euonymus europaeus* $\times \frac{2}{3}$

735 *Malva alcea* × ½

72

740 *Lavatera arborea* × 1

747 *Althaea officinalis* × ⅓

748 *Alcea pallida* × ⅕

742 *Lavatera trimestris* × ½

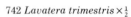

738 *Malva nicaeensis* × ⅓

752 *Hibiscus trionum* × ⅘

750 *Gossypium herbaceum* × ½

73

753 *Thymelaea hirsuta* × ½

754 *Thymelaea tartonraira* × 1

759 *Daphne mezereum* × $1\frac{1}{4}$

756 *Daphne cneorum* × $\frac{3}{4}$

74

760 *Daphne laureola* × $\frac{1}{2}$

761 *Hippophaë rhamnoides* × $\frac{4}{5}$

763 *Hypericum androsaemum* × $\frac{1}{2}$

(763) *Hypericum calycinum* $\times \frac{1}{2}$

768 *Hypericum perforatum* $\times \frac{1}{2}$

765 *Hypericum hirsutum* $\times \frac{4}{5}$

764 *Hypericum montanum* $\times \frac{1}{3}$

75

766 *Hypericum elodes* $\times \frac{1}{2}$

783 *Viola tricolor* $\times \frac{3}{5}$

782 *Viola biflora* × ½

773 *Viola palustris* × 3

774 *Viola odorata* × ½

777 *Viola riviniana* × 3

76

786 *Viola calcarata* × ⅔

785 *Viola lutea* × ⅘

790 *Cistus salvifolius* $\times \frac{3}{4}$

791 *Cistus monspeliensis* $\times \frac{1}{2}$

787 *Cistus incanus* $\times \frac{1}{4}$

793 *Cistus ladanifer* $\times \frac{2}{3}$

77

794 *Cistus laurifolius* $\times \frac{1}{2}$

788 *Cistus albidus* $\times \frac{3}{5}$

798 *Tuberaria guttata* × 1

803 *Helianthemum apenninum* × $\frac{4}{5}$

802 *Helianthemum nummularium* × $\frac{4}{5}$

796 *Halimium commutatum* × 1

78

807 *Tamarix africana* × $\frac{1}{2}$

(812) *Citrullus colocynthis* × $\frac{1}{4}$

811 *Ecballium elaterium* $\times \frac{1}{4}$

818 *Opuntia ficus-indica* $\times \frac{1}{10}$

79

820 *Lythrum salicaria* $\times \frac{1}{3}$

823 *Trapa natans* $\times \frac{1}{3}$

815 *Bryonia cretica* $\times \frac{4}{5}$

821 *Lythrum virgatum* $\times \frac{1}{2}$ 829 *Circaea lutetiana* $\times 1\frac{1}{4}$ 835 *Epilobium angustifolium* $\times \frac{3}{5}$

80

827 *Punica granatum* $\times \frac{1}{2}$

819 *Lythrum portula* × ¼

824 *Myrtus communis* × ⅔

831 *Oenothera biennis* × ⅓

834 *Oenothera rosea* × ⅔

81

837 *Epilobium hirsutum* × ½

839 *Epilobium montanum* × ⅘

845 *Cornus mas* $\times \frac{4}{5}$

82

848 *Hedera helix* $\times \frac{1}{4}$

847 *Cornus suecica* $\times 1$

846 *Cornus sanguinea* $\times \frac{2}{5}$

(843) *Myriophyllum spicatum* $\times \frac{3}{5}$

(847) *Aucuba japonica* × ½

851 *Hacquetia epipactis* × ½

852 *Astrantia major* × ½

855 *Eryngium campestre* × ⅘

83

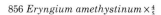
856 *Eryngium amethystinum* × ⅘

853 *Eryngium maritimum* × 1

863 *Myrrhis odorata* × ¼

866 *Smyrnium perfoliatum* × ⅕

871 *Crithmum maritimum* × ⅙

872 *Oenanthe crocata* × ⅙

84

873 *Oenanthe fistulosa* × ½

(883) *Bupleurum stellatum* × 1

886 *Apium nodiflorum* $\times \frac{1}{2}$

897 *Ferula communis* $\times \frac{1}{10}$

85

895 *Angelica archangelica* $\times \frac{1}{12}$

879 *Conium maculatum* $\times \frac{1}{12}$

892 *Ligusticum scoticum* $\times \frac{1}{8}$

902 *Heracleum mantegazzianum* $\times \frac{1}{10}$

900 *Pastinaca sativa* $\times \frac{1}{10}$

904 *Tordylium apulum* $\times 1$

917 *Monotropa hypopitys* $\times \frac{2}{3}$

86

914 *Orthilia secunda* $\times \frac{2}{3}$

912 *Pyrola minor* $\times \frac{2}{3}$

910 *Daucus carota* $\times \frac{1}{2}$

(912) *Pyrola media* $\times \frac{3}{4}$

911 *Diapensia lapponica* $\times 1\frac{3}{4}$

915 *Moneses uniflora* $\times \frac{2}{3}$

919 *Rhododendron ferrugineum* $\times \frac{1}{3}$

87

(919) *Rhododendron hirsutum* $\times \frac{1}{10}$

921 *Loiseleuria procumbens* $\times 2$

922 *Daboecia cantabrica* × ⅘   (922) *Phyllodoce caerulea* × ⅘   (925) *Arctostaphylos alpina* × 2

920 *Rhododendron ponticum* × ⅙

923 *Andromeda polifolia* × ⅘

88

925 *Arctostaphylos uva-ursi* × ⅘

937 *Vaccinium oxycoccos* × 1

935 *Vaccinium vitis-idaea* × ⅔

936 *Vaccinium myrtillus* × ⅘

(927) *Erica lusitanica* × 1

927 *Erica arborea* × 1/20

89

933 *Erica herbacea* × ⅗

934 *Erica multiflora* × ½

945 *Primula viscosa* × ½

924 *Arbutus unedo* × ½

929 *Erica ciliaris* × ⅔

940 *Primula elatior* × ½

942 *Primula vulgaris* × ½

90

941 *Primula auricula* × ½

953 *Androsace carnea* × ⅔

943 *Primula farinosa* × 1⅓   (944) *Primula minima* × ½

(945) *Primula hirsuta* × ⅔   946 *Primula integrifolia* × ⅔

(956) *Soldanella pusilla* × ¾   949 *Androsace alpina* × 1   955 *Cortusa matthioli* × 1

957 *Hottonia palustris* × $\frac{1}{8}$

960 *Cyclamen repandum* × $\frac{2}{3}$

(959) *Cyclamen graecum* × $\frac{1}{2}$

958 *Cyclamen hederifolium* × $\frac{2}{3}$

92

959 *Cyclamen purpurascens* × $\frac{4}{5}$

968 *Glaux maritima* × 1

965 *Trientalis europaea* × 1

(963) *Lysimachia punctata* × ⅓

93

(967) *Anagallis linifolia* × 1½

962 *Lysimachia nemorum* × 1

961 *Lysimachia nummularia* × ⅓

972 *Limonium sinuatum* $\times \frac{1}{3}$

975 *Armeria maritima* $\times \frac{1}{8}$

(976) *Armeria fasciculata* $\times \frac{1}{6}$

985 *Olea europaea* $\times \frac{1}{2}$

94

979 *Fraxinus ornus* $\times \frac{1}{3}$

988 *Blackstonia perfoliata* $\times 1\frac{1}{4}$

986 *Centaurium erythraea* × ⅘

984 *Jasminum officinale* × ⅘

1002 *Swertia perennis* × ⅔

95

996 *Gentiana lutea* × ⅖

983 *Jasminum fruticans* × 1

(992) *Gentiana clusii* $\times \frac{1}{3}$ 992 *Gentiana kochiana* $\times \frac{2}{3}$

991 *Gentiana verna* $\times 1$ (990) *Gentiana utriculosa* $\times \frac{2}{3}$ 989 *Gentiana cruciata* $\times \frac{1}{2}$

96

993 *Gentiana pneumonanthe* $\times \frac{2}{3}$ 999 *Gentianella campestris* $\times \frac{3}{4}$

998 *Gentiana purpurea* × ½

1004 *Nymphoides peltata* × ⅓

997 *Gentiana punctata* × ½

994 *Gentiana asclepiadea* × ⅔

1005 *Vinca minor* × ½

97

(1005) *Vinca herbacea* × ⅖

1009 *Periploca graeca* × ⅔

1003 *Menyanthes trifoliata* × ½

98

1007 *Nerium oleander* × ⅙

1013 *Putoria calabrica* × ⅖

1030 *Valantia hispida* × 1½

1031 *Rubia peregrina* × ⅔

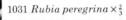

1011 *Asclepias syriaca* × $\frac{2}{3}$ 1010 *Vincetoxicum hirundinaria* × $\frac{1}{2}$ 1014 *Sherardia arvensis* × 1

1027 *Galium odoratum* × $\frac{3}{5}$ 1029 *Cruciata laevipes* × $1\frac{1}{4}$ (1022) *Galium palustre* × $\frac{1}{2}$

1039 *Convolvulus althaeoides* × 1

1032 *Polemonium caeruleum* × ½

(1039) *Convolvulus elegantissimus* × ½

1035 *Convolvulus tricolor* × ½

100

1041 *Calystegia soldanella* × 1

(1033) *Ipomoea hederacea* × ½

1044 *Cuscuta epithymum* × $1\frac{1}{4}$

1045 *Heliotropium europaeum* × $\frac{1}{3}$

101

1047 *Omphalodes verna* × $\frac{4}{5}$

(1053) *Symphytum orientale* × $\frac{1}{3}$

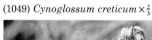

(1049) *Cynoglossum creticum* × $\frac{2}{3}$

1053 *Symphytum tuberosum* × ⅖

(1052) *Symphytum* × *uplandicum* × ⅗

102

1058 *Pentaglottis sempervirens* × ⅗

1049 *Cynoglossum officinale* × ½

1056 *Anchusa azurea* × ⅙

1054 *Borago officinalis* × ¼

1055 *Anchusa officinalis* × ½

(1054) *Trachystemon orientalis* × ½

1061 *Alkanna tinctoria* × 1¼

1067 *Myosotis alpestris* × 1

1062 *Mertensia maritima* × ⅗

1065 *Myosotis scorpioides* × 1

1064 *Pulmonaria longifolia* × 4/5

104

1070 *Eritrichium nanum* × 1

1074 *Lithospermum diffusum* × 1

1073 *Lithospermum purpurocaeruleum* × 2/3

1082 *Echium vulgare* × 1/8

1072 *Lithospermum officinale* × $\frac{2}{5}$

1076 *Onosma echioides* × $\frac{3}{4}$

1081 *Echium italicum* × 1

105

1090 *Ajuga genevensis* × $\frac{1}{2}$

1079 *Cerinthe major* × $\frac{1}{3}$

1078 *Cerinthe minor* × $\frac{1}{2}$

1088 *Callitriche stagnalis* × 1

1083 *Echium lycopsis* × ½

106

1087 *Vitex agnus-castus* × ⅓

1091 *Ajuga pyramidalis* × ½

1094 *Ajuga chamaepitys* × ⅔

1089 *Ajuga reptans* × 1¼

1099 *Teucrium chamaedrys* × ½

1103 *Teucrium montanum* × ⅔

1101 *Teucrium fruticans* × ⅘

1105 *Rosmarinus officinalis* × ⅓

107

1096 *Teucrium pseudochamaepitys* × ¾

1106 *Prasium majus* × 1⅔

1107 *Scutellaria alpina* × ⅔ (1107) *Scutellaria orientalis* × 1 1108 *Scutellaria galericulata* × ⅔

108

1110 *Lavandula staechas* × $\frac{1}{15}$

1116 *Nepeta cataria* × $\frac{1}{3}$

1118 *Glechoma hederacea* × $\frac{2}{3}$

1115 *Sideritis hyssopifolia* × 1

109

1121 *Melittis melissophyllum* × $\frac{1}{2}$

1112 *Marrubium vulgare* × $\frac{1}{3}$

1123 *Phlomis lychnitis* × ⅓   (1122) *Phlomis tuberosa* × ½   1120 *Prunella laciniata* × ⅔

110

1124 *Phlomis fruticosa* × 1/10   (1120) *Prunella grandiflora* × ½

(1126) *Galeopsis speciosa* $\times \frac{2}{3}$

(1131) *Lamium orvala* $\times \frac{3}{4}$

1126 *Galeopsis tetrahit* $\times 1$

1130 *Lamium maculatum* $\times \frac{4}{5}$

111

1127 *Lamium amplexicaule* $\times \frac{3}{4}$

1132 *Galeobdolon luteum* × ⅔   1133 *Leonurus cardiaca* × ⅖   1138 *Stachys palustris* × 1

112

1134 *Ballota nigra* × ⅓

1140 *Stachys germanica* × ⅕

1146 *Salvia glutinosa* × 1

1136 *Stachys recta* × $1\frac{1}{3}$

(1143) *Salvia triloba* × $\frac{1}{5}$

(1144) *Salvia argentea* × $1\frac{1}{2}$

113

1144 *Salvia sclarea* × $\frac{1}{2}$

1151 *Horminum pyrenaicum* $\times \frac{1}{2}$

114

1147 *Salvia pratensis* $\times 1\frac{1}{4}$

1156 *Calamintha nepeta* $\times \frac{2}{3}$

1154 *Calamintha grandiflora* $\times 1$

(1157) *Acinos alpinus* $\times \frac{4}{5}$

1149 *Salvia horminum* × $\frac{4}{5}$

1158 *Clinopodium vulgare* × $\frac{1}{3}$

1159 *Hyssopus officinalis* × $\frac{2}{5}$

1160 *Origanum vulgare* × $\frac{4}{5}$

115

1163 *Thymus vulgaris* × $\frac{1}{2}$

1164 *Thymus serpyllum* × $\frac{4}{5}$

1169 *Mentha aquatica* $\times \frac{1}{2}$

1171 *Mentha longifolia* $\times \frac{1}{3}$

116

1166 *Lycopus europaeus* $\times \frac{1}{5}$

1167 *Mentha pulegium* $\times \frac{1}{3}$

1175 *Scopolia carniolica* $\times \frac{3}{5}$

1176 *Hyoscyamus niger* × $1\frac{1}{3}$

1177 *Hyoscyamus albus* × $\frac{3}{4}$

1174 *Atropa bella-donna* × $\frac{3}{5}$

(1177) *Hyoscyamus aureus* × $\frac{1}{2}$

117

1174 *Atropa bella-donna* × $\frac{4}{5}$

1185 *Mandragora officinarum* $\times \frac{3}{4}$

(1186) *Datura metel* $\times \frac{1}{3}$

118

1186 *Datura stramonium* $\times \frac{4}{5}$

1180 *Solanum sodomeum* $\times \frac{2}{5}$

1186 *Datura stramonium* $\times \frac{2}{3}$

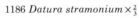

1178 *Physalis alkekengi* × ⅕

1187 *Nicotiana glauca* × ⅔

1188 *Nicotiana rustica* × ½

(1188) *Nicotiana tabacum* × ⅓

119

1179 *Nicandra physalodes* × ⅘

1181 *Solanum dulcamara* × ½

1192 *Verbascum blattaria* × ½   1196 *Verbascum creticum* × ¼   1190 *Verbascum nigrum* × 1

120

(1193) *Verbascum thapsiforme* × $\frac{1}{10}$ (1195) *Verbascum undulatum* × $\frac{1}{5}$

1200 *Asarina procumbens* × 1¼

1198 *Antirrhinum latifolium* × ½

(1192) *Verbascum phoeniceum* × ⅗

1189 *Buddleja davidii* × ¼

121

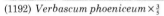

1198 *Antirrhinum latifolium* × ⅓   (1205) *Linaria dalmatica* × ⅔

1202 *Linaria repens* × 1¼

1199 *Antirrhinum orontium* × ⅔

1204 *Linaria alpina* × 1

(1204) *Linaria triornithophora* × ½₂

122

1210 *Cymbalaria muralis* × ½

1208 *Linaria triphylla* × ⅔

1215 *Scrophularia scorodonia* × 4/5

(1216) *Scrophularia hoppii* × 1/4

123

(1217) *Mimulus moschatus* × 4/5

1211 *Anarrhinum bellidifolium* × 1/3

1217 *Mimulus guttatus* × 1/2

1218 *Gratiola officinalis* × $\frac{4}{5}$

124

(1220) *Veronica fruticans* × 2

1223 *Veronica persica* × $\frac{1}{2}$

1224 *Veronica filiformis* × $1\frac{1}{4}$

1225 *Veronica beccabunga* × $\frac{2}{3}$

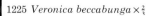

1230 *Digitalis ferruginea* × $\frac{1}{2}$

1232 *Digitalis grandiflora* × $\frac{4}{5}$

1233 *Digitalis lutea* × $\frac{2}{3}$

125

(1233) *Digitalis obscura* × $\frac{2}{3}$

1234 *Digitalis purpurea* × $\frac{1}{4}$

1238 *Parentucellia viscosa* × 1   1237 *Bellardia trixago* × ¾   1241 *Odontites verna* × ⅓

126

1236 *Bartsia alpina* × 1   1239 *Parentucellia latifolia* × 1¼

1235 *Erinus alpinus* × $\frac{1}{2}$

1242 *Euphrasia rostkoviana* × $\frac{2}{3}$

(1244) *Euphrasia minima* × $1\frac{1}{3}$

1247 *Rhinanthus minor* × $1\frac{1}{4}$

127

1252 *Pedicularis foliosa* × $\frac{1}{2}$

1250 *Pedicularis tuberosa* × $\frac{1}{3}$

1258 *Melampyrum arvense* × 1

128

1259 *Melampyrum nemorosum* × ¾

1254 *Pedicularis recutita* × ⅓

1255 *Pedicularis kerneri* × ⅔

(1256) *Pedicularis sylvatica* × ⅘

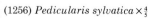

1261 *Tozzia alpina* × 2

1262 *Globularia alypum* × ⅕

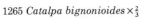

1264 *Globularia vulgaris* × ⅔

129

1265 *Catalpa bignonioides* × ⅔

1263 *Globularia cordifolia* × ½

1266 *Acanthus mollis* × ¼

1267 *Acanthus spinosus* × ⅙

130

1269 *Lathraea squamaria* × ⅔

(1269) *Lathraea clandestina* × 1

1268 *Ramonda myconi* × 1

1274 *Orobanche crenata* × ¼

1275 *Orobanche caryophyllacea?* × 1

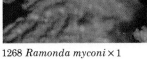

1277 *Orobanche minor* × ⅔

1272 *Orobanche rapum-genistae* × ⅓

(1280) *Pinguicula grandiflora* × $\frac{1}{2}$

132

1281 *Utricularia vulgaris* × $1\frac{1}{4}$

1279 *Pinguicula alpina* × 1

1280 *Pinguicula vulgaris* × 1

1287 *Plantago maritima* × $\frac{1}{6}$

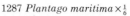

1292 *Plantago media* × 1¼

1285 *Plantago coronopus* × ¼

1298 *Viburnum opulus* × ½

1283 *Plantago indica* × 1

133

1295 *Sambucus ebulus* × ½

1295 *Sambucus ebulus* × ⅓

1299 *Viburnum lantana* × $\frac{1}{2}$

134

1300 *Viburnum tinus* × 1

1299 *Viburnum lantana* × $\frac{1}{2}$

1297 *Sambucus racemosa* × $\frac{1}{3}$

1297 *Sambucus racemosa* × $\frac{1}{3}$

1305 *Lonicera etrusca* × $\frac{2}{3}$

(1303) *Lonicera caerulea* × 1

1302 *Lonicera xylosteum* × $\frac{1}{2}$

1307 *Linnaea borealis* × $1\frac{1}{3}$

1301 *Symphoricarpos rivularis* × $\frac{2}{3}$ 1304 *Lonicera periclymenum* × $\frac{2}{3}$

(1315) *Valeriana montana* × 1

136

1316 *Centranthus ruber* × ⅔

1309 *Valerianella locusta* × ⅖

1315 *Valeriana tripteris* × ⅓

1313 *Valeriana officinalis* × ½

(1316) *Centranthus angustifolius* × $\frac{4}{5}$

(1318) *Dipsacus laciniatus* × $\frac{2}{3}$

137

1327 *Scabiosa ochroleuca* × $\frac{4}{5}$

1308 *Adoxa moschatellina* × $1\frac{1}{3}$

1312 *Fedia cornucopiae* × $1\frac{1}{5}$

1322 *Knautia arvensis* × 1⅓

1326 *Scabiosa atropurpurea* × 1

(1325) *Scabiosa lucida* × 1

1323 *Knautia sylvatica* × ⅓

138

1333 *Campanula glomerata* × ⅔

(1332) *Campanula spicata* × ⅓

1330 *Campanula barbata* × ¾

1332 *Campanula thyrsoides* × ¼

1340 *Campanula trachelium* × ⅘

1336 *Campanula persicifolia* × ⅔

(1339) *Campanula scheuchzeri* × ⅔

1338 *Campanula cochleariifolia* × ⅘

1342 *Campanula rapunculoides* × ⅔

1343 *Campanula bononiensis* × 1

140

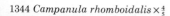

1344 *Campanula rhomboidalis* × ⅘

(1352) *Phyteuma comosum* × 1

(1350) *Phyteuma betonicifolium* × 1

1350 *Phyteuma spicatum* × $\frac{1}{10}$

1347 *Legousia speculum-veneris* × $\frac{4}{5}$

1352 *Phyteuma orbiculare* × $1\frac{3}{4}$

141

1355 *Jasione montana* × 1

1357 *Eupatorium cannabinum* × $\frac{1}{5}$

(1359) *Solidago gigantea* × ⅓

142

1365 *Aster tripolium* × 1

1363 *Aster alpinus* × ⅖

1362 *Bellidastrum michelii* × ⅓

1364 *Aster amellus* × ½

1374 *Evax pygmaea* × $\frac{4}{5}$

1358 *Solidago virgaurea* × $\frac{2}{3}$

143

1375 *Filago vulgaris* × 1

1369 *Erigeron acer* × $\frac{2}{3}$

1378 *Antennaria dioica* × $\frac{1}{2}$

1388 *Inula helenium* × $\frac{1}{3}$

144

1387 *Inula conyza* × $\frac{1}{3}$

1379 *Leontopodium alpinum* × $\frac{4}{5}$

1385 *Helichrysum stoechas* × $\frac{1}{6}$

1380 *Gnaphalium uliginosum* × $\frac{2}{3}$

1390 *Inula crithmoides* × $\frac{1}{10}$

(1391) *Inula britannica* × $\frac{1}{3}$

1386 *Phagnalon rupestre* × $\frac{1}{3}$

1393 *Pulicaria dysenterica* × $\frac{1}{2}$

145

1395 *Pallenis spinosa* × $\frac{1}{2}$

1397 *Telekia speciosa* × $\frac{1}{3}$

1398 *Asteriscus maritimus* × ⅔

1399 *Asteriscus aquaticus* × ½

1402 *Xanthium spinosum* × ⅔

1401 *Xanthium strumarium* × ½

146

1403 *Rudbeckia laciniata* × ⅓

1406 *Bidens cernua var. radiata* × ½

1419 *Achillea nana* × 1

1413 *Chamaemelum nobile* × ½

1407 *Galinsoga parviflora* × ⅘

147

1409 *Anthemis tinctoria* × ⅓

1421 *Achillea tomentosa* × ½

1426 *Chrysanthemum vulgare* × ⅕

(1424) *Chrysanthemum myconis* × $\frac{1}{3}$

1425 *Chrysanthemum coronarium* × $\frac{1}{12}$

148

1427 *Chrysanthemum leucanthemum* × $\frac{1}{4}$

1429 *Chrysanthemum parthenium* × $\frac{1}{6}$

1432 *Matricaria matricarioides* × $\frac{1}{2}$

1433 *Cotula coronopifolia* × 1

1439 *Tussilago farfara* × ½

1436 *Artemisia absinthium* × ⅓

1440 *Petasites hybridus* × ½

149

1441 *Petasites albus* × ⅔

1443 *Homogyne alpina* × 1⅕

1445 *Adenostyles alliariae* × ½

150

(1456) *Senecio nemorensis* × ¼

1448 *Doronicum grandiflorum* × ¼

(1452) *Senecio incanus* × ⅖

1447 *Doronicum pardalianches* × ⅓

1460 *Calendula arvensis* × ⅓

1464 *Xeranthemum annuum* × ½

1452 *Senecio vernalis* × ⅔

1446 *Arnica montana* × ¼

151

1455 *Senecio doronicum* × ¼

1467 *Carlina acaulis* × ⅓

1462 *Echinops ritro* × ⅓

(1475) *Saussurea alpina* × 1

152

1472 *Arctium tomentosum* × ⅔

1469 *Carlina corymbosa* × ⅓

1478 *Carduus nutans* × ½

1479 *Carduus personata* × ¼

1481 *Notobasis syriaca* × ⅔

(1484) *Cirsium candelabrum* × ⅛

153

1488 *Cirsium acarna* × ½

(1489) *Cirsium tuberosum* × ⅔

1483 *Cirsium spinosissimum* × ⅕

1482 *Cirsium oleraceum* × ⅓

(1491) *Cynara scolymus* × ¼

1485 *Cirsium eriophorum* × ⅓

154

1490 *Cirsium acaulon* × ½

1493 *Galactites tomentosa* × ⅔

1494 *Onopordum acanthium* × $\frac{4}{5}$

1492 *Silybum marianum* × $\frac{3}{4}$

1499 *Centaurea solstitialis* × $\frac{4}{5}$

1495 *Onopordum illyricum* × $\frac{2}{3}$

155

1497 *Serratula tinctoria* × $1\frac{1}{5}$

1501 *Centaurea cyanus* × 2

1500 *Centaurea calcitrapa* × $\frac{2}{3}$

1503 *Centaurea salonitana* × $\frac{1}{2}$

1506 *Centaurea rhapontica* × $\frac{4}{5}$

156

(1504) *Centaurea phrygia* × $\frac{1}{2}$

1504 *Centaurea nervosa* × $\frac{2}{5}$

(1526) *Urospermum dalechampii* × $\frac{2}{3}$ 1507 *Centaurea conifera* × $\frac{1}{2}$

1508 *Carthamus lanatus* × $\frac{2}{5}$ (1510) *Scolymus maculatus* × $\frac{1}{2}$

157

1511 *Catananche coerulea* × $\frac{2}{3}$ 1515 *Tolpis barbata* × $\frac{4}{5}$

1512 *Cichorium intybus* × ½

158

1521 *Hypochoeris uniflora* × ⅓

1519 *Hedypnois rhagadioloides* × ½

1528 *Tragopogon porrifolius* × 1¼

1524 *Leontodon hispidus* × 1⅓

1538 *Sonchus arvensis* × ½

1533 *Andryala integrifolia* × ⅘

1535 *Taraxacum officinale* × ⅓

1539 *Sonchus asper* × ⅓

159

1537 *Cicerbita alpina* × ⅙

1541 *Lactuca perennis* $\times \frac{2}{3}$

1545 *Prenanthes purpurea* $\times \frac{1}{4}$

1509 *Cnicus benedictus* $\times \frac{3}{4}$

1549 *Crepis aurea* $\times \frac{4}{5}$

160

1550 *Hieracium pilosella* $\times \frac{1}{3}$

1546 *Crepis vesicaria* $\times \frac{1}{5}$

1565 *Stratiotes aloides* $\times \frac{1}{4}$

1566 *Hydrocharis morsus-ranae* $\times \frac{1}{3}$

1580 *Posidonia oceanica* $\times \frac{1}{10}$

1564 *Butomus umbellatus* $\times \frac{1}{3}$

161

1570 *Potamogeton natans* $\times \frac{4}{5}$

(1583) *Tofieldia calyculata* × 1

1586 *Veratrum album* × $\frac{1}{4}$

162

1593 *Aphodeline lutea* × $\frac{3}{5}$

1585 *Aphyllanthes monspeliensis* × $\frac{3}{5}$

(1586) *Veratrum nigrum* × $\frac{2}{3}$

1588 *Colchicum autumnale* × $\frac{3}{5}$

1587 *Merendera montana* × $\frac{3}{5}$

1595 *Paradisea liliastrum* × $\frac{1}{3}$

1591 *Asphodelus aestivus* × $\frac{1}{4}$

163

1592 *Asphodelus fistulosus* × $\frac{1}{10}$

1589 *Bulbocodium vernum* × $\frac{2}{3}$

(1601) *Gagea graeca* × 1

164

(1597) *Hemerocallis lilioasphodelus* × $\frac{4}{5}$

1596 *Anthericum liliago* × $\frac{1}{2}$

1601 *Gagea fistulosa* × 1

1629 *Lloydia serotina* × $\frac{4}{5}$

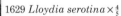

1603 *Allium schoenoprasum* $\times \frac{1}{2}$

(1604) *Allium sphaerocephalon* $\times 1\frac{1}{5}$

1606 *Allium flavum* $\times \frac{4}{5}$

1607 *Allium carinatum* $\times 1$

165

1609 *Allium neapolitanum* $\times \frac{1}{2}$

1610 *Allium triquetrum* $\times \frac{1}{3}$

1616 *Allium ursinum* × ½

166

(1619) *Lilium pomponium* × ⅓

1615 *Allium ampeloprasum* × ⅓

1611 *Allium roseum* × ⅔

1620 *Lilium bulbiferum* × ¾

1622 *Fritillaria meleagris* × ⅔   1619 *Lilium pyrenaicum* × ½

167

1618 *Lilium martagon* × ⅓

1650 *Asparagus acutifolius* × ¾

1628 *Erythronium dens-canis* × ⅔

1626 *Tulipa oculus-solis* × ⅓

1625 *Tulipa australis* × ⅓

168

1632 *Scilla peruviana* × ⅔

1636 *Scilla autumnalis* × 1

1630 *Urginea maritima* × $\frac{2}{5}$ 1640 *Ornithogalum pyrenaicum* × $\frac{4}{5}$ (1638) *Endymion hispanicus* × $\frac{1}{4}$

1646 *Muscari commutatum* × 1   1645 *Muscari comosum* × $\frac{1}{5}$

1658 *Paris quadrifolia* × ½

1648 *Muscari botryoides* × ⅘

1639 *Ornithogalum umbellatum* × ½

1635 *Scilla bifolia* × ⅘

170

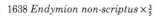

1638 *Endymion non-scriptus* × ⅗

(1647) *Muscari neglectum* × ⅘

1643 *Hyacinthus orientalis* × ½

1660 *Agave americana* × $\frac{1}{100}$

171

1655 *Polygonatum multiflorum* × $\frac{2}{3}$

1657 *Convallaria majalis* × ½

1654 *Polygonatum odoratum* × ½

1661 *Leucojum vernum* × 1

172

1662 *Leucojum aestivum* × $\frac{4}{5}$

(1662) *Leucojum trichophyllum* × $1\frac{1}{2}$

1664 *Sternbergia lutea* × $\frac{1}{2}$

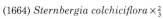

(1664) *Sternbergia colchiciflora* × $\frac{2}{3}$

1668 *Narcissus requienii* × ½

1663 *Galanthus nivalis* × ½

1666 *Narcissus bulbocodium* × 1

1671 *Narcissus poeticus* × ½

173

1672 *Narcissus serotinus* × ¾

(1666) *Narcissus cantabricus* × ⅔

1683 *Hermodactylus tuberosus* × 1

174

1676 *Crocus nudiflorus* × ½

1690 *Iris pseudacorus* × ½

1673 *Pancratium maritimum* × ⅕

1678 *Crocus albiflorus* × ⅔

1684 *Iris sisyrinchium* × $\frac{2}{3}$

(1685) *Iris xiphioides* × $\frac{1}{10}$

1686 *Iris graminea* × 1

1693 *Iris germanica* × $\frac{1}{2}$

175

1685 *Iris xiphium* × $\frac{1}{5}$

1692 *Iris chamaeiris* × $\frac{1}{2}$

1695 *Gladiolus segetum* × $\frac{2}{3}$

176

(1695) *Gladiolus byzantinus* × $\frac{1}{2}$

1691 *Iris pumila* × $\frac{2}{3}$

1718 *Phoenix canariensis* × $\frac{1}{60}$

1719 *Chamaerops humilis* × $\frac{1}{10}$

(1707) *Juncus articulatus* × $\frac{2}{3}$

1701 *Juncus jacquinii* × $\frac{1}{3}$

1705 *Juncus acutus* × $\frac{1}{3}$

177

1698 *Juncus inflexus* × 1

1700 *Juncus subuliflorus* × $\frac{2}{3}$

1699 *Juncus effusus* × $\frac{1}{3}$

1702 *Juncus squarrosus* × $\frac{1}{5}$

1712 *Luzula nivea* × $\frac{1}{2}$

1735 *Hordeum murinum* × $1\frac{1}{3}$

178

1727 *Agropyron junceiforme* × $\frac{4}{5}$

1714 *Luzula campestris* × $\frac{1}{3}$

1720 *Bromus erectus* × $\frac{1}{4}$

1730 *Aegilops ovata* × $1\frac{1}{4}$

1722 *Bromus ramosus* × $\frac{1}{3}$

1728 *Agropyron repens* × $\frac{1}{4}$

(1732) *Triticum durum* × 1

1736 *Elymus arenarius* × $\frac{1}{20}$

1737 *Sesleria caerulea* × $\frac{1}{4}$

1740 *Phragmites communis* × $\frac{1}{15}$

179

1745 *Melica ciliata* × 1

1742 *Molinia caerulea* × $\frac{1}{4}$

1746 *Briza maxima* × $\frac{1}{2}$

1748 *Cynosurus echinatus* × $\frac{2}{3}$

1751 *Poa alpina* × $\frac{2}{3}$

1765 *Avena sterilis* × ⅔

180

(1760) *Lolium multiflorum* × ½

1755 *Glyceria maxima* × ¼

(1756) *Glyceria declinata* × ½

(1757) *Festuca arundinacea* × ⅘

1759 *Catapodium rigidum* × ⅔

1772 *Deschampsia flexuosa* × ⅛

1764 *Trisetum flavescens* × ½

1770 *Holcus lanatus* × ½

1779 *Lagurus ovatus* × ½

1775 *Ammophila arenaria* × ¼

181

1796 *Spartina* × *townsendii* × ⅗

1780 *Alopecurus pratensis* × ½

1784 *Phleum pratense* × ½

1791 *Milium effusum* × ¼

1795 *Phalaris canariensis* × ⅔

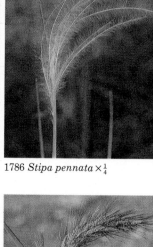

1786 *Stipa pennata* × ¼

(1803) *Panicum miliaceum* × ⅓

1807 *Erianthus ravennae* × 1/15

182

1741 *Ampelodesma mauritanica* × 1/10

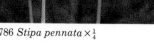

1804 *Echinochloa crus-galli* × ½

(1805) *Setaria italica* × ½

1809 *Hyparrhenia hirta* × ⅖

1808 *Bothriochloa ischaemum* × ⅔

1816 *Acorus calamus* $\times \frac{4}{5}$

1817 *Calla palustris* $\times \frac{1}{4}$

(1818) *Arum italicum* $\times \frac{1}{3}$

1819 *Dracunculus vulgaris* $\times \frac{1}{5}$

183

1820 *Biarum tenuifolium* $\times \frac{1}{2}$

1821 *Arisarum vulgare* $\times 1$

1827 *Typha latifolia* $\times \frac{1}{5}$

184

1825 *Sparganium erectum* $\times \frac{1}{3}$

1834 *Eriophorum latifolium* $\times \frac{2}{3}$

1835 *Eriophorum vaginatum* $\times \frac{1}{2}$

1822, 1823 *Lemna polyrhiza, L. minor* $\times \frac{4}{5}$

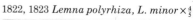

1840 *Scirpus sylvaticus* × ⅓

1839 *Scirpus maritimus* × ⅓

1841 *Scirpus holoschoenus* × 1½

185

(1843) *Scirpus tabernaemontani* × 1/15

1846 *Eleocharis palustris* × ¼

1847 *Schoenus nigricans* × ⅔

1851 *Carex extensa* × ½

1853 *Carex pseudocyperus* × ⅓

1858 *Carex riparia* × $\frac{2}{5}$

186

1861 *Carex nigra* × 1

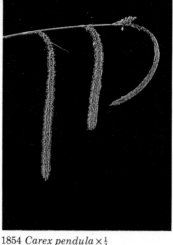

1854 *Carex pendula* × $\frac{1}{3}$

1857 *Carex rostrata* × $\frac{1}{4}$

1859 *Carex flacca* × $\frac{3}{4}$

1860 *Carex hirta* × $\frac{2}{5}$

1867 *Carex ovalis* × $\frac{2}{5}$

1869 *Carex atrata* × 1

1884 *Orchis papilionacea* × ½

1886 *Orchis coriophora* × ¼

1898 *Dactylorhiza majalis* × 1¼

187

1888 *Orchis ustulata* × 1

1892 *Orchis purpurea* × ½

1889 *Orchis tridentata* × ⅔

1893 *Orchis laxiflora* × ½

(1890) *Orchis italica* × 1

1882 *Ophrys
tenthredinifera* × 1

188

1878 *Ophrys apifera* × 1

1873 *Ophrys fusca* × 1¾

1874 *Ophrys lutea* × ⅔

1877 *Ophrys insectifera* × 2

1876 *Ophrys scolopax* × 1¼

1881 *Ophrys bertolonii* × 1¼

1879 *Ophrys bombyliflora* × 1

1895 *Orchis quadripunctata* × $\frac{2}{5}$

1891 *Orchis militaris* × 1

(1900) *Dactylorhiza fuchsii* × $1\frac{1}{2}$

189

1905 *Aceras anthropophorum* × $1\frac{2}{3}$

1899 *Dactylorhiza sambucina* × 1

1901 *Nigritella nigra* × $\frac{4}{5}$

(1903) *Serapias neglecta* × 1

1904 *Serapias vomeracea* × 1

1907 *Himantoglossum hircinum* × $\frac{2}{3}$

190

1906 *Himantoglossum longibracteatum* × 1

1909 *Herminium monorchis* × $\frac{1}{2}$

1908 *Anacamptis pyramidalis* × 2

1912 *Gymnadenia conopsea* × $1\frac{1}{4}$

1914 *Platanthera bifolia* × 1⅔

1915 *Platanthera chlorantha* × 1⅓

1921 *Limodorum abortivum* × 1¼

1920 *Cephalanthera rubra* × ⅘

1916 *Epipactis palustris* × ½

(1919) *Cephalanthera longifolia* × ⅔

1913 *Leucorchis albida* × 2

1925 *Goodyera repens* × 1

1922 *Spiranthes spiralis* × 1

1924 *Neottia nidus-avis* × ⅓

192

(1923) *Listera cordata* × 1¼

1926 *Corallorhiza trifida* × ¾

ENGLISH INDEX

Numbers given are running numbers, *except* for colour plates (as pl. 54), line drawings (as *p. 54*), and page numbers (as p. 54)

Calamint,
  Alpine, (1157), pl. 114
  Common, (1155)
  Large-flowered, 1154, pl. 114
  Lesser, 1156, pl. 114
  Wood, 1155, *p. 365*
Caltrops,
  Small, 655, pl. 64
Campion,
  Bladder, 169
  Alpine, 169
  Moss, 170, pl. 16
  Red, 165, pl. 15
  Rock, 171, *p. 90*
  Rose, 158
  Sea, 169, pl. 14
  White, 164
Canary-grass, 1795, pl. 181
  Reed, 1794
Candytuft,
  Annual, 348, pl. 36
  Burnt, 346, pl. 36
  Evergreen, 347
Cane, 1739
Cantaloupe, 817
Canterbury Bell, 1331
Cape Gooseberry, (1178)
Caper, 283, pl. 31
Capsicum, (1179)
Caraway, 890
  Corn, (887)
  Whorled, (890)
Cardoon, 1491
Carline Thistle, 1468
  Flat-topped, 1469, pl. 152
  Purple, (1469)
  Stemless, 1467, pl. 151
Carnation, (194)
Carob, 486, pl. 49
Carrot,
  Wild, 910, pl. 87
Cassie, 491
Castor Oil Plant, 668, pl.
  65
Catalpa, 1265, pl. 129
Catchfly,
  Berry, 177, pl. 15
  Italian, 166, *p. 90*
  Night-flowering, 172
  Nottingham, 167, pl. 15
  Red Alpine, 162, pl. 14
  Red German, 161, pl. 14
  Small-flowered, 174, pl. 16
  Spanish, 168, pl. 15
  Sticky, 173
  Striated, 175
  Sweet-william, 176, *p. 90*
Catmint, 1116, pl. 109
  Hairless, 1117, *p. 352*
  Lesser, (1116)
Catsear, 1520
  Giant, 1521, pl. 158
  Smooth, (1520)
  Spotted, 1522, *p. 469*
Cat's-foot, 1378, pl. 143
  Carpathian, (1378)
Cat's-tail, 1827, pl. 184
  Alpine, (1784)

Purple-stem, (1784)
Sand, 1785
Cedar,
  Western Red, (11)
Celandine,
  Greater, 274
  Lesser, 233, pl. 25
Celery,
  Wild, 885
Centaury,
  Common, 986, pl. 95
  Yellow, 987, *p. 315*
Century Plant, 1660, pl. 171
Chaffweed, (966)
Chamomile, 1413, pl. 147
  Chian, (1410)
  Corn, 1411
  Wild, 1431
  Woolly, (1410)
  Yellow, 1409, pl. 147
Charlock, 362
Chaste Tree, 1087, pl. 106
Cherry,
  Bird-, 482, pl. 49
  St Lucie's, 484, pl. 49
  Sour, 481
  Wild, 480, pl. 49
Cherry-plum, (477), pl. 48
Cherry Woodbine, (1303), *p. 399*
Chervil, 861
  Bur, (861)
  Hairy, (859)
  Rough, 859
Chestnut,
  Spanish, 46, pl. 5
  Sweet, 46, pl. 5
Chick-pea, 541, *p. 192*
Chickweed, 135
  Greater, (135)
  Jagged, 139, *p. 81*
  Mouse-ear,
    Alpine, 142, pl. 12
    Common, 144, pl. 12
    Field, 141
    Little, (144)
    Sticky, (144)
  Upright, (144)
  Water, 145, pl. 12
Chicory, 1512, pl. 158
  Spiny, (1512)
Chillies, (1179)
Chives, 1603, pl. 165
Christmas Rose, 201, pl. 19
Christ's Thorn, 726
Cineraria, 1453
Cinquefoil,
  Alpine, (455), *p. 161*
  Creeping, 450
  Golden, 455
  Hoary, 452
  Marsh, 446, pl. 46
  Norwegian, 454, *p. 161*
  Pink, (445), pl. 46
  Rock, 448, pl. 46
  Shrubby, 457, pl. 46
  Shrubby White, (447)
  Sulphur, 453

Spring, 451, pl. 46
White, 447, *p. 161*
Cistus,
  Grey-leaved, 788, pl. 77
  Gum, 793, pl. 77
  Laurel-leaved, 794, pl. 77
  Narrow-leaved, 791, pl. 77
  Poplar-leaved, 792
  Sage-leaved, 790, pl. 77
Citron, 692
Clary, 1144, pl. 113
  Meadow, 1147, pl. 114
  Whorled, 1145
  Wild, 1148
  Woolly, (1144), *p. 360*
Claytonia,
  Perfoliate, 124, pl. 11
  Pink, (124), pl. 11
Cleavers, 1023
  False, (1023)
Clematis,
  Alpine, 227, pl. 23
  Fragrant, 225, pl. 24
Cloudberry, 426, pl. 44
Clover,
  Alpine, 603, pl. 60
  Alsike, 601, pl. 59
  Crimson, 595, pl. 59
    Narrow-leaved, 596
  Dutch, 600
  Hare's-foot, 594, pl. 60
  Mountain, (601), pl. 59
  Purple, (596)
  Red, 605
  Star, 602, pl. 60
  Strawberry, 604
  Subterranean, (598)
  Suckling, 590
  Suffocated, (598)
  Sulphur, 606
  White, 600
  Zigzag, 607, pl. 59
    Mountain, 608
Club-rush,
  Round-headed, 1841, pl. 185
  Sea, 1839, pl. 185
  Wood, 1840, pl. 185
Cob-nut, 44
Cocklebur, 1401, pl. 146
  Spiny, 1402, pl. 146
  Stinking, (1401)
Cockscomb, 118
Cock's-foot, 1747
Codlins and Cream, 837, pl. 81
Coltsfoot, 1439, pl. 149
  Alpine, 1443, pl. 150
Columbine, 253
  Alpine, 254, pl. 27
  Pyrenean, (254)
Comfrey, 1052
  Blue, (1052), pl. 102
  Eastern, (1053), pl. 101
  Rough, (1052)
  Tuberous, 1053, pl. 102
Cone Flower, 1403, pl. 146
Convolvulus,
  Dwarf, 1035, pl. 100
  Pink, 1036, *p. 326*

Medick—*cont.*
  Spotted, 586
  Tree, 585
Medlar, 460
  False, 468, *p. 171*
  Mediterranean, 462, *p. 171*
Melick,
  Hairy, 1745, pl. 179
  Mountain, 1744
  Nodding, 1744
  Wood, (1744)
Melilot,
  Common, 575
  Furrowed, (575)
  Small-flowered, 574
  Tall, 576, pl. 57
  White, 573, pl. 57
Melon, 817
  Water, 812
Mercury,
  Annual, 667
  Dogs, 666, pl. 65
  Hairy, (666)
Meu, 878, *p. 287*
Mezereon, 759, pl. 74
  Alpine, 757, *p. 257*
  Mediterranean, 758
Michaelmas Daisy, 1367
  European, 1364, pl. 142
Midsummer-men, 397, pl. 40
Mignonette, (374)
  Rampion, 374, pl. 38
  Upright, 373, pl. 38
  Wild, 372, pl. 38
Milfoil, 1417
  Dark, 1420
  Dwarf, 1419, pl. 147
  Musk, (1420)
  Yellow, 1421, pl. 147
Milk-thistle, 1492, pl. 155
Milkwort,
  Box-leaved, 696, pl. 68
  Chalk, 701
  Common, 698, pl. 68
  Large, 699, *p. 220*
  Montpellier, 697, *p. 220*
  Sea, 968, pl. 92
  Shrubby, 696, pl. 68
  Tufted, 700
Milk-vetch, 527, pl. 54
  Alpine, 530, pl. 54
    Yellow, 529, *p. 192*
  Beaked, p. 191
    Meadow, 533, *p. 192*
    Mountain, 535
    Northern, (535)
    Purple, 534
  Iberian, (527)
  Montpellier, 526, *p. 192*
  Purple, 532
  Sainfoin, (532)
Millet, (1812)
  Creeping, 1803, *p. 553*
  Foxtail, (1805), pl. 182
  Italian, (1803), pl. 182; (1805),
    pl. 182
  Red, (1802)
  Wood, 1791, pl. 181

Mimosa, 490
Mind-Your-Own-Business, (69)
Mint,
  Apple-scented, 1172
  Corn, 1168
  Water, 1169, pl. 116
Mirror-of-Venus, 1875
Mistletoe, 73, pl. 7
Monkey-flower, 1217, pl. 123
Monkshood,
  Branched, (210)
  Common, 210, pl. 19
  Yellow, 209
Monk's Rhubarb, 94, pl. 9
Moon-daisy, 1427, pl. 148
  Alpine, 1428
Moor-grass,
  Blue, 1737, pl. 179
  Purple, 1742, pl. 179
Moor-king, 1251
Morning Glory, 1033
Moschatel, 1308, pl. 137
Mother of Thousands, (69)
Motherwort, 1133, pl. 112
  False, (1133)
Mountain Ash, 466, pl. 48
Mountain Tassel Flower, (956)
Mournful Widow, 1326, pl. 138
Mourning Widow, 646, pl. 64
Mouse-tail, 252, pl. 26
Mugwort, 1435
Mulberry,
  Common, 61
  White, (61), pl. 6
Mullein,
  Dark, 1190, pl. 120
  Hoary, (1195)
  Moth, 1192, pl. 120
  Purple, (1192), pl. 121
  Twiggy, (1192)
  White, 1195
Musk, (1217), pl. 123
Mustard,
  Ball,
    White, 369, *p. 139*
    Yellow, 339, *p. 135*
  Black, 361
  Buckler, 349
  Garlic, 289, pl. 31
  Hedge, 285
  Hoary, 365, *p. 128*
  Tower, 315
  Treacle, 295
    Swiss, 294
  Wall, 357
  White, (362), pl. 37
Myrtle, 824, pl. 81

Naked Ladies, 1588, pl. 163
Narcissus,
  Paper-white, (1667)
  Pheasant's-eye, 1671, pl. 173
  Rush-leaved, 1668, pl. 173
Nasturtium, (654)
Navelwort, 379, pl. 39
Nettle, 65
  Roman, 67, pl. 6
  Small, 66, pl. 6

Nettle Tree, 60, *p. 51*
Nightshade,
  Black, 1182
  Deadly, 1174, pl. 117
  Woody, 1181, pl. 119
Nipplewort, 1513
Northern Shorewort, 1062, pl. 103

Oak,
  American Red, (51)
  Common, 52, pl. 5
  Cork, 49
  Durmast, 53
  Holly, 47, pl. 5
  Holm, 48
  Kermes, 47, pl. 5
  Macedonian, (50), *p. 54*
  Pedunculate, 52, pl. 5
  Pyrenean, 55, *p. 54*
  Sessile, 53
  Turkey, 51, *p. 54*
  Valonia, 50, pl. 5
  White, 54, *p. 54*
Oat, 1766
  Animated, 1765, pl. 180
  Bristle, (1765)
  Common Wild, (1765)
  French, 1769
  Small, (1765)
  Yellow, 1764, pl. 180
Oat-grass, 1768
  Hairy, 1767
  Meadow, (1767)
Old Man's Beard, 224, pl. 24
Oleander, 1007, pl. 98
Oleaster, 762, *p. 264*
Olive, 985, pl. 94
Onion, 1602
  Welsh, (1602)
  Yellow, 1606, pl. 165
Opopanax, 491
Orache, 109, *p. 71*
  Common, 108
  Frosted, (111)
  Hastate, 110
  Shore, (108), *p. 71*
  Shrubby, 107
Orange,
  Seville, 689
  Sweet, 690, pl. 67
Oregon Grape, 262, pl. 27
Orchid,
  Bee, 1878, pl. 188
    Brown, 1873, pl. 188
    Bumble, 1879, pl. 188
    Yellow, 1874, pl. 188
  Bird's-nest, 1924, pl. 192
  Bog, (1893)
  Bug, 1886, pl. 187
  Burnt, 1888, pl. 187
  Butterfly,
    Greater, 1915, pl. 191
    Lesser, 1914, pl. 191
    Pink, 1884, pl. 187
  Coral-root, 1926, pl. 192
  Dark-winged, 1888, pl. 187
  Early Purple, 1894
  Elder-flowered, 1899, pl. 189

Spiderwort,
  Common, 1717
Spignel, 878, *p. 287*
Spike-heath, (934)
Spike-rush,
  Common, 1846, pl. 185
  Many-stemmed, (1846)
  Slender, (1846)
Spinach, 106, *p. 73*
Spindle-tree, 718, pl. 71
Spiraea,
  Willow, 419, pl. 44
Spruce,
  Norway, 2, pl. 1
  Sitka, (2)
Spurge,
  Broad, 673
  Caper, 676, pl. 67
  Cypress, 680
  Dwarf, (675)
  Glaucous,
    Broad-leaved, 682
    Narrow-leaved, 683
  Large Mediterranean, 678, pl. 66
  Laurel, 760, pl. 74
  Marsh, 674, *p. 225*
  Petty, 675
  Saw-leaved, 679
  Sea, 684, pl. 66
  Sickle, (675), *p. 225*
  Spiny, 670
    Greek, (670), pl. 66
  Sun, 671, pl. 67
  Tree, 669, pl. 66
  Whorled, (684), *p. 225*
  Wood, 677, pl. 66
Spurrey,
  Cliff, 156, pl. 13
  Corn, 154
  Sand, 157, pl. 13
  Sea,
    Greater, 155
    Lesser, (156)
Squill,
  Alpine, 1635, pl. 170
  Autumn, 1636, pl. 168
  Pyrenean, (1634)
  Sea, 1630, pl. 169
  Spring, 1633
Squinancy Wort, 1017
Star-of-Bethlehem, 1639, pl. 170
  Drooping, 1642
  Yellow, 1599
Starwort,
  Common, 1088, pl. 105
Stavesacre, 212
Sternbergia,
  Common, 1664, pl. 172
  Slender, (1664), pl. 172
Stitchwort,
  Greater, 136, pl. 12
  Lesser, 138, pl. 12
  Marsh, 137
  Wood, 134
Stock, 300, pl. 32
  Sad, (301), pl. 33
  Sea, 301, pl. 33

Three-horned, (301)
  Virginia, 298, pl. 32
Stonecrop,
  Alpine, (389)
  Annual, 390
  Blue, (394), pl. 40
  Dark, 292
  English, (391)
  Hairy, 394, pl. 40
  Rock, 388, pl. 40
  Starry, (396)
  Thick-leaved, (391)
  White, 391, pl. 39
Storax, 977
Storksbill,
  Common, 653
  Long-beaked, (652), pl. 64
  Musk, 654
  Sea, (652)
  Soft, 652, pl. 64
Stranglewort, 1008, *p. 315*
Strapwort, (148)
Strawberry,
  Barren, 445
    Pink, (445)
  Garden, (459)
  Hautbois, (459)
  Wild, 459
Strawberry Tree, 924, pl. 90
  Eastern, (924)
Succory,
  Lamb's, (1516)
  Swine's, (1516)
  Wild, 1512, pl. 158
Sumach, 707
  Staghorn, (707)
Sundew,
  Common, 376
  Great, 377, pl. 38
  Long-leaved, 378, pl. 39
  Yellow, (378), pl. 38
Sunflower, 1404
Sweep's Brush, 1714, pl. 178
Sweet Alison, 329, pl. 37
Sweet Bay, 263
Sweet Briar, 433
Sweet Cicely, 863, pl. 84
Sweet Gale, 34, pl. 4
Sweet-grass,
  Floating, 1756
  Glaucous, (1756), pl. 180
  Plicate, (1756)
  Reed, 1755, pl. 180
Sweet Scabious, 1371, *p. 432*
Sweet Vernal-grass, 1793
Sweet William, 185, *p. 90*
Swine-cress, 354
  Lesser, (354), pl. 36

Tamarisk, 806
  German, 808, *p. 264*
Tangerine, 691
Tansy, 1426, pl. 147
Tare,
  Hairy, 542
  Smooth, (542)
Tarragon, (1436)

Teasel, 1318
  Fuller's, 1318
  Small, 1319
Tea-tree, 1173, *p. 384*
Terebinth, 704, pl. 69
Thistle,
  Alpine, (1478)
  Blessed, 1509, pl. 160
  Cabbage, 1482, pl. 154
  Candelabra, (1484), pl. 153
  Cotton, 1494, pl. 155
  Creeping, 1487
  Holy, 1492, pl. 155
  Marsh, 1486
    Great, 1479, pl. 153
  Marsh Plume, (1489)
  Meadow, (1489)
  Melancholy, 1489
    Yellow, 1484
  Musk, 1478, pl. 152
  St. Barnaby's, 1499, pl. 155
  Scotch, 1494, pl. 155
  Slender, 1477
  Spear, (1485)
  Spiniest, 1483, pl. 154
  Star, 1500, pl. 156
    Maltese, (1499)
  Stemless, 1490, pl. 154
  Syrian, 1481, pl. 153
  Tuberous, (1489), pl. 153
  Welted, (1480)
  Woolly, 1485, pl. 154
Thorn-apple, 1186, pl. 118
Thorow-wax, 881
Thrift, 975, pl. 94
  Alpine, (975)
  Jersey, 976
  Plantain-leaved, 976
  Spiny, (976), pl. 94
Throatwort, 1348
Thrumwort, 1562, *p. 480*
Thyme, 1163, pl. 115
  Cat, 1100
  Round-head, (1165)
  Wild, 1164, pl. 115
    Larger, 1165
Toadflax,
  Alpine, 1204, pl. 122
  Bastard, p. 59
    Alpine, 71, pl. 6
  Common, 1205
  Italian, (1205)
  Ivy-leaved, 1210, pl. 122
  Pale, 1202, pl. 122
  Small, (1209)
  Three-leaved, 1208, pl. 122
  White, 1207
Tobacco,
  Large, (1188), pl. 119
  Shrub, 1187, pl. 119
  Small, 1188, pl. 119
Tomato, 1184
Toothwort, 1269, pl. 130
  Purple, (1269), pl. 130
Tormentil,
  Common, 449
Touch-Me-Not, 714, pl. 70
Town Hall Clock, 1308, pl. 137

LATIN INDEX

Numbers given are running numbers, *except* for colour plates (as pl. 54), line drawings (as *p. 54*), and page numbers (as p. 54). Synonyms are shown in italic.

Ajuga, 1089–1094
  chamaepitys, 1094, pl. 106
  chia, (1094)
  genevensis, 1090, pl. 105
  iva, 1093
  laxmannii, 1092
  orientalis, (1091)
  pyramidalis, 1091, pl. 106
  reptans, 1089, pl. 106
Albizia, 493
  julibrissin, 493, pl. 50
Alcea, 748
  pallida, 748, pl. 72
  rosea, (748)
Alchemilla, 423–424
  alpina, 424, *p. 161*
  arvensis, 425, *p. 161*
  pentaphyllea, (424), *p. 161*
  vulgaris, 423
Alisma, 1560–1561
  lanceolatum, 1561
  plantago-aquatica, 1560
Alismataceae, p. 477
Alkanna, 1061
  tinctoria, 1061, pl. 103
Alliaria, 289
  officinalis, 289, pl. 31
  petiolata, 289, pl. 31
Allium, 1602–1617
  ampeloprasum, 1615, pl. 166
  carinatum, 1607, pl. 165
  cepa, 1602
  fistulosum, (1602)
  flavum, 1606, pl. 165
  neapolitanum, 1609, pl. 165
  nigrum, 1612
  oleraceum, 1605
  porrum, (1615)
  roseum, 1611, pl. 166
  rotundum, 1614
  sativum, (1612)
  schoenoprasum, 1603, pl. 165
  scorodoprasum, 1613
  sibiricum, 1603
  sphaerocephalon, 1604, pl. 165
  subhirsutum, 1608
  triquetrum, 1610, pl. 165
  ursinum, 1616, pl. 166
  victorialis, 1617
  vineale, 1604
Alnus, 39–41
  cordata, (41)
  glutinosa, 40, pl. 4
  incana, 41, *p. 51*, pl. 4
  viridis, 39
Aloe, 1598
  vera, 1598
Alopecurus, 1780–1782
  arundinaceus, (1780)
  geniculatus, (1782)
  myosuroides, 1782
  pratensis, 1780, pl. 181
  utriculatus, 1781
Althaea, 745–747
  cannabina, 746, *p. 246*
  hirsuta, 745
  officinalis, 747, pl. 72

pallida, 748, pl. 72
rosea, (748)
Alyssoides, 323
  utriculata, 323
Alyssum, 324–326
  alyssoides, 326, *p. 135*
  calycinum, 326, *p. 135*
  incanum, 328
  montanum, 325
  saxatile, 324
Amaranthaceae, p. 74
Amaranthus, 117
  albus, (117)
  blitoides, (117)
  cruentus, (117)
  graecizans, (117)
  hybridus, (117)
  lividus, (117)
  retroflexus, 117, pl. 10
Amaryllidaceae, p. 507
Ambrosia, 1400
  artemisiifolia, 1400
Amelanchier, 471–472
  confusa, 472
  grandiflora, 472
  laevis, 472
  ovalis, 471, pl. 47
Ammi, 888
  majus, 888, *p. 287*
  visnaga, (888)
Ammophila, 1775
  arenaria, 1775, pl. 181
Ampelodesma, 1741
  mauritanica, 1741, pl. 182
  tenax, 1741, pl. 182
Amsinckia, 1046
  angustifolia, 1046
Anacamptis, 1908
  pyramidalis, 1908, pl. 190
Anacardiaceae, p. 231
Anacyclus, 1414
  clavatus, 1414
  radiatus, (1414)
  tomentosus, 1414
Anagallis, 966–967
  arvensis, 967
  foemina, (967)
  linifolia, (967), pl. 93
  minima, (966)
  tenella, 966
Anagyris, 495
  foetida, 495
Anarrhinum, 1211
  bellidifolium, 1211, pl. 123
Anchusa, 1055–1057
  arvensis, 1059
  azurea, 1056, pl. 102
  barrelieri, 1057
  hybrida, (1057)
  officinalis, 1055, pl. 103
  sempervirens, 1058, pl. 102
  undulata, (1057)
Andrachne, 685
  telephioides, 685, *p. 220*
Andromeda, 923
  polifolia, 923, pl. 88
Andropogon
  gryllus, 1811

hirtus, 1809, pl. 182
  ischaemum, 1808, pl. 182
Androsace, 948–954
  alpina, 949, pl. 91
  carnea, 953, pl. 90
  chamaejasme, 954
  glacialis, 949, pl. 91
  helvetica, 948, *p. 304*
  imbricata, (948)
  lactea, 952
  maxima, 950
  multiflora, (948)
  septentrionalis, 951, *p. 304*
  vandellii, (948)
  villosa, (954), *p. 304*
Androsaemum
  hircinum, (763)
Andryala, 1533
  integrifolia, 1533, pl. 159
Anemone, 214–218
  alpina, 220, pl. 23
  apennina, (214), pl. 22
  blanda, (214), pl. 21
  coronaria, 216, pl. 21
  hepatica, 219, pl. 20
  hortensis, (216), pl. 21
  narcissiflora, 218, pl. 22
  nemorosa, 214, pl. 22
  palmata, 217, pl. 21
  pavonina, (216), pl. 21
  pratensis, 222
  pulsatilla, 223, pl. 22
  ranunculoides, (214)
  stellata, (216), pl. 21
  sylvestris, 215
  vernalis, 221, pl. 22
Anethum, 876
  graveolens, 876
Angelica, 894–895
  archangelica, 895, pl. 85
  sylvestris, 894
Anisantha
  madritensis, (1723)
  sterilis, (1724)
  tectorum, 1724
Antennaria, 1378
  carpatica, (1378)
  dioica, 1378, pl. 143
Anthemis, 1409–1412
  altissima, 1410
  arvensis, 1411
  chia, (1410)
  cota, 1410
  cotula, 1412
  nobile, 1413, pl. 147
  tinctoria, 1409, pl. 147
  tomentosa, (1410)
Anthericum, 1596
  liliago, 1596, pl. 164
  ramosum, (1596)
Anthoxanthum, 1793
  odoratum, 1793
Anthriscus, 860–861
  caucalis, (861)
  cerefolium, 861
  sylvestris, 860
  vulgaris, (861)

Erinacea, (515)
  anthyllis, (515), pl. 52
  pungens, (515), pl. 52
Erinus, 1235
  alpinus, 1235, pl. 127
Eriobotrya, 470
  japonica, 470, pl. 48
Eriophorum, 1833–1836
  angustifolium, 1833
  latifolium, 1834, pl. 184
  polystachion, 1833
  scheuchzeri, 1836
  vaginatum, 1835, pl. 184
Eritrichium, 1070
  nanum, 1070, pl. 104
Erodium, 652–654
  ciconium, (654)
  cicutarium, 653
  gruinum, (652), pl. 64
  malacoides, 652, pl. 64
  maritimum, (652)
  moschatum, 654
Erophila, 335
  verna, 335
Eruca, 363
  vesicaria, 363, pl. 37
Erucastrum, 364
  gallicum, 364, p. 128
  nasturtiifolium, (364)
Ervum
  lens, 552, *p. 198*
  nigricans, (552)
Eucalyptus, 825–826
  amygdalinus, 826
  globulus, 825
  viminalis, (826)
Euonymus, 718
  europaeus, 718, pl. 71
  latifolius, (718), p. 239
Eupatorium, 1357
  cannabinum, 1357, pl. 141
Euphorbia, 669–684
  acanthothamnos, (670), pl. 66
  amygdaloides, 677, pl. 66
  biglandulosa, (683)
  biumbellata, (684), *p. 225*
  characias, 678, pl. 66
    subsp. wulfenii, 678, pl. 66
  cyparissias, 680
  dendroides, 669, pl. 66
  esula, 681
  exigua, (675)
  falcata, (675), *p. 225*
  helioscopia, 671, pl. 67
  lathyris, 676, pl. 67
  myrsinites, 682
  palustris, 674, *p. 225*
  paralias, 684, pl. 66
  peplus, 675
  pinea, (680), *p. 225*
  platyphyllos, 673
  rigida, 683
  serrata, 679
  spinosa, 670
  villosa, 672, *p. 225*
Euphorbiaceae, p. 222
Euphrasia, 1242–1246
  brevipila, 1243

ericetorum, 1246
  minima, (1244), pl. 127
  nemorosa, 1245
  nitidula, 1245
  rostkoviana, 1242, pl. 127
  salisburgensis, 1244
  stricta, 1246
Evax, 1374
  pygmaea, 1374, pl. 143
Eryngium, 853–856
  alpinum, 854
  amethystinum, 856, pl. 83
  campestre, 855, pl. 83
  creticum, (856)
  maritimum, 853, pl. 83
  planum, (856)
Erysimum, 294–295
  cheiranthoides, 295
  decumbens, (294)
  helveticum, 294
  ochroleucum, (294)
Erythronium, 1628
  dens-canis, 1628, pl. 168
Eschscholzia, 275
  californica, 275, pl. 29

Faba
  bona, 546
Fagaceae, p. 50
Fagopyrum, 89
  esculentum, 89, pl. 9
  tataricum, (89), *p. 64*
Fagus, 45
  orientalis, (45)
  sylvatica, 45
Falcaria, 889
  vulgaris, 889, *p. 287*
Farsetia
  clypeata, 327, pl. 35
Fedia, 1312
  cornucopiae, 1312, pl. 137
Ferula, 897
  communis, 897, pl. 85
Ferulago, 898
  campestris, 898
Festuca, 1757
  arundinacea, (1757), pl. 180
  bromoides, (1758)
  gigantea, 1757
  myuros, (1758)
  ovina, (1757)
  pratensis, (1757)
  rubra, (1757)
Fibigia, 327
  clypeata, 327, pl. 35
Ficaria
  verna, 233, pl. 25
Ficus, 62
  carica, 62
Filago, 1375–1377
  apiculata, (1375)
  arvensis, 1377
  gallica, 1376, *p. 432*
  minima, (1377)
  pygmaea, 1374, pl. 143
  vulgaris, 1375, pl. 143
Filipendula, 421–422

  ulmaria, 421, pl. 44
  vulgaris, 422, pl. 44
Foeniculum, 875
  officinale, 875
  vulgare, 875
    var, azoricum, 875
Fragaria, 459
  × ananassa, (459)
  chiloensis × virginiana, (459)
  moschata, (459)
  vesca, 459
  viridis, (459)
Frangula, 724
  alnus, 724
Frankenia, 809–810
  laevis, 810
  pulverulenta, 809
Frankeniaceae, p. 259
Fraxinus, 978–979
  excelsior, 978
  ornus, 979, pl. 94
Freesia, (1697)
  refracta (1697)
Fritillaria, 1622–1623
  meleagris, 1622, pl. 167
  pyrenaica, 1623
  tenella, (1623)
Fumana, 804–805
  ericoides, (805)
  procumbens, 805, p. 257
  thymifolia, 804
Fumaria, 280–281
  bastardii, (281)
  capreolata, 280, pl. 30
  densiflora, (281)
  muralis, (281)
  officinalis, 281
  parviflora, (281)
  spicata, 282
  vaillantii, (281)

Gagea, 1599–1601
  arvensis, 1600
  fistulosa, 1601, pl. 164
  graeca, (1601), pl. 164
  lutea, 1599
Galactites, 1493
  tomentosa, 1493
Galanthus, 1663
  nivalis, 1663, pl. 173
Galega, 524
  officinalis, 524, pl. 53
Galeobdolon, 1132
  luteum, 1132, pl. 112
Galeopsis, 1125–1126
  angustifolia, 1125
  segetum, (1125), *p. 360*
  speciosa, (1126), pl. 111
  tetrahit, 1126, pl. 111
Galinsoga, 1407
  parviflora, 1407, pl. 147
Galium, 1020–1028
  aparine, 1023
  boreale, 1020
  cruciata, 1029, pl. 99
  glabra, (1029)
  glaucum, 1028, *p. 326*
  mollugo, 1025

odoratum, 1027, pl. 99
palustre, (1022), pl. 99
pumilum, (1026)
rotundifolium, 1021, *p. 326*
saxatile, 1026
spurium, (1023)
sylvaticum, (1025)
tricorne, (1023)
tricornutum, (1023)
uliginosum, 1022
vernum, (1029)
verum, 1024
Gasoul
nodiflorum, 121
Gaudinia, 1769
fragilis, 1769
Genista, 508–512
anglica, 512
germanica, 511, *p. 185*
hispanica, (511), pl. 52
pilosa, 508
prostrata, 500
radiata, 510, *p. 185*
sagittalis, 513, pl. 52
tinctoria, 509
Genistella
sagittalis, 513, pl. 52
Gentiana, 989–998
asclepiadea, 994, pl. 97
bavarica, (991)
clusii, (992), pl. 96
cruciata, 989, pl. 96
kochiana, 992, pl. 96
lutea, 996, pl. 95
nivalis, 990
pannonica, (998)
pneumonanthe, 993, pl. 96
punctata, 997, pl. 97
purpurea, 998, pl. 97
pyrenaica, 995
utriculosa, (990), pl. 96
verna, 991, pl. 96
Gentianaceae, p. 313
Gentianella, 999–1001
amarella, (999)
campestris, 999, pl. 96
ciliata, 1001
germanica, (999)
tenella, 1000
Geraniaceae, p. 215
Geranium, 641–651
columbinum, (648)
dissectum, 648
lucidum, 651, pl. 63
macrorrhizum, 647, pl. 64
molle, (649)
nodosum, (642), pl. 63
phaeum, 646, pl. 64
pratense, 644
purpureum, (650)
pusillum, (649)
pyrenaicum, 642, pl. 63
robertianum, 650
rotundifolium, 649
sanguineum, 641, pl. 63
striatum, (642)
sylvaticum, 645, pl. 63
  subsp, rivulare, 645

tuberosum, 643
versicolor, (642)
Geropogon
glaber, (1528)
Gesneriaceae, p. 394
Geum, 442–444
montanum, 444
reptans, (444)
rivale, 443, pl. 45
urbanum, 442
Gladiolus, 1695–1697
byzantinus, (1695), pl. 176
communis, 1696
illyricus, 1697
palustris, (1696)
segetum, 1695, pl. 176
Glaucium, 272–273
corniculatum, 273, pl. 29
flavum, 272, pl. 29
Glaux, 968
maritima, 968, pl. 92
Glechoma, 1118
hederacea, 118, pl. 109
Gleditsia, 487
triacanthos, 487
Globularia, 1262–1264
alypum, 1262, pl. 129
cordifolia, 1263, pl. 129
nudicaulis, (1264)
vulgaris, 1264, pl. 129
Globulariaceae, p. 392
Glyceria, 1755–1756
declinata, (1756), pl. 180
fluitans, 1756
maxima, 1755, pl. 180
plicata, (1756)
Glycine, 540
hispida, 540
max, 540
Glycyrrhiza, 536
echinata, (536)
glabra, 536, *p. 192*
Gnaphalium, 1380–1383
luteoalbum, 1381
norvegicum, (1383), *p. 432*
supinum, 1382, *p. 432*
sylvaticum, 1383
uliginosum, 1380, pl. 144
Gomphocarpus, 1012
fruticosus, 1012, *p. 315*
Goodyera, 1925
repens, 1925, pl. 192
Gossypium, 750
herbaceum, 750, pl. 73
Gramineae, p. 522
Gratiola, 1218
officinalis, 1218, pl. 124
Groenlandia, 1577
densa, 1577, *p. 484*
Grossulariaceae, p. 155
Guttiferae, p. 249
Gymnadenia, 1912
albida, 1913, pl. 192
conopsea, 1912, pl. 190
odoratissima, (1912)
Gynandriris
sisyrinchium, 1684, pl. 175

Gypsophila, 178–179
fastigiata, (178), *p. 90*
muralis, 179
repens, 178, pl. 16

Hackelia, (1050)
deflexa, (1050)
Hacquetia, 851
epipactis, 851, pl. 83
Halimione, 112
portulacoides, 112, *p. 73*
Halimium, 795–797
commutatum, 796, pl. 78
halimifolium, 797
libanotis, 796, pl. 78
umbellatum, 795, *p. 257*
Haloragaceae, p. 269
Hedera, 848
helix, 848, pl. 82
Hedypnois, 1519
rhagadioloides, 1519, pl. 158
Hedysarum, 634–635
capitatum, (634), pl. 62
coronarium, 634, pl. 62
glomeratum, (634), pl. 62
hedysaroides, 635, pl. 61
spinosissimum, (634)
Helechloa
schoenoides, (1783)
Helianthemum, 800–803
apenninum, 803, pl. 78
canum, 801
ledifolium, (800)
nummularium, 802, pl. 78
pulverulentum, 803
salicifolium, 800
Helianthus, 1404–1405
annuus, 1404
tuberosus, 1405
Helichrysum, 1384–1385
arenarium, 1384, *p. 432*
italicum, (1385)
stoechas, 1385, pl. 144
Helictotrichon, 1767
pratense, (1767)
pubescens, 1767
Heliotropium, 1045
europaeum, 1045, pl. 101
supinum, (1045)
Helleborus, 199–201
cyclophyllus, (200), pl. 18
foetidus, 199, pl. 19
lividus, (199)
niger, 201, pl. 19
viridis, 200
Helxine
soleirolii, (69)
Hemerocallis, 1597
flava, (1597), pl. 164
fulva, 1597
lilioasphodelus, (1597), pl. 164
Hepatica, 219
nobilis, 219, pl. 20
Heracleum, 901–902
mantegazzianum, 902, pl. 86
sphondylium, 901
Herminium, 1909
monorchis, 1909, pl. 190

circinatus, (251)
ficaria, 233, pl. 25
flammula, 243
fluitans, (251)
glacialis, (247), pl. 26
gramineus, 245
hederaceus, 249
lanuginosus, 240, *p. 105*
lingua, 244, pl. 25
montanus, 241, pl. 25
muricatus, 232, pl. 26
ophioglossifolius, (243), *p. 105*
parnassifolius, 248, pl. 25
parviflorus, (232), *p. 105*
peltatus, 250, pl. 26
platanifolius, (246)
polyanthemos, (238), *p. 105*
pseudofluitans, (251), pl. 26
pyrenaeus, (248)
repens, 238
reptans, (243)
sardous, (237), *p. 105*
sceleratus, 236
thora, 234, pl. 25
trichophyllus, 251
Raphanus, 370
raphanistrum, 370
subsp. maritimus, 370
sativus, (370)
Rapistrum, 367
perenne, 367, *p. 135*
rugosum, (367), *p. 135*
Reichardia, 1536
picroides, 1536
Reseda, 371–374
alba, 373, pl. 38
lutea, 372, pl. 38
luteola, 371
odorata, (374)
phyteuma, 374, pl. 38
sesamoides, 375
Resedaceae, p. 144
Retama
monosperma, 514, pl. 52
Reynoutria, 88
japonica, 88, pl. 8
sachalinensis, (88)
Rhagadiolus, 1518
stellatus, 1518
Rhamnaceae, p. 237
Rhamnus, 720–723
alaternus, 720, pl. 71
alpinus, 722, *p. 239*
catharticus, 723, pl. 71
frangula, 724
pumilus, 721
saxatilis, (723)
Rhinanthus, 1247–1248
minor, 1247, pl. 127
serotinus, 1248
Rhodiola, 397
rosea, 397, pl. 40
Rhododendron, 919–920
ferrugineum, 919, pl. 87
hirsutum, (919), pl. 87
ponticum, 920, pl. 88
Rhus, 706–707
coriaria, 707

cotinus, 706, pl. 69
hirta, (707)
typhina, (707)
Ribes, 414–416
alpinum, (414), pl. 43
grossularia, 416, pl. 43
nigrum, 415
petraeum, (415)
rubrum, 414, pl. 43
spicatum, (414)
sylvestre, 414, pl. 43
uva-crispa, 416, pl. 43
Ricinus, 668
communis, 668, pl. 65
Robinia, 523
pseudacacia, 523, pl. 53
Roemeria, 271
hybrida, 271, pl. 28
Romulea, 1681
bulbocodium, 1681
columnae, (1681)
parviflora, (1681)
Rorippa, 303–305
amphibia, 303, pl. 33
austriaca, (303), *p. 128*
islandica, 305
microphylla, (307)
nasturtium-aquaticum, 307
sylvestris, 304
Rosa, 430–435
alpina, 435, pl. 45
arvensis, 430, pl. 45
canina, 432
eglanteria, (433)
gallica, 431
pendulina, 435, pl. 45
pimpinellifolia, 434, pl. 45
rubiginosa, 433
sempervirens, (430)
spinosissima, 434, pl. 45
tomentosa, (432)
villosa, (432)
Rosaceae, p. 157
Rosmarinus, 1105
officinalis, 1105, pl. 107
Rubia, 1031
peregrina, 1031, pl. 98
tinctorum, (1031)
Rubiaceae, p. 321
Rubus, 426–429
arcticus, (427), pl. 43
caesius, (429), pl. 44
chamaemorus, 426, pl. 44
fruticosus, 429
idaeus, 428
saxatilis, 427
Rudbeckia, 1403
laciniata, 1403, pl. 146
Rumex, 91–100
acetosa, 93
acetosella, 91
alpinus, 94, pl. 9
aquaticus, (95), *p. 67*
bucephalophorus, 95
conglomeratus, 98
crispus, 97, pl. 9
hydrolapathum, 96, pl. 9
maritimus, 100, *p. 67*

obtusifolius, (99)
palustris, (100), *p. 67*
pulcher, 99, *p. 67*
sanguineus, (98), pl. 9
scutatus, 92, *p. 67*
Ruppia, 1578
maritima, 1578
spiralis, 1578
Ruscus, 1651
aculeatus, 1651
hypoglossum, (1651), *p. 504*
Ruta, 686–687
chalepensis, (687)
graveolens, 686, pl. 67
montana, (686)
Rutaceae, p. 227
Rynchospora, 1848
alba, 1848, *p. 564*

Sagina, 146–147
apetala, 146
maritima, (146)
nodosa, 147, *p. 81*
procumbens, (146), pl. 13
saginoides, (146)
subulata, (146)
Sagittaria, 1563
sagittifolia, 1563
Salicaceae, p. 43
Salicornia, 114
europaea, 114
fruticosa, (113)
herbacea, 114
perennis, 113
Salix, 19–29
alba, 25, pl. 3
aurita, (28), *p. 45*
babylonica, (25)
caprea, 29, pl. 3
cinerea, 28
fragilis, 24
hastata, 22, *p. 45*, pl. 3
helvetica, (21)
herbacea, (20), *p. 45*
lapponum, (21)
myrsinites, 21, *p. 45*
pentandra, (24), *p. 45*
purpurea, 27, *p. 45*
repens, 23, *p. 45*
reticulata, 19, pl. 3
retusa, 20, pl. 3
triandra, (24), *p. 45*
viminalis, 26
Salsola, 116
kali, 116, *p. 73*
soda, (116), *p. 73*
Salvia, 1143–1149
aethiopis, (1144), *p. 360*
argentea, (1144), pl. 113
glutinosa, 1146, pl. 113
horminoides, 1148
horminum, 1144, pl. 115
nemorosa, (1147)
officinalis, 1143
pratensis, 1147, pl. 114
sclarea, 1144, pl. 113
triloba, (1143), pl. 113

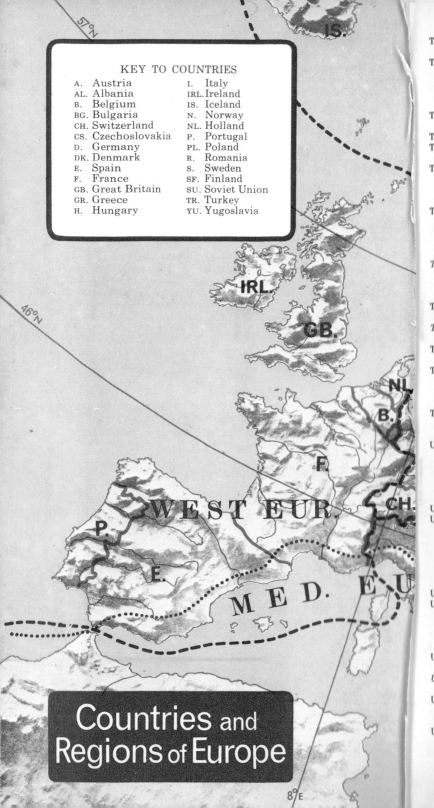

KEY TO COUNTRIES

A. Austria
AL. Albania
B. Belgium
BG. Bulgaria
CH. Switzerland
CS. Czechoslovakia
D. Germany
DK. Denmark
E. Spain
F. France
GB. Great Britain
GR. Greece
H. Hungary
I. Italy
IRL. Ireland
IS. Iceland
N. Norway
NL. Holland
P. Portugal
PL. Poland
R. Romania
S. Sweden
SF. Finland
SU. Soviet Union
TR. Turkey
YU. Yugoslavia

Countries and Regions of Europe

Trisetum, 1764
  flavescens, 1764, pl. 180
Triticum, 1732
  aestivum, (1732)
  durum, (1732), pl. 178
  polonicum, (1732)
  turgidum, 1732
Trollius, 204
  europaeus, 204, pl. 19
Tropaeolaceae, p. 218
Tropaeolum, (654)
  majus, (654)
Tuberaria, 798–799
  guttata, 798, pl. 78
  lignosa, 799
  vulgaris, 799
Tulipa, 1624–1627
  australis, 1625, pl. 168
  clusiana, 1627
  oculus-solis, 1626, pl. 168
  sylvestris, 1624
Tunica
  prolifera, 184
  saxifraga, 183
  velutina, (184)
Turgenia, 908
  latifolia, 908
Turritis
  glabra, 315
Tussilago, 1439
  farfara, 1439, pl. 149
Typha, 1827–1828
  angustifolia, 1828
  latifolia, 1827, pl. 184
  minima, (1828)
Typhaceae, p. 559

Ulex, 516–517
  europaeus, 516
  gallii, (517)
  minor, 517, pl. 53
  nanus, 517, pl. 53
  parviflorus, (517)
Ulmaceae, p. 55
Ulmus, 56–59
  carpinifolia, 58
  effusa, 59, *p. 51*
  glabra, 56, pl. 5
  laevis, 59, *p. 51*
  minor, 58
  procera, 57
Umbelliferae, p. 271
Umbilicus, 379–380
  erectus, 380
  horizontalis, (379)
  pendulinus, 379, pl. 39
  rupestris, 379, pl. 39
Urginea, 1630
  maritima, 1630, pl. 169
Uropetalum
  serotinum, 1637,
Urospermum, 1526
  dalechampii, (1526), pl. 157
  picroides, 1526, *p. 469*
Urtica, 65–67
  dioica, 65
  dubia, (66), pl. 6

pilulifera, 67, pl. 6
  urens, 66
Urticaceae, p. 57
Utricularia, 1281–1282
  intermedia, 1282
  minor, (1282)
  neglecta, (1281)
  vulgaris, 1281, pl. 132

Vaccaria, 182
  pyramidata, 182
Vaccinium, 935–937
  myrtillus, 936, pl. 89
  oxycoccos, 937, pl. 88
  uliginosum, (936)
  vitis-idaea, 935, pl. 89
Valantia, 1030
  hispida, 1030, pl. 98
Valeriana, 1313–1315
  dioica, 1314
  montana, (1315), pl. 136
  officinalis, 1313, pl. 136
  tripteris, 1315, pl. 136
  tuberosa, (1313)
Valerianaceae, p. 404
Valerianella, 1309–1311
  carinata, (1311)
  coronata, 1311, *p. 399*
  dentata, (1311)
  discoidea, (1311)
  echinata, 1310
  eriocarpa, (1311)
  locusta, 1309, pl. 136
  olitoria, 1309, pl. 136
  rimosa, (1311)
Vallisneria, 1567
  spiralis, 1567, *p. 480*
Velezia, 195
  rigida, 195, *p. 90*
Veratrum, 1586
  album, 1586, pl. 162
  nigrum, (1586), pl. 162
Verbascum, 1190–1196
  blattaria, 1192, pl. 120
  creticum, 1196, pl. 120
  longifolium, (1194)
  lychnitis, 1195
  nigrum, 1190, pl. 120
  phlomoides, 1194
  phoeniceum, (1192), pl. 121
  pulverulentum, (1195)
  sinuatum, 1191
  speciosum, (1195)
  thapsiforme, (1193), pl. 120
  thapsus, 1193
  undulatum, (1195), pl. 120
  virgatum, (1192)
Verbena, 1084
  officinalis, 1084
  supina, (1084)
Verbenaceae, p. 342
Veronica, 1219–1229
  agrestis, (1223)
  alpina, (1220)
  anagallis-aquatica, (1225)
  arvensis, (1220)
  beccabunga, 1225, pl. 124
  chamaedrys, 1228

cymbalaria, 1222
  filiformis, 1224, pl. 124
  fruticans, (1220), pl. 124
  hederifolia, (1222)
  latifolia, 1229, *p. 384*
  montana, (1227)
  officinalis, 1227
  persica, 1223, pl. 124
  polita, (1223)
  scutellata, 1226
  serpyllifolia, 1220
  spicata, 1219
  spuria, (1219)
  teucrium, 1229, *p. 384*
  triphyllos, 1221, *p. 384*
  verna, (1221)
Vesicaria
  utriculata, 323
Viburnum, 1298–1300
  lantana, 1299, pl. 134
  opulus, 1298, pl. 133
  tinus, 1300, pl. 134
Vicia, 542–551
  atropurpurea, (545), pl. 55
  benghalensis, (545), pl. 55
  bithynica, 551
  cracca, 545, pl. 55
  ervilia, (542)
  faba, 546
  grandiflora, (547)
  hirsuta, 542
  hybrida, 547
  lutea, (547)
  melanops, (547)
  narbonensis, 550
  onobrychioides, (544), *p. 198*
  orobus, 543
  sativa, 549, pl. 55
    subsp. nigra, 549
    subsp. sativa, 549
  sepium, 548
  sylvatica, 544
  tenuifolia, (545)
  tetrasperma, (542)
  villosa, (545), pl. 55
Vinca, 1005–1006
  difformis, (1005)
  herbacea, (1005), pl. 97
  major, 1006
  media, (1005)
  minor, 1005, pl. 97
Vincetoxicum, 1010
  hirundinaria, 1010, pl. 99
  nigrum, (1010)
Viola, 773–786
  arborescens, 781
  arvensis, 784
  biflora, 782, pl. 76
  calcarata, 786, pl. 76
  canina, 776
  cenisia, (786)
  elatior, 780, *p. 257*
  hirta, 775
  lutea, 785, pl. 76
  odorata, 774, pl. 76
  palustris, 773, pl. 76
  persicifolia, 779
  pinnata, (773)

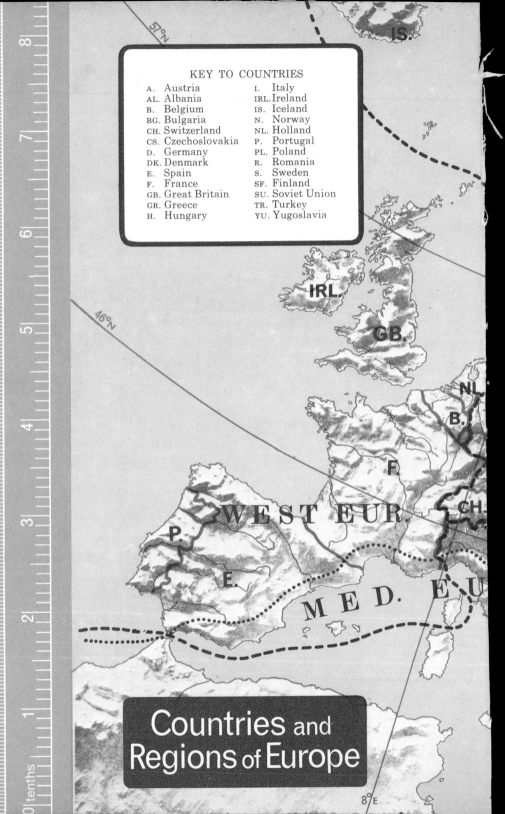

KEY TO COUNTRIES

A.	Austria	I.	Italy
AL.	Albania	IRL.	Ireland
B.	Belgium	IS.	Iceland
BG.	Bulgaria	N.	Norway
CH.	Switzerland	NL.	Holland
CS.	Czechoslovakia	P.	Portugal
D.	Germany	PL.	Poland
DK.	Denmark	R.	Romania
E.	Spain	S.	Sweden
F.	France	SF.	Finland
GB.	Great Britain	SU.	Soviet Union
GR.	Greece	TR.	Turkey
H.	Hungary	YU.	Yugoslavia

Countries and Regions of Europe